Table 16-1 **The Planets**

Planet	Symbol	Mean Distance from Sun, Earth = 1[a]	Diameter, Thousands of km	Mass, Earth = 1[b]	Mean Density, Water = 1[c]	Surface Gravity, Earth = 1[d]	Escape Speed, km/s[e]	Period of Rotation on Axis	Period of Revolution around Sun	Eccentricity of Orbit[h]	Inclination of Orbit to Ecliptic[i]	Known Satellites[j]
Mercury	☿	0.39	4.9	0.055	5.4	0.38	4.3	59 days	88 days	0.21	7°00′	0
Venus	♀	0.72	12.1	0.82	5.25	0.90	10.4	243 days[f]	225 days	0.01	3°34′	0
Earth	⊕	1.00	12.7	1.00	5.52	1.00	11.2	24 h	365 days	0.02	—	1
Mars	♂	1.52	6.8	0.11	3.93	0.38	5.0	24.5 h	687 days	0.09	1°51′	2
Jupiter	♃	5.20	143	318	1.33	2.6	60	10 h	11.9 yr	0.05	1°18′	63
Saturn	♄	9.54	120	95	0.71	1.2	36	10 h	29.5 yr	0.06	2°29′	60
Uranus	⛢	19.2	51	15	1.27	1.1	22	16 h[g]	84 yr	0.05	0°46′	27
Neptune	♆	30.1	50	17	1.70	1.2	24	16 h	165 yr	0.01	1°46′	13

[a] *The mean earth-sun distance is called the astronomical unit, where 1 AU = 1.496×10^8 km.*

[b] *The earth's mass is 5.98×10^{24} kg.*

[c] *The density of water is 1 g/cm^3 = 10^3 kg/m^3.*

[d] *The acceleration of gravity at the earth's surface is 9.8 m/s^2.*

[e] *Speed needed for permanent escape from the planet's gravitational field.*

[f] *Venus rotates in the opposite direction from the other planets.*

[g] *The axis of rotation of Uranus is only 8° from the plane of its orbit.*

[h] *The difference between the minimum and maximum distances from the sun divided by the average distance.*

[i] *The ecliptic is the plane of the earth's orbit.*

[j] *Probably more small ones around Jupiter, Saturn and Uranus.*

The Physical Universe

Thirteenth Edition

Konrad B. Krauskopf
Late Professor Emeritus of Geochemistry, Stanford University

Arthur Beiser

Higher Education

Boston Burr Ridge, IL Dubuque, IA New York San Francisco St. Louis
Bangkok Bogotá Caracas Kuala Lumpur Lisbon London Madrid Mexico City
Milan Montreal New Delhi Santiago Seoul Singapore Sydney Taipei Toronto

THE PHYSICAL UNIVERSE, THIRTEENTH EDITION

Published by McGraw-Hill, a business unit of The McGraw-Hill Companies, Inc., 1221 Avenue of the Americas, New York, NY 10020.

This book is printed on acid-free paper.

1 2 3 4 5 6 7 8 9 0 VNH/VNH 0 9

ISBN 978-0-07-351212-9

MHID 0-07-351212-5

Publisher: *Thomas D. Timp*
Sponsoring Editor: *Debra B. Hash*
Director of Development: *Kristine Tibbetts*
Senior Developmental Editor: *Mary E. Hurley*
Senior Marketing Manager: *Lisa Nicks*
Senior Project Manager: *April R. Southwood*
Senior Production Supervisor: *Kara Kudronowicz*
Lead Media Project Manager: *Stacy A. Patch*
Designer: *Laurie B. Janssen*
Cover Designer: *Ron Bissell*
(USE) Cover Image: *©Alaska Stock LLC/Alamy*
Senior Photo Research Coordinator: *John C. Leland*
Photo Research: *Mary Reeg*
Supplement Producer: *Mary Jane Lampe*
Compositor: *Laserwords Private Limited*
Typeface: *10/12 New Aster*
Printer: *R. R. Donnelley, Jefferson City, MO*

The credits section for this book begins on page C-1 and is considered an extension of the copyright page.

Library of Congress Cataloging-in-Publication Data

Krauskopf, Konrad Bates, 1910-
The physical universe/Konrad B. Krauskopf, Arthur Beiser.—13th ed.
p. cm.
Includes index.
ISBN 978-0-07-351212-9—ISBN 0-07-351212-5 (hard copy : alk. paper) 1. Physical sciences—Textbooks. I. Beiser, Arthur. II. Title.

Q161.2.K7 2010
500.2—dc22

2008031858

www.mhhe.com

Brief Contents

Contents

14 Atmosphere and Hydrosphere 499

15 The Rock Cycle 541

16 The Evolving Earth 587

17 The Solar System 639

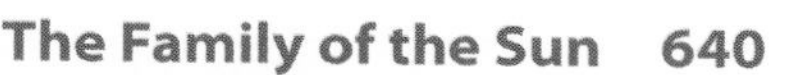

18 The Stars 687

19 The Universe 721

Preface

Creating Informed Citizens

The aim of *The Physical Universe* is to present, as simply and clearly as possible, the essentials of physics, chemistry, earth science, and astronomy to students whose main interests lie elsewhere.

Because of the scope of these sciences and because we assume minimal preparation on the part of the reader, our choice of topics and how far to develop them had to be limited. The emphasis throughout is on the basic concepts of each discipline. We also try to show how scientists approach problems and why science is a never-ending quest rather than a fixed set of facts.

The book concentrates on those aspects of the physical sciences most relevant to a nonscientist who wants to understand how the universe works and to know something about the connections between science and everyday life. We hope to equip readers to appreciate major developments in science as they arrive and to be able to act as informed citizens on matters that involve science and public policy. In particular, there are serious questions today concerning energy supply and use and the contribution of carbon dioxide emissions to global warming. Debates on these questions require a certain amount of scientific literacy, which this book is intended to provide, in order that sensible choices be made that will determine the welfare of generations to come. Past choices have not always benefited our planet and its inhabitants: it is up to us to see that future choices do.

> *"[Krauskopf/Beiser's* The Physical Universe *provides] a good coverage of the basic physical sciences. It gives the basic principles of the different physical sciences and builds real content knowledge. (In contrast, so many texts now tend to "discuss" topics rather than developing an understanding of basic principles.) It has sufficient real-world applications to make the text interesting to the students. I like the way timely, controversial topics are discussed."*
>
> —Linda Arney Wilson, *Middle Tennessee State University*

Scope and Organization

There are many possible ways to organize a book of this kind. We chose the one that provides the most logical progression of ideas, so that each new subject builds on the ones that came before.

> *"This textbook has more of what we teach in Physical Science than any other book on the market. The improvement of each revision is great. . . . I have used 4 editions of this textbook because it is well written; the multiple choice questions and the end-of-the-chapter questions and problems provide numerous opportunities for the student to apply concepts discussed in each chapter. Numerous illustrations are provided for the more visual learner."*
>
> —Etta C. Gravely, *NC A&T University*

Our first concern in *The Physical Universe* is the scientific method, using as illustration the steps that led to today's picture of the universe and the earth's place in it. Next we consider motion and the influences that affect moving bodies. Gravity, energy, and momentum are examined, and the theory of relativity is introduced. Then we examine the various problems associated with energy supply and use in today's world. Matter in its three states now draws our attention, and we pursue this theme from the kinetic-molecular model to the laws of thermodynamics and the significance of entropy. A grounding in electricity and magnetism follows, and then an exploration of wave phenomena that includes the electromagnetic theory of light. We go on from there to the atomic nucleus and elementary particles, followed by a discussion of the quantum theories of light and of matter that lead to the modern view of atomic structure.

> *"This was my favorite chapter [Chapter 1]. It was also my students' favorite. It generated a great deal of discussion and it motivated the students. . . . I was extremely impressed with how this text introduced the scientific method and then used that methodology to discuss one of the "Great Debates" in scientific history, Geocentric vs. Heliocentric. My students not only learned how the method is applied, but they enjoyed the banter of which view made sense. In fact, I received a number of emails where students went out on their own to do further investigation. . . . It also set the stage for more engaging conversation about the world around them."*
>
> —Leroy Salary, Jr., *Norfolk State University*

The transition from physics to chemistry is made via the periodic table. A look at chemical bonds and how they act to hold together molecules, solids, and liquids is followed by a survey of chemical reactions, organic chemistry, and the chemistry of life.

> *"The authors do a great job of explaining the historical relevance of the periodic table and they give an excellent introduction of the definition of what is chemistry by painting a clear picture of how to relate atoms and elements to compounds and chemical reactions."*
>
> —Antonie H. Rice, *University of Arkansas at Pine Bluff*

Our concern next shifts to the planet on which we live, and we begin by inquiring into the oceans of air and water that cover it. From there we proceed to the materials of the earth, to its ever-evolving crust, and to its no-longer-mysterious interior. After a brief narrative of the earth's geological history we go on to what we know about our nearest neighbors in space—planets and satellites, asteroids, meteoroids, and comets.

Now the sun, the monarch of the solar system and the provider of nearly all our energy, claims our notice. We go on to broaden our astronomical sights to include the other stars, both individually and as members of the immense assemblies called galaxies. The evolution of the universe starting from the big bang is the last major subject, and we end with the origin of the earth and the likelihood that other inhabited planets exist in the universe and how we might communicate with them.

> *"This is one of the best chapters [Chapter 18] on stars in a text of this level that I have read. It addresses the various aspects of the stars (size,*

distance, evolution, etc.) in an easy to understand manner. It also provides information concerning the history of the current knowledge of stars."

—Wilda Pounds, *Northeast Mississippi Community College*

Mathematical Level

The physical sciences are quantitative, which has both advantages and disadvantages. On the plus side, the use of mathematics allows many concepts to be put in the form of clear, definite statements that can be carried further by reasoning and whose predictions can be tested objectively. Less welcome is the discomfort many of us feel when faced with mathematical discussions.

The mathematical level of *The Physical Universe* follows Albert Einstein's prescription for physical theories: "Everything should be as simple as possible, but not simpler." A modest amount of mathematics enables the book to show how science makes sense of the natural world and how its findings led to the technological world of today. To give two examples, the formula $\frac{1}{2}mv^2$ for kinetic energy does not have to be pulled out of a hat here, and how the unit called the mole connects chemical ideas with actual measurements can be explored. In general, the more complicated material supplements rather than dominates the presentation, and full mastery is not needed to understand the rest of the book. The basic algebra needed is reviewed in the Math Refresher. Powers-of-ten notation for small and large numbers is carefully explained there. This section is self-contained and can provide all the math background needed.

How much mathematics is appropriate for a given classroom is for each individual instructor to decide. To this end, a section is included in the thirteenth edition Instructor's Manual that lists the slightly more difficult computational material in the text. This material can be covered as wished or omitted without affecting the continuity or conceptual coverage of a course.

"The author has done a wonderful job balancing the verbal and mathematical explanations. The clear, well-labeled diagrams included to assist understanding mathematical expressions are excellent."

—Paul A. Withey, *Northwestern State University of Louisiana*

New To This Edition

Because the organization of the previous edition worked well in the classroom, it was not altered. The principal changes for this edition were these:

Text Revision The entire book was brought up to date with earth history and various aspects of astronomy receiving particular attention. A number of sections were rewritten for greater clarity and to incorporate additional information. New sidebars introduce such topics as energy use in walking and running, the effects of solar ultraviolet radiation on health, atomic sizes, and the desalination of seawater. There are more illustrations, both drawings and photographs, than before, and many of the earlier photographs were replaced with better ones for their purposes.

New Chapter 4 A new chapter, "Energy and the Future," is a largely nontechnical overview of world energy demand and population pressure, global warming and greenhouse-gas emissions, the chief present nonrenewable and renewable energy sources, future energy sources, energy conservation, and strategies for reducing fuel consumption and carbon dioxide emissions. This material is cross-referenced to subsequent chapters for further details where appropriate. Previous editions contained much of what is now in this chapter, but it was distributed among seven chapters according to where the concepts fit into the book's organization. Now, to quote from the chapter introduction, "the energy problem in all its complexity is considered in one place so how its various parts fit together is clear. Even though the scientific elements that cannot be properly discussed this early in the book are left for later chapters, the basic ideas are all here so that those who do not cover the entire text can view the situation as a whole and appreciate how it affects them (and how they affect it)."

> *"Chapter 4 is a now-necessary application of what these students are learning about energy, chemistry, and the evolving earth and our responsible use of scientific knowledge and technological enhancements to human life."*
>
> —Roxanne R. Lane, *Northwestern State University of Louisiana, Natchitoches*

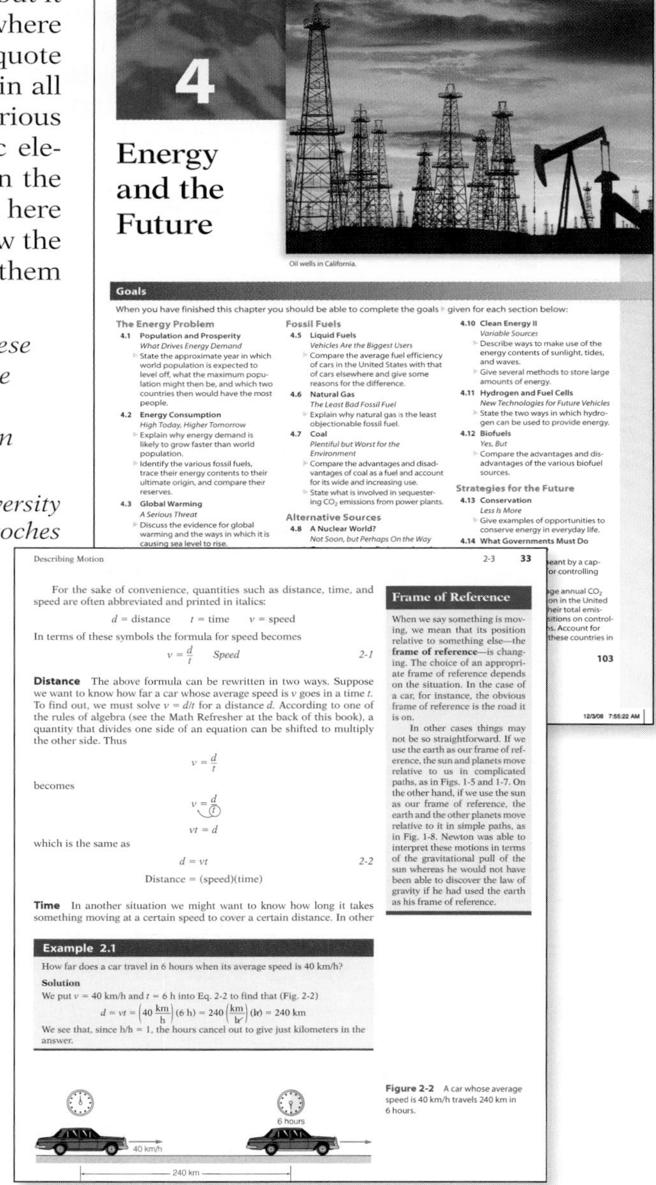

4

Energy and the Future

Oil wells in California.

Goals

When you have finished this chapter you should be able to complete the goals given for each section below:

The Energy Problem

4.1 Population and Prosperity
What Drives Energy Demand
- State the approximate year in which world population is expected to level off, what the maximum population might then be, and which two countries then would have the most people.

4.2 Energy Consumption
High Today, Higher Tomorrow
- Explain why energy demand is likely to grow faster than world population.
- Identify the various fossil fuels, trace their energy contents to their ultimate origin, and compare their reserves.

4.3 Global Warming
A Serious Threat
- Discuss the evidence for global warming and the ways in which it is causing sea level to rise.

Fossil Fuels

4.5 Liquid Fuels
Vehicles Are the Biggest Users
- Compare the average fuel efficiency of cars in the United States with that of cars elsewhere and give some reasons for the difference.

4.6 Natural Gas
The Least Bad Fossil Fuel
- Explain why natural gas is the least objectionable fossil fuel.

4.7 Coal
Plentiful but Worst for the Environment
- Compare the advantages and disadvantages of coal as a fuel and account for its wide and increasing use.
- State what is involved in sequestering CO_2 emissions from power plants.

Alternative Sources

4.8 A Nuclear World?
Not Soon, but Perhaps On the Way

4.10 Clean Energy II
Variable Sources
- Describe ways to make use of the energy contents of sunlight, tides, and waves.
- Give several methods to store large amounts of energy.

4.11 Hydrogen and Fuel Cells
New Technologies for Future Vehicles
- State the two ways in which hydrogen can be used to provide energy.

4.12 Biofuels
Yes, But
- Compare the advantages and disadvantages of the various biofuel sources.

Strategies for the Future

4.13 Conservation
Less Is More
- Give examples of opportunities to conserve energy in everyday life.

4.14 What Governments Must Do

103

Describing Motion 2-3 33

For the sake of convenience, quantities such as distance, time, and speed are often abbreviated and printed in italics:

d = distance t = time v = speed

In terms of these symbols the formula for speed becomes

$$v = \frac{d}{t} \quad \text{Speed} \qquad 2\text{-}1$$

Distance The above formula can be rewritten in two ways. Suppose we want to know how far a car whose average speed is v goes in a time t. To find out, we must solve $v = d/t$ for a distance d. According to one of the rules of algebra (see the Math Refresher at the back of this book), a quantity that divides one side of an equation can be shifted to multiply the other side. Thus

$$v = \frac{d}{t}$$

becomes

$$v = \frac{d}{t}$$

$$vt = d$$

which is the same as

$$d = vt \qquad 2\text{-}2$$

$$\text{Distance} = (\text{speed})(\text{time})$$

Time In another situation we might want to know how long it takes something moving at a certain speed to cover a certain distance. In other

Frame of Reference

When we say something is moving, we mean that its position relative to something else—the **frame of reference**—is changing. The choice of an appropriate frame of reference depends on the situation. In the case of a car, for instance, the obvious frame of reference is the road it is on.

In other cases things may not be so straightforward. If we use the earth as our frame of reference, the sun and planets move relative to us in complicated paths, as in Figs. 1-5 and 1-7. On the other hand, if we use the sun as our frame of reference, the earth and the other planets move relative to it in simple paths, as in Fig. 1-8. Newton was able to interpret these motions in terms of the gravitational pull of the sun whereas he would not have been able to discover the law of gravity if he had used the earth as his frame of reference.

Example 2.1

How far does a car travel in 6 hours when its average speed is 40 km/h?

Solution

We put v = 40 km/h and t = 6 h into Eq. 2-2 to find that (Fig. 2-2)

$$d = vt = \left(40\,\frac{\text{km}}{\text{h}}\right)(6\ \text{h}) = 240\left(\frac{\text{km}}{\text{h}}\right)(\text{h}) = 240\ \text{km}$$

We see that, since h/h = 1, the hours cancel out to give just kilometers in the answer.

Figure 2-2 A car whose average speed is 40 km/h travels 240 km in 6 hours.

The Learning System

A variety of aids are provided in *The Physical Universe* to help the reader master the text.

Chapter Opener An outline provides a preview of major topics, showing at a glance what the chapter covers. A list of goals, in order by section, helps to focus the reader on what is most important in the chapter.

Illustrations The illustrations, both line drawings and photographs, are full partners to the text and provide a visual pathway to understanding scientific observations and principles for students unaccustomed to abstract argument.

Worked Examples A full grasp of physical and chemical ideas includes an ability to solve problems based on these ideas. Some students, although able to follow the discussions in the book, nevertheless may have trouble putting their knowledge to use in this way. To help them, detailed solutions of typical problems are provided that show how to apply formulas and equations to real-world situations. Besides the worked examples, outline solutions for half the end-of-chapter exercises are given at the end of the text, which include over 200 model problem solutions. Thinking through these solutions should bring the unsolved even-numbered problems within reach. In addition to its role in reinforcing the understanding of physical and chemical ideas, solving problems can provide great pleasure, and it would be a shame to miss out on this pleasure. The worked examples in the text are not limited to problems—nearly half of them show how basic ideas can be used to answer serious questions that do not involve calculations.

The Solar System 1-13 13

BIOGRAPHY Nicolaus Copernicus (1473–1543)

When Columbus made his first voyage to the New World Copernicus was a student in his native Poland. In the years that followed intellectual as well as geographical horizons receded before eager explorers. In 1496 Copernicus went to Italy to learn medicine, theology, and astronomy. Italy was then an exciting place to be, a place of business expansion and conflicts between rival cities, great fortunes and corrupt governments, brilliant thinkers and inspired artists such as Leonardo da Vinci and Michelangelo. After 10 years in Italy Copernicus returned to Poland where he practiced medicine, served as a canon in the cathedral of which his uncle was the bishop, and became involved in currency reform, but much of his time was devoted to developing the idea that the planets move around the sun rather than around the earth. The idea was not new—the ancient Greeks were aware of it—but Copernicus went further and worked out the planetary orbits and speeds in detail. Although a summary of his results had been circulated in manuscript form earlier, not until a few weeks before his death was Copernicus's *De Revolutionibus Orbium Coelestium* published in book form. Today *De Revolutionibus* is recognized as one of the foundation stones of modern science, but soon after its appearance it was condemned by the Catholic Church (which did not lift its ban until 1835) and had little

impact on astronomy until Kepler further developed its concepts a half century later.

The apparent shifting of the sun among the stars is due to the earth's motion in its orbit. As the earth swings around the sun, we see the sun changing its position against the background of the stars. The moon's gradual eastward drift is mainly due to its orbital motion. Apparently irregular movements of the planets are really just combinations of their motions with our own shifts of position as the earth moves.

The **copernican system** offended both Protestant and Catholic religious leaders, who did not want to see the earth taken from its place at the hub of the universe. The publication of Copernicus's manuscript began a long and bitter argument. To us, growing up with the knowledge that the earth moves, it seems odd that this straightforward idea was so long and so violently opposed. But in the sixteenth century good arguments were available to both sides.

Consider, said supporters of Ptolemy, how fast the earth's surface must loose from produ does are co groun much stars, a fixed

Leap Years

A day is the time needed for the earth to make a complete turn on its axis, and a year is the time it needs to complete an orbit around the sun. The length of the year is slightly less than 365 days and 6 hours, so adding an extra day to February every 4 years (namely those years evenly divisible by 4, which are accordingly called **leap years**) enables the seasons

Elementary Particles 8-33 321

A PHYSICIST AT WORK Michelangelo D'Agostino

University of California, Berkeley, California

Starting your day at 10 P.M. is never easy. Slipping on several layers of long underwear, overalls, and a bulky red parka, I walk out the door for another day (or technically, another night) in the harshest physics lab on earth. I'm here at the geographic South Pole with 250 or so other scientists and staff. We're taking advantage of the 24-hour sunlight during the summer season to work on a wide variety of experiments spanning the fields of astrophysics, particle physics, seismology, climatology, and atmospheric science.

Antarctica is like no other place on the planet. In the summer, temperatures can reach −50°F. A two-mile thick icesheet stretches as far as the eye can see in all directions. A coworker once compared being here at the pole to standing on a piece of paper. Since we live on top of that icesheet, the altitude makes it difficult to breathe, and the air is so cold that it holds less moisture than the Sahara. All passengers and supplies are flown in from New Zealand by U.S. Air National Guard cargo planes that land on skis in the soft snow. With the extreme isolation (we have internet and phone access for only 8 hours a day and no television at all), working here is quite a challenge.

As a graduate student, I'm lucky enough to have the opportunity to tackle these challenges because I work on the IceCube experiment. IceCube is a giant particle detector being constructed under the ice here at the South Pole. In a way, it's a new, strange type of telescope. Instead of detecting light as a normal telescope does, IceCube is trying to detect another kind of fundamental particle that may come from astrophysical objects—the ghostly neutrino. Neutrinos are very light, electrically neutral particles that interact extremely weakly with normal matter. Trillions of them pass straight through your body every second like light through a piece of glass.

For this reason, neutrinos should make good astronomical messengers. While light can be obscured by dust or other matter that blocks our view, neutrinos should travel unimpeded from their source directly to us. For this same reason though, neutrinos are difficult to detect. Doing so requires a giant target—a target so large that it can't be contained inside of a normal laboratory.

That's where the South Pole comes in. When IceCube is completed, it will monitor a very special target—a cubic kilometer chunk of crystal-clear glacial ice. Using hot water, we melt holes into the ice to install very sensitive light detectors. Occasionally, a neutrino passing through half of the earth will hit a molecule of the ice, giving off a faint blue flash of light. By checking which of our sensors saw this light and comparing the timing, we can "connect the dots" to tell where the neutrino came from within the universe. By studying such neutrinos, we hope to learn about the properties of the extremely violent astrophysical objects that we think may produce them. These include explosions from dying stars (supernovas and gamma ray bursts) and eruptions of hot material from the neighborhood of giant black holes at the center of galaxies.

Doing research here is completely unlike studying physics in college, though the problem-solving skills I learned there continue to serve me well. My experiment involves a multitude of different tasks that require a variety of skills. One day, I may find myself shoveling snow or carrying crates from place to place, doing manual labor to help those who melt the holes in which we install our instruments. Another day, I may end up writing computer programs to analyze the volumes of data that tell us how our detector is performing. On still another day, I may work to design or fix electronics. In all these jobs, I work closely with fascinating people from all over the world. That is why I find this kind of physics so fulfilling—working in a team to solve difficult problems that may yield answers to some of our fundamental questions about the universe.

Bringing Science to Life

Biographies Brief biographies of 40 major figures in the development of the physical sciences appear where appropriate throughout the text. The biographies provide human and historical perspectives by attaching faces and stories to milestones in these sciences.

Sidebars These are brief accounts of topics related to the main text. A sidebar may provide additional information on a particular subject, comment on its significance, describe its applications, consider its historical background, or present recent findings. Twenty-five new ones have been added for this edition.

> *"The textbook does a nice job of covering contemporary topics, which is a way to keep non-science majors interested. I also like the biographies and have sometimes assigned students to go deeper into someone's biography and report to class. Somehow, making these scientists become 'real' for the students makes them enjoy the course more. I also like to show them how some of the topics covered in the course are fairly recent, how there are things we still do not understand. It is important for them to know that science is alive and continues to develop its body of knowledge."*
>
> —Ana Ciereszko, *Miami Dade College*

At Work Essays Four scientists who are carrying out important research in their respective fields have contributed accounts of how their days at work are spent. The enthusiasm and dedication they bring to their probes of the physical universe show clearly in these essays.

End of Chapter Features

Important Terms and Ideas Important terms introduced in the chapter are listed together with their meanings, which serves as a chapter summary. A list of the **Important Formulas** needed to solve problems based on the chapter material is also given where appropriate.

Exercises An average of over a hundred exercises on all levels of difficulty follow each chapter. They are of three kinds, multiple choice, questions, and problems:

- **Multiple Choice** An average chapter has 41 Multiple-Choice exercises (with answers) that act as a quick, painless check on understanding. Correct answers provide reinforcement and encouragement; incorrect ones identify areas of weakness.
- **Exercises** Exercises consist of both questions and problems arranged according to the corresponding text section. Each group begins with questions and goes on to problems. Some of the questions are meant to find out how well the reader has understood the chapter material. Others ask the reader to apply what he or she has learned to new situations. Answers to the odd-numbered questions are given at the back of the book. The physics and chemistry chapters include problems that range from quite easy to moderately challenging. The ability to work out such problems signifies a real understanding of these subjects. Outline solutions (not just answers) for the odd-numbered problems are given at the back of the book.

150 4-48 **Chapter 4** Energy and the Future

c. sugar cane
d. natural gas

39. Compact fluorescent lightbulbs are
 a. more efficient than incandescent bulbs but less efficient than LEDs
 b. more efficient than incandescent bulbs and about as efficient as LEDs
 c. more efficient than both incandescent bulbs and LEDs
 d. about the same in efficiency as incandescent bulbs and LEDs
40. Of the following, the strategy for coping with future energy shortages with the most in its favor is to
 a. burn more coal
 b. produce more oil from tar sands
 c. divert more agricultural land to making biofuels
 d. increase energy efficiency and energy conservation

Exercises

4.1 Population and Prosperity

1. What are the three main factors that will require changes in today's patterns of energy production and consumption?

4.2 Energy Consumption

2. Even if the developed countries stabilize or reduce their energy consumption in years to come, worldwide energy consumption will increase. What are the two main reasons for this?
3. The average rate of energy consumption per person in the United States is about how many times the world average: twice, three times, four times, over four times?
4. List the fossil fuels in the order in which they will probably be used up.
5. Explain how sunlight is responsible for these energy sources: food, wood, water power, wind power, fossil fuels.
6. What energy sources cannot be traced to sunlight falling on the earth?

4.3 Global Warming

7. Approximately what proportion of the world's population lives on or near coasts and so may be under future threat from rising sea level?
8. Give two reasons why global warming is causing sea level to rise.
9. Once the polar ice sheets have melted beyond a certain amount, melting will continue even if CO_2 emissions stop rising. Why?
10. The oceans as well as the atmosphere are growing warmer. What does this imply for tropical storms such as hurricanes?
11. When was the last time world temperatures were as high as they are likely to be in 2100 if current rates of CO_2 emission continue: hundreds of years ago, thousands of years ago, millions of

4.4 Carbon Dioxide and the Greenhouse Effect

12. Every body of matter radiates light. What is characteristic of light radiated by something very hot, such as the sun? Of light radiated by something at ordinary temperatures, such as the earth's surface?
13. What is the nature of the greenhouse effect in the earth's atmosphere?
14. List the chief greenhouse gases in the atmosphere. What property do they share?
15. About half the CO_2 from burning fossil fuels enters the atmosphere. What becomes of the rest?
16. Why is deforestation so important in global warming?
17. List the fossil fuels in the order in which they contribute to world CO_2 emissions.

4.5 Liquid Fuels

18. What fuel liberates the most energy per gram when it burns? What is produced when it burns?
19. Most of the world's oil is used as a fuel for what purpose?
20. How do the oil reserves in tar sands compare with the reserves of ordinary crude oil? What are some of the disadvantages of tar sand oil?
21. What methods are available to reduce pollution by car exhausts?
22. What are some of the reasons why the average fuel efficiency of cars in the United States is the lowest in the world?

4.6 Natural Gas

23. The amount of CO_2 emitted per kilowatt-hour of electricity by a gas-fired power plant is about half that emitted by a coal-fired plant. What do you think is the reason that coal-fired plants are much more common?

"The multiple-choice exercises and the questions and problems are a very nice feature of this book and it is definitely above average. . . . There is a good balance between the conceptual versus computational questions."

—Omar Franco Guerrero, *University of Delaware*

Online Homework and Resources

McGraw-Hill's *The Physical Universe* website offers online electronic homework along with a myriad of resources for both instructors and students. Instructors can create homework with easy-to-assign algorithmically generated Exercises from the text and the simplicity of automatic grading and reporting. *The Physical Universe's* end-of-chapter Exercises appear in the online homework system in a variety of open-ended and multiple-choice formats.

Instructors also have access to PowerPoint lecture outlines, the Instructor's Manual, PowerPoint files with electronic images from the text, clicker questions, quizzes, animations, suggested web links and experiments, and many other resources directly tied to text-specific materials in *The Physical Universe.* Students have access to a variety of self-quizzes, key term matching exercises, animations, web links, and more.

See www.mhhe.com/krauskopf to learn more and register.

Complete Set of Electronic Book Images and Assets for Instructors

Build instructional materials wherever, whenever, and however you want!

Accessed from your textbook's website, an online digital library containing photos, artwork, animations, and other media types can be used to create customized lectures, visually enhanced tests and quizzes, compelling course websites, or attractive printed support materials. All assets are copyrighted by McGraw-Hill Higher Education, but can be used by instructors for classroom purposes. The visual resources in this collection include:

- **Art** Full-color digital files of all illustrations in the book can be readily incorporated into lecture presentations, exams, or custom-made classroom materials. In addition, all files are pre-inserted into PowerPoint slides for ease of lecture preparation.
- **Photos** The photos collection contains digital files of photographs from the text, which can be reproduced for multiple classroom uses.
- **Tables and Worked Examples** Tables and Worked Examples that appear in the text have been saved in electronic form for use in classroom presentations and/or quizzes.
- **Animations** Numerous full-color animations illustrating important processes are also provided. Harness the visual impact of concepts in motion by importing these files into classroom presentations or online course materials.

Also residing on your textbook's website are:

- **PowerPoint Lecture Outlines** Ready-made presentations that combine art, animation, and lecture notes are provided for each chapter of the text.
- **PowerPoint Slides** For instructors who prefer to create their lectures from scratch, illustrations, photos, tables, and worked examples from the text are pre-inserted by chapter into blank PowerPoint slides.

Computerized Test Bank Online

A comprehensive bank of test questions is provided within a computerized test bank powered by McGraw-Hill's flexible electronic testing program EZ Test Online (www.eztestonline.com). EZ Test Online allows you to create paper and online tests or quizzes in this easy to use program!

Imagine being able to create and access your test or quiz anywhere, at any time without installing the testing software. Now, with EZ Test Online, instructors can select questions from multiple McGraw-Hill test banks or author their own, and then either print the test for paper distribution or give it online.

Test Creation

- Author/edit questions online using the 14 different question type templates.
- Create printed tests or deliver online to get instant scoring and feedback.

- Create questions pools to offer multiple versions online—great for practice.
- Export your tests for use in WebCT, Blackboard, PageOut, and Apple's iQuiz.
- Compatible with EZ Test Desktop tests you've already created.
- Sharing tests with colleagues, adjuncts, TAs is easy.

Online Test Management

- Set availability dates and time limits for your quiz or test.
- Control how your test will be presented.
- Assign points by question or question type with drop down menu.
- Provide immediate feedback to students or delay until all finish the test.
- Create practice tests online to enable student mastery.
- Your roster can be uploaded to enable student self-registration.

Online Scoring and Reporting

- Automated scoring for most of EZ Test's numerous question types.
- Allows manual scoring for essay and other open-response questions.
- Manual rescoring and feedback is also available.
- EZ Test's grade book is designed to easily export to your grade book.
- View basic statistical reports.

Support and Help

- User's Guide and built-in page-specific help.
- Flash tutorials for getting started on the support site.
- Support Website—www.mhhe.com/eztest.
- Product specialist available at 1-800-331-5094.
- Online Training: http://auth.mhhe.com/mpss/workshops/.

Personal Response Systems

Personal Response Systems ("Clickers") can bring interactivity into the classroom or lecture hall. Wireless response systems give the instructor and students immediate feedback from the entire class. Wireless response pads are essentially remotes that are easy to use and engage students, allowing instructors to motivate student preparation, interactivity, and active learning. Instructors receive immediate feedback to assess which concepts students understand. Questions covering the content of *The Physical Universe* text and formatted in PowerPoint are available on *The Physical Universe* website.

> *"I require students to use eInstruction's remotes. I use the remotes to measure whether students have mastered some of the important concepts. . . . I find them very useful—they give me immediate feedback—they allow daily attendance to be taken quickly and rather painlessly."*
>
> —Robert J. Backes, *Pittsburg State University*

"I use the website materials to prepare all my lectures in PowerPoint, since I develop my lectures to my liking. All tests are prepared using the test bank provided by the publisher. . . . These materials have tremendously lightened my workload. I haven't had to re-invent the wheel so to speak. I can change my exams every semester and shuffle my answers from class to class. Well done."

—Colley Baldwin, *Medgar Evers College, CUNY*

Student Study Guide

Another helpful resource can be found in *The Physical Universe* Student Study Guide. With this study guide, students will maximize their use of *The Physical Universe* text package. It supplements the text with additional, self-directed activities and complements the text by focusing on the important concepts, theories, facts, and processes presented by the authors.

Primis Online

This text can be customized in print or in an electronic format to meet exact course needs. McGraw-Hill's Primis Online allows instructors to select desired chapters and preferred sequence and to choose supplements from the many science items on our database. Visit **http://www.primiscontentcenter.com/** to begin today.

Electronic Books

If you or your students are ready for an alternative version of the traditional textbook, McGraw-Hill brings you innovative and inexpensive electronic textbooks. By purchasing E-books from McGraw-Hill, students can save as much as 50% on selected titles delivered on the most advanced E-book platforms available.

E-books from McGraw-Hill are smart, interactive, searchable, and portable with powerful tools that allow detailed searching, highlighting, and note taking. E-books from McGraw-Hill will help students study smarter and quickly find the information they need, and they will save money. Contact your McGraw-Hill sales representative to discuss E-book packaging options.

Acknowledgments

Comments from users have always been of much help in revising *The Physical Universe.* Detailed reviews of its twelfth edition and the new Chapter 4 by the following teachers were especially valuable and are much appreciated:

Miah Muhammad Adel, *University of Arkansas at Pine Bluff*

William K. Adeniyi, *North Carolina Agricultural and Technical State University, Greensboro*

Ignacio Birriel, *Morehead State University*

Reggie Blake, *New York City College of Technology, City University of New York*

Debra L. Burris, *University of Central Arkansas*
Peter E. Busher, *Boston University*
Brian D. Campbell, *Southwestern Oklahoma State University*
Jason A. Carr, *Pasco-Hernando Community College*
Ana A. Ciereszko, *Miami Dade College*
Paul J. Dolan, Jr., *Northeastern Illinois University*
Milton W. Ferguson, *Norfolk State University*
Brent Gutierrez, *Augusta State University*
Mahmoud Khalili, *Northeastern Illinois University*
Linda C. Kondrick, *Arkansas Tech University*
Roxanne R. Lane, *Northwestern State University of Louisiana, Natchitoches*
Kari L. Lavalli, *College of General Studies, Boston University*
Rahul Mehta, *University of Central Arkansas*
Tchao Podona, *Miami Dade College*
Jeff W. Robertson, *Arkansas Tech University*
Toni D. Sauncy, *Angelo State University*
Bruce Schulte, *Pulaski Technical College*
Sam Subramania, *Miles College*
Matthew F. Ware, *Grambling State University*
Paul A. Withey, *Northwestern State University of Louisiana*

Many constructive suggestions, new ideas, and invaluable advice were also provided by reviewers of earlier editions. Special thanks are owed to those who reviewed the text in the past:

Paul E. Adams, *Fort Hays State University*
William K. Adeniyi, *North Carolina A&T State University*
Adedoyin M. Adeyiga, *Cheyney University of Pennsylvania*
Brian Adrian, *Bethany College*
Z. Altounian, *McGill University*
I. J. Aluka, *Prairie View A&M University*
Louis G. Arnold, *Otterbein College*
Jose D'Arruda, *University of North Carolina—Pembroke*
Robert J. Backes, *Pittsburg State University*
Colley Baldwin, *Medgar Evers College, CUNY*
Rama Bansil, *Boston University*
Sharon Barnewall, *Columbus State Community College*
Mohammad Bhatti, *The University of Texas Pan American*
Ignacio Birriel, *Morehead State University*
Jennifer J. Birriel, *Morehead State University*
Charles C. Blatchley, *Pittsburg State University*
Claude E. Bolze, *Tulsa Community College*
Robert Boram, *Morehead State University*
Terry Bradfield, *Northeastern State University*
Art Braundmeier, Jr., *Southern Illinois University, Edwardsville*
Jeanne M. Buccigross, *College of Mount St. Joseph*
Peter E. Busher, *Boston University*
Korey Champe, *Bakersfield Junior College*
Edward S. Chang, *University of Massachusetts, Amherst*
Tu-nan Chang, *University of Southern California*
Ana A. Ciereszko, *Miami Dade College*
Francis Cobbina, *Columbus State Community College*
Michael R. Cohen, *Shippensburg University*
Whitman Cross II, *Towson State University*

William C. Culver, *St. Petersburg Junior College*
Guillermina Damas, *Miami Dade College*
Paul J. Dolan, Jr., *Northeastern Illinois University*
Timothy T. Ehler, *Buena Vista University*
John Encarnación, *Saint Louis University*
Bernd Enders, *College of Marin*
Terry Engelder, *Pennsylvania State University*
Walter A. Flomer, *Northwestern State University of Louisiana*
Frederick Floodstrand, *Arkansas State University*
Carl K. Frederickson, *University of Central Arkansas*
Andrea M. Gorman Gelder, *University of Maine at Presque Isle*
Etta C. Gravely, *North Carolina Agricultural and Technical State University*
William Gregg, *Louisiana State University*
Omar Franco Guerrero, *University of Delaware*
Austin F. Gulliver, *Brandon University*
Peter Hamlet, *Pittsburg State University*
Kenny Hebert, *Carl Albert State College*
Timothy G. Heil, *University of Georgia*
Esmail Hejazifar, *Wilmington College*
Jean-Francois Henry, *Northern Virginia Community College*
J. Horacio Hoyos, *St. Petersburg Junior College*
C. A. Hughes, *University of Central Oklahoma*
Eric Jerde, *Morehead State University*
Booker Juma, *Fayetteville State University*
Sher S. Kannar, *Stillman College*
Linda C. Kondrick, *Arkansas Tech University*
Richard Lahti, *Minnesota State University, Moorhead*
Eric T. Lane, *The University of Tennessee at Chattanooga*
Lauree Lane, *Tennessee State University*
Andrei Ludu, *Northwestern State University of Louisiana*
Kingshuk Majumdar, *Berea College*
Benjamin K. Malphrus, *Morehead State University*
Alan Marscher, *Boston University*
T. Ted Morishige, *University of Central Oklahoma*
Garry Noe, *Virginia Wesleyan College*
John Oakes, *Marian College*
Jennifer M. K. O'Keefe, *Morehead State University*
Patrick Owens, *Winthrop University*
Patrick Papin, *San Diego State University*
James D. Patterson, *Florida Institute of Technology*
James L. Pazun, *Pfeiffer University*
A. G. Unil Perera, *Georgia State University*
Brian L. Pickering, *North Central Michigan College*
Jerry Polson, *Southeastern Oklahoma State University*
Wilda Pounds, *Northeast Mississippi Community College*
Kent J. Price, *Morehead State University*
G. S. Rahi, *Fayetteville State University*
Antonie H. Rice, *University of Arkansas at Pine Bluff*
Jeff Robertson, *Arkansas Technical University*
Lee Roecker, *Berea College*
Charles Rop, *University of Toledo*
Klaus Rossberg, *Oklahoma City University*
Jack Ryan, *South Arkansas Community College*

Leroy Salary, Jr., *Norfolk State University*
T. D. Sauncy, *Angelo State University*
Robert M. Schoch, *Boston University*
Eric Schulman, *University of Virginia*
Regina Schulte-Ladbeck, *University of Pittsburgh*
William M. Scott, *Fort Hays State University*
Tom Shoberg, *Pittsburg State University*
David A. Slimmer, *Lander University*
Frederick R. Smith, *Memorial University of Newfoundland*
Pam Smith, *Cowley County Community College*
Steve Storm, *Arkansas State University, Heber Springs*
Sam Subramaniam, *Miles College*
John H. Summerfield, *Missouri Southern State College*
Arjun Tan, *Alabama A&M University*
Sergio E. Ulloa, *Ohio University*
Wytse van Dijk, *Redeemer College*
Peter van Keken, *University of Michigan*
Daniel A. Veith, *Nicholls State University*
Rita K. Voltmer, *Miami University*
William J. Wallace, *San Diego State University*
Ling Jun Wang, *University of Tennessee at Chattanooga*
Sylvia Washington, *Elgin Community College*
Linda Arney Wilson, *Middle Tennessee State University*
Paul A. Withey, *Northwestern State University of Louisiana*
Heather L. Woolverton, *University of Central Arkansas*
Capp Yess, *Morehead State University*

I would also like to thank Ana A. Ciereszko and the students at Miami Dade College and Andrei Ludu and the students at Northwestern State University of Louisiana for the many student evaluations, which provided very beneficial feedback on how students use the text and what they find most helpful.

Nancy Woods of Des Moines Area Community College compiled the Videolists in the Instructor's Manual/Test Bank for *The Physical Universe.* Steven Carey of the University of Mobile helped prepare the Goals for each chapter. Linda Kondrick of Arkansas Tech University was of great help in checking the exercises and their answers while preparing the online homework versions of the exercises. I am grateful to all of them.

I would also like to thank the various ancillary authors. Steven Carey of the University of Mobile wrote the Student Study Guide to accompany the text. The following contributed to the many online resources: Charles Hughes of the University of Central Oklahoma wrote the daily concept quizzes; Robert Schoch of Boston University authored the multiple-choice quizzes; S. Raj Chaudhury of Christopher Newport University contributed the clicker questions; and Toni Sauncy of Angelo State University authored the Powerpoint lecture outlines.

Finally, I want to thank my friends at McGraw-Hill, especially Mary Hurley and April Southwood, for their skilled and dedicated help in producing this edition.

Arthur Beiser

Meet the Authors

Konrad B. Krauskopf was born and raised in Madison, Wisconsin and earned a B.S. in chemistry from University of Wisconsin in 1931. He then earned a Ph.D. in chemistry at the University of California in Berkeley. When the Great Depression made jobs in chemistry scarce, Professor Krauskopf decided to study geology, which had long fascinated him. Through additional graduate work at Stanford University, he earned a second Ph.D. and eventually a position on the Stanford faculty. He remained at Stanford until his retirement in 1976. During his tenure, Professor Krauskopf also worked at various times with the U.S. Geological Survey, served with the U.S. army in occupied Japan, and traveled to Norway, France, and Germany on sabbatical leaves. His research interests included field work on granites and metamorphic rocks and laboratory study on applications of chemistry to geologic problems, especially the formation of ore deposits. In later years, Professor Krauskopf spent time working with various government agencies on the problem of radioactive waste disposal. Professor Krauskopf passed away on May 8, 2003.

Arthur Beiser, a native of New York City, received B.S., M.S., and Ph.D. degrees in physics from New York University, where he later served as Associate Professor of Physics. He then was Senior Research Scientist at the Lamont Geological Observatory of Columbia University. His research interests were chiefly in cosmic rays and in magnetohydrodynamics as applied to geophysics and astrophysics. In addition to theoretical work, he participated in a cosmic-ray expedition to an Alaskan peak and directed a search for magnetohydrodynamic waves from space in various Pacific locations. A Fellow of The Explorers Club, Dr. Beiser was the first chairman of its Committee on Space Exploration. He is the author or coauthor of 36 books, mostly college texts on physics and mathematics, 14 of which have been translated into a total of 26 languages. Two of his books are on sailing, *The Proper Yacht* and *The Sailor's World.* Figure 13-18 is a photograph of Dr. Beiser at the helm of his 58-ft sloop; he and his wife Germaine have sailed over 130,000 miles, including two Atlantic crossings and a rounding of Cape Horn. Germaine Beiser, who has degrees in physics from the Massachusetts Institute of Technology and New York University, is the author or coauthor of 7 books on various aspects of physics and has contributed to *The Physical Universe.*

1 The Scientific Method

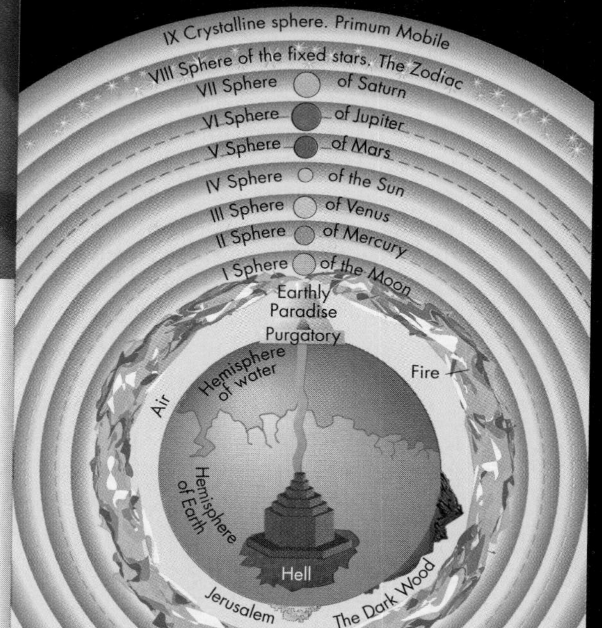

Medieval picture of the universe.

Goals

When you have finished this chapter you should be able to complete the goals ▶ given for each section below:

How Scientists Study Nature

1.1 The Scientific Method
Four Steps
- ▶ Outline the scientific method.
- ▶ Distinguish between a law and a theory.
- ▶ Discuss the role of a model in formulating a scientific theory.

1.2 Why Science Is Successful
Science Is a Living Body of Knowledge, Not a Set of Frozen Ideas
- ▶ Explain why the scientific method has been more successful than other approaches to understanding the natural world.

The Solar System

1.3 A Survey of the Sky
Everything Seems to Circle the North Star
- ▶ Give the reason why Polaris is the heavenly body that remains most nearly stationary in the sky.
- ▶ Define constellation.
- ▶ Tell how to distinguish planets from stars by observations of the night sky made several weeks or months apart.

1.4 The Ptolemaic System
The Earth as the Center of the Universe

1.5 The Copernican System
A Spinning Earth That Circles the Sun
- ▶ Compare how the ptolemaic and copernican systems account for the observed motions of the sun, moon, planets, and stars across the sky.
- ▶ Define day and year.

1.6 Kepler's Laws
How the Planets Actually Move
- ▶ Explain the significance of Kepler's laws.

1.7 Why Copernicus Was Right
Evidence Was Needed That Supported His Model While Contradicting Ptolemy's Model
- ▶ State why the copernican system is considered correct.

Universal Gravitation

1.8 What Is Gravity?
A Fundamental Force
- ▶ Define fundamental force.

1.9 Why the Earth Is Round
The Big Squeeze
- ▶ Explain why the earth is round but not a perfect sphere.

1.10 The Tides
Up and Down Twice a Day
- ▶ Explain the origin of tides.

1.11 The Discovery of Neptune
Another Triumph for the Law of Gravity
- ▶ Explain in terms of the scientific method why the discovery of Neptune was so important in confirming the law of gravity.

How Many of What

1.12 The SI System
All Scientists Use These Units
- ▶ Change the units in which a quantity is expressed from those of one system of units to those of another system.
- ▶ Use metric prefixes for small and large numbers.
- ▶ Use significant figures correctly in a calculation.

All of us belong to two worlds, the world of people and the world of nature. As members of the world of people, we take an interest in human events of the past and present and find such matters as politics and economics worth knowing about. As members of the world of nature, we also owe ourselves some knowledge of the sciences that seek to understand this world. It is not idle curiosity to ask why the sun shines, why the sky is blue, how old the earth is, why things fall down. These are serious questions, and to know their answers adds an important dimension to our personal lives.

We are made of atoms linked together into molecules, and we live on a planet circling a star—the sun—that is a member of one of the many galaxies of stars in the universe. It is the purpose of this book to survey what physics, chemistry, geology, and astronomy have to tell us about atoms and molecules, stars and galaxies, and everything in between. No single volume can cover all that is significant in this vast span, but the basic ideas of each science can be summarized along with the raw material of observation and reasoning that led to them.

Like any other voyage into the unknown, the exploration of nature is an adventure. This book records that adventure and contains many tales of wonder and discovery. The search for knowledge is far from over, with no end of exciting things still to be found. What some of these things might be and where they are being looked for are part of the story in the chapters to come.

How Scientists Study Nature

Every scientist dreams of lighting up some dark corner of the natural world—or, almost as good, of finding a dark corner where none had been suspected. The most careful observations, the most elaborate calculations will not be fruitful unless the right questions are asked. Here is where creative imagination enters science, which is why most of the greatest scientific advances have been made by young, nimble minds.

Scientists study nature in a variety of ways. Some approaches are quite direct: a geologist takes a rock sample to a laboratory and, by inspection and analysis, finds out what it is made of and how and when it was probably formed. Other approaches are indirect: nobody has ever visited the center of the earth or ever will, but by combining a lot of thought with clues from different sources, a geologist can say with near certainty that the earth has a core of molten iron. No matter what the approaches to particular problems may be, however, the work scientists do always fits into a certain pattern of steps. This pattern, a general scheme for gaining reliable information about the universe, has become known as the **scientific method.**

1.1 The Scientific Method

Four Steps

We can think of the scientific method in terms of four steps: (1) formulating a problem, (2) observation and experiment, (3) interpreting the

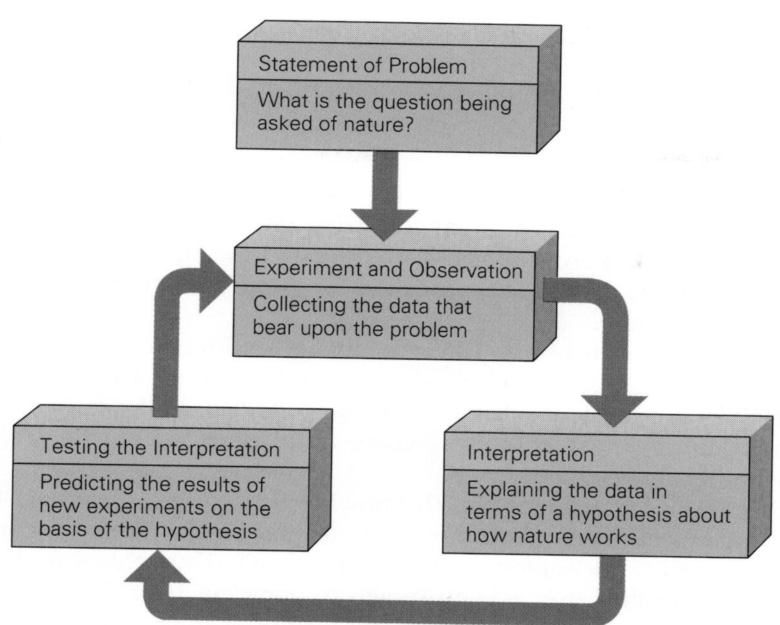

Figure 1-1 The scientific method. No hypothesis is ever final because future data may show that it is incorrect or incomplete. Unless it turns out to be wrong, a hypothesis never leaves the loop of experiment, interpretation, testing. Of course, the more times the hypothesis goes around the loop successfully, the more likely it is to be a valid interpretation of nature. Experiment and hypothesis thus evolve together, with experiment having the final word. Although a hypothesis may occur to a scientist as he or she studies experimental results, as shown here, often the hypothesis comes first and relevant data are sought afterward to test it.

data, and (4) testing the interpretation by further observation and experiment to check its predictions. These steps are often carried out by different scientists, sometimes many years apart and not always in this order. Whatever way it is carried out, though, the scientific method is not a mechanical process but a human activity that needs creative thinking in all its steps. Looking at the natural world is at the heart of the scientific method, because the results of observation and experiment serve not only as the foundations on which scientists build their ideas but also as the means by which these ideas are checked (Fig. 1-1).

1. **Formulating a problem** may mean no more than choosing a certain field to work in, but more often a scientist has in mind some specific idea he or she wishes to investigate. In many cases formulating a problem and interpreting the data overlap. The scientist has a speculation, perhaps only a hunch, perhaps a fully developed concept, about some aspect of nature but cannot come to a definite conclusion without further study.
2. **Observation and experiment** are carried out with great care. Facts about nature are the building blocks of science and the ultimate test of its results. This insistence on accurate, objective data is what sets science apart from other modes of intellectual endeavor.
3. **Interpretation** may lead to a general rule or **law** to which the data seem to conform. Or it may be a **theory,** which is a more ambitious attempt to account for what has been found in terms of how nature works. In any case, the interpretation must be able to cover new data obtained under different circumstances. As put forward orginally, a scientific interpretation is usually called a **hypothesis.**
4. **Testing the interpretation** involves making new observations or performing new experiments to see whether the interpretation correctly

Finding the Royal Road

Hermann von Helmholtz, a German physicist and biologist of a century ago, summed up his experience of scientific research in these words: "I would compare myself to a mountain climber who, not knowing the way, ascends slowly and toilsomely and is often compelled to retrace his steps because his progress is blocked; who, sometimes by reasoning and sometimes by accident, hits upon signs of a fresh path, which leads him a little farther; and who, finally, when he has reached his goal, discovers to his annoyance a royal road which he might have followed if he had been clever enough to find the right starting point at the beginning."

Experiment Is the Test

A master of several sciences, Michael Faraday, who lived a century and a half ago, is best remembered for his discoveries in electricity and magnetism (see biography in Sec. 6.18). This statement appears in the entry for March 19, 1849 in his laboratory notebook: "Nothing is too wonderful to be true if it be consistent with the laws of nature, and . . . experiment is the best test of such consistency."

predicts the results. If the results agree with the predictions, the scientist is clearly on the right track. The new data may well lead to refinements of the original idea, which in turn must be checked, and so on indefinitely.

The Laws of Nature The laws of a country tell its citizens how they are supposed to behave. Different countries have different laws, and even in one country laws are changed from time to time. Furthermore, though he or she may be caught and punished for doing so, anybody can break any law at any time.

The laws of nature are different. Everything in the universe, from atoms to galaxies of stars, behaves in certain regular ways, and these regularities are the laws of nature. To be considered a law of nature, a given regularity must hold everywhere at all times within its range of applicability.

The laws of nature are worth knowing for two reasons apart from satisfying our curiosity about how the universe works. First, we can use them to predict phenomena not yet discovered. Thus Newton's law of gravity was applied over a century ago to apparent irregularities in the motion of the planet Uranus, then the farthest known planet from the sun. Calculations not only showed that another, more distant planet should exist but also indicated where in the sky to look for it. Astronomers who looked there found a new planet, which was named Neptune.

Second, the laws of nature can give us an idea of what goes on in places we cannot examine directly. We will never visit the sun's interior (much too hot) or the interior of an atom (much too small), but we know a lot about both regions. The evidence is indirect but persuasive.

Theories and Models A **law** tells us *what;* a **theory** tells us *why.* A theory explains why certain events take place and, if they obey a particular law, how that law originates in terms of broader considerations. For example, Albert Einstein's general theory of relativity interprets gravity as a distortion in the properties of space and time around a body of matter. This theory not only accounts for Newton's law of gravity but goes

Theory

The word *theory* has two meanings. In one sense, a theory is a suggestion, a proposal, what in the scientific method is called a hypothesis: "My theory is that the dog ran away because the children pulled her tail." In science, though, a theory refers to a fully developed logical structure based on general principles that ties together a variety of observations and experimental findings and permits as-yet-unknown phenomena and connections to be predicted. A theory may be more or less speculative when proposed, but the point is that it is a large-scale framework of ideas and relationships whose validity has nothing to do with the other meaning of the word.

People ignorant of science are sometimes confused about which meaning of theory is appropriate in a given case. For instance, believers in creationism, the unsupported notion that all living things simultaneously appeared on earth a few thousand years ago, scorn Darwin's theory of evolution (see Sec. 16.8) as "just a theory" despite the wealth of evidence in its favor and its bedrock position in modern biology. In fact, few aspects of our knowledge of the natural world are as solidly established as the theory of evolution.

further, including the prediction—later confirmed—that light should be affected by gravity.

As the French mathematician Henri Poincaré once remarked, "Science is built with facts just as a house is built with bricks, but a collection of facts is not a science any more than a pile of bricks is a house."

It may not be easy to get a firm intellectual grip on some aspect of nature. Therefore a **model**—a simplified version of reality—is often part of a hypothesis or theory. In developing the law of gravity, Newton considered the earth to be perfectly round, even though it is actually more like a grapefruit than like a billiard ball. Newton regarded the path of the earth around the sun as an oval called an **ellipse,** but the actual orbit has wiggles no ellipse ever had. By choosing a sphere as a model for the earth and an ellipse as a model for its orbit, Newton isolated the most important features of the earth and its path and used them to arrive at the law of gravity. If he had started with a more realistic model—a somewhat squashed earth moving somewhat irregularly around the sun—he probably would have made little progress. Once he had formulated the law of gravity, Newton was then able to explain how the spinning of the earth causes it to become distorted into the shape of a grapefruit and how the attractions of the other planets cause the earth's orbit to differ from a perfect ellipse.

1.2 Why Science Is Successful

Science Is a Living Body of Knowledge, Not a Set of Frozen Ideas

What has made science such a powerful tool for investigating nature is the constant testing and retesting of its findings. As a result, science is a living body of information and not a collection of dogmas. The laws and theories of science are not necessarily the final word on a subject: they are valid only as long as no contrary evidence comes to light. If such contrary evidence does turn up, the law or theory must be modified or even discarded. To rock the boat is part of the game; to overturn it is one way to win. Thus science is a self-correcting search for better understanding of the natural world, a search with no end in sight.

Scientists are open about the details of their work, so that others can follow their thinking and repeat their experiments and observations. Nothing is accepted on anybody's word alone, or because it is part of a religious or political doctrine. "Common sense" is not a valid argument, either. What counts are definite measurements and clear reasoning, not vague notions that vary from person to person.

The power of the scientific approach is shown not only by its success in understanding the natural world but also by the success of the technology based on science. It is hard to think of any aspect of life today untouched in some way by science. The synthetic clothing we wear, the medicines that lengthen our lives, the cars and airplanes we travel in, the telephone, radio, and television by which we communicate—all are ultimately the products of a certain way of thinking. Curiosity and imagination are part of that way of thinking, but the most important part is that nothing is ever taken for granted but is always subject to test and change.

Degrees of Doubt

Although in principle everything in science is open to question, in practice many ideas are not really in doubt. The earth is certainly round, for instance, and the planets certainly revolve around the sun. Even though the earth is not a perfect sphere and the planetary orbits are not perfect ellipses, the basic models will always be valid.

Other beliefs are less firm. An example is the current picture of the future of the universe. Quite convincing data suggest that the universe has been expanding since its start in a "big bang" about 13.7 billion years ago. What about the future? It seems likely from the latest measurements that the expansion will continue forever, but this conclusion is still tentative and is under active study by astronomers today.

Religion and Science In the past, scientists were sometimes punished for daring to make their own interpretations of what they saw. Galileo, the first modern scientist, was forced by the Roman Catholic Church in 1633 under threat of torture to deny that the earth moves about the sun. Even today, attempts are being made to compel the teaching of religious beliefs—for instance, the story of the Creation as given in the Bible—under the name of science. But "creation science" is a contradiction in terms. The essence of science is that its results are open to change in the light of new evidence, whereas the essence of creationism is that it is a fixed doctrine with no basis in observation. The scientific method has been the means of liberating the world from ignorance and superstition. To discard this method in favor of taking at face value every word in the Bible is to replace the inquiring mind with a closed mind.

Those who wish to believe that the entire universe came into being in 6 days a few thousand years ago are free to do so. What is not proper is for certain politicians (whom Galileo would recognize if he were alive today) to try to turn back the intellectual clock and compel such matters of faith to be taught in schools alongside or even in place of scientific concepts, such as evolution, that have abundant support in the world around us. To anyone with an open mind, the evidence that the universe and its inhabitants have developed over time and continue to do so is overwhelming, as we shall see in later chapters. Nothing stands still. The ongoing evolution of living things is central to biology; the ongoing evolution of the earth is central to geology; the ongoing evolution of the universe is central to astronomy.

Many people find religious beliefs important in their lives, but such beliefs are not part of science because they are matters of faith with ideas that are meant to be accepted without question. Skepticism, on the other hand, is at the heart of science. Science follows where evidence leads; religion has fixed principles. It is entirely possible—and indeed most religious people do this—to consult sacred texts for inspiration and guidance while accepting that observation and reason represent the path to another kind of understanding. But religion and science are not interchangeable because their routes and destinations are different—which means that science classrooms are not the place to teach religion. To mix the religious and the scientific ways of looking at the world is good for neither, particularly if compulsion is involved.

Intelligent Design The founders of the United States of America insisted on the separation of church and state, a separation that is part of the Constitution. What happens in countries with no such separation, in the past and in the present, testifies to the wisdom of the founders. In 1987 the U.S. Supreme Court ruled that teaching creationism in the public schools is illegal because it is a purely religious doctrine. In response, the believers in creationism changed its name to "intelligent design" (ID) without specifying who the designer was or how the design was put into effect. Their sole argument is that life is too complex and diverse to be explained by evolution, when in fact this is precisely what evolution does with overwhelming success (see Sec. 16.8).

The supporters of ID demand that it be taught as a science—despite ID's not being based on evidence or subject to test, as all science must be—and call for "balance" in its presentation. But putting evolution and

Intelligent Design in Court

In 2004 the school board of Dover, Pennsylvania, required that ID be introduced in the science classes of its high school. A lawsuit was then filed by 11 alarmed parents who accused the board of violating the First Amendment to the U.S. Constitution, which has been ruled to prohibit public officials from pursuing religious agendas in their work. After a 6-week trial, federal judge John E. Jones III agreed that the board's action was indeed illegal. "We conclude that the religious nature of intelligent design would be readily apparent to an objective observer, adult or child. The writings of leading ID proponents reveal that the designer postulated by their arguments is the God of Christianity." Here are some further excerpts from his 137-page opinion [U.S. District Court for the Middle District of Pennsylvania, *Kitzmiller, et al. v. Dover Area School District, et al.*, Case no. O4cv2688]:

"In making this determination, we have addressed the seminal question of whether ID is science. We have concluded that it is not, and moreover that ID cannot uncouple itself from its creationist, and thus religious, antecedents.

"Both defendants and many of the leading proponents of ID make a bedrock assumption which is utterly false. Their presupposition is that evolutionary theory is antithetical to a belief in the existence of a supreme being and to religion in general. Repeatedly in this trial, plaintiffs' scientific experts testified that the theory of evolution represents good science, is overwhelmingly accepted by the scientific community, and that it in no way conflicts with, nor does it deny, the existence of a divine creator.

"To be sure, Darwin's theory of evolution is imperfect. However, the fact that a scientific theory cannot yet render an explanation on every point should not be used as a pretext to thrust an untestable alternative hypothesis grounded in religion into the science classroom or to misrepresent well-established scientific propositions.

"The citizens of the Dover area were poorly served by the members of the board who voted for the ID policy. It is ironic that several of these individuals, who so staunchly and proudly touted their religious convictions in public, would time and again lie to cover their tracks and disguise the real purpose behind the ID policy. . . .

"The breathtaking inanity of the board's decision is evident when considered against the factual backdrop which has now been fully revealed through this trial. The students, parents, and teachers of the Dover Area School District deserved better than to be dragged into this legal maelstrom, with its resulting utter waste of monetary and personal resources."

Dover is not the only place in the United States where science teaching is in danger. In order to impose their personal religious beliefs, such as ID, on its school system, the members of the Kansas State Board of Education recently rewrote its official definition of science. The former definition called science "the human activity of seeking natural explanations for what we observe in the world around us." The new definition, called "lunacy" by *The New York Times*, omits the word "natural." Supernatural explanations were to be acceptable in Kansas science classrooms. Did the parents of Kansas really want their children to believe that "what we observe in the world around us" is the work of witches and warlocks, ghosts and goblins, pixies and the Tooth Fairy? They did not: a later Board rebottled the genie of unreason.

ID on the same footing is absurd because it suggests that they have the same intellectual status. Evolution opens windows to further knowledge; ID shuts them. It would be like equating an encyclopedia with a book whose title is No! and whose pages are blank. To teach or even debate ID in a science classroom is to give it a respectability it is not entitled to.

Advocates of ID assert that evolution is an atheistic concept. Yet Charles Darwin, the father of evolution, was a Christian. The United Methodist Church, the Episcopalian Church, the Lutheran World Federation, Popes Pius XII and John Paul II, and the Central Conference of American Rabbis have all stated that evolution does not conflict with religious belief. In 2005 Cardinal Paul Poupard, head of the Roman Catholic Church's Pontifical Council for Culture, said, "we . . . know the dangers of a religion that severs its links with reason and becomes prey to fundamentalism. The faithful have the obligation to listen to that which secular modern science has to offer."

The Solar System

Each day the sun rises in the east, sweeps across the sky, and sets in the west. The moon, planets, and most stars do the same. These heavenly bodies also move relative to one another, though more slowly.

There are two ways to explain the general east-to-west motion. The most obvious is that the earth is stationary and all that we see in the sky revolves around it. The other possibility is that the earth itself turns once a day, so that the heavenly bodies only appear to circle it. How the second alternative came to be seen as correct and how this finding led to the discovery of the law of gravity are important chapters in the history of the scientific method.

1.3 A Survey of the Sky

Everything Seems to Circle the North Star

One star in the northern sky seems barely to move at all. This is the North Star, or **Polaris,** long used as a guide by travelers because of its nearly unchanging position. Stars near Polaris do not rise or set but instead move around it in circles (Fig. 1-2). These circles carry the stars under Polaris from west to east and over it from east to west. Farther from Polaris the circles get larger and larger, until eventually they dip below the horizon. Sun, moon, and stars rise and set because their circles lie partly below the horizon. Thus, to an observer north of the equator, the whole sky appears to revolve once a day about this otherwise ordinary star.

Why does Polaris occupy such a central position? The earth rotates once a day on its axis, and Polaris happens by chance to lie almost

Figure 1-2 Time exposure of stars in the northern sky. The trail of Polaris is the bright arc slightly to the left of the center of the larger arcs. The dome in the foreground houses one of the many telescopes on the summit of Mauna Kea, Hawaii. This location is favored by astronomers because observing conditions are excellent there. The lights of cars that moved during the exposure are responsible for the yellow traces near the dome.

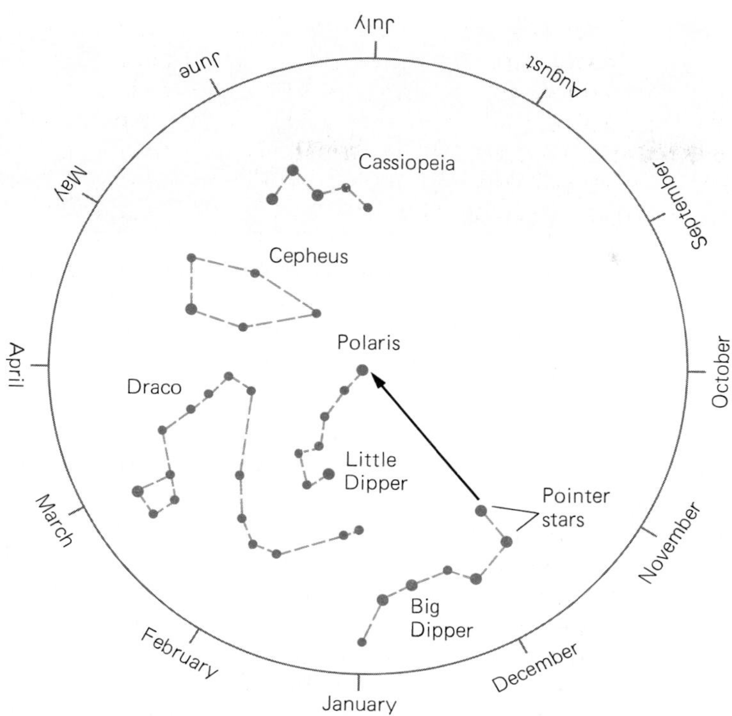

Figure 1-3 Constellations near Polaris as they appear in the early evening to an observer who faces north with the figure turned so that the current month is at the bottom. Polaris is located on an imaginary line drawn through the two "pointer" stars at the end of the bowl of the Big Dipper. The brighter stars are shown larger in size.

directly over the North Pole. As the earth turns, everything else around it seems to be moving. Except for their circular motion around Polaris, the stars appear fixed in their positions with respect to one another. Stars of the Big Dipper move halfway around Polaris between every sunset and sunrise, but the shape of the Dipper itself remains unaltered. (Actually, as discussed later, the stars *do* change their relative positions, but the stars are so far away that these changes are not easy to detect.)

Easily recognized groups of stars, like those that form the Big Dipper, are called **constellations** (Fig. 1-3). Near the Big Dipper is the less conspicuous Little Dipper with Polaris at the end of its handle. On the other side of Polaris from the Big Dipper are Cepheus and the W-shaped Cassiopeia, named for an ancient king and queen of Ethiopia. Next to Cepheus is Draco, which means dragon.

Elsewhere in the sky are dozens of other constellations that represent animals, heroes, and beautiful women. An especially easy one to recognize on winter evenings is Orion, the mighty hunter of legend. Orion has four stars, three of them quite bright, at the corners of a warped rectangle with a belt of three stars in line across its middle (Fig. 1-4). Except for the Dippers, a lot of imagination is needed to connect a given star pattern with its corresponding figure, but the constellations nevertheless are useful as convenient labels for regions of the sky.

Figure 1-4 Orion, the mighty hunter. Betelgeuse is a bright red star, and Bellatrix and Rigel are bright blue stars. Stars that seem near one another in the sky may actually be far apart in space. The three stars in Orion's belt, for instance, are actually at very different distances from us.

Sun, Moon, and Planets In their daily east-west crossing of the sky, the sun and moon move more slowly than the stars and so appear to drift eastward relative to the constellations. In the same way, a person on a

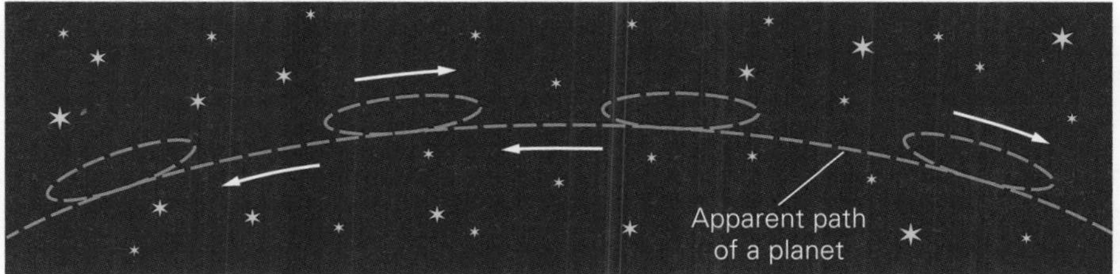

Figure 1-5 Apparent path of a planet in the sky looking south from the northern hemisphere of the earth. The planets seem to move eastward relative to the stars most of the time, but at intervals they reverse their motion and briefly move westward.

train traveling west who walks toward the rear car is moving east relative to the train although still moving west relative to the ground. In the sky, the apparent eastward motion is most easily observed for the moon. If the moon is seen near a bright star on one evening, by the next evening it will be some distance east of that star, and on later nights it will be farther and farther to the east. In about 4 weeks the moon drifts eastward completely around the sky and returns to its starting point.

The sun's relative motion is less easy to follow because we cannot observe directly which stars it is near. But if we note which constellations appear where the sun has just set, we can estimate the sun's location among the stars and follow it from day to day. We find that the sun drifts eastward more slowly than the moon, so slowly that the day-to-day change is scarcely noticeable. Because of the sun's motion each constellation appears to rise about 4 min earlier each night, and so, after a few weeks or months, the appearance of the night sky becomes quite different from what it was when we started our observations.

By the time the sun has migrated eastward completely around the sky, a year has gone by. In fact, the year is defined as the time needed for the sun to make such an apparent circuit of the stars.

Five other celestial objects visible to the naked eye also shift their positions with respect to the stars. These objects, which themselves resemble stars, are **planets** (Greek for "wanderer") and are named for the Roman gods Mercury, Venus, Mars, Jupiter, and Saturn. Like the sun and moon, the planets shift their positions so slowly that their day-to-day motion is hard to detect. Unlike the sun, they move in complex paths. In general, each planet drifts eastward among the stars, but its relative speed varies and at times the planet even reverses its relative direction to head westward briefly. Thus the path of a planet appears to consist of loops that recur regularly, as in Fig. 1-5.

1.4 The Ptolemaic System

The Earth as the Center of the Universe

Although the philosophers of ancient Greece knew that the apparent daily rotation of the sky could be explained by a rotation of the earth, most of them preferred to regard the earth as stationary. The scheme most widely accepted was originally the work of Hipparchus. Ptolemy of Alexandria

(Fig. 1-6), later included Hipparchus's ideas into his *Almagest,* a survey of astronomy that was to be the standard reference on the subject for over a thousand years. This model of the universe became known as the **ptolemaic system.**

The model was intricate and ingenious (Fig. 1-7). Our earth stands at the center, motionless, with everything else in the universe moving about it either in circles or in combinations of circles. (To the Greeks, the circle was the only "perfect" curve, hence the only possible path for a celestial object.) The fixed stars are embedded in a huge crystal sphere that makes a little more than a complete turn around the earth each day. Inside the crystal sphere is the sun, which moves around the earth exactly once a day. The difference in speed between sun and stars is just enough so that the sun appears to move eastward past the stars, returning to a given point among them once a year. Near the earth in a small orbit is the moon, revolving more slowly than the sun. The planets Venus and Mercury come between moon and sun, the other planets between sun and stars.

To account for irregularities in the motions of the planets, Ptolemy imagined that each planet moves in a small circle about a point that in turn follows a large circle about the earth. By a combination of these circular motions a planet travels in a series of loops. Since we observe

Figure 1-6 Ptolemy (A.D. 100–170).

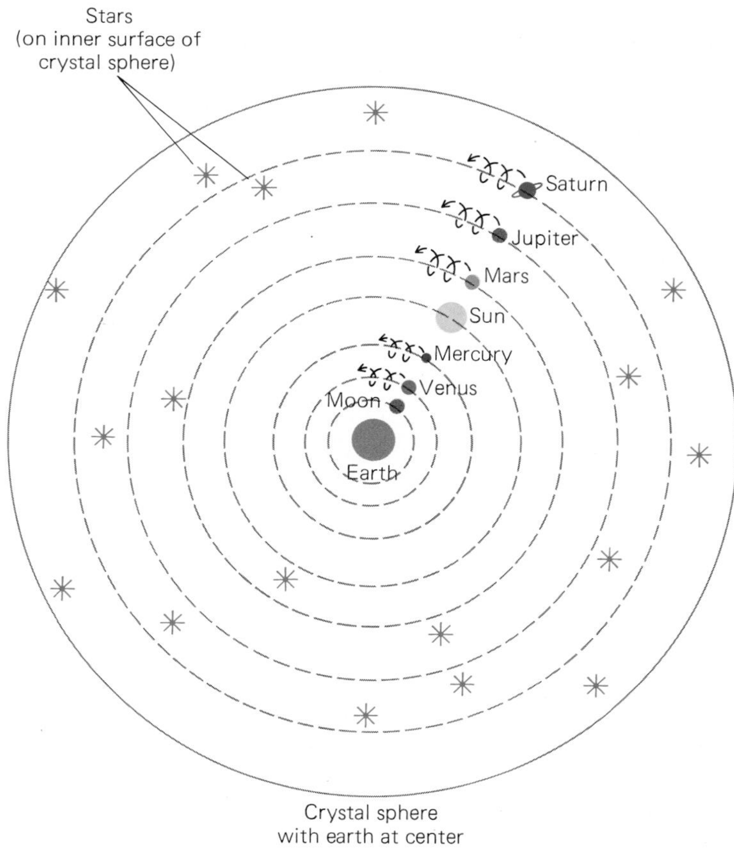

Figure 1-7 The ptolemaic system, showing the assumed arrangement of the members of the solar system within the celestial sphere. Each planet is supposed to travel around the earth in a series of loops, while the orbits of the sun and moon are circular. Only the planets known in Ptolemy's time are shown. The stars are all supposed to be at the same distance from the earth.

these loops edgewise, it appears to us as if the planets move with variable speeds and sometimes even reverse their directions of motion in the sky.

From observations made by himself and by others, Ptolemy calculated the speed of each celestial object in its assumed orbit. Using these speeds he could then figure out the location in the sky of any object at any time, past or future. These calculated positions checked fairly well, though not perfectly, with positions that had been recorded centuries earlier, and the predictions also agreed at first with observations made in later years. So Ptolemy's system fulfilled all the requirements of a scientific theory: it was based on observation, it accounted for the celestial motions known in his time, and it made predictions that could be tested in the future.

The Temple of the Sun

Here is how Copernicus summed up his picture of the solar system: "Of the moving bodies first comes Saturn, who completes his circuit in 30 years. After him Jupiter, moving in a twelve-year revolution. Then Mars, who revolves biennially. Fourth in order an annual cycle takes place, in which we have said is contained the earth, with the lunar orbit as an epicycle, that is, with the moon moving in a circle around the earth. In the fifth place Venus is carried around in 9 months. Then Mercury holds the sixth place, circulating in the space of 80 days. In the middle of all dwells the Sun. Who indeed in this most beautiful temple would place the torch in any other or better place than one whence it can illuminate the whole at the same time?"

1.5 The Copernican System

A Spinning Earth That Circles the Sun

By the sixteenth century it had become clear that something was seriously wrong with the ptolemaic model. The planets were simply not in the positions in the sky predicted for them. The errors could be removed in two ways: either the ptolemaic system could be made still more complicated, or it could be replaced by a different model of the universe.

Nicolaus Copernicus, a versatile and energetic Pole of the early sixteenth century, chose the second approach. Let us consider the earth, said Copernicus, as one of the planets, a sphere rotating once a day on its axis. Let us imagine that all the planets, including the earth, circle the sun (Fig. 1-8), that the moon circles the earth, and that the stars are all far away. In this model, it is the earth's rotation that explains the daily rising and setting of celestial objects, not the motions of these objects.

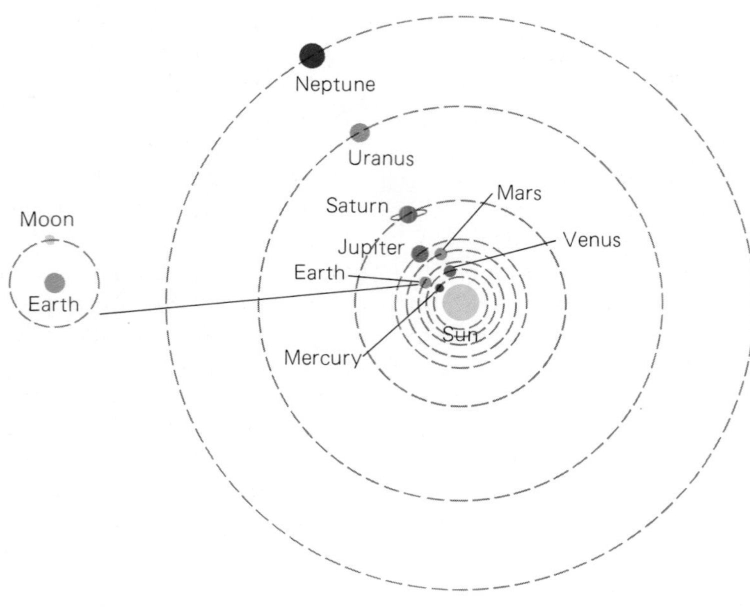

Figure 1-8 The copernican system. The planets, including the earth, are supposed to travel around the sun in circular orbits. The earth rotates daily on its axis, the moon revolves around the earth, and the stars are far away. All planets in the solar system are shown here. There are also a number of dwarf planets, such as Pluto; see Sec. 17.11. The actual orbits are ellipses and are not spaced as shown here, though they do lie in approximately the same plane.

BIOGRAPHY Nicolaus Copernicus (1473–1543)

When Columbus made his first voyage to the New World Copernicus was a student in his native Poland. In the years that followed intellectual as well as geographical horizons receded before eager explorers. In 1496 Copernicus went to Italy to learn medicine, theology, and astronomy. Italy was then an exciting place to be, a place of business expansion and conflicts between rival cities, great fortunes and corrupt governments, brilliant thinkers and inspired artists such as Leonardo da Vinci and Michelangelo. After 10 years in Italy Copernicus returned to Poland where he practiced medicine, served as a canon in the cathedral of which his uncle was the bishop, and became involved in currency reform, but much of his time was devoted to developing the idea that the planets move around the sun rather than around the earth. The idea was not new—the ancient Greeks were aware of it—but Copernicus went further and worked out the planetary orbits and speeds in detail. Although a summary of his results had been circulated in manuscript form earlier, not until a few weeks before his death was Copernicus's *De Revolutionibus Orbium Coelestium* published in book form. Today *De Revolutionibus* is recognized as one of the foundation stones of modern science, but soon after its appearance it was condemned by the Catholic Church (which did not lift its ban until 1835) and had little impact on astronomy until Kepler further developed its concepts a half century later.

The apparent shifting of the sun among the stars is due to the earth's motion in its orbit. As the earth swings around the sun, we see the sun changing its position against the background of the stars. The moon's gradual eastward drift is mainly due to its orbital motion. Apparently irregular movements of the planets are really just combinations of their motions with our own shifts of position as the earth moves.

The **copernican system** offended both Protestant and Catholic religious leaders, who did not want to see the earth taken from its place at the hub of the universe. The publication of Copernicus's manuscript began a long and bitter argument. To us, growing up with the knowledge that the earth moves, it seems odd that this straightforward idea was so long and so violently opposed. But in the sixteenth century good arguments were available to both sides.

Consider, said supporters of Ptolemy, how fast the earth's surface must move to complete a full turn every 24 h. Would not everything loose be flung into space by this whirling ball, just as mud is thrown from the rim of a carriage wheel? And would not such dizzying speeds produce a great wind to blow down buildings, trees, plants? The earth does spin rapidly, replied the followers of Copernicus, but the effects are counterbalanced by whatever force it is that holds our feet to the ground. Besides, if the speed of the earth's rotation is a problem, how much more of a problem would be the tremendous speeds of the sun, stars, and planets if they revolve, as Ptolemy thought, once a day around a fixed earth?

Leap Years

A day is the time needed for the earth to make a complete turn on its axis, and a year is the time it needs to complete an orbit around the sun. The length of the year is slightly less than 365 days and 6 hours, so adding an extra day to February every 4 years (namely those years evenly divisible by 4, which are accordingly called **leap years**) enables the seasons to recur at very nearly the same dates every year. The remaining discrepancy adds up to a full day too much every 128 years. To take care of most of this discrepancy, century years not divisible by 400 will not be leap years; thus 2000 was a leap year but 2100 will not be one.

1.6 Kepler's Laws

How the Planets Actually Move

Fortunately, improvements in astronomical measurements—the first since the time of the Greeks—were not long in coming. Tycho Brahe (1546–1601), an astronomer working for the Danish king, built an observatory on the island of Hven near Copenhagen in which the instruments were remarkably precise (Fig. 1-9). With the help of these instruments, Tycho, blessed with exceptional eyesight and patience, made thousands of measurements, a labor that occupied much of his life. Even without the telescope, which had not yet been invented, Tycho's observatory was able to determine celestial angles to better than $\frac{1}{100}$ of a degree.

At his death in 1601, Tycho left behind his own somewhat peculiar model of the solar system, a body of superb data extending over many years, and an assistant named Johannes Kepler. Kepler regarded the copernican scheme "with incredible and ravishing delight," in his words, and fully expected that Tycho's improved figures would prove Copernicus correct once and for all. But this was not the case; after 4 years of work on the orbit of Mars alone, Kepler could not get Tycho's data to fit any of the models of the solar system that had by then been proposed. If the facts do not agree with the theory, then the scientific method requires that the theory, no matter how attractive, must be discarded. Kepler then began to look for a new cosmic design that would fit Tycho's observations better.

Figure 1-9 A 1598 portrait of Tycho Brahe in his observatory. The man at the right is determining the position of a celestial body by shifting a sighting vane along a giant protractor until the body is visible through the aperture at upper left. There were four of each kind of instrument in the observatory, which were used simultaneously for reliable measurements.

BIOGRAPHY Johannes Kepler (1571–1630)

As a child, Kepler, who was born in Germany, was much impressed by seeing a comet and a total eclipse of the moon. In college, where astronomy was his worst subject, Kepler concentrated on theology, but his first job was as a teacher of mathematics and science in Graz, Austria. There he pondered the copernican system and concluded that the sun must exert a force (which he later thought was magnetic) on the planets to keep them in their orbits. Kepler also devised a geometrical scheme to account for the spacing of the planetary orbits and put all his ideas into a book called *The Cosmic Mystery.* Tycho Brahe, the Danish astronomer, read the book and took Kepler on as an assistant in his new observatory in Prague in what was then Bohemia. Upon Brahe's death (the result of drinking too much at a party given by the Emperor of Bohemia), Kepler replaced him at the observatory and gained access to all of Brahe's data, the most complete and accurate set then in existence.

Kepler felt that the copernican model of the solar system was not only capable of better agreement with the data than had yet been achieved but also contained within it yet-undiscovered regularities. Many years of labor resulted in three laws of planetary motion that fulfilled Kepler's vision and were to bear their ultimate fruit in Newton's law of gravity. Kepler also found time to prepare new tables of planetary positions, to explain how telescopes produce magnified images, to father 13 children, and to prepare horoscopes for the Emperor of Bohemia, the main reason for his employment (as it had been for Brahe). In 1620 Kepler's mother was accused of being a witch, but he was able to get her acquitted.

The First Law After considering every possibility, which meant years of drudgery in making calculations by hand, Kepler found that circular orbits for the planets were out of the question even when modified in various ways. He abandoned circular orbits reluctantly, for he was something of a mystic and believed, like Copernicus and the Greeks, that circles were the only fitting type of path for celestial bodies. Kepler then examined other geometrical figures, and here he found the key to the puzzle (Fig. 1-10). According to **Kepler's first law:**

> The paths of the planets around the sun are ellipses with the sun at one focus.

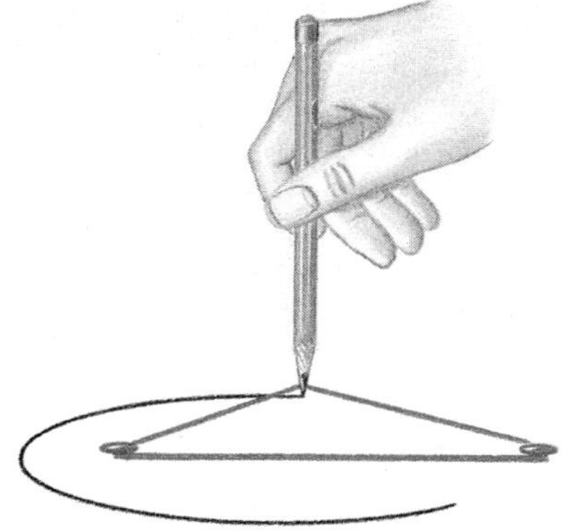

Figure 1-10 To draw an ellipse, place a loop of string over two tacks a short distance apart. Then move the pencil as shown, keeping the string taut. By varying the length of the string, ellipses of different shapes can be drawn. The points in an ellipse corresponding to the positions of the tacks are called **focuses;** the orbits of the planets are ellipses with the sun at one focus, which is Kepler's first law.

The Second Law Even this crucial discovery was not enough, as Kepler realized, to establish the courses of the planets through the sky. What was needed next was a way to relate the speeds of the planets to their positions in their elliptical orbits. Kepler could not be sure a general relationship of this kind even existed, and he was overjoyed when he had figured out the answer, known today as **Kepler's second law:**

> A planet moves so that its radius vector sweeps out equal areas in equal times.

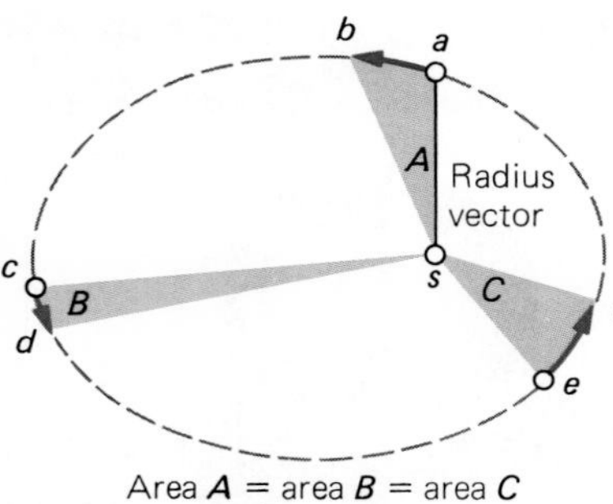

Figure 1-11 Kepler's second law. As a planet goes from *a* to *b* in its orbit, its radius vector (an imaginary line joining it with the sun) sweeps out the area *A*. In the same amount of time the planet can go from *c* to *d*, with its radius vector sweeping out the area *B*, or from *e* to *f*, with its radius vector sweeping out the area *C*. The three areas *A*, *B*, and *C* are equal.

The radius vector of a planet is an imaginary line between it and the sun. Thus in Fig. 1-11 each of the shaded areas is covered in the same period of time. This means that each planet travels faster when it is near the sun than when it is far away. The earth, for instance, has a speed of 30 km/s when it is nearest the sun and 29 km/s when it is farthest away, a difference of over 3 percent.

The Third Law A great achievement, but Kepler was not satisfied. He was obsessed with the idea of order and regularity in the universe, and spent 10 more years making calculations. It was already known that, the farther a planet is from the sun, the longer it takes to orbit the sun. **Kepler's third law** of planetary motion gives the exact relationship:

> The ratio between the square of the time needed by a planet to make a revolution around the sun and the cube of its average distance from the sun is the same for all the planets.

In equation form, this law states that

$$\frac{(\text{Period of planet})^2}{(\text{Average orbit radius})^3} = \text{same value for all the planets}$$

The period of a planet is the time needed for it to go once around the sun; in the case of the earth, the period is 1 year. Figure 1-12 illustrates Kepler's third law. Table 17-1 gives the values of the periods and average orbit radii for the planets.

At last the solar system could be interpreted in terms of simple motions. Planetary positions computed from Kepler's ellipses agreed not only with Tycho's data but also with observations made thousands of years earlier. Predictions could be made of positions of the planets in the future—accurate predictions this time, no longer approximations. Furthermore, Kepler's laws showed that the speed of a planet in different parts of its orbit was governed by a simple rule and that the speed was related to the size of the orbit.

Occam's Razor

In science, as a general rule, the simplest explanation for a phenomenon is most likely to be correct: less is more. This principle was first clearly expressed by the medieval philosopher William of Occam (or Ockham), who was born in England in 1280. In 1746 the French philosopher Etienne de Condillac called the principle **Occam's razor,** an elegant metaphor that suggests cutting away unnecessary complications to get at the heart of the matter. Copernicus was one of many successful users of Occam's razor. To be sure, as when shaving with an actual razor, it is possible to go too far; as the mathematician Alfred Whitehead said, "Seek simplicity, and distrust it."

1.7 Why Copernicus Was Right

Evidence Was Needed That Supported His Model While Contradicting Ptolemy's Model

It is often said that Kepler proved that Copernicus was "right" and that Ptolemy was "wrong." True enough, the copernican system, by having the planets move around the sun rather than around the earth, was simpler than the ptolemaic system. As modified by Kepler, the copernican system was also more accurate. However, the ptolemaic system could also be modified to be just as accurate, though in a very much more complicated way. Astronomers of the time squared themselves both with the practical needs of their profession and with the Church by using the copernican system for calculations while asserting the truth of the ptolemaic system.

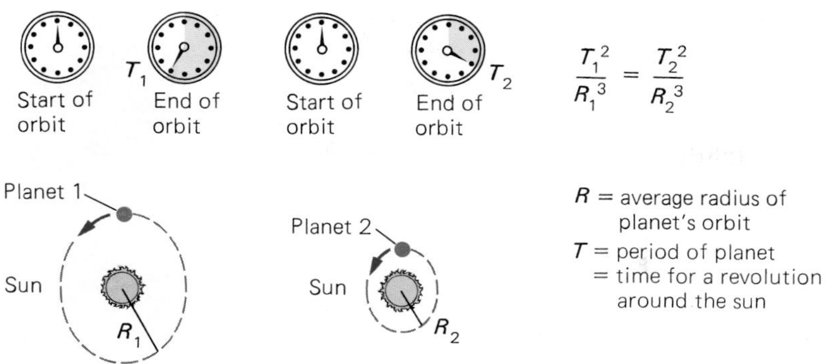

Figure 1-12 Kepler's third law states that the ratio T^2/R^3 is the same for all the planets.

Example 1.1

Kepler's laws should be obeyed by all satellite systems, not just the solar system. In the seventeenth century the French astronomer Cassini discovered four of Saturn's satellites (more have been discovered since). The names, periods, and orbit radii of these satellites are as follows:

Tethys	1.89 days	Rhea	4.52 days
	2.95×10^5 km		5.27×10^5 km
Dione	2.74 days	Iapetus	79.30 days
	3.77×10^5 km		35.60×10^5 km

Verify that Kepler's third law holds for these satellites.

Solution

What we must do is calculate the ratio T^2/R^3 for each satellite. The result for Tethys is

$$\frac{(1.89 \text{ days})^2}{(2.95 \times 10^5 \text{ km})^3} = 1.40 \times 10^{-16} \text{ days}^2/\text{km}^3$$

The ratio turns out to be the same for the other satellites as well, so we conclude that Kepler's third law holds for this satellite system. [Calculations that involve powers of ten are discussed in the Math Refresher at the end of this book. We note that $(10^5)^3 = 10^{3(5)} = 10^{15}$ and $1/10^{15} = 10^{-15}$.]

The copernican system is attractive because it accounts in a straightforward way for many aspects of what we see in the sky. However, only observations that contradict the ptolemaic system can prove it wrong. The copernican system is today considered correct because there is direct evidence of various kinds for the motions of the planets around the sun and for the rotation of the earth. An example of such evidence is the change in apparent position of nearby stars relative to the background of distant ones as the earth revolves around the sun (Fig. 1-13). Shifts of this kind are small because all stars are far away, but they have been found.

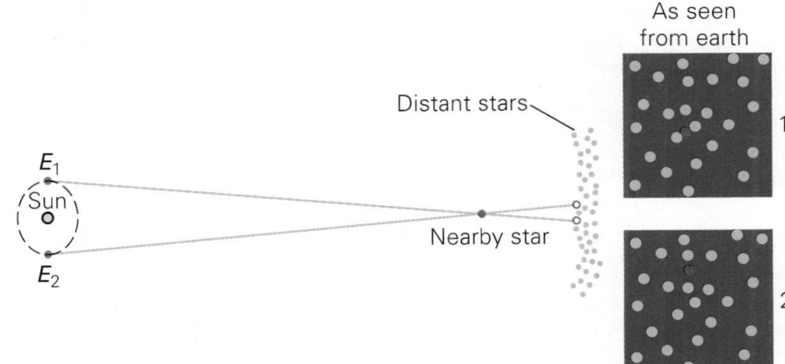

Figure 1-13 As a consequence of the earth's motion around the sun, nearby stars shift in apparent position relative to distant stars.

Astrology

To our ancestors of thousands of years ago, things happened in the world because gods caused them to happen. Famine and war, earthquake and eclipse—any conceivable catastrophe—all occurred under divine control. In time the chief gods were identified with the sun, the moon, and the five planets visible to the naked eye: Mercury, Venus, Mars, Jupiter, and Saturn. Early observers of the sky were primarily interested in finding links between celestial events and earthly ones, a study that became known as **astrology.**

Until only a few hundred years ago, astronomy was almost entirely in the service of astrology. The wealth of precise astronomical measurements that ancient civilizations compiled had as their purpose interpreting the ways of the gods.

Almost nobody today takes seriously the mythology of old. Although the basis of the connection has disappeared, however, some people still believe that the position in the sky of various celestial bodies at certain times controls the world we live in and our individual destinies as well.

It does not seem very gracious for contemporary science to dismiss astrology in view of the great debt astronomy owes its practitioners of long ago. However, it is hard to have confidence in a doctrine that, for all its internal consistency and often delightful notions, nevertheless lacks any basis in scientific theory or observation and has proved no more useful in predicting the future than a crystal ball.

Universal Gravitation

As we know from everyday experience, and as we shall learn in a more precise way in Chap. 2, a force is needed to cause something to move in a curved path (Fig. 1-14). The planets are no exception to this rule: a force of some kind must be acting to hold them in their orbits around the sun. Three centuries ago Isaac Newton had the inspired idea that this force must have the same character as the familiar force of gravity that pulls things to the earth's surface.

1.8 What Is Gravity?

A Fundamental Force

Perhaps, thought Newton, the moon revolves around the earth much as the ball in Fig. 1-14 revolves around the hand holding the string, with gravity taking the place of the pull of the string. In other words, perhaps the moon is a falling object, pulled to the earth just as we are, but moving so fast in its orbit that the earth's pull is just enough to keep the moon from flying off (Fig. 1-15). The earth and its sister planets might well be held in their orbits by a stronger gravitational pull from the sun. These notions turned out to be true, and Newton was able to show that his detailed theory of gravity accounts for Kepler's laws.

It is worth noting that Newton's discovery of the **law of gravity** depended on the copernican model of the solar system. "Common sense" tells us that the earth is the stationary center of the universe, and people were once severely punished for believing otherwise. Clearly the progress of our knowledge about the world we live in depends upon people, like Copernicus, who are able to look behind the screen of appearances that make up everyday life and who are willing to think for themselves.

Figure 1-14 An inward force is needed to keep an object moving in a curved path. The force here is provided by the string. If no force acts on it, a moving object will continue moving in a straight line at constant speed. (This is Newton's first law of motion and is discussed in Sec. 2.7.)

Gravity is a **fundamental force** in the sense that it cannot be explained in terms of any other force. Only four fundamental forces are known: gravitational, electromagnetic, weak, and strong. These forces are responsible for everything that happens in the universe. Gravitational forces act between all bodies everywhere and hold together planets, stars, and the giant groups of stars called galaxies. Electromagnetic forces, which (like gravity) are unlimited in range, act between electrically charged particles and govern the structures and behavior of atoms, molecules, solids, and liquids. When a bat hits a ball, the interaction between them can be traced to electromagnetic forces. The weak and strong forces have very short ranges and act inside atomic nuclei.

The Law of Gravity Is the Same Everywhere How can we be sure that Newton's law of gravity, which fits data on the solar system, also holds throughout the rest of the universe? The evidence for this generalization is indirect but persuasive. For instance, many double stars are known in which each member of the pair revolves around the other, which means some force holds them together. Throughout the universe stars occur in galaxies, and only gravity could keep them assembled in this way.

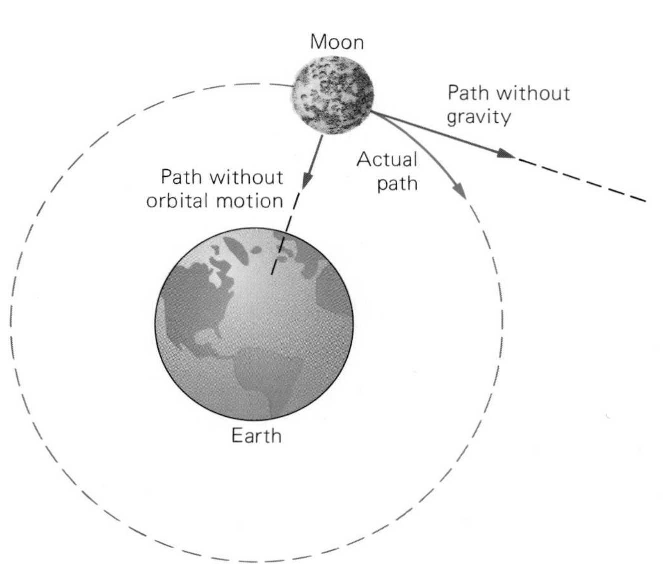

Figure 1-15 The gravitational pull of the earth on the moon causes the moon to move in an orbit around the earth. If the earth exerted no force on the moon, the moon would fly off into space. If the moon had no orbital motion, it would fall directly to the earth.

BIOGRAPHY Isaac Newton (1642–1727)

Although his mother wanted him to stay on the family farm in England, the young Newton showed a talent for science and went to Cambridge University for further study. An outbreak of plague led the university to close in 1665, the year Newton graduated, and he returned home for 18 months. In that period Newton came up with the binomial theorem of algebra; invented calculus, which gave science and engineering a new and powerful mathematical tool; discovered the law of gravity, thereby not only showing why the planets move as they do but also providing the key to understanding much else about the universe; and demonstrated that white light is a composite of light of all colors—an amazing list. As Newton later wrote, "In those days I was in the prime of my age for invention, and minded mathematics and philosophy more than at any time since."

When Cambridge University reopened, Newton went back and 2 years later became professor of mathematics there. He lived quietly and never married, carrying out experimental as well as theoretical research in many areas of physics; a reflecting telescope he made with his own hands was widely admired. Especially significant was Newton's development of the laws of motion (see Chap. 2), which showed exactly how force and motion are related, and his application of them to a variety of problems. Newton collected the results of his work on mechanics in the *Principia,* a scientific classic that was published in 1687. A later book, *Opticks,* summarized his efforts in this field. Newton also spent much time on chemistry, though here with little success.

After writing the *Principia,* Newton began to drift away from science. He became a member of Parliament in 1689 and later an official, eventually the Master, of the British Mint. At the Mint Newton helped reform the currency (one of Kepler's interests, too) and fought counterfeiters. Newton's spare time in his last 30 years was mainly spent in trying to date events in the Bible. He died at 85, a figure of honor whose stature remains great to this day.

But is the gravity that acts between stars the same as the gravity that acts in the solar system? Analyzing the light and radio waves that reach us from space shows that the matter in the rest of the universe is the same as the matter found on the earth. If we are to believe that the universe contains objects that do not obey Newton's law of gravity, we must have evidence for such a belief—and there is none. This line of thought may not seem as positive as we might prefer, but taken together with various theoretical arguments, it has convinced nearly all scientists that gravity is the same everywhere.

Figure 1-16 Astronauts in the Apollo 11 spacecraft saw this view of the earth as they orbited the moon, part of whose bleak landscape appears in the foreground. The earth is indeed round.

1.9 Why the Earth Is Round

The Big Squeeze

A sign of success of any scientific theory is its ability to account for previously mysterious findings. One such finding is the roundness of the earth (Fig. 1-16), which was known by the Greeks as long ago as the fifth century B.C. (Fig. 1-17). Early thinkers believed the earth was round because a sphere is the only "perfect" shape, a vague idea that actually explains nothing. In fact, the earth is round because gravity squeezes it into this shape.

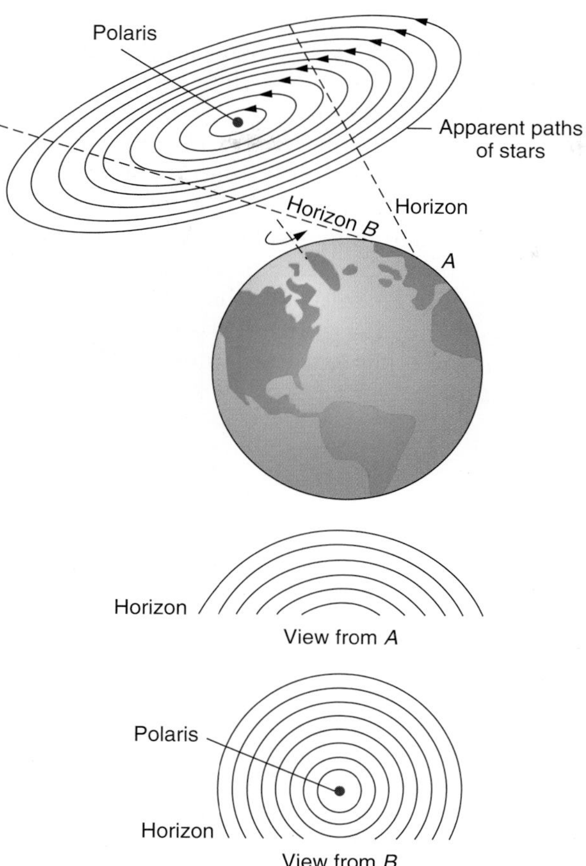

Figure 1-17 In the distant past evidence for the spherical shape of the earth came from travelers who found that, when they went north, more stars stayed above the horizon all night, and that, when they went south, additional stars became visible. Eratosthenes (276–194 B.C.) determined the earth's size with remarkable accuracy by comparing the length of the sun's shadow at noon on the same day in two places on the same north-south line.

As shown in Fig. 1-18, if any part of the earth were to stick out very much, the gravitational attraction of the rest of the earth would pull downward on the projection. The material underneath would then flow out sideways until the projection became level or nearly so. The downward forces around the rim of a deep hole would similarly cause the surrounding material to flow into it. The same argument applies to the moon, the sun, and the stars.

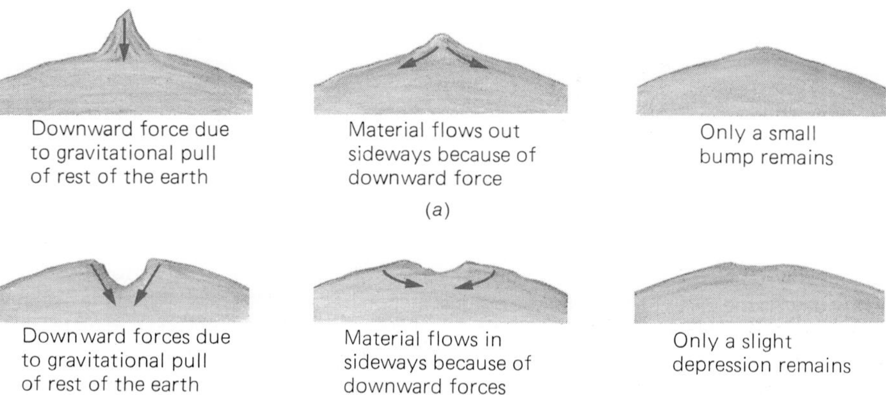

Figure 1-18 Gravity forces the earth to be round. *(a)* How a large bump would be pulled down. *(b)* How a large hole would be filled in.

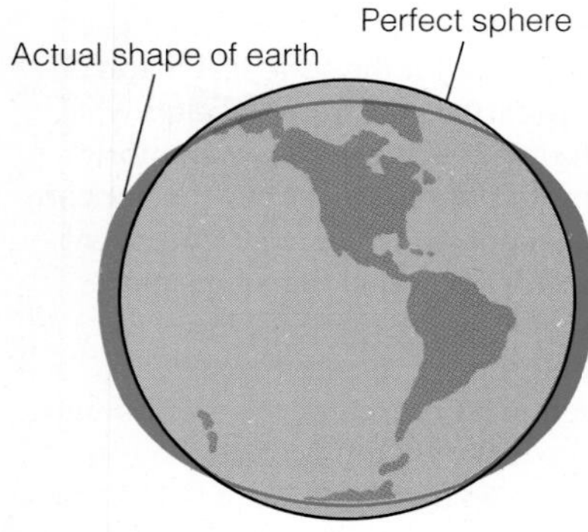

Figure 1-19 The influence of its rotation distorts the earth. The effect is greatly exaggerated in the figure; the equatorial diameter of the earth is actually only 43 km (27 mi) more than its polar diameter.

Such irregularities as mountains and ocean basins are on a very small scale compared with the earth's size. The total range from the Pacific depths to the summit of Everest is less than 20 km, not much compared with the earth's radius of 6400 km.

The earth is not a perfect sphere. The reason was apparent to Newton: since the earth is spinning rapidly, inertia causes the equatorial portion to swing outward, just as a ball on a string does when it is whirled around. As a result the earth bulges slightly at the equator and is slightly flattened at the poles, much like a grapefruit. The total distortion is not great, for the earth is only 43 km wider than it is high (Fig. 1-19). Venus, whose "day" is 243 of our days, turns so slowly that it has almost no distortion. Saturn, at the other extreme, spins so rapidly that it is almost 10 percent out of round.

1.10 The Tides

Up and Down Twice a Day

Those of us who live near an ocean know well the rhythm of the tides, the twice-daily rise and fall of water level. Usually the change in height is no more than a few meters, but in some regions—the Bay of Fundy in eastern Canada is one—the total range can be over 20 m. What causes the advance and retreat of the oceans on such a grand scale?

One factor is that the moon gravitationally attracts different parts of the earth to different extents. In Fig. 1-20 the moon's tug is strongest at *A*, which is closest, and weakest at *B*, which is farthest away. Also, the rotation of the moon around the earth is too simple a picture—what actually happens is that both bodies rotate around the center of mass (CM) of the earth-moon system. (Think of the earth and the moon as opposite ends of a dumbbell. The CM is the balance point of the dumbbell; it is inside the earth 4700 km from its center.)

As it wobbles around the CM, the solid earth is pulled away from the water at *B*, where the moon's tug is weakest, to leave the water there heaped up in a tidal bulge. At *A*, the greater tug of the moon dominates to cause a tidal bulge there as well. The bulges stay in place as the earth

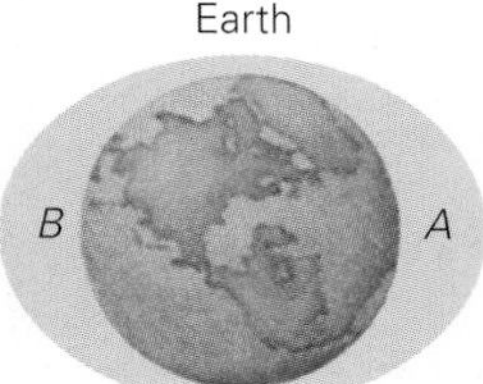

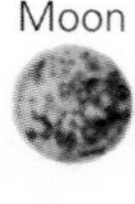

Figure 1-20 The origin of the tides. The moon's attraction for the waters of the earth is greatest at *A*, least at *B*. As the earth and moon rotate around the center of mass of the earth-moon system, which is located inside the earth, water is heaped up at *A* and *B*. The water bulges stay in place as the earth turns on its axis to produce two high and two low tides every day. As the earth turns under the bulges, friction between the oceans and the ocean floors slows down the earth's rotation. As a result the tidal bulges lag slightly behind the earth-moon line. The effect of tidal friction is thus to lengthen the day. The rate of increase is a mere 1 s per day in every 43,500 years, but it adds up. Measurements of the daily growth markings on fossil corals show that the day was only 22 h long 380 million years ago.

Figure 1-21 High and low water in the Bay of Fundy at Blacks Harbour in New Brunswick, Canada. Two tidal cycles occur daily.

revolves under them to produce two high tides and two low tides at a given place every day (Fig. 1-21).

There is more to the story. The sun also affects the waters of the earth, but to a smaller extent than the moon even though the gravitational tug of the sun exceeds that of the moon. The reason is that what is involved in the tides is the *difference* between the attractions on the near and far sides of the earth, and this difference is greater for the moon because it is closer to the earth than the sun. About twice a month—when the sun, moon, and earth are in a straight line—solar tides add to lunar tides to give the especially high (and low) **spring tides;** see Fig. 1-22. When the line between moon and earth is perpendicular to that between sun and earth, the tide-raising forces partly cancel to give **neap tides,** whose range is smaller than average.

1.11 The Discovery of Neptune

Another Triumph for the Law of Gravity

In Newton's time, as in Ptolemy's, only six planets were known: Mercury, Venus, Earth, Mars, Jupiter, and Saturn. In 1781 a seventh, Uranus, was identified. Measurements during the next few years enabled astronomers to work out details of the new planet's orbit and to predict its future positions in the sky. To make these predictions, not only the sun's attraction but also the smaller attractions of the nearby planets Jupiter and Saturn had to be considered. For 40 years, about half the time needed for Uranus to make one complete revolution around the sun, calculated positions of the planet agreed well with observed positions.

Then a discrepancy crept in. Little by little Uranus moved away from its predicted path among the stars. The calculations were checked and rechecked, but no mistake could be found. There were two possibilities: either the law of gravity, on which the calculations were based, was wrong, or else some unknown body was pulling Uranus away from its predicted path.

So firmly established was the law of gravitation that two young men, Urbain Leverrier in France and John Couch Adams in England,

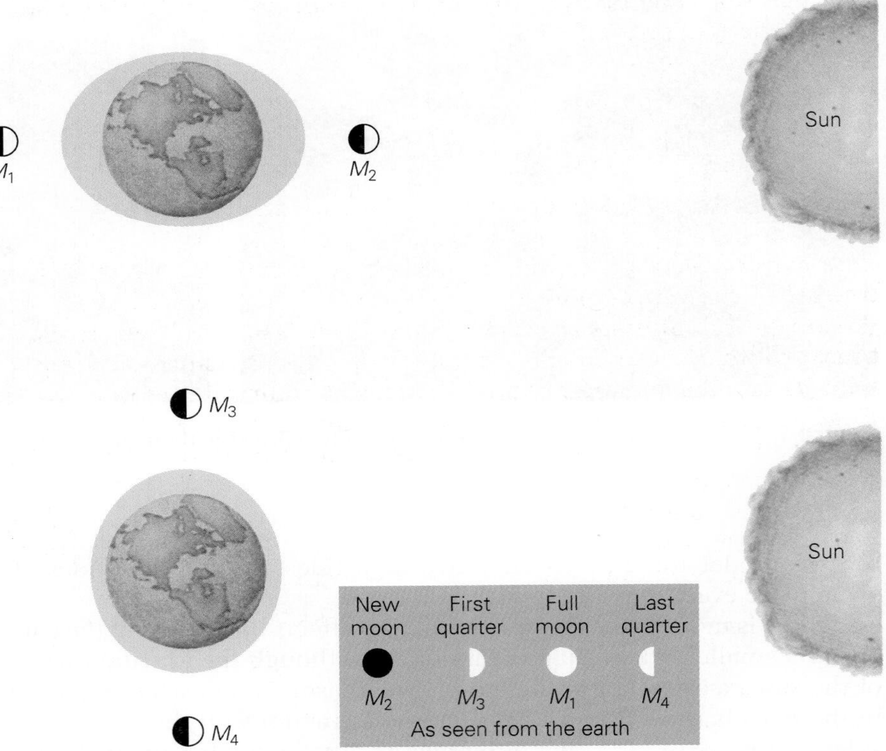

Figure 1-22 Variation of the tides. Spring tides are produced when the moon is at M_1 or M_2, neap tides when the moon is at M_3 or M_4. The range between high and low water is greatest for spring tides.

set themselves the task of calculating the orbit of an unknown body that might be responsible for the discrepancies in Uranus's position. Adams sent a sketchy account of his studies to George Airy, England's Astronomer Royal. Because the calculations were incomplete, although later found to be correct as far as they went, Airy asked for further details. Adams (who later blamed habitual lateness and a dislike of writing) did not respond. A year later, in 1846, Leverrier, with no knowledge of Adams's work, went further and proposed an actual position in the sky where the new planet should be found. He sent his result to a German astronomer, Johan Gottlieb Galle, who turned his telescope to the part of the sky where the new planet should appear. Very close to the position predicted by Leverrier, Galle found a faint object, which had moved slightly by the following night. This was indeed the eighth member of the sun's family and was called Neptune. The theory of gravity had again successfully gone around the loop of the scientific method shown in Fig. 1-1.

How Many of What

When we say that the distance between Chicago and Minneapolis is 405 miles, what we are really doing is comparing this distance with a certain standard length called the mile. Standard quantities such as the mile are known as **units.** The result of every measurement thus has two parts. One is a number (405 for the Chicago-Minneapolis distance) to answer the question "How many?" The other is a unit (the mile in this case) to answer the question "Of what?"

1.12 The SI System

All Scientists Use These Units

The most widely used units today are those of the International System, abbreviated **SI** after its French name Système International d'Unités. Examples of SI units are the **meter** (m) for length, the **second** (s) for time, the **kilogram** (kg) for mass, the **joule** (J) for energy, and the **watt** (W) for power. SI units are used universally by scientists and in most of the world in everyday life as well. Although the British system of units, with its familiar foot and pound, remains in common use only in a few English-speaking countries, it is on the way out and eventually will be replaced by the SI. Since this is a book about science, only SI units will be used from here on.

The great advantage of SI units is that their subdivisions and multiples are in steps of 10, 100, 1000, and so on, in contrast to the irregularity of British units. In the case of lengths, for instance (Fig. 1-23),

$$1 \text{ meter (m)} = 100 \text{ centimeters (cm)}$$
$$1 \text{ kilometer (km)} = 1000 \text{ meters}$$

whereas

$$1 \text{ foot (ft)} = 12 \text{ inches (in.)}$$
$$1 \text{ mile (mi)} = 5280 \text{ feet}$$

Table 1-1 lists the most common subdivisions and multiples of SI units. Each is designated by a prefix according to the corresponding power of 10. (Powers of 10 are reviewed in the Math Refresher at the back of the book.)

Meter, Kilogram, Second

SI units are derived from the units of the older **metric system.** This system was introduced in France two centuries ago to replace the hodgepodge of traditional units, often different in different countries and even in different parts of the same country, that was making commerce and industry difficult.

The **meter,** the standard of length, was originally defined as one ten-millionth of the distance from the equator to the North Pole. The **gram,** the standard of mass, was defined as the mass of 1 cubic centimeter (cm^3) of water; 1 cm^3 is the volume of a cube 1 cm (0.01 m) on each edge, and 1 kilogram = 1000 grams. The meter and gram were new units. The ancient division of a day into 24 hours, an hour into 60 minutes, and a minute into 60 seconds was kept for the definition of the second as 1/(24)(60)(60) = 1/86,400 of a day.

As more and more precision became needed, these definitions were modified several times. Today the second is specified in terms of the microwave radiation given off under certain circumstances by one type of cesium atom, ^{133}Cs: 1 s equals the time needed for 9,192,631,770 cycles of this radiation to be emitted. The meter, which for convenience had become the distance between two scratches on a platinum-iridium bar kept at Sévres, France, is now the distance traveled in 1/299,792,458 s by light in a vacuum. There are approximately 3.28 feet in a meter.

The kilogram is the mass of a platinum-iridium cylinder 39 mm in diameter and 39 mm high at Sévres. Despite much effort, a unit of mass based on a physical property measurable anywhere has not proved practical as yet. As discussed in Sec. 2.10, mass and weight are not the same. The weight of a given mass is the force with which gravity attracts it to the earth; the weight of 1 kg is 2.2 pounds on the earth's surface and decreases with altitude (see Fig. 2-39).

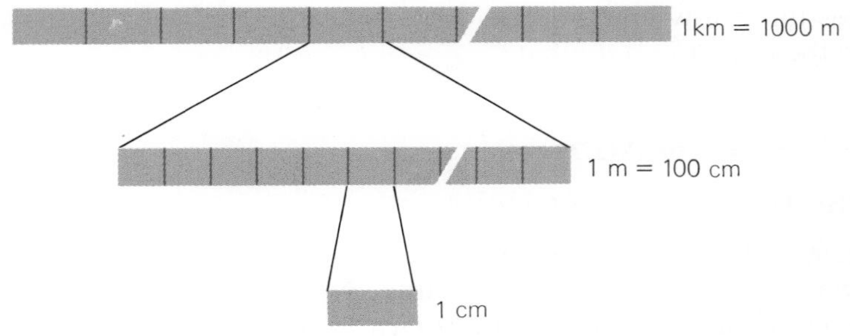

Figure 1-23 There are 1000 meters in a kilometer and 100 centimeters in a meter.

Table 1-1 Subdivisions and Multiples of SI Units (The Symbol μ Is the Greek Letter "mu")

Prefix	Power of 10	Abbreviation	Pronunciation	Common Name
Pico-	10^{-12}	p	pee' koe	Trillionth
Nano-	10^{-9}	n	nan' oe	Billionth
Micro-	10^{-6}	μ	my' kroe	Millionth
Milli-	10^{-3}	m	mil' i	Thousandth
Centi-	10^{-2}	c	sen' ti	Hundredth
Hecto-	10^{2}	h	hec' toe	Hundred
Kilo-	10^{3}	k	kil' oe	Thousand
Mega-	10^{6}	M	meg' a	Million
Giga-	10^{9}	G	ji' ga	Billion
Tera-	10^{12}	T	ter' a	Trillion

Example 1.2

How many nanometers are in a kilometer?

Solution

A nanometer is a billionth (10^{-9}) of a meter and a kilometer is a thousand (10^3) meters. Hence

$$\frac{\text{kilometer}}{\text{nanometer}} = \frac{10^3\ \text{m}}{10^{-9}\ \text{m}} = 10^{12}$$

There are 10^{12}—a trillion—nanometers in a kilometer. [We note from the Math Refresher at the end of this book that $10^n/10^m = 10^{n-m}$, so here $10^3/10^{-9} = 10^{3-(-9)} = 10^{3+9} = 10^{12}$.]

Table 1-2 contains conversion factors for changing a length expressed in one system to its equivalent in the other. (More conversion factors are given inside the back cover of this book.) We note from the table that there are about $2\frac{1}{2}$ centimeters in an inch, so a centimeter is roughly the width of a shirt button; a meter is a few inches longer than 3 feet; and a kilometer is nearly $\frac{2}{3}$ mile.

Significant Figures In Example 1-3 the distance expressed in centimeters is 8.78×10^4 cm. Does this mean that $d = 87{,}800$ cm exactly?

The answer is, not necessarily. We are only sure of the three digits 8, 7, and 8, which are the **significant figures** here. By writing $d = 8.78 \times 10^4$ cm we can see just how precisely the distance is being

Table 1-2 Conversion Factors for Length

Multiply a Length Expressed in	By	To Get the Same Length Expressed in
Centimeters	$0.394\ \frac{\text{in.}}{\text{cm}}$	Inches
Meters	$39.4\ \frac{\text{in.}}{\text{m}}$	Inches
Meters	$3.28\ \frac{\text{ft}}{\text{m}}$	Feet
Kilometers	$0.621\ \frac{\text{mi}}{\text{km}}$	Miles
Inches	$2.54\ \frac{\text{cm}}{\text{in.}}$	Centimeters
Feet	$30.5\ \frac{\text{cm}}{\text{ft}}$	Centimeters
Feet	$0.305\ \frac{\text{m}}{\text{ft}}$	Meters
Miles	$1.61\ \frac{\text{km}}{\text{mi}}$	Kilometers

expressed. If we needed less precision, we could round off the value of d to 8.8×10^4 cm. However, we could *not* write $d = 8.780 \times 10^4$ cm, which contains four significant figures, because the original distance was given as 878 m, which has only three such figures.

In part (b) of Example 1-3, the actual result of the calculation is

$$d = (0.878\ \text{km})(0.621\ \text{mi/km}) = 0.545238\ \text{mi}$$

Example 1.3

A few years ago a NASA official quoted a distance of 878 m to a reporter and added, "I don't know what this is in terms of kilometers or miles." Let us help him.

Solution

(a) Since $1\ \text{km} = 10^3\ \text{m} = 1000\ \text{m}$, the distance in kilometers is

$$d = \frac{878\ \text{m}}{1000\ \text{m/km}} = 0.878\ \text{km}$$

We note that

$$\frac{1}{\text{m/km}} = \frac{\text{km}}{\text{m}}$$

and therefore

$$\frac{\text{m}}{\text{m/km}} = \frac{(\cancel{\text{m}})(\text{km})}{\cancel{\text{m}}} = \text{km}$$

If instead we wanted this distance in centimeters, we would proceed in this way:

$$d = (878\ \text{m})(10^2\ \text{cm/m}) = 878 \times 10^2\ \text{cm} = (8.78 \times 10^2)(10^2)\ \text{cm}$$
$$= 8.78 \times 10^{2+2}\ \text{cm} = 8.78 \times 10^4\ \text{cm}$$

This is the usual way such a quantity would be expressed. The Math Refresher at the end of the book might come in handy here.

(b) From Table 1-2 the conversion factor we need is 0.621 mi/km, so

$$d = (0.878\ \cancel{\text{km}})\left(0.621\ \frac{\text{mi}}{\cancel{\text{km}}}\right) = 0.545\ \text{mi}$$

Because both the initial numbers had only three significant figures, the result can only have three also, so it was given as $d = 0.545$ mi.

When a calculation has several steps, it is a good idea to keep an extra digit in the intermediate steps. Then, at the end, the final result can be rounded off to the correct number of significant figures.

For simplicity, in this book zeros after the decimal point have usually been omitted from values given in problems. For instance, it should be assumed that when a length of 7 m is stated, what is really meant is 7.000 . . . m.

Important Terms and Ideas

The **scientific method** of studying nature has four steps: (1) formulating a problem; (2) observation and experiment; (3) interpreting the results; (4) testing the interpretation by further observation and experiment. When first proposed, a scientific interpretation is called a **hypothesis.** After thorough checking, it becomes a **law** if it states a regularity or relationship, or a **theory** if it uses general considerations to account for specific phenomena.

Polaris, the North Star, lies almost directly above the North Pole. A **constellation** is a group of stars that form a pattern in the sky. The **planets** are heavenly bodies that shift their positions regularly with respect to the stars.

In the **ptolemaic system,** the earth is stationary at the center of the universe. In the **copernican system,** the earth rotates on its axis and, with the other planets, revolves around the sun. Observational evidence supports the copernican system.

Kepler's laws are three regularities that the planets obey as they move around the sun.

Newton's **law of gravity** describes the attraction all bodies in the universe have for one another. The gravitational forces the sun exerts on the planets are what hold them in their orbits. Kepler's laws are explained by the law of gravity.

The **tides** are periodic rises and falls of sea level caused by differences in the gravitational pulls of the moon and sun. Water facing the moon is attracted to it more than the earth itself is, and the earth moves away from water on its far side. The corresponding effect of the sun is smaller than that of the moon and acts to increase or decrease tidal ranges, depending on the relative positions of the moon and sun.

To measure something means to compare it with a standard quantity of the same kind called a **unit.** The **SI** system of units is used everywhere by scientists and in most of the world in everyday life as well. The SI unit of length is the **meter** (m).

The **significant figures** in a number are its accurately known digits. When numbers are combined arithmetically, the result has as many significant figures as those in the number with the fewest of them.

Multiple Choice

1. The "scientific method" is
- **a.** a continuing process
- **b.** a way to arrive at ultimate truth
- **c.** a laboratory technique
- **d.** based on accepted laws and theories

2. A scientific law or theory is valid
- **a.** forever
- **b.** for a certain number of years, after which it is retested
- **c.** as long as a committee of scientists says so
- **d.** as long as it is not contradicted by new experimental findings

3. A hypothesis is
- **a.** a new scientific idea
- **b.** a scientific idea that has been confirmed by further experiment and observation
- **c.** a scientific idea that has been discarded because it disagrees with further experiment and observation
- **d.** a group of linked scientific ideas

4. The object in the sky that apparently moves least in the course of time is
- **a.** Polaris
- **b.** Venus
- **c.** the sun
- **d.** the moon

5. The stars in a constellation are
- **a.** about the same age
- **b.** about the same distance from the earth
- **c.** members of the solar system
- **d.** unrelated except for proximity in the sky as seen from the earth

6. Which of the following is no longer considered valid?
- **a.** the ptolemaic system
- **b.** the copernican system
- **c.** Kepler's laws of planetary motion
- **d.** Newton's law of gravity

7. A planet not visible to the naked eye is
- **a.** Mercury
- **b.** Saturn
- **c.** Neptune
- **d.** Jupiter

8. Arrange the following planets in the order of their distance from the sun:
   a. Mars
   b. Jupiter
   c. Uranus
   d. Saturn
9. The planet closest to the sun is
   a. earth
   b. Venus
   c. Mars
   d. Mercury
10. The length of the year is
    a. less than 365 days
    b. exactly 365 days
    c. more than 365 days
    d. any of the above, depending on the year
11. Kepler modified the copernican system by showing that the planetary orbits are
    a. ellipses
    b. circles
    c. combinations of circles forming looped orbits
    d. the same distance apart from one another
12. The speed of a planet in its elliptical orbit around the sun
    a. is constant
    b. is highest when the planet is closest to the sun
    c. is lowest when the planet is closest to the sun
    d. varies, but not with respect to the planet's distance from the sun
13. According to Kepler's third law, the time needed for a planet to complete an orbit around the sun
    a. is the same for all the planets
    b. depends on the planet's size
    c. depends on the planet's distance from the sun
    d. depends on how fast the planet spins on its axis
14. The law of gravity
    a. applies only to large bodies such as planets and stars
    b. accounts for all known forces
    c. holds only in the solar system
    d. holds everywhere in the universe
15. The earth bulges slightly at the equator and is flattened at the poles because
    a. it spins on its axis
    b. it revolves around the sun
    c. of the sun's gravitational pull
    d. of the moon's gravitational pull
16. The usual tidal pattern in most parts of the world consists of
    a. a high tide one day and a low tide on the next
    b. one high tide and one low tide daily
    c. two high tides and two low tides daily
    d. three high tides and three low tides daily
17. Tides are caused
    a. only by the sun
    b. only by the moon
    c. by both the sun and the moon
    d. sometimes by the sun and sometimes by the moon
18. High tide occurs at a given place
    a. only when the moon faces the place
    b. only when the moon is on the opposite side of the earth from the place
    c. both when the moon faces the place and when the moon is on the opposite side of the earth from the place
    d. when the place is halfway between facing the moon and being on the opposite side of the earth from the moon
19. The prefix micro stands for
    a. 1/10
    b. 1/100
    c. 1/1000
    d. 1/1,000,000
20. A centimeter is
    a. 0.001 m
    b. 0.01 m
    c. 0.1 m
    d. 10 m
21. Of the following, the shortest is
    a. 1 mm
    b. 0.01 in.
    c. 0.001 m
    d. 0.001 ft
22. Of the following, the longest is
    a. 1000 ft
    b. 500 m
    c. 1 km
    d. 1 mi
23. A person is 180 cm tall. This is equivalent to
    a. 4 ft 6 in.
    b. 5 ft 9 in.
    c. 5 ft 11 in.
    d. 7 ft 1 in.

Exercises

1.2 Why Science Is Successful

1. What role does "common sense" play in the scientific method?
2. What is the basic distinction between the scientific method and other ways of looking at the natural world?
3. What is the difference between a hypothesis and a law? Between a law and a theory?
4. Scientific models do not correspond exactly to reality. Why are they nevertheless so useful?
5. According to the physicist Richard Feynman, "Science is the culture of doubt." Does this mean that science is an unreliable guide to the natural world?

1.3 A Survey of the Sky

6. You are lost in the northern hemisphere in the middle of nowhere on a clear night. How

could you tell the direction of north by looking at the sky?

7. What must be your location if the stars move across the sky in circles centered directly overhead?

8. In terms of what you would actually observe, what does it mean to say that the moon apparently moves eastward among the stars?

1.5 The Copernican System

9. From observations of the moon, why would you conclude that it is a relatively small body revolving around the earth rather than another planet revolving around the sun?

10. The sun, moon, and planets all follow approximately the same path from east to west across the sky. What does this suggest about the arrangement of these members of the solar system in space?

11. What is the basic difference between the ptolemaic and copernican models? Why is the ptolemaic model considered incorrect?

12. Ancient astronomers were troubled by variations in the brightnesses of the various planets with time. Does the ptolemaic or the copernican model account better for these variations?

13. Compare the ptolemaic and copernican explanations for (a) the rising and setting of the sun; (b) the eastward drift of the sun relative to the stars that takes a year for a complete circuit; (c) the eastward drift of the moon relative to the stars that takes about 4 weeks for a complete circuit.

14. What do you think is the reason scientists use an ellipse rather than a circle as the model for a planetary orbit?

15. The average distance from the earth to the sun is called the astronomical unit (AU). If an asteroid is 4 AU from the sun and its period of revolution around the sun is 8 years, does it obey Kepler's third law?

1.9 Why The Earth Is Round

16. What, if anything, would happen to the shape of the earth if it were to rotate on its axis faster than it does today?

1.10 The Tides

17. What is the difference between spring and neap tides? Under what circumstances does each occur?

18. The length of the day has varied. When did the longest day thus far occur?

19. The earth takes almost exactly 24 h to make a complete turn on its axis, so we might expect each high tide to occur 12 h after the one before. However, the actual time between high tides is 12 h 25 min. Can you account for the difference?

20. Does the sun or the moon have the greater influence in causing tides?

1.12 The SI System

21. How many microphones are there in a megaphone?

22. The speedometer of a European car gives its speed in kilometers per hour. What is the car's speed in miles per hour when the speedometer reads 80?

23. Express the 405-mi distance between Chicago and Minneapolis in kilometers.

24. The diameter of an atom is roughly 10^4 times the diameter of its nucleus. If the nucleus of an atom were 1 mm across, how many feet across would the atom be?

25. How many square feet are there in an area of 1.00 square meters? Use the proper number of significant figures in the answer.

26. Use the proper number of significant figures to express the values of

 a. $93.2 + 8.56 - 12$

 b. $(4.6 \times 10^5)(8.75 \times 10^3)$

 c. $\dfrac{32.4 \times 10^4}{5.11 \times 10^{-2}} + 2.58 \times 10^2$

 d. $\sqrt{43}$

Answers to Multiple Choice

1. a	4. a	7. c	10. c	13. c	16. c	19. d	22. d
2. d	5. d	8. a, b, d, c	11. a	14. d	17. c	20. b	23. c
3. a	6. a	9. d	12. b	15. a	18. c	21. b	

Motion

Astronaut Bruce McCandless near the orbiting Space Shuttle *Challenger*.

Goals

When you have finished this chapter you should be able to complete the goals ▸ given for each section below:

Describing Motion

2.1 Speed
How Fast Is Fast
- Distinguish between instantaneous and average speeds.
- Use the formula $v = d/t$ to solve problems that involve distance, time, and speed.

2.2 Vectors
Which Way as Well as How Much
- Distinguish between scalar and vector quantities and give several examples of each.
- Use the Pythagorean theorem to add two vector quantities of the same kind that act at right angles to each other.

2.3 Acceleration
Vroom!
- Define acceleration and find the acceleration of an object whose speed is changing.
- Use the formula $v_2 = v_1 + at$ to solve problems that involve speed, acceleration, and time.

2.4 Distance, Time, and Acceleration
How Far?
- Use the formula $d = v_1t + \frac{1}{2}at^2$ to solve problems that involve distance, time, speed, and acceleration.

Acceleration of Gravity

2.5 Free Fall
What Goes Up Must Come Down
- Explain what is meant by the acceleration of gravity.
- Separate the velocity of an object into vertical and horizontal components in order to determine its motion.

2.6 Air Resistance
Why Raindrops Don't Kill
- Describe the effect of air resistance on falling objects.

Force and Motion

2.7 First Law of Motion
Constant Velocity Is as Natural as Being at Rest
- Define force and indicate its relationship to the first law of motion.

2.8 Mass
A Measure of Inertia

2.9 Second Law of Motion
Force and Acceleration
- Discuss the significance of the second law of motion, $F = ma$.

2.10 Mass and Weight
Weight Is a Force
- Distinguish between mass and weight and find the weight of an object of given mass.

2.11 Third Law of Motion
Action and Reaction
- Use the third law of motion to relate action and reaction forces.

Gravitation

2.12 Circular Motion
A Curved Path Requires an Inward Pull
- Explain the significance of centripetal force in motion along a curved path.
- Relate the centripetal force on an object moving in a circle to its mass, speed, and the radius of the circle.

2.13 Newton's Law of Gravity
What Holds the Solar System Together
- State Newton's law of gravity and describe how gravitational forces vary with distance.

2.14 Artificial Satellites
Thousands Circle the Earth
- Account for the ability of a satellite to orbit the earth without either falling to the ground or flying off into space.
- Define escape speed.

Everything in the universe is in nonstop movement. Whatever the scale of size, from the tiny particles inside atoms to the huge galaxies of stars far away in space, motion is the rule, not the exception. In order to understand the universe, we must begin by understanding motion and the laws it obeys.

The laws of motion that govern the behavior of atoms and stars apply just as well to the objects of our daily lives. Engineers need these laws to design cars and airplanes, machines of all kinds, even roads—how steeply to bank a highway curve is calculated from the same basic formula that Newton combined with Kepler's findings to arrive at the law of gravity. Terms such as speed and acceleration, force and weight, are used by everyone. Let us now see exactly what these terms mean and how the quantities they refer to are related.

Describing Motion

When an object goes from one place to another, we say it moves. If the object gets there quickly, we say it moves fast; if the object takes a long time, we say it moves slowly. The first step in analyzing motion is to be able to say just how fast is fast and how slow is slow.

2.1 Speed

How Fast Is Fast

The **speed** of something is the rate at which it covers distance. The higher the speed, the faster it travels and the more distance it covers in a given period of time.

If a car goes through a distance of 40 kilometers in a time of 1 hour, its speed is 40 kilometers per hour, usually written 40 km/h.

What if the time interval is not exactly 1 hour? For instance, the car might travel 60 km in 2 hours on another trip. The general formula for speed is distance divided by time:

$$\text{Speed} = \frac{\text{distance}}{\text{time}}$$

Hence the car's speed in the second case is

$$\text{Speed} = \frac{\text{distance}}{\text{time}} = \frac{60\text{ km}}{2\text{ h}} = 30\text{ km/h}$$

The same formula works for times of less than a full hour. The speed of a car that covers 24 km in half an hour is, since $\frac{1}{2}$ h = 0.5 h,

$$\text{Speed} = \frac{\text{distance}}{\text{time}} = \frac{24\text{ km}}{0.5\text{ h}} = 48\text{ km/h}$$

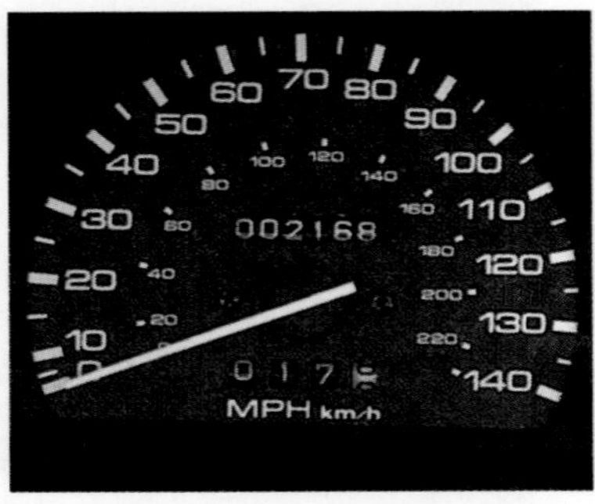

Figure 2-1 The speedometer of a car shows its instantaneous speed. This speedometer is calibrated in both mi/h (here MPH) and km/h.

These speeds are all **average speeds,** because we do not know the details of how the cars moved during their trips. They probably went slower than the average during some periods, faster at others, and even came to a stop now and then at traffic lights. What the speedometer of a car shows is the car's **instantaneous speed** at any moment, that is, how fast it is going at that moment (Fig. 2-1).

For the sake of convenience, quantities such as distance, time, and speed are often abbreviated and printed in italics:

$$d = \text{distance} \qquad t = \text{time} \qquad v = \text{speed}$$

In terms of these symbols the formula for speed becomes

$$v = \frac{d}{t} \qquad \textit{Speed} \qquad \textit{2-1}$$

Distance The above formula can be rewritten in two ways. Suppose we want to know how far a car whose average speed is v goes in a time t. To find out, we must solve $v = d/t$ for a distance d. According to one of the rules of algebra (see the Math Refresher at the back of this book), a quantity that divides one side of an equation can be shifted to multiply the other side. Thus

$$v = \frac{d}{t}$$

becomes

$$v = \frac{d}{t}$$

$$vt = d$$

which is the same as

$$d = vt \qquad \textit{2-2}$$

$$\text{Distance} = (\text{speed})(\text{time})$$

Time In another situation we might want to know how long it takes something moving at a certain speed to cover a certain distance. In other

Frame of Reference

When we say something is moving, we mean that its position relative to something else—the **frame of reference**—is changing. The choice of an appropriate frame of reference depends on the situation. In the case of a car, for instance, the obvious frame of reference is the road it is on.

In other cases things may not be so straightforward. If we use the earth as our frame of reference, the sun and planets move relative to us in complicated paths, as in Figs. 1-5 and 1-7. On the other hand, if we use the sun as our frame of reference, the earth and the other planets move relative to it in simple paths, as in Fig. 1-8. Newton was able to interpret these motions in terms of the gravitational pull of the sun whereas he would not have been able to discover the law of gravity if he had used the earth as his frame of reference.

Example 2.1

How far does a car travel in 6 hours when its average speed is 40 km/h?

Solution

We put v = 40 km/h and t = 6 h into Eq. 2-2 to find that (Fig. 2-2)

$$d = vt = \left(40\,\frac{\text{km}}{\text{h}}\right)(6\text{ h}) = 240\left(\frac{\text{km}}{\cancel{\text{h}}}\right)(\cancel{\text{h}}) = 240\text{ km}$$

We see that, since h/h = 1, the hours cancel out to give just kilometers in the answer.

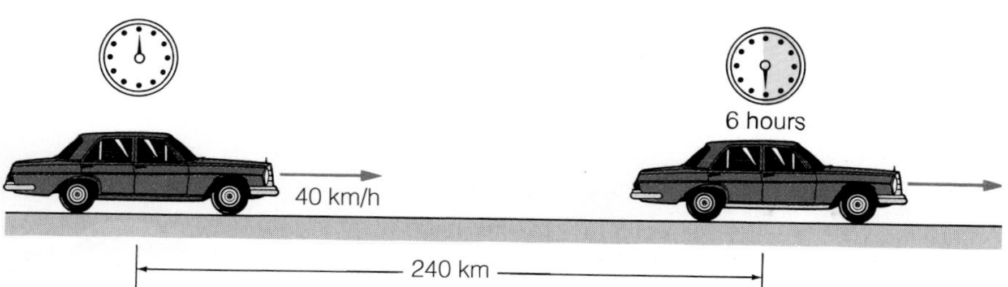

Figure 2-2 A car whose average speed is 40 km/h travels 240 km in 6 hours.

words, we know v and d and want to find the time t. What we do here is solve $d = vt$ for the time t. From basic algebra we know that something that multiplies one side of an equation can be shifted to divide the other side. What we do, then, is shift the v in the formula $d = vt$ to divide the d:

$$d = v\,t$$

$$\frac{d}{v} = t$$

which is the same as

$$t = \frac{d}{v} \qquad \text{2-3}$$

$$\text{Time} = \frac{\text{distance}}{\text{speed}}$$

Example 2.2

You are standing 100 m north of your car when an alligator appears 20 m north of you and begins to run toward you at 8 m/s, as in Fig. 2-3. At the same moment you start to run toward your car at 5 m/s. Will you reach the car before the alligator reaches you?

Solution

You are 100 m from your car and so would need

$$t_1 = \frac{d_1}{v_1} = \frac{100\ \text{m}}{5\ \text{m/s}} = 20\ \text{s}$$

to reach it. The alligator is 120 m from the car but would need only

$$t_2 = \frac{d_2}{v_2} = \frac{120\ \text{m}}{8\ \text{m/s}} = 15\ \text{s}$$

to reach it. Hence the alligator would overtake you before you get to the car. Too bad.

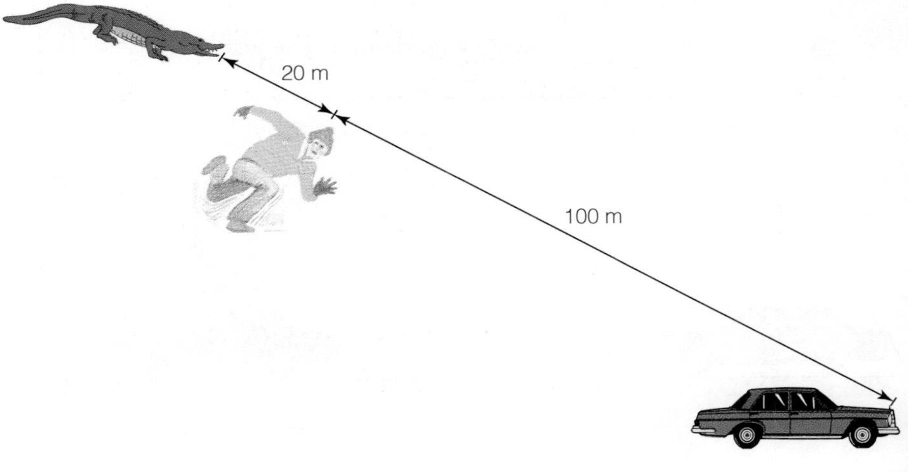

Figure 2-3 Watch out for alligators.

2.2 Vectors

Which Way as Well as How Much

Some quantities need only a number and a unit to be completely specified. It is enough to say that the area of a farm is 600 acres, that the frequency of a sound wave is 440 cycles per second, that a lightbulb uses electric energy at the rate of 75 watts. These are examples of **scalar quantities.** The **magnitude** of a quantity refers to how large it is. Thus the magnitudes of the scalar quantities given above are respectively 600 acres, 440 cycles per second, and 75 watts.

A **vector quantity,** on the other hand, has a direction as well as a magnitude associated with it, and this direction can be important. Displacement (change in position) is an example of a vector quantity. If we drive 1000 km north from Denver, we will end up in Canada; if we drive 1000 km south, we will end up in Mexico. Force is another example of a vector quantity. Applying enough upward force to this book will lift it from the table; applying a force of the same magnitude downward on the book will press it harder against the table, but the book will not move.

Speed and Velocity The **speed** of a moving object tells us only how fast the object is going, regardless of its direction. Speed is therefore a scalar quantity. If we are told that a car has a speed of 40 km/h, we do not know where it is headed, or even if it is moving in a straight line—it might well be going in a circle. The vector quantity that includes both speed and direction is called **velocity.** If we are told that a car has a constant velocity of 40 km/h toward the west, we know all there is to know about its motion and can easily figure out where it will be in an hour, or 2 hours, or at any other time.

A handy way to represent a vector quantity on a drawing is to use a straight line called a **vector** that has an arrowhead at one end to show the direction of the quantity. The length of the line is scaled according to the magnitude of the quantity. Figure 2-4 shows how a velocity of 40 km/h to the right is represented by a vector on a scale of 1 cm = 10 km/h. All other vector quantities can be pictured in a similar way.

Vector quantities are usually printed in boldface type (**F** for force, **v** for velocity). Italic type is used for scalar quantities (f for frequency, V for volume). Italic type is also used for the magnitudes of vector quantities: F is the magnitude of the force **F;** v is the magnitude of the velocity **v.** For instance, the magnitude of a velocity **v** of 40 km/h to the west is the speed v = 40 km/h. A vector quantity is usually indicated in handwriting by an arrow over its symbol, so that $\vec{F}$ means the same thing as **F.**

Adding Vectors To add scalar quantities of the same kind, we just use ordinary arithmetic. For example, 5 kg of onions plus 3 kg of onions equals

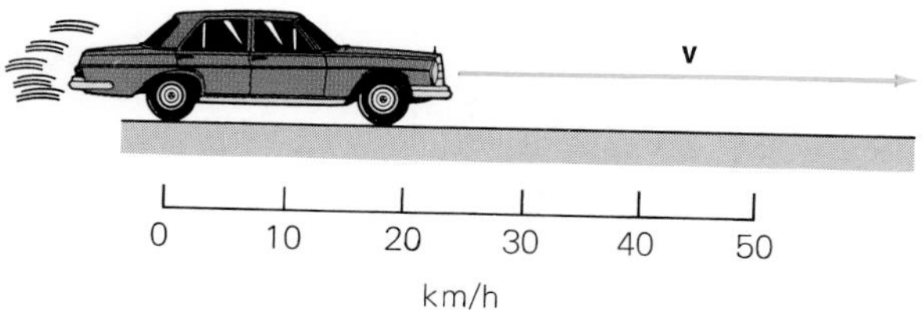

Figure 2-4 The vector **v** represents a velocity of 40 km/h to the right. The scale is 1 cm = 10 km/h.

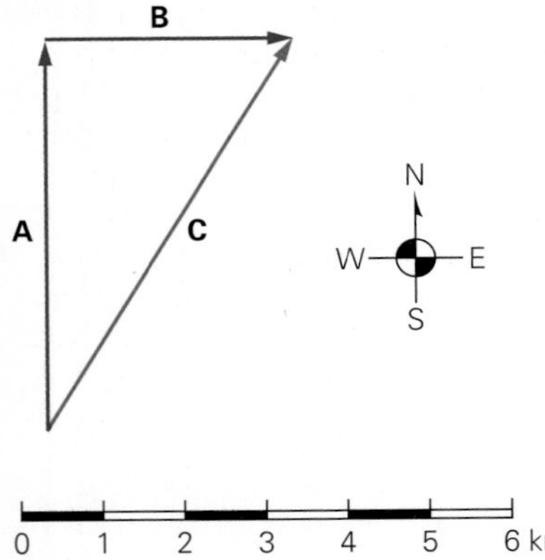

Figure 2-5 Adding vector **B** (3 km east) to vector **A** (5 km north) gives vector **C** whose length corresponds to 5.83 km. According to the Pythagorean theorem, $A^2 + B^2 = C^2$ in any right triangle.

8 kg of onions. The same method holds for vector quantities of the same kind whose directions are the same. If we drive north for 5 km and then continue north for another 3 km, we will go a total of 8 km to the north.

What if the directions are different? If we drive north for 5 km and then east for 3 km, we will not end up 8 km from our starting point. The vector diagram of Fig. 2-5 provides the answer. To add the vectors **A** and **B,** we draw **B** with its tail at the head of **A.** Connecting the tail of **A** with the head of **B** gives us the vector **C,** which corresponds to our net displacement from the start of our trip to its finish. The length of **C** tells us that our displacement was slightly less than 6 km. Any number of vectors of the same kind can be added in this way by stringing them together tail to head and then joining the tail of the first with the head of the last one.

Pythagorean Theorem A **right triangle** is one in which two of its sides are perpendicular, that is, meet at a 90° angle. The **Pythagorean theorem** is a useful relationship that holds in such a triangle. This theorem states that the sum of the squares of the short sides of a right triangle is equal to the square of its hypotenuse (longest side). For the triangle of Fig. 2-5,

$$A^2 + B^2 = C^2 \qquad \textit{Pythagorean theorem} \qquad 2\text{-}4$$

where A, B, and C are the respective magnitudes of the vectors **A, B,** and **C.**

We can therefore express the length of any of the sides of a right triangle in terms of the other sides by solving Eq. 2-4 accordingly:

$$A = \sqrt{C^2 - B^2} \qquad 2\text{-}5$$

$$B = \sqrt{C^2 - A^2} \qquad 2\text{-}6$$

$$C = \sqrt{A^2 + B^2} \qquad 2\text{-}7$$

Example 2.3

Use the Pythagorean theorem to find the displacement of a car that goes north for 5 km and then east for 3 km, as in Fig. 2-5.

Solution

Here we let A = 5 km and B = 3 km. From Eq. 2-7 we have

$$C = \sqrt{A^2 + B^2} = \sqrt{(5\text{ km})^2 + (3\text{ km})^2} = \sqrt{(25 + 9)\text{ km}^2}$$
$$= \sqrt{34\text{ km}^2} = 5.83\text{ km}$$

This method evidently gives a more accurate result than using a scale drawing.

2.3 Acceleration

Vroom!

An **accelerated** object is one whose velocity is changing. As in Fig. 2-6, the change can be an increase or a decrease in speed—the object can be going faster and faster, or slower and slower (Fig. 2-7). A change in direction, too, is an acceleration, as discussed later. Acceleration in general is

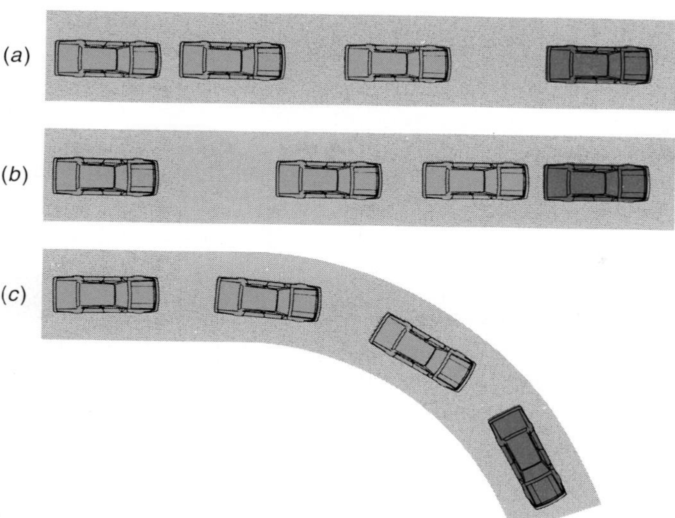

Figure 2-6 Three cases of accelerated motion, showing successive positions of a body after equal periods of time. (*a*) The intervals between the positions of the body increase in length because the body is traveling faster and faster. (*b*) The intervals decrease in length because the body is slowing down. (*c*) Here the intervals are the same in length because the speed is constant, but the direction of motion is constantly changing.

a vector quantity. For the moment, though, we will stick to straight-line motion, where acceleration is the rate of change of speed. That is,

$$a = \frac{v_2 - v_1}{t} \qquad \textit{Straight-line motion} \qquad 2\text{-}8$$

$$\text{Acceleration} = \frac{\text{change in speed}}{\text{time interval}}$$

where the symbols mean the following:

a = acceleration
t = time interval
v_1 = speed at start of time interval = initial speed
v_2 = speed at end of time interval = final speed

Not all accelerations are constant, but a great many are very nearly so. In what follows all accelerations are assumed to be constant. In Sec. 2-7 we will see why acceleration is such an important quantity in physics.

Figure 2-7 Express elevators in tall buildings have accelerations of no more than 1 m/s^2 to prevent passenger discomfort. One of the world's fastest elevators, in the Yokohama Landmark Tower (Japan's highest building), climbs 69 floors in 40 s, but only 5 s is spent at its top speed of 12.5 m/s (28 mi/h). The elevator reaches this speed at the 27th floor and then begins to slow down at the 42nd floor.

Example 2.4

The speed of a car changes from 15 m/s (about 34 mi/h) to 25 m/s (about 56 mi/h) in 20 s when its gas pedal is pressed hard (Fig. 2-8). Find its acceleration.

Solution

Here

$$v_i = 15 \text{ m/s} \quad v_f = 25 \text{ m/s} \quad t = 20 \text{ s}$$

and so the car's acceleration is

$$a = \frac{v_f - v_i}{t} = \frac{25 \text{ m/s} - 15 \text{ m/s}}{20 \text{ s}} = \frac{10 \text{ m/s}}{20 \text{ s}} = \frac{0.5 \text{ m/s}}{\text{s}} = 0.5 \text{ m/s}^2$$

This result means that the speed of the car increases by 0.5 m/s during each second the acceleration continues. It is customary to write (m/s)/s (meters per second per second) as just m/s^2 (meters per second squared) since

$$\frac{\text{m/s}}{\text{s}} = \frac{\text{m}}{(\text{s})(\text{s})} = \frac{\text{m}}{\text{s}^2}$$

Figure 2-8 A car whose speed increases from 15 m/s to 25 m/s in 20 s has an acceleration of 0.5 m/s^2.

Example 2.5

A car whose brakes can produce an acceleration of -6 m/s^2 is traveling at 30 m/s when its brakes are applied. (a) What is the car's speed 2 s later? (b) What is the total time needed for the car to come to a stop?

Solution

(a) From Eq. 2-9,

$$v_2 = v_1 + at = 30 \text{ m/s} + (-6 \text{ m/s}^2)(2 \text{ s})$$
$$= (30 - 12) \text{ m/s} = 18 \text{ m/s}$$

(b) Now $v_2 = 0 = v_1 + at$, so $at = -v_1$ and

$$t = -\frac{v_1}{a} = -\frac{30 \text{ m/s}}{-6 \text{ m/s}^2} = 5 \text{ s}$$

Suppose we know the acceleration of a car (or anything else) and want to know its speed after it has been accelerated for a time t. What we do is first rewrite Eq. 2-8 in the form

$$at = v_2 - v_1$$

which gives us what we want,

$$v_2 = v_1 + at \qquad \textit{Final speed} \qquad 2\text{-}9$$

$$\text{Final speed} = \text{initial speed} + \text{change in speed}$$

Not all accelerations increase speed. Something whose speed is decreasing is said to have a **negative acceleration.** For instance, when the brakes of a car are applied, its acceleration might be -6 m/s^2, which means that its speed drops by 6 m/s in each second that the acceleration continues. (Sometimes a negative acceleration is called a **deceleration.**)

2.4 Distance, Time, and Acceleration

How Far?

An interesting question is, how far does something, say a car, go when it is accelerated from speed v_1 to speed v_2 in the time t?

To find out, we begin by noting that the car's *average* speed $\overline{v}$ during the acceleration (assumed uniform) is

$$\overline{v} = \frac{v_1 + v_2}{2} \qquad \textit{Average speed} \qquad 2\text{-}10$$

The car moves exactly as far in the time t as if it had the constant speed v equal to its average speed $\bar{v}$. Therefore the distance the car covers in the time t is

$$d = \bar{v}t = \left(\frac{v_1 + v_2}{2}\right)t = \frac{v_1 t}{2} + \frac{v_2 t}{2}$$

The value of v_2, the car's final speed, is given by Eq. 2-9, which means that

$$d = \frac{v_1 t}{2} + \left(\frac{v_1 + at}{2}\right)t = \frac{v_1 t}{2} + \frac{v_1 t}{2} + \frac{at^2}{2}$$

$$d = v_1 t + \tfrac{1}{2}at^2 \qquad \textit{Distance under constant acceleration} \qquad 2\text{-}11$$

If the car is stationary at the start of the acceleration, $v_1 = 0$, and

$$d = \tfrac{1}{2}at^2 \qquad \textit{Distance starting from rest} \qquad 2\text{-}12$$

Example 2.6

How far did the car of Example 2-5 go while coming to a stop?

Solution

Here $v_1 = 30$ m/s, $a = -6$ m/s^2, and $t = 5$ s, so

$$d = v_1 t + \tfrac{1}{2}at^2 = (30 \text{ m/s})(5 \text{ s}) + \tfrac{1}{2}(-6 \text{ m/s}^2)(5 \text{ s})^2$$
$$= 150 \text{ m} - 75 \text{ m} = 75 \text{ m}$$

Example 2.7

An airplane needed 20 s to take off from a runway 500 m long. What was its acceleration? Its final speed?

Solution

Since the airplane started from rest, $v_1 = 0$ and $d = \frac{1}{2}at^2$. Therefore its acceleration was

$$a = \frac{2d}{t^2} = \frac{2\,(500 \text{ m})}{(20 \text{ s})^2} = 2.5 \text{ m/s}^2$$

The airplane's final speed was

$$v_2 = v_1 + at = 0 + (2.5 \text{ m/s}^2)(20 \text{ s}) = 50 \text{ m/s}$$

As we can see, by defining certain quantities (here speed and acceleration) and relating them to each other and to directly measurable quantities (here distance and time) we can build up a structure of equations that enables us to answer questions with a pencil and paper that otherwise would need separate, perhaps difficult, observations on real objects.

Acceleration of Gravity

Drop a stone, and it falls. Does the stone fall at a constant speed, or does it go faster and faster? Does the stone's motion depend on its weight, or its size, or its shape?

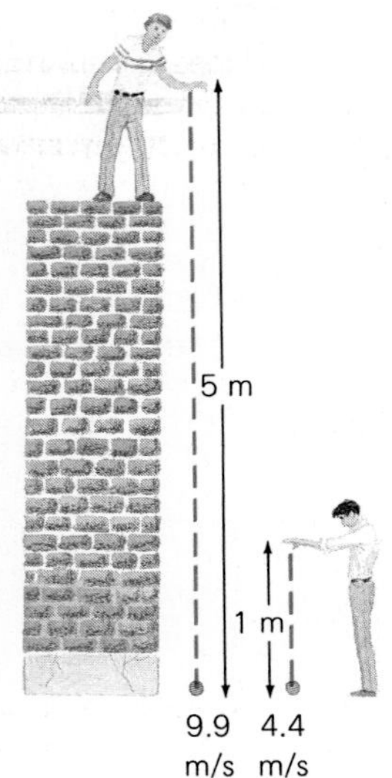

Figure 2-9 Falling bodies are accelerated downward. A stone dropped from a height of 5 m strikes the ground with a speed more than double that of a stone dropped from a height of 1 m.

Before Galileo, philosophers tried to answer such questions in terms of supposedly self-evident principles, concepts so seemingly obvious that there was no need to test them. This was the way in which Aristotle (384–322 B.C.), the famous thinker of ancient Greece, approached the subject of falling bodies. To Aristotle, every kind of material had a "natural" place where it belonged and toward which it tried to move. Thus fire rose "naturally" toward the sun and stars, whereas stones were "earthy" and so fell downward toward their home in the earth. A big stone was more earthy than a small one and so, Aristotle thought, ought to fall faster. The trouble with these ideas, and many others like them, is that they are wrong—only the scientific method, not unsupported speculation, can provide reliable information about how the universe works.

2.5 Free Fall

What Goes Up Must Come Down

Almost two thousand years later the Italian physicist Galileo, the first modern scientist, found that the higher a stone is when it is dropped, the greater its speed when it reaches the ground (Figs. 2-9 and 2-10). This means the stone is accelerated. Furthermore, the acceleration is the same for *all* stones, big and small. For more accuracy with the primitive instruments of his time, Galileo measured the accelerations of balls rolling down an inclined plane rather than their accelerations in free fall, but his conclusions were perfectly general; modern experiments have verified them to at least 1 part in 10^{12} (a trillion!).

Galileo's experiments showed that, if there were no air for them to push their way through, all falling objects near the earth's surface would have the same acceleration of 9.8 m/s^2. This acceleration is usually abbreviated *g:*

$$\textbf{Acceleration due to gravity} = g = 9.8 \text{ m/s}^2$$

Ignoring for the moment the effect of air resistance, something that drops from rest has a speed of 9.8 m/s at the end of the first second, a speed of (9.8 m/s^2)(2 s) = 19.6 m/s at the end of the next second, and so on (Fig. 2-11). In general, under these circumstances

$$v_{\text{downward}} = gt \qquad \textit{Object falling from rest} \qquad \text{2-13}$$

How Far Does a Falling Object Fall? Equation 2-9 tells us the speed of a falling object at any time t after it has been dropped from rest (and before it hits the ground, of course). To find out how far h the object has fallen in the time t, we refer back to Eq. 2-12 for accelerated motion starting from rest. Here the distance is $d = h$ and the acceleration is that of gravity, so $a = g$, which gives

$$h = \tfrac{1}{2}gt^2 \qquad \textit{Object falling from rest} \qquad \text{2-14}$$

The t^2 factor means that h increases with time much faster than the object's speed v, which is given by $v = gt$. Figure 2-12 shows h and v for various times of fall. At $t = 10$ s, the object's speed is 10 times its speed at $t = 1$ s, but the distance it has fallen is 100 times the distance it fell during the first second.

Thrown Objects The downward acceleration g is the same whether an object is just dropped or is thrown upward, downward, or sideways. If a ball is held in the air and dropped, it goes faster and faster until it hits the ground. If the ball is thrown horizontally, we can imagine its velocity as having two parts, a horizontal one that stays constant and a vertical

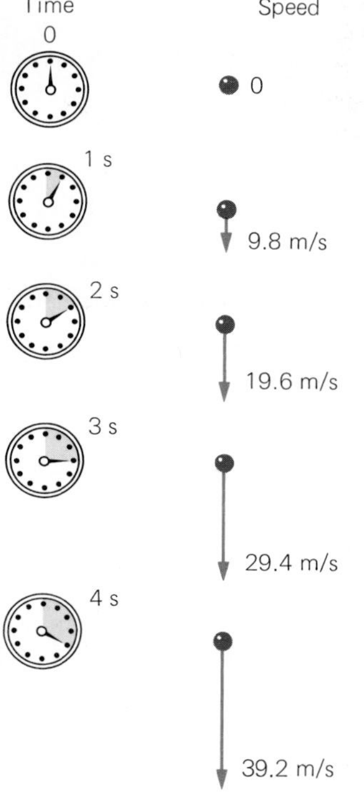

Figure 2-11 All falling objects near the earth's surface have a downward acceleration of 9.8 m/s^2. (The distance an object will have fallen in each time interval is not shown to scale here.)

Figure 2-10 The Leaning Tower of Pisa, which is 58 m high, was begun in 1174 and took over two centuries to complete. During its construction the tower started to sink into the clay soil under its south side, and corrections were made to the upper floors to try to make them level. Today the tilt is 5.5° and continues to increase, leaving the tower in danger of collapse if efforts to stabilize the ground under it fail. According to legend, Galileo dropped a bullet and a cannonball from the tower to show that all objects fall with the same acceleration.

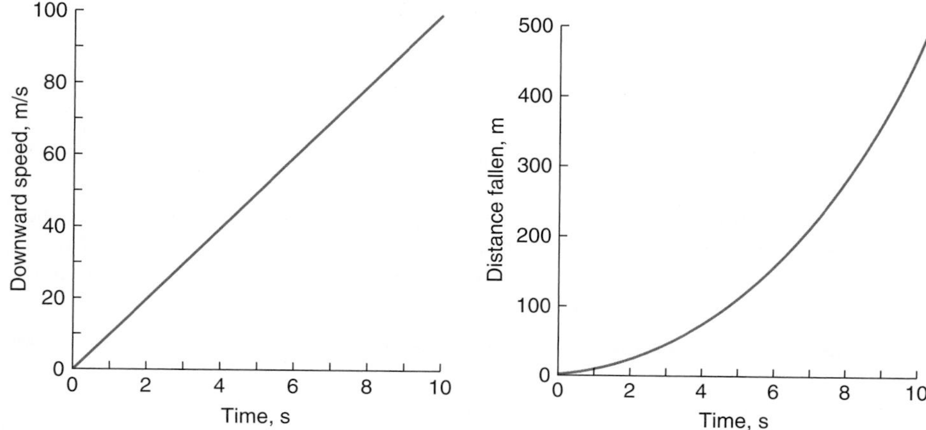

Figure 2-12 Downward speed and distance fallen in the first 10 seconds after an object is dropped from rest. Air resistance (Sec. 2.6) is ignored here.

BIOGRAPHY Galileo Galilei (1564–1642)

Galileo fathered modern science by clearly stating the central idea of the scientific method: the study of nature must be based on observation and experiment. He also pioneered the use of mathematical reasoning to interpret and generalize his findings.

Galileo was born in Pisa, Italy; following a local custom, his given name was a variation of his family name. Although his father thought that medicine would be a more sensible career choice, Galileo studied physics and mathematics and soon became a professor at Pisa and afterward at Padua. His early work was on accelerated motion, falling bodies, and the paths taken by projectiles. Later, with a telescope he had built, Galileo was the first person to see sunspots, the phases of Venus, the four largest satellites of Jupiter, and the mountains of the moon. Turning his telescope to the Milky Way, he found that it consisted of individual stars. To Galileo, as to his contemporaries, these discoveries were "infinitely stupendous."

By then famous, in 1610 Galileo went to Florence as court mathematician to Cosimo, Duke of Tuscany. Here Galileo expounded the copernican model of the universe and pointed out that his astronomical observations supported this model. Galileo discovered Jupiter's four largest satellites in 1610 and concluded from his observations, shown here, that they revolve around Jupiter.

Because the ptolemaic model with the earth as the stationary center of the universe was part of the doctrine of the Catholic Church, the Holy Office told Galileo to stop advocating the contrary, and he obeyed. Then, when a friend of his became pope in 1631, Galileo resumed teaching the merits of the copernican system. But the pope turned against him, and in 1633, when he was 70, Galileo was convicted of heresy by the Inquisition. Although he escaped being burnt at the stake, the fate of other heretics, Galileo was sentenced to house arrest for the remainder of his life and was forced to publicly deny that

the earth moves. (According to legend, Galileo then muttered, "Yet it does move.")

Even in 1992, the 350th anniversary of Galileo's death, a Vatican commission did not mention, let alone regret, the Church's role in silencing Galileo. Pope John Paul II spoke of Galileo's condemnation, which almost extinguished Italian science for over a century, as merely a matter of "mutual incomprehension."

It would be nice to think that today, four centuries after Galileo's time, the age of reason has become firmly established. But religious fundamentalists who seek to replace the findings of science by their own particular interpretations of the Bible are again on the march. Do we really want a return to the blind ignorance of the past? Evolution in biology, geology, and astronomy is now under attack, and a movement to return to "Biblical astronomy," with the earth at the center of the universe, has actually started. The struggle to keep reason in the life of the mind is never ending, but it is an essential struggle.

one that is affected by gravity. The result, as in Fig. 2-13, is a curved path that becomes steeper as the downward speed increases.

When a ball is thrown upward, as in Fig. 2-14, the effect of the downward acceleration of gravity is at first to reduce the ball's upward speed. The upward speed decreases steadily until finally it is zero. The ball is

Example 2.8

A stone dropped from a bridge strikes the water 2.2 s later. How high is the bridge above the water?

Solution

Substituting in Eq. 2-14 gives

$$h = \frac{1}{2}gt^2 = \frac{1}{2}(9.8 \text{ m/s}^2)(2.2 \text{ s})^2 = 24 \text{ m}$$

Example 2.9

An apple is dropped from a window 20 m above the ground. (a) How long does it take the apple to reach the ground? (b) What is its final speed?

Solution

(a) Since $h = \frac{1}{2}gt^2$,

$$t = \sqrt{\frac{2h}{g}} = \sqrt{\frac{(2)(20 \text{ m})}{9.8 \text{ m/s}^2}} = 2.0 \text{ s}$$

(b) From Eq. 2-13 the ball's final speed is

$$v = gt = 19.8 \text{ m/s}$$

Example 2.10

An airplane is in level flight at a velocity of 150 m/s and an altitude of 1500 m when a wheel falls off. What horizontal distance will the wheel travel before it strikes the ground?

Solution

The horizontal velocity of the wheel does not affect its vertical motion. The wheel therefore reaches the ground at the same time as a wheel dropped from rest at an altitude of 1500 m, which is

$$t = \sqrt{\frac{2h}{g}} = \sqrt{\frac{(2)(1500 \text{ m})}{9.8 \text{ m/s}^2}} = 17.5 \text{ s}$$

In this time the wheel will travel a horizontal distance of

$$d = v_{\text{horiz}}t = (150 \text{ m/s})(17.5 \text{ s}) = 2625 \text{ m} = 2.63 \text{ km}$$

then at the top of its path, when the ball is at rest for an instant. The ball then begins to fall at ever-increasing speed, exactly as though it had been dropped from the highest point. Interestingly enough, something thrown upward at a certain speed will return to its starting point with the same speed, although the object is now moving in the opposite direction.

What happens when a ball is thrown downward? Now the ball's original speed is steadily increased by the downward acceleration of gravity. When the ball reaches the ground, its final speed will be the sum of its original speed and the speed increase due to the acceleration.

When a ball is thrown upward at an angle to the ground, the result is a curved path called a **parabola** (Fig. 2-15). The maximum range (horizontal distance) for a given initial speed occurs when the ball is thrown at an angle of 45° above the ground. At higher and lower angles, the range

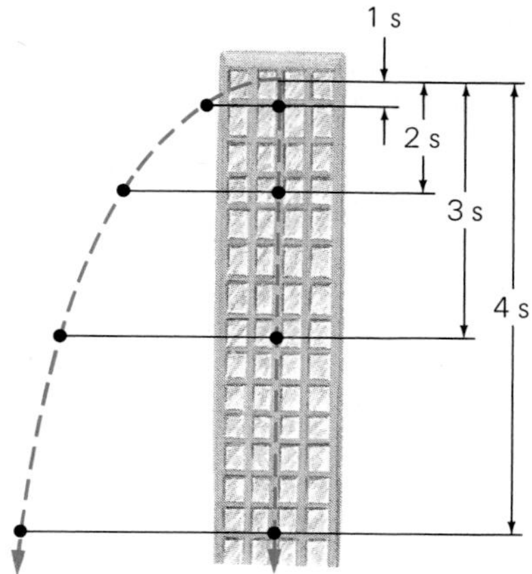

Figure 2-13 The acceleration of gravity does not depend upon horizontal motion. When one ball is thrown horizontally from a building at the same time that a second ball is dropped vertically, the two reach the ground at the same time because both have the same downward acceleration.

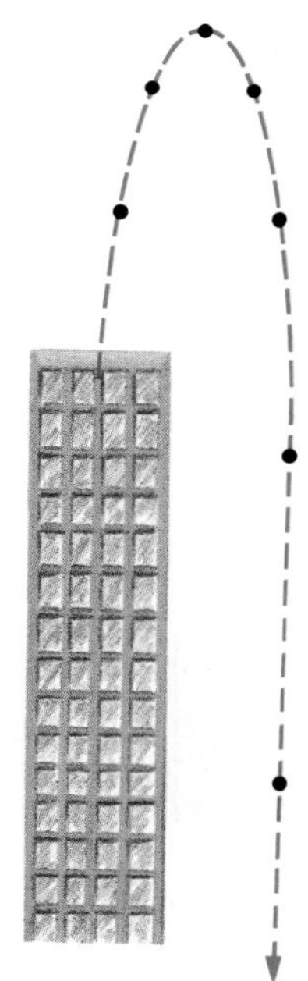

Figure 2-14 When a ball is thrown upward, its downward acceleration reduces its original speed until it comes to a momentary stop. At this time the ball is at the top of its path, and it then begins to fall as if it had been dropped from there. The ball is shown after equal time intervals.

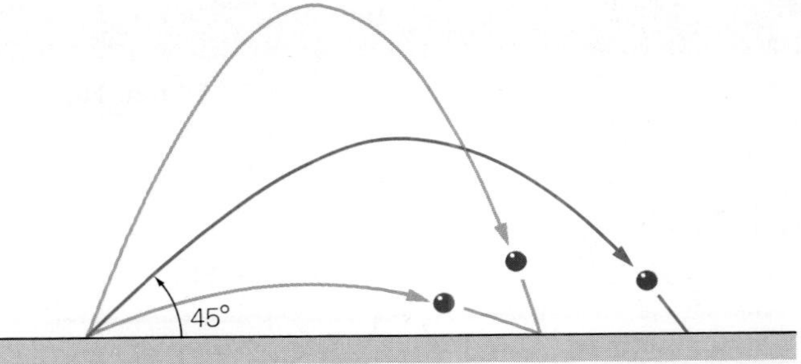

Figure 2-15 In the absence of air resistance, a ball travels farthest when it is thrown at an angle of 45°.

Example 2.11

A stone thrown upward reaches its highest point 2.2 s later. (a) How high did it go? (b) What was its initial speed?

Solution

(a) The height h is the same as that from which the stone would have been dropped to reach the ground in 2.2 s. Hence

$$h = \frac{1}{2}gt^2 = \frac{(9.8\ \text{m/s}^2)(2.2\ \text{s})}{2} = 23.7\ \text{m}$$

(b) The upward speed was the same as the downward speed the stone would have had 2.2 s after being dropped, so

$$v = gt = (9.8\ \text{m/s}^2)(2.2\ \text{s}) = 21.6\ \text{m/s}$$

will be shorter. As the figure shows, for every range up to the maximum there are two angles at which the ball can be thrown and land in the same place.

2.6 Air Resistance

Why Raindrops Don't Kill

Air resistance keeps falling things from developing the full acceleration of gravity. Without this resistance raindrops would reach the ground with bulletlike speeds and even a light shower would be dangerous.

In air, a stone falls faster than a feather because air resistance affects the stone less. In a vacuum, however, there is no air, and the stone and feather fall with the same acceleration of 9.8 m/s^2 (Fig. 2-16).

The faster something moves, the more the air in its path resists its motion. At 100 km/h (62 mi/h), the drag on a car due to air resistance is about 5 times as great as the drag at 50 km/h (31 mi/h). In the case of a falling object, the air resistance increases with speed until it equals the force of gravity on the object. The object then continues to drop at a constant **terminal speed** that depends on its size and shape and on how heavy it is (Table 2-1). A person in free fall has a terminal speed of about 54 m/s (120 mi/h), but with an open parachute the terminal speed of only about 6.3 m/s (14 mi/h) permits a safe landing (Fig. 2-17).

Air resistance reduces the range of a projectile. Figure 2-18 shows how the path of a ball is affected. In a vacuum, as we saw in Fig. 2-15, the ball

Table 2-1 Some Terminal Speeds (1 m/s = 2.2 mi/h)

Object	Terminal Speed
16-lb shot	145 m/s
Baseball	42
Golf ball	40
Tennis ball	30
Basketball	20
Large raindrop	10
Ping-Pong ball	9

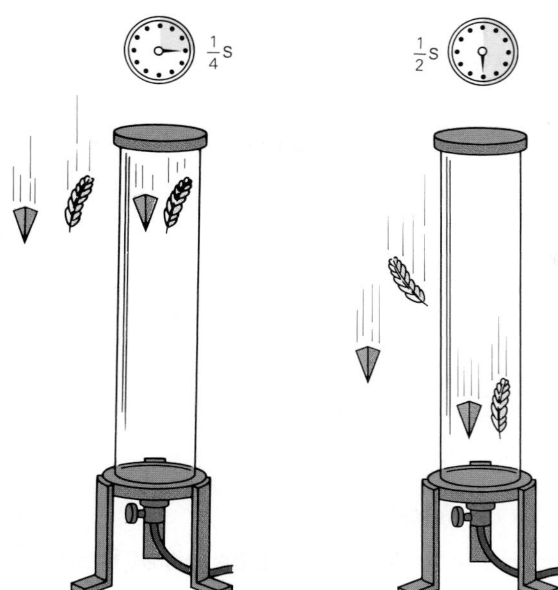

Figure 2-16 In a vacuum all bodies fall with the same acceleration.

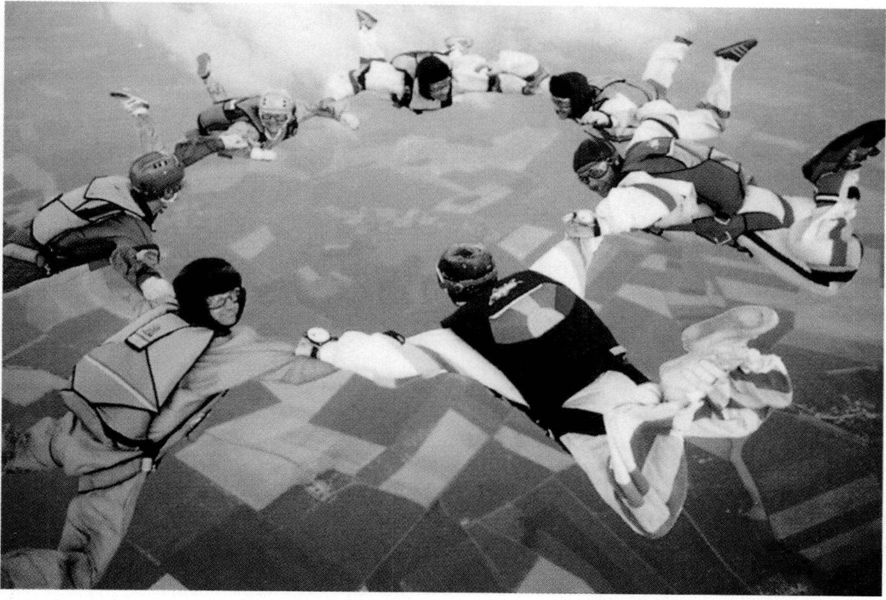

Figure 2-17 The terminal speeds of sky divers are greatly reduced when their parachutes open, which permits them to land safely.

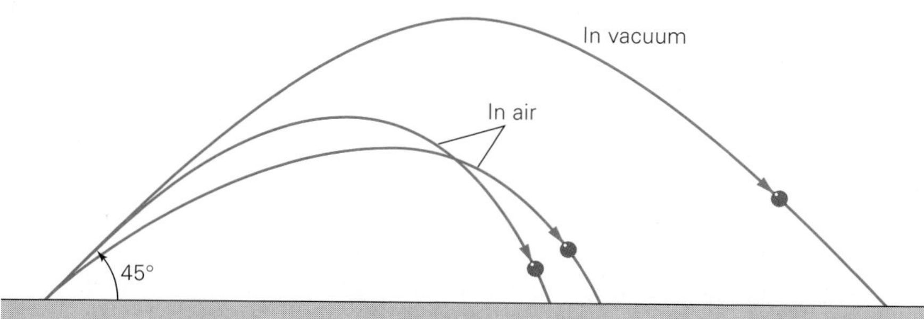

Figure 2-18 Effect of air resistance on the path of a thrown ball. An angle of less than 45° now gives the greatest range.

goes farthest when it is thrown at an angle of 45°, but in air (that is, in real life), the maximum range occurs for an angle of less than 45°. For a baseball struck hard by a bat, an angle of 40° will take it the greatest distance.

Force and Motion

What can make something originally at rest begin to move? Why do some things move faster than others? Why are some accelerated and others not? Questions like these led Isaac Newton to formulate three principles that summarize so much of the behavior of moving bodies that they have become known as the **laws of motion.** Based on observations made by Newton and others, these laws are as valid today as they were when they were set down about three centuries ago.

2.7 First Law of Motion

Constant Velocity Is as Natural as Being at Rest

Imagine a ball lying on a level floor. Left alone, the ball stays where it is. If you give it a push, the ball rolls a short way and then comes to a stop. The smoother the floor, the farther the ball rolls before stopping. With a perfectly round ball and a perfectly smooth and level floor, and no air to slow down its motion, would the ball ever stop rolling?

There will never be a perfect ball and a perfect surface for it to roll on, of course. But we can come close. The result is that, as the resistance to its motion becomes less and less, the ball goes farther and farther for the same push. We can reasonably expect that, under ideal conditions, the ball would keep rolling forever.

This conclusion was first reached by Galileo. Later it was stated by Newton as his **first law of motion:**

> If no net force acts on it, an object at rest remains at rest and an object in motion remains in motion at constant velocity (that is, at constant speed in a straight line).

According to this law, an object at rest never begins to move all by itself—a force is needed to start it off. If it is moving, the object will continue going at constant velocity unless a force acts to slow it down, to speed it up, or to change its direction. Motion at constant velocity is just as "natural" as staying at rest.

Force In thinking about **force,** most of us think of a car pulling a trailer or a person pushing a lawn mower or lifting a crate. Also familiar are the force of gravity, which pulls us and things about us downward, the pull of a magnet on a piece of iron, and the force of air pushing against the sails of a boat. In these examples the central idea is one of pushing, pulling, or lifting. Newton's first law gives us a more precise definition:

> A force is any influence that can change the speed or direction of motion of an object.

When we see something accelerated, we know that a force must be acting upon it.

Although it seems simple and straightforward, the first law of motion has far-reaching implications. For instance, until Newton's time most people believed that the orbits of the moon around the earth and of the planets around the sun were "natural" ones, with no forces needed to keep these bodies moving as they do. However, because the moon and planets move in curved paths, their velocities are not constant, so according to the first law, forces of some kind *must* be acting on them. The search for these forces led Newton to the law of gravitation.

It is worth emphasizing that just applying a force to an object at rest will not necessarily set the object in motion. If we push against a stone wall, the wall is not accelerated as a result. Only if a force is applied to an object that is able to respond will the state of rest or motion of the object change. On the other hand, *every* acceleration can be traced to the action of a force.

An object continues to be accelerated only as long as a **net force**—a force not balanced out by one or more other forces—acts upon it (Fig. 2-19). An ideal car on a level road would therefore need its engine only to be accelerated to a particular speed, after which it would keep moving at this speed forever with the engine turned off. Actual cars are not so cooperative because of the retarding forces of friction and air resistance, which require counteracting by a force applied by the engine to the wheels.

Figure 2-19 When several forces act on an object, they may cancel one another out to leave no net force. Only a net (or **unbalanced**) force can accelerate an object.

2.8 Mass

A Measure of Inertia

The reluctance of an object to change its state of rest or of uniform motion in a straight line is called **inertia.**

When you are in a car that starts to move, you feel yourself pushed back in your seat (Fig. 2-20). What is actually happening is that your inertia tends to keep your body where it was before the car started moving. When the car stops, on the other hand, you feel yourself pushed forward. What is actually happening now is that your inertia tends to keep your body moving while the car comes to a halt.

The name **mass** is given to the property of matter that shows itself as inertia. The inertia of a bowling ball exceeds that of a basketball, as you can tell by kicking them in turn, so the mass of the bowling ball exceeds that of the basketball. Mass may be thought of as quantity of matter: the more mass something has, the greater its inertia and the more matter it contains (Fig. 2-21).

The SI unit of mass is the **kilogram** (kg). A liter of water, which is a little more than a quart, has a mass of 1 kg (Fig. 2-22). Table 2-2 lists a range of mass values.

Table 2-2 Some Approximate Mass Values

The sun	2×10^{30} kg
The earth	6×10^{24}
Large tanker	4×10^{8}
747 airliner (at takeoff)	4×10^{5}
Large car	2×10^{3}
165-lb person	75
This book	1.4
Pencil	3×10^{-3}
Postage stamp	3×10^{-5}
Smallest known bacterium	1×10^{-19}
Oxygen molecule	5×10^{-26}
Electron	9×10^{-31}

2.9 Second Law of Motion

Force and Acceleration

Throw a baseball hard, and it leaves your hand going faster than if you toss it gently. This suggests that the greater the force, the greater the

Friction

Friction is a force that acts to oppose the motion of an object past another object with which it is in contact. The harder the objects are pressed together, the stronger the frictional force. Friction is an actual force, unlike inertia. Even a small net force can accelerate an object despite its inertia, but friction may prevent a small force from pushing one object across another.

Friction has two chief causes. One is the interlocking of irregularities in the two surfaces, which prevents one surface from sliding smoothly past the other. The second cause is the tendency for materials in very close contact to stick together because of attractive forces between their respective atoms and molecules, as described in Sec. 11.3.

Sometimes friction is welcome. The fastening ability of nails and screws and the resistive action of brakes depend on friction, and walking would be impossible without it. In other cases friction means wasted effort, and to reduce it lubricants (oil and grease) and rollers or wheels are commonly used. About half the power of a car's engine is lost to friction in the engine itself and in its drive train. The joints of the human body are lubricated by a substance called synovial fluid, which resembles blood plasma.

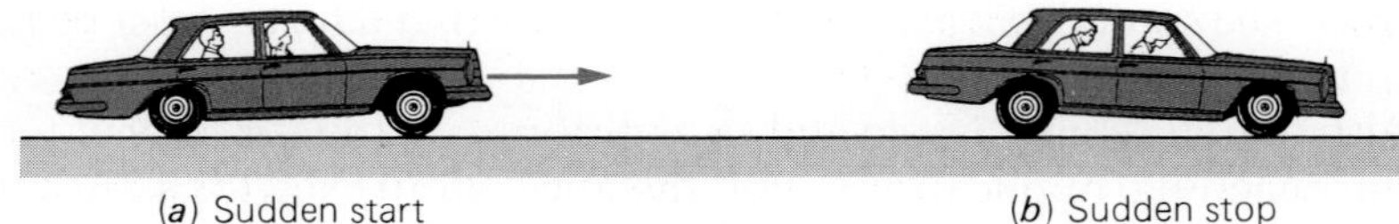

Figure 2-20 (*a*) When a car suddenly starts to move, the inertia of the passengers tends to keep them at rest relative to the earth, and so their heads move backward relative to the car. (*b*) When the car comes to a sudden stop, inertia tends to keep the passengers moving, and so their heads move forward relative to the car.

Figure 2-21 The more mass an object has, the greater its resistance to a change in its state of motion, as this shot-putter knows.

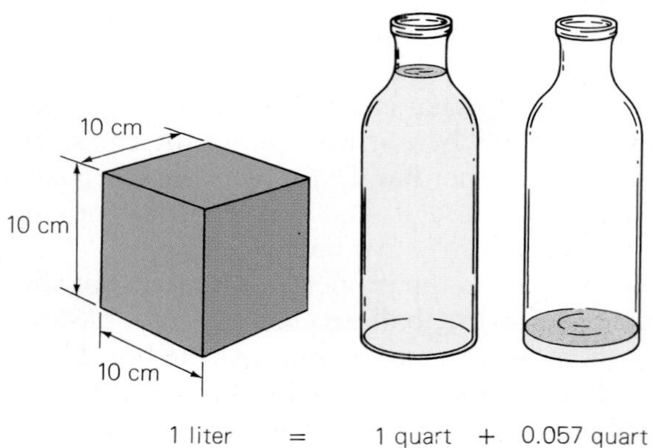

Figure 2-22 A liter, which is equal to 1.057 quarts, represents a volume of 1000 cubic centimeters (cm^3). One liter of water has a mass of 1 kg.

acceleration while the force acts. Experiments show that doubling the net force doubles the acceleration, tripling the net force triples the acceleration, and so on (Fig. 2-23).

Do all balls you throw with the same force leave your hand with the same speed? Heave an iron shot instead, and it is clear that the more mass something has, the less its acceleration for a given force.

Experiments make the relationship precise: for the same net force, doubling the mass cuts the acceleration in half, tripling the mass cuts the acceleration to one-third its original value, and so on.

Newton's **second law of motion** is a statement of these findings. If we let F = net force and m = mass, this law states that

$$a = \frac{F}{m} \qquad \textit{Second law of motion} \qquad 2\text{-}15$$

$$\text{Acceleration} = \frac{\text{force}}{\text{mass}}$$

Another way to express the second law of motion is in the form of a definition of force:

$$F = ma \qquad 2\text{-}16$$

$$\text{Force} = (\text{mass})(\text{acceleration})$$

The second law of motion was experimentally verified down to an acceleration of 5×10^{-14} m/s^2 in 2007.

An important aspect of the second law concerns direction. The direction of the acceleration is always the same as the direction of the net force. A car is going faster and faster—therefore the net force on it is in the same direction as that in which the car is headed. The car then slows down—therefore the net force on it is now in the direction *opposite* that in which it is headed (Fig. 2-24).

Thus we can say that

> The net force on an object equals the product of the mass and the acceleration of the object. The direction of the force is the same as that of the acceleration.

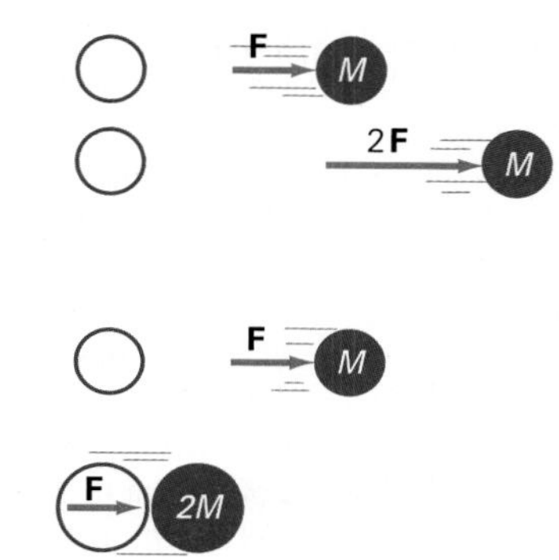

Figure 2-23 Newton's second law of motion. When different forces act upon identical masses, the greater force produces the greater acceleration. When the same force acts upon different masses, the greater mass receives the smaller acceleration.

Muscular Forces

The forces an animal exerts result from contractions of its skeletal muscles, which occur when the muscles are electrically stimulated by nerves. The maximum force a muscle can exert is proportional to its cross-sectional area and can be as much as 70 N/cm^2 (100 lb/in.^{2}). An athlete might have a biceps muscle in his arm 8 cm across, so it could produce up to 3500 N (790 lb) of force. This is a lot, but the geometry of an animal's skeleton and muscles favors range of motion over force. As a result the actual force a person's arm can exert is much smaller than the forces exerted by the arm muscles themselves, but the person's arm can move through a much greater distance than the amount the muscles contract.

An animal whose length is L has muscles that have cross-sectional areas and hence strengths roughly proportional to L^2. But the mass of the animal depends on its volume, which is roughly proportional to L^3. Therefore the larger an animal is, in general, the weaker it is relative to its mass. This is obvious in nature. For instance, even though insect muscles are intrinsically weaker than human muscles, many insects can carry loads several times their weights, whereas animals the size of humans are limited to loads comparable with their weights.

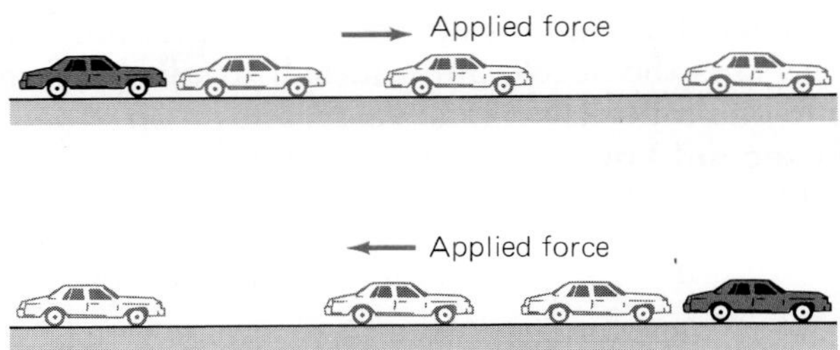

Figure 2-24 The direction of a force is significant. A force applied to a car in the direction in which it is moving (for instance by giving more fuel to its engine) produces a positive acceleration, which increases the speed of the car. A force applied opposite to the direction of motion (for instance by using the brakes) produces a negative acceleration, which decreases the speed of the car until it comes to a stop. An acceleration that reduces the speed of a moving object is sometimes called a deceleration.

The second law of motion is the key to understanding the behavior of moving objects because it links cause (force) and effect (acceleration) in a definite way (Fig. 2-25). When we speak of force from now on, we know exactly what we mean, and we know exactly how an object free to move will respond when a given force acts on it.

Figure 2-25 The relationship between force and acceleration means that the less acceleration something has, the smaller the net force on it. If you drop to the ground from a height, as this jumper has, you can reduce the force of the impact by bending your knees as you hit the ground so you come to a stop gradually instead of suddenly. The same reasoning can be applied to make cars safer. If a car's body is built to crumple progressively in a crash, the forces acting on the passengers will be smaller than if the car were rigid.

The Newton The second law of motion shows us how to define a unit for force. If we express mass m in kilograms and acceleration a in m/s^2, force F is given in terms of (kg)(m/s^2). This unit is given a special name, the **newton** (N). Thus

$$1 \text{ newton} = 1 \text{ N} = 1 \text{ (kg)(m/s}^2\text{)}$$

When a force of 1 N is applied to a 1-kg mass, the mass is given an acceleration of 1 m/s^2 (Fig. 2-26).

In the British system, the unit of force is the **pound** (lb). The pound and the newton are related as follows:

$$1 \text{ N} = 0.225 \text{ lb}$$
$$1 \text{ lb} = 4.45 \text{ N}$$

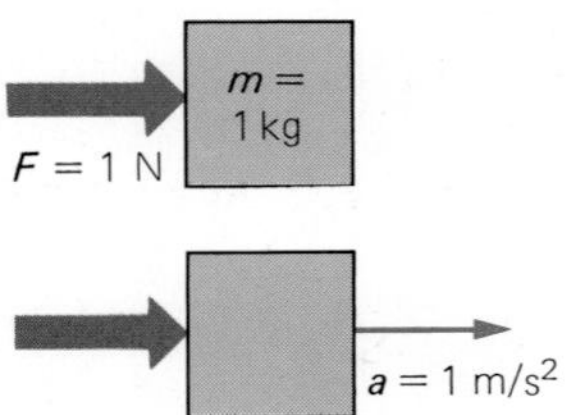

Figure 2-26 A force of 1 newton gives a mass of 1 kilogram an acceleration of 1 m/s^2.

Example 2.12

When a tennis ball is served, it is in contact with the racket for a time that is typically 0.005 s, which is 5 thousandths of a second. Find the force needed to serve a 60-g tennis ball at 30 m/s.

Solution

Since the ball starts from rest, $v_1 = 0$, and its acceleration when struck by the racket is, from Eq. 2-8,

$$a = \frac{v_2 - v_1}{t} = \frac{30 \text{ m/s} - 0}{0.005 \text{ s}} = 6000 \text{ m/s}^2$$

Because the ball's mass is 60 g = 0.06 kg, the force the racket must exert on it is

$$F = ma = (0.06 \text{ kg})(6000 \text{ m/s}^2) = 360 \text{ N}$$

In more familiar units, this force is 81 lb. Of course, it does not seem so great to the person serving because the duration of the impact is so brief (Fig. 2-27).

Figure 2-27 A person serving a tennis ball must exert a force of 360 N on it for the ball to have a speed of 30 m/s if the racket is in contact with the ball for 0.005 s.

2.10 Mass and Weight

Weight Is a Force

The **weight** of an object is the force with which it is attracted by the earth's gravitational pull. If you weigh 150 lb (668 N), the earth is pulling you down with a force of 150 lb. Weight is different from mass, which refers to how much matter something contains. There is a very close relationship between weight and mass, however.

Let us look at the situation in the following way. Whenever a net force F is applied to a mass m, Newton's second law of motion tells us that the acceleration a of the mass will be in accord with the formula

$$F = ma$$

$$\text{Force} = (\text{mass})(\text{acceleration})$$

In the case of an object at the earth's surface, the force gravity exerts on it is its weight w. This is the force that causes the object to fall with the constant acceleration $g = 9.8\ \text{m/s}^2$ when no other force acts. We may therefore substitute w for F and g for a in the formula $F = ma$ to give

$$w = mg \qquad \textit{Weight and mass} \qquad \textit{2-17}$$

$$\text{Weight} = (\text{mass})(\text{acceleration of gravity})$$

The weight w of an object and its mass m are always proportional to each other: twice the mass means twice the weight, and half the mass means half the weight.

In the SI system, mass rather than weight is normally specified. A customer in a French grocery might ask for a kilogram of bread or 5 kg of potatoes. To find the weight in newtons of something whose mass in kilograms is known, we simply turn to $w = mg$ and set $g = 9.8\ \text{m/s}^2$. Thus the weight of 5 kg of potatoes is

$$w = mg = (5\ \text{kg})(9.8\ \text{m/s}^2) = 49\ \text{N}$$

This is the force with which the earth attracts a mass of 5 kg.

At the earth's surface, the weight of a 1-kg mass in British units is 2.2 lb. The weight in pounds of 5 kg of potatoes is therefore 5(2.2 lb) = 11 lb. The mass that corresponds to a weight of 1 lb is 454 g.

Your Weight Elsewhere in the Solar System

The more mass a planet has and the smaller it is, the greater the acceleration of gravity g at its surface. The values of g for the various planets are listed in Table 17-1. From these values, if you know your mass you can figure out your weight on any planet using the formula $w = mg$. In familiar units, if you weigh 150 lb on the earth, here is what you would weigh on the planets and the moon:

Mercury	57 lb	Saturn	180 lb
Venus	135	Uranus	165
Earth	150	Neptune	180
Mars	57	Moon	25
Jupiter	390		

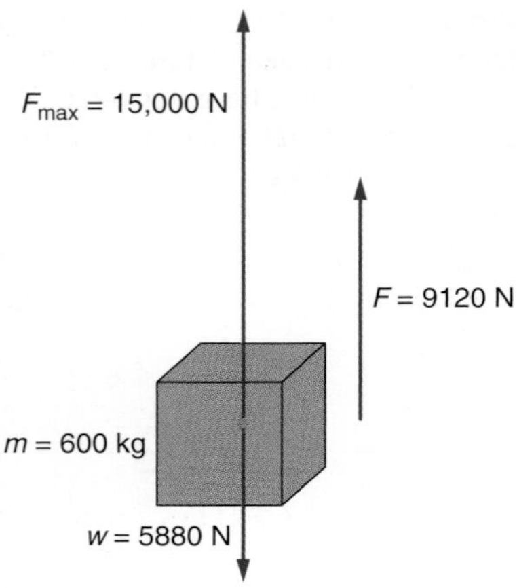

Figure 2-28 The net upward force on an elevator of mass 600 kg is 9120 N when its supporting cable exerts a total upward force of 15,000 N.

Example 2.13

An elevator whose total mass is 600 kg is suspended by a cable that can exert a maximum upward force of $F_{max} = 15,000$ N. What is the greatest upward acceleration the elevator can have? The greatest downward acceleration?

Solution

When the elevator is stationary (or moving at constant speed) the upward force the cable exerts is just the elevator's weight of

$$w = mg = (600 \text{ kg})(9.8 \text{ m/s}^2) = 5880 \text{ N}$$

To accelerate the elevator upward, an additional upward force F is needed (Fig. 2-28), where

$$F = F_{max} - w = 15,000 \text{ N} - 5880 \text{ N} = 9120 \text{ N}$$

The elevator's upward acceleration when this net force acts on it is

$$a = \frac{F}{m} = \frac{9120 \text{ N}}{600 \text{ kg}} = 15.2 \text{ m/s}$$

For the elevator to have a downward acceleration of more than the acceleration of gravity g, a downward force besides its own weight is needed. The cable cannot push the elevator downward, so its greatest downward acceleration is $g = 9.8$ m/s.

Figure 2-29 Action and reaction forces act on different bodies. Pushing a table on a frozen lake results in person and table moving apart in opposite directions.

The mass of something is a more basic property than its weight because the pull of gravity on it is not the same everywhere. This pull is less on a mountaintop than at sea level and less at the equator than near the poles because the earth bulges slightly at the equator. A person who weighs 200 lb in Lima, Peru, would weigh nearly 201 lb in Oslo, Norway. On the surface of Mars the same person would weigh only 76 lb, and he or she would be able to jump much higher than on the earth. However, the person would not be able to throw a ball any faster: because the force F the person exerts on the ball and the ball's mass m are the same on both planets, the acceleration a would be the same, too.

2.11 Third Law of Motion

Action and Reaction

Suppose you push against a heavy table and it does not move. This must mean that the table is resisting your push on it. The table stays in place because your force on it is matched by the opposing force of friction between the table legs and the floor. You don't move because the force of the table on you is matched by a similar opposing force between your shoes and the floor.

Now imagine that you and the table are on a frozen lake whose surface is so slippery on a warm day that there is no friction. Again you push on the table, which this time moves away as a result (Fig. 2-29). But you can stick to the ice no better than the table can, and you find yourself sliding backward. No matter what you do, pushing on the table always means that the table pushes back on you.

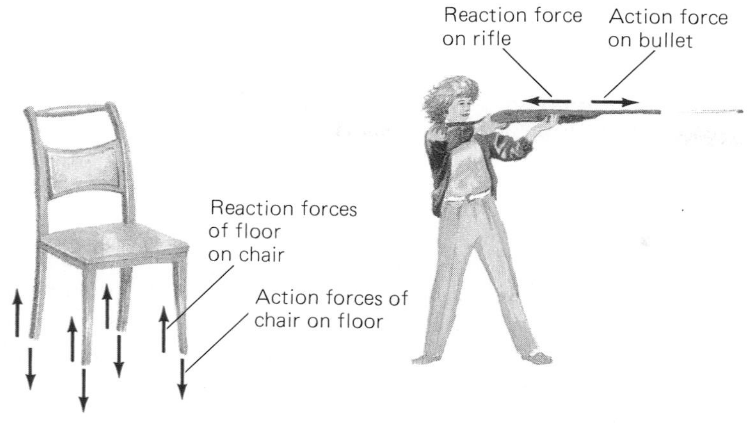

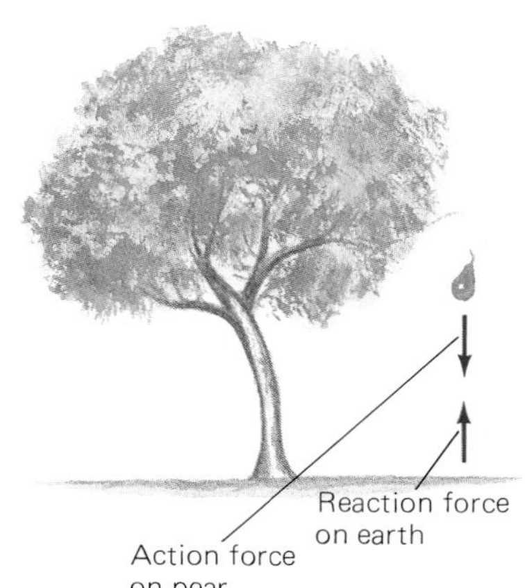

Figure 2-30 Some examples of action-reaction pairs of forces.

Considerations of this kind led Newton to his **third law of motion:**

> When one object exerts a force on a second object, the second object exerts an equal force in the opposite direction on the first object.

No force ever occurs singly. A chair pushes downward on the floor; the floor presses upward on the chair (Fig. 2-30). The firing of a rifle exerts a force on the bullet; at the same time the firing exerts a backward push (recoil) on the rifle. A pear falls from a tree because of the earth's pull on the pear; there is an equal upward pull on the earth by the pear that is not apparent because the earth has so much more mass than the pear, but this upward force is nevertheless present.

Newton's third law always applies to two different forces on two different objects—the **action force** that the first object exerts on the second, and the opposite **reaction force** the second exerts on the first.

The third law of motion permits us to walk. When you walk, what is actually pushing you forward is not your own push on the ground but instead the reaction force of the ground on you (Fig. 2-31). As you move forward, the earth itself moves backward, though by too small an amount (by virtue of its enormous mass) to be detected.

Figure 2-31 When these people push backward on the ground with their feet, the ground pushes forward on them. The latter reaction force is what leads to their forward motion.

Sometimes the origin of the reaction force is not obvious. A book lying on a table exerts the downward force of its weight; but how can an apparently rigid object like the table exert an upward force on the book? If the tabletop were made of rubber, we would see the book push it down, and the upward force would result from the elasticity of the rubber. A similar explanation actually holds for tabletops of wood or metal, which are never perfectly rigid, although the depressions made in them may be extremely small.

It is sometimes arbitrary which force of an action-reaction pair to consider action and which reaction. For instance, we can't really say that the gravitational pull of the earth on a pear is the action force and the pull of the pear on the earth is the reaction force, or the other way around. When you push on the ground when walking, however, it is legitimate to call this force the action force and the force with which the earth pushes back on you the reaction force.

Gravitation

Left to itself, a moving object travels in a straight line at constant speed. Because the moon circles the earth and the planets circle the sun, forces must be acting on the moon and planets. As we learned in Chap. 1, Newton discovered that these forces are the same in nature as the gravitational force that holds us to the earth. Before we consider how gravity works, we must look into exactly how curved paths come about.

2.12 Circular Motion

A Curved Path Requires an Inward Pull

Tie a ball to the end of a string and whirl the ball around your head, as in Fig. 2-32. What you will find is that your hand must pull on the string to keep the ball moving in a circle. If you let go of the string, there is no longer an inward force on the ball, and it flies off to the side.

The force that has to be applied to make something move in a curved path is called **centripetal** ("toward the center") **force:**

> Centripetal force = inward force on an object moving in a curved path

The centripetal force always points toward the center of curvature of the object's path, which means the force is at right angles to the object's direction of motion at each moment. In Fig. 2-32 the ball is moving in a circle, so its velocity vector **v** is always tangent to the circle and the centripetal force vector $\mathbf{F}_c$ is always directed toward the center of the circle.

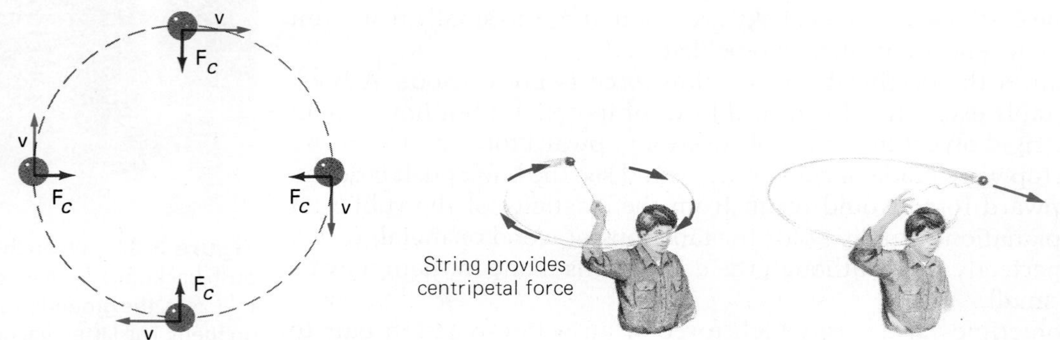

Figure 2-32 A centripetal force is necessary for circular motion. An inward centripetal force $\mathbf{F}_c$ acts upon every object that moves in a curved path. If the force is removed, the object continues moving in a straight line tangent to its original path.

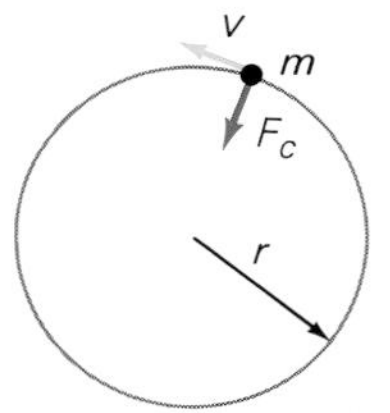

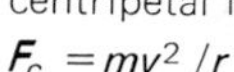

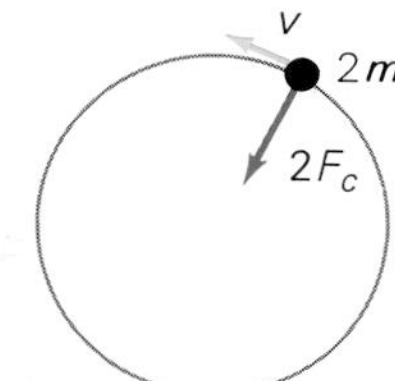

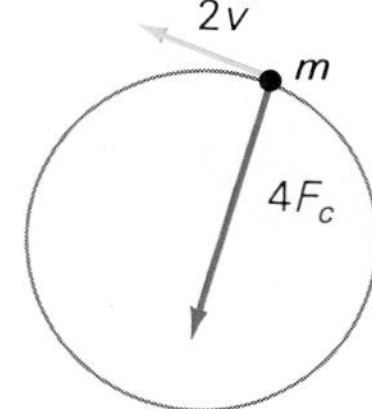

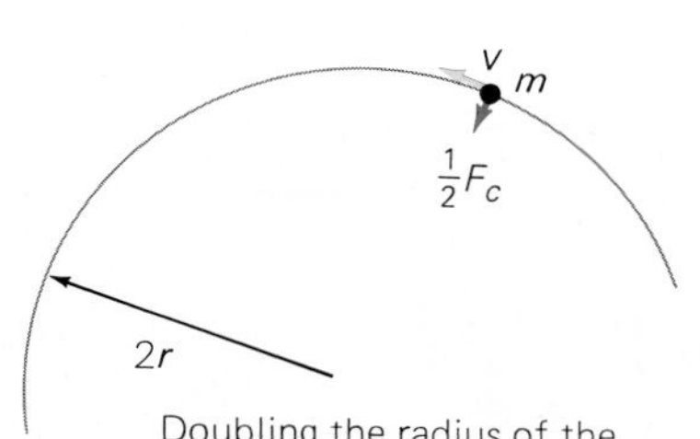

Figure 2-33 The centripetal force needed to keep an object moving in a circle depends upon the mass and speed of the object and upon the radius of the circle. The direction of the force is always toward the center of the circle. (Centripetal force is the name given to any force that is always directed toward a center of motion. It is not a distinct type of force, such as gravity or friction.)

A detailed calculation shows that the centripetal force $\mathbf{F}_c$ needed for something of mass m and speed v to travel in a circle of radius $\boldsymbol{r}$ has the magnitude

$$F_c = \frac{mv^2}{r} \qquad \textit{Centripetal force} \qquad \textit{2-18}$$

This formula tells us three things about the force needed to cause an object to move in a circular path: (1) the greater the object's mass, the greater the force; (2) the faster the object, the greater the force; and (3) the smaller the circle, the greater the force (Fig. 2-33).

Example 2.14

Find the centripetal force needed by a 1000-kg car moving at 5 m/s to go around a curve 30 m in radius, as in Fig. 2-34.

Solution

The centripetal force needed to make the turn is

$$F_c = \frac{mv^2}{r} = \frac{(1000\text{ kg})(5\text{ m/s})^2}{30\text{ m}} = 833\text{ N}$$

This force is easily transferred from the road to the car's tires if the road is dry and in good condition. However, if the car's speed were 20 m/s, the force needed would be 16 times as great, and the car would probably skid outward.

To reduce the chance of skids, particularly when the road is wet and therefore slippery, highway curves are often **banked** so that the roadbed tilts inward. A car going around a banked curve has an inward reaction force on it provided by the road itself, apart from friction (Fig. 2-35).

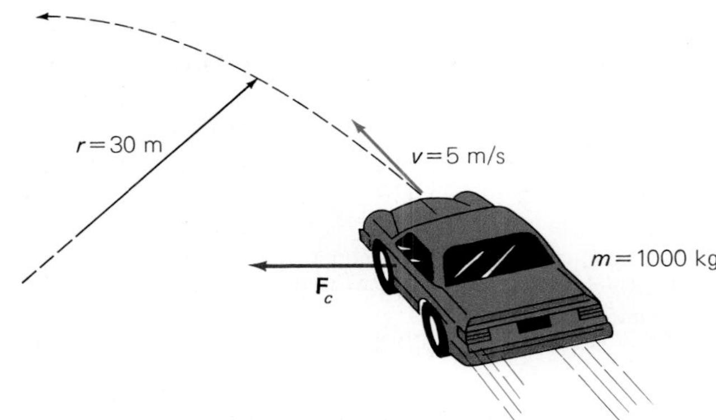

Figure 2-34 A centripetal force of 833 N is needed by this car to make the turn shown.

Figure 2-35 A wall of snow provides this bobsled with the centripetal force it needs to round the turn.

Example 2.15

A road has a hump 12 m in radius. What is the minimum speed at which a car will leave the road at the top of the hump?

Solution

The car will leave the hump when the required centripetal force mv^2/r is more than the car's weight of mg, since it is this weight that is providing the centripetal force. Hence

$$mg = \frac{mv^2}{r}, \qquad g = \frac{v^2}{r}, \qquad v = \sqrt{rg}$$

We note that the mass of the car does not matter here. (We will find the same formula later as the orbital speed a satellite of the earth must have.) Substituting for r and g gives

$$v = \sqrt{(12 \text{ m})(9.8 \text{ m/s}^2)} = 10.8 \text{ m/s}$$

This is about 24 mi/h. Driving faster over the hump is not recommended.

From the formula for F_c we can see why cars rounding a curve are so difficult to steer when the curve is sharp (small r) or the speed is high (a large value for v means a very large value for v^2). On a level road, the centripetal force is supplied by friction between the car's tires and the road. If the force needed to make a particular turn at a certain speed is more than friction can supply, the car skids outward.

2.13 Newton's Law of Gravity

What Holds the Solar System Together

Newton used Kepler's laws of planetary motion and Galileo's findings about falling bodies to establish how the gravitational force between two objects depends on their masses and on the distance between them. His conclusion was this:

> Every object in the universe attracts every other object with a force proportional to both of their masses and inversely proportional to the square of the distance between them.

In equation form, **Newton's law of gravity** states that the force F that acts between two objects whose masses are m_1 and m_2 is

$$\text{Gravitational force} = F = \frac{Gm_1m_2}{R^2} \qquad \textit{Law of gravity} \qquad \textit{2-19}$$

Here R is the distance between the objects and G is a constant of nature, the same number everywhere in the universe. The value of G is $6.670 \times 10^{-11}\ \text{N} \cdot \text{m}^2/\text{kg}^2$.

The point in an object from which R is to be measured depends on the object's shape and on the way in which its mass is distributed. The **center of mass** of a uniform sphere is its geometric center (Fig. 2-36).

The inverse square—$1/R^2$—variation of gravitational force with distance R means that this force drops off rapidly with increasing R (Fig. 2-37).

Figure 2-38 shows how this variation affects the weight of a 61-kg astronaut who leaves the earth on a spacecraft. At the earth's surface she weighs 600 N (135 lb); that is, the gravitational attraction of the earth on her is 600 N. When she is 100 times farther from the center of the earth, her weight is $1/100^2$ or $\frac{1}{10{,}000}$ as great, only 0.06 N—the weight of a cigar on the earth's surface.

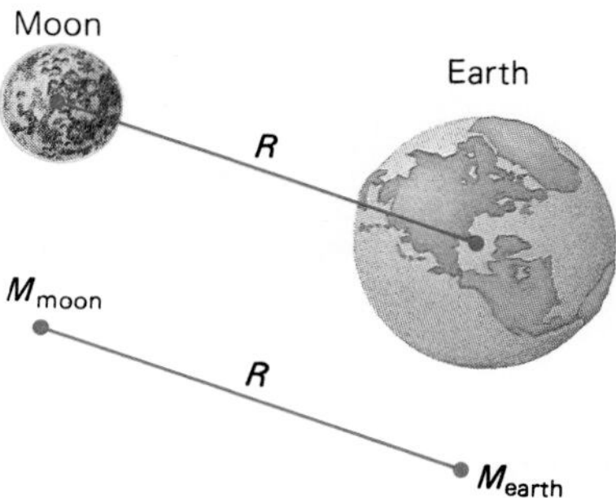

Figure 2-36 For computing gravitational effects, spherical bodies (such as the earth and moon) may be regarded as though their masses are located at their geometrical centers, provided that they are uniform spheres or consist of concentric uniform spherical shells.

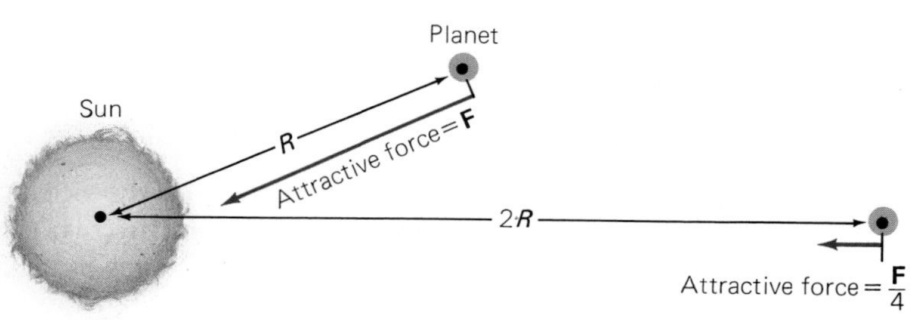

Figure 2-37 The gravitational force between two bodies depends upon the square of the distance between them. The gravitational force on a planet would drop to one-fourth its usual amount if the distance of the planet from the sun were to be doubled. If the distance is halved, the force would increase to 4 times its usual amount.

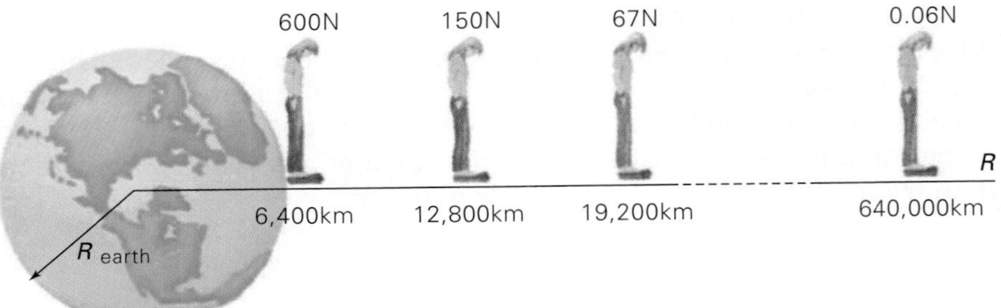

Figure 2-38 The weight of a person near the earth is the gravitational force the earth exerts upon her. As she goes farther and farther away from the earth's surface, her weight decreases inversely as the square of her distance from the earth's center. The mass of the person here is 61 kg.

Example 2.16

On the basis of what we know already, we can find the mass of the earth. This sounds, perhaps, like a formidable job, but it is really fairly easy to do. It is worth following as an example of the indirect way in which scientists go about performing such seemingly impossible feats as "weighing" the earth, the sun, other planets, and even distant stars.

Solution

Let us focus our attention on an apple of mass m on the earth's surface. The downward force of gravity on the apple is its weight of mg:

$$\text{Weight of apple} = F = mg$$

We can also use Newton's law of gravity to find F, with the result

$$\text{Gravitational force on apple} = F = \frac{GmM}{R^2}$$

Here M is the earth's mass and R is the distance between the apple and the center of the earth, which is the earth's radius of 6400 km $= 6.4 \times 10^6$ m (Fig. 2-39). The two ways to find F must give the same result, so

$$\text{Gravitational force on apple} = \text{weight of apple}$$

$$\frac{GmM}{R^2} = mg$$

We note that the apple's mass m appears on both sides of this equation, hence it cancels out. Solving for the earth's mass M gives

$$M = \frac{gR^2}{G} = \frac{(9.8\ m/s^2)(6.4 \times 10^6\ m)^2}{6.67 \times 10^{-11}\ N \cdot m^2/kg^2} = 6 \times 10^{24}\ \text{kg}$$

The number 6×10^{24} is 6 followed by 24 zeros! Enormous as it is, the earth is one of the least massive planets: Saturn has 95 times as much mass, and Jupiter 318 times as much. The sun's mass is more than 300,000 times that of the earth.

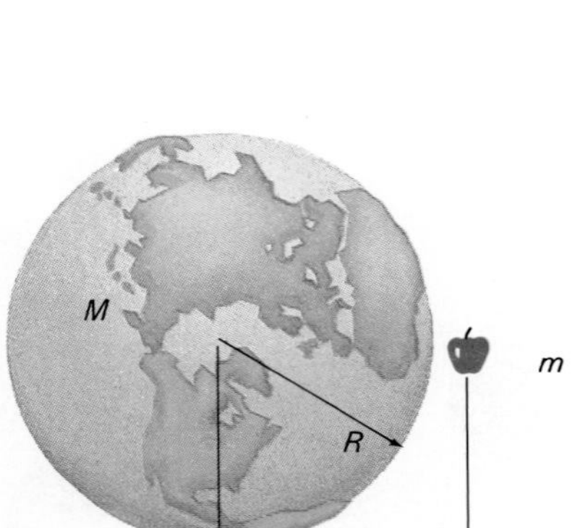

Figure 2-39 The gravitational force of the earth on an apple at the earth's surface is the same as the force between masses M and m the distance R apart. This force equals the weight of the apple.

2.14 Artificial Satellites

Thousands Circle the Earth

The first artificial satellite, Sputnik I, was launched by the Soviet Union in 1957. Since then thousands of others have been put into orbits around the earth, most of them by the United States and the former Soviet Union. Men and women have been in orbit regularly since 1961, when a Soviet cosmonaut circled the globe at an average height of 240 km (Fig. 2-40).

Figure 2-40 An earth satellite is always falling toward the earth. As a result, an astronaut inside feels "weightless," just as a person who jumps off a diving board feels "weightless." But a gravitational force does act on both people—what is missing is the upward reaction force of the ground, the diving board, the floor of a room, the seat of a chair, or whatever each person would otherwise be pressing on. In the case of an astronaut, the floor of the satellite falls just as fast as he or she does instead of pushing back.

Almost a thousand active satellites are now in orbit. (Over 2000 more are inactive.) About half belong to the United States, 10 percent to Russia, and 4 percent to China, with the rest distributed among two dozen other countries. The closest satellites, from 80 to 2000 km above the earth, are mostly "eyes in the sky" that map the earth's surface; survey it for military purposes; provide information on weather and resources such as mineral deposits, crops, and water; and carry out various scientific studies. A group of 66 satellites in low-earth orbit supports the worldwide Iridium telephone system. Twenty-seven satellites (3 of them spares) at an altitude of 20,360 km are used in the Global Positioning System (GPS) developed by the United States (Fig. 2-41). GPS receivers,

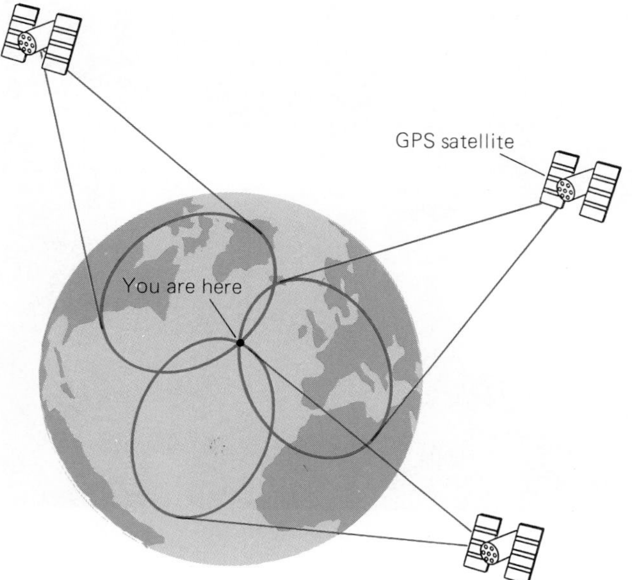

Figure 2-41 In the Global Positioning System (GPS), each of a fleet of orbiting satellites sends out coded radio signals that enable a receiver on the earth to determine both the exact position of the satellite in space and its exact distance from the receiver. Given this information, a computer in the receiver then calculates the circle on the earth's surface on which the receiver must lie. Data from three satellites give three circles, and the receiver must be located at the one point where all three intersect.

Space Junk

A remarkable amount of debris is in orbit around the earth, relics of the 5000 or so rockets that have thus far gone into space. The U.S. Space Surveillance Network uses radar to track more than 10,000 objects over 10 cm across, which range from dropped astronaut tools all the way up to discarded rocket stages and "dead" satellites, that might collide catastrophically with spacecraft. Most of the debris orbits are 700 to 1000 km above the earth. In addition, millions of smaller bits and pieces are out there, many of them able to damage spacecraft windows, solar cells, and other relatively fragile components. Even encountering something the size of a pea would be an unwelcome event when its impact speed is 15 to 45 times the speed of a bullet. This is no idle worry: in 2006 a Russian broadcasting satellite was crippled by a piece of space junk. The windows of the space shuttles have to be replaced regularly due to damage from small bits of debris.

Seeing Satellites

On a dark night many satellites, even quite small ones, are visible to the naked eye because of the sunlight they reflect. The best times to look are an hour or so before sunrise and an hour or so after sunset. High-altitude satellites can be seen for longer periods because they spend more time outside the earth's shadow. When conditions are just right, the as-yet-incomplete International Space Station is about as bright as the brightest stars. When it is finished, the ISS may rival Venus, except for the moon the most brilliant object in the night sky.

some hardly larger than a wristwatch, enable users to find their positions, including altitude, anywhere in the world at any time with uncertainties of only a few meters (Fig. 2-42). A similar system, called Glonass, is operated by Russia; China has begun work on a system called Baldu ("Big Dipper"); and the European Union is developing yet another system, called Galileo.

The most distant satellites, nearly half the total, circle the equator exactly once a day, so they remain in place indefinitely over a particular location on the earth. A satellite in such a geostationary orbit can "see"

Figure 2-42 An automobile navigation system uses GPS data to show the location of a car on a displayed map and can give visual or verbal directions to a destination. The map can be a normal top-down view or, in some models, a perspective view. Portable GPS receivers can be as small as a large wristwatch.

a large area of the earth's surface. Most satellites in geostationary orbits are used to relay communications of all kinds from one place to another, which is often cheaper than using cables between them.

Why Satellites Don't Fall Down What keeps all these satellites up there? The answer is that a satellite *is* actually falling down but, like the moon (which is a natural satellite), at exactly such a rate as to circle the earth in a stable orbit. "Stable" is a relative term, to be sure, since friction due to the extremely thin atmosphere present at the altitudes of actual satellites will eventually bring them down. Satellite lifetimes in orbit range from a matter of days to hundreds of years.

Let us think about a satellite in a circular orbit. The gravitational force on the satellite is its weight mg, where g is the acceleration of gravity at the satellite's altitude (the value of g decreases with increasing altitude). The centripetal force a satellite of speed v needs to circle the earth at the distance r from the earth's center is mv^2/r. Since the earth's gravity is providing this centripetal force,

$$\text{Centripetal force} = \text{gravitational force}$$

$$\frac{mv^2}{r} = mg$$

$$v^2 = rg$$

$$\text{Satellite speed} = v = \sqrt{rg}$$

The mass of the satellite does not matter.

For an orbit a few kilometers above the earth's surface, the satellite speed turns out to be about 28,400 km/h. Anything sent off around the earth at this speed will become a satellite of the earth. (Of course, at such a low altitude air resistance will soon bring it down.) At a lower speed than this an object sent into space would simply fall to the earth, while at a higher speed it would have an elliptical rather than a circular orbit (Fig. 2-43). A satellite initially in an elliptical orbit can be given a circular orbit if it has a small rocket motor to give it a further push at the required distance from the earth (Fig. 2-44).

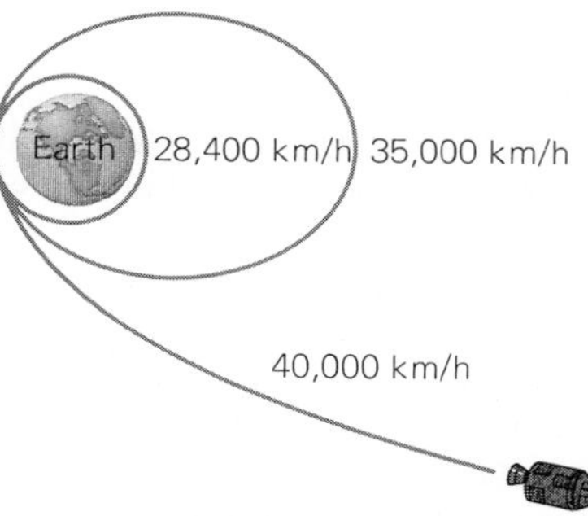

Figure 2-43 The minimum speed an earth satellite can have is 28,400 km/h. The escape speed from the earth is 40,000 km/h.

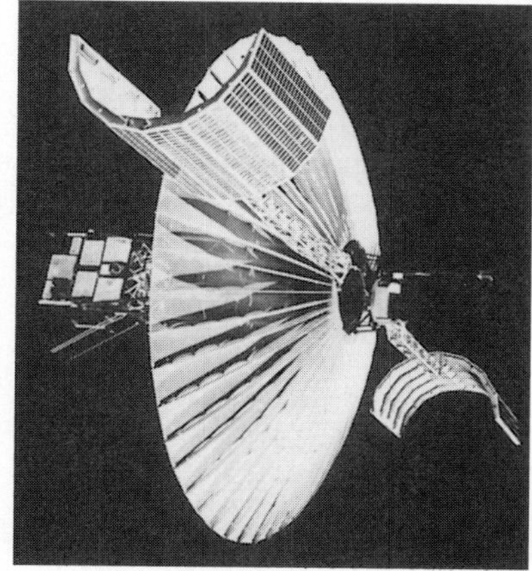

Figure 2-44 This Landsat satellite circles the earth at an altitude of 915 km. The satellite carries a television camera and a scanner system that provides images of the earth's surface in four color bands. The data radioed back provide information valuable in geology, water supply, agriculture, and land-use planning. Today the cost of placing a satellite in earth orbit is between about $10,000 and $20,000 per kg of payload, but new rocket designs are hoped to do this for as little as $2000 per kg.

Escape Speed If its original speed is high enough, at least 40,000 km/h, a spacecraft can escape entirely from the earth. The speed required for something to leave the gravitational influence of an astronomical body permanently is called the **escape speed.** Readers of *Through the Looking Glass* may recall the Red Queen's remark, "Now, here, you see, it takes all the running you can do to stay in the same place. If you want to get somewhere else, you must run at least twice as fast as that!" The ratio between escape speed and minimum orbital speed is actually $\sqrt{2}$, about 1.41. Escape speeds for the planets are listed in Table 16-1.

It is worth keeping in mind that escape speed is the initial speed needed for something to leave a planet (or other astronomical body, such as a star) permanently when there is no further propulsion. A spacecraft whose motor runs continuously can go out into space without ever reaching the escape speed—even a snail's pace would be enough, given sufficient time.

Important Terms and Ideas

When we say something is moving, we mean that its position relative to something else—the **frame of reference**—is changing. The choice of an appropriate frame of reference depends on the situation.

The **speed** of an object is the rate at which it covers distance relative to a frame of reference. The object's **velocity** specifies both its speed and the direction in which it is moving. The **acceleration** of an object is the rate at which its speed changes. Changes in direction are also accelerations.

A **scalar quantity** has magnitude only; mass and speed are examples. A **vector quantity** has both magnitude and direction; force and velocity are examples. An arrowed line that represents the magnitude and direction of a quantity is called a **vector.**

The **acceleration of gravity** is the downward acceleration of a freely falling object near the earth's surface. Its value is $g = 9.8\ \text{m/s}^2$.

The **inertia** of an object is the resistance the object offers to any change in its state of rest or motion. The property of matter that shows itself as inertia is called **mass;** mass may be thought of as quantity of matter. The unit of mass is the **kilogram** (kg).

A **force** is any influence that can cause an object to be accelerated. The unit of force is the **newton** (N). The **weight** of an object is the gravitational force with which the earth attracts it.

Newton's **first law of motion** states that, if no net force acts on it, every object continues in its state of rest or uniform motion in a straight line. Newton's **second law of motion** states that when a net force F acts on an object of mass m, the object is given an acceleration of F/m in the same direction as that of the force. Newton's **third law of motion** states that when one object exerts a force on a second object, the second object exerts an equal but opposite force on the first. Thus for every **action force** there is an equal but opposite **reaction force.**

The **centripetal force** on an object moving along a curved path is the inward force needed to cause this motion. Centripetal force acts toward the center of curvature of the path.

Newton's law of gravity states that every object in the universe attracts every other object with a force directly proportional to both their masses and inversely proportional to the square of the distance separating them.

Important Formulas

Pythagorean theorem for right triangle (C is longest side): $A^2 + B^2 = C^2$

Speed: $v = \dfrac{d}{t}$

Acceleration: $a = \dfrac{v_2 - v_1}{t}$

$$d = v_1 t + \tfrac{1}{2}at^2$$

$$h = \tfrac{1}{2}gt^2$$

Second law of motion: $F = ma$

Weight: $w = mg$

Centripetal force: $F_c = \dfrac{mv^2}{r}$

Law of gravity: $F = \dfrac{Gm_1m_2}{R^2}$

Multiple Choice

1. Which of the following quantities is not a vector quantity?
   - **a.** velocity
   - **b.** acceleration
   - **c.** mass
   - **d.** force
2. Which of the following units could be associated with a vector quantity?
   - **a.** kilogram
   - **b.** hour
   - **c.** liter
   - **d.** meter/second
3. A box suspended by a rope is pulled to one side by a horizontal force. The tension in the rope
   - **a.** is less than before
   - **b.** is unchanged
   - **c.** is greater than before
   - **d.** may be any of the above, depending on how strong the force is
4. The sum of two vectors is a minimum when the angle between them is
   - **a.** 0
   - **b.** 45°
   - **c.** 90°
   - **d.** 180°
5. In which of the following examples is the motion of the car not accelerated?
   - **a.** A car turns a corner at the constant speed of 20 km/h
   - **b.** A car climbs a steep hill with its speed dropping from 60 km/h at the bottom to 15 km/h at the top
   - **c.** A car climbs a steep hill at the constant speed of 40 km/h
   - **d.** A car climbs a steep hill and goes over the crest and down on the other side, all at the same speed of 40 km/h

6. Two objects have the same size and shape but one of them is twice as heavy as the other. They are dropped simultaneously from a tower. If air resistance is negligible,

a. the heavy object strikes the ground before the light one
b. they strike the ground at the same time, but the heavy object has the higher speed
c. they strike the ground at the same time and have the same speed
d. they strike the ground at the same time, but the heavy object has the lower acceleration because it has more mass

7. The acceleration of a stone thrown upward is

a. greater than that of a stone thrown downward
b. the same as that of a stone thrown downward
c. less than that of a stone thrown downward
d. zero until it reaches the highest point in its path

8. You are riding a bicycle at constant speed when you throw a ball vertically upward. It will land

a. in front of you
b. on your head
c. behind you
d. any of the above, depending on the ball's speed

9. When an object is accelerated,

a. its direction never changes
b. its speed always increases
c. it always falls toward the earth
d. a net force always acts on it

10. If we know the magnitude and direction of the net force on an object of known mass, Newton's second law of motion lets us find its

a. position
b. speed
c. acceleration
d. weight

11. The weight of an object

a. is the quantity of matter it contains
b. is the force with which it is attracted to the earth
c. is basically the same quantity as its mass but is expressed in different units
d. refers to its inertia

12. Compared with her mass and weight on the earth, an astronaut on Venus, where the acceleration of gravity is 8.8 m/s^2, has

a. less mass and less weight
b. less mass and the same weight
c. less mass and more weight
d. the same mass and less weight

13. In Newton's third law of motion, the action and reaction forces

a. act on the same object
b. act on different objects
c. need not be equal in strength but must act in opposite directions
d. must be equal in strength but need not act in opposite directions

14. A car that is towing a trailer is accelerating on a level road. The magnitude of the force the car exerts on the trailer is

a. equal to the force the trailer exerts on the car
b. greater than the force the trailer exerts on the car
c. equal to the force the trailer exerts on the road
d. equal to the force the road exerts on the trailer

15. When a boy pulls a cart, the force that causes him to move forward is

a. the force the cart exerts on him
b. the force he exerts on the cart
c. the force he exerts on the ground with his feet
d. the force the ground exerts on his feet

16. In order to cause something to move in a circular path, it is necessary to provide

a. a reaction force
b. an inertial force
c. a centripetal force
d. a gravitational force

17. A body moving in a circle at constant speed is accelerated

a. in the direction of motion
b. toward the center of the circle
c. away from the center of the circle
d. any of these, depending upon the circumstances

18. A car rounds a curve on a level road. The centripetal force on the car is provided by

a. inertia
b. gravity
c. friction between the tires and the road
d. the force applied to the steering wheel

19. The centripetal force that keeps the earth in its orbit around the sun is provided

a. by inertia
b. by the earth's rotation on its axis
c. partly by the gravitational pull of the sun
d. entirely by the gravitational pull of the sun

20. The gravitational force with which the earth attracts the moon

a. is less than the force with which the moon attracts the earth
b. is the same as the force with which the moon attracts the earth
c. is more than the force with which the moon attracts the earth
d. varies with the phase of the moon

21. The speed needed to put a satellite in orbit does not depend on
a. the mass of the satellite
b. the radius of the orbit
c. the shape of the orbit
d. the value of *g* at the orbit

22. An astronaut inside an orbiting satellite feels weightless because
a. he or she is wearing a space suit
b. the satellite is falling toward the earth just as fast as the astronaut is, so there is no upward reaction force on him or her
c. there is no gravitational pull from the earth so far away
d. the sun's gravitational pull balances out the earth's gravitational pull

23. A bicycle travels 12 km in 40 min. Its average speed is
a. 0.3 km/h **c.** 18 km/h
b. 8 km/h **d.** 48 km/h

24. Which one or more of the following sets of displacements might be able to return a car to its starting point?
a. 5, 5, and 5 km **c.** 5, 10, and 10 km
b. 5, 5, and 10 km **d.** 5, 5, and 20 km

25. An airplane whose airspeed is 200 km/h is flying in a wind of 80 km/h. The airplane's speed relative to the ground is between
a. 80 and 200 km/h **c.** 120 and 200 km/h
b. 80 and 280 km/h **d.** 120 and 280 km/h

26. A ship travels 200 km to the south and then 400 km to the west. The ship's displacement from its starting point is
a. 200 km **c.** 450 km
b. 400 km **d.** 600 km

27. How long does a car whose acceleration is 2 m/s^2 need to go from 10 m/s to 30 m/s?
a. 10 s **c.** 40 s
b. 20 s **d.** 400 s

28. A ball is thrown upward at a speed of 12 m/s. It will reach the top of its path in about
a. 0.6 s **c.** 1.8 s
b. 1.2 s **d.** 2.4 s

29. A car that starts from rest has a constant acceleration of 4 m/s^2. In the first 3 s the car travels
a. 6 m **c.** 18 m
b. 12 m **d.** 172 m

30. A car traveling at 10 m/s begins to be accelerated at 12 m/s^2. The distance the car covers in the first 5 s after the acceleration begins is
a. 15 m **c.** 53 m
b. 25 m **d.** 65 m

31. A car with its brakes applied has an acceleration of $-1.2\ m/s^2$. If its initial speed is 10 m/s, the distance the car covers in the first 5 s after the acceleration begins is
a. 15 m **c.** 35 m
b. 32 m **d.** 47 m

32. The distance the car of Multiple Choice 31 travels before it comes to a stop is
a. 6.5 m **c.** 21 m
b. 8.3 m **d.** 42 m

33. A bottle falls from a blimp whose altitude is 1200 m. If there was no air resistance, the bottle would reach the ground in
a. 5 s **c.** 16 s
b. 11 s **d.** 245 s

34. When a net force of 1 N acts on a 1-kg body, the body receives
a. a speed of 1 m/s
b. an acceleration of 0.1 m/s^2
c. an acceleration of 1 m/s^2
d. an acceleration of 9.8 m/s^2

35. When a net force of 1 N acts on a 1-N body, the body receives
a. a speed of 1 m/s
b. an acceleration of 0.1 m/s^2
c. an acceleration of 1 m/s^2
d. an acceleration of 9.8 m/s^2

36. A car whose mass is 1600 kg (including the driver) has a maximum acceleration of 1.2 m/s^2. If three 80-kg passengers are also in the car, its maximum acceleration will be
a. 0.5 m/s^2 **c.** 1.04 m/s^2
b. 0.72 m/s^2 **d.** 1.2 m/s^2

37. A 300-g ball is struck with a bat with a force of 150 N. If the bat was in contact with the ball for 0.020 s, the ball's speed is
a. 0.01 m/s **c.** 2.5 m/s
b. 0.1 m/s **d.** 10 m/s

38. A bicycle and its rider together have a mass of 80 kg. If the bicycle's speed is 6 m/s, the force needed to bring it to a stop in 4 s is
a. 12 N **c.** 120 N
b. 53 N **d.** 1176 N

39. The weight of 400 g of onions is
a. 0.041 N **c.** 3.9 N
b. 0.4 N **d.** 3920 N

40. A salami weighs 3 lb. Its mass is
a. 0.31 kg **c.** 6.6 kg
b. 1.36 kg **d.** 29.4 kg

41. An upward force of 600 N acts on a 50-kg dumbwaiter. The dumbwaiter's acceleration is

a. 0.82 m/s^2 **c.** 11 m/s^2
b. 2.2 m/s^2 **d.** 12 m/s^2

42. The upward force the rope of a hoist must exert to raise a 400-kg load of bricks with an acceleration of 0.4 m/s^2 is

a. 160 N **c.** 3760 N
b. 1568 N **d.** 4080 N

43. The radius of the circle in which an object is moving at constant speed is doubled. The required centripetal force is

a. one-quarter as great as before
b. one-half as great as before
c. twice as great as before
d. 4 times as great as before

44. A car rounds a curve at 20 km/h. If it rounds the curve at 40 km/h, its tendency to overturn is

a. halved
b. doubled
c. tripled
d. quadrupled

45. A 1200-kg car whose speed is 6 m/s rounds a turn whose radius is 30 m. The centripetal force on the car is

a. 48 N **c.** 240 N
b. 147 N **d.** 1440 N

46. If the earth were 3 times as far from the sun as it is now, the gravitational force exerted on it by the sun would be

a. 3 times as large as it is now
b. 9 times as large as it is now
c. one-third as large as it is now
d. one-ninth as large as it is now

47. A woman whose mass is 60 kg on the earth's surface is in a spacecraft at an altitude of one earth's radius above the surface. Her mass there is

a. 15 kg **c.** 60 kg
b. 30 kg **d.** 120 kg

48. A man whose weight is 800 N on the earth's surface is also in the spacecraft of Multiple Choice 47. His weight there is

a. 200 N **c.** 800 N
b. 400 N **d.** 1600 N

Exercises

2.1 Speed

1. A woman standing before a cliff claps her hands, and 2.8 s later she hears the echo. How far away is the cliff? The speed of sound in air at ordinary temperatures is 343 m/s.

2. How many seconds are needed by a car whose speed is 40 km/h to cover a distance of 800 m?

3. In 1977 Steve Weldon ate 91 m of spaghetti in 29 s. At the same speed, how long would it take Mr. Weldon to eat 5 m of spaghetti?

4. How many minutes would you save by making a 100-km trip at 125 km/h instead of at 80 km/h?

5. The speed of light is 3.0×10^8 m/s. How many minutes does it take for light to reach the earth from the sun, which is 1.5×10^{11} m away?

2.2 Vectors

6. Is it correct to say that scalar quantities are abstract, idealized quantities with no precise counterparts in the physical world, whereas vector quantities properly represent reality because they take directions into account?

7. What is the minimum number of unequal forces whose vector sum can equal zero?

8. An airplane whose speed through the air is 500 km/h covers a distance over the ground of 1000 km in 2.5 h. How fast was the wind against it?

9. A woman walks 70 m to an elevator and then rises upward 40 m. What is her displacement from her starting point?

10. Two cars leave a crossroads at the same time. One heads north at 50 km/h and the other heads east at 70 km/h. How far apart are the cars after 0.5 h? After 2.0 h?

2.3 Acceleration

11. Can a rapidly moving object have the same acceleration as a slowly moving one?

12. Can anything have an acceleration in the opposite direction to its velocity? If so, give some examples.

13. A car whose acceleration is constant reaches a speed of 80 km/h in 20 s starting from rest. How much more time is required for it to reach a speed of 130 km/h?

14. The tires of a car begin to lose their grip on the road at an acceleration of 5 m/s^2. At this acceleration, how long does the car need to reach a speed of 25 m/s starting from 10 m/s?

15. A car starts from rest and reaches a speed of 40 m/s in 10 s. If its acceleration remains the same, how fast will it be moving 5 s later?

16. The brakes of a car moving at 14 m/s are applied, and the car comes to a stop in 4 s. (a) What was the car's acceleration? (b) How long would the car take to come to a stop starting from 20 m/s with the same acceleration? (c) How long would the car take to slow down from 20 m/s to 10 m/s with the same acceleration?

2.4 Distance, Time, and Acceleration

17. A car is moving at 10 m/s when it begins to be accelerated at 2.5 m/s^2. (a) How long does the car take to reach a speed of 25 m/s? (b) How far does it go during this period?

18. An airplane has a takeoff speed of 80 m/s which it reaches 35 s after starting from rest. What is the minimum length of the runway?

19. A car starts from rest and covers 400 m (very nearly $\frac{1}{4}$ mi) in 20 s. Find the average acceleration of the car and its final speed.

2.5 Free Fall

20. Can anything ever have a downward acceleration greater than g? If so, how can this be accomplished?

21. Suppose you are in a barrel going over Niagara Falls and during the fall you drop an apple inside the barrel. Would the apple appear to move toward the top of the barrel or toward the bottom, or would it remain stationary within the barrel?

22. A rifle is aimed directly at a squirrel in a tree. Should the squirrel drop from the tree at the instant the rifle is fired or should it remain where it is? Why?

23. When a football is thrown, it follows a curved path through the air like the ones shown in Fig. 2-18. Where in its path is the ball's speed greatest? Where is it least?

24. A crate is dropped from an airplane flying horizontally at constant speed. How does the path of the crate appear to somebody on the airplane? To somebody on the ground?

25. A stone is thrown horizontally from a cliff and another, identical stone is dropped from there at the same time. Do the stones reach the ground at the same time? How do their speeds compare when they reach the ground? Their accelerations?

26. (a) Imagine that Charlotte drops a ball from a window on the twentieth floor of a building while at the same time Fred drops another ball from a window on the nineteenth floor of that building. As the balls fall, what happens to the distance between them (assuming no air resistance)? (b) Next imagine that Charlotte and Fred are at the same window on the twentieth floor and that Fred drops his ball a few seconds after Charlotte drops hers. As the balls fall, what happens to the distance between them now (again assuming no air resistance)?

27. A person in a stationary elevator drops a coin and the coin reaches the floor of the elevator 0.6 s later. Would the coin reach the floor in less time, the same time, or more time if it were dropped when the elevator was (a) falling at a constant speed? (b) falling at a constant acceleration? (c) rising at a constant speed? (d) rising at a constant acceleration?

28. A movie seems to show a ball falling down. From what appears on the screen, could you tell whether the movie is actually showing a ball being thrown upward but the film is being run backward in the projector?

29. A person dives off the edge of a cliff 33 m above the surface of the sea below. Assuming that air resistance is negligible, how long does the dive last and with what speed does the person enter the water?

30. If there were no air resistance, a quarter dropped from the top of New York's Empire State Building would reach the ground 9.6 s later. (a) What would its speed be? (b) How high is the building?

31. A ball is thrown upward from the edge of a cliff with an initial speed of 6 m/s. (a) How fast is it moving 0.5 s later? In what direction? (b) How fast is it moving 2 s later? In what direction? (Consider upward as + and downward as −; then $v_1 = +6$ m/s and $g = -9.8$ m/s^2)

32. A ball thrown upward reaches a height of 13 m. How much time did it take to get this high? How much time will it take to fall to the ground?

33. A ball is thrown downward at 12 m/s. What is its speed 1.0 s later?

34. A ball is thrown upward at 12 m/s. What is its speed 1.0 s later?

35. A ball is thrown vertically upward with an initial speed of 30 m/s. (a) How long will it take the ball to reach the highest point in its path? (b) How long will it take the ball to return to its starting place? (c) What will the ball's speed be there?

36. A rescue line is to be thrown horizontally from the deck of a ship 20 m above sea level to a lifeboat 40 m away. What should the speed of the line be? (Hint: First calculate the time of fall.)

37. An airplane is in level flight at a speed of 100 m/s and an altitude of 1200 m when a windshield wiper falls off. What will the wiper's speed be when it reaches the ground? (Hint: A vector calculation is needed.)

38. A ball is thrown horizontally from a cliff at 8 m/s. How fast is it moving 2 s later?

39. A bullet is fired horizontally from a rifle at 200 m/s from a cliff above a plain below. The bullet reaches the plain 5 s later. (a) How high was the cliff? (b) How far from the cliff did the bullet reach the plain? (c) What was the bullet's speed when it reached the plain?

40. A stone is thrown horizontally from a vertical cliff 25 m high and lands on the ground below, 22 m from the foot of the cliff. What was the original horizontal speed of the stone?

41. A person at the masthead of a sailboat moving at constant speed in a straight line drops a wrench. The masthead is 20 m above the boat's deck and the stern of the boat is 10 m behind the mast. Is there a minimum speed the sailboat can have so that the wrench will not land on the deck? If there is such a speed, what is it?

2.9 Second Law of Motion

42. Compare the tension in the coupling between the first two cars of a train with the tension in the coupling between the last two cars when (a) the train's speed is constant and (b) the train is accelerating.

43. In accelerating from a standing start to a speed of 300 km/h (186 mi/h—not its top speed!), the 1900-kg Bugatti Veyron sports car exerts an average force on the road of 9.4 kN. How long does the car take to reach 300 km/h?

44. A 1200-kg car goes from 10 m/s to 24 m/s in 16 s. What is the average force acting upon it?

45. The brakes of the car from Exercise 44 exert a force of 4000 N. How long will it take for them to slow the car to a stop from an initial speed of 24 m/s?

46. The brakes of an 800-kg car can exert a force of 2.4 kN. (a) How long will the brakes take to slow the car to a stop from a speed of 30 m/s? (b) How far will the car travel in this time?

47. A bicycle and its rider together have a mass of 80 kg. If the bicycle's speed is 6 m/s, how much force is needed to bring it to a stop in 4 s?

48. A 430-g soccer ball at rest on the ground is kicked with a force of 600 N and flies off at 15 m/s. How long was the toe of the person kicking the ball in contact with it?

49. A force of 20 N gives a brick an acceleration of 5 m/s^2. (a) What force would be needed to give the brick an acceleration of 1 m/s^2? (b) An acceleration of 10 m/s^2?

50. A car has a maximum acceleration of 4 m/s^2. What is its maximum acceleration when it is towing another car with the same mass?

2.10 Mass and Weight

51. Distinguish between mass and weight.

52. Consider the statement: Sara weighs 55 kg. What is wrong with the statement? Give two ways to correct it.

53. Albert is standing still on the ground. Does this mean that there is no gravitational force acting on him? If such a force is acting on him, why is he not moving?

54. The acceleration of gravity on Jupiter is about 25 m/s^2 and your mass is 70 kg. What is your weight on Jupiter? What would your mass have to be in order to weigh this amount on the earth?

55. A person weighs 85 N on the surface of the moon and 490 N on the surface of the earth. What is the acceleration of gravity on the surface of the moon?

56. An 80-kg man slides down a rope at constant speed. (a) What is the minimum breaking strength the rope must have? (b) If the rope has precisely this strength, will it support the man if he tries to climb back up?

57. A 60-g tennis ball approaches a racket at 15 m/s, is in contact with the racket for 0.005 s, and then rebounds at 20 m/s. What was the average force the racket exerted on the ball?

58. A parachutist whose total mass is 100 kg is falling at 50 m/s when her parachute opens. Her speed drops to 6 m/s in 2 s. What is the total force her harness had to withstand? How many times her weight is this force?

59. A woman whose mass is 60 kg is riding in an elevator whose upward acceleration is 2 m/s^2. What force does she exert on the floor of the elevator?

60. A person stands on a scale in an elevator. When the elevator is at rest, the scale reads 700 N. When the elevator starts to move, the scale reads 600 N. (a) Is the elevator going up or down? (b) Is it accelerated? If so, what is the acceleration?

2.11 Third Law of Motion

61. Since the opposite forces of the third law of motion are equal in magnitude, how can anything ever be accelerated?

62. What is the relationship, if any, between the first and second laws of motion? Between the second and third laws of motion?

63. A book rests on a table. (a) What is the reaction force to the force the book exerts on the table? (b) To the force gravity exerts on the book?

64. Two children wish to break a string. Are they more likely to succeed if each takes one end of the string and they pull against each other, or if they tie one end of the string to a tree and both pull on the free end? Why?

65. An engineer designs a propeller-driven spacecraft. Because there is no air in space, the engineer includes a supply of oxygen as well as a supply of fuel for the motor. What do you think of the idea?

66. A 55-kg cheetah takes 3 s to reach a speed of 21 m/s from a standing start. What is its acceleration? What is its final speed in km/h? What average force does it exert on the ground to reach that speed? Is this the force that actually causes the acceleration?

67. When a 5-kg rifle is fired, the 9-g bullet is given an acceleration of 30 km/s while it is in the barrel. (a) How much force acts on the bullet? (b) Does any force act on the rifle? If so, how much and in what direction?

2.12 Circular Motion

68. Show from its defining formula that the unit of centripetal force is the newton.

69. Under what circumstances, if any, can something move in a circular path without a centripetal force acting on it?

70. When you whirl a ball at the end of a string, the ball seems to be pulling outward away from your hand. When you let the string go, however, the ball moves along a straight path perpendicular to the direction of the string at the moment you let go. Explain each of these effects.

71. What centripetal force is needed to keep a 1-kg ball moving in a circle of radius 2 m at a speed of 5 m/s?

72. The 200-g head of a golf club moves at 40 m/s in a circular arc of 1.2 m radius. How much force must the player exert on the handle of the club to prevent it from flying out of her hands at the bottom of the swing? Ignore the mass of the club's shaft.

73. The string of a certain yo-yo is 80 cm long and will break when the force on it is 10 N. What is the highest speed the 200-g yo-yo can have when it is being whirled in a circle? Ignore the gravitational pull of the earth on the yo-yo.

74. The greatest force a road can exert on the tires of a certain 1400-kg car moving at 25 m/s is 8 kN. What is the minimum radius of a turn the car can make without skidding?

75. A 40-kg crate is lying on the flat floor of the rear of a station wagon moving at 15 m/s. A force of 150 N is needed to slide the crate against the friction between the bottom of the crate and the floor. What is the minimum radius of a turn the station wagon can make if the box is not to slip?

76. Some people believe that aliens from elsewhere in the universe visit the earth in spacecraft that travel faster than jet airplanes and can turn in their own lengths. Calculate the centripetal force on a 50-kg alien in a spacecraft moving at 500 m/s (1120 mi/h) while it is making a turn of radius 30 m. How many times the weight of the alien is this force? Do you think such stories can be believed?

77. Find the minimum radius at which an airplane flying at 300 m/s can make a U-turn if the centripetal force on it is not to exceed 4 times the airplane's weight.

78. An airplane flying at a constant speed of 160 m/s pulls out of a dive in a circular arc. The 80-kg pilot presses down on his seat with a force of 4000 N at the bottom of the arc. What is the radius of the arc?

2.13 Newtion's Law of Gravity

79. A track team on the moon could set new records for the high jump or pole vault (if they did not need space suits, of course) because of the smaller gravitational force. Could sprinters also improve their times for the 100-m dash?

80. If the moon were twice as far from the earth as it is today, how would the gravitational force it exerts on the earth compare with the force it exerts today?

81. Compare the weight and mass of an object at the earth's surface with what they would be at an altitude of two earth's radii.

82. A hole is bored to the center of the earth and a stone is dropped into it. How do the mass and weight of the stone at the earth's center compare with their values at the earth's surface?

83. Is the sun's gravitational pull on the earth the same at all seasons of the year? Explain.

84. According to the theory of gravitation, the earth must be continually "falling" toward the sun. If this is true, why does the average distance between earth and sun not grow smaller?

85. According to Kepler's second law, the earth travels fastest when it is closest to the sun. Is this consistent with the law of gravitation? Explain.

86. A 2-kg mass is 20 cm away from a 5-kg mass. What is the gravitational force (a) that the 5-kg mass exerts on the 2-kg mass, and (b) that the 2-kg mass exerts on the 5-kg mass? (c) If both masses are free to move, what are their respective accelerations if no other forces are acting?

87. A dishonest grocer installs a 100-kg lead block under the pan of his scale. How much gravitational force does the lead exert on 2 kg of cheese placed on the pan if the centers of mass of the lead and cheese are 0.3 m apart? Compare this force with the weight of 1 g of cheese to see if putting the lead under the scale was worth doing.

88. A bull and a cow elephant, each of mass 2000 kg, attract each other gravitationally with a force of 1×10^{-5} N. How far apart are they?

2.14 Artificial Satellites

89. An airplane makes a vertical circle in which it is upside down at the top of the loop. Will the passengers fall out of their seats if there is no belt to hold them in place?

90. Is an astronaut in an orbiting spacecraft actually "weightless"?

91. With the help of the data in Table 17-1, find the minimum speed artificial satellites must have to pursue stable orbits about Jupiter.

92. Find the speed of a satellite that orbits the earth in a circle of radius 8000 km. At that distance from the earth g = 6.2 m/s.

Answers to Multiple Choice

1. c	**7.** b	**13.** b	**19.** d	**25.** d	**31.** c	**37.** d	**43.** b
2. d	**8.** b	**14.** a	**20.** b	**26.** c	**32.** d	**38.** c	**44.** d
3. c	**9.** d	**15.** d	**21.** a	**27.** a	**33.** c	**39.** c	**45.** d
4. d	**10.** c	**16.** c	**22.** b	**28.** b	**34.** c	**40.** b	**46.** d
5. c	**11.** b	**17.** b	**23.** c	**29.** c	**35.** d	**41.** b	**47.** c
6. c	**12.** d	**18.** c	**24.** a, b, c	**30.** d	**36.** c	**42.** d	**48.** a

3 Energy

In a pole vault, the athlete's energy of motion while running is first transformed into energy of position at the top of the vault, then back into energy of motion while falling, and finally into work done when landing.

Goals

When you have finished this chapter you should be able to complete the goals ▸ given for each section below:

Work

3.1 The Meaning of Work
A Measure of the Change a Force Produces
- ▸ Explain why work is an important quantity.

3.2 Power
The Rate of Doing Work
- ▸ Relate work, power, and time.

Energy

3.3 Kinetic Energy
The Energy of Motion

3.4 Potential Energy
The Energy of Position
- ▸ Distinguish between kinetic energy and potential energy.
- ▸ Give several examples of potential energy.

3.5 Energy Transformations
Easy Come, Easy Go

3.6 Conservation of Energy
A Fundamental Law of Nature
- ▸ State the law of conservation of energy and give several examples of energy transformations.
- ▸ Use the principle of conservation of energy to analyze events in which work and different forms of energy are transformed into one another.

3.7 The Nature of Heat
The Downfall of Caloric
- ▸ Discuss why heat is today regarded as a form of energy rather than as an actual substance.

Momentum

3.8 Linear Momentum
Another Conservation Law
- ▸ Define linear momentum and discuss its significance.

3.9 Rockets
Momentum Conservation Is the Basis of Space Travel
- ▸ Use the principle of conservation of linear momentum to analyze the motion of objects that collide with each other or push each other apart, for instance, when a rocket is fired.

3.10 Angular Momentum
A Measure of the Tendency of a Spinning Object to Continue to Spin
- ▸ Explain how conservation of angular momentum is used by skaters to spin faster and by footballs to travel farther.

Relativity

3.11 Special Relativity
Things Are Seldom What They Seem
- ▸ Describe several relativistic effects and indicate why they are not conspicuous in the everyday world.

3.12 Rest Energy
Matter Is a Form of Energy
- ▸ Explain what is meant by rest energy and be able to calculate the rest energy of an object of given mass.

3.13 General Relativity
Gravity Is a Warping of Spacetime
- ▸ Describe how gravity is interpreted in Einstein's general theory of relativity.

The word energy has become part of everyday life. We say that an active person is energetic. We hear a candy bar described as being full of energy. We complain about the cost of the electric energy that lights our lamps and turns our motors. We worry about some day running out of the energy stored in coal and oil. We argue about whether nuclear energy is a blessing or a curse. Exactly what is meant by energy?

In general, energy refers to an ability to accomplish change. When almost anything happens in the physical world, energy is somehow involved. But "change" is not a very precise notion, and we must be sure of exactly what we are talking about in order to go further. Our procedure will be to begin with the simpler idea of work and then use it to relate change and energy in the orderly way of science.

Work

Changes that take place in the physical world are the result of forces. Forces are needed to pick things up, to move things from one place to another, to squeeze things, to stretch things, and so on. However, not all forces act to produce changes, and it is the distinction between forces that accomplish change and forces that do not that is central to the idea of work.

No work done

Work done

Figure 3-1 Work is done by a force when the object it acts on moves while the force is applied. No work is done by pushing against a stationary wall. Work is done when throwing a ball because the ball moves while being pushed during the throw.

3.1 The Meaning of Work

A Measure of the Change a Force Produces

Suppose we push against a wall. When we stop, nothing has happened even though we exerted a force on the wall. But if we apply the same force to a stone, the stone flies through the air when we let it go (Fig. 3-1). The difference is that the wall did not move during our push but the stone did. A physicist would say that we have done work on the stone, and as a result it was accelerated and moved away from our hand.

Or we might try to lift a heavy barbell. If we fail, the world is exactly the same afterward. If we succeed, though, the barbell is now up in the air, which represents a change (Fig. 3-2). As before, the difference is that in the second case an object moved while we exerted a force on it, which means that work was done on the object.

To make our ideas definite, **work** is defined in this way:

> The work done by a force acting on an object is equal to the magnitude of the force multiplied by the distance through which the force acts when both are in the same direction.

If nothing moves, no work is done, no matter how great the force. And even if something moves, work is not done on it unless a force is acting on it.

What we usually think of as work agrees with this definition. However, we must be careful not to confuse becoming tired with the amount of work done. Pushing against a wall for an afternoon in the hot sun is certainly tiring, but we have done no work because the wall didn't move.

In equation form,

$$W = Fd \qquad \text{Work} \qquad \text{3-1}$$

Work done = (applied force)(distance through which force acts)

The direction of the force **F** is assumed to be the same as the direction of the displacement **d.** If not, for example in the case of a child pulling a wagon with a rope not parallel to the ground, we must use for F the magnitude F_d of the projection of the applied force **F** that acts in the direction of motion (Fig. 3-3).

A force that is perpendicular to the direction of motion of an object can do no work on the object. Thus gravity, which results in a downward force on everything near the earth, does no work on objects moving horizontally along the earth's surface. However, if we drop an object, work is definitely done on it as it falls to the ground.

Figure 3-2 Work is done when a barbell is lifted, but no work is done while it is being held in the air even though this can be very tiring.

The Joule The SI unit of work is the **joule** (J), where one joule is the amount of work done by a force of one newton when it acts through a distance of one meter. That is,

$$1 \text{ joule (J)} = 1 \text{ newton-meter (N} \cdot \text{m)}$$

The joule is named after the English scientist James Joule and is pronounced "jool." To raise an apple from your waist to your mouth takes about 1 J of work. Since $1 \text{ N} = 1 \text{ kg} \cdot \text{m/s}^2$, the joule can also be expressed as $1 \text{ J} = 1 \text{ N m} = 1 \text{ kg} \cdot \text{m}^2/\text{s}^2$, which is more convenient in some problems.

Work Done Against Gravity It is easy to find the work done in lifting an object against gravity. The force of gravity on the object is its weight of mg. In order to raise the object to a height h above its original position (Fig. 3-4a), we need to apply an upward force of $F = mg$. With $F = mg$ and $d = h$, Eq. 3-1 becomes

$$W = mgh \qquad \textit{Work done against gravity} \qquad \text{3-2}$$

Work = (weight)(height)

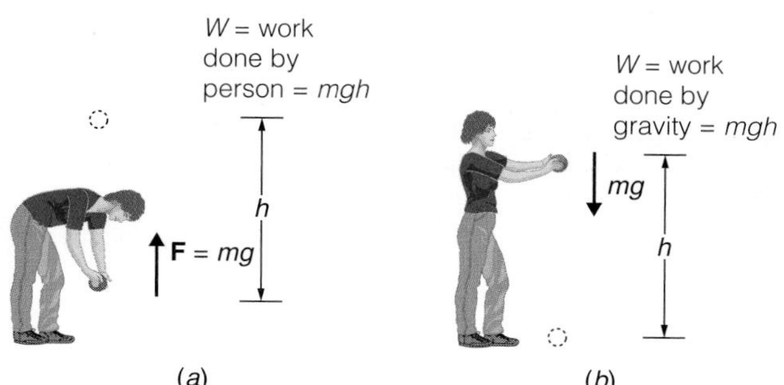

Figure 3-4 (a) The work a person does to lift an object to a height h is mgh. (b) If the object falls through the same height, the force of gravity does the work mgh.

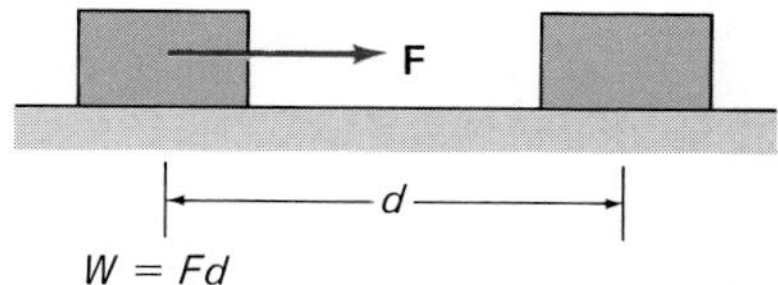

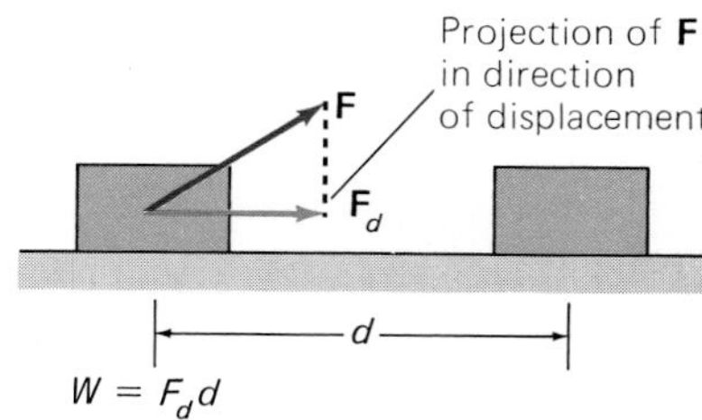

Figure 3-3 When a force and the distance through which it acts are parallel, the work done is equal to the product of F and d. When they are not in the same direction, the work done is equal to the product of d and the magnitude F_d of the projection of **F** in the direction of **d.**

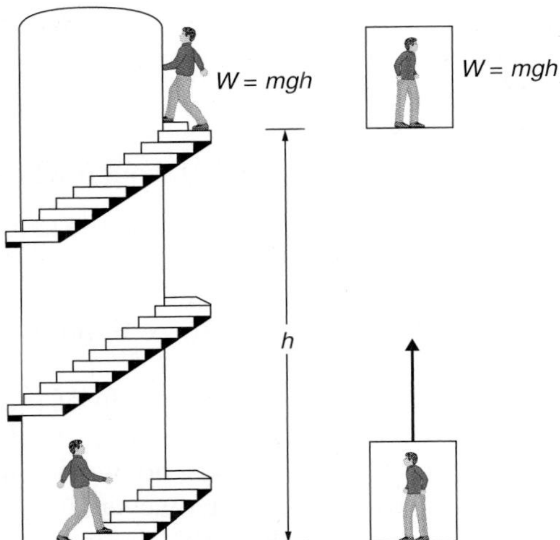

Figure 3-5 Neglecting friction, the work needed to raise a person to a height h is the same regardless of the path taken.

Only the total height h is involved here: the particular route upward taken by the object is not significant. Excluding friction, exactly as much work must be done when you climb a flight of stairs as when you go up to the same floor in an elevator (Fig. 3-5)—though the source of the work is not the same, to be sure.

If an object of mass m at the height h falls, the amount of work done *by* gravity on it is given by the same formula, $W = mgh$ (Fig. 3-4*b*).

Example 3.1

(a) A horizontal force of 100 N is used to push a 20-kg box across a level floor for 10 m. How much work is done? (b) How much work is needed to raise the same box by 10 m?

Solution

(a) The work done in pushing the box is

$$W = Fd = (100 \text{ N})(10 \text{ m}) = 1000 \text{ J}$$

The mass of the box does not matter here. What counts is the applied force, the distance through which it acts, and the relative directions of the force and the displacement of the box.

(b) Now the work done is

$$W = mgh = (20 \text{ kg})(9.8 \text{ m/s}^2)(10 \text{ m}) = 1960 \text{ J}$$

The work done in this case does depend on the mass of the box.

3.2 Power

The Rate of Doing Work

The time needed to carry out a job is often as important as the amount of work needed. If we have enough time, even the tiny motor of a toy train can lift an elevator as high as we like. However, if we want the elevator to take us up fairly quickly, we must use a motor whose output of work is rapid in terms of the total work needed. Thus the rate at which work is being done is significant. This rate is called **power:** The more powerful something is, the faster it can do work.

If the amount of work W is done in a period of time t, the power involved is

$$P = \frac{W}{t} \qquad \textit{Power} \qquad \textit{3-3}$$

$$\text{Power} = \frac{\text{work done}}{\text{time interval}}$$

The SI unit of power is the **watt** (W), where

$$1 \text{ watt (W)} = 1 \text{ joule/second (J/s)}$$

Example 3.2

A 15-kW electric motor provides power for the elevator of a building. What is the minimum time needed for the elevator to rise 30 m to the sixth floor when its total mass when loaded is 900 kg?

Solution

The work that must be done to raise the elevator is $W = mgh$. Since $P = W/t$, the time needed is

$$t = \frac{W}{P} = \frac{mgh}{P} = \frac{(900 \text{ kg})(9.8 \text{ m/s}^2)(30 \text{ m})}{15 \times 10^3 \text{ W}} = 17.6 \text{ s}$$

Thus a motor with a power output of 500 W is capable of doing 500 J of work per second. The same motor can do 250 J of work in 0.5 s, 1000 J of work in 2 s, 5000 J of work in 10 s, and so on. The watt is quite a small unit, and often the **kilowatt** (kW) is used instead, where 1 kW = 1000 W.

The Horsepower

The **horsepower** (hp) is the traditional unit of power in engineering. The origin of this unit is interesting. In order to sell the steam engines he had perfected two centuries ago, James Watt had to compare their power outputs with that of a horse, a source of work his customers were familiar with. After various tests he found that a typical horse could perform work at a rate of 497 W for as much as 10 hours per day. To avoid any disputes, Watt increased this figure by one-half to establish the unit he called the horsepower. Watt's horsepower therefore represents a rate of doing work of 746 W:

$$1 \text{ horsepower (hp)} = 746 \text{ W} = 0.746 \text{ kW}$$

$$1 \text{ kilowatt (kW)} = 1.34 \text{ hp}$$

Few horses can develop this much power for very long. The early steam engines ranged from 4 to 100 hp, with the 20-hp model being the most popular.

A person in good physical condition is usually capable of a continuous power output of about 75 W, which is 0.1 horsepower. A runner or swimmer during a distance event may have a power output 2 or 3 times greater. What limits the power output of a trained athlete is not muscular development but the supply of oxygen from the lungs through the bloodstream to the muscles, where oxygen is used in the metabolic processes that enable the muscles to do work. However, for a period of less than a second, an athlete's power output may exceed 5 kW, which accounts for the feats of weightlifters and jumpers.

Energy

We now go from the straightforward idea of work to the complex and many-sided idea of **energy:**

> Energy is that property something has that enables it to do work.

When we say that something has energy, we mean it is able, directly or indirectly, to exert a force on something else and perform work. When work is done on something, energy is added to it. Energy is measured in the same unit as work, the joule.

3.3 Kinetic Energy

The Energy of Motion

Energy occurs in several forms. One of them is the energy a moving object has because of its motion. Every moving object has the capacity to do work. By striking something else, the moving object can exert a force and cause the second object to shift its position, to break apart, or to otherwise show the effects of having work done on it. It is this property that defines energy, so we conclude that all moving things have energy by virtue of their motion. The energy of a moving object is called **kinetic energy** (KE). ("Kinetic" is a word of Greek origin that suggests motion is involved.)

The kinetic energy of a moving thing depends upon its mass and its speed. The greater the mass and the greater the speed, the more the KE. A train going at 30 km/h has more energy than a horse galloping at the same speed and more energy than a similar train going at 10 km/h. The exact way KE varies with mass m and speed v is given by the formula

$$\text{KE} = \tfrac{1}{2}mv^2 \qquad \textit{Kinetic energy} \qquad 3\text{-}4$$

The v^2 factor means the kinetic energy increases very rapidly with increasing speed. At 30 m/s a car has 9 times as much KE as at 10 m/s—

Deriving the Kinetic Energy Formula

Here is a simple derivation of the formula $KE = \frac{1}{2}mv^2$ for the kinetic energy of a moving object.

When we throw a ball, the work we do on it becomes its kinetic energy KE when it leaves our hand. Suppose we apply a constant force F for a distance d while the ball is in our hand, as in Fig. 3-6*a*. The work we do is $W = Fd$, so the ball's kinetic energy is

$$KE = Fd \qquad \textit{Work done on ball} \qquad 3\text{-}5$$

According to the second law of motion,

$$F = ma \qquad \textit{Force applied to ball} \qquad 3\text{-}6$$

where a is the ball's acceleration while the force acts on it. If the time during which the force was applied is t, as in Fig. 3-6*b*, Eq. 2-12 gives us the distance d as

$$d = \tfrac{1}{2}at^2 \qquad \textit{Distance moved during the acceleration} \qquad 3\text{-}7$$

Next we substitute the formulas for F and d into Eq. 3-5 to give

$$KE = Fd = (ma)\left(\tfrac{1}{2}at^2\right) = \tfrac{1}{2}m(at)^2$$

But at is the ball's speed v when it leaves our hand at the end of the acceleration, as in Fig. 3-6*c*, so that

$$KE = \tfrac{1}{2}mv^2 \qquad \textit{Kinetic energy of moving ball}$$

which is Eq. 3-4.

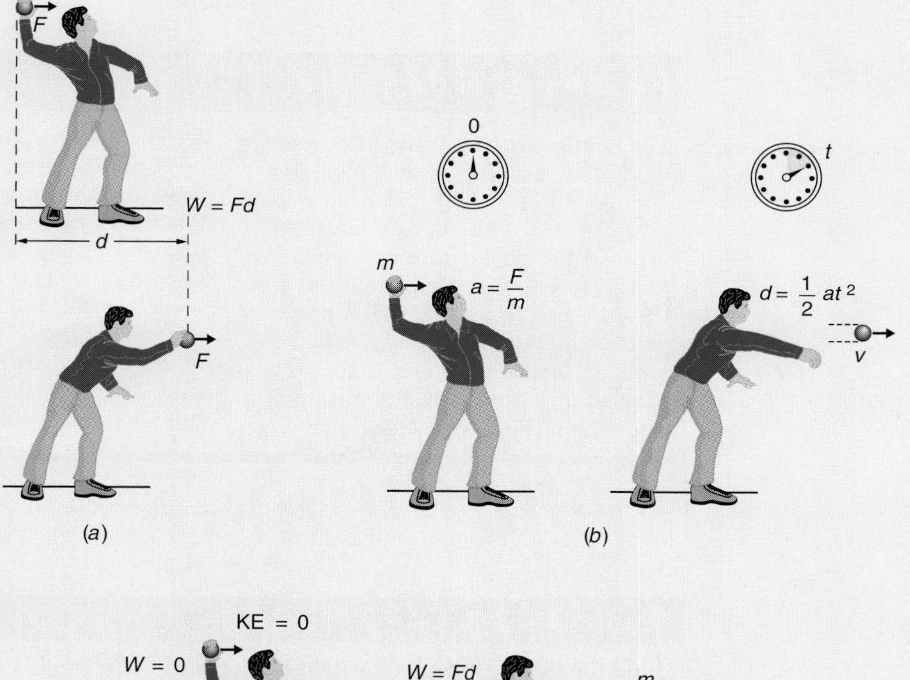

Figure 3-6 How the formula $KE = \frac{1}{2}mv^2$ for the kinetic energy of a moving object can be derived.

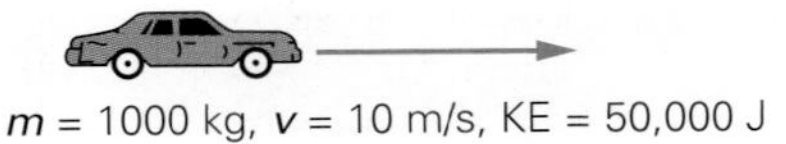

Figure 3-7 Kinetic energy is proportional to the square of the speed. A car traveling at 30 m/s has 9 times the KE of the same car traveling at 10 m/s.

and requires 9 times as much force to bring to a stop in the same distance (Fig. 3-7). The fact that KE, and hence the ability to do work (in this case, damage), depends upon the square of the speed is what is responsible for the severity of automobile accidents at high speeds. The variation of KE with mass is less marked: a 2000-kg car going at 10 m/s has just twice the KE of a 1000-kg car with the same speed.

Running Speeds

The relationship $Fd = \frac{1}{2}mv^2$ between work done and the resulting kinetic energy can be solved for speed v to give $v = \sqrt{2Fd/m}$. Let us interpret v as an animal's running speed, F as the force its muscles exert over the distance d, and m as its mass. As mentioned in Sec. 2-9, if L is the animal's length, its mass is roughly proportional to L^3 and its muscular forces are roughly proportional to L^2. The distance through which corresponding muscles act is roughly proportional to L. This means that the quantity Fd/m in the formula for v varies with L as $(L^2)(L)/L^3 = 1$, so that in general v should not vary with L at all! And, in fact, although different animals have different running speeds, there is little correlation with size over a wide span. A fox can run about as fast as a horse.

Example 3.3

Find the kinetic energy of a 1000-kg car when its speed is 10 m/s.

Solution

From Eq. 3-4 we have

$$\text{KE} = \tfrac{1}{2}mv^2 = (\tfrac{1}{2})(1000\text{ kg})(10\text{ m/s})^2$$
$$= (\tfrac{1}{2})(1000\text{ kg})(10\text{ m/s})(10\text{ m/s}) = 50{,}000\text{ J} = 50\text{ kJ}$$

In order to bring the car to this speed from rest, 50 kJ of work had to be done by its engine. To stop the car from this speed, the same amount of work must be done by its brakes.

Example 3.4

Have you ever wondered how much force a hammer exerts on a nail? Suppose you hit a nail with a hammer and drive the nail 5 mm into a wooden board (Fig. 3-8). If the hammer's head has a mass of 0.6 kg and it is moving at 4 m/s when it strikes the nail, what is the average force on the nail?

Solution

The KE of the hammer head is $\frac{1}{2}mv^2$, and this amount of energy becomes the work Fd done in driving the nail the distance $d = 5\text{ mm} = 0.005\text{ m}$ into the board. Hence

$$\text{KE of hammer head} = \text{work done on nail}$$

$$\tfrac{1}{2}mv^2 = Fd$$

and

$$F = \frac{mv^2}{2d} = \frac{(0.6\text{ kg})(4\text{ m/s})^2}{2(0.005\text{ m})} = 960\text{ N}$$

This is 216 lb—watch your fingers!

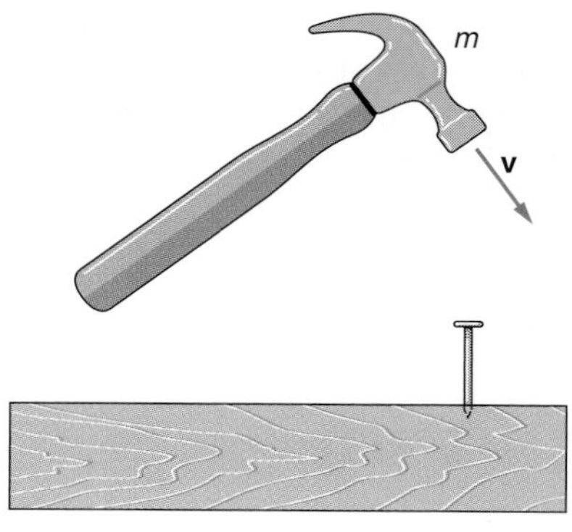

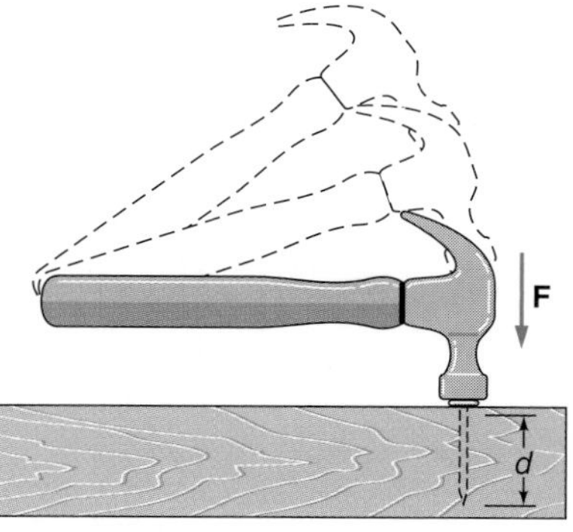

Figure 3-8 When a hammer strikes this nail, the hammer's kinetic energy is converted into the work done to push the nail into the wooden board.

3.4 Potential Energy

The Energy of Position

When we drop a stone, it falls faster and faster and finally strikes the ground. If we lift the stone afterward, we see that it has done work by making a shallow hole in the ground. In its original raised position, the stone must have had the capacity to do work even though it was not moving at the time and therefore had no KE.

The amount of work the stone could do by falling to the ground is called its **potential energy** (PE). Just as kinetic energy may be thought of as energy of motion, potential energy may be thought of as energy of position (Fig. 3-9).

Examples of potential energy are everywhere. A book on a table has PE since it can fall to the floor. A skier at the top of a slope, water at the top of a waterfall, a car at the top of the hill, anything able to move toward the earth under the influence of gravity has PE because of its position. Nor is the earth's gravity necessary: a stretched spring has PE since it can do work when it is let go, and a nail near a magnet has PE since it can do work in moving to the magnet (Fig. 3-10).

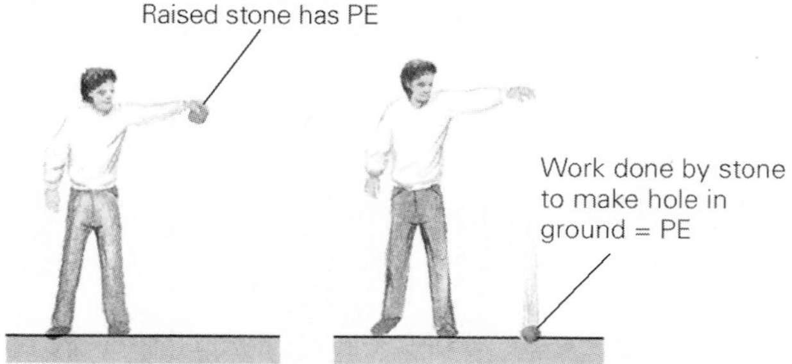

Figure 3-9 A raised stone has potential energy because it can do work on the ground when dropped.

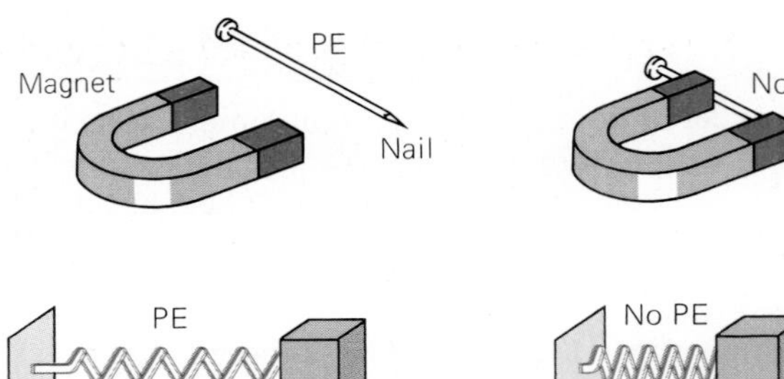

Figure 3-10 Two examples of potential energy.

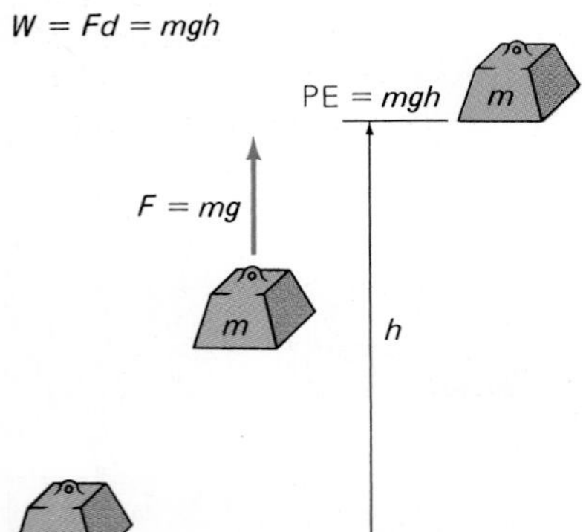

Figure 3-11 The increase in the potential energy of a raised object is equal to the work *mgh* used to lift it.

Gravitational Potential Energy When an object of mass m is raised to a height h above its original position, its gravitational potential energy is equal to the work that was done against gravity to bring it to that height (Fig. 3-11). According to Eq. 3-2 this work is $W = mgh$, and so

$$\mathrm{PE} = mgh \qquad \textit{Gravitational potential energy} \qquad 3\text{-}8$$

Example 3.5

Find the potential energy of a 1000-kg car when it is on top of a 45-m cliff.

Solution

From Eq. 3-8 the car's potential energy is

$$\mathrm{PE} = mgh = (1000\ \mathrm{kg})(9.8\ \mathrm{m/s^2})(45\ \mathrm{m}) = 441{,}000\ \mathrm{J} = 441\ \mathrm{kJ}$$

This is less than the KE of the same car when it moves at 30 m/s (Fig. 3-7). Thus a crash at 30 m/s into a wall or tree will yield more work—that is, do more damage—than dropping the car from a cliff 45 m high.

Figure 3-12 In the operation of a pile driver, the gravitational potential energy of the raised hammer becomes kinetic energy as it falls. The kinetic energy in turn becomes work as the pile is pushed into the ground.

This result for PE agrees with our experience. Consider a pile driver (Fig. 3-12), a simple machine that lifts a heavy weight (the "hammer") and allows it to fall on the head of a pile, which is a wooden or steel post, to drive the pile into the ground. From the formula PE = *mgh* we would expect the effectiveness of a pile driver to depend on the mass *m* of its hammer and the height *h* from which it is dropped, which is exactly what experience shows.

PE Is Relative It is worth noting that the gravitational PE of an object depends on the level from which it is reckoned. Often the earth's surface is convenient, but sometimes other references are more appropriate.

Suppose you lift this book as high as you can above the table while remaining seated. It will then have a PE *relative to the table* of about 12 J. But the book will have a PE *relative to the floor* of about twice that, or 24 J. And if the floor of your room is, say, 50 m above the ground, the book's PE *relative to the ground* will be about 760 J.

What is the book's true PE? The answer is that there is no such thing as "true" PE. Gravitational PE is a relative quantity. However, the *difference* between the PEs of an object at two points *is* significant, since it is this difference that can be changed into work or KE.

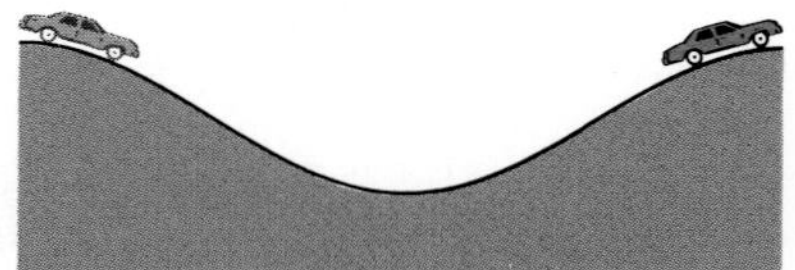

Figure 3-13 In the absence of friction, a car can coast from the top of one hill into a valley and then up to the top of another hill of the same height as the first. During the trip the initial potential energy of the car is converted into kinetic energy as the car goes downhill, and this kinetic energy then turns into potential energy as the car climbs the next hill. The total amount of energy (KE + PE) remains unchanged.

3.5 Energy Transformations

Easy Come, Easy Go

Nearly all familiar mechanical processes involve interchanges among KE, PE, and work. Thus when the car of Fig. 3-13 is driven to the top of a hill, its engine must do work in order to raise the car. At the top, the car has an amount of PE equal to the work done in getting it up there (neglecting friction). If the engine is turned off, the car can still coast down the hill, and its KE at the bottom of the hill will be the same as its PE at the top.

Changes of a similar nature, from kinetic energy to potential and back, occur in the motion of a planet in its orbit around the sun (Fig. 3-14) and in the motion of a pendulum (Fig. 3-15). The orbits of the planets are ellipses with the sun at one focus (Fig. 1-10), and each planet is therefore at a constantly varying distance from the sun. At all times the total of its potential and kinetic energies remains the same. When close to the sun, the PE of a planet is low and its KE is high. The additional speed due to increased KE keeps the planet from being pulled into the sun by the greater gravitational force on it at this point in its path. When the planet is far from the sun, its PE is higher and its KE lower, with the reduced speed exactly keeping pace with the reduced gravitational force.

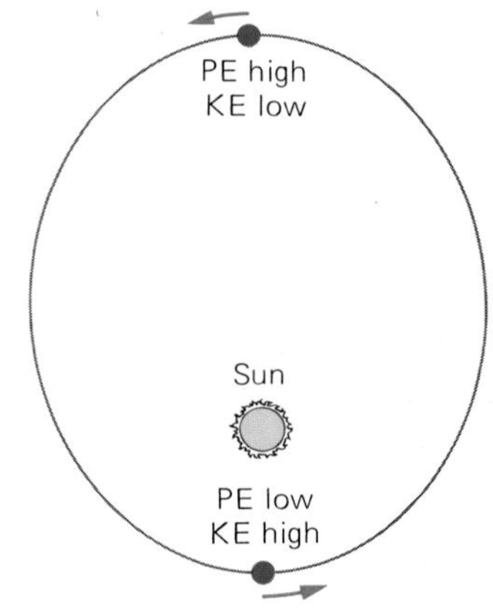

Figure 3-14 Energy transformations in planetary motion. The total energy (KE + PE) of the planet is the same at all points in its orbit. (Planetary orbits are much more nearly circular than shown here.)

A pendulum (Fig. 3-15) consists of a ball suspended by a string. When the ball is pulled to one side with its string taut and then released, it swings back and forth. When it is released, the ball has a PE relative to the bottom of its path of *mgh*. At its lowest point all this PE has become kinetic energy $\frac{1}{2}mv^2$. After reaching the bottom, the ball continues in its motion until it rises to the same height *h* on the opposite side from its initial position. Then, momentarily at rest since all its KE is now PE, the ball begins to retrace its path back through the bottom to its initial position.

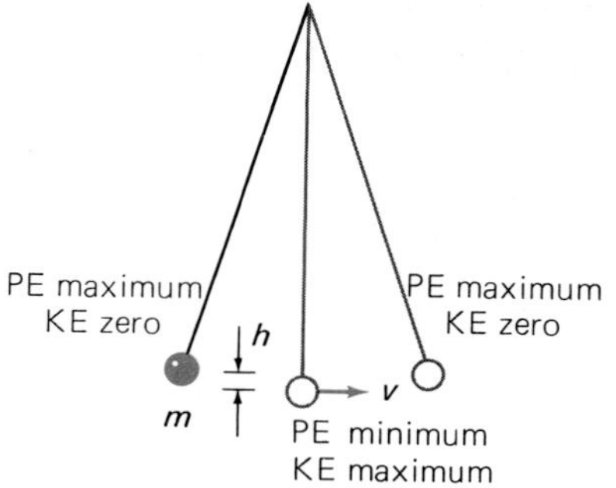

Figure 3-15 Energy transformations in pendulum motion. The total energy of the ball stays the same but is continuously exchanged between kinetic and potential forms.

Example 3.6

A girl on a swing is 2.2 m above the ground at the ends of her motion and 1.0 m above the ground at the lowest point. What is the girl's maximum speed?

Solution

The maximum speed v will occur at the lowest point where her potential energy above this point has been entirely converted to kinetic energy. If the difference in height is $h = (2.2 \text{ m}) - (1.0 \text{ m}) = 1.2 \text{ m}$ and the girl's mass is m, then

$$\text{Kinetic energy} = \text{change in potential energy}$$

$$\tfrac{1}{2}mv^2 = mgh$$

$$v = \sqrt{2gh} = \sqrt{2(9.8 \text{ m/s}^2)(1.2 \text{ m})} = 4.8 \text{ m/s}$$

The girl's mass does not matter here.

Transformations to and from kinetic energy may involve potential energies other than gravitational. An example is the elastic potential energy of a bent bow, as in Fig. 3-16.

Other Forms of Energy Energy can exist in a variety of forms besides kinetic and potential. The *chemical energy* of gasoline is used to propel our cars and the chemical energy of food enables our bodies to perform work. *Heat energy* from burning coal or oil is used to form the steam that drives the turbines of power stations. *Electric energy* turns motors in home and factory. *Radiant energy* from the sun performs work in causing water from the earth's surface to rise and form clouds, in producing differences in air temperature that cause winds, and in promoting chemical reactions in plants that produce foods.

Just as kinetic energy can be converted to potential energy and potential to kinetic, so other forms of energy can readily be transformed. In the cylinders of a car engine, for example, chemical energy stored in gasoline and air is changed first to heat energy when the mixture is ignited by the spark plugs, then to kinetic energy as the expanding gases push down on the pistons. This kinetic energy is in large part transmitted to the wheels,

Figure 3-16 The elastic potential energy of the bent bow becomes kinetic energy of the arrow when the bowstring is released.

but some is used to turn the generator and thus produce electric energy for charging the battery, and some is changed to heat by friction in bearings. Energy transformations go on constantly, all about us.

3.6 Conservation of Energy

A Fundamental Law of Nature

A skier slides down a hill and comes to rest at the bottom. What became of the potential energy he or she had at the top? The engine of a car is shut off while the car is allowed to coast along a level road. Eventually the car slows down and comes to a stop. What became of its original kinetic energy?

All of us can give similar examples of the apparent disappearance of kinetic or potential energy. What these examples have in common is that heat is always produced in an amount just equivalent to the "lost" energy (Fig. 3-17). One kind of energy is simply being converted to another; no energy is lost, nor is any new energy created. Exactly the same is true when electric, magnetic, radiant, and chemical energies are changed into one another or into heat. Thus we have a law from which no deviations have ever been found:

Figure 3-17 The potential energy of these skiers at the top of the slope turns into kinetic energy and eventually into heat as they slide downhill.

> Energy cannot be created or destroyed, although it can be changed from one form to another.

This generalization is the **law of conservation of energy.** It is the principle with the widest application in science, applying equally to distant stars and to biological processes in living cells.

We shall learn later in this chapter that matter can be transformed into energy and energy into matter. The law of conservation of energy still applies, however, with matter considered as a form of energy.

3.7 The Nature of Heat

The Downfall of Caloric

Although it comes as little surprise to us today to learn that heat is a form of energy, in earlier times this was not so clear. Less than two centuries ago most scientists regarded heat as an actual substance called **caloric.** Absorbing caloric caused an object to become warmer; the escape of caloric caused it to become cooler. Because the weight of an object does not change when the object is heated or cooled, caloric was considered to be weightless. It was also supposed to be invisible, odorless, and tasteless, properties that, of course, were why it could not be observed directly.

Actually, the idea of heat as a substance was fairly satisfactory for materials heated over a flame, but it could not account for the unlimited heat that could be generated by friction. One of the first to appreciate this difficulty was the American Benjamin Thompson (Fig. 3-18), who had supported the British during the Revolutionary War and thought it wise to move to Europe afterward, where he became Count Rumford.

Figure 3-18 Count Rumford (1753–1814).

One of Rumford's many occupations was supervising the making of cannon for a German prince, and he was impressed by the large amounts

Figure 3-19 James Prescott Joule (1818–1889).

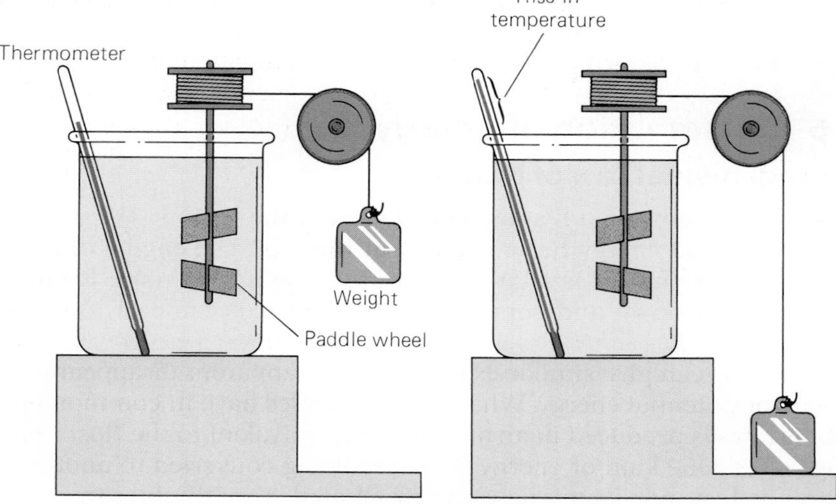

Figure 3-20 Joule's experimental demonstration that heat is a form of energy. As the weight falls, it turns the paddle wheel, which heats the water by friction. The potential energy of the weight is converted first into the kinetic energy of the paddle wheel and then into heat.

What is Heat?

As we shall learn in Chap. 5, the heat content of a body of matter consists of the KE of random motion of the atoms and molecules of which the body consists.

of heat given off by friction in the boring process. He showed that the heat could be used to boil water and that heat could be produced again and again from the same piece of metal. If heat was a fluid, it was not unreasonable that boring a hole in a piece of metal should allow it to escape. However, even a dull drill that cut no metal produced a great deal of heat. Also, it was hard to imagine a piece of metal as containing an infinite amount of caloric, and Rumford accordingly regarded heat as a form of energy.

James Prescott Joule (Fig. 3-19) was an English brewer who performed a classic experiment that settled the nature of heat once and for all. Joule's experiment used a small paddle wheel inside a container of water (Fig. 3-20). Work was done to turn the paddle wheel against the resistance of the water, and Joule measured exactly how much heat was supplied to the water by friction in this process. He found that a given amount of work always produced exactly the same amount of heat. This was a clear demonstration that heat is energy and not something else.

Joule also carried out chemical and electrical experiments that agreed with his mechanical ones, and the result was his announcement of the law of conservation of energy in 1847, when he was 29. Although Joule was a modest man ("I have done two or three little things, but nothing to make a fuss about," he later wrote), many honors came his way, including naming the SI unit of energy after him.

Momentum

Because the universe is so complex, a variety of different quantities besides the basic ones of length, time, and mass are useful to help us understand its many aspects. We have already found velocity, acceleration, force, work, and energy to be valuable, and more are to come. The idea behind defining each of these quantities is to single out something that is involved in a wide range of observations. Then we can boil down a great many separate findings about nature into a brief, clear statement, for example, the law of conservation of energy. Now we shall learn how the concepts of linear and angular momenta can give us further insights into the behavior of moving things.

3.8 Linear Momentum

Another Conservation Law

As we know (Sec. 2.7), a moving object tends to continue moving at constant speed along a straight path. The **linear momentum** of such an object is a measure of this tendency. The more linear momentum something has, the more effort is needed to slow it down or to change its direction. Another kind of momentum is **angular momentum,** which reflects the tendency of a spinning body to continue to spin. When there is no question as to which is meant, linear momentum is usually referred to simply as momentum.

The linear momentum **p** of an object of mass *m* and velocity **v** (we recall that velocity includes both speed and direction) is defined as

$$\mathbf{p} = m\mathbf{v} \qquad \textit{Linear momentum} \qquad 3\text{-}9$$

Linear momentum = (mass)(velocity)

The greater *m* and **v** are, the more difficult it is to change the object's speed or direction.

This definition of momentum is in accord with our experience. A baseball hit squarely by a bat (large **v**) is more difficult to stop than a baseball thrown gently (small **v**). The heavy iron ball used for the shotput (large *m*) is more difficult to stop than a baseball (small *m*) when their speeds are the same (Fig. 3-21).

Conservation of Momentum Momentum considerations are most useful in situations that involve explosions and collisions. When outside forces do not act on the objects involved, their combined momentum (taking directions into account) is conserved, that is, does not change:

> In the absence of outside forces, the total momentum of a set of objects remains the same no matter how the objects interact with one another.

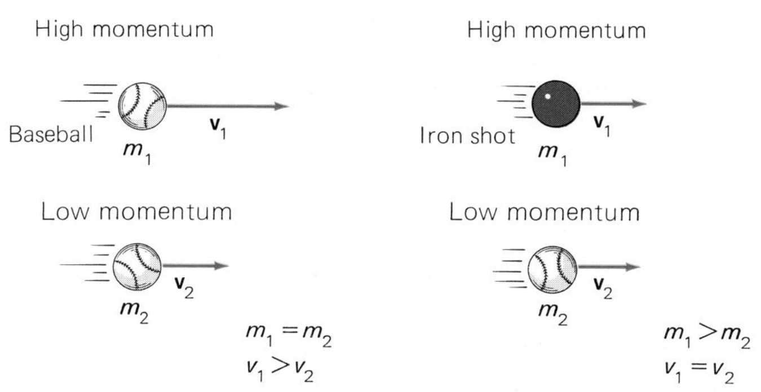

Figure 3-21 The linear momentum *m***v** of a moving object is a measure of its tendency to continue in motion at constant velocity. The symbol > means "greater than."

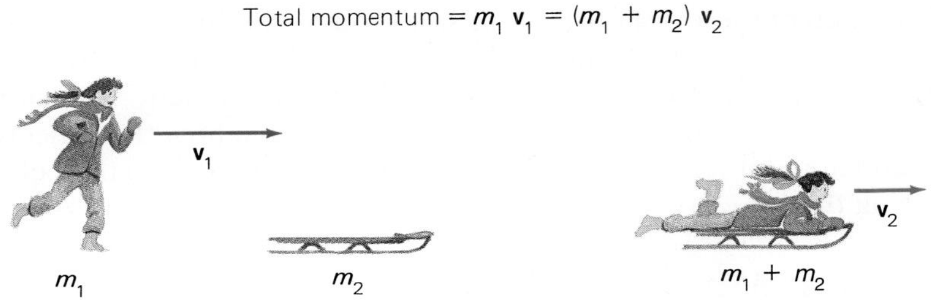

Total momentum = $m_1 \mathbf{v}_1 = (m_1 + m_2)\,\mathbf{v}_2$

$\mathbf{v}_1$ $\mathbf{v}_2$

m_1 m_2 $m_1 + m_2$

Figure 3-22 When a running girl jumps on a stationary sled, the combination moves off more slowly than the girl's original speed. The total momentum of girl + sled is the same before and after she jumps on it.

This statement is called the **law of conservation of momentum.** What it means is that, if the objects interact only with one another, each object can have its momentum changed in the interaction, provided that the total momentum after it occurs is the same as it was before.

Momentum is conserved when a running girl jumps on a stationary sled, as in Fig. 3-22. Even if there is no friction between the sled and the snow, the combination of girl and sled moves off more slowly than the girl's running speed. The original momentum, which is that of the girl alone, had to be shared between her and the sled when she jumped on it. Now that the sled is also moving, the new speed must be less than before in order that the total momentum stays the same.

Example 3.7

Let us see what happens when an object breaks up into two parts. Suppose that an astronaut outside a space station throws away a 0.5-kg camera in disgust when it jams (Fig. 3-23). The mass of the spacesuited astronaut is 100 kg, and the camera moves off at 6 m/s. What happens to the astronaut?

Solution

The total momentum of the astronaut and camera was zero originally. According to the law of conservation of momentum, their total momentum must therefore be zero afterward as well. If we call the astronaut *A* and the camera *C*, then

$$\text{Momentum before} = \text{momentum afterward}$$
$$0 = m_A v_A + m_C v_C$$

Hence

$$m_A v_A = -m_C v_C$$

where the minus sign signifies that $\mathbf{v}_A$ is opposite in direction to $\mathbf{v}_C$. Throwing the camera away therefore sets the astronaut in motion as well, with camera and astronaut moving in opposite directions. Newton's third law of motion (action-reaction) tells us the same thing, but conservation of momentum enables us to find the astronaut's speed at once:

$$v_A = -\frac{m_C v_C}{m_A} = -\frac{(0.5\text{ kg})(6\text{ m/s})}{100\text{ kg}} = -0.03\text{ m/s}$$

After an hour, which is 3600 s, the camera will have traveled $v_C t$ = 21,600 m = 21.6 km, and the astronaut will have traveled $v_A t$ = 108 m in the opposite direction if not tethered to the space station.

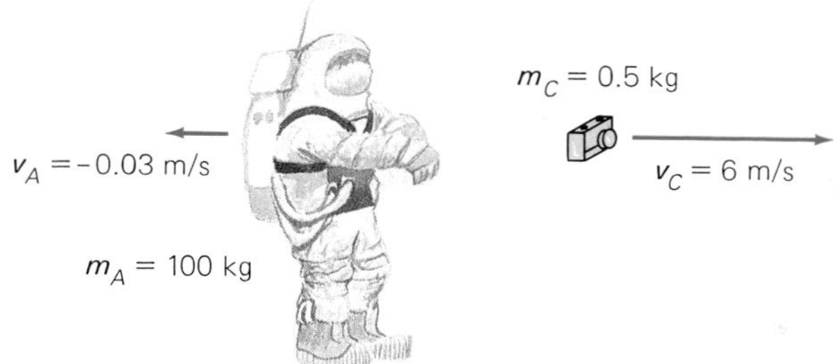

Figure 3-23 The momentum $m_C v_C$ to the right of the thrown camera is equal in magnitude to the momentum $m_A v_A$ to the left of the astronaut who threw it away.

Collisions

Applying the law of conservation of momentum to collisions gives some interesting results. These are shown in Fig. 3-24 for an object of mass m and speed v that strikes a stationary object of mass M and does not stick to it. Three situations are possible:

1. The target object has more mass, so that $M > m$. What happens here is that the incoming object bounces off the heavier target one and they move apart in opposite directions.
2. The two objects have the same mass, so that $M = m$. Now the incoming object stops and the target object moves off with the same speed v the incoming one had.
3. The target object has less mass, so that $m > M$. In this case the incoming object continues in its original direction after the impact but with reduced speed while the target object moves ahead of it at a faster pace. The greater m is compared with M, the closer the target object's final speed is to $2v$.

The third case corresponds to a golf club striking a golf ball (Fig. 3-25). This suggests that the more mass the clubhead has for a given speed, the faster the ball will fly off when struck. However, a heavy golf club is harder to swing fast than a light one, so a compromise is necessary. Experience has led golfers to use clubheads with masses about 4 times the 46-g mass of a golf ball when they want maximum distance. A good golfer can swing a clubhead at over 50 m/s.

Figure 3-25 The speed of a golf ball is greater than the speed of the clubhead that struck it because the mass of the ball is smaller than that of the clubhead.

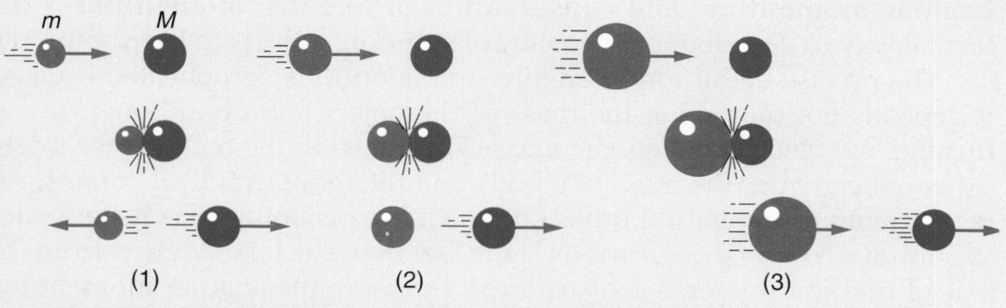

Figure 3-24 How the effects of a head-on collision with a stationary target object depend on the relative masses of the two objects.

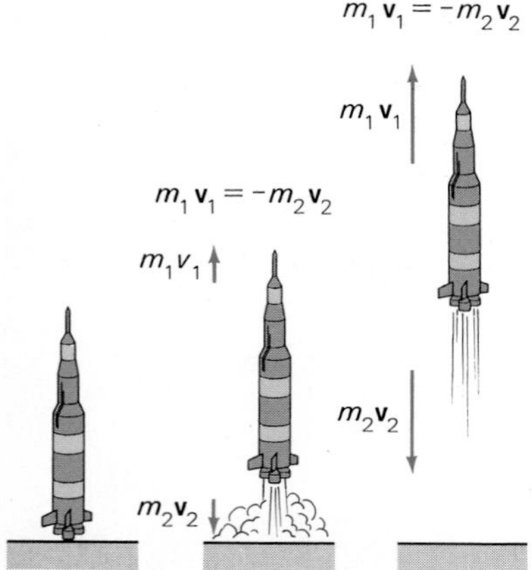

Figure 3-26 Rocket propulsion is based upon conservation of momentum. If gravity is absent, the downward momentum of the exhaust gases is equal in magnitude and opposite in direction to the upward momentum of the rocket at all times.

3.9 Rockets

Momentum Conservation Is the Basis of Space Travel

The operation of a rocket is based on conservation of linear momentum. When the rocket stands on its launching pad, its momentum is zero. When it is fired, the momentum of the exhaust gases that rush downward is balanced by the momentum in the other direction of the rocket moving upward. The total momentum of the entire system, gases and rocket, remains zero, because momentum is a vector quantity and the upward and downward momenta cancel (Fig. 3-26).

Thus a rocket does not work by "pushing" against its launching pad, the air, or anything else. In fact, rockets function best in space where no atmosphere is present to interfere with their motion.

The ultimate speed a rocket can reach is governed by the amount of fuel it can carry and by the speed of its exhaust gases. Because both these quantities are limited, **multistage rockets** are used in the exploration of space. The first stage is a large rocket that has a smaller one mounted in front of it. When the fuel of the first stage has burnt up, its motor and empty fuel tanks are cast off. Then the second stage is fired. Since the second stage is already moving rapidly and does not have to carry the motor and empty fuel tanks of the first stage, it can reach a much higher final speed than would otherwise be possible.

Depending upon the final speed needed for a given mission, three or even four stages may be required. The Saturn V launch vehicle that carried the Apollo 11 spacecraft to the moon in July 1969 had three stages. Just before takeoff the entire assembly was 111 m long and had a mass of nearly 3 million kg (Fig. 3-27).

Figure 3-27 Apollo 11 lifts off its pad to begin the first human visit to the moon. The spacecraft's final speed was 10.8 km/s, which is equivalent to 6.7 mi/s. Conservation of linear momentum underlies rocket propulsion.

3.10 Angular Momentum

A Measure of the Tendency of a Spinning Object to Continue to Spin

We have all noticed the tendency of rotating objects to continue to spin unless they are slowed down by an outside agency. A top would spin indefinitely but for friction between its tip and the ground. Another example is the earth, which has been turning for billions of years and is likely to continue doing so for many more to come.

The rotational quantity that corresponds to linear momentum is called **angular momentum,** and **conservation of angular momentum** is the formal way to describe the tendency of spinning objects to keep spinning.

The precise definition of angular momentum is complicated because it depends not only upon the mass of the object and upon how fast it is turning, but also upon how the mass is arranged in the body. As we might expect, the greater the mass of a body and the more rapidly it rotates, the more angular momentum it has and the more pronounced is its tendency to continue to spin. Less obvious is the fact that, the farther away from the axis of rotation the mass is distributed, the more the angular momentum.

An illustration of both the latter fact and the conservation of angular momentum is a skater doing a spin (Fig. 3-29). When the skater starts the spin, she pushes against the ice with one skate to start turning. Initially

Conservation Principles

The conservation principles of energy, linear momentum, and angular momentum are useful because they are obeyed in all known processes. They are significant for another reason as well. In 1917 the German mathematician Emmy Noether (Fig. 3-28) proved that:

1. If the laws of nature are the same at all times, past, present, and future, then energy must be conserved.
2. If the laws of nature are the same everywhere in the universe, then linear momentum must be conserved.
3. If the laws of nature do not depend on direction, then angular momentum must be conserved.

Figure 3-28 Emmy Noether (1882–1935).

All other conservation principles in physics, for instance conservation of electric charge (Sec. 6.2), can also be traced to similar general regularities in the universe. Thus the existence of these principles testifies to a profound order in the universe, despite the irregularities and randomness of many aspects of it, a truly remarkable finding. In 1933 Noether moved to the United States where, after a period at the Institute for Advanced Study in Princeton, she became a professor at Bryn Mawr.

both arms and one leg are extended, so that her mass is spread as far as possible from the axis of rotation. Then she brings her arms and the outstretched leg in tightly against her body, so that now all her mass is as close as possible to the axis of rotation. As a result, she spins faster. To make up for the change in the mass distribution, the speed must change as well to conserve angular momentum.

Planetary Motion

Kepler's second law of planetary motion (Fig. 1-11) has an origin similar to that of the changing spin rate of a skater. A planet moving around the sun has angular momentum, which must be the same everywhere in its orbit. As a result the planet's speed is greatest when it is close to the sun, least when it is far away.

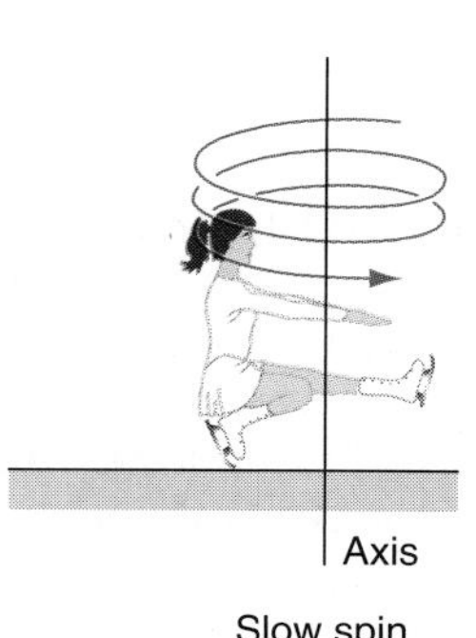

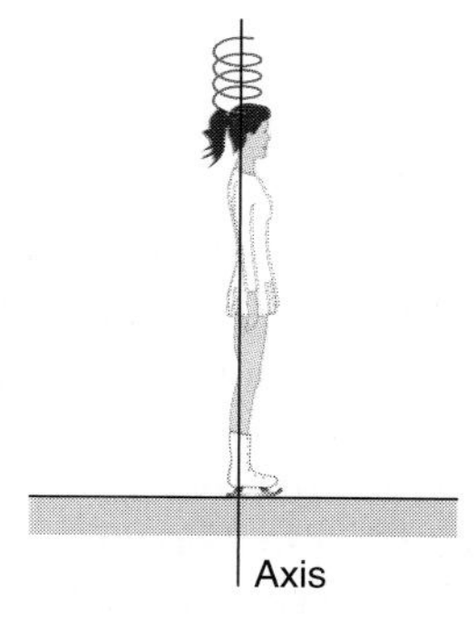

Figure 3-29 Conservation of angular momentum. Angular momentum depends upon both the speed of turning and the distribution of mass. When the skater pulls in her arms and extended leg, she spins faster to compensate for the change in the way her mass is distributed.

Figure 3-30 The faster a top spins, the more stable it is. When all its angular momentum has been lost through friction, the top falls over.

Figure 3-31 Conservation of angular momentum keeps a spinning football from tumbling end-over-end, which would slow it down and reduce its range.

Spin Stabilization Like linear momentum, angular momentum is a vector quantity with direction as well as magnitude. Conservation of angular momentum therefore means that a spinning body tends to maintain the *direction* of its spin axis in addition to the amount of angular momentum it has. A stationary top falls over at once, but a rapidly spinning top stays upright because its tendency to keep its axis in the same orientation by virtue of its angular momentum is greater than its tendency to fall over (Fig. 3-30). Footballs and rifle bullets are sent off spinning to prevent them from tumbling during flight, which would increase air resistance and hence shorten their range (Fig. 3-31).

Relativity

In 1905 a young physicist of 26 named Albert Einstein published an analysis of how measurements of time and space are affected by motion between an observer and what he or she is studying. To say that Einstein's **theory of relativity** revolutionized science is no exaggeration.

Relativity links not only time and space but also energy and matter. From it have come a host of remarkable predictions, all of which have been confirmed by experiment. Eleven years later Einstein took relativity a step further by interpreting gravity as a distortion in the structure of space and time, again predicting extraordinary effects that were verified in detail.

3.11 Special Relativity

Things Are Seldom What They Seem

Thus far in this book no special point has been made about how such quantities as length, time, and mass are measured. In particular, who makes a certain measurement would not seem to matter—everybody ought to get the same result. Suppose we want to find the length of an airplane when we are on board. All we have to do is put one end of a tape measure at the airplane's nose and look at the number on the tape at the airplane's tail.

But what if we are standing on the ground and the airplane is in flight? Now things become more complicated because the light that carries information to our instruments travels at a definite speed. According to Einstein, our measurements from the ground of length, time, and mass in the airplane would differ from those made by somebody moving with the airplane.

Einstein began with two postulates. The first concerns **frames of reference,** which were mentioned in Sec. 2.1. Motion always implies a frame of reference relative to which the location of something is changing. A passenger walking down the aisle moves relative to an airplane, the airplane moves relative to the earth, the earth moves relative to the sun, and so on (Fig. 3-32).

If we are in the windowless cabin of a cargo airplane, we cannot tell whether the airplane is in flight at constant velocity or is at rest on the ground, since without an external frame of reference the question has no meaning. To say that something is moving always requires a frame of reference. From this follows Einstein's first postulate:

> The laws of physics are the same in all frames of reference moving at constant velocity with respect to one another.

Figure 3-32 All motion is relative to a chosen frame of reference. Here the photographer has turned the camera to keep pace with one of the cyclists. Relative to him, both the road and the other cyclists are moving. There is no fixed frame of reference in nature, and therefore no such thing as "absolute motion"; all motion is relative.

If the laws of physics were different for different observers in relative motion, the observers could find from these differences which of them were "stationary" in space and which were "moving." But such a distinction does not exist, hence the first postulate.

The second postulate, which follows from the results of a great many experiments, states that

> The speed of light in free space has the same value for all observers.

The speed of light in free space is always $c = 3 \times 10^8$ m/s, about 186,000 mi/s.

Length, Time, and Kinetic Energy Let us suppose I am in an airplane moving at the constant velocity **v** relative to you on the ground. I find that the airplane is L_0 long, that it has a mass of m, and that a certain time interval (say an hour on my watch) is t_0. Einstein showed from the above postulates that you, on the ground, would find that

1. The length L you measure is shorter than L_0.
2. The time interval t you measure is longer than t_0.
3. The kinetic energy KE you determine is greater than $\frac{1}{2}mv^2$.

That is, to you on the ground, the airplane appears shorter than to me and to have more KE, and to you, my watch appears to tick more slowly.

The differences between L and L_0, t and t_0, and KE and $\frac{1}{2}mv^2$ depend on the ratio v/c between the relative speed v of the frames of reference (here the speed of the airplane relative to the ground) and the speed of light c. Because c is so great, these differences are too small to detect at speeds like those of airplanes. However, they must be taken into account in spacecraft flight. And, at speeds near c, which often occur in the subatomic world of such tiny particles as electrons and protons, relativistic effects are conspicuous. Although at speeds much less than c the formula $\frac{1}{2}mv^2$ for kinetic energy is still valid, at high speeds the theory of relativity shows that the KE of a moving object is higher than $\frac{1}{2}mv^2$ (Fig. 3-33).

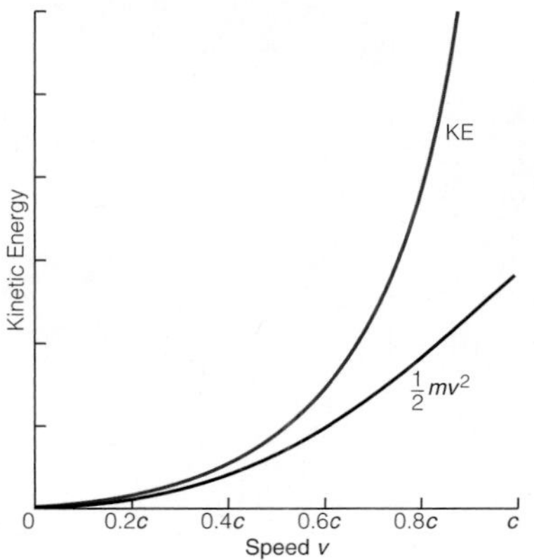

Figure 3-33 The faster an object moves relative to an observer, the more the object's kinetic energy KE exceeds $\frac{1}{2}mv^2$. This effect is only conspicuous at speeds near the speed of light $c = 3 \times 10^8$ m/s, which is about 186,000 mi/s. Because an object would have an infinite KE if $v = c$, nothing with mass can ever move that fast or faster.

As we can see from the graph, the closer v gets to c, the closer KE gets to infinity. Since an infinite kinetic energy is impossible, this conclusion means that nothing can travel as fast as light or faster: c is the absolute speed limit in the universe. The implications of this limit for space travel are discussed in Chap. 19.

Einstein's 1905 theory, which led to the above results among others, is called **special relativity** because it is restricted to constant velocities. His later theory of **general relativity,** which deals with gravity, includes accelerations.

3.12 Rest Energy

Matter Is a Form of Energy

The most far-reaching conclusion of special relativity is that mass and energy are related to each other so closely that matter can be converted into energy and energy into matter. The **rest energy** of a body is the energy equivalent of its mass. If a body has the mass m, its rest energy is

$$E_0 = mc^2 \qquad \textit{Rest energy} \qquad \textit{3-10}$$

$$\text{Rest energy} = (\text{mass})(\text{speed of light})^2$$

Experiments show that this formula is accurate to at least 0.00004 percent.

The rest energy of a 1.5-kg object, such as this book, is

$$E_0 = mc^2 = (1.5\ \text{kg})(3 \times 10^8\ \text{m/s})^2 = 1.35 \times 10^{17}\ \text{J}$$

quite apart from any kinetic or potential energy it might have. If liberated, this energy would be more than enough to send a million tons to the moon. By contrast, the PE of this book on top of Mt. Everest, which is 8850 m high, relative to its sea-level PE is less than 10^4 J.

Example 3.8

How much mass is converted into energy per day in a 100-MW nuclear power plant?

Solution

There are (60)(60)(24) = 86,400 s/day, so the energy liberated per day is

$$E_0 = Pt = (10^2)(10^6\ \text{W})(8.64 \times 10^4\ \text{s}) = 8.64 \times 10^{12}\ \text{J}$$

From Eq. 3-10 the corresponding mass is

$$m = \frac{E_0}{c^2} = \frac{8.64 \times 10^{12}\ \text{J}}{(3 \times 10^8\ \text{m/s})} = 9.6 \times 10^{-5}\ \text{kg}$$

This is less than a tenth of a gram—not much. To liberate the same amount of energy from coal, about 270 tons would have to be burned.

BIOGRAPHY Albert Einstein (1879–1955)

Figure 3-34 The irregular path of a microscopic particle bombarded by molecules. The line joins the positions of a single particle observed at constant intervals. This phenomenon is called brownian movement and is direct evidence of the reality of molecules and their random motions. It was discovered in 1827 by the British botanist Robert Brown.

Bitterly unhappy with the rigid discipline of the schools of his native Germany, Einstein went to Switzerland at 16 to complete his education and later got a job examining patent applications at the Swiss Patent Office in Berne. Then, in 1905, ideas that had been in his mind for years when he should have been paying attention to other matters (one of his math teachers called Einstein a "lazy dog") blossomed into three short papers that were to change decisively the course of not only physics but modern civilization as well.

The first paper proposed that light has a dual character with particle as well as wave properties. This work is described in Chap. 9 together with the quantum theory of the atom that flowed from it. The subject of the second paper was brownian motion, the irregular zigzag motion of tiny bits of suspended matter such as pollen grains in water (Fig. 3-34). Einstein arrived at a formula that related brownian motion to the bombardment of the particles by randomly moving molecules of the fluid in which they were suspended. Although the molecular theory of matter had been proposed many years before, this formula was the long-awaited definite link with experiment that convinced the remaining doubters that molecules actually exist. The third paper introduced the theory of relativity.

Although much of the world of physics was originally either indifferent or skeptical, even the most unexpected of Einstein's conclusions were soon confirmed and the development of what is now called modern physics began in earnest. After university posts in Switzerland and Czechoslovakia, in 1913 Einstein took up an appointment at the Kaiser Wilhelm Institute in Berlin that left him able to do research free of financial worries and routine duties. His interest was now mainly in gravity, and he began where Newton had left off more than 200 years earlier.

The general theory of relativity that resulted from Einstein's work provided a deep understanding of gravity, but his name remained unknown to the general public. This changed in 1919 with the dramatic discovery that gravity affects light exactly as Einstein had predicted. He immediately became a world celebrity, but his well-earned fame did not provide security when Hitler and the Nazis came to power in Germany in the early 1930s. Einstein left in 1933 and spent the rest of his life at the Institute for Advanced Study in Princeton, New Jersey, thereby escaping the fate of millions of other European Jews at the hands of the Germans. Einstein's last years were spent in a fruitless search for a "unified field theory" that would bring together gravitation and electromagnetism in a single picture. The problem was worthy of his gifts, but it remains unsolved to this day although progress is being made.

How is it possible that so much energy can be bottled up in even a little bit of matter without anybody having known about it until Einstein's work? In fact, we do see matter being converted into energy around us all the time. We just do not normally think about what we find in these terms. All the energy-producing reactions of chemistry and physics, from the lighting of a match to the nuclear fusion that powers

Table 3-1 Energy, Power, and Momentum

Quantity	Type	Symbol	Unit	Meaning	Formula
Work	Scalar	W	Joule (J)	A measure of the change produced by a force that acts on something	$W = Fd$
Power	Scalar	P	Watt (W)	The rate at which work is being done	$P = W/t$
Kinetic energy	Scalar	KE	Joule (J)	Energy of motion	$\text{KE} = \frac{1}{2}mv^2$
Potential energy	Scalar	PE	Joule (J)	Energy of position	$\text{PE}_{\text{gravitational}} = mgh$
Rest energy	Scalar	E_0	Joule (J)	Energy equivalent of the mass of an object	$E_0 = mc^2$
Linear momentum	Vector	$\mathbf{p}$	Kg · m/s	A measure of the tendency of a moving object to continue moving in the same straight line at the same speed	$\mathbf{p} = m\mathbf{v}$
Angular momentum	Vector	—	—	A measure of the tendency of a rotating object to continue rotating about the same axis at the same speed	—

the sun and stars, involve the disappearance of a small amount of matter and its reappearance as energy. The simple formula $E_0 = mc^2$ has led not only to a better understanding of how nature works but also to the nuclear power plants—and nuclear weapons—that are so important in today's world.

The discovery that matter and energy can be converted into each other does not affect the law of conservation of energy provided we include mass as a form of energy. Table 3-1 lists the basic features of the various quantities introduced in this chapter.

3.13 General Relativity

Gravity Is a Warping of Spacetime

Einstein's general theory of relativity, published in 1916, related gravitation to the structure of space and time. What is meant by "the structure of space and time" can be given a quite precise meaning mathematically, but unfortunately no such precision is possible using ordinary language. All the same, we can legitimately think of the force of gravity as arising from a warping of spacetime around a body of matter so that a nearby mass tends to move toward the body, much as a marble rolls toward the bottom of a saucer-shaped hole (Fig. 3-35). In an apt formulation, "Matter tells spacetime how to curve, and spacetime tells matter how to move."

It may seem as though one abstract concept is merely replacing another, but in fact the new point of view led Einstein and other scientists to a variety of remarkable discoveries that could not have come from the older way of thinking.

Perhaps the most spectacular of Einstein's results was that light ought to be subject to gravity. The effect is very small, so a large mass, such as that of the sun, is needed to detect the influence of its gravity on light. If Einstein was right, light rays that pass near the sun should be bent toward

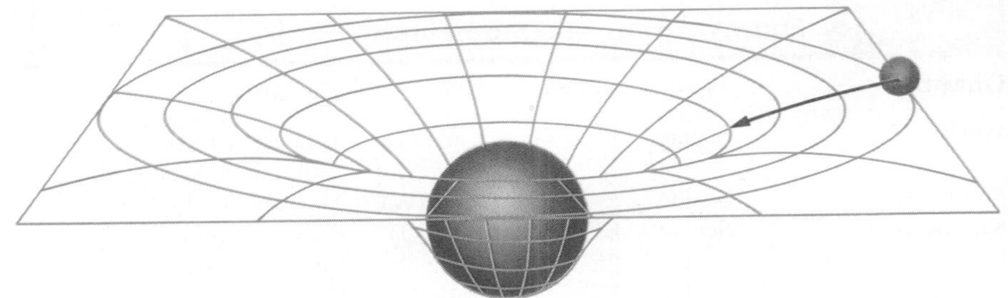

Figure 3-35 General relativity pictures gravity as a warping of the structure of space and time due to the presence of a body of matter. An object nearby experiences an attractive force as a result of this distortion in spacetime, much as a marble rolls toward the bottom of a saucer-shaped hole in the ground.

it by 0.0005°—the diameter of a dime seen from a mile away. To check this prediction, photographs were taken of stars that appeared in the sky near the sun during an eclipse in 1919, when they could be seen because the moon obscured the sun's disk (see Chap. 17). These photographs were then compared with photographs of the same region of the sky taken when the sun was far away (Fig. 3-36), and the observed changes in the apparent positions of the stars matched Einstein's calculations. Other predictions based on general relativity have also been verified, and the theory remains today without serious rival.

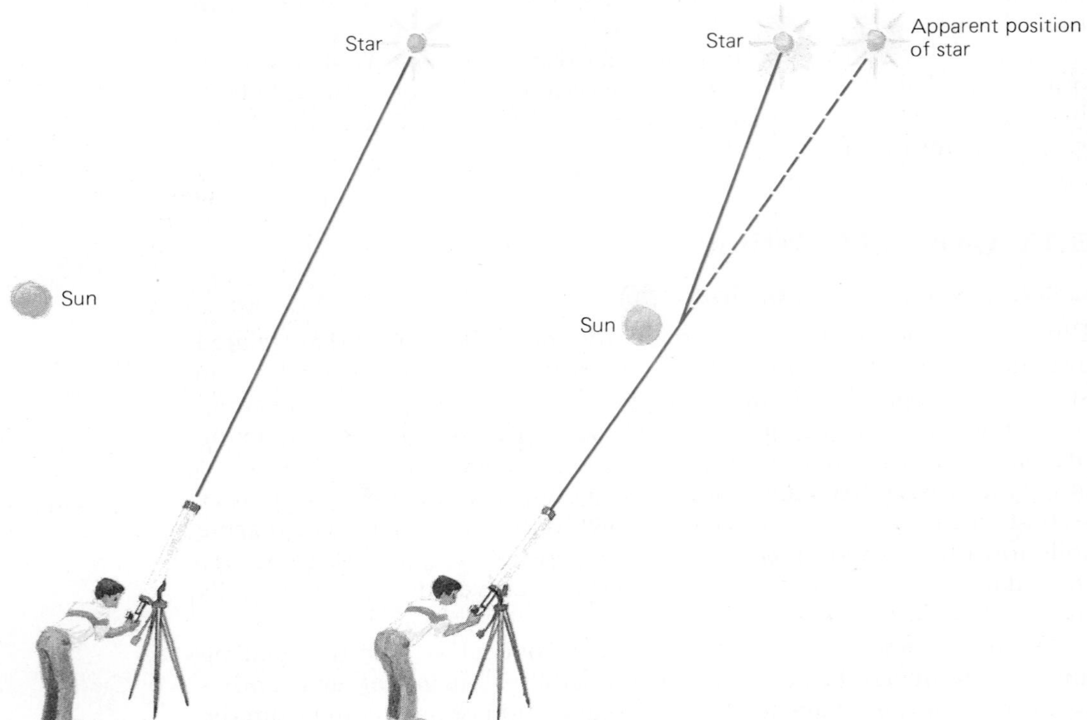

Figure 3-36 Starlight that passes near the sun is deflected by its strong gravitational pull. The deflection, which is very small, can be measured during a solar eclipse when the sun's disk is obscured by the moon.

Gravitational Waves

The existence of **gravitational waves** that travel with the speed of light was the prediction of general relativity that had to wait longest for experimental evidence. To visualize such waves, we can think in terms of the two-dimensional model of Fig. 3-35 by imagining spacetime as a rubber sheet distorted by masses lying on it. If one of the masses vibrates, waves will be sent out in the sheet (like waves on a water surface) that set other masses in vibration. Gravitational waves—"ripples in spacetime"—are expected to be extremely weak, and none has yet been directly detected.

However, in 1974 indirect but strong evidence for their existence was discovered in the behavior of a pair of close-together stars that revolve around each other. A system of this kind gives off gravitational waves and slows down as it loses energy to them. This slowing down was indeed observed and agrees well with the theoretical expectation. Gravitational waves should have been produced in abundance early in the history of the universe and may be present throughout today's universe. Ultrasensitive instruments are now operating that may be able to pick up such waves directly.

Important Terms and Ideas

Work is a measure of the change, in a general sense, that a force causes when it acts upon something. The work done by a force acting on an object is the product of the magnitude of the force and the distance through which the object moves while the force acts on it. If the direction of the force is not the same as the direction of motion, the projection of the force in the direction of motion must be used. The unit of work is the **joule** (J).

Power is the rate at which work is being done. Its unit is the **watt** (W).

Energy is the property that something has that enables it to do work. The unit of energy is the joule. The three broad categories of energy are **kinetic energy,** which is the energy something has by virtue of its motion, **potential energy,** which is the energy something has by virtue of its position, and **rest energy,** which is the energy something has by virtue of its mass. According to the **law of conservation of energy,** energy cannot be created or destroyed, although it can be changed from one form to another (including mass).

Linear momentum is a measure of the tendency of a moving object to continue in motion along a straight line. **Angular momentum** is a measure of the tendency of a rotating object to continue spinning about the same axis. Both are vector quantities. If no outside forces act on a set of objects, then their linear and angular momenta are **conserved,** that is, remain the same regardless of how the objects interact with one another.

According to the **special theory of relativity,** when there is relative motion between an observer and what is being observed, lengths are shorter than when at rest, time intervals are longer, and kinetic energies are greater. Nothing can travel faster than the speed of light.

The **general theory of relativity,** which relates gravitation to the structure of space and time, correctly predicts that light should be subject to gravity.

Important Formulas

Work: $W = Fd$

Power: $P = \frac{W}{t}$

Kinetic energy: $\mathrm{KE} = \frac{1}{2}mv^2$

Gravitational potential energy: $\mathrm{PE} = mgh$

Linear momentum: $\mathbf{p} = m\mathbf{v}$

Rest energy: $E_0 = mc^2$

Multiple Choice

1. Which of the following is not a unit of power?
   a. joule-second
   b. watt
   c. newton-meter/second
   d. horsepower
2. An object at rest may have
   a. velocity
   b. momentum
   c. kinetic energy
   d. potential energy
3. A moving object does not necessarily have
   a. velocity
   b. momentum
   c. kinetic energy
   d. potential energy
4. An object that has linear momentum must also have
   a. acceleration
   b. angular momentum
   c. kinetic energy
   d. potential energy
5. The total amount of energy (including the rest energy of matter) in the universe
   a. cannot change
   b. can decrease but not increase
   c. can increase but not decrease
   d. can either increase or decrease
6. When the speed of a body is doubled,
   a. its kinetic energy is doubled
   b. its potential energy is doubled
   c. its rest energy is doubled
   d. its momentum is doubled
7. Two balls, one of mass 5 kg and the other of mass 10 kg, are dropped simultaneously from a window. When they are 1 m above the ground, the balls have the same
   a. kinetic energy
   b. potential energy
   c. momentum
   d. acceleration
8. A bomb dropped from an airplane explodes in midair.
   a. Its total kinetic energy increases
   b. Its total kinetic energy decreases
   c. Its total momentum increases
   d. Its total momentum decreases
9. The operation of a rocket is based upon
   a. pushing against its launching pad
   b. pushing against the air
   c. conservation of linear momentum
   d. conservation of angular momentum
10. When a spinning skater pulls in her arms to turn faster,
    a. her angular momentum increases
    b. her angular momentum decreases
    c. her angular momentum remains the same
    d. any of these, depending on the circumstances
11. According to the principle of relativity, the laws of physics are the same in all frames of reference
    a. at rest with respect to one another
    b. moving toward or away from one another at constant velocity
    c. moving parallel to one another at constant velocity
    d. all of these
12. When the speed v of an object of mass m approaches the speed of light c, its kinetic energy
    a. is less than $\frac{1}{2}mv^2$
    b. equals $\frac{1}{2}mv^2$
    c. is more than $\frac{1}{2}mv^2$ but less than $\frac{1}{2}mc^2$
    d. is more than $\frac{1}{2}mv^2$ and can exceed $\frac{1}{2}mc^2$
13. A spacecraft has left the earth and is moving toward Mars. An observer on the earth finds that, relative to measurements made when the spacecraft was at rest, its
    a. length is shorter
    b. KE is less than $\frac{1}{2}mv^2$
    c. clocks tick faster
    d. rest energy is greater
14. In the formula $E_0 = mc^2$, the symbol c represents
    a. the speed of the body
    b. the speed of the observer
    c. the speed of sound
    d. the speed of light
15. It is not true that
    a. light is affected by gravity
    b. the mass of a moving object depends upon its speed
    c. the maximum speed anything can have is the speed of light
    d. momentum is a form of energy
16. Albert Einstein did not discover that
    a. the length of a moving object is less than its length at rest
    b. the acceleration of gravity g is a universal constant

c. light is affected by gravity
d. gravity is a warping of spacetime

17. The work done in holding a 50-kg object at a height of 2 m above the floor for 10 s is
a. 0
b. 250 J
c. 1000 J
d. 98,000 J

18. The work done in lifting 30 kg of bricks to a height of 20 m is
a. 61 J
b. 600 J
c. 2940 J
d. 5880 J

19. A total of 4900 J is used to lift a 50-kg mass. The mass is raised to a height of
a. 10 m
b. 98 m
c. 960 m
d. 245 km

20. The work a 300-W electric grinder can do in 5.0 min is
a. 1 kJ
b. 1.5 kJ
c. 25 kJ
d. 90 kJ

21. A 150-kg yak has an average power output of 120 W. The yak can climb a mountain 1.2 km high in
a. 25 min
b. 4.1 h
c. 13.3 h
d. 14.7 h

22. A 40-kg boy runs up a flight of stairs 4 m high in 4 s. His power output is
a. 160 W
b. 392 W
c. 40 W
d. 1568 W

23. Car A has a mass of 1000 kg and is moving at 60 km/h. Car B has a mass of 2000 kg and is moving at 30 km/h. The kinetic energy of car A is
a. half that of car B
b. equal to that of car B
c. twice that of car B
d. 4 times that of car B

24. A 1-kg object has a potential energy of 1 J relative to the ground when it is at a height of
a. 0.102 m
b. 1 m
c. 9.8 m
d. 98 m

25. A 1-kg object has kinetic energy of 1 J when its speed is
a. 0.45 m/s
b. 1 m/s
c. 1.4 m/s
d. 4.4 m/s

26. The 2-kg blade of an ax is moving at 60 m/s when it strikes a log. If the blade penetrates 2 cm into the log as its KE is turned into work, the average force it exerts is
a. 3 kN
b. 90 kN
c. 72 kN
d. 180 kN

27. A 1-kg ball is thrown in the air. When it is 10 m above the ground, its speed is 3 m/s. At this time most of the ball's total energy is in the form of
a. kinetic energy
b. potential energy relative to the ground
c. rest energy
d. momentum

28. A 10,000-kg freight car moving at 2 m/s collides with a stationary 15,000-kg freight car. The two cars couple together and move off at
a. 0.8 m/s
b. 1 m/s
c. 1.3 m/s
d. 2 m/s

29. A 30-kg girl and a 25-kg boy are standing on frictionless roller skates. The girl pushes the boy, who moves off at 1.0 m/s. The girl's speed is
a. 0.45 m/s
b. 0.55 m/s
c. 0.83 m/s
d. 1.2 m/s

30. An object has a rest energy of 1 J when its mass is
a. 1.1×10^{-17} kg
b. 3.3×10^{-9} kg
c. 1 kg
d. 9×10^{16} kg

31. The smallest part of the total energy of the ball of Multiple Choice 27 is
a. kinetic energy
b. potential energy relative to the ground
c. rest energy
d. momentum

32. The lightest particle in an atom is an electron, whose rest mass is 9.1×10^{-31} kg. The energy equivalent of this mass is approximately
a. 10^{-13} J
b. 10^{-15} J
c. 3×10^{-23} J
d. 10^{-47} J

Exercises

3.1 The Meaning of Work

1. Is it correct to say that all changes in the physical world involve energy transformations of some sort? Why?

2. Under what circumstances (if any) is no work done on a moving object even though a net force acts upon it?

3. A horizontal force of 80 N is used to move a 20-kg crate across a level floor. How much work is done when the crate is moved 5 m? How much work would have been done if the crate's mass were 30 kg?
4. How much work is needed to raise a 110-kg load of bricks 12 m above the ground to a building under construction?
5. The sun exerts a gravitational force of 4.0×10^{28} N on the earth, and the earth travels 9.4×10^{11} m in its yearly orbit around the sun. How much work is done by the sun on the earth each year?
6. The acceleration of gravity on the surface of Mars is 37 m/s^2. If an astronaut in a space suit can jump upward 20 cm on the earth's surface, how high could he jump on the surface of Mars?
7. A total of 490 J of work is needed to lift a body of unknown mass through a height of 10 m. What is its mass?

3.2 Power

8. The kilowatt-hour is a unit of what physical quantity or quantities?
9. A motorboat that develops 60 kW covers a distance of 1 km in 4 min. With how much force did the water resist the motion of the boat?
10. How much power must the legs of a 50-kg woman develop in order to run up a staircase 5 m high in 7 s?
11. An 80-kg mountaineer climbs a 3000-m mountain in 10 h. What is the average power output during the climb?
12. A weightlifter raises a 90-kg barbell from the floor to a height of 2.2 m in 0.6 s. What was his average power output during the lift?
13. A 700-kg horse whose power output is 1.0 hp is pulling a sled over the snow at 3.5 m/s. Find the force the horse exerts on the sled.
14. What should be the minimum horsepower rating of the motor of an elevator that is required to raise a total mass of 1200 kg through a height of 80 m in 15 s?
15. A crane whose motor has a power input of 5.0 kW lifts a 1200-kg load of bricks through a height of 30 m in 90 s. Find the efficiency of the crane, which is the ratio between its output power and its input power.
16. A total of 10^4 kg of water per second flows over a waterfall 25 m high. If half of the power this flow represents could be converted into electricity, how many 100-W lightbulbs could be supplied?

3.3 Kinetic Energy

17. A moving object whose initial KE is 10 J is subject to a frictional force of 2 N that acts in the opposite direction. How far will the object move before coming to a stop?
18. What is the speed of an 800-kg car whose KE is 250 kJ?
19. Is the work needed to bring a car's speed from 0 to 10 km/h less than, equal to, or more than the work needed to bring its speed from 10 to 20 km/h?
20. Which of these energies might correspond to the KE of a person riding a bicycle on a road? 10 J; 1 kJ; 100 kJ.
21. A 1-kg salmon is hooked by a fisherman and it swims off at 2 m/s. The fisherman stops the salmon in 50 cm by braking his reel. How much force does the fishing line exert on the fish?
22. During a circus performance, John Tailor was fired from a compressed-air cannon whose barrel was 20 m long. Mr. Tailor emerged from the cannon (twice on weekdays, three times on Saturdays and Sundays) at 40 m/s. If Mr. Tailor's mass was 70 kg, what was the average force on him when he was inside the cannon's barrel?
23. How long will it take a 1000-kg car with a power output of 20 kW to go from 10 m/s to 20 m/s?

3.4 Potential Energy

24. Does every moving body possess kinetic energy? Does every stationary body possess potential energy?
25. As we will learn in Chap. 6, electric charges of the same kind (both positive or both negative) repel each other, whereas charges of opposite sign (one positive and the other negative) attract each other. (a) What happens to the PE of a positive charge when it is brought near another positive charge? (b) When it is brought near a negative charge?
26. A 50-kg woman jumps off a wall 80 cm high and lands on a concrete road with her knees stiff. Her body is compressed by 6 cm at the moment of impact. (a) What was the average force the road exerted on her body? (b) If the woman bent her knees on impact so that she came to a stop over a distance of 50 cm, what would the average force on her body be?

3.5 Energy Transformations

27. In what part of its orbit is the earth's potential energy greatest with respect to the sun? In what

part of its orbit is the earth's kinetic energy greatest? Explain your answers.

28. A ball is dropped from a height of 100 m. At what height will half of its energy be potential and half kinetic?

29. Two identical balls move down a tilted board. Ball A slides down without friction and ball B rolls down. Which ball reaches the bottom first? Why?

30. A yo-yo is swung in a vertical circle in such a way that its total energy KE + PE is constant. At what point in the circle is its speed a maximum? A minimum? Why?

31. If the yo-yo in Exercise 30 has a speed of 3 m/s at the top of the circle, whose radius is 80 cm, what is its speed at the bottom?

32. A person sitting under a coconut palm is struck by a 2-kg coconut that fell from a height of 20 m. (a) Find the kinetic energy of the coconut when it reaches the person. (b) Find the average force exerted by the coconut if its impact is absorbed over a distance of 5 cm. (c) What is this force in pounds? Is it a good idea to sit under a coconut palm?

33. A skier is sliding downhill at 8 m/s when she reaches an icy patch on which her skis move freely with negligible friction. The difference in altitude between the top of the icy patch and its bottom is 10 m. What is the speed of the skier at the bottom of the icy patch? Do you have to know her mass?

34. A force of 20 N is used to lift a 600-g ball from the ground to a height of 1.8 m, when it is let go. What is the speed of the ball when it is let go?

3.6 Conservation of Energy

3.7 The Nature of Heat

35. Why does a nail become hot when it is hammered into a piece of wood?

36. A man skis down a slope 120 m high. If 80 percent of his initial potential energy is lost to friction and air resistance, what is his speed at the bottom of the slope?

37. In an effort to lose weight, a person runs 5 km per day at a speed of 4 m/s. While running, the person's body processes consume energy at a rate of 1.4 kW. Fat has an energy content of about 40 kJ/g. How many grams of fat are metabolized during each run?

38. A man drinks a bottle of beer and proposes to work off its 460 kJ by exercising with a 20-kg barbell. If each lift of the barbell from chest height to over his head is through 60 cm and the efficiency of his body is 10 percent under these circumstances, how many times must he lift the barbell?

39. An 80-kg crate is raised 2 m from the ground by a man who uses a rope and a system of pulleys. He exerts a force of 220 N on the rope and pulls a total of 8 m of rope through the pulleys while lifting the crate, which is at rest afterward. (a) How much work does the man do? (b) What is the change in the potential energy of the crate? (c) If the answers to these questions are different, explain why.

3.8 Linear Momentum

40. An 800-kg car coasts down a hill 40 m high with its engine off and the driver's foot pressing on the brake pedal. At the top of the hill the car's speed is 6 m/s and at the bottom it is 20 m/s. How much energy was converted into heat on the way down?

41. A golf ball and a Ping-Pong ball are dropped in a vacuum chamber. When they have fallen halfway to the bottom, how do their speeds compare? Their kinetic energies? Their potential energies? Their momenta?

42. Is it possible for an object to have more kinetic energy but less momentum than another object? Less kinetic energy but more momentum?

43. What happens to the momentum of a car when it comes to a stop?

44. The speed of an airplane doubles in flight. (a) How is the law of conservation of momentum obeyed in this situation? (b) The law of conservation of energy?

45. When the kinetic energy of an object is doubled, what happens to its momentum?

46. What, if anything, happens to the speed of a fighter plane when it fires a cannon at an enemy plane in front of it?

47. A ball of mass m rolling on a smooth surface collides with a stationary ball of mass M. (a) Under what circumstances will the first ball come to a stop while the second ball moves off? (b) Under what circumstances will the first ball reverse its direction while the second ball moves off in the original direction of the first ball? (c) Under what circumstances will both balls move off in the original direction of the first ball?

48. A railway car is at rest on a frictionless track. A man at one end of the car walks to the other end. (a) Does the car move while he is walking? (b) If so, in which direction? (c) What happens when the man comes to a stop?

49. An empty dump truck coasts freely with its engine off along a level road. (a) What happens to the truck's speed if it starts to rain and water collects in it? (b) The rain stops and the accumulated water leaks out. What happens to the truck's speed now?

50. A boy throws a 4-kg pumpkin at 8 m/s to a 40-kg girl on roller skates, who catches it. At what speed does the girl then move backward?

51. A 70-kg person dives horizontally from a 200-kg boat with a speed of 2 m/s. What is the recoil speed of the boat?

52. A 30-kg girl who is running at 3 m/s jumps on a stationary 10-kg sled on a frozen lake. How fast does the sled with the girl on it then move?

53. The 176-g head of a golf club is moving at 45 m/s when it strikes a 46-g golf ball and sends it off at 65 m/s. Find the final speed of the clubhead after the impact, assuming that the mass of the club's shaft can be neglected.

54. A 1000-kg car moving north at 20 m/s collides head-on with an 800-kg car moving south at 30 m/s. If the cars stick together, in what direction and at what speed does the wreckage begin to move?

55. A 40-kg skater moving at 4 m/s overtakes a 60-kg skater moving at 2 m/s in the same direction and collides with her. The two skaters stick together. (a) What is their final speed? (b) How much kinetic energy is lost?

56. The two skaters of Exercise 55 are moving in opposite directions when they collide and stick together. Answer the same questions for this case.

3.10 Angular Momentum

57. If the polar ice caps melt, the length of the day will increase. Why?

58. All helicopters have two rotors. Some have both rotors on vertical axes but rotating in opposite directions, and the rest have one rotor on a horizontal axis perpendicular to the helicopter body at the tail. Why is a single rotor never used?

59. The earthquake that caused the Indian Ocean tsunami of 2004 (see Fig. 14.38) led to changes in the earth's crust that reduced its diameter slightly. What effect, if any, do you think this reduction had on the length of the day?

3.11 Special Relativity

60. What are the two postulates from which Einstein developed the special theory of relativity?

61. The theory of relativity predicts a variety of effects that disagree with our everyday experience. Why do you think this theory is universally accepted by scientists?

62. What physical quantity will all observers always find the same value for?

63. The length of a rod is measured by several observers, one of whom is stationary with respect to the rod. What must be true of the value obtained by the stationary observer?

64. If the speed of light were smaller than it is, would relativistic phenomena be more or less conspicious than they are now?

65. Why is it impossible for an object to move faster than the speed of light?

3.12 Rest Energy

66. The potential energy of a golf ball in a hole is negative with respect to the ground. Under what circumstances (if any) is the ball's kinetic energy negative? Its rest energy?

67. What is the effect on the law of conservation of energy of the discovery that matter and energy can be converted into each other?

68. A certain sedentary person uses energy at an average rate of 70 W. All of this energy has its ultimate origin in the sun. How much matter is converted to energy in the sun per day to supply this person?

69. One kilogram of water at 0°C contains 335 kJ of energy more than 1 kg of ice at 0°C What is the mass equivalent of this amount of energy?

70. When 1 g of natural gas is burned in a stove or furnace, about 56 kJ of heat is produced. How much mass is lost in the process? Do you think this mass change could be directly measured?

71. Approximately 5.4×10^6 J of chemical energy is released when 1 kg of dynamite explodes. What fraction of the total energy of the dynamite is this?

72. Approximately 4×10^9 kg of matter is converted into energy in the sun per second. Express the power output of the sun in watts.

Answers to Multiple Choice

1. a	**5.** a	**9.** c	**13.** a	**17.** a	**21.** b	**25.** c	**29.** c
2. d	**6.** d	**10.** c	**14.** d	**18.** d	**22.** b	**26.** d	**30.** a
3. d	**7.** d	**11.** d	**15.** d	**19.** a	**23.** c	**27.** c	**31.** a
4. c	**8.** a	**12.** d	**16.** b	**20.** d	**24.** a	**28.** a	**32.** a

4

Energy and the Future

Oil wells in California.

Goals

When you have finished this chapter you should be able to complete the goals ▶ given for each section below:

The Energy Problem

4.1 Population and Prosperity
What Drives Energy Demand
- ▶ State the approximate year in which world population is expected to level off, what the maximum population might then be, and which two countries then would have the most people.

4.2 Energy Consumption
High Today, Higher Tomorrow
- ▶ Explain why energy demand is likely to grow faster than world population.
- ▶ Identify the various fossil fuels, trace their energy contents to their ultimate origin, and compare their reserves.

4.3 Global Warming
A Serious Threat
- ▶ Discuss the evidence for global warming and the ways in which it is causing sea level to rise.

4.4 Carbon Dioxide and the Greenhouse Effect
The Cause of Global Warming
- ▶ Explain the greenhouse effect and how it acts to heat the atmosphere.
- ▶ State the role of carbon dioxide in global warming and give examples of other greenhouse gases.
- ▶ Outline the role of deforestation in global warming.

Fossil Fuels

4.5 Liquid Fuels
Vehicles Are the Biggest Users
- ▶ Compare the average fuel efficiency of cars in the United States with that of cars elsewhere and give some reasons for the difference.

4.6 Natural Gas
The Least Bad Fossil Fuel
- ▶ Explain why natural gas is the least objectionable fossil fuel.

4.7 Coal
Plentiful but Worst for the Environment
- ▶ Compare the advantages and disadvantages of coal as a fuel and account for its wide and increasing use.
- ▶ State what is involved in sequestering CO_2 emissions from power plants.

Alternative Sources

4.8 A Nuclear World?
Not Soon, but Perhaps On the Way
- ▶ Compare nuclear fission and nuclear fusion as energy sources.
- ▶ State the approximate percentage of electricity in the United States that comes from nuclear energy and explain why no new nuclear plants have been built for many years.

4.9 Clean Energy I
Continuous Sources
- ▶ Define geothermal energy and give some methods of using it.

4.10 Clean Energy II
Variable Sources
- ▶ Describe ways to make use of the energy contents of sunlight, tides, and waves.
- ▶ Give several methods to store large amounts of energy.

4.11 Hydrogen and Fuel Cells
New Technologies for Future Vehicles
- ▶ State the two ways in which hydrogen can be used to provide energy.

4.12 Biofuels
Yes, But
- ▶ Compare the advantages and disadvantages of the various biofuel sources.

Strategies for the Future

4.13 Conservation
Less Is More
- ▶ Give examples of opportunities to conserve energy in everyday life.

4.14 What Governments Must Do
Their Role Is Crucial
- ▶ Describe what is meant by a cap-and-trade system for controlling CO_2 emissions.
- ▶ Compare the average annual CO_2 emissions per person in the United States and China, their total emissions, and their positions on controlling these emissions. Account for the importance of these countries in CO_2 control.

The rise of modern civilization would have been impossible without the discovery of vast resources of energy and the development of ways to transform it into useful forms. All that we do requires energy. The more energy we have at our command, the better we can satisfy our desires for food, clothing, shelter, warmth, light, transport, communication, and manufactured goods.

Unfortunately oil and natural gas, the most convenient fuels, although currently abundant, have become expensive, have limited reserves, and, together with the more plentiful coal, are largely responsible for global warming through the carbon dioxide their burning produces. Other energy sources have handicaps of one kind or another, some serious, as well as good features. Nuclear fusion, the ultimate energy source, remains a technology of the future at best. At the same time, world population is increasing, people everywhere seek better lives, and both factors bring a need for more and more energy. The Industrial Revolution of the nineteenth century was powered largely by coal; in the twentieth century oil and gas became the leading fuels. Nothing is more important today than the choice and implementation of an appropriate energy strategy for the twenty-first century.

This is an unusual chapter for this book both because it is nontechnical and because it covers some of the essential social, economic, and even political dimensions of its subject, which does not exist in a vacuum. The decisions that businesses and governments make now and in the future are critical, and it is essential that they be made in full view of an informed public. In this chapter the energy problem in all its complexity is considered in one place so how its various parts fit together is clear. Even though scientific elements that cannot be properly discussed this early in the book are left for later chapters, the basic ideas are all here so that those who do not cover the entire text can view the situation as a whole and appreciate how it affects them (and how they affect it).

The Energy Problem

The energy problem has three elements:

1. Ever-increasing demand for energy driven by an expanding world population and its growing prosperity.
2. Inevitable decline in the economical supply of fossil fuels, which now furnish about 85 percent of the world's energy.
3. Carbon dioxide from the burning of fossil fuels is the chief contributor to the global warming that affects life on earth.

We will look at these matters in turn before going on to examine present and future sources of energy and then considering how best to secure the future while continuing to meet the most essential of our needs.

4.1 Population and Prosperity

What Drives World Energy Demand

For most of the hundred thousand or so years that modern humans have existed there were too few of them to have much effect on their resources or environments. Ten thousand years ago, when agriculture began, the world's human population was probably about 5 million. It was perhaps 500 million in 1650, 1 billion in 1850, and 1.6 billion as recently as 1900. It is over 6.7 billion today and is climbing rapidly (Fig. 4-1). The current rate of population increase is 230,000 per day—another United States every $3\frac{1}{2}$ years, another China every 15 years.

It is obvious that the world's population cannot keep growing at the present rate, and indeed must decrease. Already, according to the U.N. Environment Program, "the human population is now so large that the amount of resources needed to sustain it exceeds what is available at current consumption patterns." An average fertility rate of 2.1 children per woman (the 0.1 takes into account girls and women who do not live long enough to reproduce) means a constant population. Over 60 countries have already reached this rate or even less; it is 2.1 in the United States and only 1.2 in Spain and Italy. In other countries, however, fertility rates are higher, with six of them—India, Pakistan, Nigeria, Indonesia, Bangladesh, and China—responsible for half the world's current population growth. In sub-Saharan Africa the overall fertility rate is 5.4. Africa's population is expected to quadruple before stabilizing, the last region to do so.

Poverty, low status of women, ignorance, tradition, and religion are factors that contribute to rising populations by obstructing access to the safe and efficient family planning methods already used in much of the world. Nevertheless a worldwide shift to smaller families has begun with some developing countries already having reduced their rates of population growth. In China, whose population is now 1.3 billion, fertility halved between 1970 and 1996; in Bangladesh, with 150 million people, a similar reduction took even less time.

As Fig. 4-1 shows, plausible estimates of world population suggest a leveling off at around 9 billion by 2050. (If current fertility rates were unchanged, there would be over a billion more of us than that.) By 2050 India is likely to have passed China as the most populous nation and might then have more people than there were in the entire world in 1900. After peaking in 2050 the curve of Fig. 4-1 is projected to turn downward. If that happens, an eventual population fall to a stable size that permits comfortable lives for everybody might then occur. What would a sustainable population count for the earth be? Half of today's figure? A third? Even less? Nobody knows because any estimate must involve a wide variety of factors, many without reliable numbers attached. What is clear, however, is that there are too many people today for the world to support for much longer both their demands on natural resources and their assaults on the environment in which they live.

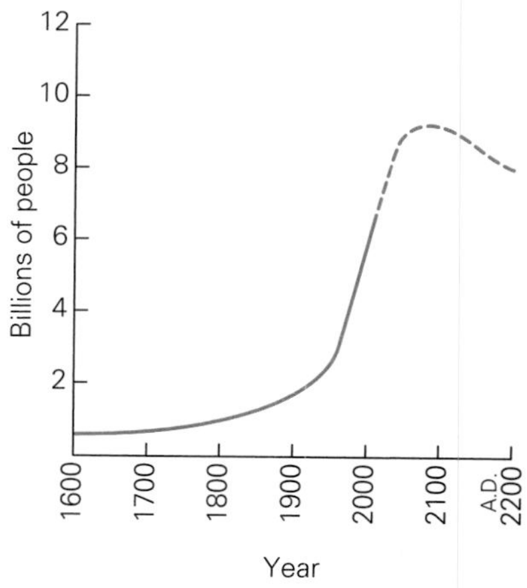

Figure 4-1 World population according to United Nations figures—currently 6.7 billion. Estimates for future population vary widely. Shown is an optimistic projection with a peak of about 9 billion in 2050. An eventual decrease in population would be the first since the Black Death of the fourteenth century. Unless the decrease occurs in time, resources of all kinds will give out and the environment, already under threat, will turn hostile.

Prosperity In parallel with a ballooning population is a broad rise in prosperity. In 1990 nearly 30 percent of the world's people lived on less than the equivalent of \$1/day; today the proportion (still a disgrace) is down to half that, which means more energy use per person. Higher up the economic ladder, life is also getting better, which has the same effect. As an example of what this implies for the future, we can compare car ownership in China (1.3 billion people) with that in the United States (300 million people). In 2006 the average income per person in each country was respectively \$1,740 and \$36,800 (three centuries ago the Chinese were ahead), with car ownership respectively 9 and 450 cars/1000 people. Incomes in China are increasing by 9 to 10 percent/year (China expects to expand its middle class to at least half its population by 2020)—and, not surprisingly, it is the world's fastest-growing car market. Even if China never catches up with the car ownership rate of the United States, in time there will be hundreds of millions more cars there gobbling up fuel of some kind. In India (1.1 billion largely poor people), whose population and economy are both growing steadily, car sales are increasing by over 20 percent each year; some new cars there cost as little as \$2500. The world's automotive fleet, now around 880 million cars, is expected to grow to 2 billion by 2050.

4.2 Energy Consumption

High Today, Higher Tomorrow

As the world economy expands, so does its consumption of energy. In the advanced countries, the standard of living is already high and their populations are stable, so their need for energy is unlikely to grow very much. Indeed, this need may even decline as their energy use becomes more efficient. Elsewhere, rates of energy consumption are still low, less than 1 kW per person for more than half the people of the world compared with about 11 kW per person in the United States (Fig. 4-2). These people seek better lives, which means more energy, and their numbers are swelling, which means still more energy. About a tenth of all spending in the United States goes to energy.

Figure 4-3 shows world energy consumption from 1980 to the present together with three projections for the future. The middle curve assumes an average annual rise in energy use of 1 percent in the advanced countries and a 3 percent rise in other countries, figures thought to be realistic. The bottom curve corresponds to a lower rate of economic growth than anticipated and the top curve to a higher rate. The midrange estimate for 2030 is nearly one and a half times today's energy consumption.

Almost all the energy available to us today has a single source—the sun. Light and heat reach us directly from the sun; food and wood owe their energy content to photosynthesis (Sec. 13.12) powered by sunlight

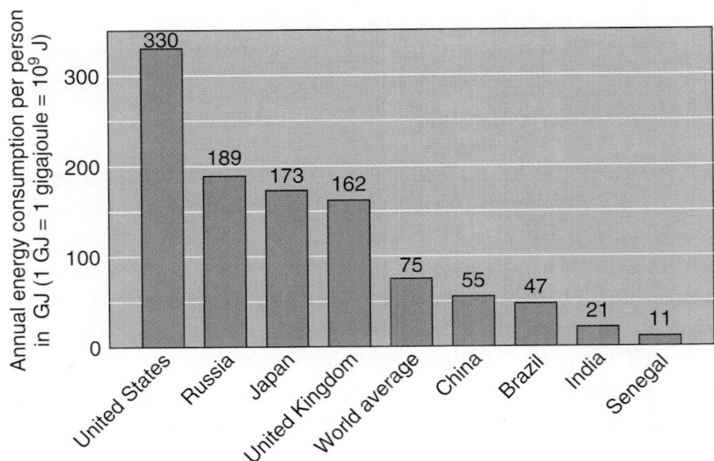

Figure 4-2 Energy use per person in various countries. The energy needs of the huge populations at the lower end of the list are increasing. China alone has a population of 1.3 billion and its energy consumption is growing at over four times the world average. The United Kingdom consists of England, Scotland, Wales, and Northern Ireland. Senegal is a fairly typical African country.

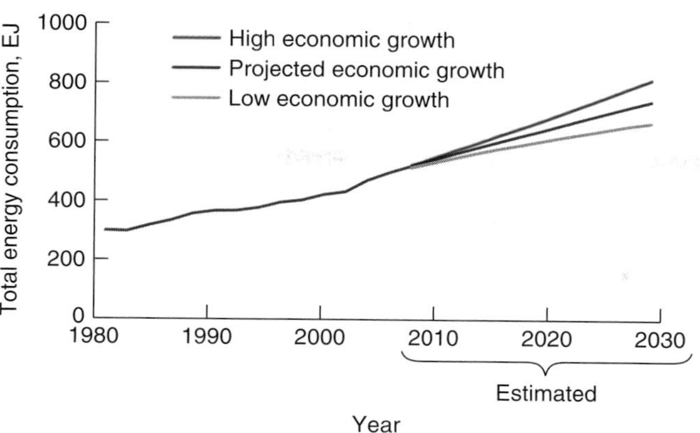

Figure 4-3 Annual world energy consumption 1980–2030. The energy unit is the exojoule (EJ), where 1 EJ = 10^{18} J.

falling on plants; water power exists because the sun's heat evaporates water from the oceans that falls later as rain and snow on high ground; wind power comes from motions in the atmosphere due to unequal heating of the earth's surface by the sun. The fossil fuels coal, oil, and natural gas were formed from the remains of plants and animals that contain energy derived from sunlight millions of years ago. Only nuclear energy, tidal energy, and heat from sources inside the earth cannot be traced to the sun's rays (Figs. 4-4 and 4-5).

The Future Fossil fuels, which today furnish by far the greatest part of the world's energy, cannot last forever. As their reserves decline, their prices will go up accordingly, which is already happening. The increased cost of energy will burden all economies, especially those of developing countries, which use energy less efficiently. For instance, China needs over twice as much energy as the United States does per unit of output. Eventually renewable and nuclear (fission and, perhaps, fusion) energies will become the principal energy sources.

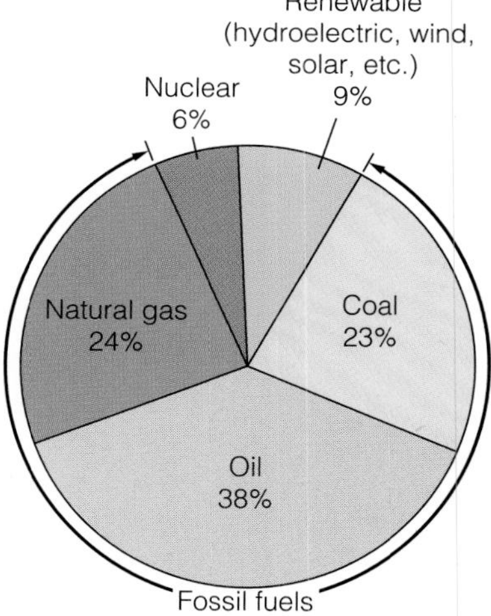

Figure 4-4 Sources of commercial energy production worldwide. Fossil fuels are responsible for 85 percent of the world's energy consumption (apart from firewood, still widely used, which is not included here). The percentages for energy sources in the United States are not very different from those of the world as a whole. The total is over 10 billion tons of oil equivalent. (An average family car burns about 1 ton of oil equivalent per year.)

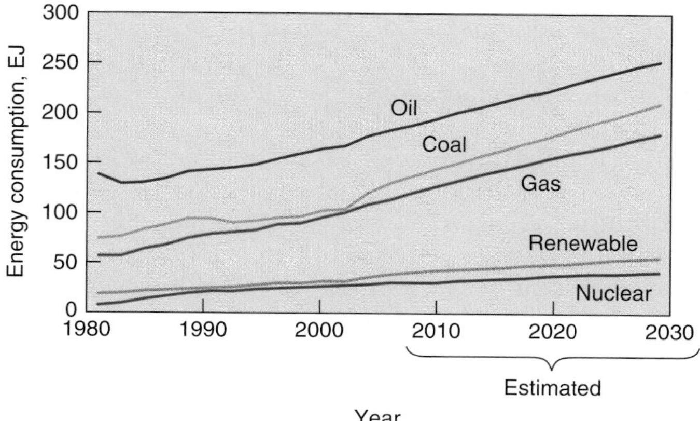

Figure 4-5 Annual world energy consumption from various sources. Serious international action to moderate global warming would reduce the projected rises in fossil-fuel use and increase the others.

Petroleum—more familiarly, oil—will be the first fossil fuel to be exhausted. At present the world uses over 85 million barrels of oil per day and demand is predicted to grow to 120 million by 2030. Where will all the oil come from? More oil will certainly be found and better technology will increase the yield from existing wells. However, the last year in which more oil was discovered than consumed was more than a quarter of a century ago, and by now two barrels of oil are burned for every barrel discovered. Sooner or later—probably before 2030, possibly well before then—oil production will reach a peak and start to decline. The flow of oil will not stop then, of course, but its price will soar. This will bring about a drastic change in the world's patterns of energy use that will be hard to adjust to because oil burns efficiently and is easy to extract, process, store, and transport. Seventy percent of the oil used today goes into fuels that power ships, trains, aircraft, cars, and trucks, and oil is a valued feedstock for synthetic material of all kinds.

Natural gas is an exceptionally clean fuel that supplies more and more power stations in the United States and already heats more than half its homes. Reserves of natural gas, too, will not last forever; demand for gas is increasing faster than it is for oil. Although liquid and gas fuels can be made from coal and coal itself can serve as the raw material for synthetics, these technologies involve greater expense.

Even though the coal we consume every year took about 2 million years to accumulate, apparently—the data are not entirely reliable—enough remains for perhaps a century at the present rate of consumption. Coal reserves are equivalent in energy content to several times oil reserves. Before 1941, coal was the world's chief fuel, and it is likely to return to first place when oil and gas run out. As with them, coal prices are headed upward, more than tripling in the past decade, and will increase even faster as production declines.

Nuclear fuel reserves exceed those of fossil fuels. Besides having an abundant fuel supply, a properly built and properly operating nuclear plant is in many respects an excellent energy source. Nuclear energy is already responsible for about a fifth of the electricity generated in the United States, and in a number of other countries the proportion is even higher; in France it is three-quarters. After a period of being largely out of favor, nuclear energy is about to come into wider use, as discussed in Sec. 4.8.

What about the energy of direct sunlight, of winds and tides, of falling water, of trees and plants, of the earth's own internal heat? After all, the technologies needed to exploit these renewable resources already exist and are steadily improving. A close look shows that it will not be easy—though far from impossible—for such sources to supply a really large part of future needs. In every case the required installation is either expensive (though decreasingly so) for the energy obtained, or practical only in favorable locations, or both. Some cannot provide energy reliably all the time, and most of them need a lot of space.

A city of medium size might use 1 GW (= 1 gigawatt = 1000 MW = 1 billion watts) of power. Less than 150 acres is enough for a 1 GW nuclear plant, whereas solar collectors of the same capacity might need 5000 acres (including rooftops), and wind turbines over 10,000 acres. To grow crops for conversion to fuel might require 200 square miles of farmland to give 1 GW averaged over a year. This doesn't mean that such energy sources are without value, particularly where local conditions are suitable, only that in the foreseeable future they are unlikely to satisfy by themselves the world's swelling energy appetite.

So, although fossil fuels clearly cannot cope indefinitely with expected energy demand, with the help of nuclear and renewable technologies there are sufficient reserves of fossil fuels to last at least much of the rest of this century. But there is another consideration: the billions of tons of carbon dioxide produced each year by the burning of fossil fuels are mainly responsible for the warming of the atmosphere that is going on today, a warming that seems sure to have serious consequences for our planet. To continue burning fossil fuels at the ever-increasing rates of Fig. 4-5 is a recipe for disaster, as we shall see next.

4.3 Global Warming

A Serious Threat

The average temperature of the earth's surface and the atmosphere just above it has varied throughout the earth's history. Warm spells and cold ones have alternated, including ice ages in which immense sheets of ice blanketed much of the globe, but the changes back and forth occurred over relatively long periods of time (see Chap. 14). In recent years a totally new pattern of change has begun in which the earth is warming up much faster than it ever has before (Fig. 4-6).

The earth's atmosphere is not heated directly by sunlight but indirectly through the greenhouse effect described in Secs. 4.4 and 14.4. The chief agent responsible for the greenhouse effect in the atmosphere is the gas carbon dioxide (CO_2), and global warming is mainly due to its growing CO_2 content. (The symbol CO_2 means that each carbon dioxide molecule consists of two oxygen atoms bonded to a carbon atom.)

Some consequences of increasing world temperatures are already obvious. Sea ice in the Arctic is melting steadily and in 20 or 30 years the

Signs of Warming

Increasing air temperatures, shrinking glaciers, and rising seas are not the only signs that the world is getting hotter. The oceans store far more heat than the atmosphere, so changes in their temperatures are more significant—and they, too, are climbing. Hurricanes and other tropical storms, whose energy comes from warm ocean water, are becoming stronger. Storms elsewhere are becoming more frequent and more severe. Climate patterns are changing, with record rainfalls in some areas and record droughts in others. Deserts in Africa and central Asia are spreading. Wildfires are more and more common worldwide. In the United States, the Atlantic and Pacific coastal regions are becoming wetter while some of the central states can expect to be increasingly starved of water.

Not all is bad for now: spring comes earlier every year, which lengthens the growing season in the high latitudes to increase food production there. But some plants and animals are already having trouble keeping up with their new environments, and a quarter of all species may die out by 2100. Overall crop yields sooner or later will decline as temperatures go up and droughts occur more frequently, unwelcome tidings for the still-swelling world population.

Also ominous is the effect of rising temperatures on the spread of disease. Previously safe parts of the Mediterranean Sea now host toxic warm-water algae. Milder winters have allowed the ticks that carry Lyme disease to spread farther across North America and Scandinavia. Mosquitoes, which are vectors of such maladies as malaria and dengue fever, range over a larger part of the world than before, and because their metabolism goes up with temperature, they also feed more often. There are many more examples.

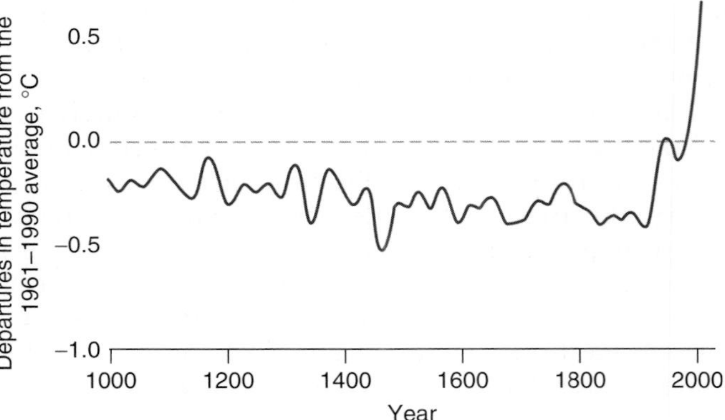

Figure 4-6 Average global surface temperatures for the past thousand years. Temperatures are continuing to rise sharply.

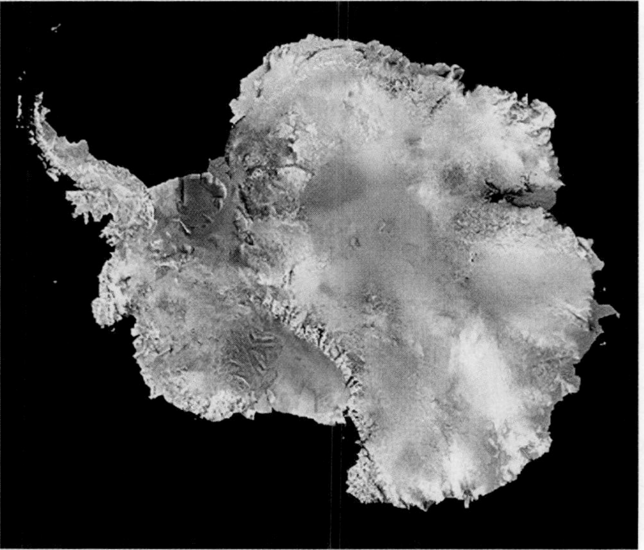

Figure 4-7 Much of Antarctica is surrounded by giant ice shelves fed by glaciers on shore. Global warming has led to the breakup of large sections of the ice shelves, which drift out to sea as icebergs and eventually melt. Because the shelves are floating to start with, their melting does not raise sea level, but meltwater from the Antarctic ice cap itself will continue to do so as global warming proceeds. If the Antarctic and Greenland ice caps were to melt, sea level would rise by at least 10 m, which would drastically change the map of the world's land areas. Complete melting would take a long time, but once well under way it would be irreversible because seawater and bare ground absorb sunlight more efficiently than ice, an excellent reflector, does.

North Pole is likely to be free of ice in the summer, for the first time in 3 million years. Similar melting is taking place around the Antarctic continent (Fig. 4-7). The melting of sea ice does not affect sea level, just as the melting of an ice cube in a glass of water does not change the water level, but the melting of ice on land is another story. Global warming is causing sea level to rise at 3.2 mm/year today, almost twice as fast as a decade ago and still accelerating. Although much of the rise is due to water expanding as it is heated, as most substances do, the melting of the vast glaciers of Greenland and Antarctica is responsible for a growing proportion.

If things go on as they are, the polar icecaps will eventually melt on a scale large enough to inundate coastal regions everywhere. This will leave huge numbers of people to be resettled on higher ground; nearly half the world's population now live on or near coasts, a proportion expected to be three-quarters in the not-too-distant future. Even before gondolas fill the flooded streets of coastal cities, the underground water deposits that supply most of their freshwater will have been contaminated by seawater.

How much higher will sea level go? A U.N. panel in 2007 predicted somewhere between 18 and 59 cm by 2100. This estimate, adopted even though most of the participating scientists believed it improbably low, assumes that the ice sheets of Greenland and Antarctica melt gradually. However, ice sheets crack as they melt, which allows water to flow down through the ice. This water hastens the melting of a glacier's interior and also lubricates the glacier's downhill slide oceanward. Lakes and river systems have been found under Antarctic glaciers, similarly helping them move. The observed result is that Greenland and Antarctic ice is making its way into the oceans considerably faster than simple melting from the top downward would suggest, leading to a sea level rise possibly as much as 5 meters by 2100. Even a rise of just 1 m, considered a minimum by many scientists, would displace at least 130 million coastal dwellers. In any case the melting will not stop in 2100 but will persist as long as elevated temperatures do, raising sea level yet more in centuries to come.

The average surface temperature since 1900 has gone up by about 0.76°C. Sunlight is reflected back into space by ice, and the loss of ice means much more solar energy absorbed by the land and sea that it once covered. The result is a feedback loop that accelerates global warming. The 2007 U.N. panel considered several scenarios for estimating temperature rises through 2100. The most optimistic and most pessimistic ones gave these predictions:

Optimistic. No further population growth and a fall in carbon dioxide emissions. Temperature increases 1.1°–2.9°C with 1.8°C most likely.

Pessimistic. Present trends in population and carbon dioxide emissions per person continue. Temperature increases 2.4°–6.4°C with 4.0°C most likely.

A rise of 1.8°C does not seem like much, but between +1 and +2°C lies a threshold beyond which major impacts on human (and all other) life will occur. Three million years ago, when temperatures were 2°–3°C above those of today, sea level was over 15 m higher. A jump of 4.0°C would result in a planet warmer than it has been for millions of years with environmental changes on a scale beyond the reach of any of today's crystal balls. All one can say for sure is that the world would become an unrecognizably different place, with most living things gone and any human survivors struggling to survive in the polar regions.

Conflicts to Come

The large-scale disruptions of normal life that global warming seems to be on the way to bringing about are not likely to be met passively by those involved. It is entirely possible that within the lifetimes of many people alive today, tens of millions of Latin Americans, hundreds of millions of Africans, and a billion Asians will run short of freshwater and food. Rising seas made worse by storm surges will flood out hundreds of millions, eventually billions of people. How will the rest of the world, by then already near or at the limit of its ability to support its existing population, react to the refugees swarming in, desperate to survive? We all know the answer. An alarmist view? Not according to the U.N. Panel on Climate Change, the U.N. Security Council, the Pentagon, and other worried observers. Only if both global population and CO_2 emissions begin to fall very soon is there any chance of a peaceful world of thriving people to come.

4.4 Carbon Dioxide and the Greenhouse Effect

The Cause of Global Warming

Every body of matter radiates light regardless of its temperature; the hotter it is, the more it gives off (see Fig. 5-6). The radiation from something very hot, such as the sun, is obvious because its glow is mainly visible light. The radiation from something at room temperature, however, is chiefly infrared light to which the eye is not sensitive. The interior of a greenhouse is warmer than the outside air because the glass of its windows is transparent to visible light from the sun whereas the infrared light given off by its contents is absorbed by the glass, so that the incoming energy is trapped.

As discussed in Sec. 14.4, this **greenhouse effect** is largely responsible for heating the earth's atmosphere and its surface. The visible light from the sun that reaches the surface is reradiated as infrared light that is readily absorbed by several gases in the atmosphere. One of the most important of these gases is carbon dioxide. As a result the atmosphere is heated mainly from below by the earth and only to a smaller extent from above by the sun, as shown in Fig. 14-12. Without the greenhouse effect the earth's surface would average −18°C instead of its current average of 15°C.

In the past the total energy that the earth and its atmosphere reradiated back into space equalled the total energy they received from the sun. However, the CO_2 content of the atmosphere is steadily increasing due to the burning of fossil fuels, which means that the earth and its atmosphere are absorbing energy at a greater rate than before and are heating up. The result is global warming.

When the CO_2 content of the atmosphere stops rising (as it eventually will), there will still be a time lag until the massive earth reaches a final temperature at which the energy input and output flows are in balance. What this temperature will be depends on future emissions of CO_2, but even if they were to cease right now, global warming would continue for decades to come. The temperature surge of recent years shown in Fig. 4-6 is only about half of what is needed to balance the energy flows. Time is not on our side.

Analyzing air bubbles trapped in Greenland and Antarctic ice shows that the CO_2 concentration in the atmosphere is currently 27 percent higher than at any time in the past 650,000 years. The chief cause in recent times is the burning of fossil fuels to generate electricity; heat our homes; propel our cars, trains, ships, and airplanes; and power various industrial processes. Most of the world's energy by far comes from carbon-based fuels. Each kilogram of carbon burned yields 3.7 kg of CO_2, and at present our chimneys and exhaust pipes pour out about 30 billion tons of CO_2 per year; the United States and China are by far the largest contributors (see Figs. 4-8 and 4-9). Some years ago China was expected to overtake the United States as the world's champion CO_2 emitter around 2025; it actually did so in 2007.

As we can see from Fig. 4-10, the CO_2 content of the atmosphere has gone up by over 20 percent since 1860 and is today increasing faster than ever. This increase represents about half the CO_2 from burning fossil fuels;

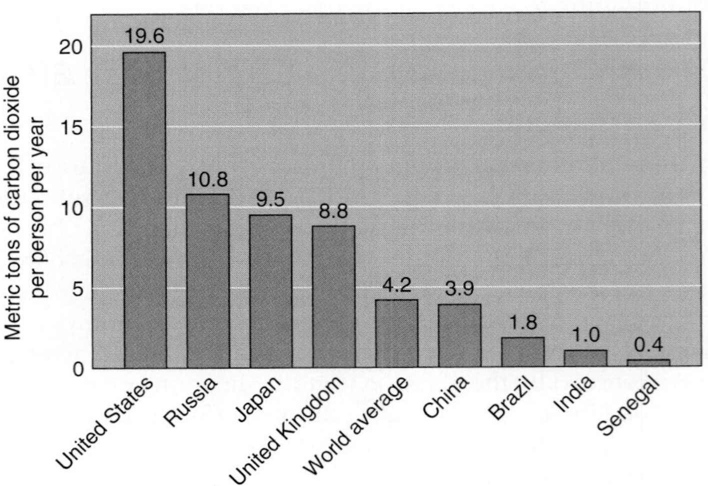

Figure 4-8 Annual carbon dioxide emissions from fossil fuels per person in various countries in 2005. Senegal is a fairly typical African country. (1 metric ton = 1000 kg)

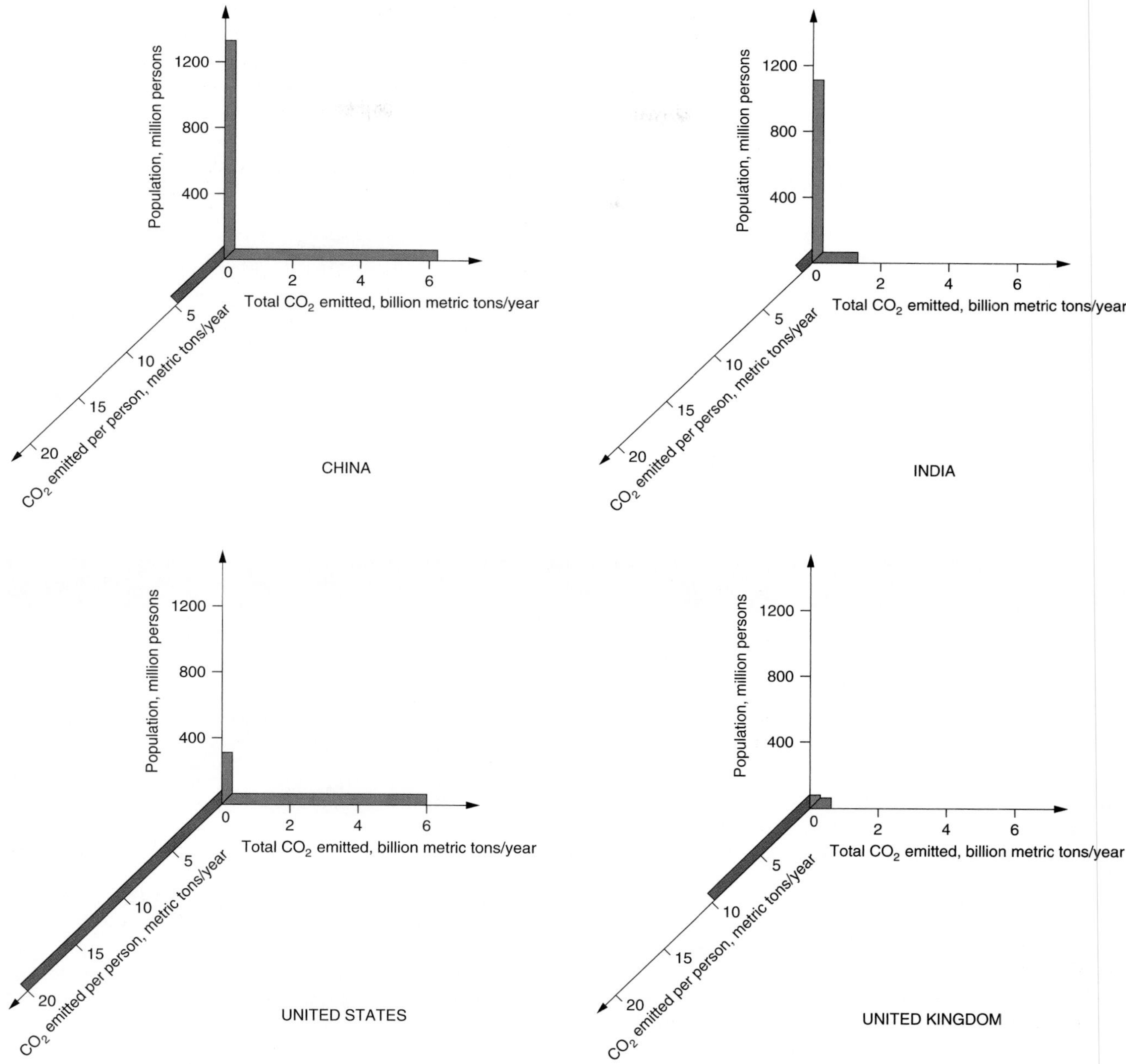

Figure 4-9 Population and annual CO_2 emissions from fossil fuels of four countries in 2007. China and India have the largest populations, with three out of eight of the world's people between them. For now India emits less than a fifth as much CO_2 per person as China does. China and the United States each account for nearly a quarter of the world's CO_2 emissions, but there are over four times as many Chinese. The United Kingdom, whose CO_2 emissions per person are about twice China's and half those of the United States, is a fairly typical industrialized country. CO_2 emissions from most less-developed countries are very small, only about 400 kg per person per year in Africa's Senegal, for example, but because such countries are largely agricultural they will suffer most from global warming.

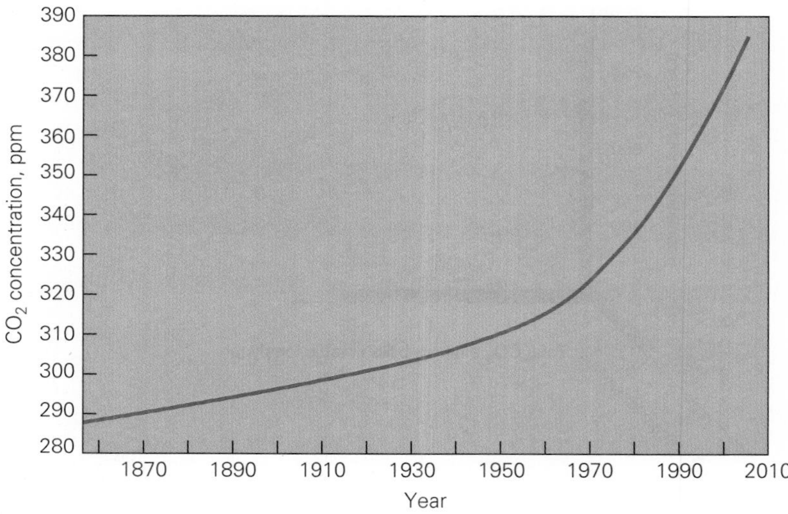

Figure 4-10 Carbon dioxide concentration in the atmosphere since 1860, in parts per million (ppm). The total today is nearly 3 trillion tons. There was little change in CO_2 concentration in the 10 thousand years before 1860. If CO_2 emissions do not fall significantly, their concentration is expected to climb to at least 500 ppm in this century. The resulting enhanced greenhouse effect will then push global temperatures past the threshold for severe and long-lasting environmental damage.

Other Greenhouse Gases

Although CO_2 is the most important of the **greenhouse gases** human activities are responsible for, it is not the only one (Fig. 4-11). Following it in significance are the CFCs and HCFCs, a group of artificially made gases mainly used in refrigeration and air-conditioning (Sec. 14.1). They leak into the atmosphere in much smaller amounts than Fig. 4-11 suggests, but are highly efficient as greenhouse gases—1 kg of most of them is equivalent to several tons of CO_2—and remain active for several decades. The quoted figure of 24 percent corresponds to their contribution to global warming.

Next in its impact on global warming is **methane,** the chief constituent of natural gas. A methane molecule consists of four hydrogen atoms bonded to a carbon atom, so its chemical formula is CH_4. Methane is 23 times as efficient as CO_2 in trapping heat but has a shorter lifetime of about a dozen years. About 600 million tons of methane are released annually into the atmosphere from wetlands, in the production of fossil fuels, in the decay of organic matter (for instance in landfills), in rice growing, and, in surprising quantities, as by-products of the digestion of food by cattle, sheep, and termites. A cow produces 100–200 liters of methane every day.

Methane in vast quantities—perhaps 50 billion tons—from the decomposition of organic remains has been locked into the frozen lands of Siberia and northern Canada for thousands of years. Now global warming is melting the permafrost and methane is bubbling out—an estimated 100,000 tons every summer day from Siberia's peat bogs alone. Like the increased absorption of sunlight by newly ice-free areas of polar lands and sea, this is another feedback loop that accelerates global warming. The methane concentration in the atmosphere today from all sources is over twice what it was in preindustrial times.

Nitrous oxide, N_2O, is 310 times as potent a greenhouse gas as CO_2 and has a similar average lifetime of 120 or so years. It is given off when fossil fuels and organic matter are burned, when fertilizers are used (3 to 5 percent of the nitrogen added by them to the soil ends up as N_2O), and in various industrial processes. Rainforests and the oceans also emit some N_2O. The N_2O concentration in the atmosphere, now 18 percent over its preindustrial level, is also on the way up.

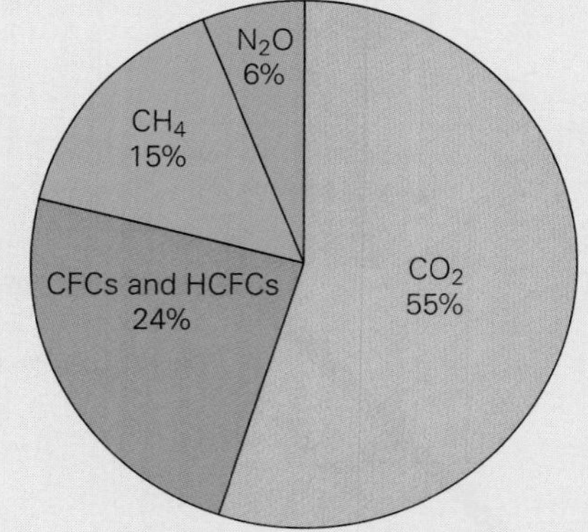

Figure 4-11 The contributions to global warming of the chief greenhouse gases.

the rest is absorbed by the oceans, soils, and forests. As fossil fuels continue to be burned at a high rate, the greenhouse "window" of CO_2 becomes a better trap for heat and the atmosphere will continue to warm up.

Interestingly enough, the warming due to increased CO_2 has been moderated in some regions—notably North America, Europe, and China—by the presence of smoke particles of various kinds, which are also given off when fossil fuels are burned. These particles form a thin haze that partly blocks sunlight. But the particles stay aloft for only a few days before falling to the ground and hence do not build up in the atmosphere. On the other hand, the lifetime in the atmosphere of emitted CO_2 is a century or more, and as the CO_2 continues to accumulate, its warming effect will further overwhelm the cooling due to the haze.

It is clear that the steady rise in CO_2 levels is increasing global temperatures and thereby putting in peril the ability of our planet to support our present ways of life, perhaps eventually human life itself. After examining the principal energy sources of today and those proposed for that status tomorrow, we will go on to see what strategies are available to cut CO_2 emissions and thereby to permit not merely life but civilized life to continue to flourish.

Deforestation

Every year 50,000 square miles of rainforest are destroyed, for the most part to create farmland. About half the tropical forests that the world once had are now gone, which has reduced the diversity of living things. As a rule, the soil under a tropical forest is poor and wears out after only a few crops, and additional forest is then cleared. Trees are about half carbon, and cutting them down to rot or be burned adds an estimated 3 billion tons of CO_2 annually into the atmosphere, almost a fifth of total CO_2 emissions. Another unfortunate aspect of deforestation comes from the fact that CO_2 and water are the raw materials from which trees manufacture carbohydrates with the help of sunlight (Sec. 13.12). A typical rainforest tree removes 22 kg of CO_2 from the atmosphere every year, a process that stops when it is cut down. Indonesia releases more CO_2 through deforestation than any other country, which puts it into third place, behind the United States and China, as a source of CO_2 emissions due to human activity. Large-scale deforestation continues in Brazil, which gains it fourth place in this list. Many tropical countries have lost most of their rainforests: the Philippines, 90 percent; Madagascar, 95 percent; Haiti, 99 percent. Figure 4-12 shows how large the contribution of deforestation is to overall greenhouse gas emissions.

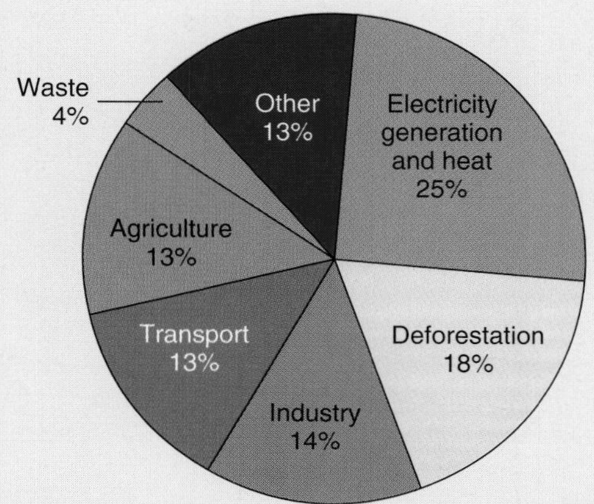

Figure 4-12 Origins of greenhouse gas emissions.

Fossil Fuels

Coal, oil, and natural gas are called **fossil fuels** because they were formed millions of years ago by the partial decay of the remains of swamp plants (coal) and marine organisms such as algae and plankton (oil and natural gas); see Sec. 16.15. Coal consists mainly of carbon, oil and natural gas consist of both carbon and hydrogen. Burning coal liberates energy as its carbon combines chemically with oxygen from the air to form carbon dioxide. Burning natural gas and fuels such as gasoline liberates more energy per gram (Fig. 4-13) as their carbon and hydrogen combine with oxygen to form carbon dioxide and water vapor, respectively. Figure 4-14 shows the recent and projected CO_2 emissions traceable to the various fossil fuels.

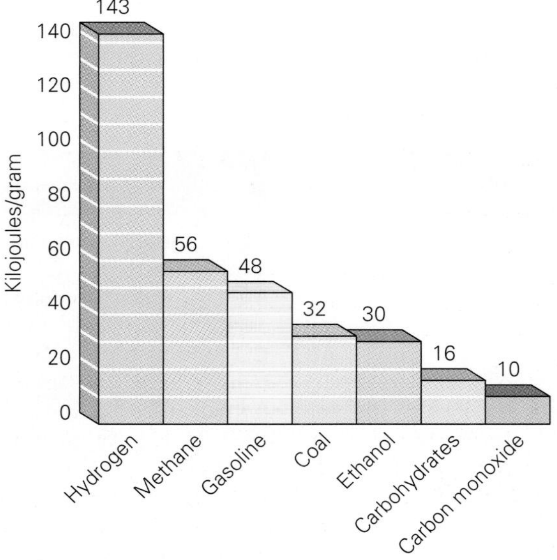

Figure 4-13 Energy contents of various fuels. Shown are the number of kilojoules of energy liberated when 1 g of each fuel is burned. Carbohydrates provide much of the energy in our diets (see Sec. 13.11). These fuels produce carbon dioxide (coal), water (hydrogen), or both when burned.

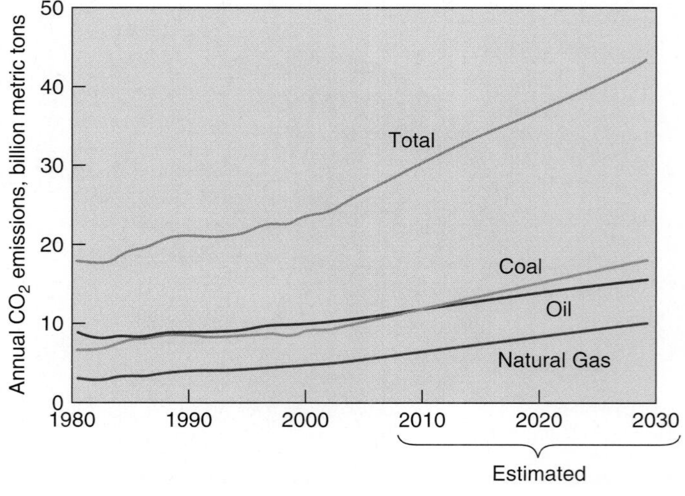

Figure 4-14 World carbon dioxide emissions from fossil fuels (1 metric ton = 1000 kg). The projections are based on current trends continuing and predict a total rise of 2.1 percent per year to give 44 billion metric tons of CO_2 given off in 2030, $1\frac{1}{2}$ times as much as today. To reverse the upward sweep of these curves and keep the earth habitable will require immediate effective action to keep population growth down, to use energy more efficiently, and to replace fossil fuels with nuclear and renewable sources.

Fossil fuels today provide about 85 percent of world energy consumption (Fig. 4-4). Oil and natural gas are versatile and convenient to transport and use, but they are growing increasingly expensive as they become harder to extract from declining reserves. Oil prices have more than quadrupled since 2003. Coal is relatively cheap and its widely distributed reserves are greater, but burning coal does the most damage to the environment. Various schemes have been proposed to utilize fossil fuels more efficiently and cleanly. Some of these schemes are more practical than others, but in the long run the role of fossil fuels in energy production will have to decline.

Another consideration in the case of the United States is that, because it imports two-thirds of the oil it uses (and has only 3 percent of world reserves), it is vulnerable to disruptions in its oil supply such as the 1973 Arab oil embargo. Making do with less oil would make the country more self-sufficient in energy supply as well as benefitting the environment.

4.5 Liquid Fuels

Vehicles Are the Biggest Users

The world's largest producers of crude oil are Saudi Arabia, the United States, Russia, and Iran, in that order. Smaller but still substantial amounts come from a number of other places, as shown in Fig. 4-15. Most new oil wells nowadays are located offshore (Fig. 4-16). The United States, which used 6.9 billion barrels in 2007, is by far the largest importer of crude oil.

As oil reserves continue to decline, previously uneconomical oil sources are likely to be exploited, but none are ideal. An example is the tar sands (mixtures of tar, sand, and clay) found over a vast area—the size of Florida—in Canada's province of Alberta; other large deposits are found in Venezuela with smaller ones elsewhere. Tar sands worldwide contain as much oil as there is in reserves of ordinary crude oil. Tar sands oil costs many times as much as crude oil per barrel to produce, in part because converting tar sands to usable oil needs a lot of energy, usually from the burning of natural gas. Even so, it is a profitable business in Canada. Unfortunately the accompanying CO_2 emissions are enormous: to obtain

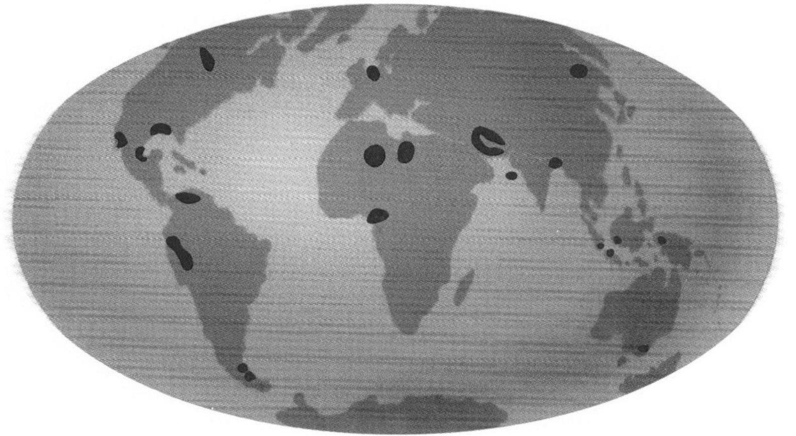

Figure 4-15 The chief oil and gas deposits in the world.

Figure 4-16 Oil drilling rig in the Gulf of Mexico off the Louisiana coast. It is more and more difficult to find new oil deposits to satisfy the world's increasing appetite.

a barrel of oil from a well involves the release of about 30 kg of CO_2, whereas the amount released is about 125 kg for a barrel of tar sands oil. There is more bad news—three to five barrels of water goes into the steam needed to produce each barrel of tar sands oil, and the waste water is too contaminated for further use. Nevertheless billions of dollars are currently being invested in tar sands projects and the current ouput from them is over a million barrels per day, expected to triple by 2015. U.S. law bans federal agencies from buying fuel from alternative sources, such as tar sands, if their production and use result in more greenhouse-gas emissions than in the case of ordinary sources. But nongovernmental consumers are under no such restriction.

As described in Chap. 13, gasoline and diesel fuel are mixtures of various hydrocarbons—chemical compounds of carbon and hydrogen—derived from oil. In a vehicle engine, the gaseous products of burning fuel expand rapidly because of the intense heat generated by the reaction. This expansion forces down the pistons of the engine, which in turn causes the crankshaft to rotate and provide power to the vehicle wheels (see Sec. 5-13). For every gallon of gasoline burned, over 8 kg of CO_2 are produced. A typical SUV can put 20 kg of CO_2 into the atmosphere in a 40-mile commute.

Most American cars use gasoline engines in which each gram of gasoline requires 15 g of air to burn completely. The CO_2 and H_2O (water) that result are odorless and nonpoisonous. A "rich" mixture of gasoline and air contains a greater proportion of gasoline than this, and a "lean" mixture contains a smaller proportion. If the burning is not complete, which occurs a little even under the best circumstances but much more when the gasoline-air mixture is rich, carbon monoxide (CO) and various hydrocarbons, all harmful substances, are formed as well as CO_2 and H_2O. The hazard of CO comes from its tendency to combine permanently with the hemoglobin in the blood in place of oxygen. This deprives the body of some of the oxygen it needs and leads to brain damage or death if too much CO is inhaled; no other poison injures or kills as many people as CO.

Saving Fuel Increasing the fuel efficiency of cars and trucks will save money, reduce pollution and global warming, and postpone the day

Reducing Pollution

We might think that the way to minimize pollution by car exhausts is to use a lean gasoline-air mixture, which increases the chance of complete burning by providing more than enough oxygen. But lean mixtures are hard to ignite. Worse, the leaner the mixture, the hotter it burns, and the higher temperatures promote the oxidation (combination with oxygen) of the nitrogen present in the air. Nitrogen oxides such as NO and N_2O react with water to form corrosive nitric acid and with unburned hydrocarbons to form a variety of toxic compounds, including some that cause cancer (Fig. 4-17).

There are several ways to reduce the amount of pollutants in vehicle exhausts. One is to build an engine in which just the right mixture of gasoline and air (to minimize carbon monoxide and unburned hydrocarbons) is burned at just the right temperature (to minimize nitrogen oxides) throughout all the variations in speed and power output involved in driving a vehicle. Such an engine will obviously be very efficient as well. But it is extremely hard to design an engine with these ideal properties, though much progress has been made by using computer control of its operation.

Another approach is to pass the exhaust gases of a car through a device called a catalytic converter, which contains an assembly of small tubes lined with a porous ceramic that contains particles of platinum and rhodium. These metals promote chemical reactions that change polluting gases into harmless ones without themselves being permanently affected. About 95 percent of the polluting gases are eliminated in this way.

A third method is to change the composition of the fuel. For example, a common additive to gasoline is ethanol, a type of alcohol (see Sec. 13-9). The mixture contains more oxygen than pure gasoline, so its burning is more efficient and leaves fewer pollutants. Ordinary gasoline engines can run on gasoline with up to about 15–20 percent of added ethanol; modified engines are needed for a higher proportion, and they are being installed in more and more new cars. Using ethanol as an additive or even as a fuel by itself in suitable engines conserves oil since it can be made from renewable crops and agricultural waste, as we shall see in Sec. 4-12, but it does not in every case reduce overall energy consumption or CO_2 emissions.

Figure 4-17 Traffic jam on multilane highway. Car exhausts contain carbon monoxide and various hydrocarbons, which are poisonous, and nitrogen oxides, which contribute to acid rain through the nitric acid they form. More efficient engines, better fuels, and catalytic converters reduce such emissions. Cars also produce carbon dioxide, which contributes to global warming through the greenhouse effect.

when oil runs out. One approach is to minimize air resistance by designing more streamlined vehicle shapes. Sport utility vehicles (SUVs) are especially bad in this respect. Engines can be improved; one method is to have several cylinders shut down when less power is needed, others involve direct fuel injection and variable valve timing. Better transmissions would also help, and reducing vehicle weight would pay big dividends.

Another approach is the hybrid car (Fig. 4-18), which has both a gasoline engine and one or more electric motors. When not much power is needed, the engine stops and the motors take over. The motors obtain their energy from a large battery that is charged both by the engine and by "regenerative braking"—slowing the car is done by using its motors as generators to convert its KE of motion to electric energy for storage. A smaller than usual gasoline engine, which is more economical with

Figure 4-18 The Toyota Prius is a hybrid car that has both electric motors and a gasoline engine, which under computer control are used separately or together as driving circumstances require. With regenerative braking, the result is a vehicle with twice the average mileage of ordinary cars and extremely low pollution. If all cars in the United States were as efficient as the Prius, 1.5 billion barrels of imported oil would be saved each year.

fuel, can be used because the electric motors can supplement it for extra power when accelerating. Carbon dioxide emissions from cars in the United States average 5.6 metric tons per person per year, which adds up to about 20 percent of the country's CO_2 emissions. Using hybrid cars could cut those figures perhaps in half.

In most of the world, gasoline and diesel fuel are expensive (in Europe, around twice American prices), so fuel economy is prized and efficient cars are normal. In the United States, where fuel until recently was relatively cheap, fuel economy was ignored by car makers until legislation in 1975 required a minimum of 27.5 mi/gal averaged over a maker's range of ordinary cars. However, SUVs, vans, and pickups—over half the cars on the road—were not covered by this requirement, and 15 mi/gal is not unusual for them. As a result, the actual average mileage in the United States is today the lowest in the world, little over half the current European average of about 40 mi/gal (expected to reach 50 mi/gal in a few years; in Japan it is nearly that high already). Forty percent of the oil consumed in the United States is used by its passenger cars. Hence improving their mileage, fought bitterly by car makers (just as they did seat belts and pollution controls earlier) but finally required by a 2007 law to reach a minimum average of 35 mi/gal, will make a big difference some years after it takes full effect in 2020.

The fuel economy of American cars is low not only because of their unnecessary size and weight and gas-guzzling engines but also because 97 percent of them use gasoline engines whereas the more efficient diesel cars are in the majority elsewhere. The latest diesel engines (unlike older ones) are quiet, produce little pollution, and are 20 to 40 percent more efficient than gasoline engines. If only a third of American cars and light trucks were diesels (the heavy trucks are already), the savings would amount to the equivalent of all the oil imported from Saudi Arabia. Performance is not an issue: a diesel car won the classic 24-h Le Mans race in France recently.

4.6 Natural Gas

The Least Bad Fossil Fuel

When burned, natural gas combines with oxygen to give carbon dioxide and water vapor, as liquid fuels derived from oil do, but with less pollution. Natural gas consists of the lighter hydrocarbons, chiefly methane. Natural gas is cheaper than oil even though its price has tripled in the past decade. It is more efficient than other fossil fuels in producing electricity, and its share of world generating capacity, now over 30 percent (20 percent in the United States) is steadily increasing. A gas-fired power plant typically gives off only about half as much CO_2 per kilowatt-hour (kWh) as a conventional coal-fired plant. Also widely used for heating, natural gas is expected to eventually replace oil as the world's chief energy source until it, too, begins to run out. Natural gas is an important feedstock for manufacturing chemicals of many kinds.

The largest producers of natural gas are Russia, the United States, and Canada, in that order; the United States also imports a great deal from Canada. Natural gas reserves exceed those of oil, but only 4 percent of

Figure 4-19 Natural gas, which is mainly methane, is carried in liquid form at low temperature (methane boils at −161°C) in tankers such as this as well as in pipelines. Liquid methane occupies about 600 times less volume than methane gas. The tanks are spherical to minimize heat flow into them: a sphere has the least surface area for a given volume. The liquid natural gas is kept cold by being allowed to evaporate continuously, which absorbs the heat that passes through the tank walls. The gas that comes off is used to power the ship's engines, so it is not wasted.

them are in North America compared with 60 percent in the three countries of Russia, Iran, and Qatar (a small Arab state in the Middle East). Expanding the use of natural gas in the United States inevitably means shipping more of it by sea from abroad (Fig. 4-19) as LNG—liquified natural gas—to special terminals where it is regasified and then carrying it by pipeline to consumers. At present there are only four such terminals in the United States—in Massachusetts, Maryland, Georgia, and Louisiana—that supply a few percent of its natural gas needs. A dozen more terminals have been approved for construction, mainly on the coast of the Gulf of Mexico. They should increase LNG imports to the country to 10 percent of its needs soon and to 25 percent by 2020.

Because natural gas is fairly plentiful and liquids are easier than gases to store and transport, a number of plants are being built to convert natural gas to diesel fuel. This is not economically practical when oil is cheap, but it is unlikely that the era of cheap oil will ever return.

4.7 Coal

Plentiful but Worst for the Environment

Coal was once the chief energy source under human control but has since been overtaken by oil. Since coal is cheap and its reserves exceed those of the other fossil fuels, it is likely to eventually return to the lead. (The energy in a dollar's worth of coal would, at 2007 prices, cost \$3.49 in the form of natural gas and \$6.32 in the form of fuel oil.) Coal is widely distributed as well as abundant; the United States has a quarter of the world's reserves—the coal in Illinois alone contains more energy than all

Underground Coal Fires

Stubborn fires in underground coal seams are surprisingly common around the world, pouring huge amounts of heat-trapping carbon dioxide plus various pollutants into the atmosphere each year and baking the land above them. Most have proved impossible to extinguish; one has been burning in India since 1916. The only remedy that works is to excavate coal around each fire to form a fire break, an immense task.

These fires are nothing new, and were started in the past by forest and grassland fires, by lightning, and also by the heat produced when certain minerals in coal react with oxygen. Today human activities are sometimes responsible. An example is the fire in the coal remaining in disused mine tunnels under Centralia, Pennsylvania that was ignited by burning trash. The fire started in 1961 and may continue to burn for centuries to come; the entire town had to be abandoned.

The several dozen such persistent fires in the United States are greatly outnumbered by those in Asia. In both China and Indonesia underground fires are widespread with hundreds of millions of tons of a valuable natural resource literally going up in smoke every year.

Figure 4-20 About half the coal extracted in the United States comes from underground mines such as this one. The rest is gouged in the open from deposits that lie near the surface after the overlying soil has been stripped away. Most of the underground mines are in the eastern part of the country; most of the surface mines are in the western part. The coal currently consumed in the world each year took about 2 million years to accumulate.

the oil in Saudi Arabia (Fig. 4-20). The chief producers of coal are China (40 percent of the world's total), the United States, India, and Australia (which exports most of the output of its mines).

If present trends in its energy supply continue, by 2030 the United States may consume as much as 50 percent more coal than the 1.2 billion tons per year it does today. China, which relies on coal for two-thirds of its energy, already burns more coal than the United States, Europe, and Japan combined and by 2030 will have more than doubled its current usage. Every week one or two new coal-fired power plants open somewhere in China. In all, over a thousand such plants are either newly built or under construction around the world. About 150 of them are in the United States and will add to the 600 currently operating there (Fig. 4-21), which generate more than half its electricity, and more are expected in the future.

Unfortunately coal is not an ideal fuel. For the same energy output, coal produces nearly a third more CO_2 than oil fuels and about twice as much as natural gas. Burning coal is responsible for 40 percent of the 30 billion tons of CO_2 released into the atmosphere per year by human activities (see Fig. 4-14), a proportion that will grow as coal use widens. During the normal 60-year lifespan of a 500 MW coal-fired power plant, it will emit 200 million tons of CO_2.

As discussed in Sec. 5.14, a basic physical principle called the second law of thermodynamics states that it is imposible to take heat from a source (such as a furnace or a nuclear reactor) and convert all of it to mechanical energy or work (for instance in a steam turbine connected to an electric generator). Some heat, usually a lot, must go to waste. In the case of an electric power station, the actual efficiency is less than half—only about 3–4 J of every 10 J of heat input becomes electric energy. Older power stations just discharge the leftover heat into nearby bodies of water or into the atmosphere via cooling towers. Nowadays combined

Figure 4-21 Coal, shown here piled next to an electric power plant in Newark, New Jersey, is the most abundant fossil fuel and is used to produce over half the electric energy generated in the United States. Coal is responsible for more CO_2 and other pollutants per unit of energy released than any other source.

heat and power stations are being built that capture the excess heat and use it for domestic heat and in various industrial applications. Such cogeneration conserves fuel and thereby cuts CO_2 emissions as well. This is not a technology of the future but a practical way of getting the most out of every ton of fossil fuel burned in a power station and every ton of CO_2 it releases.

Carbon Capture and Storage Because coal is going to remain a major energy source for a long time, it is essential to find ways to eliminate, or at least severely reduce, the CO_2 that using coal dumps into the atmosphere. A straightforward method is to pump the CO_2 from coal-burning plants deep underground for permanent burial. Suitable geological formations are widely available, and if necessary the captured CO_2 can be carried by pipeline as much as hundreds of kilometers from its source to a storage site. This is now being done on a small scale with CO_2 liberated in other processes. In Europe's North Sea, where natural gas from Norwegian wells is contaminated with excessive CO_2, a million tons of CO_2 each year are stripped from the natural gas and injected into porous rocks a kilometer below the seabed where it displaces seawater. No CO_2 has been found to be leaking out in a decade of operation. A similar carbon capture and storage (CCS), or **sequestration,** is being carried out at a gas field in Algeria (Fig. 4-22). In Canada's province of Saskatchewan, CO_2 is used to flush out the remaining oil from old wells; the CO_2, more than a million tons each year, remains behind. (A million tons of CO_2 a year is modest compared with the 30 billion tons now emitted annually, but it helps.) Various pilot operations are under way to further develop the technology involved.

Unfortunately separating CO_2 from the other flue gases spewed out by coal-fired power plants and then burying it would substantially raise the cost of the electricity generated, possibly by over 50 percent. Even so,

Figure 4-22 At this facility in Algeria in the Sahara Desert, carbon dioxide found mixed with natural gas is separated out and then pumped 2 km underground. Such sequestration keeps the carbon dioxide from entering the atmosphere where it would contribute to global warming by enhancing the greenhouse effect.

if CO_2 emissions were taxed or limited by cap-and-trade systems (see Sec. 4.14), sequestration could become an attractive option. CCS projects are being considered in a number of countries, usually to be helped by subsidies or by special circumstances. In the United States, Duke Energy has announced that it will no longer build coal-fired plants without carbon sequestration, but widespread adoption of the technology will not occur without either economic inducements or government requirements, and even then not soon.

Coal Gasification A more economical approach to sequestering the CO_2 from coal use is the integrated gasification combined cycle (IGCC), in which coal is first turned into a mixture of gases. An artificial gas fuel—syngas (for "synthesis gas")—can be made by passing very hot steam over coal to yield a mixture of carbon monoxide and hydrogen. Contaminants such as sulfur and mercury are readily removed, and the result is a gas fuel that can be burned in a power plant as cleanly as natural gas. Hydrogen could be separated out for use as the energy source for vehicles whose only emissions would be water vapor. The CO_2 from burning syngas is easier to capture than the CO_2 from burning coal directly, which makes sequestering it underground a more practical proposition. Several syngas power plants have been built and more are on the way, though not yet with provision for CCS.

Syngas can be the starting point for a variety of products. One is methane, the chief constituent of natural gas, and indeed a plant in North Dakota has been making methane for use as a fuel from coal since 1984 (Fig. 4-23). Syngas can also be used to create liquid fuels such as gasoline and diesel fuel. This was done on a large scale in oil-short Germany during World War II. However, manufacturing and using these artificial fuels in place of ordinary gasoline and diesel fuel doubles the overall amount of CO_2 produced. With global warming a reality, coal-to-liquid fuels do not seem the way to go.

Pollution Even apart from its role in global warming, coal is far from being a desirable fuel. Not only is mining it dangerous and usually leaves large tracts of land unfit for further use, but also the air pollution due to coal burning adversely affects the health of millions of people. Most estimates put the number of deaths in the United States from cancer and respiratory diseases caused by coal burning at over 10,000 per year; the number in China is in the hundreds of thousands. Interestingly enough,

Figure 4-23 The Great Plains Synfuels plant near Beulah, North Dakota, in operation since 1984, produces 4.5 million m^3 of syngas per day from 18,000 tons of coal. A by-product is CO_2, which is sent through a pipeline to Canadian oil fields where it is buried in old wells after helping to recover oil from them. Other profitable by-products include ingredients for fertilizers and raw materials for plastics.

coal-fired power plants expose people living around them to more radioactivity—from traces of uranium, thorium, and radium in their smoke—than do normally operating nuclear plants.

Mercury, which attacks the nervous system and is particularly harmful to unborn children, is an especially unfortunate component of coal smoke. Coal-fired plants in the United States discharge 48 tons of mercury each year, part of the reason why one in seven women of childbearing age in this country have enough mercury in their bloodstreams to put a fetus at risk of developmental damage. Nearly all states warn their residents about mercury contamination in their waters and in fish caught there. Although some years ago the Environmental Protection Agency had considered requiring an early cut in mercury emissions by 90 percent, industry objections (despite mercury's being the cheapest pollutant to control) led to lowering the target to a 25 percent reduction by 2010 and a 70 percent reduction by 2018. Needless nervous system damage to children will therefore continue for years to come.

Coal contains several percent of sulfur, and when coal is burned, the sulfur combines with oxygen to form sulfur dioxide, SO_2. Every year 50 to 60 million tons of SO_2 are released into the atmosphere from this source. Some nitrogen from the air also combines with oxygen in furnaces to form nitrogen oxides. The sulfur and nitrogen oxides react with atmospheric moisture to give sulfuric and nitric acids. The result is acid rain (and acid snow) that can be as much as 60 times more acidic than normal rainwater.

Acid rain has two main effects on soils. One is to dissolve and carry away valuable plant nutrients. The other is to convert ordinarily harmless aluminum compounds, abundant in many soils, to toxic varieties. As a result, forests are dying (Fig. 4-24) and fish have disappeared from many lakes and rivers due to aluminum washed into them. Drinking water has been contaminated in a number of regions by metals released by acidified water, such as cadmium and copper besides aluminum. The technology exists for "scrubbing" SO_2 from exhaust gases, and 1990 legislation requires power plants in the United States to limit their SO_2 emissions. However, the acceptable level of SO_2 was set so high that the acidity of many lakes and rivers has actually not fallen since 1990. In China, by far the largest emitter of sulfur dioxide, acid rain falls on a third of the country, with serious ecological consequences.

Figure 4-24 Acid rain, together with atmospheric sulfur dioxide (which attacks chlorophyll), led to the destruction of this forest in North Carolina. Healthy and abundant forests are needed not only for timber but also because they absorb CO_2 from the atmosphere, protect soil from erosion, help prevent floods, furnish habitats for most kinds of land plants and animals, and participate in the water cycle.

Alternative Sources

We now look at the sources responsible for the 15 percent of commercial energy production that does not involve burning fossil fuels. Although each of these sources has limitations of various kinds, it may be a good choice in certain situations. If the full potential of these sources is realized, the world will depend much less, perhaps very little, on fossil fuels with all their shortcomings. Of the available alternatives, the only one that can replace fossil fuels on a major scale in the relatively near future is nuclear energy. But other technologies are rapidly advancing, and eventually it will become clear which paths are the best to follow toward a sustainable energy supply for the world.

Today energy derived from fossil fuels is cheaper—in the case of coal, much cheaper—than energy from most alternative sources. There are two reasons. The first is that damage to the environment is not reflected in the prices of fossil fuels. If this factor is taken into account, the present cost advantage of fossil fuels disappears. The second reason is that the technologies based on fossil fuels benefit from long experience with them together with economies of scale. As alternative sources mature, these advantages will fade away.

4.8 A Nuclear World?

Perhaps on the Way

A nuclear reactor obtains its energy from the **fission** (breaking apart) of the nuclei of a certain kind of uranium atoms, as described in Chap. 8. In a nuclear power plant, steam from boilers heated by such a reactor runs turbines connected to electric generators. In 1951, in Idaho, electricity was produced for the first time from a nuclear plant.

Today 443 reactors in 31 countries generate about 450 GW of electric power, a sixth of the world total. Without them over 20 million barrels of oil (or their equivalent in coal or natural gas) would have to be burned every day. France, Belgium, and Taiwan obtain more than half their electricity from nuclear plants, with several other countries close behind (Fig. 4-25). In the United States, nuclear energy is responsible for about

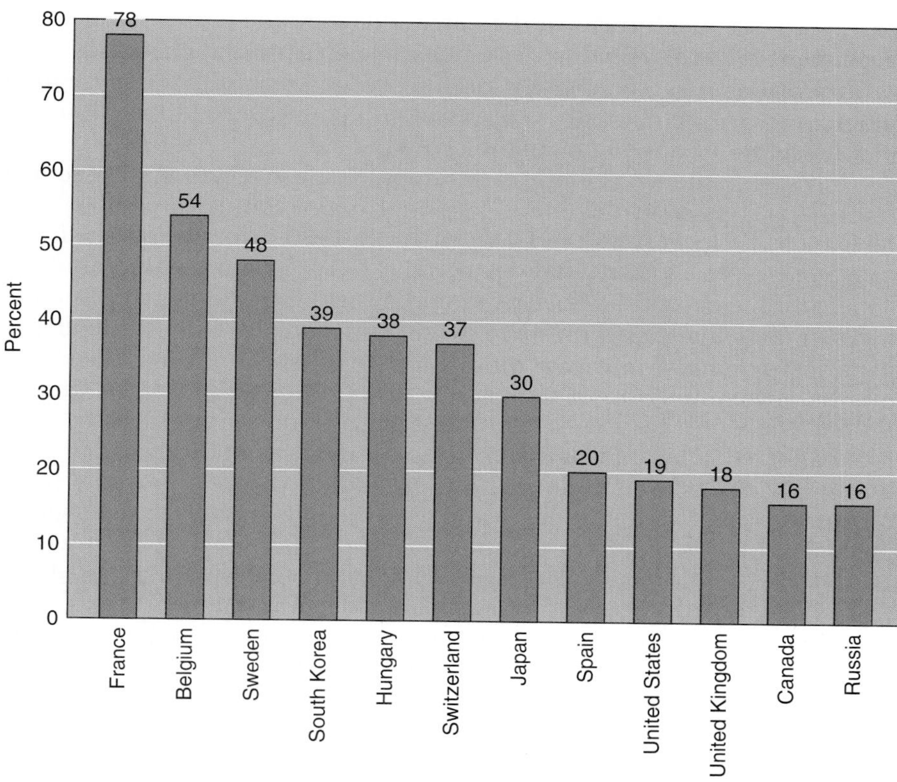

Figure 4-25 Percentage of electric energy in various countries that comes from nuclear power stations. France is more dependent on nuclear energy than any other country; it has 59 nuclear power plants. The United States has 104.

21 percent of its electricity, somewhat more than the world average of 16 percent; there are 104 reactors in 31 states. The uranium that fuels nuclear plants, fairly abundant in the earth's crust, is apparently able to support the anticipated expansion of nuclear energy in the decades to come, and is not unduly expensive to mine and purify. Nuclear plants do not emit CO_2 and so do not contribute to global warming; if the present ones were fossil-fuel plants, nearly 3 billion more tons of CO_2 would be released each year into the atmosphere. Yet for all the success of nuclear technology, construction has not begun on any new nuclear power stations in this country since 1979. Why not?

Three Mile Island and Chernobyl In March 1979 failures in its cooling system disabled one of the reactors at Three Mile Island in Pennsylvania, and a certain amount of radioactive material escaped. Although a reactor cannot explode in the way an atomic bomb does, breakdowns due to poor design, shoddy construction, inadequate maintenance, and errors in operation—all present at Three Mile Island—can occur that put large populations at risk. Although a true catstrophe was narrowly avoided, the Three Mile Island incident made it clear that the hazards associated with nuclear energy are real, and the lack of candor, then and later, by industry and government about these hazards was even more worrying.

After 1979 it was inevitable that greater safety would have to be built into new reactors, adding to their already high cost. In addition, the demand for electricity in the United States was not increasing as fast as expected,

Nuclear Weapons

Natural uranium consists of two varieties, ^{238}U and ^{235}U, of which only ^{235}U can undergo fission (see Secs. 8.9–8.11). Natural uranium contains only 0.7 percent of ^{235}U and must have this proportion increased to about 3 percent to make reactor fuel. The process by which natural uranium is enriched in ^{235}U can be continued further until the proportion is over 90 percent, and the result is the active ingredient of one kind of nuclear weapon, or "atomic bomb." Furthermore, in its operation a nuclear reactor produces another element, plutonium, that can be separated out from used fuel rods. Like ^{235}U, plutonium can also be used in nuclear weapons. At present nine countries are known to possess a total of nearly 26,000 of these weapons of mass destruction, sufficient to wipe out all human life many times over, and other countries could develop them if they wished. The threat of nuclear weapons proliferation is one of the reasons why not everybody welcomes the expansion of nuclear energy.

partly because of efforts toward greater efficiency and partly because of a decline in some of the industries (such as steel, cars, and chemicals) that are heavy users. As a result, new reactors made less economic sense than before, which together with widespread public unease led to a halt in the expansion of nuclear energy in the United States.

Elsewhere the situation was different. Nuclear reactors still seemed the best way to meet the energy needs of many countries without adequate fossil fuel resources. Then, in April 1986, a badly planned test caused a severe accident that destroyed a 1-GW reactor at Chernobyl in Ukraine, then part of the Soviet Union. This was the worst environmental disaster of technological origin in history and contributed to the collapse of the Soviet Union. The lack of a containment shell, normal elsewhere, allowed nearly 200 tons of radioactive material to escape and be carried around the world by winds. Ukraine, Belarus, and parts of Russia were most affected. The radioactivity in the fallout was 400 times that produced by the Hiroshima atomic bomb. Radiation levels in many parts of Europe rose well above usual and are still high enough to represent a hazard in a large area that was downwind of Chernobyl. Even in Great Britain, 1500 miles away, 375 farms remain so contaminated that sheep from them must be checked for radioactivity before sale.

About 350,000 residents of the Chernobyl vicinity were permanently evacuated from their homes, leaving behind ghost towns and villages. Thousands of people became ill, including about 4000 children who developed thyroid cancer. Because cancer and leukemia also have causes other than radiation, the death toll from these maladies due to Chernobyl will never be known; estimates range from thousands to tens of thousands. As in the United States after Three Mile Island, public anxiety over the safety of nuclear programs grew abroad after Chernobyl. Some countries, for instance Italy, then abandoned plans for new reactors and closed down some existing ones. (Italy today has the most expensive electricity in Europe and is converting oil-fired power stations to burn cheaper coal instead; it also expects to build reactors again.) In other countries, for instance France, the logic behind their nuclear programs remained strong enough for them to continue despite Chernobyl.

Nuclear Energy Today The latest designs for nuclear plants promise major improvements in efficiency and reliability over previous ones, which makes modern plants cheaper to run per unit of output than their fossil-fuel cousins. They are also safer than before. Together with increasing demand for energy, the result is an international boom in nuclear plants with new ones currently being built in 13 countries. China intends to add 32 such plants to its current 11 and India to add 28 to its current 14; Russia, South Korea, and Japan also expect to move in the same direction. In 2005 Finland began building a 1.6-GW plant, which will be the world's largest. The International Atomic Energy Agency thinks global nuclear capacity will quadruple by 2050. In the United States, licenses for 27 new nuclear plants (to cost several billion dollars each) are being sought; if approved, the plants may begin to enter service in 2015 or 2016. The ultimate limit to the speed of nuclear expansion may be set by shortages of skilled workers, of the needed construction materials, and of manufacturing capacity.

Nuclear Wastes

Quite apart from the safety of the reactors themselves is the issue of what to do with the radioactive wastes they produce. Although a lot of the radioactivity is gone in a few months and much of the rest in a few hundred years, some will continue for millions of years. At present over 36,000 tons of spent nuclear fuel are being stored on a temporary basis in the United States alone in cooling ponds (to prevent overheating) at reactor installations where they may leak and are vulnerable to terrorist attack.

Burying nuclear wastes deep underground currently seems to be the best long-term way to dispose of them. The right location is easy to specify but not easy to find: it must be stable geologically with no earthquakes likely, no nearby population centers, a type of rock that does not disintegrate in the presence of heat and radiation but is easy to drill into, and little groundwater that might become contaminated. In 2002 the United States government chose Yucca Mountain in Nevada as the most suitable site it could find for storing nuclear wastes indefinitely, but further studies were required and the fears of the state of Nevada have yet to be overcome (Fig. 4-26). It is by no means sure that Yucca Mountain will ever see a shipment of nuclear waste, or even that any burial site anywhere can be found free from serious objections. However, a method of dry storage in casks surrounded by an inert gas is thought to be reasonably safe for at least a century, giving time to figure out a more permanent solution.

Figure 4-26 Tunnel being bored deep into Yucca Mountain in Nevada to study its suitability for nuclear waste storage. Around $6 billion has already been spent in trying to estimate how much radioactive material might escape in the future under various possible circumstances, with no definitive conclusion. The mountain is composed of the stable volcanic rock tuff, and rainfall there is light, which reduces (but does not eliminate) the risk that wastes might leak from corroded containers and contaminate water supplies. Nevada is in an earthquake-prone region, and some time ago a small quake occurred not far from Yucca Mountain.

Nuclear Fusion Enormous as the energy produced by splitting a large atomic nucleus into smaller ones is, the joining together of small nuclei to form larger ones gives off even more energy for the same amount of starting materials (see Sec. 8.13). Such **nuclear fusion** is the energy source of the sun and stars. Here on earth, there are realistic hopes that fusion will take over the lead as a source of energy at some time in the future—safe, no greenhouse gases, very little radioactive waste, and abundant fuel, much of it from the oceans.

In laboratories, fusion reactors have been built that liberate energy for short periods as predicted by theory. In order to operate continuously and to yield energy on a commercial scale, the reactors must be much larger (a planned experimental reactor called ITER will weigh 32,000 tons), but no fundamental reason is known why such reactors should not be successful. Of course, a technical success is not necessarily an economic success, but if it becomes one before environmental disaster intervenes, fusion energy may be the ultimate solution to our energy problems.

4.9 Clean Energy I

Continuous Sources

An ideal energy source should not deplete resources or harm the environment. A number of sources meet these criteria and, despite issues of cost and location in some cases, electricity is starting to come from them in serious amounts: about 5 percent of total electricity in the United States, twice that in Europe. Just what proportion of global energy demand these sources will eventually provide remains to be seen, as does the timescale of their adoption, but their increasing popularity is a sign of hope.

"Clean" energy sources fall into two categories that depend on whether they can supply energy continuously (hydroelectric, geothermal) or only at rates that vary with the time of day (solar, tidal) or with weather conditions (wind, waves). We will first look at the two main continuous sources, moving water and geothermal heat.

Hydroelectricity The kinetic energy of moving water has been used by mills and factories for centuries, and has powered electric generators since 1870 (Fig. 4-27). Hydropower now provides 2.2 percent of the world's energy with capacity up by over 40 percent since 1980. Norway obtains 99 percent of its electricity from falling water, Brazil 84 percent, Canada 58 percent, and 13 African countries 60 percent or more each.

Figure 4-27 This hydroelectric installation on the Niagara River in New York State has a capacity of 2.2 GW of electric power. The ultimate source of this power is sunlight, which evaporates water that later falls as rain or snow that drains into the river upstream of the dam.

The United States has a total hydroelectric capacity of 96 GW; its largest installation, at the Grand Coulee Dam, produces 6.8 GW. The largest hydroelectric plant now operating anywhere is on the Brazil-Paraguay border and is rated at 14 GW. When completed, the Three Gorges Dam in China, on which work began in 1994, will take the lead with a capacity of 22.4 GW. This dam is the world's biggest civil engineering project and, together with other such projects there, will increase China's total hydroelectric output to half again its current 129 GW. Major hydroelectric installations are also under construction or having plans finalized elsewhere, mainly in Asia (a total of 14 GW in India alone) and South America (6.5 GW in Brazil). In Africa, too, more such installations are projected—with a sixth of the world's population it produces only 4 percent of the world's electricity.

Even when the new dams are completed, only about a third of the world's hydroelectric potential will have been utilized. But many of the remaining sites cannot be exploited economically for a variety of reasons. Furthermore, an increasingly significant problem with hydropower installations is the social and environmental damage they may cause, for instance by flooding wide areas and turning once fertile river valleys into wastelands unfit for agriculture. The Three Gorges Dam, which created a lake 643 km (about 400 miles) long, has already displaced 1.4 million people. The Chinese government expects that 3–4 million more people will have to be relocated in years to come as the Three Gorges project causes the bed of the Yangtze River to silt up, its banks to erode, and its waters and those of the lake to become polluted. Environmental concerns have even led to the dismantling of a number of existing hydropower dams—nearly a hundred in the United States in the past few years.

Geothermal Temperature increases with depth in the earth, and in many places water below the surface is hot enough for useful energy to be extracted. One such place is at The Geysers north of San Francisco where turbines powered by natural steam drive generators that produce 750 MW of electricity (Fig. 4-28). Even where suitable hot water or steam is not present underground, water from the surface can be pumped into cracks in deep rock formations that can be recovered as hot water or steam from wells drilled nearby. Carbon dioxide under pressure can also

Figure 4-28 This power station at The Geysers, California, runs on geothermal energy.

be used to extract such **geothermal heat.** Over their respective lifetimes, geothermal power plants produce electricity at less cost than coal-fired plants, the cheapest conventional sources.

At present 24 countries have geothermal power plants with a total capacity of almost 90 GW, and more are being built. Iceland and the Philippines obtain over a quarter of their electricity from such plants. Indonesia hopes to achieve a similar proportion by 2025. Although its current share of the world's energy supply is only 0.4 percent, a recent study found that geothermal energy has more potential than any other renewable source. In the United States, hot rocks less than 10 km underground could satisfy all the country's electrical needs at their current level for the foreseeable future.

Besides a role in generating electricity, hot subsurface water is widely used for heating purposes, mainly in buildings (all of Iceland's buildings are heated in this way) but also in agriculture to lengthen the growing season for crops.

4.10 Clean Energy II

Variable Sources

Now we consider clean energy sources whose output is not constant. This is not necessarily a major disadvantage because, when available, their electricity can replace that from fossil-fuel sources even if only intermittently. And, as mentioned later, several methods exist for storing the energy of variable sources until needed.

Solar Bright sunlight can deliver over 1 kW of power to each square meter on which it falls. At this rate, an area the size of a tennis court receives solar energy equivalent to that in a gallon of gasoline every 10 min or so. The earth receives more energy from the sun each hour than the world uses in a year.

Photovoltaic cells are available that convert the energy in sunlight directly to electricity. Although the supply of sunlight varies with location, time of day, season, and weather, such **solar cells** have the advantage of no moving parts and almost no maintenance. For a given power output solar cells are more expensive than fossil-fuel plants, but they have no fuel or operating costs. Improving technology is steadily increasing the efficiency of solar cells, now as much as 20 percent for commercial cells and over 40 percent for experimental ones, and dropping their price.

Although solar cells today provide only 0.15 percent of the world's electricity, their production is doubling about every 2 years. Arrays that produce 10 MW (enough for 10,000 homes) or more are operating in a number of countries; in China, a 100 MW array is planned for 2011. Germany, despite its often cloudy skies, has 3 GW of solar-cell capacity, more than half the world total, on the roofs of 300,000 homes and businesses. Although electricity from solar cells in the United States amounts to less than 1 GW now, it is expected to exceed 15 GW by 2020.

A big advantage of solar cells is that they can be installed close to where their electricity is to be used, for instance on rooftops (Fig. 4-29). This can mean a major saving because it eliminates distribution costs in

Solar Water Heating

Exposing pipe arrays filled with water to sunlight is a simple and cheap way to capture solar energy for household hot water and space heating. About 90 GW of solar energy is exploited in this way every year worldwide. Two-thirds of this energy is collected in China, where by replacing fossil-fuel burning, CO_2 emissions are reduced by several hundred million tons annually. In other countries such direct solar water heating is less common, only 1.8 percent of the world total in the United States, for instance.

Figure 4-29 Array of solar cells being installed over the back porch of a house in California.

rural areas where power lines would otherwise have to be built. In Kenya, more households get their electricity from solar cells than from power plants. Even when electricity grids exist, rooftop solar cells are becoming common. As part of California's efforts to have more of its energy needs come from renewable sources, the "One Million Solar Roofs" program is providing $2.9 billion in subsidies to businesses and households that install solar panels.

In another approach, concentrated solar power (CSP), mirrors direct sunlight on pipes filled with oil that becomes very hot as a result. The hot oil is then used to produce steam for turbines that drive electric generators. CSP installations are practical only where there is open land with reliable sunshine. Nine arrays in the Mohave Desert in California have been furnishing 354 MW of electricity for over 20 years (Fig. 4-30). A 2007 array in Nevada uses 19,300 4-m pipes to produce 64 MW. Other CSP projects totaling 4 GW are under way in the United States, Spain, Algeria, and Australia.

Wind Windmills are nothing new and were once widely used for such tasks as grinding grain and pumping water. Holland alone had 9000 of them. Now windmills are back in fashion for generating electricity, and although they are practical only where winds are powerful and reliable, such winds are found in many parts of the world (Fig. 4-31). Today just over 1 percent of the world's electricity comes from wind, about the same proportion as in the United States where wind electricity already supplies the equivalent of 4.5 million homes.

A typical large modern turbine has three fiberglass- or carbon fiber-reinforced blades 40–45 m long that make up to 20 revolutions per minute to generate up to 2 MW. The actual output depends on wind conditions at the time; an average of 40 percent of the maximum is considered good. The cost of windpower is on the way to becoming competitive with

A Solar Future?

Using only a modest part of the vast area of barren land in America's Southwest, installations of solar cells and CSP plus provision for storing energy overnight could provide a large fraction of the energy needs of the entire country. One ambitious but not impossible scheme envisions no less than 3500 GW of solar power in the United States by 2050 that would come from collectors of both kinds covering 46,000 square miles. (More land per GW is needed for coal-fueled electricity when the land used for mining is included.) This would provide 69 percent of the country's electricity and 35 percent of its total energy at today's prices. The cost? Plenty, but probably less per year in subsidies than current farm subsidies. A tax of 5 percent or so on the price of fossil-fuel electricity might cover it as well as discouraging fossil-fuel use. Adding energy from wind, biomass, and geothermal sources would, in principle, allow all of the electricity used in the United States and 90 percent of its total energy to come from renewable domestic sources by 2100.

Figure 4-30 This concentrated solar power (CSP) installation in California's Mohave Desert uses curved mirrors to direct sunlight to heat oil that then generates steam to power turbine generators. The array produces its maximum power on sunny summer days when electricity demand for air-conditioning is at its peak.

Figure 4-31 Wind turbine "farm" near Palm Springs, California. Such farms consist of as many as several hundred turbines and can supply energy to tens of thousands of homes and businesses. About 1 percent of the electricity used in the United States in 2003 had wind as its source. The windpower potential in the United States will exceed its energy consumption for the foreseeable future.

electricity generated by fossil-fuel or nuclear plants but without their disadvantages. Much larger turbines are being developed that should give cheaper electricity than most fossil-fuel plants and could be ready for installation by 2012.

Wind is the world's fastest-growing source of clean energy (over 30 percent annual increase in the United States) with its potential barely tapped. In 2008 the global total of wind energy capacity was over 100 GW, about five times what it was in 2000. Of that, 21 GW was installed in Germany, the leader in wind turbines, with the United States (17 GW) and Spain (15 GW) not far behind. Wind turbines have been installed in 26 countries in all. Wind-generated electricity supplies 1.8 million households in the

United States, where Texas is the leading producer, followed by California, Iowa, and Minnesota. The U.S. government has proposed a goal of 80 GW of wind turbine capacity in the country by 2020; China's target for that year is 30 GW, but it is building turbines so fast that the target will proably be reached much earlier.

More and more turbine farms are being sited in shallow offshore waters where they have minimal environmental impact and can take advantage of the stronger and steadier winds there. Denmark (the largest builder of wind turbines) expects to generate half of its electricity by 2025 from offshore turbines. A 1-GW offshore wind farm, which would be the world's largest, is planned for English waters south of London; it could supply 25 percent of London's energy needs. Other wind farms expected to be installed off the English coast have target capacities of 300 and 500 MW. In the United States, a wind farm of 130 turbines with a total capacity of 468 MW has been proposed for an offshore location south of Cape Cod in Massachusetts and may be operating by 2011. In Europe's North Sea, which has a great many oil and gas production platforms already in place, wind turbines are beginning to sprout atop them to power their operations; using existing platforms saves a third of the cost.

Tides The twice daily rise and fall of the tides (Sec. 1.10) is accompanied by corresponding flows of water into and out of bays and river mouths. Harnessing the considerable energy involved is another old idea: in Europe, tide mills go back to the twelfth century. Tidal power is reliable and has low operating costs. On the other hand, the tidal cycle means that there is no energy output for two periods of about 6 h each per day, which leaves a large investment idle for half the time.

There are two main approaches to extracting energy from the tides. One of them involves spanning a narrow inlet on a coast that has a large—over 5 m—tidal range with a dam that traps water on a rising tide and then directs it to turbine generators when the water level outside has dropped. An installation of this kind in the Rance River in northern France has supplied 240 MW of peak electric power since 1966 (Fig. 4-32). In South Korea a new 254 MW tidal power installation, for the time being the world's largest, will be followed in 2014 by another one whose capacity will be 812 MW and cost $1.9 billion. Even bigger tidal plants are being contemplated elsewhere, for instance in Canada (5 GW), India (7.4 GW), and England (8.6 GW—4.4 percent of the country's electricity use), but apart from the expense ($30 billion for the English project) there is the risk that altering tidal flow patterns may harm local ecosystems.

The other approach is to use submerged turbines to drive generators as tidal currents run back and forth past their blades. An undersea tidal farm of this kind can be on a small scale that avoids the cost and environmental issues of a dam. Such farms have been installed off the Norwegian coast and in New York's East River; locations for others elsewhere in the United States and in Europe are being studied.

Figure 4-32 This 1966 barrage across the Rance River in France uses tidal flows to drive turbine generators that supply 240 MW of peak electric power.

Waves As anybody who has stood in the surf or watched waves dash against a rocky shore knows, waves carry energy in abundance. A number of schemes have been thought up to capture this energy. In one of them ocean waves run up a sloping funnel-like channel to a reservoir above.

Water from the reservoir then powers a turbine generator on its way back down to the ocean. But this simple system is feasible only where the seabed is so shaped that wave energy is focused on a particular spot on a coast and where the winds that drive the waves are usually onshore.

A wave energy converter that can be used anywhere, called Pelamis, employs a series of cylindrical sections, about the size of railway cars, that are hinged together. The sections swing back and forth relative to one another when waves pass by, and these motions drive pumps that force oil at high pressure to hydraulic motors coupled to electric generators. Several Pelamis machines are already operating off the British and Portugeuse coasts and larger ones are being built. If all goes well, the future may see 30-MW arrays that would each be spread over a square kilometer of ocean near many of the world's coasts. Other wave energy conversion schemes are also being studied.

Storage Effective ways to store energy for later use from variable sources are clearly desirable. An obvious method is to use storage batteries in which electric energy is converted into chemical energy that can subsequently be converted back again, but until recently they had been too expensive on the required scale. However, flow batteries have been developed in which the energy-rich chemicals of a charged battery do not have to remain there but can be pumped out into separate tanks while fresh starting chemicals replace them. Reversing the flow allows the stored energy to return to its original electric form in the battery. Flow batteries are more complex than conventional batteries (Sec. 12.16) and their technology is still being improved. Still, they are already sufficiently advanced for a 7.2-GJ flow battery to have been installed in Utah and a 43-GJ one being built at an Irish wind farm.

In another approach, to be discussed in Sec. 4.11, electric energy is used to produce hydrogen, an excellent fuel, by the electrolysis of water. The hydrogen can then be burned to give heat and water or used in a fuel cell to produce electricity (no carbon dioxide in either case). Both flow-battery and hydrogen storage systems seem to have promising futures.

Electricity that is not needed at a particular time can also be used to pump water up to a high reservoir. Then, at night or when the wind stops or when the tide is not running, the water is allowed to fall through turbine generators. An arrangement of this kind has been built in Wales using a difference in water level of 440 m. Denmark now sends surplus electricity from its wind farms to Norway to replace electricity from Norwegian hydroelectric plants, thus saving water in the reservoirs for use when Danish winds are light. Norway has so many hydroelectric installations that they have been suggested for temporary storage of surplus wind electric energy from all over Europe by using the electricity to pump additional water up to the reservoirs.

On a smaller scale, air can be pumped into a sealed underground cavern, an abandoned mine, or an exhausted natural gas well and the compressed air later released to power generators. An advantage here is that suitable caverns are more common than elevated sites for water reservoirs, and costs are less than for energy storage in batteries. Compressed air energy storage facilities have been operating in Germany since 1978 and in Alabama since 1991.

4.11 Hydrogen and Fuel Cells

New Technologies for Future Vehicles

As we can see in Fig. 4-13, mass for mass, hydrogen liberates more energy when it combines with oxygen than any of the other fuels listed—three times as much as gasoline, for instance. It is a clean fuel as well: the only product of its use is water, with no carbon dioxide or harmful pollutants. Unfortunately, although hydrogen is by far the most abundant element in the universe, on earth it is found only in compounds with other elements (notably with oxygen in water, H_2O) and must be separated out from them. Therefore hydrogen is really a storage and delivery medium and not a primary fuel in itself. Nowadays most hydrogen is produced by reacting natural gas with steam, with carbon dioxide as a by-product.

Hydrogen can be used to provide energy in two ways. One is simply by burning it, which is done in welding torches and in spacecraft propulsion engines. Under development are cars that use hydrogen in place of gasoline in similar engines.

The other approach employs **fuel cells,** devices in which hydrogen and oxygen react directly to produce water and electricity rather than water and heat. Unlike batteries, which also obtain electric current directly from chemical reactions, fuel cells can provide current indefinitely without having to be replaced or recharged because the working substances are fed in continuously. Fuel cells are already employed in a variety of applications, as described in Sec. 12.16, including vehicles (Fig. 4-33) and small power stations, and in time may replace batteries in cell phones and laptop computers.

Despite currently being relatively expensive for their output, fuel cells are attractive as ultrareliable energy sources that are silent, vibrationless, more efficient than gasoline and diesel engines, nonpolluting, and that do not give off heat or carbon dioxide—a formidable list of virtues. But as long as natural gas is the chief source of hydrogen, these virtues do not necessar-

Not a New Idea

In his 1874 novel *The Mysterious Island,* one of Jules Verne's characters predicts that "water one day will be employed as fuel, that hydrogen and oxygen which constitute it . . . will furnish an inexhaustible source of heat and light . . . Water will be the coal of the future." Another character responds (as do we all), "I should like to see that."

Figure 4-33 This prototype bus is powered by fuel cells that operate on hydrogen. Similar technology is used in cars from other manufacturers, which are undergoing road tests, but commercial models are not expected for some time. Such vehicles are several times as efficient as gasoline-fueled vehicles, and in their operation only water vapor is given off.

ily compensate for the consequent depletion of resources and contribution to global warming. However, the picture changes if the CO_2 also produced is sequestered with CSS technology. A $2 billion plant to produce hydrogen in this way, the first of its kind, is planned for the United Arab Emirates.

Another way to produce hydrogen is to pass an electric current through water to break apart the H_2O molecules into their hydrogen and oxygen components. Because this process, called electrolysis (see Sec. 10.17), requires as much energy as that liberated when the hydrogen and oxygen are recombined, using electricity from fossil-fuel power plants to obtain hydrogen will do nothing to conserve resources, curb pollution, or reduce CO_2 emissions. On average, 374 g of CO_2 are released into the atmosphere for each mile an ordinary American car is driven (for a typical annual total of 6 tons), whereas if the energy used to produce the hydrogen for a fuel-cell car comes from a fossil-fuel plant, the CO_2 released is 436 g per mile. Such a comparison gives a very different answer, however, with no CO_2 given off if the electricity for the electrolysis of water comes from nuclear plants, from renewable sources such as solar cells and wind turbines, or from natural gas or even coal using CSS, which are clearly good paths to follow toward wide use of hydrogen.

Even better would be biological methods of producing hydrogen, which do not seem farfetched. Bacteria are known that liberate hydrogen when they digest cellulose in plant materials such as crop waste. Algae liberate hydrogen during photosynthesis, but with a rather low energy conversion efficiency. If this efficiency can be increased to around 10 percent by genetic engineering, which seems possible, inexpensive algae farms could provide cheap hydrogen wherever there is dependable sunshine.

But if an energy economy in which hydrogen plays an important role has many desirable aspects, the necessary replacement, even if only partial, of present energy systems would be expensive and take time to put into effect. One of the harder problems comes from the fact that gaseous hydrogen occupies a great deal of volume for the energy it can provide. One solution is to store and transport hydrogen in liquid form—which means at temperatures below its boiling point of $-253°C$. And liquifying hydrogen takes up to 40 percent of its energy content. Nevertheless methods of coping with liquid hydrogen exist and it is used in some experimental vehicles. A more practical method may be to squeeze hydrogen gas into a smaller volume under a pressure about 700 times atmospheric pressure, which is also being done in some prototype vehicles.

Even when hydrogen fuel-cell technology is perfected, though, it will be decades before fuel-cell cars fill the roads. One reason is the 15-year average lifetime of existing cars. Another is the chicken-and-egg situation of having to install an extensive (and expensive) supply system with many thousands of filling stations before people will buy fuel-cell cars in volume. More urgent are measures to reduce CO_2 emissions from fossil fuels that can take effect sooner than that.

Reformers

To avoid the problems involved in producing and distributing hydrogen on a large scale and then storing it in vehicles, the vehicles themselves can be equipped with **reformers** that extract hydrogen from such common and easy-to-handle fuels as natural gas, methanol (a type of alcohol), or gasoline. Unfortunately today's reformers are expensive, heavy, bulky, and complex, and in use they emit air pollutants, for instance carbon monoxide. Perhaps in a few years improved technology will minimize these drawbacks.

4.12 Biofuels

Yes, But

Biofuels made from crops are obviously renewable and have the advantage that the CO_2 given off when they are burned will be absorbed afterward during photosynthesis by the next crop to be grown (Sec. 13.12). Under the right circumstances they could bring energy independence closer in many countries. The simplest and cheapest way to exploit crops for energy is just to burn them in power plants alongside coal. On a small scale this is being done with an increasing number of power plants using up to 20 percent of plant matter in their fuel. On a larger scale, though, atmospheric pollution and ash disposal become problems and there is the issue of the vast amount of land needed for suitable crops. An English power company expects to plant such crops on between 0.5 and 1 million acres of land in order to replace the coal used to generate only 0.8 percent of the country's electricity.

Biofuels for vehicles are a different story. Ethanol—the alcohol in beverages such as wine, beer, and whiskey—yields almost two-thirds as much energy when burned as gasoline and can be added to or even be used instead of gasoline in car engines. Diesel engines cannot run on ethanol, but "biodiesel" made from various plant oils and animal fats can similarly supplement or replace diesel fuel derived from petroleum. Vehicle biofuels are receiving more and more attention with over 40 countries encouraging their use by various subsidies and requirements. Worldwide ethanol production has more than doubled since 2000 and biodiesel more than tripled. The United States, the European Union, and China have set ambitious targets for vehicle biofuels, in the case of the United States a fivefold increase in alternatives to gasoline by 2022. Currently biofuels provide 1 percent of the energy used globally for transport.

But even if all the corn and soybeans grown today in the United States went into biofuels, not much—10 percent?—of the country's demand for vehicle fuels would be met. Plenty of fossil-fuel energy, accompanied by CO_2 emissions, is needed to go from seeds in the ground to biofuel ready to use, the nitrous oxide from fertilizers used to grow crops is a more powerful greenhouse gas than CO_2, and a great deal of land is needed. So the biofuel picture is in shades of gray, not black and white.

Ethanol Sugar cane and corn are currently the chief raw materials for fuel ethanol. Eventually cellulose from grasses, wood, and agricultural waste is likely to enter the picture in a big way. Henry Ford's first Model T cars, introduced in 1908, used ethanol for fuel until gasoline became cheaper.

Sugar cane is about 10 times as efficient as most other plants in utilizing solar energy. The sugar in its juice can be directly fermented by yeasts into ethanol, which is then extracted by distillation. In Brazil, where large areas of suitable land benefit from ample sunshine and rainfall, ethanol costs half as much as gasoline. Most cars there use ethanol, either by itself or blended with gasoline, which has cut Brazil's gasoline use by 40 percent and lowered its CO_2 emissions by about 50 million tons per year. There is now less air pollution in Brazilian cities as well. China is a customer for Brazilian ethanol in the hope of similarly reducing the pollution of its urban air, among the world's worst.

In the United States, the government maintains the price of sugar too high for it to be an economical source of ethanol, and a tariff of \$0.54 per gallon keeps out cheap ethanol from elsewhere. As a result all fuel ethanol in this country comes from corn. Ethanol production is set to more than double as 77 new distilleries join the 113 already operating, and may triple by 2015. Only a modest amount of energy is involved in producing ethanol from sugar cane: for each joule of energy invested, the resulting ethanol can provide around 4.7 J when burned. However, so much more energy is needed to grow corn and process it into ethanol—the starch in corn must first be converted to sugars—that each joule invested gives back less than 1.25 J—not much of a return. Many studies actually indicate a return of under 1 J, a net loss of energy. Since the energy that goes into making corn ethanol, including manufacturing the fertilizer needed, comes from fossil fuels, it is not surprising that it is more expensive to produce than gasoline.

Under the best circumstances, using corn ethanol instead of gasoline does little or nothing to reduce net CO_2 emissions, as opposed to an 80 percent decrease for cane ethanol. If coal is one of the fossil fuels used in the production of corn ethanol, the result is an increase in CO_2 emissions over those from gasoline. If forests and grasslands are cleared to grow corn, more CO_2 is released than can be made up in many years, even centuries, of using the ethanol produced. Furthermore, large amounts of water are needed to produce corn ethanol—about 5 liters of water per liter of corn ethanol—and water is becoming scarce in corn-growing regions of the United States. The real attraction of corn ethanol in the United States lies in generous subsidies to corn farmers and a \$0.51 per gallon tax credit for converting corn to ethanol, which together add up to over \$6 billion per year. Elsewhere a number of countries have offered similar inducements, but many are reducing or eliminating them as the problematic aspects of corn ethanol (and, indeed, of some other biofuels also) come into focus. The 2007 United States law that calls for a large increase in ethanol production includes the welcome provision that, taking indirect as well as direct emissions into account, using ethanol from any source must give at least a 20 percent reduction in greenhouse gases relative to the use of gasoline.

Increasingly worrying is the use of agricultural land for fuel rather than for food in a time of expanding population (see Fig. 4-1): 230,000 new mouths every day. The corn needed for the ethanol to fill an SUV's fuel tank would feed a person for a year. Over a fifth of the corn harvest in the United States—which produces 40 percent of the world's corn—already goes to make ethanol, a proportion still climbing. The diversion of corn to fuel has driven its price sharply upward everywhere, as has happened with other crops that farmers are replacing with corn. With animal feed more expensive, meat and dairy products also cost more. For the world's poor, many of them in Africa, this is bad news that is getting worse as corn ethanol production rises. Both of the U.N.'s food agencies regard biofuel production warily; a report to one of them concluded that replacing food crops with biofuels is "a crime against humanity."

Fortunately the drawbacks to corn ethanol apply with much less force to cellulosic ethanol. Cellulose is the main constituent of all plants (Sec. 13.11) and agricultural waste, wood, and certain grasses are cheap and abundant sources; municipal waste contains a great deal of cellulose. Some grasses yield several times as much ethanol per acre as corn and can grow on poor agricultural land. Furthermore obtaining ethanol from cellulose involves far less fossil-fuel energy than in the case of corn ethanol, and a 90 percent reduction in CO_2 emissions. Starting from sugar cane is even better, but there is far more cellulose available and at less cost. The trouble is that, while going from cellulose to ethanol has been done in the laboratory, it is not quite practical to do so on an industrial scale. But sooner or later a workable method will inevitably be found, and cellulosic ethanol is likely to push aside corn ethanol on its way to becoming a major vehicle fuel.

Biodiesel A century ago some of Rudolf Diesel's first engines ran on peanut oil. Vegetable oils, now in processed form, are once again powering such engines either by themselves or in blends with ordinary diesel fuel. Soy, palm, rapeseed (canola), cottonseed, and sunflower oils are all feedstocks for biodiesel. Soybeans are responsible for most of the biodiesel produced in the United States, with an energy yield of 2 J for each joule of input. Other oils have even better yields and need considerably less land. On an overall basis, biodiesel use involves less CO_2 emission—the amount of reduction depends on the source—than the use of conventional diesel fuel or of ethanol from corn. Biodiesel is at present more expensive to produce than conventional diesel fuel, but, as with ethanol, its sale is subsidized or required, or both, in many countries, and its share of the market is going up every year. From about 1 million tons in 2008, biodiesel production capacity is expected to increase to 6.8 million tons in 2015.

Although biodiesel is a relatively green fuel in itself, growing the crops from which it comes may not be. For instance, the rapeseed that is the primary oil source in Europe often needs so much fertilizer made using natural gas that overall CO_2 emissions remain high. Even worse are destructive farming practices in some countries. In a notorious case, Indonesian swamps are being drained and the peat under them burned to make room for palm-oil plantations. The result is 2 billion tons of CO_2 entering the atmosphere annually, 8 percent of global CO_2 emissions from burning fossil fuels—far more CO_2 than using the palm oil grown there could ever save in the future. The European Union now bans the import of biofuels whose production involves degrading the environment in ways such as this.

Waste animal fats are cheaper than vegetable oils and are also suited for transformation into diesel fuel. A useful amount of these fats is available: one chicken processing company alone has over a million tons of chicken fat to dispose of every year. This company and other large meatpackers have their eyes on a share of the biodiesel market.

Strategies for the Future

Clearly no simple solution to the problem of providing safe, clean, cheap, and abundant energy is possible in the near future. But there is much that can be done, first of all by improving the efficiency of energy use, which gives a much better return on investment than any form of energy generation. A serious effort could probably save the United States at least half of the electricity it now consumes, for instance. This would mean changes in how we live: technology alone could not do the job. Also essential is to sensibly utilize the various available renewable sources and to expand the production of fission nuclear energy, all the while

Algae to the Rescue?

As mentioned in Sec. 4.11, certain algae can produce hydrogen that can be collected and used as fuel. As it happens, other algae can produce an oil whose conversion to diesel fuel is straightforward. This approach to biodiesel was studied in the 1970s and 1980s but experiments stopped when crude oil became cheaper. In new work to carry the idea further, suitable algae are being grown in Hawaii on 6 acres of ponds that have CO_2 bubbled up through them for the algae to use with water and sunlight in the photosynthesis that nourishes them. If all goes well, the next step would be scaling up the ponds to cover first 2500 and eventually 50,000 acres. An acre of soybeans can yield 227 liters of biofuel per year and an acre of corn a little over 1000 liters, whereas in principle an acre of algae could yield 19,000 liters per year without using agricultural land. If using algae to produce fuel proves practical on the scale required, all the transport fuel needs of the United States could perhaps be supplied from ponds whose area would correspond to only 2–3 percent of the area used for the country's agriculture.

Another scheme being tried uses plastic cylinders exposed to sunlight through which CO_2 from a nearby fossil-fuel plant is circulated (Fig. 4-34). This raises the prospect that CO_2 emissions from such plants could be used productively, a better alternative to burying them underground. In both cases it seems possible for sewage to provide the other nutrients the algae need, another benefit.

Figure 4-34 Reactors at the 1.04-GW Redhawk gas-fired power plant in Arizona use algae to convert some of its CO_2 emissions to biodiesel with the help of water and sunlight.

trying to make fusion energy practical as soon as possible. If the world's population also stabilizes or, better, decreases, social disaster (starvation, war) and environmental catastrophe (a planet unfit for life) may well be avoided even if fusion never becomes practical.

14.13 Conservation

Less Is More

Our children and grandchildren will have to live with the results of what is done (and not done) today about the energy problem and the global warming and resource depletion that are part of it. Is there anything we, as users of energy in our personal lives and in our work, can do that will earn their respect? The answer is yes, taking conservation seriously can make a real difference if enough of us participate, and our pocketbooks will benefit as well. As the saying goes, not to be part of the solution is to be part of the problem.

Major opportunities to save energy during their years of use come in the intelligent design and careful construction of new buildings, both residential and commercial. The best new buildings need as little as 20 percent as much energy as older ones for heating, cooling, and lighting.

In everyday life, just loading dishwashers full and not using hot water in clothes washers saves a lot of energy, as does setting thermostats for heating lower than usual in winter and for air-conditioning higher than

Compact Fluorescent Lightbulbs

In an ordinary incandescent lightbulb a tungsten filament glows when heated by the passage of an electric current. Such bulbs are very inefficient; about 95 percent of the energy they consume becomes heat, and there are billions of them—perhaps 4 billion in the United States alone, using 9 percent of all its electricity. They last around 1000 hours.

In a fluorescent tube an electric discharge causes the atoms in a mercury vapor to emit invisible ultraviolet light. A coating of a material called a phosphor on the inside of the glass tube absorbs the ultraviolet light and, in a process called fluorescence, gives off visible light. Fluorescent lamps are about 5 times as efficient as incandescent ones and last up to 10 times longer. A compact fluorescent lightbulb (CFL), like the ones shown in Fig. 4-35, that replaces an ordinary bulb of the same brightness saves several times its additional cost each year in electricity bills and hundreds of pounds of CO_2 emitted by a fossil-fuel power plant. A 2007 law calls for phasing out 100-W incandescent bulbs in the United States starting in 2012 in favor of energy-saving bulbs such as CFLs. The next year will be the turn of 75-W bulbs and the year after of 60-W bulbs. The ultimate result will be to cut $18 billion per year from electricity charges and reduce CO_2 emissions by about 160 million tons, equivalent to taking tens of millions of cars off the roads or shutting down as many as 80 coal-fired power plants. Australia requires a complete change to CFLs even earlier, by 2010, and other countries, including China and members of the European Union, have similar plans.

Eventually light-emitting diodes (LEDs) may well take over the lighting market. LEDs employ microchip technology like that used in modern electronic devices and are compact, rugged, and versatile. The best current ones are two to three times as efficient as CFLs and last 50,000 hours, with room for improvement on both counts. LEDs are widely used in indicator lamps, instrument lighting, traffic signals, and highway and advertising signs—a few among an increasing list of applications—but are still too expensive for general lighting.

Figure 4-35 A compact fluorescent bulb uses much less power for a given light output than an ordinary incandescent bulb.

usual in summer. Buying fuel-efficient cars, driving them at moderate speeds, and sharing rides instead of going alone will help, as will taking public transport whenever possible. Replacing ordinary lightbulbs with energy-saving ones; installing better home insulation (including double-pane windows filled with low-conductivity gas); upgrading to more efficient space and water heaters, kitchen appliances, and so forth add up to a big reduction in energy use per household. Everything counts: just switching computers, audio and video equipment, cell-phone chargers, coffeemakers, and other devices off instead of leaving them on standby at night would eliminate an estimated 5 percent of residential energy use in the United States, equivalent to the output of 18 typical power stations.

Recycling can help: to recycle aluminum uses less than 9 percent as much energy as to refine it from its ores, and billions of aluminum cans are discarded every year. Recycling other metals, glass, plastic bottles, paper, and cardboard also conserves energy and raw materials and is kinder to the environment than burial in landfills or burning in incinerators. San Francisco's recycling rate of 70 percent, well above the U.S. average, shows what can be done.

Industry, too, cannot continue with business as usual. As with individuals, those companies that have adopted better practices have often found them to save money as well as contributing to a healthy planet. Thus DuPont has cut CO_2 emissions by 72 percent in the last decade while saving \$2 billion in energy costs through greater efficiency. General Electric is another convert and plans to bring its CO_2 emissions down despite an expansion that otherwise would increase them by 40 percent. GE is sure that clean technologies are its future. These are not isolated examples: environmental awareness is now more and more accepted as part of good corporate citizenship. A recent survey of business leaders around the world put environmental concerns at the top of a list of the issues that will be most important to their companies in the near future. The U.S. Climate Action Partnership consists of several dozen major firms in a variety of fields that find global warming no idle threat and intend to work together to help combat it. They have called for "strong" federal action. But plenty of business interests remain in opposition.

4.14 What Governments Must Do

Their Role Is Crucial

Governments everywhere have become aware of the gravity of the energy problem and of the need for them to respond, though few are acting with the urgency required. An obvious step is to impose the highest feasible efficiency standards for appliances, buildings, and vehicles. Thirty years ago California began to introduce regulations that required greater efficiencies in energy use, with the result that average energy consumption per person there has changed little since then although in the rest of the United States it has increased by 50 percent.

Another step is to use both incentives and regulations to promote solar, wind, geothermal, cellulosic ethanol, and other renewable clean energy sources while avoiding such blind alleys as corn ethanol. Nuclear energy should similarly be encouraged to expand. Above all, every effort should be made to phase out fossil fuels, especially coal. Because coal will nevertheless continue to be burned in quantity for a long time to come, efforts to capture and bury the resulting CO_2 must be accelerated. A different facet of the energy problem is deforestation, which as we saw in Fig. 4-12 gives rise to 18 percent of worldwide greenhouse gas emissions. Deforestation can be tackled immediately at relatively modest cost simply by having rich countries pay poor countries not to cut down their forests, a procedure endorsed at the 2007 Bali conference on climate change.

Although subsidizing alternative energy sources to reduce the gap between their costs and those of fossil fuels is certainly useful, subsidies are not by themselves sufficient because governments have a poor record of choosing winners. Many economists think a direct approach to CO_2 emissions would be more effective by using the market to decide which sources are best at achieving the combined objectives of low cost and minimum emissions. Such an effort can proceed in two ways. The simplest and fairest is to levy a tax on CO_2 emissions—have polluters pay for the consequences of their actions. The tax rate can be adjusted from time to time to achieve the desired total CO_2 reduction. But as a practical matter politicians are allergic to taxes, however beneficial.

The other way is to use a **cap-and-trade system** in which a region-wide total (the cap) is set for annual CO_2 emissions. The government then auctions or gives away permits to emit CO_2 that add up to the overall ceiling. Companies that do not use their entire quotas can sell the left-over permits to companies whose emissions exceed their quotas. A proper choice of the cap would make the price of traded permits high enough to serve as an incentive to big emitters such as power companies to invest in greater efficiency, carbon capture and storage, and clean technologies. If the permits are auctioned, the government receives money that can be used to help ease the transition to clean alternative energy. If permits are free, there is the problem of distributing them fairly. This can easily result in a large cap being set so that all emitters are satisfied with their quotas—and emissions are reduced by little or nothing. Either way the price of traded permits, unlike a tax, would vary with general economic conditions and other unpredictable factors, which would add a new element of uncertainty for businesses planning future investments.

The European Union set targets for 2020 that consist of a cut in CO_2 emissions by 20 percent below 1990 levels and an increase in the share of renewable energy to 20 percent of total energy production. Free permits were issued, but because the quotas were too generous and some sectors of industry were left out, there was not the hoped-for effect on emissions. To remedy this disappointing result, from 2013 most permits will instead be auctioned off to make fossil-fuel use sufficiently expensive to encourage new investment in methods that reduce CO_2 emissions. Meeting the 2020 targets may increase electricity bills by 10–15 percent, considered a fair price for the expected benefits. Today under 4 percent of global CO_2 emissions are controlled by cap-and-trade schemes and even less by taxation.

Replacing words with deeds on a scale large enough to make a real difference is proving difficult in the two countries that are by far the largest emitters of CO_2, the United States and China (see Fig. 4-9). China is in the midst of rapid industrialization with coal as its main fuel: it has plenty of coal and coal-fired plants are cheap. China's position is that while going ahead as fast as it can with renewable energy sources (twice as much investment in them per year than the United States, with a goal of 15 percent of its energy coming from such sources by 2020), it must continue to press on with coal in order to keep up its rate of development. Because China has a long way to go to catch up with the advanced nations of North America and Europe, its CO_2 emissions per person are still modest, on average less than a quarter per Chinese as per American. There are so many Chinese, however, that total CO_2 emissions now exceed those of the United States, which was until 2007 the world leader. China has hinted it will consider joining future international programs to minimize these emissions.

For its part, the United States, until the change in the Administration at the beginning of 2009, claimed that reducing its CO_2 emissions would cripple its $13 trillion economy, the world's largest. It vowed it would never require emission cuts. After secretly consulting fossil-fuel producers, the previous Administration increased subsidies to them and decreased funding for research in alternative sources. It consistently downplayed global warming and employed an oil-industry lobbyist to alter hundreds of scientific reports to cast doubt on the occurrence of such warming, and then, when that became impossible, on its connection with the greenhouse effect. NASA tried to prevent its scientists from calling attention to global warming, and the White House stopped the director of the Centers for Disease Control and Prevention from reporting the implications of global warming for the spread of disease. At international conferences, the U.S. delegates always tried to block any concerted action to reduce greenhouse gas emissions or even to monitor them.

Why recall all this now? The reason is that, if we forget it happened, such refusal to face reality can happen again. Thus far living things have altered conditions for life on a planetwide scale only once, billions of years ago, when photosynthesis by primitive bacteria provided the atmosphere with its free oxygen, which was formerly absent (Sec. 16.13). Climate scientists believe that, if the CO_2 content of the atmosphere exceeds perhaps 450 million parts per million, another such long-lasting environmental change will occur in which feedback loops will cause the earth to continue warming up regardless of what we do afterward. So there is no time to waste (see Figs. 4-6 and 4-10).

Fortunately over 30 states plus hundreds of cities stepped into the vacuum created by Washington with energy-efficiency standards for buildings and plans to reduce the emission of greenhouse gases, often obliging their electricity suppliers to diversify away from fossil fuels to renewable clean sources. Three groups of states, 10 in the Northeast, 5 in the Midwest, and 6 in the West, plus adjacent Canadian provinces intend to set up regional cap-and-trade systems that will cover half the U.S. population. In 2007 Kansas denied a permit for a new coal-fired plant on the basis that its CO_2 emissions threatened health and the environment, the first time CO_2 was cited in rejecting such a permit. According to the Kansas governor, "We have an obligation to be good stewards of this state."

As responsible citizens, we should make sure the enlightened view of the Kansas governor informs decisions by governments everywhere and not allow our voices to be ignored, as they were not so long ago. A recent survey asked people in various countries if they thought their politicians were doing enough against global warming. Most said "No"—in the

United States, 75 percent. Asked if "the polluter should pay," a still larger majority said "Yes"—in the United States, 82 percent. Governments will only listen to us if they hear us. There is plenty of competition for their ears, so we should not be shy about letting them know what we think. Former vice-president Al Gore, who received a Nobel Peace Prize for his efforts to increase awareness of global warming, made these suggestions in his documentary film *An Inconvenient Truth:* "Vote for leaders who pledge to solve this crisis. Write to Congress. If they don't listen, run for Congress."

Important Terms and Ideas

The **fossil fuels** coal, oil, and natural gas were formed by the partial decay of the remains of plants and marine organisms that lived millions of years ago.

Methane, the main constituent of natural gas, is a compound of carbon and hydrogen with the chemical formula CH_4.

The **greenhouse effect** refers to the process by which a greenhouse is heated: sunlight can enter through its windows, but the infrared radiation the warm interior gives off is absorbed by glass, so the incoming energy is trapped. The earth's atmosphere is heated in a similar way by absorbing infrared radiation from the warm earth. **Greenhouse gases** are gases that absorb infrared radiation; the chief ones in the atmosphere are carbon dioxide (CO_2), methane, nitrous oxide (N_2O), and a group of gases used in refrigeration called CFCs and HCFCs.

Sequestration is a method of carbon capture and storage that involves pumping CO_2 emitted by power plants or other sources into underground reservoirs.

In **nuclear fission,** a large atomic nucleus (notably a nucleus of one kind of uranium atom) splits into smaller ones, a process that gives off considerable energy. A nuclear reactor produces energy from nuclear fissions that occur at a controlled rate.

In **nuclear fusion,** two small nuclei unite to form a larger one, a process that also gives off considerable energy. The sun and stars obtain their energy from nuclear fusion, but fusion technology for power plants is still under development.

Geothermal energy comes from the heat of the earth's interior.

A **photovoltaic cell,** also called a **solar cell,** converts the energy in sunlight directly to electric energy.

In a **fuel cell,** electric current is produced by means of chemical reactions.

A **reformer** is a device that extracts hydrogen, usually for a fuel cell, from fuels such as natural gas, ethanol, or gasoline.

In a **cap-and-trade** system for controlling CO_2 emissions, an overall cap on them is set for a region and companies there are given or buy at auction permits to emit CO_2 whose total equals the cap. Companies that do not use their full quotas can sell the leftover permits to companies that exceed their quotas.

Multiple Choice

1. The number of people in the world may reach a maximum in 2050 of about
- **a.** 1 billion
- **b.** 2.5 billion
- **c.** 6.7 billion
- **d.** 9 billion

2. Arrange these sources in the order of the energy they supply to the world today, starting with the source of the most energy.
- **a.** coal
- **b.** oil
- **c.** natural gas
- **d.** nuclear

3. Of the following, the energy source likely to be used up first is
- **a.** coal
- **b.** oil
- **c.** natural gas
- **d.** nuclear

4. Of the following, the energy source likely to last the longest is
- **a.** coal
- **b.** oil
- **c.** natural gas
- **d.** nuclear

5. In which one or more of the following countries is the energy consumption per person below the world average?
   a. Russia
   b. Japan
   c. China
   d. India

6. The midrange estimate for the increase in world energy demand in 2030 over that of today is
   a. 25 percent
   b. 50 percent
   c. 75 percent
   d. 100 percent

7. A city of medium size might use energy at a rate of
   a. 1 million watts (1 MW)
   b. 1 billion watts (1 GW)
   c. 1 trillion watts (1 TW)
   d. 1 quadrillion watts (1 PW)

8. If present trends continue, the most likely average global temperature increase by 2100 is about
   a. 1°C
   b. 2°C
   c. 4°C
   d. 10°C

9. The source that produces the most carbon dioxide per joule of energy liberated is
   a. coal
   b. oil
   c. natural gas
   d. nuclear

10. The average amount of CO_2 emitted each year per person in the United States is about
   a. 1 ton
   b. 2 tons
   c. 5 tons
   d. 20 tons

11. The radiation from an object at room temperature is mainly in the form of
   a. infrared light
   b. visible light
   c. ultraviolet light
   d. any of the above, depending on its color

12. The earth's atmosphere is primarily heated by
   a. direct sunlight
   b. sunlight reflected by the earth's surface
   c. infrared light radiated by the earth's surface
   d. carbon dioxide emissions

13. A gas that does not contribute to global warming is
   a. methane
   b. nitrogen
   c. nitrous oxide
   d. carbon dioxide

14. Arrange these countries in increasing order of their CO_2 emissions per person.
   a. China
   b. United States
   c. United Kingdom
   d. India

15. The country or countries each responsible for about one-fourth of the total of the world's CO_2 emissions is (are)
   a. China
   b. India
   c. Russia
   d. United States

16. Of the following fuels, the one that gives off the most heat per gram when burned is
   a. hydrogen
   b. methane
   c. gasoline
   d. coal

17. Of the following fuels, the one that gives off the least heat per gram when burned is
   a. hydrogen
   b. methane
   c. gasoline
   d. coal

18. Which of the following is not a fossil fuel?
   a. hydrogen
   b. natural gas
   c. oil
   d. coal

19. The contribution of fossil fuels to world energy supply is about
   a. 25 percent
   b. 60 percent
   c. 85 percent
   d. 95 percent

20. The proportion of oil used by the United States that is imported is about
   a. 1/10
   b. 1/5
   c. 1/2
   d. 2/3

21. Most oil today is used for
   a. transportation
   b. heating
   c. electricity
   d. lubrication

22. Which one or more of the following fuels produce(s) both water and CO_2 when burned?
    a. hydrogen
    b. methane
    c. gasoline
    d. coal
23. Natural gas consists largely of
    a. hydrogen
    b. oxygen
    c. nitrogen
    d. methane
24. The least polluting of the following fuels is
    a. coal
    b. gasoline
    c. diesel fuel
    d. natural gas
25. The number of coal-fired power plants either newly built or under construction in the world is roughly
    a. 100
    b. 1000
    c. 10,000
    d. 100,000
26. The impurity in coal that contributes to acid rain is
    a. nitrogen
    b. sulfur
    c. carbon
    d. chlorine
27. The worst emitters of mercury, which damages the nervous system, are power plants that use
    a. coal
    b. oil
    c. natural gas
    d. nuclear energy
28. Syngas is made from
    a. coal
    b. oil
    c. natural gas
    d. carbon dioxide
29. Of the following countries, the one that obtains the largest proportion of its electricity from nuclear energy is
    a. France
    b. China
    c. Japan
    d. United States
30. The proportion of electricity generated in the United States that comes from nuclear energy is roughly
    a. 1 percent
    b. 5 percent
    c. 20 percent
    d. 50 percent
31. In the relatively near future, the technology most able to replace fossil fuels on a large scale is
    a. nuclear
    b. wind
    c. solar
    d. biofuels
32. A major unsolved problem for nuclear energy is
    a. greenhouse gas emissions
    b. fuel supply
    c. danger of explosion
    d. permanent waste disposal
33. Bright sunlight delivers energy to an area of 1 square meter at a rate of about
    a. 1 W
    b. 10 W
    c. 100 W
    d. 1000 W
34. The output of which of the following renewable energy sources varies least?
    a. wind
    b. waves
    c. geothermal
    d. solar
35. The renewable energy source that seems to have the most potential for the future is
    a. wind
    b. waves
    c. geothermal
    d. solar
36. Of the following technologies, the one that may eventually become the chief energy source in the world involves the use of
    a. nuclear fission
    b. nuclear fusion
    c. fuel cells
    d. biofuels
37. Of the following technologies, the one farthest from being a commercial energy source in the near future involves the use of
    a. nuclear fission
    b. nuclear fusion
    c. fuel cells
    d. biofuels
38. Ethanol made from which of the following sources seems on an overall basis to have the most advantages for the future?
    a. cellulose
    b. corn

c. sugar cane
d. natural gas

39. Compact fluorescent lightbulbs are
    a. more efficient than incandescent bulbs but less efficient than LEDs
    b. more efficient than incandescent bulbs and about as efficient as LEDs
    c. more efficient than both incandescent bulbs and LEDs
    d. about the same in efficiency as incandescent bulbs and LEDs

40. Of the following, the strategy for coping with future energy shortages with the most in its favor is to
    a. burn more coal
    b. produce more oil from tar sands
    c. divert more agricultural land to making biofuels
    d. increase energy efficiency and energy conservation

Exercises

4.1 Population and Prosperity

1. What are the three main factors that will require changes in today's patterns of energy production and consumption?

4.2 Energy Consumption

2. Even if the developed countries stabilize or reduce their energy consumption in years to come, word-wide energy consumption will increase. What are the two main reasons for this?
3. The average rate of energy consumption per person in the United States is about how many times the world average: twice, three times, four times, over four times?
4. List the fossil fuels in the order in which they will probably be used up.
5. Explain how sunlight is responsible for these energy sources: food, wood, water power, wind power, fossil fuels.
6. What energy sources cannot be traced to sunlight falling on the earth?

4.3 Global Warming

7. Approximately what proportion of the world's population lives on or near coasts and so may be under future threat from rising sea level?
8. Give two reasons why global warming is causing sea level to rise.
9. Once the polar ice sheets have melted beyond a certain amount, melting will continue even if CO_2 emissions stop rising. Why?
10. The oceans as well as the atmosphere are growing warmer. What does this imply for tropical storms such as hurricanes?
11. When was the last time world temperatures were as high as they are likely to be in 2100 if current rates of CO_2 emission continue: hundreds of years ago, thousands of years ago, millions of years ago?

4.4 Carbon Dioxide and the Greenhouse Effect

12. Every body of matter radiates light. What is characteristic of light radiated by something very hot, such as the sun? Of light radiated by something at ordinary temperatures, such as the earth's surface?
13. What is the nature of the greenhouse effect in the earth's atmosphere?
14. List the chief greenhouse gases in the atmosphere. What property do they share?
15. About half the CO_2 from burning fossil fuels enters the atmosphere. What becomes of the rest?
16. Why is deforestation so important in global warming?
17. List the fossil fuels in the order in which they contribute to world CO_2 emissions.

4.5 Liquid Fuels

18. What fuel liberates the most energy per gram when it burns? What is produced when it burns?
19. Most of the world's oil is used as a fuel for what purpose?
20. How do the oil reserves in tar sands compare with the reserves of ordinary crude oil? What are some of the disadvantages of tar sand oil?
21. What methods are available to reduce pollution by car exhausts?
22. What are some of the reasons why the average fuel efficiency of cars in the United States is the lowest in the world?

4.6 Natural Gas

23. The amount of CO_2 emitted per kilowatt-hour of electricity by a gas-fired power plant is about half that emitted by a coal-fired plant. What do you think is the reason that coal-fired plants are much more common?
24. Why is natural gas rarely used as a vehicle fuel?

4.7 Coal

25. Which fossil fuel does the United States have the greatest reserves of?

26. What are the chief advantages of coal as a fuel? The chief disadvantage?

27. Coal is responsible for approximately which proportion of the electricity generated in the United States: one-quarter, one-half, three-quarters?

28. Coal smoke contains sulfur and mercury. Why are they harmful?

29. Why do you think that, per joule of energy liberated when they are burned, coal produces more carbon dioxide than the other fossil fuels?

30. List the desirable aspects of coal gasification, the process in which coal is turned into a mixture of gases called syngas.

4.8 A Nuclear World?

31. What is the basic difference between nuclear fission and nuclear fusion? In what way are they similar?

32. What role does uranium play in nuclear energy production? What is the uranium supply situation?

33. How does a nuclear power plant produce electricity?

34. Explain why no nuclear power plants were planned in the United States between 1979 and now but are currently being considered for construction.

35. List the potential advantages of fusion energy.

36. What stands in the way of the immediate use of nuclear fusion as a commercial energy source?

37. Is there anywhere outside of laboratories where fusion energy is produced today?

4.9 Clean Energy I

4.10 Clean Energy II

38. Give examples of clean sources that can supply energy continuously and examples of others whose output varies with time of day and weather conditions.

39. Of the various clean energy sources, which provides the most energy worldwide today?

40. Give several reasons why fossil-fuel energy is cheaper than energy from most renewable sources.

41. What does a photovoltaic cell do? What is another name for it?

42. What advantages do solar cells have for installation in remote regions?

43. Instead of a new 500-MW coal-fired power plant, a wind farm of turbines rated at 2 MW maximum output each is to be installed. If the average turbine output is 40 percent of the maximum, how many turbines are needed?

44. Explain how tide and wave energies can be captured.

45. What major advantage does geothermal energy have over solar, wind, tidal, and wave energies?

46. List four practical ways to store energy from noncontinuous energy sources.

4.11 Hydrogen and Fuel Cells

47. When hydrogen combines with oxygen, a great deal of energy is liberated with only water as the product. What are the two main factors that hold back wider use of hydrogen as a fuel?

48. What are the advantages and disadvantages of hydrogen fuel cells?

4.12 Biofuels

49. (a) Why is corn not regarded as an ideal choice for producing ethanol? (b) Cellulose is apparently a better choice. Why is it not in wide use?

50. Why are algae so interesting as a way of producing biodiesel fuel?

4.13 Conservation

51. A long-term goal for energy efficiency envisions an average use of 65 GJ per person per year. To what continuous power in kilowatts does this correspond?

4.14 What Governments Must Do

52. (a) In round numbers, which proportion of worldwide greenhouse gas emissions is due to deforestation: 5 percent, 10 percent, 20 percent, 40 percent? (b) What seems to be the most practical way to reduce deforestation?

53. Explain the cap-and-trade system for controlling CO_2 emissions. Is there an alternative means of control?

Answers to Multiple Choice

1. d
2. b, a, c, d
3. b
4. d
5. c,d
6. b
7. b
8. c
9. a
10. d
11. a
12. c
13. b
14. d, a, c, b
15. a,d
16. a
17. d
18. a
19. c
20. d
21. a
22. b,c
23. d
24. d
25. b
26. b
27. a
28. a
29. a
30. c
31. a
32. d
33. d
34. c
35. c
36. b
37. b
38. a
39. a
40. d

5 Matter and Heat

In 1999 a hot-air balloon circled the earth nonstop.

Goals

When you have finished this chapter you should be able to complete the goals ▸ given for each section below:

Temperature and Heat

5.1 Temperature
Putting Numbers to Hot and Cold
- ▸ Distinguish between temperature and heat.
- ▸ Describe how various thermometers work.
- ▸ Convert temperatures from the celsius to the fahrenheit scale and vice versa.

5.2 Heat
Different Substances Need Different Amounts of Heat for the Same Temperature Change
- ▸ Define the specific heat capacity of a substance and use it to relate the heat added to or removed from a given mass of the substance to a temperature change it undergoes.
- ▸ Describe the three ways heat can be transferred from one place to another.

5.3 Metabolic Energy
The Energy of People and Animals
- ▸ Discuss the significance of the metabolic rate of an animal and how to convert between kilocalories and kilojoules.

Fluids

5.4 Density
A Characteristic Property of Every Material
- ▸ Define density and calculate the mass of a body of matter given its density and volume.

5.5 Pressure
How Much of a Squeeze
- ▸ Define pressure and account for the increase in pressure with depth in a liquid or gas.

5.6 Buoyancy
Sink or Swim
- ▸ State Archimedes' principle and explain its origin.

5.7 The Gas Laws
Ideal Gases Obey Them
- ▸ Use Boyle's law to relate pressure and volume changes in a gas at constant temperature.
- ▸ Use Charles's law to relate temperature and volume changes in a gas at constant pressure.
- ▸ Show how the ideal gas law is related to Boyle's law and Charles's law.

Kinetic Theory of Matter

5.8 Kinetic Theory of Gases
Why Gases Behave as They Do
- ▸ State the three basic assumptions of the kinetic theory of gases.

5.9 Molecular Motion and Temperature
The Faster the Molecules, the Higher the Temperature
- ▸ Discuss the connection between temperature and molecular motion.
- ▸ Explain the significance of the absolute temperature scale and the meaning of absolute zero.

Changes of State

5.10 Liquids and Solids
Intermolecular Forces Hold Them Together
- ▸ Account for the differences of gases, liquids, and solids in terms of the forces between their molecules.

5.11 Evaporation and Boiling
Liquid into Gas
- ▸ Distinguish between evaporation and boiling.

5.12 Melting
Solid into Liquid
- ▸ Explain what is meant by heat of vaporization and by heat of fusion.

Energy Transformations

5.13 Heat Engines
Turning Heat into Work
- ▸ Discuss why heat engines cannot be perfectly efficient.
- ▸ Compare heat engines and refrigerators.

5.14 Thermodynamics
You Can't Win
- ▸ State the two laws of thermodynamics.

5.15 Fate of the Universe
Order into Disorder

5.16 Entropy
The Arrow of Time
- ▸ Relate entropy to the second law of thermodynamics.

Suppose our microscopes had no limit to their power, so that we could examine a drop of water at any magnification we like. What would we find if the drop were enlarged a million or more times? Would we still see a clear, structureless liquid? If not, what else?

The answer is that, on a very small scale of size, our drop of water consists of billions of tiny separate particles. Indeed, *all* matter does, whether in the form of a solid, a liquid, or a gas. This much was suspected over 2000 years ago in ancient Greece. Modern science has not only confirmed this suspicion but extended it: the particles that make up all matter are in constant random motion, and the kinetic energy of this motion is what constitutes heat. In everyday life, matter shows no direct sign of either the particles or their motion. However, plenty of indirect signs support this picture, and we shall consider some of them in this chapter.

Temperature and Heat

Temperature and heat are easy to confuse. Certainly the higher its temperature, the more heat something contains. But we cannot say that an object at one temperature contains more heat than another object at a lower temperature just because of the temperature difference. A cup of boiling water is at a higher temperature than a pailful of cool water, but the pailful of cool water would melt more ice (Fig. 5-1). And the same masses of different substances at the same temperature contain different amounts of heat. A kilogram of boiling water can melt 32 times as much ice as 1 kg of gold at the temperature of the water, for instance.

5.1 Temperature

Putting Numbers to Hot and Cold

Temperature, like force, is a physical quantity that means something to us in terms of our sense impressions. And, as with force, a certain amount of discussion is needed before a statement of exactly what temperature signifies can be given. Such a discussion appears later in this chapter. For

Figure 5-1 The heat content of a given substance depends upon both its mass and its temperature. A pail of cool water contains more heat than a cup of boiling water.

Figure 5-2 Allowance must be made in the design of a bridge for its expansion and contraction as the temperature changes. The Golden Gate Bridge in San Francisco varies by over a meter in length between summer and winter.

the time being, we can simply regard temperature as that which gives rise to sensations of hot and cold.

A **thermometer** is a device that measures temperature. Most substances expand when heated and shrink when cooled (Fig. 5-2), and the thermometers we use in everyday life are designed around this property of matter. More precisely, they are based upon the fact that different materials react to a given temperature change to different extents. The familiar liquid-in-glass thermometer (Fig. 5-3) works because a liquid expands more than glass when heated and contracts more than glass when cooled. Thus the length of the liquid column in the glass tube provides a measure of the temperature around the bulb.

Another common thermometer used for high temperatures, such as in ovens and furnaces, makes use of the different rates of expansion of different kinds of metals. Two straight strips of dissimilar metals are joined together at a particular temperature (Fig. 5-4). At higher temperatures the bimetallic strip bends so that the metal with the greater expansion is on the outside of the curve, and at lower temperatures it bends in the opposite direction. In each case the exact amount of bending depends

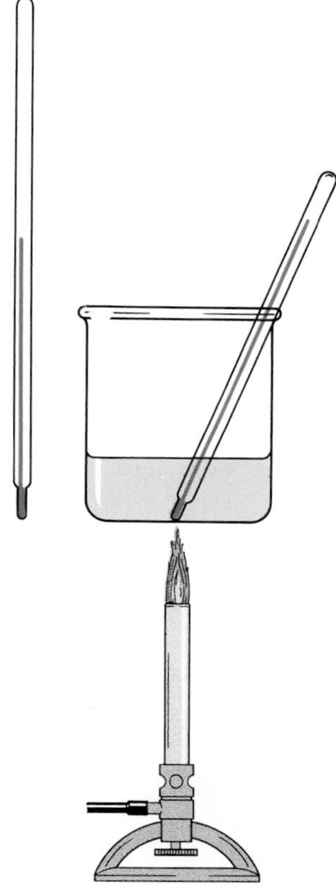

Figure 5-3 A liquid-in-glass thermometer. Mercury or a colored alcohol solution responds to temperature changes to a greater extent than glass does, and so the length of the liquid column is a measure of the temperature of the thermometer bulb.

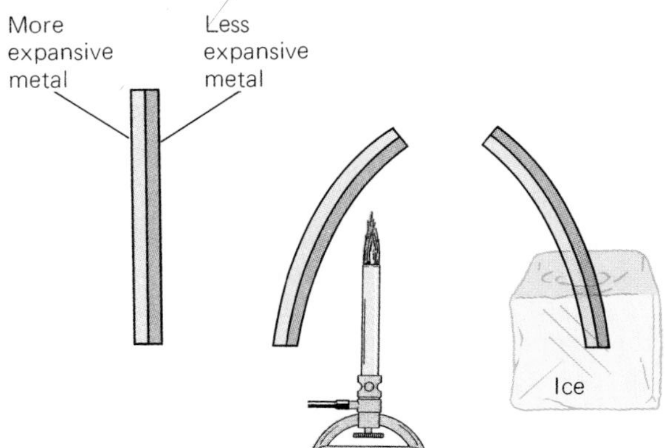

Figure 5-4 A bimetallic strip thermometer. No matter on which side the heat is applied, the bend is away from the more expansive metal. The higher the temperature, the greater the deflection. At low temperatures the deflection is in the opposite direction. Steel and copper are often used in bimetallic strips; the steel expands less when heated.

upon the temperature. Bimetallic strips of this kind are used in the **thermostats** that switch on and off heating systems, refrigerators, and freezers at preset temperatures.

Thermal expansion is not the only property of matter that can be used to make a thermometer. As another example, the color and amount of light emitted by an object vary with its temperature. A poker thrust into a fire first glows dull red, then successively bright red, orange, and yellow. Finally, if the poker achieves a high enough temperature, it becomes "white hot." The color of the light given off by a glowing object is thus a measure of its temperature (Figs. 5-5 and 5-6). This property is used by astronomers to determine the temperatures of stars.

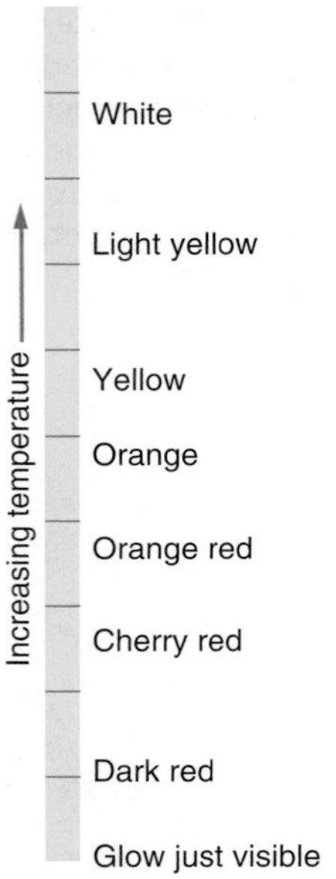

White
Light yellow
Yellow
Orange
Orange red
Cherry red
Dark red
Glow just visible
Increasing temperature

Figure 5-5 The color of an object hot enough to glow varies with its temperature as shown here.

Temperature Scales Two temperature scales are used in the United States. On the **Fahrenheit scale** the freezing point of water is 32° and the boiling point of water is 212°. On the **Celsius scale** these points are 0° and 100° (Fig. 5-7). The Fahrenheit scale is used only in a few English-speaking countries that cling to it with the same obstinacy that preserves the equally awkward British system of units. The rest of the world, and all scientists, use the more convenient Celsius (or centigrade) scale.

To go from a Fahrenheit temperature T_F to a Celsius temperature T_C, and vice versa, we note that 180°F separates the freezing and boiling points of water on the Fahrenheit scale. On the Celsius scale, however, the difference is 100°C. Therefore Fahrenheit degrees are $\frac{100}{180}$, or $\frac{5}{9}$, as large as Celsius degrees:

$$5 \text{ Celsius degrees} = 9 \text{ Fahrenheit degrees}$$

Taking into account that the freezing point of water is 0°C = 32°F, we see that

$$\text{Fahrenheit temperature} = T_F = \tfrac{9}{5}T_C + 32° \qquad 5\text{-}1$$

$$\text{Celsius temperature} = T_C = \tfrac{5}{9}(T_F - 32°) \qquad 5\text{-}2$$

Thus the Celsius equivalent of the normal body temperature of 98.6°F is

$$T_C = \tfrac{5}{9}(98.6° - 32.0°) = \tfrac{5}{9}(66.6°) = 37.0°\text{C}$$

Figure 5-6 The color and brightness of an object heated until it glows, such as this steel beam, depends on its temperature. An object that glows white is hotter than one that glows red and gives off more light as well.

Example 5.1

In 1983 a temperature of −89°C was measured at a Russian research station in Antarctica, a record low for the earth's surface. What is the Fahrenheit equivalent of this temperature?

Solution

From Eq. 5-1

$$T_F = \tfrac{9}{5}(-89°\text{C}) + 32° = -160° + 32° = -128°\text{F}$$

On both scales, brrr.

5.2 Heat

Different Substances Need Different Amounts of Heat for the Same Temperature Change

The **heat** in a body of matter is the sum of the kinetic energies of all the separate particles that make up the body. The more energy these particles have, the more heat the body contains, and the higher its temperature. Thus the heat content of a body can also be called its **internal energy.**

Since heat is a form of energy, the joule is the proper unit for it. The amount of heat needed to raise or lower the temperature of 1 kg of a substance by 1°C depends on the nature of the substance. In the case of water, 4.2 kJ of heat is required per kilogram per °C (Fig. 5-8). To heat 1 kg of water from, say, 20°C to 60°C means raising its temperature by 40°C. The amount of heat that must be added is therefore (1 kg)(4.2 kJ/kg · °C) (40°C) = 168 kJ.

For a given temperature change, liquid water must have more heat added to or taken away from it per kilogram than nearly all other materials. For instance, to change the temperature of 1 kg of ice by 1°C, we must transfer to or from it 2.1 kJ, about half as much as for water. To do the same for 1 kg of gold takes only 0.13 kJ.

The **specific heat capacity** (or just **specific heat**) of a substance is the amount of heat that must be added to or removed from 1 kg of the substance in order to change its temperature by 1°C. The symbol of specific heat is c and its unit is the kJ/kg · °C. Thus the specific heat of water is c_{water} = 4.2 kJ/kg · °C. Table 5-1 gives the specific heats of various substances. We note that metals have fairly low specific heats, as we saw for gold, which means that relatively little heat is needed to change their temperatures by a given amount compared with other materials of the same mass.

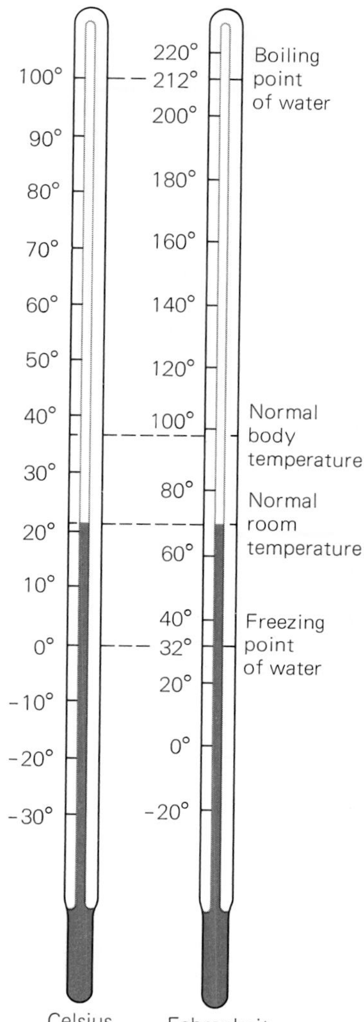

Figure 5-7 Comparison of the Celsius and Fahrenheit temperature scales.

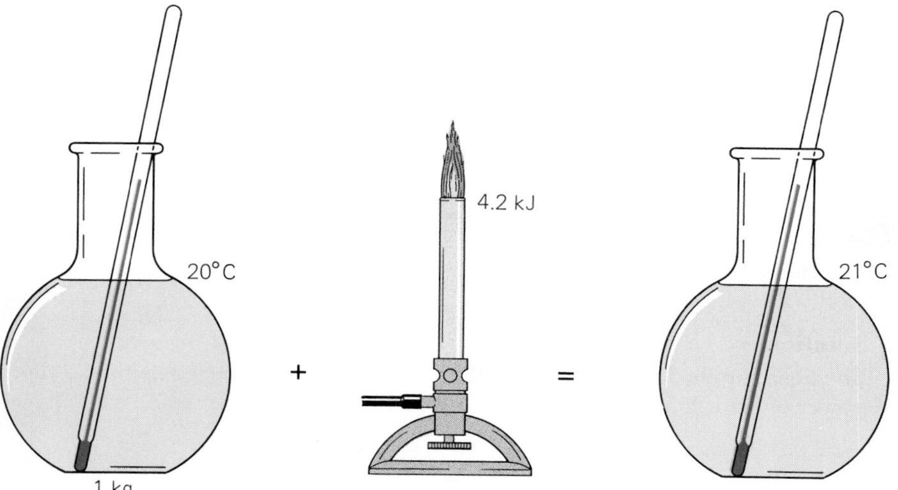

Figure 5-8 To raise the temperature of 1 kg of water by 1°C, 4.2 kJ of heat must be added to it. The same amount of heat must be removed to cool the water by 1°C.

Btu

In the United States, though no longer in Great Britain, the **British Thermal Unit** (Btu) is sometimes used as a unit of heat. The Btu is defined as the heat needed to raise the temperature of 1 lb of water by 1°F; 1 Btu = 1.054 kJ. Another common heat unit, the kilocalorie, is discussed in Sec. 5.3.

Table 5–1 Some Specific Heats

Substance	Specific Heat, kJ/kg · °C
Alcohol (ethyl)	2.4
Aluminum	0.92
Concrete	2.9
Copper	0.39
Glass	0.84
Gold	0.13
Human body	3.5
Ice	2.1
Iron	0.46
Steam	2.0
Water	4.2
Wood	1.8

When an amount of heat Q is added to or removed from a mass m of a substance whose specific heat is c, the resulting temperature change ΔT (Δ is the Greek capital letter "delta" and in physics usually means "change in") is related to Q, m, and ΔT by the formula

$$Q = mc\,\Delta T \qquad \text{5-3}$$

Heat transferred = (mass) (specific heat) (temperature change)

The examples show how Eq. 5-3 can be used.

Example 5.2

A person decides to lose weight by eating only cold food. A 100-g piece of apple pie yields about 1500 kJ of energy when eaten. If its specific heat is 1.7 kJ/kg · °C, how much less is its energy content at 5°C than at 25°C?

Solution

Since ΔT here is 25°C − 5°C = 20°C, the energy difference is

$$Q = mc\Delta T = (0.1\text{ kg})(1.7\text{ kJ/kg}\cdot°\text{C})(20°\text{C}) = 3.4\text{ kJ}$$

The saving amounts to (3.4 kJ)/(1500 kJ) = 0.0023 = 0.23 percent of the total energy provided by eating the pie—this is not really a practical path to becoming thin.

Example 5.3

Find the difference in temperature between the water at the top and at the bottom of a waterfall 50 m high. Assume that all the potential energy lost by the water in falling goes into heat.

Solution

The potential energy of a mass m of water at the top of the waterfall is PE = mgh, where h = 50 m is the height of the waterfall. If this energy is converted into the heat Q, then, since for water c = 4.2 kJ/kg · °C = 4200 J/kg · °C,

$$\text{PE} = Q$$
$$mgh = mc\,\Delta T$$
$$\Delta T = \frac{gh}{c} = \frac{(9.8\text{ m/s}^2)(50\text{ m})}{4200\text{ J/kg}\cdot°\text{C}} = 0.12°\text{C}$$

Example 5.4

How long will a 1.2-kW electric kettle take to heat 0.8 kg of water from 20°C to 100°C?

Solution

The heat supplied by the kettle in the time t is $Pt = Q$, where P is the kettle's power of 1.2 kW. We proceed as follows, with ΔT = 100°C − 20°C = 80°C:

$$Pt = Q = mc\,\Delta T$$
$$t = \frac{mc\,\Delta T}{P} = \frac{(0.8\text{ kg})(4.2\text{ kJ/kg}\cdot°\text{C})(80°\text{C})}{1.2\text{ kW}} = 224\text{ s} = 3.7\text{ min}$$

Heat Transfer Heat can be transferred from one place to another in three ways (Fig. 5-9). If we put one end of a poker in a fire, the other end becomes warm as heat flows through the poker. Such **conduction** is most efficient by far in metals (see Sec. 5.10). Conduction is inefficient in air, and a heater warms a room mainly through the actual movement of hot air. When a portion of a fluid (either a gas or a liquid) is heated, it expands so that it becomes lighter than the surrounding, cooler fluid and as a result rises upward ("heat rises"). This process is called **convection** (see Sec. 14.6). In space, which is virtually empty, neither conduction nor convection can occur to any real extent. Instead, the earth receives heat from the sun in the form of **radiation,** which consists of **electromagnetic waves** (light and radio waves are examples of such radiation; see Secs. 7.8 and 7.9). The earth in turn radiates energy to space; see Sec. 14.4.

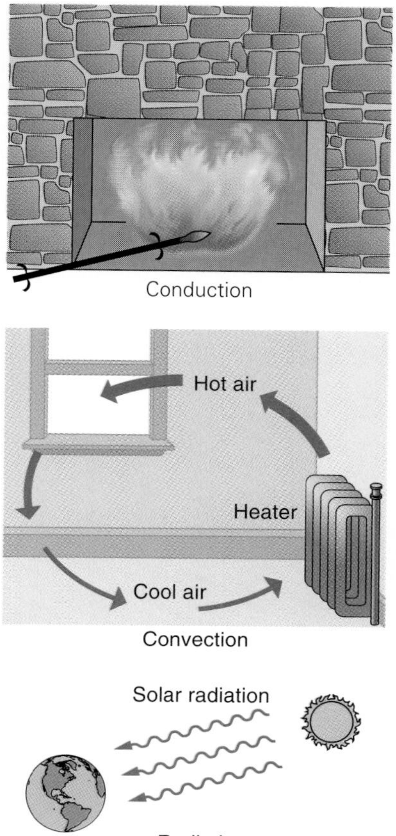

Conduction

Hot air
Heater
Cool air
Convection

Solar radiation
Radiation

Figure 5-9 The three mechanisms of heat transfer.

5.3 Metabolic Energy

The Energy of People and Animals

Metabolism refers to the biochemical processes by which the energy content of the food an animal eats is liberated. Table 5-2 lists the energy contents of some common foods. The unit is the **kilocalorie** (kcal), which is the amount of heat needed to change the temperature of 1 kg of water by 1°C. Thus 1 kcal = 4.2 kJ. The "calorie" used by dieticians is actually the kilocalorie.

Table 5–2 Energy Contents of Some Common Foods (1 kcal = 4.2 kJ)

Food	kcal
1 raw onion	5
1 dill pickle	15
6 asparagus	20
1 gum drop	35
1 poached egg	75
8 raw oysters	100
1 banana	120
1 cupcake	130
1 broiled hamburger patty	150
1 glass milk	165
1 cup bean soup	190
$\frac{1}{2}$ cup tuna salad	220
1 ice cream soda	325
$\frac{1}{2}$ broiled chicken	350
1 lamb chop	420

Example 5.5

A 50-kg girl eats a banana and she then proposes to work off its 120 kcal energy content by climbing a hill. If her body is 10 percent efficient, how high must she climb?

Solution

From Sec. 3-1 we know that the amount of work needed to raise something whose mass is m through the height h is $W = mgh$. Here the available energy is 10 percent of (120 kcal)(4.2 kJ/kcal) = 504 kJ, so $W = (0.10)(504 \text{ kJ}) = 50.4 \text{ kJ} = 50.4 \times 10^3 \text{ J}$, and the height is

$$h = \frac{W}{mg} = \frac{50.4 \times 10^3 \text{ J}}{(50 \text{ kg})(9.8 \text{ m/s}^2)} = 102 \text{ m}$$

The proportion of metabolic energy that is converted to mechanical work by muscular activity is not very high, only 10 or 20 percent. (An electric motor with the same power output as a person is typically 50 percent efficient, and larger electric motors are more efficient still.) The rest of the energy goes into heat, most of which escapes through the animal's skin. The maximum power output an animal is capable of depends upon its maximum metabolic rate, which in turn depends upon its ability to dissipate the resulting heat, and therefore upon its surface area.

A large animal has more surface area than a small one and so is capable of a higher power output. However, a large animal also has more mass than a small one, and because an animal's mass goes up faster with its size than its skin area does, its metabolic rate per kilogram decreases. Typical basal metabolic rates, which correspond to an animal resting, are 5.2 W/kg for a pigeon, 1.2 W/kg for a person, and 0.67 W/kg for a cow. African elephants partly overcome the limitation of the small surface/mass ratio of their huge bodies by their enormous ears, which help them get rid of metabolic heat. Most birds are small because, with increasing size, a bird's metabolic rate (and hence power output) per kilogram decreases while the work it must perform per kilogram to fly stays the same. Large birds such as ostriches and emus are not notable for their flying ability (Fig. 5-10).

Figure 5-10 Egret coming in for a landing in the Florida Everglades. Birds are limited in size because the larger an animal is, the less is its ability to dissipate waste metabolic energy per unit of body mass.

When an animal is active, its metabolic rate may be much greater than its basal rate. A 70-kg person, to give an example, has a basal metabolic rate of around 80 W. When the person is reading or doing light work while sitting, the rate will go up to perhaps 125 W. Walking means a metabolic rate of 300 W or so, and running hard increases it to as much as 1200 W. The highest known metabolic rates per kilogram are those of the flight muscles of insects.

An animal's brain as well as its muscles uses energy to function. Your brain, with only 2 percent of your body's mass, uses 20 to 25 percent of your metabolic energy—thinking is hard work! The proportion is only about 9 percent for monkeys and 5 percent for cats and dogs.

When the intake of energy from food exceeds a person's metabolic needs, the excess goes into additional tissue: muscle if there is enough physical activity, otherwise fat. The energy stored in fat is available should metabolic needs not be provided by food at a later time.

Walking Versus Running

Why does running a mile use more energy than walking the same distance? A major reason walking is more efficient is that a walker's legs swing forward easily at each step at about the same speed as a freely swinging pendulum of that length. A runner's legs need more muscular force to drive them forward faster than their natural speed, which means more work done and so more energy used (and muscles that are more tired afterward). In both cases all the work the muscles do ends up as heat.

People, horses, and dogs, whose legs stay relatively straight, use about half as much energy to walk than animals such as cats, which walk in a crouch with bent legs. Even though it is inefficient, a cat's way of walking presumably evolved because it enables stealthy hunting in the wild.

Fluids

The particles of matter in a solid vibrate around fixed positions, so the solid has a definite size and shape (Fig. 5-11). In a liquid, the particles are about as far apart as those in a solid but are able to move about. Hence a liquid sample has a definite volume but flows to fit its container. In a gas, the particles can move freely, so a gas has neither a definite volume nor shape but fills whatever container it is in.

Liquids and gases are together called **fluids** because they flow readily. This ability has some interesting consequences. One is buoyancy, which permits objects to float in a fluid under certain conditions—a balloon in air, a ship in water. In order to understand how buoyancy comes about, we must first look into the ideas of density and pressure.

5.4 Density

A Characteristic Property of Every Material

The **density** of a material is its mass per unit volume:

$$d = \frac{m}{V} \qquad \textit{Density} \qquad \text{5-4}$$

$$\text{Density} = \frac{\text{mass}}{\text{volume}}$$

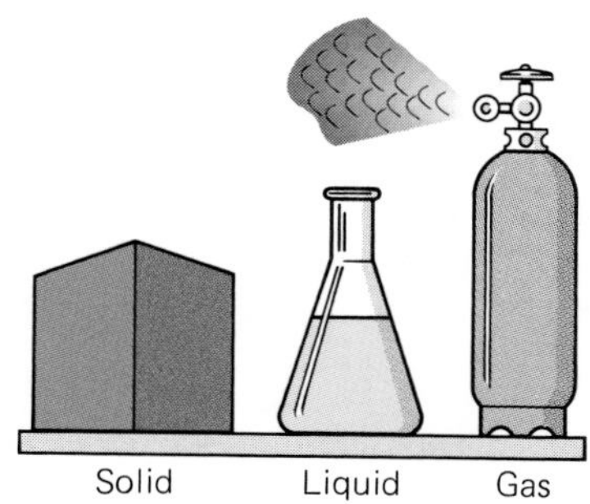

Figure 5-11 Solids, liquids, and gases. A solid maintains its shape and volume no matter where it is placed; a liquid assumes the shape of its container while maintaining its volume; a gas expands indefinitely unless stopped by the walls of a container.

When we say that lead is a "heavy" metal and aluminum a "light" one, what we really mean is that lead has a higher density than aluminum: the density of lead is 11,300 kg per cubic meter (kg/m^3) whereas that of aluminum is only 2700 kg/m^3, a quarter as much.

Although the proper SI unit of density is the kg/m^3, densities are often given instead in g/cm^3 (grams per cubic centimeter), where 1 g/cm^3 = 1000 kg/m^3 = 10^3 kg/m^3. Thus the density of lead can also be expressed as 11.3 g/cm^3. Table 5-3 lists the densities of some common substances.

Table 5-3 Densities of Various Substances at Room Temperature and Atmospheric Pressure

Substance	Density, kg/m^3	Substance	Density, kg/m^3
Air	1.3	Hydrogen	0.09
Alcohol (ethyl)	7.9×10^2	Ice	9.2×10^2
Aluminum	2.7×10^3	Iron	7.8×10^3
Balsa wood	1.3×10^2	Lead	1.1×10^4
Concrete	2.3×10^3	Mercury	1.4×10^4
Gasoline	6.8×10^2	Oak	7.2×10^2
Gold	1.9×10^4	Water, pure	1.00×10^3
Helium	0.18	Water, sea	1.03×10^3

Example 5.6

Find the mass of the water in a bathtub whose interior is 1.300 m long and 0.600 m wide and that is filled to a height of 0.300 m (Fig. 5-12).

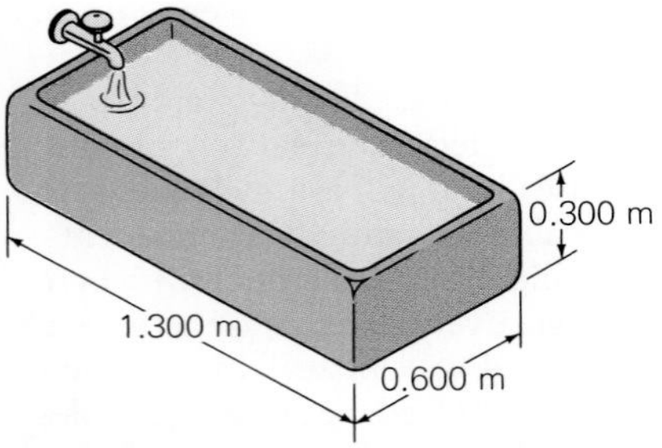

Solution

The water's volume is

$$V = \text{(length)(width)(height)}$$
$$= (1.300\text{ m})(0.600\text{ m})(0.300\text{ m}) = 0.234\text{ m}^3$$

According to Table 5-3 the density of water is 1000 kg/m^3. Rewriting $d = m/V$ as $m = dV$ gives

$$\text{Mass} = \text{(density)(volume)}$$
$$m = dV = \left(1000\frac{\text{kg}}{\text{m}^3}\right)(0.234\text{ m}^3) = 234\text{ kg}$$

The weight of this amount of water is a little over 500 lb.

0.300 m
1.300 m
0.600 m

Figure 5-12 The volume of water in this bathtub is equal to the product (length) (width) (height).

5.5 Pressure

How Much of a Squeeze

We next look into what is meant by **pressure.** When a force F acts perpendicular to a surface whose area is A, the pressure acting on the surface is the ratio between the force and the area:

$$p = \frac{F}{A} \qquad \textit{Pressure} \qquad 5\text{-}5$$

$$\text{Pressure} = \frac{\text{force}}{\text{area}}$$

The SI unit of pressure is the **pascal** (Pa), where

$$1\text{ pascal} = 1\text{ Pa} = 1\text{ newton/meter}^2$$

This unit honors the French scientist and philosopher Blaise Pascal (1623–1662). The pascal is a very small unit: the pressure exerted by pushing really hard on a table with your thumb is about a million pascals. For this reason the **kilopascal** (kPa) is often used, where 1 kPa = 1000 Pa = 10^3 Pa = 0.145 lb/in^2.

Example 5.7

A 60-kg woman balances on the heel of one shoe, whose area is 0.5 cm^2 (Fig. 5-13). How much pressure does she exert on the floor?

Solution

The force exerted by the heel of the woman's shoe is her weight, so $F = mg = (60\text{ kg})(9.8\text{ m/s}^2) = 588\text{ N}$. Since 1 cm = 10^{-2} m, the area of the heel is $A = (0.5\text{ cm}^2)(10^{-4}\text{ m}^2/\text{cm}^2) = 5 \times 10^{-5}\text{ m}^2$, and the pressure on the floor is

$$p = \frac{F}{A} = \frac{588\text{ N}}{5 \times 10^{-5}\text{ m}^2} = 1.2 \times 10^7\text{ N/m}^2 = 12\text{ MPa}$$

This is 40 times the pressure the feet of the 35-ton dinosaur apatosaurus are estimated to have exerted on the ground.

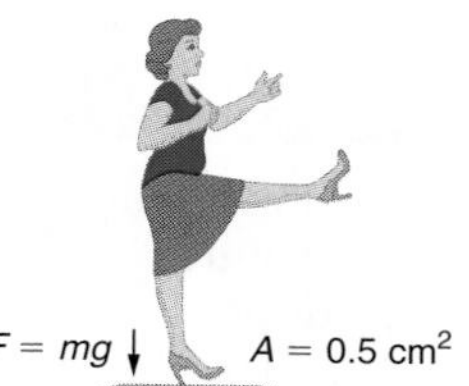

Figure 5-13 Pressure is force per unit area. The heel of this 60-kg woman exerts a pressure of 12 MPa on the floor. Being stepped on by such a heel is not recommended.

Pressure in a Fluid Pressure is a useful quantity where fluids are concerned for these reasons:

1. The forces that a fluid exerts on the walls of its container, and those that the walls exert on the fluid, always act perpendicular to the walls.
2. The force exerted by the pressure in a fluid is the same in all directions at a given depth.
3. An external pressure exerted on a fluid is transmitted uniformly throughout the fluid.

These properties mean that we can transmit a force from one place to another by applying pressure with a pump to a fluid at one end of a tube and then allowing the fluid at the other end of the tube to push against a movable piston. Machines in which forces are transmitted by liquids are

Example 5.8

Before a storm people in a house closed its doors and windows so tightly that the air pressure inside remained at 101 kPa even when the outside pressure fell to 98 kPa. How much force acted on a window 80 cm high and 120 cm wide?

Solution

The area of the window is $A = (0.80\text{ m})(1.2\text{ m}) = 0.96\text{ m}^2$. The net pressure on the window was the difference Δp between the inside pressure (which gives rise to an outward force) of 101 kPa and the outside pressure (which gives rise to an inward force) of 98 kPa, so $\Delta p = 3\text{ kPa} = 3 \times 10^3\text{ Pa}$. Therefore the outward force on the window was

$$F = \Delta p A = (3 \times 10^3\text{ Pa})(0.96\text{ m}^2) = 2880\text{ N}$$

which corresponds to 636 lb! This may well have been enough to break the window or push it out of its frame. Evidently a building should not be sealed when a large change in pressure due to a storm is forecast.

Blood Pressure

The pressures that force blood through the lungs and the rest of the body are produced by the heart (Fig. 5-14). The heart consists of two pumps, called **ventricles.** The contraction and relaxation of the muscular walls of the ventricles take the place of the piston strokes of an ordinary pump. The right ventricle pumps blood from the veins through the lungs, where it absorbs oxygen from the air that has been breathed in and gives up carbon dioxide. The oxygenated blood then goes to the more powerful left ventricle, which pumps it via the aorta and the arteries to the rest of the body. A typical rate of flow of blood in a resting person is 6 L/min.

Arterial blood pressures are measured by using an inflatable cuff that is pumped up until the flow of blood stops, as monitored by a stethoscope (Fig. 5-15). Air is then let out of the cuff until the flow just begins again, which is recognized by a gurgling sound in the stethoscope. The pressure at this point is called **systolic** and corresponds to the maximum pressure the heart produces in the arteries. Next, more air is let out until the gurgling stops, which corresponds to normal blood flow. The pressure now, called **diastolic,** corresponds to the arterial pressure between strokes of the heart.

Blood pressures are usually expressed in *torr,* which is the pressure exerted by a column of mercury 1 mm high; 1 torr = 133 Pa. In a healthy person the systolic and diastolic blood pressures are about 120 and 80 torr, respectively.

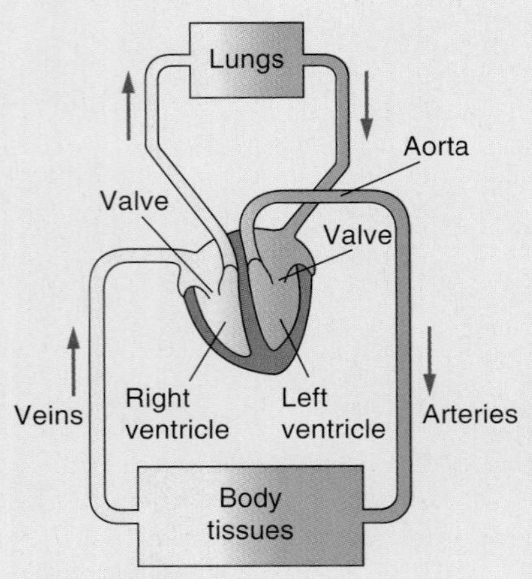

Figure 5-14 The human circulatory system.

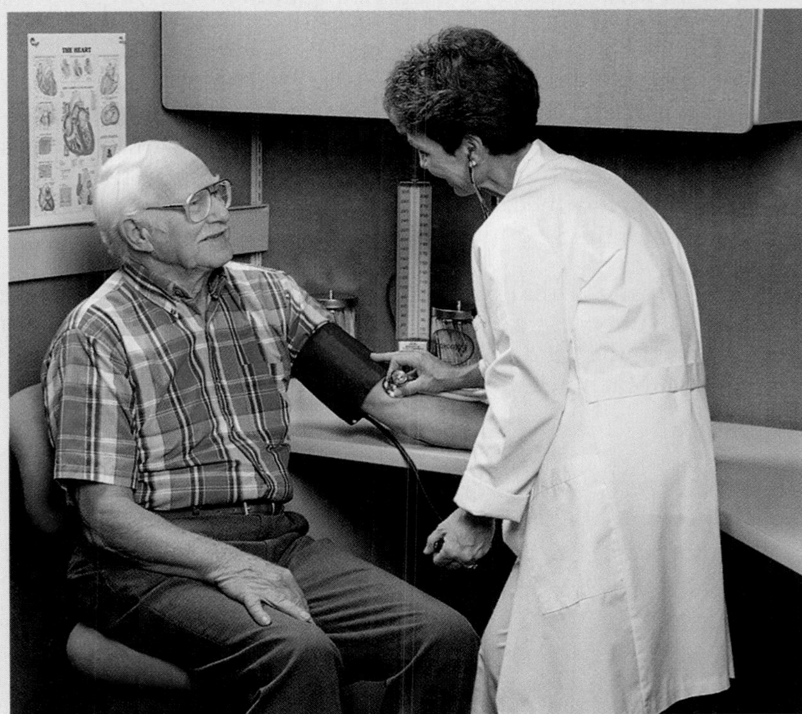

Figure 5-15 Measuring blood pressure.

called **hydraulic** (Fig. 5-16), and those that use compressed air are called **pneumatic.**

Pressure and Depth The pressure in a fluid increases with depth because of the weight of the overlying fluid. The amount of the increase is proportional to both the depth and the fluid density: the greater these are, the greater the pressure. In a tire pump, the air inside is under greater pressure at the bottom of the cylinder than at the top because of the weight of the air in the cylinder. In the case of a tire pump, the pressure difference is very small, of course, but elsewhere it can be significant. At sea level on the earth, for instance, the pressure due to the weight of the air above us averages 101 kPa, nearly 15 lb/in^2. This corresponds to a force of 10.1 N on every square centimeter of our bodies. We are not aware of this pressure because the pressures inside our bodies are the same. Atmospheric pressures are measured with instruments called **barometers,** one type of which is shown in Fig. 5-17.

Because pressure increases with depth, most submarines cannot go down more than a few hundred meters without the danger of collapsing. At a depth of 10 km in the ocean, the pressure is about 1000 times sea-level atmospheric pressure, enough to compress water by 3 percent of its volume. Fish that live at such depths are not crushed for the same reason we can survive the pressure at the bottom of our ocean of air: pressures inside their bodies are kept equal to pressures outside.

Scuba divers carry tanks of compressed air with regulator valves that provide air at the same pressure as the water around them (Fig. 5-18). ("Scuba" stands for *s*elf-*c*ontained *u*nderwater *b*reathing *a*pparatus.) When a diver returns to the surface, he or she must breathe out continuously to allow the air pressure in the lungs to decrease at the same rate as the external pressure decreases. If this is not done, the pressure difference may burst the lungs. When brought to the surface quickly, deep-sea fish sometimes explode because of their high internal pressures.

Figure 5-16 A hydraulic ram converts pressure in a liquid into an applied force. The pressure is provided by an engine-driven pump.

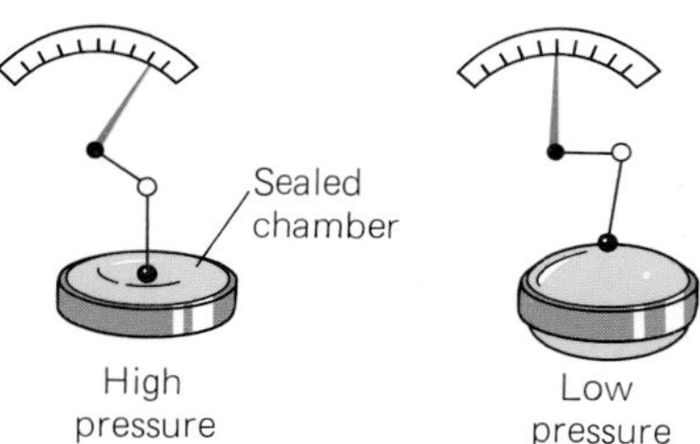

Figure 5-17 An aneroid barometer. The flexible ends of a sealed metal chamber are pushed in by a high atmospheric pressure. Under low atmospheric pressure, the air inside the chamber pushes the ends out.

Figure 5-18 Air at high pressure in the tank of a scuba diver is reduced by a regulator valve to the pressure at the depth of the water. The diver must wear lead weights to overcome his or her buoyancy. The deeper the diver goes, the greater the water pressure, and the faster the air in the tank is used up.

5.6 Buoyancy

Sink or Swim

An object immersed in a fluid is acted upon by an upward force that arises because pressures in a fluid increase with depth. Hence the upward force on the bottom of the object is greater than the downward force on its top. The difference between the two forces is the **buoyant force.**

Buoyancy enables balloons to float in the air and ships to float in the sea. If the buoyant force on an immersed object is greater than its weight, the object floats; if the force is less than its weight, the object sinks.

Imagine a solid object of any kind whose volume is V that is in a tank of water (Fig. 5-19). A body of water of the same size and shape in the tank is supported by a buoyant force F_b equal to its weight of $w_{water} = dVg$. The buoyant force is the result of all of the forces that the rest of the water in the tank exert on this particular body of water. This force is always upward because the pressure underneath the body of water is greater than the pressure above it. The pressures on the sides cancel one another out, as shown in the figure.

If we now replace the body of water by the solid object, the forces on the object are the same as before. The buoyant force therefore remains dVg so that

$$F_b = dVg \qquad \textit{Buoyant force} \qquad 5\text{-}6$$

In this formula, d is the density of the fluid (which need not be water, of course), V is the volume of fluid displaced by the solid object, and g is the acceleration of gravity, 9.8 m/s^2.

Thus we have **Archimedes' principle:**

> Buoyant force on an object in a fluid = weight of fluid displaced by the object.

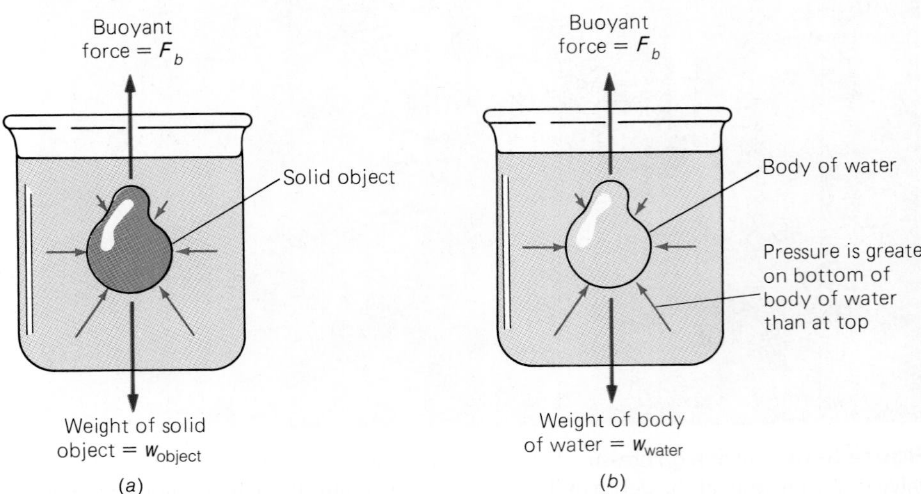

Figure 5-19 Archimedes' principle. The buoyant force F_b on an object immersed in water (or other fluid) is equal to the weight of the body of water displaced by the object.

Example 5.9

The density of a person is slightly less than that of pure water (which is why people can float). Assuming that these densities are the same, find the buoyant force of the atmosphere on a 60-kg person.

Solution

The first step is to find the person's volume. Since $d = m/V$ (density = mass/volume) and $d_{water} = 1.00 \times 10^3$ kg/m^3 from Table 5-3, we have

$$V = \frac{m}{d_{water}} = \frac{60 \text{ kg}}{1.00 \times 10^3 \text{ kg/m}^3} = 0.06 \text{ m}^3$$

The buoyant force is equal to the weight $w = m_{air}g = d_{air}Vg$ of the air displaced by the person, so

$$F_b = d_{air}Vg = (1.3 \text{ kg/m}^3)(0.06 \text{ m}^3)(9.8 \text{ m/s}^2) = 0.76 \text{ N}$$

This is about 2.7 ounces.

Archimedes' principle holds whether the object floats or sinks. If the object's weight w_{object} is greater than the buoyant force F_b, it sinks. If its weight is less than the buoyant force, it floats, in which case the volume V refers only to the part of the object that is in the fluid.

The condition for an object to float in a given fluid is that its average density be lower than the density of the fluid (Fig. 5-20). Why does a steel ship float when the density of steel is nearly 8 times that of water? The answer is that the ship is a hollow shell, so its average density is less than that of water even when loaded with cargo. If the ship springs a leak and fills with water, its average density goes up, and the ship sinks. The purpose of a life jacket is to reduce the average density of a person in the water so that she floats higher and is less likely to get water in her lungs and drown as a result.

Figure 5-20 Because water expands when it freezes, ice floats. Nearly 90 percent of the volume of this iceberg near Greenland lies below the surface. An iceberg is a chunk of freshwater ice that has broken off from an ice cap, such as those that cover Greenland and Antarctica, or from a glacier at the edge of the sea.

BIOGRAPHY Archimedes (287?–212 B.C.)

The pre-eminent scientist/mathematician of the ancient world, Archimedes was born in Syracuse, Sicily, at that time a Greek colony. He went to Alexandria, Egypt, to study under a former pupil of Euclid and then returned to Syracuse where he spent the rest of his life. Many stories have come down through the ages about Archimedes, some of them possibly true. The principle named after him, that the buoyant force on a submerged object equals the weight of the fluid it displaces, is supposed to have come to him while in his bath, whereupon he rushed naked into the street crying "Eureka!" ("I've got it!"). He had been trying to think of a way to determine whether a new crown made for King Hieron of Sicily was pure gold without damaging it and used his discovery to show that it was not; the goldsmith was then executed.

Archimedes worked out the theory of the lever and remarked (so it is said), "Give me a place to stand and I can move the world." King Hieron then challenged him to move something really large, even if not the world, and Archimedes responded by using a system of pulleys, which are developments of the lever, to haul a laden ship overland all by himself. Archimedes was an able mathematician, and among his other accomplishments he calculated that the value of π lay between 223/71 and 220/70, an extremely good approximation. Archimedes is given credit for keeping the Roman fleet that attacked Syracuse in 215 B.C. at bay for 3 years by a variety of clever devices, including giant

lenses that focused sunlight on the ships and set them afire. Finally the Romans did conquer Syracuse, and a Roman soldier found Archimedes bent over a mathematical problem scratched in the sand. When Archimedes would not stop work immediately, the soldier killed him.

5.7 The Gas Laws

Ideal Gases Obey Them

In many ways the gaseous state is the one whose behavior is the easiest to describe and account for. As an important example, the pressures, volumes, and temperatures of gas samples are related by simple formulas that have no counterpart in the cases of liquids and solids. The discovery of these formulas led to a search for their explanation in terms of the basic nature of gases, a search that resulted in the kinetic theory of matter.

Boyle's Law Suppose that a sample of some gas is placed in the cylinder of Fig. 5-21, and a pressure of 100 kPa is applied. The final volume of the sample is 1 m^3. If we double the pressure to 200 kPa, the piston will move down until the gas volume is 0.5 m^3, half its original amount, provided the gas temperature is kept unchanged. If the pressure is made 10 times greater, the piston will move down farther, until the gas occupies a volume of 0.1 m^3, again if the gas temperature is kept unchanged.

These findings can be summarized by saying that the volume of a given quantity of a gas at constant temperature is inversely proportional to the pressure applied to it. (By "inversely proportional" is meant that as the pressure increases, the volume decreases by the same proportion.) If the volume of the gas is V_1 when the pressure is p_1 and the volume changes to V_2 when the pressure is changed to p_2, the relationship among the various quantities is

$$\frac{p_1}{p_2} = \frac{V_2}{V_1} \quad \text{(at constant temperature)} \qquad \textit{Boyle's law} \qquad 5\text{-}7$$

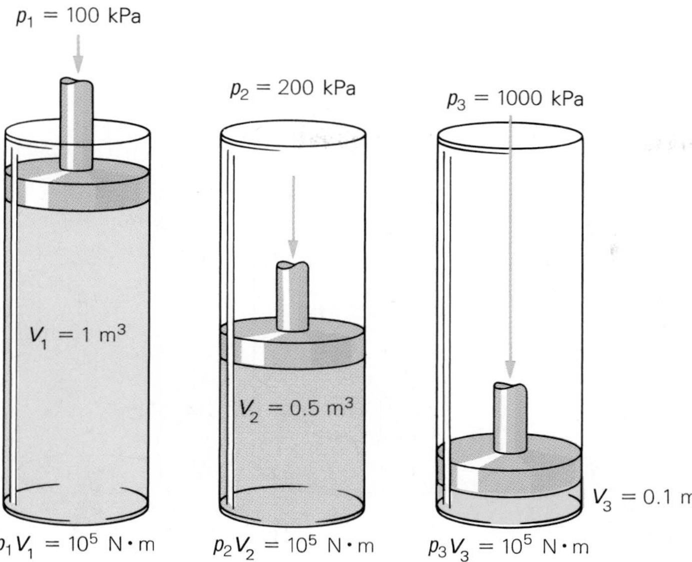

Figure 5-21 Boyle's law: At constant temperature, the volume of a sample of any gas is inversely proportional to the pressure applied to it. Here $p_1V_1 = p_2V_2 = p_3V_3$.

This relationship is called **Boyle's law,** in honor of the English physicist who discovered it. It is often written in the equivalent form $p_1V_1 = p_2V_2$.

Example 5.10

When water is boiled in a pot, the bubbles of steam increase in size as they rise through the water. Why?

Solution

The pressure in the water decreases toward the top, hence the steam bubbles expand.

Example 5.11

A scuba diver whose 12-liter (L) tank is filled with air at a pressure of 150 atmospheres (atm) is swimming at a depth of 15 m where the water pressure is 2.5 atm. If she uses 30 L of air per minute at the same pressure as the water pressure, how long can she stay at that depth?

Solution

We begin by using Boyle's law to determine the volume V_2 of air available at a pressure of $p_2 = 2.5$ atm:

$$V_2 = \frac{p_1V_1}{p_2} = \frac{(150 \text{ atm})(12 \text{ L})}{2.5 \text{ atm}}$$
$$= 720 \text{ L}$$

However, 12 L of air remains in the tank, so only 708 L is usable and will last for

$$\frac{708 \text{ L}}{30 \text{ L/min}} = 23.6 \text{ min}$$

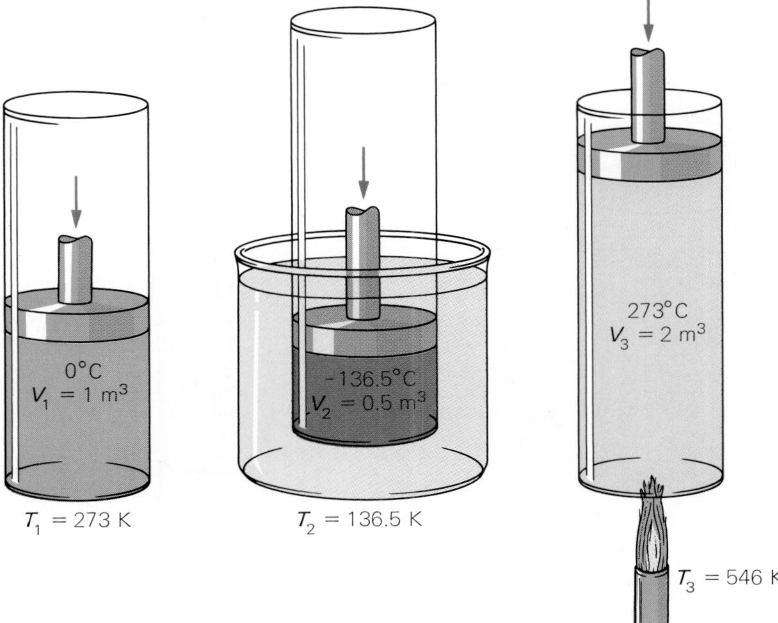

Figure 5-22 Charles's law: At constant pressure, the volume of a gas sample is directly proportional to its absolute temperature T_K, where $T_K = T_C + 273°$. Here $V_1/T_1 = V_2/T_2 = V_3/T_3$.

Figure 5-23 At constant pressure, heating a gas causes it to expand. The density of hot air therefore is less than the density of cool air at the same pressure, which is why a hot-air balloon is buoyant. These balloons have propane burners in their gondolas to supply the needed heat.

Charles's Law Changes in the volume of a gas sample are also related to temperature changes in a simple way. If a gas is cooled steadily, starting at 0°C, while its pressure is maintained constant, its volume decreases by $\frac{1}{273}$ of its volume at 0°C for every degree the temperature falls. If the gas is heated, its volume increases by the same fraction (Figs. 5-22 and 5-23). If volume rather than pressure is kept fixed, the pressure increases with rising temperature and decreases with falling temperature, again by the fraction $\frac{1}{273}$ of its 0°C value for every degree change.

These figures suggest an obvious question: What would happen to a gas if we could lower its temperature to −273°C? If we kept the gas at constant volume, the pressure at this temperature ought to fall to zero. If the pressure stayed constant, the volume ought to fall to zero.

It is hardly likely, however, that our experiments would have such results. In the first place, we should find it impossible to reach quite so low a temperature. In the second place, all known gases turn into liquids before that temperature is reached. Nevertheless, a temperature of −273°C has a special significance, a significance that will become clearer shortly. This temperature is called **absolute zero.**

For many scientific purposes it is convenient to begin the temperature scale at absolute zero. Temperatures on such a scale, given as degrees Celsius above absolute zero, are called **absolute temperatures.** Thus the freezing point of water is 273° absolute, written as 273 K in honor of the English physicist Lord Kelvin (Fig. 5-24), and the boiling point of water is 373 K. Any Celsius temperature T_C can be changed to its equivalent absolute temperature T_K by adding 273 (Fig. 5-25):

$$T_K = T_C + 273 \qquad 5\text{-}8$$

Absolute temperature = Celsius temperature + 273

Using the absolute scale, we can express the relationship between gas volumes and temperatures quite simply: the volume of a gas is directly proportional to its absolute temperature (Fig. 5-26). This relation may be expressed in the form

$$\frac{V_1}{V_2} = \frac{T_1}{T_2} \quad \text{(at constant pressure)} \qquad \textit{Charles's law} \qquad 5\text{-}9$$

where the T's are absolute temperatures. Discovered by two eighteenth-century French physicists, Jacques Alexandre Charles and Joseph Gay-Lussac, this relation is commonly known as **Charles's law.** Charles's law can also be written $V_1/T_1 = V_2/T_2$.

Figure 5-24 Lord Kelvin (1824–1907). A notable physicist but a poor prophet, he announced in 1900 that "there is nothing new to be discovered in physics now".

Ideal Gas Law Boyle's and Charles's laws can be combined in a single formula known as the **ideal gas law:**

$$\frac{p_1V_1}{T_1} = \frac{p_2V_2}{T_2} \qquad \textit{Ideal gas law} \qquad 5\text{-}10$$

At constant temperature, $T_1 = T_2$ and we have Boyle's law. At constant pressure, $p_1 = p_2$ and we have Charles's law. Another way to write the ideal gas law is

$$\frac{pV}{T} = \text{constant} \qquad \textit{Ideal gas law} \qquad 5\text{-}11$$

since this particular combination of quantities does not change in value for a gas sample even though the individual quantities p, V, and T may vary.

The ideal gas law is obeyed approximately by all gases. The significant thing is not that the agreement with experiment is never quite perfect but that *all* gases behave almost identically.

An **ideal gas** is defined as one that obeys Eq. 5-11 exactly. Although no ideal gases actually exist, they do provide a target for a theory of the gaseous state to aim at. As we shall see, the kinetic theory of gases is indeed able to explain the ideal gas law, which means that it is a valid guide to the essential nature of gases.

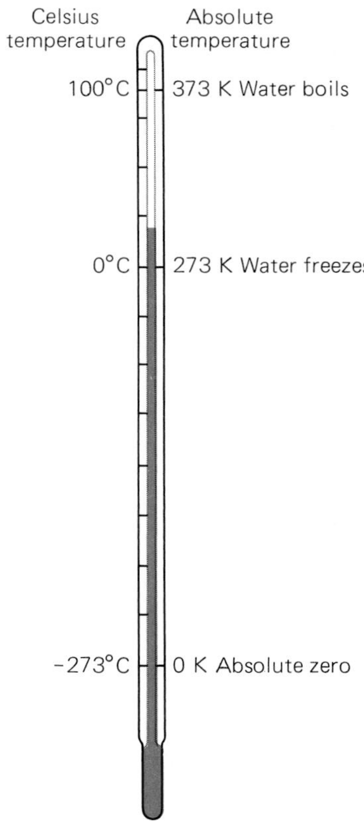

Figure 5-25 The absolute temperature scale.

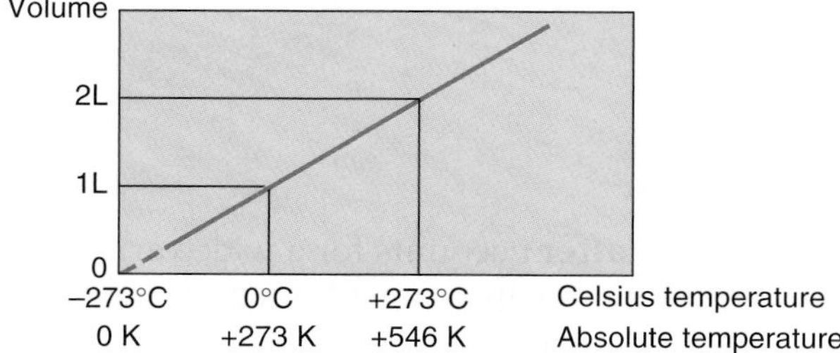

Figure 5-26 Graphic representation of Charles's law, showing the proportionality between volume and absolute temperature for a gas at constant pressure. If the temperature of the gas could be reduced to absolute zero, its volume would fall to zero. Actual gases liquefy at temperatures above absolute zero.

Example 5.12

A tank whose capacity is 0.1 m^3 contains helium at a pressure of 1000 kPa and a temperature of 20°C. A rubber weather balloon is inflated from this tank (Fig. 5-27). The helium cools as it expands, and when the pressure in the balloon is 100 kPa, which is atmospheric pressure at that time, the temperature of the helium is −40°C. (The helium has done work in expanding at the expense of its heat content, and the cooling reflects this loss of heat.) Find the volume of the inflated balloon.

Solution

Here $T_1 = 20°C = 293$ K, $T_2 = -40°C = 233$ K, $V_1 = 0.1\ m^3$, $p_1 = 1000$ kPa, and $p_2 = 100$ kPa. From Eq. 5-10,

$$V_2 = \frac{T_2 p_1 V_1}{T_1 p_1} = \frac{(233\ \text{K})(1000\ \text{kPa})(0.1\ \text{m}^3)}{(293\ \text{K})(100\ \text{kPa})} = 0.8\ \text{m}^3$$

Because the tank's capacity is 0.1 m^3, the balloon's volume is $(0.8 - 0.1)m^3 = 0.7\ m^3$. As the balloon warms up to the outside temperature, it will continue to expand until the helium and air temperatures are the same.

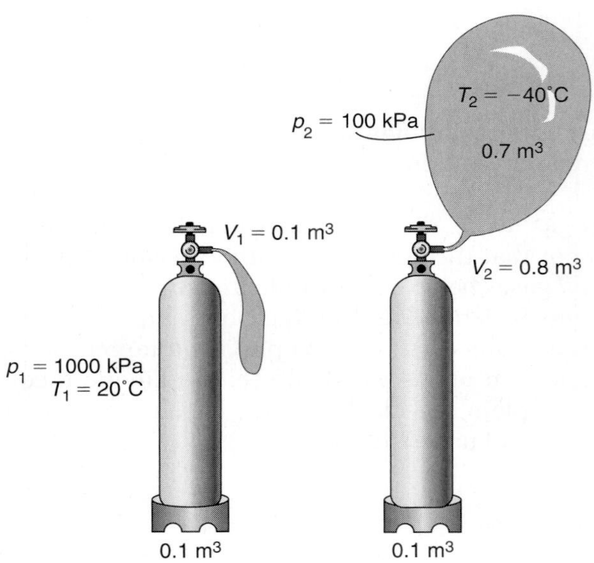

Figure 5-27 Inflating a weather balloon.

Kinetic Theory of Matter

The **kinetic theory of matter** accounts for a wide variety of physical and chemical properties of matter in terms of a simple model. According to this model, all matter is composed of tiny particles. In the case of a gas, the particles are usually **molecules** that consist of two or more atoms. In liquids and solids, the particles may be molecules, atoms, or ions, as discussed in Chap. 10. In this chapter, for simplicity, the basic particles characteristic of any substance will be called molecules.

5.8 Kinetic Theory of Gases

Why Gases Behave as They Do

Today we know a great deal about the sizes, speeds, even shapes of the molecules in various kinds of matter.

For example, a molecule of nitrogen, the chief constituent of air, is about 0.18 billionth of a meter (1.8×10^{-10} m) across and has a mass of 4.7×10^{-26} kg. It travels (at 0°C) at an average speed of 500 m/s, about the speed of a rifle bullet, and in each second collides with more than a billion other molecules. Of similar dimensions and moving with similar speeds in each cubic centimeter of air are 2.7×10^{19} other molecules. If all the molecules in such a thimbleful of air were divided equally among the 6.7 billion people on the earth, each person would receive several billion molecules.

The three basic assumptions of the kinetic theory for gas molecules, which have been verified by experiment, are these:

1. Gas molecules are small compared with the average distance between them.
2. Gas molecules collide without loss of kinetic energy.
3. Gas molecules exert almost no forces on one another, except when they collide.

A gas, then, is mostly empty space, with its isolated molecules moving helter-skelter like a swarm of angry bees in a closed room (Fig. 5-28). Each molecule collides with others perhaps billions of times a second, changing its speed and direction at each collision but unaffected by its neighbors between collisions. If a series of collisions brings it momentarily to a stop, new collisions will set it in motion. If its speed becomes greater than the average, successive collisions will slow it down. There is no order in the motion, no uniformity of speed or direction. All we can say is that the molecules have a certain average speed and that at any instant as many molecules are moving in one direction as in another.

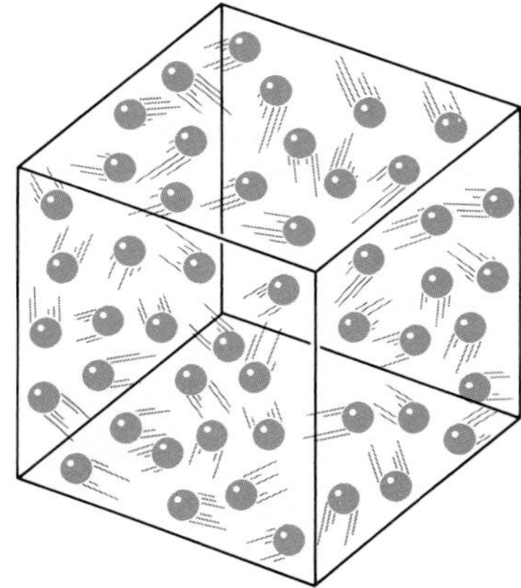

Figure 5-28 The molecules of a gas are in constant random motion.

This animated picture explains the more obvious properties of gases. The ability of a gas to expand and to leak through small openings follows from the rapid motion of its molecules and their lack of attraction for one another. Gases are easily compressed because the molecules are, on the average, widely separated. One gas mixes with another because the spaces between molecules leave plenty of room for others. Because a given volume of a gas consists mainly of empty space, the mass of that volume is much less than that of the same volume of a liquid or a solid.

Origin of Boyle's Law The pressure of a gas on the walls of its container is the result of bombardment by billions and billions of molecules, the same bombardment that causes brownian movement (Fig. 5-29). The many tiny, separate blows affect our senses and our measuring instruments as a continuous force.

The kinetic theory accounts nicely for Boyle's law, $p_1V_1 = p_2V_2$ (at constant temperature). As in Fig. 5-30, we can think of the molecules of a gas in a cylinder as moving in a regular manner, some of them vertically between the piston and the base of the cylinder and the others horizontally between the cylinder walls. If the piston is raised so that

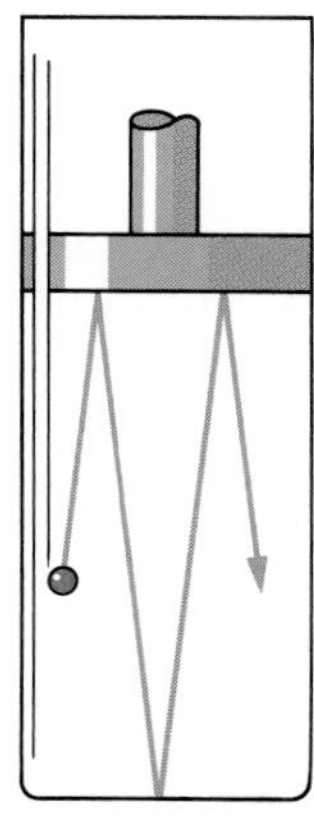

Figure 5-29 Gas pressure is the result of molecular bombardment. For simplicity, only vertical molecular motions are shown.

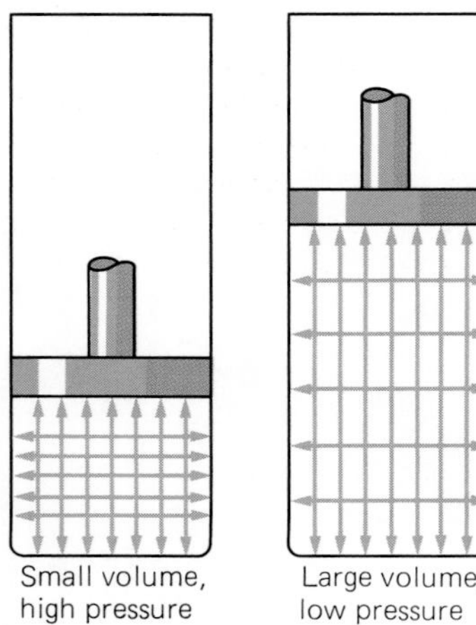

Figure 5-30 Origin of Boyle's law. Expanding a gas sample means that its molecules must travel farther between successive impacts on the container wall and that their blows are spread over a larger area, so the gas pressure drops.

the gas volume is doubled, the vertically moving molecules have twice as far to go between collisions with top and bottom and hence will strike only half as often. The horizontally moving molecules must spread their blows over twice as great an area, and hence the number of impacts per unit area will be cut in half. Thus the pressure in all parts of the cylinder is exactly halved, as Boyle's law predicts. It is not hard to extend this reasoning to a real gas whose molecules move at random.

5.9 Molecular Motion and Temperature

The Faster the Molecules, the Higher the Temperature

To account for the effect of a temperature change on a gas, the kinetic theory requires a further concept:

4. The absolute temperature of a gas is proportional to the average kinetic energy of its molecules.

This concept was used by Einstein to explain brownian motion (see Einstein biography in Sec. 3.12).

That temperature should be related to molecular energies and thus to molecular speeds follows from the increase in the pressure of a confined gas as its temperature rises. Increases in pressure must mean that the molecules are striking the walls of their container more forcefully and so must be moving faster.

Earlier in this chapter we learned that the pressure of a gas approaches zero as its temperature falls toward 0 K, which is −273°C. For the pressure to become zero, molecular bombardment must stop. Thus absolute zero is interpreted as the temperature at which gas molecules would lose their kinetic energies completely, as shown in Fig. 5-31. (This is a simplification of the actual situation: in reality, even at 0 K a molecule will have a very small amount of KE that cannot be reduced.) There can be no lower temperature, simply because there can be no smaller amount of energy. The regular increase of gas pressure with absolute temperature if the volume is constant and the similar increase

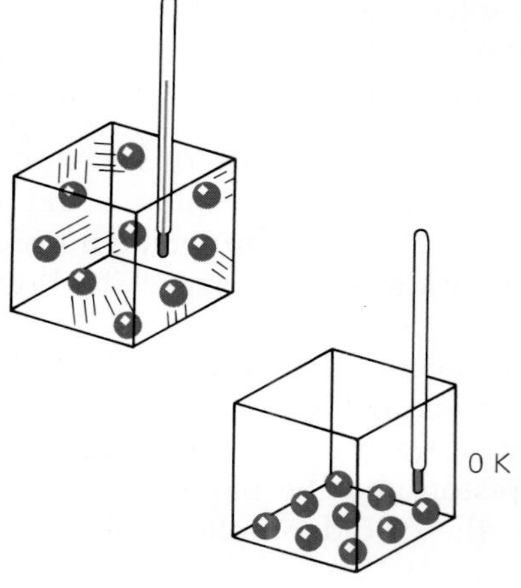

Figure 5-31 According to the kinetic theory of gases, at absolute zero the molecules of a gas would not move. More advanced theories show that even at 0 K a very slight movement will persist.

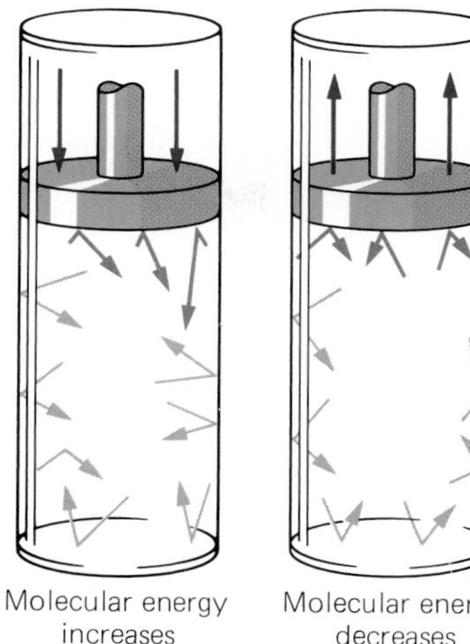

Figure 5-32 Compressing a gas causes its temperature to rise because molecules rebound from the piston with more energy. Expanding a gas causes its temperature to drop because molecules rebound from the piston with less energy.

of volume if the pressure is constant (Charles's law) follow from this definition of absolute zero.

Origin of Charles's Law If temperature is a measure of average molecular energy, then compressing a gas in a cylinder ought to cause its temperature to rise. While the piston of Fig. 5-32 is moving down, molecules rebound from it with increased energy just as a baseball rebounds with increased energy from a moving bat. To verify this prediction, all we have to do is pump up a bicycle tire and notice how hot the pump becomes after the air in it has been compressed a few times. On the other hand, if a gas is expanded by pulling a piston outward, its temperature falls, since each molecule that strikes the retreating piston gives up some of its kinetic energy (Fig. 5-33).

The cooling effect of gas expansion explains the formation of clouds from rising moist air, as discussed in Chap. 14. Atmospheric pressure decreases with altitude, and the water vapor in the moist air cools as it moves upward until it condenses into the water droplets that constitute clouds.

Figure 5-33 In the operation of a snowmaking machine, a mixture of compressed air and water is blown through a set of nozzles. The expansion of the air cools the mixture sufficiently to freeze the water into the ice crystals of snow.

Changes of State

The kinetic theory of matter is clearly a success in explaining the behavior of gases. Let us now see what this theory has to say about liquids and solids. In particular, changes of state between gas and liquid and between liquid and solid are extremely interesting in molecular terms.

5.10 Liquids and Solids

Intermolecular Forces Hold Them Together

If a gas is like a swarm of angry bees, the molecules in a liquid are more like bees in a hive, crawling over one another constantly. Liquids flow because their molecules slide past one another easily, but they flow less readily than gases because of intermolecular attractions that act only over short distances.

The forces between the molecules of a solid are stronger than those in a liquid, so strong that the molecules are not free to move about (Fig. 5-34). They are hardly at rest, however. Held in position as if by springs attached to its neighbors, each molecule vibrates back and forth rapidly (Fig. 5-35). Each spring represents a bond between two adjacent molecules. Such bonds are electrical in nature. A solid is elastic because its molecules return to their normal separations after being pulled apart or pushed together when a moderate force is applied. If the force is great, the solid may be permanently deformed. In this process, the molecules shift to new positions and find new partners for their attractive forces. Too much applied force, of course, may break the solid apart.

Heat Conduction

Heat conduction is a consequence of the kinetic behavior of matter. Molecules at the hot end of a solid object vibrate faster and faster as the temperature there increases, and when these molecules collide with their less energetic neighbors, some KE transfers to them. Heat flows through a liquid by conduction in a similar way. Gases are poor conductors because their molecules are relatively far apart and so do not collide as often.

Metals are exceptionally good at conducting heat (copper is 3000 times better than wood) because some of the electrons in them are able to move freely instead of being bound to particular atoms (see Sec. 11.2). Picking up KE at the hot end of a metal object, the free electrons travel past many atoms before finally giving up their added KE in collisions. Heat conduction by free electrons in a metal compares with heat conduction in other materials as travel by express train compares with travel by local train.

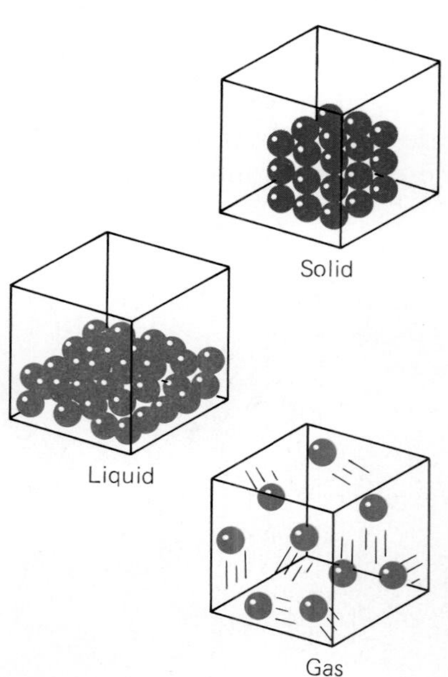

Figure 5-34 Molecular models of a solid, a liquid, and a gas. The molecules of a solid are attached to one another; those of a liquid can slide past one another; those of a gas move freely except when they collide with one another or with the walls of a container.

5.11 Evaporation and Boiling

Liquid into Gas

Suppose we have two liquids, water and alcohol, in open dishes (Fig. 5-36). Molecules in each dish are moving in all directions, with a variety of speeds. At any instant some molecules are moving fast enough upward to escape into the air in spite of the attractions of their slower neighbors. By this loss of its faster molecules each liquid gradually evaporates. Since the remaining molecules are the slower ones, evaporation leaves cool liquids behind. The alcohol evaporates more quickly than the water and cools itself more noticeably because the attraction of its particles for one another is smaller and a greater number can escape.

When we add heat to a liquid, eventually a temperature is reached at which even molecules of average speed can overcome the forces binding them together. Now bubbles of gas form throughout the liquid, and it begins to boil. This temperature is accordingly called the **boiling point** of the liquid. As we would expect, the boiling point of alcohol (78°C) is lower than that of water (100°C).

Thus evaporation differs from boiling in two ways:

1. Evaporation occurs only at a liquid surface; boiling occurs in the entire volume of liquid.
2. Evaporation occurs at all temperatures; boiling occurs only at the boiling point or higher temperatures.

Heat of Vaporization Whether evaporation takes place by itself from an open dish or is aided by heating, forming a gas from a liquid requires

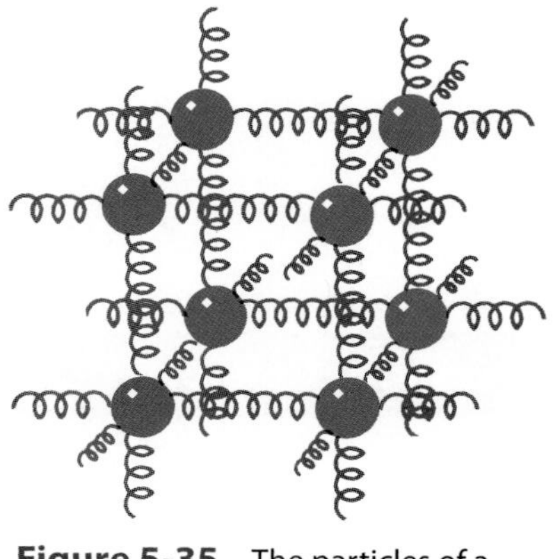

Figure 5-35 The particles of a solid can be imagined as being held together by tiny springs that permit them to vibrate back and forth. The higher the temperature, the more energetic the vibrations. When a solid is squeezed, the springs are (so to speak) pushed together; when it is stretched, the springs are pulled apart.

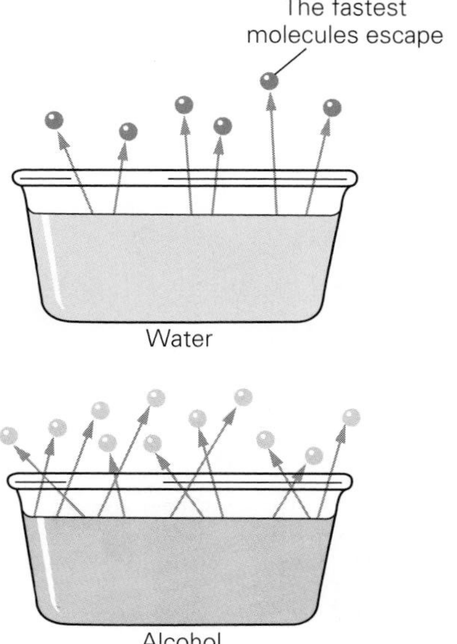

Figure 5-36 Evaporation. Alcohol evaporates more rapidly than water because the attractive forces between its molecules are smaller. In each case, the faster molecules escape. Hence the average kinetic energy of the remaining molecules is lower and the liquid temperature drops.

Gas and Vapor

A gas is a substance whose molecules are too far apart to attract one another. As a result, a gas will expand indefinitely unless stopped by the walls of a container or, in the case of a planet's atmosphere, is prevented from leaving the planet by gravity. But even a substance that is normally a liquid or solid at a certain temperature may lose molecules from its surface. These molecules make up a **vapor.** Under ordinary conditions water is a liquid, but enough water molecules escape (mainly from the oceans) for water vapor to comprise up to 4 percent of the earth's atmosphere. When hot enough, water becomes steam, which is a gas.

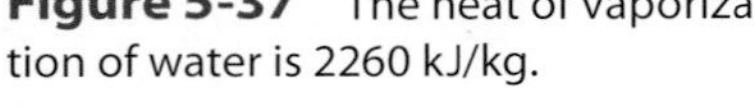

Figure 5-37 The heat of vaporization of water is 2260 kJ/kg.

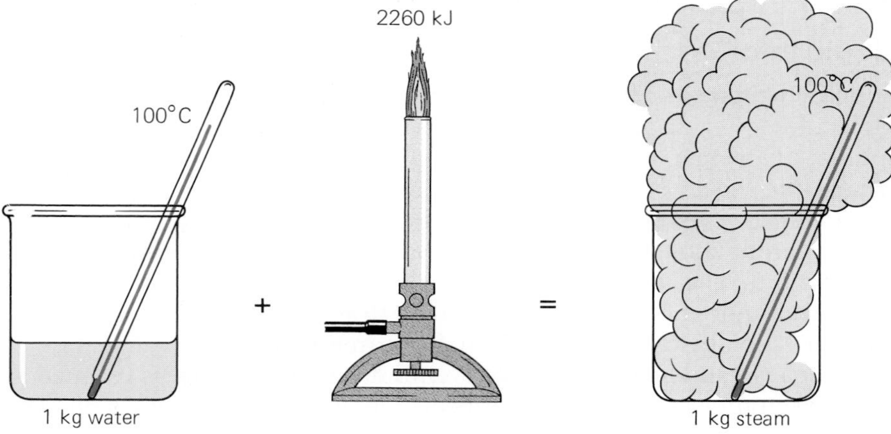

Pressure Cookers

Boiling points vary with pressure: the higher the pressure, the higher the boiling point. At twice sea-level atmospheric pressure, for instance, water boils at 120°C. The pressure cooker is based on this observation (Fig. 5-38). When water is heated in a closed container, the pressure in the container increases, and the temperature at which the water inside boils will be correspondingly higher than 100°C. In this way food can be cooked faster than in an open pan. Similarly, lowering the pressure reduces the boiling point. Atmospheric pressure decreases with altitude (see Fig. 14-1), so the boiling point of water in Denver, which is well above sea level, is 96°C; on top of the even higher Pikes Peak it is only 88°C.

energy. In the first case energy is supplied from the heat content of the liquid itself (since the liquid grows cooler), in the other case from the outside source of heat. For water at its boiling point of 100°C, 2260 kJ (the **heat of vaporization**) is needed to change each kilogram of liquid into gas (Fig. 5-37). With no difference in temperature between liquid and gas, there is no difference in their average molecular kinetic energies. If not into kinetic energy, into what form of molecular energy does the 2260 kJ of heat go?

Intermolecular forces provide the answer. In a liquid these forces are strong because the molecules are close together. To tear the molecules apart, to separate them by the wide distances that exist in the gas, requires that these strong forces be overcome. Each molecule must be moved, against the pull of its neighbors, to a new position in which their attraction for it is very small. Just as a stone thrown upward against the earth's gravity gains potential energy, so molecules moved apart in this way gain potential energy—potential energy with respect to intermolecular forces. When a gas becomes a liquid, the process is reversed. The molecules

Figure 5-38 A pressure cooker.

"fall" toward one another under the influence of their mutual attractions, and their potential energy is taken up as heat by the surroundings.

Example 5.13

The high heat of vaporization of water is what makes steam dangerous (Fig. 5-39). Compare the heat given to a person's skin when 1 g (about a third of a teaspoonful) of water at 100°C falls on it with the heat given by 1 g of steam at 100°C. Assume that the skin is at the normal body temperature of 37°C.

Solution

Since the specific heat of water is $c = 4.2$ kJ/kg · °C and here $\Delta T = (100°\text{C} - 37°\text{C}) = 63°\text{C}$, the heat given up by the hot water in cooling is

$$Q_1 = mc\,\Delta T = (0.001\ \text{kg})(4.2\ \text{kJ/kg}\cdot°\text{C})(63°\text{C}) = 0.26\ \text{kJ}$$

The heat given up by the steam is this amount plus the heat given up as the steam at 100°C condenses into water at 100°C, which is

$$Q_2 = (0.001\ \text{kg})(2260\ \text{kJ/kg}\cdot°\text{C}) = 2.26\ \text{kJ}$$

The total is $Q_1 + Q_2 = 2.52$ kJ—nearly 10 times as much as in the case of the hot water.

5.12 Melting

Solid into Liquid

Just as heat must be added at its boiling point to turn a liquid into a gas, so heat at its melting point is needed to turn a solid into a liquid. The heat required to change 1 kg of a solid at its melting point into a liquid is called the **heat of fusion** of the substance. The same amount of heat must be given off by 1 kg of the substance when it is a liquid at its melting point for it to harden into a solid. The heat of fusion of water is 335 kJ/kg (Fig. 5-40). Most other substances have lower heats of fusion (Table 5-4).

The heat of fusion of a substance is always much smaller than its heat of vaporization. The molecules of a solid are arranged in a fixed

Figure 5-39 Great care must always be taken near steam because the heat of vaporization of water is so great.

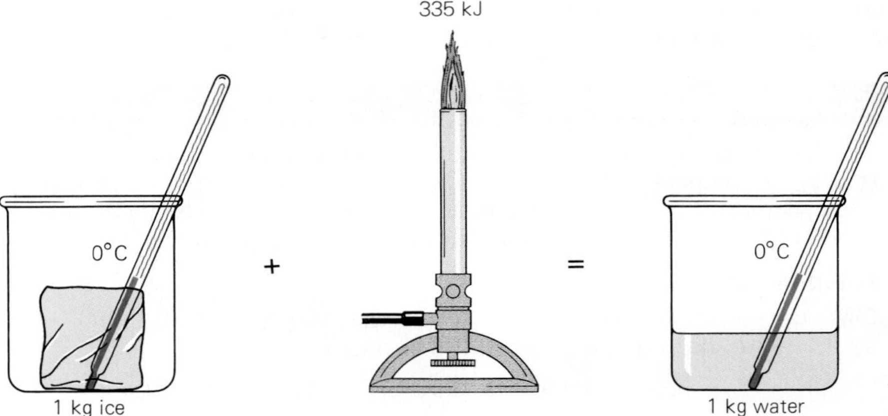

Figure 5-40 The heat of fusion of water is 335 kJ/kg.

Table 5–4 Some Heats of Fusion and Vaporization

	Melting point, °C	Heat of fusion, kJ/kg	Boiling point, °C	Heat of vaporization, kJ/kg
Copper	1083	134	1187	5069
Ethanol[1]	−114	105	78	854
Lead	330	25	1170	870
Mercury	−39	12	358	297
Nitrogen	−210	26	−196	201
Oxygen	−219	14	−183	213
Water	0	335	100	2260

[1]Ethanol is also known as ethyl alcohol.

pattern such that the forces holding each one to its neighbors are as large as possible. To overcome these forces and give the molecules the random, constantly shifting arrangement of a liquid, additional energy must be given to them (Fig. 5-41). However, the molecules still stick together sufficiently to give a liquid sample a definite volume. To turn a liquid into a gas, enough energy must be added to pull the molecules permanently apart, a much harder job. The molecules of the gas can then move

Example 5.14

An ice cube at 0°C is dropped to the ground and melts into water at 0°C. If all the original potential energy of the ice above the ground went into melting the ice, from what height did the ice cube fall?

Solution

If L is the heat of fusion of water, $Q = mL = \text{PE} = mgh$ and

$$h = \frac{\text{L}}{\text{g}} = \frac{335 \text{ kJ/kg}}{9.8 \text{ m/s}^2} = 34 \text{ km}$$

We did not have to know the mass of the ice cube.

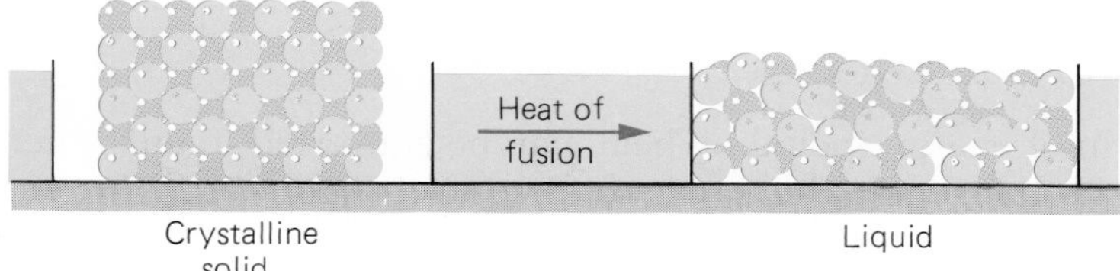

Figure 5-41 The orderly arrangement of particles in a crystalline solid changes to the random arrangement of particles in a liquid when enough energy is supplied to the solid to overcome the bonding forces within it.

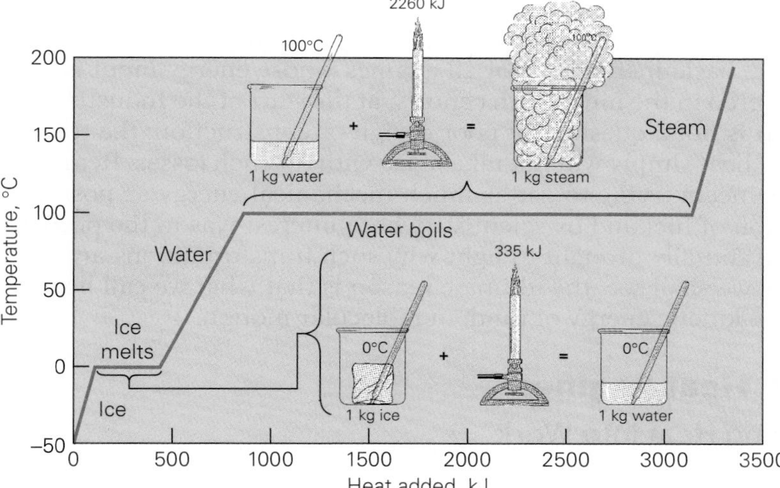

Figure 5-42 A graph of the temperature of 1 kg of water, originally ice at −50°C, as heat is added to it.

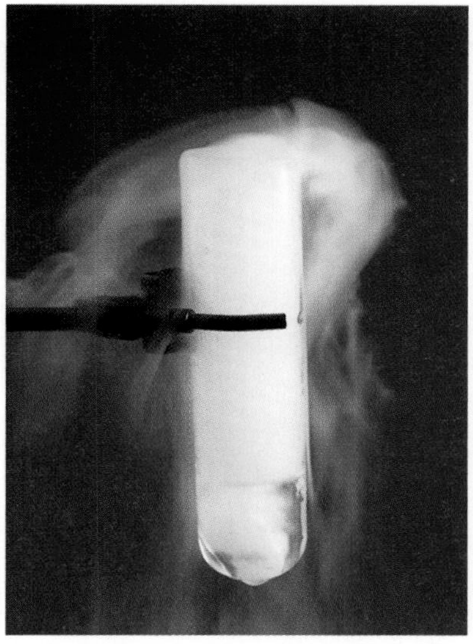

Figure 5-43 Solid carbon dioxide ("dry ice") vaporizes into a gas at atmospheric pressure, a process called sublimation.

about freely, and the gas expands. At atmospheric pressure 1 L of water becomes 1680 L of steam; in a vacuum, it would expand indefinitely.

Figure 5-42 shows what happens as we supply heat to 1 kg of ice originally at −50°C. The ice warms up until it reaches 0°C, when it begins to melt. Then the temperature remains steady at 0°C until all the ice has melted. When all the ice has become water, the temperature rises again. When the water reaches 100°C, it starts to turn into steam, which takes a lot of heat, much more than the ice needed to melt. Finally all the water has become steam at 100°C and the temperature of the steam now increases as more heat is added.

Sublimation Most substances change directly from the solid to the vapor state, a process called **sublimation,** under the right conditions of temperature and pressure. Usually pressures well under atmospheric pressure are needed for sublimation. A familiar exception is solid carbon dioxide ("dry ice"), which turns into a gas without first becoming a liquid, at temperatures above −79°C, even at atmospheric pressure (Fig. 5-43). Figure 5-44 summarizes all the changes of state we have been discussing.

The best instant coffee is prepared with the help of sublimation. The brewed coffee is first frozen and then put in a vacuum chamber. The ice in the frozen coffee sublimes to water vapor, which is pumped away. Freeze drying affects the flavor of coffee much less than drying it by heating. The same process is also used to preserve other materials of biological origin, such as blood plasma.

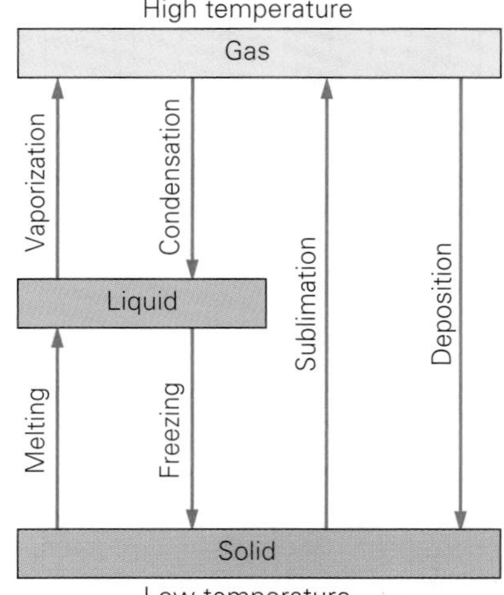

Figure 5-44 The various changes of state.

Energy Transformations

Any form of energy, including heat, can be converted to any other form. But heat is unusual in that it cannot be converted *efficiently*. We routinely obtain mechanical energy from the heat given off by burning coal and oil in engines of various types, but a large part of the heat is always wasted—about two-thirds in the case of electric power stations, for example. The losses are serious because nearly all the "raw" energy available to modern civilization is liberated from its sources as heat.

The basic inefficiency of all engines whose energy input is heat was discovered in the nineteenth century, at the start of the Industrial Revolution. It is not a question of poor design or construction; the transformation of heat simply will not take place without such losses. Research both by engineers trying to get as much mechanical energy as possible from each ton of fuel and by scientists whose interest was in the properties of heat eventually brought to light why such transformations are so wasteful. As we shall see, the ultimate reason is that what we call heat is actually the kinetic energy of random molecular motion.

5.13 Heat Engines

Turning Heat into Work

Heat is the easiest and cheapest form of energy to obtain, since all we have to do to obtain it is to burn an appropriate fuel. A device that turns heat into mechanical energy is called a **heat engine.** Examples are the gasoline and diesel engines of cars, the jet engines of aircraft, and the steam turbines of ships and power stations. All these engines operate in the same basic way: a gas is heated and allowed to expand against a piston or the blades of a turbine.

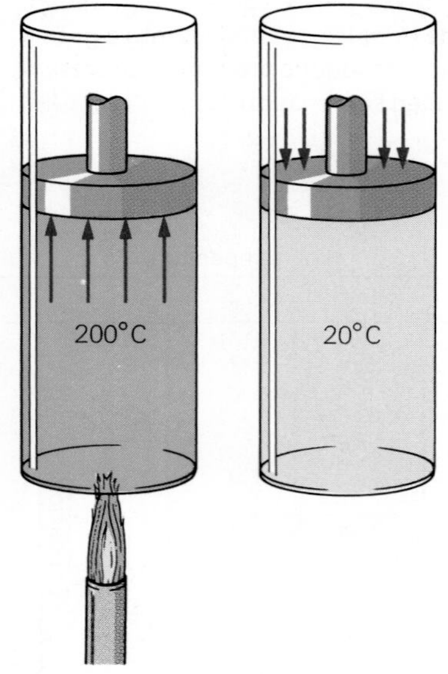

Figure 5-45 An idealized heat engine. A gas at 200°C gives out more energy in expanding than is required to compress the gas at 20°C. This excess energy is available for doing work.

Figure 5-45 shows a gas being heated in a cylinder. As the temperature of the gas increases, its pressure increases as well, and the piston is pushed upward. The energy of the upward-moving piston can be used to propel a car or turn a generator or for any other purpose we wish. When the piston reaches the top of the cylinder, however, the conversion of heat into mechanical energy stops. In order to keep the engine working, we must now push the piston back down again in order to begin another energy-producing expansion.

If we push the piston down while the gas in the cylinder is still hot, we will find that we have to do exactly as much work as the energy provided by the expansion. Thus there will be no net work done at all. To make the engine perform a net amount of work in each cycle, we must now cool the gas so that less work is needed to compress it. It is in this cooling process that heat is lost. There is no way to avoid throwing away some of the heat added to the gas in the expansion if the engine is to continue to work. The wasted heat usually ends up in the atmosphere around the engine, in the water of a nearby river, or in the ocean.

What happens in a complete cycle, then, is that heat flows in and out of the engine, and during the flow we manage to change some of the heat into mechanical energy, as in Fig. 5-46. Heat flows by itself from a hot reservoir to a cold one, so we need *both* reservoirs for a heat engine

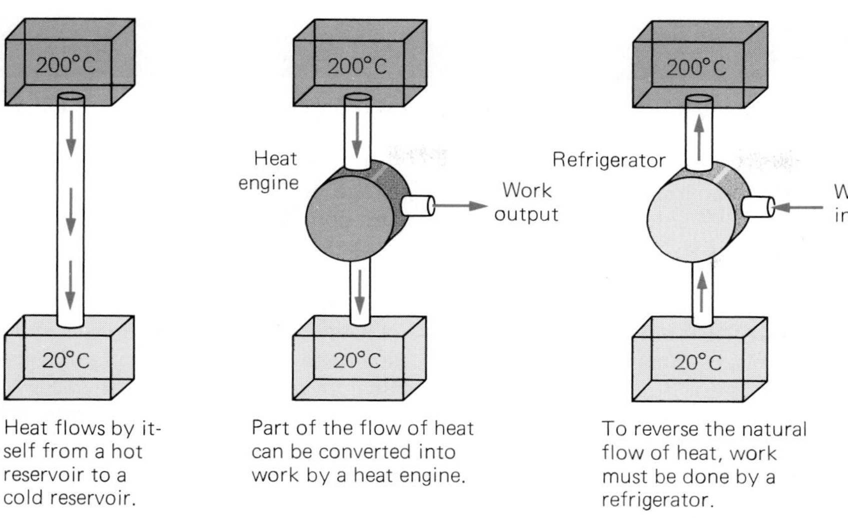

Figure 5-46 A heat engine converts part of the heat flowing from a hot reservoir to a cold one into work. A refrigerator extracts heat from a cold reservoir and delivers it to a hot one by doing work that is converted into heat.

to operate. In a gasoline or diesel engine the hot reservoir is the burning gases of the power stroke and the cold reservoir is the atmosphere.

Gasoline and Diesel Engines

The operating cycle of a four-stroke gasoline engine is shown in Fig. 5-47. In the intake stroke, a mixture of gasoline and air from the carburetor is sucked into the cylinder as the piston moves downward. In the compression stroke, the fuel-air mixture is compressed to one-seventh or one-eighth of its original volume. At the end of the compression stroke the spark plug is fired, which ignites the fuel-air mixture. The expanding gases force the piston downward in the power stroke. Finally the piston moves upward again to force the spent gases out through the exhaust valve.

In a diesel engine, only air is drawn into the cylinder in the intake stroke. At the end of the compression stroke, diesel fuel is injected into the cylinder and is ignited by the high temperature of the compressed air. No spark plug is needed. Diesel engines are more efficient because they have higher compression ratios than gasoline engines but are also heavier and more expensive.

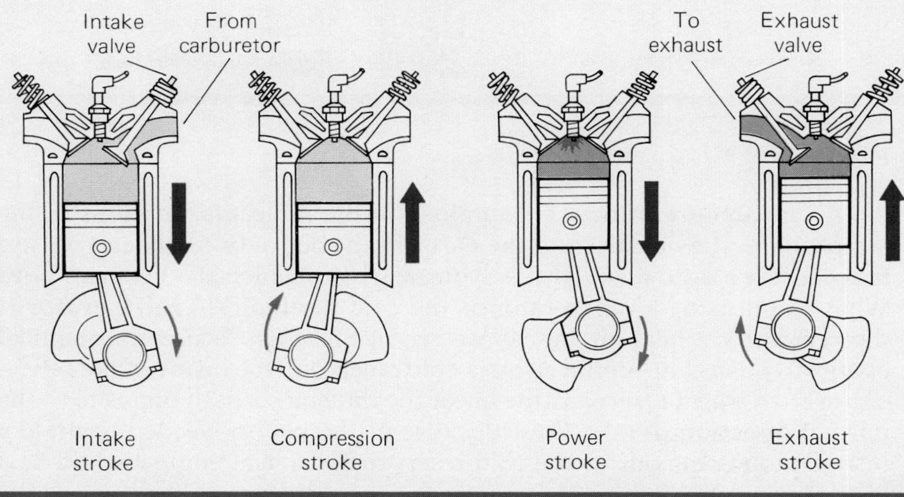

Figure 5-47 A four-stroke gasoline engine.

How a Refrigerator Works

A refrigerator takes in heat at a low temperature and exhausts it at a higher one. Figure 5-48 shows a refrigerator whose working substance is an easily liquefied gas called a **refrigerant.** The operation of the refrigeration system of Fig. 5-48 proceeds as follows:

1. The **compressor,** usually driven by an electric motor, brings the refrigerant to a high pressure, which raises its temperature as well.
2. The hot refrigerant passes through the **condenser,** an array of thin tubes that give off heat from the refrigerant to the atmosphere. The condenser is on the back of most household refrigerators. As it cools, the refrigerant becomes a liquid under high pressure.
3. The liquid refrigerant now goes into the **expansion valve,** from which it emerges at a lower pressure and temperature.
4. In the **evaporator** the cool liquid refrigerant absorbs heat from the storage chamber and vaporizes. Farther along in the evaporator the refrigerant vapor absorbs more heat and becomes warmer. The warm vapor then goes back to the compressor to begin another cycle.

In step 4 of this cycle, heat is extracted from the storage chamber by the refrigerant. In step 1, work is done on the refrigerant by the compressor. In step 2, heat from the refrigerant leaves the system. A refrigerator might remove two or more times as much heat from its storage chamber as the amount of work done.

A **heat pump** is a refrigeration system that takes heat from the cold outdoors in winter and delivers it to the interior of a house. The advantage of a heat pump is that it transfers more heat than the energy it uses in its operation. Thus a heat pump may be several times more efficient than an ordinary furnace. In summer the same heat pump can be used in reverse to serve as an air conditioner to take heat from the house and exhaust it to the warmer outdoors.

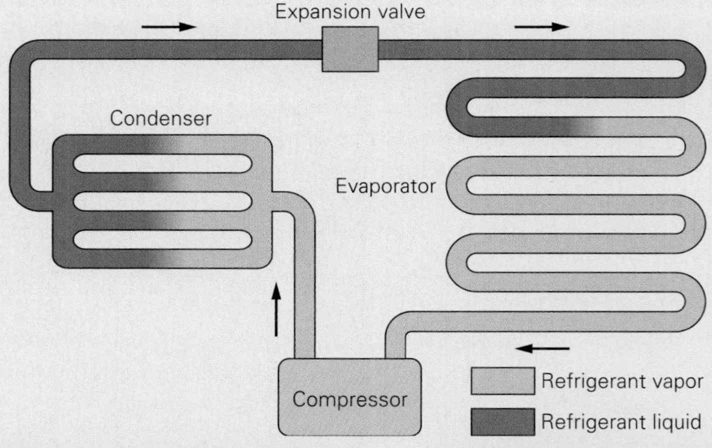

Figure 5-48 A typical refrigeration system. Heat is absorbed by the refrigerant from the storage chamber in the evaporator and is given up by the refrigerant in the condenser.

A vast amount of heat is contained in the molecular motions of the atmosphere, the oceans, and the earth itself, but only rarely can we use it because we need a colder reservoir nearby to which the heat can flow. What about using a refrigerator as the cold reservoir? A **refrigerator** is the reverse of a heat engine, as we see in Fig. 5-46. It uses mechanical energy to push heat "uphill" from a cold reservoir (the inside of the refrigerator) to a warm reservoir (the air of the kitchen), a path opposite to the normal direction of heat flow. Because of the energy needed to drive a refrigerator, using one as the cold reservoir for a heat engine would be a losing proposition.

5.14 Thermodynamics

You Can't Win

Thermodynamics is the science of heat transformation, and it has two fundamental laws:

1. Energy cannot be created or destroyed, but it can be converted from one form to another.
2. It is impossible to take heat from a source and change all of it to mechanical energy or work; some heat must be wasted.

The first law of thermodynamics is the same as the law of conservation of energy discussed in Chap. 3. What it means is that we can't get something for nothing. The second law singles out heat from other kinds of energy and recognizes that all conversions of heat into any of the others must be inefficient.

Thermodynamics is able to specify the maximum efficiency of a heat engine, ignoring losses to friction and other practical difficulties. The maximum efficiency turns out to depend only on the absolute temperatures T_{hot} and T_{cold} of the hot and cold reservoirs between which the engine operates:

$$\text{Maximum efficiency} = \left(\frac{\text{work output}}{\text{energy input}}\right)_{\text{maximum}}$$

$$\text{Eff(max)} = 1 - \frac{T_{\text{cold}}}{T_{\text{hot}}} \qquad \textit{Engine efficiency} \qquad 5\text{-}12$$

The greater the ratio between the two temperatures, the less heat is wasted and the more efficient the engine.

Figure 5-49 shows the basic design of a steam turbine. In a power station, the steam comes from a boiler heated by a coal, oil, or gas furnace or by a nuclear reactor, and the turbine shaft is connected to an electric generator (Fig. 5-50). In a typical power station, steam enters a turbine at about 570°C and leaves at about 95°C into a partial vacuum.

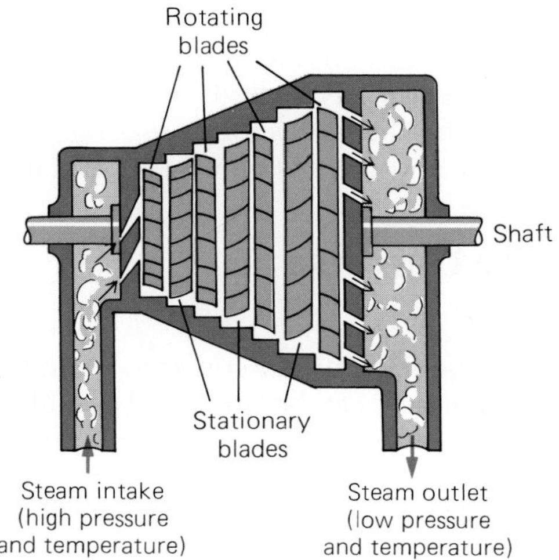

Figure 5-49 In a steam turbine, steam moves past several sets of rotating blades on the same shaft to obtain as much power as possible. The stationary blades direct the flow of steam in the most effective way.

Figure 5-50 The rotor of a steam turbine. Such a rotor might turn at 3600 revolutions per minute to produce perhaps 200 MW of power.

The corresponding absolute temperatures are 843 K and 368 K, so the maximum efficiency of such a turbine is

$$\text{Eff(max)} = 1 - \frac{T_{\text{cold}}}{T_{\text{hot}}} = 1 - \frac{368\text{ K}}{843\text{ K}} = 0.56$$

which is 56 percent. The actual efficiency is less than 40 percent because of friction and other sources of energy loss (Fig. 5-51).

Why a Heat Engine Must Be Inefficient On a molecular level, it is not hard to see why heat resists being changed into other forms of energy. When heat is added to the gas of a heat engine, its molecules increase their average speeds. But the molecules are moving in random directions, whereas the engine can draw upon the increased energies of only those molecules that are moving in more or less the same direction as the piston or turbine blades. If we could line up the molecules and aim them all, like miniature bullets, right at the piston or turbine blades, all the added energy could be turned into mechanical energy. Because this is impossible, only a fraction of any heat given to a gas can be extracted as energy of orderly motion. The nature of heat is responsible for the inefficiency of heat engines, and there is no way around it.

5.15 Fate of the Universe

Order into Disorder

Other kinds of energy can be entirely converted into heat, whereas only part of a given amount of heat can be converted the other way. As a result, there is an overall tendency toward an increase in the heat energy of the universe at the expense of the other kinds it contains. We see this tendency all around us in everyday life. When coal or oil burns in an engine, much of its chemical energy becomes heat; when any kind of machine is

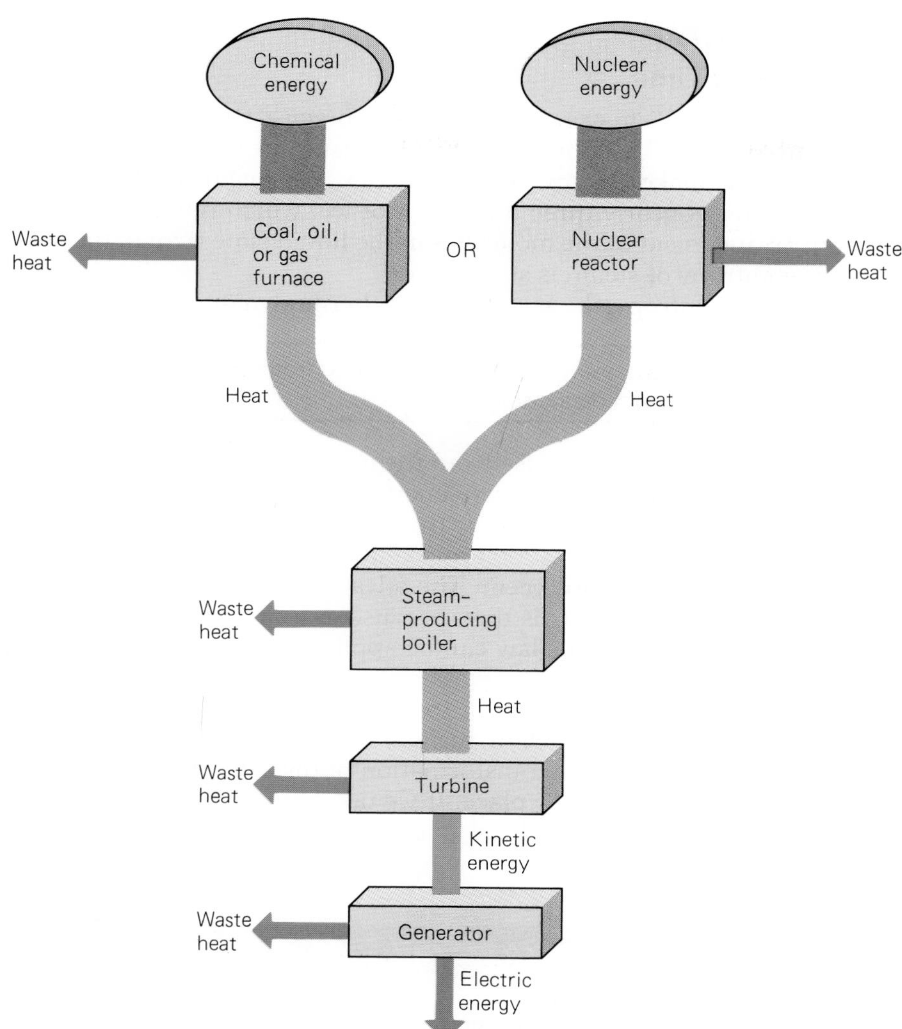

Figure 5-51 Energy flow in electric generating plants. A large part of the energy waste is due to the unavoidable thermodynamic inefficiency of the turbine.

operated, friction turns some of its energy into heat; an electric lightbulb emits heat as well as light; and so on. Most of the lost energy is dissipated in the atmosphere, the oceans, and the earth itself where it is largely unavailable for recovery.

In the world of nature a similar steady degradation of energy into unusable heat occurs. In the universe as a whole, the stars (for instance, the sun) constitute the hot reservoir and everything else (for instance, the earth) constitutes the cold reservoir from a thermodynamic point of view. As time goes on, the stars will grow cooler and the rest of the universe will grow warmer, so that less and less energy will be available to power the further evolution of the universe. On a molecular level, order will become disorder. If this process continues indefinitely, the entire universe will be at the same temperature and all its particles will have the same average energy. This condition is sometimes called the "heat death" of the universe. However, as we shall see in Sec. 19.9, this is not the only possible fate of the universe.

5.16 Entropy

The Arrow of Time

Although "disorder" may not seem a very precise concept, a quantity called **entropy** can be defined that is a measure of the disorder of the molecules that make up any body of matter. For instance, the entropy of liquid water is nearly three times that of ice, which reflects the more random arrangement of the molecules in the liquid state than in the solid state. The entropy of steam is still greater.

In terms of entropy, the second law of thermodynamics becomes

> The entropy of a system of some kind isolated from the rest of the universe cannot decrease.

If a puddle of water were to rise from the ground by itself while turning into ice, the process could conserve energy (and so obey the first law of thermodynamics) as the heat lost by the water becomes kinetic energy of the ice (Fig. 5-52). However, such an event would decrease the entropy of the water, and so cannot occur. The advantage of expressing the second law in terms of entropy is that, because entropy can be determined for a variety of systems, this law can be applied in an exact way to such systems.

Biological systems might seem to violate the second law. Certainly entropy decreases when a plant turns carbon dioxide and water into leaves and flowers. But this transformation of disorder into order needs the energy of sunlight to take place. If we take into account the increase in the entropy of the sun as it produces the required sunlight, the net

BIOGRAPHY **Ludwig Boltzmann (1844–1906)**

A native of Vienna, Boltzmann attended the university there. He then taught and carried out research at a number of institutions in Austria and Germany, moving from one to another every few years. Boltzmann was interested in poetry, music, and travel, and visited the United States three times, something unusual in those days. Of Boltzmann's many contributions to physics, the most important were to the kinetic theory of gases and to the foundations of thermodynamics. The constant k in the formula $KE_{av} = \frac{3}{2}kT$ for the average molecular energy in a gas at the absolute temperature T is called Boltzmann's constant in honor of his work. The mathematical relationship between molecular disorder and entropy was developed by Boltzmann; a monument to him in Vienna is inscribed with this relationship.

Boltzmann was a champion of the atomic theory of matter, still controversial in the late nineteenth century because there was then only indirect evidence for the existence of atoms and molecules. Battles with nonbelieving scientists deeply upset Boltzmann, and in his later years asthma, headaches, and poor eyesight further depressed his spirits. He committed suicide in 1906, not long after Albert Einstein published a paper on brownian motion that was to convince the remaining doubters of the correctness of the atomic theory.

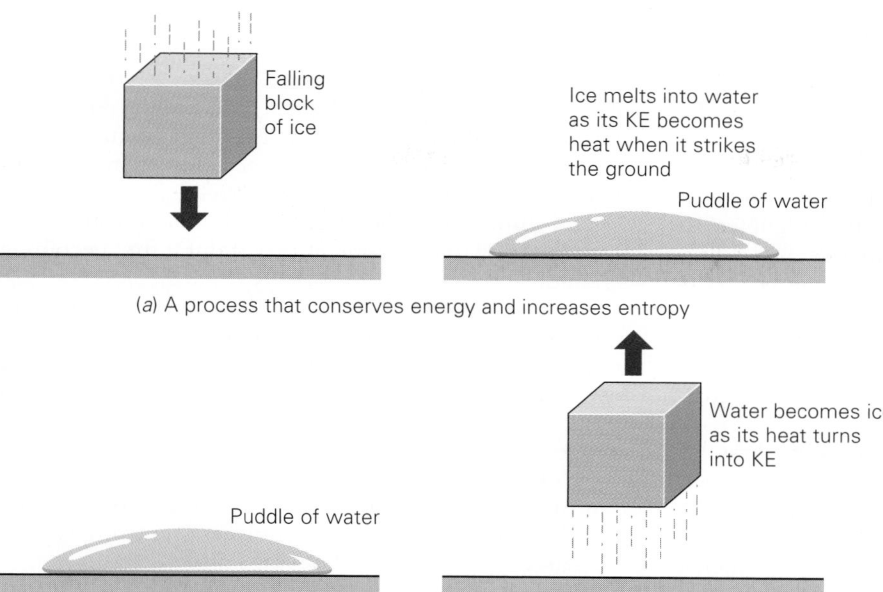

Figure 5-52 The second law of thermodynamics provides a way to distinguish between processes that conserve energy and (*a*) increase entropy, hence are possible, and those that (*b*) decrease entropy, hence are impossible.

result is an increase in the entropy of the universe. There is no way to avoid the second law.

The second law is an unusual physical principle in several respects. It does not apply to individual particles, only to assemblies of many particles. It does not say what can occur, only what cannot occur. And it is unique in that it is closely tied to the direction of time.

Events that involve only a few individual particles are always reversible. Nobody can tell whether a film of a billiard ball bouncing around is being run forward or backward. But events that involve many-particle systems are not always reversible. The film of an egg breaking when dropped makes no sense when run backward. The arrow of time always points in the direction of entropy increase. Time never runs backward; a broken egg never reassembles itself; in the universe as a whole, entropy—and the disorder it mirrors—marches on.

Important Terms and Ideas

Temperature is that property of a body of matter that gives rise to sensations of hot and cold; it is a measure of average molecular kinetic energy. **Heat** is molecular kinetic energy.

A **thermometer** is a device for measuring temperature. In the **Celsius scale,** the freezing point of water is given the value 0°C and the boiling point of water the value 100°C. In the **Fahrenheit scale,** these temperatures are given the values 212°F.

The **specific heat** of a substance is the amount of heat needed to change the temperature of 1 kg of the substance by 1°C.

In **conduction,** heat is carried from one place to another by molecular collisions. In **convection,** the transport is by the motion of a volume of hot fluid. **Radiation** transfers heat by means of **electromagnetic waves,** which require no material medium for their passage.

The **density** of a substance is its mass per unit volume.

The **pressure** on a surface is the perpendicular force per unit area acting on the surface. The unit of pressure is the **pascal,** which is equal to the newton/meter^2. The pressure at any point in a fluid depends on the weight of the fluid above the point as well as on any applied pressure.

According to **Archimedes' principle,** the upward **buoyant force** on an object immersed in a fluid is equal to the weight of fluid displaced by the object.

Boyle's law states that, at constant temperature, the volume of a gas sample is inversely proportional to its pressure.

The **absolute temperature scale** has its zero point at $-273°C$; temperatures in this scale are designated K. **Absolute zero** is $0\ K = -273°C$. **Charles's law** states that, at constant pressure, the volume of a gas sample is directly proportional to its absolute temperature.

The **ideal gas law**—which states that $pV/T =$ constant for a gas sample regardless of changes in p, V, and T—is a combination of Boyle's and Charles's laws and is approximately obeyed by all gases.

According to the **kinetic theory of matter,** all matter consists of tiny individual **molecules** that are in constant random motion. The ideal gas law can be explained by the kinetic theory on the basis that the absolute temperature of a gas is proportional to the average kinetic energy of its molecules. At absolute zero, gas molecules would have no kinetic energy.

The **heat of vaporization** of a substance is the amount of heat needed to change 1 kg of it at its boiling point from the liquid to the gaseous state. The **heat of fusion** of a substance is the amount of heat needed to change 1 kg of it at its melting point from the solid to the liquid state.

Sublimation is the direct conversion of a substance from the solid to the vapor state without it first becoming a liquid.

A **heat engine** is a device that converts heat into mechanical energy or work. The **first law of thermodynamics** is the law of conservation of energy. **The second law of thermodynamics** states that some of the heat input to a heat engine must be wasted in order for the engine to operate.

Entropy is a measure of the disorder of the particles that make up a body of matter. In a system of any kind isolated from the rest of the universe, entropy cannot decrease.

Important Formulas

Temperature scales: $T_F = \frac{9}{5}T_C + 32°$

$T_C = \frac{5}{9}(T_F - 32°)$

Amount of heat: $Q = mc\,\Delta T$

Density: $d = \dfrac{m}{V}$

Pressure: $p = \dfrac{F}{A}$

Boyle's law: $\dfrac{p_1}{p_2} = \dfrac{V_2}{V_1}$ (at constant temperature)

Absolute temperature scale: $T_K = T_C + 273$

Charles's law: $\dfrac{V_1}{V_2} = \dfrac{T_1}{T_2}$ (at constant pressure; temperature on absolute scale)

Ideal gas law: $\dfrac{p_1V_1}{T_1} = \dfrac{p_2V_2}{T_2}$ (temperature on absolute scale)

Maximum efficiency of heat engine:

$\text{Eff(max)} = 1 - \dfrac{T_{\text{cold}}}{T_{\text{hot}}}$ (temperature on absolute scale)

Multiple Choice

1. Two thermometers, one calibrated in °F and the other in °C, are used to measure the same temperature. The numerical reading on the Fahrenheit thermometer
  a. is less than that on the Celsius thermometer
  b. is equal to that on the Celsius thermometer
  c. is greater than that on the Celsius thermometer
  d. may be any of these, depending on the temperature

2. One gram of steam at 100°C causes a more serious burn than 1 g of water at 100°C because the steam
  a. is less dense
  b. strikes the skin with greater force
  c. has a higher specific heat
  d. contains more energy

3. Heat transfer in a gas can occur by
  a. radiation only
  b. convection only
  c. radiation and convection only
  d. radiation, convection, and conduction

4. Heat transfer in a vacuum can occur by
  a. radiation only
  b. convection only
  c. radiation and convection only
  d. radiation, convection, and conduction

5. The fluid at the bottom of a container is
  a. under less pressure than the fluid at the top
  b. under the same pressure as the fluid at the top
  c. under more pressure than the fluid at the top
  d. any of these, depending upon the circumstances

6. The pressure of the earth's atmosphere at sea level is due to
   a. the gravitational attraction of the earth for the atmosphere
   b. the heating of the atmosphere by the sun
   c. the fact that most living things constantly breathe air
   d. evaporation of water from the seas and oceans
7. A cake of soap placed in a bathtub of water sinks. The buoyant force on the soap is
   a. 0
   b. less than its weight
   c. equal to its weight
   d. more than its weight
8. The density of freshwater is 1.00 g/cm^3 and that of seawater is 1.03 g/cm^3. A ship will float
   a. higher in freshwater than in seawater
   b. lower in freshwater than in seawater
   c. at the same level in freshwater and seawater
   d. any of the above, depending on the shape of its hull
9. An ice cube whose center consists of liquid water is floating in a glass of water. When the ice melts, the level of water in the glass
   a. rises
   b. remains the same
   c. falls
   d. any of the above, depending on the relative volume of water inside the ice cube
10. A person stands on a very sensitive scale and inhales deeply. The reading on the scale
    a. increases
    b. does not change
    c. decreases
    d. any of the above, depending on how the expansion of the person's chest compares with the volume of air inhaled
11. At constant pressure, the volume of a gas sample is directly proportional to
    a. the size of its molecules
    b. its Fahrenheit temperature
    c. its Celsius temperature
    d. its absolute temperature
12. Which of the following statements is not correct?
    a. Matter is composed of tiny particles called molecules.
    b. These molecules are in constant motion, even in solids.
    c. All molecules have the same size and mass.
    d. The differences between the solid, liquid, and gaseous states of matter lie in the relative freedom of motion of their respective molecules.
13. Molecular motion is not responsible for
    a. the pressure exerted by a gas
    b. Boyle's law
    c. evaporation
    d. buoyancy
14. Absolute zero may be regarded as that temperature at which
    a. water freezes
    b. all gases become liquids
    c. all substances become solid
    d. molecular motion in a gas would be the minimum possible
15. On the molecular level, heat is
    a. kinetic energy
    b. potential energy
    c. rest energy
    d. all of these, in proportions that depend on the circumstances
16. At a given temperature
    a. the molecules in a gas all have the same average velocity
    b. the molecules in a gas all have the same average energy
    c. light gas molecules have lower average energies than heavy gas molecules
    d. heavy gas molecules have lower average energies than light gas molecules
17. The temperature of a gas sample in a container of fixed volume is raised. The gas exerts a higher pressure on the walls of its container because its molecules
    a. lose more PE when they strike the walls
    b. lose more KE when they strike the walls
    c. are in contact with the walls for a shorter time
    d. have higher average velocities and strike the walls more often
18. The volume of a gas sample is increased while its temperature is held constant. The gas exerts a lower pressure on the walls of its container because its molecules strike the walls
    a. less often
    b. with lower velocities
    c. with less energy
    d. with less force
19. When evaporation occurs, the liquid that remains is cooler because
    a. the pressure on the liquid decreases
    b. the volume of the liquid decreases
    c. the slowest molecules remain behind
    d. the fastest molecules remain behind

20. When a vapor condenses into a liquid,
a. its temperature rises
b. its temperature falls
c. it absorbs heat
d. it gives off heat

21. Food cooks more rapidly in a pressure cooker than in an ordinary pot with a loose lid because
a. the pressure forces heat into the food
b. the high pressure lowers the boiling point of water
c. the high pressure raises the boiling point of water
d. the tight lid keeps the heat inside the cooker

22. A heat engine takes in heat at one temperature and turns
a. all of it into work
b. some of it into work and rejects the rest at a lower temperature
c. some of it into work and rejects the rest at the same temperature
d. some of it into work and rejects the rest at a higher temperature

23. In any process, the maximum amount of heat that can be converted to mechanical energy
a. depends on the amount of friction present
b. depends on the intake and exhaust temperatures
c. depends on whether kinetic or potential energy is involved
d. is 100 percent

24. In any process, the maximum amount of mechanical energy that can be converted to heat
a. depends on the amount of friction present
b. depends on the intake and exhaust temperatures
c. depends on whether kinetic or potential energy is involved
d. is 100 percent

25. A frictionless heat engine can be 100 percent efficient only if its exhaust temperature is
a. equal to its input temperature
b. less than its input temperature
c. 0°C
d. 0 K

26. The physics of a refrigerator most closely resembles the physics of
a. a heat engine
b. the melting of ice
c. the freezing of water
d. the evaporation of water

27. The working substance (or refrigerant) used in most refrigerators is a
a. gas that is easy to liquify
b. gas that is hard to liquify
c. liquid that is easy to solidify
d. liquid that is hard to solidify

28. Heat is absorbed by the refrigerant in a refrigerator when it
a. melts
b. vaporizes
c. condenses
d. is compressed

29. The heat a refrigerator absorbs from its contents is
a. less than it gives off
b. the same amount it gives off
c. more than it gives off
d. any of these, depending on its design

30. The second law of thermodynamics does not lead to the conclusion that
a. on a molecular level, order will eventually become disorder in the universe
b. all the matter in the universe will eventually end up at the same temperature
c. no heat engine can convert heat into work with 100 percent efficiency
d. the total amount of energy in the universe, including rest energy, is constant

31. The greater the entropy of a system of particles,
a. the less the energy of the system
b. the more the energy of the system
c. the less the order of the system
d. the more the order of the system

32. Ethyl alcohol boils at 172°F. The Celsius equivalent of this temperature is
a. 64°C
b. 78°C
c. 140°C
d. 278°C

33. A temperature of 20°C is the same as
a. −20.9°F
b. −6.4°F
c. 68°F
d. 94°F

34. The heat needed to warm 8 kg of water from 20°C to 70°C is
a. 400 kJ
b. 420 kJ
c. 1680 kJ
d. 2016 kJ

35. When 1 kg of steam at 200°C loses 3 MJ of heat, the result is
a. ice
b. water and ice

c. water
d. water and steam

36. Fifty kilojoules of heat are removed from 2 kg of ice initially at −5°C. The final temperature of the ice is
a. −6°C
b. −11°C
c. −12°C
d. −17°C

37. A 400-kg concrete block has the dimensions 1 m × 0.6 m × 0.3 m. Its density is
a. 72 kg/m^3
b. 222 kg/m^3
c. 667 kg/m^3
d. 2222 kg/m^3

38. The block of Multiple Choice 37 can exert three different pressures on a horizontal surface, depending upon which face it rests on. The highest pressure is
a. 0.7 kPa
b. 2.2 kPa
c. 13.1 kPa
d. 21.8 kPa

39. An object suspended from a spring scale is lowered into a pail filled to the brim with water, and 4 N of water overflows. The scale shows that the object weighs 6 N in the water. The weight in air of the object is
a. 2 N
b. 4 N
c. 6 N
d. 10 N

40. A wooden plank 200 cm long, 30 cm wide, and 40 mm thick floats in water with 10 mm of its thickness above the surface. The mass of the board is
a. 1.8 kg
b. 18 kg
c. 24 kg
d. 176 kg

41. The pressure on 100 liters of helium is increased from 100 kPa to 400 kPa. The new volume of the helium is
a. 25 liters
b. 50 liters
c. 400 liters
d. 1600 liters

42. Lead melts at 330°C. On the absolute scale this temperature corresponds to
a. 57 K
b. 362 K
c. 571 K
d. 603 K

43. At which of the following temperatures would the molecules of a gas have twice the average kinetic energy they have at room temperature, 20°C?
a. 40°C
b. 80°C
c. 313°C
d. 586°C

44. A heat engine absorbs heat at a temperature of 127°C and exhausts heat at a temperature of 77°C. Its maximum efficiency is
a. 13 percent
b. 39 percent
c. 61 percent
d. 88 percent

45. If it is to be 40 percent efficient, a heat engine that exhausts heat at 350 K must absorb heat at a temperature no less than
a. 210 K
b. 583 K
c. 875 K
d. 1038 K

Exercises

5.1 Temperature

1. Running hot water over the metal lid of a glass jar makes it easier to open the jar. Why?
2. When a mercury-in-glass thermometer is heated, its mercury column goes down briefly before rising. Why?
3. Three iron bars are heated in a furnace to different temperatures. One of them glows white, another yellow, and the third red. Which is at the highest temperature? The lowest?
4. Why do you think the Celsius temperature scale is sometimes called the centigrade scale?
5. The normal temperature of the human body is 37°C. What is this temperature on the Fahrenheit scale?
6. What is the Celsius equivalent of a temperature of 100°F?
7. Dry ice (solid carbon dioxide) starts to vaporize into a gas at −112°F. What is this temperature on the Celsius scale?
8. You have a Fahrenheit thermometer in your left hand and a Celsius thermometer in your right hand. Both thermometers show the same reading in degrees. What is the temperature?

5.2 Heat

9. Why is a piece of ice at 0°C more effective in cooling a drink than the same mass of cold water at 0°C?

10. Would it be more efficient to warm your bed on a cold night with a hot water bottle that contains 1 kg of water at 50°C or with a 1-kg gold bar at 50°C? Why?

11. A cup of hot coffee can be cooled by placing a cold spoon in it. A spoon of which of the following materials would be most effective for this purpose? Assume the spoons all have the same mass: aluminum, copper, iron, glass.

12. You can safely put your hand inside a hot oven for a short time, but even a momentary contact with the metal walls of the oven will cause a burn. Explain.

13. Outdoors in winter, why does the steel blade of a shovel feel colder than its wooden handle?

14. A thermos bottle consists of two glass vessels, one inside the other, with air removed from the space between them. The vessels are both coated with thin metal films. Why is this device so effective in keeping its contents at a constant temperature?

15. How many kJ of heat are needed to raise the temperature of 200 g of water from 20°C to 100°C in preparing a cup of coffee?

16. The specific heat of granite is 0.80 kJ/kg · °C. If 2 MJ of heat are added to a 100-kg granite statue of James Prescott Joule that is originally at 20°C, what is the final temperature of the statue?

17. The diet of a 60-kg person provides 12,000 kJ daily. If this amount of energy were added to 60 kg of water, by how much would its temperature be increased?

18. Forty kilojoules of heat were added to a 500-g piece of wood whose temperature then rose from 20°C to 68°C. What was the specific heat of the wood?

19. The average specific heat of a certain 25-kg storage battery is 0.84 kJ/kg · °C. When it is fully charged, the battery contains 1.4 MJ of electric energy. If all of this energy were dissipated inside the battery, by how much would its temperature increase?

20. An essential part of a home solar heating system is a way to store heat for use at night and on cloudy days. In a certain system, the water used for storage is initially at 70°C and it is required to provide an average of 7.5 kW to keep the house at an average of 18°C for 3 days. How much water is needed?

21. A 10-kg stone is dropped into a pool of water from a height of 100 m. How much energy in joules does the stone have when it strikes the water? If all this energy goes into heat and if the pool contains 10 m^3 of water, by how much is its temperature raised? (The mass of 1 m^3 of water is 10^3 kg.)

5.4 Density

22. Why do tables of densities always include the temperature for which the listed values hold? What would be true of the densities of most solids and liquids at a temperature higher than the quoted one?

23. A room is 5 m long, 4 m wide, and 3 m high. What is the mass of the air it contains?

24. A 2.0-kg brass monkey has a volume of 240 cm^3. What is the density of brass?

25. A 50-g bracelet is suspected of being gold-plated lead instead of pure gold. It is dropped into a full glass of water and 4 cm^3 of water overflows. Is the bracelet pure gold?

26. A 1200-kg concrete slab that measures 2 m × 1 m × 20 cm is delivered to a building under construction. Does the slab contain steel reinforcing rods or is it plain concrete?

27. Mammals have approximately the same density as freshwater. (a) Find the volume in liters of a 55-kg woman and (b) the volume in cubic meters of a 140,000-kg blue whale. (Note: 1 liter $= 10^{-3}$ m^3 $= 0.001$ m^3.)

28. A cube of gold 30 mm long on each edge (the size of an ice cube) is worth about $10,000. How much is a gram of gold worth?

29. The radius of the earth is 6.37×10^6 m and its mass is 5.98×10^{24} kg. (a) Find the average density of the earth. (b) The average density of the rocks at the earth's surface is 2.7×10^3 kg/m^3. What must be true of the matter of which the earth's interior is composed? Is it likely that the earth is hollow and peopled by another species, as the ancients believed? (Note: The volume of a sphere of radius R is $\frac{4}{3}\pi R^3$.)

5.5 Pressure

30. Some water is boiled briefly in an open metal can. The can is then sealed while still hot. Why does the can collapse when it cools?

31. When a person drinks a soda through a straw, where does the force come from that causes the soda to move upward?

32. A U-shaped tube contains water and an unknown liquid separated by mercury, as in Fig. 5-53. How does the density of the liquid compare with the density of water? How do the pressures at A and B compare?

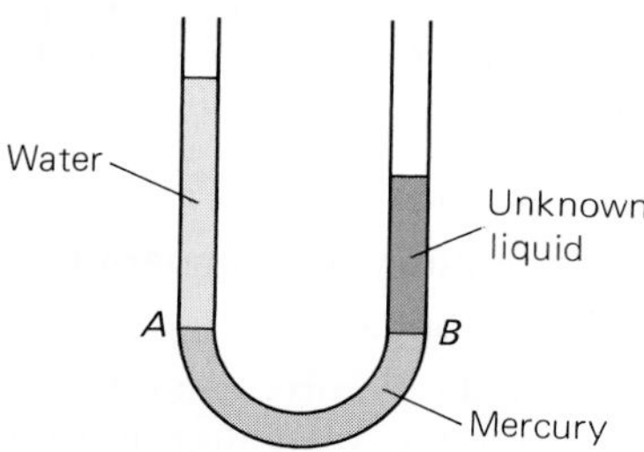

Figure 5-53

33. The three containers shown in Fig. 5-54 are filled with water to the same height. Compare the pressures at the bottoms of the containers.

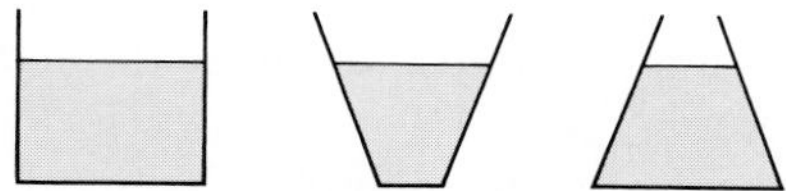

Figure 5-54

34. The two upper front teeth of a person exert a total force of 30 N on a piece of steak. If the biting edge of each tooth measures 10 mm $\times$ 1 mm, find the pressure on the steak.

35. A tire pump has a piston whose cross-sectional area is 0.001 m^2. If a force of 150 N is applied to the piston, find the pressure on the air in the pump.

36. A nail whose cross-sectional area is 3 mm^2 is embedded in a tire in which the air pressure is 1.8 bar. How much force tends to push the nail out?

37. A 1200-lb car is equally supported by its 4 tires, which are inflated to a pressure of 2 bar. What area of each tire is in contact with the road?

38. The smallest bone in the index finger of a 75-kg circus acrobat has a cross-sectional area of 0.5 cm^2 and breaks under a pressure of 1.7×10^8 Pa. Is it safe for the acrobat to balance his entire weight on this finger?

39. A hypodermic syringe whose cylinder has a cross-sectional area of 60 mm^2 is used to inject a liquid medicine into a patient's vein in which the blood pressure is 2 kPa. (a) What is the minimum force needed on the plunger of the syringe? (b) Why is the cross-sectional area of the needle irrelevant?

40. At the distance of the earth from the sun, the pressure exerted by sunlight is about 10^{-5} Pa. How much force does this pressure exert on a 10 m^2 array of solar cells on a satellite when the array is perpendicular to the direction of the sunlight? If the satellite has a mass of 200 kg, what acceleration does this force give it?

5.6 Buoyancy

41. Why does buoyancy occur? Under what circumstances does an object float in a fluid?

42. Suppose the pressure in a liquid did not increase with depth. Would anything float in such a liquid?

43. A wooden block is submerged in a tank of water and pressed down against the bottom of the tank so that there is no water underneath it. The block is released. Will it rise to the surface or stay where it is?

44. A jar is filled to the top with water, and a piece of cardboard is slid over the opening so that there is only water in the jar. If the jar is turned over, will the cardboard fall off? What will happen if there is any air in the jar?

45. A ship catches fire, and its steel hull expands as it heats up. What happens to the volume of water the ship displaces? What happens to the height of the ship's deck above the water?

46. A bridge in Sweden carries the Göta Canal over a highway. What, if anything, happens to the load on the bridge when a boat passes across it in the canal?

47. As Table 5-3 shows, ice has a lower density than water, which is why ice floats. What will happen when an ice cube floating in a glass of water filled to the brim begins to melt?

48. An ice cube with an air bubble inside it is floating in a glass of water. Compare what happens to the water level in the glass when the ice melts with what would happen if the cube had no bubble in it.

49. A sailboat has a lead or iron keel to keep it upright despite the force wind exerts on its sails. What difference, if any, is there between the stability of a sailboat in freshwater and in seawater?

50. An aluminum canoe is floating in a swimming pool. After a while it begins to leak and sinks to the bottom of the pool. What, if anything, happens to the water level in the pool?

51. How much force is needed to support a 100-kg iron anchor when it is submerged in seawater? (Hint: First find the anchor's volume.)

52. A 50-kg girl dives off a raft 2 m square floating in a freshwater lake. By how much does the raft rise?

53. A raft 3 m long, 2 m wide, and 30 cm thick, made from solid balsa wood, is floating in a freshwater lake. How many 65-kg people can it support?

54. An iceberg is a body of freshwater ice that is floating in seawater. What proportion of the volume of an iceberg is submerged? (Hint: Set equal the weights of the iceberg and of the volume of seawater it displaces and then solve for $V_{submerged}/V_{iceberg}$.)

55. A 200-L iron tank has a mass of 36 kg. (a) Will it float in seawater when empty? (b) When filled with freshwater? (c) When filled with gasoline?

5.7 The Gas Laws

56. What are the equivalents of 0 K, 0°C, and 0°F in the other temperature scales?

57. A certain quantity of hydrogen occupies a volume of 1000 cm^3 at 0°C (273 K) and ordinary atmospheric pressure. (a) If the pressure is tripled but the temperature is held constant, what will the volume of the hydrogen be? (b) If the temperature is increased to 273°C but the pressure is held constant, what will the volume of the hydrogen be?

58. A tire contains air at a pressure of 2 bar at 15°C. If the tire's volume is unchanged, what will the air pressure in it be when the tire warms up to 40°C as the car is driven?

59. An oxygen cylinder used for welding contains 40 L of oxygen at 20°C and a pressure of 17 MPa. The density of oxygen is 1.4 kg/m^3 at 20°C and atmospheric pressure of 101 kPa. Find the mass of the oxygen in the cylinder.

60. A weather balloon carries instruments that measure temperature, pressure, and humidity as it rises through the atmosphere. Suppose such a balloon has a volume of 1.2 m^3 at sea level where the pressure is 1 atm and the temperature is 20°C. When the balloon is at an altitude of 11 km (36,000 ft) the pressure is down to 0.5 atm and the temperature is about −55°C. What is the volume of the balloon then?

61. To what Celsius temperature must a gas sample initially at 20°C be heated if its volume is to double while its pressure remains the same?

62. The propellant gas that remains in an empty can of spray paint is at atmospheric pressure. If such a can at 20°C is thrown into a fire and is heated to 600°C, how many times atmospheric pressure is the new pressure inside the can? (The higher pressure may burst the can, which is why such cans should not be thrown into a fire.)

63. An air tank used for scuba diving has a safety valve set to open at a pressure of 28 MPa. The normal pressure of the full tank at 20°C is 20 MPa. If the tank is heated after being filled to 20 MPa, at what temperature will the safety valve open?

5.8 Kinetic Theory of Gases

5.9 Molecular Motion and Temperature

5.10 Liquids and Solids

64. A glass of water is stirred and then allowed to stand until the water stops moving. What becomes of the KE of the moving water?

65. Is it meaningful to say that an object at a temperature of 200°C is twice as hot as one at 100°C?

66. Gas molecules have speeds comparable with those of rifle bullets, yet it is observed that a gas with a strong odor (ammonia, for instance) takes a few minutes to diffuse through a room. Why?

67. At absolute zero, a sample of an ideal gas would have zero volume. Why would this not be true of an actual gas at absolute zero?

68. The pressure on a sample of hydrogen is doubled, while its temperature is kept unchanged. What happens to the average speed of the hydrogen molecules?

69. When they are close together, molecules attract one another slightly. As a result of this attraction, are gas pressures higher or lower than expected from the ideal gas law?

70. Temperatures in both the Celsius and Fahrenheit scales can be negative. Why is a negative temperature impossible on the absolute scale?

71. How can the conclusion of kinetic theory that molecular motion occurs in solids be reconciled with the observation that solids have definite shapes and volumes?

72. A 1-L tank holds 1 g of hydrogen at 0°C and another 1-L tank holds 1 g of oxygen at 0°C. The mass of an oxygen molecule is 16 times the mass of a hydrogen molecule. (a) Do the tanks hold the same number of molecules? If not, which holds more? (b) Do the gases exert the same pressure? If not, which exerts the greater pressure? (c) Do the molecules in the tanks have the same average energy? If not, in which are the average energies greater? (d) Do the molecules in the tanks have the same average speeds? If not, in which are the average speeds greater?

73. To what temperature must a gas sample initially at 27°C be raised in order for the average energy of its molecules to double?

74. A tank holds 1 kg of nitrogen at 25°C and a pressure of 100 kPa. What happens to the pressure when 3 kg of nitrogen are added to the tank at the same temperature? Explain your answer in terms of kinetic theory of gases.

75. A mixture of the gases hydrogen and carbon dioxide (CO_2) is at 20°C. A CO_2 molecule has about 22 times the mass of a hydrogen molecule, whose average speed at 20°C is about 1.6 km/s. What is the average speed of a CO_2 molecule in the mixture?

5.11 Evaporation and Boiling

76. Why does evaporation cool a liquid?

77. Why does blowing across hot coffee cool it down?

78. If you wish to speed up the rate at which potatoes are cooking in a pan of boiling water, would it be better to turn up the gas flame or use a pressure cooker?

79. Give as many methods as you can think of that will increase the rate of evaporation of a liquid sample. Explain why each method will have this effect.

80. How much heat is given off when 1 kg of steam at 100°C condenses and cools to water at 20°C?

81. A total of 500 kJ of heat are added to 1 kg of water at 20°C. What is the final temperature of the water? If it is 100°C, how much steam, if any, is produced?

82. Many power stations get rid of their waste heat by using it to boil water and allowing the resulting steam to escape into the atmosphere via a cooling tower. How much water would a power station need per second to dispose of 1000 MW of waste heat? Consider only the vaporization of the water; the heat used to raise the water to the boiling point is much less.

83. Solar energy arrives at a rate of 1.2 kW at a reflecting dish with an area of 1 m^2 that focuses the light on a jar that contains 100 g of water initially at 20°C. How long will it take the water to reach 100°C and then boil away if no energy is lost?

5.12 Melting

84. If all the heat lost by 1 kg of water at 0°C when it turns into ice at 0°C could be turned into kinetic energy, what would the speed of the ice be?

85. Water at 50°C can be obtained by mixing together which one or more of the following? Which of the others would have final temperatures higher than 50°C and which lower than 50°C?

   a. 1 kg of ice at 0°C and 1 kg of steam at 100°C
   b. 1 kg of ice at 0°C and 1 kg of water at 100°C
   c. 1 kg of water at 0°C and 1 kg of steam at 100°C
   d. 1 kg of water at 0°C and 1 kg of water at 100°C

5.13 Heat Engines

5.14 Thermodynamics

86. Why are both a hot and a cold reservoir needed for a heat engine to operate?

87. The oceans contain an immense amount of heat energy. Why can a submarine not make use of this energy for propulsion?

88. A person tries to cool a kitchen by switching on an electric fan and closing the kitchen door and windows. What will happen?

89. In another attempt to cool the kitchen, the person leaves the refrigerator door open, again with the kitchen door and windows closed. Now what will happen?

90. Is it correct to say that a refrigerator "produces cold"? If not, why not?

91. The first law of thermodynamics is the same as the law of conservation of energy. Is there any law of nature that is the same as the second law of thermodynamics?

92. An engine that operates at the maximum efficiency possible takes in 1.0 MJ of heat at 327°C and exhausts waste heat at 127°C. How much work does it perform?

93. An engine that operates between 2000 K and 700 K has an efficiency of 40 percent. What percentage of its maximum possible efficiency is this?

94. An engine is proposed that is to operate between 200°C and 50°C with an efficiency of 35 percent. Will the engine perform as predicted? If not, what would its maximum efficiency be?

95. Surface water in a tropical ocean is typically at 27°C whereas at a depth of a kilometer or more it is at only about 5°C. It has been proposed to operate heat engines using surface water as the hot reservoir and deep water (pumped to the surface) as the cold reservoir. What would the maximum efficiency of such an engine be? Why

might such an engine eventually be a practical proposition even with so low an efficiency?

96. Three designs for an engine to operate between 450 K and 300 K are proposed. Design *A* is claimed to require a heat input of 800 J for each 1000 J of work output, design *B* a heat input of 2500 J, and design *C* a heat input of 3500 J. Which design would you choose and why?

5.16 Entropy

97. The evolution of today's animals from their primitive ancestors of billions of years ago represents a large increase in order, which corresponds to a decrease in entropy. How can this be reconciled with the second law of thermodynamics?

98. When salt is dissolved in water, do you think the entropy of the system of salt + water increases or decreases? Why?

Answers to Multiple Choice

1. d	**6.** a	**11.** d	**16.** b	**21.** c	**26.** a	**31.** c	**36.** d	**41.** a
2. d	**7.** b	**12.** c	**17.** d	**22.** b	**27.** a	**32.** b	**37.** d	**42.** d
3. d	**8.** b	**13.** d	**18.** a	**23.** b	**28.** b	**33.** c	**38.** d	**43.** c
4. a	**9.** b	**14.** d	**19.** c	**24.** d	**29.** a	**34.** c	**39.** d	**44.** a
5. c	**10.** d	**15.** a	**20.** d	**25.** d	**30.** d	**35.** b	**40.** b	**45.** b

6 Electricity and Magnetism

Lightning is an electrical discharge in the atmosphere.

Goals

When you have finished this chapter you should be able to complete the goals ▶ given for each section below:

Electric Charge

6.1 Positive and Negative Charge
Opposites Attract

6.2 What Is Charge?
Protons, Electrons, and Neutrons
- ▶ Discuss what is meant by electric charge.
- ▶ Describe the structure of an atom.

6.3 Coulomb's Law
The Law of Force for Electric Charges
- ▶ State Coulomb's law for electric force and compare it with Newton's law of gravity.

6.4 Force on an Uncharged Object
Why a Comb Can Attract Bits of Paper
- ▶ Account for the attraction between a charged object and an uncharged one.

Electricity and Matter

6.5 Matter in Bulk
Gravity versus Electricity

6.6 Conductors and Insulators
How Charge Flows from One Place to Another
- ▶ Distinguish among conductors, semiconductors, and insulators.
- ▶ Define ion and give several ways of producing ionization.

6.7 Superconductivity
A Revolution in Technology May Be Near
- ▶ Define superconductivity and discuss its potential importance.

Electric Current

6.8 The Ampere
The Unit of Electric Current

6.9 Potential Difference
The Push behind a Current
- ▶ Describe electric current and potential difference (voltage) by analogy with the flow of water in a pipe.

6.10 Ohm's Law
Current, Voltage, and Resistance
- ▶ Use Ohm's law to solve problems that involve the current in a circuit, the resistance of the circuit, and the voltage across the circuit.

6.11 Electric Power
Current Times Voltage
- ▶ Relate the power consumed by an electrical appliance to the current in it and the voltage across it.

Magnetism

6.12 Magnets
Attraction and Repulsion

6.13 Magnetic Field
How Magnetic Forces Act
- ▶ Describe what is meant by a magnetic field and discuss how it can be pictured by field lines.

6.14 Oersted's Experiment
Magnetic Fields Originate in Moving Electric Charges
- ▶ State the connection between electric charges and magnetic fields.
- ▶ Use the right-hand rule to find the direction of the magnetic field around an electric current.

6.15 Electromagnets
How to Create a Strong Magnetic Field
- ▶ Explain how an electromagnet works.

Using Magnetism

6.16 Magnetic Force on a Current
A Sidewise Push
- ▶ Describe the force a magnetic field exerts on an electric current.

6.17 Electric Motors
Mechanical Energy from Electric Energy
- ▶ Discuss the operation of an electric motor.

6.18 Electromagnetic Induction
Electric Energy from Mechanical Energy
- ▶ Describe electromagnetic induction and explain how a generator makes use of it to produce an electric current.

6.19 Transformers
Stepping Voltage Up or Down
- ▶ Explain how a transformer changes the voltage of an alternating current and why this is useful.

We have now learned about force and motion, mass and energy, the law of gravity, and the concept of matter as being made up of tiny moving molecules. With the help of these ideas we have been able to make sense of a wide variety of observations, from the paths of the planets across the sky to the melting of ice and the boiling of water. Is this enough for us to understand how the entire physical universe works?

For an answer all we need do is run a hard rubber comb through our hair on a dry day. Little sparks occur, and the comb then can pick up small bits of dust and paper. What is revealed in this way is an **electrical** phenomenon, something that neither gravity nor the kinetic theory of matter can account for.

In everyday life electricity is familiar as that which causes our lightbulbs to glow, many of our motors to turn, our telephones and radios to bring us sounds, our television screens to bring us images. But there is more to electricity than its ability to transport energy and information. All matter turns out to be electrical in nature, and electric forces are what bind electrons to nuclei to form atoms and what hold atoms together to form molecules, liquids, and solids. Most of the properties of the ordinary matter around us—an exception is mass—can be traced to electrical forces.

Electric Charge

The first recorded studies of electricity were made in Greece by Thales of Miletus about 2500 years ago. Thales experimented with amber, called *electron* in Greek, and fur. The name **electric charge** is today given to whatever it is that a piece of amber (or hard rubber) possesses as a result of being rubbed with fur. It is this charge that causes sparks to occur and that attracts light objects such as bits of paper.

6.1 Positive and Negative Charge

Opposites Attract

Let us begin by hanging a small plastic ball from a thread, as in Fig. 6-1. We touch the ball with a hard rubber rod and find that nothing happens.

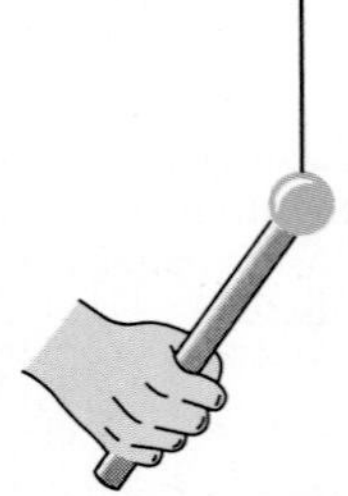

1. A plastic ball held by a string is touched by a hard rubber rod. Nothing happens.

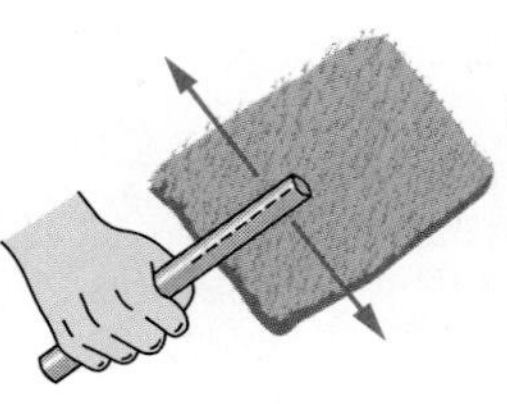

2. The rubber rod is stroked against a piece of fur.

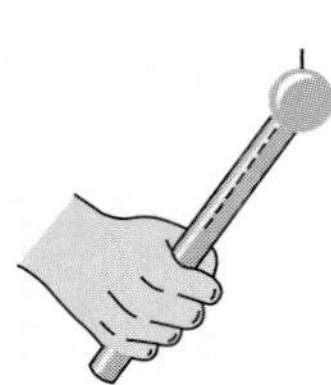

3. The plastic ball is again touched by the rubber rod.

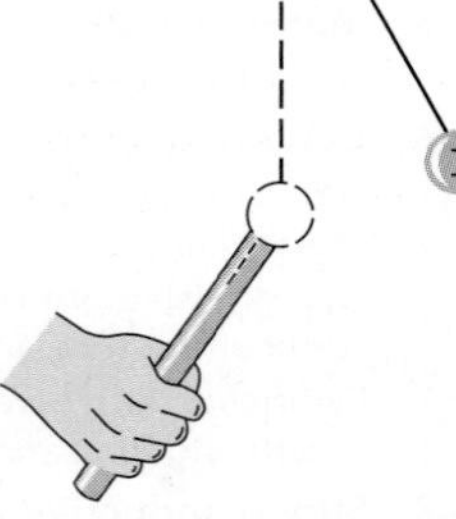

4. After the touch, the plastic ball flies away from the rod.

Figure 6-1 A rubber rod stroked with fur becomes negatively charged. When it is touched against a plastic ball, some of the negative charge flows to the ball. The plastic ball then flies away because like charges repel each other.

BIOGRAPHY Benjamin Franklin (1706–1790)

Although best known for his role in establishing the United States as an independent country (he helped draft the Declaration of Independence and the Constitution), Franklin was also the first notable American scientist and inventor. The Franklin stove, the lightning rod, and bifocal eyeglasses were products of his ingenuity. By analyzing the records of Atlantic voyages in the sailing ships of his time, he inferred the existence of the Gulf Stream (see Sec. 14.14).

Franklin was especially interested in electricity, and he interpreted what was then known about it in terms of a fluid: a positive charge meant an excess of the fluid, and a negative charge a deficiency. In this picture an electric current involves the motion of the fluid from a positively charged object to a negatively charged object where it is absorbed. If the excess in one object matches the deficiency in the other, both become neutral as a result. Even today electricians regard electric currents in this way, although it is now known that what moves are negatively charged electrons in the opposite direction. In the most famous of his experiments, Franklin produced sparks from the end of the wet string of a kite flown in a thunderstorm, from which he concluded that lightning is an electrical phenomenon. (He was careful not to touch the string directly; an imitator who did was electrocuted.)

Next we stroke the rod with a piece of fur and again touch the plastic ball with the rod. This time the ball flies away from the rod. What must have happened is that some of the electric charge on the rod has flowed to the ball, and the fact that the ball then flies away from the rod means that charges of the same kind repel each other (Fig. 6-2).

Is there only one kind of electric charge? To find out, we try other combinations of materials and see what happens when the various charged plastic balls are near each other. Figure 6-3 shows the result when one ball has been charged by a rubber rod stroked with fur and the other ball has been charged by a glass rod stroked with silk: the two balls fly together. We conclude that the charges on the rods are somehow different and that different charges attract each other.

Comprehensive experiments show that *all* electric charges fall into one of these two types. Regardless of origin, charges always behave as though they came either from a rubber rod rubbed with fur or from a glass rod rubbed with silk. Benjamin Franklin suggested names for these two basic kinds of electricity. He called the charge produced on the rubber rod **negative charge** and the charge produced on the glass rod **positive charge.** These definitions are still used today.

These experiments can be summarized very simply:

> All electric charges are either positive or negative. Like charges repel one another; unlike charges attract one another.

Figure 6-2 This girl has been given an electric charge by touching the terminal of a static electric generator. Because all her hairs have charge of the same sign, they repel one another.

Charge Separation We have thus far been paying attention to the positive charge of the glass and the negative charge of the rubber. However,

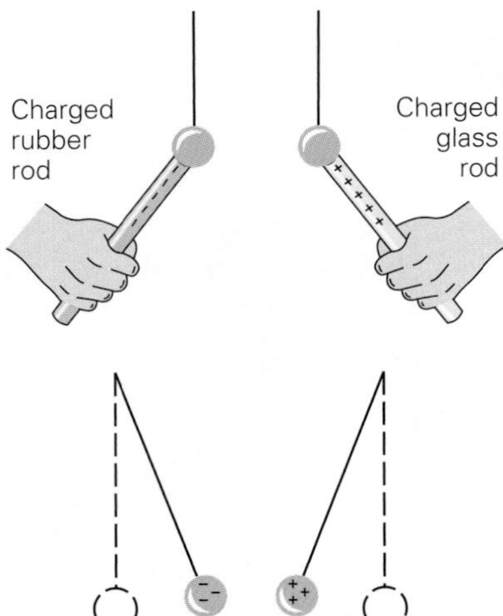

Figure 6-3 A glass rod stroked with silk becomes positively charged. When one plastic ball is touched with a negatively charged rubber rod and another plastic ball is touched with a positively charged glass rod, the two balls fly together because unlike charges attract each other.

Figure 6-4 When a rubber rod is stroked against a piece of fur, charges that were originally mixed together evenly become separated so that the rod becomes negatively charged and the fur becomes positively charged.

Conservation of Charge

We have already met the laws of conservation of energy (with mass considered as a form of energy), momentum, and angular momentum. Here is another, the **law of conservation of charge:**

> The net electric charge in an isolated system remains constant.

"Net charge" means the algebraic sum of the charges present in the system, that is, the total with negative charges canceling out positive charges of the same magnitude (and vice versa if there is more negative charge present). Net charge can be positive, negative, or zero. Every known physical process in the universe conserves electric charge. Separating or bringing together charges does not affect their magnitudes, so such rearrangements leave the net charge the same.

we do not produce only positive charge by rubbing glass with silk or only negative charge by stroking rubber with fur. If the fur used with the rubber is brought near a negatively charged plastic ball, the ball is attracted. Thus the fur must have a positive charge (Fig. 6-4). Similarly the silk used with the glass turns out to have a negative charge. Whenever electric charge is produced by contact between two objects of different materials, one of them ends up with a positive charge and the other a negative charge. Which is which depends on the particular materials used.

The rubbing process does not *create* the electric charges that appear as a result. All "uncharged" objects actually contain equal amounts of positive and negative charge. For some pairs of materials, as we have seen, mere rubbing is enough to separate some of the charges from each other. In most cases, however, the charges are firmly held in place and more elaborate treatment is needed to pull them apart.

An object whose positive and negative charges exactly balance out is said to be electrically **neutral.**

6.2 What Is Charge?

Protons, Electrons, and Neutrons

In our own experience, matter and electric charge seem continuous, so that we can imagine dividing them into smaller and smaller portions without limit. But there is another level beyond reach of our senses, though not beyond reach of our instruments, on which every substance is revealed as being composed of tiny bits of matter called **atoms.** (Atoms are discussed in detail in later chapters along with how they join together to form molecules, liquids, and solids.) Over 100 varieties of atoms are known, but each of them is made up of just three kinds of **elementary particles;** nature is very economical. Two of the particles carry electric charges, so that charge, like mass, comes in small parcels of definite size. The third particle has no charge.

The three elementary particles found in atoms are

1. The **proton,** which has a mass of 1.673×10^{-27} kg and is positively charged
2. The **electron,** which has a mass of 9.11×10^{-31} kg and is negatively charged
3. The **neutron,** which has a mass of 1.675×10^{-27} kg and is uncharged

The proton and electron have exactly the same amounts of charge, although of opposite sign. Protons and neutrons have almost equal masses, which are nearly 2000 times greater than the electron mass.

Every atom has a small, central **nucleus** of protons and neutrons with its electrons moving about the nucleus some distance away (Fig. 6-5). Different types of atoms have different combinations of protons and neutrons in their nuclei. For instance, the most common carbon atom has a nucleus that contains six protons and six neutrons; the most common uranium atom has a nucleus that contains 92 protons and 146 neutrons. The electrons in an atom are normally equal in number to the protons, so the atom is electrically neutral unless disturbed in some way.

What actually *is* charge? All that can be said is that electric charge, like mass, is a fundamental property of certain elementary particles of which all matter is composed. Their masses give rise to gravitational forces; their charges give rise to electric forces. Elementary particles have other attributes, too, that affect their behavior. Much about elementary particles is understood at present, as we shall learn in Secs. 8.14 to 8.16. However, just why the properties of these particles exist and have the magnitudes they do is still the subject of much study. One thing is clear, though: there is no simple answer to the question, "What is charge?"

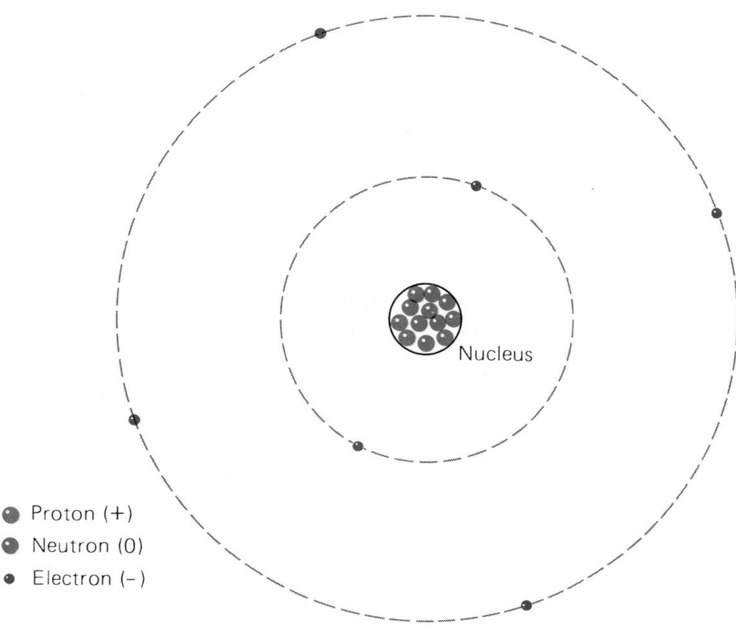

Figure 6-5 An atom consists of a central nucleus of protons and neutrons with electrons moving around it some distance away. Shown is a schematic diagram (not to scale) of the most common type of carbon atom, which has six protons, six neutrons, and six electrons. Two of the electrons are relatively near the nucleus, the others are farther away. A more realistic way to think of atomic structure is discussed in Sec. 9.13.

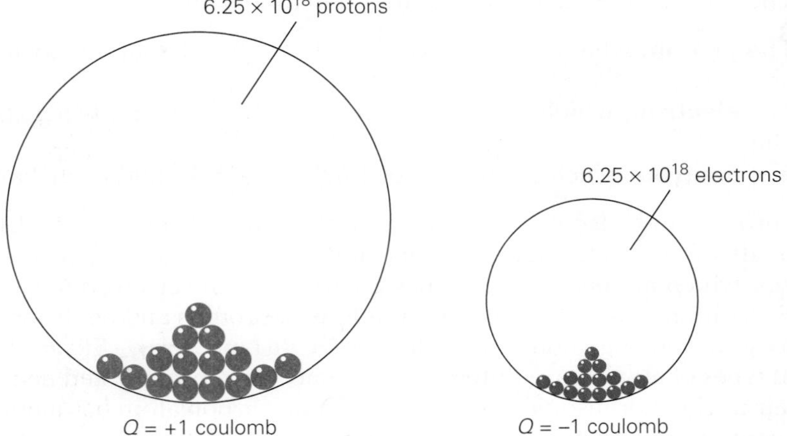

Figure 6-6 Electric charge is not continuous but occurs in multiples of $\pm e = \pm 1.6 \times 10^{-19}$ coulomb. (These pictures are not to be taken literally, of course: the mutual repulsion of the particles would prevent their assembly in this way.)

The Coulomb The unit of electric charge is the **coulomb** (C). The proton has a charge of $+1.6 \times 10^{-19}$ C and the electron has a charge of -1.6×10^{-19} C. All charges, both positive and negative, are therefore found only in multiples of 1.6×10^{-19} C. This basic quantity of charge is abbreviated e:

$$e = 1.6 \times 10^{-19} \text{ C} \qquad \textit{Basic unit of charge in nature}$$

Electric charge appears continuous outside the laboratory because e is such a small quantity. A charge of -1 C, for example, corresponds to more than 6 billion billion electrons (Fig. 6-6). Atoms are small, too: coal is almost pure carbon, and 6 billion billion carbon atoms would make a piece of coal only about the size of a pea.

6.3 Coulomb's Law

The Law of Force for Electric Charges

The forces between electric charges can be studied in rather simple experiments, such as that shown in Fig. 6-7. What we find is that the force between a charged rod and a charged plastic ball depends on two things: how close the rod is to the ball, and how much charge each one has.

Precise measurements show that the force between charges follows the same inverse-square variation with distance that the gravitational force between two masses does (see Sec. 2.13). For instance, when the charges are 2 cm apart, the force between them is $\frac{1}{4}$ as great as the force when they are 1 cm apart; it is 4 times greater when they are $\frac{1}{2}$ cm apart (Fig. 6-8*a*). If R is the distance between the charges, we can say that the force between them is proportional to $1/R^2$.

The force also depends on the magnitude of each charge: if either charge is doubled, the force doubles too, and if both charges are doubled, the force increases fourfold (Fig. 6-8*b*). If the charges have the respective magnitudes Q_1 and Q_2, then the force varies as their product Q_1Q_2.

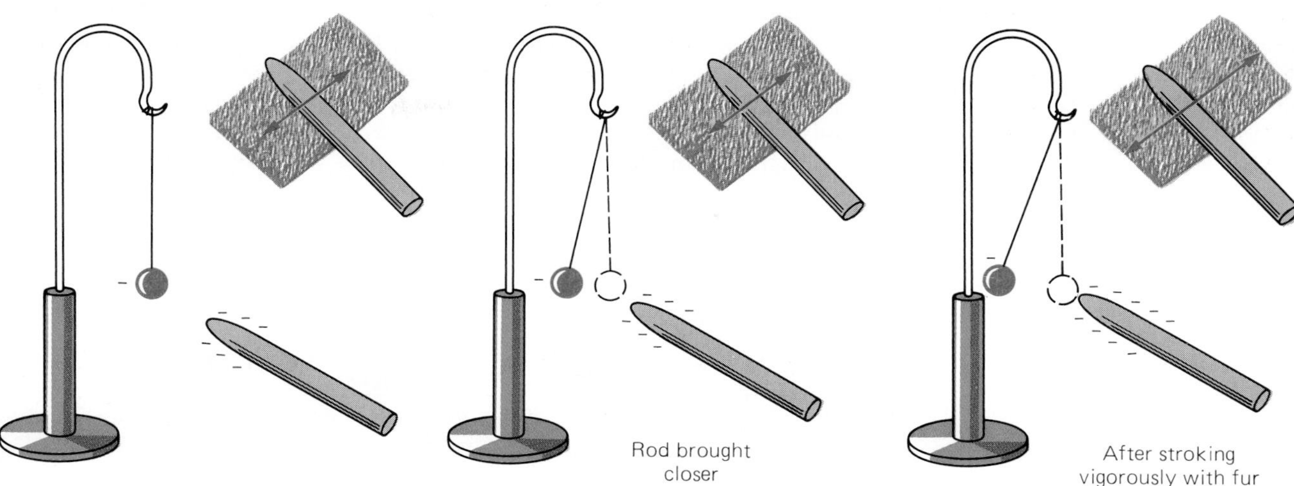

Figure 6-7 The forces between electric charges. When a rubber rod that has been stroked with fur is brought near a negatively charged plastic ball, the force on the ball is greater when the rod is held close to it and also greater when the rod has been vigorously stroked.

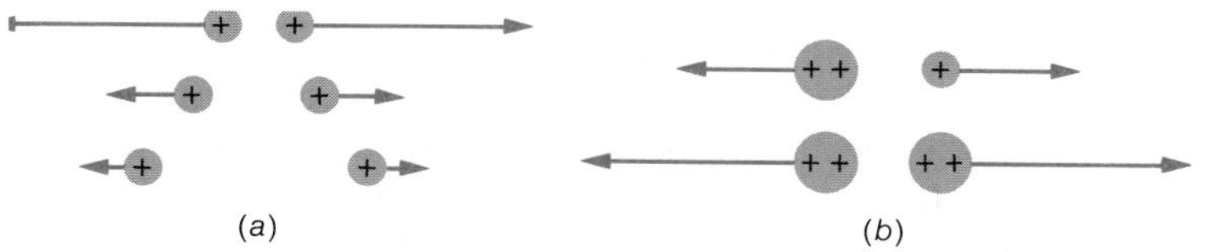

Figure 6-8 (*a*) The force between two charges varies inversely as the square of their separation; increasing the distance reduces the force. (*b*) The force is proportional to the product of the charges. Attractive forces behave the same way.

These results are summarized in **Coulomb's law:**

$$F = \frac{KQ_1Q_2}{R^2} \qquad \textit{Electric force} \qquad 6\text{-}1$$

which is named in honor of Charles Coulomb, who helped develop it (Fig. 6-9). The quantity K is a constant whose value is almost exactly

$$K = 9 \times 10^9 \text{ N} \cdot \text{m}^2/\text{C}^2 \qquad \textit{Electric force constant}$$

Just by looking at this formula we can see that the force between two charges of 1 C each that are separated by 1 m is 9×10^9 N, 9 billion newtons. This is an enormous force, equal to about 2 billion lb! We conclude that the coulomb is a very large unit indeed, and that even the most highly charged objects that can be produced cannot contain more than a small fraction of a coulomb of net charge of either sign.

The concept of a force field, described in Sec. 6.13, is a useful way to think about how electric charges interact with one another.

Figure 6-9 Charles Coulomb (1736–1806).

Example 6.1

How far apart should two electrons be if the force each exerts on the other is to equal the weight of an electron at sea level?

Solution

The weight of an electron at sea level is $m_e g$, $Q_1 = Q_2 = e$ here, so from Eq. 6-1

$$m_e g = \frac{KQ_1Q_2}{R^2} = \frac{Ke^2}{R^2}$$

Solving for R gives

$$R = \sqrt{\frac{Ke^2}{m_e g}} = \sqrt{\frac{(9 \times 10^9\ \text{N} \cdot \text{m}^2/\text{C}^2)(1.6 \times 10^{-19}\ \text{C})^2}{(9.1 \times 10^{-31}\ \text{kg})(9.8\ \text{m/s}^2)}} = 5.1\ \text{m}$$

6.4 Force on an Uncharged Object

Why a Comb Can Attract Bits of Paper

One sign that a body has an electric charge is that it causes small, uncharged objects such as dust particles, bits of paper, and suspended plastic balls to move toward it. Where does the force come from?

The explanation comes from the fact that the electrons in a solid have some freedom of movement. In a metal this freedom is considerable, but even in other substances the electrons can shift around a little without leaving their parent atoms or molecules. When a comb is given a negative charge by being run through our hair, electrons in a nearby bit of paper are repelled by the negative charge and move away as far as they can (Fig. 6-10). The side of the paper near the comb is left with a positive charge, and the paper is accordingly attracted to the comb. If the comb is removed without actually touching the paper, the disturbed electrons resume their normal positions. Only a small amount of charge separation actually occurs, and so, with little force available, only very light things can be picked up this way.

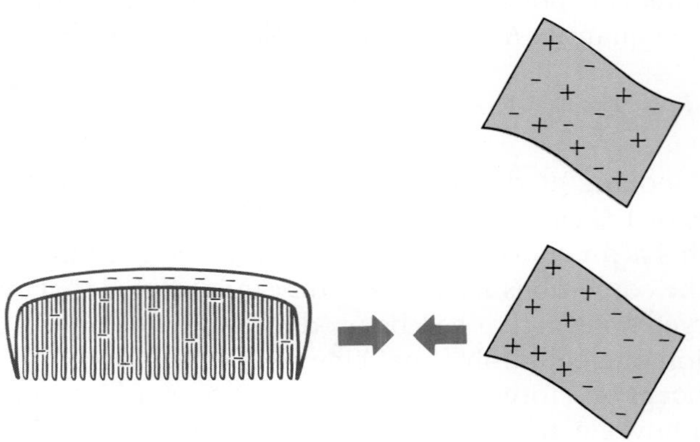

Figure 6-10 A charged object attracts an uncharged one by first causing a separation of charge in the latter. The effect is much exaggerated here; except in metals, the charge displacement takes place within individual atoms and molecules.

Electricity and Matter

Let us now look into some aspects of the electrical behavior of matter.

6.5 Matter in Bulk

Gravity Versus Electricity

Coulomb's law for the force between charges is one of the fundamental laws of physics, in the same category as Newton's law of gravity. The latter, as we know, is written in equation form as

$$F = \frac{Gm_1m_2}{R^2} \qquad \textit{Gravitational force} \qquad 2\text{-}19$$

Coulomb's law resembles the law of gravity with the difference that gravitational forces are always attractive but electric forces may be either attractive or repulsive.

This last fact has an important consequence. Because one lump of matter always attracts another lump gravitationally, matter in the universe tends to come together into large masses. Even though dispersive influences of various kinds exist, they must fight against this steady attraction. Galaxies, stars, and planets, which condensed from matter that was originally spread out in space, bear witness to this cosmic herd instinct.

To collect much electric charge of either sign, however, is far more of a feat. Charges of opposite sign attract each other strongly, so it is hard to separate neutral matter into differently charged portions. And charges of the same sign repel each other, so putting together a large amount of charge of one sign is difficult.

To sum up, we can say that a system of electrically neutral particles is most stable (that is, has a minimum potential energy) when the particles make up a single body, while a system of electric charges is most stable when charges of opposite signs pair off to cancel each other out. Hence on a cosmic scale gravitational forces are significant and electric ones are not. On an atomic scale, however, the reverse is true. The masses of subatomic particles are too small for them to interact gravitationally to any appreciable extent, whereas their electric charges are large enough for electric forces to exert marked effects.

Example 6.2

The hydrogen atom, the simplest of all, consists of a single proton as its nucleus with an electron an average of 5.3×10^{-11} m away. Compare the electric and gravitational forces between the proton and the electron in this atom.

Solution

The electric force between the proton and electron is, from Eq. 6-1,

$$F_{\text{elec}} = \frac{KQ_1Q_2}{R^2} = \frac{Ke^2}{R^2}$$

(continued)

Example 6.2 (Continued)

and the gravitational force between them is, from Eq. 2-19,

$$F_{\text{grav}} = \frac{Gm_1m_2}{R^2} = \frac{Gm_pm_e}{R^2}$$

Since both forces vary with distance as $1/R^2$, their ratio will be the same regardless of how far apart the proton and electron are. This ratio is

$$\frac{F_{\text{elec}}}{F_{\text{grav}}} = \frac{Ke^2/R^2}{Gm_pm_e/R^2} = \frac{Ke^2}{Gm_pm_e}$$

$$= \frac{(9 \times 10^9 \text{ N}\cdot\text{m/C}^2)(1.6 \times 10^{-19} \text{ C})^2}{(6.7 \times 10^{-11} \text{ N}\cdot\text{m}^2/\text{kg}^2)(1.7 \times 10^{-27} \text{ kg})(9.1 \times 10^{-31} \text{ kg})}$$

$$= 2.2 \times 10^{39}$$

The electric force is over 10^{39} times the gravitational force! Clearly, gravitational effects are negligible within atoms compared with electric effects.

6.6 Conductors and Insulators

How Charge Flows from One Place to Another

A substance through which electric charge can flow readily is called a **conductor.** Metals are the only solid conductors at room temperature, copper being an especially good one. In a metal, each atom gives up one or more electrons to a "gas" of electrons that can move relatively freely inside the metal. The atoms themselves stay in place and are not involved in the movement of charge. Metals are also the most efficient conductors of heat.

In an **insulator,** charge can flow only with great difficulty. Nonmetallic solids are insulators because all their electrons are tightly bound to particular atoms or groups of atoms. Glass, rubber, and plastics are good insulators.

A few substances, called **semiconductors,** are between conductors and insulators in their ability to let charge move through them. Semiconductors have made possible devices called **transistors,** whose ability to transmit charge can be changed at will. Transistors are widely used in modern electronics, notably in portable telephones and in radio and television receivers. A computer contains millions of transistors that act as miniature switches to perform arithmetic and carry out logical operations. Semiconductor memories are also used in computers, with huge numbers of memory elements built into a "chip" smaller than a fingernail (Fig. 6-11).

Ions The conduction of electricity through gases and liquids—in a neon sign, for instance, or in the acid of a storage battery—involves the movement of charged atoms and molecules called **ions.** An atom or molecule gains a positive charge (becomes a positive ion) when it loses one or more electrons, and it gains a negative charge (becomes a negative ion) when electrons in excess of its normal number become attached to it.

Electrical Grounding

The earth as a whole, at least that part of it beneath the outer dry soil, is a fairly good electrical conductor. Hence if a charged object is connected with the earth by a piece of metal, the charge is conducted away from the object to the earth. This convenient method of removing the charge from an object is called **grounding** the object. As a safety measure, the metal shells of electrical appliances are grounded through special wires that give electric charges in the shells paths to the earth. The round post in the familiar three-prong electric plug is the ground connection.

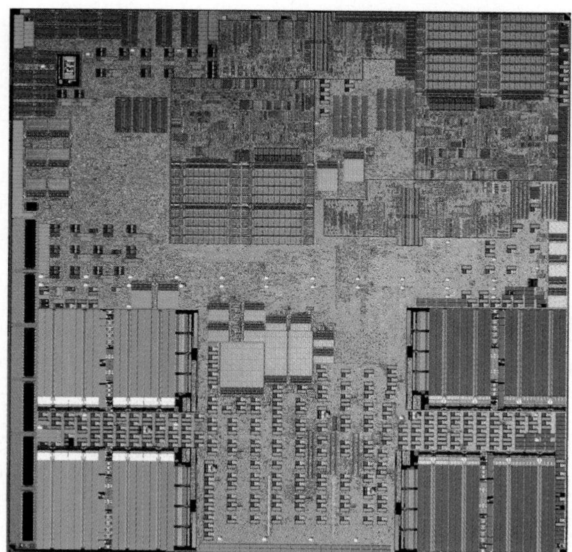

Figure 6-11 Enlargement of a semiconductor "chip" that contains the millions of circuit elements that make up a microprocessor used in a supercomputer.

The process of forming ions, or **ionization,** can take place in a number of ways. A gas like ordinary air, which is normally a poor conductor, becomes ionized when x-rays, ultraviolet light, or radiation from a radioactive material pass through it, when an electric spark is produced, or even when a flame burns in it. Air molecules are sufficiently disturbed by these processes that electrons are torn loose from some of them. The electrons thus set free may attach themselves to adjacent molecules, so both positive and negative ions are formed (Fig. 6-12). Eventually oppositely charged ions come together, whereupon the extra electrons on negative ions shift to positive ions to give neutral molecules again. At normal atmospheric pressure and temperature the ions last no more than a few seconds.

In the upper part of the earth's atmosphere, air molecules are so far apart on the average that the ionization produced by x-rays and ultraviolet

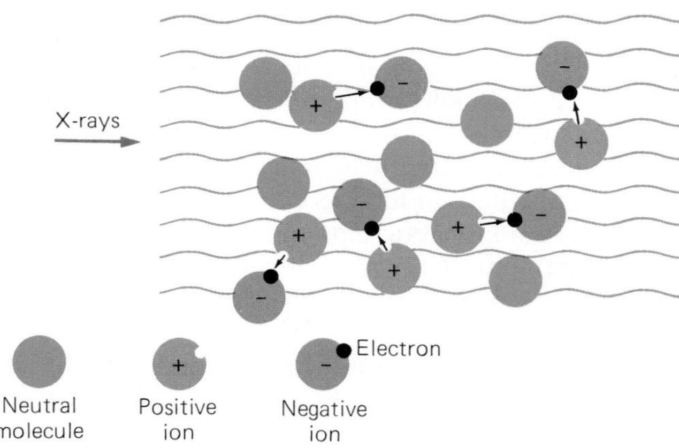

Figure 6-12 A gas such as air becomes ionized when x-rays disrupt its molecules. A molecule losing an electron becomes a positive ion; a molecule gaining an electron becomes a negative ion. Ultraviolet light, radiation from radioactive substances, sparks, and flames also cause ionization to occur.

light from the sun tends to persist. The ability of these ions to reflect radio waves makes possible long-range radio communication (see Sec. 7-9).

In contrast with gases, certain liquids may be permanently ionized to a greater or lesser extent (see Chap. 11). The conductivity of pure water itself is extremely small, but even traces of some impurities increase its conductivity enormously. Since most of the water we use in daily life is somewhat impure, it is usually considered a fair conductor of electricity.

6.7 Superconductivity

A Revolution in Technology May Be Near

Even the best conductors resist to some extent the flow of charge through them at ordinary temperatures. However, when extremely cold, some substances lose all electrical resistance. This phenomenon, called **superconductivity,** was discovered by Kamerlingh Onnes in the Netherlands in 1911. For example, aluminum is a superconductor below 1.2 K, which is −272°C. Such temperatures are difficult and expensive to reach, and as a result superconductivity has not been exploited commercially to any great extent.

If electrons are set in motion in a closed wire loop at room temperature, they will come to a stop in less than a second even in a good conductor such as copper. In a superconducting wire loop, on the other hand, electrons have circulated for years with no outside help.

Superconductivity is important because electric currents—flows of charge—are the means by which electric energy is carried from one place to another. Electric currents are also the means by which magnetic fields are produced in such devices as electric motors, as we shall learn later in this chapter. In an ordinary conductor, some of the energy of a current is lost as heat. Where long distances or large currents are involved, quite a bit of energy can be wasted in this way. About 8 percent of the electric energy generated in the United States is lost as heat in transmission lines.

High-Temperature Superconductors Despite much effort, until 1986 no substance was known that was superconducting above 23 K. In that year Alex Müller and Georg Bednorz, working in Switzerland, discovered a ceramic that was superconducting up to 35 K. Soon afterward others extended their approach to produce superconductivity at temperatures as high as 164 K. Although still extremely cold (−109°C) by everyday standards, such temperatures are above the 77 K boiling point of liquid nitrogen, which is cheap (cheaper than milk) and readily available. Despite being difficult to manufacture, more complicated to install, and costlier, superconducting cables cooled with liquid nitrogen can be attractive when large currents must be transmitted. For instance, in many places existing underground ducts are already filled with copper wires, leaving no room to expand electricity supply. Using superconducting cables in such situations may be less expensive than putting in new ducts. Electromagnets in MRI medical scanners and in some particle accelerators use superconducting wires, and electric motors, generators, and transformers using them are being developed.

BIOGRAPHY John Bardeen (1908–1991)

In 1945 John Bardeen, who was born in Wisconsin and educated there and at Princeton, joined a research group at Bell Telephone Laboratories led by William Shockley. In 1948 the group produced the first transistor, for which Shockley, Bardeen, and their collaborator Walter Brattain received a Nobel Prize in 1956. Bardeen later said, "I knew the transistor was important, but I never foresaw the revolution in electronics it would bring."

In 1951 Bardeen left Bell Labs for the University of Illinois where, together with Leon Cooper and J. Robert Schrieffer, he developed the theory of superconductivity. Compared with his earlier work on the transistor, "Superconductivity was more difficult to solve, and it required some radically new concepts." According to the theory, the motions of two electrons can become correlated through their interactions with the atoms in a crystal, which enables the pair to move with complete freedom through the crystal. Bardeen received his second Nobel Prize in 1972 for this theory along with Cooper and Schrieffer; he was the first person to receive two such prizes in the same field.

A room-temperature superconductor would truly revolutionize the world's technology. As just one example, all trains would run suspended above the ground by magnetic forces, resulting in better fuel efficiency and high speeds. In general, less waste of electric energy would mean a lower rate of depletion of fuel resources and reduced pollution. A room-temperature superconductor, though still a dream, does not seem quite so hopeless a dream as it did not so long ago.

Electric Current

A flow of charge from one place to another constitutes an **electric current.** Currents and not stationary charges are involved in nearly all the practical applications of electricity.

6.8 The Ampere

The Unit of Electric Current

A battery changes chemical energy into electric energy. If we connect a wire between the terminals of a battery to make a complete conducting path, or **circuit,** electrons will flow into the wire from the negative terminal and out of the wire into the positive terminal. Chemical reactions in the battery keep the electrons moving. We do not say, "The electrons carry the current," or "The motion of the electrons produces a current"; the moving electrons *are* the current.

The flow of electricity along a wire is a lot like the flow of water in a pipe (Fig. 6-13). When we describe the rate at which water moves through

Direction of Current

It is customary to regard an electric current as consisting of positive charges in motion. Thus a current is assumed to go from the positive terminal of a battery or generator to its negative terminal through the external circuit, as in Fig. 6-13.

In reality, an electric current in a metal is a flow of electrons, which are negatively charged. However, a current of negative charges going one way is electrically the same as a current of positive charges going the other way. Since it makes no difference in practice, this book will follow the usual convention of considering current as a flow of positive charge.

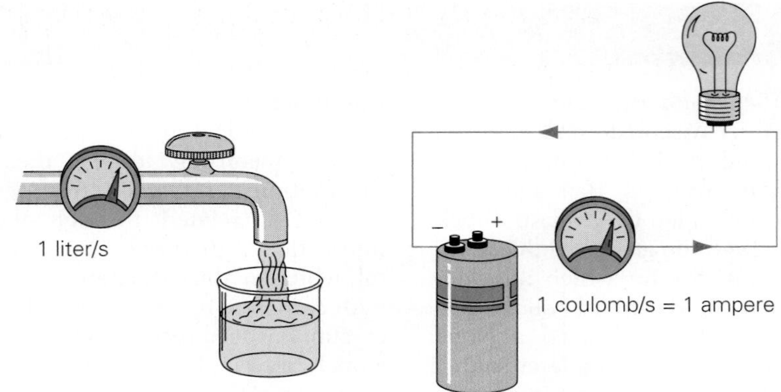

Figure 6-13 The ampere is the unit of electric current. The flow of charge in a circuit is like the flow of water in a pipe except that a return wire is needed in order to have a complete conducting path. An electric current is always assumed to go from the + terminal of a battery or generator to its − terminal through the external circuit.

a pipe, we give the flow in terms of, say, liters per second. If 5 liters of water passes through a given pipe each second, the flow is 5 liters/s.

The description of electric current, symbol *I*, follows the same pattern. As we know, quantity of electric charge is measured in coulombs, as quantity of water is measured in liters. The natural way to refer to a flow of charge in a wire, then, is in terms of the number of coulombs per second that go past any point in the wire:

$$I = \frac{Q}{t} \qquad \textit{Electric current} \qquad 6\text{-}2$$

$$\text{Electric current} = \frac{\text{charge transferred}}{\text{time interval}}$$

The unit of electric current is called the **ampere** (A), after the French physicist André Marie Ampère. That is,

$$1 \text{ ampere} = 1 \frac{\text{coulomb}}{\text{second}}$$

$$1 \text{ A} = 1 \text{ C/s}$$

The current in the lightbulb of most desk lamps is a little less than 1 A.

Example 6.3

Since electric current is a flow of charge, why are two wires, not just one, needed to connect a battery or generator to an appliance such as a motor or lightbulb?

Solution

If a single wire were used, charge of one sign or the other (depending on the situation) would be permanently transferred from the source of current to the appliance. In a short time so much charge would have been transferred that the source would be unable to shift further charge against the repulsive force of the charge piled up at the appliance. Thus a single wire cannot carry a current continuously. With two wires, however, charge can be circulated from source to appliance and back, which permits a steady flow of energy to the appliance.

Example 6.4

How many electrons per second flow past a point in a wire carrying a current of 1 A?

Solution

The electron charge is $Q = e = 1.6 \times 10^{-19}$ C, and so a current of 1 A = 1 C/s corresponds to a flow of

$$\frac{\text{Electrons}}{\text{time}} = \frac{Q/t}{Q/\text{electron}} = \frac{1 \text{ C/s}}{1.6 \times 10^{-19} \text{ C/electron}}$$
$$= 6.3 \times 10^{18} \text{ electrons/s}$$

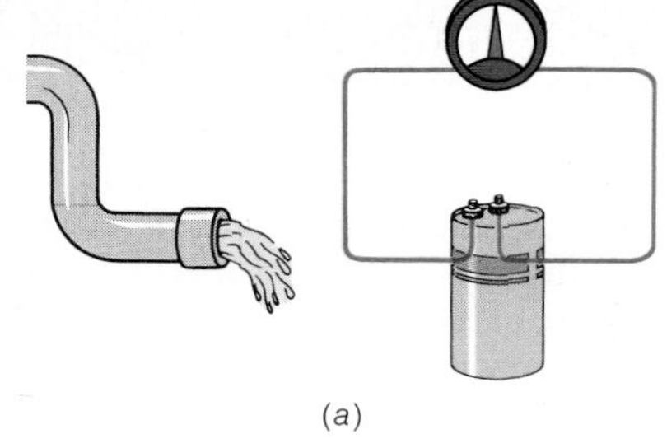
(a)

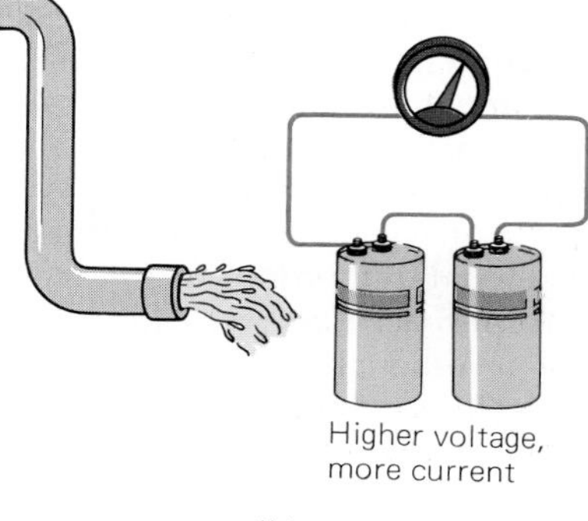

(b)

Figure 6-14 The flow of electric charge in a wire is analogous to the flow of water in a pipe. Thus having the water fall through a greater height at (*b*) than at (*a*) yields a greater flow of water, which corresponds to using two batteries to obtain a higher potential difference and thereby a greater current.

6.9 Potential Difference

The Push behind a Current

Consider a liter of water at the top of a waterfall. The water has potential energy there, since it can move downward under the pull of gravity. When the water drops, its PE decreases. As we learned in Chap. 3, the work that can be obtained from the liter of water during its fall is equal to its decrease in PE.

Now consider a coulomb of negative charge on the − terminal of a battery. It is repelled by the − terminal and attracted by the + terminal, and so it has a certain amount of PE. When the coulomb of charge has moved along a wire to the + terminal, its PE is gone. The work the coulomb of charge can perform while flowing from the − to the + terminal of the battery is equal to this decrease in PE.

The decrease in its PE brought about by the motion of 1 C of charge from the − to the + terminal is called the **potential difference** between the two terminals. It is analogous to the difference in height in the case of water (Fig. 6-14). The potential difference V between two points is equal to the corresponding energy difference W per coulomb:

$$V = \frac{W}{Q} \qquad \textit{Potential difference} \qquad 6\text{-}3$$

$$\text{Potential difference} = \frac{\text{potential energy difference}}{\text{charge transferred}}$$
$$= \frac{\text{work done to transfer charge } Q}{\text{charge transferred}}$$

We measure difference of height in meters; we measure difference of potential in **volts,** named for the Italian physicist Alessandro Volta (Fig. 6-15). When 1 coulomb of charge travels through 1 volt of potential difference, the work that it does is equal to 1 joule. By definition:

$$1 \text{ volt} = 1 \frac{\text{joule}}{\text{coulomb}}$$
$$1 \text{ V} = 1 \text{ J/C}$$

Potential difference is often called simply **voltage.**

Figure 6-15 Alessandro Volta (1745–1827).

Lightning

Charge separation occurs in a thunderstorm when water droplets and ice crystals in a cloud collide, and the separation grows in scale through convectional updrafts and downdrafts. The result is a region of positive charge in the upper part of the cloud with a region of negative charge below it. By the mechanism shown in Fig. 6-10, the latter region induces a charge separation in the ground under the cloud, with positive charge uppermost facing the negative charge in the cloud (Fig. 6-16). Potential differences inside the cloud and between the cloud and the ground range up to millions of volts and result in electric discharges (which are really giant sparks) that appear as lightning strokes along the paths where charges flow (Fig. 6-17). Exactly what triggers a lightning stroke is not completely understood, but a promising idea is that cosmic-ray particles from space (see Sec. 19.5) are responsible.

Although cloud-to-ground strokes are the most familiar, lightning is actually much more frequent inside thunderclouds, where it is hidden from our view by the clouds although we can hear the thunder it causes. Thunder is the result of the heating of the air along the path of a lightning stroke, whose sudden expansion gives rise to intense sound waves. "Sheet lightning" is a luminosity in the sky due to a lightning stroke that is behind a cloud or occurs below the horizon.

Lightning usually strikes the highest point in an area, such as a tall tree or a ship offshore. The safest place to be in a thunderstorm is inside a building or a car. If caught in the open, one should crouch down but not lie flat, because strong horizontal currents flow in the ground near where lightning strikes.

Airplanes are regularly struck by lightning, and indeed often trigger lightning when flying through charged clouds. Most airplanes have aluminum skins that conduct the currents of lightning strokes harmlessly around their interiors. The reinforced plastic skins of other aircraft incorporate metal fibers for such shielding.

Figure 6-16 The negative charge on the bottom of a thundercloud induces a positive charge in the ground under it.

Figure 6-17 About 20 million lightning strokes reach the ground in the United States each year, killing several hundred people and causing much property damage.

Batteries The normal potential difference between the terminals of a car's storage battery is about 12 V, in the case of a dry cell about 1.5 V. Every coulomb of electricity at the negative terminal of the storage battery can do 8 times as much work as a coulomb at the negative terminal of a dry cell—just as a liter of water at the top of a waterfall 12 m high can do 8 times as much work as a liter at the top of a 1.5-m fall. If a storage battery and a dry cell are connected in identical circuits, the battery will push 8 times as many electrons around its circuit in a given time as the dry cell, giving a current 8 times as great. We may think of the potential difference between two points as the amount of "push" available to move charge between the points.

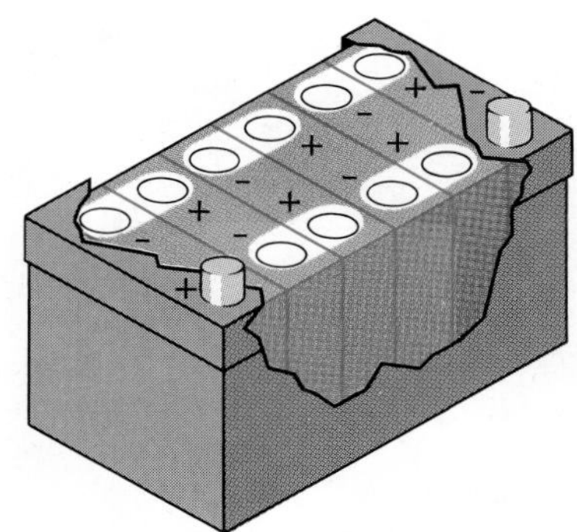

Figure 6-18 A 12-V storage battery consists of six 2-V cells connected in series.

If we connect two or more batteries together as shown in Fig. 6-14*b*, the available voltage is increased. The method of connection, called **series,** is − terminal to + terminal, so that each battery in turn supplies its "push" to electrons flowing through the set. The voltage of a particular cell depends on the chemical reactions that take place in it. In the case of the lead-acid storage battery of a car, each cell has a voltage of 2 V, and six of them are connected together to give the 12 V needed to run the car's electrical equipment (Fig. 6-18). As its name suggests, a storage battery can be **recharged** when the energy it contains is used up. Ordinary dry-cell batteries, such as those used in flashlights and portable radios, have voltages of 1.5 V and cannot be recharged.

A battery is rated according to the total amount of charge it can transfer from one terminal to the other, expressed in ampere-hours (A · h). A typical car battery has a capacity of 60 A · h, which means it can supply a current of 60 A for 1 h, a current of 30 A for 2 h, a current of 1 A for 60 h, and so on. The less the current, the longer the battery can supply it.

Example 6.5

How much energy is stored in a 12-V, 60 A · h battery when it is fully charged? If the mass of the battery is 20 kg and all the energy stored in it is used to raise it from the ground, how high would it go?

Solution

Since 1 A = 1 C/s and 1 h = 3600 s, the amount of charge the battery can transfer from one of its terminals to the other is

$$Q = (60 \text{ A} \cdot \text{h})\left(1 \frac{\text{C/s}}{\text{h}}\right)\left(3600 \frac{\text{s}}{\text{h}}\right) = 2.16 \times 10^5 \text{ C}$$

and the stored energy is, from Eq. 6-3,

$$W = QV = (2.16 \times 10^5 \text{ C})(12 \text{ V}) = 2.59 \times 10^6 \text{ J}$$

The potential energy of the battery at the height h is PE = mgh. Hence $W = mgh$ and

$$h = \frac{W}{mg} = \frac{2.59 \times 10^6 \text{ J}}{(20 \text{ kg})(9.8 \text{ m/s}^2)} = 1.32 \times 10^4 \text{ m} = 13.2 \text{ km}$$

Mt. Everest is only 8.85 km high—a charged battery can contain a lot of energy.

Figure 6-19 Georg Ohm (1787–1854).

6.10 Ohm's Law

Current, Voltage, and Resistance

When different voltages are applied to the ends of the same piece of wire, we find that the current in the wire is proportional to the potential difference. Doubling the voltage doubles the current. This generalization is called **Ohm's law** after its discoverer, the German physicist Georg Ohm (Fig. 6-19).

The property of a conductor that opposes the flow of charge in it is called **resistance.** We can think of resistance as a kind of friction. The more the resistance in a circuit, the less the current for a given applied voltage (Fig. 6-20). If we write I for current, V for voltage, and R for resistance, Ohm's law says that

$$I = \frac{V}{R} \qquad \textit{Ohm's law} \qquad 6\text{-}4$$

$$\text{Current} = \frac{\text{voltage}}{\text{resistance}}$$

Circuit Faults

An electric circuit basically consists of a source of electric energy, such as a battery or generator, a load—for instance, a lightbulb or motor—and wires that connect them. There are two common modes of failure in actual circuits. In one, a wire breaks or comes loose from its terminal. The resulting **open circuit** does not provide a complete conducting path and no current flows.

In a **short circuit,** the connecting wires either accidentally touch each other or are joined by a stray conductor such as a misplaced screwdriver. The current now has an alternate path of very low resistance and, because $I = V/R$, a high current flows. A short circuit is dangerous because the high current produces a lot of heat, which can start a fire or even melt the wires involved.

Fuses and circuit breakers are designed to open electric circuits whenever unsafe amounts of current pass through them, and so they provide protection from the results of a short circuit. All power lines have fuses or circuit breakers for this reason, and many individual electric appliances are so equipped as well.

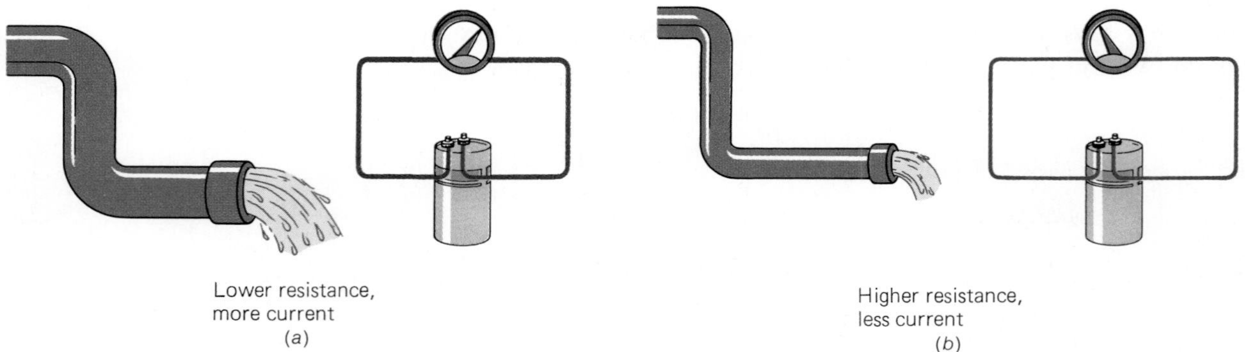

Figure 6-20 (*a*) A short, wide pipe yields a large flow of water, which corresponds to using a short, thick wire that offers less resistance to the flow of charge. (*b*) A long, narrow pipe yields a small flow of water, which corresponds to using a long, thin wire that offers more resistance to the flow of charge.

The unit of resistance is the **ohm,** whose abbreviation is Ω, the Greek capital letter "omega." Hence 1 A = 1 V/Ω and

$$1 \text{ ohm} = 1 \frac{\text{volt}}{\text{ampere}}$$

$$1\ \Omega = 1 \text{ V/A}$$

The resistance of a wire or other metallic conductor depends on the material it is made of (an iron wire has 7 times the resistance of a copper wire of the same size); its length (the longer the wire, the more its resistance); its cross-sectional area (the greater this area, the less the resistance); and the temperature (the higher the temperature, the more the resistance).

Despite its name, Ohm's law is not a basic physical principle such as the law of conservation of energy. Ohm's law is obeyed only by metallic conductors, not by gaseous or liquid conductors and not by such electronic devices as transistors.

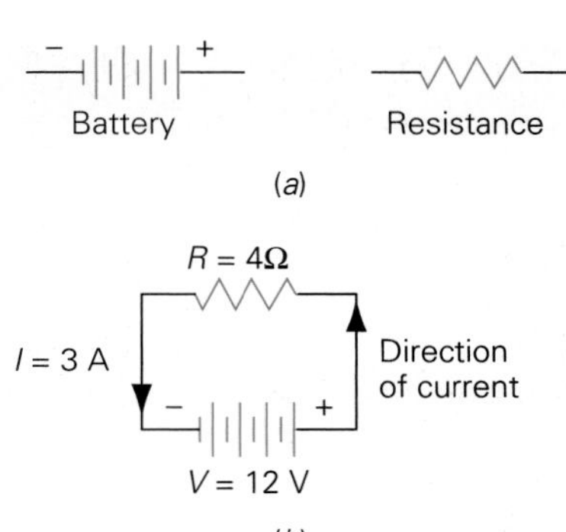

Figure 6-21 (*a*) Symbols for a battery and a resistance. (*b*) A current of 3 A flows in a circuit whose resistance is 4 Ω when a potential difference of 12 V is applied. The current direction is from the + terminal of the battery to the − terminal.

Example 6.6

A car has a 12-V battery whose capacity is 60 A · h. If the car's headlights and taillights have a total resistance of 4 Ω (Fig. 6-21), how long can they be left on before the battery runs down? Assume that the car's engine is not running, so its generator is not recharging the battery.

Solution

The first step is to find the current. From Eq. 6-4,

$$I = \frac{V}{R} = \frac{12\ V}{4\ \Omega} = 3 \text{ A}$$

Because the battery's capacity is 60 A · h, the lights can be left on for

$$t = \frac{60 \text{ A} \cdot \text{h}}{3 \text{ A}} = 20 \text{ h}$$

before the battery runs down.

Example 6.7

Find the resistance of a 120-V electric toaster that draws a current of 4 A.

Solution

To find the resistance, we rewrite Ohm's law, Eq. 6-4, in the form $R = V/I$ and substitute the given values:

$$R = \frac{V}{I} = \frac{120 \text{ V}}{4 \text{ A}} = 30\ \Omega$$

The resistance of the toaster is 30 Ω.

6.11 Electric Power

Current Times Voltage

Electric energy is so useful both because it is conveniently carried by wires and because it is easily converted into other kinds of energy.

Electrical Safety

Body tissue is a fairly good electrical conductor because it contains ions in solution. Dry skin has the most resistance and can protect the rest of the body in case of accidental exposure to a high voltage. This protection disappears when the skin is wet. An electric current in body tissue stimulates nerves and muscles and produces heat. Most people can feel a current as small as 0.0005 A, one of 0.005 A is painful, and one of 0.01 A or more leads to muscle contractions that may prevent a person from letting go of the source of the current. Breathing becomes impossible when the current is greater than about 0.018 A.

Touching a single "live" conductor has no effect if the body is isolated since a complete conducting path is necessary for a current to occur. However, if a person is at the same time grounded by being in contact with a water pipe, by standing on wet soil, or in some other way, a current will pass through his or her body. The human body's resistance is in the neighborhood of 1000 Ω, so if a potential difference of 120 V is applied via wet skin, the resulting current will be somewhere near $I = V/R = 120$ V/1000 Ω = 0.12 A. Such a current is exceedingly dangerous because it causes the heart muscles to contract rapidly and irregularly, which is fatal if allowed to continue.

Electrical devices in bathrooms and kitchens are potential sources of danger because the moisture on a wet finger may be enough to provide a conducting path to the interior of the devices. If a person is in a bathtub and thus is grounded through the tub's water to its drainpipe, or the person has one hand on a faucet, even touching a switch with a wet finger is risky.

Today new electrical installations in bathrooms, kitchens, garages, and outdoors always include **ground fault circuit interruptors** (Fig. 6-22) that trip when the current (white wire) returning from an outlet connected to an appliance is less than the current (black wire) going to the outlet. The difference means that some current is leaking to ground, perhaps through a person's body. Such devices are sensitive to leaks of as little as 0.005 A and can break the circuit in 0.025 s.

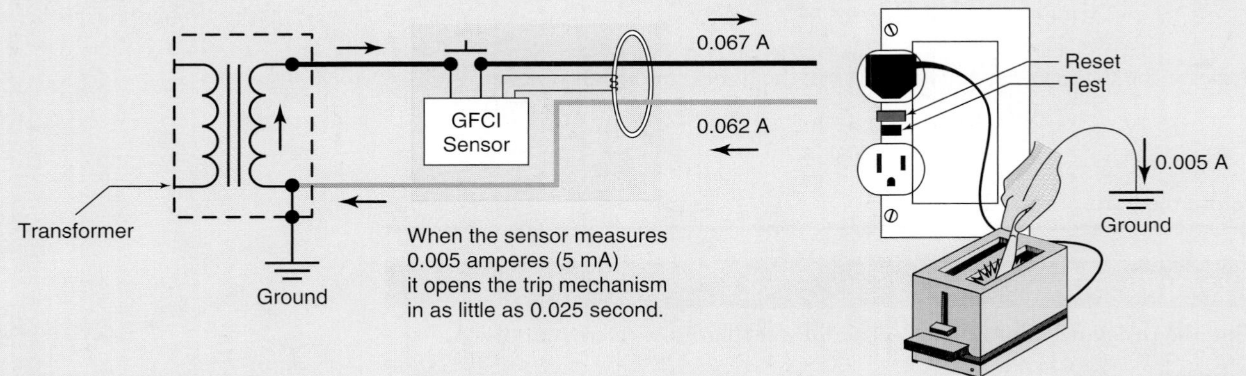

Figure 6-22 A ground fault circuit interruptor (GFCI) opens an electric circuit when even a small amount of current leaks from it to ground, which is especially important if a person's body is accidentally providing the path. In this way a GFCI protects people from possibly fatal electric shocks. (Transformers are discussed in Sec. 6.19.)

Electric energy in the form of electric current becomes radiant energy in a lightbulb, chemical energy when a storage battery is charged, kinetic energy in an electric motor, heat in an electric oven. In each case the current performs work on the device it passes through, and the device then turns this work into another kind of energy.

As in the case of the energy lost due to friction, the energy lost due to the resistance of a conductor becomes heat. This is the basis of electric heaters and stoves. In a lightbulb, the filament is so hot that it glows white. In electric circuits it is obviously important to use wires large enough in diameter, and hence small enough in resistance, to prevent the wires becoming so hot that they melt their insulation and start fires. A thin extension cord suitable for a lamp or radio might well be dangerous used for a heater or power tool.

An important quantity in any discussion of electric current is the rate at which a current is doing work—in other words, the **power** of the current. From Eq. 6-3 we know that $V = W/Q$, where W is the work done in transferring the charge Q through a potential difference of V. Since $W = Pt$ and $Q = It$, we have

$$V = \frac{W}{Q} = \frac{Pt}{It} = \frac{P}{I}$$

Thus we have the useful result that electric power is given by the product of the current and the voltage of the circuit:

$$P = IV \qquad \textit{Electric power} \qquad 6\text{-}5$$

$$\text{Power} = (\text{current})(\text{voltage})$$

Now we can see why electrical appliances are rated in watts; as we learned in Sec. 3.2, the watt is the unit of power. A 60-W lightbulb uses twice the power of a 30-W bulb, and one-tenth the power of a 600-W electric drill (Table 6-1).

A fuse or circuit breaker interrupts a power line if the current exceeds a safe limit. Many of the fuses normally used in homes are rated at 15 A. Since the power-line voltage is 120 V, the greatest power a 15-A line can provide without blowing the fuse is

$$P = IV = (15\text{ A})(120\text{ V}) = 1800\text{ W} = 1.8\text{ kW}$$

Because $P = IV$, it is easy to find how much current is needed by an appliance rated in watts when connected to a power line of given voltage. For instance, a 60-W bulb connected to a 120-V line needs a current of

$$I = \frac{P}{V} = \frac{60\text{ W}}{120\text{ V}} = 0.5\text{ A}$$

Table 6-1 Typical Power Ratings of Various Appliances

Appliance	Power, W
Charger for electric toothbrush	1
Clothes dryer	5,000
Coffeemaker	700
Dishwasher	1,600
Fan	150
Fax transmitter/ receiver	65
Heater	2,000
Iron	1,000
Personal computer	150
Portable sander	200
Refrigerator	400
Stove	12,000
TV receiver	120
Vacuum cleaner	750

Example 6.8

Can the energy of lightning strokes be harnessed for human needs? A lightning stroke might be driven by a potential difference of 2 million volts, involve a current of 5000 A, and last for 1 ms (0.001 s). How much energy is released in this stroke?

Solution

The power of the lightning stroke is

$$P = IV = (5 \times 10^3\text{ A})(2 \times 10^6\text{ V}) = 1 \times 10^{10}\text{ W}$$

This is 10 billion watts! However, although this is a lot of *power*, the *energy* released is modest:

$$E = Pt = (1 \times 10^{10}\text{ W})(1 \times 10^{-3}\text{ s}) = 1 \times 10^7\text{ J} = 10\text{ MJ}$$

Burning a cup of gasoline liberates more energy than this. Hence the response to the question of how to utilize the energy of lightning is, why bother? The reason lightning strokes are so destructive despite their relatively minor energy content is that the release of this energy takes place very quickly in a fairly small volume, which produces an explosive effect.

Figure 6-23 A kilowatthour meter registers the electric energy that has been supplied to a house or other user of electricity.

Table 6-2 Electrical Quantities

Quantity	Symbol	Unit	Meaning	Formula
Charge	Q	Coulomb (C)	A basic property of most elementary particles. The electron has a charge of -1.6×10^{-19} C.	
Current	I	Ampere (A) (1 A = 1 C/s)	Rate of flow of charge.	$I = \frac{Q}{t} = \frac{P}{V}$
Potential difference (voltage)	V	Volt (V) (1 V = 1 J/C)	Potential energy difference per coulomb of charge between two points; corresponds to pressure in water flow.	$V = \frac{W}{Q} = IR = \frac{P}{I}$
Resistance	R	Ohm (Ω) (1Ω = 1 V/A)	A measure of the opposition to the flow of charge in a particular circuit. For a given voltage, the higher the resistance, the lower the current.	$R = \frac{V}{I}$
Power	P	Watt (W) (1 W = 1 V · A)	Rate of energy flow.	$P = \frac{W}{t} = IV$

The Kilowatthour Users of electricity pay for the amount of energy they consume. The usual commercial unit of electric energy is the **kilowatthour** (kWh), which is the energy supplied per hour when the power level is 1 kilowatt (Fig. 6-23). If electricity is sold at \$0.12 per kilowatthour, the cost of operating a 1.5-kW electric heater for 7 h would be

$$\begin{aligned}\text{Cost} &= (\text{price per unit of energy})(\text{energy used})\\ &= (\text{price per unit of energy})(\text{power})(\text{time})\\ &= (\$0.12/\text{kWh})(1.5\ \text{kW})(7\ \text{h}) = \$1.26\end{aligned}$$

Table 6-2 summarizes the various electrical quantities we have been discussing.

Magnetism

Electricity and magnetism were once considered as completely separate phenomena. One of the great achievements of nineteenth-century science was the realization that they are really very closely related, a realization that led to the discovery of the electromagnetic nature of light. And one of the great achievements of nineteenth-century technology was the invention of electric motors and generators, whose operation depends upon the connection between electricity and magnetism.

Series and Parallel

There are two basic ways to connect circuit elements. The resistors in Fig. 6-24*a* are joined end-to-end in **series** with the same current flowing through all of them. In (*b*) the resistors are in **parallel** and the total current is split up among them. This is the arrangement used in household wiring, so that each lamp or other appliance has the same voltage (usually 120 V) across it. If the appliances were in series, all would have to be switched on for any of them to work, and each would have a smaller voltage than 120 V.

When batteries are connected in series, as in Fig. 6-24*c*, their voltages add: three 12-V batteries in series gives 36 V for the set. In parallel, as in (*d*), the voltage of the set is still 12 V but the set can provide three times as much current as each battery alone. When a car's battery is too weak to start the car, the remedy is another battery connected in parallel to it (+ terminal to +, − terminal to −) using jumper cables.

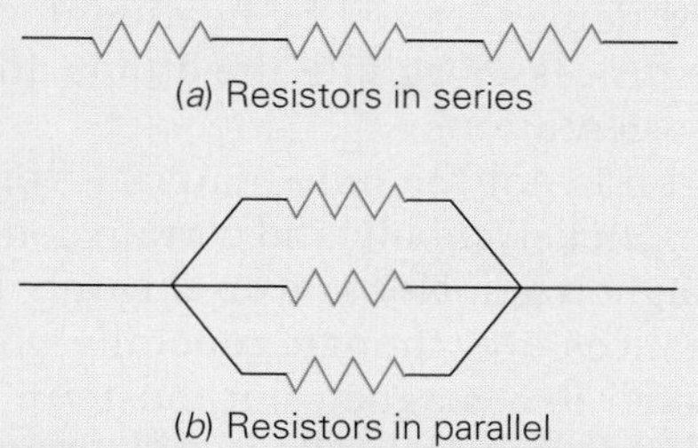

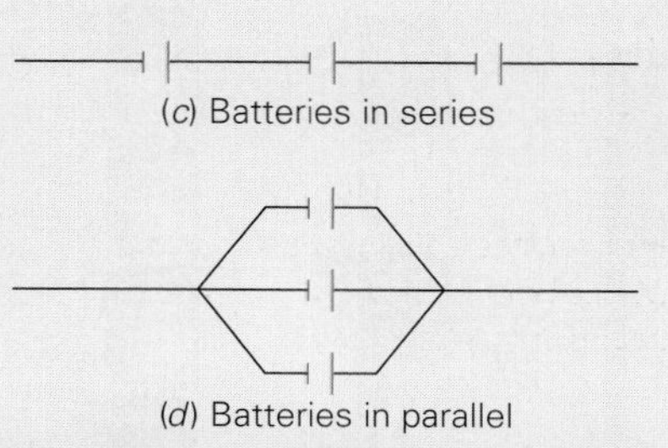

Figure 6-24 Series and parallel connections. Household electric outlets are always connected in parallel so each one has the same applied voltage of 120 V.

6.12 Magnets

Attraction and Repulsion

Ordinary magnets are familiar to everybody. The simplest is a bar of iron that has been magnetized in one way or another, say, by having been stroked by another magnet. A magnetized iron bar is recognized, of course, by its ability to attract and hold other pieces of iron to itself. Most of the force a magnet exerts comes from its ends, as we can see by testing the attraction of different parts of a bar magnet for iron nails.

If we pivot a magnet at its center so that it can swing freely, we will find that it turns so that one end points north and the other south. The north-pointing end is called the **north pole** of the magnet, and the south-pointing end is called its **south pole.** The tendency of a magnet to line up with the earth's axis is the basis of the compass, whose needle is a small magnet (Fig. 6-25). (In Chap. 15 we shall find that the reason for this behavior is that the earth itself is a giant magnet.)

If the north poles of two magnets are brought near each other, the magnets repel. If the north pole of one magnet is brought near the south pole of the other, the magnets attract (Fig. 6-26). This gives us a simple rule like that for electric charges:

> Like magnetic poles repel one another, unlike poles attract one another.

Figure 6-25 A magnetic compass uses the earth's magnetic field to establish direction. The magnetic axis of the earth is not quite aligned with its axis of rotation, so the needle of a magnetic compass does not point exactly to true north.

Poles Always Come in Pairs Positive and negative charges in neutral matter can be separated from each other. Can the north and south poles of a magnet also be separated? It would seem that all we have to do is

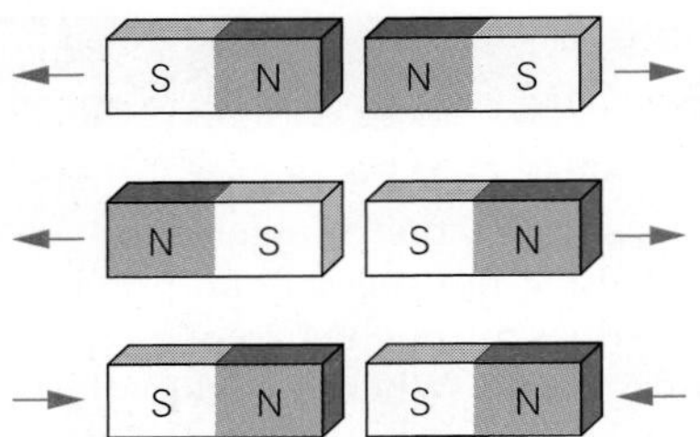

Figure 6-26 Like magnetic poles repel each other; unlike magnetic poles attract.

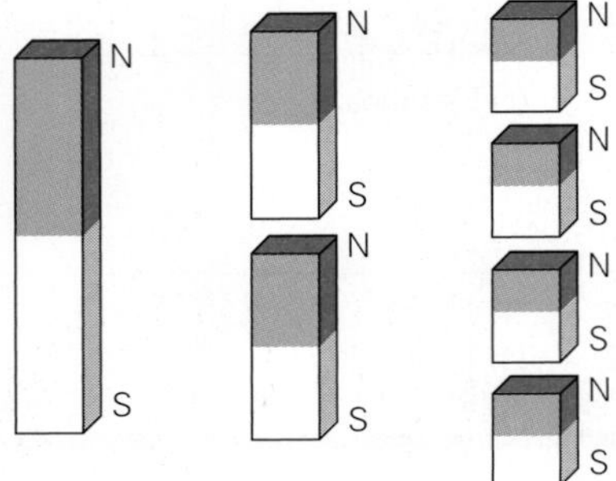

Figure 6-27 Cutting a magnet in half produces two other magnets. There is no such thing as a single free magnetic pole.

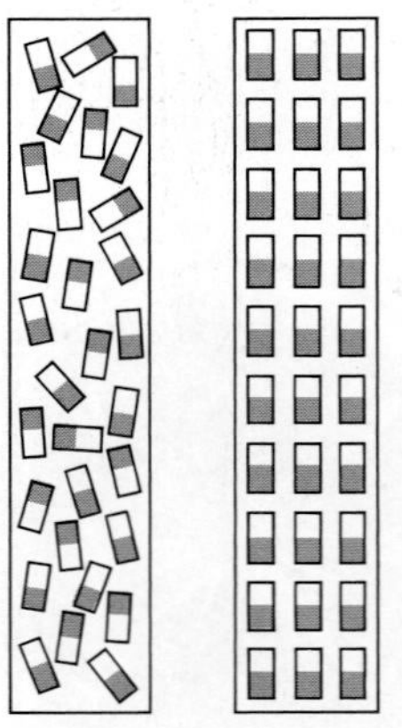

Figure 6-28 The iron atoms in an unmagnetized iron bar are randomly oriented, whereas in a magnetized bar they are aligned with their north poles pointing in the same direction. The ability of iron atoms to remain aligned in this way is responsible for the magnetic properties of iron.

to saw the magnet in half. But if we do this, as in Fig. 6-27, we find that the resulting pieces each have an N pole and an S pole. We may cut the resulting magnets in two again and continue as long as we like, but each piece, however small, will still have both an N pole and an S pole. There is no such thing as a single free magnetic pole.

Since a magnet can be cut into smaller and smaller pieces indefinitely with each piece a small magnet in itself, we conclude that magnetism is a property of the iron atoms themselves. Each atom of iron behaves as if it has an N pole and an S pole. In ordinary iron the atoms have their poles randomly arranged, and nearby N and S poles cancel out each other's effect. When a bar of iron is magnetized, many or all of the atoms are aligned with the N poles in the same direction, so that the strengths of all the tiny magnets are added together (Fig. 6-28). A "permanent" magnet can be demagnetized by heating it strongly or by hammering it. Both of these processes agitate the atoms and restore them to their normal random orientations.

Iron is not the only material from which permanent magnets can be made. Nickel, cobalt, and certain combinations of other elements can also be magnetized. Nor is iron the only material affected by magnetism—*all* substances are, though generally only to a very slight extent. Some are attracted to a magnet, but most are repelled. In the case of mercury the repulsion, though weak, is still enough to be easily observed.

6.13 Magnetic Field

How Magnetic Forces Act

We are so familiar with gravitational, electric, and magnetic forces that we take them for granted. However, if we think about them, it is clear that something remarkable is going on: these forces act without the objects involved touching each other. We cannot move a book from a table by just waving our hand at it, and a golf ball will not fly off until a golf club actually strikes it. An iron nail, however, does not wait until a magnet touches it, but is pulled to the magnet when the two are some distance apart. The properties of space near the magnet are somehow altered by the magnet's presence, just as a mass or an electric charge alters the space around itself, in each case in a different way.

The altered space around a mass, an electric charge, or a magnet is called a **force field.** A physicist describes a force field in terms of what it does, which is to exert a force on appropriate objects. Although we cannot see a force field, we can detect its presence by its effects. In fact, even the forces we think of as being exerted by direct contact turn out to involve force fields. For instance, when a golf club strikes a ball, it is the action of electric forces on the molecular level that leads to the observed transfer of energy and momentum to the ball. There is actually no such thing as "direct contact" since the atoms involved never touch each other.

When iron filings are scattered on a card held over a magnet, they form a pattern that suggests the form of the magnet's field. At each point on the card, the filings line up in the direction in which a piece of iron would move if put there, and the filings gather most thickly where the

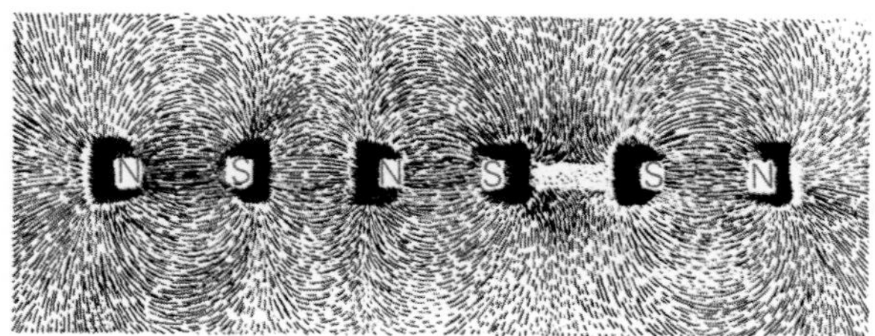

Figure 6-29 Patterns formed by iron filings sprinkled on a card held over three bar magnets. The filings align themselves in the direction of the magnetic field. It is convenient to think of the pattern in terms of "field lines," but such lines do not actually exist since the field is a continuous property of the region of space it occupies.

force on the iron would be greatest. Figure 6-29 shows the patterns of iron filings near three bar magnets.

Field Lines It is traditional, and convenient, to think of a **magnetic field** in terms of imaginary **field lines** that correspond to the patterns formed by iron filings. A magnetic field line traces the path that would be taken by a small iron object if placed in the field, with the lines close together where the field is strong and far apart where the field is weak. Although the notion of field lines is helpful in illustrating a number of magnetic effects, we must keep in mind that they are imaginary—a force field is a continuous property of the region of space where it is present, not a collection of strings. Figure 15-35 shows the field lines of the earth's magnetic field.

Figure 6-30 Hans Christian Oersted (1777–1851).

6.14 Oersted's Experiment

Magnetic Fields Originate in Moving Electric Charges

Electric currents may not be familiar to us as sources of magnetic fields, yet every current has such a field around it. To repeat a famous experiment first performed in 1820 by the Danish physicist Hans Christian Oersted (Fig. 6-30), we can connect a horizontal wire to a battery and hold under the wire a small compass needle (Fig. 6-31). The needle at once swings into a position at right angles to the wire. When the compass is placed just above the wire, the needle swings around until it is again perpendicular to the wire but pointing in the opposite direction.

We can use iron filings to study the magnetic field pattern around a wire carrying a current. When we do this, we find that the field lines

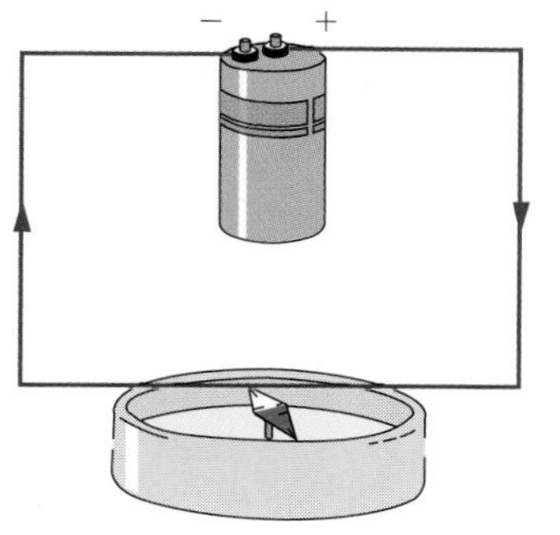

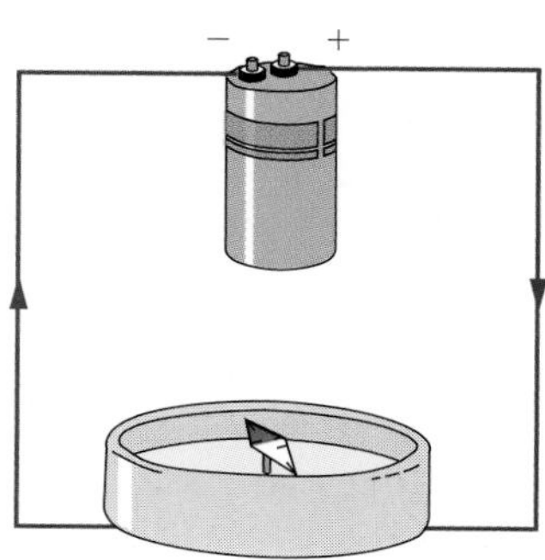

Figure 6-31 Oersted's experiment showed that a magnetic field surrounds every electric current. The field direction above the wire is opposite to that below the wire.

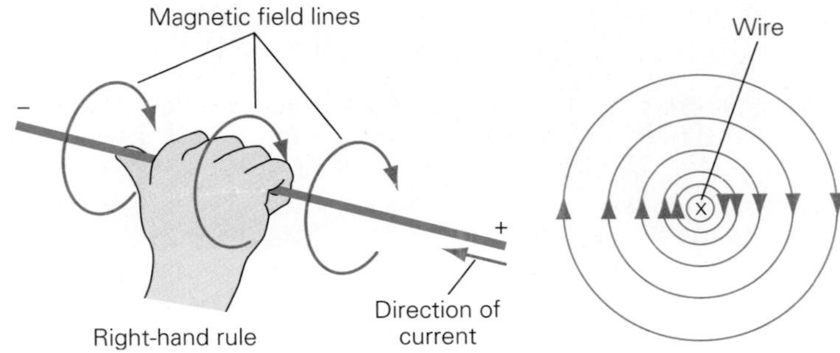

Figure 6-32 Magnetic field lines around a wire carrying an electric current. The direction of the lines may be found by placing the thumb of the right hand in the direction of the current; the curled fingers then point in the direction of the field lines. In the right-hand diagram the current flows into the paper. (The × represents the tail feathers of an arrow.)

near the wire consist of circles, as in Fig. 6-32. The direction of the field lines (that is, the direction in which the N pole of the compass points) depends on the direction of flow of electrons through the wire. When one is reversed, the other reverses also.

In general, the direction of the magnetic field around a wire can be found by encircling the wire with the fingers of the right hand so that the extended thumb points along the wire in the direction of the current. According to this **right-hand rule,** the current and the field are perpendicular to each other.

Oersted's discovery showed for the first time that a connection exists between electricity and magnetism. It was also the first demonstration of the principle on which the electric motor is based. Magnetism and electricity are related, but only through moving charges. An electric charge *at rest* has no magnetic properties. A magnet is not influenced by a stationary electric charge near it, and vice versa.

When a current passes through a wire bent into a circle, the resulting magnetic field, shown in Fig. 6-33, is the same as that produced by a bar magnet. One side of the loop acts as a north pole, the other as a south pole. If free to turn, the loop swings to a north-south position. A current loop attracts pieces of iron just as a bar magnet does.

The results of Oersted's experiment and of many others allow us to say that

> All moving electric charges give rise to magnetic fields.

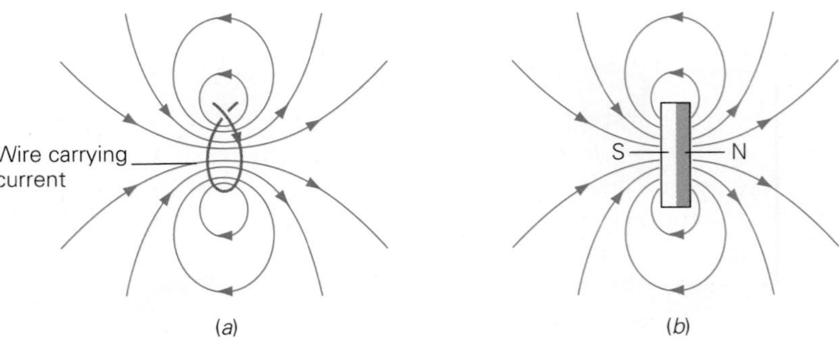

Figure 6-33 The magnetic field of a loop of electric current is the same as that of a bar magnet.

BIOGRAPHY André Marie Ampère (1775–1836)

Ampère was largely self-taught and had mastered advanced mathematics by his early teens. Starting as a teacher in local schools near Lyon, he went on to a series of professorships in Paris and was appointed by Napoleon as inspector-general of the French university system. Ampère's personal life was one misfortune after another: the execution of his father during the French Revolution, the early death of his much-loved first wife, a disastrous second marriage, financial problems.

In contrast were the successes of his scientific career. Upon learning of Oersted's discovery that a magnetic field surrounds every electric current, Ampère carried out experiments of his own that resulted in "a great theory of these phenomena and of all others known for magnets," as he wrote to his son only two weeks later. Ampère's results included the law that describes the magnetic force between two currents and the observation that the magnetic field around a current loop is the same as the field around a bar magnet. He went on to speculate that the magnetism of a material such as iron is due to loops of electric current in its atoms, a concept very much ahead of his time. The unit of electric current is called the ampere because he was the first to distinguish clearly between current and potential difference.

The Electromagnetic Field An electric charge at rest is surrounded only by an electric field, and when the charge is moving it is surrounded by a magnetic field as well. Suppose we travel alongside a moving charge, in the same direction and at the same speed. All we find now is an electric field—the magnetic field has disappeared. But if we move past a stationary charge with our instruments, we find both an electric and a magnetic field! Clearly the *relative motion* between charge and observer is needed to produce a magnetic field: no relative motion, no magnetic field.

According to the theory of relativity, whatever it is in nature that shows itself as an electric force between charges at rest *must* also show itself as a magnetic force between moving charges. One effect is not possible without the other. Thus the proper way to regard what we think of as separate electric and magnetic fields is that they are both aspects of a single electromagnetic field that surrounds every electric charge. The electric field is always there, but the magnetic field appears only when relative motion is present.

In the case of a wire that carries an electric current, there is only a magnetic field because the wire itself is electrically neutral. The electric field of the electrons is canceled out by the opposite electric field of the positive ions in the wire. However, the positive ions are stationary and therefore have no magnetic field to cancel the magnetic field of the moving electrons. If we simply move a wire that has no current in it, the electric and magnetic fields of the electrons are canceled by the electric and magnetic fields of the positive ions.

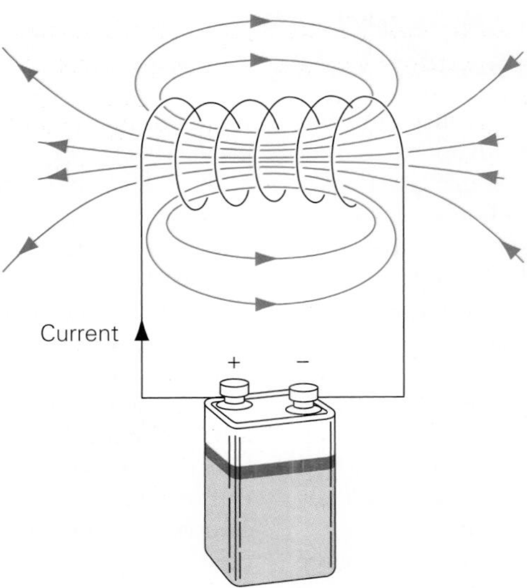

Figure 6-34 The magnetic field of a coil is like that of a single loop but is stronger.

6.15 Electromagnets

How to Create a Strong Magnetic Field

When several wires that carry currents in the same direction are side by side, their magnetic fields add together to give a stronger total magnetic field. This effect is often used to increase the magnetic field of a current loop. Instead of one loop, many loops of wire are wound into a coil, as in Fig. 6-34, and the resulting magnetic field is as many times stronger than the field of one turn as there are turns in the coil. A coil with 50 turns produces a field 50 times greater than a coil with just one turn.

The magnetic field of the coil is enormously increased if a rod of iron is placed inside it (Fig. 6-35). This combination of coil and iron core is called an **electromagnet.** An electromagnet exerts magnetic force only

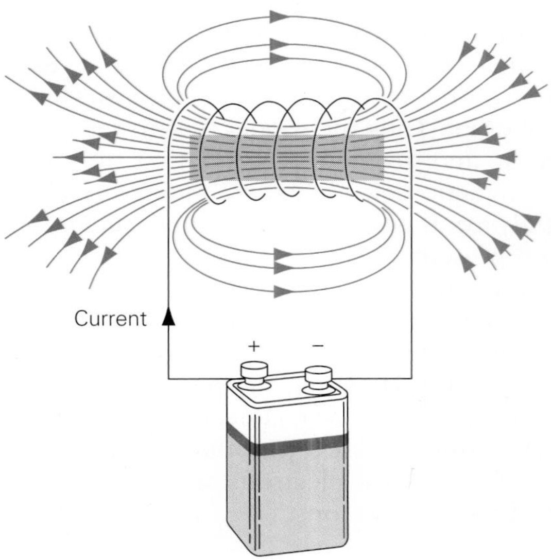

Figure 6-35 An electromagnet consists of a coil with an iron core, which considerably enhances the magnetic field produced.

when current flows through its turns, and so its action can be turned on and off. Also, by using many turns and enough current, an electromagnet can be made far more powerful than a permanent magnet. Electromagnets are widely used and range in size from the tiny coils in telephone receivers to the huge ones that load and unload scrap iron (Fig. 6-36).

Figure 6-36 Electromagnet loading scrap iron and steel.

Using Magnetism

An electric motor uses magnetic fields to turn electric energy into mechanical energy, and a generator uses magnetic fields to turn mechanical energy into electric energy. As we shall find, magnetic fields also play essential roles in television picture tubes, in sound and video recording, and in the transformers used to distribute electric power over large areas.

6.16 Magnetic Force on a Current

A Sidewise Push

Suppose that a horizontal wire connected to a battery is suspended as in Fig. 6-37, so that it is free to move from side to side, and the N pole of a bar magnet is then placed directly under it. This arrangement is the reverse of Oersted's experiment. Oersted placed a movable magnet near a wire fixed in position, whereas here we have a movable wire near a fixed magnet. We might predict, from Oersted's results and Newton's third law of motion, that in this case the wire will move. It does indeed, swinging out to one side as soon as the current is on. The direction of the wire's motion is perpendicular to the bar magnet's field. Whether the

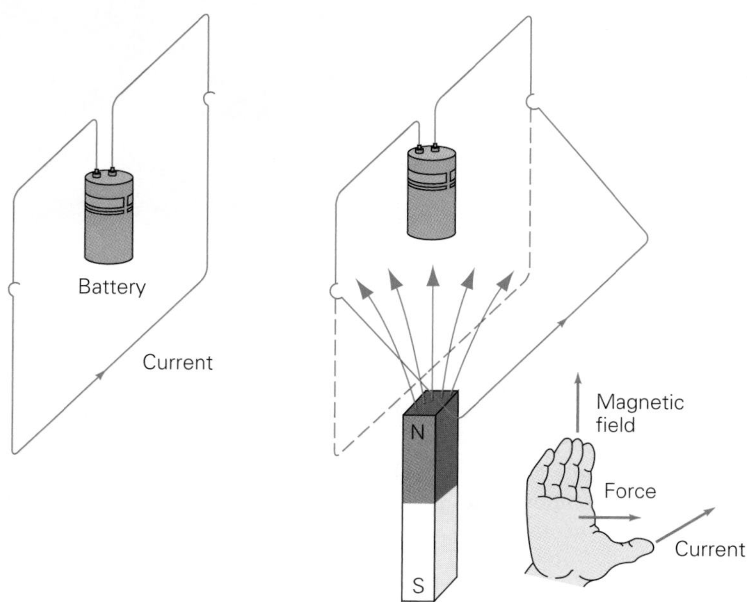

Figure 6-37 A magnetic field exerts a sidewise push on an electric current. In this arrangement, the wire moves to the side in a direction perpendicular to both the magnetic field and the current. A handy way to figure out the direction of the force is to open your right hand so that the fingers are together and the thumb sticks out. When your thumb is in the direction of the current and your fingers are in the direction of the magnetic field, your palm faces in the direction of the force. To remember this rule, think of your thumb in terms of hitchhiking and so as the current, of your parallel fingers as magnetic field lines, and of your palm as pushing on something. The same rule holds for the force on a moving positive charge. The force on a moving negative charge is in the opposite direction.

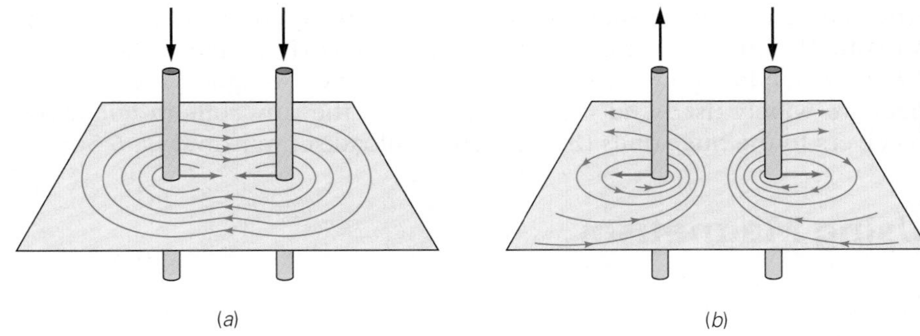

Figure 6-38 Equal and opposite forces are exerted by parallel currents on each other. The forces are attractive when the currents are in the same direction, repulsive when they are in opposite directions.

wire swings to one side or the other depends on the direction of flow of electrons in the wire and on which pole of the magnet is used.

Thus the force a magnetic field exerts on an electric current is not a simple attraction or repulsion but a *sidewise push.* The maximum

Maglev Trains

Electromagnets can provide enough force to support—"levitate"—trains and so eliminate friction (although air resistance remains). Magnetic forces also propel such **maglev** trains. Experimental maglev trains based on pioneering work done in the United States 40 years ago were first built in Germany and Japan. In Germany, attractive forces are developed by conventional electromagnets that curl under a T-shaped guideway. A maglev system of this kind was built in China that connects Shanghai with its new airport 30 km away. The trains carry 600 passengers at 430 km/h (267 mi/h). A proposed 175-km maglev link between Shanghai and Hangzhou that would have cut the journey time between these cities to 30 min from the 140 min that conventional trains take was canceled because of its expected cost of over $5 billion.

In Japan, where an 8.9-km maglev system with nine stations came into service near Nagoya in 2005, the upward force is provided by magnetic repulsion using superconducting coils for the magnetic fields (Fig. 6-39). The Japanese system holds the speed record of 581 km/h (361 mi/h). In both systems the magnetic field of an alternating current passed through electromagnets along the guideway creates attractive forces that pull the train's own magnets forward and repulsive forces that push them from behind. The higher the frequency of the alternating current, the higher the train's speed. Such a **linear motor** is like an ordinary electric motor cut open and unrolled.

A third type of maglev train, called Inductrak, would use powerful permanent magnets in the train cars; a small-scale working model suggests important advantages of efficiency and safety over the German and Japanese approaches.

When perfected, maglev trains are expected to use less than half the energy per person per kilometer than jet aircraft do and ought to be competitive in total trip times for distances of up to perhaps 1000 km, as well as being more convenient. At present, construction costs—$1.25 billion for the Shanghai airport link—are the chief obstacle to wider use of maglev trains. Even so, a number of maglev systems are being considered around the world. In the United States high-speed maglev links have been proposed to connect Pittsburgh with its airport, Las Vegas with Los Angeles, and Baltimore with Washington, DC. In Japan maglev trains may someday speed at 500 km/h between Tokyo and both Nagoya and Osaka.

Figure 6-39 This experimental Japanese train uses magnetic forces for both support and propulsion.

sidewise push occurs when the current is perpendicular to the magnetic field, as in Fig. 6-37. At other angles the push is less, and it disappears when the current is parallel to the magnetic field.

Every current has a magnetic field around it, and as a result nearby currents exert magnetic forces on each other. When parallel currents are in the same direction, as in Fig. 6-38*a*, the forces are attractive; when the currents are in opposite directions, as in Fig. 6-38*b*, the forces are repulsive.

6.17 Electric Motors

Mechanical Energy from Electric Energy

The sidewise push of a magnetic field on a current-carrying wire can be used to produce continuous motion in an arrangement like that shown in Fig. 6-40. A magnet gives rise to a magnetic field inside which a wire loop is free to turn. When the plane of the loop is parallel to the magnetic field, there is no force on the two sides of the loop that lie along the magnetic field. The side of the loop at left in the diagram, however, receives a downward push, and the side at right receives an upward one. Thus the loop is turned counterclockwise.

To produce a continuous movement, the direction of the current in the loop must be reversed when the loop is vertical. The reversed current then interacts with the magnetic field to continue to rotate the loop through 180°. Now the current must have its direction reversed once more, whereupon the loop will again swing around through a half-turn. The device used to automatically change the current direction is called a **commutator;** it is visible on the shaft of a direct-current motor as a copper sleeve divided into segments. Normally more than two loops and commutator segments are used in order to yield the maximum turning force.

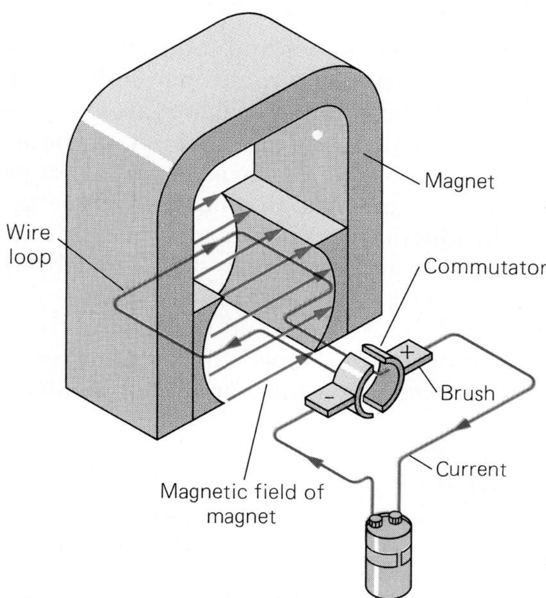

Figure 6-40 A simple direct-current electric motor. The commutator reverses the current in the loop periodically so that the loop always rotates in the same direction.

Figure 6-41 The stationary windings of a large electric motor. Magnetic forces underlie the operation of such motors.

Actual direct-current motors, such as the starter motor of a car, are more complicated than the one shown in Fig. 6-40, but their basic operating principle is the same. Usually electromagnets are employed rather than permanent ones to create the field (Fig. 6-41), and in some motors the coil is fixed in place and the magnet or magnets rotate inside it. Motors built for alternating rather than direct current do not need commutators because the current direction changes back and forth many times per second.

6.18 Electromagnetic Induction

Electric Energy from Mechanical Energy

The electric energy that our homes and industries use in such quantity comes from generators driven by turbines powered by running water or, more often, by steam. In the latter case, as we saw in Fig. 5-50, the boilers that supply the steam obtain heat from coal, oil, or natural gas, or from nuclear reactors. Ships and isolated farms have smaller generators operated by gasoline or diesel engines. In all cases the energy that is turned into electricity is the kinetic energy of moving machinery.

The principle of the generator was discovered by the nineteenth-century English physicist Michael Faraday. Faraday's curiosity was aroused by the research of Ampère and Oersted on the magnetic fields around electric currents. He reasoned that, if a current can produce a magnetic field, then somehow a magnet should be able to generate an electric current.

A wire placed in a magnetic field and connected to a meter shows no sign of a current. What Faraday found instead is that

> A current is produced in a wire when there is relative motion between the wire and a magnetic field.

BIOGRAPHY Michael Faraday (1791–1867)

The son of a blacksmith, Faraday was apprenticed to a bookbinder at 13 and taught himself chemistry and physics from the books he was learning to bind. At 21 he became bottle washer for Humphrey Davy at that noted chemist's laboratory in the Royal Institution in London. Within 20 years Faraday succeeded Davy as head of the Institution. During that period Faraday had, among other things, liquefied a number of gases for the first time and formulated what are today called Faraday's laws of electrolysis. In the later years of his life came the remarkable work in electricity and magnetism whose results include building the first electric motor and the discovery of electromagnetic induction.

Faraday realized at once the implications of electromagnetic induction and soon had generators and transformers working in his laboratory. Asked by a politician what use these devices were, Faraday replied, "At present I do not know, but one day you will be able to tax them." To make sense of the electric and magnetic fields he could not see or feel or represent mathematically (he was poor at math), Faraday invented field lines—lines that do not exist but help us to picture what is going on. At his death he left behind notebooks with over 16,000 entries that testify to his originality, intuition, skill, and diligence.

As long as the wire continues to move across magnetic field lines, the current continues. When the motion stops, the current stops. Because it is produced by motion through a magnetic field, this sort of current is called an **induced current.** The entire effect is known as **electromagnetic induction.**

Let us repeat Faraday's experiment. Suppose that the wire of Fig. 6-42 is moved back and forth across the field lines of force of the bar magnet. The meter will indicate a current first in one direction, then in the other. The direction of the induced current through the wire depends on the relative directions of the wire's motion and of the field lines. Reverse the motion, or use the opposite magnetic pole, and the current is reversed. The strength of the current depends on the strength of the magnetic field and on how rapid the wire's motion is.

Electromagnetic induction is related to the sidewise force a magnetic field exerts on electrons flowing along a wire. In Faraday's experiment electrons are again moved through a field, but now by moving the wire as a whole. The electrons are pushed sidewise as before and, in response to the push, move along the wire as an electric current.

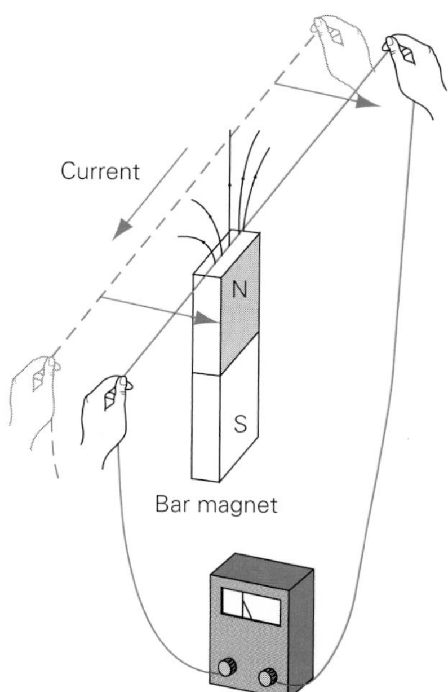

Figure 6-42 Electromagnetic induction. The direction of the induced current is perpendicular both to the magnetic lines of force and to the direction in which the wire is moving. No current is induced when the wire is at rest.

Alternating and Direct Currents In order to obtain a large induced current, an actual generator uses several coils rather than a single wire and several electromagnets instead of a bar magnet. Turned rapidly between the electromagnets, wires of the coil cut lines of force first one way, then the other. How a generator works is shown in Fig. 6-43, where a coil is shown turning between two magnets. During one part of each turn, each side of the coil cuts the field in one direction. Then, during the other part of the turn, each side of the coil cuts the field in the opposite

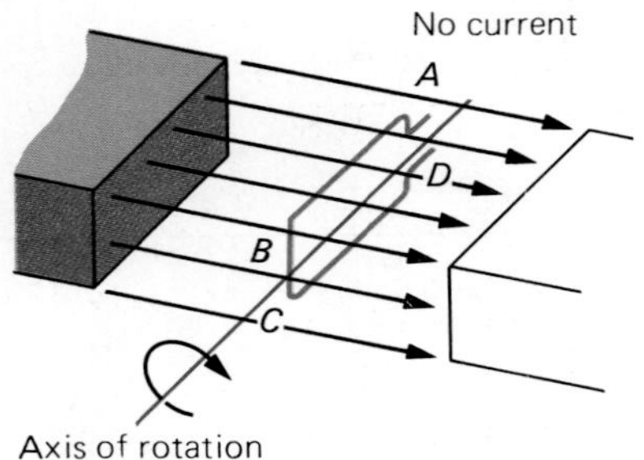

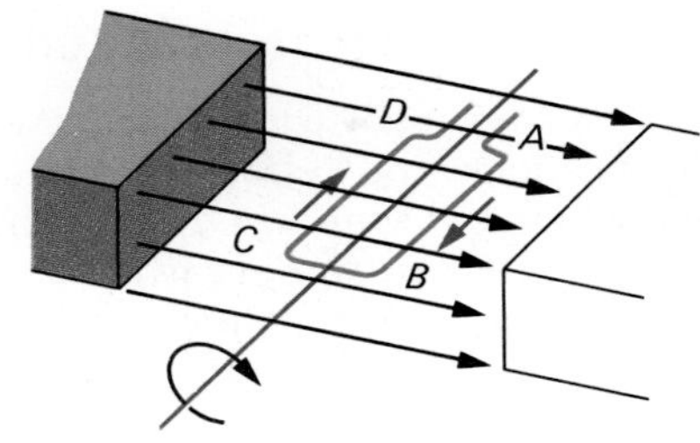

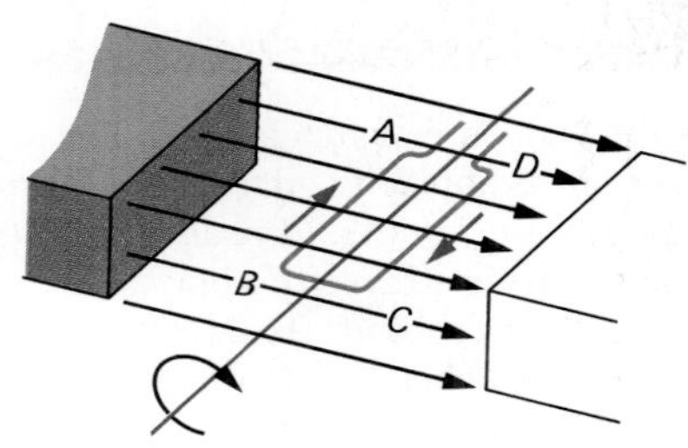

Figure 6-43 An alternating-current generator. As the loop rotates, current is induced in it first in one direction (*ABCD*) and then in the other (*DCBA*). No current flows at those times when the loop is moving parallel to the magnetic field.

Magnetic Navigation by Animals

A number of animals—notably certain birds, turtles, fish, and insects—use the earth's magnetic field to help them find their way on journeys that may cover thousands of kilometers. Sharks detect the field by means of sense organs in their snouts that respond to tiny electric currents induced by their motion through the field (Fig. 6-44). The amount of current depends on the angle between the field and the direction in which the shark is swimming.

Figure 6-44 Sharks navigate with the help of the earth's magnetic field. They detect the field using electromagnetic induction.

Other animals have bits of magnetite, a mineral that contains iron and is affected by magnetic fields, in their brains. Exactly how animals employ magnetite to sense direction is not known, but experiments leave no doubt that they do. Some bacteria that contain magnetite even use the earth's field to distinguish between up and down when floating in a pond, which enables them to find their preferred habitat in the ooze at the bottom. The human brain also contains magnetite: does that mean we have built-in compasses? Nobody knows for sure.

A third sensing mechanism is based on the presence of certain pigments in the eyes of some animals that become weakly magnetic when light falls on them. The signals such eyes send to the brain seem to be affected by a magnetic field.

Magnetic sensing is not the only way that migratory birds navigate on their voyages—as far as 19,000 km for the Arctic tern, which goes from Canada to Antarctica and then back as the seasons change. Such birds also depend for direction on the sun, which traces an east-west arc in the sky with true north (or south in the southern hemisphere) at its highest point, and on the stars, which appear to circle a point in the sky that is always true north (or south). Apparently birds use clues from the sun and stars to calibrate their magnetic compasses as well as directly for navigation.

direction. Hence the induced current flows first one way and then the other. Such a back-and-forth current is an **alternating current.**

The pressure variations of a sound wave (described in detail in Chap. 7) are changed into an alternating current by a microphone, of which there are several kinds. One of them, shown in Fig. 6-45, makes use of electromagnetic induction. A loudspeaker, which changes an alternating current into sound waves, resembles this type of microphone. Its operation is based on the force exerted on a current-carrying wire in a magnetic field.

The electric currents that come from such sources as batteries and photoelectric cells are always one-way **direct currents** that can be reversed only by changing the connections. In the 60-Hz (1 Hz = 1 hertz = 1 cycle/second) alternating current that we ordinarily use in our homes, electrons change their direction 120 times each second (Fig. 6-46). The usual abbreviation for alternating current is ac and that for direct current is dc.

By using commutators like those used on dc motors, generators can be built that produce direct current. Another way to obtain direct current from an ac generator (or **alternator**) is to use a **rectifier,** a device that permits current to pass through it in only one direction. Because alternators are simpler to make and more reliable than dc generators, they are often used together with rectifiers to give the direct current needed to charge the batteries of cars and trucks.

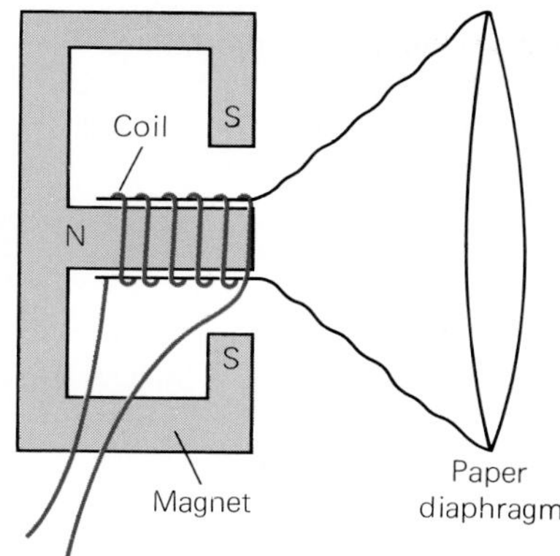

Figure 6-45 A moving-coil microphone. When sound waves reach the diaphragm, it vibrates accordingly. The motion of the coil through the magnetic field of the magnet induces an alternating current in the coil that corresponds to the original sound. A loudspeaker is similar in construction except that an alternating current in its coil causes the diaphragm to vibrate and thereby produce sound waves.

6.19 Transformers

Stepping Voltage Up or Down

To induce a current requires that magnetic field lines move across a conductor. As in Fig. 6-42, one way to do this is to move a wire past a magnet. Another way is to hold the wire stationary while the magnet is moved. We come now to a third, less obvious, method, which involves no visible motion at all.

Let us connect coil *A* in Fig. 6-47 to a switch and a battery and connect the separate coil *B* to a meter. When the switch is closed, a current flows through *A*, building up a magnetic field around it. The current and field do not reach their full strengths at once. A fraction of a second is needed for the current to increase from zero to its final value, and the magnetic field increases along with the current. As this happens, the field lines from coil *A* spread outward across the wires of coil *B*. This motion of the lines across coil *B* produces in it a momentary current. Once the current in *A* reaches its normal, steady value, the magnetic field becomes stationary and the induced current in *B* stops.

Next let us open the switch to break the circuit. In a fraction of a second the current in *A* drops to zero, and its magnetic field collapses. Once more field lines cut across *B* to induce a current, this time in the opposite direction since the field lines are now moving the other way past *B*. Thus starting and stopping the current in *A* has the same effect as moving a magnet in and out of *B*. An induced current is generated whenever the switch is opened or closed.

Suppose *A* is connected not to a battery but to a 60-Hz alternating current. Now we need no switch. Automatically, 120 times each second,

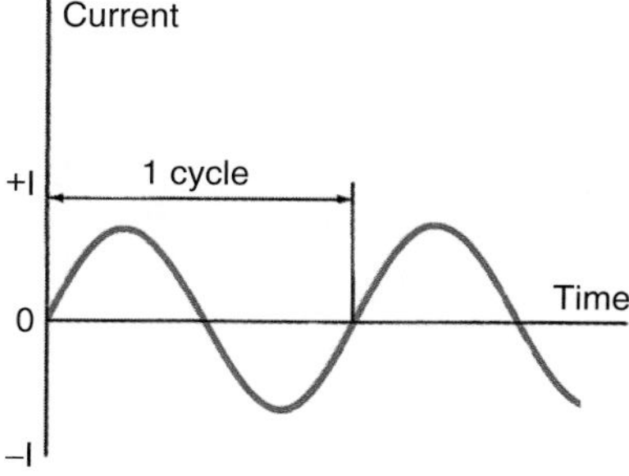

Figure 6-46 How a 60-Hz (60 cycles/s) alternating current varies with time. The frequency of the current is the number of cycles that occur per second. Here each complete cycle takes $\frac{1}{60}$ s. If a current in one direction in a circuit is considered +, a current in the opposite direction is considered −.

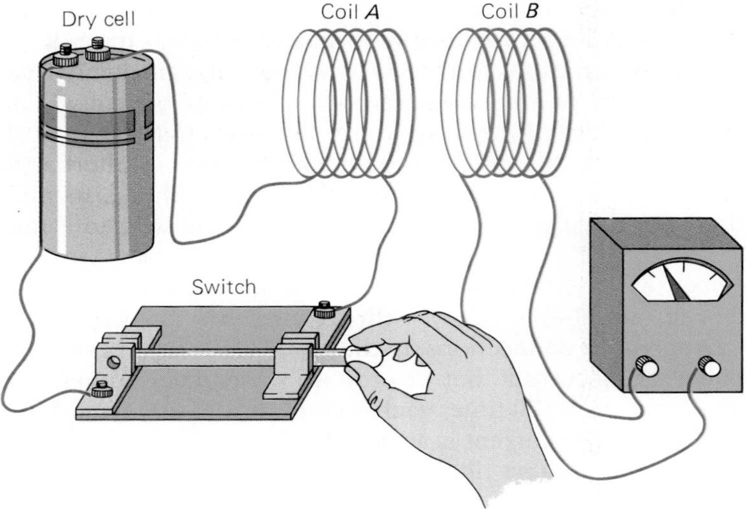

Figure 6-47 A simple transformer. Momentary currents are detected by the meter when the current in coil *A* is started or stopped.

the current comes to a complete stop and starts off again in the other direction. Its magnetic field expands and contracts at the same rate, and the field lines cutting *B*, first in one direction and then in the other, induce an alternating current similar to that in *A*. An ordinary meter will not respond to these rapid alterations, but an instrument meant for ac will show the induced current.

Thus an alternating current in one coil produces an alternating current in a nearby (but unconnected) coil. Such a combination of two coils and an iron core is a **transformer.** To generate an induced current most efficiently, the two coils should be close together and wound on a core of soft iron (Fig. 6-48). The coil into which electricity is fed from an outside source is the primary coil, and the coil in which an induced current is generated is the secondary coil.

Why Transformers Are Useful Transformers are useful because the voltage of the induced current can be raised or lowered by suitable windings of the coils. If the secondary coil has the same number of turns as the primary, the induced voltage will be the same as the primary voltage. If the secondary has twice as many turns, its voltage is twice that of the primary; if it has one-third as many turns, its voltage is one-third that of the primary; and so on. By using a suitable transformer, we can obtain any voltage we like, high or low, from a given alternating current.

When the secondary coil of a transformer has a higher voltage than the primary coil, its current is lower than that in the primary (and vice versa), so that the power $P = IV$ is the same in both coils. Thus

$$\frac{N_1}{N_2} = \frac{V_1}{V_2} = \frac{I_2}{I_1} \qquad \textit{Transformer} \qquad 6\text{-}6$$

$$\frac{\text{Primary turns}}{\text{Secondary turns}} = \frac{\text{primary voltage}}{\text{secondary voltage}} = \frac{\text{secondary current}}{\text{primary current}}$$

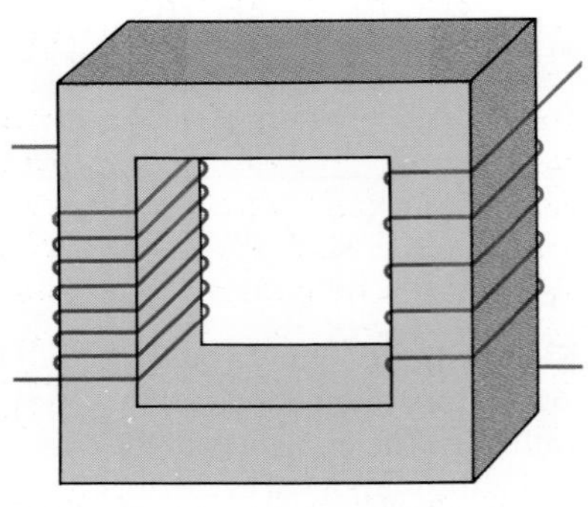

Figure 6-48 Actual transformers usually have iron cores. The winding with the greater number of turns has the higher voltage across it and carries the lower current. The power in both windings is the same.

For many purposes in homes, factories, and laboratories it is desirable to change the voltage of alternating currents. But most valuable of all, transformers permit the efficient long-distance transmission of

Example 6.9

A transformer connected to a 120-V ac power line has 200 turns in its primary coil and 50 turns in its secondary coil. The secondary is connected to a 100-Ω lightbulb. How much current is drawn from the 120-V power line?

Solution
The voltage across the secondary coil is

$$V_2 = \frac{N_2}{N_1} V_1 = \frac{50 \text{ turns}}{200 \text{ turns}}(120 \text{ V}) = 30 \text{ V}$$

Ohm's law, Eq. 6-4, then gives us the current I_2 in the secondary coil, which is what passes through the lightbulb:

$$I_2 = \frac{V_2}{R} = \frac{30 \text{ V}}{100 \ \Omega} = 0.3 \text{ A}$$

This means that the current in the primary coil is

$$I_1 = \frac{N_2}{N_1} I_2 = \frac{50 \text{ turns}}{200 \text{ turns}}(0.3 \text{ A}) = 0.075 \text{ A}$$

Figure 6-49 Transformers such as this at a power station step up the voltage of the electric power generated there for transmission over long distances. The higher the voltage V, the lower the current I for the same power P, since $P = IV$. The advantage of a low current is that less energy is lost as heat in the transmission lines. Other transformers step down the voltage for the consumer to the usual 220–240 V or 110–120 V.

power. Currents in long-distance transmission must be as small as possible, since large currents mean energy lost in heating the transmission wires. Hence at a power plant electricity from the generator is led into a "step-up" transformer, which increases the voltage and decreases the current, each by several hundred times (Fig. 6-49). On high-voltage lines (sometimes carrying currents at voltages exceeding 1 million V) this current is carried to local substations, where other transformers "step down" its voltage to make it safe for local transmission and use.

Important Terms and Ideas

Electric charge is a fundamental property of certain elementary particles of which all matter is composed. The two kinds of charge are called **positive** and **negative.** Charges of the same sign repel each other; charges of opposite sign attract each other. The unit of charge is the **coulomb** (C). All charges, of either sign, occur in multiples of $e = 1.6 \times 10^{-19}$ C.

Atoms are composed of **electrons,** whose charge is $-e$; **protons,** whose charge is $+e$; and **neutrons,** which have no charge. Protons and neutrons have almost equal masses, which are nearly 2000 times greater than the electron mass. Every atom has a small, central **nucleus** of protons and neutrons with its electrons moving about the nucleus some distance away. The number of protons and electrons is equal in a normal atom, which is therefore electrically neutral. An atom that has lost one or more electrons is a **positive ion,** and an atom that has picked up one or more electrons in excess of its usual number is a **negative ion.**

A flow of charge from one place to another is an **electric current.** The unit of electric current is the **ampere** (A), which is equal to a flow of 1 coulomb/second. Charge flows easily through a **conductor,** with some difficulty through a **semiconductor,** and only with great difficulty through an **insulator.** A **superconductor** offers no resistance at all to the flow of charge.

The **potential difference** (or **voltage**) between two points is the work needed to take a charge of 1 C from one of the points to the other. The unit of potential difference is the **volt** (V), which is equal to 1 joule/coulomb.

According to **Ohm's law,** the current in a metal conductor is proportional to the potential difference between its ends and inversely proportional to its **resistance.** The unit of resistance is the ohm (Ω), which is equal to 1 volt/ampere.

The **power** of an electric current is the rate at which it does work.

Every electric current (and moving charge) has a **magnetic field** around it that exerts a sidewise force on any other electric current (or moving charge) in its presence. All atoms contain moving electrons, and **permanent magnets** are made from substances, notably iron, whose atomic magnetic fields can be lined up instead of being randomly oriented.

Electromagnetic induction refers to the production of a current in a wire when there is relative motion between the wire and a magnetic field.

The direction of an **alternating current** reverses itself at regular intervals. In a **transformer** an alternating current in one coil of wire induces an alternating current in another nearby coil. Depending on the ratio of turns of the coils, the induced current can have a voltage that is larger, smaller, or the same as that of the primary current.

Important Formulas

Columb's law: $F = \dfrac{KQ_1Q_2}{R^2}$

Electric current $= \dfrac{Q}{t}$

Potential difference $= \dfrac{W}{Q}$

Ohm's law: $I = \dfrac{V}{R}$

Electric power: $P = IV$

Transformer: $\dfrac{N_1}{N_2} = \dfrac{V_1}{V_2} = \dfrac{I_2}{I_1}$

Multiple Choice

1. Electric charge
   - **a.** can be subdivided indefinitely
   - **b.** occurs only in separate parcels of $\pm 1.6 \times 10^{-19}$ C
   - **c.** occurs only in separate parcels of ± 1 C
   - **d.** occurs only in separate parcels whose value depends on the particle carrying the charge
2. A negative electric charge
   - **a.** interacts only with positive charges
   - **b.** interacts only with negative charges
   - **c.** interacts with both positive and negative charges
   - **d.** may interact with either positive or negative charges, depending on circumstances
3. A positively charged rod is brought near an isolated metal ball. Which of the sketches in Fig. 6-50 best illustrates the arrangement of charges on the ball?
   - **a.** *a*
   - **b.** *b*
   - **c.** *c*
   - **d.** *d*

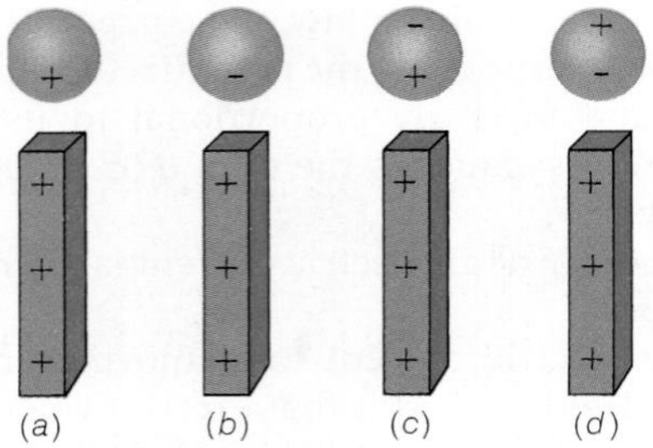

(a) (b) (c) (d)

Figure 6-50

4. Which of the following statements is not true?
   - **a.** The positive charge in an atomic nucleus is due to the protons it contains.
   - **b.** All protons have the same charge.
   - **c.** Protons and electrons have charges equal in magnitude although opposite in sign.
   - **d.** Protons and electrons have equal masses.
5. Coulomb's law for the force between electric charges belongs in the same general category as
   - **a.** the law of conservation of energy
   - **b.** Newton's second law of motion
   - **c.** Newton's law of gravitation
   - **d.** the second law of thermodynamics
6. The electric force between a proton and an electron
   - **a.** is weaker than the gravitational force between them
   - **b.** is equal in strength to the gravitational force between them
   - **c.** is stronger than the gravitational force between them
   - **d.** is any of these, depending on the distance between the proton and the electron
7. The electrons in an atom
   - **a.** are bound to it permanently
   - **b.** are some distance away from the nucleus
   - **c.** have more mass than the nucleus
   - **d.** may be positively or negatively charged
8. Atoms and molecules are normally
   - **a.** electrically neutral
   - **b.** negatively charged

c. positively charged
d. ionized

9. An object has a positive electric charge whenever
   a. it has an excess of electrons
   b. it has a deficiency of electrons
   c. the nuclei of its atoms are positively charged
   d. the electrons of its atoms are positively charged
10. Which of the following statements is correct?
   a. Electrons carry electric current.
   b. The motion of electrons produces an electric current.
   c. Moving electrons constitute an electric current.
   d. Electric currents are carried by conductors and insulators only.
11. Superconductivity occurs in certain substances
   a. only at very low temperatures
   b. only at very high temperatures
   c. at all temperatures
   d. any of the above, depending on the substance
12. Match each of the electrical qualities listed below with the appropriate unit from the list on the right:

a. resistance	volt
b. current	ampere
c. potential difference	ohm
d. power	watt

13. Electric power is equal to
   a. (current)(voltage)
   b. current/voltage
   c. voltage/current
   d. (resistance)(voltage)
14. The electric energy lost when a current passes through a resistance
   a. becomes magnetic energy
   b. becomes potential energy
   c. becomes heat
   d. disappears completely
15. When a magnetized bar of iron is strongly heated, its magnetization
   a. becomes weaker
   b. becomes stronger
   c. reverses its direction
   d. is unchanged
16. All magnetic fields originate in
   a. iron atoms
   b. permanent magnets
   c. stationary electric charges
   d. moving electric charges
17. The force on an electron that moves in a curved path must be
   a. gravitational
   b. electrical
   c. magnetic
   d. one or more of these
18. A drawing of the field lines of a magnetic field provides information on
   a. the direction of the field only
   b. the strength of the field only
   c. both the direction and the strength of the field
   d. the source of the field
19. Magnetic field lines provide a convenient way to visualize a magnetic field. Which of the following statements is not true?
   a. The path followed by an iron particle released near a magnet corresponds to a field line.
   b. The path followed by an electric charge released near a magnet corresponds to a field line.
   c. A compass needle in a magnetic field turns until it is parallel to the field lines around it.
   d. Magnetic field lines do not actually exist.
20. A moving electric charge produces
   a. only an electric field
   b. only a magnetic field
   c. both an electric and a magnetic field
   d. any of these, depending on its speed
21. The magnetic field of a bar magnet resembles most closely the magnetic field of
   a. a straight wire carrying a direct current
   b. a straight wire carrying an alternating current
   c. a wire loop carrying a direct current
   d. a wire loop carrying an alternating current
22. The magnetic field shown in Fig. 6-51 is produced by
   a. two north poles
   b. two south poles
   c. a north pole and a south pole
   d. a south pole and an unmagnetized iron bar

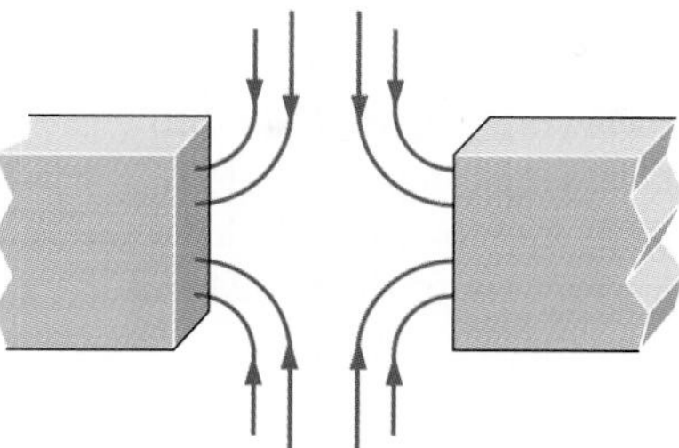

Figure 6-51

23. The magnetic field lines around a long, straight current are
   a. straight lines parallel to the current
   b. straight lines that radiate from the current like spokes of a wheel
   c. concentric circles around the current
   d. concentric helixes around the current

24. A magnet does not exert a force on
- **a.** an unmagnetized iron bar
- **b.** a magnetized iron bar
- **c.** a stationary electric charge
- **d.** a moving electric charge

25. A current-carrying wire is in a magnetic field with the direction of the current the same as that of the field.
- **a.** The wire tends to move parallel to the field.
- **b.** The wire tends to move perpendicular to the field.
- **c.** The wire tends to turn until it is perpendicular to the field.
- **d.** The wire has no tendency to move or to turn.

26. An electromagnet
- **a.** uses an electric current to produce a magnetic field
- **b.** uses a magnetic field to produce an electric current
- **c.** is a magnet that has an electric charge
- **d.** operates only on alternating current

27. The nature of the force that is responsible for the operation of an electric motor is
- **a.** electric
- **b.** magnetic
- **c.** a combination of electric and magnetic
- **d.** either electric or magnetic, depending on the design of the motor

28. A generator is said to "generate electricity." What it actually does is act as a source of
- **a.** electric charge
- **b.** electrons
- **c.** magnetism
- **d.** electric energy

29. The alternating current in the secondary coil of a transformer is induced by
- **a.** the varying electric field of the primary coil
- **b.** the varying magnetic field of the primary coil
- **c.** the varying magnetic field of the secondary coil
- **d.** the iron core of the transformer

30. A transformer can change
- **a.** the voltage of an alternating current
- **b.** the power of an alternating current
- **c.** alternating current to direct current
- **d.** direct current to alternating current

31. If 10^5 electrons are added to a neutral object, its charge will be
- **a.** -1.6×10^{-24} C
- **b.** -1.6×10^{-14} C
- **c.** $+1.6 \times 10^{-24}$ C
- **d.** $+1.6 \times 10^{-14}$ C

32. A positive and a negative charge are initially 4 cm apart. When they are moved closer together so that they are now only 1 cm apart, the force between them is
- **a.** 4 times smaller than before
- **b.** 4 times larger than before
- **c.** 8 times larger than before
- **d.** 16 times larger than before

33. The force between two charges of -3×10^{-9} C that are 5 cm apart is
- **a.** 1.8×10^{-16} N
- **b.** 3.6×10^{-15} N
- **c.** 1.6×10^{-6} N
- **d.** 3.2×10^{-5} N

34. Five joules of work are needed to shift 10 C of charge from one place to another. The potential difference between the places is
- **a.** 0.5 V
- **b.** 2 V
- **c.** 5 V
- **d.** 10 V

35. The current in a 12-Ω toaster operated at 120 V is
- **a.** 0.1 A
- **b.** 10 A
- **c.** 12 A
- **d.** 1440 A

36. The voltage needed to produce a current of 5 A in a resistance of 40 Ω is
- **a.** 0.125 V
- **b.** 5 V
- **c.** 8 V
- **d.** 200 V

37. The resistance of a lightbulb that draws a current of 2 A when connected to a 12-V battery is
- **a.** 1.67 Ω
- **b.** 2 Ω
- **c.** 6 Ω
- **d.** 24 Ω

38. The current in a 40-W, 120-V electric lightbulb is
- **a.** $\frac{1}{3}$ A
- **b.** 3 A
- **c.** 80 A
- **d.** 4800 A

39. A car's storage battery is being charged at a rate of 75 W. If the potential difference across the battery's terminals is 13.6 V, charge is being transferred between its plates at
- **a.** 0.18 C/s
- **b.** 2.8 C/s
- **c.** 5.5 C/s
- **d.** 1020 C/s

40. A 120-V, 1-kW electric heater is mistakenly connected to a 240-V power line that has a 15-A fuse. The heater will
   a. give off less than 1 kW of heat
   b. give off 1 kW of heat
   c. give off more than 1 kW of heat
   d. blow the fuse

41. A 240-V, 1-kW electric heater is mistakenly connected to a 120-V power line that has a 15-A fuse. The heater will
   a. give off less than 1 kW of heat
   b. give off 1 kW of heat
   c. give off more than 1 kW of heat
   d. blow the fuse

42. A transformer whose primary winding has twice as many turns as its secondary winding is used to convert 240-V ac to 120-V ac. If the current in the secondary circuit is 4 A, the primary current is
   a. 1 A
   b. 2 A
   c. 4 A
   d. 8 A

Exercises

6.1 Positive and Negative Charge

6.2 What Is Charge?

1. What reasons might there be for the universal belief among scientists that there are only two kinds of electric charge?
2. Electricity was once thought to be a weightless fluid, an excess of which was "positive" and a deficiency of which was "negative." What phenomena can this hypothesis still explain? What phenomena can it not explain?
3. A plastic ball has a charge of $+10^{-12}$ C. (a) Does the ball have more electrons or fewer electrons than when it is electrically neutral? (b) How many such electrons?
4. Why does the production of electricity by friction always yield equal amounts of positive and negative charge?
5. Nearly all the mass of an atom is concentrated in its nucleus. Where is its charge located?
6. Compare the basic characters of electric and gravitational forces.
7. Find the total charge of 1 g of protons.

6.3 Coulomb's Law

6.4 Force on an Uncharged Object

6.5 Matter in Bulk

8. Is there any distance at which the gravitational force between two electrons is greater than the electric force between them?
9. When two objects attract each other electrically, must both of them be charged? When two objects repel each other electrically, must both of them be charged?
10. How do we know that the force holding the earth in its orbit about the sun is not an electric force, since both gravitational and electric forces vary inversely with the square of the distance between centers of force?
11. A hydrogen molecule consists of two hydrogen atoms whose nuclei are single protons. Find the force between the two protons in a hydrogen molecule whose distance apart is 7.42×10^{-11} m. (The two electrons in the molecule spend more time between the protons than outside them, which leads to attractive forces that balance the repulsion of the protons and permit a stable molecule; see Chap. 10.)
12. A charge of $+2 \times 10^{-6}$ C is 20 cm from a charge of -3×10^{-6} C. Find the magnitude and direction of the force on each charge.
13. A charge of $+3 \times 10^{-9}$ C is 50 cm from a charge of -5×10^{-9} C. Find the magnitude and direction of the force on each charge.
14. Two charges repel each other with a force of 1×10^{-6} N when they are 10 cm apart. Find the forces between the same charges when they are 2 cm and 20 cm apart.
15. Two charges originally 80 mm apart are brought together until the force between them is 16 times greater. How far apart are they now?
16. Two small spheres are given identical positive charges. When they are 1 cm apart, the repulsive force on each of them is 0.002 N. What would the force be if (a) the distance is increased to 3 cm? (b) one charge is doubled? (c) both charges are tripled? (d) one charge is doubled and the distance is increased to 2 cm?
17. (a) A metal sphere with a charge of $+1 \times 10^{-5}$ C is 10 cm from another metal sphere with a charge of -2×10^{-5} C. Find the magnitude of the attractive force on each sphere. (b) The two spheres are brought in contact and again separated by 10 cm. Find the magnitude of the new force on each sphere.

18. How far apart are two charges of $+1 \times 10^{-8}$ C that repel each other with a force of 0.1 N?

19. How much positive charge must be added to the earth and the moon so that the resulting electrical repulsion balances the gravitational attraction between them? Assume equal amounts of charge are added to each body. (The mass of the earth is 6.0×10^{24} kg and that of the moon is 7.3×10^{22} kg.)

6.6 Conductors and Insulators

6.7 Superconductivity

20. How is the movement of electricity through air different from its movement through a copper wire?

21. One terminal of a battery is connected to a lightbulb. What, if anything, happens?

22. Why do you think bending a wire does not affect its electrical resistance, even though a bent pipe offers more resistance to the flow of water than a straight one?

23. What basic aspect of superconductivity has prevented its large-scale application thus far?

6.8 The Ampere

6.9 Potential Difference

24. Sensitive instruments can detect the passage of as few as 60 electrons/s. To what current does this correspond?

25. How much energy is stored in a 6-V battery rated at 20 A · h?

26. The energy stored in a certain 12-V battery is 2 MJ. (a) How much charge was transferred from one of its terminals to the other when it was charged? (b) How long would a 10-A charger take to charge the battery, assuming a constant current? (Actually, the current decreases as a battery nears a full charge.)

27. The potential difference between a cloud and the ground is 4 MV. A charge of 200 C is transferred in a lightning stroke between the cloud and the ground. How much energy is dissipated?

6.10 Ohm's Law

28. A person can be electrocuted while taking a bath if he or she touches a poorly insulated light switch. Why is the electric shock received under these conditions so much more dangerous than usual?

29. A 240-V water heater has a resistance of 24 Ω. What must be the minimum rating of the fuse in the electric circuit to which the water heater is connected?

30. A 120-V electric coffeepot draws a current of 0.6 A. What is the resistance of its heating element?

31. What potential difference must be applied across a 1500-Ω resistance in order that the resulting current be 50 mA? (1 mA = 1 milliampere = 0.001 A)

6.11 Electric Power

32. An electrical appliance is sometimes said to "use up" electricity. What does it actually use in its operation?

33. A fuse prevents more than a certain amount of current from flowing in a particular circuit. What might happen if too much current were to flow? What determines how much is too much?

34. Heavy users of electric power, such as large electric stoves and clothes dryers, are sometimes designed to operate on 240 V rather than 120 V. What advantage do you think the higher voltage has in these applications?

35. How are the terminals of a set of batteries connected when the batteries are in series? In parallel? What is the advantage of each combination?

36. Wire *A* has a potential difference of 50 V across it and carries a current of 2 A. Wire *B* has a potential difference of 100 V across it and also carries a current of 2 A. Compare the resistances, rates of flow of charge, and rates of flow of energy in the two wires.

37. (a) If a 75-W lightbulb is connected to a 120-V power line, how much current flows through it? (b) What is the resistance of the bulb? (c) How much power does the bulb consume?

38. A solar cell whose area is 80 cm^2 produces a current of 2.1 A at 0.5 V in bright sunlight whose intensity is 0.1 W/cm^2. Find the efficiency with which the cell turns solar energy into electric energy.

39. An electric drill rated at 400 W is connected to a 240-V power line. How much current does it draw?

40. A power of 1 horsepower (hp) is equivalent to 746 W. What is the power output in hp of an electric motor that draws a current of 4 A at 120 V and is 80 percent efficient?

41. If your home has a 120-V power line, how much power in watts can you draw from the line before a 30-A fuse will burn out? How many 100-W lightbulbs can you put in the circuit before the fuse will burn out?

42. A 120-V electric motor draws 2.5 A. (a) How many coulombs of charge pass through it in 15 min? (b) How many joules of energy does it use in 15 min?

43. A 240-V clothes dryer draws a current of 15 A. How much energy, in kilowatthours and in joules, does it use in 45 min of operation?

44. A fully-charged 12-V storage battery contains 3.5 MJ of energy. What is its capacity in ampere-hours?

45. A 1.35-V mercury cell with a capacity of 1.5 A · h is used to power a cardiac pacemaker. (a) If the power required is 0.1 mW (1 mW = 1 milliwatt = 0.001 W), what is the average current? (b) How long will the cell last?

46. When a certain 1.5-V battery is used to power a 3-W flashlight bulb, it is dead after an hour's use. If the battery costs $0.50, what is the cost of a kilowatthour of electric energy obtained in this way? How does this compare with the cost of electric energy supplied to your home?

47. A hot-water heater employs a 2000-W resistance element. If all the heat from the resistance element is absorbed by the water in the heater, how much water per hour can be warmed from 10 to 70°C?

48. A trolley bus whose mass is 10^4 kg takes 10 s to reach a speed of 8 m/s starting from rest. It operates from a 5-kV overhead power line and is 50 percent efficient. What is the average current drawn by the bus during the acceleration? (Hint: First calculate the final KE of the bus.)

6.12 Magnets

6.13 Magnetic Field

49. Why is a piece of iron attracted to either pole of a magnet?

50. The magnetic poles of the earth are called geomagnetic poles. Is the north geomagnetic pole a north magnetic pole or a south magnetic pole?

51. Explain why lines of force can never cross one another.

6.14 Oersted's Experiment

52. A current flows west through a power line. Find the directions of the magnetic field above and below the power line; ignore the earth's magnetic field.

53. Figure 6-52 shows a current-carrying wire and a compass. In which direction will the compass needle point?

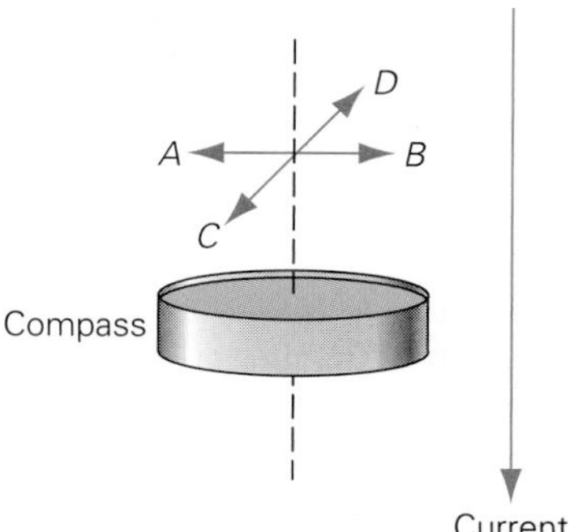

Figure 6-52

54. In a TV picture tube, a beam of electrons perpendicular to a fluorescent screen scans across the screen, which glows where the electrons strike it. When you face the screen, in what direction is the magnetic field of the electron beam?

6.15 Electromagnets

6.16 Magnetic Force on a Current

6.17 Electric Motors

55. Two parallel wires carry currents in the same direction. Do they attract each other, repel each other, or not affect each other? What happens when the currents are in opposite directions?

56. A physicist is equipped to measure electric, magnetic, and gravitational fields. Which will she detect when a proton moves past her? When she moves past a proton?

57. A current-carrying wire is in a magnetic field. What angle should the wire make with the direction of the field for the force on it to be zero? What should the angle be for the force to be a maximum?

58. A current is passed through a helical (corkscrew-shaped) spring. What, if anything, do you think happens to the length of the spring?

59. A length of copper wire *AB* rests across a pair of parallel copper wires that are connected to a battery through a switch, as in Fig. 6-53. The arrangement is placed between the poles of a magnet and the switch is closed. In what direction does the wire *AB* move?

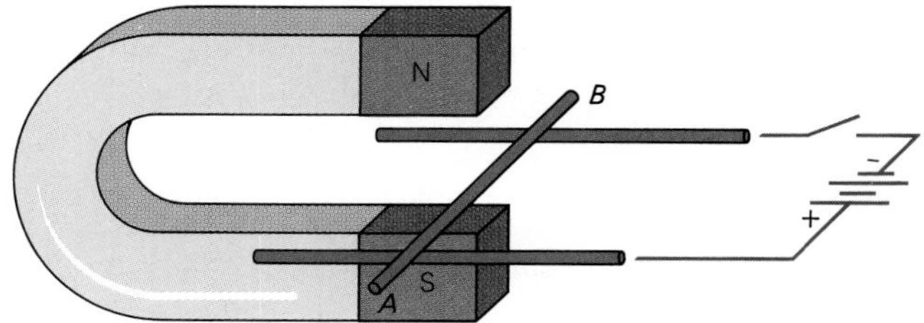

Figure 6-53

60. A beam of protons, at first moving slowly, is accelerated to higher and higher speeds. When the protons are moving slowly, the beam spreads out, but when they are moving fast, the beam diameter decreases. What do you think is the reason for this behavior?

61. When a wire loop is rotated in a magnetic field, the direction of the current induced in the loop reverses itself twice per rotation. Why?

6.18 Electromagnetic Induction

62. Would you expect to find direct or alternating current in (a) the filament of a lightbulb in your home? (b) the filament of a lightbulb in a car? (c) the secondary coil of a transformer? (d) the output of a battery charger?

63. The shaft of a generator is much easier to turn when the generator is not connected to an outside circuit than when such a connection is made. Why?

64. A generator driven by a diesel engine that develops 12 hp delivers 30 A at 240 V. What is the efficiency of the generator?

6.19 Transformers

65. What acts on the secondary winding of a transformer to cause an alternating voltage to occur across its ends even though the primary and secondary windings are not connected?

66. Given a coil of wire and a small lightbulb, how can you tell whether the current in another coil is direct or alternating without touching the second coil or its connecting wires?

67. What would happen if the primary winding of a transformer were connected to a battery?

68. An electric welding machine employs a current of 400 A. The device uses a transformer whose primary coil has 400 turns and that draws 4 A from a 220-V power line. How many turns are there in the secondary coil of the transformer? What is the potential difference across the secondary coil?

69. A transformer has 50 turns in its primary coil and 100 turns in its secondary. What are the nature and magnitude of the current in the secondary when (a) a 60-Hz, 3-A alternating current passes through the primary, and (b) when a 3-A direct current passes through the primary?

70. A transformer has a 600-turn primary coil and a 200-turn secondary coil. If the secondary voltage is 80 V, what is the primary voltage? If the transformer delivers 300 W, what is the primary current?

71. A transformer rated at a maximum power of 10 kW is used to couple a 5000-V transmission line to a 240-V circuit. (a) What is the ratio of turns in the transformer? (b) What is the maximum current in the 240-V circuit?

Answers to Multiple Choice

1. b
2. c
3. d
4. d
5. c
6. c
7. b
8. a
9. b
10. c
11. a
12. **a.** ohm
b. ampere
c. volt
d. watt
13. a
14. c
15. a
16. d
17. d
18. c
19. b
20. c
21. c
22. b
23. c
24. c
25. d
26. a
27. b
28. d
29. b
30. a
31. b
32. d
33. d
34. a
35. b
36. d
37. c
38. a
39. c
40. d
41. a
42. b

7

Waves

The dispersion of sunlight by water droplets produces rainbows.

Goals

When you have finished this chapter you should be able to complete the goals ▸ given for each section below:

Wave Motion

7.1 Water Waves
Crests and Troughs

7.2 Transverse and Longitudinal Waves
Across and Back; Toward and Away
- ▸ State what a wave is and give examples of different kinds of waves.
- ▸ Distinguish between transverse and longitudinal waves.

7.3 Describing Waves
Wavelength, Frequency, and Speed Are Related
- ▸ Use the formula $v = \lambda f$ to relate the frequency and wavelength of a wave to its speed.

7.4 Standing Waves
They Generate Most Musical Sounds
- ▸ Describe what a standing wave is and how musical instruments make use of them.

Sound Waves

7.5 Sound
Pressure Waves in a Solid, Liquid, or Gas
- ▸ Discuss the nature of sound.

7.6 Doppler Effect
Higher Pitch When Approaching; Lower Pitch When Receding
- ▸ State what the doppler effect is and explain its origin.

7.7 Musical Sounds
Fundamentals and Overtones

Electromagnetic Waves

7.8 Electromagnetic Waves
Waves without Matter
- ▸ Discuss the nature of electromagnetic waves and describe the difference between polarized and unpolarized light.

7.9 Types of EM Waves
They Carry Information as Well as Energy
- ▸ Distinguish between amplitude and frequency modulation.
- ▸ Explain how a radar works.

7.10 Light "Rays"
The Paths Light Takes
- ▸ Describe what is meant by a light ray.

Wave Behavior

7.11 Reflection
Mirror, Mirror on the Wall

7.12 Refraction
A Change in Direction Caused by a Change in Speed
- ▸ Describe how reflection and refraction occur.
- ▸ Explain how a mirror produces an image.
- ▸ Explain how refraction makes a body of water seem shallower than it actually is.
- ▸ Explain what is meant by internal reflection.

7.13 Lenses
Bending Light to Form an Image
- ▸ Define lens and distinguish between converging and diverging lenses.
- ▸ Use ray tracing to find the properties of the image a converging lens produces of an object.

7.14 The Eye
A Remarkable Optical Instrument
- ▸ Describe the differences between farsightedness, nearsightedness, and astigmatism.

7.15 Color
Each Frequency of Light Produces the Sensation of a Different Color
- ▸ Account for the dispersion of white light into a spectrum when it is refracted.
- ▸ Discuss the origin of rainbows and why the sky is blue.

7.16 Interference
Waves in Step and out of Step
- ▸ Distinguish between constructive and destructive interference.
- ▸ Explain why thin films of soap or oil are brightly colored.

7.17 Diffraction
Why Shadows Are Never Completely Dark
- ▸ Describe the diffraction of waves at the edge of an obstacle.
- ▸ Discuss the factors that determine the sharpness of the image produced by an optical instrument.

A **wave** is a periodic disturbance—a back-and-forth change of some kind that is repeated regularly as time goes on—that spreads out from a source and carries energy with it. Throw a stone into a lake: water waves move out from the splash. Clap your hands: sound waves carry the noise around you. Switch on a lamp: light waves flood the room. Water waves, sound waves, and light waves are very different from one another in various ways, but all have in common some basic properties that are explored in this chapter.

Wave Motion

Two important categories of waves are **mechanical waves** and **electromagnetic waves.** Mechanical waves, such as water waves and sound waves, travel only through matter and involve the motion of particles of the matter they pass through. Electromagnetic waves, such as light waves and radio waves, consist of varying electric and magnetic fields and can travel through a vacuum as well as through matter. Because mechanical waves are the easier of the two kinds to understand, we shall begin by looking at what they are, how they are described, and how they behave.

7.1 Water Waves

Crests and Troughs

If we stand on an ocean beach and watch the waves roll in and break one after the other, we might guess that water is moving bodily toward the shore (Fig. 7-1). After a few minutes, though, we see that this cannot be true. Between the breakers, water rushes back out to sea, and there is no piling up of water on the beach. The overall motion is really an endless movement of water to and fro.

Figure 7-1 Water molecules in deep water move in circular orbits as waves pass by (see Fig. 7-2). When the waves reach shallow water, the molecules in the lower parts of their orbits touch the bottom, which slows them down. The wave crests, however, continue to move forward as before, which causes the fronts of the waves to become steeper and steeper as the water gets shallower. Finally the wave crests topple over, or "break," in a shower of foam that spills down the wave front. At this stage the water depth is about 1.3 times the wave height, and the wave crests may be moving toward the shore twice as fast as the waves themselves.

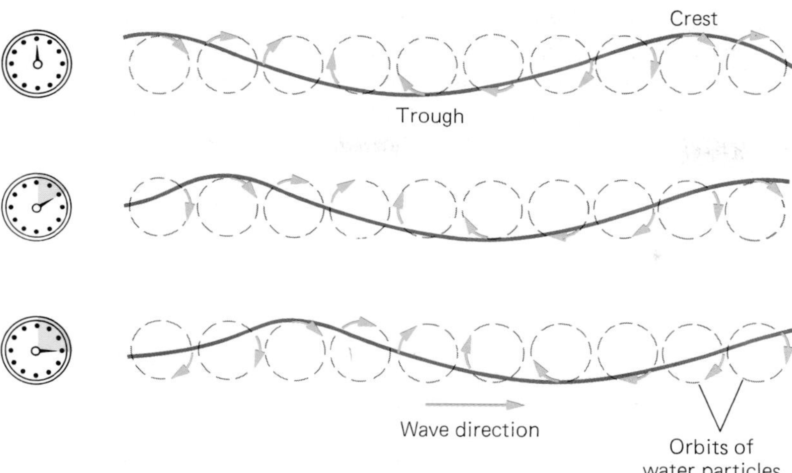

Figure 7-2 Nature of a water wave in deep water. Each water molecule performs a periodic motion in a small circle. Because successive molecules reach the tops of their circles at slightly later times, their combination appears as a series of crests and troughs moving along the surface of the water. There is no net transfer of water by the wave.

We can see what is happening better by moving out beyond the breakers, say to the end of a pier. If we study a piece of seaweed floating on the water, we find little change in its position. As the crest of a wave passes, the seaweed rises and moves shoreward. In the trough that follows the crest, the seaweed falls and moves the same distance seaward. On the whole the seaweed moves in a roughly circular path perpendicular to the water surface.

The illusion of overall movement toward the shore comes about because each molecule of water undergoes its circular motion a moment later than the molecule behind it (Fig. 7-2). At the crest of a wave the molecules move in the direction of the wave, while in a trough the molecules move in the backward direction.

What *does* move shoreward is not water but energy. Ocean waves are produced by wind, and it is the energy from the wind out at sea that is carried by means of wave motion to the shore. All mechanical waves behave the same way: they transfer energy from place to place by a series of periodic motions of individual particles but cause no permanent shift in the position of matter.

7.2 Transverse and Longitudinal Waves

Across and Back; Toward and Away

Water waves are familiar but complicated. Simpler are the waves set up when we shake one end of a rope whose other end is fixed in place (Fig. 7-3*a*). Here the rope particles move perpendicular to the direction in which the wave moves. Such waves are said to be **transverse.**

Another type of wave can occur in a long coil spring (Fig. 7-3*b*). If the left-hand end of the spring is moved back and forth, a series of **compressions** and **rarefactions** move along the spring. The compressions are places where the loops of the spring are pressed together; the rarefactions are places where the loops are stretched apart. Any one loop simply moves back and forth, transmitting its motion to the next in line, and the regular series of back-and-forth movements gives rise to the

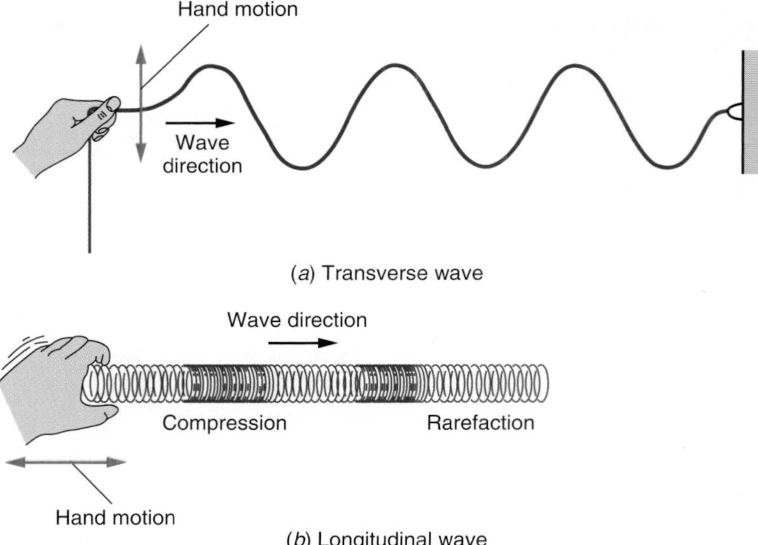

Figure 7-3 Transverse and longitudinal waves. (*a*) Transverse waves travel along the rope in the direction of the black arrow. The individual particles of the rope move back and forth (red arrows) perpendicular to the direction of the waves. (*b*) In longitudinal waves, successive regions of compression and rarefaction move along the spring. The particles of the spring move back and forth parallel to the spring.

compressions and rarefactions. Waves of this kind, in which the motion of individual units is along the same line that the wave travels, are called **longitudinal waves.**

Water waves are a combination of transverse and longitudinal waves, as Fig. 7-2 shows. Pure transverse mechanical waves can occur only in solids, whereas longitudinal waves can travel in any medium, solid or fluid. Transverse motion requires that each particle, as it moves, drag with it adjacent particles to which it is tightly bound. This is impossible in a fluid, where molecules easily slide past their neighbors. Longitudinal motion, on the other hand, merely requires that each particle push on its neighbors, which can happen as easily in a gas or liquid as in a solid. (Surface waves on water—in fact any waves at the boundary between two fluids—are an exception to this rule, for in part they involve transverse motion.) The fact that longitudinal waves that originate in earthquakes pass through the center of the earth while transverse earthquake waves cannot is one of the reasons the earth is thought to have a liquid core (Chap. 15).

7.3 Describing Waves

Wavelength, Frequency, and Speed Are Related

All waves can be represented by a curve like that in Fig. 7-4. The resemblance to transverse wave motion is easiest to see; in fact the curve is an idealized picture of continuous waves in a rope like that of Fig. 7-3*a*. As the wave moves to the right, each point on the curve can be thought of as moving up or down just as any point on the rope would move. In the case of a longitudinal wave, the high points of the curve represent the maximum shifts of particles in one direction and the low points represent their maximum shifts in the other direction.

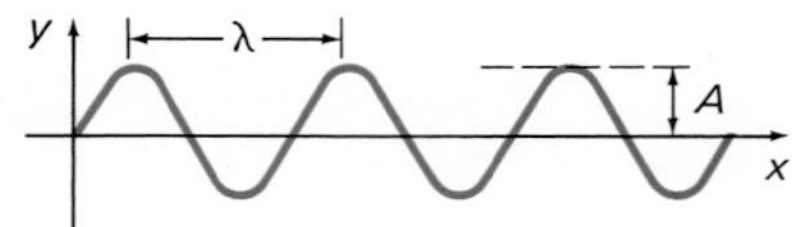

Figure 7-4 A transverse wave moving in the *x* direction whose displacements are in the *y* direction. The wavelength is λ and the amplitude is *A*.

With the help of Fig. 7-4 we can assign numbers to certain key properties of a wave, so that different waves can be compared. The distance from crest to crest (or trough to trough) is called the **wavelength,** usually

Table 7-1 Wave Quantities

Quantity	Symbol	Formula	Meaning
Speed	v	$v = f\lambda$	Distance through which each wave moves per second
Wavelength	λ	$\lambda = v/f$	Distance between adjacent crests or troughs
Frequency	f	$f = v/\lambda$	Number of waves that pass a given point per second
Period	T	$T = 1/f$	Time needed for a wave to pass a given point
Amplitude	A		Maximum displacement of oscillating particle from its normal position

symbolized by the Greek letter λ (lambda). The **speed** v of the waves is the rate at which each crest moves, and the **frequency** f is the number of crests that pass a given point each second. The **period** T is the time needed for a complete wave (crest + trough) to pass a given point (Table 7-1).

The **amplitude** A of a wave is the height of the crests above the undisturbed level (or the depth of the troughs below this level). Not surprisingly, the energy carried by waves depends on amplitude and frequency, that is, on the violence of the waves and the number of them per second. It turns out that the energy is proportional to the square of each of these quantities.

The unit of frequency is the cycle per second (c/s). As mentioned in Chap. 6, this unit is usually called the **hertz** (Hz), after Heinrich Hertz, a pioneer in the study of electromagnetic waves.

Basic Wave Formula The number of waves that pass a point per second multiplied by the length of each wave gives the speed with which the waves travel (Fig. 7-5). Thus frequency f times wavelength λ gives speed v:

$$v = f\lambda \qquad \textit{Wave speed} \qquad 7\text{-}1$$

Wave speed = (frequency)(wavelength)

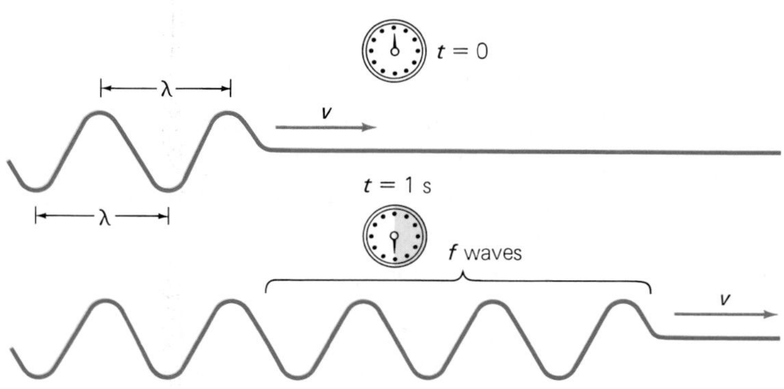

Figure 7-5 Wave speed equals frequency times wavelength.

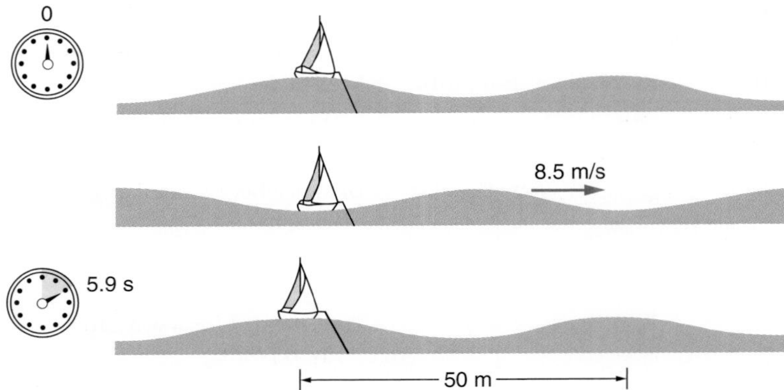

Figure 7-6 Waves whose speed is 8.5 m/s and whose wavelength is 50 m have a frequency of 0.17 Hz. This means that such waves pass an anchored boat once every 5.9 s.

If 10 waves, each 2 m long, pass in a second, then each wave must travel 20 m during that second to give a speed of 20 m/s. This formula applies to waves of all kinds.

Example 7.1

Waves in the open sea whose wavelength is 50 m travel at about 8.5 m/s (Fig. 7-6). Find their frequency and period.

Solution

The frequency of these waves is

$$f = \frac{v}{\lambda} = \frac{8.5 \text{ m/s}}{50 \text{ m}} = 0.17 \text{ Hz}$$

The period of the waves is

$$T = \frac{1}{f} = \frac{1}{0.17 \text{ Hz}} = 5.9 \text{ s}$$

so a wave passes a given point every 5.9 s.

Example 7.2

A tuning fork vibrating at 300 Hz is placed in a tank of water. (a) Find the frequency and wavelength of the sound waves in the water. (b) Find the frequency and wavelength of the sound waves produced in the air above the tank by the vibrations of the water surface. The speed of sound is 1498 m/s in water and 343 m/s in air at room temperature.

Solution

(a) In water, the frequency of the sound waves is the 300 Hz of their source. Their wavelength is, from Eq. 7-1,

$$\lambda_1 = \frac{v_1}{f} = \frac{1498 \text{ m/s}}{300 \text{ Hz}} = 5.0 \text{ m}$$

(b) In air, the frequency of the sound waves is the same as the frequency of their source, which is the vibrating water surface. Hence $f = 300$ Hz again, but the wavelength is different:

$$\lambda_2 = \frac{v_2}{f} = \frac{343 \text{ m/s}}{300 \text{ Hz}} = 1.1 \text{ m}$$

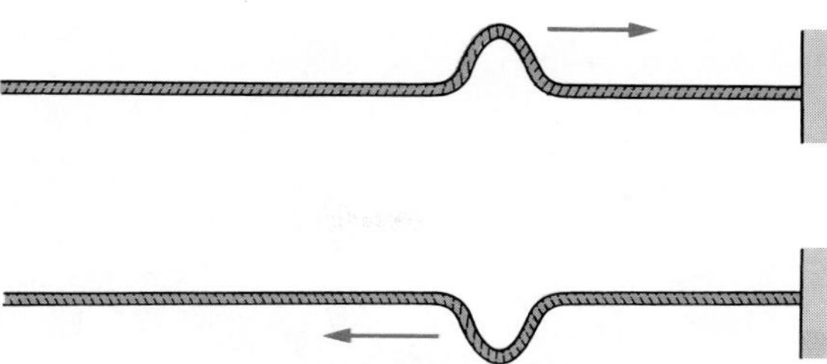

Figure 7-7 A wave in a stretched rope is reflected when it reaches a fixed end. The reflected wave is inverted.

7.4 Standing Waves

They Generate Most Musical Sounds

When a single wave sent down a stretched rope by a shake of the free end meets the attached end, it re-forms itself and travels back along the rope (Fig. 7-7). When a series of waves is sent along the rope, the reflected waves will meet the forward-moving waves head on. Each point on the rope must then respond to two different impulses at the same time. The two impulses add together. If the point on the rope is being pushed in the same direction by both waves, it will move in that direction with an amplitude equal to the sum of the amplitudes of the two waves. If the wave impulses at a point on the rope are in opposite directions, that point will have an amplitude equal to the difference of the two wave amplitudes.

With the timing just right, the two motions may cancel out completely for some points of the rope while other points move with twice the normal amplitude. In this situation the waves appear not to travel at all. Some parts of the rope simply move up and down, and other parts remain at rest (Fig. 7-8). Waves of this sort are called **standing waves.**

Vibrating strings in musical instruments are the most familiar examples of standing waves (Fig. 7-9). Longitudinal waves traveling in opposite directions over the same path may also set up standing waves, as in the vibrating air columns of whistles, organ pipes, flutes, and clarinets (Fig. 7-10). Standing waves set up in structures, for instance bridges (Fig. 7-11), sometimes lead to severe damage.

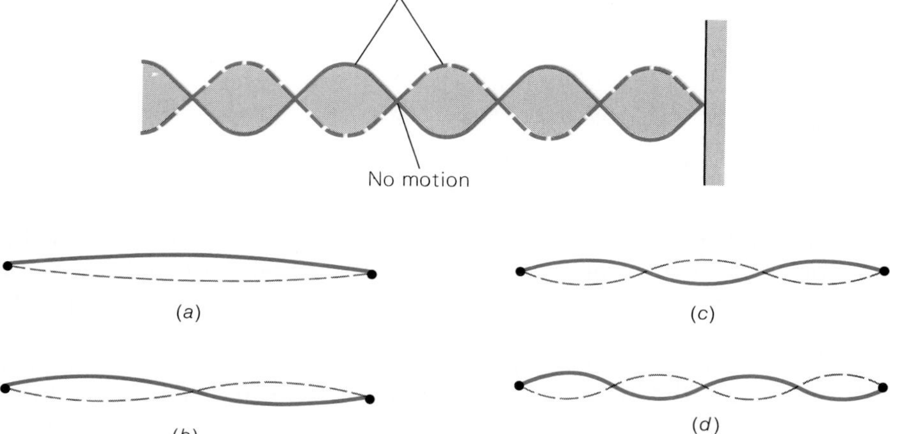

Figure 7-8 Standing waves in a stretched rope.

Figure 7-9 A few of the possible standing waves in a stretched string, such as a violin string.

Figure 7-10 Sound waves can be generated in various ways, as by the vibrating strings of a violin, the vibrating air column of a clarinet, and the vibrating membrane of a drum.

Figure 7-11 Strong winds set up standing waves in the Tacoma Narrows Bridge in Washington State soon after its completion in 1940. The bridge collapsed as a result. Today bridges are stiffened to prevent such disasters.

Sound Waves

7.5 Sound

Pressure Waves in a Solid, Liquid, or Gas

Most sounds are produced by a vibrating object, such as the cone of a loudspeaker (Fig. 7-12). When it moves outward, the cone pushes the air molecules in front of it together to form a region of high pressure that spreads outward. The cone then moves backward, which expands the space available to nearby air molecules. Some of these molecules now flow toward the cone, leaving a region of low pressure that spreads outward behind the high-pressure region. The repeated vibrations of the loudspeaker cone thus send out a series of compressions and rarefactions that constitute sound waves.

Sound waves are longitudinal, because the molecules in their paths move back and forth in the same direction as that of the waves. The air (or other material) in the path of a sound wave becomes alternately denser and rarer, and the resulting pressure changes cause our eardrums to vibrate, which produces the sensation of sound.

The great majority of sounds consist of waves of this type, but a few—the crack of a rifle, the first sharp sound of a thunderclap—are single, sudden compressions of the air rather than periodic phenomena.

The speed of sound is about 343 m/s (767 mi/h) in sea-level air at ordinary temperatures. Since the particles in liquids and solids are closer together than those in gases and therefore respond more quickly to one another's motions, sound travels faster in liquids and solids than in gases: about 1500 m/s in water and 5000 m/s in iron.

Our ears are most sensitive to sounds whose frequencies are between 3000 and 4000 Hz. Almost nobody can hear sounds with frequencies below about 20 Hz **(infrasound)** and above about 20,000 Hz **(ultrasound),** although for many animals the upper limit is higher. Dogs, for instance, respond to sound frequencies up to about 45,000 Hz. Hearing deteriorates with age, most noticeably at the higher frequencies.

Ultrasound has a number of applications, notably in medical imaging and in determining water depths (Fig. 7-13). The latter technique, called **sonar,** is also used to detect submarines and, by bats in air, to detect prey.

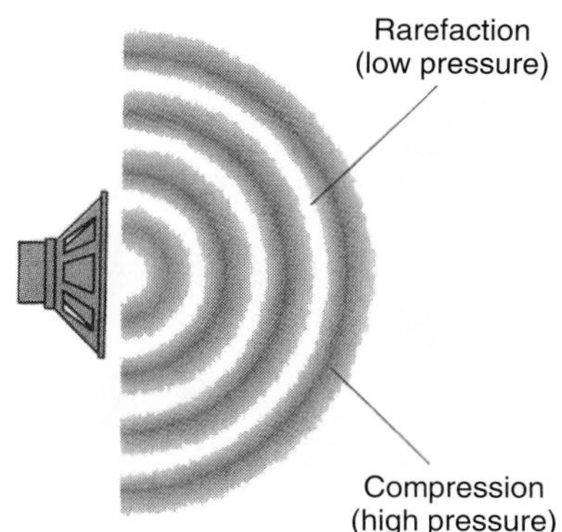

Figure 7-12 Sound waves produced by a loudspeaker. Alternate regions of compression and rarefaction move outward from the vibrating cone of the loudspeaker.

Ultrasound Imaging

Ultrasound is used in medicine to produce images of internal parts of the body, including unborn babies. Ultrasound is better able than x-rays to distinguish between soft tissues and liquids, and is far less harmful. What is done is to send a pulse of ultrasound waves in a narrow beam into a patient's body through the skin; the reflections of the pulse from interfaces between different materials return to a detector at different times. A picture of the internal structure of the body can be built up by moving the ultrasound beam in a scanning pattern (Fig. 7-14).

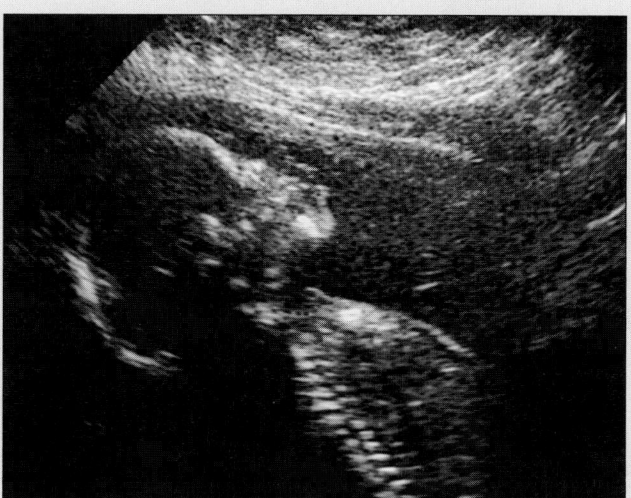

Figure 7-14 Ultrasound image of a fetus 20 weeks old.

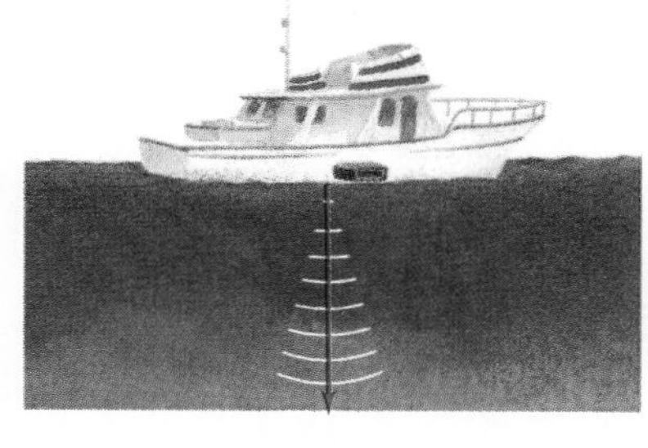

(*a*)

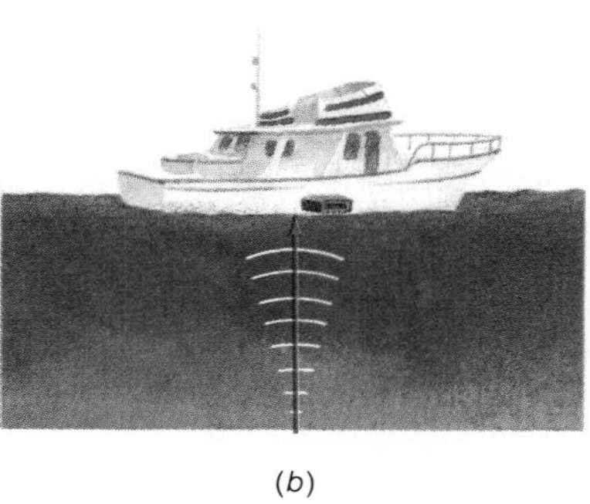

(*b*)

Figure 7-13 The principle of echo sounding. (*a*) A pulse of high-frequency sound waves is sent out by a suitable device on a ship. (*b*) The time at which the pulse returns to the ship is a measure of the sea depth.

Figure 7-15 Decibel scale for sounds.

The Decibel The more energy a sound wave carries, the louder it sounds. However, our ears respond to sound waves in a peculiar way. Doubling the rate of energy flow of a particular sound gives the sensation of an only slightly louder sound, not nearly twice as loud. This is why a solo instrument can be heard in a concerto even though a full orchestra is playing at the same time, and why you can carry on a conversation at a party even though many others are talking at the same time.

A special scale, whose unit is the **decibel** (dB), is therefore used to describe how powerful a sound is (Fig. 7-15). A sound that can barely be heard by a normal person is given the value 0 dB. Each 10-dB change corresponds to a 10-fold change in sound energy. Thus a 50-dB sound is 10 times stronger than a 40-dB sound and 100 times stronger than a 30-dB sound. The sound of ordinary conversation is usually about 60 dB, which is 10^6—a million—times more intense than the faintest sound that can be heard.

Exposure to sounds of 85 dB or more can lead to permanent hearing damage. Rock concerts (as much as 125 dB) have left many attendees with significant hearing loss, and millions of young people in the United States have damaged hearing due to daily doses of overamplified music. Three-quarters of the hearing loss of a typical older person in the United States is due to exposure to loud sounds.

7.6 Doppler Effect

Higher Pitch When Approaching; Lower Pitch When Receding

We all know that sounds produced by vehicles moving toward us seem higher pitched than usual, whereas sounds produced by vehicles moving away from us seem lower pitched than usual. Anybody who has listened to the siren of a police car as it passes by at high speed is aware of these changes in frequency, called the **doppler effect.**

The doppler effect arises from the relative motion of the listener and the source of the sound. Either or both may be moving. When the motion reduces the distance between source and listener, as in Fig. 7-16*b*, the wavelength decreases to make the frequency higher. When the motion takes source and listener farther away from each other, as in Fig. 7-16*c*, the wavelength increases to make the frequency lower.

To get an idea of the magnitude of such frequency shifts, the frequency of the sound heard by someone when a fire engine with a 500-Hz siren approaches at 60 km/h (37 mi/h) will be 526 Hz, and it will be 477 Hz when the fire engine has passed and is moving away. The ear can easily pick up frequency changes like these.

A simple way to visualize the doppler effect is to imagine traveling in a boat on a windy day. If we head into the wind, waves strike the boat more often than when the boat is at rest, and the ride may be very choppy. On the other hand, if we head away from the wind, waves catch up with us more slowly than when the boat is at rest, so their apparent frequency is less.

An interesting use of the doppler effect is to measure the speed of blood in an artery. When an ultrasound beam is directed at an artery, the waves reflected from the moving blood cells show a doppler shift in frequency because the cells then act as moving wave sources. From this shift the speed of the blood can be calculated. It is a few centimeters per second in the main arteries, less in the smaller ones.

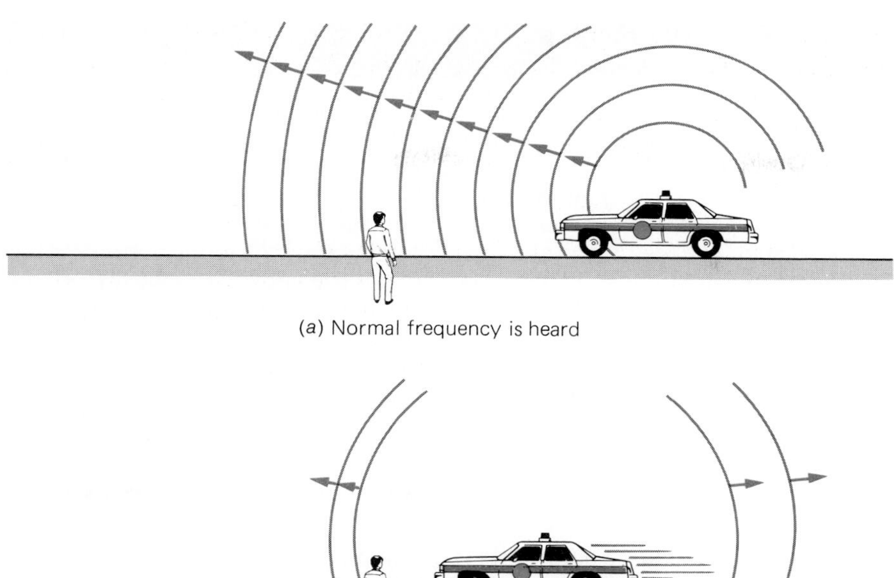

(*a*) Normal frequency is heard

(*b*) Higher frequency (shorter wavelength) is heard

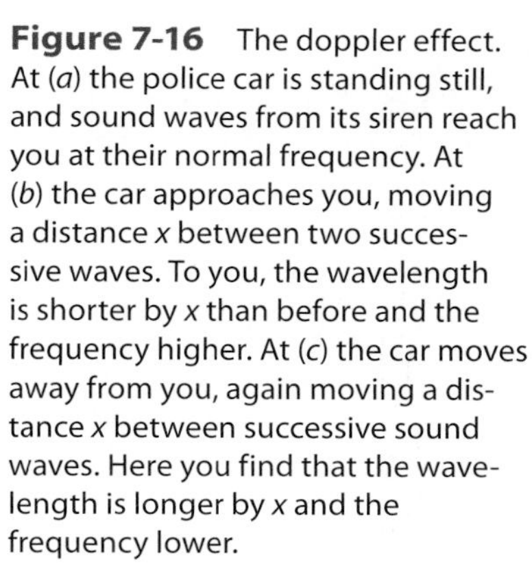

Figure 7-16 The doppler effect. At (*a*) the police car is standing still, and sound waves from its siren reach you at their normal frequency. At (*b*) the car approaches you, moving a distance *x* between two successive waves. To you, the wavelength is shorter by *x* than before and the frequency higher. At (*c*) the car moves away from you, again moving a distance *x* between successive sound waves. Here you find that the wavelength is longer by *x* and the frequency lower.

(*c*) Lower frequency (longer wavelength) is heard

The doppler effect occurs in light waves and is one of the ways by which astronomers detect and measure motions of the stars. Stars emit light that has only certain characteristic wavelengths. When a star moves either toward or away from the earth, these wavelengths appear, respectively, shorter or longer than usual. From the amount of the shift, it is possible to calculate the speed with which the star is approaching or receding. As we shall learn in Chap. 19, this is how the expansion of the universe was discovered. There are still other applications of the doppler effect (Fig. 7-17).

Figure 7-17 Doppler shifts in radar waves are widely used by police to determine vehicle speeds.

7.7 Musical Sounds

Fundamentals and Overtones

Musical sounds are produced by vibrating objects—stretched gut or wire in stringed instruments, vocal cords in the throat, membranes in drums, air columns in wind instruments. The simplest vibration of a stretched string is one in which a single standing wave takes up the entire length of the string, as in Fig. 7-9*a*. The frequency may be varied by changing the

Flute

Violin

Clarinet

Human voice:
vowel sound "e"

Figure 7-18 The waveforms of sounds can be analyzed electronically with the help of an oscilloscope, a device that displays electric signals on a screen like that of a television set. A microphone is used to convert sound waves into electric signals, and these in turn can be displayed on the oscilloscope screen. "Pure" tones, like those produced by a tuning fork, have simple waveforms like that of Fig. 7-4, while musical instruments and the human voice produce complex waveforms. Ordinary nonmusical noises consist of waves with complex and rapidly changing forms.

Harmony and Discord

Certain mixtures of frequencies are pleasing to the ear. For instance, the combination of a tone and its first overtone, whose frequency is twice as great, appears harmonious to a listener. Such an interval is called an **octave** in music because it includes eight notes. Also agreeable are tones with a frequency ratio of 2:3, such as C (262 Hz) and G (392 Hz). This interval is called a **fifth** because it includes five notes, here C, D, E, F, G. Somewhat less agreeable are tones whose frequencies are in the ratio 4:5, such as C and E; the interval here spans three notes and is called a **third.** The larger the numbers that express their frequency ratio, the less attractive a combination of tones appears. Thus C and D (ratio 8:9) seems discordant, and E and F (ratio 15:16) more discordant still. Ordinary sounds are mixtures of frequencies that have no special relationship with one another. If the mixture seems particularly harsh, we consider it noise.

tension on the string, as a violinist does when tuning the instrument. The tighter the string, the higher the frequency. For a given tension, the frequency can be varied by changing the length of the string, as the violinist does with the pressure of a finger on the string.

Depending on where the string is plucked, bowed, or struck, more complex vibrations may be set up: standing waves may form with two, three, or even more crests (Fig. 7-9*b*, *c*, *d*). Sound waves set up by these shorter standing waves have higher frequencies, and the frequencies are related to the frequency of the longest wave by simple ratios—2:1, 3:1, and so on. The tone produced by the string vibrating as in Fig. 7-9*a* is called the **fundamental,** and the higher frequencies produced when it vibrates in segments are called **overtones.**

Resonance In practice, the strings of a musical instrument give not just the fundamental or a single overtone but a combination of the fundamental plus several overtones. The motion of the string, and so the form of the sound wave, may be very complex (Fig. 7-18). To the ear a fundamental tone by itself seems flat and uninteresting. As overtones are added the tone becomes richer, with the quality, or timbre, of the tone depending on which particular overtones are emphasized. The emphasis largely depends on the shape of the instrument, which enables it to **resonate** at particular frequencies. The sounding part of the instrument—the belly of the violin or the soundboard of the piano—has certain natural frequencies of vibration, and it is more readily set vibrating at these frequencies than at others. The resulting sounds may include a large number of overtones, but the greater emphasis on certain overtones provides the musical quality characteristic of the instrument (Fig. 7-19).

Wind instruments produce sounds by means of vibrating air columns. In an organ, there is a separate pipe for each note. The shorter the pipe, the higher the pitch. Woodwinds, such as flutes and clarinets, use a single tube with holes whose opening and closing controls the effective length of the air column. Most brass instruments have valves connected to loops of tubing. Opening a valve adds to the length of the air column and thus produces a note of lower pitch. In a slide trombone, the length of the air column is

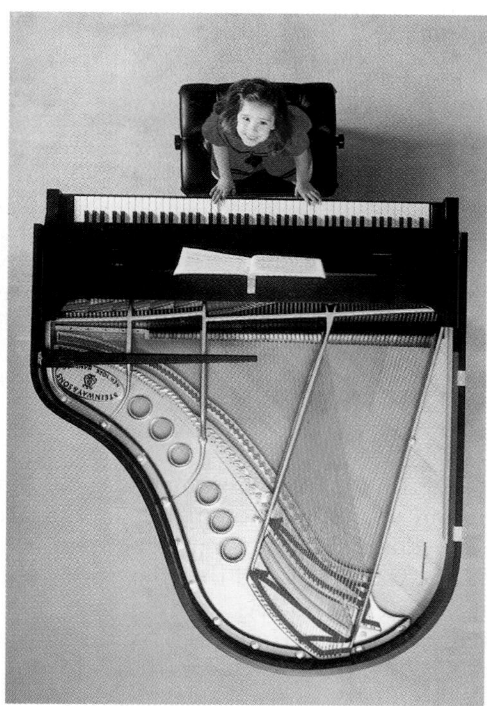

Figure 7-19 The 88 notes of a piano are produced by 220 metal wires (two or three per note to give louder sounds) whose fundamental frequencies range from 27 Hz for the lowest bass note to 4186 Hz for the highest treble note. The total tension of all the wires is about 18 tons and is borne by a heavy cast iron frame.

varied by sliding in or out a telescoping U tube. A bugle has neither holes nor valves, and a bugler obtains different notes with the lips alone.

The fundamental frequencies of the speaking voice average about 145 Hz in men and about 230 Hz in women. Even considering the overtones present, the frequencies in ordinary speech are mostly below 1000 Hz. In singing, the first and second overtones may be louder than their fundamentals, and even higher overtones add to the beauty of the sound (Fig. 7-20). Because our ears are more sensitive to higher sound frequencies, the presence of these overtones also helps a singer to be heard over the typically lower sound frequencies of an instrumental background.

Earthquake Damage

Earthquakes send out waves that cause the earth's surface to vibrate. A Mexican earthquake in 1985 was especially destructive in Mexico City because the frequency of the waves happened to match the natural frequencies of many buildings there, particularly narrow ones 8 to 15 stories high. The acceleration of the ground due to the waves was about 0.2*g*, but resonance amplified the building vibrations to produce accelerations of over 1*g*. This was more than the buildings could withstand, and they collapsed as a result.

Figure 7-20 The human voice is produced when the vocal cords in a person's throat set in vibration an air column that extends from the throat to the mouth and the nasal cavity above it. The shape of this column, which we adjust while speaking or singing by manipulating the mouth and tongue, determines the different vowel sounds by emphasizing some overtones and suppressing others. This shape also gives rise to the subtle differences in sound quality that enable us to tell one person's voice from another's. Vocal power is typically less than 1 W.

Electromagnetic Waves

In 1864 the British physicist James Clerk Maxwell suggested that an accelerated electric charge generates combined electrical and magnetic disturbances able to travel indefinitely through empty space. These disturbances are called **electromagnetic waves.** Such waves are hard to visualize because they represent fluctuations in fields that are themselves difficult to form mental images of. But they certainly exist—light, radio waves, and x-rays are examples of electromagnetic waves, and like all types of waves, they carry energy from one place to another.

7.8 Electromagnetic Waves

Waves without Matter

We learned in Chap. 6 that a changing magnetic field gives rise by electromagnetic induction to an electric current in a nearby wire. We can reasonably conclude that a changing magnetic field has an electric field associated with it. Maxwell proposed that the opposite effect also exists, so that a changing electric field has an associated magnetic field. The electric fields produced by electromagnetic induction are easy to measure because metals offer little resistance to the flow of electrons. There is no such thing as a magnetic current, however, and it was impossible in Maxwell's time to detect the weak magnetic fields he had predicted. But there is another way to check Maxwell's idea.

If Maxwell was right, then electromagnetic (em) waves must occur in which changing electric and magnetic fields are coupled together by both electromagnetic induction and the mechanism he proposed. The linked

BIOGRAPHY James Clerk Maxwell (1831–1879)

Maxwell was born in Scotland shortly before Michael Faraday discovered electromagnetic induction. While still a student, he used his ideas on color vision to make the first color photograph. At 24 he showed that the rings of Saturn could not be solid or liquid but must consist of separate small bodies. At about this time Maxwell became interested in electricity and magnetism and soon was convinced that these were not separate phenomena but had an underlying unity of some kind.

Starting from the results of Faraday and others, Maxwell created a single comprehensive theory of electricity and magnetism that remains the foundation of the subject today. From his equations Maxwell predicted that electromagnetic waves should exist that travel with the speed of light and surmised that light consisted of such waves. Sadly, he did not live to see his work confirmed in the experiments of Heinrich Hertz. Maxwell died of cancer at 48 in 1879, the year in which Albert Einstein was born. Maxwell had been the greatest theoretical physicist of the nineteenth century; Einstein was to be the greatest theoretical physicist of the twentieth century. (By a similar coincidence, Newton was born in the year of Galileo's death.)

fields spread out in space much as ripples spread out when a stone is dropped into a body of water. The energy carried by an em wave is constantly being exchanged between its fluctuating electric and magnetic fields. Calculations show that the wave speed in empty space should have the same value as the speed of light, which is 3×10^8 m/s (186,000 mi/s), regardless of frequency or amplitude. The symbol for the speed of light in empty space is c, so that

$$\textbf{Speed of light} = c = 3 \times 10^8 \text{ m/s}$$

Figure 7-21 shows the relationship between the electric and magnetic fields in an electromagnetic wave. Here the fields are represented by a

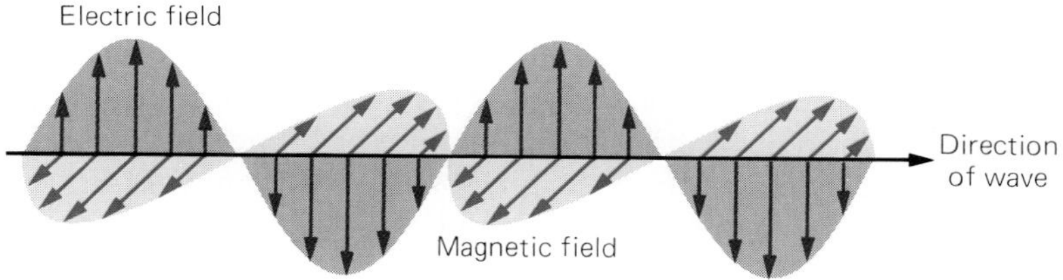

Figure 7-21 The magnitudes of the electric and magnetic fields in an electromagnetic wave vary together. The fields are perpendicular to each other and to the direction of the wave.

Solar Sails

Em waves carry momentum as well as energy, and their impact on a surface gives it a tiny push called **radiation pressure.** A spacecraft could use sunlight from the sun striking a suitable sail to propel it far out into the solar system. Anything bouncing off a surface exerts twice as much force as it would if it were simply absorbed, hence a solar sail should have a reflective surface. The sail should also be large: for a thrust of only 1 N (0.225 lb) an area of about 10^5 m^2 (nearly 25 acres!) would be needed at the distance of the earth's orbit. But there is no air resistance in space, so even a very small thrust would eventually lead to speeds well beyond those possible for a conventional spacecraft with a fixed fuel supply. Solar sails are a serious possibility for the future. Unfortunately, trials in 2001, 2005, and 2008 were inconclusive because in each case the launch vehicle failed.

Polarized Light

A polarized beam of light is one in which the electric fields of the waves are all in the same direction. (For simplicity, only the electric fields are considered here; the magnetic fields are perpendicular to them, as in Fig. 7-21.) If the electric fields are in random directions (though, of course, always in a plane perpendicular to the direction of the light beam), the beam is unpolarized. Figure 7-22*a* corresponds to a polarized beam and *b* corresponds to an unpolarized beam. Materials can be made that permit only light polarized in a certain direction to pass through them. Polaroid is a material of this kind. Because skylight is partly polarized, sunglasses with correctly oriented Polaroid lenses reduce glare from the sky while affecting unpolarized light from elsewhere to a lesser extent. Among other applications of polarized light are the liquid-crystal displays used for digital readouts, for instance in watches and some TV screens. These displays are based on the response of certain substances to polarized light when subjected to electric fields.

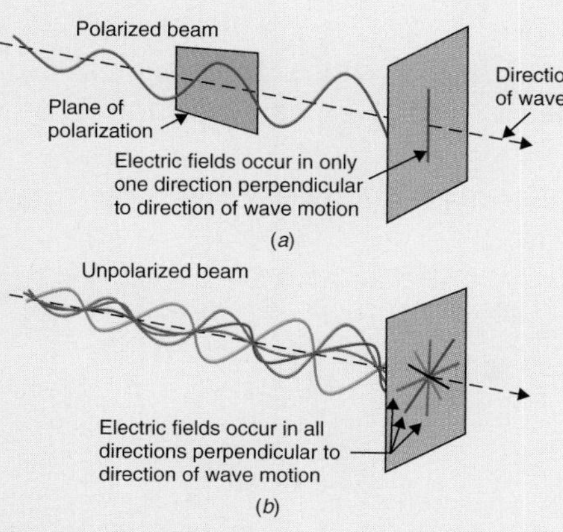

Figure 7-22 *(a)* A polarized and *(b)* an unpolarized beam of electromagnetic waves.

series of vectors (not field lines) that indicate the magnitude and direction of the fields in the path of the wave. The fields are perpendicular to each other and to the direction of the wave, and they remain in step as they periodically reverse their directions.

Figure 7-23 Heinrich Hertz (1857–1894).

7.9 Types of EM Waves

They Carry Information as Well as Energy

During Maxwell's lifetime em waves remained an unproven idea. Finally, in 1887, the German physicist Heinrich Hertz (Fig. 7-23) showed experimentally that em waves indeed exist and behave exactly as Maxwell expected them to.

Hertz was not concerned with the commercial possibilities of em waves, and other scientists and engineers developed what we now call radio. A radio signal is sent by means of em waves produced by electrons that move back and forth hundreds of thousands to millions of times per second in the antenna of the sending station. The signal resides either in variations in the strength of the waves (**amplitude modulation,** or AM) or in variations in the frequency of the waves (**frequency modulation,** or FM), as in Fig. 7-24. Frequency modulation is less subject to random disturbances ("static").

When the waves reach the antenna of a receiver, the electrons there vibrate in step with the waves. The receiver can be tuned to respond only to a narrow frequency band. Since transmitters operate on different frequencies, a receiver can pick up the signals sent out by whatever station we wish. The currents set up in the receiving antenna are very weak, but they are strong enough for electronic circuits in the receiver to extract the signal from them and turn it into sounds from a loudspeaker.

The frequencies of ordinary radio waves extend up to about 2 MHz (1 MHz = 1 megahertz = 10^6 Hz) and those of waves used in long-range short-wave communication extend up to about 30 MHz. Still higher frequencies have found widespread use in television and radar. Such extremely short waves are not reflected by the ionosphere (see opposite page), so direct reception of television is limited by the horizon unless rebroadcast by a satellite station.

Waves whose frequencies are around 10 GHz (1 GHz = 1 gigahertz = 10^9 Hz), corresponding to wavelengths of a few centimeters, can

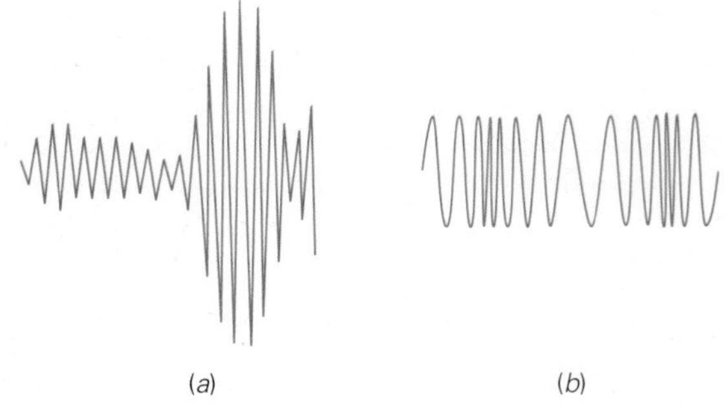

Figure 7-24 *(a)* In amplitude modulation (AM), variations in the amplitude of a constant-frequency radio wave constitute the signal being sent out. *(b)* In frequency modulation (FM), variations in the frequency of a constant-amplitude wave constitute the signal.

The Ionosphere

Not long after Hertz's experiments with em waves, the Italian engineer Guglielmo Marconi (Fig. 7-25) thought of using them for communication. In Hertz's work the transmitter and receiver were only a few meters apart. Marconi was able to extend the range to many kilometers, and in 1899 he sent a radio message across the English Channel. Two years later he sent signals across the Atlantic Ocean from England to Newfoundland, using kites to raise his antennas.

Radio waves, like light waves, tend to travel in straight lines, and the curvature of the earth should therefore prevent radio communication over long distances. For this reason Marconi's achievements came as a surprise. The mystery was solved with the discovery of a region of ionized gas called the **ionosphere** that extends from about 70 km to several hundred km above the earth's surface (Fig. 7-26). The ions are produced by the action of high-energy ultraviolet rays and x-rays from the sun. The ionosphere behaves like a mirror to high-frequency radio waves, which can bounce one or more times between the ionosphere and the earth's surface (Fig. 7-26). Low-frequency radio waves are absorbed by the ionosphere, and very-high frequency (VHF) and ultrahigh frequency (UHF) waves pass through it and so can be used to communicate with satellites and spacecraft.

Figure 7-25 Guglielmo Marconi at his laboratory in Newfoundland with the instruments that detected the first radio transmission across the Atlantic Ocean. The radio waves were reflected by the ionosphere.

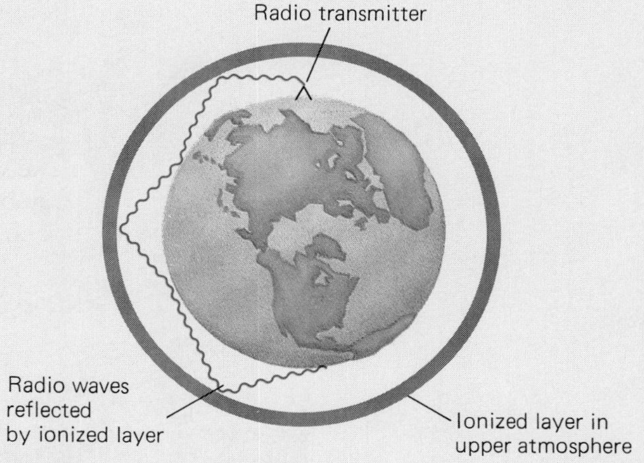

Figure 7-26 The ionosphere is a region in the upper atmosphere whose ionized layers make possible long-range radio communication by their ability to reflect short-wavelength radio waves.

readily be formed into narrow beams. Such beams are reflected by solid objects such as ships and airplanes, which is the basis of **radar** (from *ra*dio *d*etection *a*nd *r*anging) (Fig. 7-27). A rotating antenna is used to send out a pulsed beam, and the distance of a particular target is found from the time needed for the echo to return to the antenna. The direction of the target is the direction in which the antenna is then pointing.

Figure 7-28 shows the range (or **spectrum**) of em waves. The human eye can detect only light waves in a very short frequency band, from about 4.3×10^{14} Hz for red light to about 7.5×10^{14} Hz for violet light. Infrared radiation has lower frequencies than those in visible light, and ultraviolet radiation has higher frequencies. Still higher are the frequencies of x-rays and of the gamma radiation from atomic nuclei.

Figure 7-27 *(a)* As a radar scanner rotates, it sends out pulses of high-frequency radio waves in a narrow beam and receives reflections of them from objects around it. The reflections are displayed on a screen. The wider the antenna, the narrower the beam and the more detail in the image. *(b)* Long- and short-range radar images of the surroundings of a ship whose location corresponds in each case to the center of the circles that indicate distance from the antenna. The heading of the ship corresponds to upward in these images.

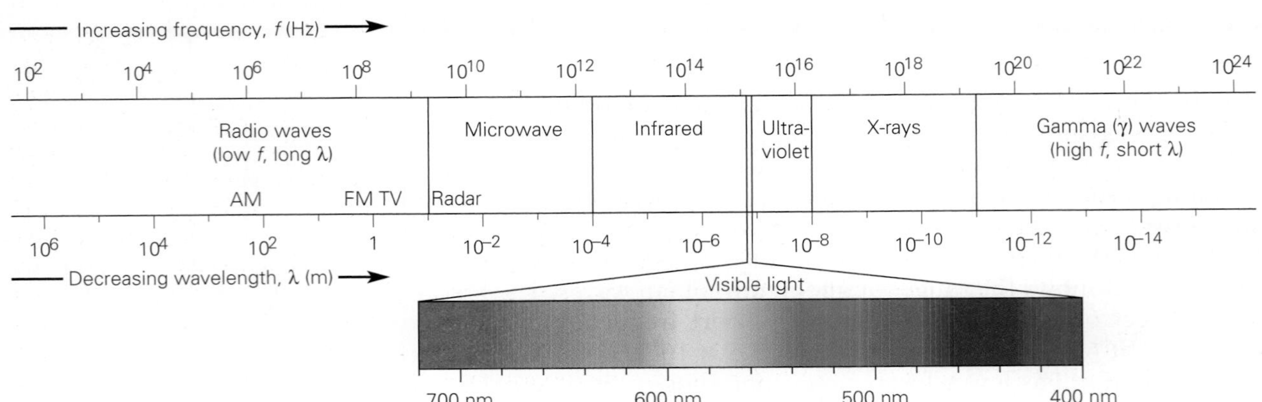

Figure 7-28 The electromagnetic spectrum. All em waves have the same fundamental character and the same speed in a vacuum, but how they interact with matter depends on their frequency (1 nm = 1 nanometer = 10^{-9} m = 1 billionth of a meter).

Example 7.3

A certain radar emits 9.4 GHz radio waves in groups 0.08 μs in duration. Find (a) the wavelength of these waves, (b) the length of each wave group, which governs how precisely the radar can measure distances, and (c) the number of waves in each group.

Solution

(a) Since 1 GHz = 1 gigahertz = 10^9 Hz, 9.4 GHz = 9.4×10^9 Hz, and the wavelength is, from Eq. 7-1,

$$\lambda = \frac{c}{f} = \frac{3 \times 10^8 \text{ m/s}}{9.4 \times 10^9 \text{ Hz}} = 3.2 \times 10^{-2} \text{ m} = 3.2 \text{ cm}$$

(b) Since 1 μs = 1 microsecond = 10^{-6} s, 0.08 μs = 8×10^{-8} s, and the length of each group is

$$d = ct = (3 \times 10^8 \text{ m/s})(8 \times 10^{-8} \text{ s}) = 24 \text{ m}$$

(c) The number N of waves in each group is

$$N = \frac{d}{\lambda} = \frac{24 \text{ m}}{3.2 \times 10^{-2} \text{ m}} = 750 \text{ waves}$$

We can also find N from $N = ft$.

7.10 Light "Rays"

The Paths Light Takes

Many aspects of the behavior of light waves can be understood without reference to their electromagnetic nature. In fact, we can even use a yet simpler model in many (but far from all) situations by thinking of light in terms of "rays" rather than waves.

Early in life we become aware that light travels in straight lines. A simple piece of evidence is the beam of a flashlight on a foggy night. Actually,

Ultraviolet and the Skin

Solar radiation at the longer-wavelength (320–400 nm) end of the ultraviolet part of the spectrum is called UVA and acts on the skin to produce a protective tan. Shorter-wavelength (290–320 nm) ultraviolet radiation is called UVB and causes sunburn. Repeated exposure to both kinds of solar UV ages the skin and, in the case of UVB especially, may lead to skin cancer. Sunscreen products usually block UVB while permitting some UVA to get through, which allows tanning. The sun protection factor (SPF) of a sunscreen is a measure of its effectiveness against UVB. An SPF of 10, for instance, means that 10 hours in the sun with the sunscreen on is equivalent to 1 hour with bare skin. In practice, to achieve its stated SPF value a sunscreen must be applied thickly and fairly often, say every few hours.

There is another consideration here. The action of UVB on skin produces vitamin D, which has turned out to be important in many aspects of health as well as its long-known role in bone formation. Vitamin D is involved in helping the body combat cancers of various kinds, infectious diseases, and autoimmune conditions such as multiple sclerosis. Many people do not receive enough of it in their food. It is now believed that one of the reasons for the general increase in sickness in winter is the reduced exposure of people's skin to strong sunlight then. Evidently, although too much UVB is bad, so is not enough. For most people in North America, 5 to 15 minutes daily of unprotected exposure to sunlight when the sun is high is sufficient and will not harm the skin. In winter, supplementary vitamin D would be a good idea to accompany diets low in that vitamin.

our entire orientation to the world about us, our sense of the location of things in space, depends on *assuming* that light follows straight-line paths.

Just as familiar, however, is the fact that light does not always follow straight lines. We see most objects by reflected light, light that has been turned sharply on striking a surface. The distorted appearance of things seen in water or through the heated air rising above a flame further testifies to the ability of light to be bent from a straight path. In these latter cases the light is said to be **refracted,** and we note that this occurs when light moves from one transparent material to another.

Although the conscious part of our minds recognizes that light can be reflected and refracted, it is easy to be deceived about the true positions of things. When we look in a mirror, for example, we are seeing light that travels to the mirror and then from the mirror to our eyes, but our eyes seem to tell us that the light comes from an image behind the mirror. When we look at the legs of someone standing in shallow water, they appear shorter than they do in air because light going from water into air is bent. Our eyes and brains have no way to take this into account and so we register the illusion rather than the reality.

Much can be learned about the behavior of light by studying the paths that light follows under various circumstances. Since light appears to travel in a straight path in a uniform medium, we can represent its motion by straight lines called **rays.** Rays are a convenient abstraction, and we can visualize what we mean by thinking of a narrow pencil of light in a darkened room.

Wave Behavior

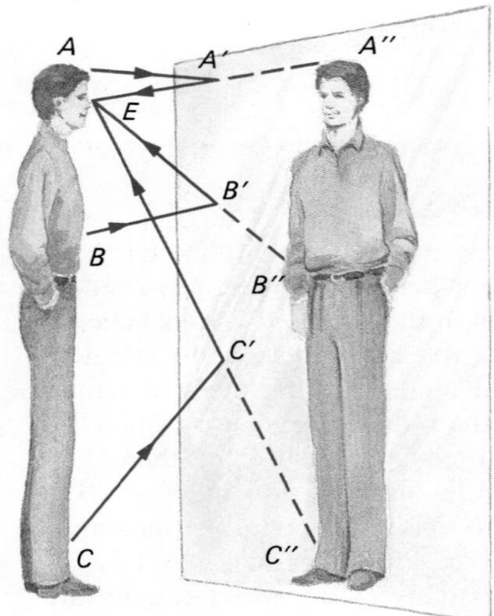

Figure 7-29 Formation of an image in a mirror. The image appears to be behind the mirror because we instinctively respond to light as though it travels in straight lines.

7.11 Reflection

Mirror, Mirror on the Wall

When we look at ourselves in a mirror, light from all parts of the body (this is reflected light, of course, but we may treat it as if it originated in the body) is reflected from the mirror back to our eyes, as in Fig. 7-29. Light from the foot, for example, follows the path $CC'E$. Our eyes, which see the ray $C'E$ and automatically project it in a straight line, register the foot at the proper distance but apparently behind the mirror at C''. A ray from the top of the head is reflected at A', and our eyes see the point A as if it were behind the mirror at A''. Rays from other points of the body are similarly reflected, and in this manner a complete **image** is formed that appears to be behind the mirror.

Left and right are interchanged in a mirror image because front and back have been reversed by the reflection. Thus a printed page appears backward in a mirror, and what seems to be one's left hand is really one's right hand.

Why do we not see images of ourselves in walls and furniture as well as in mirrors? This is simply a question of the relative roughness of surfaces. Rays of light are reflected from walls just as they are from mirrors, but the reflected rays are scattered in all directions by the many surface irregularities (Fig. 7-30). We see the wall by the scattered light reflected from it.

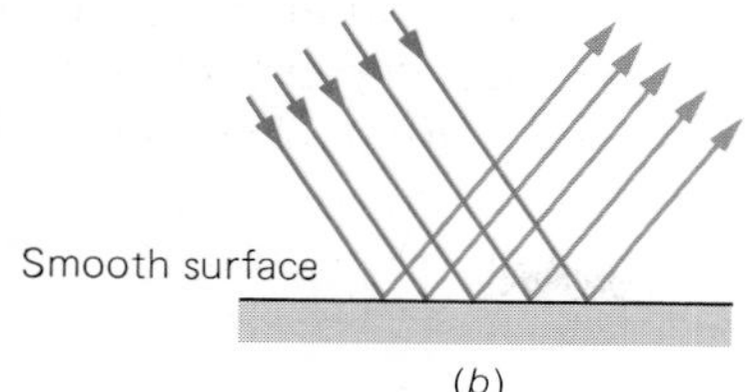

Figure 7-30 (*a*) Light that strikes an irregular surface is scattered randomly and cannot form an image. (*b*) Light that strikes a smooth, flat surface is reflected at an angle equal to the angle of incidence. Such a surface acts as a mirror.

7.12 Refraction

A Change in Direction Caused by a Change in Speed

No matter from where the wind is blowing, waves always approach a sloping beach very nearly at right angles to the shore (Fig. 7-31). Farther out in open water, the wave direction may be oblique (that is, at a slanting angle) to the shore, but the waves swing around as they move in so that their crests become roughly parallel to the shoreline. This is an example of refraction.

The explanation is straightforward. As a wave moves obliquely shoreward, its near-shore end encounters shallow water before its outer end does, and friction between the near-shore end and the sea bottom slows down that part of the wave. More and more of the wave is slowed as it continues to move toward the shore, and the slowing becomes more pronounced as the water gets shallower. As a result the whole wave turns until it is moving almost directly shoreward. The wave has turned because part of it was forced to move more slowly than the rest. Thus refraction is caused by differences in speed across the wave. Figure 7-32 shows an analogy to refraction in the motion of a tracked vehicle.

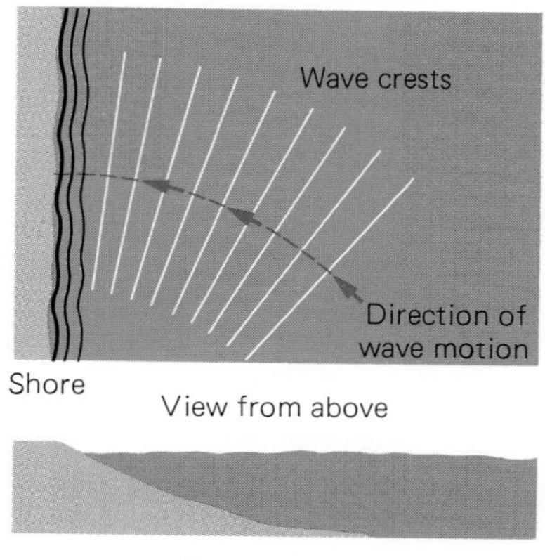

Figure 7-31 Refraction of water waves. Waves approaching shore obliquely are turned because they move more slowly in shallow water near shore.

Figure 7-32 Something similar to refraction occurs when a tracked vehicle such as a bulldozer makes a turn. The right-hand track of this bulldozer was slowed down, and the greater speed of its left-hand track then swung the bulldozer around to the right.

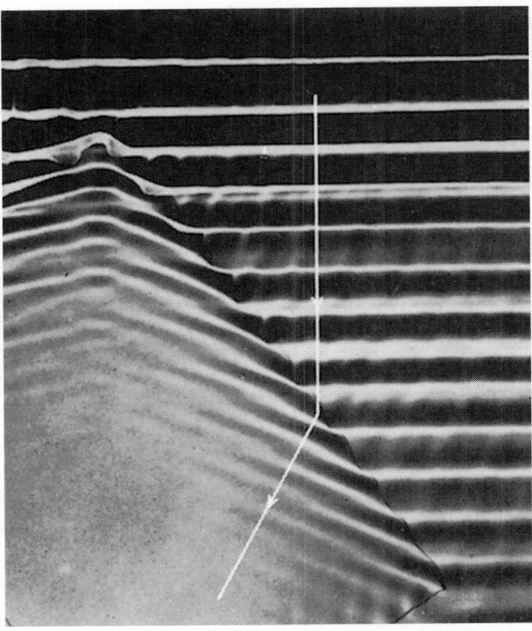

Figure 7-33 The refraction of water waves in a tank. In the left side of the tank is a glass plate over which the water is shallower than elsewhere. Waves move more slowly in shallow water than in deeper water, and hence refraction occurs at the edge of the plate. The arrows show the direction of movement of the waves.

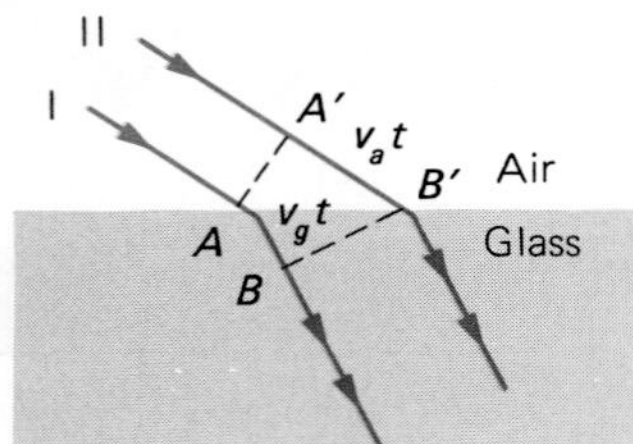

Figure 7-34 Refraction occurs whenever light passes from one medium to another in which its speed is different. Here two rays of light, I and II, pass from air, in which their speed is v_a, to glass, in which their speed is v_g. Because v_g is less than v_a, $A'B'$ is longer than AB, and the beam of which I and II are part changes direction when it enters the glass.

The bending will be at a sharp angle if the waves cross a definite boundary between regions in which they move at different speeds. Figure 7-33 shows this effect in ripples that move obliquely from deep water to shallow water in a tank. If the waves approach the boundary at right angles, no refraction occurs because the change in wave speed takes place across each wave at the same time.

Bending Light We all know that the water in a bathtub, a swimming pool, and even a puddle is always deeper than it seems to be. The reason is that light is refracted—changes direction—when it goes from one medium into another medium in which the speed of light is different (Fig. 7-34). The effect is similar to the refraction of water waves. A ray of light from the stone in Fig. 7-35 follows the bent path *ABE* to our eyes, but our brain registers that the segment *BE* is part of the straight-line path starting at *A*′.

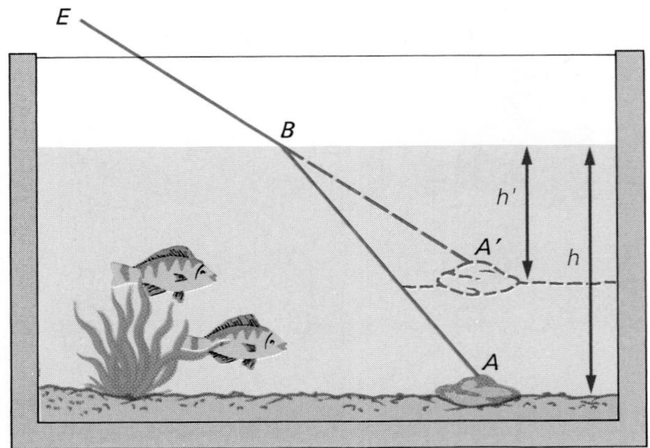

Figure 7-35 Light is refracted when it travels obliquely from one medium to another. Here the effect of refraction is to make the water appear shallower than it actually is.

The ratio between the speed of light c in free space and its speed v in a medium is called the **index of refraction** of the medium:

$$n = \frac{c}{v} \quad \textit{Index of refraction} \qquad 7\text{-}2$$

$$\text{Index of refraction} = \frac{\text{speed of light in free space}}{\text{speed of light in medium}}$$

The greater the value of n, the more a light ray is deflected when it enters a medium from air at an oblique angle. Here are some examples of indexes of refraction:

Water	1.33
Ordinary glass and clear plastics	1.52
Diamond	2.42

Because the speed of light in air is so close to c, for most purposes we can assume that they are the same with $n = 1$ for air.

The apparent depth h' of the submerged stone in Fig. 7-35 is related to its actual depth h and the index of refraction of the water (or other medium in the tank) by the formula

$$h' = \frac{h}{n} \quad \textit{Apparent depth} \qquad 7\text{-}3$$

$$\text{Apparent depth} = \frac{\text{actual depth}}{\text{index of refraction}}$$

Example 7.4

The water in a swimming pool is 2.0 m deep. How deep does it appear to someone looking into it from its rim?

Solution

For water, $n = 1.33$, so from Eq. 7-3

$$h' = \frac{h}{n} = \frac{2.0 \text{ m}}{1.33} = 1.5 \text{ m}$$

The pool seems only three-quarters as deep as it really is.

In general, light rays that go obliquely from one medium to another are bent toward a perpendicular to the surface between them if light in the second medium travels more slowly than in the first (Fig. 7-36). If light in the second medium travels faster there, the rays are bent *away*

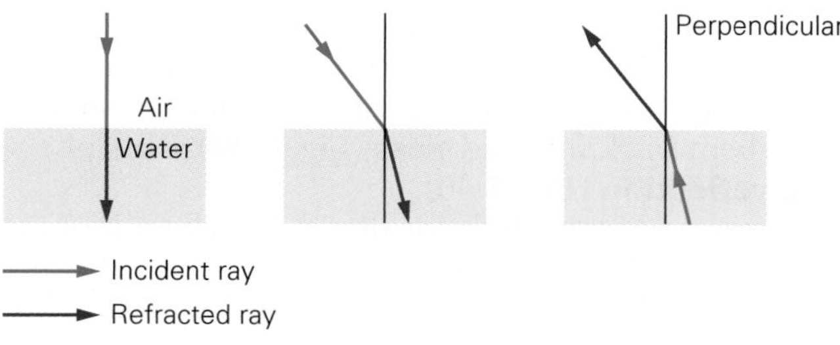

Figure 7-36 Light rays are bent toward the perpendicular when they enter an optically denser medium, away from the perpendicular when they enter an optically less dense medium. A ray moving along the perpendicular is not bent. The paths taken by light rays are always reversible.

Fiber Optics

Internal reflection makes it possible to "pipe" light by means of a series of reflections from the wall of a glass rod, as in Fig. 7-37. If a cluster of thin glass fibers is used instead of a single rod, an image can be transferred from one end to the other with each fiber carrying a part of the image. Because a fiber cluster is flexible, it can be used for such purposes as examining a person's stomach by being passed in through the mouth. Some of the fibers provide light for illumination, and the rest carry the reflected light back outside for viewing.

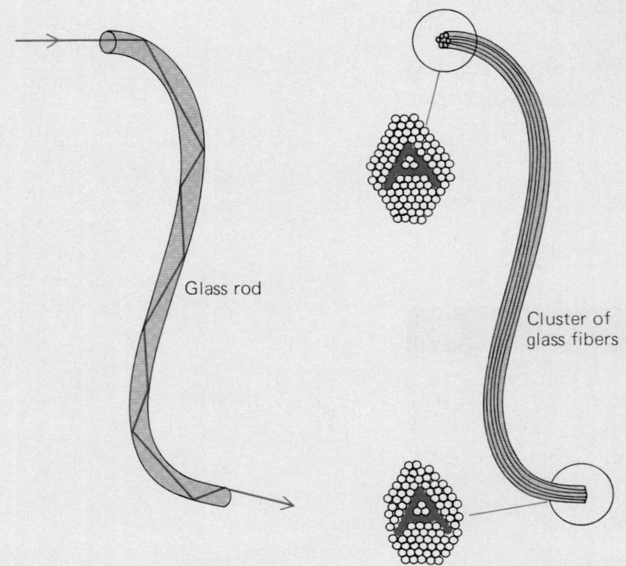

Figure 7-37 Light can be "piped" from one place to another by means of internal reflections in a glass rod. Using a cluster of glass fibers permits an image to be carried in this way.

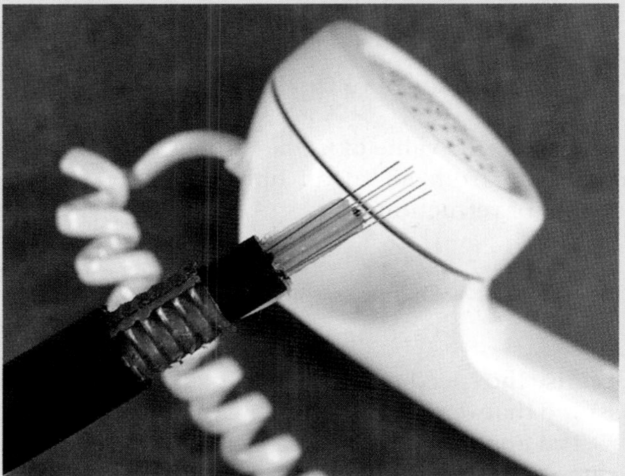

Figure 7-38 Each of the thin fibers in this cable can carry millions of telephone conversations in the form of coded flashes of light distributed among 100 or so different frequencies.

Glass fibers have been used in telephone systems since 1977. The electric signals that would otherwise be sent along copper wires are converted to a series of pulses according to a standard code and then sent as flashes of infrared light down a hair-thin glass fiber. At the other end the flashes are converted back to electric signals (Fig. 7-38). Modern electronic methods allow at most 32 telephone conversations to be carried at the same time by a pair of copper wires, but over a million can be carried by a single fiber with no problems of electrical interference. Telephone fiberoptic systems today link many cities and exchanges within cities everywhere, and fiberoptic cables span the world's seas and oceans.

from the perpendicular. Light that enters another medium perpendicular to the surface between them does not change direction.

Internal Reflection Since light travels more slowly in glass than in air, light that goes from glass to air at a slanting angle is refracted away from the perpendicular to the glass surface. If the angle is shallow enough, the light will be bent back into the glass (Fig. 7-39). This phenomenon is called **internal reflection** (Fig. 7-40).

What does an underwater fish (or diver) see when looking upward? The paths taken by light rays are always reversible. By reversing the rays of light in Fig. 7-39 we can see that light from above the water's

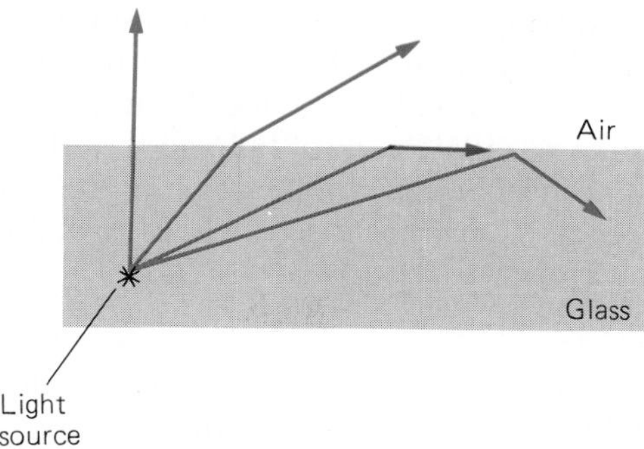

Figure 7-39 Total internal reflection occurs when the angle of refraction for a light ray going from one medium to a less optically dense medium would be more than 90° from the perpendicular to the surface between the media.

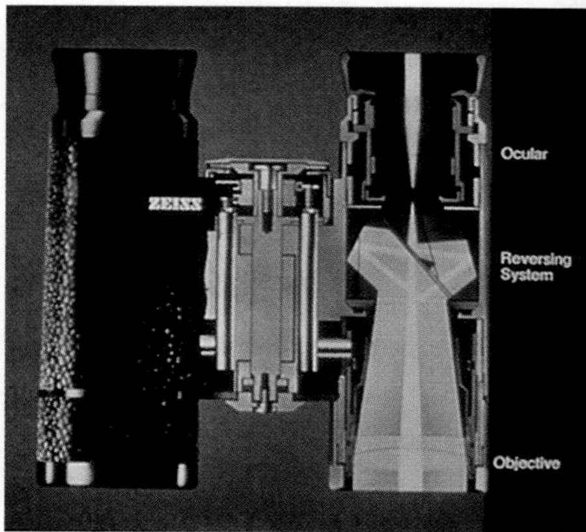

Figure 7-40 The sharpness and brightness of a light beam are better preserved by internal reflection than by reflection from an ordinary mirror. Because of this, optical devices such as binoculars use prisms instead of mirrors when light is to have its direction changed. The prisms used in binoculars have two purposes: they invert the magnified image so that it is right-side-up, and by reversing the optical path twice, they shorten the length of the instrument.

surface can reach the fish's eyes only through a circle on the surface. The rays from above are all brought together in a cone whose angular width turns out to be 98° (Fig. 7-41). Outside the cone is a reflection of the underwater scene.

7.13 Lenses

Bending Light to Form an Image

A **lens** is a piece of glass or other transparent material shaped so that it can produce an image by refracting light that comes from an object. Lenses are used for many purposes: in eyeglasses to improve vision, in cameras to record scenes, in projectors to show images on a screen, in microscopes to enable small things to be seen, in telescopes to enable distant things to be seen, and so forth.

Lenses are of two kinds, **converging** and **diverging.** A converging lens is thicker in the middle than at its rim; a diverging lens is thinner in the middle. As in Fig. 7-42*a*, a converging lens brings a parallel beam of light to a single focal point *F*. Here *F* is called a **real focal point** because

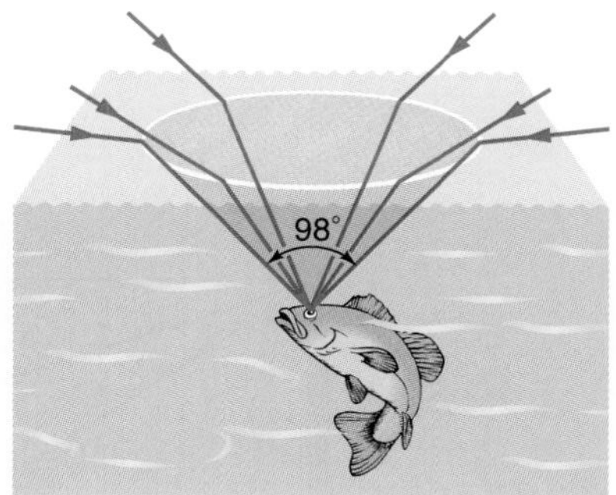

Figure 7-41 All the light reaching an underwater observer from above the surface is concentrated in a cone 98° wide, so that the observer sees a circle of light at the surface when looking upward.

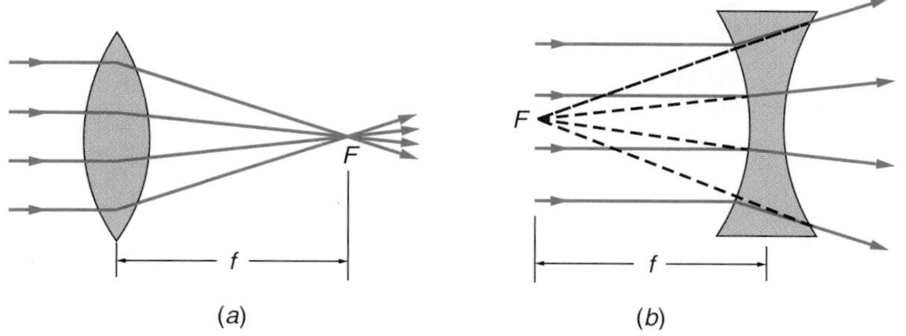

Figure 7-42 (*a*) A converging lens brings parallel rays of light together to a focal point *F*. (*b*) A diverging lens spreads out parallel rays of light so that they seem to originate at a focal point *F*. In both cases the distance between *F* and the lens is the focal length *f* of the lens.

the light rays pass through it. If sunlight is used, the concentration of radiant energy may be enough to burn a hole in a piece of paper. The distance from the lens to F is called the **focal length** of the lens. A diverging lens spreads out a parallel beam of light so that the rays seem to have come from a focal point F behind the lens, as in Fig. 7-42*b*. In this case F is called a **virtual focal point** because the light rays do not actually pass through it but only appear to.

Ray Tracing A scale drawing gives us an easy way to find the properties of the image of an object formed by a lens. What we do is trace the paths of two different light rays from a point of interest on the object to where they (or their extensions, in the case of a virtual image) come together again after passing through the lens. We should note that a lens has two focal points, one on each side, the same distance f from its center.

The three simplest rays to trace are shown in Fig. 7-43 for a converging lens:

1. A ray that leaves the object parallel to the lens axis. The lens deviates this ray so that it passes through the focal point of the lens on the far side;
2. A ray that passes through the center of the lens. This ray is not deviated;
3. A ray that passes through the nearest focal point of the lens. The lens deviates the ray so that it continues parallel to the lens axis.

In Fig. 7-43 the object is the distance $2f$ from the lens, and we can see that the image is $2f$ on the other side of the lens. The image is real (the rays actually pass through it), is the same size as the object, and is inverted. This corresponds to the optical system of a photocopier that produces a copy the same size as an original.

A diverging lens deviates rays that enter parallel to the axis away from the axis, as we saw in Fig. 7-42*b*, instead of toward the axis as a converging

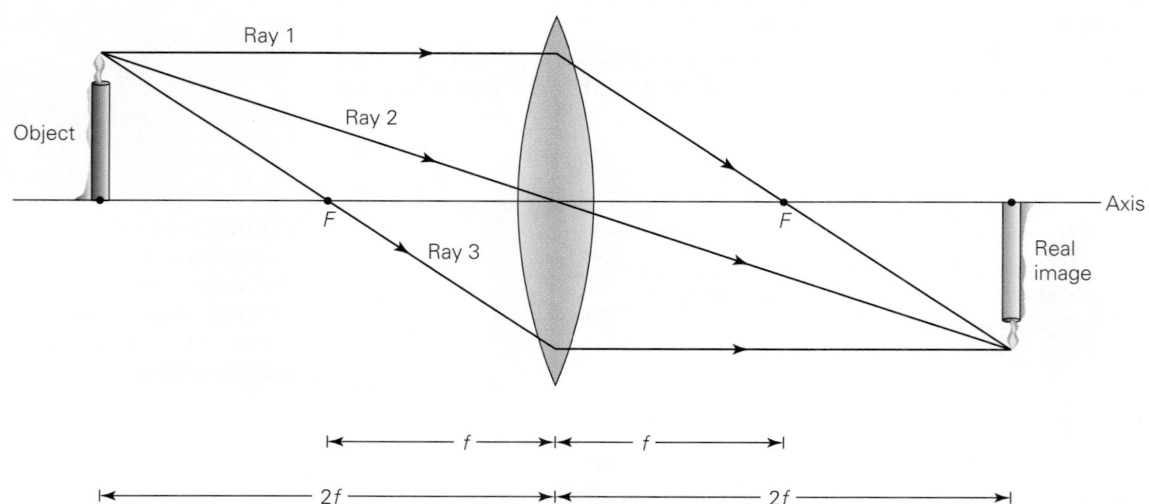

Figure 7-43 Tracing any two of the rays shown will indicate the position and size of the image formed by a converging lens and whether it is erect (right-side-up) or inverted (upside down). The image will be found where the rays come together after being refracted by the lens.

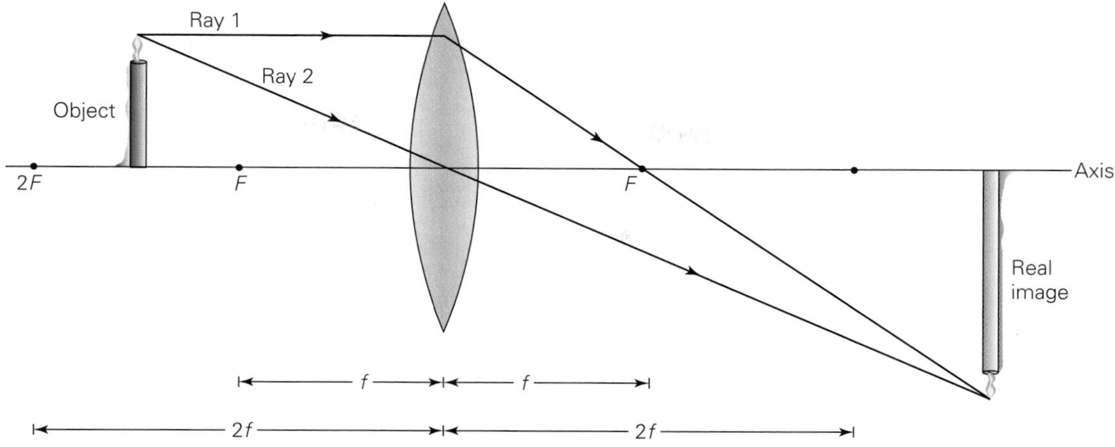

Figure 7-44 Ray diagram for an object a distance between f and $2f$ from a converging lens. This corresponds to the optical system of a projector where a slide is the object and the image appears on a screen.

Example 7.5

A slide projector uses a converging lens to produce an enlarged image of a transparent slide on a screen. Use a ray diagram to find the properties of the image when the slide is between f and $2f$ from the lens.

Solution

As we see in Fig. 7-44 by tracing rays 1 and 2, the image is real, farther than $2f$ from the lens, larger than the object on the slide, and inverted. This is why slides have to be put upside down in a projector.

Example 7.6

A camera uses a converging lens to produce a reduced image on the sensitive surface (film or electronic sensor). Use a ray diagram to find the properties of the image when the object being photographed is more than $2f$ from the lens.

Solution

As we see in Fig. 7-45, again by tracing rays 1 and 2, the image is real, between f and $2f$ from the lens, smaller than the object, and inverted.

Example 7.7

A "magnifying glass" uses a converging lens to produce an enlarged image of an object closer than f from the lens. Use a ray diagram to find the properties of the image.

Solution

As we see in Fig. 7-46, the image here seems to be behind the lens because the refracted rays diverge as though coming from a point behind it. This is a **virtual image:** although it can be seen by the eye, it cannot appear on a screen because no rays actually pass through it. The image is farther from the lens than the object, is larger than the object, and is erect.

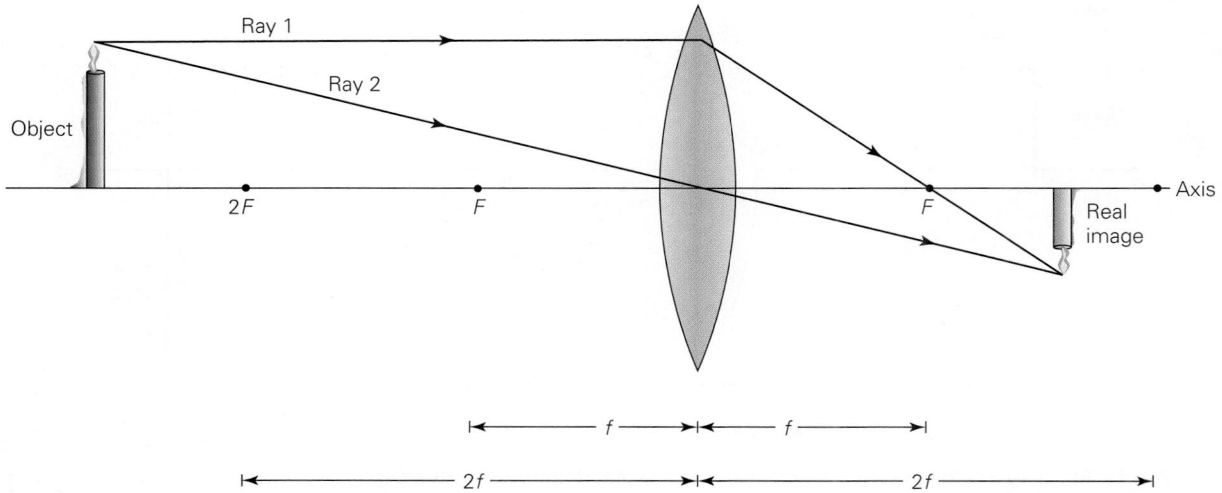

Figure 7-45 Ray diagram for an object farther than 2*f* from a converging lens. This corresponds to the optical system of a camera. The closer the object is to the lens, the farther the sensor screen or film must be from the lens. The appropriate focal length of a camera lens depends upon the desired angle of view. A "normal" lens gives an angle of view of about 45°. A lens of shorter focal length is a wide-angle lens that captures more of a given scene, though at the expense of reducing the sizes of details in the scene. A telephoto lens has a long focal length to give larger images of distant objects, though less of the scene is included. Adjustable "zoom" lenses can cover a wide range of focal lengths.

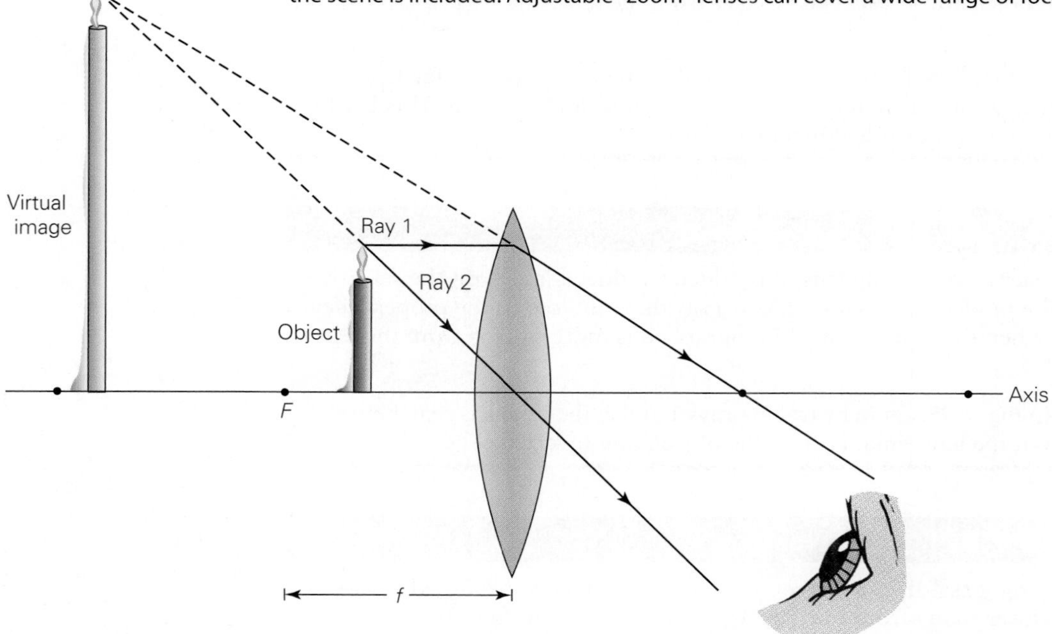

Figure 7-46 Ray diagram for an object closer than *f* from a converging lens. The image is erect and virtual and larger than the object. This is how a "magnifying glass" works.

lens does, as in Fig. 7-42*a*. As a result, the image of a real object produced by a diverging lens is always virtual, erect, smaller than the object, and closer to the lens than the object is (Fig. 7-47).

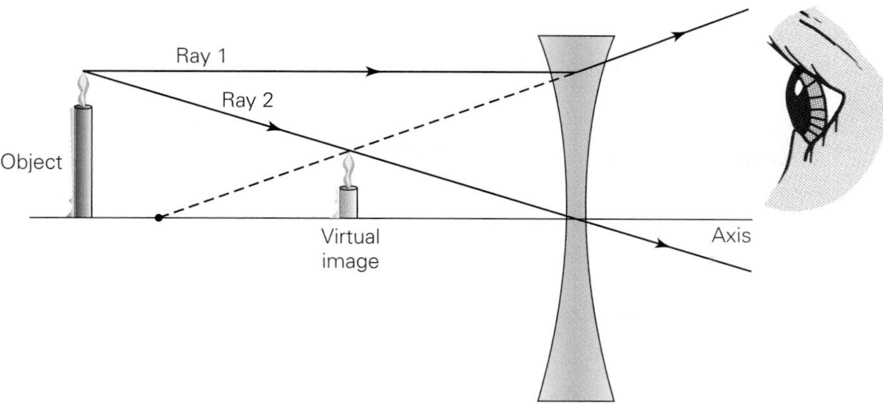

Figure 7-47 A diverging lens produces a virtual image of a real object that is always erect and smaller than the object.

7.14 The Eye

A Remarkable Optical Instrument

The structure of the human eye is shown in Fig. 7-48. The **cornea,** the transparent outer membrane, and the jellylike **lens** together focus incoming light on the sensitive **retina,** which converts what is seen into nerve impulses that are carried to the brain by the **optic nerve** (Fig. 7-49). Focusing on objects different distances away is done when the **ciliary muscle** changes the shape and hence the focal length of the lens. The colored **iris** acts like the diaphragm of a camera to control the amount of light entering the **pupil,** which is the opening of the iris. In bright light the pupil is small, in dim light it is large. A fully opened pupil lets in about 16 times as much light as a fully contracted one. The retina itself can also cope with a wide range of brightnesses.

Figure 7-49 Refraction at the cornea gives rise to most of the focusing power of the human eye. When the eye is immersed in water rather than air, the amount of refraction is much less. As a result, underwater objects cannot be brought to a sharp focus unless goggles or a face mask are used to keep water away from the cornea. Beavers cope with this problem by having transparent eyelids that adjust the optical systems of their eyes for underwater vision. Fish avoid the problem by having eyes in which most of the focusing is done by very thick lenses, with the cornea playing only a minor part.

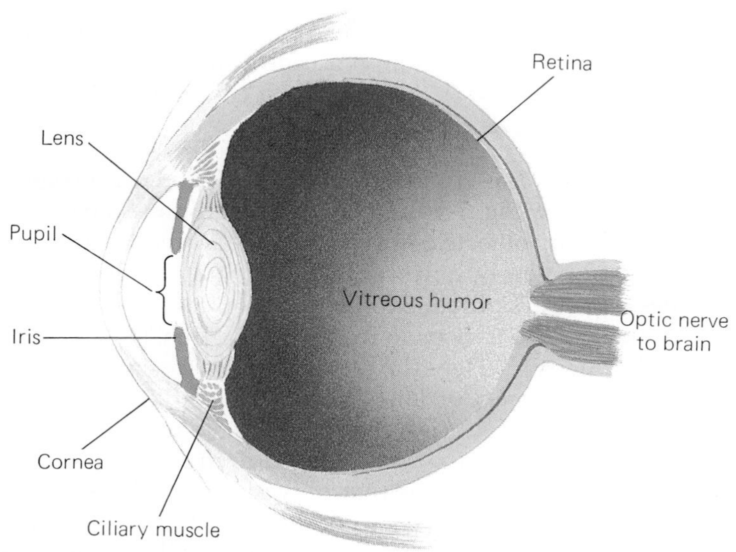

Figure 7-48 The human eye, shown larger than life size. In dim light the iris opens wide to let enough light enter through the pupil for good vision.

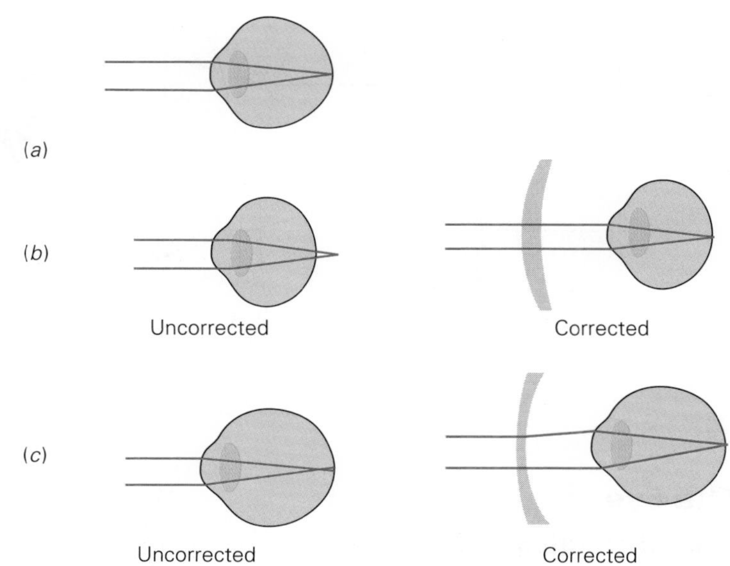

Figure 7-50 (*a*) A normal eye. (*b*) Farsightedness can be corrected with a converging lens. (*c*) Nearsightedness can be corrected with a diverging lens.

The retina has millions of tiny structures called **cones** and **rods** that are sensitive to light. The cones are specialized for color vision and occur in three types that respond respectively to red, green, and blue light. (Mammals other than primates have only two types of cone, so they have poorer color vision than we have; birds have four types, which gives them better color vision, including part of the ultraviolet.) Rods need much less light to be activated than cones do, but rods do not distinguish colors. In poor illumination, then, what we see is in shades of gray, like a black-and-white photograph. Because the central region of the retina contains only cones, it is easier to see something in a dim light by looking a bit to one side instead of directly at it. This central sensitive region, called the *fovea,* covers a region of the visual field only about the size of the sun or moon in the sky. Because the fovea is so small, each eye constantly darts about a few times a second, stopping each time for a brief moment to register what is in view and then moving on. The brain knits these flashes of information into a seamless picture.

Defects of Vision Two common defects of vision are **farsightedness** and **nearsightedness** (Fig. 7-50). In farsightedness the eyeball is too short, and light from nearby objects comes to a focus behind the retina. (Distant objects can be seen clearly, however.) A converging eyeglass lens corrects farsightedness. In nearsightedness the eyeball is too long, and light from distant objects comes to a focus in front of the retina. (Nearby objects can be seen clearly, however.) Here the correction is a diverging eyeglass lens.

Sometimes the cornea or lens of an eye has different curvatures in different planes. When light rays that lie in one plane are in focus on the retina of such an eye, rays in other planes are in focus either in front of or behind the retina. This means that only one of the bars of a cross can be in focus at any time (Fig. 7-52), a condition called **astigmatism.** Astigmatism causes eyestrain because the eye continually varies the focus of the lens as it tries to produce a completely sharp image of what it sees. A cylindrical corrective lens (Fig. 7-53) is the remedy for astigmatism.

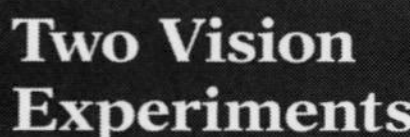

Two Vision Experiments

The retinas of vertebrate animals, like us, are inside-out (not a very intelligent design, but the result of the unplanned way evolution proceeded): the light-sensitive cells are on the outside, behind the nerves and blood vessels connected to them. A **blind spot** occurs where the nerves and blood vessels pass through the retina. To experience the blind spot, close your left eye and look directly at the cross in Fig. 7-51 with your right eye. When the cross is about 20 cm from your eye, the dot should disappear. We are not usually aware of the blind spot for two reasons: the blind spots of the two eyes obscure different fields of vision, and the eyes are in constant scanning motion.

Figure 7-51 Locating the blind spot.

To verify that cones, which are responsible for color vision, are concentrated near the center of the retina, hold something with a bright color off to one side and slowly move it around until it is in front of you. As you do this, you will find the sensation of its color going from weak to strong.

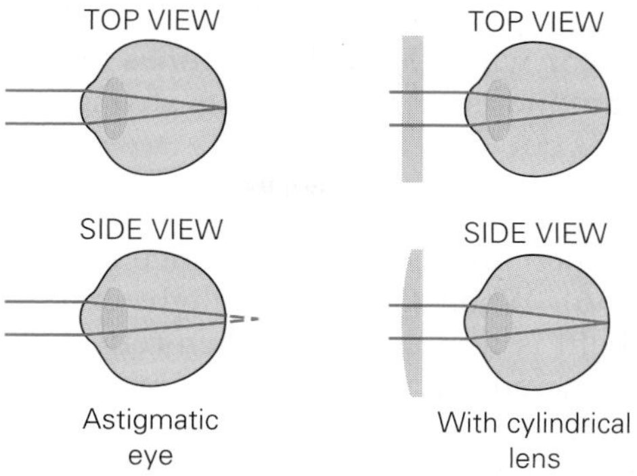

TOP VIEW
TOP VIEW
SIDE VIEW
SIDE VIEW
Astigmatic eye
With cylindrical lens

Figure 7-53 A cylindrical lens can improve the image formed by an astigmatic eye.

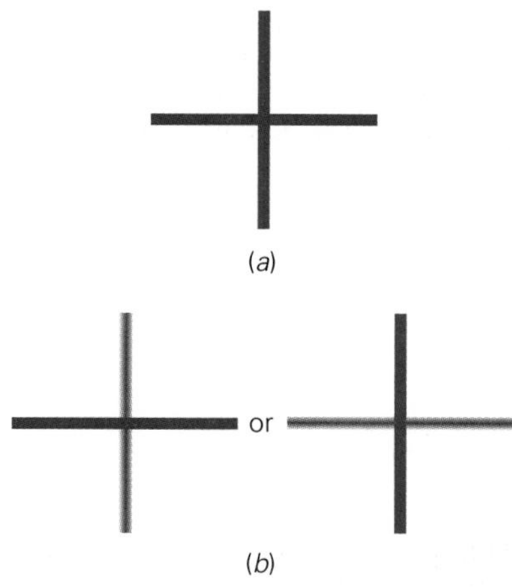

Figure 7-52 How a cross is seen (*a*) by a normal eye and (*b*) by an astigmatic eye.

7.15 Color

Each Frequency of Light Produces the Sensation of a Different Color

White light is a mixture of light waves of different frequencies, each of which produces the visual sensation of a particular color. To show this, we can direct a narrow beam of white light at a glass prism (Fig. 7-54). Because the speed of light in glass varies slightly with frequency, light of each color is refracted to a different extent. The effect is called **dispersion.** The result is that the original beam is separated by the prism into beams of various colors, with red light bent the least and violet light bent the most.

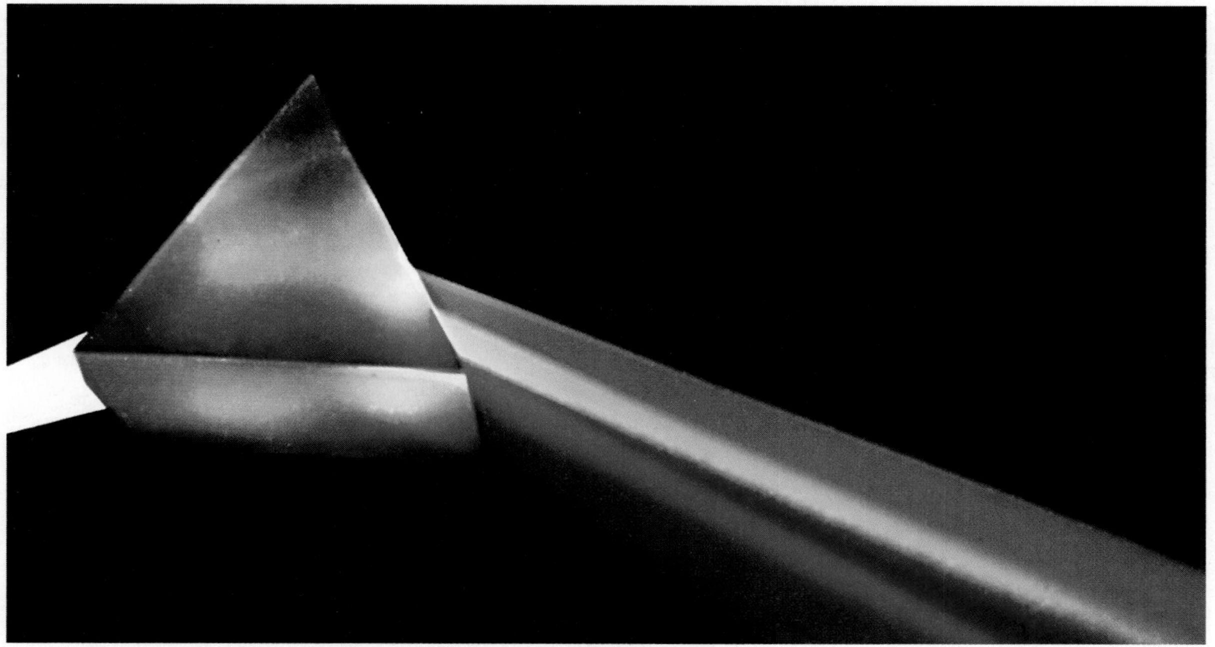

Figure 7-54 A beam of white light is separated into its component wavelengths, each of which appears to the eye as a different color, by dispersion in a glass prism.

Dispersion is especially marked in diamond, which is the reason for the vivid play of color when white light shines on a cut diamond. The sparkle of a cut diamond is partly due to its strong refractive power and partly to the way it is cut (Fig. 7-55).

Rainbows The dispersion of sunlight by water droplets is responsible for rainbows, which are seen when we face falling rain with the sun behind us (Fig. 7-56). When a ray of sunlight enters a raindrop, as in Fig. 7-57, the sunlight is first refracted, then reflected at the back of the drop, and finally refracted again when it goes back out. Dispersion occurs at each refraction. With the sky full of raindrops, the result is a colored arc that has red light on the outside and violet light on the inside. Someone

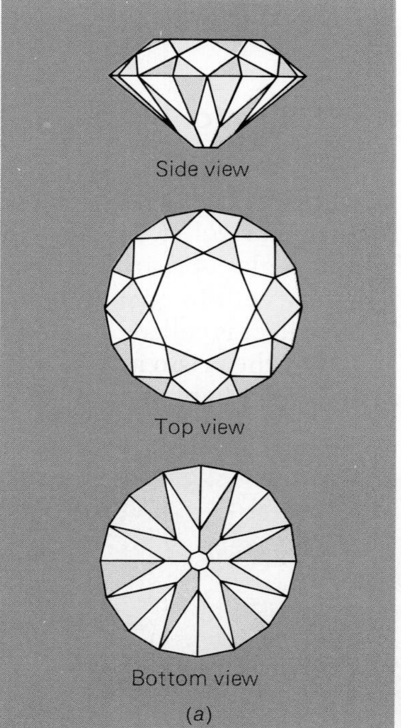

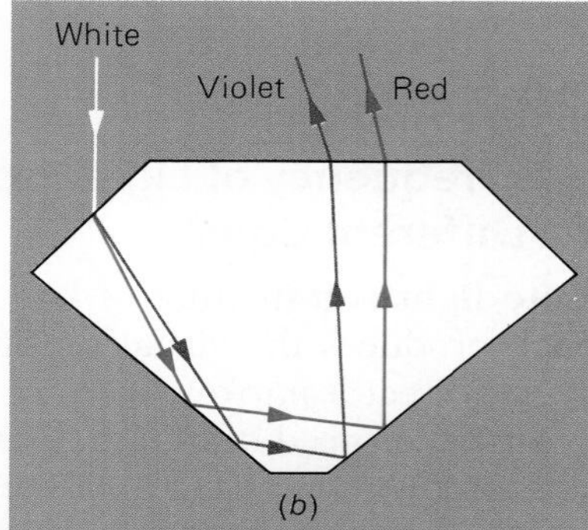

Figure 7-55 (*a*) A diamond cut in the "brilliant" style has 33 facets in its upper part and 25 in its lower part. The proportions of the facets are critical in giving the maximum of sparkle. (*b*) Dispersion gives a cut diamond its fire.

Figure 7-56 A rainbow is caused by the dispersion of sunlight into its component colors when refracted by water droplets in the atmosphere.

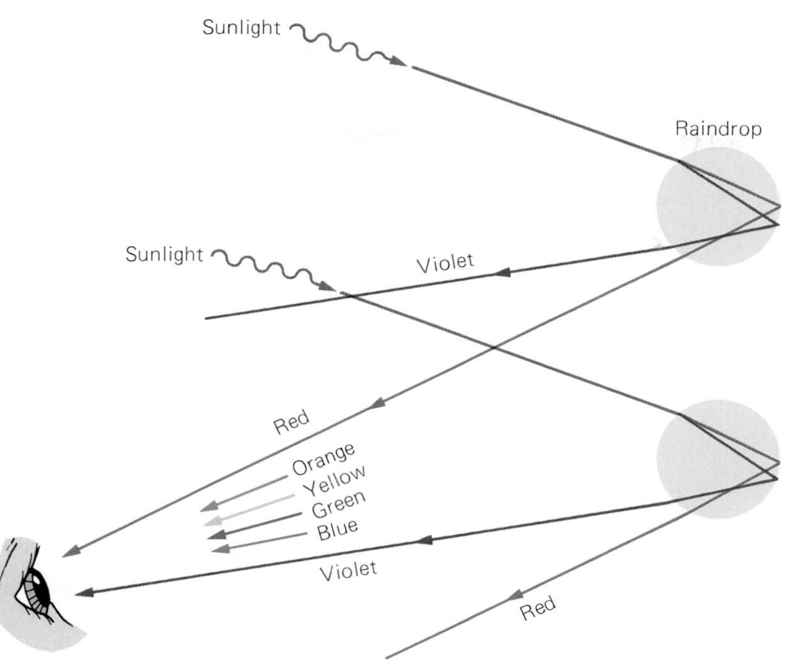

Figure 7-57 Rainbows are created by the dispersion of sunlight by raindrops. Red light arrives at the eye of the observer from the upper drop shown here, violet light from the lower drop. Other raindrops yield the other colors and produce a continuous arc in the sky.

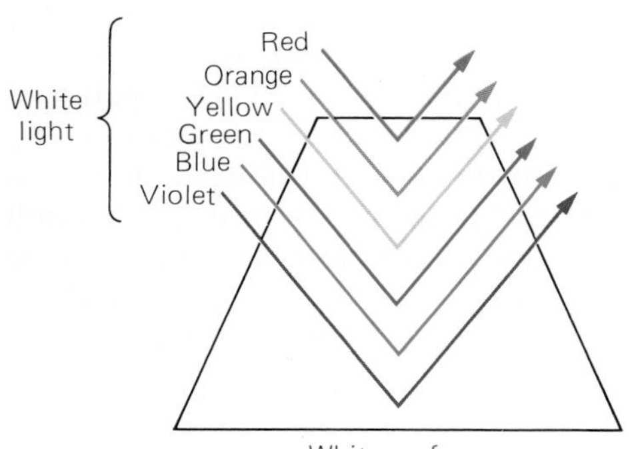

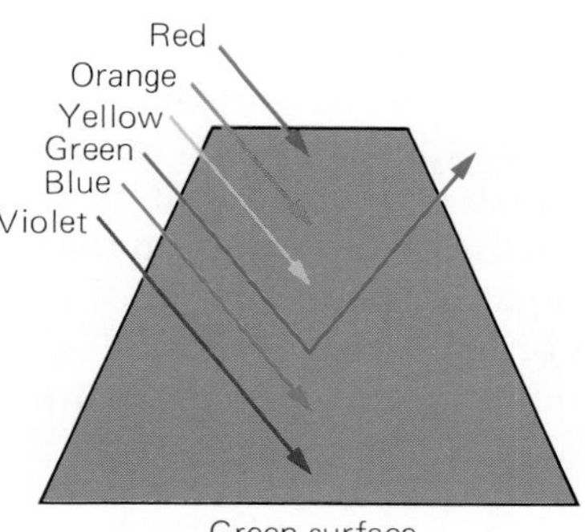

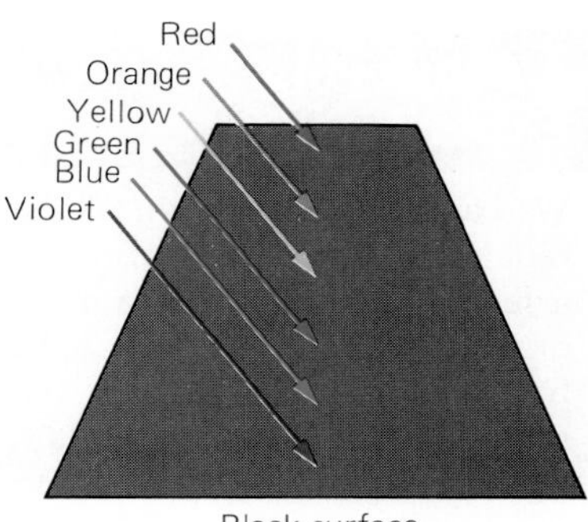

Figure 7-58 A white surface reflects all light that falls on it. A green surface reflects only green light and absorbs the rest. A black surface absorbs all light that falls on it.

in an airplane can see the entire ring of color, but from the ground only the upper part is visible.

The color of an object depends on the kind of light that falls on the object and on the nature of its surface. If a surface reflects all light that falls on it, the color of the surface will be white when white light illuminates it (Fig. 7-58), red when red light illuminates it, and so on. A surface that reflects only, say, green light will appear green only when the light illuminating it contains green; otherwise it will appear black. A surface that absorbs all light that falls on it appears black.

Blue Sky The blue color of the sky is due to scattering of the sun's light by molecules and dust particles in the atmosphere. Blue light is scattered

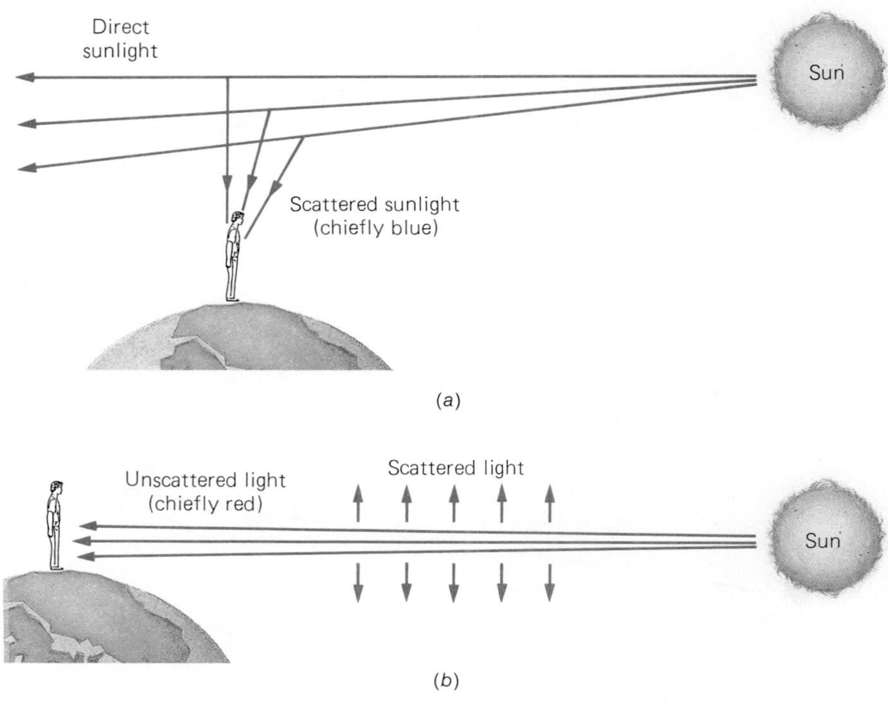

Figure 7-59 (*a*) The preferential scattering of blue light in the atmosphere is responsible for the blue color of the sky. (*b*) The remaining direct sunlight is reddish, which is the reason for the red color of the sun at sunrise and sunset.

more effectively than red light. When we look at the sky, what we see is light from the sun that has been scattered out of the direct beam and hence appears blue (Fig. 7-59). The sun itself is therefore a little more yellowish or reddish than it would appear if there were no atmosphere. At sunrise or sunset, when the sun's light must make a long passage through the atmosphere, much of its blue light is scattered out. The sun may be a brilliant red as a result. Above the atmosphere the sky is black, and the moon, stars, and planets are visible to astronauts in the daytime.

A Scattering Experiment

We can verify the origin of the blue color of the sky by adding a few drops of milk to a glass of water and then shining light from a flashlight through the mixture. If we look at the glass from the side, perpendicular to the light beam, the light scattered by the milk droplets will appear blue. If we look at the glass opposite the flashlight toward the beam, it will have an orange tint.

7.16 Interference

Waves in Step and out of Step

Interference refers to the adding together of two or more waves of the same kind that pass by the same point at the same time. The formation of standing waves is an example of interference.

How interference works is shown in Fig. 7-60. Let us shake the stretched strings *AC* and *BC* at the ends *A* and *B*. The single string *CD* is then affected by *both* sets of waves. Each portion of *CD* must respond to two impulses at the same time, and its motion is therefore the total of the effects of the two original waves. Suppose we shake *A* and *B* in step with each other so that, at *C*, crest meets crest and trough meets trough. Then the crests in *CD* are twice as high and the troughs twice as deep as those in *AB* and *BC*. This situation is called **constructive interference.**

On the other hand, if we shake *A* and *B* exactly out of step with each other, wave crests in *AC* will arrive at *C* just when troughs get there from *BC*. As a result, crest matches trough, the wave impulses cancel each

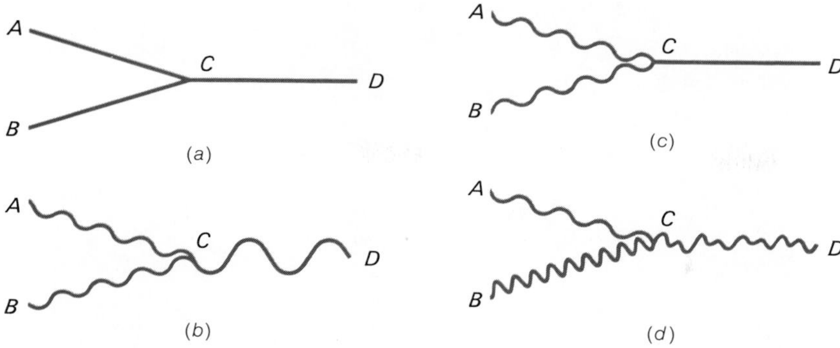

Figure 7-60 Interference. (*a*) Waves started along stretched strings *AC* and *BC* will interfere at *C*. (*b*) Constructive interference. (*c*) Destructive interference. (*d*) A mixture of constructive and destructive interference.

Figure 7-61 Destructive interference provides a way to reduce or eliminate noise. The procedure is to use an electronic device to analyze a particular noise and then produce mirror-image sound waves that cancel it out. If the device is appropriately programmed, wanted sounds such as conversation or music come through clearly. Powerful antinoise systems are already in use to counter the noise of suction grain unloaders, which are as loud as jet engines, and other systems are being developed to make the cabins of turboprop aircraft noise-free. On a smaller scale, the Noise Buster shown here feeds a pair of headphones with antinoise for personal use.

other out, if their amplitudes are the same, and *CD* remains at rest. This situation is called **destructive interference** (Fig. 7-61).

As another possibility, if *A* is shaken through a smaller range than and half as rapidly *B*, the two waves add together to give the complex waveform of Fig. 7-60*d*. The variations are endless, and the resulting waveforms depend upon the amplitudes, wavelengths, and timing of the incoming waves.

The interference of water waves is shown by ripples in Fig. 7-62. Ripples that spread out from the two vibrating rods affect the same water molecules. In some directions, crests from one source arrive at the same time as crests from the other source and the ripples are reinforced. Between these regions of vigorous motion are narrow lanes where the water is quiet. These lanes represent directions in which crests from one source arrive together with troughs from the other so that the wave motions cancel.

Why Thin Films Are Brightly Colored All of us have seen the brilliant colors that appear in soap bubbles and thin oil films. This effect can be traced to a combination of reflection and interference.

Let us consider what happens when light of only one color, and hence only one wavelength, strikes an oil film. As in Fig. 7-63, part of the light passes right through the film, but some is reflected from the upper

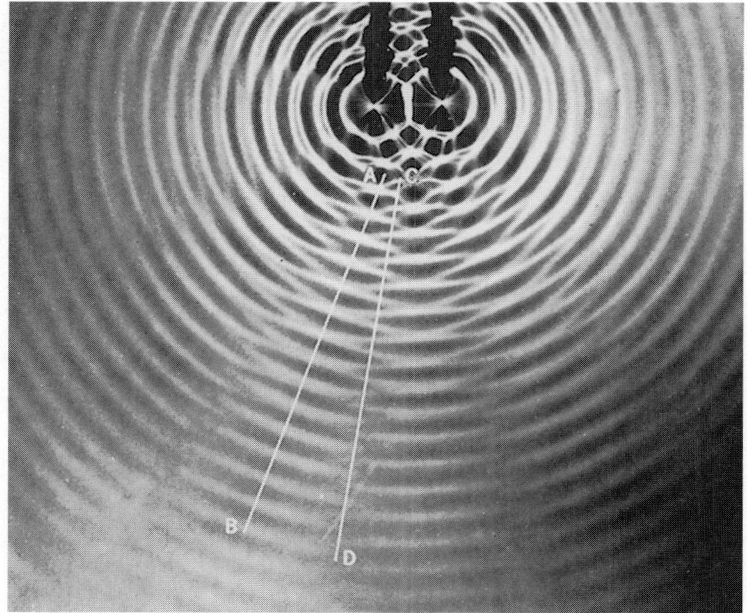

Figure 7-62 The interference of water waves. Ripples spread out across the surface of a shallow tank of water from the two sources at the top. In some directions (for instance *AB*) the ripples reinforce each other and the waves are more prominent. In other directions (for instance *CD*) the ripples are out of step and cancel each other, so that the waves are small or absent.

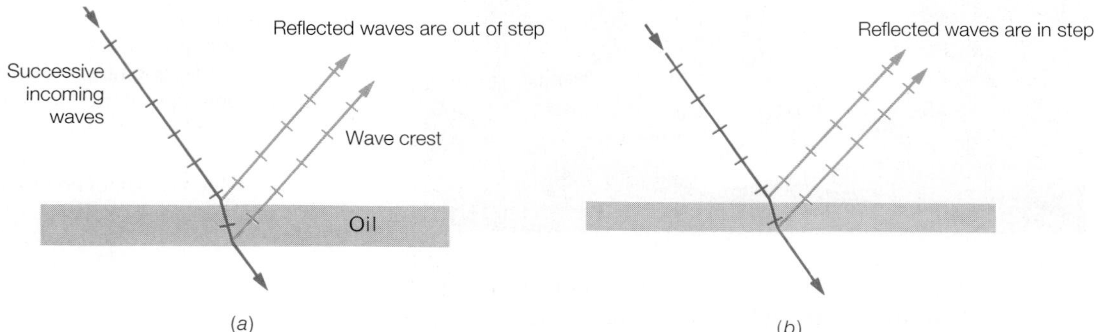

Figure 7-63 (*a*) Destructive and (*b*) constructive interference in a thin film for light of a particular wavelength. When the film has the thickness in (*a*), it appears dark; when it has the thickness in (*b*), it appears bright. Light of other wavelengths undergoes destructive and constructive interference at different film thicknesses.

Coated Lenses

A portion of the light that strikes an air-glass or glass-air interface is reflected, just as in the case of the oil film of Fig. 7-63. Because there are many such interfaces in most optical instruments (10 in the case of each half of a pair of binoculars), the total amount of light lost may be considerable, and the reflections also blur the image.

To reduce reflection at an interface, the glass can be coated with a thin layer of a transparent substance whose thickness is just right for the reflected rays from the top and bottom of the layer to interfere destructively, as in Fig. 7-63*a*. Of course, the cancellation is exact only for a particular wavelength. What is done is to choose a wavelength in the middle of the visible spectrum, which corresponds to green light, so that partial cancellation occurs over a wide range of colors. The red and violet ends of the spectrum are accordingly least affected, and the light reflected from a coated lens is a mixture of these colors, a purplish hue. Good-quality optical instruments always use coated lenses.

BIOGRAPHY Thomas Young (1773–1829)

While still a medical student in London, Young discovered that the eye focuses on objects at different distances by changing the shape of its lens, and that astigmatism (from which he suffered) is due to an irregularly shaped cornea. Although a practicing physician all his life, an inherited fortune allowed Young the time to make important discoveries in a number of areas of physics. He was the first to use the word energy in its modern sense and to relate the work done on a body to the change in its kinetic energy.

Young's most notable achievement was to demonstrate the wave nature of light, which until then had been uncertain: many contemporaries still held to Newton's view that light consisted of a stream of particles. What Young did was to show that light from two adjacent narrow slits interferes to produce alternate bright and dark fringes on a screen, the optical equivalent of the interference of water waves shown in Fig. 7-62. From this experiment Young could determine the wavelengths found in light. Physics and medicine were not all that interested Young: he was a pioneer in deciphering the hieroglyphics of ancient Egypt.

surface of the film and some is reflected from the lower surface. The two reflected waves interfere with each other. At some places in the film its thickness is just right for the reflected waves to be exactly out of step (crest-to-trough), as in Fig. 7-63*a*. This is the same effect as the destructive interference shown in Fig. 7-60*c*. Little or no reflection can take place in this part of the oil film, and nearly all the incoming light simply passes right through it. The oil film accordingly seems black in this region.

Where the film is slightly thicker or thinner than in Fig. 7-63*a*, the reflected waves may be exactly in step and therefore reinforce each other, as in *b*. This corresponds to the constructive interference shown in Fig. 7-60*b*. Here the film is a good reflector and appears bright. Shining light of one color on a thin oil film gives rise in this way to areas of light and dark whose pattern depends on the varying thickness of the film.

When white light is used, the reflected waves of only one color will be in step at a particular place while waves of other colors will not. The result is a series of brilliant colors. This is the reason for the rainbow effects we see in soap bubbles and in oil films (Fig. 7-64).

Figure 7-64 A colored pattern occurs when an uneven thin film of oil floats on water, because light of each wavelength undergoes constructive interference at a different oil thickness.

7.17 Diffraction

Why Shadows Are Never Completely Dark

An important property of all waves is their ability to bend around the edge of an obstacle in their path. This property is called **diffraction.**

A simple example of diffraction occurs when we hear the noise of a car horn around the corner of a building. The noise could not have reached us through the building, and refraction is not involved since the speed of sound does not change between the source of noise and our ears. What happens is that the sound waves spread out from the corner of the building into the "shadow" as though they come from the corner (Fig. 7-65). The diffracted waves are not as loud as those that proceed directly to a listener, but they go around the corner in a way that a stream of particles, for example, cannot.

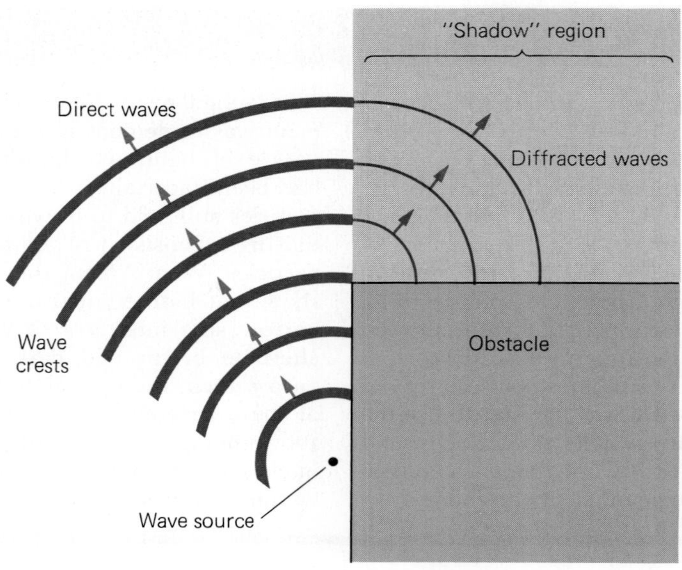

Figure 7-65 Diffraction causes waves to bend around the corner of an obstacle into the "shadow" region. The diffracted waves spread out as though they originated at the corner of the obstacle and are weaker than the direct waves. The waves shown here could be of any kind, for instance, water waves, sound waves, or light waves.

The Larger the Lens Diameter, the Sharper the Image Diffraction, too, occurs in light waves. As mentioned, diffraction refers to the "bending" of waves around the edges of an obstacle in their path. Because of diffraction, a shadow is never completely dark, although the wavelengths of light waves are so short that the effects of diffraction are largely limited to the border of the shadow region.

Diffraction limits the useful magnifications of microscopes and telescopes. The larger the diameter of a lens (or of a curved mirror that acts like a lens), the less significant is diffraction. For this reason a small telescope cannot be used at a high magnification, since the result would be a blurred image instead of a sharp one. The huge Hale telescope at Mt. Palomar in California has a mirror 5 m in diameter, but even so it can resolve objects only 50 m or more across on the moon's surface.

The **resolving power** of a telescope depends upon the wavelength of the light that enters it divided by the diameter of the lens or mirror; the smaller the resolving power, the sharper the image (Fig. 7-66). This relationship poses a severe problem for radio telescopes, which are antennas designed to receive radio waves from space (Chap. 19). Because such

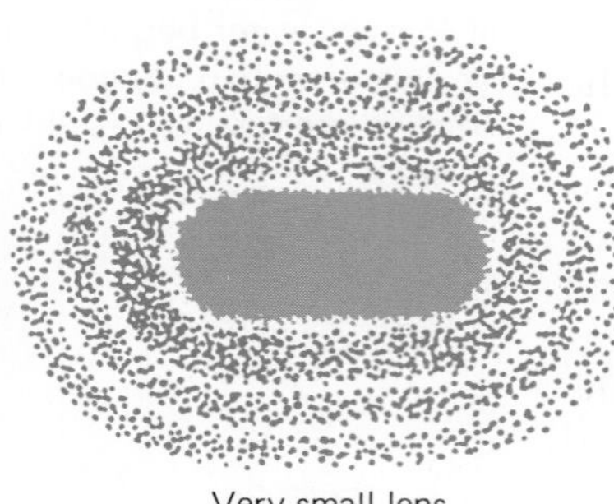

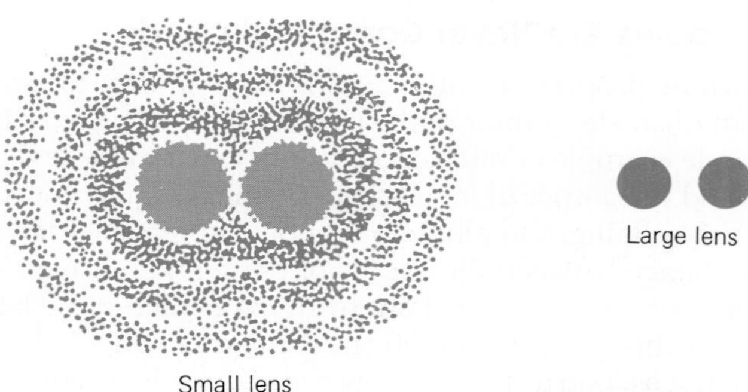

Figure 7-66 A large lens or mirror is better able to resolve nearby objects than a small one.

Resolving Power

No matter how perfect an optical system is, the image of a point source of light it produces is always a tiny disk of light with bright and dark interference fringes around it. If we are looking at two objects a distance d apart through an optical system whose diameter is D using light of wavelength λ and we are L away from the objects, we will see a single blob rather than separate objects if their separation is less than

$$d_0 = 1.22\,\frac{\lambda L}{D}$$

This minimum distance, its **resolving power,** is a measure of the ability of the optical system to keep distinct the images of two objects that are close together.

The pupils of a person's eyes under ordinary conditions of illumination are about 3 mm in diameter, and for most people the distance of most distinct vision is 25 cm. Let us use $\lambda = 550$ nm, which is in the middle of the visible spectrum of light, to find the resolving power d_0 of the eye. The above formula gives

$$d_0 = 1.22\,\frac{\lambda L}{D} = \frac{(1.22)(5.5 \times 10^{-7}\text{ m})(0.25\text{ m})}{3 \times 10^{-3}\text{ m}}$$
$$= 6 \times 10^{-5}\text{ m} = 0.06\text{ mm}$$

In fact, the photoreceptors in the retina are not quite close enough together to permit such sharp vision, and 0.1 mm is a more realistic resolving power.

The original generation of DVDs are "read" with lasers that emit red light. Since resolving power is proportional to wavelength, the invention of lasers that emit blue light, whose wavelength is much shorter (see Fig. 7-28), means that DVDs that squeeze more information on each disc can be manufactured and read. Using blue lasers together with improved software enables the second generation of DVDs to each store as much as 13 h of normal video or 2 h of high-definition video.

waves might have wavelengths as much as a million times greater than those in visible light, an antenna would have to be tens or hundreds of kilometers across to be able to separate sources as close together in the sky as optical telescopes can. In fact, modern electronics permits sufficient resolution to be achieved by combining the signals received by a series of widely spaced antennas. Radio astronomers have established such an antenna array extending from the Virgin Islands across the United States to Hawaii, a span of about 8000 km—nearly the diameter of the earth (Fig. 7-67). This array provides better resolution than even the largest optical telescope.

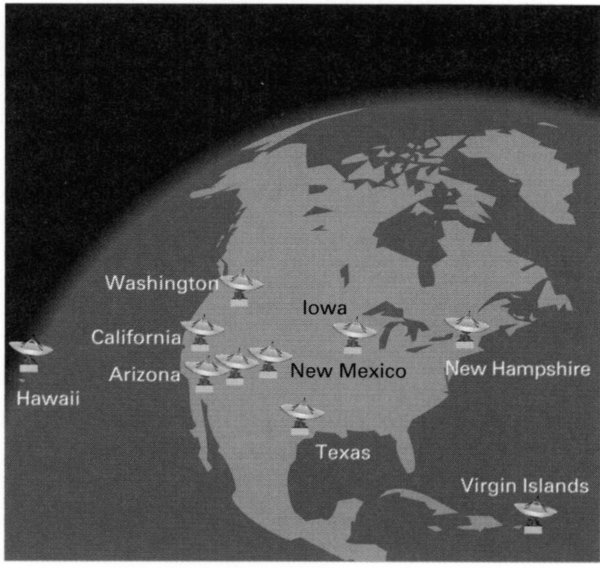

Figure 7-67 The ten radio telescopes whose locations are shown here operate together to give an angular resolving power of less than a millionth of a degree.

Important Terms and Ideas

Waves carry energy from one place to another by a series of periodic motions of the individual particles of the medium in which the waves occur. (Electromagnetic waves are an exception.) There is no net transfer of matter in wave motion.

In a **longitudinal wave** the particles of the medium vibrate back and forth in the direction in which the waves travel. In a **transverse wave** the particles vibrate from side to side perpendicular to the wave direction. Sound waves are longitudinal; waves in a stretched string are transverse; water waves are a combination of both since water molecules move in circular orbits when a wave passes.

The **frequency** of waves is the number of wave crests that pass a particular point per second. The **period** is the time needed for a complete wave to pass a given point. **Wavelength** is the distance between adjacent crests or troughs. The **amplitude** of a wave is the maximum displacement of a particle of the medium on either side of its normal position when the wave passes.

The **doppler effect** refers to the change in frequency of a wave when there is relative motion between its source and an observer.

Resonance occurs when an object (such as a musical instrument) vibrates at a frequency equal to one of its natural frequencies of vibration.

Electromagnetic waves consist of coupled electric and magnetic field oscillations. Radio waves, light waves, and x-rays are all electromagnetic waves that differ only in their frequency.

In **amplitude modulation** (AM), information is contained in variations in the amplitude of a constant-frequency wave. In **frequency modulation** (FM), information is contained in variations in the frequency of the wave.

The change in direction of waves when they enter a region in which their speed changes is called **refraction.** In **reflection,** waves strike an obstacle and rebound from it.

The **index of refraction** of a medium is the ratio between the speed of light in free space and its speed in the medium. In **internal reflection,** light that arrives at a medium of lower index of refraction (for instance, glass to air) at a large enough angle is reflected back.

A **lens** is a piece of glass or other transparent material shaped to produce an image by refracting light that comes from an object. A **converging lens** brings parallel light to a single point at a distance called the **focal length** of the lens. A **diverging lens** spreads out parallel light so that it seems to come from a point behind the lens.

A **real image** of an object is formed by light rays that pass through it; the image would therefore appear on a screen. A **virtual image** can only be seen by the eye because the light rays that seem to come from the image do not actually pass through it.

White light is a mixture of different frequencies, each of which produces the visual sensation of a particular color. Because the speed of light in a medium is slightly different for different frequencies, white light is **dispersed** into its separate colors when refracted in a glass prism or a water droplet.

Interference refers to the adding together of two or more waves of the same kind that pass by the same point at the same time. In **constructive interference** the new wave has a greater amplitude than any of the original ones; in **destructive interference** the new wave has a smaller amplitude.

The ability of waves to bend around the edge of an obstacle in their path is called **diffraction.** The **resolving power** of an optical system is a measure of its ability to produce sharp images, which is limited by diffraction.

Important Formulas

Wave speed: $v = f\lambda$

Wave Period: $T = \frac{1}{f}$

Index of refraction: $n = \frac{c}{v}$

Apparent depth: $h' = \frac{h}{n}$

Multiple Choice

1. The distance from crest to crest of any wave is called its

a. frequency
b. wavelength
c. speed
d. amplitude

2. Of the following properties of a wave, the one that is independent of the others is its

a. frequency
b. wavelength
c. speed
d. amplitude

3. When waves go from one place to another, they transport
   a. amplitude c. wavelength
   b. frequency d. energy
4. Water waves are
   a. longitudinal
   b. transverse
   c. a mixture of longitudinal and transverse
   d. sometimes longitudinal and sometimes transverse
5. Sound waves are
   a. longitudinal
   b. transverse
   c. a mixture of longitudinal and transverse
   d. sometimes longitudinal and sometimes transverse
6. Sound cannot travel through
   a. a solid c. a gas
   b. a liquid d. a vacuum
7. Sound travels fastest in
   a. air c. iron
   b. water d. a vacuum
8. The amplitude of a sound wave determines its
   a. loudness c. wavelength
   b. pitch d. overtones
9. The higher the frequency of a wave,
   a. the lower its speed
   b. the shorter its wavelength
   c. the smaller its amplitude
   d. the lower its pitch
10. Six flutes playing together produce a 60-dB sound. The number of flutes needed to produce a 70-dB sound is
    a. 7 c. 70
    b. 60 d. 120
11. The doppler effect occurs
    a. only in sound waves
    b. only in longitudinal waves
    c. only in transverse waves
    d. in all types of waves
12. A spacecraft is approaching the earth. Relative to the radio signals it sends out, the signals received on the earth have
    a. a lower frequency c. a higher speed
    b. a shorter wavelength d. all of the above
13. Maxwell based his theory of electromagnetic waves on the hypothesis that a changing electric field gives rise to
    a. an electric current
    b. a stream of electrons
    c. a magnetic field
    d. longitudinal waves
14. In a vacuum, the speed of an electromagnetic wave
    a. depends upon its frequency
    b. depends upon its wavelength
    c. depends upon the strength of its electric and magnetic fields
    d. is a universal constant
15. Electromagnetic waves transport
    a. frequency c. charge
    b. wavelength d. energy
16. Which of the following does not consist of electromagnetic waves?
    a. x-rays c. sound waves
    b. radar waves d. infrared waves
17. The energy of an electromagnetic wave resides in its
    a. frequency
    b. wavelength
    c. speed
    d. electric and magnetic fields
18. Light waves
    a. require air or another gas to travel through
    b. require some kind of matter to travel through
    c. require electric and magnetic fields to travel through
    d. can travel through a perfect vacuum
19. A beam of transverse waves whose variations occur in all directions perpendicular to their direction of motion is
    a. resolved c. polarized
    b. diffracted d. unpolarized
20. The ionosphere is a region of ionized gas in the upper atmosphere. The ionosphere is responsible for
    a. the blue color of the sky
    b. rainbows
    c. long-distance radio communication
    d. the ability of satellites to orbit the earth
21. A pencil in a glass of water appears bent. This is an example of
    a. reflection c. diffraction
    b. refraction d. interference
22. A fish that you are looking at from a boat seems to be 60 cm below the water surface. The actual depth of the fish
    a. is less than 60 cm
    b. is 60 cm
    c. is more than 60 cm
    d. may be any of these, depending on the angle of view

23. The index of refraction of a transparent substance is always
    a. less than 1
    b. 1
    c. greater than 1
    d. any of the above
24. A real image formed by any lens is always which one or more of the following?
    a. smaller than the object
    b. larger than the object
    c. erect
    d. inverted
25. The image of a real object farther from a converging lens than f is always which one or more of the following?
    a. smaller than the object
    b. the same size as the object
    c. virtual
    d. inverted
26. The image of a real object closer to a converging lens than f is always which one or more of the following?
    a. smaller than the object
    b. the same size as the object
    c. virtual
    d. inverted
27. The image formed by a diverging lens of a real object is never which one or more of the following?
    a. real
    b. virtual
    c. erect
    d. smaller than the object
28. A converging lens of focal length f is being used as a magnifying glass. The distance of an object from the lens must be
    a. less than f
    b. f
    c. between f and $2f$
    d. more than f
29. The quality in sound that corresponds to color in light is
    a. amplitude
    b. resonance
    c. waveform
    d. pitch
30. Light of which color has the lowest frequency?
    a. red
    b. blue
    c. yellow
    d. green
31. Light of which color has the shortest wavelength?
    a. red
    b. blue
    c. yellow
    d. green
32. The Spanish flag is yellow and red. When viewed with yellow light it appears
    a. all yellow
    b. yellow and red
    c. yellow and black
    d. white and red
33. The Danish flag is red and white. When viewed with red light it appears
    a. all red
    b. all white
    c. red and white
    d. red and black
34. Thin films of oil and soapy water owe their brilliant colors to a combination of reflection, refraction, and
    a. scattering
    b. interference
    c. diffraction
    d. doppler effect
35. The sky is blue because
    a. air molecules are blue
    b. the lens of the eye is blue
    c. the scattering of light is more efficient the shorter its wavelength
    d. the scattering of light is more efficient the longer its wavelength
36. Diffraction refers to
    a. the splitting of a beam of white light into its component colors
    b. the interference of light that produces bright colors in thin oil films
    c. the bending of waves around the edge of an obstacle in their path
    d. the increase in frequency due to motion of a wave source toward an observer
37. The useful magnification of a telescope is limited by
    a. the speed of light
    b. the doppler effect
    c. interference
    d. diffraction
38. Which one or more of the following will improve the resolving power of a lens?
    a. Increase the wavelength of the light used
    b. Increase the frequency of the light used
    c. Decrease the diameter of the lens used
    d. Increase the diameter of the lens used
39. The speed of sound waves having a frequency of 256 Hz compared with the speed of sound waves having a frequency of 512 Hz is
    a. half as great
    b. the same
    c. twice as great
    d. 4 times as great
40. The wavelength of sound waves having a frequency of 256 Hz compared with the wavelength of sound waves having a frequency of 512 Hz is
    a. half as great
    b. the same
    c. twice as great
    d. 4 times as great
41. Waves in a lake are observed to be 5 m in length and to pass an anchored boat 1.25 s apart. The speed of the waves is
    a. 0.25 m/s
    b. 4 m/s

c. 6.25 m/s
d. impossible to find from the information given

42. A boat at anchor is rocked by waves whose crests are 20 m apart and whose speed is 5 m/s. These waves reach the boat with a frequency of
a. 0.25 Hz c. 20 Hz
b. 4 Hz d. 100 Hz

43. One kHz (kilohertz) is equal to 10^3 Hz. What is the wavelength of the electromagnetic waves sent out by a radio station whose frequency is 660 kHz? The speed of light is 3×10^8 m/s.
a. 2.2×10^{-3} m c. 4.55×10^3 m
b. 4.55×10^2 m d. 1.98×10^{14} m

44. The speed of light in diamond is 1.24×10^8 m/s. The index of refraction of diamond is
a. 1.24 c. 2.42
b. 2.31 d. 3.72

45. A medal in a plastic cube appears to be 12 mm below the top of the cube. If $n = 1.5$ for the plastic, the depth of the medal is actually
a. 6 mm c. 18 mm
b. 8 mm d. 24 mm

Exercises

7.1 Water Waves

7.2 Tranverse and Longitudinal Waves

7.3 Describing Waves

1. (a) Distinguish between longitudinal and transverse waves. (b) Do all waves fall into one or the other of these categories? If not, give an example of one that does not.
2. Does increasing the frequency of a wave also increase its wavelength? If not, how are these quantities related?
3. Water waves whose crests are 6 m apart reach the shore every 1.2 s. Find the frequency and speed of the waves.
4. A certain groove in a phonograph record moves past the needle at 30 cm/s. If the wiggles in the groove are 0.1 mm apart, what is the frequency of the sound that is produced?
5. At one end of a ripple tank 90 cm across, a 6-Hz vibrator produces waves whose wavelength is 50 mm. Find the time the waves need to cross the tank.
6. A 1.2-MHz ultrasonic beam is used to scan body tissue. If the speed of sound in a certain tissue is 1540 m/s and the limit of resolution is one wavelength, what size is the smallest detail that can be resolved?

7.5 Sound

7. Why does sound travel fastest in solids and slowest in gases?
8. The speed of sound in a gas depends upon the average speed of the gas molecules. Why is such a relationship reasonable?
9. Even if astronauts on the moon's surface did not need to be enclosed in space suits, they could not speak directly to each other but would have to communicate by radio. Can you think of the reason?
10. What eventually becomes of the energy of sound waves?
11. A person is watching as spikes are being driven to hold a steel rail in place. The sound of each sledgehammer blow arrives 0.14 s through the rail and 2 s through the air after the person sees the hammer strike the spike. Find the speed of sound in the rail.
12. An airplane is flying at 600 km/h at an altitude of 2.0 km. When the sound of the airplane's engines seems to somebody on the ground to be coming from directly overhead, how far away is she from being directly below the airplane?
13. Find the frequency of sound waves in air whose wavelength is 25 cm.
14. A person determines the direction from which a sound comes by means of two mechanisms. One compares the loudness of the sound in one ear with that in the other ear, which is most effective at low frequencies. The other compares the phases of the waves that arrive at the two ears, which is most effective at high frequencies. (The phase of a wave is the part of its cycle it is in at a particular time and place.) The crossover point of equal effectiveness occurs at about 1200 Hz, and as a result, sound with frequencies in the vicinity of 1200 Hz are difficult to locate. How does the wavelength of a 1200-Hz sound compare with the distance between your ears?
15. How many times stronger than the 60-dB sound of a person talking loudly is the 100-dB sound of a power lawn mower?
16. A violin string vibrates 1044 times per second. How many vibrations does it make while its sound travels 10 m?

7.6 Doppler Effect

17. In what kinds of waves can the doppler effect occur?

18. A "double star" consists of two nearby stars that revolve around their center of mass (see Sec. 1.10). How can an astronomer recognize a double star from the characteristic frequencies of the light that reaches him from its member stars?

19. The characteristic wavelengths of light emitted by a distant star are observed to be shifted toward the red end of the spectrum. What does this suggest about the motion of the star relative to the earth?

7.8 Electromagnetic Waves

7.9 Types of EM Waves

20. Why are light waves able to travel through a vacuum whereas sound waves cannot?

21. How could you show that light carries energy?

22. Why was electromagnetic induction discovered much earlier than its converse, the production of a magnetic field by a changing electric field?

23. Light is said to be a transverse wave. What is it that varies at right angles to the direction in which a light wave travels?

24. How are the directions of the electric and magnetic fields of an em wave related to each other and to the direction in which the wave is moving?

25. Light waves carry both energy and momentum. Why doesn't the momentum of the sun diminish with time as its energy content does?

26. Give as many similarities and differences as you can between sound and light waves.

27. Visible light of which color has the lowest frequency? The highest frequency? The shortest wavelength? The longest wavelength?

28. A radar signal takes 2.7 s to go to the moon and return. How far away was the moon at that time?

29. An opera performance is being broadcast by radio. Who will hear a certain sound first, a member of the audience 30 m from the stage or a listener to a radio receiver 5000 km away?

30. Radio waves of very long wavelength can penetrate farther into seawater than those of shorter wavelength. The U.S. Navy communicates with submerged submarines using 76-Hz radio waves. What is their wavelength in air?

31. A nanosecond is 10^{-9} s. (a) What is the frequency of an em wave whose period is 1 ns? (b) What is its wavelength? (c) To what class of em waves does it belong?

32. A radar sends out $0.05 - \mu$s pulses of microwaves whose wavelength is 25 mm. What is the frequency of these microwaves? How many waves does each pulse contain?

7.10 Light "Rays"

7.11 Reflection

7.12 Refraction

33. When a light ray is reflected, which of the following quantities, if any, is unchanged? The direction of the ray; its speed; its frequency; its wavelength.

34. When a light ray goes from one medium to another, which of the following quantities, if any, is unchanged? The direction of the ray; its speed; its frequency; its wavelength.

35. What is the height of the smallest mirror in which you could see yourself at full length? Use a diagram to explain your answer. Does it matter how far away you are?

36. What types of waves can be refracted? Under what circumstances does refraction occur?

37. Can the index of refraction of a substance be less than 1? If not, why not?

38. (a) You are standing on a pier and want to spear a fish that swims by. Should you aim above, below, or exactly where the fish seems to be? (b) What if you are swimming underwater and want to spear a fish?

39. When a fish looks up through the water surface at an object in the air, will the object appear to be its normal size and distance above the water? Use a diagram to explain your answer, and assume that the fish's eye and brain, like the human eye and brain, are accustomed to interpreting light rays as straight lines.

40. A flashlight at the bottom of a swimming pool shines upward at an angle with the surface, as in Fig. 7-68. Which path does the light follow?

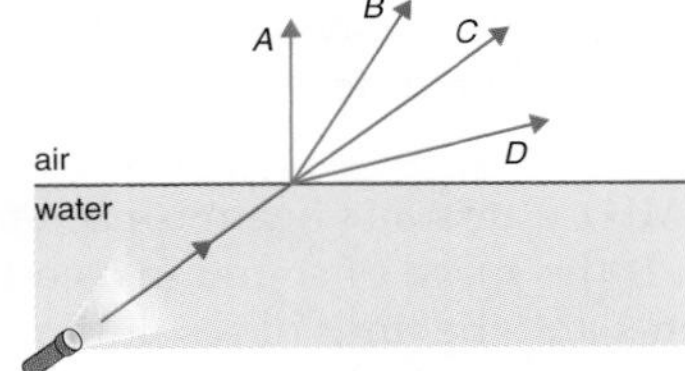

Figure 7-68

41. Which of the paths shown in Fig. 7-69 could represent a ray of light passing through a glass block in air?

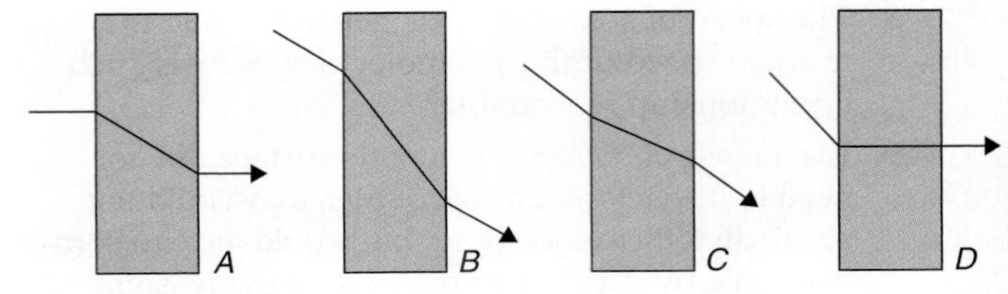

Figure 7-69

42. The olive in a cocktail ($n = 1.35$) seems to be 37 mm below the surface. What is the actual depth of the olive?

43. A leaf frozen into a pond in winter is 40 mm below the surface of the ice. How far down does the leaf seem to be? ($n_{ice} = 1.31$)

7.13 Lenses

44. Does a spherical bubble in a pane of glass act to converge or diverge light passing through it?

45. What is the difference between a real image and a virtual image?

46. A coin is placed at a focal point of a converging lens. Is an image formed? If yes, is it real or virtual, erect or inverted, larger or smaller than the object?

47. Under what circumstances, if any, will a light ray that passes through a converging lens not be deviated? Under what circumstances, if any, will a light ray that passes through a diverging lens not be deviated?

48. Under what circumstances, if any, will a converging lens form an inverted image of a real object? Under what circumstances, if any, will a diverging lens form an erect image of a real object?

49. Is the mercury column in a thermometer wider or narrower than it seems to be?

50. Is there any way in which a converging lens, by itself, can form a virtual image of a real object? Is there any way in which a diverging lens, by itself, can form a real image of a real object?

51. When it leaves the lens, which path will the incoming ray in Fig. 7-70 follow?

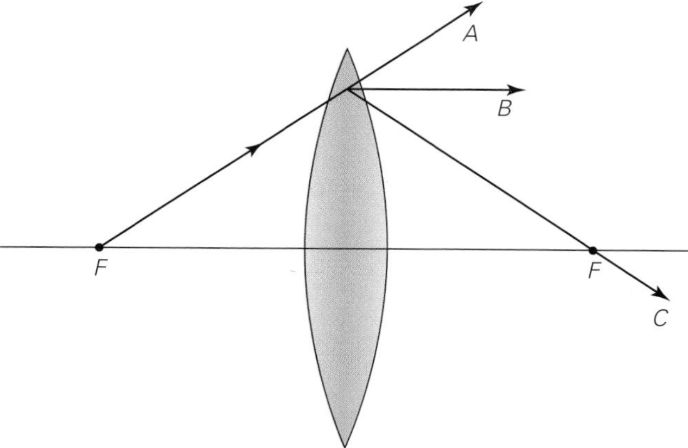

Figure 7-70

52. If the screen is moved closer to a slide projector, how should the projector's lens be moved to restore the image to a sharp focus?

53. A birthday candle 4 cm high is 10 cm from a converging lens whose focal length is 15 cm. Use a ray diagram on a suitable scale (say, $\frac{1}{3}$ full size) to find the location of the image, its height, and whether it is erect or inverted, real or virtual.

54. The candle of Exercise 53 is 15 cm from the lens. Answer the same questions for this situation.

55. The candle of Exercise 53 is 25 cm from the lens. Answer the same questions for this situation.

56. The candle of Exercise 53 is 30 cm from the lens. Answer the same questions for this situation.

57. The candle of Exercise 53 is 50 cm from the lens. Answer the same questions for this situation.

7.14 The Eye

58. Describe the image the lens of the eye forms on the retina.

59. (a) What is the name of the defect of vision in which an eye can see nearby objects clearly but not distant ones? (b) What is the name of the defect of vision in which either bar of a cross can be seen clearly but not both at the same time?

7.15 Color

60. When a beam of white light passes perpendicularly through a flat pane of glass, it is not dispersed into a spectrum. Why not?

61. When white light is dispersed by a glass prism, red light is bent least and violet light is bent most. What does this tell you about the relative speeds of red and violet light in glass?

62. What color would red cloth appear if it were illuminated by (a) white light? (b) red light? (c) green light?

63. (a) What would the American flag look like when viewed with red light? (b) With blue light?

64. If the earth had no atmosphere, what would the color of the sky be during the day?

65. Light of what color is scattered most in the atmosphere? Least?

7.16 Interference

66. How can constructive and destructive interference be reconciled with the principle of conservation of energy?

7.17 Diffraction

67. Which of the following can occur in (a) transverse waves and (b) longitudinal waves? Reflection, interference, diffraction, polarization.

68. Give two advantages that a telescope lens or mirror of large diameter has over one of small diameter.

69. Radio waves are able to diffract readily around buildings, as anybody with a portable radio receiver can verify. However, light waves, which are also electromagnetic waves, undergo no discernible diffraction around buildings. Why not?

70. A radar operating at a wavelength of 3 cm is to have a resolving power of 30 m at a range of 1 km. Find the minimum width its antenna must have.

71. Suppose you are an astronaut orbiting the earth at an altitude of 200 km. The pupils of your eyes are 3 mm in diameter and the average wavelength of the light reaching you from the earth is 500 nm. What is the length of the smallest structure you could make out on the earth with your naked eye, assuming that turbulence in the earth's atmosphere does not smear out the image?

Answers to Multiple Choice

1. b
2. d
3. d
4. c
5. a
6. d
7. c
8. a
9. b
10. b
11. d
12. b
13. c
14. d
15. d
16. c
17. d
18. d
19. d
20. c
21. b
22. c
23. c
24. d
25. d
26. c
27. a
28. a
29. d
30. a
31. b
32. c
33. a
34. b
35. c
36. c
37. d
38. b, d
39. b
40. c
41. b
42. a
43. b
44. c
45. c

8

The Nucleus

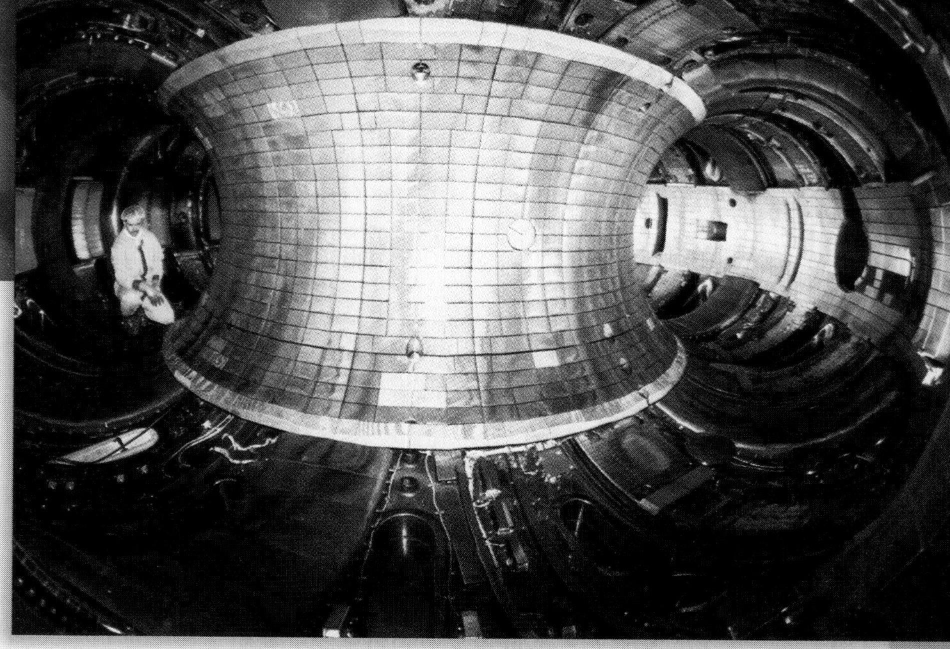

Interior of the Tokamak nuclear fusion test reactor at Princeton.

Goals

When you have finished this chapter you should be able to complete the goals ▶ given for each section below:

Atom and Nucleus

8.1 Rutherford Model of the Atom
An Atom Is Mostly Empty Space
▶ Discuss how the Rutherford experiment led to the modern picture of atomic structure.

8.2 Nuclear Structure
Protons and Neutrons
▶ Distinguish between nucleon and nuclide and between atomic number and mass number.
▶ State in what ways the isotopes of an element are similar and in what ways they are different.

Radioactivity

8.3 Radioactive Decay
How Unstable Nuclei Change into Stable Ones
▶ Describe the various kinds of radioactive decay and explain why each occurs.

8.4 Half-Life
Less and Less, but Always Some Left
▶ Define half-life.

8.5 Radiation Hazards
Invisible but Dangerous
▶ Discuss the sources and hazards of the ionizing radiation we are exposed to in daily life.

Nuclear Energy

8.6 Units of Mass and Energy
The Atomic Mass Unit and the Electronvolt
▶ Define atomic mass unit and electronvolt and use them in calculations.

8.7 Binding Energy
The Missing Energy That Keeps a Nucleus Together
▶ Explain the significance of the binding energy of a nucleus.

8.8 Binding Energy per Nucleon
Why Fission and Fusion Liberate Energy
▶ Sketch a graph of binding energy per nucleon versus mass number and indicate on it the location of the most stable nucleus and the range of mass numbers in which fusion and fission can occur.

Fission and Fusion

8.9 Nuclear Fission
Divide and Conquer
▶ Discuss nuclear fission and the conditions needed for a chain reaction to occur.

8.10 How a Reactor Works
From Uranium to Heat to Electricity
▶ Describe how a nuclear reactor works.

8.11 Plutonium
Another Fissionable Material
▶ Discuss what plutonium is, how it is made, and why it is important.

8.12 Nuclear Fusion
The Energy Source of the Future?
▶ Describe nuclear fusion and identify the conditions needed for a successful fusion reactor.

Elementary Particles

8.13 Antiparticles
The Same but Different
▶ Compare a particle with its antiparticle.
▶ Describe the processes of annihilation and pair production.

8.14 Fundamental Interactions
Only Four Give Rise to All Physical Processes
▶ List the four fundamental interactions and identify the aspects of the universe that each governs.

8.15 Leptons and Hadrons
Ultimate Matter
▶ Distinguish between leptons and hadrons and discuss the quark model of hadrons.

Atoms are the smallest particles of ordinary matter. Every atom has a central core, or **nucleus,** of protons and neutrons that provide nearly all the atom's mass. Outside the nucleus are the much lighter electrons, the same in number as the number of protons in the nucleus so that the atom as a whole is electrically neutral.

The chief properties (except mass) of atoms, molecules, solids, and liquids can be traced to the behavior of atomic electrons. But the atomic nucleus is also significant in the grand scheme of things. The continuing evolution of the universe is powered by energy that comes from nuclear reactions and transformations. Like other stars, the sun obtains its energy in this way. In turn, the coal, oil, and natural gas of the earth, as well as its winds and falling water, owe their energy contents to the sun's rays. Nuclear processes are responsible for the heat of the earth's interior and for the energy produced by nuclear reactors. Thus *all* the energy at our command has a nuclear origin, except for the energy of the tides, which are the result of the gravitational pull of the moon and sun on the waters of the world.

Atom and Nucleus

Until 1911 little was known about atoms except that they exist and contain electrons. Since electrons carry negative charges but atoms are neutral, scientists agreed that positively charged matter of some kind must be present in atoms. But what kind? And arranged in what way?

One suggestion, made by the British physicist J. J. Thomson in 1898, was that atoms are simply positively charged lumps of matter with electrons embedded in them, like raisins in a fruitcake (Fig. 8-1). Because Thomson had played an important part in discovering the electron, his idea was taken seriously. But atomic structure turned out to be very different.

8.1 Rutherford Model of the Atom

An Atom Is Mostly Empty Space

The most direct way to find out what is inside a fruitcake is to poke a finger into it. A similar method was used in 1911 in an experiment suggested by the British physicist Ernest Rutherford to find out what is inside an atom. Alpha particles were used as probes. (As discussed later in this chapter, alpha particles are emitted by certain substances. For now, all we need to know about these particles is that they are almost 8000 times heavier than electrons and each one has a charge of $+2e$.) A sample of an alpha-emitting substance was placed behind a lead screen with a small hole in it, as in Fig. 8-2, so that a narrow beam of alpha particles was produced. This beam was aimed at a thin gold foil. A zinc sulfide screen, which gives off a visible flash of light when struck by an alpha particle, was set on the other side of the foil.

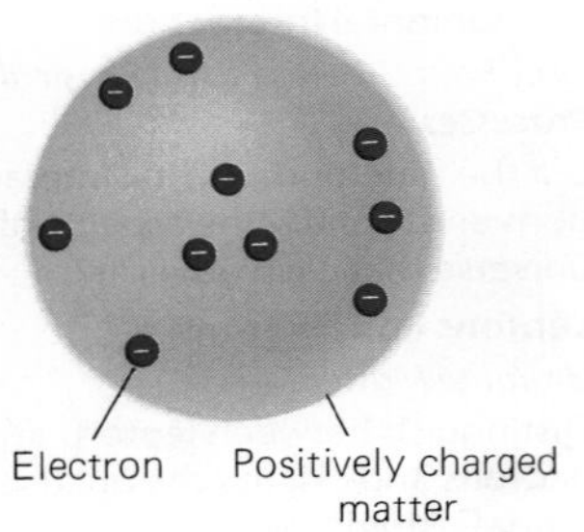

Figure 8-1 The Thomson model of the atom. Experiment shows it to be incorrect.

Rutherford expected the alpha particles to go right through the foil with hardly any deflection. This follows from the Thomson model, in which the electric charge inside an atom is assumed to be uniformly

BIOGRAPHY Ernest Rutherford (1871–1937)

He was digging potatoes on his family's farm in New Zealand when Rutherford learned that he had won a scholarship for graduate study at Cambridge University in England. "This is the last potato I will ever dig," he said, throwing down his spade. Thirteen years later he received the Nobel Prize for his work on radioactivity, which included discovering that alpha particles are the nuclei of helium atoms and that the radioactive decay of an element can give rise to a different element.

In 1911 Rutherford showed that the nuclear model of the atom was the only one that could explain the observed scattering of alpha particles by thin metal foils. Using alpha particles to bombard nitrogen nuclei, Rutherford was the first to artificially transmute one element into another. Notable achievements made in his laboratory by his associates include the discovery of the neutron in 1932 and the construction of the first high-energy particle accelerator for nuclear research. But Rutherford was not infallible: only a few years before the first nuclear reactor was built, he dismissed the idea of practical uses for nuclear energy as "moonshine." Rutherford is buried near Newton in Westminster Abbey.

spread through its volume. With only weak electric forces exerted on them, alpha particles that pass through a thin foil ought to be deflected only slightly, 1° or less.

What was found instead was that, although most of the alpha particles indeed were not deviated by much, a few were scattered through very large angles. Some were even scattered in the backward direction. As Rutherford remarked, "It was as incredible as if you fired a 15-inch shell at a piece of tissue paper and it came back and hit you."

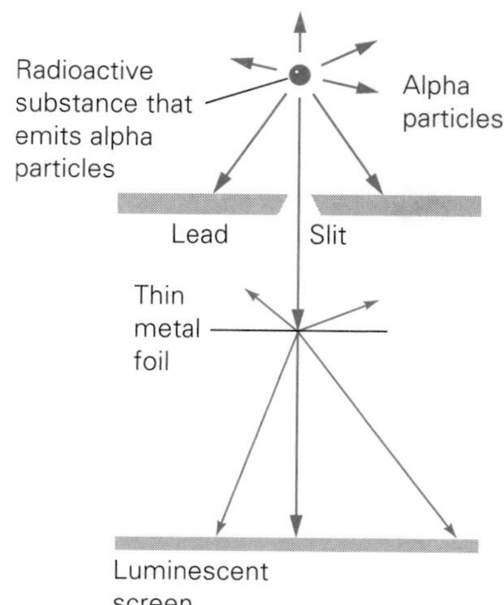

Figure 8-2 Principle of the Rutherford experiment. Nearly all the alpha particles pass through the foil with little or no deflection, but a few of the particles are scattered through large angles, even in the backward direction. This result means that strong electric fields must act on the particles, and such fields can arise only if atoms have very small nuclei in which their positive charge is concentrated.

Why the Nucleus Must Be Small Since alpha particles are relatively heavy and since those used in this experiment had high speeds, it was clear that strong forces had to be exerted upon them to cause such marked deflections. The only way to explain the results, Rutherford found, was to picture an atom as having a tiny nucleus in which the positive charge and nearly all the mass of the atom are concentrated. The electrons are some distance away, as in Fig. 8-3.

With an atom that is largely empty space, it is easy to see why most alpha particles go right through a thin foil. However, when an alpha particle happens to come near a nucleus, the strong electric field there causes the particle to be deflected through a large angle. The atomic electrons, being so light, have little effect on the alpha particles.

Suppose, as an analogy, that a star approaches the solar system from space at great speed. The chances are good that the star will not be deflected. Even a collision with a planet would not change the star's path to any great extent. Only if the star came near the great mass of the sun would the star's direction change by much. Similarly, said Rutherford, an alpha particle plows straight through an atom, unaffected by striking an electron now and then. Only a close approach to the heavy central nucleus of an atom can turn the alpha particle aside.

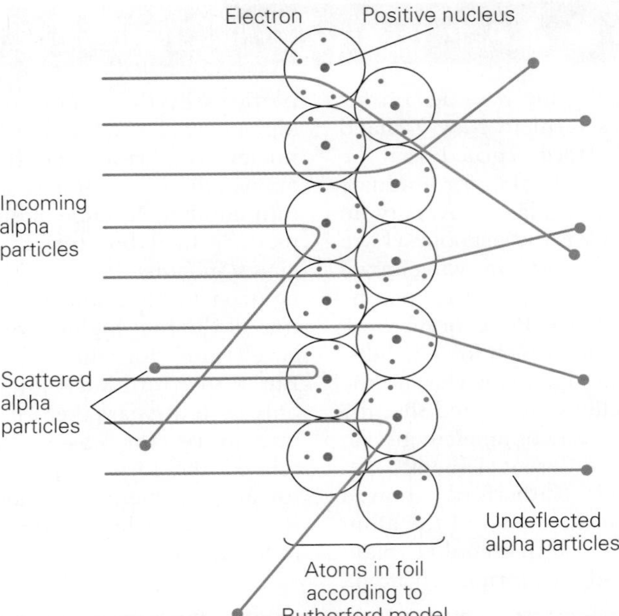

Figure 8-3 In the Rutherford model of the atom, the positive charge is concentrated in a central nucleus with the electrons some distance away. This model correctly predicts that some alpha particles striking a thin metal foil will be scattered through large angles by the strong electric fields of the nuclei.

Ordinary matter, then, is mostly empty space. The solid wood of a table, the steel that supports a bridge, the hard rock underfoot, all are just collections of electric charges, comparatively farther away from one another than the planets are from the sun. If all the electrons and nuclei in our bodies could somehow be packed closely together, we would be no larger than specks just visible with a microscope.

8.2 Nuclear Structure

Protons and Neutrons

What gives a nucleus its mass and charge? The simplest nucleus, that of the hydrogen atom, usually consists of a single **proton.** As mentioned in Chap. 6, the proton is a particle whose charge is $+e$ and whose mass is 1836 times the electron mass. Nuclei more complex than that of hydrogen contain **neutrons** as well as protons. The neutron has no charge, and its mass, 1839 times that of the electron, is slightly more than that of the proton. The compositions of several atoms are illustrated in Fig. 8-4.

Elements are the simplest substances in the bulk matter around us. Over 100 elements are known, of which 11 are gases, 2 are liquids, and the rest are solids at room temperature and atmospheric pressure. Hydrogen, helium, oxygen, chlorine, and neon are gaseous elements, and bromine and mercury are the two liquids. Most of the solid elements are metals. Elements are discussed in more detail in Chap. 10.

In a neutral atom of any element, the number of protons equals the number of electrons. This number is called the **atomic number** of the element. Thus the atomic number of hydrogen is 1, of helium 2, of lithium 3, and of beryllium 4. The atomic number of an element is its most

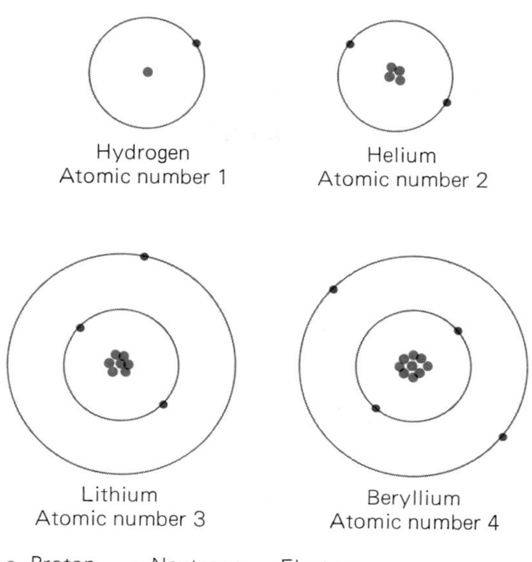

Figure 8-4 The elements that correspond to the atomic numbers 1, 2, 3, and 4 are hydrogen, helium, lithium, and beryllium. The various particles are actually far too small to be seen even on this scale.

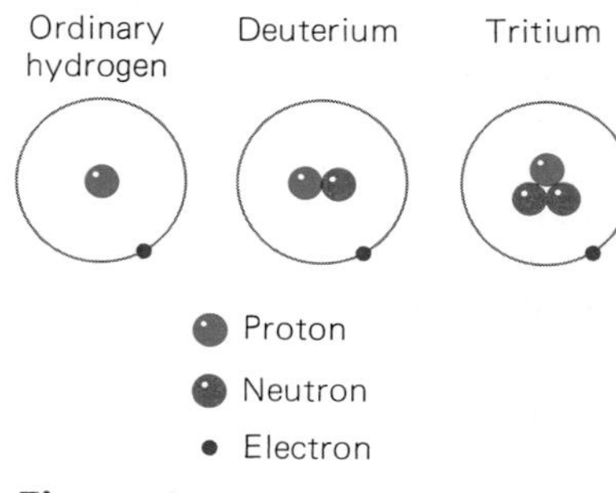

Figure 8-5 The isotopes of hydrogen.

basic property since this determines how many electrons its atoms have and how they are arranged, which in turn govern the physical and chemical behavior of the element. Atomic numbers and symbols of all the elements are given in Table 10-6, which is reproduced inside the back cover of the book.

Isotopes All atoms of a given element have nuclei with the same number of protons but not necessarily the same number of neutrons. For instance, even though more than 99.9 percent of hydrogen nuclei are just single protons, a few also contain a neutron as well, and a very few contain two neutrons along with the single proton (Fig. 8-5). The different kinds of hydrogen atom are called **isotopes.** All elements have isotopes.

A nucleus with a particular composition is called a **nuclide.** Symbols for nuclides follow the pattern

$$^{A}_{Z}X \qquad \textit{Nuclide symbol}$$

where X = chemical symbol of element

Z = atomic number of element

= number of protons in nucleus

A = **mass number** of nucleus

= number of protons and neutrons in nucleus

Thus the nucleus of the chlorine isotope that contains 17 protons and 18 neutrons has the atomic number $Z = 17$ and the mass number $A = 17 + 18 = 35$. Its symbol is accordingly

$$^{35}_{17}\text{Cl}$$

Deuterium and Tritium

About one in every 7000 hydrogen atoms is a deuterium ($^{2}_{1}$H) atom (Fig. 8-5). The proportion of tritium ($^{3}_{1}$H) atoms is even smaller—only about 2 kg of tritium of natural origin is present on the earth, nearly all of it in the oceans. Tritium is radioactive and decays to a helium isotope. Nuclear reactions in the atmosphere caused by cosmic rays from space (see Sec. 19.5) continually replenish the earth's tritium. **Heavy water** is water in which deuterium atoms instead of ordinary hydrogen ($^{1}_{1}$H) atoms are combined with oxygen atoms to form H_2O. Because $^{2}_{1}$H atoms have about twice the mass of $^{1}_{1}$H atoms, heavy water is 10 percent denser than ordinary water. Deuterium does not combine with neutrons as readily as ordinary hydrogen and for this reason heavy water is used in certain types of nuclear reactors.

The symbol of this nuclide is sometimes shortened to ^{35}Cl or Cl-35. The symbols of the nuclides of Fig. 8-4 are ^1_1H, ^4_2He, ^7_3Li, and ^9_4Be.

The term **nucleon** refers to both protons and neutrons, so that the mass number A is the number of nucleons in a particular nucleus.

Radioactivity

In 1896 Henri Becquerel accidentally discovered in his Paris laboratory that the element uranium can expose covered photographic film, can ionize gases, and can cause certain materials (such as the zinc sulfide used in the Rutherford experiment) to glow in the dark. Becquerel concluded that uranium gives off some kind of invisible but penetrating radiation, a property soon called **radioactivity.**

Not long afterward, Pierre and Marie Curie, in the course of extracting uranium from the ore pitchblende at the same laboratory, found two other elements that are also radioactive. They named one polonium, after Marie Curie's native Poland. The other, which turned out to be thousands of times more radioactive than uranium, was called radium.

Chemical reactions do not change the ability of a radioactive material to emit radiation, nor does heating it in an electric arc or cooling

BIOGRAPHY Marie Sklodowska Curie (1867–1934)

After high school in her native Poland, Marie Sklodowska worked as a governess until she was 24 so that she could study science in Paris, where she had barely enough money to survive. In 1894 Marie married Pierre Curie, who was 8 years older than she and already a noted physicist. In 1897, just after the birth of her daughter Irene (who, like her mother, was to win a Nobel Prize in physics), Marie began to investigate the newly discovered phenomenon of radioactivity—her word—for her doctoral thesis.

After a search of all the known elements, Marie learned that thorium as well as uranium was radioactive. She then examined various minerals for radioactivity and found that the uranium ore pitchblende was far more radioactive than its uranium content would suggest. Marie and Pierre together went on to identify first polonium, named for her native Poland, and then radium as the sources of the additional activity. With the primitive facilities that were all they could afford (they had to use their own money), they had succeeded by 1902 in purifying a tenth of a gram of radium from several tons of ore, a task that involved immense physical as well as intellectual labor.

Together with Becquerel, the Curies shared the 1903 Nobel Prize in physics. Pierre ended his acceptance speech with these words: "One may also imagine that in criminal hands radium might become very dangerous,

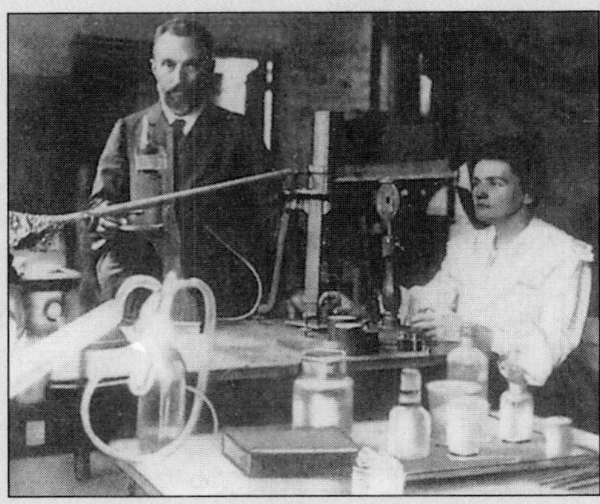

and here one may ask if humanity has anything to gain by learning the secrets of nature, if it is ready to profit from them, or if this knowledge is not harmful. . . . I am among those who think . . . that humanity will obtain more good than evil from the new discoveries."

In 1906 Pierre was struck and killed by a horse-drawn carriage in a Paris street. Marie continued work on radioactivity and became world-famous. Even before Pierre's death both Curies had suffered from ill health because of their exposure to radiation, and much of Marie's later life was marred by radiation-induced ailments, including the leukemia from which she died.

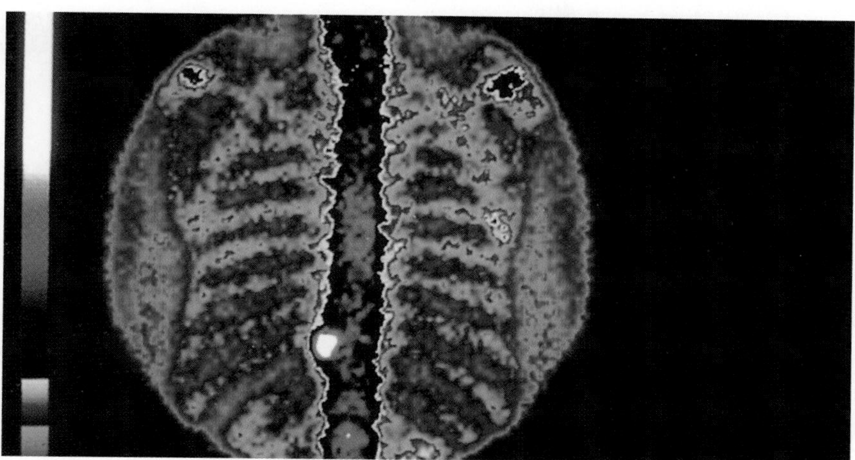

Figure 8-6 Substances that contain a radionuclide can be traced in living tissue by the radiation they emit. In this image, different levels of radiation intensity are shown in different colors. The gamma-emitting radionuclide used here is absorbed more readily by cancerous bone than by normal bone. The white area in the spine corresponds to a high rate of gamma emission and indicates a tumor there.

it in liquid air. Radioactivity must therefore be associated with atomic nuclei because these are the only parts of atoms not affected by such treatment.

The radioactivity of an element is due to the radioactivity of one or more of its isotopes. Most elements in nature have no radioactive isotopes, though such isotopes can be prepared artificially and are useful in biological and medical research as "tracers." The procedure is to incorporate a radionuclide in a chemical compound and follow what happens to the compound in a living organism by monitoring the radiation from the isotope (Fig. 8-6). Other elements, such as potassium, have some stable isotopes and some radioactive ones. A few, such as uranium, have only radioactive isotopes.

Of the 7000 or so nuclides that might possibly exist, about 2000 have either been found in nature or created in the laboratory, and of those only 256 are stable and do not undergo radioactive decay.

8.3 Radioactive Decay

How Unstable Nuclei Change into Stable Ones

Early experimenters found that a magnetic field splits the radiation from a radioactive material such as radium into three parts (Fig. 8-7). One part is deflected as though it consists of positively charged particles. Called **alpha particles,** these turned out to be the nuclei of helium atoms. Such nuclei contain two protons and two neutrons, so their symbol is ^4_2He. (These were the probes used in Rutherford's discovery of the nucleus.)

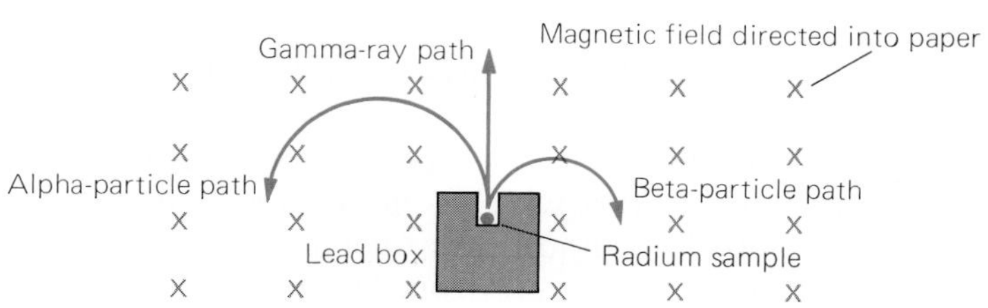

Figure 8-7 The radiations from a radium sample may be analyzed with the help of a magnetic field. Alpha particles are deflected to the left, hence they are positively charged; beta particles are deflected to the right, hence they are negatively charged; and gamma rays are not affected, hence they are uncharged.

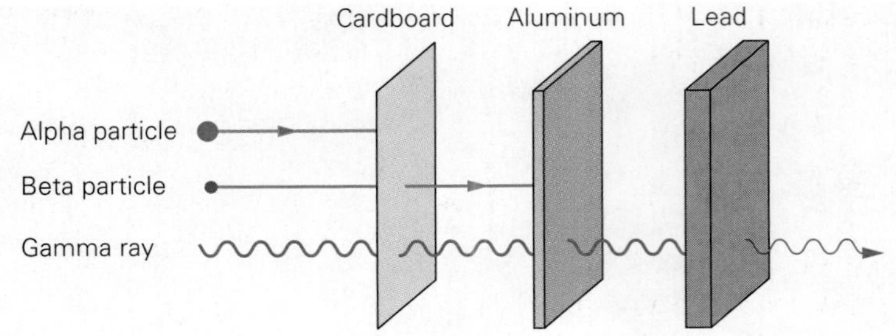

Figure 8-8 Alpha particles from radioactive materials are stopped by a piece of cardboard. Beta particles penetrate the cardboard but are stopped by a sheet of aluminum. Even a thick slab of lead may not stop all the gamma rays.

Another part of the radiation is deflected as though it consists of negatively charged particles. Called **beta particles,** these are electrons.

The rest of the radiation, which is not affected by a magnetic field, consists of **gamma rays.** Today these are known to be electromagnetic waves whose frequencies are higher than those of x-rays. A gamma ray is emitted by a nucleus that, for one reason or another, has more than its normal amount of energy. The composition of the nucleus does not change in gamma decay, unlike the cases of alpha and beta decay. Gamma rays are the most penetrating of the three kinds of radiation, alpha particles the least (Fig. 8-8).

Why Decays Occur A nucleus is said to **decay** when it emits an alpha or beta particle or a gamma ray. Alpha decay occurs in nuclei too large to be stable (Fig. 8-9). The forces that hold protons and neutrons together in a nucleus act only over short distances. As a result these particles interact strongly only with their nearest neighbors in a nucleus. Because the electrical repulsion of the protons is strong throughout the entire nucleus, there is a limit to the ability of neutrons to hold together a large nucleus. This limit is represented by the bismuth isotope $^{209}_{83}$Bi, which is the heaviest stable (that is, nonradioactive) nucleus. All larger nuclei become smaller ones by alpha decay.

Another cause of radioactive decay is a ratio of neutrons to protons that is too large or too small. A small nucleus is stable with equal numbers of neutrons and protons. However, larger nuclei need more neutrons than protons in order to overcome the electrical repulsion of the protons. In beta decay, one of the neutrons in a nucleus with too many of them spontaneously turns into a proton with the emission of an electron, as in Fig. 8-9.

In a nucleus with too few neutrons for stability, one of the protons may become a neutron with the emission of a **positron,** which is an electron that has a positive charge rather than a negative one. Alternatively one of the electrons in the atom may be absorbed by one of the protons to form a neutron. This process, which rarely occurs, is called **electron capture.** (An uncharged particle called a **neutrino,** whose mass is extremely small, is also emitted during beta decay, positron emission, and electron capture. See Sec. 8.15.)

Sometimes a certain nuclide requires a number of radioactive decays before it reaches a stable form. The uranium isotope $^{238}_{92}$U, for instance, undergoes eight alpha decays and six beta decays before it eventually becomes the lead isotope $^{206}_{82}$Pb, which is not radioactive.

Free neutrons outside of nuclei are unstable and undergo radioactive decay into a proton and an electron. Nevertheless it is not correct

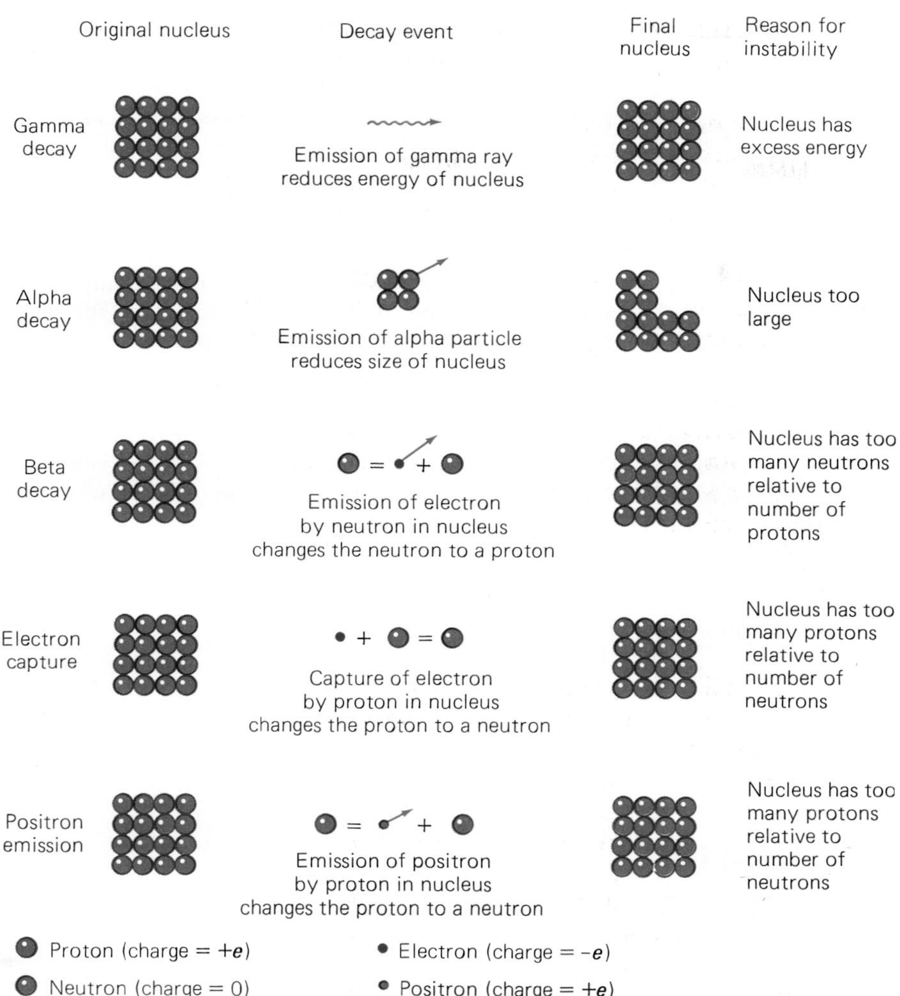

Figure 8-9 Five kinds of radioactive decay. The neutrinos emitted during beta decay, electron capture, and positron emission are not shown; they are uncharged and have very little mass.

to think of a neutron as a combination of a proton and an electron: a neutron is a separate particle with unique properties. If we were to try to create a neutron by bringing together a proton and an electron, we would merely get a hydrogen atom as the result, not a neutron. Protons outside a nucleus are apparently stable.

Example 8.1

Find the symbol of the nuclide into which the uranium isotope ${}^{232}_{92}\text{U}$ is transformed when it undergoes alpha decay.

Solution

An alpha particle is the ${}^{4}_{2}\text{He}$ nucleus, so when a nucleus undergoes alpha decay its atomic number decreases by 2 (corresponding to a loss of 2 protons) and its mass number decreases by 4 (corresponding to a loss of 4 nucleons). Hence the product of the alpha decay of ${}^{232}_{92}\text{U}$ has the atomic number 90 and the mass number 228. From Table 10-6 (see the inside back cover), thorium, Th, has the atomic number 90, so the result of the decay is ${}^{228}_{90}\text{Th}$.

Example 8.2

The bromine isotope ${}^{80}_{35}\text{Br}$ can decay by emitting an electron or a positron or by capturing an electron. What is the equation of the process in each case?

Solution

(a) When a nucleus emits an electron, its charge increases by $+e$ (corresponding to a loss of $-e$), which means that its atomic number increases by 1. Thus the new nucleus has an atomic number of $Z = 35 + 1 = 36$. There is no change in mass number since the number of nucleons (protons + neutrons) stays the same. According to Table 10-6 the element with $Z = 36$ is krypton, Kr, so the equation of the process is

$${}^{80}_{35}\text{Br} \rightarrow {}^{80}_{36}\text{Kr} + e^-$$

(b) When a nucleus emits a positron, its charge decreases by $-e$ (corresponding to a loss of $+e$), which means that Z decreases by 1. The new value of Z is therefore $35 - 1 = 34$, which is the atomic number of selenium, Se. Hence

$${}^{80}_{35}\text{Br} \rightarrow {}^{80}_{34}\text{Se} + e^+$$

(c) When a nucleus captures an electron, its charge also decreases by $-e$, so the result is again $Z = 34$. In this case

$${}^{80}_{35}\text{Br} + e^- \rightarrow {}^{80}_{34}\text{Se}$$

In each case we can see that the net electric charge does not change in the decay, so charge is conserved (as it must be).

8.4 Half-Life

Less and Less, But Always Some Left

The **half-life** of a radionuclide is the period of time needed for half of an initial amount of the nuclide to decay. As time goes on, the undecayed amount becomes smaller, but there is some left for many half-lives.

Suppose we start with 1 milligram (mg) of the radium isotope ${}^{226}_{88}\text{Ra}$, which alpha decays to the radon isotope ${}^{222}_{88}\text{Rn}$, with a half-life of about 1600 years. After 1600 years, 0.5 mg of radium will remain, with the rest having turned into radon (which, by the way, is a gas; radium is a metal). During the next 1600 years, half the 0.5 mg of radium that is left will decay, to leave 0.25 mg of radium (Fig. 8-10). After a further 1600 years, which means a total of 4800 years or 3 half-lives, 0.125 mg of radium will be left—still a fair amount. Even after 6 half-lives, more than 1 percent of an original sample will remain undecayed.

Every radionuclide has a characteristic and unchanging half-life. Some half-lives are only a millionth of a second; others are billions of years. Radon, for example, is an alpha emitter like its parent radium, but the half-life of radon is only 3.8 days instead of 1600 years. One of the biggest problems faced by nuclear power plants is the safe disposal of radioactive wastes since some of the isotopes present have long half-lives. The beta decay of neutrons when they are outside nuclei has a half-life of 14.5 min.

The dating of archaeological specimens and rock samples (including those brought back from the moon) by methods based on radioactive decay is described in Chap. 16.

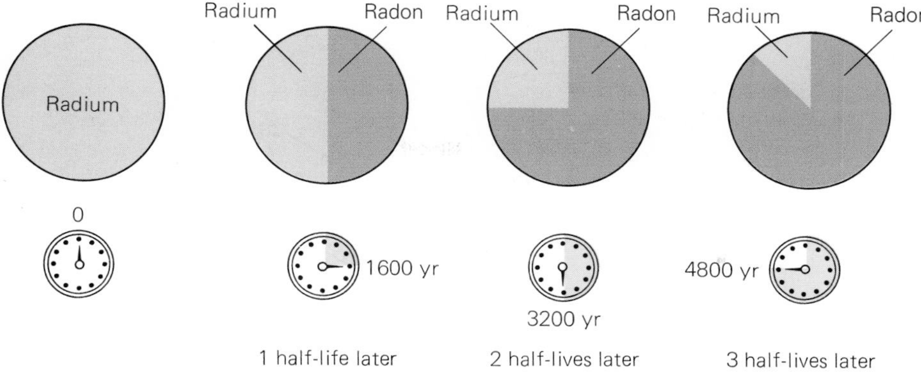

Figure 8-10 The decay of the radium isotope $^{226}_{88}$Ra. The number of undecayed radium atoms in a sample decreases by one-half in each 1600-year period. This time span is accordingly known as the "half-life" of radium. The radium alpha decays into the radon isotope $^{222}_{86}$Rn, whose own half-life is 3.8 days.

Example 8.3

Potassium contains a small proportion of the radioactive isotope $^{40}_{19}$K, which decays into the isotope $^{40}_{18}$Ar of the gas argon with a half-life of 1.3 billion years. When a rock whose minerals contain potassium is formed, no $^{40}_{18}$Ar is present. As time goes on, the $^{40}_{19}$K gradually decays into $^{40}_{18}$Ar, which is trapped in the rock. Comparing the amounts of $^{40}_{19}$K and $^{40}_{18}$Ar in a rock therefore lets us calculate how long ago the rock was formed. How old is a rock that is found to contain 3 times as much $^{40}_{18}$Ar as $^{40}_{19}$K?

Solution

One-quarter of the original $^{40}_{19}$K is left. Since $\frac{1}{4} = \frac{1}{2} \times \frac{1}{2}$, the rock is 2 half-lives old, which is 2.6 billion years.

8.5 Radiation Hazards

Invisible but Dangerous

The various radiations from radionuclides ionize matter through which they pass. X-rays ionize matter, too. All ionizing radiation is harmful to living tissue, although if the damage is slight, the tissue can often repair itself with no permanent effect. Radiation hazards are easy to underestimate because there is usually a delay, sometimes of many years, between an exposure and some of its possible consequences. These consequences include cancer, leukemia, and changes in reproductive cells that lead to children with physical deformities and mental handicaps. The em radiation emitted by power lines, cell phones, and the various electronic devices in the home do not ionize matter.

Radiation dosage is measured in **sieverts** (Sv), where 1 Sv is the amount of any radiation that has the same biological effects as those produced when 1 kg of body tissue absorbs 1 joule of x-rays or gamma rays. (A related unit sometimes used is the rem, equal to 0.01 Sv.) Although radiobiologists disagree about the exact relationship between radiation exposure and the likelihood of developing cancer or leukemia, there is no question that such a link exists. Natural sources of radiation lead to a

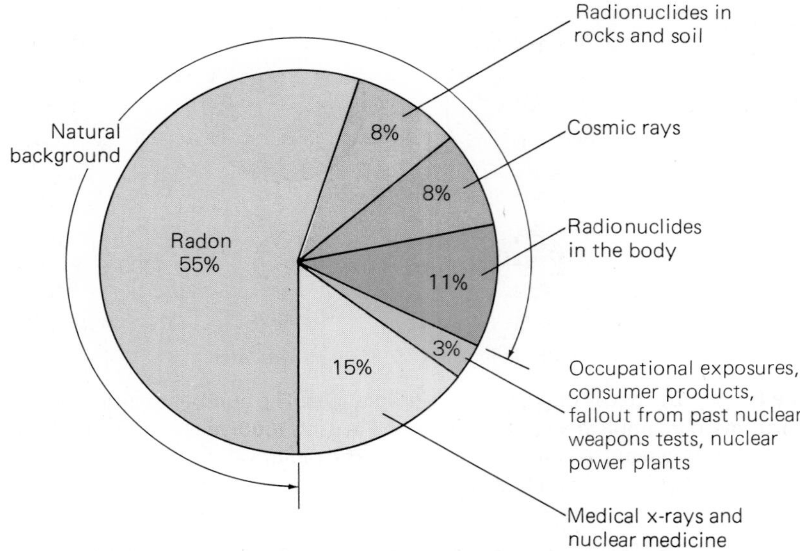

Figure 8-11 Sources of radiation dosage for an average person in the United States. Actual dosages vary widely. For instance, radon concentrations are not the same everywhere, some people receive more medical x-rays than others; cosmic rays are more intense at high altitudes; and so on. Nuclear power stations are responsible for 0.08 percent of the total, although accidents can raise the amount in affected areas to dangerous levels.

dosage rate per person of about 3 mSv/y averaged over the U.S. population (1 mSv = 0.001 Sv). Other sources of radiation add 0.6 mSv/y, with medical x-rays contributing the largest amount; a typical mammogram involves a dose of 0.7 mSv. The total per person thus averages about 3.6 mSv/y.

Natural Sources Figure 8-11 shows the relative contributions to the radiation dosage received by an average person in the United States. The most important single source is the radioactive gas radon, a decay product of radium whose own origin traces back to the decay of uranium. Uranium is found in many common rocks, notably granite. Hence radon, colorless and odorless, is present nearly everywhere, though usually in amounts too small to endanger health. Problems arise when houses are built in uranium-rich regions, since it is impossible to prevent radon from entering such houses from the ground under them. Surveys show that millions of American homes have radon concentrations high enough to pose a small but definite cancer risk. As a cause of lung cancer, radon is second only to cigarette smoking. The most effective method of reducing radon levels in an existing house in a hazardous region seems to be to extract air from underneath the ground floor and disperse it into the atmosphere before it can enter the house.

Other natural sources of radiation dosage include cosmic rays (see Chap. 19) and radionuclides present in rocks and soil. The human body itself contains tiny amounts of radionuclides of such elements as potassium and carbon. The cosmic-ray dosage depends on altitude because they are gradually absorbed by the atmosphere. Near sea level the dosage is around 0.3 mSv per year, more than that for aircrews and frequent fliers, 2 mSv per year in La Paz, Bolivia, 3700 m above sea level, and as much as 1 mSv per day for astronauts in orbit.

Dose Limits

Many useful processes involve ionizing radiation. Some employ such radiation directly, as in the x-rays and gamma rays used in medicine and industry. In other cases the radiation is an unwanted but inescapable by-product, notably in the operation of nuclear reactors and in the disposal of their wastes. An estimated 9 million people around the world are exposed to radiation at work. The radiation dosage limit for such people in the United States is 50 mSv per year. The maximum dose to the general public (who have no choice in the matter) from artificial sources has been set internationally at 1 mSv per year. By comparison, smoking 10 cigarettes a day gives a cancer death risk 100 times greater.

X-Rays It is not always easy to find an appropriate balance between risk and benefit for medical x-ray exposures, many of which are made for no strong reason and may do more harm than good. In this category are many "routine" x-rays. Particularly dangerous is the x-raying of pregnant women, until not long ago another "routine" procedure, which dramatically increases the chance of cancer in their children. Children themselves are extremely susceptible to harm from x-rays.

Of course, x-rays have many valuable applications in medicine. The point is that every x-ray exposure should have a definite justification that outweighs the risk involved, and patients should never hesitate to question the strength of such justification. Unfortunately not all doctors appreciate the very real hazards of x-rays. This is especially true of CT scans (Sec. 9.4); a typical abdominal CT scan delivers an x-ray dosage of 10 mSv, 500 times that of an ordinary chest x-ray. In the United States over 60 million CT scans are made every year, and the need for so many has not been established. Worst of all are whole-body CT scans of symptomless people to look for possible hidden abnormalities, which the Food and Drug Administration, among other authorities, feels are never justified.

Nuclear Energy

The atomic nucleus is the energy source of the reactors that produce more and more of the world's electricity. It is also the energy source of the most destructive weapons ever invented. But there is more to nuclear energy than these applications: nearly all the energy that keeps the sun and stars shining comes from the nucleus as well. Before considering what nuclear energy does, let us look into exactly what it is.

8.6 Units of Mass and Energy

The Atomic Mass Unit and the Electronvolt

Until now we have been using the kilogram as the unit of mass and the joule as the unit of energy. These units are far too large in the atomic world, and physicists find it more convenient to use smaller units for mass and energy in this world.

The **atomic mass unit** (u) has the value

$$1 \text{ atomic mass unit} = 1 \text{ u} = 1.66 \times 10^{-27} \text{ kg}$$

This mass is approximately equal to the mass of the hydrogen atom, whose actual mass is 1.008 u.

The energy unit used in atomic physics is the **electronvolt** (eV), which is the energy gained by an electron accelerated by a potential difference of 1 volt. The joule equivalent of the electronvolt is

$$1 \text{ electronvolt} = 1 \text{ eV} = 1.60 \times 10^{-19} \text{ J}$$

A typical quantity expressed in electronvolts is the energy needed to remove an electron from an atom. In the case of a nitrogen atom this energy is 14.5 eV, for example.

In nuclear physics the electronvolt is too small, and its multiple the **megaelectronvolt** (MeV) is more suitable:

$$1 \text{ megaelectronvolt} = 1 \text{ MeV} = 10^{6} \text{ eV} = 1.60 \times 10^{-13} \text{ J}$$

(*Mega* is the prefix for million.) A typical quantity expressed in megaelectronvolts is the energy of the radiation emitted by a radionuclide. The alpha particle emitted by a nucleus of the radium isotope $^{226}_{88}\text{Ra}$ has an energy of 4.9 MeV, for example.

The energy equivalent ($E_0 = mc^2$) of a rest mass of 1 u is 931 MeV.

8.7 Binding Energy

The Missing Energy That Keeps a Nucleus Together

An ordinary hydrogen atom has a nucleus that consists of a single proton, as its symbol ^1_1H indicates. The isotope of hydrogen called deuterium, ^2_1H, has a neutron as well as a proton in its nucleus. Thus we expect the mass of the deuterium atom to equal the mass of a ^1_1H hydrogen atom plus the mass of a neutron:

Mass of ^1_1H atom	1.0078 u
+Mass of neutron	+1.0087 u
Expected mass of ^2_1H atom	2.0165 u

However, the measured mass of the ^2_1H atom is only 2.0141 u, which is 0.0024 u *less* than the combined masses of a ^1_1H atom and a neutron (Fig. 8-12).

Deuterium atoms are not the only ones that have less mass than the combined masses of the particles they are composed of—*all* atoms (except ^1_1H) are like that. We conclude that nuclei are stable because they lack enough mass to break up into separate nucleons.

What happens when a nucleus is formed is that a certain amount of energy is given off due to the action of the forces that hold the neutrons and protons together. Energy is similarly given off due to the action of gravity when a stone strikes the ground or due to the action of intermolecular forces when water freezes into ice. In the case of a nucleus, the energy comes from the mass of the particles that join together. The resulting nucleus therefore has less mass than the total mass of the particles before they interact.

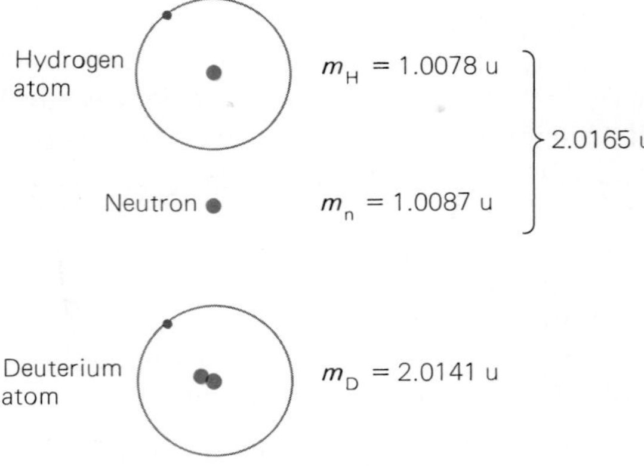

Figure 8-12 The mass of a deuterium atom (^2_1H) is less than the sum of the masses of a hydrogen atom (^1_1H) and a neutron. The energy equivalent of the missing mass is called the binding energy of the nucleus.

Since the energy equivalent of 1 u of mass is 931 MeV, the energy that corresponds to the missing deuterium mass of 0.0024 u is

$$\text{Missing energy} = (0.0024\ \text{u})(931\ \text{MeV/u}) = 2.2\ \text{MeV}$$

To test the above interpretation of the missing mass, we can perform experiments to see how much energy is needed to break apart a deuterium nucleus into a separate neutron and proton. The required energy turns out to be 2.2 MeV, as we expect (Fig. 8-13). When less energy than 2.2 MeV is given to a ${}^{2}_{1}\text{H}$ nucleus, the nucleus stays together. When the added energy is more than 2.2 MeV, the extra energy goes into kinetic energy of the neutron and proton as they fly apart.

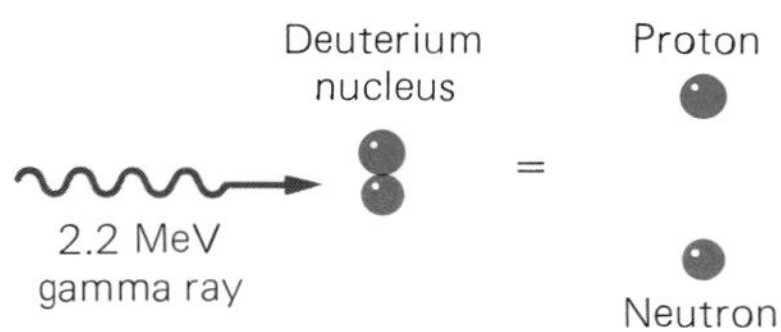

Figure 8-13 The binding energy of the deuterium nucleus is 2.2 MeV. A gamma ray whose energy is 2.2 MeV or more can split a deuterium nucleus into a proton and neutron. A gamma ray whose energy is less than 2.2 MeV cannot do this.

> The energy equivalent of the missing mass of a nucleus is called the **binding energy** of the nucleus. The greater its binding energy, the more the energy that must be supplied to break up the nucleus.

Nuclear binding energies are strikingly high. The range for stable nuclei is from 2.2 MeV for ${}^{2}_{1}\text{H}$ (deuterium) to 1640 MeV for ${}^{209}_{83}\text{Bi}$ (an isotope of the metal bismuth). Larger nuclei are all unstable and decay radioactively. To appreciate how high binding energies are, we can compare them with more familiar energies in terms of kilojoules of energy per kilogram of mass. In these units, a typical binding energy is 8×10^{11} kJ/kg—800 billion kJ/kg. By contrast, to boil water involves a heat of vaporization of a mere 2260 kJ/kg, and even the heat given off by burning gasoline is only 4.7×10^{4} kJ/kg, 17 million times smaller.

8.8 Binding Energy per Nucleon

Why Fission and Fusion Liberate Energy

For a given nucleus, the **binding energy per nucleon** is found by dividing the total binding energy of the nucleus by the number of

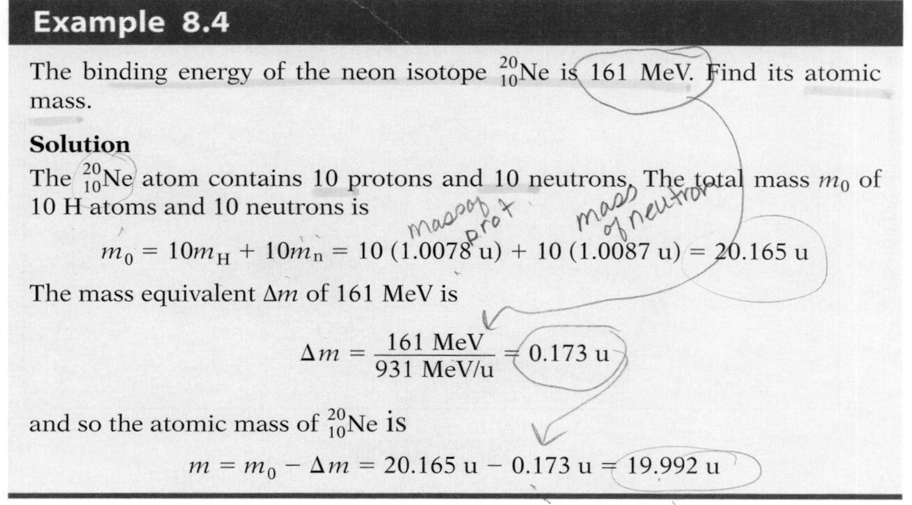

Example 8.4

The binding energy of the neon isotope ${}^{20}_{10}\text{Ne}$ is 161 MeV. Find its atomic mass.

Solution

The ${}^{20}_{10}\text{Ne}$ atom contains 10 protons and 10 neutrons. The total mass m_0 of 10 H atoms and 10 neutrons is

$$m_0 = 10m_H + 10m_n = 10\,(1.0078\ \text{u}) + 10\,(1.0087\ \text{u}) = 20.165\ \text{u}$$

The mass equivalent Δm of 161 MeV is

$$\Delta m = \frac{161\ \text{MeV}}{931\ \text{MeV/u}} = 0.173\ \text{u}$$

and so the atomic mass of ${}^{20}_{10}\text{Ne}$ is

$$m = m_0 - \Delta m = 20.165\ \text{u} - 0.173\ \text{u} = 19.992\ \text{u}$$

nucleons (protons and neutrons) it contains. Thus the binding energy per nucleon for ${}^{2}_{1}H$ is 2.2 MeV/2 = 1.1 MeV/nucleon, and for ${}^{209}_{83}Bi$ it is 1640 MeV/209 = 7.8 MeV/nucleon.

Figure 8-14 shows binding energy per nucleon plotted against mass number (number of nucleons). The greater the binding energy per nucleon, the more stable the nucleus. The graph has its maximum of 8.8 MeV/nucleon when the number of nucleons is 56. The nucleus that has 56 protons and neutrons is ${}^{56}_{26}Fe$, an iron isotope. This is the most stable nucleus of them all, since the most energy is needed to pull a nucleon away from it. Larger and smaller nuclei are less stable.

Fission and Fusion Two remarkable conclusions can be drawn from the curve of Fig. 8-14. The first is that, if we somehow split a heavy nucleus into two medium-size ones, each of the new nuclei will have *more* binding energy per nucleon (and hence less mass per nucleon) than the original nucleus did. The extra energy will be given off, and it can be a lot.

As an example, if the uranium nucleus ${}^{235}_{92}U$ is broken into two smaller nuclei, the difference in binding energy per nucleon is about 0.8 MeV. Since ${}^{235}_{92}U$ contains 235 nucleons, the total energy given off is

$$\left(0.8\ \frac{\text{MeV}}{\text{nucleon}}\right)(235\ \text{nucleons}) = 188\ \text{MeV}$$

This is a truly enormous amount of energy to come from a single atomic event. For comparison, ordinary chemical reactions involve only a few

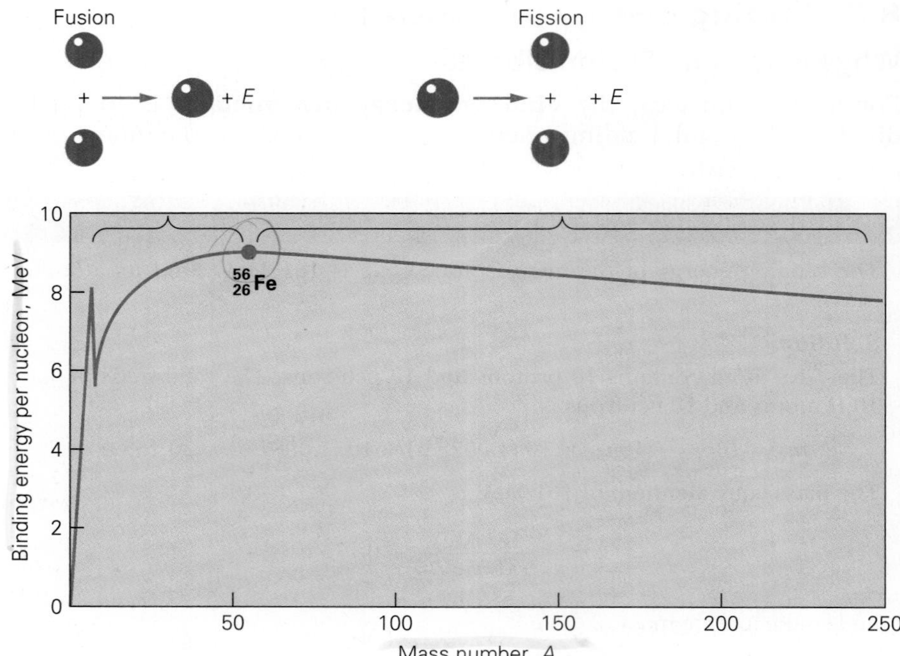

Figure 8-14 The binding energy per nucleon is a maximum for nuclei of mass number $A = 56$. Such nuclei are the most stable. When two light nuclei join to form a heavier one, a process called *fusion,* the greater binding energy of the product nucleus causes energy to be given off. When a heavy nucleus is split into two lighter ones, a process called *fission,* the greater binding energy of the product nuclei also causes energy to be given off.

eV per atom. Splitting a large nucleus, which is called **nuclear fission,** thus involves a hundred million times more energy per atom than, say, burning coal or oil.

The other notable conclusion from Fig. 8-14 is that joining two light nuclei together to give a single nucleus of medium size also means more binding energy per nucleon in the new nucleus. For instance, if two ${}^{2}_{1}H$ deuterium nuclei combine to form a ${}^{4}_{2}He$ helium nucleus, over 23 MeV is released. Such a process, called **nuclear fusion,** is also a very effective way to obtain energy. In fact, nuclear fusion is the main energy source of the sun and other stars, as described in Chap. 18.

The graph of Fig. 8-14 is extremely significant because it is the key to understanding energy production in the universe. The fact that binding energy exists at all means that nuclei more complex than the single proton of hydrogen can be stable. This stability in turn accounts for the existence of the various elements and consequently for the existence of the many and diverse forms of matter we see around us. Because the curve peaks in the middle, we have the explanation for the energy that powers, directly or indirectly, the evolution of much of the universe: this energy comes from the fusion of protons and light nuclei to form heavier nuclei. And the harnessing of nuclear fission in reactors and weapons has irreversibly changed modern civilization.

Fission and Fusion

The words "nuclear energy" bring to mind two images. One is of a huge building in which a mysterious thing called a nuclear reactor turns an absurdly small amount of uranium into an absurdly large amount of energy. The other image is of a mushroom-shaped cloud rising from the explosion of a nuclear bomb, an explosion that can level the largest city and kill millions of people.

The first image is a picture of hope, hope for a future of plentiful, cheap, pollution-free energy—a hope only partly fulfilled. The second image is a picture of horror—but such bombs have not been used in war for over 60 years. So nuclear energy has not turned out as yet to be either the overwhelming blessing or the overwhelming curse it might have been.

8.9 Nuclear Fission

Divide and Conquer

As we have seen, a lot of energy will be released if we can break a large nucleus into smaller ones. But nuclei are ordinarily not at all easy to break up. What we need is a way to split a heavy nucleus without using more energy than we get back from the process.

The answer came in 1939 with the discovery that a nucleus of the uranium isotope ${}^{235}_{92}U$ undergoes fission when struck by a neutron. It is not the impact of the neutron that has this effect. Instead, the ${}^{235}_{92}U$ nucleus absorbs the neutron to become ${}^{236}_{92}U$, and the new nucleus is so unstable

BIOGRAPHY Lise Meitner (1878–1968)

The daughter of a Viennese lawyer, Meitner became interested in science when she read about the Curies and radium. She earned her Ph.D. in physics in 1905 at the University of Vienna, only the second woman to obtain a doctorate there. She then went to Berlin where she began research on radioactivity with the chemist Otto Hahn. Their supervisor refused to have a woman in his laboratory, so they started their work in a carpentry shop. Ten years later she was a professor, a department head, and, with Hahn, the discoverer of a new element, protactinium.

In the 1930s the Italian physicist Enrico Fermi found that bombarding heavy elements with neutrons led to the production of other elements. What happened in the case of uranium was puzzling, and Meitner and Hahn tried to find the answer. At the time the German persecution of Jews had begun, but Meitner, who was Jewish, was protected by her Austrian citizenship. In 1938 Germany annexed Austria, and Meitner fled to Sweden but kept in touch with Hahn and their younger colleague Fritz Strassmann. Hahn and Strassmann finally concluded that neutrons interact with uranium to produce radium, but Meitner's calculations showed that this was impossible and she urged them to persist in their experiments. They did, and found to their surprise that the lighter element barium in fact had been created. Meitner surmised that the neutrons had caused the uranium nuclei to split apart and, with her nephew Otto Frisch, developed the theoretical picture of what they called fission.

In January 1939, Hahn and Strassmann published the discovery of fission in a German journal; because Meitner was Jewish, they thought it safer to ignore her contribution. Meitner and Frisch later published their own paper on fission in an English journal, but it was too late: Hahn disgracefully claimed full credit, and not once in the years that followed acknowledged her role. Hahn alone received the Nobel Prize in physics for discovering fission. Unfortunately Meitner did not live to see a measure of justice: the element of atomic number 109 is called meitnerium in her honor, while the tentative name of hahnium for element 105 was changed in 1997 to dubnium, after the Russian nuclear research center in Dubna.

that almost at once it splits into two pieces (Fig. 8-15). Isotopes of several elements besides uranium were later found to be fissionable by neutrons in similar processes.

Most of the energy set free in fission goes into kinetic energy of the new nuclei. These nuclei are usually radioactive, some with long half-lives. Hence the products of fission, which are found in reactor fuel rods and in the fallout from a nuclear weapon explosion, are extremely dangerous and remain so for many generations.

Chain Reaction When a nucleus breaks apart, two or three neutrons are set free at the same time. This suggests a remarkable possibility. Perhaps, under the right conditions, the neutrons emitted by one uranium nucleus as it undergoes fission can cause other uranium nuclei to split; the neutrons from these other fissions might then go on to split still more uranium nuclei; and so on, with a series of fission reactions spreading through a mass of uranium. A **chain reaction** of this kind was first demonstrated in Chicago in 1942 under the direction of Enrico Fermi, an Italian physicist who had not long before taken refuge in the United

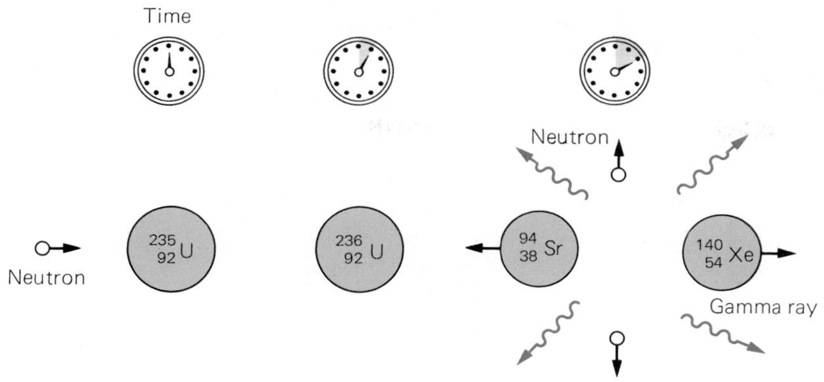

Figure 8-15 In nuclear fission an absorbed neutron causes a heavy nucleus to split into two parts. Several neutrons and gamma rays are emitted in the process. The smaller nuclei shown here are typical of those produced in the fission of $^{235}_{92}U$.

States (Fig. 8-16). Figure 8-17 is a sketch of the events that occur in a chain reaction.

For a chain reaction to occur, at least one neutron produced by each fission must, on the average, lead to another fission and not either escape or be absorbed without producing fission. If too few neutrons cause fissions, the reaction slows down and stops. If precisely one neutron per fission causes another fission, energy is released at a steady rate. This is the case in a nuclear reactor, which is an arrangement for producing controlled power from nuclear fission.

Figure 8-16 Enrico Fermi (1901–1954).

Nuclear Weapons What happens if more than one neutron from each fission causes other fissions? Then the chain reaction speeds up and the energy release is so fast that an explosion results. An "atomic" bomb makes use of this effect. The destructive power of nuclear weapons does not stop with their detonation but continues long afterward through the radioactive debris that is produced and widely dispersed. A single modern nuclear bomb set off above New York City would leave about 4 million people dead by the next day with millions more fatalities later from radiation exposure.

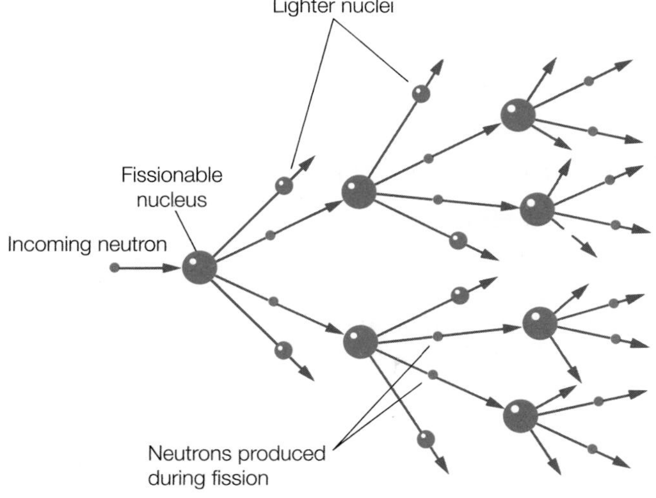

Figure 8-17 Sketch of a chain reaction. The reaction continues if at least one neutron from each fission event on the average induces another fission event. If more than one neutron per fission on the average induces another fission, the reaction is explosive.

Table 8–1 Estimated Stockpiles of Nuclear Weapons

Country	Nuclear Warheads
Russia	15,000
United States	9900
France	350
China	200
Great Britain	200
Israel	80
Pakistan	60
India	50
North Korea	<10

Note: About 5800 of Russia's weapons and 5700 of those of the United States are operational. A treaty binds these countries to reduce their stocks of "operationally deployed" nuclear weapons to 2200 each by 2012.

The discovery of fission became known in the United States in 1939, just before the start of World War II. Its military possibilities were immediately recognized. Expecting that German physicists would come to the same conclusion and would start work on a nuclear bomb, the United States began such a program in earnest. By the time it had succeeded, in 1945, Germany had been defeated, and two nuclear bombs exploded over Hiroshima and Nagasaki then ended the war with Japan. It was later learned that the German effort had amounted to very little.

Not long afterward the Soviet Union, Great Britain, and France also developed nuclear weapons, and later China, Israel, South Africa, India, Pakistan, and North Korea did so as well. South Africa voluntarily abandoned its nuclear weapons program, the only country to do so. The present nuclear powers have among themselves about 26,000 weapons (Table 8.1), many times more than enough to destroy all human life. It is impossible to ascribe a rational purpose to having so many of these weapons, or to plans by the U.S. government to develop new ones.

Unfortunately it is not a giant step from a nuclear reactor program for energy generation to a nuclear weapons program. At least 20 countries besides those listed above already have the skills and materials to develop nuclear weapons, and a few may have begun the process. Minimizing the nuclear threat is a continuing task.

8.10 How a Reactor Works

From Uranium to Heat to Electricity

For every gram of uranium that undergoes fission in a reactor, 2.6 tons of coal must be burned in an ordinary power plant of the same rating. The energy given off in a nuclear reactor becomes heat, which is removed by a liquid or gas coolant. The hot coolant is then used to boil water, and the resulting steam is fed to a turbine that can power an electric generator, a ship, or a submarine.

In order for a chain reaction to occur at a steady rate, one neutron from each fission must cause another fission to take place. Since each fission in ^{235}U liberates an average of 2.5 neutrons, no more than 1.5 neutrons per fission can be lost on the average. However, natural uranium contains only 0.7 percent of the fissionable isotope ^{235}U. The rest is ^{238}U, an isotope that captures the rapidly moving neutrons emitted during the fission of ^{235}U but usually does not undergo fission afterward. The neutrons absorbed by ^{238}U are therefore wasted, and since 99.3 percent of natural uranium is ^{238}U, too many disappear for a chain reaction to occur in a solid lump of natural uranium.

Fast and Slow Neutrons There is an ingenious way around this problem. As it happens, ^{238}U tends to pick up only fast neutrons, not slow ones. In addition, slow neutrons are more apt to induce fission in ^{235}U than fast ones. If the fast neutrons from fission are slowed down, then many more will produce further fissions despite the small proportion of ^{235}U present.

To slow down fission neutrons, the uranium fuel in a reactor is mixed with a **moderator,** a substance whose nuclei absorb energy from fast neutrons that collide with them. In general, the more nearly equal in mass colliding particles are, the more energy is transferred. A ball bounces off a wall with little loss of energy, but it can lose all its energy when it strikes another ball (Fig. 3-24). Since hydrogen nuclei are protons with nearly the same mass as neutrons, hydrogen is widely used as a moderator in the form of water, H_2O, each of whose molecules contains two hydrogen atoms along with an oxygen atom.

Unfortunately a neutron striking a proton has a certain tendency to stick to it to form a deuterium nucleus, $^{2}_{1}$H. As a result, a reactor whose moderator is water cannot use ordinary uranium as fuel but must instead use **enriched** uranium whose ^{235}U content has been increased to 3 to 5 percent. Uranium enriched to about 90 percent of ^{235}U is used in one type of nuclear weapon.

A Natural Reactor

Because ^{235}U has a shorter half-life than ^{238}U, in the past the ^{235}U/^{238}U ratio in uranium was higher than the 0.7 percent of today. This raises the possibility that, in former times when the ^{235}U/^{238}U ratio was 3 percent or more, a uranium deposit with nearby water to act as moderator could have sustained a chain reaction. In fact, such a natural reactor seems to have existed 2 billion years ago in West Africa in what is today Gabon, which developed about 100 kW. Studies of rock samples from the area indicate that the reactor operated until the heat it produced boiled away water in the rocks around the uranium deposit, which typically took 30 min. Then the lack of water stopped the chain reaction for typically 2.5 h until enough water had seeped back to restart it. This cycle continued for perhaps 150,000 years. There once may well have been other natural reactors in addition to the one in Gabon.

8.11 Plutonium

Another Fissionable Material

Some nonfissionable nuclides can be changed into fissionable ones by absorbing neutrons. A notable example is ^{238}U, which becomes ^{239}U when it captures a fast neutron. The latter uranium isotope beta-decays soon after its creation into the neptunium isotope $^{239}_{93}$Np. In turn $^{239}_{93}$Np beta-decays into the plutonium isotope $^{239}_{94}$Pu (Fig. 8-20). Like ^{235}U, ^{239}Pu undergoes fission when it absorbs a neutron and can support a chain reaction.

Both neptunium and plutonium are called **transuranium elements** because their atomic numbers are greater than the 92 of uranium. Other transuranium elements have been created in the laboratory up to atomic number 118 in high-energy collisions of lighter nuclei. Element 110, to give an example, was first produced in 1994 by bombarding lead nuclei with nickel nuclei. No transuranium elements of natural origin are found on the earth because all of them decay too fast to have survived even if they had been present when the earth came into being 4.5 billion years ago.

A certain amount of plutonium is produced in the normal operation of a uranium-fueled reactor, and its fission adds to the energy produced by the reactor. Plutonium separated from the uranium that remains in a used fuel rod can serve as a reactor fuel itself and also, like highly enriched uranium, as the active ingredient in nuclear weapons. A **breeder reactor** is one especially designed to produce more plutonium than the

Nuclear Power Plants

The fuel for a nuclear reactor consists of uranium oxide pellets sealed in long, thin tubes (Fig. 8-18). Control rods of cadmium or boron, which are good absorbers of slow neutrons, can be slid in and out of the reactor core to adjust the rate of the chain reaction. In the most common type of reactor, water under pressure (to prevent boiling) circulates around the fuel in the core where it acts as both moderator and coolant. As in Fig. 8-19, the pressurized water transfers heat from the chain reaction in the fuel rods to a steam generator. The resulting steam then passes out of the containment shell, which serves as a barrier to protect the outside world from accidents to the reactor, and is piped to a turbine that drives an electric generator. Various aspects of nuclear reactors as commercial energy sources were discussed in Sec. 4.8.

In a typical plant, the steel reactor vessel is 13.5 m high and 4.4 m in diameter and weighs 385 tons. It contains 90 tons of uranium oxide in the form of 50,952 fuel rods, each 3.85 m long and 9.5 mm in diameter. Four steam generators are used, instead of the single one shown in Fig. 8-19, as well as a number of turbine-generators. The reactor operates at 3400 MW and yields 1100 MW of electric power, enough for the needs of over a million people. The fuel must be replaced every few years as the concentration of neutron-absorbing fission products builds up.

Current designs for nuclear reactors incorporate major improvements in efficiency and reliability. An especially interesting new concept is the pebble-bed reactor. In such a reactor the fuel elements are "pebbles" the size of tennis balls that each contain thousands of tiny uranium oxide particles. Each particle has a tough ceramic coating that acts as a miniature pressure vessel to seal in the fission products. A pebble-bed reactor would have a core of several hundred thousand such pebbles and operate at a much higher temperature than today's reactors for higher efficiency (Sec. 5.14). It would be cooled by helium, which would go directly to a turbine connected to an electric generator. Unlike conventional reactors, the pebble-bed type would not have to be shut down for refueling, which would be done merely by adding new pebbles at the top while removing used ones at the bottom. Pebble-bed reactors, besides being simple and safe, can be economically built in small sizes in factories and then shipped to their places of use rather than be constructed there. China and South Africa are prominent among the countries developing such reactors. China expects to have 30 pebble-bed reactors operating by 2020 that would develop a total of 6 GW. Still other new designs are being studied.

Figure 8-18 Loading fuel rods into a reactor at the Comanche Peak power station in Texas. The rods are metal tubes filled with pellets of uranium oxide.

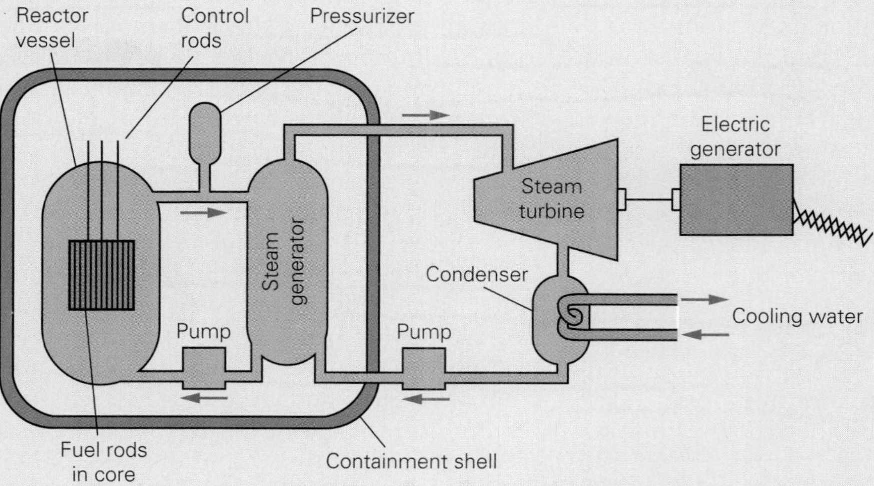

Figure 8-19 Basic design of a typical nuclear power plant.

^{235}U it consumes. Because the otherwise useless ^{238}U is 140 times more abundant than the fissionable ^{235}U, using breeder reactors would allow reserves of uranium to last much longer. In the past, breeder reactors were expensive and unreliable and few were built. However, new technologies have been proposed that could extract energy from spent reactor fuel while minimizing radioactive waste. If they succeed, uranium reserves would last for many centuries to come, but serious problems remain to be overcome and economic solutions to them may not be possible.

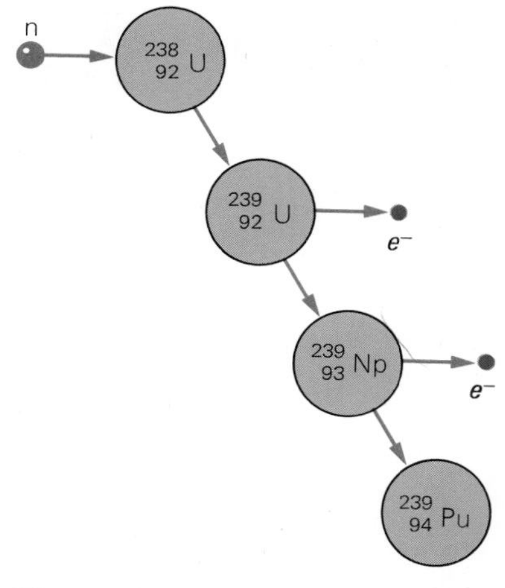

Figure 8-20 The nonfissionable uranium isotope ^{238}U, which makes up 99.3 percent of natural uranium, becomes the fissionable plutonium isotope ^{239}Pu by absorbing a neutron and beta-decaying twice. This transformation is the basis of the breeder reactor, which produces many times more nuclear fuel in the form of plutonium than it uses up in the form of ^{235}U.

8.12 Nuclear Fusion

The Energy Source of the Future?

For all the energy produced by fission, the fusion of small nuclei to form larger ones can yield even more energy per kilogram of starting materials. Nuclear fusion is the energy source of the sun and stars, as discussed in Chap. 18. On the earth, it is possible that fusion will become the ultimate source of energy: safe, almost nonpolluting, and with the oceans supplying limitless fuel.

Three conditions must be met by a successful fusion reactor. The first is a high temperature—100 million °C or more—so that the nuclei are moving fast enough to collide despite the repulsion of their positive electric charges. The second condition is a high concentration of the nuclei to ensure that such collisions are frequent. Third, the reacting nuclei must remain together for a long enough time to give off more energy than the reactor's operation uses. The last two conditions are related, since the more nuclei there are in a given volume, the shorter the minimum confinement time for a net energy output.

The fusion reaction that is the basis of current research involves the combination of a deuterium nucleus and a tritium nucleus to form a helium nucleus (Fig. 8-21):

$$\underset{\text{deuterium}}{{}^{2}_{1}\text{H}} + \underset{\text{tritium}}{{}^{3}_{1}\text{H}} \rightarrow \underset{\text{helium}}{{}^{4}_{2}\text{He}} + \underset{\text{neutron}}{{}^{1}_{0}n} + \underset{\text{energy}}{17.6\ \text{MeV}} \qquad \textit{Fusion reaction}$$

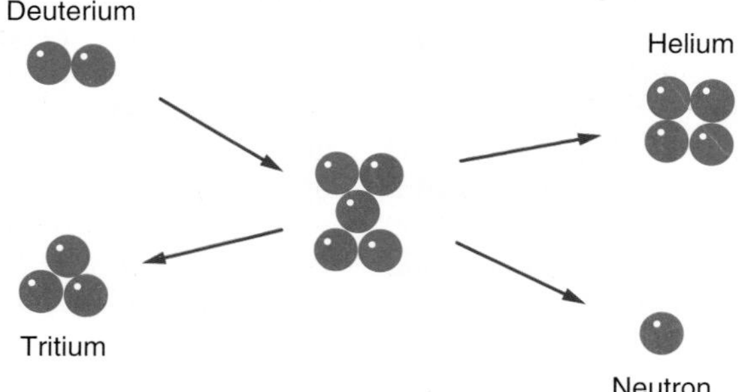

Figure 8-21 The deuterium-tritium fusion reaction liberates a great deal of energy, most of which is carried off by the neutron as kinetic energy.

Most of the energy given off is carried by the neutron that is emitted. To recover this energy, one proposal is to surround the reactor chamber with lithium to absorb the neutrons. The resulting hot lithium would then act as the heat source for a conventional electric generating system.

About 0.015 percent of the waters of the world is deuterium, which adds up to a total of over 10^{15} tons of ${}^2_1H$—no scarcity there. A gallon of seawater has the potential for fusion energy equivalent to the chemical energy in 600 gallons of gasoline. Seawater contains too little tritium for economic recovery, but as it happens, neutrons react with lithium nuclei to yield tritium and helium. Thus, once a fusion reactor is given an initial charge of tritium, it will make enough additional tritium from the surrounding lithium for its further operation.

Several Approaches The big problem in making fusion energy practical is to achieve the necessary combination of temperature, density, and confinement time. A number of approaches are being explored. In one, strong magnetic fields are used to keep the reacting nuclei close together. Five decades of research have led to larger and larger experimental magnetic fusion reactors that have brought to light no reasons why eventual success should not be possible (Fig. 8-22). In an encouraging experiment, 6.2 MW of power was produced for 4 s. The larger the reactor, the better its contents can maintain the required high temperature. The current record for size is held by the Joint European Torus (a torus has the shape of a doughtnut) in England, which is 15 m across and 20 m high. An international effort to build a similar reactor with a volume almost 6 times greater is in progress. If all goes well, commercial fusion reactors could be operating by 2050.

Another approach to practical fusion energy involves using energetic beams to both heat and compress tiny fuel pellets. Laser beams (see Chap. 9) are being tried for this purpose at the Lawrence Livermore National Laboratory in California. An initial laser beam will be split into 192 separate beams that will each be amplified 3×10^{15} times to give pulses whose total energy is 1.8 MJ. The corresponding power is 500 times the output of all the power stations in the United States, but only

ITER

The International Thermonuclear Experimental Reactor (ITER) now under construction in France represents what is hoped to be the final step before practical fusion power by magnetic confinement becomes a reality. ITER is sponsored by the United States, Japan, China, Russia, South Korea, India, and the European Union, which together represent over half the world's population; in 2008 the United States suspended its support as part of a general reduction in funding for scientific research. The reactor is expected to generate 500 MW (10 times the input power) from deuterium-tritium reactions, to weigh 32,000 tons, to cost perhaps \$14.6 billion (including operation for 20 years), and be finished by 2016. Superconducting magnets will keep the reacting ions in a doughnut-shaped region whose volume is that of a large house. About 80 percent of the energy released will be carried off by the neutrons that are produced, and these neutrons will be absorbed by lithium pellets in tubes that surround the reaction chamber. Circulating water will carry away the resulting heat; this is the heat that could be used in a working reactor to power turbines connected to electric generators.

Figure 8-22 An experimental fusion reactor at Princeton University. Based on a Soviet design called a tokamak, the reactor uses powerful magnetic fields to confine a hot ionized gas of the hydrogen isotopes deuterium and tritium. When nuclei of these isotopes react, they form a helium nucleus and a neutron in a process that gives off a great deal of energy. If practical fusion reactors can be developed, they might well supply much of the world's future energy needs.

for 3×10^{-9} s—3 billionths of a second. These beams will strike a fuel pellet from all sides to keep it in place as the conditions inside a star are created. If 10 pellets the size of a grain of sand are ignited every second, the energy output would be enough to provide electric power to a city of 175,000 people. Other laser fusion projects are under way in France and Japan.

A third method, the **Z-pinch,** would heat and compress fuel pellets in yet another way. In promising trials, thin parallel tungsten wires were arranged in cylindrical cages the size of a spool of thread. When momentary currents of millions of amperes were sent through the wires, they were heated and vaporized as magnetic forces squeezed them together (see Fig. 6-38a). The results were ionized gases at temperatures as high as 2.7 million °C. There seems to be no basic reason why further work should not be able to create conditions suitable for energy-producing fusion reaction in deuterium-tritium fuel pellets. The Z-pinch method has the potential advantages of being simpler and more energy efficient than the others.

There is a big gap between laboratory experiments, however encouraging, and large-scale production at an economic cost. For all the promise of fusion energy, it is by no means sure that the gap will eventually be bridged and fusion will end up as the ultimate solution to the world's energy needs. An old joke has it that unlimited fusion energy is just 40 years away—and always will be. Still, if fusion energy does prove a success, a realistic projection has it providing 40 percent of global energy needs by 2100 with no fuel supply problems or greenhouse gas emissions.

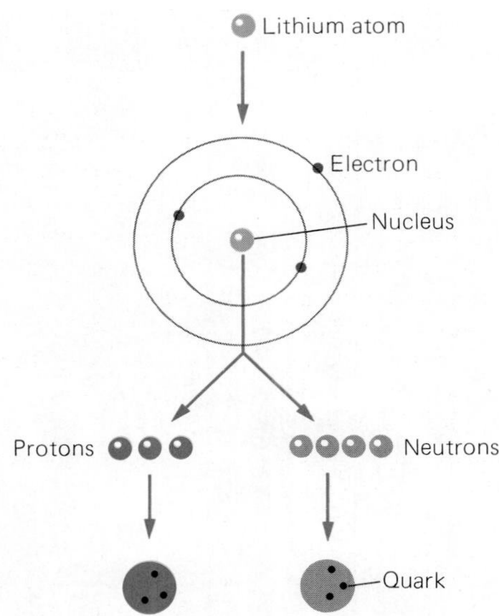

Figure 8-23 The search for truly elementary particles has led to the discovery of particles within particles. Today all ordinary matter seems to be made up of electrons and quarks. Shown are the various levels of organization of a lithium $^7_3$Li atom.

Elementary Particles

The electrons, protons, and neutrons of which atoms are composed are **elementary particles** in the sense that they cannot be broken down into anything else. Electrons are simply bits of electrically charged matter, but experiments show that nucleons (protons and neutrons) consist of still smaller particles called **quarks** (Fig. 8-23). The quarks in a nucleon stick together too tightly to permit the nucleon to be split apart, so nucleons are regarded as elementary particles despite their inner structures.

A great many other elementary particles besides electrons and nucleons are known, some composed of quarks and some not. Few of these other particles seem to have anything to do with ordinary matter, although their discovery has helped physicists in their study of how nature works.

8.13 Antiparticles

The Same but Different

Nearly all elementary particles have **antiparticles.** The antiparticle of a given particle has the same mass as the particle and behaves similarly in most respects, but its electric charge is opposite in sign. Thus the **positron** e^+ is the antiparticle of the electron e^-, and the negatively charged **antiproton** p^- is the antiparticle of the proton p^+. Certain uncharged elementary particles, such as the neutron, have antiparticles because they have properties other than charge that are different in the particle and its antiparticle.

Antiparticles are not easy to find for a very basic reason. When a particle and its antiparticle happen to come together, they destroy each other in a process called **annihilation.** The lost mass reappears as energy in the form of gamma rays when electrons and positrons are annihilated

Antimatter

There seems to be no reason why atoms could not be composed of antiprotons, antineutrons, and positrons. Indeed, hydrogenlike atoms that consist of antiprotons and positrons have already been created in the laboratory. Such **antimatter** ought to behave like ordinary matter. Of course, if antimatter comes in contact with ordinary matter, the same amount of both will disappear in a burst of energy. A postage stamp of antimatter reacting with a similar stamp of matter would release enough energy to send the space shuttle into orbit. But we might imagine that, when the universe was formed, equal quantities of matter and antimatter came into being that became separate galaxies of stars. If this were true, elsewhere in the universe would be stars, planets, and living things made entirely of antimatter.

The idea that the universe consists of both matter and antimatter is an attractive one, but unfortunately it does not seem to be the case. Although galaxies are far apart on the average, now and then two of them collide. A collision between a matter galaxy and an antimatter galaxy would be a violent explosion giving rise to a flood of gamma rays with characteristic energies. Very few such gamma rays are observed, from which astronomers conclude that there cannot be much antimatter in the universe. Current theories of elementary particles suggest that matter and antimatter are not exactly mirror images of each other, and as a result slightly more matter than antimatter was created when the universe came into being in the big bang described in Chap. 19. After all the antimatter had been annihilated, the excess of matter remained to become today's universe.

(Figs. 8-24 and 8-25). Unstable particles of various kinds may be produced instead of gamma rays when protons and antiprotons (or neutrons and antineutrons) are annihilated.

The reverse of annihilation can also take place, with energy becoming matter and electric charge being created where none existed before. In the remarkable process of **pair production,** a particle and its antiparticle materialize when a high-energy gamma ray passes near an atomic nucleus (Figs. 8-26 and 8-27). According to Einstein's formula $E_0 = mc^2$, the energy equivalent of the electron mass is 0.51 MeV. To produce an electron-positron pair therefore requires a gamma ray whose energy is at least 1.02 MeV. If the gamma ray has more energy than 1.02 MeV, the excess goes into the kinetic energies of the electron and positron. The minimum energy needed for a proton-antiproton or neutron-antineutron

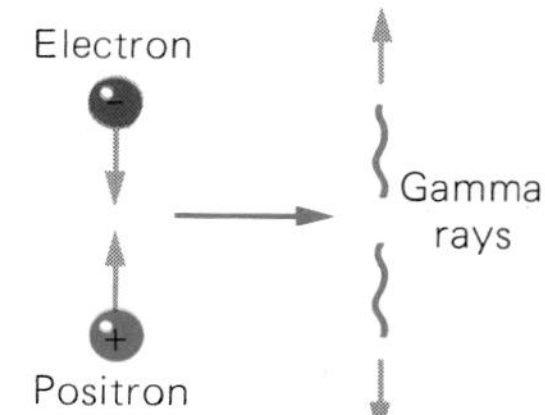

Figure 8-24 The mutual annihilation of an electron and a positron results in a pair of gamma rays whose total energy is equal to mc^2, where m is the total mass of the electron and positron.

BIOGRAPHY Paul A. M. Dirac (1902–1984)

Born in Bristol, England, Dirac originally studied electrical engineering. He then switched to physics and obtained his Ph.D. from Cambridge University in 1926. A new and revolutionary theory of the atom called quantum mechanics (see Sec. 9.12) was just then coming into being, and Dirac made a number of major contributions to it. Soon Dirac had joined special relativity to quantum mechanics to give a theory of the electron that predicted the existence of positively charged electrons, or positrons, which were then unknown. At first he thought that protons were the positive antiparticles of electrons despite their much greater mass and the fact that they are not annihilated by electrons. Then, in 1932, the American physicist Carl Anderson found that positrons do exist and have the same mass as electrons. In the same year Dirac became Lucasian Professor of Mathematics at Cambridge, the post Newton had held two-and-a-half centuries earlier. Dirac remained active in physics for the rest of his life, after 1969 in the warmer climate of Florida, but as is often the case in science he will be remembered for the brilliant achievements of his youth.

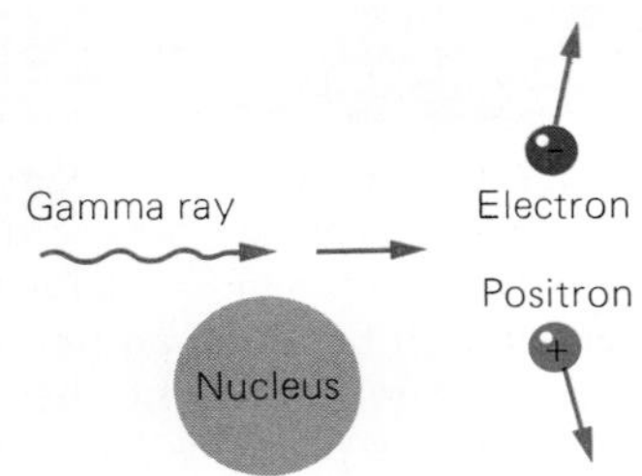

Figure 8-26 Pair production. (The presence of a nucleus is required in order that both momentum and energy be conserved.) Proton-antiproton and neutron-antineutron pairs can also be produced if the gamma ray has enough energy.

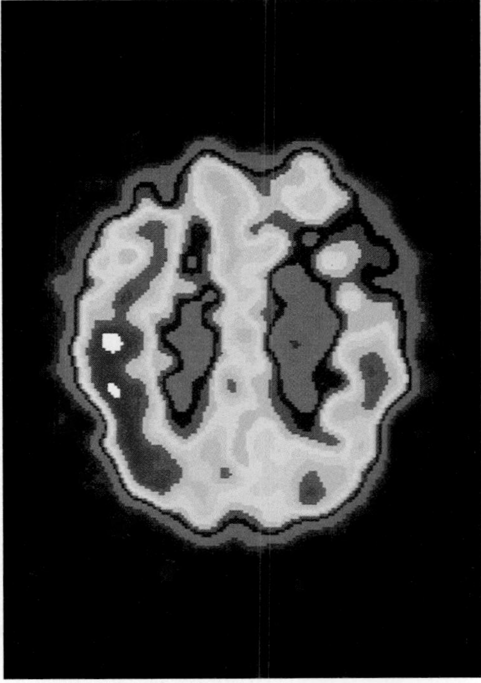

Figure 8-25 A positron emission tomography (PET) scan of the brain of a patient with Alzheimer's disease. Different colors correspond to different rates of metabolic activity. In PET, a positron-emitting isotope of an element appropriate to the condition being studied (here an oxygen isotope) is injected and allowed to circulate in a patient's body. When an emitted positron encounters an electron, which it does almost immediately, both are annihilated and a pair of gamma rays is created. Tracing back the directions of the gamma rays gives the location of the annihilation, which is very close to that of the emitting nucleus. In this way, a map of the concentration of the radionuclide can be built up. In a normal brain, metabolic activity produces a similar PET pattern in each side. Here the irregular appearance of the scan indicates that brain tissue has degenerated.

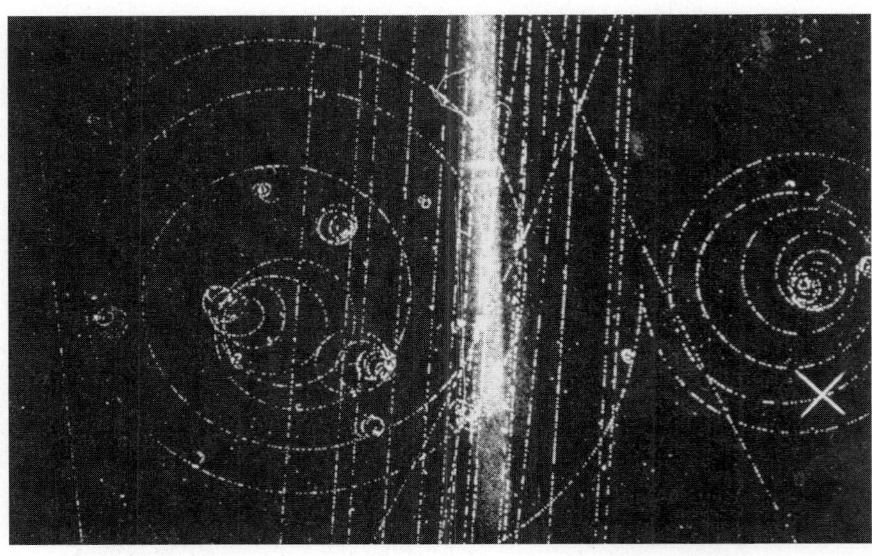

Figure 8-27 A bubble chamber contains liquid hydrogen under high pressure. When the pressure is suddenly released, tiny bubbles form along tracks of fast electrically charged particles. This photograph shows an electron-positron pair that formed in a bubble chamber in which there was a magnetic field perpendicular to the page. The field deflected the electron and positron into oppositely curved paths. These paths are spirals because the particles lost energy as they moved through the hydrogen in the chamber and so were deflected more and more by the field.

pair is nearly 2 GeV. The antiparticles formed in pair production exist for only a short time before they meet up with their particle counterparts in ordinary matter and are annihilated.

8.14 Fundamental Interactions

Only Four Give Rise to All Physical Processes

Elementary particles interact with each other in only four ways. These fundamental interactions seem able to account for all the physical processes and structures in the universe on all scales of size from atomic nuclei to galaxies of stars (Fig. 8-28). In order of decreasing strength these interactions are

1. The **strong interaction,** which holds protons and neutrons together to form atomic nuclei despite the mutual repulsion of the protons. The forces produced by this interaction have short ranges, only about 10^{-15} m, which is why nuclei are limited in size. Because the strong interaction is what its name suggests, nuclear binding energies are high. Electrons are not affected by the strong interaction.
2. The **electromagnetic interaction,** which gives rise to electric and magnetic forces between charged particles. This interaction is responsible for the structures of atoms, molecules, liquids, and solids. The force exerted when a bat hits a ball is electromagnetic. Although the electromagnetic interaction is about 100 times weaker than the strong interaction at short distances, electromagnetic forces are unlimited in range and, unlike strong forces, act on electrons.
3. The **weak interaction,** which affects all particles. By causing beta decay this interaction helps determine the compositions of atomic

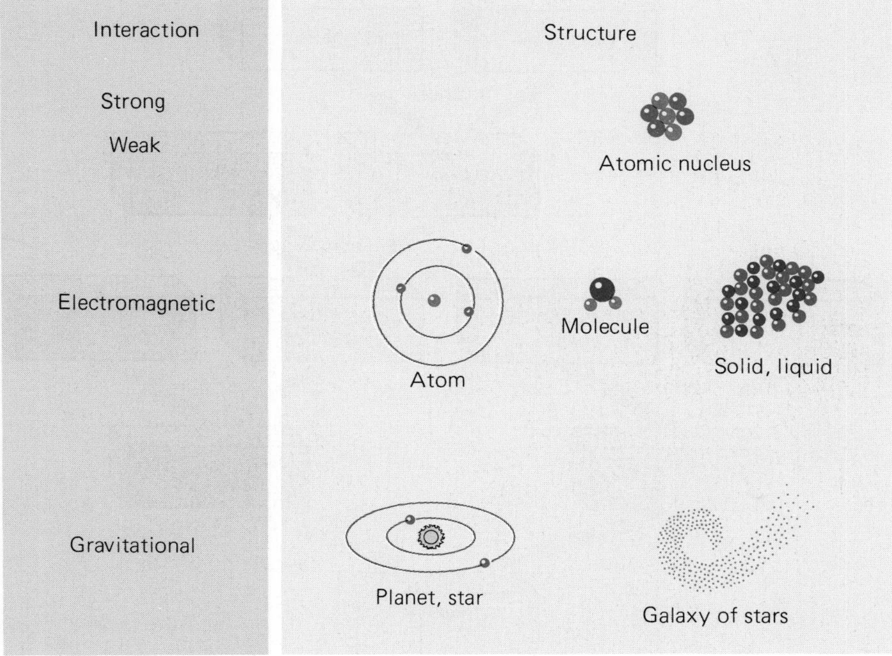

Figure 8-28 The four fundamental interactions determine how matter comes together to form the characteristic structures of the universe.

nuclei. The range of the weak interaction is even shorter than that of the strong interaction, and 10 trillion times less powerful.

4. The **gravitational interaction,** which is responsible for the attractive force one mass exerts on another. Because the strong and weak forces are severely limited in range and because matter in bulk is electrically neutral, the gravitational interaction dominates on a large scale. Gravitation is what pulls matter together into the planets, stars, and galaxies that populate space. This interaction is nevertheless extremely feeble on a small scale; the gravitational pull of one electron on another is 10^{43} times weaker than their electric repulsion.

Before Newton, it was not clear that the gravity that pulls things down to the earth—which we might call terrestrial gravity—is the same as the gravity that holds the planets in their orbits around the sun. One of Newton's great accomplishments was to show that both terrestrial and astronomical gravity have the same nature. Another notable unification was made by Maxwell when he demonstrated that electric and magnetic forces can both be traced to a single interaction between charged particles.

Unifying the Interactions What about the four fundamental interactions listed above? Are they all truly fundamental or are any of them, too, related in some way?

Studies made independently by Steven Weinberg and Abdus Salam in the 1960s indicated that the weak and electromagnetic interactions are really different aspects of the same basic phenomenon, a conclusion supported by experiment (Fig. 8-29). Later work has linked the electroweak

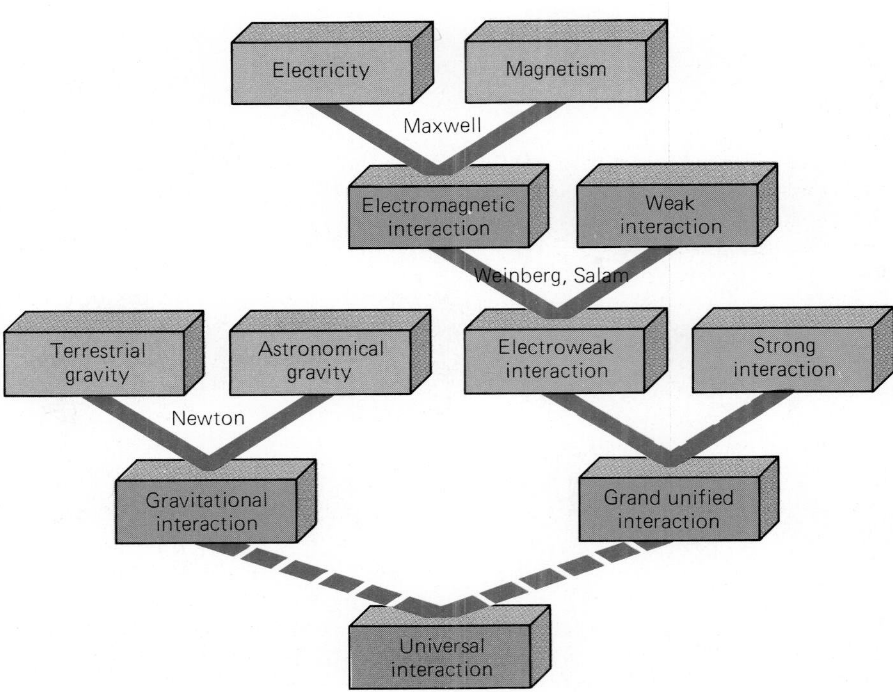

Figure 8-29 One of the goals of physics is a single theoretical picture that unites all the ways in which particles of matter interact with each other. Much progress has been made, but the task is not finished.

and strong interactions as part of the current, very successful Standard Model of elementary particles. One of the merits of this model is that it can explain why the proton and the electron, which are very different kinds of particle (as discussed in the next section), have electric charges of exactly the same size.

What about gravitation? The final step in understanding how nature works is a single theory that ties together all the particles and interactions that are known—a "theory of everything." There are hints that this supreme goal is not beyond reach.

8.15 Leptons and Hadrons

Ultimate Matter

All elementary particles fall into two broad categories that depend on their response to the strong interaction. **Leptons** (Greek for "light" or "swift") are not affected by this interaction and seem to be point particles with no size or internal structure. The electron is a lepton. **Hadrons**

New Accelerators

The recently completed Large Hadron Collider at the CERN laboratory on the border between France and Switzerland is the most powerful particle accelerator in the world (Fig. 8-30); it uses 120 MW in its operation, as much as a small city. Giant superconducting magnets keep beams of lead ions circulating in opposite directions in a ring-shaped tube 27 km around as electric fields bring the energies of the protons in the ions up to 7 TeV each. Millions of ions per second are then brought together in head-on collisions that give rise to a variety of elementary particles for study. One of the particles the machine was designed to produce is the long-sought Higgs boson, which current theories suggest is the key to understanding the nature of mass. Hitherto unknown particles and perhaps new physical laws that apply only at high energies, such as those found in the early universe, may be found as well. No new particles have been discovered since 1995—an unusually long gap in this branch of physics.

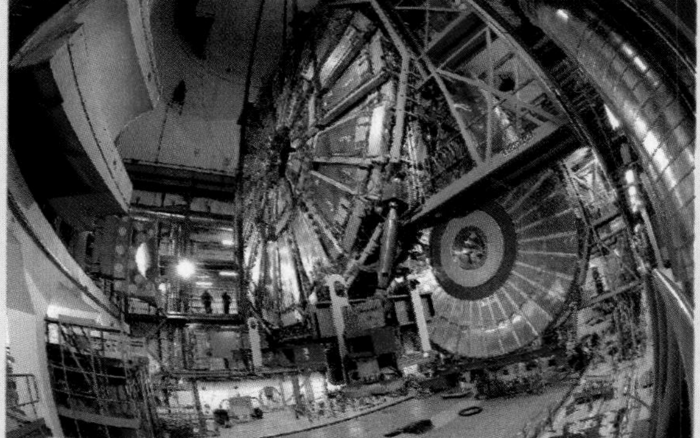

Figure 8-30 One of the four particle detectors, shown under construction, of the Large Hadron Collider at CERN, which is pushing back the frontier of knowledge. Seven thousand magnets, their superconducting windings cooled by liquid helium, guide and focus two beams of 7 TeV protons (the most energetic yet accelerated on the earth) around the 27-km circumference of the collider. At four points on the ring the beams cross to give over 600 million collisions per second. The particles produced in these events are tracked and measured by the detectors; the raw data would fill hundreds of thousands of CDs every second, but only the information about the 100 most interesting events each second is kept for analysis.

Planning has already begun on the next major accelerator, the International Linear Collider, which will be 31 km long and will smash together beams of high-energy electrons and positrons. Collaborating on the design of the projected machine, which is expected to cost \$8 billion and be completed possibly in 2019, are about 1000 scientists and engineers from 100 countries.

(Greek for "heavy" or "strong") are subject to the strong interaction and have definite sizes—they are about 10^{-15} m across—and internal structures. The proton and neutron are hadrons.

A very interesting lepton is the **neutrino,** which has no charge and very little mass (a millionth of the electron's mass, so little that until recently it was thought to have no mass at all). The neutrino is associated with the weak interaction. Whenever a nucleus undergoes beta decay, a neutrino as well as an electron (or positron) is emitted. A neutrino can pass through vast amounts of matter—over 100 *light-years* of solid iron on the average—before interacting. (A light-year is the distance light travels in empty space in a year.) A vast number of neutrinos are produced in the sun in the course of the nuclear reactions that occur within it, and these neutrinos carry into space 6 to 8 percent of all the energy the sun generates. About 65 billion neutrinos, mostly from the sun, pass through each square centimeter (the area of a fingernail) of your body per second. The energy the neutrinos from the sun and other stars carry is apparently lost forever in the sense that it cannot be changed into any other form. Neutrinos outnumber protons in the universe by about a billion to one, but they represent less than 1 percent of the total energy of the universe.

Besides the proton and the neutron, the hadron family includes several hundred particles with extremely short lifetimes, less than a billionth of a second for some. These particles seem to play no role in the behavior of ordinary matter. They decay in various ways, often in a series of steps, and usually end up as protons, neutrons, or electrons; a few become gamma rays. Some of these decays involve the emission of neutrinos that are different in certain respects (mass, for one thing) than those emitted in the beta decays of atomic nuclei.

Quarks The discovery that hadrons have internal structures was made with the help of experiments in which fast electrons were scattered by collisions with protons and neutrons. (We recall that the internal structure of the atom was revealed by the similar Rutherford experiment that used alpha particles as probes.) The particles that make up hadrons are the **quarks** mentioned earlier.

Only six kinds of quark are needed to account for all known hadrons. Those hadrons that are lighter than the proton consist of a quark and an antiquark. The proton, neutron, and heavier hadrons consist of three quarks. This is a welcome simplification, but quarks turn out to have two unprecedented properties. The first is that, unlike any other particle known, their electric charge is less than $\pm e$. Some quarks have a charge of $\pm\frac{1}{3}e$; others have a charge of $\pm\frac{2}{3}e$.

The second unusual aspect of quarks is that they do not seem able to exist outside of hadrons. No quark has ever been found by itself, even in experiments that ought to have been able to set them free. Thus there is no direct way to confirm that quarks have fractional charges, or even that they really exist.

Nevertheless there is a great deal of indirect evidence strongly in favor of quarks. For instance, every known hadron matches up with a particular arrangement of quarks, and predictions of hitherto unknown hadrons made on the basis of the quark model have turned out correct. Furthermore, a theory of the strong interaction based on quarks has been quite successful. This is the theory mentioned in the previous section that

A PHYSICIST AT WORK

Michelangelo D'Agostino
University of California, Berkeley, California

Starting your day at 10 P.M. is never easy. Slipping on several layers of long underwear, overalls, and a bulky red parka, I walk out the door for another day (or technically, another night) in the harshest physics lab on earth. I'm here at the geographic South Pole with 250 or so other scientists and staff. We're taking advantage of the 24-hour sunlight during the summer season to work on a wide variety of experiments spanning the fields of astrophysics, particle physics, seismology, climatology, and atmospheric science.

Antarctica is like no other place on the planet. In the summer, temperatures can reach −50°F. A two-mile thick icesheet stretches as far as the eye can see in all directions. A coworker once compared being here at the pole to standing on a piece of paper. Since we live on top of that icesheet, the altitude makes it difficult to breathe, and the air is so cold that it holds less moisture than the Sahara. All passengers and supplies are flown in from New Zealand by U.S. Air National Guard cargo planes that land on skis in the soft snow. With the extreme isolation (we have internet and phone access for only 8 hours a day and no television at all), working here is quite a challenge.

As a graduate student, I'm lucky enough to have the opportunity to tackle these challenges because I work on the IceCube experiment. IceCube is a giant particle detector being constructed under the ice here at the South Pole. In a way, it's a new, strange type of telescope. Instead of detecting light as a normal telescope does, IceCube is trying to detect another kind of fundamental particle that may come from astrophysical objects—the ghostly neutrino. Neutrinos are very light, electrically neutral particles that interact extremely weakly with normal matter. Trillions of them pass straight through your body every second like light through a piece of glass. For this reason, neutrinos should make good astronomical messengers. While light can be obscured by dust or other matter that blocks our view, neutrinos should travel unimpeded from their source directly to us. For this same reason though, neutrinos are difficult to detect. Doing so requires a giant target—a target so large that it can't be contained inside of a normal laboratory.

That's where the South Pole comes in. When IceCube is completed, it will monitor a very special target—a cubic kilometer chunk of crystal-clear glacial ice. Using hot water, we melt holes into the ice to install very sensitive light detectors. Occasionally, a neutrino passing through half of the earth will hit a molecule of the ice, giving off a faint blue flash of light. By checking which of our sensors saw this light and comparing the timing, we can "connect the dots" to tell where the neutrino came from within the universe. By studying such neutrinos, we hope to learn about the properties of the extremely violent astrophysical objects that we think may produce them. These include explosions from dying stars (supernovas and gamma ray bursts) and eruptions of hot material from the neighborhood of giant black holes at the center of galaxies.

Doing research here is completely unlike studying physics in college, though the problem-solving skills I learned there continue to serve me well. My experiment involves a multitude of different tasks that require a variety of skills. One day, I may find myself shoveling snow or carrying crates from place to place, doing manual labor to help those who melt the holes in which we install our instruments. Another day, I may end up writing computer programs to analyze the volumes of data that tell us how our detector is performing. On still another day, I may work to design or fix electronics. In all these jobs, I work closely with fascinating people from all over the world. That is why I find this kind of physics so fulfilling—working in a team to solve difficult problems that may yield answers to some of our fundamental questions about the universe.

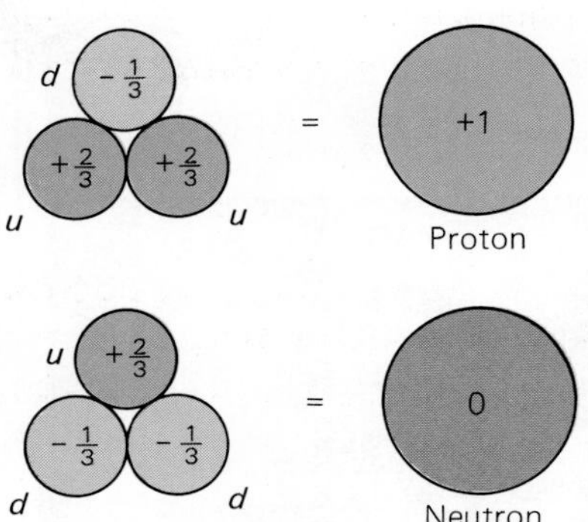

Figure 8-31 Quark models of the proton and neutron. Electric charges are given in units of *e*.

has been linked to the theory of the electromagnetic-weak interaction to make a unified picture that accounts for all aspects of the behavior of matter except—thus far—gravitation.

The two quarks that make up the proton and neutron are called *u* (charge $+\frac{2}{3}e$) and *d* (charge $-\frac{1}{3}e$). As shown in Fig. 8-31, a proton consists of one *d* quark and two *u* quarks, and a neutron consists of one *u* quark and two *d* quarks. Thus all the properties of ordinary matter can be understood on the basis of just two leptons, the electron and the neutrino, and two quarks, *u* and *d*. Considering how diverse these properties are, this is an astonishing achievement. The other leptons and quarks are connected only with unstable particles created in high-energy collisions and seem to have nothing to do with ordinary matter. Sensitive experiments, backed up by otherwise successful theories, suggest that leptons and quarks represent the limit of simplification and are not themselves composed of still other, more fundamental particles.

Important Terms and Ideas

An **element** is a substance all of whose atoms have the same number of protons in their nuclei. This number is the **atomic number** of the element and equals the number of electrons that surround each nucleus of the element's atoms. The **isotopes** of an element have different numbers of neutrons in their nuclei. A **nucleon** is a neutron or proton; the **mass number** of a nucleus is the number of nucleons it contains. A **nuclide** is an atom whose nucleus has particular atomic and mass numbers.

In **radioactive decay,** certain atomic nuclei spontaneously emit **alpha particles** (helium nuclei), **beta particles** (electrons), or **gamma rays** (high-frequency electromagnetic waves). A **positron** is a positively charged electron emitted in some beta decays.

The **half-life** of a radionuclide is the time needed for half of an original sample to decay.

The mass of every nucleus is slightly less than the total mass of the same number of free neutrons and protons. The **binding energy** of a nucleus is the energy equivalent of the missing mass and must be supplied to the nucleus to break it up. Nuclei of intermediate size have the highest **binding energies per nucleon.** Hence the **fusion** of light nuclei to form heavier ones and the **fission** of heavy nuclei into lighter ones are both processes that liberate energy.

An **elementary particle** cannot be separated into other particles. The **antiparticle** of an elementary particle has the same mass and general behavior, but it has a charge of opposite sign and differs in certain other respects. A particle and its antiparticle can **annihilate** each other, with their masses turning entirely into energy. In the opposite process of **pair production,** a particle-antiparticle pair materializes from energy.

The four **fundamental interactions** are, in order of decreasing strength, the strong, electromagnetic, weak, and gravitational.

A **lepton** is an elementary particle that is not affected by the strong interaction and has no internal structure; the electron is a lepton. A **hadron** is an elementary particle that is affected by the strong interaction and is composed of *quarks,* particles with electric charges of $\pm\frac{1}{3}e$ or $\pm\frac{2}{3}e$ that have not been found outside hadrons as yet. Protons and neutrons are hadrons.

Multiple Choice

1. The basic idea of the Rutherford atomic model is that the positive charge in an atom is
   a. spread uniformly throughout its volume
   b. concentrated at its center
   c. readily deflected by an incoming alpha particle
   d. the same for all atoms

2. Nearly all the volume occupied by matter consists of
   a. electrons **c.** neutrons
   b. protons **d.** nothing

3. The atomic number of an element is the number of
   a. protons in its nucleus
   b. neutrons in its nucleus
   c. electrons in its nucleus
   d. protons and neutrons in its nucleus

4. The nuclei of the isotopes of an element all contain the same number of
   a. neutrons **c.** nucleons
   b. protons **d.** electrons

5. Which of the following is not an isotope of hydrogen?
   a. ${}^{1}_{0}H$
   b. ${}^{1}_{1}H$
   c. ${}^{2}_{1}H$
   d. ${}^{3}_{1}H$
6. The chemical behavior of an atom is determined by its
   a. atomic number
   b. mass number
   c. binding energy
   d. number of isotopes
7. An alpha particle consists of
   a. two protons
   b. two protons and two electrons
   c. two protons and two neutrons
   d. four protons
8. An electron is emitted by an atomic nucleus in the process of
   a. alpha decay
   b. beta decay
   c. gamma decay
   d. nuclear fission
9. Gamma rays have the same basic nature as
   a. alpha particles
   b. beta particles
   c. x-rays
   d. sound waves
10. Radioactive materials do not emit
    a. electrons
    b. protons
    c. alpha particles
    d. gamma rays
11. Which of these types of radiation has the least ability to penetrate matter?
    a. alpha particles
    b. beta particles
    c. gamma rays
    d. x-rays
12. Which of these types of radiation has the greatest ability to penetrate matter?
    a. alpha particles
    b. beta particles
    c. gamma rays
    d. x-rays
13. When a nucleus undergoes radioactive decay, the number of nucleons it contains afterward is
    a. always less than the original number
    b. always more than the original number
    c. never less than the original number
    d. never more than the original number
14. Which of these particles is radioactive?
    a. electron
    b. proton
    c. neutron
    d. alpha particle
15. The largest amount of radiation received by an average person in the United States comes from
    a. medical x-rays
    b. nuclear reactors
    c. fallout from past weapons tests
    d. natural sources
16. The half-life of a radionuclide is
    a. half the time needed for a sample to decay entirely
    b. half the time a sample can be kept before it begins to decay
    c. the time needed for half a sample to decay
    d. the time needed for the remainder of a sample to decay after half of it has already decayed
17. As a sample of a radionuclide decays, its half-life
    a. decreases
    b. remains the same
    c. increases
    d. any of these, depending upon the nuclide
18. In a stable nucleus other than ${}^{1}_{1}H$ the number of neutrons is always
    a. less than the number of protons
    b. less than or equal to the number of protons
    c. equal to or more than the number of protons
    d. more than the number of protons
19. The electronvolt is a unit of
    a. charge
    b. potential difference
    c. energy
    d. momentum
20. Relative to the sum of the masses of its constituent particles, the mass of an atom is
    a. greater
    b. the same
    c. smaller
    d. any of these, depending on the element
21. The binding energy per nucleon is
    a. the same for all nuclei
    b. greater for very small nuclei
    c. greatest for nuclei of intermediate size
    d. greatest for very large nuclei
22. The splitting of an atomic nucleus, such as that of ${}^{235}U$, into two or more fragments is called
    a. fusion
    b. fission
    c. a chain reaction
    d. beta decay
23. In a chain reaction
    a. protons and neutrons join to form atomic nuclei
    b. light nuclei join to form heavy ones
    c. neutrons emitted during the fission of heavy nuclei induce fissions in other nuclei
    d. uranium is burned in a type of furnace called a reactor
24. Enriched uranium is a better fuel for nuclear reactors than natural uranium because enriched uranium has a greater proportion of
    a. slow neutrons
    b. deuterium
    c. plutonium
    d. ${}^{235}U$

25. In a nuclear power plant the nuclear reactor itself is used as a source of
a. neutrons **c.** radioactivity
b. heat **d.** electricity

26. Fusion reactions on the earth are likely to use as fuel
a. ordinary hydrogen **c.** plutonium
b. deuterium **d.** uranium

27. Of the following particles, the one that is not an elementary particle is the
a. alpha particle **c.** neutron
b. beta particle **d.** neutrino

28. An example of a particle-antiparticle pair is the
a. proton and positron
b. proton and neutron
c. neutron and neutrino
d. electron and positron

29. Atomic nuclei are stable despite the mutual repulsion of the protons they contain because of the action of the
a. gravitational interaction
b. electromagnetic interaction
c. weak interaction
d. strong interaction

30. The weakest of the four fundamental interactions is the
a. gravitational interaction
b. electromagnetic interaction
c. strong interaction
d. weak interaction

31. The mass of the neutrino is
a. equal to that of the neutron
b. equal to that of the electron
c. equal to that of a quark
d. very small

32. Quarks are particles that
a. have no mass
b. have charges whose magnitudes are less than e
c. decay into protons
d. decay into neutrinos

33. A particle that is believed to consist of quarks is the
a. electron **c.** neutron
b. positron **d.** neutrino

34. The number of protons in a nucleus of the boron isotope $^{11}_{5}B$ is
a. 5 **c.** 11
b. 6 **d.** 16

35. The number of neutrons in a nucleus of the potassium nucleus $^{40}_{19}K$ is
a. 19 **c.** 40
b. 21 **d.** 59

36. When the nitrogen isotope $^{13}_{7}N$ decays into the carbon isotope $^{13}_{6}C$, it emits
a. a gamma ray **c.** a positron
b. an electron **d.** an alpha particle

37. The product of the alpha decay of the bismuth isotope $^{214}_{83}Bi$ is
a. $^{210}_{79}Au$ **c.** $^{210}_{83}Bi$
b. $^{210}_{81}Tl$ **d.** $^{218}_{85}At$

38. The product of the gamma decay of the aluminum isotope $^{27}_{13}Al$ is
a. $^{27}_{12}Mg$
b. $^{26}_{13}Al$
c. $^{27}_{13}Al$
d. $^{27}_{14}Si$

39. After 2 h has elapsed, one-sixteenth of the original quantity of a certain radioactive substance remains undecayed. The half-life of this substance is
a. 15 min **c.** 45 min
b. 30 min **d.** 60 min

40. The half-life of tritium is 12.5 years. If we start out with 1 g of tritium, after 25 years there will be
a. no tritium left
b. $\frac{1}{4}$ g of tritium left
c. $\frac{1}{2}$ g of tritium left
d. a total of 4 g of tritium

Exercises

8.1 Rutherford Model of the Atom

8.2 Nuclear Structure

1. How do the ways in which the mass and the charge of an atom are distributed differ?

2. Alpha particle tracks through gases and thin metal foils show few deflections. What does this tell us about the atom?

3. What are the similarities and differences among the isotopes of an element?

4. Find the number of neutrons and protons in each of the following nuclei: $^{6}_{3}Li$; $^{13}_{6}C$; $^{31}_{15}C$; $^{94}_{40}Zr$.

5. Find the number of neutrons and protons in each of the following nuclei: $^{18}_{8}O$; $^{26}_{12}Mg$; $^{57}_{26}Fe$; $^{109}_{47}Ag$.

8.3 Radioactive Decay

6. The following statements were thought to be correct in the nineteenth century. Which of them are now known to be incorrect? For those that are incorrect, indicate why the statement is wrong and modify it to be in accordance with modern views. (a) Energy can be neither created nor destroyed. (b) The acceleration of an object is proportional to the force applied to it and inversely proportional to its mass. (c) Atoms are indivisible and indestructible. (d) All atoms of a particular element are identical.

7. What limits the size of a nucleus?

8. How does the number of neutrons in a stable nucleus compare with the number of protons? Why is this?

9. (a) What is an alpha particle? A beta particle? A gamma ray? (b) How do they compare in general in ability to penetrate matter?

10. Radium spontaneously decays into helium and radon. Why do you think radium is regarded as an element rather than as a chemical compound of helium and radon in the way that water, for example, is considered a chemical compound of hydrogen and oxygen?

11. What happens to the atomic number and mass number of a nucleus when it emits an alpha particle?

12. What happens to the atomic number and mass number of a nucleus when it emits (a) an electron? (b) a positron? (c) a gamma ray?

13. (a) Under what circumstances does a nucleus emit an electron? A positron? (b) The oxygen nuclei $^{14}_{8}O$ and $^{19}_{8}O$ both undergo beta decay to become stable nuclei. Which would you expect to emit a positron and which an electron?

14. The carbon isotope $^{14}_{6}C$ decays into the nitrogen isotope $^{14}_{7}N$. What kind of particle is emitted in the decay?

15. The polonium isotope $^{210}_{84}Po$ undergoes alpha decay to become an isotope of lead. Find the atomic number and mass number of this isotope.

16. The helium isotope $^{6}_{2}He$ is unstable. What kind of decay would you expect it to undergo? What would the resulting nuclide be?

17. The thorium nucleus $^{233}_{90}Th$ undergoes two successive negative beta decays. Find the atomic number, mass number, and chemical name of the resulting nucleus.

18. A $^{64}_{29}Cu$ nucleus can decay by emitting an electron or a positron and also by capturing an electron. What is the final nucleus in each case?

19. The uranium isotope $^{235}_{92}U$ decays into a lead isotope by emitting seven alpha particles and four electrons. What is the symbol of the lead isotope?

20. A reaction often used to detect neutrons occurs when a neutron is absorbed by a $^{10}_{5}B$ boron nucleus, which then emits an alpha particle. What are the atomic number, mass number, and chemical name of the remaining nucleus?

8.4 Half-Life

21. What happens to the half-life of a radionuclide as it decays?

22. If the half-life of a radionuclide is 1 month, is a sample of it completely decayed after 2 months?

23. After 10 years, 75 g of an original sample of 100 g of a certain radionuclide has decayed. What is the half-life of the nuclide?

24. One-eighth of a sample of $^{30}_{15}P$ remains undecayed after 7.5 min. What is the half-life of this phosphorus isotope?

25. If 1 kg of radium (half-life = 1600 years) is sealed into a container, how much of it will remain as radium after 1600 years? after 4800 years? If the container is opened after a period of time, what gases would you expect to find inside it?

8.6 Units of Mass and Energy

26. When the radium isotope $^{226}_{88}Ra$ undergoes alpha decay, the energy liberated is 4.87 MeV. (a) Identify the resulting nuclide. (b) The alpha particle has a KE of 4.78 MeV. Where do you think the other 0.09 MeV goes?

27. Find the kinetic energy (in eV) of an electron whose speed is 10^6 m/s.

28. Find the kinetic energy (in keV) of a $^{14}_{7}N$ atom of mass 14.0 u whose speed is 10^6 m/s.

29. Find the speed of an electron whose kinetic energy is 26 eV.

30. Find the speed of a proton whose kinetic energy is 100 eV.

8.7 Binding Energy

8.8 Binding Energy per Nucleon

31. How does the energy needed to remove an electron from an atom compare with the energy needed to remove a proton from its nucleus?

32. Why is the $^{56}_{26}Fe$ nucleus the most stable (that is, the most difficult to break apart) nucleus?

33. What property of atomic nuclei makes it possible for nuclear fission and fusion to give off energy?

34. Atomic mass always refers to the mass of a neutral atom, not the mass of its bare nucleus. With this definition in mind, determine by how much the mass of a parent atom changes when its nucleus emits (a) an electron and (b) a positron. Ignore the kinetic energy of the emitted particle.

35. The binding energy per nucleon in the iron nucleus $^{56}_{26}Fe$ is 8.8 MeV. Find its atomic mass.

36. The binding energy of ${}^{24}_{12}\text{Mg}$ is 198 MeV. Find its atomic mass.

37. The mass of ${}^{4}_{2}\text{He}$ is 4.0026 u. Find its binding energy and binding energy per nucleon.

38. The binding energy per nucleon in the chlorine isotope ${}^{35}_{17}\text{CI}$ is 8.5 MeV. What is its atomic mass?

39. The neutron decays in free space into a proton and an electron after an average lifetime of 15 min. What must be the minimum binding energy contributed by a neutron to a nucleus in order that the neutron not decay inside the nucleus? How does this figure compare with the observed binding energies per nucleon in stable nuclei?

8.9 Nuclear Fission
8.10 How a Reactor Works
8.11 Plutonium

40. Why can ordinary uranium not be used to fuel a reactor cooled by ordinary water?

41. What is the function of the moderator in a uranium-fueled nuclear reactor?

42. What fuel other than uranium can be used in a nuclear reactor?

43. (a) How much mass is lost per day by a nuclear reactor operated at a 1.0-GW power level? (b) If each fission releases 200 MeV, how many fissions occur per second to give this power level?

44. ${}^{235}_{92}\text{U}$ loses about 0.1 percent of its mass when it undergoes fission. (a) How much energy is released when 1 kg of ${}^{235}_{92}\text{U}$ undergoes fission? (b) A ton of TNT releases about 9×10^9 J when it explodes. How many tons of TNT are equivalent in destructive power to a bomb that contains 1 kg of ${}^{235}_{92}\text{U}$?

8.12 Nuclear Fusion

45. What are the differences and similarities between fusion and fission?

46. Old stars obtain part of their energy by the fusion of three alpha particles to form a ${}^{12}_{6}\text{C}$ nucleus, whose mass is 12.0000 u. How much energy is given off in each such reaction?

8.13 Antiparticles
8.14 Fundamental Interactions
8.15 Leptons and Hadrons

47. What distinguishes a charged particle from its antiparticle? What happens when they come together?

48. (a) Could a gamma ray energetic enough to materialize into a proton-antiproton pair alternatively materialize into a neutron-antineutron pair? (b) Could a gamma ray energetic enough to materialize into a neutron-antineutron pair alternatively materialize into a proton-antiproton pair? Explain.

49. Suppose the strong interaction did not exist, so there were no nuclear binding energies. If the early universe contained protons, neutrons, and electrons, what kind or kinds of matter would eventually fill the universe?

50. The gravitational interaction alone governs the motions of the planets around the sun. Why are the other fundamental interactions not significant in planetary motion?

51. Discuss the similarities and differences between the neutron and the neutrino.

52. Why can neutrinos travel immense distances through matter whereas other elementary particles cannot?

53. Leptons and hadrons are the two classes of basic particle. How do they differ?

54. Which constituents of an atom consist of quarks and which do not?

55. No particle of fractional charge has yet been observed. If none is found in the future either, does this necessarily mean that the quark hypothesis is wrong?

56. Would you expect the gravitational attractive force between two protons in a nucleus to counterbalance their electrical repulsion? Calculate the ratio between the electric and gravitational forces acting between two protons. Does this ratio depend upon how far apart the protons are?

Answers to Multiple Choice

1. b	7. c	13. d	19. c	25. b	31. d	37. b
2. d	8. b	14. c	20. c	26. b	32. b	38. c
3. a	9. c	15. d	21. c	27. a	33. c	39. b
4. b	10. b	16. c	22. b	28. d	34. a	40. b
5. a	11. a	17. b	23. c	29. d	35. b	
6. a	12. c	18. c	24. d	30. a	36. c	

9

The Atom

Laser experiment.

Goals

When you have finished this chapter you should be able to complete the goals ▶ given for each section below:

Quantum Theory of Light

9.1 Photoelectric Effect
How Can Electrons Be Set Free from Atoms by Light?
- Describe the photoelectric effect and discuss why the wave theory of light cannot account for it.

9.2 Photons
Particles of Light
- Explain how the quantum theory of light accounts for the photoelectric effect in terms of photons.

9.3 What Is Light?
Both Wave and Particle
- Compare the quantum and wave theories of light and discuss why both are needed.

9.4 X-Rays
High-Energy Photons
- Describe x-rays and interpret their production in terms of the quantum theory of light.

Matter Waves

9.5 De Broglie Waves
Matter Waves Are Significant Only in the Atomic World

9.6 Waves of What?
Waves of Probability
- Discuss what is meant by the matter wave of a moving particle.

9.7 Uncertainty Principle
We Cannot Know the Future Because We Cannot Know the Present
- State the uncertainty principle and interpret it in terms of matter waves.

The Hydrogen Atom

9.8 Atomic Spectra
Each Element Has a Characteristic Spectrum
- Distinguish between emission and absorption spectra and describe what is meant by a spectral series.

9.9 The Bohr Model
Only Certain Electron Energies Are Possible in an Atom

9.10 Electron Waves and Orbits
Standing Waves in the Atom
- Give the basic ideas of the Bohr model of the atom and show how they follow from the wave nature of moving electrons.
- Define quantum number, energy level, ground state, and excited state.
- Explain the origins of emission and absorption spectra and of spectral series.

9.11 The Laser
An Amplifier of Light That Produces Waves All in Step
- Explain how a laser works.
- List the three characteristic properties of laser light.

Quantum Theory of the Atom

9.12 Quantum Mechanics
Probabilities, not Certainties
- Compare quantum mechanics and newtonian mechanics.

9.13 Quantum Numbers
An Atomic Electron Has Four in All
- Describe what is meant by the probability cloud of an atomic electron.
- List the four quantum numbers of an atomic electron according to quantum mechanics together with the quantity each governs.

9.14 Exclusion Principle
A Different Set of Quantum Numbers for Each Electron in an Atom
- State the exclusion principle.

Every atom consists of a tiny, positively charged nucleus with negatively charged electrons some distance away. What keeps the electrons out there?

By analogy with the planets of the solar system, we might suppose that atomic electrons avoid being sucked into the nucleus by circling around it at just the right speed. This is not a bad idea, but it raises a serious problem. According to Maxwell's theory (see Chap. 6), a circling electron should lose energy all the time by giving off electromagnetic waves. Thus the electron's orbit should become smaller and smaller, and soon it should spiral into the nucleus. However, atomic electrons do not behave like this. Under ordinary conditions atoms emit no radiation, and needless to say, they never collapse.

Whenever they have been tested outside the atomic domain, the laws of motion and of electromagnetism have always agreed with experiment—yet atoms are stable. In this chapter we shall see how the strange and radical concepts of the quantum theory of light and the wave theory of moving particles are needed to understand the world of the atom.

Quantum Theory of Light

The concepts of "particle" and "wave" are clear enough to everybody. We regard a stone as a particle and the ripples in a lake as waves. A stone thrown into a lake and the ripples that spread out from where it lands seem to have in common only that both carry energy from one place to another. **Classical physics,** which refers to the physics covered in Chaps. 1 through 7, treats particles and waves as separate aspects of the reality we find in everyday life.

But the physical reality around us arises from the small-scale world of atoms and molecules, electrons and nuclei. In this world there are neither particles nor waves in our sense of these terms.

We think of electrons as particles because they have charge and mass and behave according to the laws of particle mechanics in such familiar devices as television picture tubes. However, there is plenty of evidence that makes sense only if a moving electron is a type of wave. We think of electromagnetic (em) waves as waves because they can exhibit such characteristic wave behavior as diffraction and interference. However, em waves also behave as though they consist of streams of particles. The wave-particle duality is central to an understanding of **modern physics,** which is the physics of the atomic world.

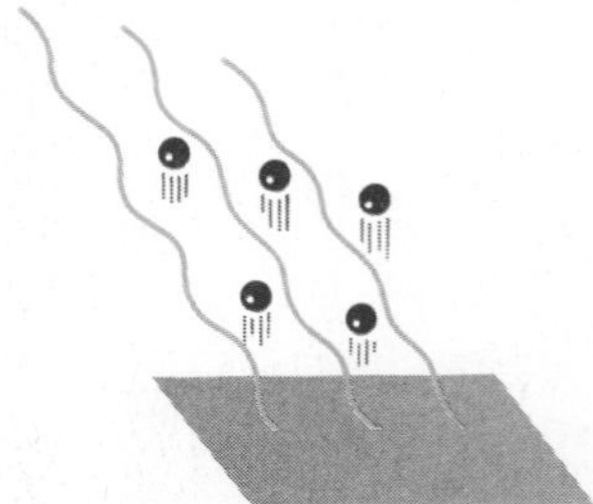

Figure 9-1 In the photoelectric effect, electrons are emitted from a metal surface when a light beam is directed on it.

9.1 Photoelectric Effect

How Can Electrons Be Set Free from Atoms by Light?

A century ago experiments showed that electrons are given off by a metal surface when light is directed onto it (Fig. 9-1). For most metals ultraviolet light is needed for this **photoelectric effect** to occur, but some metals, such as potassium and cesium, and certain other substances as well, also respond to visible light. The photosensitive screen in a digital camera, the solar cell that produces electric current when sunlight falls on it, and the television camera tube that converts the image of a scene into an electric signal are all based upon the photoelectric effect.

Since light is electromagnetic in nature and carries energy, there seems to be nothing unusual about the photoelectric effect—it should be like water waves dislodging pebbles from a beach. But three experimental findings show that no such simple explanation is possible.

1. The electrons are always emitted at once, even when a faint light is used. However, because the energy in an em wave is spread out across the wave, a certain period of time should be needed for an individual electron to gather enough energy to leave the metal. Several months ought to be needed for a really weak light beam.
2. A bright light causes more electrons to be emitted than a faint light, but the average kinetic energy of the electrons is the same. The electromagnetic theory of light, on the contrary, predicts that the stronger the light, the greater the KE of the electrons.
3. The higher the frequency of the light, the more KE the electrons have. Blue light yields faster electrons than red light (Fig. 9-2). According to the electromagnetic theory of light, the frequency should not matter.

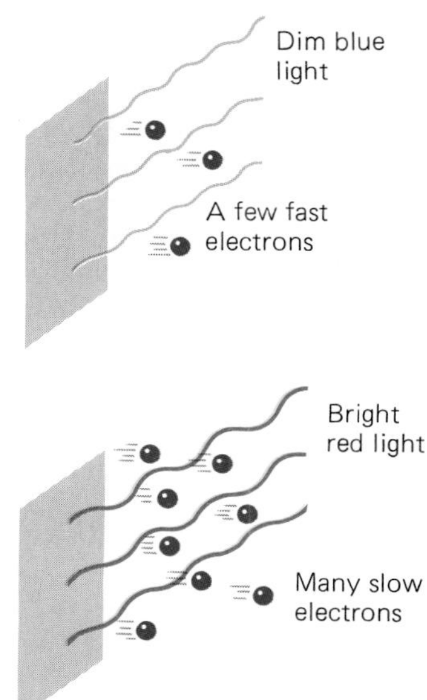

Figure 9-2 The higher the frequency of the light, the more KE the photoelectrons have. The brighter the light, the more photoelectrons are emitted. Blue light has a higher frequency than red light.

Until the discovery of the photoelectric effect, the electromagnetic theory of light had been completely successful in explaining the behavior of light. But no amount of ingenuity could bring experiment and theory together in this case. The result was the creation of the entirely new **quantum theory of light** in 1905 by Albert Einstein. The same year saw the birth of his equally revolutionary theory of relativity. All of modern physics has its roots in these two theories.

9.2 Photons

Particles of Light

Einstein proposed that light consists of tiny bursts of energy called **photons.** He began with a hypothesis suggested 5 years earlier by the German physicist Max Planck (Fig. 9-3) to account for the spectrum of the light given off by hot objects. Figure 18-4 shows how the brightness of this light varies with wavelength for objects at three different temperatures. Actually, something need not be so hot that it glows for it to radiate em waves—*all* objects radiate such waves whatever their temperatures, though which wavelength appears strongest depends on the temperature. The higher the temperature, the shorter the predominant wavelength (or, equivalently, the higher the predominant frequency); a bar of iron that glows yellow is hotter than one that glows red (Fig. 5-5). For an object at room temperature, or a person, most of the radiation is in the infrared and hence is invisible (Fig. 9-4).

Figure 9-3 Max Planck (1858–1947).

In order to explain the spectrum of emitted radiation, Planck found it necessary to assume that hot objects contribute energy in separate units, or **quanta,** to the light they give off. The higher the frequency of the light, the more the energy per quantum. All the quanta associated with a particular frequency f of light have the same energy

$$E = hf \qquad \textit{Quantum energy} \qquad 9\text{-}1$$

Quantum energy = (Planck's constant)(frequency)

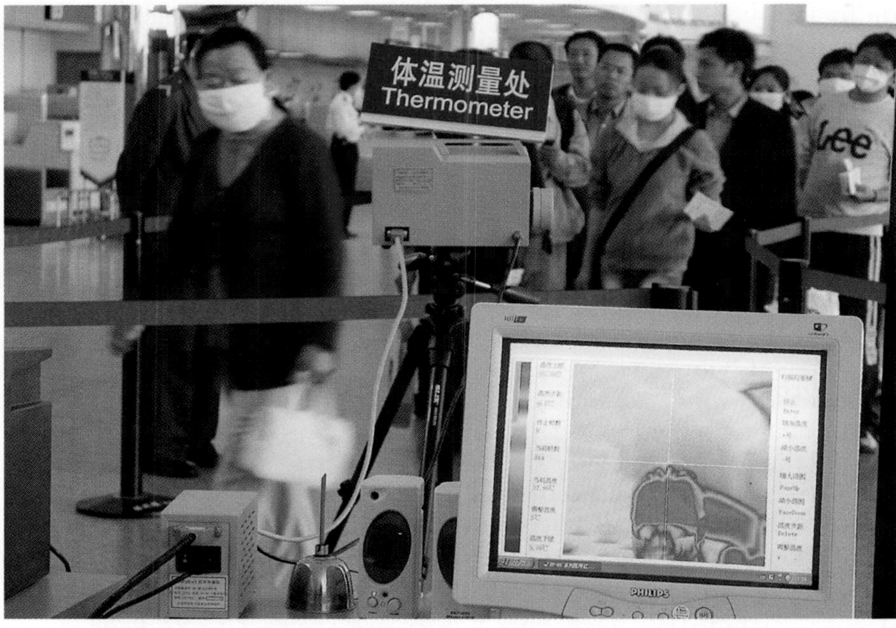

Figure 9-4 The spectrum of the electromagnetic radiation from a surface depends on the temperature of the surface. The system shown here detects people with fevers on the basis of their infrared emissions, with red indicating skin temperatures above normal. In this way people with illnesses that may be infectious can be easily identified in airports, hotels, hospitals, and other public places.

In this formula the quantity h, today known as **Planck's constant,** has the value

$$\text{Planck's constant} = h = 6.63 \times 10^{-34} \text{ joule} \cdot \text{second}$$

Planck was not happy about this assumption, which made no sense in terms of the physical theories known at that time. He took the position that, although energy apparently had to be given to the light emitted by a glowing object in small bursts, the light nevertheless traveled with its energy spread out in waves exactly as everybody thought.

Einstein's Hypothesis Einstein, however, felt that, if light is emitted in little packets, it should also travel through space and finally be absorbed in the same little packets. His idea fit the experiments on the photoelectric effect perfectly (Fig. 9-5). He supposed that some minimum energy w is needed to pull an electron away from a metal surface. If the frequency of the light is too low—so that E, the quantum energy, is less than w—no electrons can come out. When E is greater than w, a photon of light striking an electron can give the electron enough energy for it to leave the metal with a certain amount of kinetic energy (Fig. 9-6). Einstein's formula for the process is very simple:

$$hf = \text{KE} + w \qquad \textit{Photoelectric effect} \qquad 9\text{-}2$$

where hf = energy of a photon of light whose frequency is f

KE = kinetic energy of the emitted electron

w = energy needed to pull the electron from the metal

Although the photon has no mass and always moves with the speed of light, it has most of the other properties of particles—it is localized in a small region of space, it has energy and momentum, and it interacts with other particles in more or less the same way as a billiard ball interacts with other billiard balls.

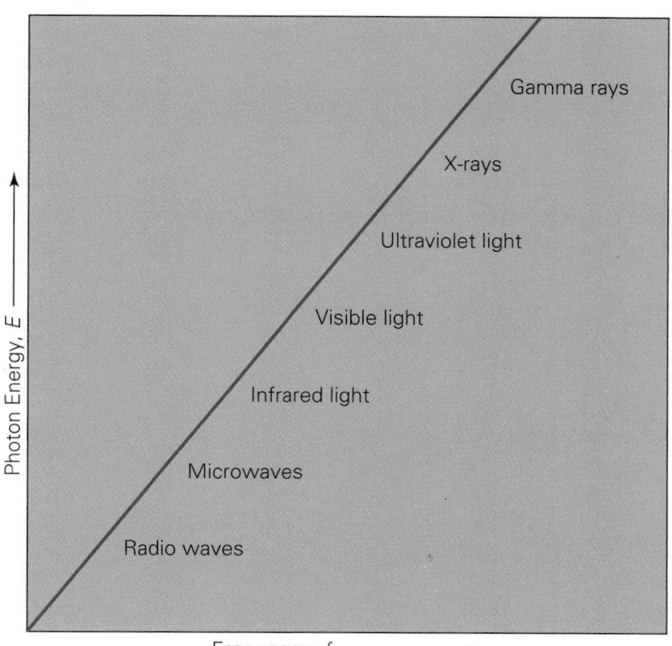

Figure 9-5 The higher the frequency of an electromagnetic wave, the greater the energy of its photons, since $E = hf$. Radio waves have the lowest frequencies, hence their photons have the least energy; gamma rays have the highest frequencies, hence their photons have the most energy. (The spacing of the categories here is not to scale.)

Figure 9-6 All light-sensitive detectors, including the eye and the one used in this digital camera, are based on the absorption of energy from photons of light by electrons in the atoms the light falls on.

Example 9.1

The average frequency of the light emitted by a 100-W lightbulb is 5.5×10^{14} Hz. How many photons per second does the lightbulb emit?

Solution

The energy of each photon is, from Eq. 9-1,

$$E = hf = (6.63 \times 10^{-34}\ \text{J}\cdot\text{s})(5.5 \times 10^{14}\ \text{Hz}) = 3.6 \times 10^{-19}\ \text{J}$$

Since 100 W = 100 J/s, the number of photons emitted per second is

$$\frac{\text{Energy/second}}{\text{energy/photon}} = \frac{100\ \text{J/s}}{3.6 \times 10^{-19}\ \text{J/photon}} = 2.8 \times 10^{20}\ \text{photons/s}$$

Such an enormous number of photons makes it impossible for us to experience light as a stream of individual particles.

9.3 What Is Light?

Both Wave and Particle

The idea that light travels as a series of little packets of energy is directly opposed to the wave theory of light. And the latter, which provides the only way to explain such optical effects as diffraction and interference, is one of the best established of physical theories. Planck's suggestion that a hot object gives energy to light in separate quanta led to no more than raised eyebrows among physicists in 1900 since it did not apparently conflict with the picture of light as a wave. Einstein's suggestion in 1905 that light travels through space in the form of distinct photons, on the other hand, astonished most of his colleagues.

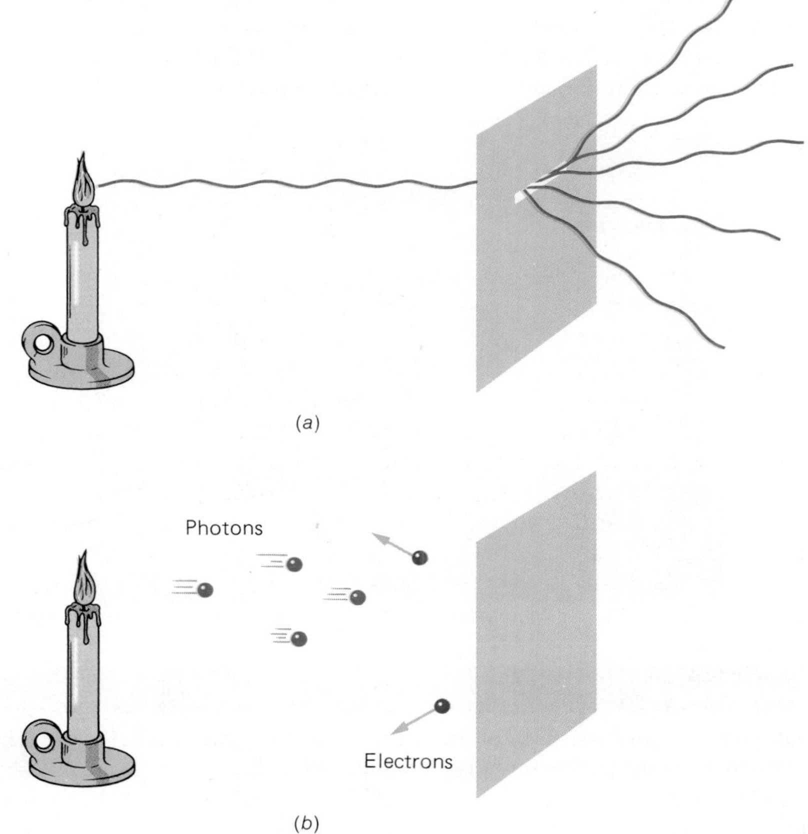

Figure 9-7 (*a*) The wave theory of light accounts for the diffraction of light into the shadow region when it passes through a narrow slit. (*b*) The quantum theory of light accounts for the photoelectric effect. Neither theory by itself can account for all aspects of the behavior of light. The two theories therefore complement each other.

Photons and Gravity

In Sec. 3.13 we saw that light is affected by gravity, which was predicted by Einstein's general theory of relativity and confirmed by measurements on starlight that passes close to the sun. This suggests that photons must be subject to gravity, which is indeed the case.

If we drop a stone, the gravitational pull of the earth accelerates the stone so that it falls faster and thus gains kinetic energy on the way to the ground. All photons travel with the speed of light and so cannot go any faster. However, a photon can manifest a gain in energy by an increase in frequency. This increase, which has been observed, is proportional to the original frequency of the photon and to the height through which it falls. For visible light the frequency change is a few hertz for a fall through 20 m.

According to wave theory, light waves spread out from a source in the way ripples spread out on the surface of a lake when a stone falls into it. The energy carried by the light in this picture is spread out through the wave pattern (Fig. 9-7). According to the quantum theory, however, light travels from a source as a series of tiny bursts of energy, each burst so small that it can be taken up by a single electron. Curiously, the quantum theory of light, which treats light as a particle phenomenon, incorporates the light frequency f, a wave concept.

Which theory are we to believe? A great many scientific ideas have had to be changed or discarded when they were found to disagree with experiment. Here, for the first time, two entirely different theories are needed to account for a single physical phenomenon.

In any particular event light exhibits *either* a wave nature or a particle nature, never both at the same time. This is an important point. The light beam that shows diffraction in passing the edge of an obstacle can also cause photoelectrons to be emitted from a metal surface, but these processes occur independently:

> The wave theory of light and the quantum theory of light complement each other.

Electromagnetic waves provide the only explanation for some experiments involving light, and photons provide the only explanation for all other experiments involving light. Light incorporates both wave and particle characters even though there is nothing in everyday life like that to help us form a mental picture of it.

Figure 9-8 Wilhelm Roentgen (1845–1923).

9.4 X-Rays

High-Energy Photons

The photoelectric effect shows that photons of light can give energy to electrons. Is the reverse process also possible? That is, can part or all of the kinetic energy of an electron be turned into a photon? As it happens, the inverse photoelectric effect not only does occur but also had been discovered (though not understood) before the work of Planck and Einstein.

In 1895, in his laboratory in Germany, Wilhelm Roentgen (Fig. 9-8) accidentally found that a screen coated with a fluorescent salt glowed every time he switched on a nearby cathode-ray tube. (A cathode-ray tube is a tube with the air pumped out in which electrons are accelerated by an electric field. A TV picture tube is a type of cathode-ray tube.) Roentgen knew that the electrons themselves could not get through the glass walls of his tube, but it was clear that some sort of invisible radiation was falling on the screen.

The radiation was very penetrating. Thick pieces of wood, glass, and even metal could be placed between tube and screen, and still the screen glowed. Soon Roentgen found that his mysterious rays would penetrate flesh and produce shadows of the bones inside. He gave them the name **x-rays** after the algebraic symbol for an unknown quantity. Roentgen refused to benefit financially from his work and died in poverty in the German inflation that followed World War I.

X-rays are given off whenever fast electrons are stopped suddenly. Figure 9-9 shows a cathode-ray tube especially designed to produce x-rays. In a television picture tube, an electron beam strikes the inside of the glass screen, which emits x-rays as a result. To prevent harm to viewers, the glass of the screen contains barium, a heavy metal that absorbs x-rays. X-rays are, of course, widely used today in medicine (Fig. 9-10) and industry (Fig. 9-11).

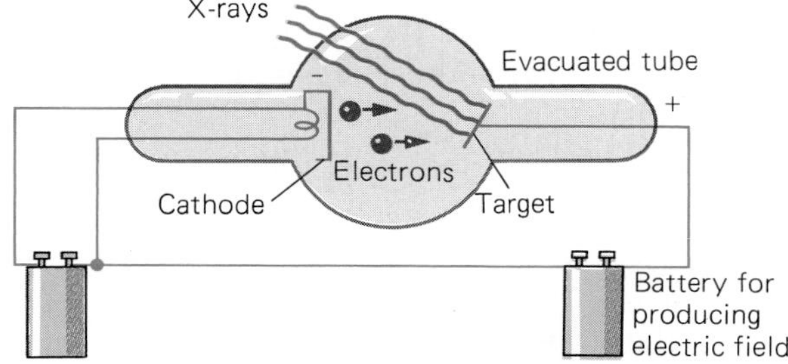

Figure 9-9 A simple x-ray tube. High-frequency electromagnetic waves called x-rays are emitted by a metal target when it is struck by fast-moving electrons. The cathode (negative electrode) is heated by a filament and emits electrons that are accelerated by an electric field between it and the positively charged target.

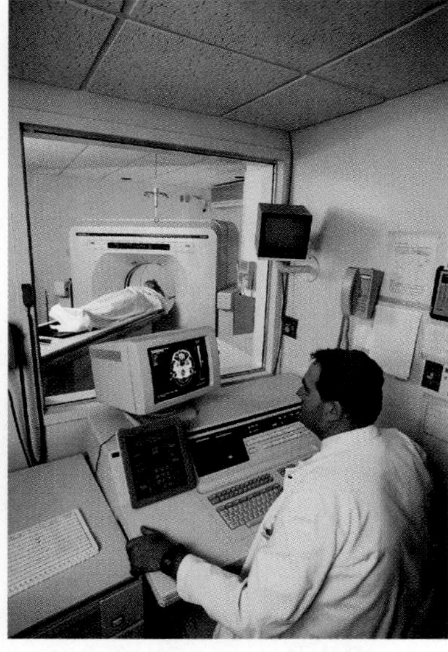

Figure 9-10 In a CT (computerized tomography) scanner, a series of x-ray exposures of a patient taken from different directions are combined by a computer to give cross-sectional images of the part of the body being examined. In effect, the tissue is sliced up by the computer on the basis of the x-ray exposures, and any desired slice can be displayed. This technique enables an abnormality to be detected and its exact location established, which might be impossible to do from an ordinary x-ray picture. The patient's head is being examined here.

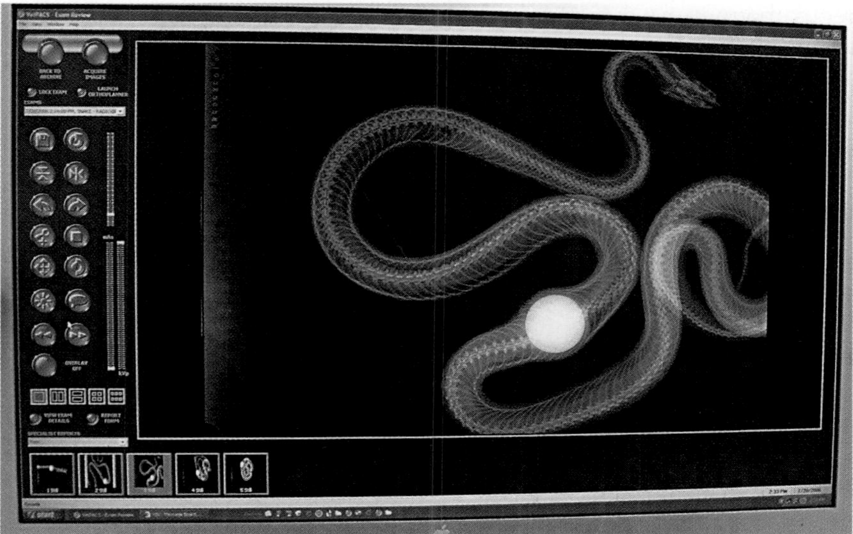

Figure 9-11 X-ray image of a yellow rat snake that had swallowed a golf ball, presumably under the impression it was an egg. The golf ball was later removed by a veterinarian.

What Are X-Rays? After many attempts had been made to determine their nature, in 1912 Max von Laue was able to show by means of an interference experiment that they are electromagnetic waves of extremely high frequency. X-ray frequencies are much higher than those of ultraviolet light, but somewhat lower than those of the gamma rays produced by radioactive atomic nuclei.

The early workers with x-rays noted that increasing the voltage applied to the tube, which means faster electrons, gave rise to x-rays of greater penetrating power. The greater the penetrating ability, the higher the x-ray frequency turned out to be. Hence high-energy electrons produce high-frequency x-rays. The more electrons in the beam, the more x-rays were produced, but their energy depended only on the electron energy.

The quantum theory of light is in complete accord with these observations. Instead of photon energy being transformed into electron KE, electron KE is transformed into photon energy. The energy of an x-ray photon of frequency f is hf, and therefore the minimum KE of the electron that produced the x-ray should be equal to hf. This prediction agrees with experimental data.

Matter Waves

As we have seen, light has both wave and particle aspects. In the topsy-turvy world of the very small, is it possible that what we normally think of as particles—electrons, for instance—have wave properties as well? So extraordinary is this question that it was not asked for two decades after Einstein's work. Soon after the question was raised came the even more extraordinary answer: yes.

BIOGRAPHY Louis de Broglie (1892–1987)

De Broglie originally pursued a career in history but later turned to physics. In his University of Paris doctoral thesis of 1924 de Broglie proposed that moving objects have wave properties that complement their particle properties: these "seemingly incompatible conceptions can each represent an aspect of the truth. . . . They may serve in turn to represent the facts without ever entering into direct conflict."

Looking back, it may seem odd that two decades passed between Einstein's 1905 discovery of the particle behavior of light waves and de Broglie's speculation. It is one thing, however, to suggest a revolutionary concept to explain otherwise mysterious data and quite another to suggest an equally revolutionary concept without such data in hand. Experiments soon showed that de Broglie's idea was correct, and it was developed by Erwin Schrödinger and others into a detailed theory, called *quantum mechanics,* that explained a wide variety of atomic phenomena. In 1929 de Broglie received the Nobel Prize.

9.5 De Broglie Waves

Matter Waves Are Significant Only in the Atomic World

In 1924 the French physicist Louis de Broglie proposed that moving objects act in some respects like waves. Reasoning by analogy with the properties of photons, he suggested that a particle of mass m and speed v behaves as though it is a wave whose wavelength is

$$\lambda = \frac{h}{mv} \qquad \textit{de Broglie waves} \qquad 9\text{-}3$$

$$\text{de Broglie wavelength} = \frac{\text{Planck's constant}}{\text{momentum}}$$

The more momentum mv a particle has, the shorter its **de Broglie wavelength** λ.

How can de Broglie's hypothesis be tested? Only waves can be diffracted and can reinforce and cancel each other by interference. A few years after de Broglie's work, experiments were performed in the United States and in England in which streams of electrons were shown to exhibit both diffraction and interference. The wavelengths of the electrons could be found from the data, and they agreed exactly with the formula $\lambda = h/mv$.

There is nothing imaginary about these **matter waves.** They are perfectly real, just as light and sound waves are. Not only electrons but also all other moving objects behave like waves. However, these waves are not necessarily evident in every situation, as Example 9.2 indicates. Only on an atomic scale are matter waves significant, and there they turn out to be the key to understanding atomic structure and behavior.

As with electromagnetic waves, the wave and particle aspects of moving bodies can never be observed at the same time. It therefore makes no sense to ask which is the "correct" description. All we can say is that in certain situations a moving body exhibits wave properties and in other situations it exhibits particle properties.

Electron Microscopes

The wave nature of moving electrons is the basis of the electron microscope. The resolving power of any optical instrument depends on the wavelength of whatever is used to illuminate the specimen being studied. In the case of a microscope that uses visible light, the highest useful magnification is about 500×. Higher magnifications give larger images but do not show more detail. Fast electrons, however, have wavelengths much shorter than those of visible light, and electron microscopes can produce useful magnifications of over 1,000,000×. X-rays also have short wavelengths, but it is not (yet?) possible to focus them adequately. In an electron microscope, an electron beam is directed at a thin specimen. Magnetic fields that act as lenses to focus the beam then produce an enlarged image of the specimen on a fluorescent screen or photographic film (Figs. 9-12 and 9-13).

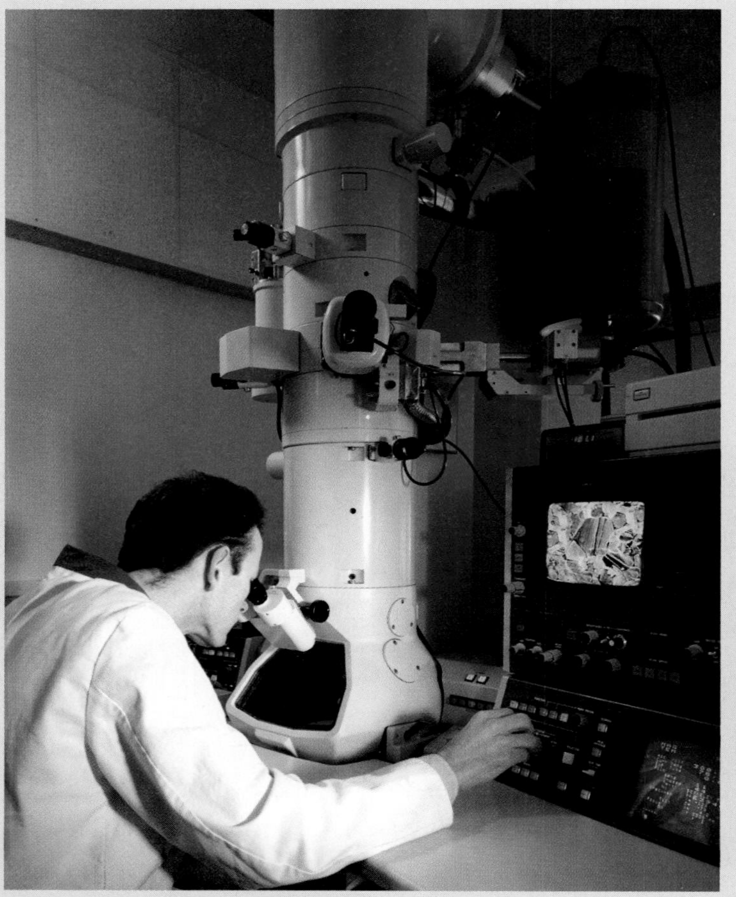

Figure 9-12 An electron microscope.

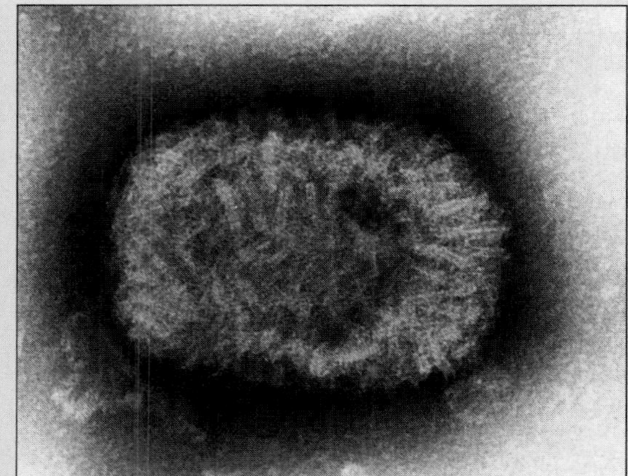

Figure 9-13 Electron micrograph of a smallpox virus particle at a magnification of 310,000×.

Example 9.2

Find the de Broglie wavelengths of (a) a 46-g golf ball whose speed is 30 m/s, and (b) an electron whose speed is 10^7 m/s.

Solution

(a) From Eq. 9-3, with $m = 46\text{ g} = 0.046\text{ kg}$,

$$\lambda = \frac{h}{mv} = \frac{6.63 \times 10^{-34}\text{ J}\cdot\text{s}}{(0.046\text{ kg})(30\text{ m/s})} = 4.8 \times 10^{-34}\text{ m}$$

(continued)

Example 9.2 Continued

The wavelength of the golf ball is so small compared with its dimensions that we would not expect to find any wave aspects in its behavior.

(b) The electron mass is 9.1×10^{-31} kg, so

$$\lambda = \frac{h}{mv} = \frac{6.63 \times 10^{-34}\ \text{J} \cdot \text{s}}{(9.1 \times 10^{-31}\ \text{kg})(10^7\ \text{m/s})} = 7.3 \times 10^{-11}\ \text{m}$$

The dimensions of atoms are comparable with this wavelength—the radius of the hydrogen atom is 5.3×10^{-11} m, for example. It is therefore not surprising that the wave character of moving electrons is so important in the world of the atom.

9.6 Waves of What?

Waves of Probability

In water waves, the quantity that varies periodically is the height of the water surface. In sound waves, it is air pressure. In light waves, electric and magnetic fields vary. What is it that varies in the case of matter waves?

The quantity whose variations make up matter waves is called the **wave function,** symbol ψ (the Greek letter *psi*). The value of ψ^2 at a given place and time for a given particle determines the probability of finding the particle there at that time. For this reason ψ^2 is called the **probability density** of the particle. A large value of ψ^2 means the strong possibility of the particle's presence; a small value of ψ^2 means its presence is unlikely.

The de Broglie waves associated with a moving particle are in the form of a group, or packet, of waves, as in Fig. 9-14. This wave packet travels with the same speed v as the particle does. Even though we cannot visualize what is meant by ψ and so cannot form a mental image of matter waves, the agreement between theory and experiment means that we must take them seriously.

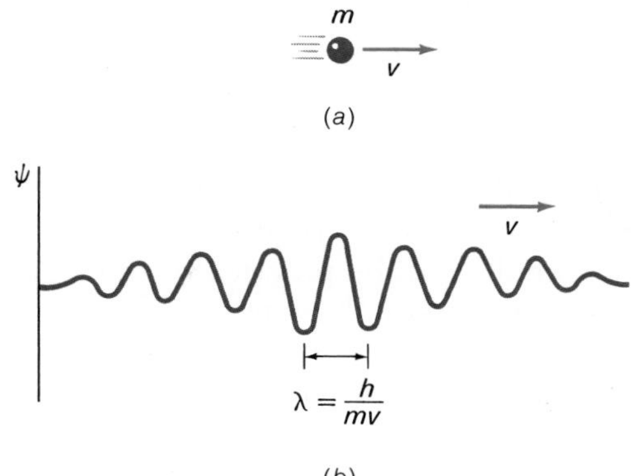

Figure 9-14 (*a*) Particle description of a moving object. (*b*) Wave description of the same moving object. The packet of matter waves that corresponds to a certain object moves with the same speed v as the object does. The waves are waves of probability.

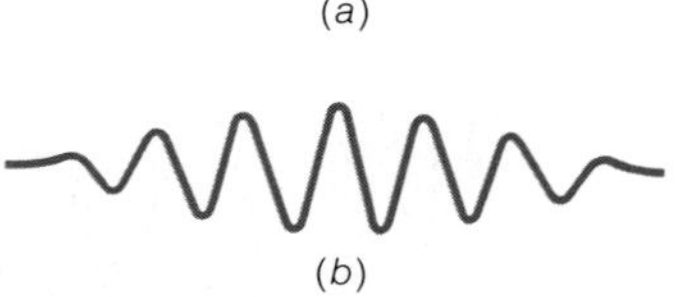

Figure 9-15 (*a*) A narrow wave packet. The position of the particle can be precisely determined, but the wavelength (and hence the particle's momentum) cannot be established because there are not enough waves to measure λ accurately. (*b*) A wide wave packet. Now the wavelength can be accurately determined, but not the position of the particle.

9.7 Uncertainty Principle

We Cannot Know the Future Because We Cannot Know the Present

To regard a moving particle as a wave packet suggests that there are limits to the accuracy with which we can measure such "particle" properties as position and speed. The particle whose wave packet is shown in Fig. 9-14 may be located anywhere within the packet at a given time. Of course, the probability density ψ^2 is a maximum in the middle of the packet, so the particle is most likely to be found there. But we may still find the particle anywhere that ψ^2 is not 0.

The narrower its wave packet, the more precisely a particle's position can be specified (Fig. 9-15*a*). However, the wavelength of the waves in a narrow packet is not well defined. There are just not enough waves to measure λ accurately. This means that, since $\lambda = h/mv$, the particle's momentum mv and hence speed v are not precise quantities. If we make a series of momentum measurements, we will find a broad range of values.

On the other hand, a wide wave packet such as that in Fig. 9-15*b* has a clearly defined wavelength. The momentum that corresponds to this wavelength is therefore a precise quantity, and a series of measurements will give a narrow range of values. But where is the particle located? The width of the packet is now too great for us to be able to say just where the particle is at a given time.

Thus we have the **uncertainty principle:**

> It is impossible to know both the exact position and the exact momentum of a particle at the same time.

This principle, which was discovered by Werner Heisenberg, is one of the most significant of physical laws.

Since we cannot know exactly both where a particle is right now and what its speed is, we cannot say anything definite about where it will be 2 s from now and how fast it will be moving then. *We cannot know the future for sure because we cannot know the present for sure.* But our ignorance is not total. We can still say that the particle is more likely to be in one place than another and that its speed is more likely to have a certain value than another.

All objects, of whatever size, are governed by the uncertainty principle, which means that their positions and motions likewise can be expressed only as probabilities. There is a chance that this book will someday defy the law of gravity and rise up in the air by itself. But for objects this large—in fact, even for objects the size of molecules—such probabilities are so small as to be practically zero. The likelihood that this book will continue to obey the law of gravity is so great that we can be quite sure it will stay where it is if left alone. Only in the behavior of electrons and other atomic particles do matter waves play an important part.

BIOGRAPHY Werner Heisenberg (1901–1976)

Heisenberg studied theoretical physics in Munich, Germany, where he also became an enthusiastic skier and mountaineer. In 1927 he showed that $\Delta x \Delta(mv)$, the product of the uncertainties in a simultaneous measurement of a particle's position and momentum, can never be less than $h/4\pi$, where h is Planck's constant. This conclusion is an inescapable consequence of the wave nature of moving particles. Because h is so small, the uncertainty principle is an important factor only in the atomic world, where it tells us what we might be able to know about events in this world and what we cannot ever know. Heisenberg distrusted mechanical models of the atom: "Any picture of the atom that our imagination is able to invent is for that very reason defective," he remarked.

Heisenberg was one of the very few distinguished scientists to remain in Germany during the Nazi period. In World War II he led research there on nuclear weapons, but little progress had been made by the war's end. Poor experimental data seems to have been the reason—there is no evidence that Heisenberg, as he later claimed, had moral qualms about creating such weapons for Hitler and deliberately dragged his feet. Alarmed by the news that Heisenberg was working on a nuclear bomb, the U.S. government sent the former Boston Red Sox catcher Moe Berg to shoot Heisenberg during a lecture in neutral Switzerland in 1944. Berg, sitting in the second row, found himself uncertain from Heisenberg's remarks about how advanced the German program was, and kept his gun in his pocket.

Using the Uncertainty Principle

The uncertainty principle is not just a negative statement; like the second law of thermodynamics, it can sometimes give valuable information in a simple way. For example, the average radius of the hydrogen atom is 5.3×10^{-11} m. If we assume that the uncertainty in the location of its single electron has the same value, the uncertainty principle at once gives a minimum value of its momentum that corresponds to a least kinetic energy of 3.4 eV. In fact, the KE of an electron in the lowest hydrogen energy level (see Sec. 9.9) is 13.6 eV, but a calculation based on a detailed theory of the atom is needed to get this figure.

We can also use the uncertainty principle to set a lower limit to the energy an electron must have if it is to be part of an atomic nucleus. A typical nuclear radius is 5×10^{-15} m, much smaller than an atomic radius, and now the minimum electron energy turns out to be 20 MeV. Experiments show that even the electrons associated with unstable nuclei never have more than a fraction of this energy, so we arrive at the important result that electrons cannot be present inside nuclei.

The Hydrogen Atom

We now have what we need to make sense of atomic structures: the Rutherford model of the atom, the quantum theory of light, and the wave theory of moving particles. When linked together, these concepts give rise to a theory of the atom that agrees with experiment. Our starting point will be the hydrogen atom—the simplest of all—with its single electron outside a nucleus that consists of a single proton.

Figure 9-16 Hong Kong at night. Gas atoms excited by electric currents in the tubes of these signs radiate light of wavelengths characteristic of the gas used.

9.8 Atomic Spectra

Each Element Has a Characteristic Spectrum

When an electric current is passed through a gas, electrons in the gas atoms absorb energy from the current. The gas is said to be **excited.** An excited neon gas gives off a bright orange-red light; other excited gases give off light of other colors. We have all seen signs based on this effect (Fig. 9-16), but not all of us are aware of how closely related the color of the light is to the way the electrons in the gas atoms are arranged.

Figure 9-17 shows an instrument called a **spectroscope** that disperses (spreads out) the light emitted by an excited gas into the different frequencies the light contains. Each frequency appears on the screen as a bright line, and the resulting series of bright lines is called an **emission spectrum** (Fig. 9-18*a*). Because some of the lines are more intense than the rest, the original undispersed light usually gives the impression of being a specific color, orange-red in the case of neon, even though other colors are present as well. An emission spectrum is different from a **continuous spectrum,** which is the rainbow band produced when light from a hot object passes through a spectroscope (see Fig. 7-55). A continuous spectrum contains all frequencies, not just a few.

Absorption Spectra Spectra of a different kind, **absorption spectra,** occur when light from a hot source passes through a cool gas before entering the spectroscope. The light source alone would give a continuous spectrum, but atoms of the gas absorb certain frequencies from the light that goes through it. Hence the original continuous spectrum is now crossed by dark lines, each line corresponding to one of the absorbed frequencies (Fig. 9-18*b*).

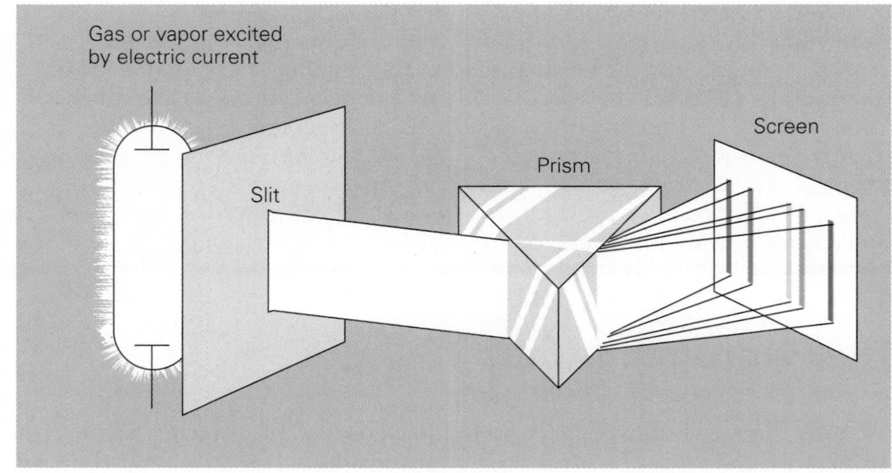

Figure 9-17 An idealized spectroscope. Dispersion in the prism separates light of different frequencies.

Figure 9-18 (*a*) Part of the emission spectrum of sodium. Each bright line represents a specific frequency in the light given off when sodium vapor is excited in an arrangement such as that shown in Figure 9-17. Every element emits a set of frequencies that is characteristic of that element. (*b*) Part of the absorption spectrum of sodium. Each dark line represents a specific frequency in the light absorbed by sodium vapor when white light passes through the vapor. Every dark absorption line corresponds to one of the bright lines in the emission spectrum of the same element.

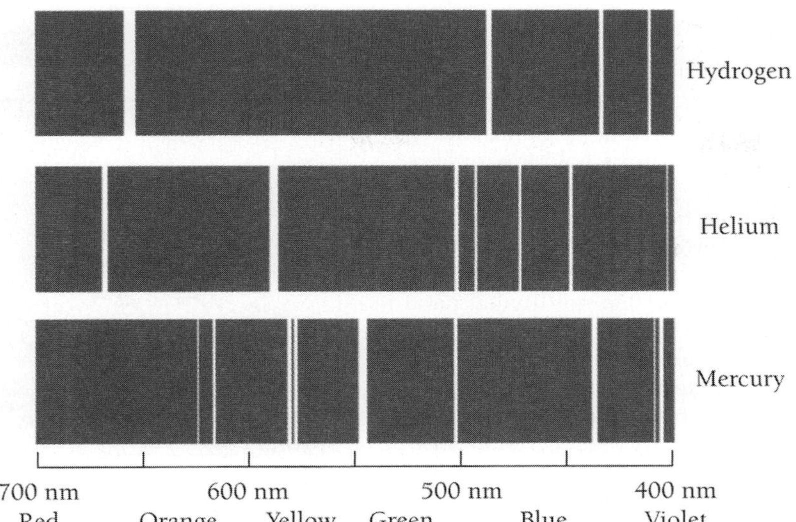

Figure 9-19 Some of the principal lines in the emission spectra of hydrogen, helium, and mercury. These were produced by passing electric currents through gaseous hydrogen and helium and through mercury vapor, which caused them to radiate light whose frequencies are characteristic of the atoms involved. One of the great triumphs of the modern theory of the atom has been its explanation of why such particular frequencies occur, and from such frequencies a great deal can be learned about the electron structures of the atoms of the elements.

If the emission spectrum of an element is compared with the absorption spectrum of the same element, the dark lines in the latter spectrum have the same frequencies as a number of the bright lines in the former spectrum. Thus a cool gas absorbs some of the frequencies of the light that it emits when excited. The spectrum of sunlight has dark lines in it because the luminous part of the sun, which radiates much like an object heated to 6000 K, has around it an envelope of cooler gas (Chap. 18).

Because the line spectrum of each element (either emission or absorption) contains frequencies that are characteristic of that element only, the spectrometer is a valuable tool in chemical analysis (Fig. 9-19). Even the smallest traces of an element can be identified by the lines in a spectrum of an unknown substance. Helium was discovered in the sun through its spectrum 17 years before it was identified on the earth in 1895 (*helios* is Greek for "sun").

A century ago it was discovered that the frequencies in the spectrum of an element fall into sets called **spectral series** (Fig. 9-20). A simple formula relates the frequencies in each series. When the foundations of the modern picture of the atom had been laid, these spectral series provided the final clues for working out the details of atomic structure.

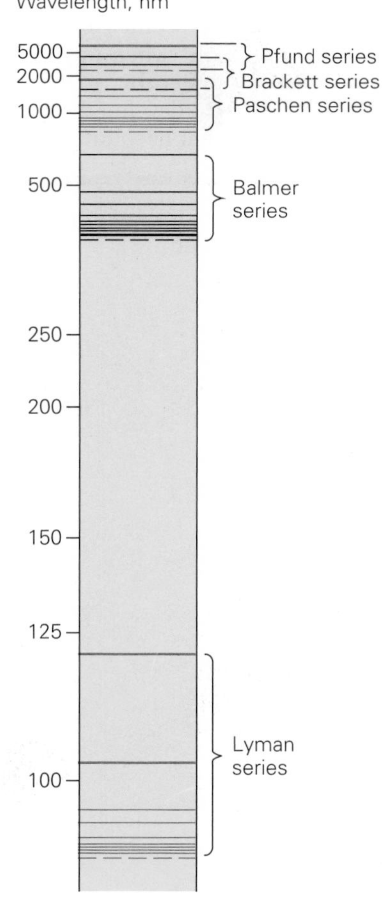

Figure 9-20 The spectral series of hydrogen. The wavelengths (and hence frequencies) in each series can be related by simple formulas (1 nm = 1 nanometer = 10^{-9} m).

9.9 The Bohr Model

Only Certain Electron Energies Are Possible in an Atom

Niels Bohr, a Dane, put forward in 1913 a theory of the hydrogen atom that could account both for its stability and for the frequencies of the spectral lines of hydrogen. Bohr applied the then-new quantum ideas to atomic structure to come up with a model that, even though later replaced by a more complex picture of greater accuracy and usefulness, is still the mental image many scientists have of the atom.

Bohr began by proposing that an electron in an atom can circle the nucleus without losing energy only in certain specific orbits. Because these orbits are each a different distance from the nucleus, the energy

BIOGRAPHY Niels Bohr (1884–1962)

The Bohrs were a distinguished Danish family: Niels's father was a professor of physiology, his brother Harald was a noted mathematician, and his son Aage, like Niels himself, would win a Nobel Prize in physics. After receiving his doctorate in 1911, Bohr visited Rutherford's laboratory in England where he was introduced to the just-discovered nuclear model of the atom.

To understand how atomic spectra are produced by such atomic structures, Bohr began with two revolutionary ideas. The first was that an atomic electron can circle its nucleus only in certain orbits, and the other was that an atom emits or absorbs a photon of light when an electron jumps from one orbit to another. Bohr was able to develop these ideas into a theory that accounted for the spectral series of hydrogen. Einstein, who was impressed by Bohr's theory, commented on its bold mix of classical and quantum concepts: "One ought to be ashamed for the successes [of the theory] because they have been earned according to the Jesuit maxim, 'Let not thy left hand know what the other doeth.'"

A decade later de Broglie clarified the basis of Bohr's theory by showing how the restriction of atomic electrons to certain orbits arises from the wave nature of the electrons. Bohr did other important work, including the explanation in 1939 of why nuclear fission, which had just been discovered, occurs in some nuclides but not in others.

Legend has it that, when asked if he really thought the horseshoe nailed above the door to his summer house brought good luck, Bohr replied, "Of course I don't believe such superstitious nonsense, but I'm told it works even if you don't believe it."

of the electron depends on which orbit it is in. Thus Bohr suggested that atomic electrons can have only certain particular energies. An analogy is a person on a ladder, who can stand only on its rungs and not in between.

An electron in the innermost orbit has the least energy. The larger the orbit, the more the electron energy. The orbits are identified by a **quantum number,** n, which is $n = 1$ for the innermost orbit, $n = 2$ for the next, and so on. Each orbit corresponds to an **energy level** of the atom.

Explaining Spectral Lines That atoms emit and absorb only light of certain frequencies, which we observe as spectral lines, fits Bohr's atomic model perfectly. An electron in a particular orbit can absorb only those photons of light whose energy will permit it to "jump" to another orbit farther out, where the electron has more energy. When an electron jumps from a particular orbit to another orbit closer to the nucleus, where it has less energy, it emits a photon of light. The difference in energy between the two orbits is hf, where f is the frequency of the absorbed or emitted light.

Figure 9-21 shows the possible orbits of the electron in a hydrogen atom. The circle nearest the nucleus represents the electron orbit under ordinary conditions, when the atom has the lowest possible energy. Such an atom is said to be in its **ground state.** The other circles represent orbits in which the electron would have more energy, since it would then be farther from the nucleus. (Similarly a stone on the roof of a building has more potential energy than it has on the ground, since it is farther from the earth's center when it is on the roof.)

Suppose an atom is in its ground state. If the atom is given energy—by strong heating, by an electric discharge, or by radiation—the electron may jump to a larger orbit (Fig. 9-22). This jump means that the atom has absorbed energy. The atom keeps the added energy as long as it is in the **excited state,** that is, as long as the electron stays in the larger orbit. Because excited states are unstable, in a fraction of a second the electron drops to a smaller orbit, emitting a photon of light as it does so.

The energy (and hence the frequency) of the photon emitted from a hydrogen atom depends on the particular jump that its electron makes. If the electron jumps from orbit $n = 4$ to orbit $n = 1$ (Fig. 9-23), the energy of the photon will be greater than if the electron jumps from 3 to 1 or 2 to 1. Starting from orbit 4, the electron may return to 1 not only by a single leap but also by stopping at 3 and 2 on the way. Corresponding to these jumps are photons with energies determined by the energy differences between 4 and 3, 3 and 2, 2 and 1.

Each electron jump gives a photon of a characteristic frequency and therefore appears in the hydrogen spectrum as a single bright line. The frequencies of the different lines are related since they correspond to different jumps in the same set of orbits. And the relations among the lines

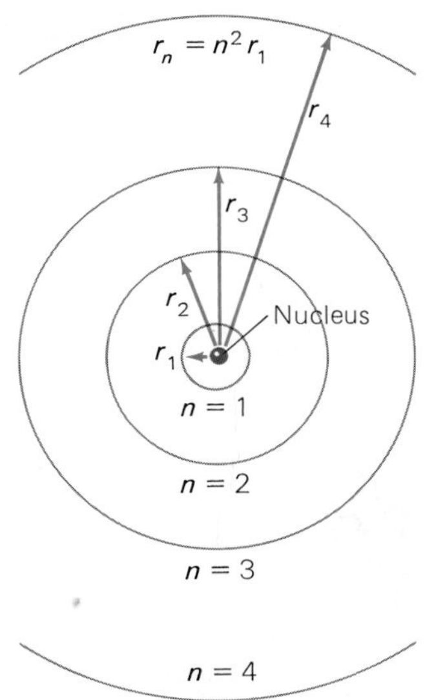

Figure 9-21 Electron orbits in the Bohr model of the hydrogen atom (not to scale). The radius of each orbit is proportional to n^2, the square of the orbit's quantum number. The inner orbit is the electron's normal path, and the outer orbits represent states of higher energy. If the electron absorbs enough energy to jump to an outer orbit, it will return to the $n = 1$ orbit by a single jump or combination of jumps. Each inward jump is accompanied by the emission of a photon.

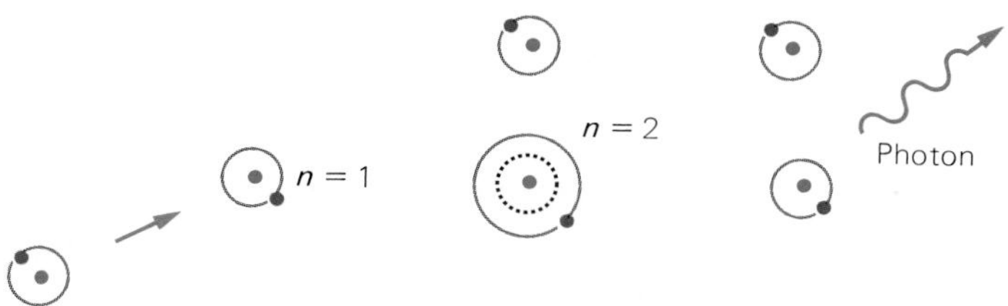

Figure 9-22 Excitation by collision. When two atoms collide, some of the available energy is absorbed by one of the atoms, which goes into an excited energy state. The atom then emits a photon in returning to its ground (normal) state.

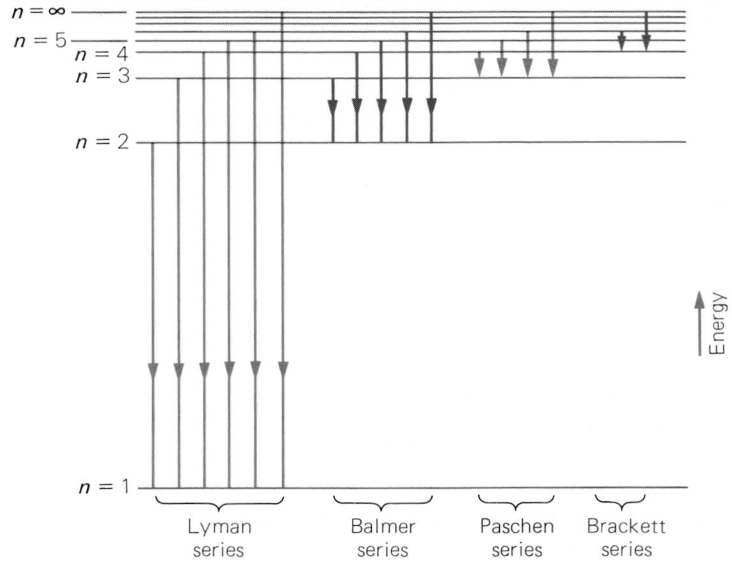

Figure 9-23 Spectral lines are the result of jumps between energy levels. The spectral series of hydrogen are shown in Figure 9-20. When $n = \infty$, the electron is free.

Origin of Absorption Spectra

Figure 9-24 shows how emission and absorption spectra arise. In (*a*) the jump of an electron from a certain excited state to the ground state of an atom involves the emission of a photon of wavelength λ. This is how an emission (bright) spectrum line originates. If a photon with the same wavelength λ is absorbed by the atom when it is in its ground state, as in (*b*), an electron in this state will jump to the same excited state as in (*a*). This is how an absorption (dark) spectrum line originates.

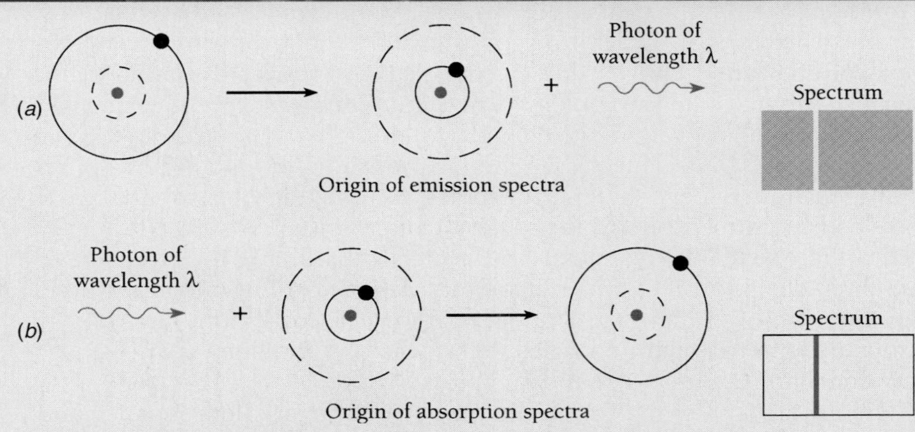

Figure 9-24 How emission and absorption spectra originate.

When white light, which contains all wavelengths, is passed through a gas or vapor, as in Fig. 9-25, photons of those wavelengths that correspond to electron jumps to higher energy levels are absorbed. The resulting excited atoms reradiate their extra energy almost at once, but these photons come off in random directions. Because a few of the reradiated photons are in the direction of the original beam, the dark lines in an absorption spectrum are never completely black. The lines in an absorption spectrum have the same wavelengths as those of the emission lines that correspond to jumps to the ground state. Hence not all of the emission lines of an element are present in its absorption spectrum, as we saw in Fig. 9-18.

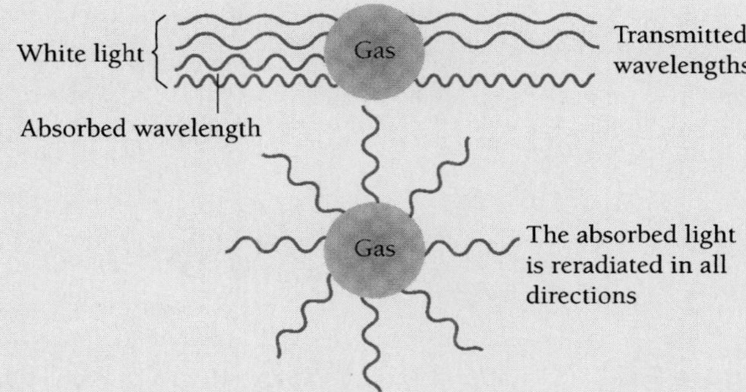

Figure 9-25 The dark lines in an absorption spectrum are never totally dark.

that Bohr predicted by this mechanism precisely matched the observed relations among the lines in the hydrogen spectrum.

9.10 Electron Waves and Orbits

Standing Waves in the Atom

Why does an atomic electron follow certain orbits only? The answer comes from an analysis of the wave properties of an electron that circles a hydrogen nucleus. It turns out that the de Broglie wavelength of the electron is exactly equal to the circumference of its ground-state (that is, innermost) orbit. Thus the $n = 1$ orbit of the electron in a hydrogen atom corresponds to one complete electron wave joined on itself (Fig. 9-26).

This fact provides us with the final clue we need for a theory of the atom. If we consider the vibrations of a wire loop (Fig. 9-27), we find that their wavelengths always fit a whole number of times into the loop's circumference, so that each wave joins smoothly with the next. These are the

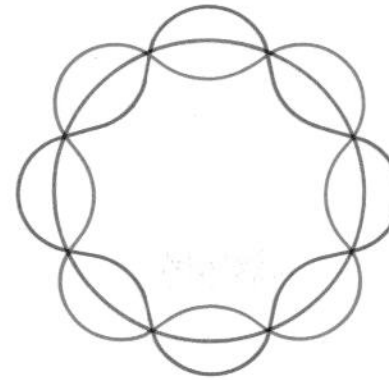

Figure 9-27 Vibrations of a wire loop. In each case a whole number of wavelengths fit into the circumference of the loop.

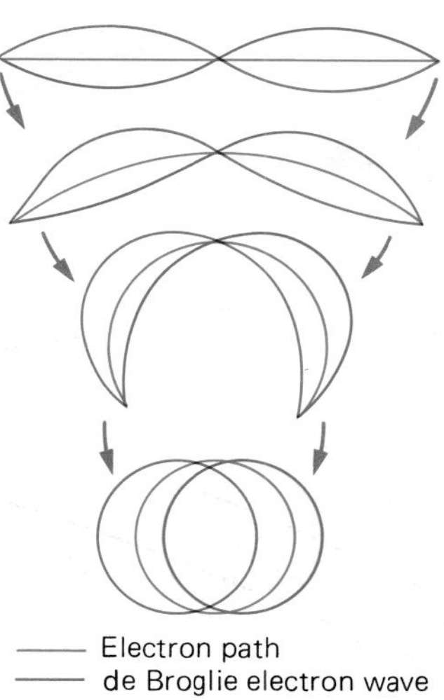

Figure 9-26 The condition for a stable electron orbit. The orbit of the electron in a hydrogen atom corresponds to a complete de Broglie electron wave joined on itself.

only vibrations possible. Regarding electron waves in an atom as analogous to standing waves in a wire loop leads to an interesting concept:

> An electron can circle a nucleus only in orbits that contain a whole number of de Broglie wavelengths.

Example 9.3

Use the energy levels shown in Fig. 9-28 to find the frequency of the photon that must be absorbed for a hydrogen atom in the ground state to be excited to the $n = 2$ state.

Solution

From Fig. 9-28 the energy difference ΔE between the two states is

$$\Delta E = E_{\text{final}} - E_{\text{initial}} = (-3.4 \text{ eV}) - (-13.6 \text{ eV})$$
$$= -3.4 \text{ eV} + 13.6 \text{ eV} = 10.2 \text{ eV}$$

Since $1 \text{ eV} = 1.6 \times 10^{-19}$ J,

$$\Delta E = (10.2 \text{ eV})(1.6 \times 10^{-19} \text{ J/eV}) = 16.3 \times 10^{-19} \text{ J}$$

From Eq. 9-1, $E = hf$, so we have here

$$f = \frac{E}{h} = \frac{16.3 \times 10^{-19} \text{ J}}{6.63 \times 10^{-34} \text{ J} \cdot \text{s}} = 2.46 \times 10^{15} \text{ Hz}$$

This is in the ultraviolet part of the spectrum.

Quantization in the Atomic World

The presence of energy levels in an atom—which is true for all atoms, not just the hydrogen atom—is a further example of the basic graininess of physical quantities. In the world of our daily lives, matter, electric charge, energy, and so forth seem to be continuous and able to be cut up (so to speak) into chunks of any size at all. But in the world of the atom matter consists of elementary particles with definite masses; charge always comes in multiples of $+e$ and $-e$; em waves of frequency f appear as streams of photons each with the energy hf; and stable systems of particles, such as atoms, can have only certain energies.

Other quantities in nature are also grainy, or *quantized.* This quantization enters into every aspect of how electrons, protons, and neutrons interact to give the matter around us (and of which we are made) its familiar properties. In the case of an atom, the quantization of energy follows from the wave nature of moving bodies: the electron waves must be standing waves, hence the electrons can have only certain energies.

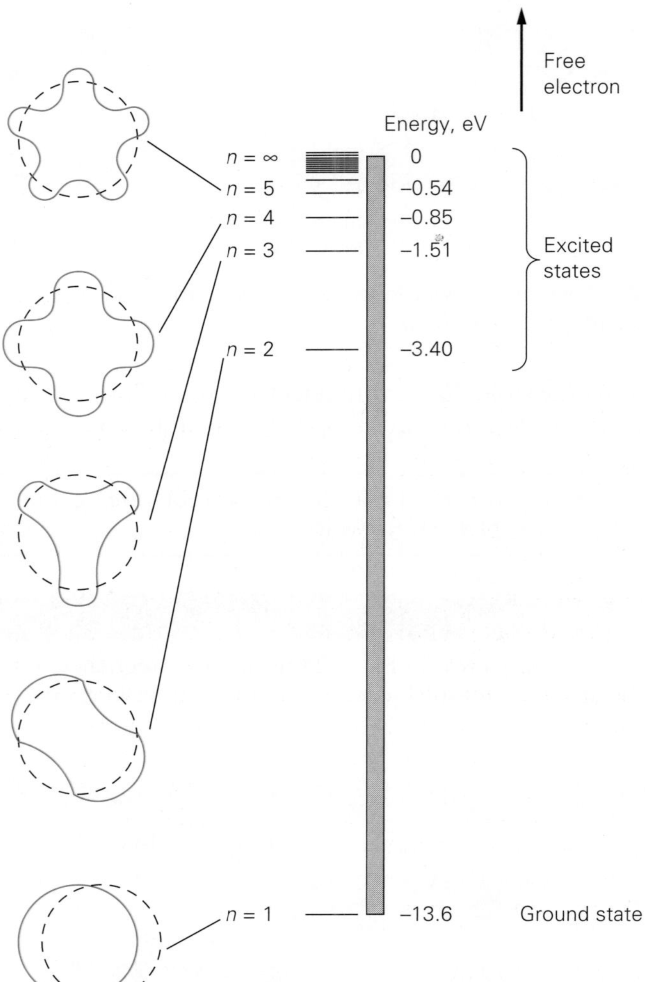

Figure 9-28 Energy levels of the hydrogen atom. The energies are negative, which signifies that the electron is bound to its nucleus.

This idea is the decisive one for understanding the atom. It combines both the particle and the wave characters of the electron into a single statement, since the electron wavelength depends upon the orbital speed needed to balance the electrical attraction of the nucleus. These contradictory characters are basic aspects of the atomic world.

Now we see what the quantum number n of an orbit means—it is the number of electron waves that fit into the orbit (Fig. 9-28).

9.11 The Laser

An Amplifier of Light That Produces Waves All in Step

A **laser** is a device that produces an intense beam of single-frequency, **coherent** light from the cooperative radiation of excited atoms. The light waves in a coherent beam are all in step with one another, as shown in

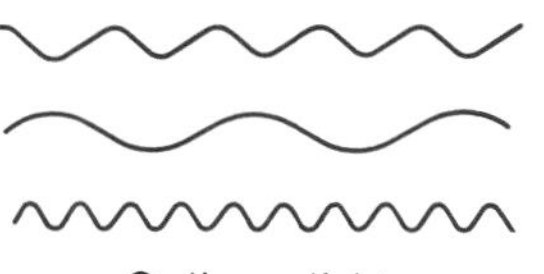

Ordinary light contains many frequencies

Single-frequency incoherent light

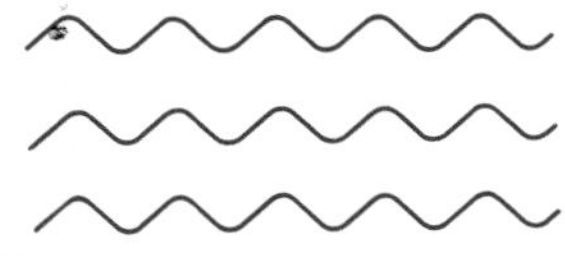

Single-frequency coherent light

Figure 9-29 A laser produces a beam of light whose waves all have the same frequency and are in step with one another (coherent). The beam is also very narrow and spreads out very little even over long distances.

Figure 9-30 Laser beams at a laboratory in Madrid, Spain.

Fig. 9-29. Ordinary light is incoherent since the atoms in light sources such as lamps and the sun emit light waves randomly.

A laser beam hardly spreads out at all (Fig. 9-30). One sent from the earth to a mirror left on the moon by the Apollo 11 expedition remained narrow enough to be detected on its return to earth, a round-trip distance of over three-quarters of a million km. A light beam produced by any other means would have spread out too much to have been detected. The word *laser* comes from *l*ight *a*mplification by *s*timulated *e*mission of *r*adiation.

The key to the laser is that many atoms have one or more excited energy levels whose lifetimes are as much as 0.001 s instead of the usual 10^{-8} s. Such relatively long lived states are called **metastable.**

A laser uses atoms whose metastable states have some excitation energy E_1 (Fig. 9-31). The first step in laser operation is to bring as many of these atoms as possible to this metastable level. Often it is necessary to raise the atoms to a still higher state E_2, from which a number of

Who Invented the Laser?

In 1951 Charles Townes, a physics professor at Columbia University, was sitting on a park bench when the idea came to him that microwaves could be amplified by a molecular mechanism similar to the atomic mechanism shown in Fig. 9-31. Two years later a device called the maser that he had built based on this idea began operating. In 1958 Townes and Arthur Schawlow attracted much attention with a paper showing that a similar scheme ought to be possible at optical wavelengths. Slightly earlier Gordon Gould, then a graduate student at Columbia, had come to the same conclusion, but he did not publish his calculations since that would prevent securing a patent. Gould tried to develop the laser—his term—in private industry, but the Defense Department classified the project (and his original notebooks) as secret and then denied him clearance to work on it.

Twenty years later Gould succeeded in establishing his priority. He received two patents on the laser and, subsequently, a third. The first working laser was built by Theodore Maiman at Hughes Research Laboratories in 1960. In 1964 Townes, along with two Russian laser pioneers, was awarded a Nobel Prize. In 1981 Schawlow shared a Nobel Prize for precision spectroscopy using lasers.

The stimulated emission of radiation, a key concept behind the laser, was first proposed by Einstein in 1917. By the early 1920s this idea, together with what had already become known about the physics of the atom, would have enabled the laser to have been invented then. Somehow nobody managed to connect the dots until over 30 years later.

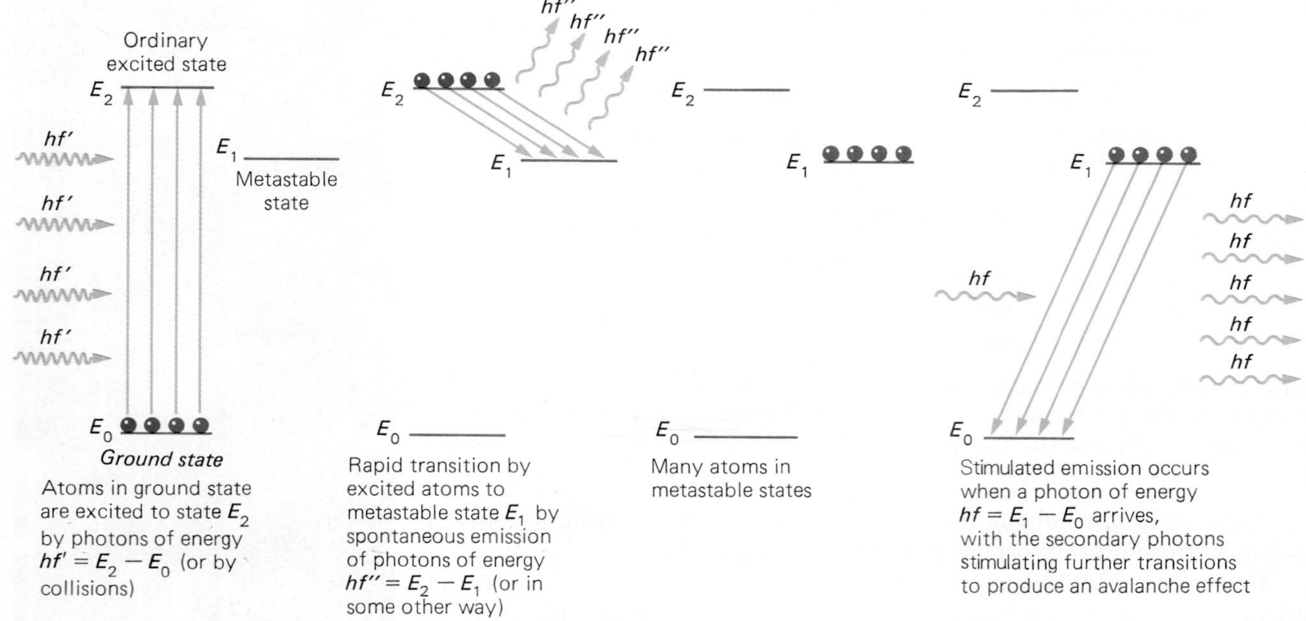

Figure 9-31 The principle of the laser. A metastable atomic state is one that lasts a much longer time than usual before a photon is emitted that brings the atom to a state of lower energy.

them fall to the metastable level by emitting a photon of energy $E_2 - E_1$. Several ways exist to do this. In one of them, an external light source provides photons with the right energy. This method was used in the first lasers, in which xenon-filled flash lamps excited chromium ions in ruby rods to the required level E_2 (Fig. 9-32).

Another method is used in the helium-neon laser. Here an electric discharge in the gas mixture produces fast electrons whose impact on the gas atoms brings them to the required energy level. The advantage is that such a laser can operate continuously, whereas a ruby laser produces separate flashes of light.

With many atoms in the metastable state E_1, a few of them are likely to spontaneously emit photons of energy $hf = E_1 - E_0$ before the others, thereby falling to the ground state E_0. A typical laser is a transparent solid (such as a ruby rod) or a gas-filled tube with mirrors at both ends, one of them only partly silvered to allow some of the light inside to get out. The distance between the mirrors is made equal to a whole number of half-wavelengths of light of frequency f, so that the trapped light forms an optical standing wave (Fig. 7-8). This standing wave stimulates the

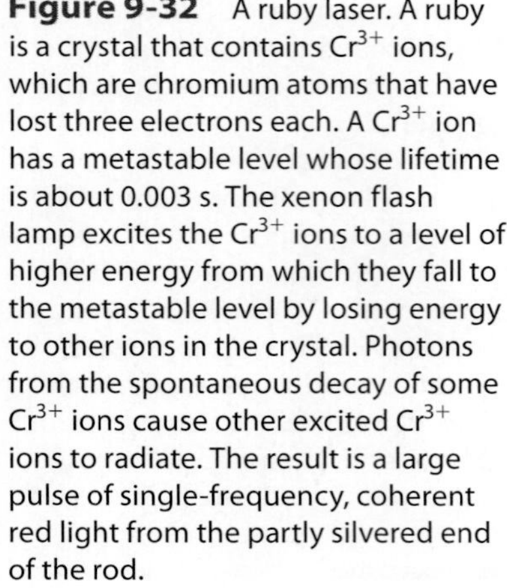

Figure 9-32 A ruby laser. A ruby is a crystal that contains Cr^{3+} ions, which are chromium atoms that have lost three electrons each. A Cr^{3+} ion has a metastable level whose lifetime is about 0.003 s. The xenon flash lamp excites the Cr^{3+} ions to a level of higher energy from which they fall to the metastable level by losing energy to other ions in the crystal. Photons from the spontaneous decay of some Cr^{3+} ions cause other excited Cr^{3+} ions to radiate. The result is a large pulse of single-frequency, coherent red light from the partly silvered end of the rod.

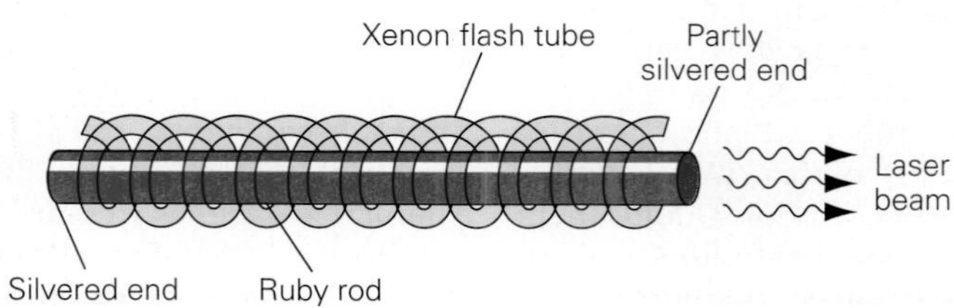

Practical Lasers

Soon after its invention, the laser was spoken of as "a solution looking for a problem" because few applications were then imagined for it. Today, of course, lasers are widely employed for a variety of purposes.

In a compact disk (CD) or digital video disk (DVD) player, the beam of a tiny solid-state semiconductor laser is focused to a spot a micrometer (a millionth of a meter) across to read data coded as pits that appear as dark spots on a reflective disk 12 cm in diameter. Similar lasers are used in transmission systems that carry telephone and television signals as flashes of infrared light along thin glass fibers.

Helium-neon gas lasers produce the narrow red beams that read bar codes in shops. More powerful carbon dioxide gas lasers are used in surgery, where a laser beam has the advantage of sealing small blood vessels it cuts through (Fig. 9-33). Lasers find numerous applications in industry, from cutting fabric for clothing and making holes in nipples for babies' bottles to welding pipelines and heat-treating the surfaces of engine crankshafts to harden them. Lasers have been built for research that deliver pulses of around a joule that last only 10^{-13} s, which means a power of 10^{13} W—more than the total output of all the power plants of the world, though only for an instant. Lasers are being developed for military use that can deliver over 100 kW for minutes at a time.

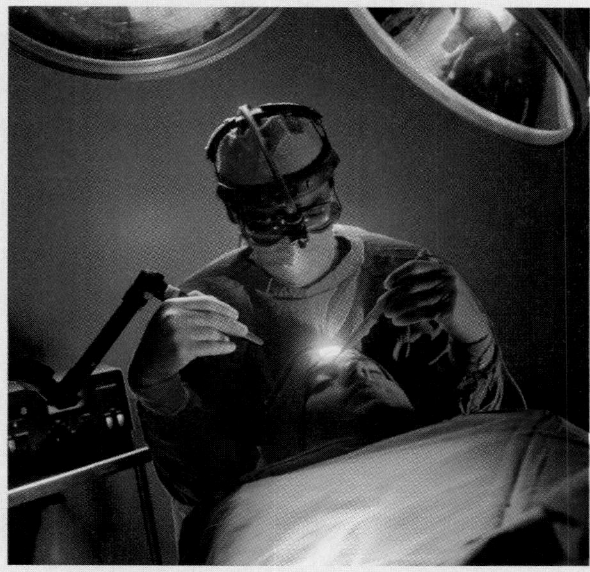

Figure 9-33 A laser produces an intense beam of single-frequency coherent light from the cooperative radiation of excited atoms or molecules. The light waves in a coherent beam are all in step, which greatly increases their effectiveness. A laser beam is here being used in plastic surgery.

other atoms in metastable states to radiate before they would normally do so. The result is an avalanche of photons, all of the same frequency f and all of whose waves are exactly in step, which greatly increases the power they can deliver.

Quantum Theory of the Atom

The preceding theory of the hydrogen atom is basically that developed by Bohr in 1913 (although he did not have de Broglie's idea of electron waves to guide his thinking). It can account for much experimental data in a convincing manner. However, it has some severe limitations. For instance, although the Bohr theory correctly predicts the spectral series of hydrogen, it cannot do the same for the spectra of atoms that have two or more electrons each. Perhaps most important of all, it does not give what a really successful theory of the atom ought to: an understanding of how individual atoms interact with one another to form molecules, solids, and liquids.

BIOGRAPHY Erwin Schrödinger (1887–1961)

Schrödinger was born in Vienna, Austria, and studied at the university there. Late in 1925, when he was a professor of physics in Zurich, Switzerland, Schrödinger wrote to a friend that he was "struggling with a new atomic theory. If only I knew more mathematics! I am very optimistic about this thing and expect that if I can only . . . solve it, it will be *very* beautiful." The struggle was successful, and early in 1926 Schrödinger published four papers on quantum mechanics that revolutionized physics and were indeed beautiful.

Later, while at Dublin's Institute for Advanced Study, Schrödinger became interested in biology, in particular the mechanism of heredity. He seems to have been the first to make definite the idea of a genetic code and to identify genes as long molecules that carry the code in the form of variations in how their atoms are arranged. Schrödinger's 1944 book *What Is Life?* was enormously influential and started James Watson and Francis Crick on their search for "the secret of the gene," which they discovered in 1953 to be the structure of the DNA molecule (see Sec. 13.16).

These objections to the Bohr theory are not meant to be unfriendly, for it was one of those historic achievements that transform scientific thought, but rather to emphasize that a more general approach to the atom is required. Such an approach was developed in 1925–1926 by Erwin Schrödinger, Werner Heisenberg, and others, under the apt name of **quantum mechanics.** By the early 1930s the application of quantum mechanics to problems involving nuclei, atoms, molecules, and matter in the solid state made it possible to understand a vast body of otherwise puzzling data and—vital for any theory—led to predictions of remarkable accuracy.

Two Views of Quantum Physics

According to Eugene Wigner, one of the early workers in the field, "The discovery of quantum mechanics was nearly a total surprise. It described the physical world in a way that was fundamentally new. It seemed to many of us a miracle." Richard Feynman, one of the most notable of the next generation of quantum physicists, brushed aside the strangeness of the ideas involved: "It is not philosophy we are after, but the behavior of real things," he remarked, and compared the agreement between theory and experiment to finding the distance between New York and Los Angeles to within the thickness of a single hair.

9.12 Quantum Mechanics

Probabilities, not Certainties

The real difference between newtonian mechanics and quantum mechanics lies in what they describe. The newtonian mechanics of Chap. 2 deals with the motion of an object under the influence of applied forces, and it takes for granted that such quantities as the object's position, mass, velocity, and acceleration can be measured. This assumption agrees completely with our everyday experience. Newtonian mechanics provides the "correct" explanation for the behavior of moving objects in the sense that the values it predicts for observable quantities agree with the measured values of those quantities.

Quantum mechanics, too, consists of relationships between observable quantities, but the uncertainty principle radically alters the meaning of "observable quantity" in the atomic realm. According to the uncertainty principle, the position and momentum of a particle cannot both be accurately known at the same time. (In newtonian physics, of course, such quantities are assumed to always have definite, measurable values.) What quantum mechanics explores are *probabilities*. Instead of saying, for example, that the electron in a normal hydrogen atom is always exactly 5.3×10^{-11} m from the nucleus, quantum mechanics holds that this is the *most probable* distance. In a suitable experiment, many trials would yield different values, but the one most likely to be found would be 5.3×10^{-11} m.

Quantum mechanics does not try to invent a mechanical model based on ideas from everyday life to represent the atom. Instead it deals only with quantities that can actually be measured. We can measure the mass of the electron and its electric charge, we can measure the frequencies of spectral lines emitted by excited atoms, and so on, and the theory must be able to relate them all. But we *cannot* measure the precise diameter of an electron's orbit or watch it jump from one orbit to another, and these notions therefore are not part of the theory.

Newtonian and Quantum Mechanics Quantum mechanics abandons the traditional approach to physics in which models we can visualize are the starting points of theories. But although quantum mechanics does not give us a look into the inner world of the atom, it does tell us everything we need to know about the measurable properties of atoms. And there is something more: *quantum mechanics includes newtonian mechanics as a special case.* The certainties of Newton do not hold on all scales of size. Their agreement with experiment is due to the fact that ordinary objects contain so many atoms that deviations from the most probable behavior are unnoticeable. Instead of two sets of physical principles, one for the world of the large and one for the world of the small, there is only a single set, and quantum mechanics represents our best effort to date at formulating it.

9.13 Quantum Numbers

An Atomic Electron Has Four in All

In the Bohr model of the hydrogen atom, the electron moves around the nucleus in a circular orbit. The only quantity that changes as the electron moves is its position on the circle. The single quantum number n is enough to specify the physical state of such an electron.

In the quantum theory of the atom, an electron has no fixed orbit but is free to move about in three dimensions. We can think of the electron as circulating in a **probability cloud** that forms a certain pattern in space. Where the cloud is most dense (that is, where ψ^2 has a high value), the electron is most likely to be found. Where the cloud is least dense (ψ^2 has a low value), the electron is least likely to be found (see Sec. 9.6). Figure 9-34 shows a cross section of the probability cloud for the ground (lowest-energy) state of the hydrogen atom.

Three quantum numbers determine the size and shape of the probability cloud of an atomic electron. One of them, the **principal quantum number,** is designated n as in the Bohr theory. This quantum number is the chief factor that governs the electron's energy (the larger n is, the greater the energy) and its average distance from the nucleus (the larger n is, the farther the electron tends to be from the nucleus).

The other two quantum numbers, l and m_l, together govern the electron's angular momentum and the form of its probability cloud. Angular momentum, as we learned in Chap. 3, is the rotational analog of linear

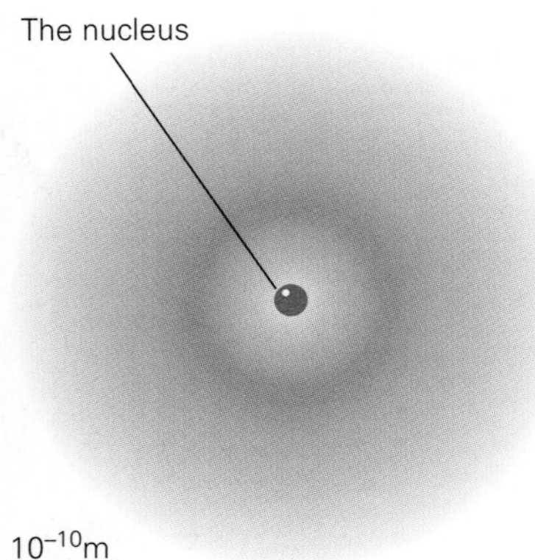

Figure 9-34 Probability cloud for the ground state of the hydrogen atom. The denser the cloud, the more likely the electron is to be found there.

momentum. According to quantum mechanics, angular momentum as well as energy is quantized—restricted to certain particular values—in an atom. The possible values of angular momentum for an atomic electron are determined by l, the **orbital quantum number.** An electron whose principal quantum number is n can have an orbital quantum number of 0 or any whole number up to $n - 1$. For instance, if $n = 3$, the values l can have are 0, 1, or 2.

The orbital quantum number l determines the *magnitude* of the electron's angular momentum. However, angular momentum, like linear momentum, is a vector quantity (see Sec. 2.2), and so to describe it completely requires that its *direction* be specified as well as its magnitude (Fig. 9-35). This is the role of the **magnetic quantum number** m_l.

What meaning can a direction in space have for an atom? The answer becomes clear when we reflect that an electron revolving about a nucleus is a current loop and has a magnetic field like that of a tiny bar magnet. In a magnetic field the potential energy of a bar magnet depends both upon how strong the magnet is and upon its orientation with respect to the field. It is the direction of the angular-momentum vector (that is, the direction of the axis about which the electron may be thought to revolve) with respect to a magnetic field that is determined by m_l.

An electron whose orbital quantum number is l can have a magnetic quantum number that is 0 or any whole number between $-l$ and $+l$. For instance, if $l = 2$, the values m_l can have are -2, -1, 0, $+1$, and $+2$.

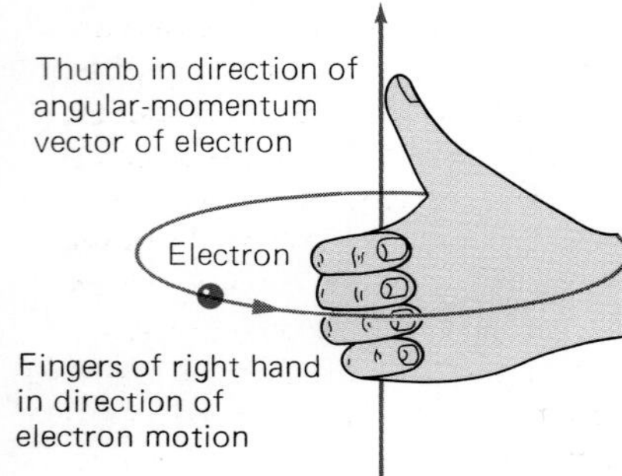

Figure 9-35 The right-hand rule for direction of angular-momentum vector.

Electron Spin There is still another quantum number needed to describe completely an atomic electron. This is the electron **spin magnetic quantum number** m_s.

Electrons behave as though they were, in themselves, little bar magnets, which we can visualize as arising from electrons spinning on their axes. If we picture an electron as a charged sphere, such spinning means a circular electric current and hence magnetic behavior. (This is only a way to help us get a mental handle on what is going on. In reality, an electron has no measurable size; its angular momentum and magnetic properties are built into it, so to speak.) Protons and neutrons also exhibit spin.

An electron can align itself so that its spin is either along a magnetic field, in which case m_s has the value $+\frac{1}{2}$, or opposite to the field, in which case $m_s = -\frac{1}{2}$, (Fig. 9-36). The concept of electron spin is essential for understanding many atomic phenomena such as permanent magnetism. Proton spin underlies the operation of MRI scanners (Fig. 9-37).

Table 9-1 lists the quantum numbers of an atomic electron, their possible values, and their significance.

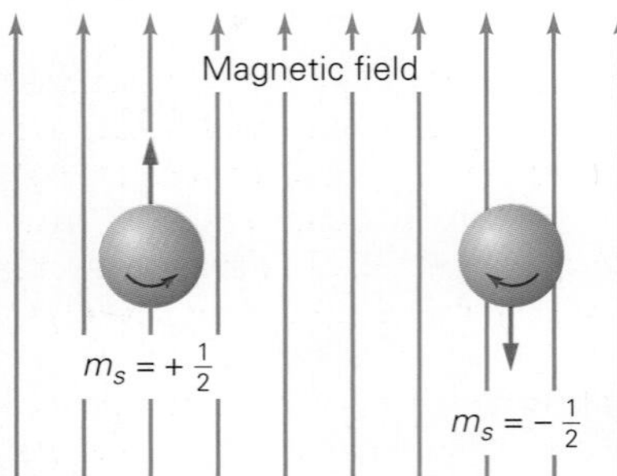

Figure 9-36 The spin magnetic quantum number m_s of an atomic electron has two possible values, $+\frac{1}{2}$ and $-\frac{1}{2}$, depending upon how the electron aligns itself with a magnetic field.

9.14 Exclusion Principle

A Different Set of Quantum Numbers for Each Electron in an Atom

In an unexcited hydrogen atom, the electron is in its quantum state of lowest energy. What about more complex atoms? Are all 92 electrons of a uranium atom in the same quantum state, jammed into a single probability cloud? Many lines of evidence make this idea unlikely.

Table 9-1 Quantum Numbers of an Atomic Electron

Name	Symbol	Possible Values	Quantity Determined
Principal	n	$1, 2, 3, \ldots$	Electron energy
Orbital	l	$0, 1, 2, \ldots, n-1$	Magnitude of angular momentum
Magnetic	m_l	$-l, \ldots, 0, \ldots, +l$	Direction of angular momentum
Spin magnetic	m_s	$-\frac{1}{2}, +\frac{1}{2}$	Direction of electron spin

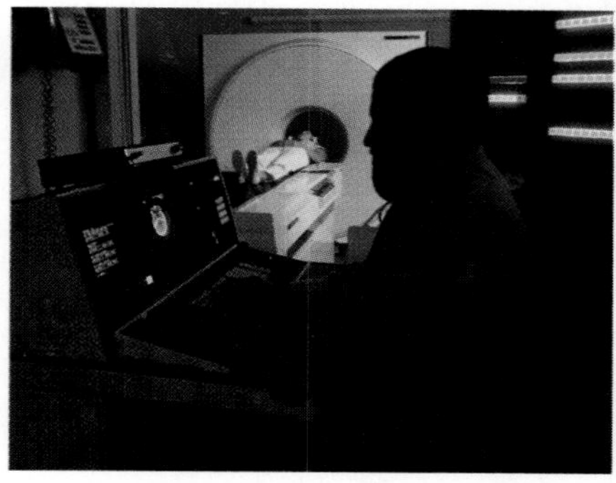

Figure 9-37 Magnetic resonance imaging (MRI) is a method for producing a series of maps of tissue density in the body that reveals details of soft tissues better than x-ray scans do with no radiation hazard. MRI is based on the magnetic behavior of proton spin. In an MRI scanner, a strong magnetic field aligns the spins of the proton nuclei of hydrogen atoms present in a tissue. A pulse of radio waves with a span of frequencies then bombards the tissue, which causes some of the proton spins to absorb the amount of energy *hf* needed to flip to the opposite direction. When the pulse stops, the proton spins realign with the magnetic field and in doing so reemit radio waves of frequency *f*. The exact value of *f* depends on the environment of the protons and can be interpreted in terms of tissue density. This photograph shows the head of a patient being scanned using MRI.

An example is the great difference in chemical behavior shown by certain elements whose atomic structures differ by just one electron. Thus the elements that have the atomic numbers 9, 10, and 11 are respectively the chemically active gas fluorine, the inert gas neon, and the metal sodium. Since the electron structure of an atom controls how the atom interacts with other atoms, it makes no sense that the chemical properties of the elements should change so sharply with a small change in atomic number if all the electrons in an atom were in the same quantum state.

In 1925 Wolfgang Pauli solved the problem of the electron arrangement in an atom that has more than one electron. His **exclusion principle** states that

> Only one electron in an atom can exist in a given quantum state.

Each electron in an atom must have a different set of quantum numbers n, l, m_l, m_s. In the next chapter we will see how the exclusion principle, together with the limits on the possible values of the various quantum numbers, determines the chemical behavior of the elements.

Important Terms and Ideas

The **photoelectric effect** is the emission of electrons from a metal surface when light shines on it. The **quantum theory of light** states that light travels in tiny bursts (or **quanta**) of energy called **photons.** The photoelectric effect can be explained only by the quantum theory of light, whereas the wave theory of light is needed to account for such other phenomena as interference; the two theories complement each other.

X-rays are high-frequency electromagnetic waves given off when matter is struck by fast electrons.

Moving objects have wave as well as particle properties; the smaller the object, the more conspicuous its wave behavior. The **matter waves** that correspond to a moving object have a **de Broglie wavelength** inversely proportional to its momentum. The quantity that varies in a matter wave is called the **wave function,** and its square is the object's **probability density.** The greater the probability density at a certain time and place, the greater the likelihood of finding the object there at that time. The **uncertainty principle** expresses the limit set by the wave nature of matter on finding both the position and the state of motion of a moving object at the same time.

An **emission spectrum** consists of the various frequencies of light given off by an excited substance. An **absorption spectrum** consists of the various frequencies absorbed by a substance when white light is passed through it.

According to the **Bohr model of the atom,** an electron can circle an atomic nucleus only if the electron's orbit is a whole number of de Broglie wavelengths in circumference. The number of wavelengths is the **quantum**

number n of the orbit. Each orbit corresponds to a specific energy, and spectral lines originate in electron shifts from one orbit, and hence **energy level,** to another. An atom in its **ground state** has the lowest possible energy; **excited states** correspond to higher energy levels.

A **laser** is a device that produces an intense beam of single-frequency light whose waves are all in step with one another, which greatly increases their effectiveness. Such light is said to be **coherent.**

Quantum mechanics is based on the wave nature of moving things; newtonian mechanics turns out to be a special case of quantum mechanics valid only on large scales of size. Quantum mechanics shows that four quantum numbers are needed to specify the physical state of each atomic electron. One of these quantum numbers governs the direction of the **spin** of the electron. According to the **exclusion principle,** no two electrons in an atom can have the same set of quantum numbers.

Important Formulas

Quantum energy of photon: $E = hf$

Photoelectric effect: $hf = \text{KE} + w$

De Broglie wavelength: $\lambda = \dfrac{h}{mv}$

Multiple Choice

1. When light is directed at a metal surface, the energies of the emitted electrons
   a. vary with the intensity of the light
   b. vary with the frequency of the light
   c. vary with the speed of the light
   d. are random
2. An increase in the brightness of the light directed at a metal surface causes an increase in the emitted electrons'
   a. wavelength
   b. speed
   c. energy
   d. number
3. The photoelectric effect can be understood on the basis of
   a. the electromagnetic theory of light
   b. the interference of light waves
   c. the special theory of relativity
   d. none of these
4. In a vacuum, all photons have the same
   a. frequency
   b. wavelength
   c. energy
   d. speed
5. The mass of a photon
   a. is 0
   b. is the same as that of an electron
   c. depends on its frequency
   d. is the size of the x-rays emitted
6. When the speed of the electrons that strike a metal surface is increased, the result is an increase in
   a. the number of x-rays emitted
   b. the frequency of the x-rays emitted
   c. the speed of the x-rays emitted
   d. the size of the x-rays emitted
7. A phenomenon that cannot be understood with the help of the quantum theory of light is
   a. the photoelectric effect
   b. x-ray production
   c. the spectrum of an element
   d. interference of light
8. According to the theories of modern physics, light
   a. is exclusively a wave phenomenon
   b. is exclusively a particle phenomenon
   c. combines wave and particle properties
   d. has neither wave nor particle properties
9. According to the theories of modern physics,
   a. only stationary particles exhibit wave behavior
   b. only moving particles exhibit wave behavior
   c. only charged particles exhibit wave behavior
   d. all particles exhibit wave behavior
10. The speed of the wave packet that corresponds to a moving particle is
    a. less than the particle's speed
    b. equal to the particle's speed
    c. more than the particle's speed
    d. any of these, depending on the circumstances
11. De Broglie waves can be regarded as waves of
    a. pressure
    b. probability
    c. electric charge
    d. momentum
12. The description of a moving body in terms of matter waves is legitimate because
    a. it is based upon common sense
    b. matter waves have actually been seen

c. the analogy with electromagnetic waves is plausible
d. theory and experiment agree

13. The narrower the wave packet of a particle is,
a. the shorter its wavelength
b. the more precisely its position can be established
c. the more precisely its momentum can be established
d. the more precisely its energy can be established

14. According to the uncertainty principle, it is impossible to precisely determine at the same time a particle's
a. position and charge
b. position and momentum
c. momentum and energy
d. charge and mass

15. If Planck's constant were larger than it is,
a. moving bodies would have shorter wavelengths
b. moving bodies would have higher energies
c. moving bodies would have higher momenta
d. the uncertainty principle would be significant on a larger scale of size

16. The emission spectrum produced by the excited atoms of an element contains frequencies that are
a. the same for all elements
b. characteristic of the particular element
c. evenly distributed throughout the entire visible spectrum
d. different from the frequencies in its absorption spectrum

17. A neon sign does not produce
a. a line spectrum
b. an emission spectrum
c. an absorption spectrum
d. photons

18. The sun's spectrum consists of a bright background crossed by dark lines. This suggests that the sun
a. is a hot object surrounded by a hot atmosphere
b. is a hot object surrounded by a cool atmosphere
c. is a cool object surrounded by a hot atmosphere
d. is a cool object surrounded by a cool atmosphere

19. The classical model of the hydrogen atom fails because
a. an accelerated electron radiates electromagnetic waves
b. a moving electron has more mass than an electron at rest
c. a moving electron has more charge than an electron at rest
d. the attractive force of the nucleus is not enough to keep an electron in orbit around it

20. An electron can revolve in an orbit around an atomic nucleus without radiating energy provided that the orbit
a. is far enough away from the nucleus
b. is less than a de Broglie wavelength in circumference
c. is a whole number of de Broglie wavelengths in circumference
d. is a perfect circle

21. According to the Bohr model of the atom, an electron in the ground state
a. radiates electromagnetic energy continuously
b. emits only spectral lines
c. remains there forever
d. can jump to another orbit if given enough energy

22. In the Bohr model of the atom, the electrons revolve around the nucleus of an atom so as to
a. emit spectral lines
b. produce x-rays
c. form energy levels that depend upon their speeds only
d. keep from falling into the nucleus

23. A hydrogen atom is said to be in its ground state when its electron
a. is at rest
b. is inside the nucleus
c. is in its lowest energy level
d. has escaped from the atom

24. An atom emits a photon when one of its orbital electrons
a. jumps from a higher to a lower energy level
b. jumps from a lower to a higher energy level
c. is removed by the photoelectric effect
d. is struck by an x-ray

25. The energy difference between adjacent energy levels in the hydrogen atom
a. is smaller for small quantum numbers
b. is the same for all quantum numbers
c. is larger for small quantum numbers
d. has no regularity

26. When an atom absorbs a photon of light, which one or more of the following can happen?
a. An electron shifts to a state of smaller quantum number
b. An electron shifts to a state of higher quantum number
c. An electron leaves the atom
d. An x-ray photon is emitted

27. Which of the following types of radiation is not emitted by the electronic structures of atoms?
    a. ultraviolet light
    b. visible light
    c. x-rays
    d. gamma rays
28. The operation of the laser is based upon
    a. the uncertainty principle
    b. the interference of de Broglie waves
    c. stimulated emission of radiation
    d. stimulated absorption of radiation
29. Which of the following properties is not characteristic of the light waves from a laser?
    a. The waves all have the same frequency
    b. The waves are all in step with one another
    c. The waves form a narrow beam
    d. The waves have higher photon energies than light waves of the same frequency from an ordinary source
30. The quantum-mechanical theory of the atom is
    a. based upon a mechanical model of the atom
    b. a theory that restricts itself to physical quantities that can be measured directly
    c. less accurate than the Bohr theory of the atom
    d. impossible to reconcile with Newton's laws of motion
31. A quantum number is not associated with an atomic electron's
    a. mass
    b. energy
    c. spin
    d. orbital angular momentum
32. Electrons behave like
    a. pure charges with no magnetic properties
    b. tiny bar magnets with different strengths that never change
    c. tiny bar magnets with strengths that may change
    d. tiny bar magnets with the same strength that never changes
33. The electrons in an atom all have the same
    a. speed
    b. spin magnitude
    c. orbit
    d. quantum numbers
34. According to the exclusion principle, no two electrons in an atom can have the same
    a. spin direction
    b. speed
    c. orbit
    d. set of quantum numbers
35. Light of wavelength 5×10^{-7} m consists of photons whose energy is
    a. 1.1×10^{-48} J
    b. 1.3×10^{-27} J
    c. 4×10^{-19} J
    d. 1.7×10^{-15} J
36. An x-ray photon has an energy of 6.6×10^{-15} J. The frequency that corresponds to this energy is
    a. 4.4×10^{-48} Hz
    b. 10^{-19} Hz
    c. 10^{15} Hz
    d. 10^{19} Hz
37. The de Broglie wavelength of an electron whose speed is 10^8 m/s is
    a. 5.9×10^{-56} m
    b. 1.5×10^{-19} m
    c. 7.3×10^{-12} m
    d. 1.4×10^{11} m
38. The speed of an electron whose de Broglie wavelength is 10^{-10} m is
    a. 6.6×10^{-24} m/s
    b. 3.8×10^3 m/s
    c. 7.3×10^6 m/s
    d. 10^{10} m/s

Exercises

9.1 Photoelectric Effect

9.2 Photons

9.3 What Is Light?

1. What differences can you think of between the photon and the electron?
2. The photon and the neutrino are both uncharged. What are the differences between them?
3. Compare the evidence for the wave nature of light with the evidence for its particle nature.
4. A certain metal surface emits electrons when light is shone on it. (a) How can the number of electrons per second be increased? (b) How can the energies of the electrons be increased?
5. Energy is carried in light by means of separate photons, yet even the faintest light we can see does not appear as a series of flashes. Explain.
6. How does the speed of a photon compare with the speed of an em wave?
7. When the speed of the electrons that strike a metal surface is increased, what happens to the speed, energy, and number per second of the x-ray photons that are emitted?
8. Why do you think the wave aspect of light was discovered earlier than its particle aspect?
9. Find the energy of a photon of ultraviolet light whose frequency is 2×10^{16} Hz. Do the same for a photon of radio waves whose frequency is 2×10^5 Hz.

10. Find the frequency and wavelength of a 100-MeV gamma-ray photon.

11. The eye can detect as little as 10^{-18} J of energy in the form of light. How many photons of frequency 5×10^{14} Hz does this amount of energy represent?

12. A 1-kW radio transmitter operates at a frequency of 880 kHz. How many photons per second does it emit?

13. The radiant energy reaching the earth from the sun is about 1400 W/m^2. If this energy is all green light of wavelength 5.5×10^{-7} m, how many photons strike each square meter per second?

14. A microwave oven operating at 2.5 GHz has a power output of 500 W. (a) What is the wavelength of the microwaves? (b) What is the energy of each photon? (c) How many photons per second does the oven produce?

15. A detached retina is being "welded" back by using 20-ms pulses from a 0.50-W laser operating at a wavelength of 643 nm. How many photons are in each pulse?

16. An electron needs an energy of 2.2 eV to escape from a potassium surface. If ultraviolet light of wavelength 350 nm falls on a potassium surface, what is the maximum KE of the emitted electrons?

17. An energy of 4×10^{-19} J is required to remove an electron from the surface of a particular metal. (a) What is the frequency of the light that will just dislodge electrons from the surface? (b) What is the maximum energy of electrons emitted through the action of light of wavelength 2×10^{-7} m?

9.4 X-Rays

18. What is the shortest wavelength present in the radiation from an x-ray machine whose operating potential difference is 50,000 V?

19. In a television picture tube, electrons are accelerated through voltages of about 10 kV. Find the highest frequencies of the em waves emitted when these electrons strike the screen of the tube. What type of waves are they (see Fig. 7-28)?

20. What voltage must be applied to an x-ray tube for it to emit x-rays with a maximum frequency of 10^{19} Hz?

9.5 De Broglie Waves

9.6 Waves of What

21. Must a particle have an electric charge in order for matter waves to be associated with its motion?

22. What kind of experiment might you use to distinguish between a gamma ray of wavelength 10^{-11} m and an electron whose de Broglie wavelength is also 10^{-11} m?

23. A photon and a proton have the same wavelength. How does the photon's energy compare with the proton's kinetic energy?

24. A proton and an electron have the same de Broglie wavelength. How do their speeds compare?

25. How does the speed of the wave packet that corresponds to a moving object compare with (a) the object's speed and (b) the speed of light?

26. An electron microscope has a much greater useful magnification than an optical microscope because it can resolve smaller details. What makes the higher resolving power possible?

27. Find the de Broglie wavelength of an electron whose speed is 2×10^7 m/s. How significant are the wave properties of such an electron likely to be?

28. Find the de Broglie wavelength of a 1500-kg car when its speed is 30 m/s. How significant are the wave properties of this car likely to be?

29. An oxygen molecule has a mass of 5.3×10^{-26} kg and an average speed in air at room temperature of about 480 m/s. How does the de Broglie wavelength of such a molecule compare with its diameter of about 4×10^{-10} m? Would you expect such a molecule to exhibit wave behavior?

30. The de Broglie wavelength of a 1-mg grain of sand being blown by the wind is 3×10^{-29} m. What is the speed of the grain of sand? How significant are its wave properties likely to be?

31. An electron microscope uses 40-keV (4×10^4 eV) electrons. Find its ultimate resolving power on the assumption that this is equal to the wavelength of the electrons.

9.7 Uncertainty Principle

32. What aspect of nature has the uncertainty principle as a consequence?

33. The uncertainty principle applies to *all* bodies, yet its consequences are significant only for such extremely small particles as electrons, protons, and neutrons. Explain.

9.8 Atomic Spectra

34. Most stars are hot objects surrounded by cooler atmospheres. What kind of spectrum does such a star give rise to?

35. What kind of spectrum is observed in
(a) light from the hot filament of a lightbulb;
(b) light from a sodium-vapor highway lamp;

(c) light from a lightbulb surrounded by cool sodium vapor?

36. Why does the hydrogen spectrum contain many lines, even though the hydrogen atom has only a single electron?

9.9 The Bohr Model

9.10 Electron Waves and Orbits

37. In the Bohr model of the atom, the electron is in constant motion. How can such an electron have a negative amount of energy?

38. Why is the Bohr theory incompatible with the uncertainty principle?

39. (a) What is meant by the ground state of an atom? (b) What is the quantum number of the ground state of a hydrogen atom in the Bohr model?

40. What is an excited atom?

41. The atoms of an excited gas are in rapid random motion. What effect do you think this has on the frequencies of the spectral lines in the emission spectrum of the gas?

42. Of the following transitions in a hydrogen atom (a) which emits the photon of highest frequency, (b) which emits the photon of lowest frequency, and (c) which absorbs the photon of highest frequency? $n = 1$ to $n = 2$, $n = 2$ to $n = 1$, $n = 2$ to $n = 6$, $n = 6$ to $n = 2$.

43. Calculate the speed of the electron in the innermost ($n = 1$) Bohr orbit of a hydrogen atom. The radius of this orbit in 5.3×10^{-11} m. (Hint: Begin by setting the centripetal force on the electron equal to the electrical attraction of the proton it circles around.)

44. With the help of Fig. 9-28 find the frequency of the photon emitted when an electron in the $n = 3$ state of hydrogen falls into the $n = 2$ state.

45. The earth's mass is 6×10^{24} kg, the circumference of its orbit around the sun is 9.4×10^{11} m, and its orbital speed is 3×10^4 m/s. (a) Find the de Broglie wavelength of the earth. (b) Find the quantum number of the earth's orbit. (c) Do you think quantum considerations play an important part in the earth's orbital motion?

9.11 The Laser

46. What is coherent light? Is the light from a lightbulb coherent? The light from the sun?

47. In what way does light from a laser differ from light from other sources?

48. Why is the optical length of a laser so important?

49. What is a metastable atomic state?

50. For laser action to occur, the medium used must have at least three energy levels. What must the nature of each of these levels be?

9.12 Quantum Mechanics

9.13 Quantum Numbers

9.14 Exclusion Principle

51. The Bohr theory permits us to visualize the structure of the atom, whereas quantum mechanics is very complex and concerned with such ideas as wave functions and probabilities. What reasons would lead to the replacement of the Bohr theory by quantum mechanics?

52. In the Bohr model of the hydrogen atom, the radius of the electron's orbit in the ground state is 5.3×10^{-11} m. What aspect of the quantum mechanical model of this atom would you expect to correspond to this figure?

53. What is the significance of a high value of the probability density ψ^2 of a particle at a certain time and place? Of a low value of ψ^2?

54. What physical quantities are governed by the quantum numbers of an atomic electron?

55. Under what circumstances do electrons exhibit spin?

56. Under what circumstances can two electrons share the same probability cloud in an atom?

Answers to Multiple Choice

1. b	**7.** d	**13.** b	**19.** a	**25.** c	**31.** a	**37.** c
2. d	**8.** c	**14.** b	**20.** c	**26.** b, c	**32.** d	**38.** c
3. d	**9.** b	**15.** d	**21.** d	**27.** d	**33.** b	
4. d	**10.** b	**16.** b	**22.** d	**28.** c	**34.** d	
5. a	**11.** b	**17.** c	**23.** c	**29.** d	**35.** c	
6. b	**12.** d	**18.** b	**24.** a	**30.** b	**36.** d	

10

The Periodic Law

These precolumbian ornaments do not show their age because gold is a relatively inactive element.

Goals

When you have finished this chapter you should be able to complete the goals ▸ given for each section below:

Elements and Compounds

10.1 Chemical Change
A Chemical Reaction Alters the Substances Involved

10.2 Three Classes of Matter
Elements, Compounds, and Mixtures
- ▸ Distinguish among the three classes of matter—elements, compounds, and mixtures—and describe how they can be told apart.
- ▸ State the law of definite proportions.

10.3 The Atomic Theory
The Building Blocks of Matter
- ▸ Explain the meanings of the letters, numbers, and parentheses in the chemical formula of a compound, for instance, $Al_2(SO_4)_3$.

The Periodic Law

10.4 Metals and Nonmetals
A Basic Distinction
- ▸ Compare the properties of metals and nonmetals.

10.5 Chemical Activity
The More Active an Element, the More Stable Its Compounds
- ▸ Discuss the relationship between the chemical activity of an element and the stability of its compounds.

10.6 Families of Elements
Members of Each Family Have a Lot in Common
- ▸ List some of the characteristic properties of the halogens, the alkali metals, and the inert gases.

10.7 The Periodic Table
A Pattern of Recurring Similarities among the Elements
- ▸ State the periodic law and describe how the periodic table is drawn up.

10.8 Groups and Periods
Elements in a Group Have Similar Properties; Elements in a Period Have Different Ones
- ▸ Distinguish between the groups and periods of the periodic table.

Atomic Structure

10.9 Shells and Subshells
They Contain Electrons with Similar Energies
- ▸ State what is meant by atomic shells and subshells.

10.10 Explaining the Periodic Table
How an Atom's Electron Structure Determines Its Chemical Behavior
- ▸ Distinguish between metal and nonmetal atoms in terms of their electron structures.
- ▸ Explain the origin of the periodic law in terms of the electron structures of atoms.

Chemical Bonds

10.11 Types of Bond
Electric Forces Hold Atoms to One Another

10.12 Covalent Bonding
Sharing Electron Pairs Produces an Attractive Force

10.13 Ionic Bonding
Electron Transfer Creates Ions That Attract Each Other
- ▸ Compare covalent and ionic bonds.
- ▸ State what is meant by a polar covalent molecule.

10.14 Ionic Compounds
Matching Up Ions
- ▸ Explain how the formula of an ionic compound can be predicted from the charges on the ions it contains.

10.15 Atom Groups
They Act as Units in Chemical Reactions
- ▸ Discuss the nature of an atom group.

10.16 Naming Compounds
The Vocabulary of Chemistry
- ▸ Establish the formula of a simple compound from its chemical name.

10.17 Chemical Equations
The Atoms on Each Side Must Balance
- ▸ Explain what a chemical equation represents and does not represent.
- ▸ Recognize whether a chemical equation is balanced or unbalanced.
- ▸ Balance an unbalanced chemical equation.

Although the line between physics and chemistry is hazy, with this chapter we are definitely across it. Chemistry began with a search for a way to change ordinary metals into gold. This fruitless task, called **alchemy** by the Arabs, was not abandoned until the seventeenth century. At that time John Mayow and Robert Boyle in England, Jean Rey in France, and Georg Stahl in Germany, among others, started to look systematically into the properties of matter and how they change in chemical reactions.

After a look at what is meant by chemical change, we go on to consider the periodic law, a natural classification of the elements into groups with similar characteristics. As we shall find, the periodic law has its roots in atomic structure. This is not surprising, since the way in which electrons are arranged in an atom is what determines how that atom interacts with other atoms—in other words, how it behaves chemically.

Elements and Compounds

The properties of matter are altered in a number of processes. When a solid melts into a liquid or a liquid vaporizes into a gas, the cause is a change in the motions and separations of the molecules of the material. In other processes, however, the changes are in the molecules themselves. Examples are the rusting of iron, the burning of wood, and the souring of milk. Such processes are called **chemical reactions.**

10.1 Chemical Change

A Chemical Reaction Alters the Substances Involved

To begin our study of chemistry, let us examine a specific chemical reaction. Suppose we mix some powdered zinc metal with a somewhat larger volume of powdered sulfur on a ceramic surface and then ignite the mixture, say with a gas flame. The result is a small explosion with light and heat given off. When the fireworks have died down, we are left with a brittle white substance that resembles neither the original zinc nor the original sulfur (Fig. 10-1). What has happened?

Further experiments would show (1) that neither zinc nor sulfur alone gives such a reaction when heated; (2) that the explosion takes place just as well in a vacuum as in air; and (3) that the ceramic surface may be replaced by a metal or asbestos one without affecting the reaction. Clearly the process involves both zinc and sulfur, but nothing else. We conclude

Zinc

Gray metal

Melts at 420°C

Density 7.1 g/cm³

Dissolves in dilute acids

Does not dissolve in carbon disulfide

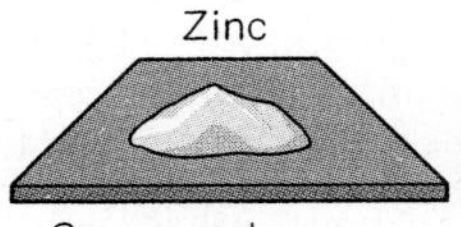

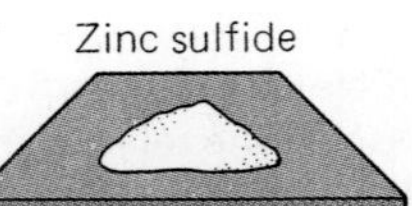

Figure 10-1 Zinc and sulfur react chemically to give zinc sulfide, a substance whose properties are different from those of zinc and sulfur.

that zinc and sulfur have joined chemically to form the new material, which is called zinc sulfide.

From Fig. 10-1 we can see that the properties of zinc, sulfur, and zinc sulfide are quite different from one another. Each material is a pure substance. Every particle of sulfur in the sulfur pile is like every other particle of sulfur, and the same is true for zinc and for zinc sulfide. However, if we simply mix zinc and sulfur together without heating them, the result is a **heterogeneous substance** whose properties vary from one particle to the next. With a microscope and tweezers we can separate particles of zinc from those of sulfur, which we cannot do in the case of zinc sulfide. There has been no change at all in the ingredients of the mixture of zinc and sulfur.

10.2 Three Classes of Matter

Elements, Compounds, and Mixtures

Although the alchemists never reached their goal of turning ordinary metals into gold, their work did have an important result. What the alchemists discovered was that certain substances—such as zinc, sulfur, and gold—could be neither broken down nor changed into one another. Slowly the belief grew that only a limited number of such **elements** exist and that all other substances are combinations of them. A new material can be formed from other materials by chemical change only if the elements of the new material are present in the original ones. This observation, little more than two centuries old, marks the beginning of the science of chemistry.

Today more than 100 elements are known, most of them solids at room temperature and atmospheric pressure. About 75 percent (by mass) of the matter in the universe is a single element, hydrogen, and nearly all the rest consists of one other element, helium. The other elements amount to less than 1 percent of the total. As for our planet, the four elements iron, oxygen, silicon, and magnesium make up 96 percent of the earth's mass. In the human body, oxygen is the most abundant element,

Naming the Elements

The names of the elements have a variety of origins. A few elements are identified by their properties: hydrogen, a constituent of water, gets its name from the Greek word for "water maker" and chlorine, a green gas, from the Greek word for green. Helium, revealed in the sun's atmosphere by its spectral lines before being found on the earth, gets its name from *helios,* Greek for the sun. Mercury, uranium, neptunium, and plutonium are named after planets; europium and americium after continents; francium after France; rhenium after the Rhine River; copper after the island of Cyprus where, according to legend, it was first discovered; and californium and berkelium after the University of California at Berkeley where they were created. Yttrium, ytterbium, terbium, and erbium were identified in ores mined near the Swedish town of Ytterby and were named accordingly. The names of several elements honor notable scientists: curium, einsteinium, fermium, and, appropriately, mendelevium.

The scarcity and inactivity of certain gases kept them from being discovered until only a century ago. Some of their names reflect these properties: neon (from the Greek for "new"); argon ("inert"); krypton ("hidden"); and xenon ("stranger").

The symbol of an element is often an abbreviation of its name. For many elements the first letter is used: O for oxygen, H for hydrogen, C for carbon. When the names of two or more elements begin with the same letter, two letters may be used: Cl for chlorine, He for helium, Zn for zinc. For some elements abbreviations of Latin names are used: Cu for copper (cuprum), Fe for iron (ferrum), Hg for mercury (hydragyrum).

followed by carbon, hydrogen, nitrogen, calcium, and phosphorus; no other element amounts to more than a fraction of a percent.

The matter around us contains elements by themselves and in a variety of combinations. Some materials consist of two or more elements joined together in chemical **compounds,** as in the case of zinc sulfide. Other materials are mixtures of elements or compounds or both. A mixture may be heterogeneous, with its components obvious to the eye and easy to separate. Wood is one example; our mixture of zinc powder and sulfur powder is another. When the components are so thoroughly mixed that the result is uniform, we have a **homogeneous mixture,** or **solution.** Thus seawater is a solution of various solids and gases dissolved in water. Figure 10-2 shows how matter is classified into its different forms.

Compound or Solution? How can we tell whether two elements have joined to make a compound or are just mixed to make a solution? A number of tests are available. Here are two:

1. See whether the new material can be separated into different substances by boiling or freezing. The changes of state we studied in Chap. 5 occur at specific temperatures for elements and compounds but not for mixtures. Air, for example, is a solution of several gases, mainly nitrogen and oxygen. Nitrogen boils at −196°C and oxygen boils at −183°C. If we heat liquid air to −196°C, most of the gas given

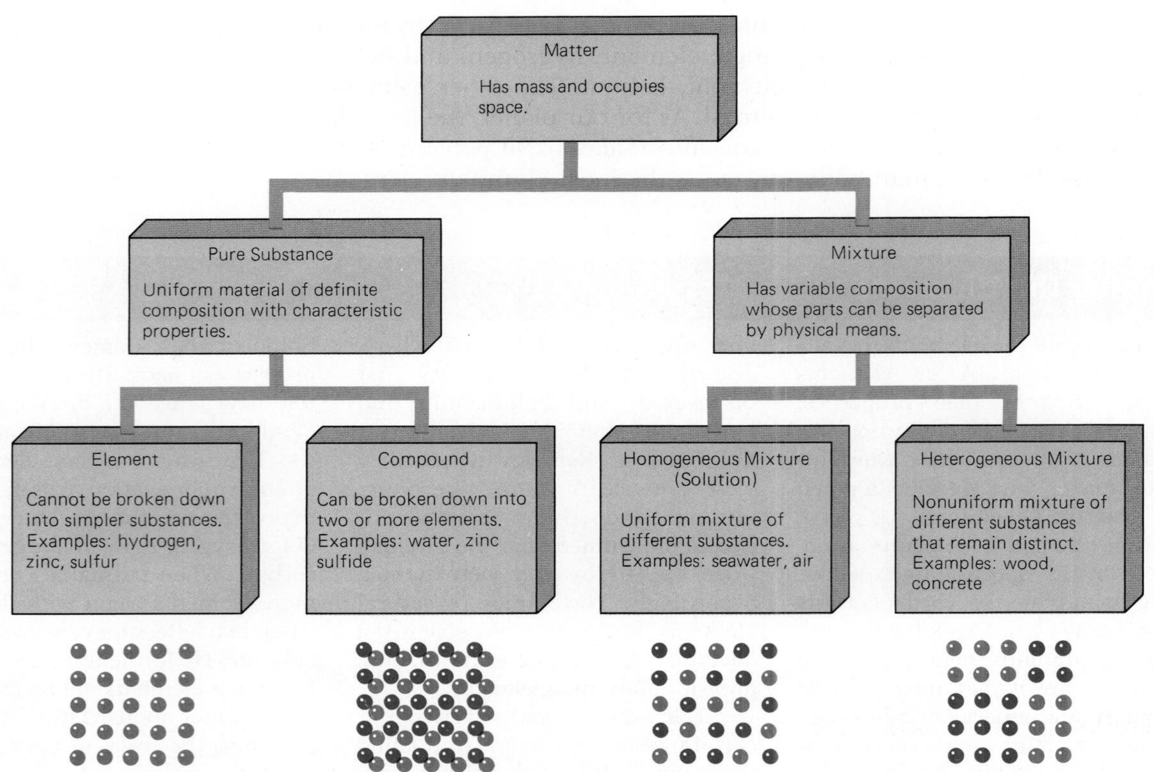

Figure 10-2 Classification of matter.

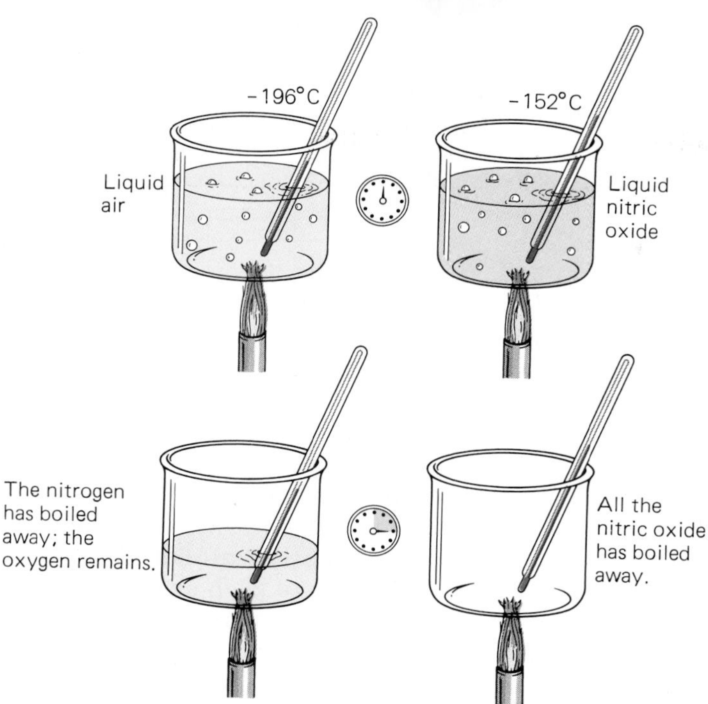

Figure 10-3 Elements and compounds have specific boiling and freezing points. A solution, here air, can therefore be separated into the elements or compounds it contains by boiling or freezing at an appropriate temperature, but a compound, here nitric oxide, cannot. Nitrogen boils at −196°C, oxygen boils at the higher temperature of −183°C, and nitric oxide boils at the still higher temperature of −152°C.

off is nitrogen. The liquid left behind is richer in oxygen than the original sample. On the other hand, nitric oxide is a compound of nitrogen and oxygen that has a boiling point of −152°C. If we heat liquid nitric oxide to −152°C, all of it will boil away at this temperature, with no change in composition (Fig. 10-3).

2. Compare the relative masses of the elements in different samples of the material. The elements in a given compound are always present in exactly the same proportions. However, the ingredients of a solution may be present in a range of proportions. At sea level, the mass ratio of the nitrogen and oxygen in air has an average value of 3.2:1, but there is more nitrogen than this at high altitudes. On the other hand, the mass ratio of these elements in nitric oxide is exactly 0.88:1 everywhere. If there is too much of either nitrogen or oxygen when nitric oxide is being made, the extra amount will not combine but will be left over and can easily be separated (Fig. 10-4). The **law of definite proportions** is as basic to chemistry as the law of conservation of momentum is to physics:

> The elements that make up a compound are always combined in the same proportions by mass.

10.3 The Atomic Theory

The Building Blocks of Matter

Two hundred years ago the structure of matter was still largely a mystery. Nobody knew what really happens when elements combine to form

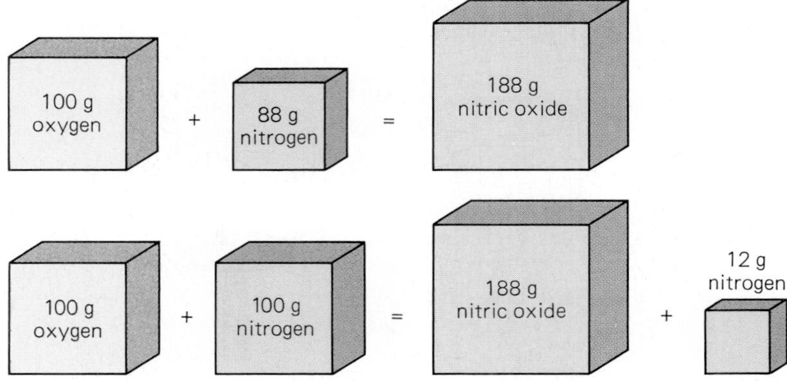

Figure 10-4 An example of the law of definite proportions. Elements combine in a specific mass ratio when they form a compound. The mass ratio between the oxygen and nitrogen in nitric oxide is always 100 : 88.

compounds. The explanation finally came from an English schoolteacher named John Dalton. Dalton began with the ancient Greek notion that all matter is built from basic particles called **atoms,** but he went much further. He proposed that all the atoms of each element were the same but were different from the atoms of other elements. Dalton was the first to establish the relative masses of atoms of the known elements, thus taking his ideas from the realm of philosophy and putting them into the realm of science. In Dalton's picture, compounds consist of atoms of different elements, with each compound having fixed ratios of the kinds of atoms present. Chemical reactions represent rearrangements of atoms, not changes in the atoms or the creation or destruction of atoms.

Molecules Our modern picture of matter grew from Dalton's work. In the case of gases, the ultimate particles of a gaseous compound are its **molecules,** which in turn are made up of atoms of the elements in the compound. Some elemental gases, such as helium and neon, consist of individual atoms. Other elemental gases consist of molecules whose atoms are all the same. Thus each molecule of gaseous oxygen consists of a pair of oxygen atoms bound together.

Atoms and most molecules are very small, and even a tiny bit of matter contains huge numbers of them. If each atom in a penny were worth 1 cent, all the money in the world would not be enough to pay for it.

The molecules of a compound have fixed compositions, as Fig. 10-5 shows. This is the reason for the law of definite proportions. Each water molecule contains two hydrogen atoms and one oxygen atom, for example, and each ammonia molecule contains three hydrogen atoms and a nitrogen atom.

Two or more atoms linked into a molecule are represented by writing the symbols for their elements side by side. Thus a carbon monoxide molecule is CO and a zinc sulfide molecule is ZnS. When a molecule contains two or more atoms of the same kind, a subscript shows the number present. The familiar H_2O means that a molecule of water contains two H atoms and one O atom. A molecule of oxygen, with two O atoms, is written O_2; a molecule of nitrogen pentoxide, with two N atoms and five O atoms, is written N_2O_5. Each subscript number applies only to the symbol just in front of it. These expressions are called **chemical formulas.**

BIOGRAPHY John Dalton (1766–1844)

The son of a Quaker weaver in England, Dalton began teaching at the age of 12, a year after his own formal education had ended. Besides having had no instruction in science, Dalton was an inept experimenter, a plodding, literal thinker, and poor at expressing his ideas. But these handicaps did not stop Dalton from placing the atomic theory of matter on a firm foundation. Indeed, by leading him to seek simple explanations for complex phenomena, they gave him an advantage over contemporaries whose minds were full of the misconceptions of the day.

Dalton's initial scientific interest was meteorology: he wrote a book on the subject and recorded weather data every day of his adult life. Studying the atmosphere started Dalton reflecting on the nature of gases and then on the nature of matter in general. He developed the concepts of atom and molecule, element and compound in detail, with numerical values derived from experiment. For instance, Dalton observed that in what is now called carbon monoxide gas the carbon:oxygen ratio by mass is 3:4 whereas in carbon dioxide it is 3:8. This led him to suggest that carbon monoxide molecules consist of one atom of carbon and one of oxygen (in symbols: CO), and that carbon dioxide molecules consist of one atom of carbon and two of oxygen (CO_2). Ratios such as these enabled Dalton to work out the relative atomic masses of many elements. Sometimes his figures were wrong (he assumed that water molecules contain one hydrogen atom for each oxygen atom, so the observed mass ratio of 1:8 means that the atomic mass of oxygen is 8 times that of hydrogen; in fact, of course, there are two H atoms per O atom in water, H_2O, so the O mass is 16 times the H mass), but on the whole he did very well. Soon after his book on this work appeared in 1808, most chemists accepted Dalton's ideas, and he became famous.

The atomic theory was not all that occupied Dalton, who always said he was too busy to marry. Among his other achievements were the first description of color blindness, from which he suffered, and the discovery that the warmer air is, the more water vapor it can hold.

Not All Compounds Consist of Molecules Elements in liquid and solid form are usually assemblies of individual atoms. Some liquid and solid compounds are also assemblies of individual molecules; others are

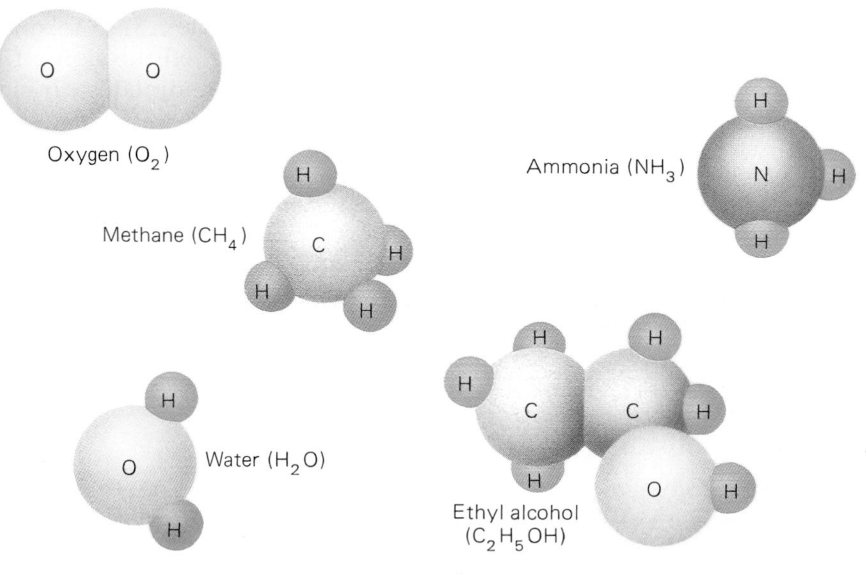

Figure 10-5 Structures of several common molecules. Ethyl alcohol is also called ethanol.

assemblies of ions. For example, crystals of table salt, which is a compound of sodium and chlorine, consist of sodium and chlorine ions rather than of neutral atoms or molecules. The sodium ions are positively charged and the chlorine ions are negatively charged, as in Fig. 10-6. For every sodium ion Na^+, there is a chlorine ion Cl^-, so that the ratio between them is fixed, and the ions are firmly held together in a definite pattern. Sodium chloride is as much a compound as water, even though it is not composed of separate molecules, and its formula is NaCl.

The Periodic Law

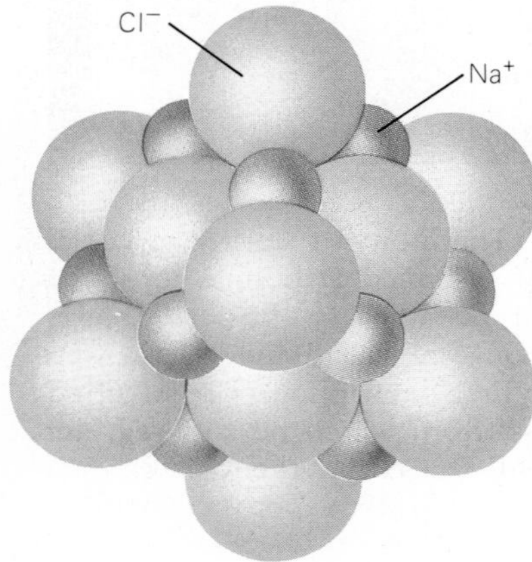

Figure 10-6 Sodium chloride crystals consist of Na^+ and Cl^- ions rather than of neutral Na and Cl atoms or individual NaCl molecules.

The periodic law, now over a century old, was a giant step for chemists along the path toward understanding the nature and behavior of the elements. There cannot be many chemistry laboratories in the world that do not have a copy of the periodic table hanging on a wall. Before we examine the periodic law, we shall look at some of the ideas behind its discovery.

10.4 Metals and Nonmetals

A Basic Distinction

The division between metals and nonmetals is a familiar one. All metals except mercury are solid at room temperature. Iron, copper, aluminum, tin, silver, and gold are examples. Nonmetals may be solid (carbon, sulfur), liquid (bromine), or gaseous (chlorine, oxygen, nitrogen) at room temperature. Metals outnumber nonmetals by more than 5:1.

A number of physical properties distinguish metals from nonmetals (Table 10-1). An obvious one is **metallic luster,** the characteristic sheen of a clean metal surface. Related to this sheen is the fact that all metals are opaque—light cannot pass through even the thinnest sheet of a metal. Solid nonmetals do not show metallic luster and nearly all are transparent in thin sheets.

Another typical property of metals is their ability to be shaped by bending or hammering. One gram of gold can be beaten into a square meter of foil (Fig. 10-7), and a copper rod can be pulled through a tiny hole in a steel plate to make a hair-thin wire. Solid nonmetals, though, are brittle and break instead of being deformed when enough force is applied. Metals are all good conductors of heat and electricity; nonmetals are insulators.

Table 10-1 Some Physical Properties of Metals and Nonmetals

Property	Metals	Nonmetals
Metallic luster	Yes	No
Opaque to light	Yes	Only a few
Can be deformed without breaking	Yes	No
Conducts heat and electricity	Yes	No

Carbon is one of a few elements whose properties put it on the borderline between metals and nonmetals. Carbon conducts heat and electricity better than other nonmetals do, and one form of it, graphite (familiar as the "lead" in pencils), is somewhat lustrous. However, all forms of carbon are brittle. Other elements intermediate between metals and nonmetals are boron, silicon, germanium, arsenic, antimony, and tellurium, which are called **semimetals** or **metalloids.** Modern electronics is largely based on the electrical properties of silicon and germanium.

Figure 10-7 Gold leaf thinner than this page was glued to this statue of Buddha. Metals differ from other solids in their ability to be rolled into thin sheets or be otherwise deformed without breaking.

10.5 Chemical Activity

The More Active an Element, the More Stable Its Compounds

Metals and nonmetals also differ in their chemical properties, but these differences are less clear-cut than the differences in physical properties because the elements in each category vary a great deal among themselves. In particular, some metals and nonmetals are very **active,** which means that they readily combine to form compounds. At the other extreme, **inactive** elements have little tendency to react chemically.

Sodium is an example of an active metal and gold is an example of an inactive one. A few seconds in the open air and sodium has lost its luster through chemical reactions, but a gold ring remains bright after a lifetime of exposure to perspiration as well as air. Sodium combines spectacularly with chlorine, giving off much heat and light. Gold combines with chlorine only sluggishly, with little energy set free. Sodium reacts with dilute acids and even with water. Gold is affected only by a mixture of concentrated hydrochloric and nitric acids.

Determining Activities The relative activities of different elements can be established by measuring the amounts of heat given off in similar chemical reactions. Suppose we combine a given mass of chlorine with sodium and then the same mass of chlorine with gold. We would find that forming sodium chloride gives off more than 15 times as much heat as forming gold chloride. The conclusion is that sodium is much more active than gold (Fig. 10-8).

Or we might start with similar compounds and ask how easily they can be separated into their component elements. In the case of gold chloride and sodium chloride, the results are that gold chloride breaks up when it is heated to about 300°C, but sodium chloride must be heated to well over 1000°C for this to happen. Gold chloride is accordingly considered to be a relatively unstable compound and sodium chloride to be a relatively stable compound. In general, the more active an element is, the more difficulty we have in decomposing its compounds.

Both metals and nonmetals can be arranged in order of their activities. In the partial listing of Table 10-2 the most active elements are at the top of each series and the least active are at the bottom.

The more active a metal is, the harder it is to extract from its ores, which are the minerals that contain compounds of the metal. Lead is more active than copper. If we heat the copper sulfide Cu_2S in air, the result is metallic copper and sulfur dioxide gas, SO_2. But if we heat lead

Table 10-2 Relative Activities of Metals and Nonmetals

Metals		Nonmetals
Potassium		Fluorine
Sodium		Chlorine
Lithium	↑ More active	Bromine
Calcium		Oxygen
Magnesium		Iodine
Aluminum		Sulfur
Zinc		
Iron		
Tin		
Lead	↓ Less active	
Copper		
Mercury		
Silver		
Gold		
Platinum		

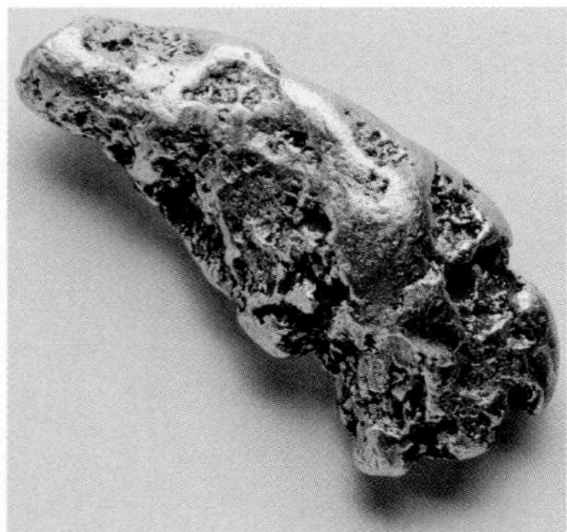

Figure 10-8 Gold occurs as the free metal because it is very inactive chemically, which is why gold objects such as rings and this nugget do not tarnish or corrode.

sulfide, PbS, in air, we get lead oxide, PbO, and sulfur dioxide. Refining lead is evidently more difficult than refining copper.

10.6 Families of Elements

Members of Each Family Have a Lot in Common

Some elements resemble one another so much that they seem to be members of the same natural family. Three examples of such families are a group of active nonmetals called the *halogens,* a group of active metals called the *alkali metals,* and a group of gases that undergo almost no chemical reactions, the *inert gases.*

Halogens The **halogens** (Table 10-3) are all highly active elements. In fact, fluorine is the most active element of all and can even corrode platinum, one of the most stable metals. The halogens are responsible for some of the worst odors (*bromos* is Greek for "stink") and most brilliant colors (*chloros* is Greek for "green") to be found in the laboratory. The name halogen means "salt former," a token of the fact that these elements combine with many metals to give white solids that resemble table salt (which is NaCl, sodium chloride).

At room temperature fluorine is a pale-yellow gas, chlorine is a greenish-yellow gas, bromine is a reddish-brown liquid, iodine is a steel-gray solid, and astatine is a radioactive solid.

What are the similarities among the halogens? For one thing, their molecules contain two atoms at ordinary temperatures: F_2, Cl_2, Br_2, I_2. (The half-life of astatine is too short for its chemical properties to be known.) Also, the compounds they form with metals have similar formulas. Here are three examples:

NaF	ZnF_2	AlF_3
NaCl	$ZnCl_2$	$AlCl_3$

Table 10-3 The Halogens

Element	Symbol	Atomic Number
Fluorine	F	9
Chlorine	Cl	17
Bromine	Br	35
Iodine	I	53
Astatine	At	85

Uses of the Halogens

Fluorine is a constituent of the nonstick plastic Teflon and is added in small quantities to water to help prevent tooth decay by making tooth enamel more resistant to acid attack. About one part of chlorine per million parts of water is enough to kill any bacteria present, which is why chlorine is added to water supplies and to swimming pools; it is also widely used as a bleach. Silver bromide, a compound of bromine and silver, is the light-sensitive material in photographic film and paper. Because the thyroid hormone thyroxine contains iodine, this halogen is essential in the diet, and accordingly table salt in the United States is usually "iodized" with a small percentage of KI or NaI. Iodine is the active ingredient in various antiseptics.

In an ordinary incandescent lightbulb a tungsten filament glows when it is heated by the passage of an electric current. As the bulb is used, the tungsten gradually evaporates and condenses on the glass envelope until the bulb eventually "burns out." To achieve a brighter light, the filament would have to be hotter and its lifetime would then be correspondingly shorter.

A way around this problem is to fill the bulb with an appropriate vapor; usually the halogen element iodine is chosen. The evaporated tungsten cools after it leaves the filament and reacts with the iodine to form tungsten iodide vapor instead of being deposited on the glass. When tungsten iodide vapor comes in contact with the hot filament, it breaks up, and the tungsten returns to the filament to leave the iodine vapor free once more. This cycle allows a halogen bulb to be operated at much higher temperatures than an ordinary bulb, and hence to be much brighter, without reducing its lifetime.

NaBr	$ZnBr_2$	$AlBr_3$
NaI	ZnI_2	AlI_3

All the halogens react with hydrogen to form HF, HCl, HBr, and so on. These compounds can be dissolved in water to form acids, of which hydrochloric acid is a familiar example. The halogens dissolve readily in a liquid called carbon tetrachloride to give solutions colored in the same way as their vapors, but the halogens are only slightly soluble in water.

Table 10-4 The Alkali Metals

Element	Symbol	Atomic Number
Lithium	Li	3
Sodium	Na	11
Potassium	K	19
Rubidium	Rb	37
Cesium	Cs	55
Francium	Fr	87

Alkali Metals The **alkali metals** (Table 10-4) are all soft and very active chemically. They lose their lusters quickly in air, liberate hydrogen from water and dilute acids, and combine with active nonmetals to form very stable compounds. Formulas for their compounds follow similar patterns, for instance,

Bromides:	LiBr	NaBr	KBr	RbBr	CsBr	FrBr
Sulfides:	Li_2S	Na_2S	K_2S	Rb_2S	Cs_2S	Fr_2S
Hydroxides:	LiOH	NaOH	KOH	RbOH	CsOH	FrOH

Sodium is quite abundant, making up about 2.5 percent of the earth's crust, but its activity prevents it from occurring free in nature. Its compounds are widely distributed in rocks, soil, and, in solution, in bodies of water.

All the alkali metals have rather low melting points for metals: cesium melts on a really hot day, and even lithium, with the highest melting point of the group, melts at only 186°C. Because the isotopes of francium are radioactive with very short half-lives, little is known about its properties. If enough of it could be gathered together, francium would probably join bromine and mercury as the only elements that are liquid at room temperature.

Inert Gases The **inert gases** (Table 10-5), in contrast with the active halogens and alkali metals, are so inactive that they form only a handful of compounds with other elements. In fact, these elements are so inactive that their atoms do not even join together into molecules as the atoms of other gaseous elements do. All the inert gases are found in small amounts in the atmosphere, with argon making up about 1 percent of the air and the others much less.

A volume of helium weighs much less than the same volume of air. Because it also cannot burn or explode, helium is ideal for lighter-than-air craft such as balloons and blimps (Fig. 10-9). We have already met radon, a radioactive product of radium decay, in Chap. 8. The other inert gases glow in various colors when excited by an electric current and are widely used in signs. Argon is often used in welding as a shield to prevent the hot metal from reacting with atmospheric oxygen.

Table 10-5 The Inert Gases

Element	Symbol	Atomic Number
Helium	He	2
Neon	Ne	10
Argon	Ar	18
Krypton	Kr	36
Xenon	Xe	54
Radon	Rn	86

10.7 The Periodic Table

A Pattern of Recurring Similarities among the Elements

A curious feature of the elements listed in Tables 10-3, 10-4, and 10-5 is that each halogen is followed in atomic number by an inert gas and

Figure 10-9 The inert gas helium is used in these blimps because helium cannot burn or explode, besides being less dense than air. Helium is found in small amounts in natural gas. To separate it out, the raw gas is cooled to a temperature at which the other gases present (mainly methane) have become liquid, leaving gaseous helium behind.

then by an alkali metal. Thus fluorine, neon, and sodium have the atomic numbers Z = 9, Z = 10, and Z = 11, a sequence that continues through astatine (85), radon (86), and francium (87). When the properties of all the elements are checked to see what other regularities occur, the result is the **periodic law:**

> When the elements are listed in order of atomic number, elements with similar chemical and physical properties appear at regular intervals.

The periodic law was first formulated in detail by the Russian chemist Dmitri Mendeleev about 1869, although the general idea was not new (Fig. 10-10). While the modern quantum theory of the atom was many years in the future, Mendeleev was fully aware of the significance of his work. As he remarked, "The periodic law, together with the revelations of spectrum analysis, have contributed to again revive an old but remarkably long-lived hope—that of discovering, if not by experiment then at least by mental effort, the *primary matter.*"

A **periodic table** is a listing of the elements according to atomic number in a series of rows such that elements with similar properties form vertical columns. Table 10-6 is a simple form of the periodic table. Let us see how it organizes our knowledge of the elements.

Building the Periodic Table The first element in the table is hydrogen, which behaves chemically much like an active metal although physically it is a nonmetal. Next comes the inert gas helium, the alkali metal lithium, and the less active metal beryllium. Then follows a series of nonmetals of increasing nonmetallic activity: boron, carbon, nitrogen, oxygen, and finally the halogen fluorine. Thus from lithium to fluorine we have a complete sequence that goes from a highly active metal to a highly active nonmetal.

Following fluorine is neon, an inert gas like helium, and after neon is sodium, an alkali metal like lithium. Clearly it makes sense to break off the rows at helium and neon and start new rows with lithium and sodium under hydrogen. In the seven elements beyond neon, we find again a transition from active metals to active nonmetals.

Atomic Mass

As we recall from Chap. 8, nearly all elements have isotopes whose nuclei have different numbers of neutrons and hence whose atomic masses are different. The atomic mass of an element that chemists use is the *average* mass of the atoms of its various isotopes in the proportion in which they occur in nature. For instance, chlorine consists of 76 percent of the $^{35}_{17}Cl$ isotope, whose atomic mass is 34.97 u, and 24 percent of the $^{37}_{17}Cl$ isotope, whose atomic mass is 36.97 u. The average atomic mass of chlorine is 35.45 u, and this is the value given in Table 10-6.

BIOGRAPHY Dmitri Mendeleev (1834 –1907)

Mendeleev was born in Siberia and grew up there, going on to Moscow and later France and Germany to study chemistry. In 1866 he became professor of chemistry at the University of St. Petersburg and 3 years later he published the first version of the periodic table. The notion of atomic number was then unknown and Mendeleev had to deviate from the strict sequence of atomic masses for some elements and to leave gaps in the table in order that the known elements (only 63 at that time) occupy places appropriate to their properties. Other chemists of the time were thinking along the same lines, but Mendeleev went further in 1871 by proposing that the gaps correspond to then-unknown elements. When his detailed predictions of the properties of these elements were fulfilled upon their discovery, Mendeleev became world-famous.

A further triumph for the periodic table came at the end of the nineteenth century when the inert gases were discovered. Here were six elements of whose existence Mendeleev (and everybody else) had been unaware, but they fit perfectly as a new group in the table. The element of atomic number 101 is called mendelevium in his honor.

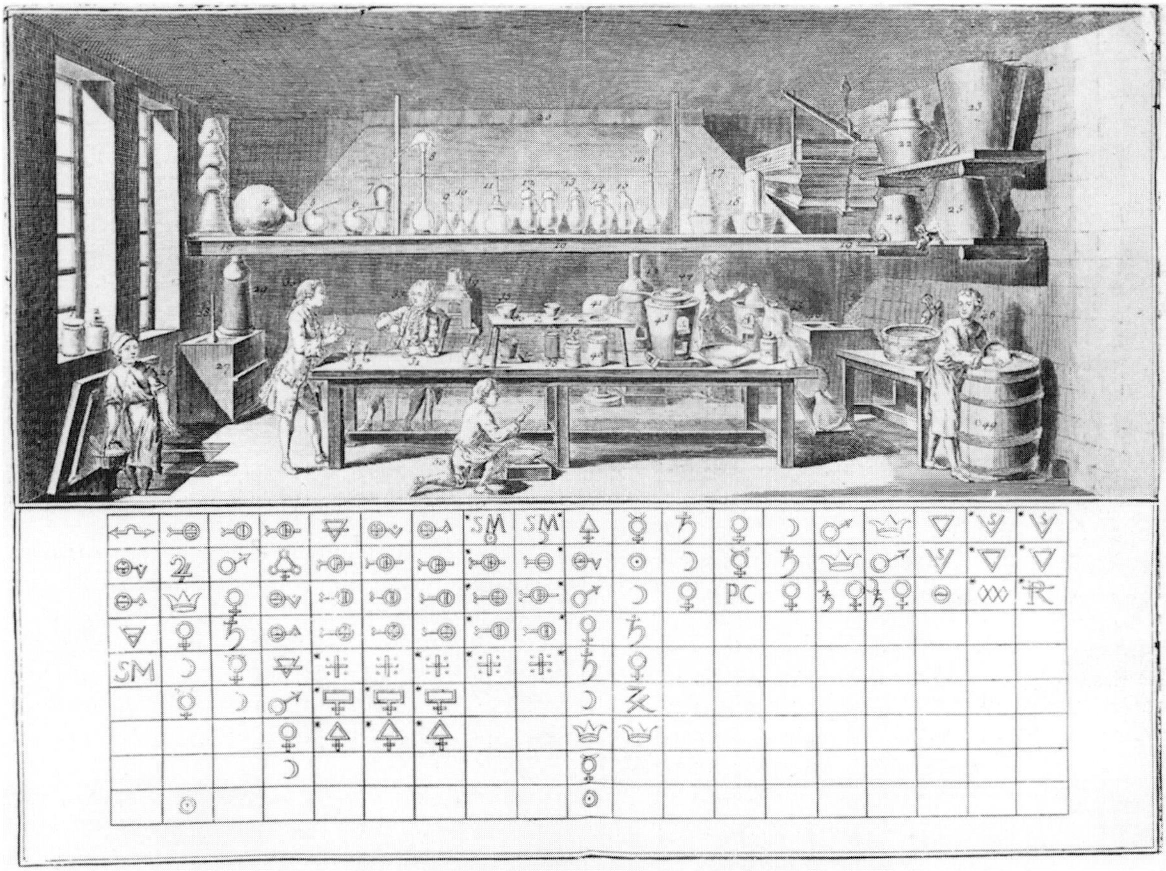

Figure 10-10 This 1780 table of chemical symbols is an early attempt to show relationships among the elements. A chemistry laboratory of the time is pictured in the engraving.

Table 10-6 The Periodic Table of the Elements

The number above the symbol of each element is its atomic number, and the number below its name is its average atomic mass. The elements whose atomic masses are given in parentheses do not occur in nature but have been created in nuclear reactions. The atomic mass in such a case is the mass number of the most long-lived radioisotope of the element. The elements with atomic numbers 113, 114, 115, 116, and 118 have also been created in the laboratory.

Group / Period	1	2	Transition metals										3	4	5	6	7	8
1	1 H Hydrogen 1.008																	2 He Helium 4.003
2	3 Li Lithium 6.941	4 Be Beryllium 9.012											5 B Boron 10.81	6 C Carbon 12.01	7 N Nitrogen 14.01	8 O Oxygen 16.00	9 F Fluorine 19.00	10 Ne Neon 20.18
3	11 Na Sodium 22.99	12 Mg Magnesium 24.31											13 Al Aluminum 26.98	14 Si Silicon 28.09	15 P Phosphorus 30.97	16 S Sulfur 32.07	17 Cl Chlorine 35.45	18 Ar Argon 39.95
4	19 K Potassium 39.10	20 Ca Calcium 40.08	21 Sc Scandium 44.96	22 Ti Titanium 47.88	23 V Vanadium 50.94	24 Cr Chromium 52.00	25 Mn Manganese 54.94	26 Fe Iron 55.8	27 Co Cobalt 58.93	28 Ni Nickel 58.69	29 Cu Copper 63.55	30 Zn Zinc 65.39	31 Ga Gallium 69.72	32 Ge Germanium 72.59	33 As Arsenic 74.92	34 Se Selenium 78.96	35 Br Bromine 79.90	36 Kr Krypton 83.80
5	37 Rb Rubidium 85.47	38 Sr Strontium 87.62	39 Y Yttrium 88.91	40 Zr Zirconium 91.22	41 Nb Niobium 92.91	42 Mo Molybdenum 95.94	43 Tc Technetium (98)	44 Ru Ruthenium 101.1	45 Rh Rhodium 102.9	46 Pd Palladium 106.4	47 Ag Silver 107.9	48 Cd Cadmium 112.4	49 In Indium 114.8	50 Sn Tin 118.7	51 Sb Antimony 121.8	52 Te Tellurium 127.6	53 I Iodine 126.9	54 Xe Xenon 131.8
6	55 Cs Cesium 132.9	56 Ba Barium 137.3		72 Hf Hafnium 178.5	73 Ta Tantalum 180.9	74 W Tungsten 183.9	75 Re Rhenium 186.2	76 Os Osmium 190.2	77 Ir Iridium 192.2	78 Pt Platinum 195.1	79 Au Gold 197.0	80 Hg Mercury 200.6	81 Tl Thallium 204.4	82 Pb Lead 207.2	83 Bi Bismuth 209.0	84 Po Polonium (209)	85 At Astatine (210)	86 Rn Radon (222)
7	87 Fr Francium (223)	88 Ra Radium 226.0		104 Rf Rutherfordium (261)	105 Db Dubnium (262)	106 Sg Seaborgium (263)	107 Bh Bohrium (262)	108 Hs Hassium (265)	109 Mt Meitnerium (266)	110 (269)	111 (272)	112 (277)						
	Alkali metals																Halogens	Inert gases

Lanthanides (rare earths)

57 La Lanthanum 138.9	58 Ce Cerium 140.1	59 Pr Praseodymium 140.9	60 Nd Neodymium 144.2	61 Pm Promethium (145)	62 Sm Samarium 150.4	63 Eu Europium 152.0	64 Gd Gadolinium 157.3	65 Tb Terbium 158.9	66 Dy Dysprosium 162.5	67 Ho Holmium 164.9	68 Er Erbium 167.3	69 Tm Thulium 168.9	70 Yb Ytterbium 173.0	71 Lu Lutetium 175.0
89 Ac Actinium (227)	90 Th Thorium 232.0	91 Pa Protactinium 231.0	92 U Uranium 238.0	93 Np Neptunium (237)	94 Pu Plutonium (244)	95 Am Americium (243)	96 Cm Curium (247)	97 Bk Berkelium (247)	98 Cf Californium (251)	99 Es Einsteinium (252)	100 Fm Fermium (257)	101 Md Mendelevium (260)	102 No Nobelium (259)	103 Lw Lawrencium (262)

Actinides

After calcium, in the fourth row, complications appear. Scandium, the next element, is similar to aluminum in some properties but different in others. Titanium (Ti) is even less like carbon and silicon. Then come 10 metals (including iron, copper, and zinc) that are quite similar among themselves but conspicuously different from the nonmetals at the end of the first three rows. Only after the 10 metals do three relatives of these nonmetals appear, arsenic (As), selenium (Se), and bromine (Br).

Between the gases helium and neon is a sequence of eight elements, and between neon and argon is another sequence of eight. However, between argon and krypton the sequence includes 18 elements. Beyond krypton is a second sequence of 18, including again a dozen metals with many properties in common. From xenon to the last inert gas, radon, is an even more complex sequence of 32 elements.

10.8 Groups and Periods

Elements in a Group Have Similar Properties; Elements in a Period Have Different Ones

The periodic table arranges families of similar elements in vertical columns called **groups.** The horizontal rows, called **periods,** contain elements with widely different properties (Fig. 10-11). Across each period is a steady change from an active metal through less active metals and weakly active nonmetals to highly active nonmetals and finally to an inert gas (Fig. 10-12). Within each column there is also a steady change in properties. Thus activity increases in the alkali metal family as we go from top to bottom down the group 1 column, and activity decreases in the halogen family as we go down the group 7 column.

Eight of the groups in Table 10-6 are numbered. The inert gases of group 8 are placed at the right since this puts them with the other nonmetals (Fig. 10-13). Each of the eight-element periods (periods 2 and 3) is broken after the second element in order to keep the members of the period in line with the most closely related elements of the long periods, which are 4 to 7.

The **transition metals** in periods 4 and 5 are metals that resemble one another in chemical behavior but do not much resemble elements in the numbered groups (Fig. 10-14). They are what we usually think of as typical metals. Iron, nickel, copper, and gold are transition metals, all of which are less reactive than the metals in groups 1 and 2. Period 6

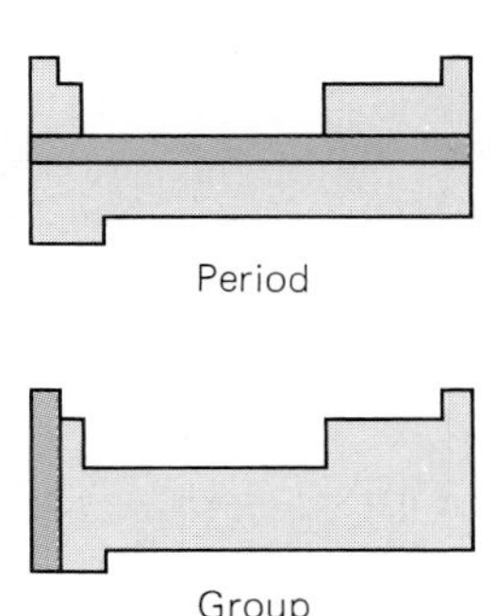

Figure 10-11 The elements in a group of the periodic table have similar properties whereas those in a period have different properties.

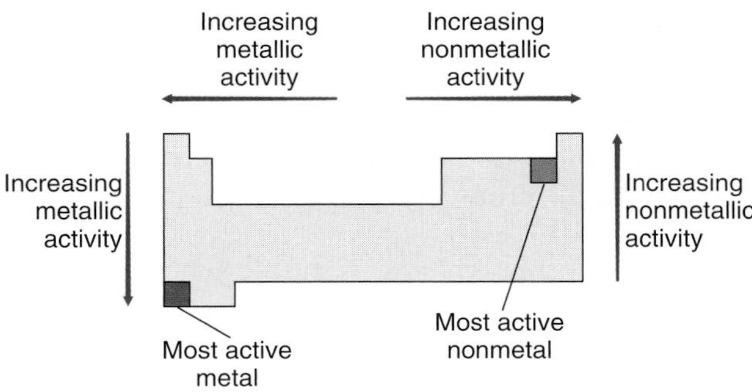

Figure 10-12 How chemical activity varies in the periodic table.

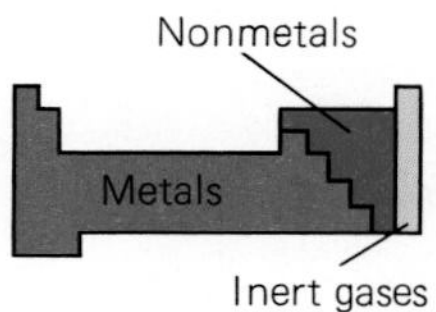

Figure 10-13 The majority of the elements are metals. The semimetals lie just to the right of the division between metals and nonmetals.

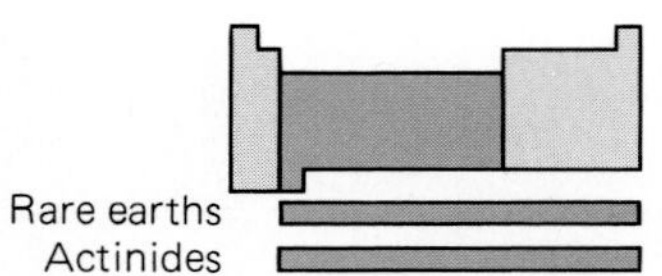

Figure 10-14 The transition elements are metals.

contains 32 elements, but 15 of them are brought out to a separate box. These **rare-earth** metals, which are so much alike that they are hard to separate chemically, are all placed together in the spot just to the right of barium, Z = 56. A similar group of closely related elements, the **actinides,** appears in the same position in period 7, and these elements are also shown in a separate box.

The relationships brought out by the periodic table are a little vague in places, but on the whole the table brings together similar elements with considerable accuracy. Mendeleev's achievement is all the more remarkable when we recall that in 1869, when the periodic law was developed, the notion of atomic number had not been discovered and only 63 elements were known. Mendeleev used average atomic mass, not atomic number, to arrange the elements in the periodic table, and he and later chemists found it necessary to deviate from the strict sequence of atomic masses for certain elements. When atomic numbers were later determined for the elements, the values of Z were found to fit the sequence in the periodic table perfectly.

Mendeleev's Predictions Because so few elements were known in his time, Mendeleev had to leave gaps in his table in order to have similar elements fall in line. Sure of the correctness of his classification, he proposed that these gaps represented undiscovered elements. From the position of each gap, from the properties of the elements around it, and from the variation of these properties across the periods and down the columns, he went on to predict the properties of the unknown elements. His predictions included not only general chemical activity but also numerical values for boiling points, melting points, and so on.

As the unknown elements were discovered one by one and as their properties were found to agree with Mendeleev's predictions, the validity and usefulness of the periodic table became firmly established. Perhaps its greatest triumph came at the end of the nineteenth century, when the inert gases were discovered. Here were six new elements whose existence Mendeleev was not aware of, but they fitted perfectly as one more family of similar elements into the periodic table. The history of the periodic table is a beautiful example of the scientific method in action.

Atomic Structure

Now we return to the atomic theory of Chap. 9 to seek the basis of the periodic law. Two basic principles determine the structure of an atom that contains more than one electron:

1. The exclusion principle, which states that only one electron can exist in each quantum state of an atom. Thus each electron in a complex atom must have a different set of the four quantum numbers n, l, m_l, and m_s (see Table 9-1).
2. An atom, like any other system, is stable when its total energy is a minimum. This means that the various electrons in a normal atom are in the quantum states of lowest energy permitted by the exclusion principle.

10.9 Shells and Subshells

They Contain Electrons with Similar Energies

Let us look into how electron energy varies with quantum state. In any atom, all the electrons with the same quantum number n are, on the average, about the same distance from the nucleus. These electrons therefore move around in nearly the same electric field and have similar energies. Such electrons are said to occupy the same atomic **shell.**

The energy of an electron in a particular shell also depends to some extent on the electron's orbital quantum number l, because l influences the shape of the electron's probability cloud and hence its average distance from the nucleus. The higher the value of l, the higher the energy. Electrons that share a certain value of l in a shell are said to occupy the same **subshell.** All the electrons in a subshell have very nearly the same energy.

The subshells in a shell of given n can have any value of l from 0 to $n - 1$. Thus the $n = 1$ shell has only the single subshell $l = 0$; the $n = 2$ shell has the subshells $l = 0$ and $l = 1$; the $n = 3$ shell has the subshells $l = 0$, $l = 1$, and $l = 2$; and so on.

Closed Shells and Subshells The exclusion principle limits the number of electrons that can occupy a given shell or subshell. A shell or subshell that contains its full quota of electrons is said to be **closed.**

The larger the orbital quantum number l, the more electrons the corresponding subshell can hold (see Sec. 9.13). When $l = 0$, the maximum number of electrons turns out to be 2; when $l = 1$, it is 6; when $l = 2$, it is 10; and so on. Adding up the electrons in its closed subshells gives the maximum number of electrons in a closed shell. Thus a closed $n = 1$ shell holds 2 $l = 0$ electrons; a closed $n = 2$ shell holds 2 $l = 0$ electrons plus 6 $l = 1$ electrons for a total of 8 electrons; a closed $n = 3$ shell holds these 8 electrons plus 10 more $l = 2$ electrons for a total of 18 electrons; and so on.

The concept of electron shells and subshells fits perfectly into the pattern of the periodic table, which turns out to mirror the atomic structures of the elements. Let us see how this pattern arises.

10.10 Explaining the Periodic Table

How an Atom's Electron Structure Determines Its Chemical Behavior

Table 10-7, which is illustrated in Fig. 10-15, shows the number of electrons in the shells of a number of elements. The table is arranged in the same manner as the periodic table to emphasize the relationship between the two tables.

Inert Gas Atoms In order to interpret Table 10-7 we note that the electrons in a closed shell are all tightly bound to the atom, since the positive nuclear charge that attracts them is large relative to the negative charge of any electrons in inner shells. An atom that contains only closed shells or subshells has its electric charge uniformly distributed, so it does not attract other electrons and its electrons cannot be easily removed. We would expect such atoms to be passive chemically, like the

Table 10-7 Simplified Table of Electron Structures of Some Atoms (Subshells Are Filled When a Shell Has 2, 8, or 18 Electrons)

Electrons in	H							He
1st shell	1							2
Electrons in	Li	Be	B	C	N	O	F	Ne
1st shell	2	2	2	2	2	2	2	2
2nd shell	1	2	3	4	5	6	7	8
Electrons in	Na	Mg	Al	Si	P	S	Cl	Ar
1st shell	2	2	2	2	2	2	2	2
2nd shell	8	8	8	8	8	8	8	8
3rd shell	1	2	3	4	5	6	7	8
Electrons in	K	Ca					Br	Kr
1st shell	2	2	. . .	. . .	. . .	. . .	2	2
2nd shell	8	8					8	8
3rd shell	8	8					18	18
4th shell	1	2					7	8
Electrons in	Rb	Sr					I	Xe
1st shell	2	2	. . .	. . .	. . .	. . .	2	2
2nd shell	8	8					8	8
3rd shell	18	18					18	18
4th shell	8	8					18	18
5th shell	1	2					7	8

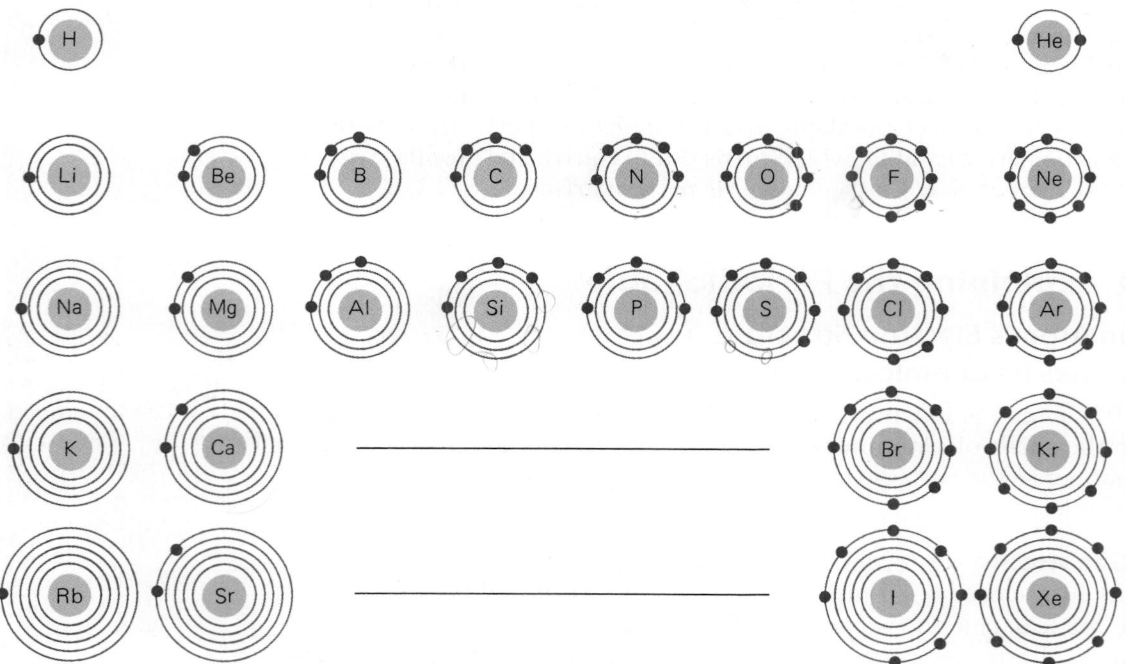

Figure 10-15 Electron structures of some atoms. In this schematic illustration of Table 10-7 the circles without dots represent closed (that is, completely filled) inner shells.

inert gases—and the inert gases all turn out to have closed-subshell electron structures!

Hydrogen and Alkali Metal Atoms Hydrogen and the alkali metals have single outer electrons. In the case of the hydrogen atom, the attractive force on the electron is due to a nuclear charge of only $+e$ and is not very great. In the case of the sodium atom, the total nuclear charge of $+11e$ acts on the two inner electrons, which are held very tightly. These two electrons shield part of the nuclear charge from the 8 electrons in the second shell, which are therefore attracted by a net charge of $+9e$. All 10 electrons in the first and second shells act to shield the outermost electron. This electron "sees" a net nuclear charge of only $+e$ and is held much less securely to the atom than any of the other electrons (Fig. 10-16*a*). This analysis also holds for the other alkali metals. As a result, atoms of hydrogen and of the alkali metals all tend to lose their outermost electrons in chemical reactions and therefore have similar chemical behavior.

As noted in Fig. 10-12, the alkali metals become more reactive going down group 1. This is because atoms become larger with increasing atomic number in group 1. As a result, their outer electrons are farther and farther from the nucleus, so the electric force on them is progressively weaker. Therefore these electrons are held less tightly and are more easily given up in a reaction.

Halogen Atoms An atom whose outer shell lacks one electron from being closed tends to pick up such an electron through the strong attraction of the poorly shielded nuclear charge. The chemical behavior of the halogens is the result. In the chlorine atom, for instance, there are 10 electrons in the inner two shells, just as in the sodium atom. However, the nuclear charge of chlorine is $+17e$ as compared with only $+11e$ for sodium (Fig. 10-16*b*). Hence the net charge "felt" by each of the 7 outer electrons in chlorine is $+7e$, not the $+e$ in the case of the single outer electron of sodium, and the attractive force on an outer electron in chlorine is 7 times greater.

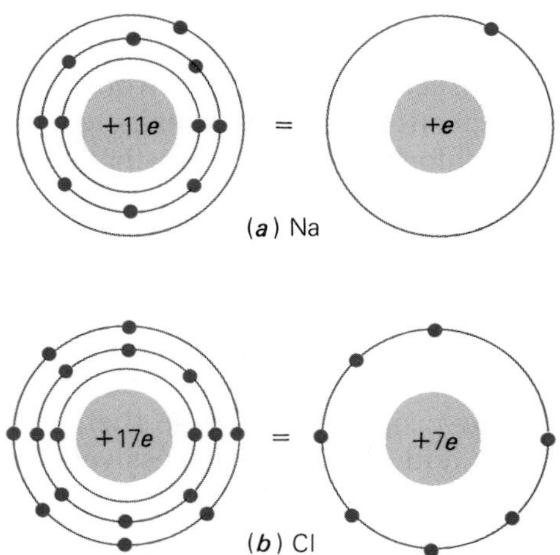

Figure 10-16 Electron shielding in sodium (*a*) and chlorine (*b*). Each outer electron in a Cl atom is acted upon by an effective nuclear charge 7 times greater than that acting upon the outer electron in a Na atom, even though the outer electrons in both cases are in the same shell.

In contrast to the alkali metals, the halogens become *less* reactive going down group 7. The reason again is the increase in atomic size with increasing atomic number. The halogens react by picking up electrons, and the larger the atom, the weaker the attractive electric force on the electrons in the outer shell. Thus the smaller the halogen atom, the more strongly it attracts an electron from elsewhere to fill its outer shell and the more reactive it is, as shown in Fig. 10-12.

Metals and Nonmetals These considerations lead us to general descriptions of metal and nonmetal atoms:

> A metal atom has one or several electrons outside closed shells or subshells. Such an atom combines chemically by losing these electrons to nonmetal atoms.

> A nonmetal atom needs one or several electrons to achieve closed shells or subshells. Such an atom combines chemically by picking up electrons from metal atoms or by sharing electrons with other nonmetal atoms.

The inert gases are exceptions to these statements, of course, since their atomic structures make it hard for them to gain or lose electrons. As a result, they have almost no ability to react chemically.

The steady change in chemical properties as we go across a period from an alkali metal on the left to a halogen on the right is easy to account for. An atom of an element in group 2, for instance magnesium (Mg), has two electrons outside closed inner shells, as we see in Fig. 10-15. These electrons "feel" an effective nuclear charge of $+2e$ and so are more tightly held than the single outer electron in sodium, which "feels" an effective nuclear charge of only $+e$. Not surprisingly, the outer electrons in an Mg atom are harder to pull away than the outer electron in Na. Hence Mg is less active as a metal than Na. Aluminum (Al), with three outer electrons, holds them still more securely, which is why Al is less active than Mg.

In a nonmetal atom, the more the gaps in its outer shell, the weaker the electric field that attracts additional electrons to complete the shell. Sulfur (S), with two electrons missing from its outer shell, is therefore less active a nonmetal than chlorine, which is missing just one electron. Phosphorus (P), with three electrons missing, is even less active. We can now see why, in any period, metallic activity (losing electrons) decreases going to the right, while nonmetallic activity (gaining electrons) increases going to the right (Fig. 10-12).

Electron shells and subshells are not always filled in consecutive order in atoms with many electrons. The transition elements in any period have similar properties because their outer electron shells are the same and they add electrons successively to inner shells (see Table 10-7).

Atomic Sizes

How big are atoms? Even though we have to think of an atom as being more like a tiny cloud, as in Fig. 9–34, than as a tiny ball, there are various ways to specify atomic sizes. For instance, we can determine how far apart the atoms in a solid composed of one element are and suppose that the atoms are spheres in contact with their nearest neighbors. This leads to a figure for what we can regard as the radius of each atom of that element.

The variations in size are what we would expect from the discussions of this section. The alkali metals have the largest atoms in each period of the periodic table, since their structures consists of a single electron outside closed inner shells that shield the electron from all but $+e$ of nuclear charge. Then there is a regular decrease in size within each period as the nuclear charge increases, which pulls the outer electrons in closer to the nucleus. Thus in period 3 sodium, magnesium, and aluminum have atomic radii of 0.19 nm, 0.16 nm, and 0.15 nm, respectively. At the end of each period there is a small increase in size due to the mutual repulsion of the outer electrons. Within a given group, atomic size increases with atomic number: in group 1, the atomic radii of lithium, sodium, and potassium are 0.15 nm, 0.19 nm, and 0.23 nm, respectively.

The sizes of ions can be found from measurements on crystals that contain them. Positive ions turn out to be smaller than the corresponding neutral atoms, negative ions are larger. As examples, the radii of Na^+ and Na are 0.10 nm and 0.19 nm, respectively, and those of S^- and S are 0.18 nm and 0.13 nm.

Chemical Bonds

What is the nature of the forces that bond atoms together when compounds are formed? This question is of basic importance to the chemist. It is also important to the physicist because the quantum theory of the atom cannot be complete unless it provides a satisfactory answer. The ability of the quantum theory to explain chemical bonding is further testimony to the power of this approach.

10.11 Types of Bond

Electric Forces Hold Atoms to One Another

Let us consider what happens when two atoms are brought closer and closer together. Three extreme situations may occur:

1. A **covalent bond** is formed. One or more pairs of electrons are shared by the two atoms. The shared electrons spend more time between the atoms than on their far sides, which produces an attractive force. An example is H_2, the hydrogen molecule, whose two electrons belong jointly to the two protons (Fig. 10-17).
2. An **ionic bond** is formed. One or more electrons from one atom shift to another atom, and the resulting positive and negative ions attract each other. An example is NaCl, where the bond exists between Na^+ and Cl^- ions and not between Na and Cl atoms (Fig. 10-18).
3. No bond is formed. The atoms do not interact to produce an attractive force.

In H_2 the bond is purely covalent and in NaCl it is purely ionic, but in many other molecules an intermediate type of bond occurs in which the atoms share electrons to an unequal extent. An example is the HCl molecule, where the Cl atom attracts the shared electrons more strongly than the H atom.

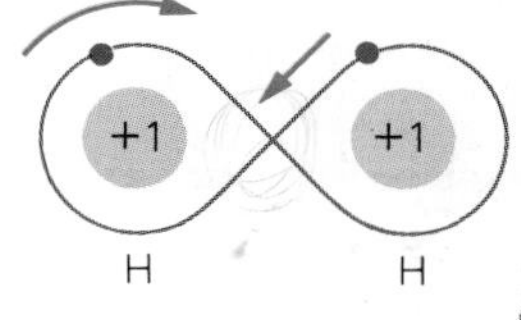

Figure 10-17 A simplified model of covalent bonding in hydrogen. The shared electrons spend more time on the average between their parent nuclei than on the far sides of the nuclei and therefore lead to an attractive internuclear force.

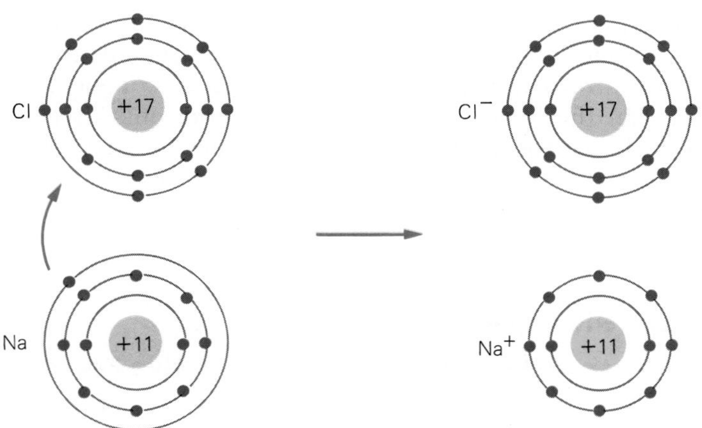

Figure 10-18 A simplified model of ionic bonding. Sodium and chlorine combine chemically by the transfer of electrons from sodium atoms to chlorine atoms. The resulting ions attract electrically.

BIOGRAPHY Linus Pauling (1901–1994)

A native of Oregon, Pauling received his Ph.D. from the California Institute of Technology and remained there for his entire scientific career except for a period in the middle 1920s when he was in Germany to study the new quantum mechanics. A pioneer in applying quantum theory to chemistry, Pauling provided many of the key insights that led to an understanding of the details of chemical bonding. He also did important work in molecular biology, in particular protein structure: with the help of x-ray diffraction, he discovered the helical and pleated sheet forms that protein molecules can have. It was Pauling who realized that sickle cell anemia is a "molecular disease" due to hemoglobin with one wrong amino acid resulting from a genetic fault. He received the Nobel Prize in chemistry in 1954.

In 1923 Pauling met Ava Helen Miller in a chemistry class and she married him despite his admission that, "if I had to choose between you and science, I'm not sure that I would choose you." She introduced him to the world outside the laboratory, and in his later years he became more and more politically active. Pauling fought to stop the atmospheric testing of nuclear weapons with its attendant radioactive fallout, a crusade that led to a 2500-page FBI file, the nuclear test ban treaty, and the Nobel Peace Prize in 1965.

Ionic bonds usually do not result in the formation of molecules. Strictly speaking, a molecule is an electrically neutral group of atoms that is held together strongly enough to be experimentally observable as a particle. Thus the individual units that constitute gaseous hydrogen each consist of two hydrogen atoms, and we are entitled to regard them as molecules.

Crystals of table salt (NaCl), however, are aggregates of sodium and chlorine ions (Fig. 10-19). Although arranged in a certain definite way, as we saw in Fig. 10-6, the ions do not pair off into individual molecules consisting of one Na^+ ion and one Cl^- ion. Salt crystals may in fact be of almost any size. There are always equal numbers of Na^+ and Cl^- ions in salt, so that the formula NaCl correctly represents its composition. Despite the absence of individual NaCl molecules in solid NaCl, the electric force between adjacent Na^+ and Cl^- ions makes NaCl as characteristic an example of chemical bonding as H_2.

Figure 10-19 Much of the world's salt was once produced by evaporating seawater, as is still done in the Canary Islands.

10.12 Covalent Bonding

Sharing Electron Pairs Produces an Attractive Force

We saw in Fig. 10-17 how two identical atoms, in this case hydrogen atoms, can bond together by sharing a pair of electrons. In some molecules more than one pair of electrons is shared. Examples are O_2, which has two shared electron pairs, and N_2, which has three. If we use a pair of dots to stand for a shared pair of electrons, the H_2, O_2, and N_2 molecules can be represented as follows:

H : H	O : : O	N : : : N
hydrogen molecule	oxygen molecule	nitrogen molecule

Substances whose atoms are joined by shared electron pairs are called **covalent.** In general, they are either nonmetallic elements or else compounds of one nonmetal with another, although some compounds that contain metals belong to this class. The shared pair or pairs of electrons between two atoms constitute a **bond.** In general, double or triple bonds are stronger than single ones. The triple bonds in nitrogen molecules are among the strongest known. As a result, nitrogen molecules are extremely stable and do not react readily; what this means for living things, which need nitrogen for their proteins, is discussed in Sec. 13.15.

In some covalent compounds the shared electron pairs are closer to one atom than to the other. Two examples are HCl (hydrochloric acid) and H_2O:

H : Cl   H :O
                  ¨
                  H

These substances are called **polar covalent compounds,** because one part of the molecule is relatively negative and another part is relatively positive. All gradations can be found between uniformly covalent molecules at one extreme, through polar covalent molecules, to ionic compounds at the other extreme. For example,

Covalent	Cl	:	Cl
Polar covalent	H	:	Cl
Ionic	Na	:	Cl

The carbon atom has four outer electrons to share with each other and with other atoms in covalent bonds. Covalent compounds that contain carbon are called **organic compounds.** Organic compounds are so important that all of Chap. 13 is devoted to them.

10.13 Ionic Bonding

Electron Transfer Creates Ions That Attract Each Other

The simplest example of a chemical reaction that involves electron transfer is the combination of a metal and a nonmetal. For a specific case, let us consider the burning of sodium in chlorine to give sodium chloride. From Fig. 10-15 it is clear that Na and Cl are perfect mates—one has an electron to lose, the other an electron to gain. In the process of combination, an electron goes from Na to Cl, as shown in Fig. 10-18.

The stability of the resulting closed electron shells in both ions is shown by the large amount of energy given off in the form of heat and light when this reaction takes place. The compound NaCl is quite unreactive because each of its ions has a stable electron structure. To break NaCl apart, which means to return the electron from Cl^- to Na^+, requires the same considerable energy that was set free when the compound was formed.

As we know, metal atoms tend to lose their outer electrons, like sodium in NaCl. Nonmetal atoms, on the other hand, tend to gain electrons so as to fill in gaps in their outer shells. In most reactions of this sort the metal loses all its outer electrons, and the nonmetal fills all the gaps in its structure. When sodium combines with sulfur, for instance, each S atom has two spaces to fill for a closed outer shell (Fig. 10-20), but each Na atom has only one electron to give. Hence two Na atoms are needed for each S atom, and the resulting compound is Na_2S. When calcium combines with oxygen, each Ca atom contributes two electrons to each O atom, and the formula of the compound is CaO.

Compounds formed by electron transfer are called **ionic compounds.** Some are simple compounds like NaCl, Na_2S, and CaO. Others have more complex formulas, such as Na_2SO_4, KNO_3, and $CaCO_3$. In these latter compounds electrons from the metal atoms have been transferred to nonmetal **atom groups** (SO_4, NO_3, CO_3) instead of to single nonmetal atoms.

Ionic compounds in general contain a metal and one or more nonmetals, and their crystal structures have alternate positive and negative ions. Most of them are crystalline solids with high melting points. We expect this, since melting involves separating the ions.

Example 10.1

Why do only the outermost electrons of an atom usually participate in bonding?

Solution

Inner electrons are much more tightly held to a nucleus both because they are closer to it and because they are shielded by fewer intervening electrons. Hence the inner electrons are unable either to transfer to another atom in an ionic bond or to be shared with another atom in a covalent bond.

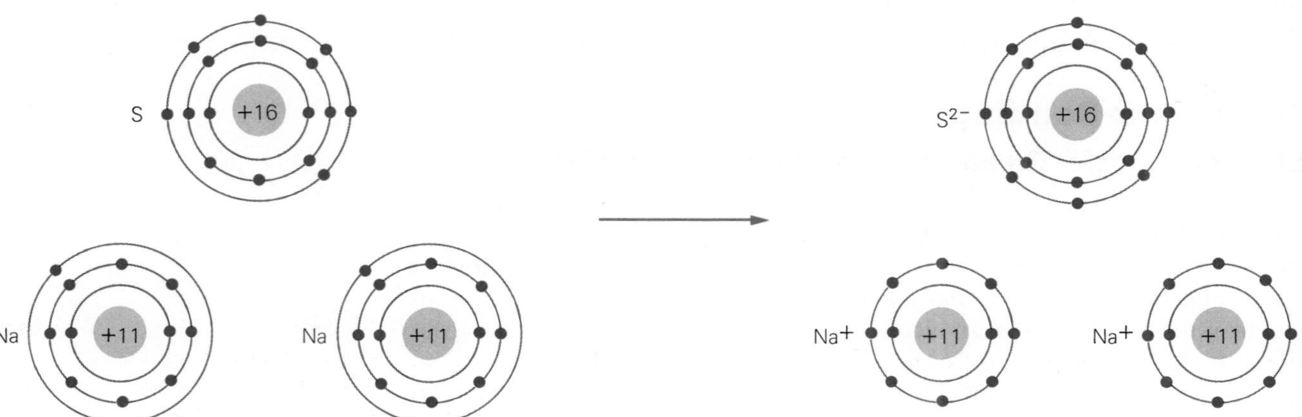

Figure 10-20 Ionic bonding in Na_2S. Each sodium atom contributes one electron to the sulfur atom, and the resulting S^{2-} ion attracts the two Na^+ atoms.

10.14 Ionic Compounds

Matching Up Ions

When a metal and a nonmetal combine to form an ionic compound, the atoms of the metal give up one or more electrons to atoms of the nonmetal. We can figure out the formula of the compound by knowing how many electrons the metal atoms tend to lose and how many electrons the nonmetal atoms tend to gain.

As we have already seen, Na tends to lose one electron to become Na^+ and Cl tends to gain one electron to become Cl^-. Hence the formula of sodium chloride is NaCl. Similarly sulfur tends to gain two electrons to become S^{2-}, so sodium sulfide must have the formula Na_2S in order that two electrons be available for each S atom. Calcium forms Ca^{2+} ions, hence calcium sulfide must have the formula CaS with two electrons shifting from each Ca atom to each S atom.

Table 10-8 shows the ions formed by some common elements when they enter into compounds. A few elements form different ions under different circumstances, for example copper (Cu^+, Cu^{2+}) and iron (Fe^{2+}, Fe^{3+}). In such cases the name of the element in its compound is followed by a Roman numeral to indicate the ionic charge. Thus $FeCl_2$ is called iron(II) chloride (in speech, "iron-two chloride") because it contains Fe^{2+} ions, and $FeCl_3$ is called iron(III) chloride because it contains Fe^{3+} ions.

With the help of Table 10-8 we can see what happens when a given metal combines with a given nonmetal. The positive and negative charges

Table 10-8 Ions of Some Common Elements

Element	Ion
Hydrogen	H^+
Lithium	Li^+
Sodium	Na^+
Potassium	K^+
Silver	Ag^+
Copper	Cu^+, Cu^{2+}
Mercury	Hg^+, Hg^{2+}
Magnesium	Mg^{2+}
Calcium	Ca^{2+}
Barium	Ba^{2+}
Zinc	Zn^{2+}
Iron	Fe^{2+}, Fe^{3+}
Aluminum	Al^{3+}
Tin	Sn^{2+}, Sn^{4+}
Lead	Pb^{2+}, Pb^{4+}
Fluorine	F^-
Chlorine	Cl^-
Bromine	Br^-
Iodine	I^-
Oxygen	O^{2-}
Sulfur	S^{2-}
Nitrogen	N^{3-}
Phosphorus	P^{3-}

Example 10.2

Find the formula of aluminum oxide, which consists of aluminum and oxygen ions.

Solution

From Table 10-8 we see that aluminum forms Al^{3+} ions, and oxygen forms O^{2-} ions. Because the ion charges are different in magnitude, the formula cannot be simply AlO. A straightforward way to arrive at the correct formula is to first write down the symbols for each ion in parentheses:

$$(Al^{3+}) \qquad (O^{2-})$$

Now we put a subscript after each parentheses equal to the magnitude of the charge on the *other* ion (that is, the amount of charge regardless of sign):

$$(Al^{3+})_2(O^{2-})_3$$

Finally we delete the ionic charges and, where possible, the parentheses:

$$Al_2O_3$$

This is the formula of aluminum oxide (Fig. 10-21).

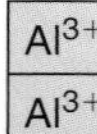

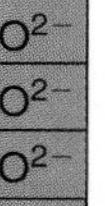

Figure 10-21 The charges in each unit of an ionic compound, such as Al_2O_3, must be in balance. Here the 6 positive charges of $2Al^{3+}$ are balanced by the 6 negative charges of $3O^{2-}$.

on the ions must always balance out, and a little thought may be needed to find the right combination.

10.15 Atom Groups

They Act as Units in Chemical Reactions

Certain groups of atoms appear as units in many compounds and remain together during chemical reactions. An example is the group SO_4, which consists of a sulfur atom joined to four oxygen atoms. This **sulfate group,** whose ion has a charge of −2, is found in a number of compounds:

Sodium sulfate	Na_2SO_4
Potassium sulfate	K_2SO_4
Copper(II) sulfate	$CuSO_4$
Magnesium sulfate	$MgSO_4$

How can we be sure that the sulfate group enters into chemical reactions as a unit? One way is to mix solutions of magnesium sulfate and barium chloride, $BaCl_2$. What happens is that a precipitate is formed, which analysis shows consists of barium sulfate, $BaSO_4$. (A **precipitate** is an insoluble solid that results from a chemical reaction in solution.) The solution left behind contains magnesium ions and chlorine ions (Fig. 10-22). The sulfate group has changed partners.

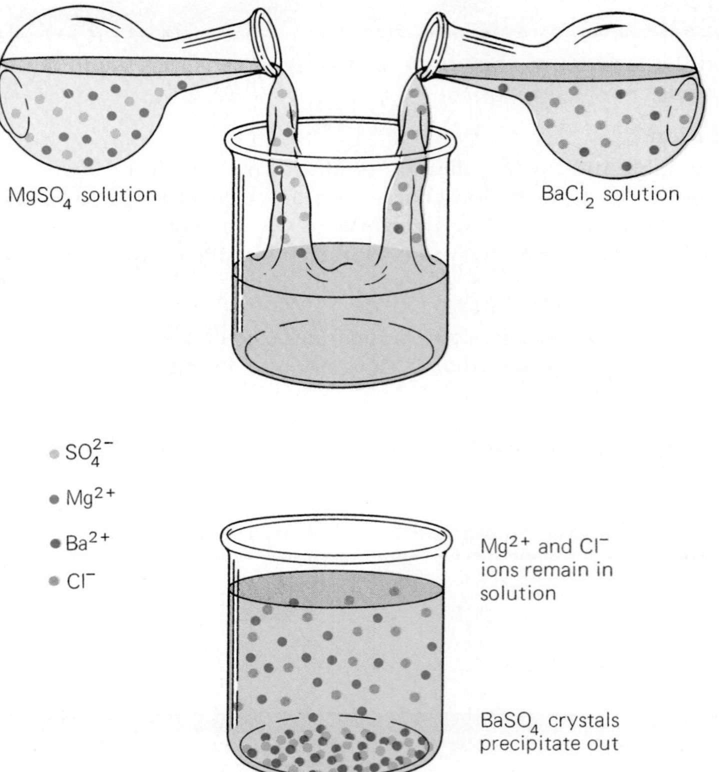

Figure 10-22 When magnesium sulfate ($MgSO_4$) and barium chloride ($BaCl_2$) are dissolved in water, a precipitate of the insoluble compound barium sulfate is produced. The magnesium and chlorine ions remain in solution.

When two or more groups of a single kind are present in each molecule of a compound, the formula is written with parentheses around the group. An example is

Calcium nitrate $Ca(NO_3)_2$

The Ca^{2+} ion needs two NO_3^- ions to combine with in order that the charges balance out. Table 10-9 is a list of common atom groups and the charges their ions have.

Table 10-9 Ions of Some Common Atom Groups

Atom Group	Ion
Ammonium	NH_4^+
Nitrate	NO_3^-
Permanganate	MnO_4^-
Chlorate	ClO_3^-
Hydroxide	OH^-
Cyanide	CN^-
Sulfate	SO_4^{2-}
Carbonate	CO_3^{2-}
Chromate	CrO_4^{2-}
Silicate	SiO_4^{2-}
Phosphate	PO_4^{3-}

10.16 Naming Compounds

The Vocabulary of Chemistry

Here are some of the rules chemists use to name compounds:

1. The ending *-ide* usually indicates a compound having only two elements:

   Sodium chloride $NaCl$
   Calcium oxide CaO

   The hydroxides, which contain the OH^- ion, are the most common exceptions to this rule:

   Barium hydroxide $Ba(OH)_2$

2. The ending *-ate* indicates a compound that contains oxygen and two or more other elements:

   Sodium sulfate Na_2SO_4
   Potassium nitrate KNO_3

3. When the same pair of elements occurs in two or more compounds, a prefix (*mono-* = 1, *di-* = 2, *tri-* = 3, *tetra-* = 4, *penta-* = 5, *hexa-* = 6, and so on) may be used to indicate the number of one or both kinds of atom in the molecule:

   Carbon monoxide CO
   Carbon dioxide CO_2

4. When one of the elements in a compound is a metal that can form different ions, the scheme mentioned in Sec. 10.14 is used. In this scheme the ionic charge of the metal is given by a roman numeral:

   Iron(II) chloride $FeCl_2$
   Iron(III) chloride $FeCl_3$

   The names of molecular compounds that contain hydrogen often follow tradition instead of a definite system. Thus

   Methane CH_4
   Water H_2O
   Ammonia NH_3

10.17 Chemical Equations

The Atoms on Each Side Must Balance

A **chemical equation** is a shorthand way to express the results of a chemical change. In a chemical equation the formulas of the **reactants** (reacting substances) appear on the left-hand side and the formulas of the products appear on the right-hand side. When charcoal (which is almost pure carbon) burns in air, for instance, what is happening is that carbon atoms are reacting with oxygen molecules in the air to form carbon dioxide molecules. The corresponding equation is therefore

$$\underset{\text{carbon atom}}{C} + \underset{\text{oxygen molecule}}{O_2} \rightarrow \underset{\text{carbon dioxide molecule}}{CO_2}$$

In order to correctly represent a chemical reaction, a chemical equation must be **balanced:** the number of atoms of each kind must be the same on both sides of the equation. Let us consider the decomposition (breaking down) of water that occurs when an electric current is passed through a water sample (Fig. 10-23). The process is called **electrolysis** and is written in words as

$$\text{Water} \rightarrow \text{hydrogen} + \text{oxygen}$$

Using the formulas for these substances, we might write

$$H_2O \rightarrow H_2 + O_2 \qquad \textit{Unbalanced equation}$$

Here two atoms of oxygen are shown on the right-hand side but only one atom of oxygen on the left. The equation is therefore **unbalanced.** We cannot just write O instead of O_2 on the right because gaseous oxygen has the formula O_2. Nor can we write a subscript "2" after the O in H_2O because H_2O_2 is the formula for hydrogen peroxide, not water.

Balancing an Equation The first step toward balancing the equation is to show two molecules of H_2O on the left:

$$2H_2O \rightarrow H_2 + O_2 \qquad \textit{Oxygen atoms balanced}$$

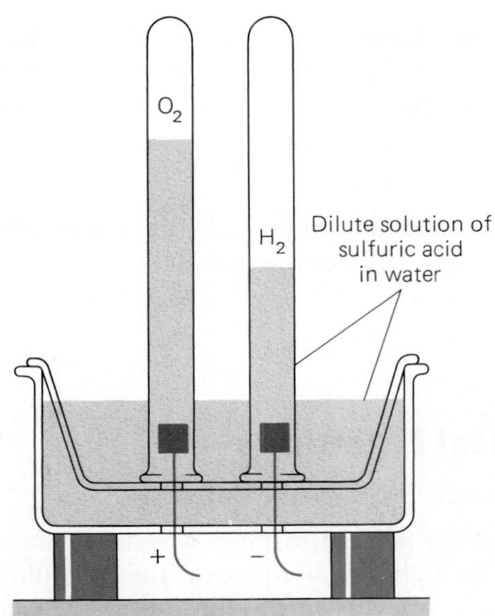

Figure 10-23 Electrolysis of water. An electric current decomposes water into gaseous hydrogen and oxygen. The volume of the hydrogen evolved is twice that of the oxygen, since water contains twice as many hydrogen atoms as oxygen atoms. A trace of sulfuric acid is used to enable the water to conduct electricity.

Now we have two O atoms on both sides of the equation, which means that these atoms are balanced. However, there are four H atoms on the left but only two H atoms on the right. The remedy is to put two H_2 molecules on the right:

$$2H_2O \rightarrow 2H_2 + O_2 \quad \textit{Balanced equation}$$

2 water molecules → 2 hydrogen molecules + 1 oxygen molecule

Since two O atoms and four H atoms appear on each side, the equation is balanced (Fig. 10-24).

An important point is that a number in front of a formula multiplies everything in the formula, whereas a subscript applies only to the symbol before it. Thus $2H_2O$ refers to two complete H_2O molecules, each one of which has two H atoms and one O atom.

It is worth noting that being able to balance the chemical equation for a certain reaction does not necessarily mean that the reaction can occur. And even if a reaction can take place, the balanced equation for the reaction does not tell us the particular conditions (of temperature and pressure, for instance) that might be needed.

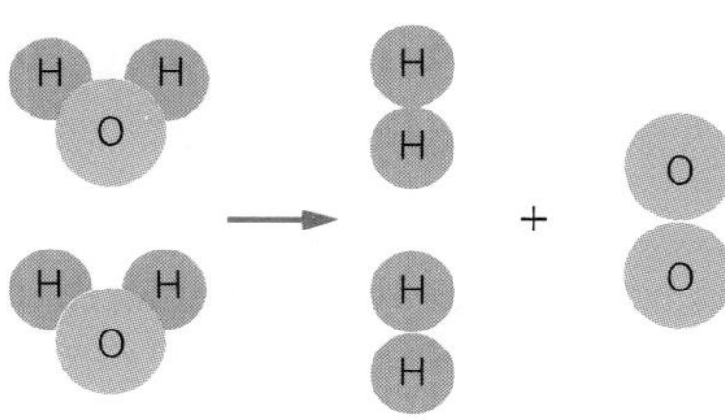

Figure 10-24 Schematic diagram of the electrolysis of water.

Example 10.3

Propane, C_3H_8, is a gas that is widely used in cooking stoves and blowtorches. When it burns, propane combines with oxygen from the air to form carbon dioxide and water vapor (Fig. 10-25). Find the balanced equation of the process.

Solution

We begin by writing the unbalanced equation of the process:

$$C_3H_8 + O_2 \rightarrow CO_2 + H_2O \quad \textit{Unbalanced equation}$$

Generally it is easiest to balance hydrogen and oxygen atoms last, so here we start with carbon atoms. To balance the three carbon atoms on the left we need three on the right:

$$C_3H_8 + O_2 \rightarrow 3CO_2 + H_2O \quad \textit{Carbon atoms balanced}$$

Eight hydrogen atoms appear on the left. Hence we must have eight on the right, which means four H_2O molecules:

$$C_3H_8 + O_2 \rightarrow 3CO_2 + 4H_2O \quad \textit{Hydrogen atoms balanced}$$

The three CO_2 molecules have six oxygen atoms and the four H_2O molecules have four, for a total of ten oxygen atoms on the right. Five O_2 molecules on the left will provide the required ten oxygen atoms there:

$$C_3H_8 + 5O_2 \rightarrow 3CO_2 + 4H_2O \quad \textit{Balanced equation}$$

This is the balanced equation for the burning of propane. Since a propane stove or torch uses large amounts of oxygen and produces carbon dioxide, good ventilation is clearly necessary.

Figure 10-25 The combustion of propane produces carbon dioxide and water vapor and releases a great deal of energy.

Important Terms and Ideas

Elements are the simplest substances present in bulk matter. An element cannot be decomposed or changed into other elements by chemical means. Two or more elements may combine chemically to form a **compound,** a new substance whose properties are different from those of the elements it contains. According to the **law of definite proportions,** the elements that make up a compound are always combined in the same proportions by mass.

Other materials are **mixtures** of elements or compounds or both. The constituents of a mixture keep their characteristic properties. A **solution** is a uniform (or **homogeneous**) mixture.

The ultimate particles of any element are called **atoms.** The ultimate particles of gaseous compounds consist of atoms of the elements they contain joined together in separate **molecules.** Some compounds in the liquid and solid state also consist of molecules, but in many others the atoms are linked in larger arrays. In a given compound, however, the ratios between its various atoms are fixed.

The **periodic law** states that if the elements are listed in order of atomic number, elements with similar chemical and physical properties appear at regular intervals. Such similar elements form **groups.** The halogens, the alkali metals, and the inert gases are examples.

The electrons in an atom that have the same principal quantum number n are said to occupy the same **shell.** Electrons in a given shell that have the same orbital quantum number l are said to occupy the same **subshell.** Shells and subshells are **closed** when they contain the maximum number of electrons permitted by the exclusion principle. Atoms that contain only closed shells and subshells are extremely stable. The concept of shells and subshells is able to account for the periodic law.

A **metal** atom has one or several electrons outside closed shells or subshells. It combines chemically by losing these electrons to nonmetal atoms. A **nonmetal** atom lacks having closed shells or subshells by one or several electrons. It combines chemically by picking up electrons from metal atoms or by sharing electrons with other nonmetal atoms. **Semimetals,** or **metalloids,** are elements whose properties are intermediate between metals and nonmetals.

In a **covalent bond** between atoms, the atoms share one or more electron pairs. In an **ionic bond,** electrons are transferred from one atom to another and the resulting ions then attract each other. Many bonds in liquids and solids are intermediate between covalent and ionic.

Atom groups, such as SO_4^{2-} (the sulfate group), appear as units in many compounds and remain together during chemical reactions.

A **chemical equation** expresses the result of a chemical change. When the equation is **balanced,** the number of each kind of atom is the same on both sides of the equation.

Multiple Choice

1. A pure substance that cannot be decomposed by chemical means is
   a. an element c. a solid
   b. a compound d. a solution
2. Elements can be distinguished unambiguously by their
   a. hardnesses c. atomic numbers
   b. colors d. electrical properties
3. The number of known elements is approximately
   a. 50 c. 200
   b. 100 d. 500
4. At room temperature and atmospheric pressure, most elements are
   a. gases c. metallic solids
   b. liquids d. nonmetallic solids
5. Which of the following substances is a homogeneous mixture?
   a. iron c. salt
   b. seawater d. paper
6. Which of the following substances is a compound?
   a. iron c. salt
   b. seawater d. paper
7. The nonmetal whose chemical behavior is most like that of typical metals is
   a. hydrogen c. chlorine
   b. helium d. carbon
8. Which one or more of the following properties are characteristic of all metals?
   a. Conducts electricity well
   b. Conducts heat well
   c. Is a solid at room temperature
   d. Is transparent to light
9. Iodine is an example of
   a. an inert gas c. a halogen
   b. an alkali metal d. a compound
10. Which of the following is (are) true for active elements?
    a. They form compounds readily
    b. They form stable compounds
    c. They never occur as gases at room temperature
    d. They liberate more heat when they react than inactive elements do

11. Of the following metals, the most active chemically is
 a. gold c. iron
 b. aluminum d. sodium
12. Of the following metals, the least active chemically is
 a. gold c. lead
 b. calcium d. sodium
13. Of the following nonmetals, the most active chemically is
 a. helium c. oxygen
 b. fluorine d. sulfur
14. At room temperature, chlorine is
 a. a colorless gas
 b. a greenish-yellow gas
 c. a reddish-brown liquid
 d. a steel-gray solid
15. Of the following nonmetals, the one not a halogen is
 a. fluorine c. sulfur
 b. bromine d. iodine
16. The place of an element in the periodic table is determined by its
 a. atomic number c. density
 b. atomic mass d. chemical activity
17. Each vertical column of the periodic table includes elements with chemical characteristics that are, in general,
 a. identical
 b. similar
 c. different
 d. sometimes similar and sometimes different
18. The periodic table of the elements does not
 a. permit us to make accurate guesses of the properties of undiscovered elements
 b. reveal regularities in the occurrence of elements with similar properties
 c. include the inert gases
 d. tell us the arrangement of the atoms in a molecule
19. The elements in group 1 of the periodic table (except for hydrogen) are
 a. all metals
 b. all nonmetals
 c. both metals and nonmetals
 d. neither metals nor nonmetals
20. Of the elements in group 8 of the periodic table, at room temperature and atmospheric pressure
 a. all are gases
 b. all are liquids
 c. some are gases and the others liquids
 d. some are liquids and the others solids
21. In each period of the periodic table, metallic activity
 a. increases to the right
 b. remains constant
 c. decreases to the right
 d. varies with no regular pattern
22. An alkali metal atom
 a. has one electron in its outer shell
 b. has two electrons in its outer shell
 c. has a filled outer shell
 d. lacks one electron of having a filled outer shell
23. A halogen atom
 a. has one electron in its outer shell
 b. has two electrons in its outer shell
 c. has a filled outer shell
 d. lacks one electron of having a filled outer shell
24. An inert gas atom
 a. has one electron in its outer shell
 b. has two electrons in its outer shell
 c. has a filled outer shell
 d. lacks one electron of having a filled outer shell
25. The most important factor in determining the chemical behavior of an atom is its
 a. nuclear structure c. atomic mass
 b. electron structure d. solubility
26. An atom that loses its outer electron or electrons readily is
 a. an active metal
 b. an active nonmetal
 c. an inactive metal
 d. an inactive nonmetal
27. When they combine chemically with metal atoms, nonmetal atoms tend to
 a. gain electrons to become negative ions
 b. lose electrons to become positive ions
 c. remain electrically neutral
 d. any of these, depending upon the circumstances
28. When atoms join to form a molecule,
 a. energy is absorbed
 b. energy is given off
 c. energy is neither absorbed nor given off
 d. any of these, depending on the circumstances
29. Relative to the number of electrons in the atoms that join to form a molecule, the number of electrons in the molecule is
 a. smaller c. larger
 b. the same d. any of the above, depending on the molecule
30. In a covalent molecule,
 a. at least one metal atom is always present
 b. one or more electrons are transferred from one atom to another
 c. adjacent atoms share one or more electrons
 d. adjacent atoms share one or more pairs of electrons

31. An element that can form an ionic compound with chlorine is
   a. carbon c. sulfur
   b. copper d. neon
32. Sodium chloride crystals consist of
   a. NaCl molecules c. Na^+ and Cl^- ions
   b. Na and Cl atoms d. Na^- and Cl^+ ions
33. A compound whose name ends in -ate always contains
   a. hydrogen c. hydrogen and oxygen
   b. oxygen d. carbon
34. The number of atoms in a molecule of ammonium sulfide, $(NH_4)_2S$, is
   a. 3 c. 10
   b. 6 d. 11
35. The number of oxygen atoms in a molecule of aluminum sulfate, $Al_2(SO_4)_3$, is
   a. 3 c. 7
   b. 4 d. 12
36. The nitrate ion has the formula NO_3^-. The formula of mercury(II) nitrate is
   a. $HgNO_3$ c. $Hg(NO_3)_2$
   b. Hg_2NO_3 d. $Hg_2 (NO_3)_2$
37. Which of the following chemical equations is balanced?
   a. $Fe_2O_3 + CO \rightarrow Fe + 2CO_2$
   b. $Na_2 SO_2 + S \rightarrow S_2O_3 + S$
   c. $3CuO + 2NH_3 \rightarrow 3Cu + 3H_2O + N_2$
   d. $4Al + 3Fe_3O_4 \rightarrow 4Al_2 O_3 + 9Fe$
38. Which of the following chemical equations is unbalanced?
   a. $2Hg + O_2 \rightarrow 2HgO$
   b. $2H_2S + 3O_2 \rightarrow 2H_2O + 2SO_2$
   c. $Na_2O + H_2O \rightarrow 2NaOH$
   d. $SO_2 + H_2O \rightarrow H_2SO_4$
39. The missing number in the equation $2Ca(NO_3)_2 \rightarrow 2CaO + [\ ]NO_2 + O_2$ is
   a. 1 c. 3
   b. 2 d. 4
40. The missing number in the equation $4NH_3 + [\ ]O_2 \rightarrow 4NO + 6H_2O$ is
   a. 1 c. 5
   b. 2 d. 10

Exercises

10.1 Chemical Change
10.2 Three Classes of Matter
10.3 The Atomic Theory

1. The conversion of water to ice is considered a physical change, whereas the conversion of iron to rust is considered a chemical change. Why?
2. How can you show that water is a compound rather than a homogeneous mixture of hydrogen and oxygen?
3. Heating is a physical process. When mercuric oxide is heated, it becomes mercury and oxygen. Does this mean that mercuric oxide is a mixture rather than a compound?
4. Which of the following substances are homogeneous and which are heterogeneous? Blood, carbon dioxide gas, solid carbon dioxide, rock, steak, iron, rust, concrete, air, oxygen, salt, milk.
5. Which of the following homogeneous liquids are elements, which are compounds, and which are solutions? Alcohol, mercury, liquid hydrogen, pure water, seawater, beer.
6. How does the law of definite proportions help to distinguish between a compound of certain elements and a mixture of the same elements?
7. What is the most abundant element in the universe? In the human body?
8. What kind(s) of particles make up (a) gaseous compounds, (b) liquid compounds, and (c) solid compounds?
9. The formula for liquid water is H_2O, for solid zinc sulfide ZnS, and for gaseous nitrogen dioxide NO_2. Precisely what information do these formulas convey? What information do they *not* convey?
10. What is the difference in meaning between C_4 and 4C?

10.4 Metals and Nonmetals

11. From what physical and chemical characteristics of iron do we conclude that it is a metal? From what physical and chemical characteristics of sulfur do we conclude that it is a nonmetal?

10.5 Chemical Activity

12. The Bronze Age got its name from the ability of people in that stage of human development to refine tin and copper from their ores; bronze is an alloy (mixture) of tin and copper and is stronger than either of these metals by itself. In the later Iron Age, the still stronger iron could be won from its ores. Nowadays metals such as aluminum and magnesium are refined electrically. Relate this sequence of metallurgical skill to the sequence of metal activity in Table 10-2.
13. Sodium never occurs in nature as the free element, and platinum seldom occurs in

combination. How are these observations related to the chemical activities of the two metals?

14. What energy change would you expect when a molecule breaks up into its constituent atoms?

10.6 Families of Elements

10.7 The Perodic Table

10.8 Groups and Periods

15. Are the chemical properties of the elements in a vertical column or in a horizontal row of the periodic table similar to one another?

16. The element astatine (At), which appears at the bottom of the halogen column in the periodic table, has been prepared artificially in minute amounts but has not been found in nature. Using the periodic law and your knowledge of the halogens, predict the properties of this element, as follows:

- **a.** At room temperature, is it solid, liquid, or gaseous?
- **b.** How many atoms does a molecule of its vapor contain?
- **c.** Is it very soluble, moderately soluble, or slightly soluble in water?
- **d.** What is the formula for its compound with hydrogen?
- **e.** What are the formulas for its compounds with potassium and calcium?
- **f.** Is its compound with potassium more or less stable than potassium iodide?

17. The following metals are listed in order of decreasing chemical activity: potassium, sodium, calcium, magnesium. How does this order agree with their positions in the periodic table? Where would you place cesium in the above list?

10.9 Shells and Subshells

10.10 Explaining the Periodic Table

18. A century ago an entirely new group of elements, the inert gases, was discovered. Is it possible that, in the future, another as yet unknown group of the periodic table might be found?

19. (a) What is characteristic about the outer electron shells of the alkali metals? (b) Of the halogens? (c) Of the inert gases?

20. Group 2 of the periodic table contains the family of elements called the **alkaline earths.** How active chemically would you expect an alkaline earth element to be compared with the alkali metal next to it? Why?

21. Why do fluorine and chlorine exhibit similar chemical behavior?

22. Why do lithium and sodium exhibit similar chemical behavior?

23. Electrons are much more readily liberated from metals than from nonmetals when irradiated with visible or ultraviolet light. Can you explain why this is true? From metals of what group would you expect electrons to be liberated most easily?

24. Would you expect magnesium or calcium to be the more active metal? Explain your answer in terms of atomic structure.

25. Why are chlorine atoms more chemically active than chlorine ions?

26. What is the difference in atomic structure between the two isotopes of chlorine? How would you account for the great chemical similarity of the two isotopes?

27. The transition elements in any period have the same or nearly the same outer electron shells and add electrons successively to inner shells. How does this bear upon their chemical similarity?

28. (a) Would you expect N or Br to differ most in its chemical properties from F? (b) B or Si from C? (c) P or O from N?

29. The rare element selenium has the following arrangement of electrons: 2 in the first shell, 8 in the second, 18 in the third, and 6 in the fourth. Would you expect selenium to be a metal or a nonmetal?

30. What is the effective nuclear charge that acts on each electron in the outer shell of the sulfur ($Z = 16$) atom? Would you think that such an electron is relatively easy or relatively hard to detach from the atom?

31. What is the effective nuclear charge that acts on each electron in the outer shell of the calcium ($Z = 20$) atom? Would you think that such an electron is relatively easy or relatively hard to detach from the atom?

10.11 Types of Bond

10.12 Convalent Bonding

10.13 Ionic Bonding

32. Illustrate with electronic diagrams (a) the reaction between a lithium atom and a fluorine atom, and (b) the reaction between a magnesium atom and a sulfur atom. Would you expect lithium fluoride and magnesium sulfide to be ionic or covalent compounds?

33. More energy is needed to remove an electron from a hydrogen molecule than from a hydrogen atom. Why do you think this is so?

34. Which of the following compounds do you expect to be ionic and which covalent? IBr, NO_2, SiF_4, Na_2S, CCl_4, RbCl, Ca_3N_2.

35. Why do the inert gas atoms almost never participate in covalent bonds?

36. Under what circumstances would you expect the shared electron pair to be equal distances on the average from each of the atoms participating in a covalent bond?

10.14 Ionic Compounds

37. What is the charge on alkali metal ions? On halogen ions? On oxygen ions?

38. With the help of Tables 10-8 and 10-9 find the formulas of the following compounds: silicon carbide; lead(II) oxide; manganese(IV) oxide; sodium nitride.

39. With the help of Tables 10-8 and 10-9 find the formulas of the following compounds: barium iodide; ammonium chlorate; tin(II) chromate; lithium phosphate.

10.15 Atom Groups

40. How many atoms of which elements are present in a molecule of $(C_2H_5)_3N$?

41. How many atoms are present in a molecule of $C_3H_5(OH)_3$? How many of them are hydrogen atoms?

10.16 Naming Compounds

42. Name these compounds: $CaMnO_2$, $CaWO_4$, $Ca_3(AsO_4)_2$.

43. Name these compounds: BaH_2, Li_3PO_4, PbO, $CuBr_2$, KOH.

44. Write the formulas of these compounds: sulfur trioxide; phosphorus pentachloride; dinitrogen tetroxide.

10.17 Chemical Equations

45. Which of the following equations are balanced?
   a. $Zn + H_2SO_4 \rightarrow H_2 + ZnSO_4$
   b. $Al + 3O_2 \rightarrow Al_2O_3$
   c. $H_2CO_3 \rightarrow H_2O + CO_2$
   d. $3CO + Fe_2O_3 \rightarrow 3CO_2 + 2Fe$

46. Which of the following equations are balanced?
   a. $6Na + Fe_2O_3 \rightarrow 2Fe + 3Na_2O$
   b. $MnO + 4HCl \rightarrow MnCl_2 + 2H_2O + Cl_2$
   c. $C_4H_{10} + 9O_2 \rightarrow 4CO_2 + 5H_2O$
   d. $3H_2S + 2HNO_3 \rightarrow 3S + 2NO + 4H_2O$

47. Insert the missing numbers in the following equations:
   a. $Ca + [\ ]H_2O \rightarrow Ca(OH)_2 + H_2$
   b. $2Al + [\ ]H_2SO_4 \rightarrow Al_2(SO_4)_3 + 3H_2$
   c. $C_7H_{16} \rightarrow 11O_2 \rightarrow 7CO_2 + [\ ]H_2O$
   d. $6H_3BO_3 \rightarrow H_4B_6O_{11} + [\ ]H_2O$

48. Insert the missing numbers in the following equations:
   a. $4NH_3 + 3O_2 \rightarrow 2N_2 + [\ ]H_2O$
   b. $4NH_3 + 5O_2 \rightarrow 4NO + [\ ]H_2O$
   c. $4FeS_2 + 11O_2 \rightarrow 2Fe_2O_3 + [\ ]SO_2$
   d. $2HNO_3 + 3H_2S \rightarrow 2NO + [\ ]H_2 + 3S$

Write balanced equations for the following reactions:

49. Sodium reacts with water to give sodium hydroxide and gaseous hydrogen.

50. Calcium hydride reacts with water to give gaseous hydrogen and calcium hydroxide.

51. Aluminum reacts with gaseous chlorine to give aluminum chloride.

52. Sulfur dioxide and carbon react to give carbon disulfide and carbon monoxide.

53. Ethane (C_2H_6) burns in air—that is, reacts with oxygen—to give carbon dioxide and water.

54. Acetylene gas (C_2H_2) burns in air to give carbon dioxide and water.

55. Butane (C_4H_{10}) burns in air to give carbon dioxide and water.

56. Lead(II) sulfide reacts with oxygen to give lead(II) oxide and sulfur dioxide.

57. Iron(III) oxide reacts with carbon monoxide in a blast furnace to give iron and carbon dioxide.

Answers to Multiple Choice

1. a	6. c	11. d	16. a	21. c	26. a	31. b	36. c
2. c	7. a	12. a	17. b	22. a	27. a	32. c	37. c
3. b	8. a, b	13. b	18. d	23. d	28. b	33. b	38. d
4. c	9. c	14. b	19. a	24. c	29. b	34. d	39. d
5. b	10. a, b, d	15. c	20. a	25. b	30. d	35. d	40. c

11

Crystals, Ions, and Solutions

Diamond is one of the crystalline forms of carbon.

Goals

When you have finished this chapter you should be able to complete the goals ▶ given for each section below:

Solids

11.1 Ionic and Covalent Crystals
Electron Transfer and Electron Sharing in Solids

11.2 The Metallic Bond
The Electron "Sea" That Bonds Metals Makes Them Good Conductors

11.3 Molecular Crystals
Van der Waals Forces Can Hold Molecules Together

- ▶ Distinguish between crystalline and amorphous solids.
- ▶ List the four classes of crystalline solids and identify the nature of the bonds in each class.
- ▶ Explain the origin of van der Waals forces.

Solutions

11.4 Solubility
Solvent and Solute

- ▶ Distinguish between solvent and solute.
- ▶ Define the solubility of a substance.
- ▶ Describe what is meant by unsaturated, saturated, and supersaturated solutions.
- ▶ Discuss how the solubilities of gases and solids in water vary with temperature and, in the case of gases, with pressure.

11.5 Polar and Nonpolar Liquids
Like Dissolves Like

- ▶ Compare polar and nonpolar liquids as solvents.

11.6 Ions in Solution
Ions Have Characteristic Properties of Their Own

11.7 Evidence for Dissociation
A Daring Idea a Century Ago

- ▶ Give some of the reasons why an ionic crystal is believed to dissociate into ions when it dissolves.
- ▶ Explain how dissociation occurs.

11.8 Water
The Most Important Liquid

11.9 Water Pollution
A Menace Hard to Eliminate

- ▶ Discuss some of the chief causes of water pollution.

Acids and Bases

11.10 Acids
Hydrogen Ions Give Acidic Solutions Their Characteristic Properties

11.11 Strong and Weak Acids
The More It Dissociates, the Stronger the Acid

- ▶ Define acid and distinguish between strong and weak acids.

11.12 Bases
Hydroxide Ions Give Basic Solutions Their Characteristic Properties

- ▶ Define base and distinguish between strong and weak bases.

11.13 The pH Scale
Less Than 7 Is Acidic; More Than 7 Is Basic

- ▶ Describe the pH scale.

11.14 Salts
An Acid Plus a Base Gives Water and a Salt

- ▶ Explain what happens when an acid and a base neutralize each other.
- ▶ Describe how to prepare a salt.
- ▶ Give some examples of acids, bases, and salts.

Solids

The modern theory of the atom provides deep insights into many properties of matter. Exactly how are atoms held together in a solid? Why do metals conduct electricity but other solids do not? Why do some substances dissolve only in water and others dissolve only in liquids like alcohol or gasoline? In this chapter we shall look into the answers to these questions and others like them.

A solid consists of atoms, ions, or molecules packed closely together and held in place by electric forces. Most solids are **crystalline,** which means that the particles they are made of are arranged in regular, repeated patterns. Every crystal of a given kind, whether large or small, has the same geometric form. The word *crystal* suggests salt and sugar grains, mineral samples, sparkling gemstones. But metals and snowflakes are crystalline, too, as are the fibers of asbestos and the clear, flat plates of mica. Clay is composed of tiny crystals that can trap water between them to give an easily shaped material.

Glass

Glass is a transparent, amorphous solid that consists of silica (silicon dioxide, SiO_2, the chief constituent of most sands) combined with other oxides (Fig. 11-1). Some glasses slowly crystallize with time and crack easily when that happens. Silica alone forms an excellent temperature-resistant glass ("quartz glass") that is transparent to ultraviolet light, unlike other glasses, but is too difficult to make for everyday use. Ordinary glass consists of about 75 percent SiO_2, 15 percent Na_2O, and 10 percent CaO. Pyrex glass, more resistant to temperature changes, is largely silica and B_2O_3 with small amounts of other oxides. Lead glass, a soft, highly refractive glass used in optical instruments and expensive glassware, is made up of SiO_2, PbO, and K_2O.

Figure 11-1 Glass is an amorphous solid that softens gradually when heated instead of melting at a specific temperature as crystalline solids do. For this reason hot glass is easily shaped.

Traces of certain metal oxides are responsible for most colored glass. The green glass of cheap bottles contains a little of the iron oxide FeO that was originally present as an impurity in its ingredients. Cobalt oxide gives glass a blue color, manganese oxide a violet color, and uranium oxide a yellow color. Red glass gets its hue from tiny particles of gold and copper.

Glass has been made for a long time; the Egyptians used glass containers 5000 years ago. The energy saved by recycling a glass bottle could power a 100-W lightbulb for nearly an hour.

Solids whose particles are irregularly arranged with no definite pattern are called **amorphous** (Greek for "without form"). Examples of amorphous solids are glass, pitch, and various plastics. One way to distinguish between the two kinds of solid is to see what happens when samples of each kind are heated. A crystalline solid melts at a specific temperature when the thermal energy of its particles is enough to break the bonds between them. An amorphous solid is really a very stiff liquid and softens gradually when heated because of the random nature of the bonds between its particles.

Crystalline solids fall into four classes, depending on how their particles are bonded together: **ionic, covalent, metallic,** and **molecular.** Let us look into how each type of bond arises.

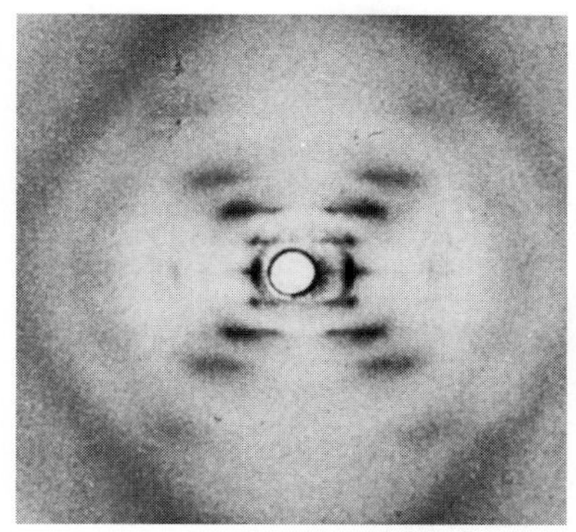

Figure 11-2 An x-ray diffraction photograph of a DNA fiber obtained in 1952 by the English crystallographer Rosalind Franklin (Fig. 11-3). From this photograph and others like it, the helical structure of DNA and its dimensions can be inferred. Franklin's work was essential to Watson and Crick in their analysis of DNA (see Sec. 13.16).

11.1 Ionic and Covalent Crystals

Electron Transfer and Electron Sharing in Solids

As we know, ionic bonds occur when metal atoms, which tend to lose electrons, interact with nonmetal atoms, which tend to pick up electrons. The result is a stable assembly of positive and negative ions. Ionic bonds are usually fairly strong and result in hard crystals with high melting points.

X-rays are often used to study the structure of a solid. The basic particles—atoms, ions, or molecules—in a solid diffract a beam of x-rays at angles that vary with the wavelength of the x-rays, the spacing of the particles, and their arrangement. From the pattern of diffracted x-rays, the corresponding pattern of particles in the solid can be found (Fig. 11-2).

Figure 11-3 Rosalind Franklin (1921–1958).

Figure 11-4 shows the arrangement of Na^+ and Cl^- ions in a sodium chloride crystal. The ions of each kind may be thought of as located at the corners and centers of the faces of a series of cubes, with the Na^+ and Cl^- cubes overlapping. Each ion thus has six nearest neighbors of the other kind.

A different structure is found in cesium chloride crystals, where each ion is located at the center of a cube at whose corners are ions of the other kind (Fig. 11-5). Here each ion has eight nearest neighbors of the other kind. Still other types of structures are found in ionic crystals.

The forces that hold covalent crystals together can be traced to electrons between adjacent atoms. Each atom involved in a covalent bond

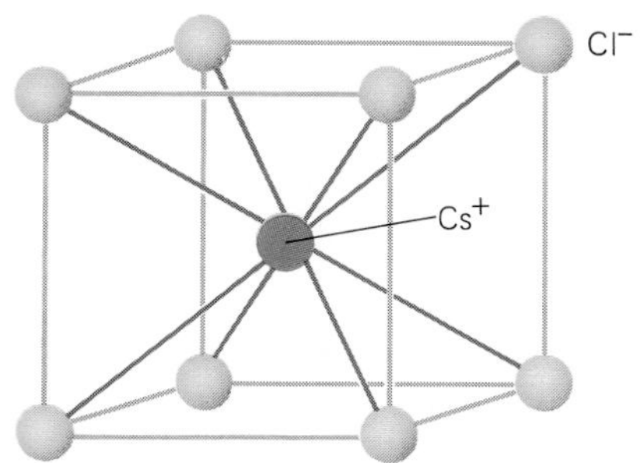

Figure 11-5 The crystal structure of cesium chloride.

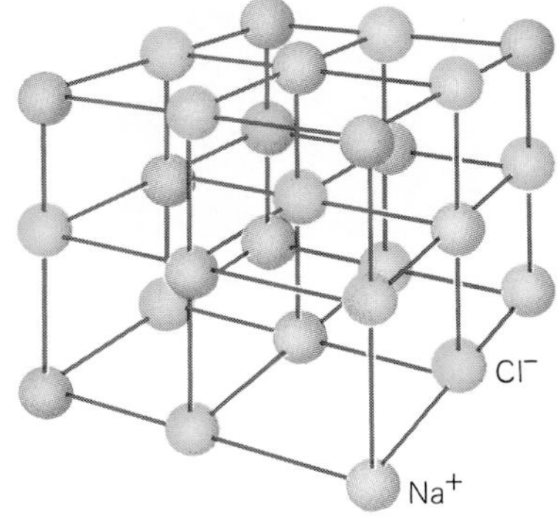

Figure 11-4 The crystal structure of sodium chloride.

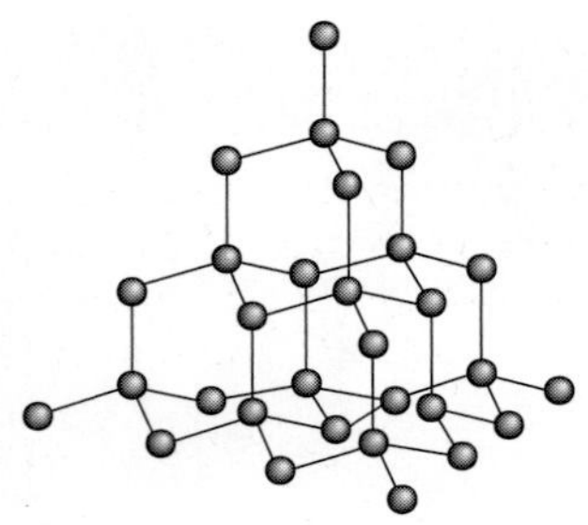

Figure 11-6 The crystal lattice of diamond. The carbon atoms are held together by covalent bonds, which are shared electron pairs.

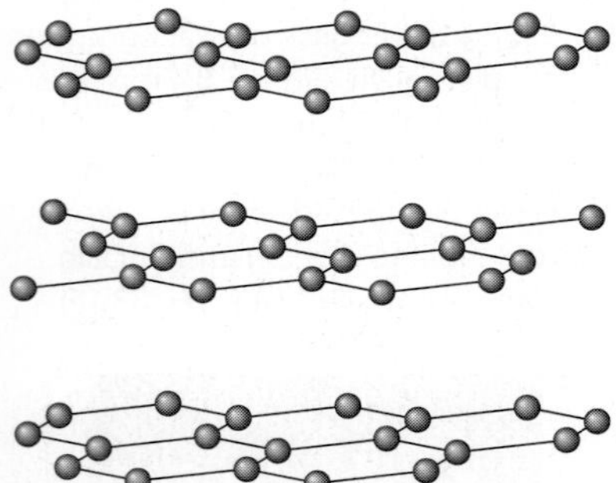

Figure 11-7 Graphite is a form of carbon that consists of layers of carbon atoms in hexagonal arrays. The layers are held together by the weak van der Waals forces described in Sec. 11.3.

Figure 11-8 Extremely high temperatures and pressures are needed to change graphite, the ordinary form of carbon, into diamond. Only relatively small diamonds, such as these, have been produced artificially.

donates an electron to the bond, and these electrons are shared by both atoms. Few purely covalent crystals are known; some examples are diamond, silicon, germanium, and silicon carbide ("Carborundum"). They are not soluble in water and do not conduct electricity.

As in the case of molecules, it is not always possible to classify a given crystal as being wholly ionic or wholly covalent. Silicon dioxide (quartz) and tungsten carbide, for instance, contain bonds of mixed character.

Diamond and Graphite Figure 11-6 shows the structure of a diamond crystal. This is the most symmetrical crystal structure possible. In diamond, each carbon atom has four nearest neighbors and shares an electron pair with each of them. Since all the electrons in the outer shells of the carbon atoms participate in the bonding, it is not surprising that diamonds are extremely hard and must be heated to over 3500°C before they melt. (The name comes from the Greek *adamas,* which means "unconquerable.")

Carbon can occur in other forms besides diamond. One is the familiar **graphite,** a soft, black, lustrous solid that is a fair conductor of electricity—all properties very different from those of diamond. Coke, charcoal, and soot are composed mainly of small graphite crystals. Figure 11-7 shows the structure of graphite, which consists of sheets of carbon atoms in hexagonal arrays in which each atom is linked to three others. Weak van der Waals forces (see Sec. 11.3) bond the layers together. The layers can slide past each other readily and are easily flaked apart, which is why graphite is so useful as a lubricant and in pencils, where it is mixed with a clay binder. Graphite does not melt but becomes a gas directly from its solid form when heated above 3000°C.

Under ordinary conditions graphite is more stable than diamond, so crystallizing carbon produces only graphite. Because graphite is less dense than diamond, high pressures favor the formation of diamond. Natural diamond originates deep in the earth where pressures are enormous. To synthesize diamonds, graphite is dissolved in molten cobalt or nickel and the mixture is compressed at 1400°C or more to about 60,000 atmospheres. The diamonds that form are less than 1 mm across and are widely used for grinding and cutting tools; a small number are of gem quality (Fig. 11-8). About 100 tons of synthetic diamonds are produced each year, 10 times the amount of natural diamonds mined.

11.2 The Metallic Bond

The Electron "Sea" That Bonds Metals Makes Them Good Conductors

A metal atom has only one or a few electrons in its outer shell, and these electrons are loosely attached. When metal atoms come together to form a solid, their outer electrons are given up to a common "sea" of

Buckyballs and Nanotubes

A form of carbon other than diamond or graphite was accidentally discovered in 1985 at Rice University in Texas. The commonest version consists of 60 carbon atoms arranged in a cage structure of 12 pentagons and 20 hexagons whose geometry is the same as that of a soccer ball (Fig. 11-9). This extraordinary molecule was called "buckminsterfullerene" in honor of the American architect R. Buckminster Fuller, whose geodesic domes it resembles; the name is usually shortened to **buckyball.**

Buckyballs can be made in the laboratory from graphite, and are present in small quantities in ordinary soot and in a carbon-rich rock found in Russia. The original C_{60} buckyball is not the only form of fullerene known: C_{28}, C_{32}, C_{50}, C_{70} (also present in the Russian rock), C_{76}, C_{78}, and still larger ones have been made. Fullerene molecules are held together to form solids by weak van der Waals forces (like those that hold together the layers of C atoms in graphite); solid C_{60} is yellowish-brown, and C_{70} is reddish-brown. Since their discovery, the fullerenes and their offshoots have shown some remarkable properties. For instance, solid C_{60} with potassium atoms in the spaces between the buckyballs ("potassium buckide") is a superconductor.

Carbon nanotubes, relatives of buckyballs, consist of tiny cylinders of carbon atoms arranged in hexagons, like rolled-up chicken wire.

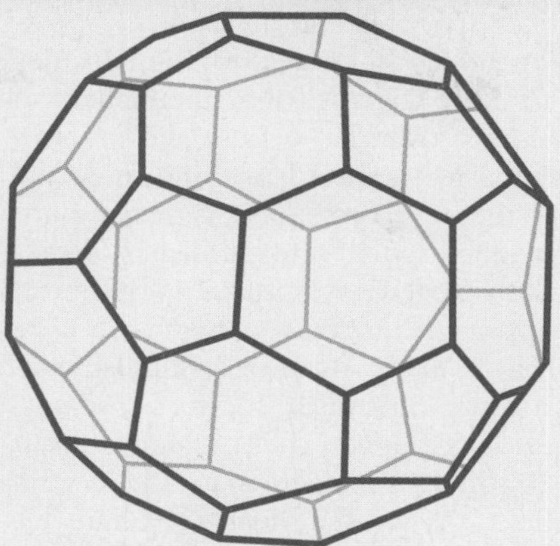

Figure 11-9 In a buckyball, carbon atoms form a closed cagelike structure in which each atom is bonded to three others. Shown here is the C_{60} buckyball that contains 60 carbon atoms. The lines represent carbon-carbon bonds; their pattern of hexagons and pentagons closely resembles the pattern made by the seams of a soccer ball. Other buckyballs have different numbers of carbon atoms.

They were first made over a century ago for use as filaments in early lightbulbs, although their structures are a much more recent discovery. Their name comes from nanometer, a billionth of a meter, which is about the diameter of the thinnest nanotubes. Depending on whether their rows of hexagons are straight or wind around in a helix, carbon nanotubes act either as electrical conductors or as semiconductors and various possible applications in electronics are under active study. Most of today's cars have carbon nanotubes embedded in their nylon fuel lines to carry away any electronic charge that may build up during fuel flow and cause dangerous sparks.

Carbon nanotubes form exceedingly strong fibers, 50 times stronger than steel wires of the same size although a quarter as dense, that are highly flexible as well. They have a remarkable ability to repair themselves when tears occur in their structures. If carbon nanotubes can be made long enough cheaply, fibers like this would be ideal in composite materials, a big improvement on the glass and graphite fibers currently used to reinforce polyester and epoxy resins in aircraft components, boat hulls, wind turbine blades, and so forth. Virtually crash-proof cars? Perhaps. Nanotubes also have promise for use in water desalination membranes and for storing the hydrogen needed for the fuel cells of future electric cars without the use of heavy steel containers.

Imagine unrolling a nanotube. The result, called **graphene,** is a one-atom-thick sheet of carbon atoms linked by bonds arranged in hexagons, like one layer of the carbon atoms in graphite (Fig. 11-7). (Graphene is not actually made this way.) Experimental elecronic devices of great promise have been made with graphene, and it can be embedded in a plastic matrix to form stiff, tough composites that could be used in lightweight structures and as protective coatings.

electrons that move relatively freely through the assembly of metal ions. This negatively charged electron sea acts to hold together the positively charged metal ions.

The electron-sea picture of the **metallic bond** accounts nicely for the properties of metals. Metals conduct heat and electricity well because

Metallic Hydrogen

Hydrogen is in group 1 of the periodic table, all the other elements of which are metals. Hydrogen is the exception, which is not surprising when it is in the gaseous state, but it does not behave as a metal—for instance, by being a good electrical conductor—even when it has been cooled to the liquid or solid states. The reason is that both liquid and solid hydrogen at atmospheric pressure consist of hydrogen molecules, H_2, and these molecules hold their electrons so tightly that none of these electrons can break loose and move about freely as in the case of the atomic electrons of metals.

However, extremely high pressures—several million times atmospheric pressure—turn hydrogen into a conducting liquid. What the pressure does is force the H_2 molecules so close together that their electron clouds overlap, which allows electrons to migrate from one molecule to the next. Pressures inside the giant planet Jupiter, which consists largely of hydrogen, are sufficient for Jupiter apparently to have a hydrogen core that is in the form of a liquid metal. Electric currents in Jupiter's core produce its magnetic field, which is about 20 times stronger than the earth's field. (As we shall learn in Chap. 15, the earth's magnetic field is due to currents in its molten iron core.)

It is not inconceivable that someday solid metallic hydrogen could be created, possibly combined with other substances, that would be stable at ordinary temperatures and pressures. The properties of solid metallic hydrogen are unknown but can be guessed at. Some guesses include superconductivity and light weight combined with mechanical strength. The energy that would be released by allowing solid hydrogen to turn into a gas could be used for propelling spacecraft—it might give five times as much thrust per kilogram as current rocket fuels. Because solid hydrogen would be much denser than ordinary hydrogen, in the form of its isotopes deuterium and tritium it would make an extremely efficient fuel for fusion reactors. All in all, wonderful prospects, but they will not necessarily ever come to pass.

Figure 11-10 The electron "sea" in a metal is responsible for the shiny surfaces of metal objects such as this silver teapot.

the free electrons can move about easily. In nonmetallic solids, all the electrons are bound firmly to particular atoms or pairs of atoms, which is why nonmetals are poor conductors (in other words, good insulators). Free electrons in metals respond readily to electromagnetic waves, which is why metals are opaque to light and have shiny surfaces (Fig. 11-10). Since neighboring atoms in a metal are not linked to each other by specific bonds, most alloys—mixtures of different metals—do not obey the law of definite proportions discussed in Sec. 10.2. The copper and zinc in brass, for instance, need not be present in any exact ratio.

Example 11.1

Lithium atoms, like hydrogen atoms, have only one electron in their outer shells, yet lithium atoms do not join together to form Li_2 molecules the way hydrogen atoms form H_2 molecules. Instead, lithium is a metal with each atom part of a crystal structure. What do you think is the reason for the difference?

Solution

At most there can be 2 electrons in the first shell of an atom, so only 2 H atoms can join at a time. A maximum of 8 electrons can exist in the second shell of an atom, so there is no limit to the number of Li atoms that can join to form an array of atoms. As a result lithium is a metallic solid under ordinary conditions, whereas hydrogen is a diatomic gas.

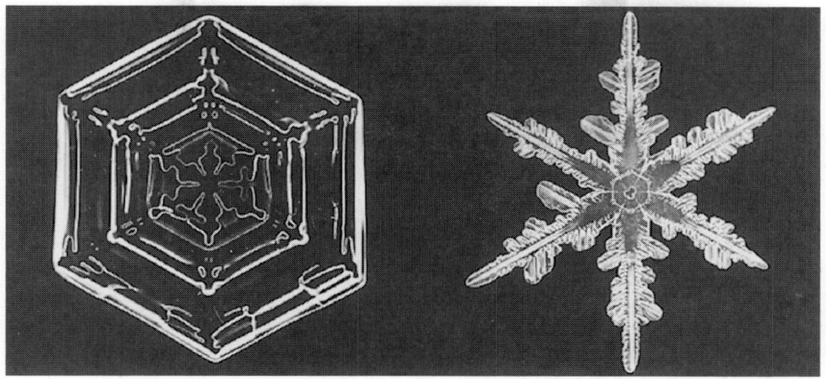

Figure 11-11 The water molecules in a snowflake are held together by van der Waals bonds.

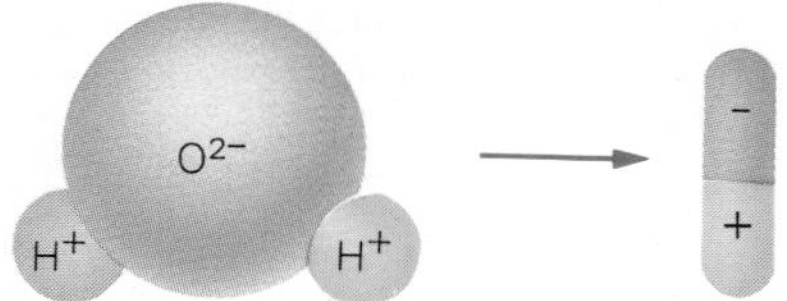

Figure 11-12 The electron distribution in a water molecule is such that the end where the H atoms are attached behaves as if positively charged and the opposite end behaves as if negatively charged. The water molecule is therefore polar.

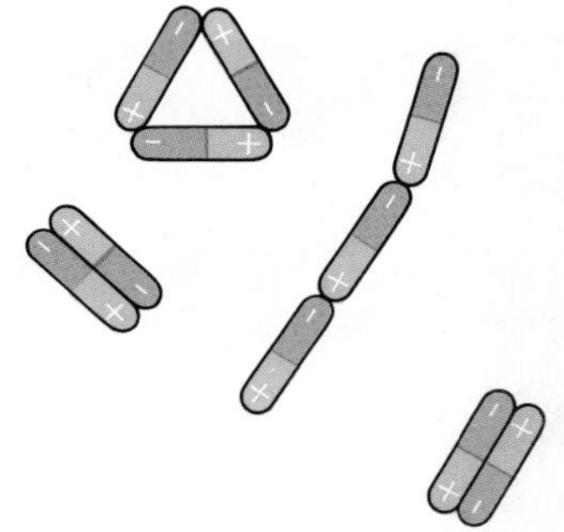

Figure 11-13 Polar molecules attract each other by means of a polar-polar interaction.

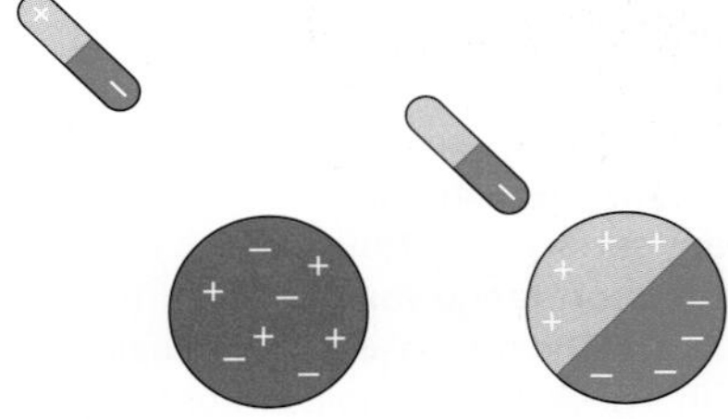

Figure 11-14 Polar molecules attract normally nonpolar molecules by means of a polar-nonpolar interaction.

11.3 Molecular Crystals

Van der Waals Forces Can Hold Molecules Together

Many molecules are so stable that they have no tendency to join together by transferring or sharing electrons. However, even these stable molecules can form liquids and solids through the action of what are called **van der Waals forces** (Fig. 11-11). These forces are named after the Dutch physicist Johannes van der Waals, who suggested their existence nearly a century ago to account for the small but definite departures of actual gases from the ideal gas law. The explanation of how the forces come into being is more recent, of course, since it is based on the quantum theory of the atom.

We recall from Sec. 10.12 that molecules held together by polar covalent bonds behave as though they are negatively charged at one end and positively charged at the other end. An example is the H_2O molecule. In this molecule, the tendency for the shared electrons to favor the O atom makes the oxygen end of the molecule more negative than the end where the hydrogen atoms are (Fig. 11-12). Such **polar molecules** line up with the ends that have opposite charges adjacent, as in Fig. 11-13. This is a **polar-polar** interaction.

A polar molecule can also attract nonpolar molecules through a **polar-nonpolar** interaction. Figure 11-14 shows a polar molecule approaching a nonpolar molecule. When the two molecules are close enough, the electric field of the polar molecule causes charges in the nonpolar molecule to separate. The two molecules now have charges of opposite sign facing each other, which produces an attractive force. The thin plastic sheets that stick so readily to whatever they touch do so because of polar molecules on their surfaces (Fig. 11-15). The polar molecules cause molecules in the other material (the glass of the bowl you cover with cling film, for instance) that were originally nonpolar to become polar, and as a result, the plastic sheet is held firmly in place.

More remarkably, two nonpolar molecules can attract each other electrically. The electrons in a **nonpolar molecule** are distributed evenly *on the average*. However, the electrons are in constant motion, and so *at any*

Figure 11-15 Cling film owes its properties to polar molecules on its surface.

Figure 11-17 Geckos are small lizards famous for their ability to scamper up walls and across ceilings. Insects do this by means of sticky secretions, but geckos use van der Waals forces. The bottom of each of their toes is covered with several hundred thousand tiny hairs, with each hair ending in a number of minute pads. These pads are so small that they come in intimate contact with any surface they touch, which enables the short-range van der Waals forces to act. The total of 6.5 million hairs on a gecko's feet could altogether support about 130 kg—a gecko could hang from a ceiling suspended from just one toe. This raises the question of how a gecko manages to lift its feet to walk. The answer is that a toe hair can act as a lever to help peel its pads away from a surface; a change in angle of only about 30° is enough to release the grip, and only a few hairs need be detached at a time. Synthetic gecko hairs have been made with promising results, but there is a long way to go before wearing a pair of the right gloves will enable us to walk across a ceiling.

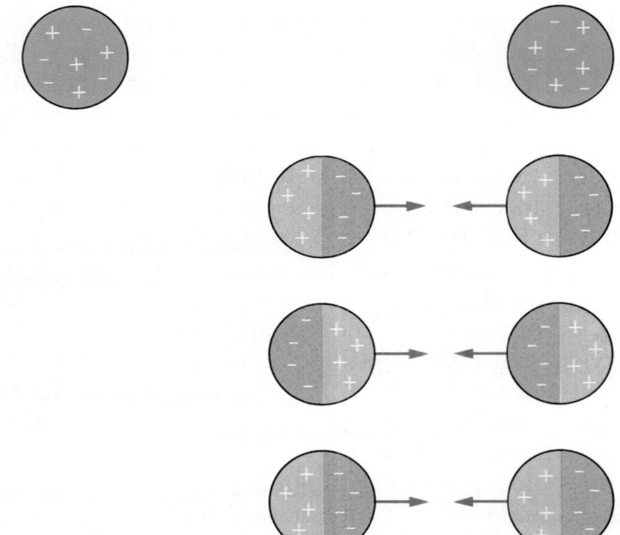

Figure 11-16 Nonpolar molecules normally have, on the average, uniform distributions of charge, but at any one moment the distributions may be uneven. When two nonpolar molecules are close together, the fluctuations in their charge distributions keep in step, which leads to an attractive force between them. This is a nonpolar-nonpolar interaction.

moment one part of the molecule has more than the usual number of electrons and the rest of the molecule has less. When two nonpolar molecules happen to get close together, their changing charge distributions tend to shift together, with adjacent ends always having opposite signs (Fig. 11-16). The result of this **nonpolar-nonpolar** interaction is an attractive force. Although individual van der Waals bonds are relatively weak, they can add up to substantial forces in some situations (Fig. 11-17).

Ice

The molecular solid ice deserves an additional word. The crystal structure of ice is very open (Fig. 11-18) because an H_2O molecule can form bonds with only four other H_2O molecules. In other solids, each atom or molecule may have as many as 12 neighbors, which gives crystals that are much more compact than ice crystals. The molecules in liquid water are closer together on the average than those in ice. Water therefore expands when it freezes, which is why water pipes may burst on a cold winter's day. The expansion also means that a given volume of ice weighs less than the same volume of water, which is why ice cubes float in a glass of water (Fig. 11-19).

Since ice floats, a body of water outdoors freezes from the top down. Ice is a fair insulator of heat, and so a layer of it on the surface of a body of water helps prevent further freezing. As a result, many lakes, rivers, and arms of the sea do not freeze solid in winter, which allows their plant and animal life to survive until the following spring.

When seawater freezes, its salt content is not incorporated in the ice crystals that form but ends up as brine (concentrated salt solution) in tiny pockets and channels around the crystals. As time goes on, the brine gradually drains away into the seawater below the ice. As a result, after a few months the upper layers of sea ice contain very little brine, and when melted, yield water that is fresh enough to drink. This was known by nineteenth century whalers and explorers, who replenished their water supplies in the Arctic by melting sea ice.

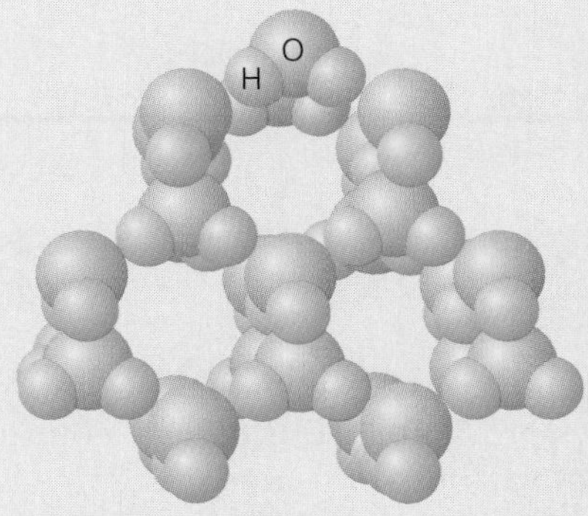

Figure 11-18 Top view of an ice crystal, showing the open hexagonal arrangement of the H_2O molecules. The molecules in liquid water are randomly arranged; hence water is denser and ice floats.

Figure 11-19 Fishing through floating ice in the Arctic.

Van der Waals forces occur not only between all molecules but also between all atoms, including those of the inert gases that do not otherwise interact. Van der Waals bonds are much weaker than ionic, covalent, and metallic bonds. As a result, molecular crystals generally have low melting and boiling points and little mechanical strength. Ordinary ice and dry ice (solid CO_2) are examples of molecular solids.

Table 11-1 summarizes the characteristics of the four kinds of crystalline solids.

Table 11-1 Crystal Types (Bonds Are Strongest in Covalent Crystals, Weakest in Molecular Crystals)

Type		Bond	Example	Properties
Covalent	Shared electrons	Shared electrons	Diamond C	Very hard; high melting point
Ionic	Negative ion; Positive ion	Electrical attraction	Sodium chloride NaCl	Hard; high melting point
Metallic	Electron sea; Metal ion	Electron sea	Copper Cu	Can be deformed; metallic luster; high electrical and thermal conductivity
Molecular	Instantaneous charge separation in molecule	Van der Waals forces	Ice H_2O	Soft; low melting point

Solutions

A solution is an intimate mixture of two or more substances. Solutions can be formed of any of the three states of matter. Thus air is a solution of several gases, seawater is a solution of various solids and gases in a liquid, and many alloys, such as brass, are "solid solutions" of two or more metals. Here our concern will be with solutions in liquids.

11.4 Solubility

Solvent and Solute

In a solution that contains two substances, the substance present in the larger amount is called the **solvent** and the other substance is called the **solute.** When solids or gases dissolve in liquids, the liquid is always considered the solvent. When sugar is stirred into water, the sugar is the solute and the water is the solvent. Water is by far the most common and most effective of all solvents.

The **concentration** of a solution is the amount of solute in a given amount of solvent. Solutions, like compounds, are homogeneous, but unlike compounds, solutions do not have fixed compositions. To a sodium chloride solution whose concentration is 10 g of NaCl in 100 g of water, for example, we can add somewhat more NaCl or as much more water as we like. The concentration of the solution is altered, but it remains uniform.

Some pairs of liquids form solutions in all proportions. Any amount of alcohol may be mixed with any amount of water to form a homogeneous liquid, for instance. In general, however, a given liquid will dissolve only a limited amount of another substance. Sodium chloride can be stirred into water at 20°C until the solution contains 36 g of the salt for every 100 g of water. More salt will not dissolve, no matter how much we stir (Fig. 11-20). This figure, 36 g per 100 g of water, is called the **solubility** of NaCl in water at 20°C:

> The solubility of a substance is the maximum amount that can be dissolved in a given quantity of a particular solvent at a given temperature and pressure.

Saturated Solutions A solution that contains the maximum amount of solute possible is said to be **saturated.** The solubilities of most solids increase with increasing temperature (Fig. 11-21). We all know that hot water is a better solvent than cold water; for example, hot tea can dissolve about twice as much sugar as iced tea. When a solution that is saturated at a high temperature is allowed to cool, some of the solute usually crystallizes out (Fig. 11-22).

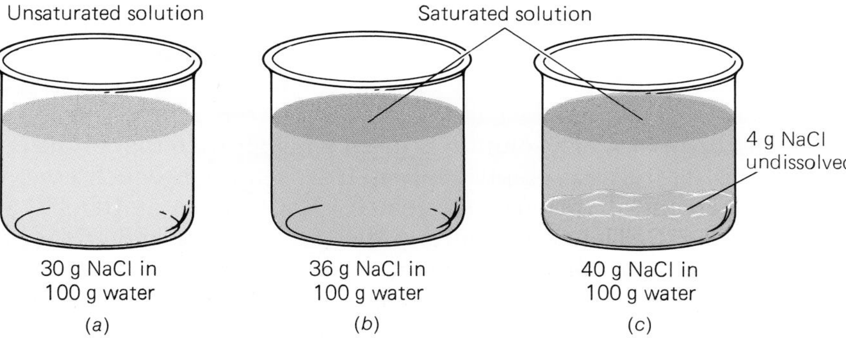

Figure 11-20 The solubility of NaCl is 36 g per 100 g of water at 20°C. (*a*) 30 g of NaCl in 100 g of water produces an unsaturated solution. (*b*) 36 g of NaCl is the maximum amount that can dissolve, and it produces a saturated solution. (*c*) If 40 g of NaCl is added to 100 g of water, 4 g will remain undissolved.

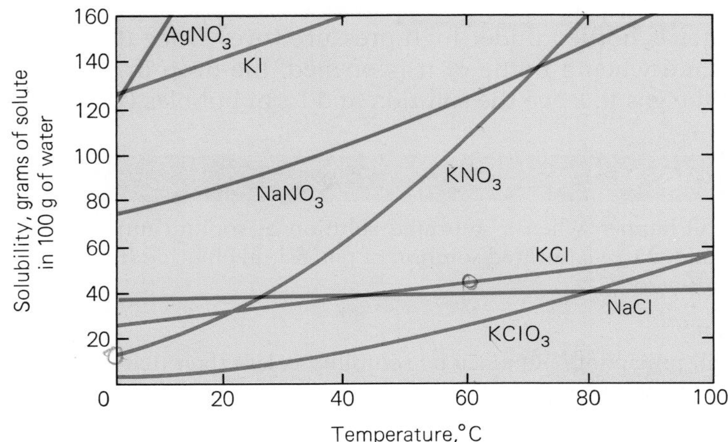

Figure 11-21 How the solubilities of various compounds in water vary with temperature. The higher the temperature, the greater the solubility.

Colloids

A **colloid** is an intimate mixture of two or more substances that is intermediate between the homogeneous and heterogeneous mixtures shown in Fig. 10-2. Oil does not dissolve in water, but when they are shaken together vigorously, tiny droplets of oil become dispersed in the water to form an **emulsion,** which is one kind of colloid. Salad dressing is an emulsion, as is ordinary (nonskim) milk. Shaving foam and whipped cream are colloids in which gas bubbles are dispersed in a liquid. A **sol** is a colloidal suspension of solid particles in a liquid; toothpaste is an example. A **gel** is a colloid in which a solid forms a continuous fine network in a liquid; gelatin is an example. An **aerosol** is a colloid in which solid particles (as in smoke) or liquid droplets (as in fog) are dispersed in a gas (in these cases, air).

Figure 11-23 The higher the pressure and the lower the temperature, the greater the solubility of a gas in water.

Figure 11-22 The solubility of potassium nitrate, KNO_3, is 136 g per 100 g of water at 70°C and 31 g at 20°C. Cooling a saturated solution of KNO_3 from 70°C to 20°C causes 136 g − 31 g = 105 g of the salt per 100 g of water to crystallize out.

Sometimes, if the cooling of a saturated solution is allowed to take place slowly and without disturbance, a solute may remain in solution even though its solubility is exceeded. The result is a **supersaturated** solution. Supersaturated solutions are often unstable, with the solute crystallizing out suddenly when the solution is jarred or otherwise disturbed.

The boiling point of a solution is usually higher than that of the pure solvent, and its freezing point is lower. Thus seawater, which contains about 3.5 percent of various salts (chiefly NaCl), boils at 100.3°C and freezes at −1.2°C. The more concentrated the solution, the greater the changes in boiling and freezing points. Ethylene glycol, $C_2H_4(OH)_2$, is often added to the water in the cooling system of a car to prevent the water from freezing in cold weather. A solution of 83 g of ethylene glycol per 100 g of water will not freeze until −25°C.

In contrast to the case of solids, the solubilities of gases in liquids *decrease* with increasing temperature. We all know that warming a glass of soda water, which is a solution of carbon dioxide gas in water, causes some of the gas to escape as bubbles. The solubility of a gas in a liquid depends on the pressure as well, increasing with increasing pressure. Soda water is bottled under high pressure (over twice that of the atmosphere), and when a bottle of it is opened, the drop in pressure causes some of the gas to leave the solution and form bubbles (Fig. 11-23).

Example 11.2

What will happen when a saturated solution of sodium nitrate ($NaNO_3$) at 50°C is added to a saturated solution of potassium chloride (KCl) at the same temperature?

Solution

NaCl will precipitate out since its solubility is less than that of KCl whereas the solubilities of $NaNO_3$ and KNO_3 are greater than that of KCl. KCl is the less soluble of the two initial compounds.

The Bends

As a diver goes down to greater and greater depths, the pressure of the air being breathed increases as the water pressure increases. Because of the higher pressure, more oxygen and nitrogen are dissolved in the blood and tissues of the diver than usual. The extra oxygen is mostly used up in the normal metabolism of the diver's body, but the inert nitrogen can build up to a high concentration.

If the diver returns to the surface slowly, the added nitrogen comes out of solution gradually and is simply breathed out. However, if the diver ascends rapidly, nitrogen bubbles form in his or her body. Nitrogen bubbles in the joints cause great pain—"the bends." More dangerous are bubbles in the blood, which can block arteries and lead to nervous system damage and even death. Apparently sperm whales also suffer from the bends from time to time when they ascend too rapidly, and this may be true for other whales as well.

The only remedy is to put the diver into a decompression chamber at a high enough pressure to redissolve the nitrogen and then to gradually reduce the pressure. Among the first things student divers are taught is the proper way to ascend after a deep dive. Professional divers often use a mixture of helium and oxygen instead of air because helium is less soluble than nitrogen and so less likely to cause trouble.

11.5 Polar And Nonpolar Liquids

Like Dissolves Like

Some liquids are better solvents for some substances than for others. Water readily dissolves salt and sugar but not fats or oil. Gasoline, on the other hand, dissolves fats and oils but not salt or sugar.

The explanation for this behavior depends upon the electrical characters of the solvent and solute. Water is a **polar liquid** since its molecules behave as if negatively charged at one end and positively charged at the other (Fig. 11-12). Gasoline is a **nonpolar liquid** since the charges in its molecules are evenly distributed. Let us see what difference this makes.

Water and other polar liquids consist of groups of molecules rather than single, freely moving molecules. The molecules join together in clumps, positive charges near negative charges, as in Fig. 11-24. Water molecules can join together in a similar way with polar molecules of other substances, such as sugar (Fig. 11-25), so water dissolves these substances with ease.

Molecules of fats and oils are nonpolar and so do not interact with water molecules. If oil is shaken with water, the strong attraction of

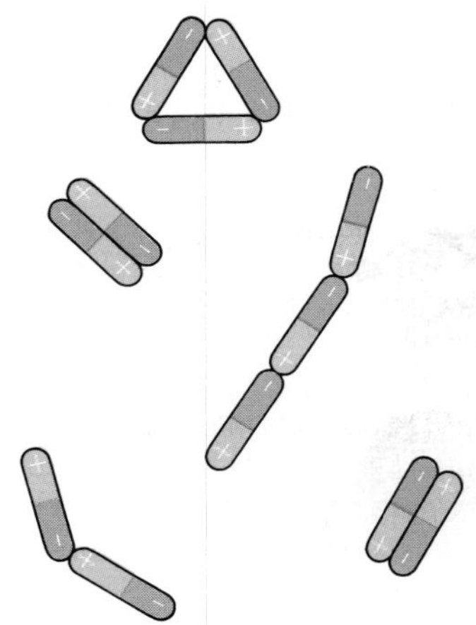

Figure 11-24 Water molecules cluster together because of electric forces that arise from their polar character. In a microwave oven, microwave radiation breaks up some of the bonds between water molecules, which increases their potential energies. When these molecules reestablish bonds, their potential energies become kinetic energies of motion, which means an increase in the water temperature. Thus food in a microwave oven is heated from the inside, so to speak, rather than from the outside as in ordinary cooking.

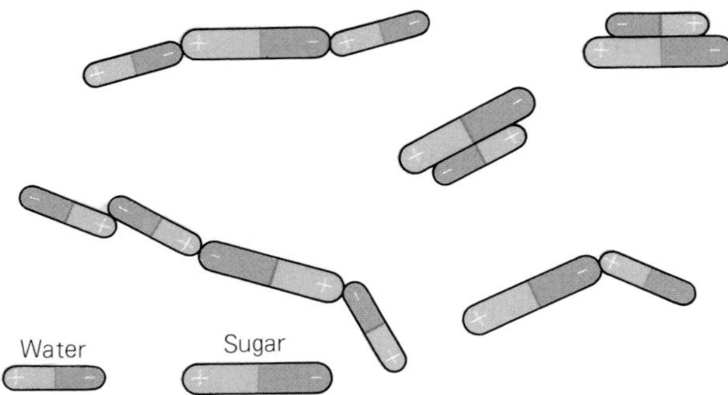

Figure 11-25 Sugar dissolved in water. Polar compounds readily dissolve in water because their molecules can link up with water molecules.

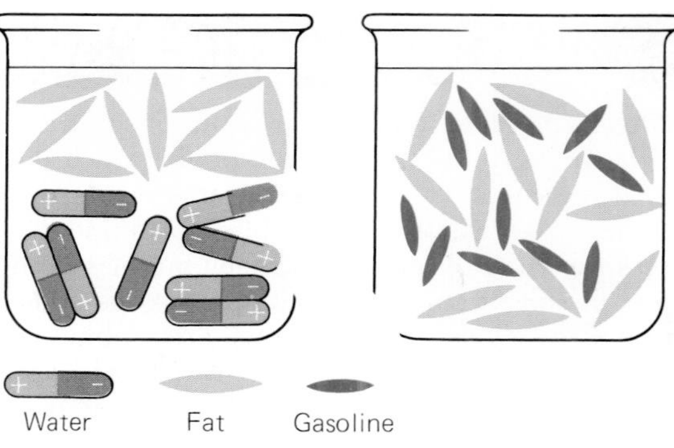

Figure 11-26 Gasoline dissolves fat; water does not. Nonpolar compounds dissolve only in nonpolar liquids.

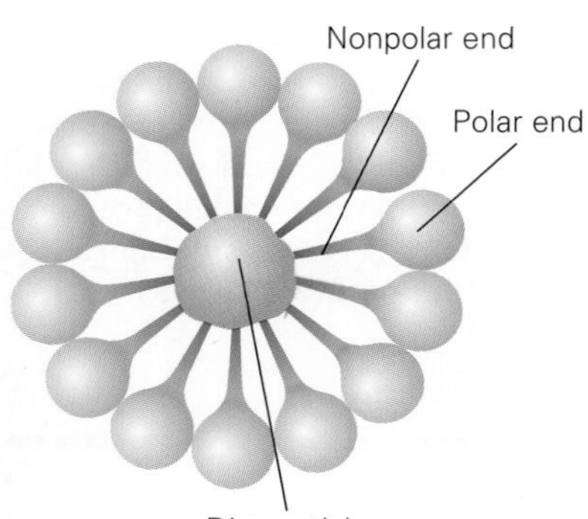

Figure 11-27 Soap and detergent molecules are polar at one end and nonpolar at the other. In water they form spherical cages, called micelles, of several dozen molecules with the nonpolar ends of the molecules on the inside. Micelles trap nonpolar dirt particles as shown. Because the outside of a micelle is polar, it moves freely in water, whereas the dirt particle inside cannot. Micelles that have picked up dirt from a surface can then be rinsed away, taking the dirt with them.

water molecules for one another squeezes out the nonpolar oil molecules from between them and the oil and water separate into layers. Oil or fat molecules mix readily, however, with the similarly nonpolar molecules of gasoline (Fig. 11-26).

A covalent substance, then, dissolves only in liquids whose molecules have similar electrical structures. In general, "like dissolves like."

Soaps and Detergents Some dirt is water soluble, but much is greasy and thus nonpolar, so plain water will not wash it away. Soap molecules are negatively charged at one end and nonpolar at the other, and in water they clump together to form clusters called **micelles** with the nonpolar ends of the molecules inside (Fig. 11-27). Nonpolar dirt particles are absorbed inside micelles, which, unlike the particles themselves, move freely through water because their outsides are polar. In this manner dirt on skin, clothing, and other surfaces is loosened by soapy water and can then be rinsed away.

Soap is not effective in "hard" water (see Sec. 11.8) because the minerals dissolved in it interfere with micelle formation. Synthetic **detergents** act in the same way as soaps even in hard water.

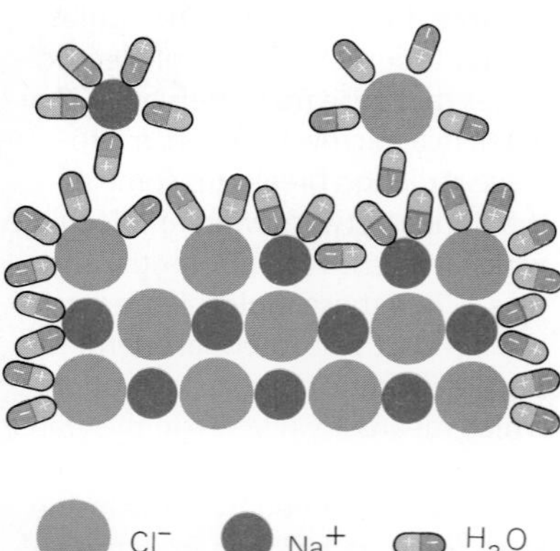

Figure 11-28 Solution of sodium chloride crystal in water. Water molecules exert electric forces on the Na^+ and Cl^- ions that are strong enough to remove them from the crystal lattice.

Dissociation Ionic compounds consist of positive and negative ions and therefore dissolve only in highly polar liquids. Figure 11-28 shows how NaCl dissolves in water. At the surface of a salt crystal, water molecules are attracted to the ions, positive ends toward negative ions and negative ends toward positive ions. The pull of several water molecules is enough to overcome the electric forces that hold an ion to the crystal. The ion then moves off into the solution with its cluster of water molecules. As each layer of ions is removed, the next is attacked, until either the salt is completely dissolved or the solution becomes saturated. The separation of a compound into ions when it dissolves is called **dissociation.**

The ions released when an ionic compound dissolves are the same as those in its crystal structure. This is true not only for such simple compounds as NaCl, which dissociates into Na^+ and Cl^- ions, but also for more complex compounds that involve atom groups. An example is potassium nitrate, KNO_3, which dissociates into K^+ and NO_3^- ions.

Substances that separate into ions when dissolved in water are called **electrolytes.** Electrolytes include all ionic compounds soluble in water and some covalent compounds containing hydrogen (for example,

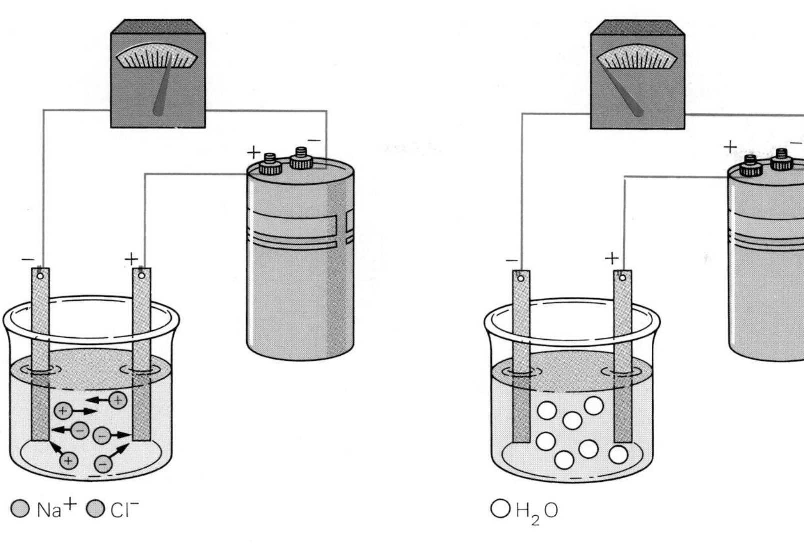

Figure 11-29 (*a*) An electrolyte such as NaCl in solution conducts electric current through the motion of its ions. (*b*) Pure water is a nonelectrolyte, as are solutions of compounds that do not dissociate.

hydrochloric acid, HCl) that form ions by reaction with water. Soluble covalent compounds, such as sugar and alcohol, that do not dissociate in solution are **nonelectrolytes.**

Electrolytes can be recognized by the ability of their solutions to conduct electric current. Hence their name. Conduction is possible because the ions are free to move, with positive ions migrating through the solution toward the negative terminal, negative ions migrating toward the positive terminal (Fig. 11-29). The electric currents in nerves are carried by ions.

11.6 Ions in Solution

Ions Have Characteristic Properties of Their Own

One of the early objections to the theory of ionic solutions was that sodium chloride was supposed to break down into separate particles of sodium and chlorine, yet the solution remains colorless. Why, if chlorine is present as free ions, should we not find the greenish-yellow color of chlorine in the solution? The answer is that chloride ion Cl^- has altogether different properties from gaseous chlorine—a different color, a different taste, different chemical reactions.

We must regard a solution of sodium chloride not as a solution of NaCl or of Na and Cl atoms but as a solution of the two ions Na^+ and Cl^-. Each of these ions in solution has its own set of properties, properties quite different from those of NaCl crystals or of the active metal Na and the poisonous gas Cl_2. Each ion in an electrolytic solution is a new and separate substance.

The "properties of an ion" are really the properties of solutions in which the ion occurs. A solution of a single kind of ion, all by itself, cannot be prepared; positive ions and negative ions must always be present together, so that the total number of charges of each sign will be the

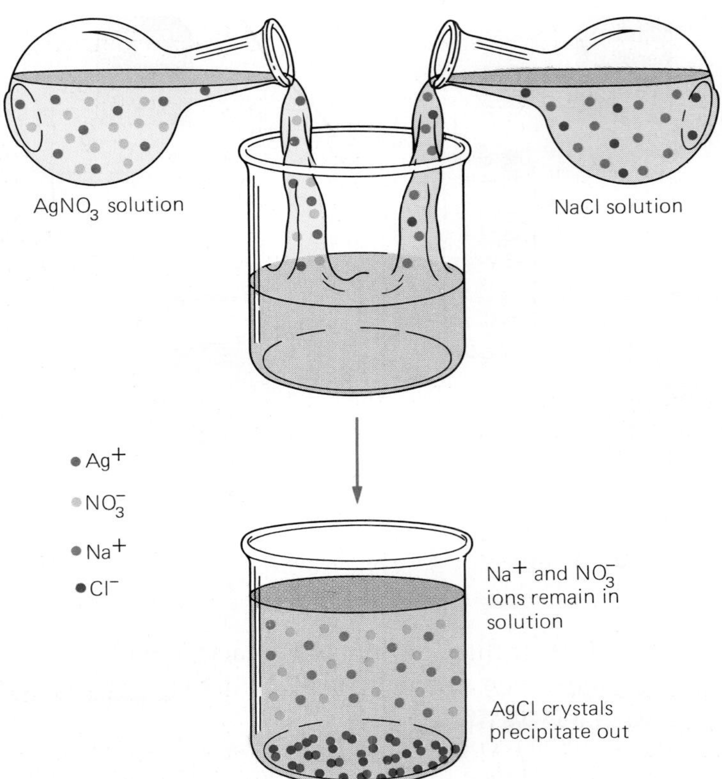

Figure 11-30 When silver nitrate ($AgNO_3$) and sodium chloride (NaCl) are dissolved in water, a precipitate of the insoluble compound silver chloride is produced. The sodium and nitrate ions remain in solution.

Breath Analyzer

The difference in color between the chromic ion (Cr^{3+}), which is green, and the dichromate ($Cr_2O_7^{2-}$) ion, which is yellow-orange, is used by police to determine the level of ethanol—ethyl alcohol—in the breath of a person suspected of drunken driving (Fig. 11-31). A sample of the breath is mixed with a solution of potassium dichromate in sulfuric acid, and any ethanol present reacts with the solution to give chromic ions. The amount of ethanol can be found from the extent of the color change, which is measured in a device called a breath analyzer. Because ethanol in the bloodstream passes into air in the lungs, the proportion of ethanol vapor in a breath sample indicates the percentage of ethanol in the person's blood.

Figure 11-31 This device uses the color difference between chromic and dichromate ions to measure the alcohol concentration in a person's breath.

same. But each ion gives its own characteristic properties to all solutions containing it, and these properties can be recognized whenever they are not masked by other ions.

Let us look at some examples. A property of the copper ion Cu^{2+} is its blue color, and all solutions of this ion are blue (unless some other ion that has a stronger color is present). A characteristic of the hydrogen ion H^+ is its sour taste, and all solutions containing this ion (namely acids) are sour. The silver ion Ag^+ forms an insoluble white precipitate of AgCl when mixed with solutions of the chloride ion Cl^-. (As we recall from Chap. 10, a precipitate is an insoluble solid that results from a chemical reaction.) Any solution of an electrolyte that contains silver will give this precipitate when mixed with a solution of any chloride (Fig. 11-30).

To emphasize the differences between the properties of an ion and those of the corresponding neutral substance, Table 11-2 compares the chloride ion, Cl^-, and molecular chlorine, Cl_2. In general, Cl_2 is much more active chemically. This is to be expected, since Cl atoms have only seven outer electrons whereas Cl^- ions have closed outer shells with eight electrons.

For any ion we can list the properties common to all its solutions. In general, the properties of a solution of an electrolyte are the sum of the properties of the ions that the solution contains. The properties of a sodium chloride solution are the properties of Na^+ plus those of Cl^-; the properties of a copper sulfate solution are the properties of Cu^{2+} plus those of SO_4^{2-}. Instead of learning the individual properties of hundreds of different

Table 11-2 The Properties of Molecular Chlorine and Chloride Ion in Solution

Cl_2	Cl^-
Greenish-yellow color	Colorless
Strong, irritating taste and odor	Mild, pleasant taste
Combines with all metals	Does not react with metals
Combines readily with hydrogen	Does not react with hydrogen
Does not react with Ag^+	Forms AgCl with Ag^+

electrolytes, we need only learn the properties of a few ions to be able to predict the behavior of any electrolytic solution that contains them.

11.7 Evidence for Dissociation

A Daring Idea a Century Ago

The hypothesis that many substances exist as ions in solution was proposed in 1887 by a young Swedish chemist, Svante Arrhenius. Today the idea of ions in solution follows naturally from our knowledge of the electrical structure of matter. We know that some compounds are formed by the shift of electrons from one kind of atom to another, so that some of the atoms become positive ions and the others negative ions. It is not hard for us to imagine that a polar liquid like water can separate these ions from a crystal. But in 1887 the modern picture of the atom was not even a dream. Without this knowledge Arrhenius's fellow chemists were

BIOGRAPHY Svante Arrhenius (1859–1927)

Born near Uppsala, Sweden, Arrhenius attended the university there. As a graduate student, he developed his theory that electrolytes dissociate into ions in solution and presented this as part of his Ph.D. thesis in 1884. His examiners found the ideas difficult to swallow—not only was atomic structure then unknown, but the electron itself was unknown as well—and gave him the lowest passing grade. In the years that followed a few other chemists agreed with Arrhenius, but most, Mendeleev among them, did not. Once the electron had been discovered (by the English physicist J. J. Thomson in 1897), though, it became clear that atoms were not indivisible particles, and in 1903 Arrhenius received the Nobel Prize in chemistry.

Arrhenius did other notable work, for instance, in proposing the concept of activation energy in chemical reactions (see Sec. 12.7). The connection between the greenhouse effect and climate was another of his contributions. He pointed out that carbon dioxide in the atmosphere absorbs infrared radiation emitted by the sunwarmed earth, just as the windows of a greenhouse absorb the infrared emitted by its interior. As a result, said Arrhenius, changes in the CO_2 content of the atmosphere would be followed by changes in its temperature and so might be responsible for such variations in climate

as the ice ages. He recognized that industrialization would lead to an increase in atmospheric CO_2 but, not foreseeing its eventual extent, thought the result would be beneficial rather than the potential catastrophe it has turned out to be.

hard to convince that neutral substances can break up into electrically charged fragments in solution.

Until the work of Arrhenius, Faraday's explanation for the ability of certain solutions to conduct electricity was generally accepted. Faraday held that the passage of a current caused the substance in solution to break up into ions. Arrhenius instead felt that ions are set free whenever an electrolyte dissolves, and he gave a number of reasons to support this notion.

One of the points Arrhenius made was that reactions between electrolytes take place almost instantaneously in solution, but occur very slowly or not at all if the electrolytes are dry. An example is the reaction between the silver nitrate and sodium chloride solutions shown in Fig. 11-30, which is very rapid. The speed with which the insoluble AgCl is formed suggests that its silver and chlorine components are already free in the original solutions and so are ready to combine at once. However, if dry $AgNO_3$ is mixed with dry NaCl, nothing happens because the components of each salt are held firmly in their respective crystals.

Freezing and Boiling Points Another piece of evidence cited by Arrhenius was the unexpectedly low freezing points of electrolyte solutions. The amount by which the freezing point of a solution is reduced (or its boiling point increased) depends upon the concentration of solute particles present, not upon their nature. Equal numbers of sugar and of alcohol molecules dissolved in the same amount of water lower its freezing point by almost exactly the same amount. But the same number of NaCl units lowers the freezing point nearly *twice as much.* This suggests that there are no NaCl molecules as such and that solid NaCl breaks up into Na^+ and Cl^- ions when it dissolves. Similarly calcium chloride, $CaCl_2$, lowers the freezing point of water by nearly three times as much as sugar or alcohol, because each $CaCl_2$ unit dissociates in solution into three particles, one Ca^{2+} ion and two Cl^- ions.

11.8 Water

The Most Important Liquid

Although a minor constituent of the earth as a whole, water covers three-quarters of its surface. Most of the earth's water was once part of the rock of its interior and was freed as a result of geological processes. As evidence, water is present in meteorites, which are leftovers from the youth of the solar system, and water vapor is common in the gases that present-day volcanoes emit (see Chap. 15). Some water has also been brought to the earth's surface by comets, whose impacts were frequent in the earth's early history. Life may have begun in or near the early oceans, and water is essential to all living things.

Seawater has a salt content (or **salinity**) that averages 3.5 percent. The composition of seawater is shown in Fig. 11-32. The ions Na^+ and Cl^- account for over 85 percent of the total salinity. Figure 11-33 shows where the ions found in seawater come from.

Waste Not, Want Not Only 3 percent of the world's water is fresh, and of that 3 percent, two-thirds is trapped as ice in the Arctic and Antarctic. Half of the rest is already being employed for various

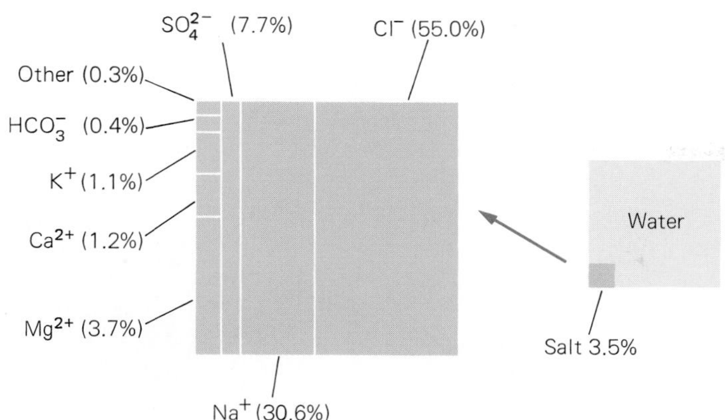

Figure 11-32 The composition of seawater. In the open ocean the total salt content varies about an average of 3.5 percent, but the relative proportions of the various ions are quite constant. (Percentages given are by mass.)

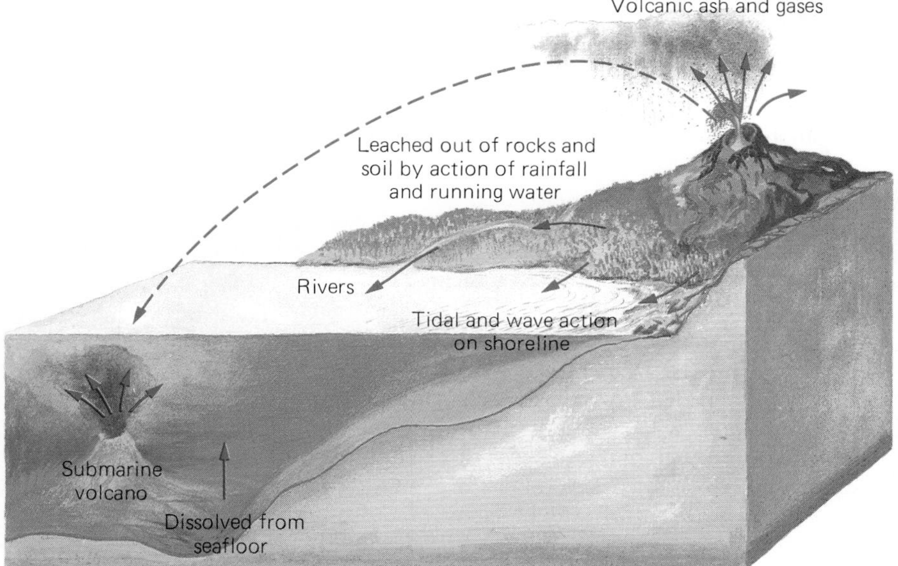

Figure 11-33 Origins of seawater salts.

human purposes: roughly 70 percent by farms and ranches (an average of about 1 liter of water is needed to produce 1 kcal of food energy); 20 percent by industry; and 10 percent for drinking, washing, and other domestic uses. In the United States, water consumption for all uses averages about 1500 gallons per day per person, three times the world average. World water demand is increasing even faster than population (Fig. 11-34). As a result readily available freshwater is well on the way to becoming a scarce commodity: over a billion people today lack reliable access to safe drinking water, and a United Nations estimate has water shortages affecting two-thirds of the world's people by 2025. China is already in a serious situation with 20 percent of the world's population but only 7 percent of the world's water resources (and most of them in its south, whereas demand is mainly in its north).

Although freshwater can be made by desalinating seawater, the process is costly in both money and energy and is practical only in a limited number of locations. The only real way to prevent large-scale water

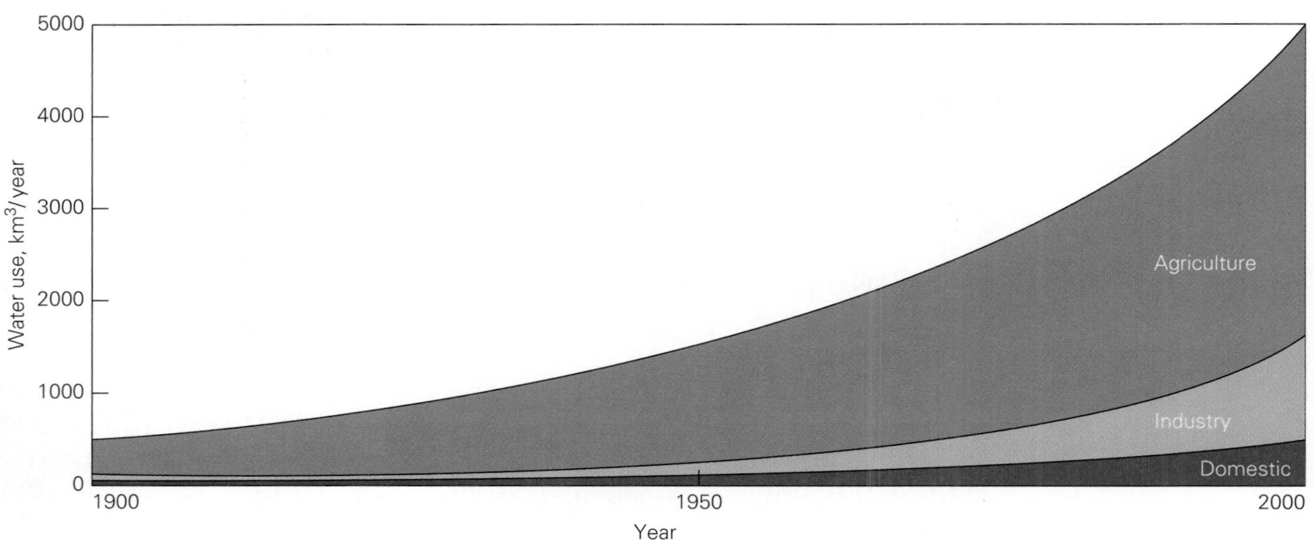

Figure 11-34 Worldwide use of freshwater in cubic kilometers per year. The rate of increase is somewhat greater than the rate of population increase. Without effective conservation, serious water shortages are sure to become widespread in years to come.

Desalination

Modern desalinators use special semipermeable membranes that allow water molecules to pass through while holding back salt ions. If pure water is on one side of such a membrane with salt water on the other side and both are under the same pressure, more water molecules will migrate through the membrane from the pure side to the salt side than the other way. This process is called **osmosis.** However, if the salt water is under higher pressure than the pure water, more water molecules will migrate from the salt side to the pure side. This is **reverse osmosis.** An extremely high pressure is needed for a reasonable output of pure water, which means expensive pumps that consume a great deal of energy, several kWh per thousand liters of water. As a result reverse osmosis desalinators are currently in use only in places where water is really scarce and energy relatively cheap. Singapore is such a place, with a giant desalination plant that provides the island with over 100 million liters per day, 10 percent of its consumption. Improved membranes are being developed, some using the carbon nanotubes mentioned in Sec. 11.2, that could reduce the energy required considerably and make desalination realistic anywhere that freshwater is in short supply and salt water is available, which is a lot of the world.

shortages is conservation. Simply charging appropriate prices for water has reduced water wastage wherever it has been tried. Repairing leaky water delivery systems and recycling industrial water are other practical strategies. As much as 100 tons of water once had to be supplied to make a ton of steel, but today less than 6 tons of new water is needed; the remainder is water that was formerly discarded. And in agriculture, by far the biggest user (and waster) of water, new irrigation techniques are being adopted. Merely burying perforated pipes in the soil to deliver water directly to plant roots minimizes water losses by evaporation and runoff, which can be substantial in traditional agriculture. So there is hope, but much remains to be done.

"Hard" Water Even freshwater is rarely free of ions in solution. "Hard" water contains dissolved minerals that prevent soap from forming suds

Figure 11-35 Hard water left a deposit of scale in this pipe.

and react with soap to produce a precipitate. When heated, hard water forms deposits called scale in boilers, water heaters, and tea kettles that decrease their efficiency by acting as heat insulators (Fig. 11-35). Hot water pipes often become partly or even completely blocked by scale. These deposits are insoluble in water but not in acids, which provides a way to remove them. Calcium and magnesium ions are usually responsible for hard water.

Water Softeners

Hard water can be softened by removing the Ca^{2+} and Mg^{2+} ions it contains, which can be done in several ways. In one common method, hard water is passed through a column filled with either a synthetic ion-exchange resin or a natural material called zeolite (Fig. 11-36). Both absorb Ca^{2+} and Mg^{2+} ions into their structures while releasing an equivalent number of Na^+ ions. Since Na^+ ions do not affect soap or form compounds that precipitate out from hot water, the water is now "soft." When the ion-exchange column has reached its capacity of Ca^{2+} and Mg^{2+} ions, it can be flushed with a concentrated solution of NaCl to reverse the process and replace the accumulated Ca^{2+} and Mg^{2+} ions with Na^{2+} ions.

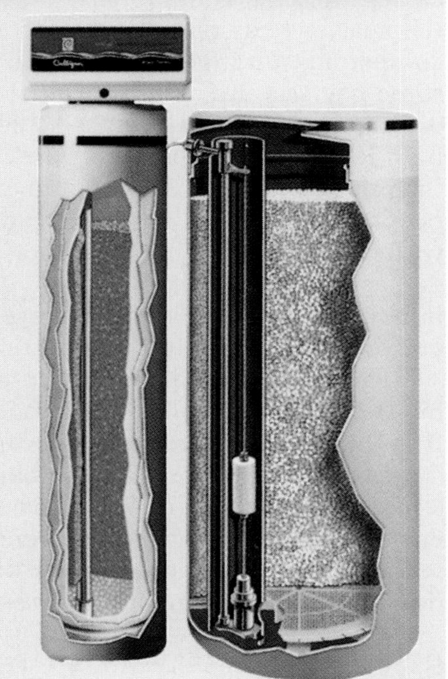

Figure 11-36 These household water softeners use ion-exchange resins.

Groundwater commonly becomes hard by flowing past limestone deposits on or near the earth's surface. Although calcium carbonate ($CaCO_3$), the main constituent of limestone, is insoluble in pure water, water with dissolved carbon dioxide (which is present in air) can react with $CaCO_3$ as follows:

$$CaCO_3 + CO_2 + H_2O \longrightarrow Ca^{2+} + 2HCO_3^-$$

The HCO_3^- ion is the *hydrogen carbonate ion,* also known as the *bicarbonate ion.* When water that contains Ca^{2+} and HCO_3^- ions is heated, this reaction occurs in reverse to precipitate solid $CaCO_3$ as scale.

11.9 Water Pollution

A Menace Hard to Eliminate

Water can be polluted—that is, rendered unsuitable for a particular purpose, not necessarily human consumption only—in a variety of ways. Now that the dangers of water pollution are widely recognized, a great deal of effort is going into minimizing it. But so widespread are the existing sources of pollution, together with new ones appearing as patterns of industry and agriculture evolve, that this is a problem that will never go away.

Industry Common sources of pollution are industry and mining. Some pollutants are especially dangerous because they are concentrated in the food chain of living things. A notorious example is mercury, which is widely used in the production of sodium hydroxide (NaOH) and chlorine gas from solutions of NaCl by electrolysis, as well as in other processes. Mercury-containing wastes have traditionally been dumped into the nearest body of water on the assumption that, since most mercury compounds are insoluble in water, no harm would result. Unfortunately some bacteria are able to convert mercury into the soluble compound dimethyl mercury, $(CH_3)_2Hg$, which is highly toxic. As lower forms of life are eaten by higher ones, the concentration of dimethyl mercury in the organisms increases. A number of years ago in Japan, thousands of people who regularly ate fish caught in Minamata Bay were afflicted with mercury poisoning due to wastes dumped by a nearby factory: brain damage, paralysis, blindness, and deformed children were the result, and hundreds died. "Minamata disease" is only one of many occurrences of poisoning due to industrial pollution.

The United States has made great progress in reducing pollution, but more remains to be done. Today nearly half of all toxic releases are from metal mines and refineries in Nevada, Utah, Arizona, and Alaska (Fig. 11-37). About 500,000 abandoned mines continue to pollute streams in the West. In China, whose industrialization has led to severe environmental damage, 70 percent of the lakes and rivers are polluted, with resulting harm to the health of people and crops. At least half a million Chinese die every year as a result of water and air pollution.

Figure 11-37 Water pollution from a mine. Public anger has forced governments to act against such abuse of the environment.

Agriculture The fertilizers and pesticides used in agriculture are another major source of water pollution. Unlike the case of industry, where proper procedures can control the dispersion of harmful wastes, there is no way to keep chemicals deposited on the soil from spreading further. What can happen is illustrated by the potent and long-lasting insecticide DDT, now banned in the United States and elsewhere. Washed from farmland into a body of water such as a lake, DDT enters the chain of life through tiny plant and animal organisms called plankton. With less than 0.1 part per million (ppm) of DDT in the lake water, the plankton may contain several ppm of DDT. The DDT concentration increases rapidly in the fish that eat plankton because DDT is retained in their fat and can reach over 1000 ppm in birds that eat the fish. One way that DDT affects birds is by weakening the shells of their eggs, which tend to break before the chicks hatch. This can wipe out whole species of birds in a DDT-polluted region.

Because pesticides are of great benefit to agriculture and, in many parts of the world, to disease control, it is not satisfactory merely to stop using them. The remedy is to develop pesticides that decompose rapidly, and much has already been done in this respect. But unfortunately no really effective alternative to DDT has yet been found for preventing the spread of malaria, whose parasites are carried by mosquitoes. More people may have died from malaria over history than from any other disease. Because its drawbacks led to a major decline in the use of DDT, malarial regions were left with the disastrous consequence that this disease, once on the way out, flourished once more and now kills or contributes to the deaths of perhaps 3 million people, mainly in Africa, each year. As a result, DDT has been reintroduced as an indoor spray for malarial control, which uses much less of the pesticide than its former agricultural employment and so affects the environment to only a minor extent.

Dissolved Oxygen Fertilizers contain the plant nutrients nitrogen, phosphorus, and potassium. (Nitrogen goes into the proteins in plant tissues; phosphorus promotes root growth and fruit ripening; potassium helps plants to cope with disease and frost and promotes seed growth.) When a body of water becomes rich in these elements as a result of pollution by fertilizers, tiny organisms called algae cover its surface (Fig. 11-38). As the algae die, aerobic bacteria (which need oxygen to live) use oxygen dissolved in the water to break down the algae into simple compounds, largely water and CO_2. Sewage and other organic wastes (such as those from food-processing plants and paper mills) are similarly attacked by these bacteria.

The oxygen needed to completely oxidize the organic debris in a given sample of water is called the **biochemical oxygen demand,** or BOD, of the water. BOD is a useful index of pollution because, when it is too great, the oxygen content of the water falls to the point where fish and other creatures in the water die out. A huge (sometimes over 8000 square miles) dead zone of this kind forms each summer in the Gulf of Mexico south of Louisiana and has significantly harmed the fishing industry there. The Mississippi River and the rivers that flow into it drain 40 percent of the continental United States, and since 1960 the amount of nitrogen in the water the Mississippi delivers to the Gulf has tripled and the amount of phosphorus doubled. Fertilizer runoff from farms in the Middle West is largely responsible and the dead zone is widening as more and more corn is being grown for conversion into ethanol. More than half the other river mouths of the United States have also been degraded by nutrient pollution, though not to the same extent as in the dead zone of the Gulf. In Europe's Black Sea a dead zone the size of Switzerland occurs off the mouth of the Danube River.

Figure 11-38 Water pollution by fertilizers caused this algal bloom on the surface of a lake in Wisconsin.

A sufficiently high BOD can lower the oxygen content of water so far that aerobic bacteria are unable to break down organic wastes. Anaerobic bacteria (which do not need oxygen) then take over to produce such gases as methane (CH_4), which is flammable, and hydrogen sulfide (H_2S), whose unpleasant odor is that of rotten eggs.

Another way in which the oxygen content of a body of water can be lowered is by heating, since the solubility of a gas in water decreases with increasing temperature. Because the heat output of an electric power plant is two or more times its electric output, such a plant has plenty of heat to get rid of. Using water from a nearby river or lake for cooling is the most common method, but the resulting thermal pollution does not benefit the local fish population.

Acids and Bases

We continue our study of ions in solution by considering the three important classes of electrolytes: acids, bases, and salts. All are familiar in everyday life.

11.10 Acids

Hydrogen Ions Give Acidic Solutions Their Characteristic Properties

Acids are hydrogen-containing substances whose water solutions taste sour and change the color of the dye litmus from blue to red. In concentrated form such strong acids as the sulfuric acid used in storage batteries are poisonous, cause painful burns if allowed to remain on skin, and damage many materials. Hydrochloric acid, another strong acid, helps to digest food in our stomachs. Weak acids—such as the acetic acid of vinegar, the citric acid of lemons, and the lactic acid of yogurt—are far from being harmful and add a pleasant sour taste to foods and drinks.

What is it that underlies the behavior of acids? We have two clues:

1. All acids consist of hydrogen in combination with one or more nonmetals.
2. Solutions of acids conduct electricity and hence must contain ions.

It is therefore tempting to think that acids such as hydrochloric acid, HCl, and sulfuric acid, H_2SO_4 dissociate into hydrogen and nonmetal ions as follows:

$$HCl \longrightarrow H^+ + Cl^- \quad \textit{Dissociation of hydrochloric acid}$$

$$H_2SO_4 \longrightarrow H^+ + HSO_4^- \quad \textit{Dissociation of sulfuric acid}$$

From this it is natural to conclude that the characteristic properties of acid solutions are the properties of the hydrogen ion H^+.

This simple picture presents difficulties, however. For one thing, pure acids in liquid form do not conduct electric current, so a pure acid cannot be made up of ions. Because acids are covalent rather than ionic substances, they form ions not by the separation of ions already present but by reacting with water.

A second problem with the above simple picture of acids is that the ion H^+ is just a single proton—the nucleus of a hydrogen atom without its electron. All other ions are particles of the same general size as atoms, particles that consist of nuclei and electron clouds. The H^+ ion would be entirely different, a naked proton a million billion times smaller than other ions. Such a particle cannot exist by itself in a liquid but must become attached immediately to some other atom or molecule.

Hydronium To avoid these difficulties, we could think in terms of reactions like this:

$$HCl + H_2O \longrightarrow H_3O^+ + Cl^-$$

Here the acid HCl is shown reacting with water instead of simply splitting up into ions, and the proton (the H^+ ion) is shown attached to a water molecule rather than free in solution. The ion H_3O^+ is a combination of H^+ with H_2O and is called the **hydronium ion** (Fig. 11-39). The characteristic properties of acids are described more correctly as properties of the hydronium ion than as properties of the simple hydrogen ion.

Nevertheless, chemists customarily write H^+ for the characteristic ion of acids rather than H_3O^+ because it is more convenient to do so. For most purposes, then, we can say that

> An acid is a substance that contains hydrogen and whose solution in water increases the number of H^+ ions present.

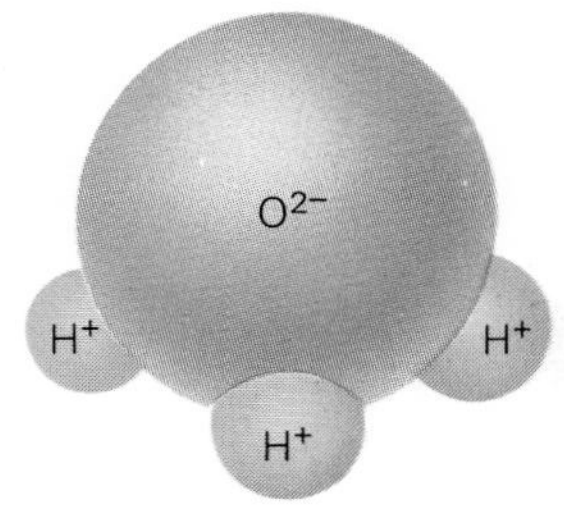

Figure 11-39 A model of the hydronium ion, H_3O^+.

Although not strictly true, it is still legitimate to think of free hydrogen ions as being present in all acid solutions and giving these solutions their common properties. When we say that acid solutions taste sour, turn litmus pink, and liberate hydrogen gas by reaction with metals, we mean that the hydrogen ion does these things.

11.11 Strong and Weak Acids

The More It Dissociates, the Stronger the Acid

Acids differ greatly in how much they dissociate. Some acids, called **strong acids,** dissociate completely. The three most common strong acids are HCl (hydrochloric), H_2SO_4 (sulfuric), and HNO_3 (nitric). Other acids, called **weak acids,** dissociate only slightly.

The stronger an acid, the weaker the attachment of hydrogen in its molecules. In a strong acid like HCl the link is so weak that all the H^+ and Cl^- ions split apart and go their separate ways in solution. In a weak acid like acetic acid, $HC_2H_3O_2$, on the other hand, the link is strong enough so that most of the molecules remain undissociated—in the case of acetic acid, 98 out of 100.

The greater dissociation of strong acids means that, in solutions of the same total concentration, a strong acid has a much larger proportion of hydrogen ions than a weak acid. It has a more sour taste, it is a better conductor of electricity, and if the two acids are poured on zinc, hydrogen gas is given off much faster from the reaction with strong acid.

Why Carbon Dioxide Solutions Are Acidic Certain substances that do not contain hydrogen still yield acidic solutions by reacting with water to liberate H^+ from H_2O. An interesting example is the gas CO_2, which when dissolved in water produces H^+ and HCO_3^- (hydrogen carbonate) ions:

$$CO_2 + H_2O \longrightarrow H^+ + HCO_3^-$$

Some HCO_3^- ions further dissociate into H^+ and CO_3^{2-} (carbonate) ions:

$$HCO_3^- \longrightarrow H^+ + CO_3^{2-}$$

It is customary to consider a solution of CO_2 in water (which is what soda water is) as containing "carbonic acid," although H_2CO_3 rarely exists as such. "Carbonic acid" is weak because relatively little of the dissolved CO_2 reacts with water to give H^+ ions. Rain and snow are slightly acidic because of the dissolved CO_2 they contain.

11.12 Bases

Hydroxide Ions Give Basic Solutions Their Characteristic Properties

Bases are familiar as substances whose solutions in water have a bitter taste, a slippery or soapy feel, and an ability to turn red litmus to blue.

Their formulas, such as NaOH for sodium hydroxide and $Ba(OH)_2$ for barium hydroxide, show that bases consist of a metal together with one or more hydroxide (OH) groups. On dissolving in water, bases dissociate into ions according to reactions such as

$$NaOH \longrightarrow Na^+ + OH^- \qquad \textit{Dissociation of sodium hydroxide}$$

$$Ba(OH)_2 \longrightarrow Ba^{2+} + 2OH^- \qquad \textit{Dissociation of barium hydroxide}$$

Just as H^+ is the characteristic ion of acidic solutions, so OH^- is the characteristic ion in water solutions of bases. The properties of bases are properties of the OH^- ion. We may therefore say that

> A base is a substance that contains hydroxide groups and whose solution in water increases the number of OH^- ions present.

The last part of the definition is needed because not all compounds that contain OH groups release them as OH^- ions in solution. An example is methanol (methyl alcohol), CH_3OH.

Like acids, bases may be classed as strong and weak according to how they dissociate in solution. Thus potassium hydroxide, KOH, is a strong base because it breaks up completely into K^+ and OH^- ions when it dissolves. The most common strong bases are KOH (caustic potash), NaOH (lye or caustic soda, used in oven cleaners), and $Ca(OH)_2$ (slaked lime, used in the mortar that holds the bricks of a building together). Widely employed in industry, these bases are all poisonous and just as destructive to flesh and clothing as the strong acids.

Why Ammonia Solutions Are Basic Bases differ from acids in that soluble weak bases are rare. However, many substances that do not contain OH in their formulas give basic solutions because they react with water to release OH^- ions from H_2O molecules. An example is the gas ammonia, NH_3, which reacts with water as follows:

$$NH_3 + H_2O \longrightarrow NH_4^+ + OH^- \qquad \textit{Solution of ammonia}$$

The process is analogous to that by which CO_2 reacts with water to give an acidic solution. Ammonia solutions are often used in household cleansers. Two other compounds that are not bases but that give basic solutions are sodium carbonate (washing soda), Na_2CO_3, and sodium tetraborate (borax), $Na_2B_4O_7$, both also used as cleansing agents.

The name **alkali** is sometimes used for a substance that dissolves in water to give a basic solution. Alkali is an old Arabic word that referred originally to a bitter extract obtained from the ashes of a desert plant. Because NaOH and KOH are strong alkalis, sodium and potassium are known as alkali metals. An alkaline solution is one that contains OH^- ions; the terms alkaline and basic mean the same thing.

Example 11.3

Acetic acid, $HC_2H_3O_2$, is a weak acid. Would you expect a solution of sodium acetate, $NaC_2H_3O_2$, to be acidic, basic, or neutral?

Solution

Sodium acetate dissolves in water to form Na^+ and $C_2H_3O_2^-$ ions:

$$NaC_2H_3O_2 \longrightarrow Na^+ + C_3H_3O_2^-$$

Some of the acetate ions then react with water to form undissociated acetic acid, a process that liberates OH^- ions:

$$C_2H_3O_2^- + H_2O \longrightarrow HC_2H_3O_2 + OH^-$$

This reaction occurs because acetic acid is weak and therefore can exist undissociated in solution. The sodium acetate solution contains $HC_2H_3O_2$ molecules together with Na^+, $C_2H_3O_2^-$, and OH^- ions and so is basic.

Gastric Fluid

Gastric fluid, produced by glands in the stomach lining, acts upon food in the stomach. This fluid contains pepsin, an enzyme that helps break down proteins in food, and hydrochloric acid, which creates the acid environment needed for pepsin to act. (Most digestion occurs in the small intestine, not the stomach.)

Gastric fluid normally has a pH of about 1.5, acidic enough to dissolve some metals. If the acid concentration is too high, the stomach lining can become inflamed (gastritis), which causes pain and sometimes bleeding. To reduce excess acidity, an antacid can be taken to neutralize some of the HCl. A number of antacids are widely used, one of which is magnesium hydroxide, $Mg(OH)_2$, commonly known as milk of magnesia. It reacts with HCl as follows:

$$Mg(OH)_2 + 2HCl \longrightarrow MgCl_2 + 2H_2O$$

Some antacids are not bases like $Mg(OH)_2$ but nevertheless can neutralize HCl; see the discussion at the end of Sec. 11.12. An example is sodium bicarbonate, $NaHCO_3$:

$$NaHCO_3 + HCl \longrightarrow NaCl + H_2O + CO_2$$

The carbon dioxide gas that is produced may cause discomfort and belching.

11.13 The pH Scale

Less Than 7 Is Acidic; More Than 7 Is Basic

Even pure water dissociates to a small extent. The reaction can be written

$$H_2O \longrightarrow H^+ + OH^- \qquad \textit{Dissociation of water}$$

The hydroxide ion OH^- attracts protons much more strongly than the neutral water molecule H_2O, and the reverse reaction

$$H^+ + OH^- \longrightarrow H_2O \qquad \textit{Recombination of water}$$

occurs readily. Thus we can write

$$H_2O \rightleftharpoons H^+ + OH^-$$

where the double arrow means that both reactions take place all the time in water.

The dissociation of water means that there are always some H^+ and OH^- ions in pure water, and the tendency for these ions to recombine keeps their concentration low. Only 0.0000002 percent of pure water is dissociated into ions on the average: 2 molecules out of every billion. In an acidic solution the concentration of H^+ is greater than in pure water, and the concentration of OH^- is lower. In a basic solution, the concentration of OH^- is greater than in pure water, and that of H^+ is lower.

The **pH scale** is a method for expressing the exact degree of acidity or basicity of a solution in terms of its H^+ ion concentration (Fig. 11-40). This scale is so widely used that an acquaintance with it is worth having, but we need not concern ourselves here with its mathematical basis.

A solution that, like pure water, is neither acidic nor basic is said to be **neutral** and has, by definition, a pH of 7. Acidic solutions have pH values of less than 7; the more strongly acidic they are, the lower the pH. Basic solutions have pH values of more than 7; the more strongly basic they are, the higher the pH. A change in pH of 1 means a change in H^+ concentration by a factor of 10. Thus a solution of pH 4 is 10 times more acidic than a solution of pH 5 and 100 times more acidic than a solution of pH 6. Figure 11-41 illustrates the pH scale and Fig. 11-42 shows typical pH values of some familiar solutions.

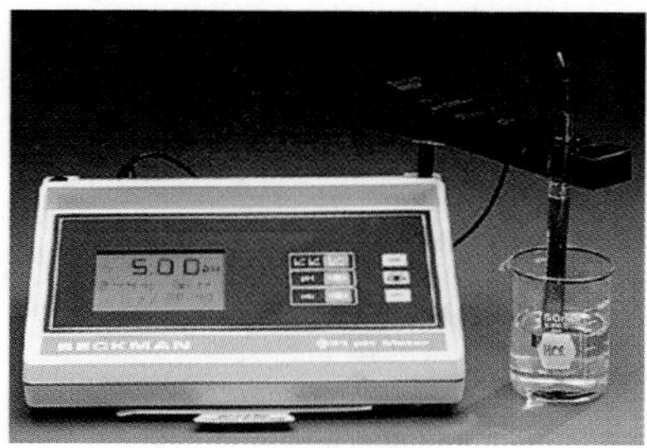

Figure 11-40 This device determines the pH of a solution electrically. Its pH is a measure of how strongly acidic or basic the solution is.

11.14 Salts

An Acid Plus a Base Gives Water and a Salt

When a sodium hydroxide solution is added slowly to hydrochloric acid, there is no visible sign that anything is happening. Both original solutions are colorless, and the resulting solution is also colorless. That a reaction does indeed occur can be shown in several ways, however. One sign is that the mixture becomes warm, which means that chemical energy is being liberated. In addition, if we measure the pH of the mixture as we add the NaOH, we would find that it gets closer and closer to 7 as the base is added—the concentration of H^+ ions is decreasing.

Evidently a base destroys, or **neutralizes,** the characteristic properties of an acid, and the reaction is accordingly called **neutralization.** In the same way the characteristic properties of a base can be neutralized by adding a strong acid.

Neutralization What is the chemical change in the neutralization of HCl by NaOH? We could write simply

$$HCl + NaOH \longrightarrow H_2O + NaCl$$

However, we gain more insight into the process by considering the ions involved. HCl, a strong acid, dissociates completely in water to give H^+ and Cl^-; NaOH, a strong base, dissociates into Na^+ and OH^-; the product NaCl, also a soluble electrolyte, remains dissociated in solution. Of the four substances shown, only water is a nonelectrolyte, so it alone should appear intact in the equation. Hence we have

$$H^+ + Cl^- + Na^+ + OH^- \longrightarrow H_2O + Na^+ + Cl^-$$

Since Na^+ and Cl^- appear on both sides, we can omit them, leaving

$$H^+ + OH^- \longrightarrow H_2O \qquad \textit{Neutralization}$$

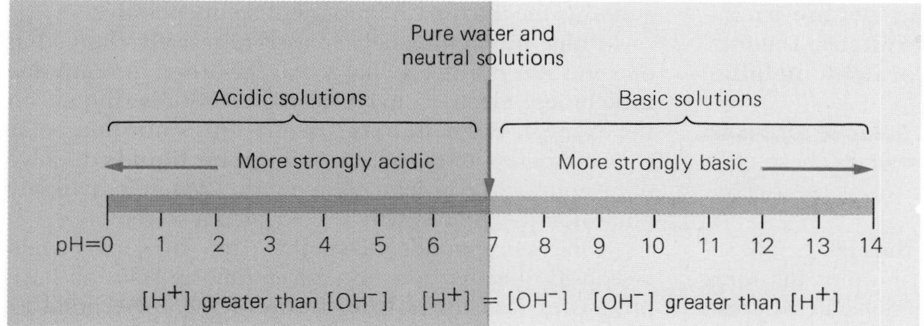

Figure 11-41 The pH scale. The concentration of hydrogen ion is symbolized by $[H^+]$ and that of the hydroxide ion by $[OH^-]$. A neutral solution has a pH of 7. Litmus paper is red in an acidic solution, blue in a basic solution. An increase of 1 in pH corresponds to a decrease of a factor of 10 in H^+ concentration.

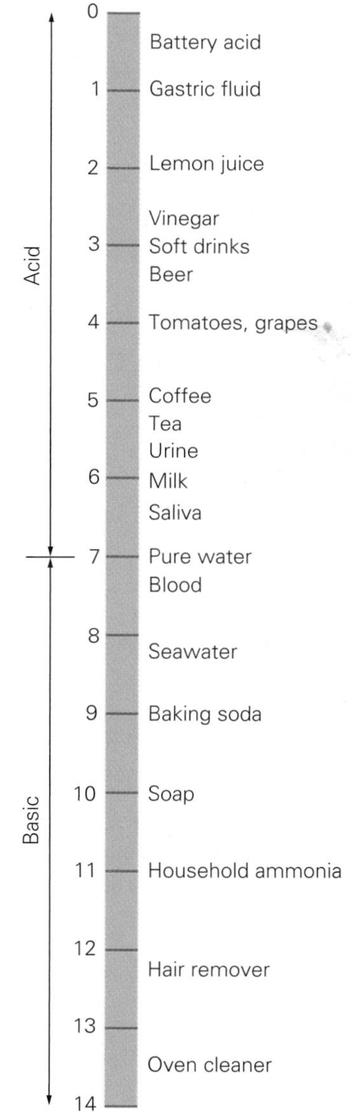

Figure 11-42 Typical pH values.

This is the chemical change stripped of all nonessentials. The neutralization of a strong acid by a strong base in water solution is really a reaction between hydrogen ions and hydroxide ions to form water.

When an NaOH solution is neutralized with HCl, the resulting solution contains only the ions Na^+ and Cl^-. If the solution is evaporated to dryness, the ions combine to form the white solid NaCl. This substance, ordinary table salt, gives its name to an important class of compounds. Most **salts** are crystalline solids at ordinary temperatures and consist of a metal combined with one or more nonmetals.

How to Prepare a Salt Any salt can be made by mixing the appropriate acid and base and evaporating the solution to dryness. Thus potassium nitrate, KNO_3, is formed when solutions of potassium hydroxide, KOH, and nitric acid, HNO_3, are mixed and evaporated; copper sulfate, $CuSO_4$, is formed when sulfuric acid, H_2SO_4, is poured on insoluble copper hydroxide, $Cu(OH)_2$, and the resulting solution is evaporated. In general, then, neutralization reactions give water and a solution of salt.

It is important to remember that a salt itself is not produced directly by a neutralization. Neutralization is a reaction between hydrogen ions and hydroxide ions. As a result of neutralization, ions may be left in solution that combine to form a salt when the solution is evaporated.

Table salt, NaCl, is not the only salt familiar to us in everyday life. Washing soda, Na_2CO_3, and borax, $Na_2B_4O_7$, have already been mentioned. Here are a few other examples. Baking soda, $NaHCO_3$, is used in baking powder, as a deodorizer, and in medicine as an antacid; it is also known as bicarbonate of soda and, to chemists, as sodium hydrogen carbonate. Epsom salt, $MgSO_4 \cdot 7H_2O$ (this formula means that seven water molecules are associated with each $MgSO_4$ unit in its crystals), has various medical uses. Saltpeter, KNO_3, is used to preserve meat and in gunpowder. Gypsum, $CaSO_4 \cdot 2H_2O$, is an ingredient of plaster. Many other salts can be added to this list.

Important Terms and Ideas

Solids that consist of particles arranged in repeated patterns are called **crystalline.** If the particles are irregularly arranged, the solid is called **amorphous.** The four types of bonds in crystals are **ionic, covalent, metallic,** and **molecular.**

The **metallic bond** arises from a "sea" of electrons that can move freely through a solid metal. These electrons are also responsible for the ability of metals to conduct heat and electricity well.

Van der Waals forces arise from the electric attraction between nonuniform charge distributions in atoms and molecules. They enable atoms and molecules to form solids without sharing or transferring electrons.

In a solution, the substance present in larger amount is the **solvent;** the other is the **solute.** When a solid or gas is dissolved in a liquid, the liquid is always considered the solvent. The **solubility** of a substance is the maximum amount that can be dissolved in a given quantity of solvent at a given temperature. A **saturated** solution is one that contains the maximum amount of solute possible.

Polar molecules behave as if negatively charged at one end and positively charged at the other; in **nonpolar molecules,** electric charge is uniformly distributed on the average. **Polar liquids** dissolve only ionic and polar covalent compounds, whereas **nonpolar liquids** dissolve only nonpolar covalent compounds. Water is a highly polar liquid, which is why it is so good a solvent.

Ionic compounds **dissociate** into free ions when dissolved in water; ions of a given kind in solution have properties that differ from those of the corresponding neutral substance. An **electrolyte** is any substance that separates into ions when dissolved in water.

Solutions of **acids** in water contain H^+ ions; solutions of **bases** in water contain OH^- ions. **Strong** acids and bases dissociate completely in solution; **weak** acids and bases only partially. The **pH** of a solution is a measure of its degree of acidity or basicity. Acid solutions have pH values of less than 7; basic solutions have pH values of more than 7. A **neutral** solution is neither acidic nor basic and has a pH of 7.

In acid-base **neutralization,** H^+ and OH^- ions join to form H_2O molecules.

Salts are usually crystalline solids that consist of positive metal ions and negative nonmetal ions. A salt can be formed by neutralizing the acid that contains the appropriate nonmetal ion with the base that contains the appropriate metal ion and then evaporating the solution to dryness.

Multiple Choice

1. An amorphous solid
- **a.** has its particles arranged in a regular pattern
- **b.** is held together by ionic bonds
- **c.** does not melt at a definite temperature but softens gradually
- **d.** consists of nonpolar molecules

2. An amorphous solid is closest in structure to
- **a.** a covalent crystal
- **b.** an ionic crystal
- **c.** a van der Waals crystal
- **d.** a liquid

3. Ionic crystals
- **a.** contain a "sea" of freely moving electrons
- **b.** consist of either positive or negative ions only
- **c.** dissolve only in polar liquids
- **d.** are soft and melt at low temperatures

4. A "sea" of freely moving electrons is present in
- **a.** ionic crystals
- **b.** covalent crystals
- **c.** molecular crystals
- **d.** metal crystals

5. A polar molecule can attract
- **a.** only ions
- **b.** only other polar molecules
- **c.** only nonpolar molecules
- **d.** all of these

6. Van der Waals forces between atoms and between molecules arise from
- **a.** uniform charge distributions
- **b.** nonuniform charge distributions
- **c.** electron transfer
- **d.** electron sharing

7. Which solids have the lowest melting points in general?
- **a.** covalent **c.** van der Waals
- **b.** ionic **d.** metallic

8. Which solids are the best electrical conductors?
- **a.** covalent **c.** van der Waals
- **b.** ionic **d.** metallic

9. Which solids are the best heat conductors?
- **a.** covalent **c.** van der Waals
- **b.** ionic **d.** metallic

10. A property of metals that is not due to the electron "sea" in them is their ability to
- **a.** conduct electricity **c.** reflect light
- **b.** conduct heat **d.** form oxides

11. Diamond and graphite are forms of solid carbon. They both
- **a.** conduct electricity well
- **b.** can be used as lubricants
- **c.** are very hard
- **d.** form CO_2 when burned in air

12. Suppose there were molecules that had no attraction whatever for one another. A collection of such molecules would form a (an)
- **a.** gas **c.** amorphous solid
- **b.** liquid **d.** crystalline solid

13. A saturated solution is a solution that
- **a.** contains the maximum amount of solute
- **b.** contains the maximum amount of solvent
- **c.** is in the process of crystallizing
- **d.** contains polar molecules

14. A gas is dissolved in a liquid. When the temperature of the solution is increased, the solubility of the gas
- **a.** increases
- **b.** decreases
- **c.** remains the same
- **d.** any of these, depending upon the nature of the solution

15. A solid is dissolved in a liquid. When the temperature of the solution is increased, the solubility of the solid
- **a.** increases
- **b.** decreases
- **c.** remains the same
- **d.** any of these, depending upon the nature of the solution

16. Dissolving a solute in a liquid
- **a.** decreases the liquid's freezing point
- **b.** increases the liquid's freezing point
- **c.** does not change the liquid's freezing point
- **d.** any of these, depending on the substances involved

17. A molecule that behaves as though positively charged at one end and negatively charged at the other is called a (an)
a. hydronium ion
b. acid molecule
c. polar molecule
d. nonpolar molecule

18. The most strongly polar liquid of the following is
a. water
b. alcohol
c. a detergent
d. gasoline

19. In general, ionic compounds
a. are insoluble in all liquids
b. dissolve best in polar liquids
c. dissolve best in nonpolar liquids
d. dissolve equally well in polar and nonpolar liquids

20. Nonpolar substances usually dissolve most readily in
a. polar liquids
b. nonpolar liquids
c. acids
d. bases

21. A substance that separates into free ions when dissolved in water is said to be
a. polar
b. saturated
c. an electrolyte
d. covalent

22. Which of the following would you expect to be a strong electrolyte in solution?
a. H_2O
b. HCl
c. sugar
d. alcohol

23. Which of the following has the least ability to conduct electric current?
a. an acid solution
b. a basic solution
c. a salt solution
d. a nonpolar liquid

24. The ions and atoms (or molecules) of an element
a. have very nearly the same properties except that the ions are electrically charged
b. may have strikingly different properties
c. always exhibit different colors
d. differ in that the ions are always more active chemically than the atoms or molecules

25. Dissociation refers to
a. the formation of a precipitate
b. the separation of a mixture of polar and nonpolar liquids (such as oil and water) into separate layers
c. the separation of a solution containing ions into separate layers of + and − ions
d. the separation of a substance into free ions

26. When an electrolyte is dissolved in water,
a. the freezing point of the water is raised
b. the solution is acidic
c. the solution contains free ions
d. the solution contains free electrons

27. The ions usually responsible for "hard" water are
a. H^+ and OH^-
b. Na^+ and Cl^-
c. Ca^{2+} and Mg^{2+}
d. SO_4^{2-} and NO_3^-

28. Acids invariably contain
a. hydrogen
b. oxygen
c. chlorine
d. water

29. The reason pure acids in the liquid state are not dissociated is that their chemical bonds are
a. ionic
b. covalent
c. metallic
d. van der Waals

30. While it is convenient to regard acidic solutions as containing H^+ ions, it is more realistic to describe them as containing
a. hydronium ions
b. hydroxide ions
c. polar molecules
d. hydrogen atoms

31. A common strong acid is
a. acetic acid
b. boric acid
c. nitric acid
d. citric acid

32. A base dissolved in water liberates
a. H^-
b. OH
c. OH^+
d. OH^-

33. A substance whose formula does not contain OH yet yields a basic solution when dissolved in water is
a. NH_3
b. CO_2
c. HCl
d. NaCl

34. The symbol of the ammonium ion is
a. NH_3^-
b. NH_3^+
c. NH_4^-
d. NH_4^+

35. The net reaction between a strong acid and a strong base is
a. $H_2O \rightarrow H^+ + OH^-$
b. $H^+ + OH^- \rightarrow H_2O$
c. $H_2O + H_2O \rightarrow H_3O^+ + OH^-$
d. $H^+ + H_2O \rightarrow H_3O^+$

36. The formula for iron(III) hydroxide is
a. FeOH
b. Fe_3OH
c. $Fe(OH)_3$
d. $Fe_3(OH)_3$

37. A strong acid or base in solution is completely
a. dissociated
b. neutralized
c. hydrolyzed
d. precipitated

38. Pure water contains
a. only H^+ ions
b. only OH^- ions
c. both H^+ and OH^- ions
d. neither H^+ nor OH^- ions

39. A pH of 7 signifies a (an)
   a. acid solution
   b. basic solution
   c. neutral solution
   d. solution of polar molecules

40. A concentrated solution of which of the following has the lowest pH?
   a. hydrochloric acid
   b. acetic acid
   c. sodium hydroxide
   d. ammonia

41. A concentrated solution of which of the following has the highest pH?
   a. hydrochloric acid
   b. acetic acid
   c. sodium hydroxide
   d. ammonia

Exercises

11.1 Ionic and Covalent Crystals
11.2 The Metallic Bond
11.3 Molecular Crystals

1. (a) State the four principal types of bonding in crystalline solids and give an example of each. (b) What is the fundamental physical origin of all of them? (c) What kind of particle is present in the crystal structure of each of them?
2. What kind of solid is ice? Why does ice float when nearly all other solids sink when they freeze?
3. Are ionic or covalent crystals more common?
4. You are given two solids that look nearly alike, one of which is held together by ionic bonds and the other by van der Waals bonds. How could you tell them apart?
5. How could you tell experimentally whether a fragment of a clear, colorless material is glass or a crystalline solid?
6. What kind of solid contains a "sea" of freely moving electrons? Does this sea include all the electrons present?
7. Van der Waals forces are strong enough to hold inert gas atoms together to form liquids at low temperatures, but these forces do not lead to inert gas molecules at higher temperatures. Why not?
8. What ions would you expect to find in the crystal structures of MgO and K_2S?
9. What ions would you expect to find in the crystal structures of CaF_2 and KI?

11.4 Solubility
11.5 Polar and Nonpolar Liquids

10. Why is the solubility of one gas in another unlimited?
11. Why do bubbles of gas form in a glass of soda water when it warms up?
12. Ordinary tap water tastes different after it has been boiled. Can you think of the reason why?
13. How do unsaturated, saturated, and supersaturated solutions differ?
14. How can an unsaturated solution of a solid in a liquid become saturated? How can a saturated solution of a solid in a liquid become supersaturated (if this is possible in a particular case)?
15. Give two ways to tell whether a sugar solution is saturated or not.
16. At 10°C, which is more concentrated, a saturated solution of potassium nitrate or a saturated solution of potassium chloride? At 60°C?
17. You have saturated solutions of silver nitrate ($NaNO_3$) and potassium nitrate (KNO_3) at 60°C. What is an easy way to tell which is which?
18. What is the difference between a molecular ion and a polar molecule?

11.6 Ions in Solution
11.7 Evidence for Dissociation

19. How could you distinguish experimentally between an electrolyte and a nonelectrolyte?
20. The ions of potassium (K) and calcium (Ca) both contain 18 electrons. Would you expect the chemical behaviors of these elements to be similar?
21. (a) You have a solution that contains Cl^- ions and another that contains NO_3^- ions. How would adding a solution that contains Ag^+ ions to these solutions enable you to tell which is which? (b) You have a solution that contains Ag^+ ions and another that contains Na^+ ions. How would adding a solution that contains Cl^- ions to these solutions enable you to tell which is which?
22. Seawater freezes at a lower temperature than pure water because of the salts dissolved in it. How does the boiling point of seawater compare with that of pure water?

11.8 Water

11.9 Water Pollution

23. When water that contains Ca^{++} and HCO_3^- ions is heated, calcium carbonate precipitates out. Carbon dioxide and something else are also produced. What is the something else? Give the equation of the process, which is the cause of the scale found in boilers, water heaters, and tea kettles.

24. What are the two chief ions found in seawater?

25. (a) Is the percentage of the world's water that is freshwater 1 percent? 3 percent? 10 percent? 50 percent? (b) Is the proportion of freshwater in the form of ice in the polar regions about one-third? One-half? Two-thirds?

26. The pesticide DDT concentrates in the fat of animals and tends to remain in the soil despite heavy rain that washes away other contaminants. What do these observations tell you about the nature of the DDT molecule?

11.10 Acids

11.11 Strong and Weak Acids

27. Do pure acids in the liquid state contain H^+ ions? If not, what do such acids consist of?

28. Which of the following are weak acids? Hydrochloric acid, nitric acid, acetic acid, sulfuric acid, citric acid.

29. Would you expect HBr to be a weak or strong acid? Why?

11.12 Bases

30. Even though ammonia is not a base because its molecules do not contain OH groups, its solution in water is basic. Why?

31. What is the difference, if any, between a basic solution and an alkaline solution?

11.13 The pH Scale

32. Is it correct to say that the only ions an acidic solution contains are H^+ ions and that the only ions a basic solution contains are OH^- ions? If not, what would correct descriptions of such solutions be?

33. Which is more strongly acidic, a solution of pH 3 or one of pH 5? Which is more strongly basic, a solution of pH 8 or one of pH 10?

34. In an acidic solution, why is the OH^- concentration lower than it is in pure water?

35. Justify the statement that water is both a weak acid and a weak base.

11.14 Salts

36. When a salt that contains the negative ion of a weak acid is dissolved in water, the solution is basic. For example, a solution of sodium acetate (the corresponding acid is acetic acid) is basic. Why?

37. Give the ionic equation for the neutralization of potassium hydroxide by nitric acid. What chemical changes does this equation show?

38. What salt is formed when a solution of calcium hydroxide is neutralized by phosphoric acid, H_3PO_4? Give the equation of the process.

39. (a) What salt is formed when a solution of calcium hydroxide is neutralized by hydrochloric acid? (b) Give the equation of the process.

40. What salt is formed when a solution of sodium hydroxide is neutralized by sulfuric acid? Give the equation of the process.

41. What salt is formed when a solution of potassium hydroxide is neutralized by acetic acid, $HC_2H_3O_2$? Give the equation of the process.

42. Give the equation of the reaction described below:
Johnny, finding life a bore,
Drank some H_2SO_4.
Johnny's father, an MD,
Gave him $CaCO_3$.
Now he's neutralized, it's true,
But he's full of CO_2.

43. The fertilizer ammonium sulfate can be made by using sulfuric acid to neutralize a solution of ammonia in water. Give the overall equation of the process.

44. Boric acid (H_3BO_3) is a very weak acid. What would happen if solutions of Na_3BO_3 (sodium borate) and HCl were mixed?

45. The Al^{3+} ion tends to form $AlOH^{2+}$ ions in water solution. Would you expect a solution of $AlCl_3$ to be acidic, basic, or neutral? Explain your answer.

Answers to Multiple Choice

1. c
2. d
3. c
4. d
5. d
6. b
7. c
8. d
9. d
10. d
11. d
12. a
13. a
14. b
15. a
16. a
17. c
18. a
19. b
20. b
21. c
22. b
23. d
24. b
25. d
26. c
27. c
28. a
29. b
30. a
31. c
32. d
33. a
34. d
35. b
36. c
37. a
38. c
39. c
40. a
41. c

Chemical Reactions

A fire is an exothermic chemical reaction.

Goals

When you have finished this chapter you should be able to complete the goals ▶ given for each section below:

Quantitative Chemistry

12.1 Phlogiston
Now It's There, Now It Isn't
▶ Discuss the phlogiston hypothesis and explain how Lavoisier's experiments showed it to be incorrect.

12.2 Oxygen
Combustion Is Rapid Oxidation
▶ Define oxide and oxidation.

12.3 The Mole
The Chemist's Unit of Quantity

12.4 Formula Units
A Mole of Anything Contains Avogadro's Number of Formula Units
▶ Define mole, Avogadro's number, and formula mass.
▶ Explain why the mole is so valuable as a unit in chemistry.

Chemical Energy

12.5 Exothermic and Endothermic Reactions
Some Reactions Liberate Energy; Others Absorb It
▶ Distinguish between exothermic and endothermic reactions.

12.6 Chemical Energy and Stability
The Less PE Its Electrons Have, the More Stable the Compound
▶ Identify the nature of chemical energy.
▶ Describe the relationship between the chemical energy absorbed or given off in a chemical change and the stabilities of the substances involved.

12.7 Activation Energy
The Initial Energy Needed to Start an Exothermic Reaction
▶ Explain what is meant by activation energy.

Reaction Rates

12.8 Temperature and Reaction Rates
Hotter Means Faster
▶ Explain why reaction rates depend strongly on temperature.

12.9 Other Factors
Concentration, Surface Area, and Catalysts
▶ List the four factors that affect the speed of a chemical reaction.

12.10 Chemical Equilibrium
One Step Forward; One Step Back
▶ Describe what is meant by a chemical equilibrium.

12.11 Altering an Equilibrium
How to Get Farther Up the Down Escalator
▶ List the three main ways in which a chemical equilibrium can be altered to favor one direction over the other.

Oxidation and Reduction

12.12 Oxidation-Reduction Reactions
They Always Go Together
▶ Distinguish between oxidation and reduction in terms of the electrons transferred in each case.
▶ Describe electrolysis.

12.13 Electrochemical Cells
Turning Chemical Energy into Electric Energy
▶ Explain the basic principle behind the operation of electrochemical cells.
▶ Compare batteries and fuel cells.

Chemical reactions have significant aspects quite apart from the changes that occur when the reactants combine to form the products. An important one concerns the quantities involved, for instance, how much of A must be added to how much of B to give a certain amount of C? Energy considerations are also relevant. After all, the energy given off in chemical reactions powers our cars, airplanes, and ships, heats our homes, cooks most of our food, and is the energy source of the generating plants that produce most of our electricity. Our own bodies obtain the energy they need from chemical reactions in which the food we eat combines with oxygen from the air we breathe.

Not all chemical reactions liberate energy—some reactions must be supplied with energy in order to occur. Even those reactions that liberate energy may not take place unless some initial energy is furnished to start the process. Another aspect of chemical reactions is that they take time to be completed: a fraction of a second to many years, depending on a number of factors. Not all reactions even go to completion. Instead, an intermediate equilibrium situation often occurs with the products undergoing reverse reactions to form the starting substances just as fast as the primary reaction proceeds. These are some of the topics considered in this chapter.

Quantitative Chemistry

The most spectacular chemical change our ancestors were familiar with was **combustion,** the process of burning. Early explanations of how wood turns into smoke and ashes amid dancing flames were based on demons and spirits. The fire god has a respected place in many religions.

The ancient Greeks made the first attempt at a nonsupernatural explanation, as recorded by Aristotle. Every flammable material was supposed to consist of "earth" and "fire." When the material burned, the fire escaped, leaving the earth behind as ashes. This idea persisted in various forms until the time of the French Revolution two centuries ago when Antoine Lavoisier gave combustion its modern explanation. Lavoisier's discovery was made possible by his use of the balance and by his insistence on the importance of mass in studying chemical reactions. This emphasis on mass marked a profound change in viewpoint and is one of Lavoisier's great contributions to chemistry. From his day to ours the balance has remained the chemist's most valuable tool.

12.1 Phlogiston

Now It's There, Now It Isn't

The notion of fire as a basic substance was developed by two Germans, Johann Becher (1635–1682) and his student Georg Stahl (1660–1734), into the **phlogiston** hypothesis. The starting point was the same as Aristotle's, but Becher and Stahl showed how it could be extended to reactions other than burning. They used the word phlogiston (from the Greek word for "flame") for the substance that supposedly escaped during combustion.

The story of the downfall of the phlogiston hypothesis and the growth of the modern picture of chemical change is a notable chapter in the history of ideas.

Today we never hear the word phlogiston, but once there was no more respected concept in chemistry. All substances that can be burned were supposed to contain phlogiston, which escapes as the burning takes place. Combustion requires air, but this is explained by assuming that phlogiston can leave a substance only when air is present to absorb it. When heated in air, many metals change slowly to soft powders: zinc and tin give white powders, mercury a reddish powder, iron a black scaly material. These changes, like the changes in ordinary burning, were ascribed to the escape of phlogiston.

A metal was assumed to be a compound of the corresponding powder plus phlogiston, and heating the metal simply caused the compound to decompose. Now, many of these powders can be changed back into metal by heating with charcoal. This observation was interpreted to mean that charcoal must be a form of phlogiston that simply reunited with the powder to form the compound (the metal). When hydrogen was discovered in 1766, its ability to burn without leaving any ash suggested that it was another form of phlogiston. One could predict, then, that heating one of these powders with hydrogen would form a metal, and this prediction was confirmed by experiment.

So far so good, but soon the phlogiston hypothesis ran into serious trouble. When wood burns, its ashes weigh less than the original wood, and the decrease in mass can reasonably be explained as due to the escape of phlogiston. But when a metal is heated until it turns into a powder, the powder weighs *more* than the original metal! The believers in phlogiston were forced to assume that it sometimes could have negative mass, so that if phlogiston left a substance, the remaining material could weigh more than before. To us this notion of negative mass is nonsense, but in the eighteenth century it was taken quite seriously.

The Downfall of Phlogiston The French chemist Antoine Lavoisier carried out a series of experiments in the latter part of the eighteenth century that overthrew the phlogiston hypothesis. Lavoisier knew that tin changed into a white powder when heated and that the powder weighed more than the original metal. To study the process in detail, he placed a piece of tin on a wooden block floating in water, as in Fig. 12-1. He covered the block with a glass jar and heated the tin by focusing sunlight on it with a magnifying glass—a common method of heating before gas burners and electric heaters were invented. The tin was partly changed into a white powder and the water level rose in the jar until only four-fifths as much air was left in the jar as there had been at the start. Further heating caused nothing more to happen.

In another experiment, Lavoisier heated tin in a sealed flask until as much as possible was turned into powder. The flask was weighed before and after heating, and the two masses were the same. Then the flask was opened, and air rushed in. With the additional air, the mass of the flask was more than it had been at the start. The increase in mass was equal to the increase in mass of the tin.

BIOGRAPHY Antoine Lavoisier (1743–1794)

The son of a wealthy French lawyer, Lavoisier studied law at first but soon became fascinated by chemistry and published his first paper, on the mineral gypsum, when he was 22. In that paper, as in his later work, Lavoisier stressed the importance of accurate measurements of mass in chemistry. This emphasis on reliable data marked the start of the modern era of chemistry: Lavoisier's influence on chemistry was much like that of Galileo on physics. With the help of a sensitive balance and a large magnifying glass for focusing sunlight, Lavoisier carried out a series of experiments that, among other achievements, demolished the notion of phlogiston and made clear the role of oxygen in combustion and other processes.

In his research Lavoisier was greatly aided by his wife, Marie-Anne, who was 14 to his 28 when they were married. Besides work in the laboratory, Marie-Anne made engravings of the apparatus for publication and translated scientific books and papers from English to French. Benjamin Franklin and Thomas Jefferson were among the visitors to the Lavoisier laboratory.

In 1787 Lavoisier and several colleagues systematized the language of chemistry in the book *Methode de Nomenclature Chimique.* They listed 55 substances they could not decompose, most of which (but not caloric or light, which were included) indeed proved to be elements, and named compounds after their constituents according to a standard scheme. Such familiar terms as oxide, nitrate, and sulfate were thereby introduced for the first time. Two years later Lavoisier published the first modern textbook on chemistry, an influential book in which the central place of the law of conservation of mass was made clear.

In order to support his research, while still a young man Lavoisier had invested in the Ferme Générale, a private company that collected taxes for the French government, and he became an official of the company, as was his father-in-law. Brutal and corrupt, the Ferme Générale was the most hated institution in the country. In 1793, after the French Revolution, Lavoisier was arrested because of his association with the Ferme Générale; when he protested that he was a scientist, he was told that "the Republic has no need of scientists." In 1794 he and his father-in-law were guillotined. Lavoisier's widow later married Count Rumford, who proved that caloric did not exist 25 years after Lavoisier had done the same for phlogiston, but it was an unhappy union.

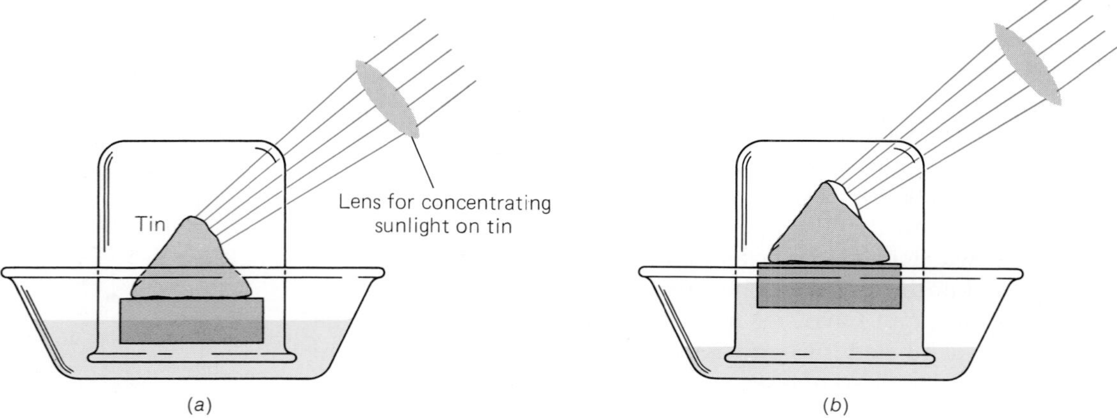

Figure 12-1 Lavoisier's experiment showed that tin, upon heating, combines with a gas from the air. (*a*) Before heating; (*b*) after heating. The tin is partly changed to a white powder, and the water level rises until only four-fifths as much air is left as there was at the start. Further heating causes no additional change.

To Lavoisier these results suggested that the tin had combined with a gas from the air. Since four-fifths of the original air was left after the reaction, he reasoned that one-fifth of air consisted of a gas that can combine with tin. Then the powder was a compound formed from this gas and tin, and the increase in mass of the powder over the tin was the mass of the gas. Water rose in the jar of Fig. 12-1 to take the place of the gas that had combined with the tin. When the sealed flask was opened in the second experiment, air rushed in to take the place of the gas that had combined with the tin. These explanations are simple and direct and involved substances that, unlike phlogiston, had definite masses.

Figure 12-2 Joseph Priestley (1733–1804).

12.2 Oxygen

Combustion Is Rapid Oxidation

At about the time he was making these experiments, Lavoisier learned that Joseph Priestley (Fig. 12-2) had prepared a new gas with remarkable properties. Priestley was the poverty-striken minister of a small church in England, with only limited time and equipment, yet his scientific talents led him to a number of significant discoveries. The gas he had found caused lighted candles to flare up brightly and glowing charcoal to burst into flames. A mouse kept in a closed jar of the gas lived longer than one kept in a closed jar of air.

Lavoisier gave Priestley's new gas its modern name, **oxygen,** and found it to be involved not only in the changes that occur in metals when heated but in the process of combustion as well. The burning of candles, wood, and coal, according to Lavoisier, involves combining their materials with oxygen. When they burn, these materials seem to lose mass only because some of the products of the reactions are gases. Actually, as experiment shows, the total mass of the products in each case is more than the original mass of the solid material.

Under ordinary conditions, oxygen is a colorless, odorless, tasteless gas. Air owes its ability to support combustion to its oxygen content. Air cannot support combustion as well as pure oxygen because air consists of only about one-fifth oxygen (Fig. 12-3). The other four-fifths is mainly nitrogen, together with small amounts of other inactive gases.

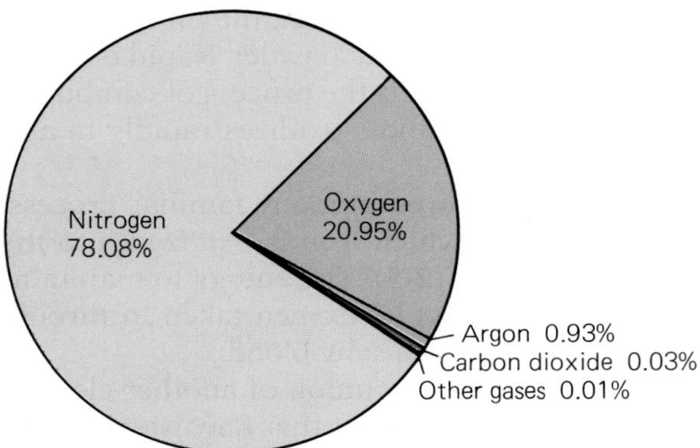

Figure 12-3 Composition of dry air near ground level. A variable amount of water vapor is usually present also.

Ozone

The molecules of ordinary oxygen consist of two oxygen atoms: O_2. A less stable form of oxygen is **ozone,** whose molecules consist of three oxygen atoms: O_3. The pungent odor of ozone is familiar near electrical discharges, such as sparks and lightning, which can produce O_3 from atmospheric O_2. Ozone is used industrially as a bleach and to purify water; it is toxic to living things and damages various materials, notably rubber.

Two important processes in the atmosphere, one bad and one good, involve the small amount of ozone it contains. At low altitudes ozone damages our lungs and contributes to the formation of smog, which is also harmful to our health. At high altitudes, however, ozone provides a vital service by absorbing dangerous ultraviolet radiation from the sun (see Sec. 14.1).

Figure 12-4 A fire can be put out in three ways. (1) Water can be sprayed on to cool the burning material below its ignition temperature. (2) The fire can be smothered by a heavier-than-air agent (such as carbon dioxide or a foam that contains carbon dioxide bubbles) that does not support combustion and keeps out air and hence oxygen. (3) A chemical that interferes with the oxidation process can be sprayed on the fire.

Figure 12-5 Rust forms when iron or steel reacts with oxygen and water. The formula for rust is often written $2Fe_2O_3 \cdot xH_2O$, which means that a variable number of water molecules are associated with every two iron(III) oxide units. Because rust is porous, oxygen and water vapor can penetrate it and continue to react with the metal underneath. A rusty object may therefore have very little strength left. On the other hand, the oxidation of aluminum to Al_2O_3, which occurs naturally in air, produces a hard, durable coating that prevents further corrosion. This is why steel cans are plated with tin to protect them, whereas aluminum cans need no treatment. For even greater protection of aluminum, an electrical process called anodization produces a much thicker layer of Al_2O_3.

Priestley discovered a number of other gases—for instance, ammonia, sulfur dioxide, hydrogen sulfide, and carbon monoxide. His political opinions—one of his books provided ideas to Thomas Jefferson—and religious beliefs—he opposed the Church of England—were unpopular, and after his house and laboratory were burnt down by a mob he left England and spent the last 10 years of his life in Pennsylvania.

Oxidation When oxygen combines chemically with another substance, the process is called **oxidation,** and the other substance is said to be **oxidized.** In the experiments of Lavoisier, the tin reacted with oxygen in the air to become oxidized to a white powder. Rapid oxidation in which a lot of heat and light are given off is the process of combustion (Fig. 12-4). As Priestley found, a lighted candle oxidizes rapidly in air, even more rapidly in pure oxygen.

Slow oxidation is involved in many familiar processes. One of them is the rusting of iron, in which iron is oxidized into the reddish-brown material we call rust (Fig. 12-5). The energy to maintain life comes from the steady oxidation of food by oxygen taken in through our lungs and carried to all parts of our bodies by blood.

A substance formed by the union of another element with oxygen is called an **oxide.** The white powder that Lavoisier obtained by heating tin is tin oxide. Rust is largely iron oxide. In general, oxides of metals are solids. Oxides of other elements may be solid, liquid, or gaseous. Thus

Stainless Steel

Chromium reacts with atmospheric oxygen to produce a thin airtight oxide coating on a chromium surface, which prevents further oxidation. If the oxide layer is scratched, new oxide rapidly covers the exposed metal. For this reason chromium surfaces are always shiny, which is why many metal objects are chrome plated. When iron and ordinary steel are oxidized, on the other hand, the resulting rust does not seal the surface and oxidation can continue underneath. With enough chromium alloyed (mixed) with steel, the chromium is able to form an oxide coating that protects the entire surface. Such an alloy is called **stainless steel.** To be sure, some "stainless" steels do not have enough expensive chromium to be completely stainless and corrode in certain environments.

one of the oxides of sulfur is the foul-smelling gas sulfur dioxide (SO_2). Carbon forms two gaseous oxides, carbon monoxide (CO) and carbon dioxide (CO_2). The oxide of silicon SiO_2 is found in nature as the solid called quartz, the chief constituent of ordinary sand and abundant in rocks (see Figs. 15-2 and 15-6). The oxide of hydrogen is water, H_2O.

Oxides of nearly all the elements can be prepared, most of them simply by heating the elements with oxygen. A few oxides (mercury oxide, lead oxide, barium peroxide) are easily decomposed by heating, which provides a convenient laboratory method for preparing oxygen. Other oxides, such as lime (calcium oxide), are not decomposed even at the temperature of an electric arc, 3000°C.

Example 12.1

The most common building material is concrete, which is often reinforced by embedded steel rods. Iron is the chief ingredient of steel, and if the iron in the steel rods rusts, the concrete may crack. Why?

Solution

Rust is a compound of iron and oxygen and so occupies more volume than the same amount of iron by itself. As their iron content rusts, the rods expand and thus may crack the concrete.

12.3 The Mole

The Chemist's Unit of Quantity

Regardless of how small may be the samples of matter involved in a chemical process in industry or the laboratory, so many atoms are present that counting them is out of the question. To measure the mass of a sample, however, is easy. What the chemist therefore needs is a way to relate the number of atoms in a chemical formula or equation to the corresponding masses of the substances.

To make clear the train of thought used to set up such a method, let us consider atoms of carbon and of oxygen. Since the mass of a carbon atom is 12 u and the mass of an oxygen atom is 16 u, the ratio of their masses is exactly 12 : 16. (We recall from Chap. 8 that u is the abbreviation of the atomic mass unit, which is equal to 1.66×10^{-27} kg.)

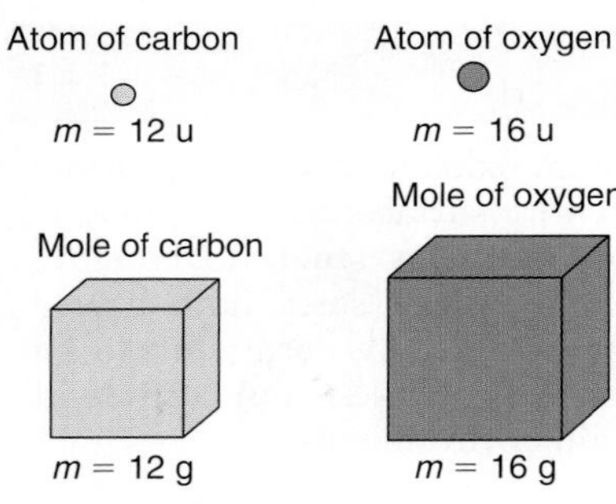

Figure 12-6 A mole of any element is equal to its atomic mass expressed in grams.

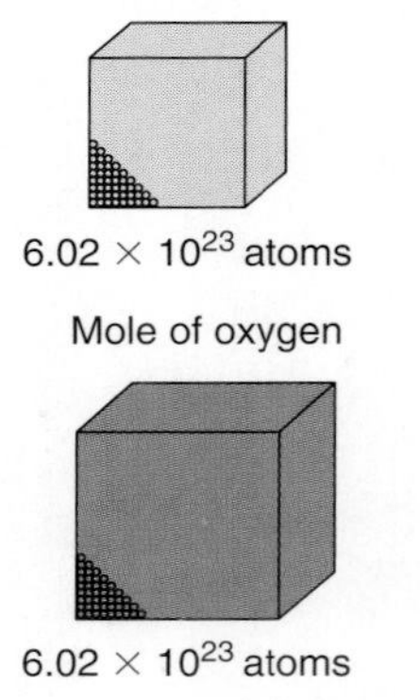

Figure 12-7 A mole of any element contains 6.02×10^{23} atoms, which is Avogadro's number.

Now suppose we have samples of carbon and of oxygen that contain many atoms. However, no matter how many atoms the samples contain, if the ratio between the carbon and oxygen masses is the same 12:16 as the ratio between their atomic masses, the samples contain the same numbers of each kind of atom.

This reasoning can be extended to the atoms of any element. For convenience a quantity called the **mole** is defined in this way:

> A mole of any element is that amount of it whose mass in grams is equal to its atomic mass expressed in u.

The abbreviation of the mole is just mol. Thus a mole of carbon is 12 g and a mole of oxygen is 16 g (Fig. 12-6).

The definition of the mole means that *a mole of any element contains the same number of atoms as a mole of any other element.* This number is a constant of nature called **Avogadro's number** (Fig. 12-7):

$$N_0 = 6.02 \times 10^{23} \text{ atoms/mol} \qquad \textit{Avogadro's number}$$

Avogadro's number = number of atoms per mole of any element

The number of atoms in a sample of any element is just the number of moles in the sample multiplied by N_0.

Example 12.2

How many atoms are present in 100 g of iron?

Solution

Since the atomic mass of iron is 55.85 u, the mass of 1 mole of iron is 55.85 g. The number of moles in 100 g of iron is

$$\text{Moles of iron} = \frac{\text{mass of Fe}}{\text{molar mass of Fe}} = \frac{100 \text{ g}}{55.85 \text{ g/mol}} = 1.79 \text{ mol}$$

The number of iron atoms is therefore (Fig. 12-8)

$$\begin{aligned}\text{Atoms of iron} &= (\text{moles of iron})(\text{atoms/mol}) \\ &= (1.79 \text{ mol})(6.02 \times 10^{23} \text{ atoms/mol}) \\ &= 1.08 \times 10^{24} \text{ atoms}\end{aligned}$$

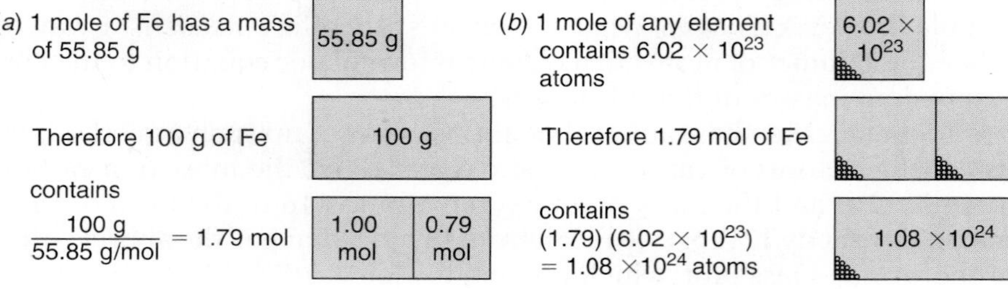

Figure 12-8 How the number of atoms in 100 g of iron can be calculated. (*a*) First the number of moles is found. (*b*) Then the total number of atoms is determined.

12.4 Formula Units

A Mole of Anything Contains Avogadro's Number of Formula Units

The concept of the mole is not limited to the elements. For instance, the gas carbon monoxide, CO, is a compound of carbon and oxygen whose molecules contain one atom of each kind. A molecule of CO therefore has a mass of 12 u + 16 u = 28 u, and a mole of CO has a mass of 28 g. There are N_0 molecules of CO in each mole of it.

Because many compounds, such as sodium chloride, NaCl, do not consist of individual molecules, it is more appropriate to deal with **formula units** rather than molecules in generalizing the definition of the mole. A formula unit of a substance is just the set of atoms given by its formula. In the case of CO, a formula unit is the same combination of one C atom and one O atom each molecule contains. For NaCl, a formula unit consists of one Na atom and one Cl atom. For the more complex compound sodium sulfate, Na_2SO_4, a formula unit consists of two Na atoms, one S atom, and four O atoms.

> The **formula mass** of a substance is the sum of the atomic masses of the elements it contains, each multiplied by the number of times it appears in the formula of the substance.

Evidently the formula mass of carbon monoxide is the same as its molecular mass of 28 u. Here is how the formula masses of sodium chloride and sodium sulfate, which do not exist in molecular form in the solid state, are found:

NaCl: 1 Na = 22.99 u	Na_2SO_4: 2Na = 2 × 22.99 = 45.98 u
1 Cl = 35.45 u	1S = 1 × 32.06 = 32.06 u
Formula mass = 58.44 u	4O = 4 × 16.00 = 64.00 u
	Formula mass = 142.04 u

Now we can give a general definition of Avogadro's number:

$$N_0 = 6.02 \times 10^{23} \text{ formula units/mol} \qquad \textit{Avogadro's number}$$

Avogadro's number = number of formula units per mole of any substance

For a grocer, the normal unit of quantity for eggs is the dozen, equal to 12. For a paper manufacturer, the normal unit of quantity for his product is the ream, equal to 500 sheets. For a chemist, the normal unit of quantity of any substance is the mole, equal to N_0 formula units:

> A mole of any substance is that amount of it whose mass in grams is equal to its formula mass expressed in u.

A mole of NaCl has a mass of 58.44 g, and a mole of Na_2SO_4 has a mass of 142.04 g. The number of formula units in a mole is Avogadro's number.

Mass Relationships in Reactions Owing to the way the mole is defined, a chemical equation can be interpreted in terms of moles as well

Molarity of a Solution

The *molarity* of a solution is the number of moles of solute per liter of solution. A solution that contains 2 mol of sulfuric acid per liter is designated 2*M* H_2SO_4. When a certain number of moles of a compound is needed for a particular reaction, it is convenient to be able just to pour out the corresponding volume of a solution of known molarity. Suppose we need 0.082 mol of sulfuric acid and have a bottle of 2*M* H_2SO_4. Since molarity = moles/liter,

$$\text{Volume needed} = \frac{\text{moles needed}}{\text{molarity}} = \frac{0.082 \text{ mol}}{2 \text{ mol/liter}} = 0.041 \text{ liter} = 41 \text{ ml}$$

as in terms of molecules or formula units. Let us consider the burning of the gas propane, which was discussed in Sec. 10.17. The process obeys the equation

$$C_3H_8 + 5O_2 \longrightarrow 3CO_2 + 4H_2O$$

which means that 1 molecule of C_3H_8 combines with 5 molecules of O_2 to yield 3 molecules of CO_2 and 4 molecules of H_2O. The equation equally correctly states that 1 mole of C_3H_8 combines with 5 moles of O_2 to yield 3 moles of CO_2 and 4 moles of H_2O:

$$\underset{\text{1 mole of propane}}{C_3H_8} + \underset{\text{5 moles of oxygen}}{5O_2} \longrightarrow \underset{\text{3 moles of carbon dioxide}}{3CO_2} + \underset{\text{4 moles of water}}{4H_2O}$$

Example 12.3

How many grams of oxygen are needed to burn 100 g of propane?

Solution

We start by finding the formula masses of oxygen and propane:

$$O_2\text{: } 2O = 2 \times 16.00 = 32.00 \text{ u} \qquad C_3H_8\text{: } 3C = 3 \times 12.00 = 36.00 \text{ u}$$
$$8H = 8 \times 1.008 = \underline{8.06 \text{ u}}$$
$$44.06 \text{ u}$$

Therefore the molar masses of oxygen and propane are, respectively, 32.00 g and 44.06 g. The number of moles in 100 g of propane is

$$\text{Moles of propane} = \frac{\text{mass of } C_3H_8}{\text{molar mass of } C_3H_8} = \frac{100 \text{ g}}{44.06 \text{ g/mol}} = 2.27 \text{ mol}$$

From the equation of the reaction, 5 moles of O_2 are needed for every mole of C_3H_8, so the number of moles of oxygen we need is

$$\text{Moles of oxygen} = \left(\frac{\text{moles of } O_2}{\text{mole of } C_3H_8}\right)(\text{moles of } C_3H_8)$$
$$= (5)(2.27 \text{ mol}) = 11.35 \text{ mol}$$

The mass of oxygen needed is

$$\text{Mass of oxygen} = (\text{moles of } O_2)(\text{molar mass of } O_2)$$
$$= (11.35 \text{ mol})(32.00 \text{ g/mol}) = 363 \text{ g}$$

A total of 363 g of oxygen is needed for the complete combustion of 100 g of propane (Fig. 12-9). Propane and other hydrocarbon gases, such as butane and methane (natural gas), need surprisingly large amounts of oxygen to burn completely.

Chemical Energy

Ever since our ancestors learned to control fire, people have been putting chemical energy to practical use. Today we transform it not only into heat and light but into mechanical energy and electric energy as well. Locked up in matter, chemical energy long remained a mystery. The

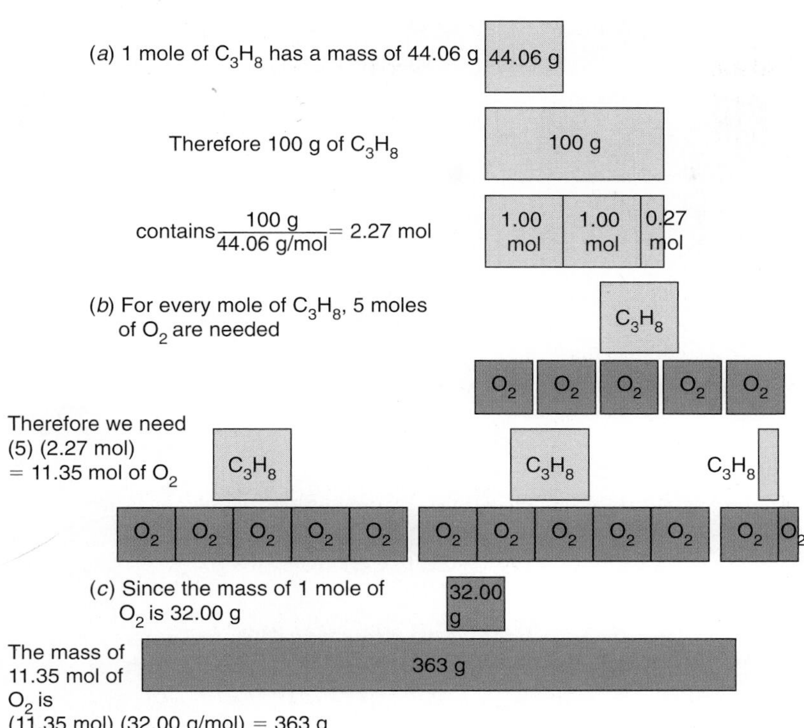

Figure 12-9 How the mass of oxygen needed to burn 100 g of propane is calculated. (*a*) First the number of moles of propane is found. (*b*) Then the number of moles of oxygen is found. (*c*) Finally the mass of oxygen is determined. The formula for the process is $C_3H_8 + 5O_2 \longrightarrow 3CO_2 + 4H_2O$.

modern picture of the atom and of the chemical bond, however, is able to explain the nature of this energy.

12.5 Exothermic and Endothermic Reactions

Some Reactions Liberate Energy; Others Absorb It

Chemical changes that *give off* energy are called **exothermic reactions.** (In Greek, *exo* means "outside.") The burning of coal, which is largely carbon, and of hydrogen are both exothermic:

$$C + O_2 \longrightarrow CO_2 + E \qquad E = 9\text{ kJ/g of }CO_2 \qquad \textit{Formation of carbon dioxide}$$

$$2H_2 + O_2 \longrightarrow 2H_2O + E \qquad E = 13.6\text{ kJ/g of }H_2O \qquad \textit{Formation of water}$$

Chemical changes that take place only when heat or some other kind of energy is *absorbed* are called **endothermic reactions.** (In Greek, *endo* means "inside.") The decomposition of water into hydrogen and oxygen requires heating to very high temperatures or the supply of electric energy during electrolysis (see Fig. 10-23), so it is endothermic:

$$2H_2O + E \longrightarrow 2H_2 + O_2 \qquad E = 13.6\text{ kJ/g of }H_2O \qquad \textit{Decomposition of water}$$

The formation of nitric oxide (NO) from the elements N_2 and O_2 is an endothermic reaction that takes place only at high temperatures (Fig. 12-11):

$$N_2 + O_2 + E \longrightarrow 2NO \qquad E = 3\text{ kJ/g of NO} \qquad \textit{Formation of nitric oxide}$$

Hot and Cold Packs

Although most salts absorb heat when they dissolve in water, some give off heat in this process. These effects are used in the hot and cold packs that are sometimes applied as first aid for minor injuries. A hot pack contains a plastic bag of water and a salt whose solution is exothermic, for example calcium chloride. In a cold pack the salt is one whose solution is endothermic, for example ammonium nitrate. Squeezing the pack breaks the water bag and, as the salt dissolves, the pack becomes hot or cold depending on what kind of salt is present.

The *Hindenburg* Fire

The German airship *Hindenburg* was filled with hydrogen. In 1937, after crossing the Atlantic, it caught fire while landing at Lakehurst, New Jersey (Fig. 12-10). The combustion of hydrogen is an exothermic reaction whose product is water. Today's airships use the inert gas helium, which is denser than hydrogen and so less buoyant but is completely safe.

Because the *Hindenburg* fire was bright whereas flames from a hydrogen fire are almost invisible, it has been proposed that the *Hindenburg* fire occurred primarily in the paint on its envelope. However, experiments on surviving pieces and replica samples of the airship's envelope show that they burn far too slowly to account for the event; instead of the observed 34 seconds, the fire would have lasted 40 hours. Of course, if the hydrogen had been ignited first, as seems probable from the available evidence, the envelope would then burn brightly and other materials in the airship would glow in the way the mantle of a gas lantern does. Both effects would have contributed to the spectacle shown in the photograph.

Figure 12-10 The hydrogen-filled airship *Hindenburg* took only 34 seconds to burn up.

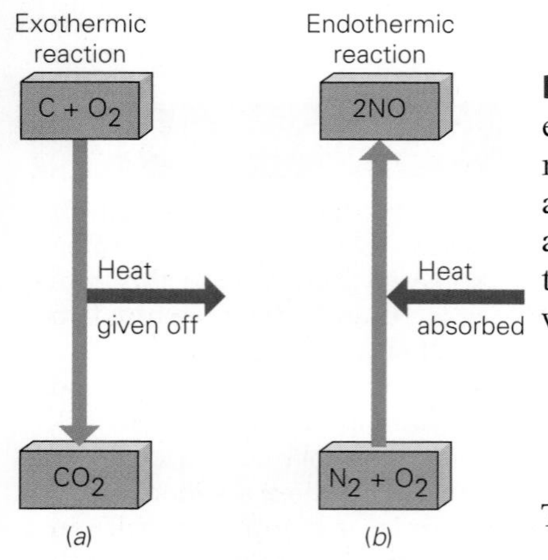

Figure 12-11 Examples of exothermic and endothermic reactions. (*a*) The burning of coal to form carbon dioxide gas gives off 9 kJ of heat per gram of CO_2. (*b*) The formation of nitric oxide requires 3 kJ of heat to be absorbed per gram of NO.

Direct and Reverse Reactions From the law of conservation of energy we can predict that, if a given reaction is exothermic, the reverse reaction will be endothermic. Furthermore, the amount of heat liberated by the exothermic reaction must be the same as the amount of heat absorbed by the endothermic reaction. This prediction is borne out in the case of water, as we just saw, and is also verified in all other reactions where it can be tested. An example is sodium reacting with chlorine:

$$2\text{Na} + \text{Cl}_2 \longrightarrow 2\text{NaCl} + E \quad E = 7 \text{ kJ/g of NaCl}$$

Formation of sodium chloride (Exothermic)

To break up NaCl takes the same amount of energy:

$$2\text{NaCl} + E \longrightarrow 2\text{Na} + \text{Cl}_2 \quad E = 7 \text{ kJ/g of NaCl}$$

Decomposition of sodium chloride (Endothermic)

Dissociation and Neutralization The dissociation of most salts is an endothermic process. For example, when KNO_3 is dissolved in water, the container becomes cold, since the dissociation of the salt requires energy:

$$KNO_3 + E \longrightarrow K^+ + NO_3^- \quad E = 0.36 \text{ kJ/g of } KNO_3$$ *Dissociation of potassium nitrate*

Neutralization, on the other hand, is an exothermic process. If concentrated solutions of NaOH and HCl are mixed, for instance, the mixture quickly becomes too hot to touch:

$$H^+ + OH^- \longrightarrow H_2O + E \quad E = 3.2 \text{ kJ/g of } H_2O$$ *Neutralization*

The Na^+ and Cl^- ions are omitted here because the principal chemical change in all neutralizations is simply the joining together of hydrogen ions and hydroxide ions. For the same reason the neutralization of any other strong acid by any other strong base liberates almost precisely the same amount of heat for each gram of water produced.

12.6 Chemical Energy and Stability

The Less PE Its Electrons Have, the More Stable the Compound

The heat given off or absorbed in a chemical change is a measure of the stabilities of the substances (or mixtures of substances) involved. If a great deal of energy is needed to decompose a substance, the substance is (with rare exceptions) relatively stable. If the decomposition is either exothermic or weakly endothermic, the substance is normally unstable.

From the reactions given in Sec. 12.5 we can see at a glance that CO_2, H_2O, and NaCl are stable compounds, since the formation of each is strongly exothermic and its decomposition is endothermic. NO, on the other hand, is unstable, since its decomposition liberates heat. The combinations H_2 and O_2, Na and Cl_2, H^+ and OH^- are relatively unstable, since they react to give off energy. On the other hand, N_2 and O_2 form a stable mixture since energy must be supplied for them to react.

Explosives

An explosive is a material in which a violent reaction can occur whose products are rapidly expanding gases. Most explosives contain nitrogen compounds because N_2 molecules, which have triple covalent bonds between their N atoms (see Sec. 10.12), are very stable and so their formation in an explosion is strongly exothermic.

The earliest explosive in wide use was gunpowder, a mixture of potassium nitrate (KNO_3), charcoal, and sulfur. (The Chinese invention of gunpowder, together with those of printing and the magnetic compass, marked the start of the modern era of human history.) When gunpowder is ignited, oxygen from the KNO_3 combines with carbon in the charcoal to give CO_2 and with sulfur to give SO_2, and the nitrogen in the KNO_3 becomes N_2. The temperature of the explosion is thought to be about 2700°C.

More recent explosives, such as nitroglycerin and trinitrotoluene (TNT), consist of molecules that each contain carbon, hydrogen, oxygen, and nitrogen whose recombination yields gaseous products. For instance, when nitroglycerin explodes,

$$4C_3H_5(NO_3)_3 \longrightarrow 12CO_2 + 10H_2O + 6N_2 + O_2$$

Nitroglycerin is dangerous to handle because it blows up at the slightest shock. Dynamite is a mixture of nitroglycerin and other explosives with inert ingredients that make the combination safe to work with. Dynamite is set off by a small explosive detonator that is activated electrically (see Fig. 13-11).

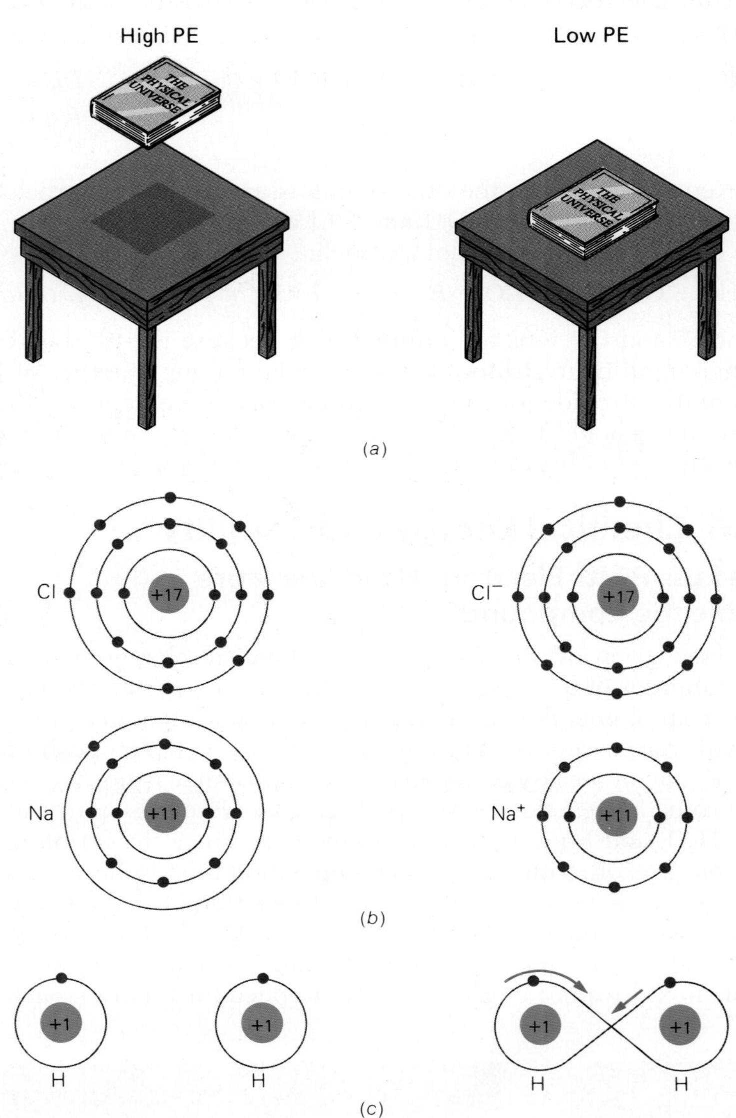

Figure 12-12 (*a*) A book raised above a table has more PE than the same book lying on the table because the attractive force the earth exerts on the book is greater when the book is closer to the earth. (*b*) The outer electron in an Na atom has more PE than the same electron has when it is attached to a Cl atom to form a Cl^- ion because the electron is more strongly attracted to the nucleus of the Cl^- ion (see Fig. 10-16). (*c*) The electron in an H atom has more PE than the same electron has when it is part of a H molecule because in the H molecule the electron is attracted by two protons rather than one proton.

We can interpret chemical-energy changes in terms of atomic structure (Fig. 12-12). When sodium reacts with chlorine, for example, the outer electron of each Na atom is transferred to the outer shell of a Cl atom. In its new position the electron has less potential energy with respect to the atomic nuclei because it is held more firmly to the chlorine nucleus than it was to the sodium nucleus. The same is true for other ionic bonds.

When two hydrogen atoms react to form a hydrogen molecule, the atoms are joined by shared electrons. Such a covalent bond involves a decrease in the PE of the electrons because each is now attracted by two nuclei instead of a single nucleus. The same is true for other covalent bonds.

The Nature of Chemical Energy Our conclusion is that

> Chemical energy is electron potential energy.

When electrons move to new locations during an exothermic reaction, some of their original PE is liberated. This released energy may show itself in faster atomic or molecular motions that correspond to a higher temperature. Or the freed energy may excite outer electrons into higher energy levels from which they return to lower levels by giving off photons of light. In endothermic reactions, energy must be supplied to the atoms involved to enable some of their electrons to form bonds in which their PEs are greater than before.

12.7 Activation Energy

The Initial Energy Needed to Start an Exothermic Reaction

Wood burns in air to give off great quantities of heat. However, we can store a pile of firewood indefinitely without its catching fire. A mixture of hydrogen and oxygen can explode violently. However, hydrogen does not explode when mixed with air unless a flame or spark sets off the reaction. Why do not all exothermic reactions take place at once of their own accord?

Clearly, in order to begin, many exothermic processes must first be supplied with energy. A mixture of hydrogen and oxygen is like the car of Fig. 12-13, whose potential energy may be converted to kinetic energy if it moves down into the valley. However, the car cannot begin to go downward unless it is first given enough energy to climb to the top of the hill. Similarly the chemical energy of a mixture of hydrogen and oxygen can be freed only if the molecules have enough energy, or are sufficiently **activated,** to make the reaction start. The energy needed for activation, corresponding to the energy required to move the car up the hill, is called the **activation energy** of the reaction.

The electron picture of chemical combination suggests the reason for activation energy. The reaction of oxygen and hydrogen involves the formation of bonds between O and H atoms, a process that gives out energy. However, before these bonds can be formed, the covalent bonds between the hydrogen atoms in H_2 molecules and the oxygen atoms in O_2 molecules must be broken. To break these bonds takes energy. Once the reaction starts, the energy already liberated can supply the needed energy, but in the beginning some outside energy must be supplied. Thus a mixture of hydrogen and oxygen need only be touched with a flame for the reaction to spread so rapidly that an explosion results. When a bed of coal is set on fire it continues to burn, since the heat liberated in one place is sufficient to ignite the coal around it.

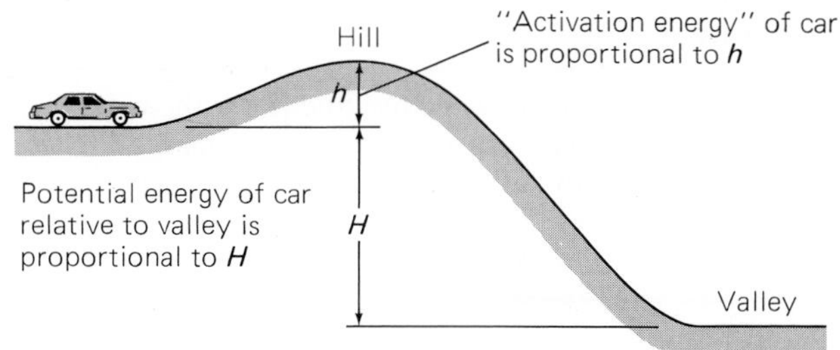

Figure 12-13 Activation energy. The potential energy of the car will be converted into kinetic energy if it moves down into the valley. However, the car requires initial kinetic energy in order to climb the hill between it and the valley, analogous to the activation energy required in many exothermic reactions.

Activated Molecules A molecule with enough energy to react is called an **activated molecule.** In reactions that take place spontaneously at room temperature (for example, the reaction between hydrogen and fluorine), enough of the initial molecules have the required KE of thermal motion for their bonds to break during collisions without further activation. Ions in solution are, so to speak, already activated and react almost instantaneously. But a large number of exothermic reactions must have the preliminary activation of some molecules before the reactions can take place in a self-sustaining way.

Reaction Rates

Some chemical changes are practically instantaneous. Thus in neutralization the acid and base react as soon as they are stirred together. Silver chloride is precipitated immediately when solutions that contain silver ions and chloride ions are mixed. The reaction involved in a dynamite explosion takes a fraction of a second. In contrast, other chemical changes, like the rusting of iron, take place slowly.

Reaction rates depend first of all on the nature of the reacting substances: iron corrodes faster than copper, for example. In any particular reaction the rate is influenced by four principal factors (Table 12-1). These are temperature, concentrations of the reacting substances, the exposed surface area in the case of reactions that involve solids, and the presence of an appropriate catalyst.

12.8 Temperature and Reaction Rates

Hotter Means Faster

Reaction rates are always increased by a rise in temperature. This is why we use hot water rather than cold for washing. Reaction rates for many common processes that occur at or near room temperature are approximately doubled for every 10°C increase in temperature.

The kinetic theory of matter suggests one obvious reason for the increase of reaction rates with temperature. Most reactions depend on collisions between particles, and the number of collisions increases with rising temperature because molecular speeds are increased. But a 10°C rise is nowhere near enough to double the number of collisions in a particular

Table 12-1 Factors That Affect Reaction Rates

Factor	Effect
Temperature	The higher the temperature, the faster the reaction
Concentration	The higher the concentration of the reactants, the faster the reaction
Surface area	The greater the surface area of a solid reactant, the faster the reaction
Catalyst	Increases the reaction rate

sample. To find the real explanation we must go back to the idea of activation energy.

Why Reaction Rates Vary with Temperature If molecules must be activated before they can react, reaction rates should depend not on the total number of collisions per second but on the number of collisions between *activated* molecules (see Sec. 12.5). Activated molecules in a fluid (liquid or gas) are produced by ordinary molecular motion as a result of exceptionally energetic collisions. Such molecules remain activated only a short time before losing energy in further collisions (unless they react in the meantime).

In any fluid, then, a certain fraction of the molecules is activated at any time. The fraction may be very small at ordinary temperatures, but it increases rapidly as the temperature rises and molecular motion speeds up. Reaction rates increase with temperature chiefly because the number of activated molecules increases.

At room temperature, for instance, a mixture of hydrogen and oxygen contains very few molecules with sufficient energy to react. The reaction is so slow that the gases may remain mixed for years without anything happening. Even at 400°C the rate is small, but at 600° enough of the molecules are activated to make the reaction fast, and at 700° so many are activated that the mixture explodes.

This kind of behavior is typical of many reactions between molecules. At low temperatures the chemical changes are so slow that for all practical purposes they do not occur; in a range of intermediate temperatures the reactions are moderately rapid; and at high temperatures they become practically instantaneous. Reactions between ions, on the other hand, occur immediately even at room temperature, since the ionic state itself is a form of activation.

12.9 Other Factors

Concentration, Surface Area, and Catalysts

The effect of concentration on reaction speed is illustrated by rates of burning in air and in pure oxygen. The pure gas has almost 5 times as many oxygen molecules per cubic centimeter as air, and combustion in pure oxygen is correspondingly faster.

As a general rule, the rate of a simple chemical reaction is proportional to the concentration of each reacting substance. The number of collisions between activated molecules, which determines the reaction speed, depends on the total number of collisions and this, in turn, depends on how many molecules each cubic centimeter contains.

Surface Area When a reaction takes place between two solids or between a fluid and a solid, the reaction speed depends markedly on the amount of solid surface exposed. A finely powdered solid presents vastly more surface than a few large chunks (Fig. 12-14), and reactions of powders are accordingly much faster. Granulated sugar dissolves more rapidly in water than lump sugar; finely divided zinc is attacked by acid quickly, larger pieces are attacked slowly; ordinary iron rusts slowly, but the oxidation of iron powder is fast enough to produce a flame.

A CHEMIST AT WORK

Judith M. Iriarte-Gross,
Middle Tennessee State University

I am a professor of chemistry at Middle Tennessee State University (MTSU) where I teach chemistry and physical science. I also direct research in the area of inorganic chemistry and in science education for nonscience majors.

I began my chemistry education in a community college and then transferred to the University of Maryland at College Park where I received my B.S. and M.S. in chemistry. I earned my Ph.D. in the laboratory of Jerry Odom at the University of South Carolina. My graduate work focused on the synthesis and characterization by nuclear magnetic resonance (NMR) of Groups 14 and 16 compounds. I was a nontraditional student and defended my dissertation three days before my son graduated from high school. Graduate school was tough but very rewarding.

I accepted a postdoctoral position in Patty Wisian-Neilson's lab at Southern Methodist University, where I studied inorganic polymers. Next, I was a Food and Drug Administration chemist, analyzing everything from pesticides in fruits and vegetables to dyes in cosmetics. I spent a few years in the plastics industry as a polymer chemist where I developed plastics for use as lenses in scanners and detectors. I still taught chemistry at night at local colleges and universities. My Ph.D. mentor once said that he had the best job in the world, where he could do research and teach and work with students. I agree with him. My favorite job as a chemist is what I am doing now—being a university professor.

In addition to my teaching, I mentor students in undergraduate research. I am interested in inorganic "bench" chemistry and science education research. We use sol-gel chemistry to synthesize glass and ceramic materials at room temperature. Sol-gel chemistry involves the formation of a sol, a colloidal suspension of solid particles in a liquid, and its subsequent transformation to a gel, a colloidal suspension in which the colloids form a continuous network in the liquid. The gel is then used as a precursor material for the synthesis of unique glasses and ceramics. Currently we are exploring how the addition of selenium and tellurium salts affects a silicate sol-gel matrix. My students find that this research is exciting since they are making new materials that have potential applications in technology. My undergraduate students are also learning to use synthetic and analytical techniques that I learned as a graduate student.

In a new research area, we are exploring a new dating method as an alternative to the well-known carbon-14 dating technique. We measure the concentration of fluoride in faunal remains, or as we say, old bones. Over time, these bones absorb fluoride from the soils in which the bones are buried. We are using a fluoride ion selective electrode (like a pH electrode) to determine the concentration of fluoride in squirrel and deer bones found at Native American archaeological sites in Tennessee. We clean and grind the bones into a fine powder. We dissolve the powder in a hydrochloric acid solution and measure the fluoride concentration using the fluoride electrode. Our fluoride concentration data will be compared to and used to corroborate the C-14 dating results. This research is exciting because of its interdisciplinary nature and because the notion of studying "old" bones appeals to many students. My science education research involves the design, development, testing, and writing of new laboratory activities for physical science. For example, a new lab titled "The 60's Lab" is about light and color and tie dying.

There is more to being a chemistry professor. Once the research is complete, we communicate our findings to others in the scientific community by presenting results at professional meetings such as those of the American Chemical Society (ACS) or publishing papers in professional journals. As a professor, I advise students on which classes to take, on graduate school admission processes, and on careers in chemistry. I serve on departmental and university committees and give back to the community by talking with middle and high school girls about the importance of science and math education and careers. As you can see, a chemist's life is never dull.

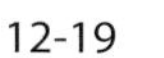

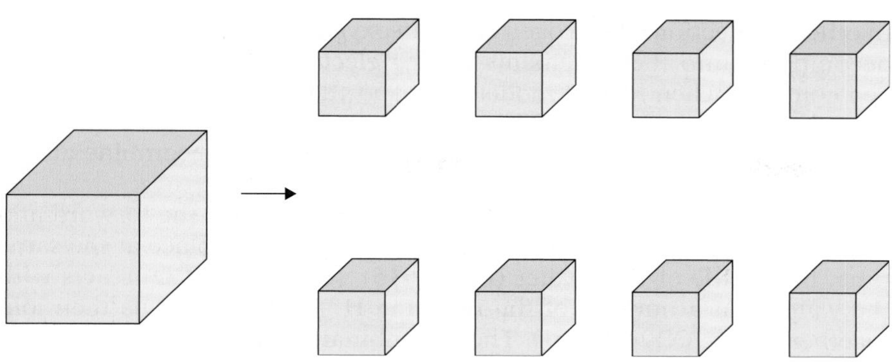

Figure 12-14 Cutting a cube into 8 smaller cubes doubles its original surface area. Cutting the smaller cubes further gives a still larger total surface area, and grinding them into a fine powder increases the area by a huge amount. When a solid undergoes a chemical reaction, the greater its surface area, the faster the reaction.

The explanation is obvious: the greater the surface, the more quickly atoms and molecules can get together to react. For a similar reason, efficient stirring speeds up reactions between fluids.

Catalysts A **catalyst** is a substance that speeds up a reaction without being permanently changed itself. As an example of catalytic action, let us consider the decomposition of hydrogen peroxide, H_2O_2. At ordinary temperatures solutions of hydrogen peroxide are unstable and slowly turn into water and oxygen:

$$2H_2O_2 \longrightarrow 2H_2O + O_2$$

If a little powdered manganese dioxide is added to the solution, the decomposition goes much faster, with oxygen bubbling up violently. At the end of the reaction the manganese dioxide remains unchanged. Commercial solutions of hydrogen peroxide usually contain a trace of the compound acetanilid, which acts to retard their decomposition.

Catalysts accelerate reactions in different ways. In some cases the catalyst forms an unstable intermediate compound with one of the reacting substances, and this compound decomposes later in the reaction. Most catalysts, notably certain metals such as platinum, increase reaction rates by producing activated molecules at their surfaces. No adequate explanation is known for the action of some catalysts. A given reaction is usually influenced only by a few catalysts, and these may or may not affect other reactions. Catalysts are essential in many industrial processes.

The many chemical processes that take place in living things, for instance the digestion of food, are controlled by catalysts called **enzymes.** An enzyme is a protein molecule whose physical structure is such that it attracts specific molecules to its surface, promotes their reaction, and then releases the products. Each enzyme catalyzes a particular reaction, and thousands of different ones are present in the human body. The reacting molecules fit into the outside of the enzyme for a process much as a key fits into a lock: if their shapes do not match, nothing will happen.

Self-cleaning Windows

The compound titanium dioxide, which is what makes most white paint white, is also an efficient photocatalyst that promotes the dissociation of water molecules by ultraviolet light. The resulting OH^- ions disrupt organic molecules of all kinds. Because water vapor is always present in the atmosphere and sunlight contains ultraviolet light, titanium dioxide coatings thin enough to be transparent are being used on window glass to make them self-cleaning. The breakdown products of dirt on the windows are easily washed away by rain. Titanium dioxide is also used in a similar way to treat polluted water and to sterilize hospital surfaces using artificial sources of ultraviolet light.

12.10 Chemical Equilibrium

One Step Forward; One Step Back

Most chemical reactions are reversible. That is, under suitable conditions the products of a chemical change can usually be made to react "backward" to give the original substances. We have seen many examples

Figure 12-15 In soda water, dissolved carbon dioxide reacts with water to give hydrogen and hydrogen carbonate ions. At the same time, hydrogen and hydrogen carbonate ions are recombining to form carbon dioxide and water. The forward and reverse reactions occur at the same rate, which is an example of chemical equilibrium.

in other connections. Hydrogen burns in oxygen to form water, and water decomposes into these elements during electrolysis. Mercury and oxygen combine when heated moderately, and the oxide decomposes when heated more strongly. Carbon dioxide reacts with water to give hydrogen and hydrogen carbonate (HCO_3^-) ions, and these ions recombine all the time to form the original CO_2 and H_2O.

There is no reason why, under the right conditions, the forward and backward processes of a chemical change cannot take place at the same time. In a bottle of soda water (Fig. 12-15), some of the CO_2 reacts with the water. But a number of the resulting H^+ and HCO_3^- ions then join together to give CO_2 and H_2O. The recombination rate increases with the ion concentration until, finally, as many ions recombine each second as are being formed. At this point the rates of the forward and backward reactions are the same, and the amounts of the various substances do not change. The situation can be represented by the single equation

$$H_2O + CO_2 \rightleftharpoons H^+ + HCO_3^-$$

The double arrow indicates that reactions in both directions occur together.

A State of Balance Such a situation is called **chemical equilibrium.** It is a state of balance determined by two opposing processes. The two processes do not reach equilibrium and stop, but instead continue indefinitely because each process constantly undoes what the other accomplishes.

As an analogy, we might imagine a person walking up an escalator while the escalator is moving down. If the person walks as fast in one direction as the escalator is moving in the other, the two motions will be in equilibrium and the person will remain in the same place indefinitely.

A great many chemical changes reach a state of equilibrium instead of going to completion. Equilibrium may be established when a reaction is nearly complete, or when it is only just starting, or when both products and reacting substances are present in comparable amounts.

The point at which equilibrium occurs depends on the rates of the opposing reactions. The initial reaction always dominates until the products are abundant enough for the reverse reaction to go at the same rate. Thus the extent to which an acid is dissociated depends on how fast its molecules break down into ions compared with how fast the ions recombine. HCl dissociates so rapidly in solution into H^+ and Cl^- that the reverse reaction has no chance to maintain a measurable amount of HCl. On the other hand, acetic acid dissociates slowly, and when only a small concentration of ions has been built up, the recombination occurs at the same rate as the dissociation.

12.11 Altering an Equilibrium

How to Get Farther Up the Down Escalator

Often a chemist wishing to prepare a compound finds that the reaction reaches equilibrium before very much of the compound has been formed. Once this happens, waiting for more of the product to form is useless, for its amount does not change after that. How can equilibrium conditions be altered to increase the yield of the product?

Since equilibrium represents a balance between two rates, what is needed to increase the yield is a way to change the speed of one reaction

Example 12.4

Under what circumstances will a liquid-phase reaction go to completion instead of an equilibrium being established?

Solution

When one of the products of a reaction leaves the system, the reaction must go to completion since the reverse reaction then cannot occur. A reaction in a liquid will go to completion if one of the products is (a) a gas that escapes; (b) an insoluble precipitate; or (c) composed of molecules that do not dissociate when the reaction involves ions.

or the other. Speeding up or slowing down one of the reactions in an equilibrium is not as simple as changing the rate of a single reaction, but the same factors that affect reaction rates also influence equilibrium. The chemist has three chief methods available for shifting an equilibrium to favor one direction or the other. These are:

1. Change the concentration of one or more substances. For example, removing the gaseous product of a reaction will retard the reverse reaction. Thus opening a soda bottle allows CO_2 to escape, which decreases the rate of formation of H^+ and HCO_3^- and so lowers the acidity of the solution.
2. Change the temperature. If one reaction in an equilibrium is exothermic (gives off energy), the other is necessarily endothermic (absorbs energy). A rise in temperature, although it makes both reactions go faster, will favor the endothermic one.
3. Change the pressure. This is most effective in gas reactions where the number of product molecules differs from the number of initial molecules. Increasing the pressure favors the reaction that gives the fewest molecules. An example is the synthesis of ammonia from nitrogen and hydrogen in the reaction

$$N_2 + 3H_2 \rightleftharpoons 2NH_3 \qquad \textit{Ammonia synthesis}$$

A rise in pressure increases the yield of ammonia because the ammonia occupies only half the volume of the gases that react to form it. Pressures of 150 to 350 atm are used in modern ammonia production plants.

Example 12.5

Hydrogen sulfide gas dissolves in water and ionizes very slightly:

$$H_2S \rightleftharpoons 2H^+ + S^{2-}$$

How would the acidity of the solution (that is, the concentration of H^+) be affected by (a) increasing the pressure of H_2S? (b) Raising the temperature? (c) Adding a solution of silver nitrate? (Silver sulfide, Ag_2S, is insoluble.)

Solution

(a) The acidity would increase because the greater the gas pressure, the more of it dissolves. (b) The acidity would decrease because the solubility of gases decreases with increasing temperature. (c) The acidity would increase because removing S^{2-} ions reduces the rate at which H_2S leaves the solution without affecting the rate at which H_2S enters it.

The Haber Process

Fritz Haber (1868–1934) was the German chemist who perfected the synthesis of ammonia from nitrogen and hydrogen. Carl Bosch adapted the process for the commercial production of ammonia, which began in 1913, just in time to provide Germany with the nitrogen compounds it needed to manufacture explosives for World War I. Without the Haber-Bosch process, German guns would have run out of ammunition in 6 months. After the war, the process was used to "fix" nitrogen for use in fertilizers; nitrogen is an essential constituent of all proteins (see Sec. 13.15). Much of the world's agriculture today depends on fertilizers based on synthetic ammonia, for whose development Haber received the Nobel Prize in chemistry in 1919. The breakthrough was hailed as *Brot aus Luft*, bread from the air.

In what he saw as a humanitarian effort to minimize overall casualties by bringing about a quick German victory in World War I, Haber pioneered the use of poison gas as a battlefield weapon. Horrified, his wife committed suicide, and the Allies labeled Haber a war criminal for the immense suffering gas warfare caused. Despite his contributions to the war effort, Haber, who was Jewish, had to flee Germany in 1933 to escape the increasing persecution of Jews there. The insecticide Zyklon B, whose development Haber had worked on in the 1920s, was later used in modified form by the Nazis to kill millions of people, mainly Jews, among whom were members of Haber's family.

Oxidation and Reduction

Until now we have used the term **oxidation** to mean the chemical combination of a substance with oxygen. A related term is **reduction,** which refers to the removal of oxygen from a compound. When oxygen reacts with another substance (except fluorine), the oxygen atoms pick up electrons donated by the atoms of that substance. When the resulting compound is reduced, the atoms of the substances that had been oxidized regain the electrons initially lost to the oxygen atoms.

12.12 Oxidation-Reduction Reactions

They Always Go Together

It is convenient to generalize oxidation and reduction to refer to *any* chemical process in which electrons are transferred from one element to another, regardless of whether or not oxygen is involved. Hence

> Oxidation refers to the loss of electrons by the atoms of an element, and reduction refers to the gain of electrons.

The oxidation of one element is always accompanied by the reduction of another. The two processes must take place together. For example, when zinc combines with chlorine, electrons are given up by the zinc atoms to the chlorine atoms. Thus the zinc is oxidized and the chlorine is reduced in the reaction:

$$\mathrm{Zn} \longrightarrow \mathrm{Zn}^{2+} + 2e^- \qquad \textit{Oxidation}$$

$$\mathrm{Cl}_2 + 2e^- \longrightarrow 2\mathrm{Cl}^- \qquad \textit{Reduction}$$

Reactions that involve electron transfer are called **oxidation-reduction reactions,** and they make up a large and important category of chemical reactions.

Example 12.6

Chlorine is a powerful oxidizing agent, which is the reason it is used as a bleach. Do you think a chlorine solution would remove a rust stain on a piece of clothing?

Solution

Because the iron in rust is already oxidized, the chlorine solution will have no effect.

Electrolysis Electrolysis is an oxidation-reduction process. Let us consider the electrolysis of molten sodium chloride, which consists of the ions Na^+ and Cl^-. An **electrode** is a conductor through which electric current enters or leaves a solution. When electrodes in the molten NaCl are connected to the terminals of a battery, Na^+ ions are attracted to the negative electrode and Cl^- ions to the positive electrode, as in Fig. 12-16.

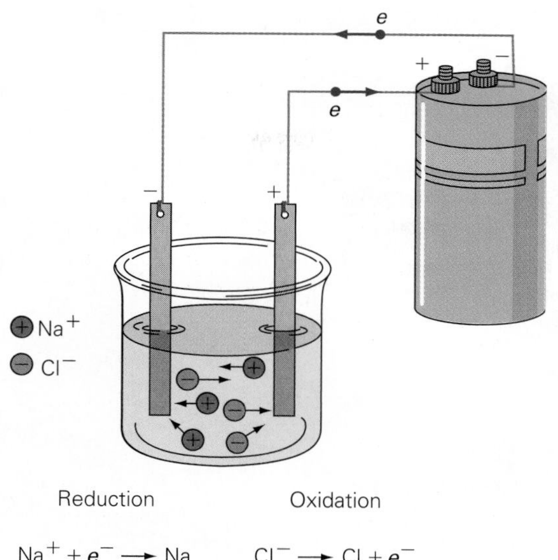

Figure 12-16 The electrolysis of molten sodium chloride. The current in the liquid consists of moving Na^+ and Cl^- ions; the current in the wires consists of moving electrons.

At the positive electrode each Cl^- is neutralized by giving up its extra electron and becomes a chlorine atom:

$$Cl^- \longrightarrow Cl + e^- \qquad \textit{Oxidation}$$

The Cl atoms pair off to form molecules of chlorine gas, Cl_2. At the negative electrode each Na^+ is neutralized by gaining an electron and becomes a sodium atom:

$$Na^+ + e^- \longrightarrow Na \qquad \textit{Reduction}$$

The net result of sending a current through molten salt, then, is to break up the compound NaCl into its constituent elements:

$$2NaCl \longrightarrow 2Na + Cl_2 \qquad \textit{Electrolysis of sodium chloride}$$

The sodium, a liquid at the temperature of molten salt, collects around the negative electrode, and chlorine gas bubbles up around the positive electrode. This procedure is commonly used to prepare metallic sodium.

Electroplating Electrolysis is used in the process of **electroplating** in which a thin layer of one metal is deposited on an object made of another metal. Sometimes this is done because the plating metal is expensive, for instance gold or silver. In other cases the object is to protect the base metal from corrosion, as in the tin or chromium plating of steel (Fig. 12-17). Nonmetallic items can be plated by first coating them with a conducting material such as graphite.

Figure 12-18 shows how a spoon can be silver-plated. The procedure uses a solution of silver nitrate, which dissociates into Ag^+ and NO_3^- ions. Silver atoms lose electrons (which flow to the battery) at the positive electrode and enter the solution as Ag^+ ions. These ions are attracted to the spoon, which acts as a negative electrode, from which they pick up electrons (supplied by the battery) to become silver atoms again. In this way silver atoms are transferred from the positive electrode to the spoon.

Aluminum

Although aluminum is the third most abundant element in the earth's outer layer and the most abundant metal there, it was not discovered until 1827. Chemical methods for preparing aluminum were so expensive then that for some time its main use was in jewelry, and attempts to refine aluminum by electrolysis were balked by the difficulty of melting or dissolving aluminum ores.

Finally, in 1886, a 22-year-old American, Charles Martin Hall, found that cryolite, a mineral abundant in Greenland, when melted would dissolve the chief aluminum ore, Al_2O_3. Passing an electric current through a solution of Al_2O_3 in molten cryolite liberates aluminum at the negative electrode and oxygen at the positive one: this is the process used today to refine aluminum.

About 11 kJ of electric energy is needed to produce each gram of aluminum metal, which is a lot of energy. To recycle aluminum, notably the aluminum cans discarded by the billion every year, basically involves heating it to its melting point and then melting it, which takes less than 1 kJ/g. Hence recycling aluminum uses less than 9 percent as much energy as refining it from its ores, a vast saving on the scale at which aluminum cans are used: about 20 percent of all the aluminum produced in the world is used for beverage cans in the United States alone.

Figure 12-17 Steel plated with tin to prevent corrosion is widely used for food and beverage containers.

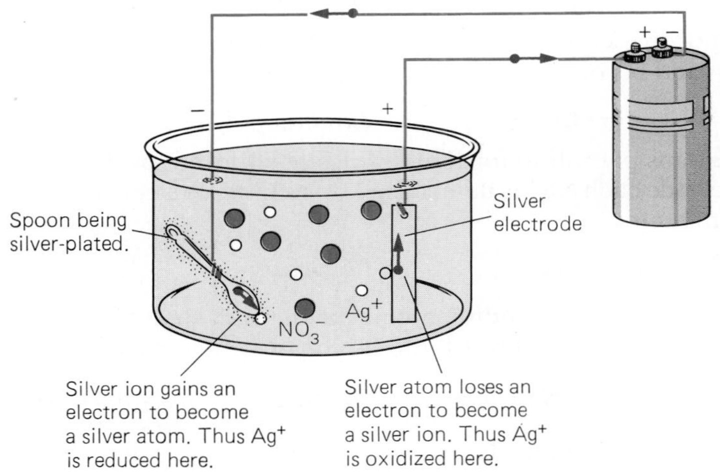

Figure 12-18 Silver plating. The bath is a solution of silver nitrate, $AgNO_3$. The nitrate ions remain in solution because Ag atoms lose electrons at the positive electrode more readily than NO_3^- ions do.

Because an Ag atom loses an electron to become Ag^+ more readily than an NO_3^- ion loses its extra electron, the NO_3^- ions stay in solution and do not participate in the plating process.

12.13 Electrochemical Cells

Turning Chemical Energy into Electric Energy

Oxidation-reduction reactions can produce electric currents. What we must do is arrange to have the electrons transferred in such a reaction pass through an external wire as they go from one reactant to the other. The dry-cell batteries of a flashlight, the storage battery of a car, and the fuel cell of a spacecraft are all based on oxidation-reduction reactions. They are known as **electrochemical cells.**

In some types of batteries, the oxidation-reduction reactions cannot be reversed; when the reactants are used up, the battery is "dead." In a

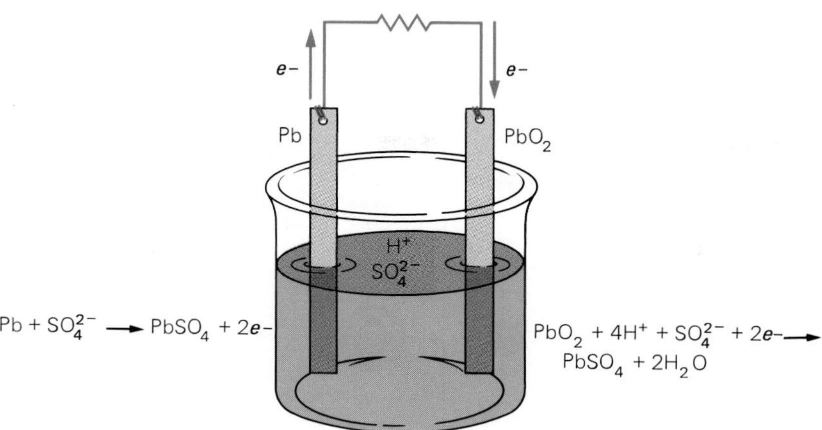

Figure 12-19 The lead-acid storage battery. The reactions shown are those that occur at each electrode when the battery provides current. These reactions are reversed when the battery is being charged.

rechargable battery, the reactions can be reversed by passing a current through it in the opposite direction. The electrode reactions then proceed backwards to restore the battery to its original state.

An example of a rechargable battery is the storage battery of a car in which plates of lead and of lead dioxide, PbO_2, are in a solution of sulfuric acid which is dissociated into H^+ and SO_4^{2-} ions. The reactions that take place at each electrode when the battery is providing current are shown in Fig. 12-19. As the battery provides current, insoluble lead sulfate, $PbSO_4$, builds up on its plates. When the reactants have been used up, the battery cannot supply any more current. Recharging the battery brings the plates and the acid bath back to their initial compositions.

The potential difference across a storage battery cell is 2.1 V; a "12-V" battery contains six cells connected together. The lower the temperature of the acid in a storage battery, the slower its ions move and the less current the battery can provide. In freezing weather, the current available for the starting motor of a car may be less than half that available on a warm day, and it may be difficult or impossible to start the car's engine. Other types of rechargable batteries have been developed. Nickel-cadmium (NiCad) batteries are widely used in small electric and electronic devices, and laptop computers often feature high-capacity lithium-ion batteries.

Fuel Cells In a **fuel cell,** the reacting substances are fed in continuously. As a result the cell can provide current indefinitely without having to be replaced or recharged. Fuel cells are used in spacecraft since they are very light in proportion to the electric power they can supply. In the future it is likely that fuel cells will be perfected to the point where they are economical sources of power for individual homes, electric cars, and large-scale electric plants.

Combining 1 kg of hydrogen and 8 kg of oxygen in a hydrogen-oxygen fuel cell produces over 2×10^8 J of electric energy, enough to power a 100-W lightbulb for 4 weeks. The overall reaction in such a cell is simply

$$2H_2 + O_2 \longrightarrow 2H_2O$$

and involves the flow of 4 electrons each time the reaction occurs (Fig. 12-20). If a mixture of two volumes of hydrogen gas and one volume of

Fuel Cell Efficiency

A fuel cell does not function in the way a heat engine does and therefore is not subject to the thermodynamic limits on efficiency of heat engines (Sec. 5.14). A hydrogen fuel cell is about twice as efficient as a gasoline or diesel engine. Because there are hundreds of millions of cars and trucks in the world, the saving in energy would be enormous if the energy used to produce and distribute the hydrogen is not too great.

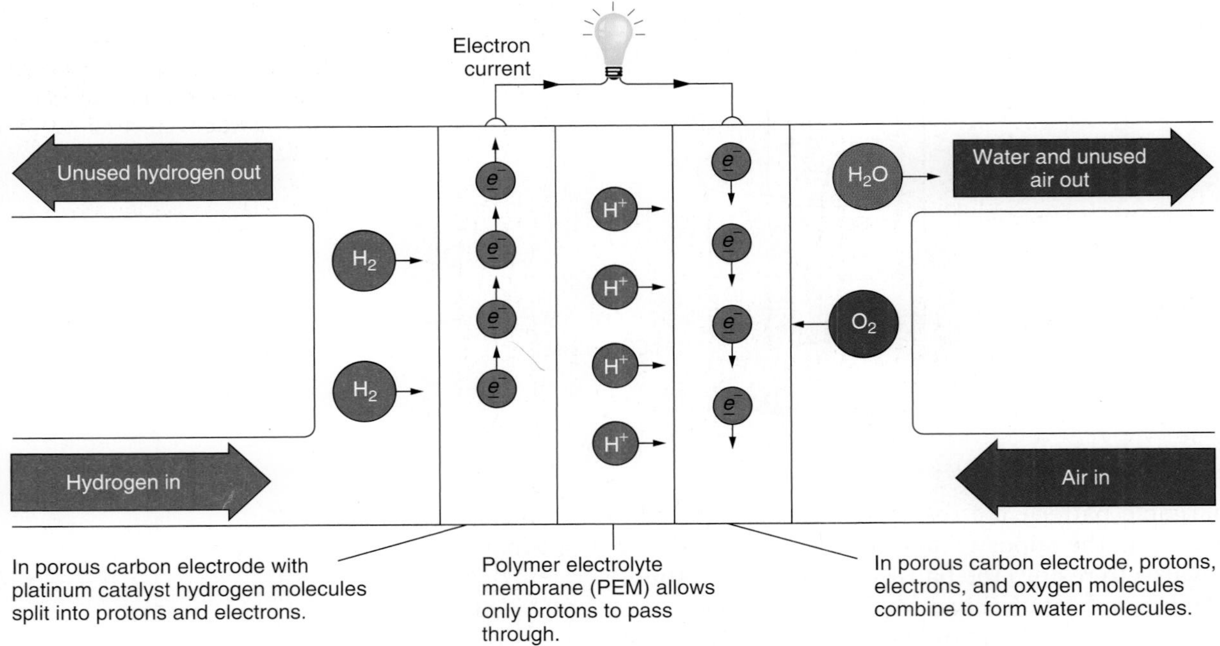

Figure 12-20 How a hydrogen PEM fuel cell operates. The overall reaction is $2H_2 + O_2 \longrightarrow 2H_2O$.

oxygen gas is ignited, the result is a violent explosion with water as the product. In a hydrogen-oxygen fuel cell the same chemical combination takes place, but the liberated energy is released in the form of electric current.

Several types of fuel cells have been developed. The fuel cells used in spacecraft are very light in weight but are extremely expensive and rely on pure oxygen, rather than the oxygen in air, because their potassium hydroxide electrolyte reacts with the CO_2 in air. More practical on earth is a cell with a phosphoric acid electrolyte that can be fed with methane or hydrogen and air. Hundreds of phosphoric-acid fuel cells, some quite large, today supply uninterruptible electricity to hospitals, banks, and computer centers. The reliability of these cells compensates for their cost, which is declining but still higher than that of conventional generators. Using fuel cells in remote locations can be cheaper and more convenient than running power lines to distant electric grids. Thus a police station in New York's Central Park was equipped with a 200-kW fuel cell because it was less expensive than digging up the park for cables. In Japan plans have been made to install phosphoric-acid cells totalling 2000 MW by the year 2010 in a number of small power stations to supply local needs.

A hydrogen fuel-cell design that employs a solid proton-exchange membrane (PEM) instead of a liquid electrolyte is better suited for vehicles. Such cells are relatively light in weight and small in size but need very pure fuel. The membrane is made of Teflon treated to permit protons but not electrons to pass through it. The membrane acts as the electrolyte and is sandwiched between two electrodes coated with platinum particles that act as catalysts. Hydrogen is fed to one electrode, where it

Figure 12-21 The power densities of PEM fuel cell stacks developed by Ballard Power Systems for use in cars have improved from 0.1 kW/liter (at left) to 1.1 kW/liter (at right). The stack at right can deliver 50 kW to a car's motors. Even greater power densities should be possible.

breaks up into protons and electrons. The protons migrate through the membrane to the other electrode where their positive charge attracts the electrons through an external circuit. (These electrons carry the energy the cell generates to the outside world.) Air is fed to this electrode, where its oxygen reacts with the protons and electrons to form water, which passes out of the cell and is the only byproduct.

A single PEM cell has a voltage of only about 0.7 V, so stacks of them must be connected in series (Fig. 6-24) to provide the higher voltages needed in practical applications; car motors, for instance, need about 300 V (Fig. 12-21). All major car makers have developed experimental cars that will use PEM cells to power electric motors (Fig. 4-33). There remains much to do, in particular figuring out how to safely store the hydrogen fuel needed for a moderately long trip; devising cheaper and more durable fuel cells; and creating a network of filling stations, estimated at about 12,000 for the continental United States—not to mention the task of supplying the hydrogen itself in the purity and amounts needed for millions of vehicles at reasonable cost and without adding pollutants, notably CO_2, to the atmosphere. Will efficient and themselves nonpolluting fuel-cell cars eventually take over with today's cars joining the dinosaurs in the pages of history? Nobody yet knows, and even if the answer is yes, the next unknown is when.

Important Terms and Ideas

Combustion is the rapid combination of oxygen with another substance during which heat and light are given off.

The **formula mass** of a substance is the sum of the atomic masses of the elements it contains, each multiplied by the number of times it appears in the formula of the substance. A **mole** of a substance is that amount of it whose mass in grams is equal to its formula mass expressed in atomic mass units (u). A mole of anything contains the same number of formula units as a mole of anything else; this number is called **Avogadro's number.** Because of the way the mole is defined, a chemical equation can be interpreted in terms of moles as well as in terms of formula units such as atoms, molecules, or ions.

Endothermic reactions absorb energy and **exothermic reactions** liberate energy. Many exothermic reactions require initial **activation energy** in order to take place.

A **catalyst** is a substance that can change the rate of a chemical reaction without itself being permanently changed.

In a **chemical equilibrium,** forward and reverse reactions occur at the same rate, so the concentrations of the reactants and products remain constant.

Oxidation involves the loss of electrons by the atoms of an element in a chemical reaction, and **reduction** involves the gain of electrons. An example of an oxidation-reduction reaction is **electrolysis,** in which free elements are liberated from a liquid by the passage of an electric current. **Batteries** and **fuel cells** produce electric current by means of oxidation-reduction reactions.

Multiple Choice

1. When something burns,
   a. it combines with phlogiston
   b. it gives off phlogiston
   c. it combines with oxygen
   d. it gives off oxygen
2. The proportion of oxygen in air is about
   a. $\frac{1}{5}$
   b. $\frac{1}{3}$
   c. $\frac{1}{2}$
   d. $\frac{4}{5}$
3. A substance of unknown composition is heated in an open container. As a result,
   a. its mass decreases
   b. its mass remains the same
   c. its mass increases
   d. any of these, depending on the nature of the substance and the temperature reached
4. When a piece of metal is oxidized, the resulting oxide is
   a. lighter than the original metal
   b. the same weight as the original metal
   c. heavier than the original metal
   d. any of these, depending on the metal
5. Chemical energy is stored within atoms, molecules, and ions as
   a. activation energy
   b. electron kinetic energy
   c. electron potential energy
   d. thermal energy
6. A chemical reaction that absorbs energy is called
   a. endothermic
   b. exothermic
   c. activated
   d. oxidation-reduction
7. An example of an endothermic reaction is
   a. the dissociation of a salt in water
   b. the neutralization of an acid by a base
   c. the freezing of water
   d. combustion
8. If a given reaction is exothermic, the reverse reaction
   a. is exothermic
   b. is endothermic
   c. may involve no energy change
   d. any of these, depending on the reaction
9. When a catalyst promotes a chemical reaction, it usually does so by
   a. providing energy
   b. providing electrons
   c. producing activated molecules
   d. becoming permanently altered
10. The neutralization of a strong acid by a strong base
    a. absorbs energy
    b. liberates energy
    c. involves no energy change
    d. requires a catalyst to occur
11. Reaction rates increase with temperature primarily because
    a. dissociation into ions is more complete
    b. more collisions occur between the molecules involved
    c. more activated molecules are formed
    d. equilibrium does not occur at high temperatures
12. At ordinary temperatures, the temperature increase needed to double the rate of many common reactions is
    a. 1°C
    b. 10°C
    c. 50°C
    d. 100°C
13. The speeds of reactions between ions in solution
    a. depend critically upon temperature
    b. are essentially independent of temperature
    c. depend upon which catalyst is used
    d. are slow in general

14. When the temperature at which a certain reversible reaction occurs is increased, the final amount of the reaction product or products
  a. increases
  b. remains the same
  c. decreases
  d. any of the above, depending on the reaction

15. At equilibrium,
  a. both forward and reverse reactions have ceased
  b. the forward and reverse reactions are proceeding at the same rate
  c. the forward reaction has come to a stop, and the reverse reaction is just about to begin
  d. the mass of reactants equals the mass of products

16. The yield of the product C in the reversible reaction A + B + energy $\rightleftharpoons$ C can be increased by
  a. decreasing the temperature
  b. increasing the temperature
  c. increasing the surface area of the reactants
  d. changing the catalyst

17. When a gas reaction involves a decrease in the total number of molecules, the equilibrium can be shifted in the direction of higher yield by
  a. increasing the pressure
  b. decreasing the pressure
  c. increasing the temperature
  d. decreasing the temperature

18. Reduction occurs when a substance
  a. loses electrons
  b. gains electrons
  c. combines with oxygen
  d. reacts with an acid

19. A catalyst affects which one or more of the following?
  a. The energy needed for a chemical reaction to occur
  b. The energy a chemical reaction gives off
  c. The speed of a chemical reaction
  d. Whether a substance is oxidized or reduced in a chemical reaction

20. When an electric current is passed through molten sodium chloride,
  a. sodium metal is deposited at the positive electrode
  b. sodium ions are deposited at the positive electrode
  c. chlorine gas is liberated at the positive electrode
  d. chlorine ions are liberated at the positive electrode

21. The quantity actually stored in a "storage battery" is
  a. electric charge
  b. electric current
  c. voltage
  d. energy

22. Batteries and fuel cells employ
  a. oxidation reactions only
  b. reduction reactions only
  c. both oxidation and reduction reactions
  d. acid-base neutralization reactions

23. A fuel cell does not require
  a. a positive electrode
  b. a negative electrode
  c. oxidation-reduction reactions
  d. recharging

24. One mole of an element has a mass equal to
  a. Avogadro's number
  b. its atomic number expressed in grams
  c. its mass number expressed in grams
  d. its atomic mass expressed in grams

25. The number of formula units in a mole of a substance
  a. depends on the formula mass of the substance
  b. depends on whether the substance is an element or a compound
  c. depends on whether the substance is in the gaseous, liquid, or solid state
  d. is the same for all substances

26. The formula mass of gaseous carbon dioxide, CO_2,
  a. is 28 u
  b. is 44 u
  c. is 56 u
  d. depends on the mass of the sample

27. An oxygen atom has a mass of 16.0 u. The number of moles of molecular oxygen, O_2, in 64 g of oxygen gas is
  a. 2
  b. 4
  c. 32
  d. 64

28. How many moles of H atoms are present in 1 mole of H_2O?
  a. $\frac{2}{3}$
  b. 1
  c. 2
  d. 3

29. The number of moles of carbon present in 3 mol of glucose, $C_6H_{12}O_6$, is
  a. 2
  b. 3

c. 6
d. 18

30. The atomic mass of helium is 4.0 u and that of carbon is 12.0 u.
 a. The mass of 1 mole of carbon is $\frac{1}{3}$ the mass of 1 mole of helium
 b. The mass of 1 mole of carbon is 3 times the mass of 1 mole of helium
 c. One mole of carbon contains $\frac{1}{3}$ as many atoms as 1 mole of helium
 d. One mole of carbon contains 3 times as many atoms as 1 mole of helium

31. One mole of which of the following compounds contains the greatest mass of bromine?
 a. HBr
 b. Br_2
 c. $AlBr_3$
 d. $SiBr_4$

32. In round numbers, the atomic mass of nitrogen is 14 u and Avogadro's number is $N_0 = 6 \times 10^{23}$ formula units/mole. One mole of molecular nitrogen, N_2, contains
 a. 6×10^{23} molecules
 b. 12×10^{23} molecules
 c. 84×10^{23} molecules
 d. 168×10^{23} molecules

33. The mass of 1 mole of molecular nitrogen is
 a. 14 g
 b. 28 g
 c. 84×10^{23} g
 d. 168×10^{23} g

34. The mass of 6×10^{23} molecules of N_2 is
 a. 14 g
 b. 28 g
 c. 84×10^{23} g
 d. 168×10^{23} g

35. Six moles of O_2 are consumed in a certain run of the reaction $2H_2S + 3O_2 \longrightarrow 2H_2O + 2SO_2$. The number of moles of water produced in the run is
 a. 1
 b. 2
 c. 4
 d. 6

Exercises

12.1 Phlogiston

12.2 Oxygen

1. What aspect of Lavoisier's work marked the beginning of chemistry as a science?
2. (a) What is the formula of ozone? (b) Ozone in the atmosphere is both harmful to and essential for our health. Explain.
3. What role does air play in combustion?
4. For a given amount of energy to be used for its propulsion, a spacecraft must have much larger tanks than an airplane. Why?

12.3 The Mole

12.4 Formula Units

5. Which of the following quantities are the same for both a mole of hydrogen molecules and a mole of oxygen molecules at the same temperature? (a) The mass of each sample; (b) the number of molecules present; (c) the average molecular energies.
6. When hydrogen is burned in oxygen, water is formed according to the reaction $2H_2 + O_2 \longrightarrow 2H_2O$. How many moles of H_2 and how many of O_2 are needed to produce 3 mol of H_2O?
7. How many moles of aluminum are present in 5 mol of $MgAl_2O_4$?
8. How many moles of propane, C_3H_8, can be prepared from 1 mol of carbon? From 1 mol of hydrogen?
9. Ammonia is produced by the reaction $N_2 + 3H_2 \longrightarrow 2NH_3$. How many moles of N_2 and how many of H_2 are needed to produce 1 mol of ammonia?
10. Find the mass of 0.4 mol of magnesium, Mg. How many atoms are present in such a sample?
11. Find the mass of 10 mol of uranium, U. How many atoms are present in such a sample?
12. Find the mass of 9.4 mol of ethylene, C_2H_4. How many carbon atoms are present in the sample?
13. Find the mass of 2 mol of iron(III) oxide, Fe_2O_3.
14. Find the mass of 85 mol of sulfuric acid, H_2SO_4.
15. How many moles are present in 500 kg of glucose, $C_6H_{12}O_6$?
16. How many moles are present in 100 g of lead nitrate, $Pb(NO_3)_2$?
17. Three compounds used as fertilizers are urea, $CO(NH_2)_2$; ammonium nitrate, NH_4NO_3; and

ammonium sulfate, $(NH_4)_2SO_4$. What is the percentage of nitrogen by mass in each compound? (Hint: Divide the formula mass of the nitrogen in each compound by the formula mass of the compound and then express the result as a percentage.)

18. When potassium chlorate, $KClO_3$, is heated, it decomposes into potassium chloride and oxygen in the reaction $2KClO_3 \longrightarrow 2KCl + 3O_2$. How much oxygen is liberated when 50 g of potassium chlorate is heated?

19. How much chlorine is needed to react with 50 g of sodium to form sodium chloride, NaCl? How much sodium chloride is produced?

20. How much sulfur is needed to react with 60 kg of aluminum to form aluminum sulfide, Al_2S_3?

21. How much sulfur is needed to react with 200 g of potassium to form potassium sulfide, K_2S?

12.5 Exothermic and Endothermic Reactions

12.6 Chemical Energy and Stability

12.7 Activation Energy

22. What is the origin of the energy liberated in an exothermic reaction?

23. In what fundamental way is the explosion of an atomic bomb different from the explosion of dynamite?

24. From the observation that the slaking of lime [addition of water to CaO to form $Ca(OH)_2$] gives out heat, would you conclude that the following reaction is endothermic or exothermic?

$$Ca(OH)_2 \longrightarrow CaO + H_2O$$

25. Which of the following are exothermic reactions and which are endothermic?

a. The explosion of dynamite

b. The burning of methane

c. The decomposition of water into its elements

d. The dissociation of water into ions

e. The burning of iron in chlorine

f. The combination of zinc and sulfur to form zinc sulfide

26. When carbon in the form of diamond is burned to produce CO_2, more heat is given off than when carbon in the form of graphite is burned. What form of carbon is more stable under ordinary conditions? What bearing does this conclusion have on the origin of diamonds?

27. What is the fundamental role of activation energy in starting an exothermic reaction?

28. Do ions in solution need activation energy to react with one another? If not, why not?

12.8 Temperature and Reaction Rates

12.9 Other Factors

29. What is the chief reason that reaction rates increase with temperature?

30. Why does an increase in temperature increase the rate of exothermic as well as endothermic reactions?

31. Suggest three ways to increase the rate at which coarse salt dissolves in a pan of water.

32. Suggest three ways to increase the rate at which zinc dissolves in sulfuric acid.

33. Give an example of a reaction that is (a) practically instantaneous at room temperatures, (b) fairly slow at room temperatures.

34. Under ordinary circumstances coal burns slowly, but the fine coal dust in mines sometimes burns so rapidly as to cause an explosion. Explain the difference in rates. Would you expect the danger from spontaneous combustion to be greater in a coal pile containing principally large chunks or in one containing finely pulverized coal? Why?

35. Why is a reaction with a high activation energy slow at room temperature?

36. To what extent does the time needed for a strong acid to neutralize a strong base in solution depend on temperature?

12.10 Chemical Equilibrium

12.11 Altering an Equilibrium

37. How common are reversible chemical reactions?

38. The solubility of a gas in a liquid decreases with increasing temperature. From this observation and what you know of how a change in temperature can affect an equilibrium, would you expect that dissolving a gas in a liquid is an exothermic or an endothermic process?

39. Changing the pressure has no effect on the equilibrium

$$CO + H_2O \rightleftharpoons CO_2 + H_2$$

in which all the substances involved are gases. Why not?

40. Methanol (CH_3OH) can be synthesized from carbon monoxide and hydrogen by the reversible reaction

$$CO + H_2 \rightleftharpoons CH_3OH + \text{energy}$$

If the temperature in the reaction chamber is increased, what happens to the yield of methanol? If the pressure is increased, what happens to the yield of methanol? Why?

41. The reaction $2SO_2 + O_2 \longrightarrow 2SO_3$ is exothermic. How will a rise in temperature affect the yield of

SO_3 in an equilibrium mixture of the three gases? Will an increase in pressure raise or lower this yield? In what possible way can the speed of the reaction be increased at moderate temperatures?

42. Ozone (O_3) and normal molecular oxygen (O_2) transform into each other according to the reversible reaction

$$3O_2 + \text{energy} \rightleftharpoons 2O_3$$

How will the equilibrium be affected by increasing the temperature? Increasing the pressure?

43. The three gases N_2, O_2, and NO are in equilibrium. The formation of NO is exothermic. Write the equation for the equilibrium. How would a decrease in temperature affect the equilibrium? A decrease in pressure? A lower concentration of N_2? A lower concentration of NO? The presence of a catalyst?

12.12 Oxidation-Reduction Reactions

44. A displacement reaction is an oxidation-reduction reaction in which one element displaces another from solution. In each of the following displacement reactions identify the element that is oxidized and the element that is reduced:

$$Zn + Cu^{2+} \longrightarrow Zn^{2+} + Cu$$
$$Fe + 2H^{+} \longrightarrow Fe^{2+} + H_2$$
$$Cl_2 + 2Br^{-} \longrightarrow 2Cl^{-} + Br_2$$

45. Which loses electrons more easily, Na or Fe? Al or Ag? I^- or Cl^-? Which gains electrons more easily, Cl or Br? Hg^{2+} or Mg^{2+}? (Hint: Reread Sec. 10.10.)

46. When magnesium is placed in an acid solution, hydrogen gas is given off. Is magnesium or hydrogen the better reducing agent?

47. Lithium reacts with water to produce lithium hydroxide. What else is produced? Write the equation of the process. Which element is reduced and which is oxidized?

48. In the refining of iron, the iron(III) oxide, Fe_2O_3, in iron ore is reduced by carbon (in the form of coke) to yield metallic iron and carbon dioxide. Write the balanced equation of the process.

49. When an electric current is passed through a solution of hydrochloric acid, what substance is liberated at the positive electrode? At the negative electrode?

50. When an electric current is passed through a solution of copper chloride, what substance is liberated at the positive electrode? At the negative electrode?

51. Which of the following metals will be deposited in the greatest mass when 1 C of charge is passed through appropriate electrolytic cells? Aluminum, nickel, silver. The ions of these metals in solution are respectively Al^{3+}, Ni^{2+}, Ag^{+}.

12.13 Electrochemical Cells

52. What becomes of the electric energy provided in electrolysis? In what device is this energy transformation reversed?

53. What do you think happens when a charging current is passed through a fully charged storage battery?

54. In what basic way is a fuel cell different from a dry cell or a storage battery?

Answers to Multiple Choice

1. c	**6.** a	**11.** c	**16.** b	**21.** d	**26.** b	**31.** d
2. a	**7.** a	**12.** b	**17.** a	**22.** c	**27.** a	**32.** a
3. d	**8.** b	**13.** b	**18.** b	**23.** d	**28.** c	**33.** b
4. c	**9.** c	**14.** d	**19.** c	**24.** d	**29.** d	**34.** b
5. c	**10.** b	**15.** b	**20.** c	**25.** d	**30.** b	**35.** c

13

Organic Chemistry

Synthetic polyester resins are used in the hull and sails of this yacht.

Goals

When you have finished this chapter you should be able to complete the goals ▸ given for each section below:

Carbon Compounds

13.1 Carbon Bonds
Carbon Atoms Can Form Covalent Bonds with Each Other
- ▸ Discuss the covalent bonding behavior of carbon atoms.

13.2 Alkanes
The Hydrocarbons in Petroleum and Natural Gas
- ▸ Define hydrocarbon and alkane and explain why the alkane series of hydrocarbons is so important.

13.3 Petroleum Products
Fractional Distillation, Catalytic Cracking, and Polymerization
- ▸ Describe how fractional distillation works.
- ▸ Compare the ways in which cracking and polymerization increase the yield of gasoline from petroleum.

Structures of Organic Molecules

13.4 Structural Formulas
They Show How Atoms Are Linked Together
- ▸ Compare molecular and structural formulas and explain why the latter are so useful in organic chemistry.

13.5 Isomers
The Same Atoms but Arranged Differently

13.6 Unsaturated Hydrocarbons
Double and Triple Carbon-Carbon Bonds
- ▸ Compare saturated and unsaturated compounds and explain why the latter are more reactive.

13.7 Benzene
Its Molecule Contains a Stable Ring of Six Carbon Atoms
- ▸ Draw the structural formula of benzene and explain the circle inside it.

Organic Compounds

13.8 Hydrocarbon Groups
A Handy Classification Scheme

13.9 Functional Groups
Atom Groups with Characteristic Behaviors
- ▸ Explain what a functional group is and list several important examples.
- ▸ Compare inorganic acids, bases, and salts with their organic equivalents.

13.10 Polymers
Molecules Linked into Giant Chains
- ▸ Distinguish between monomers and polymers and list several examples of polymers.
- ▸ Explain why Teflon is much more durable than other polymers.
- ▸ Explain why nylon is called a polyamide and dacron a polyester.

Chemistry of Life

13.11 Carbohydrates
The First Link in the Food Chain
- ▸ Identify carbohydrates, give some examples, and discuss what they are used for by living things.

13.12 Photosynthesis
How the Sun Powers the Living World
- ▸ Describe photosynthesis and give the reasons for its importance.

13.13 Lipids
Where the Calories Are
- ▸ Identify lipids and discuss what they are used for by living things.
- ▸ Identify cholesterol and discuss its role in heart disease.

13.14 Proteins
The Building Blocks of Living Matter
- ▸ Identify proteins and discuss what they are used for by living things.
- ▸ Account for the wide variety of proteins.

13.15 Soil Nitrogen
A Vital Component
- ▸ Explain the importance of soil nitrogen and list the ways in which it is replenished.

13.16 Nucleic Acids
The Genetic Code
- ▸ Describe the structure of the nucleic acid DNA and list the three fundamental attributes of life it is responsible for.
- ▸ Define gene, genome, and chromosome.

13.17 Origin of Life
An Inevitable Result of Natural Processes

In many ways carbon is the most remarkable element. Hundreds of thousands of carbon compounds are known, far more than the number of compounds that do not contain carbon. Furthermore, carbon compounds are the chief constituents of all living things—hence the name **organic chemistry** to describe the chemistry of carbon and the name **inorganic chemistry** to describe the chemistry of all the other elements.

Carbon Compounds

At one time it was thought that carbon compounds—with the exception of the carbon oxides, the carbonates, and a few others—could be produced only by plants and animals (or indirectly from other compounds produced by them). Carbon was supposed to unite with other elements only under the influence of a mysterious "life force" possessed by living things. This ancient idea was disproved in 1828 by the German chemist Friedrich Wöhler, who prepared the organic compound urea by reacting the inorganic compounds lead cyanate and ammonia. Since Wöhler's time a great number of organic compounds have been made in the laboratory from inorganic materials, but the general distinction between the chemistry of carbon compounds and inorganic chemistry nevertheless remains useful.

13.1 Carbon Bonds

Carbon Atoms Can Form Covalent Bonds with Each Other

Let us see what the periodic table can tell us about carbon. Carbon is in period 2 at the top of group 4, which means it is halfway between the active metal lithium and the active nonmetal fluorine. Active metals tend to lose their outer electrons when they react chemically, and active nonmetals tend to gain electrons. Carbon, in the middle, does neither. Instead, it forms covalent bonds in which it shares four electron pairs.

As we saw in Table 10-7 and Fig. 10-15, the carbon atom has four electrons in its outer shell. For a carbon atom to achieve a closed outer shell, it can lose these four electrons, pick up four more for a total of eight, or share its four electrons with other atoms that contribute four electrons so that eight electrons—four pairs—are shared.

The effective nuclear charge on the outer electrons in a carbon atom is $+4e$ (Fig. 13-1). The resulting force on the outer electrons is sufficient to keep them from being detached to leave a C^{4+} ion. However, the

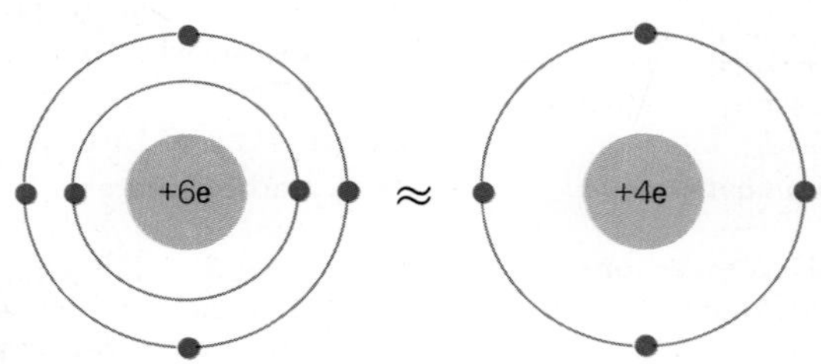

Figure 13-1 Electron shielding in carbon. Each outer electron is acted on by an effective nuclear force of $+4e$ because the inner electrons shield part of the actual nuclear charge of $+6e$.

effective nuclear charge is not enough for a carbon atom to attract and hold four more electrons to give a C^{4-} ion. (See Fig. 10-16 for the reason why Na readily becomes Na^+ and Cl readily becomes Cl^-.) The result is that carbon atoms participate in four covalent bonds each when they form molecules with other atoms.

Why Carbon Forms Many Compounds A carbon atom can bond strongly not only with many metallic and nonmetallic atoms but with *other carbon atoms* as well. This is the reason for the immense number and variety of carbon compounds, whose molecules have skeletons of linked carbon atoms. The strength of the bonds between carbon atoms is shown by the hardness of diamond, a crystalline form of carbon in which each atom is joined to four others by electron pairs (see Fig. 11-6). Atoms of a few elements near carbon in the periodic table, notably boron and silicon, are also able to bond with each other, but the range of their compounds is far more limited.

Because the bonds formed by carbon atoms are covalent, carbon compounds are mostly nonelectrolytes, and their reaction rates are usually slow. The affinity of carbon and hydrogen for oxygen makes many organic compounds subject to slow oxidation in air and to rapid oxidation if heated. Even in the absence of air most organic compounds are unstable at high temperatures; very few of them resist decomposition at temperatures over a few hundred degrees celsius.

Organic molecules are either nonpolar or nearly so, hence the van der Waals forces between them are weak and as a class, organic compounds have low melting and boiling points. Some polymers (large molecules that consist of a great many subunits joined together) are exceptions; a notable example is Teflon.

13.2 Alkanes

The Hydrocarbons in Petroleum and Natural Gas

The simplest organic compounds are the **hydrocarbons** that contain only carbon and hydrogen. A group of hydrocarbons called the **alkanes** includes the familiar gases methane (CH_4), propane (C_3H_8), and butane (C_4H_{10}), all widely used as fuels in stoves, furnaces, and even cigarette lighters. Alkane molecules have single covalent bonds between their carbon atoms. The freezing points and boiling points of the alkanes all increase regularly as the molecular size increases (Table 13-1). Other series of organic compounds show similar regular changes in properties as the number of carbon atoms per molecule increases.

Natural gas and petroleum consist mainly of alkanes. About 80 percent of natural gas is methane, 10 percent ethane, and the rest mostly propane and butane. Alkanes with five or more carbon atoms are the main ingredients of petroleum ("crude oil"), whose exact composition varies from source to source. Methane is also one of the emissions from active volcanoes and a product of the bacterial decay of plant matter in the absence of oxygen. The "marsh gas" that bubbles up from the black ooze at the bottom of stagnant pools is largely methane, as is the "fire damp" that sometimes causes explosions in coal mines.

Table 13-1 The Alkane Series of Hydrocarbon

Formula	Name	Freezing Point, °C	Boiling Point, °C	Commercial Name
CH_4	Methane	−183	−160	
C_2H_6	Ethane	−184	−89	Fuel gases
C_3H_8	Propane	−188	−42	
C_4H_{10}	Butane	−139	−1	
C_5H_{12}	Pentane	−130	36	Naphtha
C_6H_{14}	Hexane	−95	69	
C_7H_{16}	Heptane	−91	98	
C_8H_{18}	Octane	−57	126	Gasoline
C_9H_{20}	Nonane	−54	151	
$C_{10}H_{22}$	Decane	−30	174	Kerosene,
$C_{11}H_{24}$	Undecane	−27	197	jet fuel
$C_{16}H_{34}$	Hexadecane	18	287	
$C_{14}H_{30}$ to $C_{18}H_{38}$				Diesel fuel, heating oil
$C_{16}H_{34}$ to $C_{18}H_{38}$				Lubricating oil
$C_{16}H_{34}$ to $C_{32}H_{68}$				Petroleum jelly
$C_{20}H_{42}$ and up				Paraffin wax
$C_{36}H_{74}$ and up				Asphalt

Note: The data refer to the straight-chain compounds. Isomers of these hydrocarbons (see Sec. 13.5) have somewhat different properties.

13.3 Petroleum Products

Fractional Distillation, Catalytic Cracking, and Polymerization

The separation of petroleum into its different alkanes is difficult because their properties are so similar. Suppose we want to separate pentane and hexane. Because pentane boils at 36°C and hexane at 69°C, it would seem that if we heated a mixture of the two, all the pentane would boil away first to leave pure hexane. The trouble is that hexane evaporates readily at 36°C, so the vapor produced by heating the mixture to that temperature would contain a certain amount of hexane as well as pentane. This procedure would thus give a vapor rich in pentane and a remaining liquid rich in hexane, but would not separate the two compounds completely.

Usually a complete separation of the alkanes in petroleum is not necessary, however. The basic process in petroleum refining is **fractional distillation,** in which crude oil is heated and its vapors are led off and condensed at progressively higher temperatures. A diagram of a distillation tower together with some of the possible fractions is given in Fig. 13-2. The cracking process mentioned in the figure is described below. Grease is not one of the fractions because it consists of oil to which a thickening agent has been added to prevent the mixture from running out from between the surfaces being lubricated.

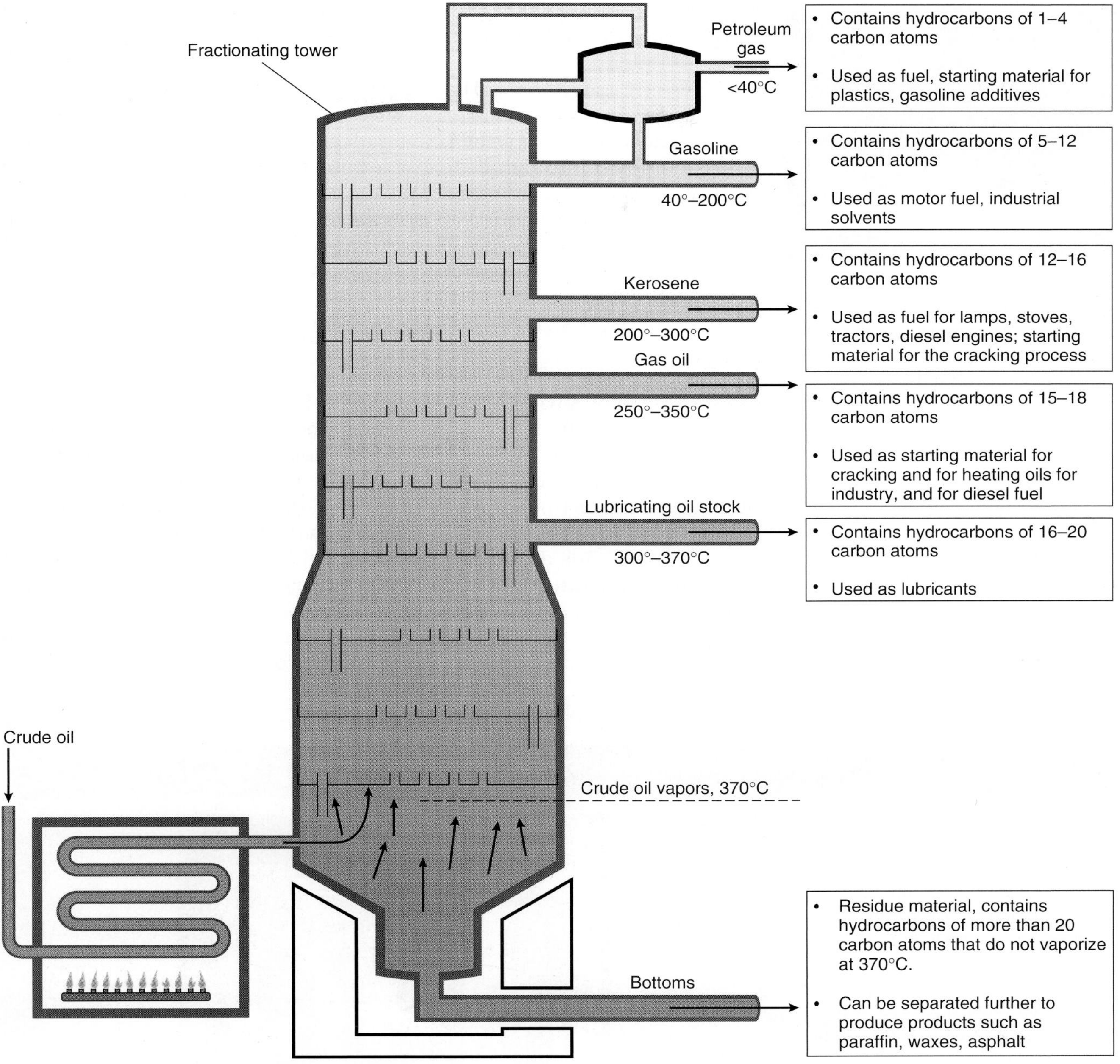

Figure 13-2 A distillation tower like the simplified one shown here separates crude oil into fractions according to their boiling points.

Gasoline Of all these products the most valuable, of course, is gasoline, needed to fuel many of the world's hundreds of millions of cars. The United States alone uses over 100 billion gallons of gasoline each year. Unfortunately the constituents of gasoline are present only to a minor extent in most petroleums. Two methods have been developed to increase the yield of gasoline. In one of them, the heavier hydrocarbons

Figure 13-3 Catalytic cracking units at an oil refinery break down complex hydrocarbons into simpler ones. There have been no new oil refineries in the United States for 30 years, mainly because of environmental concerns. Many refineries have been built recently or are under construction abroad.

are **cracked** into smaller molecules by heating them under pressure in the presence of catalysts (Fig. 13-3). A typical cracking reaction is

$$C_{16}H_{34} \longrightarrow C_8H_{18} + C_8H_{16} \qquad \textit{Cracking reaction}$$

Here hexadecane, one of the heavier alkanes in kerosene and diesel fuel, is broken down into lighter hydrocarbons that vaporize and burn more readily.

The second procedure is to **polymerize** lighter hydrocarbons, which means to join small molecules into larger ones under the influence of heat, pressure, and appropriate catalysts. An example is

$$C_3H_8 + C_4H_8 \longrightarrow C_7H_{16} \qquad \textit{Polymerization reaction}$$

in which heptane, a liquid, is formed by the polymerization of two gases. Figure 13-6 shows what a barrel of crude oil typically yields.

Alkane molecules have chains of carbon atoms linked together in line, as we shall see in Sec. 13.4. Such molecules, as we might expect, are nonpolar, with neither end much more positive or negative than the other. Because of this nonpolar character, the alkane hydrocarbons are insoluble in water. Chemically they are rather unreactive, and neither concentrated acids and bases nor most oxidizing agents affect them at moderate temperatures. Nor do biological agents such as bacteria attack them to any great extent. The combination of insolubility, relative inertness, and toxicity to living things is what makes the discharge of petroleum and its products into the sea such a serious matter.

Octane Rating

When a car's gasoline engine is under stress, for instance when climbing a hill in high gear, the gasoline-air mixture in its cylinders may ignite early, while being compressed instead of when the spark plugs fire (Fig. 5-47). The result is a rattle called "knocking." The octane rating of a gasoline (Fig. 13-4) is a measure of its ability to prevent knocking: the higher the octane rating, the less likely knocking is to occur. On the octane scale, heptane (a straight-chain hydrocarbon) is rated as 0, and isooctane (a branched-chain hydrocarbon) is rated as 100; branched-chain hydrocarbons are the more effective in reducing knocking. A compound called tetraethyl lead was once widely used to increase the octane rating of gasoline. Because the exhausts of engines using leaded gasoline discharge lead, which is toxic, into the atmosphere and the catalytic converters that reduce other pollutants are inactivated by lead, unleaded gasoline is now standard in the United States (but not everywhere else in the world, unfortunately). To obtain a high octane rating for "super" or "premium" fuel, branched-chain hydrocarbons of various kinds are incorporated in unleaded gasoline.

Figure 13-4 Gasoline with the correct octane rating is needed for optimum performance from a car engine.

Spills and Leaks

Oil spills are common. Some are accidental, the result of shipwreck or a malfunction at an offshore oil well, but a great many are deliberate, the result of tankers illegally flushing out waste oil. Although the lighter hydrocarbons soon evaporate, the heavier ones remain floating on the surface or are washed ashore on adjacent coastlines. Of every thousand tons of oil shipped around the world, one is spilled at sea.

Depending on the nature of the residues and the region, the effects of the residues on marine life may be drastic and immediate—dead plankton, dead fish, dead crustaceans, dead birds—or they may be gradual, taking the form of an altered balance of nature with declining populations. The lumps of tar that are one result of oil spills are a prominent feature on the surface of much of the world's oceans and are familiar sights on many beaches.

An especially severe oil spill occurred in 1989, when the tanker *Exxon Valdez* struck a well-marked reef outside the shipping lane in Alaska's Prince William Sound, an area of great natural beauty (Fig. 13-5). Exxon itself and both state and federal agencies reacted slowly and ineffectively to the spill, which allowed much of the oil to wash ashore. The oil killed huge numbers of birds, fish, and other wildlife, and its traces remain hazardous. Twenty years later ExxonMobil was still fighting court orders to pay damages to 33,000 Alaskans. An even greater such disaster was the breakup of the tanker *Prestige* in 2002, which poured 63,000 tons of oil in the Atlantic Ocean, much of which ended up on the Spanish coast. Again birds and marine life suffered on a large scale, and about 100,000 families were directly affected by the event in some way.

Figure 13-5 Careless operation of the tanker *Exxon Valdez* led to its grounding on a reef off the Alaskan coast. Over 40,000 tons of crude oil from the wreck devastated wildlife on a large scale; this dead otter was one of many victims. ExxonMobil has called those who want to keep our planet habitable "extremists."

What can be done? A welcome approach is to use only double-hulled tankers, which provide two layers of steel instead of just one to reduce the risk their cargoes will escape in case of an accident. For about a decade, all new tankers have been built with double hulls, and older single-hulled tankers are gradually being scrapped. But there is no substitute for responsible operation.

Spectacular as oil spills from tankers can be, even more pollution, on both land and sea, comes from the steady seepage of petroleum products from leaky tanks and pipes. Service-station tanks are particular offenders. The resulting contamination of groundwater is a threat to wildlife as well as to human health. Some leaks are on a huge scale. For instance, oil from adjacent refineries and storage tanks has been ending up for many years in Newtown Creek, which lies between the New York boroughs of Brooklyn and Queens. The oil—one and a half times as much as the *Exxon Valdez* spill—covers about 55 acres of the creek and has contaminated nearby soil and groundwater. In 1990, 12 years after the Coast Guard determined that ExxonMobil was primarily responsible for the leaks, the company agreed to clean up the creek, but the agreement lacked deadlines or penalties for noncompliance. ExxonMobil did so little thereafter that, in 2007, New York State sued to force it to act more effectively.

Pipelines, too, are at risk. In 2006 a badly maintained BP pipeline in Alaska broke open due to corrosion and an estimated 200,000 gallons of crude oil gushed out on the ground and into a nearby lake. A month later another Alaskan BP pipeline ruptured. Polluters always claim that they cannot afford to respect the environment: BP's profit the year before was \$19.3 billion; ExxonMobil's was \$36.2 billion, then the largest ever for an American company.

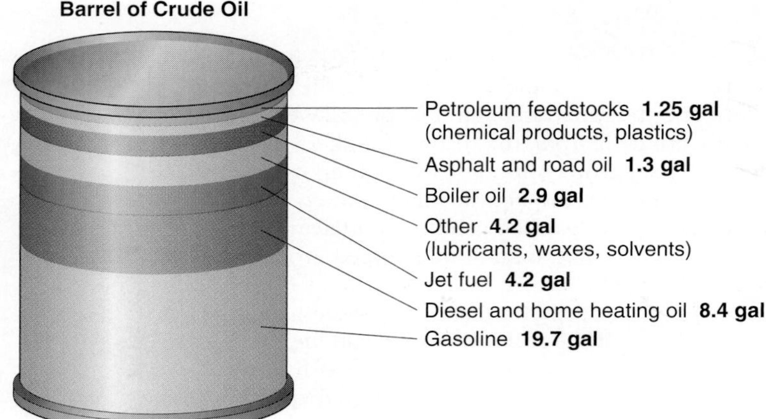

Figure 13-6 End products from the refining of a barrel (42 gallons) of crude oil. Over 80 percent is merely burnt, mostly in vehicle engines. The 3 percent that becomes feedstocks for plastics, fibers, pharmaceuticals, and other synthetic products will be hardest to replace when crude oil becomes rare. The United States, with less than 5 percent of the world's population, consumes a quarter of the world's oil production. Two-thirds of the oil used in the United States is imported.

Structures of Organic Molecules

13.4 Structural Formulas

They Show How Atoms Are Linked Together

Instead of a molecular formula such as CH_4 and C_2H_6, an organic compound is often represented by a **structural formula** in which the covalent bonds between the atoms in each molecule are shown by dashes. Each dash stands for a shared pair of electrons. Thus the structural formulas of the alkanes methane, ethane, and propane are

$$
\begin{array}{ccccc}
 & H & \\
 & | & \\
H- & C & -H \\
 & | & \\
 & H & \\
\end{array}
\qquad
\begin{array}{cccc}
 & H & H & \\
 & | & | & \\
H- & C- & C & -H \\
 & | & | & \\
 & H & H & \\
\end{array}
\qquad
\begin{array}{ccccc}
 & H & H & H & \\
 & | & | & | & \\
H- & C- & C- & C & -H \\
 & | & | & | & \\
 & H & H & H & \\
\end{array}
$$

methane ethane propane

A molecular formula tells us only how many atoms of each kind are present in each molecule of a compound. A structural formula tells us more. For instance, in the above three molecules we can see that each hydrogen atom is attached to a carbon atom and that in ethane and propane the carbon atoms are linked together. Figure 13-7 shows a three-dimensional model of the methane molecule.

The number of bonds an atom forms in an organic compound is the same as the number of electrons it has to gain or lose to achieve a closed outer shell. A carbon atom always participates in four bonds, as we have learned. A hydrogen atom always participates in a single bond, as does a chlorine atom; an oxygen atom participates in two bonds. Here are some examples:

$$
\begin{array}{ccccc}
 & H & H & H & \\
 & | & | & | & \\
H- & C- & C- & C & -Cl \\
 & | & | & | & \\
 & H & H & H & \\
\end{array}
\qquad
\begin{array}{ccccc}
 & H & & \\
 & | & & \\
H- & C & -O- & H \\
 & | & & \\
 & H & & \\
\end{array}
\qquad
\begin{array}{ccc}
 & Cl & \\
 & | & \\
Cl- & C & =O \\
\end{array}
$$

propyl chloride methanol phosgene

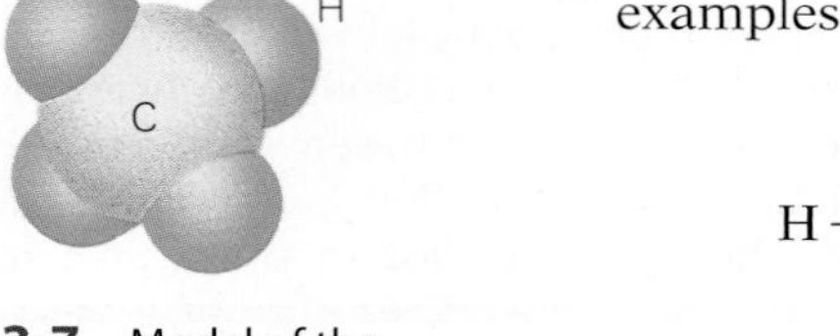

Figure 13-7 Model of the methane molecule, CH_4.

13.5 Isomers

The Same Atoms but Arranged Differently

For methane, ethane, and propane the structural formulas given earlier are the only possible arrangements of carbon and hydrogen atoms that will satisfy the combination rules. Butane, on the other hand, may have its 4 C atoms and 10 H atoms arranged in two ways:

$$\begin{array}{ccccccccc} & & H & & H & & H & & H & & \\ & & | & & | & & | & & | & & \\ H & - & C & - & C & - & C & - & C & - & H \\ & & | & & | & & | & & | & & \\ & & H & & H & & H & & H & & \end{array} \qquad \begin{array}{ccccccc} & & H & & H & & H & & \\ & & | & & | & & | & & \\ H & - & C & - & C & - & C & - & H \\ & & | & & | & & | & & \\ & & H & & C & & H & & \\ & & & / & | & \backslash & & & \\ & & H & & H & & H & & \end{array}$$

normal butane isobutane

These formulas show that there are two different compounds with the molecular formula C_4H_{10}. They differ in that one of the carbon atoms in isobutane is linked to three other carbon atoms, while in normal butane the carbon atoms are linked to only one or two others.

The physical properties of isobutane are different from those of normal butane because of this difference in molecular structure. The boiling point of isobutane, for instance, is $-12°C$, whereas that of normal butane, as listed in Table 13-1, is $-1°C$. Another difference is their densities (masses per unit volume): that of isobutane is 0.622 g/cm^3 whereas that of normal butane is 0.604 g/cm^3. Figure 13-8 shows three-dimensional models of the two kinds of butane.

Compounds that have the same molecular formulas but different structural formulas are called **isomers.** The number of possible isomers increases rapidly with the number of carbon atoms in the molecule; $C_{13}H_{28}$ has 813 theoretically possible isomers and $C_{20}H_{42}$ has 366,319. Only a few of the possible isomers have actually been prepared.

Merely flipping a structural formula end-for-end does not give the formula of an isomer. For instance, we can show the structure of methanol (methyl alcohol) in two ways:

$$\begin{array}{ccccc} & & H & & \\ & & | & & \\ H & - & C & - & OH \\ & & | & & \\ & & H & & \end{array} \qquad \begin{array}{ccccc} & & H & & \\ & & | & & \\ OH & - & C & - & H \\ & & | & & \\ & & H & & \end{array}$$

However, the molecule is exactly the same in both cases.

13.6 Unsaturated Hydrocarbons

Double and Triple Carbon-Carbon Bonds

Hydrocarbons are not limited to the alkanes. A simple example of a nonalkane hydrocarbon is **ethene,** also called **ethylene,** whose formula is C_2H_4. The alkane with two C atoms is ethane, C_2H_6, whose structural formula is

$$\begin{array}{ccccccc} & & H & & H & & \\ & & | & & | & & \\ H & - & C & - & C & - & H \\ & & | & & | & & \\ & & H & & H & & \end{array}$$

ethane

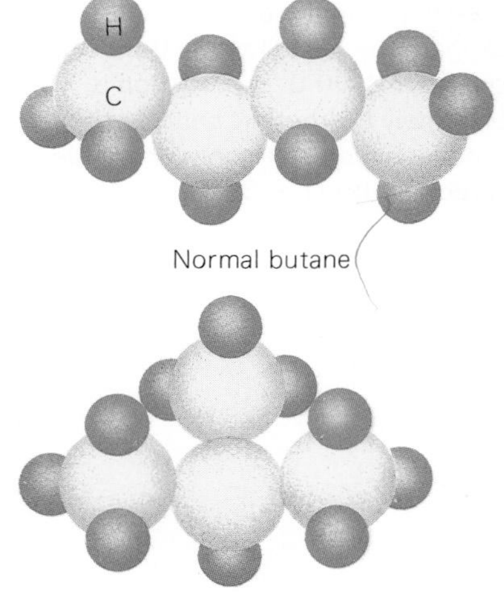

Figure 13-8 The two isomers of butane, C_4H_{10}.

How can ethene, with two fewer H atoms, still have each C atom share four electron pairs? The answer is that there are *two* covalent bonds between the C atoms in ethene:

$$\begin{array}{ccccc} H & & & & H \\ & \diagdown & & \diagup & \\ & & C = C & & \\ & \diagup & & \diagdown & \\ H & & & & H \end{array}$$
$$\text{ethene}$$

Such a link between carbon atoms is called a **double bond** and involves the sharing of two electron pairs.

Triple bonds, with carbon atoms sharing three electron pairs, are also possible. The simplest case is that of acetylene, C_2H_2, a gas widely used in welding and metal-cutting torches (Fig. 13-9). The structural formula of acetylene is

$$H - C \equiv C - H$$
$$\text{acetylene}$$

Multiple Bonds and Reactivity Compounds with double and triple bonds are much more reactive than the alkanes, which have only single bonds. Both HCl and Cl_2 combine readily with ethene, for instance:

$$\begin{array}{ccccc} H & & & & H \\ & \diagdown & & \diagup & \\ & & C = C & & \\ & \diagup & & \diagdown & \\ H & & & & H \end{array} + HCl \longrightarrow \begin{array}{ccccccc} & & H & & H & & \\ & & | & & | & & \\ H & - & C & - & C & - & H \\ & & | & & | & & \\ & & H & & Cl & & \end{array}$$

$$\begin{array}{ccccc} H & & & & H \\ & \diagdown & & \diagup & \\ & & C = C & & \\ & \diagup & & \diagdown & \\ H & & & & H \end{array} + Cl_2 \longrightarrow \begin{array}{ccccccc} & & H & & H & & \\ & & | & & | & & \\ H & - & C & - & C & - & H \\ & & | & & | & & \\ & & Cl & & Cl & & \end{array}$$

Figure 13-9 Oxyacetylene cutting torch. The short cylinder contains acetylene and the tall one contains oxygen. The flame temperature can reach 3000°C. The reaction is extremely exothermic because the acetylene molecule contains a triple carbon-carbon bond.

Example 13.1

The compound pentene has the molecular formula C_5H_{10}. Draw the structural formula(s) of pentene and its isomers, if any. Is it a saturated or unsaturated compound?

Solution

Each C atom must participate in four bonds. We begin by drawing the C atoms with single bonds between them and lines to represent the remaining bonds they can form:

$$\begin{array}{ccccccccccc} & | & & | & & | & & | & & | & \\ - & C & - & C & - & C & - & C & - & C & - \\ & | & & | & & | & & | & & | & \end{array}$$

Now we distribute the ten H atoms among these bonds:

$$\begin{array}{ccccccccccc} & H & & H & & H & & H & & H & \\ & | & & | & & | & & | & & | & \\ H- & C & - & C & - & C & - & C & - & C & - \\ & | & & | & & | & & | & & | & \\ & H & & H & & H & & H & & & \end{array}$$

Two bonds are left over, which means that one of the carbon-carbon bonds is a double bond. Thus the structural formulas of the two possible isomers of pentene are

$$\begin{array}{cccccccccc} & H & & H & & H & & H & & H \\ & | & & | & & | & & | & & / \\ H- & C & - & C & - & C & - & C & = & C \\ & | & & | & & | & & & & \backslash \\ & H & & H & & H & & & & H \end{array}$$

$$\begin{array}{ccccccccc} & H & & H & & & & H & \\ & | & & | & & & & | & \\ H- & C & - & C & - & C & - & C & -H \\ & | & & | & & \| & & | & \\ & H & & H & & C & & H & \\ & & & & H/ & & \backslash H & & \end{array}$$

(If we put the double bond between either of the other pairs of C atoms the result will not be a new isomer because it would just be a flipped version of one of the above formulas.) Pentene is evidently an unsaturated compound.

The other halogens and many other acids give similar reactions. Since compounds with multiple bonds are able to add other atoms to their molecules, they are called **unsaturated compounds.** The alkanes and similar compounds whose molecules have only single carbon-carbon bonds are called **saturated compounds** because they cannot add other atoms to their molecules.

13.7 Benzene

Its Molecule Contains a Stable Ring of Six Carbon Atoms

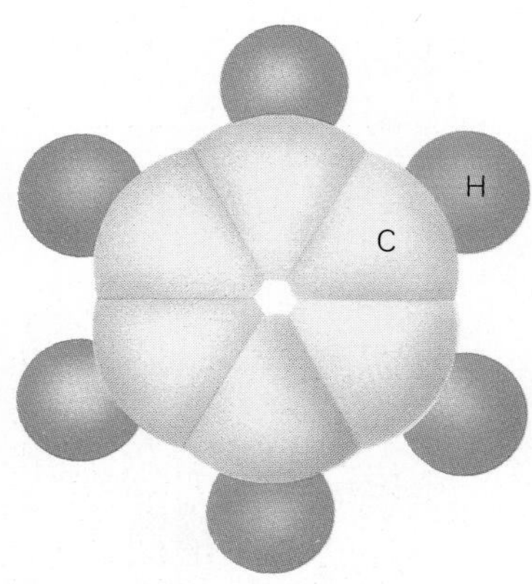

Figure 13-10 Model of the benzene molecule, C_6H_6. Over 5 million tons of benzene are produced in the United States each year.

Benzene, C_6H_6, is a clear liquid that does not mix with water and has a strong odor. Benzene is widely used as a solvent and in the manufacture of more complex organic compounds.

The six C atoms in benzene are arranged in a flat hexagonal ring, as shown in Fig. 13-10. What is especially interesting about this molecule is the manner in which its C atoms are attached to one another. In addition to single bonds between these atoms, six electrons are shared by the entire ring. The latter electrons belong to the molecule as a whole and not to any particular pair of atoms; these electrons are **delocalized.** (We recall from Chap. 11 that the outer-shell electrons in a metal are similarly delocalized.) The six delocalized electrons in benzene can be represented by an inner circle in its structural formula:

benzene

Aromatic Compounds An **aromatic** compound is defined as one that contains a ring of six carbon atoms like that in benzene. The name arose because many of these compounds have strong odors. An example is toluene, which is a common solvent and paint thinner:

toluene

The C and H atoms that are part of a benzene ring are often omitted in representing the structures of aromatic molecules, as shown above.

Some aromatic compounds contain two or more benzene rings fused together, as in the case of naphthalene:

naphthalene

BIOGRAPHY August Kekulé (1829–1896)

A native of Germany, Kekulé studied architecture before he became a chemist, and in fact his most notable work was on the architecture of molecules. He introduced the idea of structural formulas, proposed that a carbon atom forms four bonds when it combines, and suggested that carbon atoms could join with each other to form chains. All this came before the nature of chemical bonds was understood; indeed, even the electron was unknown until after Kekulé's death.

Kekulé's greatest achievement was to establish the structure of benzene, C_6H_6, a compound which had been discovered by Michael Faraday in 1826. Exactly how benzene's atoms were arranged was a mystery for the next 40 years until Kekulé (so he said) had a dream in which dancing snakes swallowed their tails. This led him to the concept of a flat ring of carbon atoms linked by alternate single and double bonds, with a hydrogen atom attached to each carbon atom:

```
          H
          |
   H      C      H
    \   //  \   /
      C       C
      |       ||
      C       C
    /   \\  /   \
   H      C      H
          |
          H
```

Although this picture accounted for most of the properties of benzene and was the key to many of the extraordinary successes of organic chemistry, it had a flaw: experiments show that all of the carbon-carbon bonds in benzene are exactly the same, not alternate single and double ones. Not until quantum

theory was applied to molecular structure in the next century did it become clear that six of the bonding electrons could be shared equally by the entire ring.

Naphthalene is familiar as the active ingredient in mothballs.

Organic compounds whose molecules do not contain ring structures are said to be **aliphatic.**

Organic Compounds

The remarkable range of organic compounds is hinted at in the hydrocarbons, which contain only carbon and hydrogen. Add just oxygen and the possibilities are multiplied many times over, giving compounds as diverse as they are numerous. Add still other elements and the result is staggering in variety and complexity. But regularities exist that permit the orderly classification of organic compounds and lead to an understanding of how their molecular structures govern their behavior. Given this understanding, the organic chemist can create compounds tailored to exhibit specific properties. Evidence of the success of this endeavor is found in the synthetic materials, from textile fibers to drugs, so widely used today.

13.8 Hydrocarbon Groups

A Handy Classification Scheme

In the classification system used for organic compounds that contain other elements besides carbon and hydrogen, the compounds are often regarded as **derivatives** of hydrocarbons—that is, as compounds obtained by substituting other atoms or atom groups for one or more of the H atoms in hydrocarbon molecules. Ordinarily such compounds are *not* prepared in this way, but their structural formulas suggest that they might be. For example, ethanol can be regarded as a derivative of ethane, with an OH group replacing an H atom:

```
    H   H                H   H
    |   |                |   |
H — C — C — H        H — C — C — OH
    |   |                |   |
    H   H                H   H
   ethane               ethanol
```

Similarly acetic acid can be regarded as a derivative of methane, with a COOH group replacing an H atom:

```
    H                H    O
    |                |   //
H — C — H        H — C — C
    |                |    \
    H                H     OH
 methane           acetic acid
```

Hydrocarbon Groups The carbon-hydrogen atom groups that appear in hydrocarbon derivatives are named from the hydrocarbons. Groups corresponding to the hydrocarbons methane, ethane, and propane are

```
    H                H   H                H   H   H
    |                |   |                |   |   |
H — C —          H — C — C —          H — C — C — C —
    |                |   |                |   |   |
    H                H   H                H   H   H
  methyl            ethyl                  propyl
  group             group                  group
```

Thus the compound CH_3Cl is methyl chloride, C_3H_7I is propyl iodide, and $CH_3C_2H_5SO_4$ is methyl ethyl sulfate.

13.9 Functional Groups

Atom Groups with Characteristic Behaviors

Inorganic compounds that contain a particular atom group, such as OH or SO_4, have important aspects of their chemical behavior in common, as we know. The chemical behavior of many organic compounds is also determined to a large extent by the presence of certain atom groups,

Example 13.2

The following compound can be called either bromoethane or ethyl bromide. Why?

```
    H   H
    |   |
H — C — C — Br
    |   |
    H   H
```

Solution

Since CH_3CH_3 is ethane, substituting a Br (bromine) atom for one of the H atoms leads to the name bromoethane. Since CH_3CH_2 is the ethyl group, the compound can also be called ethyl bromide.

called **functional groups.** Table 13-2 shows some of the main functional groups found in organic molecules.

Alcohols The hydroxyl (OH) group they contain makes many alcohol molecules somewhat polar, and the simpler alcohols are soluble in water. The polarity is not enough, however, to prevent them from mixing with many compounds less polar than water, which makes the alcohols useful as solvents. Ethanol (ethyl alcohol) is, of course, the active ingredient in wine, beer, and spirits. Ethanol is a poison that is removed from a person's blood by the liver; regularly drinking too much ethanol causes permanent liver damage.

The ethanol in beverages is produced by fermentation. In this process sugar is converted to ethanol and carbon dioxide, with yeast enzymes acting as catalysts. For example,

$$\underset{\substack{\text{glucose}\\\text{(a sugar)}}}{C_6H_{12}O_6} \xrightarrow[\text{yeast enzymes}]{} \underset{\text{ethanol}}{2C_2H_5OH} + \underset{\substack{\text{carbon}\\\text{dioxide}}}{2CO_2} \qquad \textit{Fermentation}$$

Wine is made by fermenting fruit juice, usually grape juice; traces of wine have been found in jars dating back to 5400 B.C. Beer is made by fermenting grain, usually barley, and then adding a bitter extract of a plant called hop for flavor. Since yeast cells die before the alcohol concentration reaches about 15 percent, wine and beer cannot be stronger than this. In fact, fermentation generally stops somewhat earlier because the sugar runs out.

Distillation can produce stronger liquors. The fermented liquid is heated and the alcohol-rich vapor is then led off and condensed to give brandy (starting from fruit), whiskey (grain), rum (sugar cane), vodka (traditionally potatoes), and so on. The proof of an alcoholic beverage is twice its percentage content of ethanol. Thus 80 proof whiskey contains 40 percent ethanol, and pure ethanol is 200 proof.

Ethanol for industrial purposes is usually made by reacting ethene, a byproduct of petroleum refining, with steam under high pressure. As mentioned in Sec. 4.11, ethanol for fuel is made today by fermenting the sugar in sugarcane or corn. Probably suitable catalysts will eventually be developed

Table 13-2 Common Functional Groups

Name of Group	Structural Formula	Class of Compound	Example	Formula	Comments
Hydroxyl	$-OH$	Alcohol	Ethanol	$H-\underset{H}{\overset{H}{C}}-\underset{H}{\overset{H}{C}}-OH$	Used as a solvent and in beverages; prepared by fermenting sugar solution and synthetically from ethene and water.
Ether	$-O-$	Ether	Diethyl ether	$H-\underset{H}{\overset{H}{C}}-\underset{H}{\overset{H}{C}}-O-\underset{H}{\overset{H}{C}}-\underset{H}{\overset{H}{C}}-H$	Once widely used as an anesthetic, its side effects and flammability led to its replacement by safer compounds.
Aldehyde	$-C(=O)H$	Aldehyde	Formaldehyde	$H-C(=O)H$	A gas used to preserve biological specimens and as an embalming fluid when dissolved in water ("formalin").
Carbonyl	$-\overset{O}{\overset{\Vert}{C}}-$	Ketone	Acetone	$H-\underset{H}{\overset{H}{C}}-\overset{O}{\overset{\Vert}{C}}-\underset{H}{\overset{H}{C}}-H$	A common solvent with a toxic vapor used in paints and as nail polish remover.
Carboxyl	$-C(=O)OH$	Acid	Acetic acid	$H-\underset{H}{\overset{H}{C}}-C(=O)OH$	Responsible for characteristic taste of vinegar; a weak acid like other organic acids.
Ester	$-C(=O)O-$	Ester	Methyl acetate	$H-\underset{H}{\overset{H}{C}}-C(=O)-O-\underset{H}{\overset{H}{C}}-H$	Formed by the reaction of methyl alcohol and acetic acid with water as the other product.

to enable cellulose from plant material—which is cheap, plentiful, and does not compete with food supply—to be processed efficiently into ethanol.

When an H atom is replaced by an OH group in an aromatic hydrocarbon, the result is a compound whose properties are different from those of ordinary alcohols. The simplest example is phenol ("carbolic acid"), C_6H_5OH, which was the first antiseptic and today is one of the raw materials for the plastic Bakelite and for the phenolic glues used in plywood:

$$C_6H_5-OH$$

phenol

Familiar alcohols with more than one OH group are ethylene glycol and glycerol:

```
    H   H                H   H   H
    |   |                |   |   |
H — C — C — H        H — C — C — C — H
    |   |                |   |   |
   OH  OH               OH  OH  OH
 ethylene glycol          glycerol
```

Ethylene glycol is used as an antifreeze in car engines. Glycerol, also known as glycerin, is a sweetish, viscous liquid used in many skin lotions and to prevent tobacco from drying out.

Ethers An ether has an oxygen atom bonded between two carbon atoms. Relatively inert chemically, ethers are widely used as solvents in organic processes since there is little or no danger they will interfere with the reactions.

Aldehydes and Ketones These compounds have similar chemical behavior because both contain the carbonyl atom group $>C=O$. In aldehydes the carbonyl group is at the end of a molecule with a hydrogen atom attached to the carbon atom, while in ketones the group is inside a molecule between two other carbon atoms. The double bond between C and O is highly polar, and as a result aldehydes and ketones are soluble in water.

Ethanol is oxidized in the liver into acetaldehyde:

```
    H   H                      H     O
    |   |                      |    //
H — C — C — OH + O ——→ H — C — C        + H2O
    |   |                      |    \
    H   H                      H     H
  ethanol      oxygen    acetaldehyde     water
```

Most of the acetaldehyde is oxidized further in the liver to acetic acid, which is then oxidized in the muscles to CO_2 and H_2O. The acetaldehyde that survives enters the bloodstream and is responsible for many of the ill effects of drinking too much, which include damage to most of the body's organs as well as nausea and hangovers. Acetaldehyde is also present in tobacco smoke and contributes to the harm it causes. Methanol (methyl alcohol) is oxidized in the liver to the poisonous formaldehyde, which is believed to be the reason why methanol is so toxic.

The solvent acetone is the most familiar ketone.

Organic Acids Compounds that contain the carboxyl group, —COOH, are acids because the H atom is loosely held and can detach itself as H^+. A C–H bond is stronger. Most organic acids are very weak. Familiar examples are the formic acid that causes insect bites to sting, the acetic acid of vinegar, the butyric acid of rancid butter and some cheeses, the citric acid of citrus fruits, the lactic acid of sour milk, and the acetylsalicylic acid of aspirin.

When an opened bottle of wine is stored for some time, the ethanol it contains gradually turns into acetic acid and the eventual result is vinegar.

The conversion of ethanol to acetic acid is promoted by enzymes produced by bacteria in the wine:

$$\begin{array}{ccccc} & H & H & & \\ & | & | & & \\ H- & C & - C & - OH & + O_2 \\ & | & | & & \\ & H & H & & \\ \end{array} \longrightarrow \begin{array}{cccc} & H & & O \\ & | & & /\!\!/ \\ H- & C & - C & \\ & | & & \backslash \\ & H & & O-H \\ \end{array} + H_2O$$

ethanol — oxygen — acetic acid — water

Esters Alcohols are, so to speak, organic hydroxides, but unlike their inorganic cousins they do not dissociate appreciably in water. They react slowly with acids to form compounds called esters, which are analogous to the salts of inorganic chemistry but are not electrolytes. An example is ethyl acetate, which is made by reacting ethanol with acetic acid:

$$\begin{array}{ccccccc} & H & H & & O & H & \\ & | & | & & \| & | & \\ H- & C & - C & - O - & C & - C & - H \\ & | & | & & & | & \\ & H & H & & & H & \end{array}$$

ethyl acetate

Figure 13-11 The ester nitroglycerin is the active ingredient in the explosive dynamite. Dynamite was used to demolish this building in Indianapolis.

This ester is an important commercial solvent. Several hundred thousand tons of it are used each year in the United States in manufacturing coatings of various kinds, from paint to nail polish.

Many esters have pleasant fruity or flowerlike odors and find extensive use in perfumes and flavors. Propyl acetate is responsible for the fragrance and taste of pears, octyl acetate for those of oranges, ethyl butyrate for those of apricots, and butyl butyrate for those of pineapples. The explosive nitroglycerin is an ester formed by the reaction of nitric acid with the alcohol glycerol (Fig. 13-11). Animal and vegetable fats are all esters of glycerol as well.

13.10 Polymers

Molecules Linked into Giant Chains

Polymers are giant molecules that consist of hundreds or thousands of identical (or almost identical) subunits. Proteins, starch, cellulose, and rubber are natural polymers. Polythene, polyvinyl chloride (PVC), Styrofoam, Teflon, nylon, and Dacron are synthetic polymers; the solid ones are usually called just plastics.

For a long time polymers were thought to be merely assemblies of small molecules held together by van der Waals forces. Finally, in work that began in 1926, the German chemist Hermann Staudinger (Fig. 13-12) showed that polymers are true molecules of huge size held together with covalent bonds. Nearly 60 million tons of polymers are made every year in the United States, about 0.25 percent ("bioplastics") from plant material and the rest from oil and natural gas feedstocks.

Figure 13-12 Hermann Staudinger (1881–1965).

Plastic Waste

Modern civilization produces huge amounts of plastic waste that are a problem to dispose of because ordinary polymers, unlike traditional materials such as wood, paper, and natural fibers, do not readily decompose. Dumped in landfills, where most rubbish ends up, plastic waste remains intact for a very long time. Burning plastic waste reduces its volume and provides useful heat—more than the same mass of coal or oil—but it must be carried out in expensive special incinerators because toxic gases, such as sulfur dioxide, hydrogen chloride, and hydrogen cyanide, may be given off.

Another approach is to create plastics that degrade naturally. An example is the plastic collars used to hold beer cans together, which contain polymer chains with atom groups that split the chains when exposed to light; the fragments that result decompose more readily than intact chains. Biodegradable plastics have been made that incorporate starch, which bacteria consume and thereby break down the polymer chains. Still other ways to produce degradable plastics are being explored, including the use of bacteria to convert corn sugar to a biodegradable polyester called PHA.

The best thing to do with plastic waste, of course, is to recycle it into new products. But not all polymers can be reused, and the needed careful sorting is often more expensive than starting from new raw materials. One successful recycling process starts with beverage bottles made of polyethylene terephthalate (PET). The PET is melted down and turned into fibers that are used in such items as blankets, fleece garments, and insulation. Overall, though, relatively little plastic waste is recycled today.

Of course, the problem of plastic waste can be reduced just by wasting less plastic. For instance, durable shopping bags could replace the usual plastic bags provided by stores, 100 billion or so of which are thrown away in the United States every year. To encourage such good sense, Ireland now taxes plastic bags, which has led to a 90 percent drop in their use in that country. San Francisco has simply banned its larger groceries and pharmacies from providing nonbiodegradable plastic bags at all. The city's goal is for all waste to be be biodegradable or recyclable by 2020.

Polythene We are already acquainted with the unsaturated hydrocarbon ethene:

$$\begin{array}{ccccc} H & & & & H \\ & \diagdown & & \diagup & \\ & & C=C & & \\ & \diagup & & \diagdown & \\ H & & & & H \end{array}$$

ethene

Because of the double bond, ethene molecules can, under the proper conditions of heat and pressure, polymerize to form chains thousands of units long whose formula we might write as

$$\begin{array}{ccccccccccccccc} & & H & & H & & H & & H & & H & & H & & \\ & & | & & | & & | & & | & & | & & | & & \\ \cdots & - & C & - & C & - & C & - & C & - & C & - & C & - & \cdots \\ & & | & & | & & | & & | & & | & & | & & \\ & & H & & H & & H & & H & & H & & H & & \end{array}$$

polythene

This material is polythene (or polyethylene), which is widely used as a packaging material because of its inertness and pliability. The ethene is called the **monomer** in the process, and polythene the polymer. (In Greek, *mono* means "alone" or "single," and *poly* means "many.") A train can be thought of as a polymer, with each of its cars as a monomer. Because of the large size of their molecules, polymers are usually solids.

Vinyls One of the H atoms in ethene can be replaced by another atom or atom group to form the monomer for a polymer whose properties differ from those of polythene. Because the group

$$\begin{matrix} H & & H \\ & C{=}C & \\ H & & \end{matrix}$$

vinyl group

is called the **vinyl group,** such polymers are classed as vinyls. Some familiar examples are shown in Table 13-3 with Fig. 13-13 illustrating the corresponding monomers.

The benzene rings attached to alternate C atoms in polystyrene are relatively large and project like knobs from the polymer chain. This prevents adjacent chains from sliding past one another, and as a result polystyrene is relatively stiff. If a substance that gives off a gas is added to the liquid monomer mixture, gas bubbles will form throughout the liquid as it polymerizes. The result is the familiar lightweight, rigid Styrofoam.

Table 13-3 Some Common Vinyl Polymers

Monomer	Polymer	Uses
$H_2C{=}CHCl$ **vinyl chloride**	$\cdots{-}CH_2{-}CHCl{-}\cdots$ **polyvinylchloride**	Tubing, insulation, imitation leather, rainwear (PVC, Geon, Koroseal)
$H_2C{=}CH{-}C{\equiv}N$ **acrylonitrile**	$\cdots{-}CH_2{-}CH(C{\equiv}N){-}\cdots$ **polyacrylonitrile**	Textiles, carpets (Acrilon, Orlon)
$H_2C{=}CH{-}CH_3$ **propene**	$\cdots{-}CH_2{-}CH(CH_3){-}\cdots$ **polypropylene**	Carpets, ropes, molded objects, thermal underwear
$H_2C{=}CH{-}C_6H_5$ **styrene**	$\cdots{-}CH_2{-}CH(C_6H_5){-}\cdots$ **polystyrene**	Molded objects, insulation, packing material (Styrofoam)

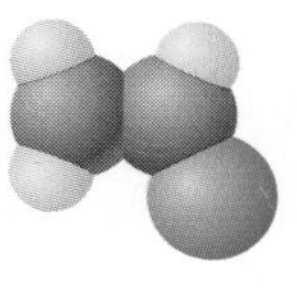
Vinyl chloride

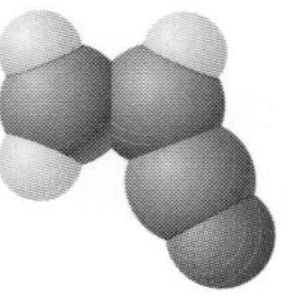
Acrylonitrile

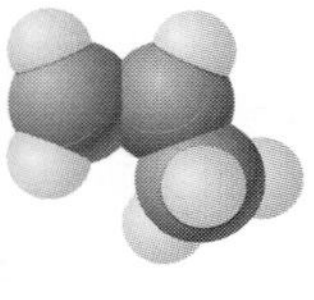
Propene

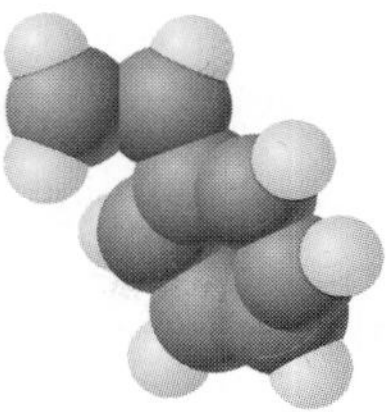
Styrene

Figure 13-13 Models of the monomers listed in Table 13-3.

Lucite and Plexiglas In some monomers, such as methyl methacrylate, two of the H atoms in ethene are replaced by atom groups. Methyl methacrylate polymerizes to form the transparent plastics whose trade names are Lucite and Plexiglas. A feature of this material is that it is **thermoplastic,** which means that it softens and can be shaped when heated but becomes rigid again upon cooling.

Teflon The monomer for Teflon is tetrafluorethene, which is ethene with all the H atoms replaced by fluorine atoms:

$$\begin{array}{c} \text{F} \qquad\quad \text{F} \\ \diagdown \quad \diagup \\ \text{C}=\text{C} \\ \diagup \quad \diagdown \\ \text{F} \qquad\quad \text{F} \end{array} \qquad\qquad \begin{array}{c} \text{F} \quad \text{F} \\ | \quad\;\; | \\ \cdots-\text{C}-\text{C}-\cdots \\ | \quad\;\; | \\ \text{F} \quad \text{F} \end{array}$$

tetrafluorethene   Teflon

The bond between fluorine and carbon is extremely strong, which makes Teflon tough and inert and able to withstand much higher temperatures than other polymers. Teflon has a very slippery surface, too.

The Cup and the Environment

Styrofoam cups have nonrenewable petroleum as their ultimate raw material, their manufacture involves the carcinogen styrene, they cannot be recycled, and they are not biodegradable. Surely paper cups, with trees as their raw material and disposable in various ways, are more friendly to the environment? And are not china cups really the greenest of all, since they can be used over and over again?

Well, it all depends (Fig. 13-14). Paper cups need more energy and are more polluting to produce than Styrofoam cups. Since paper cups weigh more than Styrofoam cups, their transportation uses more energy, too. China cups have to be washed, which involves energy, detergents, and water. Whether Styrofoam, paper, or china cups are best varies according to whether the criterion is energy use, nature and amount of raw materials, air pollution, water pollution, or volume of solid wastes. Overall, china cups win only if they are washed with the least water and detergent possible and are used several thousand times before breaking. (Dishwashing a full load by machine uses less water and less electricity for heating it than dishwashing by hand.) Between the throwaway cups, Styrofoam seems less harmful than paper, although this is not entirely certain.

Figure 13-14 Life-cycle studies of the environmental impacts of Styrofoam, paper, and china cups give no firm answer as to which is best on an overall basis.

These properties make Teflon useful industrially for seals and bearings as well as for nonstick coatings for cooking utensils (Fig. 13-15).

Figure 13-15 The strong bond between carbon and fluorine accounts for the durability and inertness of Teflon, which was used to coat this frying pan.

Copolymers Some polymers consist of two different monomers. An example of such a **copolymer** is Dynel, used among other things to make fibers for wigs, whose monomers are vinyl chloride and vinyl acetate. The kitchen wrap Saran is another copolymer.

Elastomers Certain monomers that contain two double bonds in each molecule form flexible, elastic polymers called **elastomers.** Rubber is a natural elastomer (Fig. 13-16). A widely used synthetic elastomer is neoprene, which has the valuable property that liquid hydrocarbons such as gasoline affect it less than they do natural rubber. Another elastomer, nitrile rubber, is still more resistant to hydrocarbons and is used to line gasoline hoses.

Fibers Of the various kinds of synthetic fibers that have been developed, nylon and Dacron are the most familiar. Both are composed of chains of structural elements, just like polymers, but they are produced by chemical reactions rather than by the polymerization of monomer molecules. In the case of nylon, the result is a chain whose elements can be written

$$
\begin{array}{ccccccccccccccc}
 & \mathrm{H} & \mathrm{H} & \mathrm{H} & \mathrm{H} & \mathrm{H} & \mathrm{H} & \mathrm{H} & \mathrm{H} & \mathrm{O} & \mathrm{H} & \mathrm{H} & \mathrm{H} & \mathrm{H} & \mathrm{O} \\
 & | & | & | & | & | & | & | & | & \| & | & | & | & | & \| \\
- & \mathrm{N}- & \mathrm{C}- & \mathrm{C}- & \mathrm{C}- & \mathrm{C}- & \mathrm{C}- & \mathrm{C}- & \mathrm{N}- & \mathrm{C}- & \mathrm{C}- & \mathrm{C}- & \mathrm{C}- & \mathrm{C}- & \mathrm{C}- \\
 & & | & | & | & | & | & | & & & | & | & | & | & \\
 & & \mathrm{H} & \mathrm{H} & \mathrm{H} & \mathrm{H} & \mathrm{H} & \mathrm{H} & & & \mathrm{H} & \mathrm{H} & \mathrm{H} & \mathrm{H} &
\end{array}
$$

The atom group

$$
\begin{array}{cc}
\mathrm{H} & \mathrm{O} \\
| & \| \\
-\mathrm{N}- & \mathrm{C}-
\end{array}
\qquad \textit{Amide linkage}
$$

is known as an amide linkage, so nylon is called a **polyamide.** The N—H and C$=$O groups in nylon are polar, and their mutual attraction is what holds adjacent chains of molecules firmly together (Fig. 13-17).

Dacron, whose structural elements are different from those of nylon, is a **polyester** because its elements are linked together by groups of the form

$$
\begin{array}{cc}
\mathrm{O} & \\
\| & \\
-\mathrm{C}- & \mathrm{O}-
\end{array}
\qquad \textit{Ester linkage}
$$

Silicones

Silicon is just below carbon in the periodic table and, like carbon, its atoms each participate in four covalent bonds. Polymers can be made in which silicon atoms replace some of the carbon atoms in their structures. Such polymers are called **silicones.** Some silicones are liquids, others are gels or elastomers, and some are solids with varying degrees of rigidity. Silicone surgical implants are widely used.

(*a*) Natural rubber

(*b*) Vulcanized rubber

Sulfide link

Figure 13-16 (*a*) Natural rubber is a sticky liquid whose polymer chains can slide past one another easily. (*b*) In the process of vulcanization, natural rubber is heated with sulfur, which produces links between polymer chains. The result is a material that is locally flexible but solid overall. The greater the number of sulfide links, the stiffer the result.

Figure 13-18 The hull of the yacht *Ardent Spirit* consists of fiberglass-reinforced polyester resin, and polyester fibers were woven to make the cloth for its sails.

Figure 13-17 Winding a nylon filament from liquid polyamide in the laboratory.

(see Table 13-2). Polyester resins reinforced with glass fibers are often used in boat hulls, truck bodies, and other large structures (Fig. 13-18).

The strongest synthetic fiber yet developed—10 times as strong as steel for the same weight—is a form of polythene called Spectra. The molecules in Spectra are as much as 100 times as long as those in ordinary polythene, and great care is taken when they are formed into fibers to keep them aligned in the fiber direction.

Chemistry of Life

At one time the physical and biological worlds seemed two separate realms. They interacted with each other to be sure, but nevertheless they were thought to be distinct in that an intangible "life force" was thought to be present in living things but absent everywhere else.

Nowadays it is clear there is no life force. Instead there is a continuous chain of development from simple chemical compounds through more elaborate ones through viruses (which are neither "alive" nor "dead" by conventional definitions) through primitive one-celled organisms to complex plants and animals. Given the right chemical and physical conditions and plenty of time, there is every reason to believe that life will inevitably come into being from inorganic matter, as it has on our planet.

The four chief classes of organic compounds found in living matter are carbohydrates, lipids, proteins, and nucleic acids, which we shall examine in turn.

13.11 Carbohydrates

The First Link in the Food Chain

Carbohydrates are compounds of carbon, hydrogen, and oxygen whose molecules generally contain two atoms of hydrogen for every one of oxygen. They are manufactured in the leaves of green plants from carbon dioxide and water in the process of **photosynthesis,** with energy for the reaction being provided by sunlight. Sugars, starches, and cellulose are all carbohydrates.

An important group of sugars consists of isomers that have the same molecular formula, $C_6H_{12}O_6$. These sugars exist both as straight chains of C atoms and as ring structures. The ring forms are more stable and so occur in nature most often, but a sugar molecule of this kind can shift back and forth between the two forms. Here are the straight-chain and ring forms of glucose, which is the sugar circulated by the blood to provide the body with energy:

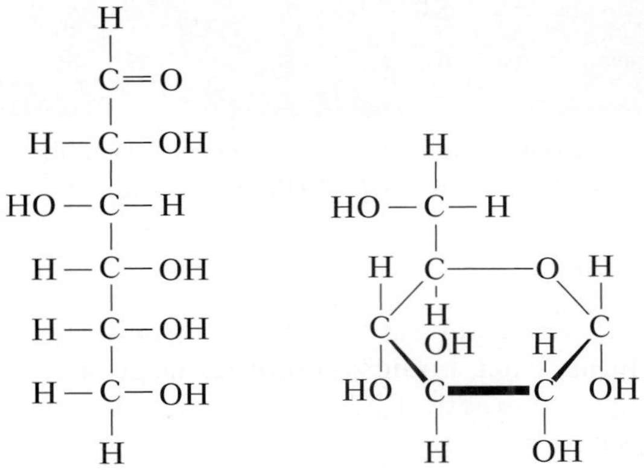

```
       H
       |
       C=O
       |
   H - C - OH
       |
  HO - C - H
       |
   H - C - OH
       |
   H - C - OH
       |
   H - C - OH
       |
       H
```

glucose
(straight-chain form)

```
          H
          |
   HO  -  C  -  H
          |
    H     C ———— O   H
    |   /  |       \  |
    C      H          C
    |      OH    H    |
   HO      C ———— C   OH
           |      |
           H      OH
```

glucose
(ring form)

The isomers of glucose—among them fructose (the sweetest-tasting of all), galactose, and mannose—have somewhat different structures and properties.

Simple sugars like those above are called **monosaccharides.** Two monosaccharide rings can link together to form a **disaccharide.** Thus sucrose (ordinary table sugar) consists of one ring of glucose and one of fructose; lactose (milk sugar) consists of one ring of glucose and one of galactose; and maltose (malt sugar) consists of two rings of glucose. The molecular formula of all three of these disaccharides is $C_{12}H_{22}O_{11}$, but of course their structures are different.

Polysaccharides **Polysaccharides** are complex sugars that consist of chains of more than two simple sugars. They are naturally occurring polymers. In living things the polysaccharides serve both as structural components and as a medium of energy storage. In plants **cellulose,** which consists of a chain of about 1500 glucose rings, is the chief constituent of cell walls. Wood is mostly cellulose, as is cotton (Fig. 13-19). In fact, cellulose is the most abundant organic compound on earth.

Figure 13-19 The cellulose in wood is extracted and made into paper at this mill in Maine.

Starch, whose 300 to 1000 glucose units are joined together in a slightly different way from those in cellulose, is a polysaccharide that plants use to store energy for later use. Starch occurs in grains that have an insoluble outer layer, and so remains in the cell in which it is formed until, when needed as fuel, it is broken down into soluble glucose molecules. (Chew a mouthful of white bread without swallowing. After a while the bread will begin to taste sweet as its starch molecules are broken up into glucose molecules by the action of enzymes in your saliva.)

A polysaccharide found in animals is **chitin,** which forms the outer shells of insects and crustaceans such as lobsters and crabs. Chitin is much like cellulose in structure and is also very abundant. Another polysaccharide is **glycogen,** which is present in the liver and muscles of animals and is released when energy is required. Glycogen, the animal equivalent of starch, is soluble, but its molecules are so large that they cannot pass readily through cell walls. When glucose is needed by an animal, its stored glycogen is split into the much smaller glucose molecules. When the stored glycogen of an endurance athlete such as a distance runner is used up, he or she "hits the wall" as extreme fatigue sets in.

Plant and Animal Energy Living things obtain the energy they need by the oxidation of nutrient molecules. Generally the nutrient molecule most directly involved is glucose, and its oxidation is an exothermic reaction that yields carbon dioxide and water as products:

$$\underset{\text{glucose}}{C_6H_{12}O_6} + \underset{\text{oxygen}}{6O_2} \longrightarrow \underset{\text{carbon dioxide}}{6CO_2} + \underset{\text{water}}{6H_2O} + \text{energy} \qquad \textit{Oxidation of glucose}$$

This reaction takes place not all at once, as this equation would indicate, but in a complex series of steps that involve a number of other substances. However, the net effect is the oxidation of glucose. The oxidation of glucose is evidently the reverse of photosynthesis and is the final process by which the energy in sunlight is turned into the energy used by living things.

Figure 13-20 Microorganisms in their digestive systems enable cattle to convert the cellulose in plants to glucose.

Usually the carbohydrates in the food we eat are in the form of disaccharides and polysaccharides. In digestion these are **hydrolyzed** with the help of water to monosaccharides. Hydrolysis is promoted by enzymes, which are the specialized protein molecules that act as catalysts in most biochemical processes.

Although many animals can hydrolyze starch to glucose, very few can hydrolyze cellulose. Some plant-eating animals, for instance cattle, have microorganisms such as yeasts, protozoa, and bacteria in their digestive tracts, and enzymes from these microorganisms hydrolyze cellulose in the plants that are eaten. The resulting glucose can then be used by the animal (Fig. 13-20).

After digestion, glucose passes into the bloodstream to be circulated throughout the body. Glucose not immediately needed by the cells is converted into glycogen in the liver and elsewhere. If there is too much glucose to be stored as glycogen, the excess is synthesized into fats.

13.12 Photosynthesis

How the Sun Powers the Living World

As just mentioned, plants combine carbon dioxide from the air with water absorbed through their roots to form carbohydrates in photosynthesis (Fig. 13-21). Photosynthesis is highly endothermic, with the necessary energy coming from sunlight:

$$6CO_2 + 6H_2O + \text{energy} \longrightarrow C_6H_{12}O_6 + 6O_2 \quad \textit{Photosynthesis}$$

The energy is absorbed not directly by the CO_2 and H_2O but instead by a substance called **chlorophyll,** which is part of the green coloring matter of leaves. Chlorophyll acts as a catalyst that passes solar energy to the reacting molecules in a complicated way.

Perhaps 70 billion tons of carbon dioxide is cycled each year, about half by land plants and the rest by free-floating plantlike organisms (algae and certain bacteria) in the upper part of the oceans (Fig. 13-22). Photosynthesis is only about 1 percent efficient on the average in utilizing the sunlight that reaches plants. A few plants have much higher efficiencies—as much as 11 percent for sugarcane, which is why the cheapest ethanol sometimes used to replace or supplement gasoline is made from this source of sugar.

Photosynthesis not only maintains the oxygen content of the atmosphere but seems to have been responsible for it in the first place. The early atmosphere of the earth, which is thought to have consisted of gases emitted during volcanic action, contained oxygen only in combination with other elements in compounds such as water (H_2O), carbon dioxide (CO_2), and sulfur dioxide (SO_2). Primitive organisms, which probably obtained their own energy originally from such sources as sulfur, iron, and methane, eventually began to produce free oxygen by photosynthesis. There is evidence that photosynthesis has been going on for at least 3.5 billion years. In time the oxygen content of the atmosphere increased to the point where more complex organisms could evolve. Besides the oxygen now in the atmosphere, photosynthesis is believed to account for much of the oxygen that is combined with other elements in the oxides, carbonates, and sulfates found in sediments and sedimentary rocks.

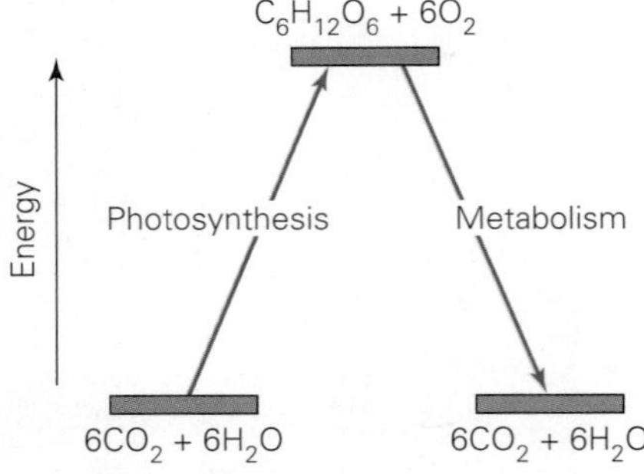

Figure 13-21 The energy needed for the photosynthesis of carbohydrates (shown is glucose) in plants comes from sunlight. The energy stored in carbohydrates is released during their metabolism by plants and animals.

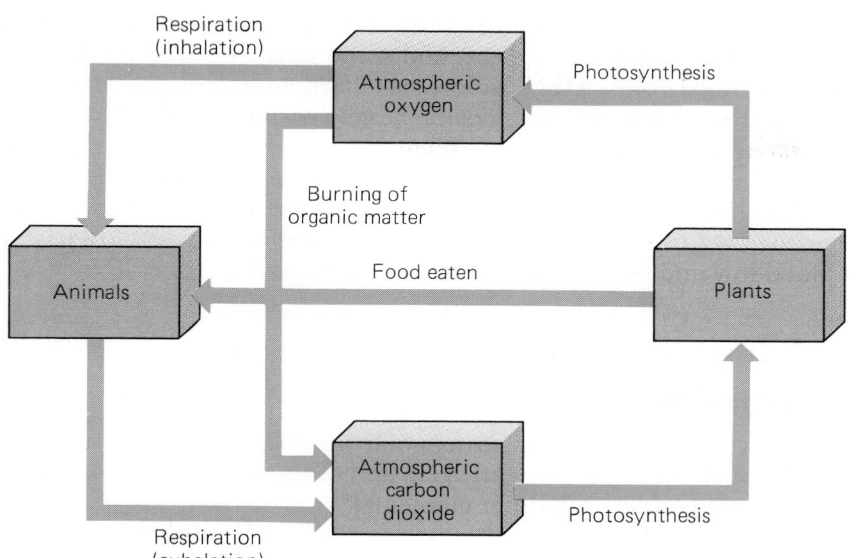

Figure 13-22 The oxygen-carbon dioxide cycle in the atmosphere.

13.13 Lipids

Where the Calories Are

Fats and such fatlike substances as oils, waxes, and sterols are collectively known as **lipids.** Like carbohydrates, lipids contain only the elements C, H, and O. This is natural since lipids are synthesized in plants and animals from carbohydrates. The proportions and arrangements of these elements are different in lipids, though.

A fat molecule consists of a glycerol molecule with three **fatty acid** molecules attached to it. The hydrocarbon chains in solid fats have only single bonds between their carbon atoms; hence they are saturated. Most animal fats are saturated. Liquid fats, such as vegetable oils, are unsaturated, with double bonds linking one or more carbon atoms. Polyunsaturated fats have more than one double bond per molecule. The double bonds introduce bends in such molecules, which prevents them from being closely packed together. As a result the interactions between nearby molecules are weaker than in the case of saturated fat molecules, and unsaturated fats are liquid at room temperature whereas saturated fats are solid.

Adding hydrogen atoms to the double-bonded carbon atoms in a liquid fat saturates the chains and gives a solid fat. Margarine is produced by such a **hydrogenation** process, with vegetable oils such as soybean and cottonseed oils being heated with hydrogen in the presence of a catalyst.

Fats are used for energy storage and other purposes, such as waterproofing, insulation against cold, and mechanical protection, in most living things. The digestion of a fat molecule involves the breaking of the ester links between its glycerol and fatty acid parts. The oxidation of the glycerol and fatty acids then proceeds in a fairly complicated way and releases nearly twice as much energy as in the case of the same mass of carbohydrate.

Plants and People

Suppose you are in a sealed room (perhaps in a space station) with plenty of light, either from the sun or from artificial sources. What area of plant life would there have to be to provide you with the oxygen you need and to absorb the carbon dioxide you exhale? If you spend all your time resting, your body will require around 1750 kcal of energy per day. To obtain this much energy by metabolizing food will take a little less than 500 g of oxygen, which can be produced by plants that cover an area between 2.5 and 18 m^2, depending on the plants used and assuming optimum illumination. The same vegetation will absorb the CO_2 you breathe out. In Russian experiments using algae to recycle air in this way, an area of 8 m^2 per person did the trick with artificial light powering the photosynthesis. A square plot of this area would be 9.3 ft on a side.

Cholesterol is a lipid found in the bloodstream. Some comes from food of animal origin that we eat, and some is synthesized by the body itself from other lipids. **Atherosclerosis** ("hardening of the arteries") is a serious condition caused by deposits, largely of cholesterol, that restrict the flow of blood. Heart attacks occur when partly blocked arteries prevent enough blood from reaching its muscles for the heart to function properly. These attacks are a leading cause of death. Eating unsaturated rather than saturated fats has been found to keep the cholesterol level in the blood low and so helps prevent atherosclerosis.

Tissue Matching

The carbohydrates and the lipids do not share the specificity of the proteins. Glucose, for instance, is a carbohydrate found in all plants and animals, but no protein is similarly widespread. Even individuals of the same species have some proteins that are not quite identical, so that tissues cannot ordinarily be transplanted because of the danger of "rejection" of the graft. The matching of blood types before a transfusion is to ensure that the proteins in the blood of the donor are the same as those in the blood of the recipient (Fig. 13-23). Drugs have been developed that can prevent transplanted organs such as kidneys and hearts from being rejected. Unfortunately such drugs also weaken the body's ability to defend itself from infection since both responses of the body to foreign proteins involve the same mechanisms.

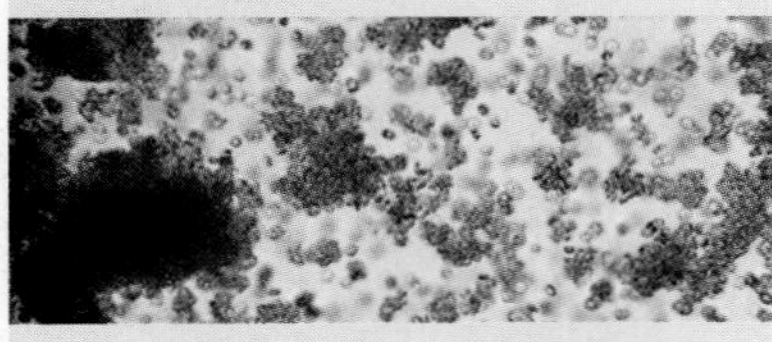

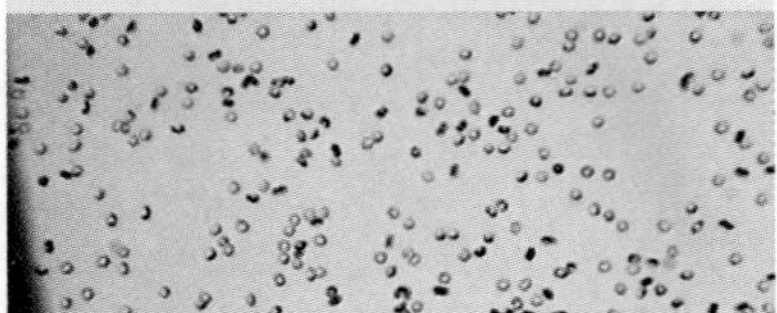

Figure 13-23 Matching blood types before a transfusion. *Top:* The blood types of donor and recipient are incompatible, which causes the red cells to clump together. *Bottom:* When the blood types are compatible, no clumping occurs.

13.14 Proteins

The Building Blocks of Living Matter

Proteins are the principal constituent of living cells. They are compounds of carbon, hydrogen, oxygen, nitrogen, and often sulfur and phosphorus; some proteins contain still other elements.

The basic chemical units of which protein molecules are composed are 20 **amino acids.** The simplest amino acid is glycine,

$$\begin{array}{ccccccc} H & & & H & & & O \\ & \diagdown & & | & & /\!\!/ & \\ & & N - & C & - C & & \\ & / & & | & & \diagdown & \\ H & & & H & & & O-H \end{array}$$

in which we recognize the characteristic carboxyl group —COOH of organic acids. Typical protein molecules consist of several hundred amino acids joined together in chains, and their structures are accordingly quite complex. The formula of one of the proteins found in milk is $C_{1864}H_{3012}O_{576}N_{468}S_{21}$, which gives an idea of the size of some protein molecules.

Plant and animal tissues contain proteins both in solution, as part of the fluid present in cells and in other fluids such as blood, and in insoluble form, such as the skin, muscles, hair, nails, horns, and so forth of animals. Silk is an almost pure protein. The human body contains thousands of different proteins, all of which it must make from the 20 amino acids it obtains from the digestion of the food proteins it takes in. One of the great successes of modern biochemistry was the discovery of how living cells build the complex arrangements of amino acids in their proteins.

In a protein molecule, the links between the amino acids consist of **peptide bonds** that are like the amide bonds in nylon. These chains of amino acids, called **polypeptide chains,** are usually coiled or folded in intricate patterns (Fig. 13-24). An important aspect of the patterns is the cross-linking that occurs between different chains and between different parts of the same chain.

The sequence of amino acids in a protein is just as important as which ones they are. The amino acid units in even a small protein molecule, such as insulin with 51 units, can be arranged in a great many different ways. However, only one arrangement has the biological effects associated with insulin. A parallel is with the formation of a word from the 26 letters of the alphabet. *Run* and *urn* have the same letters but mean

Spider Silk

A strand of spider silk is about five times as strong as a steel wire of the same size and is more elastic. Such strands consist chiefly of a protein polymer called fibroin that is similar to keratin, the protein found in hair and horns. When a spider squeezes liquid fibroin out through a tiny orifice, some of the amino acid chains in the fibroin align into highly ordered crystals whose peptide bonds are exceptionally strong. The rest of the fibroin is in the less-ordered form of helices, as in Fig. 13-24*a*. The result is a composite material in which the crystals provide strength and the helices provide elasticity. The diameter of a typical strand is 1 micrometer, a thousandth of a millimeter. Spider silk is used in medicine to make mats and sponges that serve as scaffolds on which human cells of various kinds are grown.

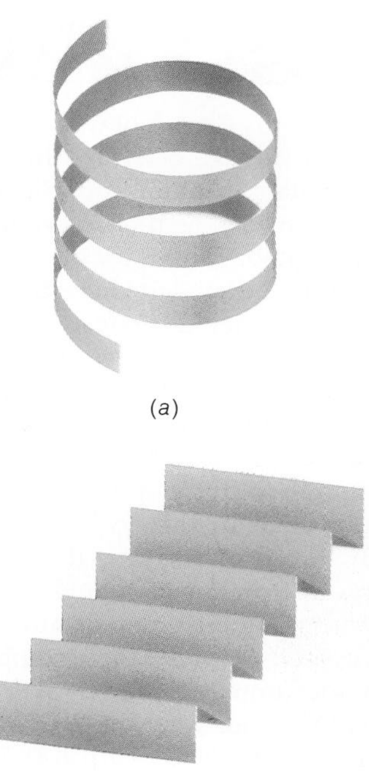

Figure 13-24 (*a*) The alpha helix form of a protein molecule. Each amino acid unit in the helix is linked by hydrogen bonds (a type of van der Waals bond) to other units above and below it. Most proteins have molecules like this. (*b*) The pleated sheet form of a protein molecule. Two or more chains of amino acid units are linked side-to-side along the sheet by hydrogen bonds. Fibrous proteins—such as those in hair, silk, cartilage, and horn—are of this kind. Some proteins have still other forms.

different things because the order of the letters is different. The alphabet of the proteins has only 20 letters, corresponding to the various amino acids, but the words may contain hundreds of letters whose relative positions in three dimensions are significant. The extraordinary number of different proteins, each serving a specific biological need in an organism, is not surprising in view of this picture of protein structure.

Every protein has a specific shape it must assume in order to carry out its function in the body. A number of serious diseases—for instance, Alzheimer's, Parkinson's, Huntington's, and several kinds of cancer—seem to be the result of incorrect protein folding.

Dietary Protein The human body can synthesize only some of the 20 amino acids it requires. The others must be present in our diets or else our bodies will not be able to manufacture the various proteins essential to life. A proper diet must therefore include not just an adequate total amount of proteins but also the right ones.

Most proteins of animal origin—such as those in meat, fish, eggs, and milk—contain all the needed amino acids, but plant proteins do not. The important amino acid lysine is missing in corn, wheat, and rice; isoleucine and valine are missing in wheat; threonine is missing in rice; and so on. Although it is certainly possible to live without eating meat or other animal products, a vegetarian diet not only must be sufficiently varied to include all the required amino acids but must provide all of these acids every day since they are not stored in the body and are needed together for protein manufacture.

13.15 Soil Nitrogen

A Vital Component

The amino acids of which proteins are composed all contain nitrogen. The ultimate source of all our protein is plants, although much of it comes to us secondhand in such animal proteins as those in meat, eggs, and milk. Plants manufacture their proteins from simpler nitrogen compounds that enter their roots from the soil in which they grow.

Green plants cannot draw upon the stable molecules of free nitrogen in the air around them. All their nitrogen, and therefore all the nitrogen that goes into animal bodies as well, comes from nitrogen compounds

BIOGRAPHY Dorothy Crowfoot Hodgkin (1910–1994)

Dorothy Crowfoot was fascinated at the age of 10 by the growth of crystals in alum and copper sulfate solutions as their solvent water evaporated. This fascination with crystals never left her. She studied chemistry at Oxford University despite the difficulties women students of science had to face in those days, and as an undergraduate had mastered x-ray crystallography well enough to have a research paper published. In this technique a narrow beam of x-rays is directed at a crystal from various angles and the resulting interference patterns are analyzed to yield the arrangement of the atoms in the crystal. Dorothy Crowfoot went on to Cambridge University to work with J. D. Bernal, who had just begun to use x-rays to investigate biological molecules. Under the right conditions many such molecules form crystals from whose structures the structures of the molecules themselves can be inferred. In particular, the structures of protein molecules are important because they are closely related to their biological functions. She and Bernal were the first to map the arrangement of the atoms in a protein, the digestive enzyme pepsin.

After two intense years at Cambridge, Dorothy Crowfoot returned to Oxford where she married Thomas Hodgkin and had three children while continuing active research. Her most notable work was on penicillin (then the most complex molecule to be successfully analyzed), vitamin B_{12}, and insulin (it took 35 years of on-and-off effort to finish the job). She was a pioneer in using computers to interpret x-ray data, an arduous task for all but the simplest molecules.

For all her achievements and their recognition in the scientific world, Hodgkin was for many years shabbily treated at Oxford: poor laboratory facilities, the lowest possible official status, half the pay of her male colleagues with continual worries about making ends meet until outside support (much of it from the Rockefeller Foundation of the United States) became available. She received the Nobel Prize in chemistry in 1964, the third woman to do so.

in the soil. The nitrogen molecules we breathe can do us no good either, for the atoms in these molecules are held together by strong triple bonds that our body processes are unable to break. Like a shipwrecked sailor surrounded by seawater but dying of thirst, we are surrounded by an ocean of nitrogen but would perish except for the combined nitrogen that plants can absorb through their roots.

The formation of plant proteins steadily removes nitrogen compounds from the soil. Just as steadily, nitrogen compounds are returned to the soil by the decay of animal wastes and of dead plants and animals. The nitrogen of proteins is converted by decay into ammonia and ammonium salts, which are then oxidized to nitrates by soil bacteria. But the replenishment is never complete. Some nitrogen is lost permanently from the soil when nitrates and ammonium salts dissolve in streams and rainwash, and when bacteria decompose nitrates into free nitrogen.

Nature makes good these losses in two ways. Another kind of soil bacteria, the "nitrogen-fixing" bacteria, have the ability to break down the stable nitrogen molecules of the air and to manufacture nitrates from the atoms. Also, lightning causes atmospheric nitrogen and oxygen to combine into nitrogen oxides, which are brought down to the earth in rainwater. So in nature nitrogen goes through a continuous cycle (Fig. 13-25) that keeps the amount of fixed nitrogen in the soil approximately constant.

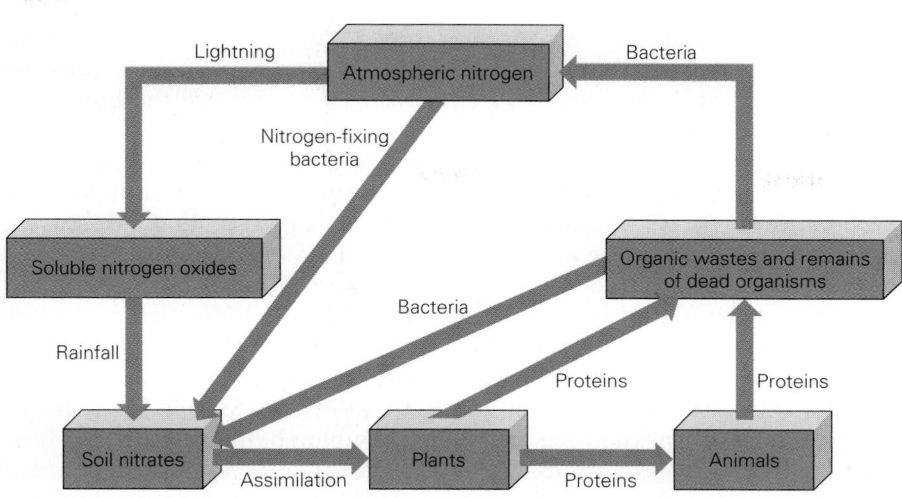

Figure 13-25 The nitrogen cycle on land.

Figure 13-26 Pure ammonia (NH_3) is sometimes applied directly to soil to supply it with the nitrogen that plants (here soybeans) need to manufacture proteins.

We have drastically disturbed this natural cycle. Much of the protein that enters our bodies is not returned to the soil but instead is dumped as sewage into various bodies of water. The use of plant material and manure for cooking fires in primitive areas further adds to the conversion of fixed nitrogen to free nitrogen. To be sure, manure is still used as fertilizer in some regions, and legumes, which are plants (such as peas and beans) on which nitrogen-fixing bacteria grow, are widely cultivated, but artificial fertilizers have become essential as sources of nitrogen for a large part of the world's agriculture (Fig. 13-26). On a global basis, a third of the protein in the human diet contains nitrogen derived from artificial fertilizers. Eighty million tons of nitrogen contained in such fertilizers is used each year, much of it ending up in groundwater, lakes, and rivers. Some of the consequences of this pollution were described in Sec. 10.9; nitrates in drinking water are also a health hazard.

13.16 Nucleic Acids

The Genetic Code

The **nucleic acids** are very minor constituents of living matter from the point of view of quantity. However, because they control the processes by which cells and organisms manufacture their proteins and reproduce themselves, these acids are extremely important. If anything may be said to be the key to the distinction between living and nonliving matter, it is the nucleic acids.

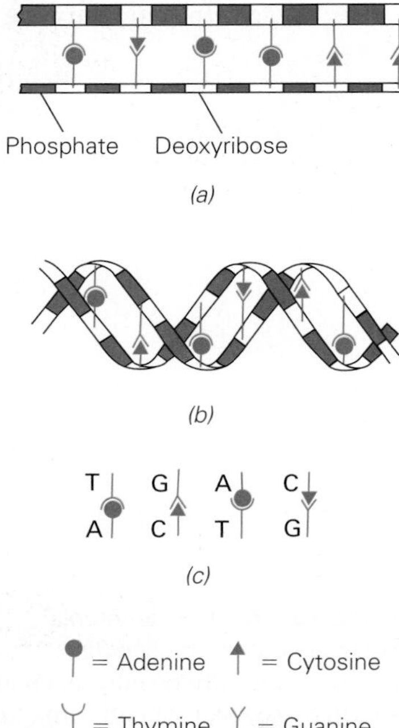

Figure 13-27 The structure of DNA. (*a*) The nitrogen bases link a double chain of alternate phosphate and deoxyribose groups. Adenine and thymine are always paired, and cytosine and guanine are always paired. (*b*) The chains are not flat but form a double helix, like a twisted ladder. (*c*) The four "letters" of the genetic code.

Nucleic acid molecules consist of long chains of units called **nucleotides.** As in the case of the amino acids in a polypeptide chain, both the kinds of nucleotide present and their arrangement govern the biological behavior of a nucleic acid.

Each nucleotide has three parts, a **phosphate group** (PO_4), **pentose sugar,** and a **nitrogen base.** A pentose sugar is one that contains five carbon atoms. In **ribonucleic acid** (RNA) the sugar is **ribose,** $C_5H_{10}O_5$, and in **deoxyribonucleic acid** (DNA) the sugar is **deoxyribose,** $C_5H_{10}O_4$, which has one O atom fewer than ribose. Nitrogen bases have characteristic ring structures of nitrogen and carbon atoms. Four nitrogen bases are found in DNA: adenine, guanine, cytosine, and thymine. The nitrogen bases in RNA are the same except that uracil replaces thymine.

The structure of a DNA molecule is shown in Fig. 13-27. Pairs of nitrogen bases form the links between a double chain of alternate phosphate and deoxyribose groups. Adenine and thymine are always coupled together, as are cytosine and guanine. The chains are not flat but spiral around each other in a double helix, as in Fig. 13-27*b*. The double helix structure of DNA was discovered in 1953 by the American biologist James D. Watson and the English physicist Francis H. C. Crick, who were working together at Cambridge University (Fig. 13-28).

Figure 13-27*c* shows the four "letters" of the genetic code. There may be hundreds of millions of such letters in a DNA molecule, and their precise sequence governs the properties of the cell in which the molecule is located. DNA molecules thus represent the biological blueprints that are translated into the processes of life.

The complexity of living things is mirrored in the complexity of DNA molecules, which are the largest known to science (Fig. 13-29). DNA molecules are normally folded and coiled into microscopic packages called **chromosomes.** If the 23 human chromosomes were stretched out, they would total about a meter in length. If DNA were as thick as a strand of spaghetti, a chromosome would be over 10 km long.

What DNA and RNA Do DNA controls the development and functioning of a cell by determining the proteins the cell makes. This is only one aspect of the role of DNA in the life process. Another follows from the ability of DNA molecules to reproduce themselves, so that when a cell divides, all the new cells have the same characteristics (that is, the same **heredity**) as the original cell (Fig. 13-30). Finally, changes in the sequence of bases in a DNA molecule can occur under certain circumstances, for example, during exposure to x-rays. These changes will be reflected in alterations in the cell containing the DNA molecule. If such a **mutation** occurs in the DNA of a reproductive cell of an organism, the result may be that the descendants of the original organism will be different in some way from their ancestor.

Thus four fundamental attributes of life can be traced to DNA: the structure of every organism, how it functions, its ability to reproduce, and its ability to evolve into different forms in later generations.

The other type of nucleic acid, RNA, differs from DNA in a number of respects. RNA molecules are much smaller than DNA molecules, for example, and usually consist of only single strands of nucleotides. One type of RNA carries instructions for the synthesis of specific proteins from the DNA in a cell's nucleus to the place where the synthesis occurs.

Figure 13-28 James D. Watson (1928–) *at left* and Francis H. C. Crick (1916–2004).

The instructions are in the form of a code in which each successive group of three nucleotides determines the particular amino acid to be added next to the protein polypeptide chain being formed. For example, the group GCA (guanine-cytosine-adenine) corresponds to the amino acid alanine, and GGA corresponds to glycine.

Figure 13-29 Model of a small part of a DNA molecule, which has the form of a double helix. The development and functioning of every living organism are ultimately controlled by the DNA in its cells. When the organism reproduces, copies of its DNA are passed on to the new generation. One cubic centimeter of DNA can hold more information than a trillion CDs.

The Human Genome Every cell in a plant or animal contains in its DNA coded instructions for making all the proteins the organism needs. The set of instructions for each protein is called a **gene;** human genes are 1000 to 1500 base pairs long. Genes make up only 1.2 percent of the human **genome,** the 3 billion or so base pairs present in our chromosomes. The rest of the genome (once dismissed as "junk DNA") contains a number of base-pair sequences that regulate the action of various genes. Some sequences that do not code for genes, amounting to 3.8 percent of the genome, are also found in other mammals, which suggests they are not random but either have an unknown function or once did. Perhaps raw material for new genes is part of the genome.

The human genome seems to contain 20,000–25,000 genes. Preparing a detailed map that shows not only each gene but also its sequence of base pairs was an immense task that was started in 1988 and is now largely complete—an extraordinary accomplishment. The result is nothing less than a blueprint for human life. It is remarkable that such a linear code is able to produce enormously complex three-dimensional living things (such as us). The genomes of over a thousand species have either been or are being decoded today, with more to come. About 500 genes seem to be universal: every known genome contains them. The human genome, with 3 billion base pairs, is not unusual in size: the range in animals is from 350 million (the pufferfish) to 130 billion (the marbled lungfish). Plants may have surprisingly large genomes. That of the mere onion, with 17 billion base pairs, is over five times the size of ours. Why? Nobody knows.

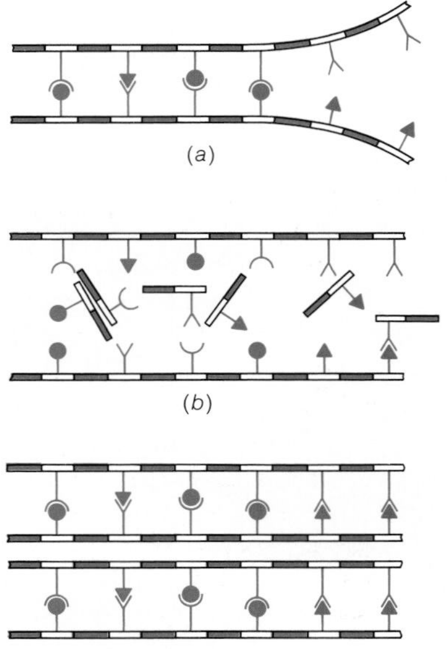

Figure 13-30 Simplified model of DNA replication. (*a*) When a cell reproduces, each double DNA chain it contains breaks into two single ones, much like a zipper opening. (*b*) The single chains then pick up from the cell material the nucleotides needed to complete their structures. (*c*) The result is two identical DNA chains.

A knowledge of the genome has practical importance apart from its intrinsic interest. Many diseases have a specifically genetic basis (for instance, cystic fibrosis and muscular dystrophy), and most, if not all, others have some genetic linkage. An example of a genetic linkage is the increased tendency members of some families have to contract a particular disease, such as cancer or Alzheimer's. The human genetic map is a guide toward better health for everybody.

13.17 Origin of Life

An Inevitable Result of Natural Processes

Whatever the earth's beginnings may have been, it is safe to assume that, at some time in the remote past, the surface was considerably warmer than it is at present. The atmosphere of the young earth almost certainly contained compounds of hydrogen, oxygen, carbon, and nitrogen; the most likely were water, methane, ammonia, and carbon dioxide (Fig. 13-31). When an electric discharge simulating lightning is passed through a mixture of water vapor, hydrogen, methane, and ammonia, amino acids and other compounds of biological importance are formed (Fig. 13-32). This result seems to suggest that the raw materials for living things could have come into being in the early atmosphere of the earth. However, it now seems unlikely that the young earth's atmosphere contained enough hydrogen and hydrogen-rich compounds, notably methane (CH_4) and ammonia (NH_3), to get very far on the journey to life. An atmosphere dominated by carbon dioxide, such as those of Venus and Mars, is more probable.

Another hypothesis is that the ingredients of life came from space, where molecules of over a hundred compounds have been identified both in clouds of matter from which stars and planetary systems condense (see Sec. 19.10) and in comets, which peppered the earth long ago. But, among other objections, the number of these compounds with biological significance is too few to fill in the blanks in any plausible picture of how life began.

Figure 13-31 **Fumaroles** are vents in the earth's surface associated with volcanic activity that give out a mixture of gases, mainly water vapor, carbon dioxide, and nitrogen and sulfur compounds. The earth's oceans and atmosphere are believed to be the result of similar outgassing on a large scale early in the earth's history. These fumaroles are in Yellowstone National Park.

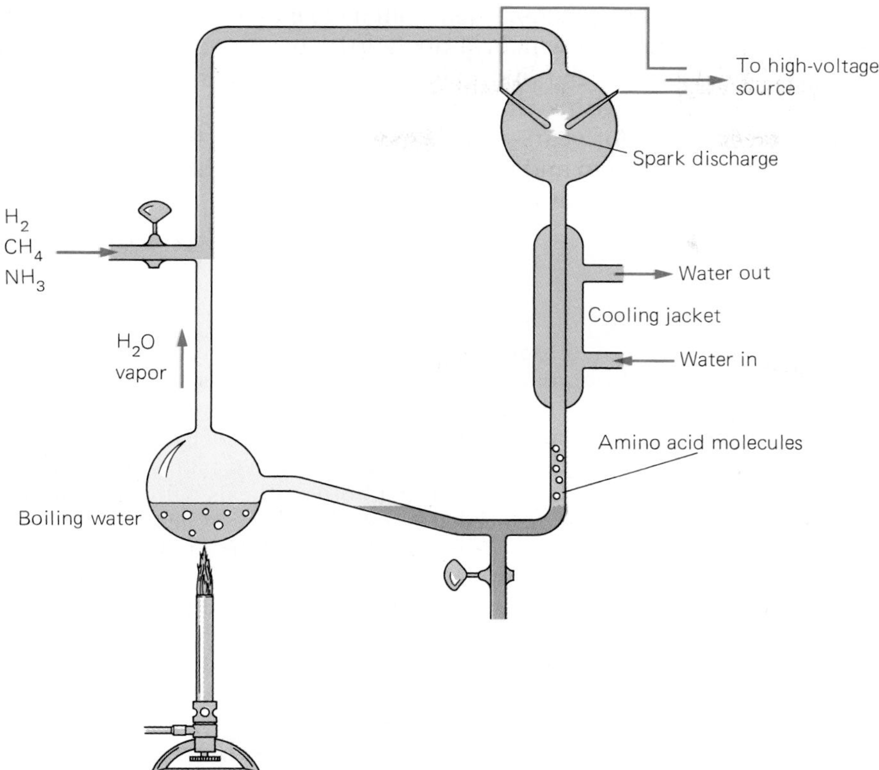

Figure 13-32 In this experiment, first performed in 1952 by Stanley Miller under the direction of Harold Urey, amino acids were created by passing sparks that simulate lightning through a mixture of water vapor, hydrogen, methane, and ammonia.

A more promising line of thought proposes that at least some of the building blocks of life were formed near hydrothermal vents, like those on the floors of modern oceans, from which scalding hot, chemical-laden water gushes out. Chemical reactions occur most readily in liquids, water is the best solvent, and there is ample energy here to power such reactions. And, indeed, not only individual amino acids but also short chains of them—miniature proteins—are created around today's vents. The nitrogen bases of RNA and DNA could perhaps also have such vents in their ancestry.

Good arguments put RNA rather than DNA as the first genetic molecule. For one thing, RNA molecules are both simpler and more stable than those of DNA; for another, RNA plays many more roles in life than just being a messenger for instructions from DNA. Where could RNA have come from? A leading idea is that certain clays acted as templates for selecting the starting molecules for RNA and as scaffolds for assembling them. In experiments, RNA molecules up to 50 letters long developed on the surface of a variety of clay called montmorillonite. Another proposal details how RNA and proteins could have appeared together, each acting as a catalyst for the production of the other. Given proteins and RNA, all that remains to make a living cell able to function and reproduce is a fatty membrane to act as a container, and it is known that such membranes can spontaneously form in certain circumstances.

Living organisms certainly did not come into existence with the neatness and dispatch with which, say, a baker combines certain ingredients, puts the mixture in an oven, and takes out a cake an hour later. Although there is no agreement on which, if any, of the proposed pathways from individual molecules to the first cells was actually followed, nothing is known that conflicts with such a transformation. And plenty of time—hundreds of millions of years—was available between the origin of the earth 4.6 billion years ago and the existence of well-established life, probably primitive bacteria, that left traces in ancient rocks.

Important Terms and Ideas

Organic chemistry concerns the chemistry of carbon compounds. The number and variety of organic compounds result from the ability of carbon atoms to form covalent bonds with each other as well as with other atoms.

Structural formulas show not only the numbers of each kind of atom in a molecule but also how the atoms are joined together. **Isomers** are compounds with the same molecular formulas but different structural formulas, corresponding to different arrangements of the same atoms.

Double bonds and **triple bonds** are possible between carbon atoms in addition to single ones. The molecules of a **saturated** organic compound contain only single carbon-carbon bonds; those of an **unsaturated** compound contain one or more double or triple bonds. Unsaturated compounds react more readily than saturated ones.

A **functional group** is a group of atoms whose presence in an organic molecule determines its chemical behavior to a large extent. Thus the hydroxyl (OH) group characterizes the alcohols, the carboxyl (COOH) group characterizes the organic acids, and so on.

A **polymer** is a long chain of simple molecules **(monomers)** linked together. Plastics, synthetic fibers, and synthetic elastomers are polymers.

Carbohydrates are compounds of carbon, hydrogen, and oxygen manufactured in green plants from carbon dioxide and water by **photosynthesis** with sunlight providing the needed energy. Sugars, starches, and cellulose are carbohydrates. **Lipids** are fats and fatlike substances such as oils and waxes synthesized from carbohydrates by plants and animals. **Proteins,** the principal constituents of living matter, consist of long chains of **amino acid** molecules. The sequence of amino acids in a protein molecule together with the shape of the molecule determines its biological role.

The **nucleic acid** molecules DNA and RNA consist of long chains of **nucleotides,** atom groups whose precise sequence governs the structure and function of cells and organisms. DNA has the form of a double helix and carries the genetic code; one type of RNA has the form of a single helix and acts as a messenger in protein synthesis.

Multiple Choice

1. The science of organic chemistry has as its subject
- **a.** compounds produced by plants and animals
- **b.** carbon compounds
- **c.** compounds with complex molecules
- **d.** the determination of structural formulas

2. Compared with inorganic compounds in general, most organic compounds
- **a.** are more readily soluble in water
- **b.** are more easily decomposed by heat
- **c.** react more rapidly
- **d.** form ions more readily in solution

3. The number of covalent bonds each carbon atom has in organic compounds is usually
- **a.** one
- **b.** two
- **c.** four
- **d.** six

4. The number of covalent bonds possible between two carbon atoms in organic molecules is
- **a.** one
- **b.** one or two
- **c.** one, two, or three
- **d.** one, two, three, or four

5. As a class, the alkanes are
- **a.** highly reactive
- **b.** soluble in water
- **c.** unaffected by acids and bases
- **d.** harmless to living things

6. In general, in the alkane series of hydrocarbons, a high molecular mass implies a (an)
- **a.** low boiling point
- **b.** high boiling point
- **c.** low freezing point
- **d.** artificial origin

7. Gasoline is a mixture of
   a. alkanes
   b. isomers of octane
   c. hydrocarbon derivatives
   d. unsaturated hydrocarbons

8. Compounds that have the same molecular formulas but different structural formulas are called
   a. hydrocarbons c. polymers
   b. isomers d. derivatives

9. Unsaturated hydrocarbon molecules are characterized by
   a. double or triple bonds between carbon atoms, so that additional atoms can be added readily
   b. the ability to absorb water
   c. the ability to dissolve in water
   d. benzene rings in their structural formulas

10. Hydrocarbon molecules with only single bonds between carbon atoms are
    a. fairly reactive c. extremely dense
    b. fairly unreactive d. unsaturated

11. Hydrocarbon molecules with double or triple bonds between carbon atoms are
    a. fairly reactive c. extremely dense
    b. fairly unreactive d. saturated

12. The number of bonds between the carbon atoms in acetylene, C_2H_2, is
    a. one c. three
    b. two d. four

13. Which of the following compounds can exist?
    a. C_2H_3 c. C_2H_5
    b. C_2H_4 d. C_2H_7

14. The benzene molecule is notable for having a
    a. ring of six carbon atoms
    b. ring that can consist of any number of carbon atoms
    c. straight chain of six carbon atoms
    d. helix of carbon atoms

15. Hydrocarbon derivatives in general are
    a. formed by burning hydrocarbons in air
    b. alcohols
    c. hydrocarbons that have had H atoms replaced by other atoms or atom groups
    d. hydrocarbons that have more than one bond between carbon atoms

16. All alcohols
    a. are safe to drink
    b. can be made only by fermentation
    c. contain just one OH group
    d. contain one or more OH groups

17. The presence of a COOH group is characteristic of an
    a. alkane c. aldehyde
    b. organic acid d. ester

18. The conversion of sugar to ethanol and carbon dioxide with enzymes acting as catalysts is called
    a. fermentation c. photosynthesis
    b. polymerization d. digestion

19. Organic acids are
    a. strong and highly corrosive
    b. rather weak
    c. not found in nature but must be artificially made
    d. characterized by simple molecules

20. The reaction of ethanol with acetic acid produces ethyl acetate, which is a (an)
    a. aldehyde c. aromatic compound
    b. ester d. polymer

21. A long chain of identical simple molecules linked together is called a (an)
    a. monomer c. isomer
    b. polymer d. elastomer

22. Dacron is an example of a
    a. monomer c. isomer
    b. polymer d. elastomer

23. Living cells consist mainly of
    a. carbohydrates c. proteins
    b. lipids d. nucleic acids

24. Living things differ most from one another in their constituent
    a. carbohydrates c. amino acids
    b. lipids d. proteins

25. The most abundant organic compound on earth is
    a. methane c. glucose
    b. benzene d. cellulose

26. Carbohydrates are produced in green plants by
    a. polymerization c. photosynthesis
    b. fermentation d. respiration

27. Animals store energy in the form of
    a. starch c. chitin
    b. cellulose d. glycogen

28. Plants store energy in the form of
    a. starch c. chitin
    b. cellulose d. glycogen

29. Fermentation can convert sugars to carbon dioxide and
    a. ethanol c. starch
    b. acetic and d. proteins

30. Cellulose is not
- **a.** a carbohydrate
- **b.** the chief constituent of wood
- **c.** present in all plants
- **d.** easily digested by most animals

31. Fats and oils are
- **a.** proteins
- **b.** carbohydrates
- **c.** nucleic acids
- **d.** lipids

32. Lipids are synthesized in plants and animals from
- **a.** proteins
- **b.** carbohydrates
- **c.** enzymes
- **d.** nucleic acids

33. A given mass of fat compared with the same mass of carbohydrate can provide the body with about
- **a.** half as much energy
- **b.** the same amount of energy
- **c.** twice as much energy
- **d.** 10 times as much energy

34. It is healthier to eat unsaturated rather than saturated fats because unsaturated fats
- **a.** are easier to digest
- **b.** contain more energy
- **c.** keep the cholesterol content of the blood low
- **d.** keep the glucose content of the blood low

35. Proteins consist of combinations of
- **a.** amino acids
- **b.** nucleic acids
- **c.** esters of glycerin with organic acids
- **d.** DNA and RNA molecules

36. The number of amino acids important to life is
- **a.** 2
- **b.** 8
- **c.** 20
- **d.** 69

37. Which of the following statements is correct?
- **a.** The human body can synthesize all the amino acids it requires.
- **b.** The human body can synthesize some of the amino acids it requires.
- **c.** The human body can synthesize none of the amino acids it requires.
- **d.** Different people require different groups of amino acids.

38. The matching of blood types before a transfusion is to ensure that the blood of donor and recipient both have the same
- **a.** carbohydrates
- **b.** proteins
- **c.** lipids
- **d.** nucleic acids

39. Most biochemical processes in living matter are catalyzed by
- **a.** enzymes
- **b.** glycogen
- **c.** lipids
- **d.** DNA

40. The structure of a DNA molecule resembles a
- **a.** single helix
- **b.** double helix
- **c.** pleated ribbon
- **d.** straight chain

41. Each three-nucleotide group in a DNA molecule corresponds to a particular
- **a.** amino acid
- **b.** nucleic acid
- **c.** enzyme
- **d.** protein

42. DNA is involved in which one or more of the following attributes of every living organism?
- **a.** Its structure.
- **b.** How it functions.
- **c.** Its ability to transmit its characteristics to its descendants.
- **d.** The ability of these characteristics to evolve, that is, to change in succeeding generations.

Exercises

13.1 Carbon Bonds

1. Why are there more carbon compounds than compounds of any other element?
2. In what ways do organic compounds, as a class, differ from inorganic compounds?
3. What is the principal bonding mechanism in organic molecules?

13.2 Alkanes

13.3 Petroleum Products

4. How can the different alkanes in petroleum be separated? What property of the alkanes makes this procedure possible?
5. Is gasoline a compound? If not, what is it?

13.4 Structural Formulas

13.5 Isomers

13.6 Unsaturated Hydrocarbons

6. Why are structural formulas more important in organic chemistry than in inorganic chemistry?
7. The isomers of a compound have the same chemical formula. In what way do they differ from one another?
8. Distinguish between unsaturated and saturated hydrocarbons, giving examples of each.
9. How many electrons are shared in a double bond between two carbon atoms?

10. What kind of carbon-carbon bonds are found in alkane molecules?

11. How many covalent bonds are present between the carbon atom and each oxygen atom in carbon dioxide, CO_2?

12. In general, how do the reactivities of hydrocarbon molecules that contain only single bonds compare with the reactivities of hydrocarbon molecules that contain double or triple bonds as well?

13. Why are substances whose molecules contain triple carbon-carbon bonds relatively rare?

14. The alkanes of Sec. 13.2 are saturated hydrocarbons with the general formula C_nH_{2n+2}. Another group of hydrocarbons called **alkenes** have molecules with two fewer H atoms than the corresponding alkane molecules, so their general formula is C_nH_{2n}. (a) Would you expect the alkenes to be saturated or unsaturated? (b) Would you expect more, less, or about the same reactivity in the alkenes as in the alkanes?

15. The structural formula of propane is given in Sec. 13.4. In bromopropane, one of the H atoms is replaced by a Br (bromine) atom. How many isomers does bromopropane have? What are their structural formulas?

16. Why does this structural formula not represent an actual molecule?

```
    H   H
    |   |
H — C — C
    |   |
    H   H
```

17. Why does this structural formula not represent an actual molecule?

```
    H   H
    |   |     H
    |   |    /
H — C — C = C
    |   |    \
    |   |     H
    H   H
```

18. In which of the compounds C_2H_2, C_2H_4, and C_4H_{10} are the carbon-carbon bonds single, in which are they double, and in which are they triple?

19. Is it possible for a molecule with the formula C_4H_2 to exist? If not, why not?

20. Is it possible for a molecule with the formula C_2H_3 to exist? If not, why not?

21. Is it possible for a molecule with the formula C_2H_6 to exist? If not, why not?

22. Each molecule of butyne, C_4H_6, has a triple bond between two of its carbon atoms. What is the structural formula of the butyne isomer in which the triple bond is in the middle of the molecule?

23. Each molecule of butene, C_4H_8, has a double bond between two of its carbon atoms. (a) Give the structural formulas for the isomers of butene in which the carbon atoms form a straight chain. (b) Another isomer is possible in which the carbon atoms form a branched chain. What is its structural formula?

24. Each molecule of propene, C_3H_6, has a double bond between two of its carbon atoms. Give the structural formula(s) for propene and its isomers, if any.

13.7 Benzene

25. What is the difference between aromatic and aliphatic compounds?

26. Why are all aromatic compounds unsaturated?

27. The carbon atoms in normal hexane, C_6H_{14}, form a straight chain. All the bonds are single. In cyclohexane the six carbon atoms are arranged in a ring. Give the structural formula of cyclohexane. Are all the bonds single?

13.8 Hydrocarbon Groups

13.9 Functional Groups

28. Ethanol can be used as an automotive fuel either by itself or added to gasoline. Give the equation for the combustion of ethanol.

29. When sugar undergoes fermentation to produce ethanol, what other compound is also formed?

30. To what class of organic compounds does the compound belong whose structure is shown below?

```
 O        H              O
  \\      |             //
    C  —  C  —  C  —  C
   /      |             \
 HO       OH             OH
```

31. To what class of organic compounds does the compound belong whose structure is shown below?

```
    H   H      O
    |   |     //
H — C — C — C
    |   |     \
    H   H      H
```

32. Which of the following (a) dissolve in water, (b) are acids, (c) react with ethyl alcohol to give esters, (d) react with acetic acid to give esters?

C_2H_5COOH	C_3H_8
C_2H_4	C_2H_5OH
HCl	$C_3H_5(OH)_3$

33. Compare the properties of a simple ester, for instance, methyl acetate, with those of a salt, for instance, sodium chloride.

34. Why do you think the compound whose structure is shown below is called dimethyl ether?

```
    H       H
    |       |
H — C — O — C — H
    |       |
    H       H
```

35. Why do you think the compound whose structure is shown below is called trichloroethene?

```
Cl        H
  \      /
   C = C
  /      \
Cl        Cl
```

36. Use structural formulas to show the reaction between methyl alcohol and acetic acid to produce methyl acetate.

37. Give structural formulas for the two isomeric propyl alcohols that share the molecular formula C_3H_7OH.

38. What is the structural formula of methyl ethyl ketone?

39. (a) Give structural formulas for the three isomers of pentane, C_5H_{12}. (b) One of these isomers is also known as methylbutane. Which is it? (c) Another of these isomers is also known as dimethyl propane. Which is it?

13.10 Polymers

40. Teflon is inert, tough, and can tolerate high temperatures because the bonds between carbon and fluorine in its structure are extremely strong. What does this suggest about the chemical activity of fluorine?

13.11 Carbohydrates

13.12 Photosynthesis

13.13 Lipids

41. How does a plant obtain its carbohydrates and fats? An animal?

42. What are the products of the oxidation of glucose? Is the process endothermic or exothermic?

43. What is believed to be the origin of atmospheric oxygen?

44. The ultimate source of the energy in food is the energy liberated when hydrogen is converted to helium in thermonuclear reactions in the sun. Trace this energy from the sun to the food you eat.

45. Can you think of any function other than energy storage that body fat might have?

13.14 Proteins

13.15 Soil Nitrogen

46. Why do plants need nitrogen? Why can they not use nitrogen from the air? Where do nitrogen compounds in the soil come from?

47. What are the basic structural units of proteins? How does the human body obtain them?

13.16 Nucleic Acids

48. How many letters are there in the genetic code by which DNA governs protein synthesis? How many letters are used to specify a particular amino acid?

49. What change in a gene is involved in a mutation? What is the significance of mutations in reproduction?

Answers to Multiple Choice

1. b	**7.** a	**13.** b	**19.** b	**25.** d	**31.** d	**37.** b
2. b	**8.** b	**14.** a	**20.** b	**26.** c	**32.** b	**38.** b
3. c	**9.** a	**15.** c	**21.** b	**27.** d	**33.** c	**39.** a
4. c	**10.** b	**16.** d	**22.** b	**28.** a	**34.** c	**40.** b
5. c	**11.** a	**17.** b	**23.** c	**29.** a	**35.** a	**41.** a
6. b	**12.** c	**18.** a	**24.** d	**30.** d	**36.** c	**42.** a, b, c, d

14

Atmosphere and Hydrosphere

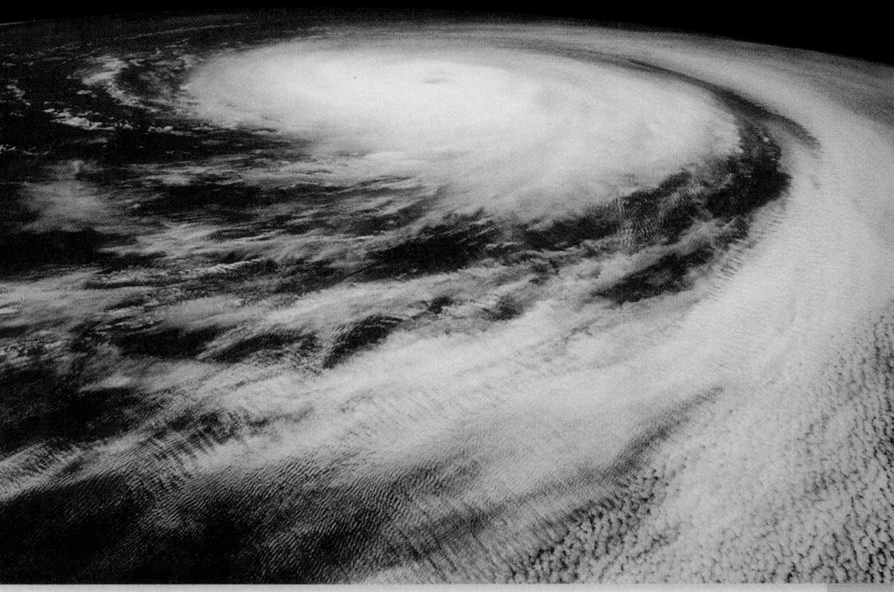

Hurricane Fefa east of Taiwan in 1991.

Goals

When you have finished this chapter you should be able to complete the goals ▶ given for each section below:

The Atmosphere

14.1 Regions of the Atmosphere
Four Layers
- ▶ List in order of abundance the four chief ingredients of dry air near ground level.
- ▶ Distinguish among the troposphere, stratosphere, mesosphere, and thermosphere.
- ▶ Define ozone and explain why the ozone layer in the upper atmosphere is so important.

14.2 Atmospheric Moisture
Another Vital Cycle
- ▶ State what is meant by saturated air and by the relative humidity of a volume of air.

14.3 Clouds
Some Are Water Droplets, Others Are Ice Crystals
- ▶ List the three principal ways in which clouds form.
- ▶ Describe what causes rain and snow to fall from a cloud.

Weather

14.4 Atmospheric Energy
A Giant Greenhouse in the Sky
- ▶ Define insolation and describe the greenhouse effect.
- ▶ Discuss why temperatures vary around the earth.

14.5 The Seasons
They Are Due to the Tilt of the Earth's Axis
- ▶ Explain how the seasons of the year originate.

14.6 Winds
Currents of Air Driven by Temperature Differences
- ▶ Describe what is meant by a convection current.
- ▶ State the influence of the coriolis effect on wind direction in the northern and southern hemispheres.

14.7 General Circulation of the Atmosphere
Alternate Belts of Wind and Calm
- ▶ Sketch on a map the main surface wind systems of the world and name them and the belts of relative calm that separate them.
- ▶ Describe what a jet stream is.
- ▶ Describe what an El Niño is.

14.8 Middle-Latitude Weather Systems
Why Our Weather Is So Fickle
- ▶ Compare cyclones and anticyclones and describe the motion of air in each of them.
- ▶ Compare warm and cold fronts and describe what happens when a cold front overtakes a warm front.

Climate

14.9 Tropical Climates
Hot and Wet or Hot and Dry
- ▶ Explain how tropical cyclones originate and where they usually occur.

14.10 Middle-Latitude Climates
Variety Is the Rule

14.11 Climate Change
An Icy Past, a Warm Future
- ▶ Describe the ice ages and the variations in the earth's motions that may be responsible for them.

The Hydrosphere

14.12 Ocean Basins
Water, Water Everywhere
- ▶ Describe what a tsunami is and how it is caused.

14.13 Ocean Currents
Four Great Whirlpools
- ▶ List the ways in which the oceans affect climates.
- ▶ Describe how the Gulf Stream affects European climate and why it has little influence on climate in the United States.

The earth's **atmosphere** is an invisible envelope of gas we hardly notice except when the wind blows or when rain or snow falls. The atmosphere is also responsible for the blue of the sky, for the colors of sunrise and sunset, and for the rainbow, as we learned in Chap. 7. Less obvious but more important is the role of the atmosphere in the living world. Its oxygen, nitrogen, and carbon dioxide are essential for life. It screens out deadly ultraviolet and x-rays from the sun. It carries energy and water over the face of the earth. And, by weathering away rocks, it helps form the soil in which plants grow.

All the water of the earth's surface is included in the **hydrosphere.** Oceans, seas, rivers, and lakes cover about three-quarters of the surface area of our planet. The oceans, of course, make up by far the greatest part of the hydrosphere, and they are a major factor in shaping the environment of life on earth.

The Atmosphere

14.1 Regions of the Atmosphere

Four Layers

The chief gases of the atmosphere and their average abundances are given in Table 14-1. Water vapor is also present but to a variable extent, from nearly none to about 4 percent. In addition, the lower atmosphere contains a great many small particles of different kinds, such as soot, bits of rock and soil, salt grains from the evaporation of seawater droplets, and spores, pollen, and bacteria.

Those of us who have been among mountains know that the higher up we go, the thinner and colder the air becomes. In the lower atmosphere, air temperature falls an average of 6.5°C per km of altitude. At an elevation of only 5 km (about 16,400 ft) the pressure is down to half what it is at sea level (Fig. 14-1) and the temperature is about −20°C. At about

Table 14-1 The Composition of Dry Air Near Ground Level

Gas	Average Percentage by Volume
Nitrogen	78.08
Oxygen	20.95
Argon	0.93
Carbon dioxide	0.03
Neon	0.0018
Helium	0.00052
Methane	0.00015
Krypton	0.00011
Hydrogen, carbon monoxide, xenon, ozone	<0.0001

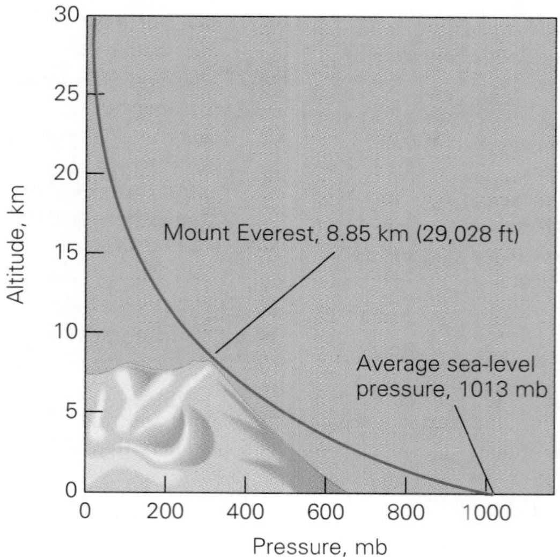

Figure 14-1 The variation of atmospheric pressure with altitude. The millibar (mb) is the pressure unit used in meteorology, where 1 mb = 100 Pa = 1 hPa. The average sea-level pressure of 1013 mb corresponds to 14.7 lb/in.^{2}. The lowest and highest surface pressures ever recorded are respectively 870 mb (in a Pacific typhoon) and 1084 mb (in a Siberian winter).

BIOGRAPHY Evangelista Torricelli (1608–1647)

Born near Ravenna, Italy, Torricelli studied mathematics in Rome. He revered Galileo and was his secretary in Florence for the last few months of Galileo's life. Torricelli then became court mathematician to the Grand Duke of Tuscany, Galileo's old position. Galileo had found it odd that, on the upstroke of its piston, a pump could lift water no more than about 10 m, and he suggested that Torricelli investigate the matter. As others did in those days, Galileo thought that pumps were able to lift water because "nature abhors a vacuum," so there seemed no reason for the observed limit of 10 m.

Torricelli concluded that the behavior of a lift pump made sense if air has mass, so that the atmosphere presses on the water feeding the pump and thereby pushes it upward in the cylinder as the piston is raised. The limit of 10 m meant that the atmosphere was finite in extent and did not fill the entire universe, despite the belief of many scientists then. To test his ideas, in 1643 Torricelli filled a glass tube, closed at one end, with mercury, whose density is nearly 14 times that of water. Holding his thumb over the open end of the tube, Torricelli turned it upside down and placed the open end in a dish of mercury. When he removed his thumb, the mercury level dropped in the tube but stopped when its height was about 760 mm above the mercury in the dish, $\frac{1}{14}$ of the maximum height that water could be lifted by a pump. This was the first barometer, with the mercury height directly proportional to the pressure of the atmosphere.

Torricelli observed that the mercury height varied slightly from day to day and interpreted this to mean that atmospheric pressure varied similarly. Three years later the French mathematician and physicist Blaise Pascal realized that, if atmospheric pressure is due to the weight of the atmosphere, then it should decrease with altitude. Himself not very robust, Pascal gave his brother-in-law two barometers to take up a mountain. The mercury columns dropped by several centimeters, which verified Torricelli's insights. A pressure unit equal to the pressure exerted by a mercury column 1 mm high is called the **torr** in honor of Torricelli; blood pressures in medicine are conventionally expressed in torr (Sec. 5.5).

---

11 km (36,000 ft) the pressure is only one-fourth its sea-level value, which means that 75 percent of the atmosphere lies below. The temperature at 11 km is about −55°C, which is cold but no colder than it sometimes is at ground level in Siberia and northern Canada. The atmosphere stays that cold for 14 km more.

A passenger in an airplane climbing past 11 km would notice a marked change in the atmosphere. Above that level there are almost no clouds, no storms, not even dust. Since the character of the atmosphere changes so abruptly at the 11-km level, this is taken as the boundary between the two lowest layers of the atmosphere. The dense part near the ground is called the **troposphere,** and the clear layer above it is called the **stratosphere.** Such features of the weather as clouds and storms, fog and haze, belong to the troposphere. Figure 14-2 shows how the atmosphere is divided into regions based on temperature behavior.

Ozone An important aspect of the stratosphere is the presence of ozone, O_3, which was described in Sec. 12.2. Ozone is produced in the stratosphere when solar radiation breaks up O_2 molecules into O atoms. Some O atoms then join O_2 molecules to give ozone molecules: $O + O_2 \rightarrow O_3$. As O_3 molecules are being formed, others are combining with O atoms to

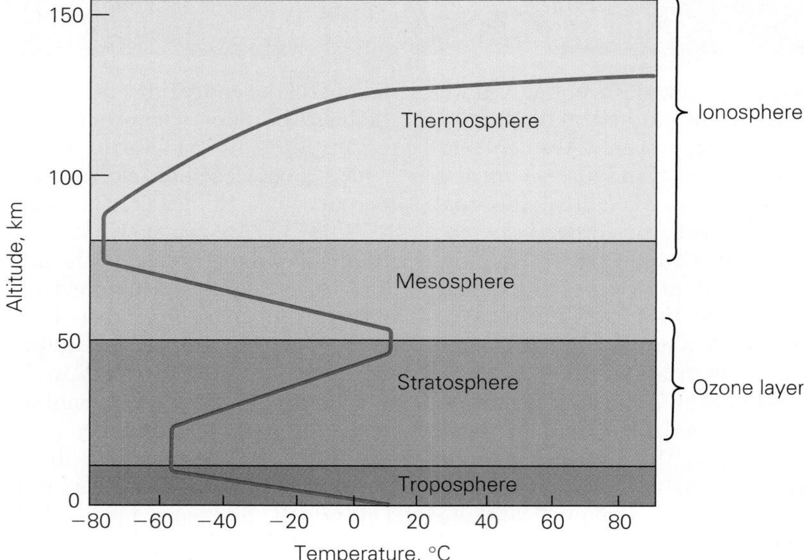

Figure 14-2 The variation of temperature with altitude. The altitudes of the boundaries between regions of the atmosphere are averages.

give O_2 molecules: $O + O_3 \rightarrow 2O_2$. Hence the situation is one of equilibrium, with ozone being formed and destroyed at the same rates.

Ozone is an excellent absorber of ultraviolet radiation. It is so good, in fact, that the relatively small amount of ozone in the stratosphere is able to filter out nearly all the dangerous short-wavelength ultraviolet radiation reaching the earth from the sun. Because this radiation is so harmful to living things, it is believed that life did not leave the protection of the sea to become established on land until the ozone layer had come into being. The maximum concentration of ozone occurs at 22 km, where less than one molecule in 4 million is O_3—hardly an impressive amount for so efficient an absorber.

Because the ozone in the stratosphere is so valuable, pollutants that attack it are highly undesirable. Among the worst such pollutants are artificially made gases called chlorofluorocarbons (CFCs) and hydrochlorofluorocarbons (HCFCs). An example is $CHClF_2$, usually referred to as HCFC-22, R-22, or Freon 22. These gases are widely used in refrigerators and air conditioners, as cleaning solvents, as the propellants in spray cans, and in making foam-plastic objects. In each application some CFCs or HCFCs escape into the atmosphere. The chlorine in them catalyzes the breakdown of O_3 molecules, and once in the stratosphere, the chlorine can remain there for a very long time. One chlorine molecule can destroy around 100,000 ozone molecules during its stay in the stratosphere.

By now the CFCs and HCFCs have measurably weakened the ozone shield. This is a serious matter since the additional ultraviolet radiation that reaches the earth as a result increases skin cancer and cataract rates, depresses immune systems, and reduces crop yields, among other effects (Sec. 7.9). Over Antarctica weather patterns have enabled the CFCs and HCFCs to punch a huge seasonal "hole" of low O_3 concentration in the

Smog in Los Angeles and Elsewhere

On an average day 9000 tons of carbon monoxide, various hydrocarbons, nitrogen and sulfur oxides, and small particles such as soot are emitted in the Los Angeles basin, mainly by vehicles and industry. (On some days, 25 percent of the particulate matter has been blown over from China, where 400,000 people die prematurely every year from diseases linked to air pollution.) Sunlight acting on this already nasty mixture adds toxic ozone to it. The result of breathing such polluted air is a high rate of respiratory diseases such as bronchitis and asthma and an increased risk of cancer. The air in Los Angeles is unhealthful on half the days each year (Fig. 14-3).

Figure 14-3 Smog in Los Angeles. When cool air from the Pacific blows into the Los Angeles basin, it forms a dense layer under warm air above it that stops polluted air from rising upward. The result of such a temperature inversion is heavy smog, made worse by nearby mountains that prevent the smog from escaping inland.

Los Angeles is not alone in having dirty air: all 20 of the world's largest cities exceed World Health Organization limits in at least one pollutant (Fig. 14-4). The situation in the Los Angeles basin is made worse by the frequent **temperature inversions** that occur there; Mexico City is another victim of this phenomenon. Ordinarily air temperature falls steadily with increasing altitude in the lower atmosphere. Sometimes, however, a temporary situation arises in which a layer of air aloft is warmer than the air below it. This is called a temperature inversion. Polluted air cannot rise past an inversion because when it reaches the inversion, the density of the polluted air is greater than that of the warm air layer on top. Hence the inversion acts to trap the polluted air, which results in a persistent smog.

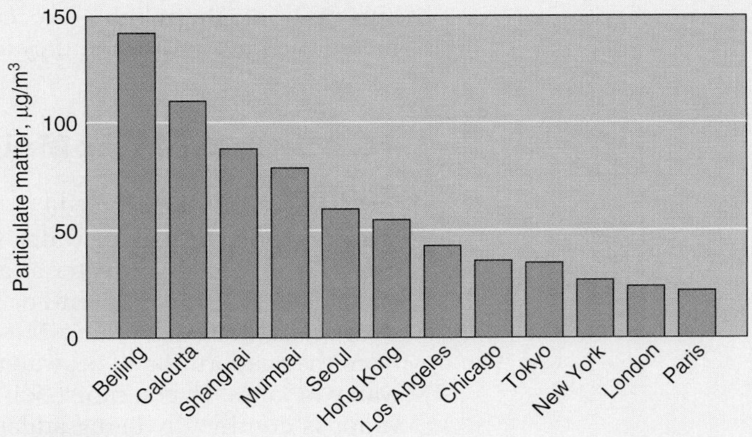

Figure 14-4 Average densities of particulate matter, mainly soot and dust, in the air of various cities. In the United States, densities of over 50 $\mu g/m^3$ (micrograms per cubic meter) are considered unhealthy; the European Union figure is 40 $\mu g/m^3$. The World Health Organization has found that deaths due to cardiopulmonary diseases and lung cancer increase above a density of 10 $\mu g/m^3$.

In the United States, a 1990 amendment to the Clean Air Act required the Environmental Protection Agency to establish regulations that would lead to reducing the emissions of 190 toxic airborne pollutants. According to a U.S. Government Accountability Office report, after 16 years less than a third of the program had been carried out, and the task of setting standards for vehicle emissions was just starting. At the same time, the EPA's own figures showed that 95 percent of the U.S. population had an increased likelihood of developing cancer due to breathing toxic substances in the air.

ozone layer larger in size than North America, and something similar seems to be happening in the Arctic.

By international agreement production of CFCs and HCFCs (and similarly destructive gases that contain bromine) is being phased out and other gases, less harmful to the ozone layer, are replacing them. In the United States, production and import of some CFCs and HCFCs are becoming increasingly restricted, with a complete ban on all of them to come in 2030. But because the changeover is taking place gradually, because air conditioners are being installed at rapid rates in China and India with their huge populations, and because these gases in the atmosphere take an average of 75 years to release their chlorine, the ozone layer will not fully recover for a long time to come.

Mesosphere and Thermosphere The ozone of the stratosphere causes a rise in temperature to 10°C or so in the vicinity of 50 km. At this altitude the atmosphere is $\frac{1}{1000}$ of its density at sea level. The temperature then falls once more to another minimum of about −75°C at 80 km. The portion of the atmosphere between 50 and 80 km is known as the **mesosphere.**

Above 80 km the properties of the atmosphere change radically, for now ions become abundant. The **thermosphere** extends from 80 km to about 600 km, with the temperature increasing to about 2000°C. (We must keep in mind that the density of the thermosphere is extremely low, so that despite the high temperatures a slowly moving object there would not get hot if shielded from direct sunlight.) The thermosphere is the home of the ionosphere, which was described in Sec. 7.9; it is where auroras—"northern lights"—occur (Sec. 18.5) and also where most small meteoroids burn up when they approach the earth (Sec. 17.3).

14.2 Atmospheric Moisture

Another Vital Cycle

Water vapor consists of water molecules that have escaped, or **evaporated,** from a body of water at a temperature below the boiling point of water. The moisture content, or **humidity,** of air refers to the amount of water vapor that it contains. Most of the atmosphere's water vapor comes from the evaporation of seawater. A little also comes from evaporation of water in lakes, rivers, moist soil, and vegetation (Fig. 14-5). Since water vapor is continually being added to air by evaporation and periodically removed by condensation as clouds, fog, rain, and snow, the humidity of the atmosphere varies a lot from day to day and from one region to another. If it were not for the ability of water to evaporate, to be carried by winds, and later to fall to the ground, all the earth's water would be in its oceans and the continents would be lifeless deserts.

We can regard air as a sort of sponge for water vapor, and like an ordinary sponge, air at a given temperature can absorb only so much water and no more. Air is said to be **saturated** when it holds this maximum amount of water vapor. (To be accurate, air has nothing to do with evaporation. Even if there were no air, vapor would still escape from bodies

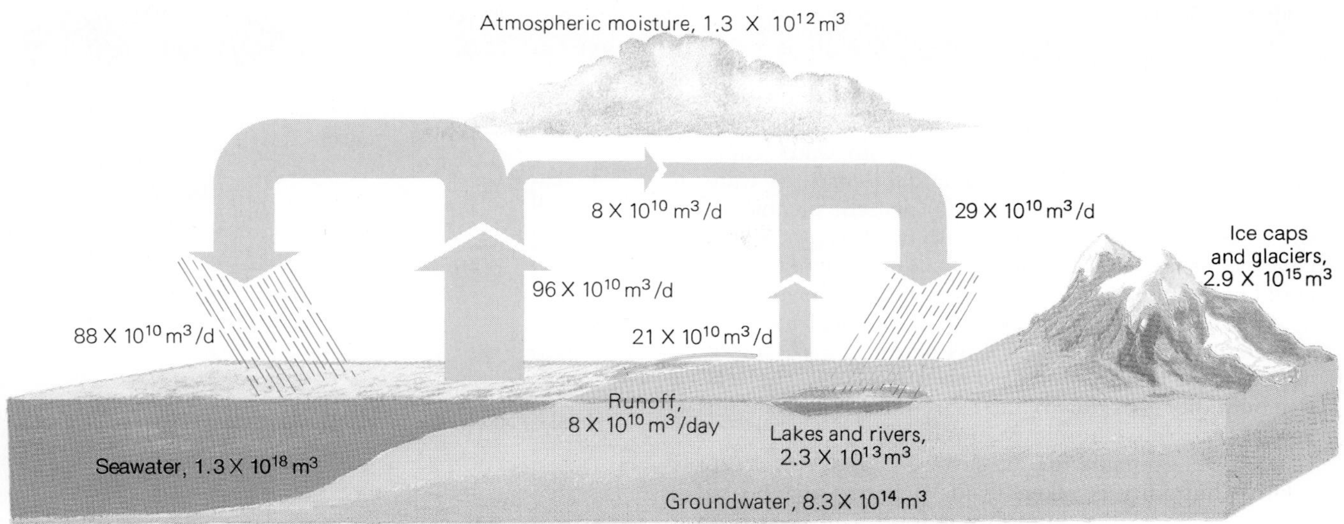

Figure 14-5 The world's water content and its daily cycle. Upward arrows indicate evaporation; downward arrows indicate precipitation. If all the water vapor in the atmosphere were condensed, it would form a liquid layer only about 2.5 cm thick. Rainfall in the United States varies from region to region; the average is about 76 cm per year.

of water. But, since moving air carries water vapor from one region to another and since air is the medium in which water vapor condenses as clouds, fog, rain, or snow, it is convenient to think of the air as "taking up" and "holding" different amounts of vapor.)

We usually describe air as humid if it is saturated or nearly saturated, and as dry if it is highly unsaturated. Humid weather is uncomfortable because little moisture can evaporate from the skin into saturated air, and so perspiration does not produce its usual cooling effect. Very dry air is harmful to the skin and mucous membranes because their moisture evaporates too rapidly to be replaced.

14.3 Clouds

Some Are Water Droplets, Others Are Ice Crystals

When air that contains water vapor is cooled past its saturation point, some of the vapor condenses to a liquid. Dew forms because the temperature of the ground falls at night, which cools the nearby air. Fogs result when large volumes of air come in contact with cold land or water. Clouds come into being when air is cooled by expansion when it rises. Clouds cover about half the earth's surface at any time.

Compressing a gas causes it to heat up, as anyone who has used a tire pump knows. The opposite effect is the cooling of a gas when it expands. When a warm, moist air mass moves upward, it expands because the pressure decreases, and it becomes cooler. A cooling rate of about 0.65°C for each 100 m of rise is normal. If the temperature drop is large enough,

Relative Humidity and Dew Point

Meteorologists express the moisture content of air in terms of **relative humidity,** a percentage that indicates the extent to which air is saturated with water vapor. (The **absolute humidity** is the actual density of water vapor.) A relative humidity of 100 percent means that the air is completely saturated with water vapor, 50 percent means that the air contains half of the maximum it could hold, and 0 percent means perfectly dry air.

The amount of moisture that air can hold increases with temperature (Fig. 14-6). As a result, a sample of air becomes less saturated when it is heated (its relative humidity goes down) and more saturated when it is cooled (its relative humidity goes up).

Suppose outside air at, say, 10°C and 70 percent relative humidity is taken inside a house and heated to 20°C. The relative humidity indoors will then drop to only about 35 percent, even though the actual vapor density stays the same. Such a low relative humidity is not desirable, so a way to add more moisture to heated air in winter should be provided.

On the other hand, if summer air at, say, 30°C and 70 percent relative humidity is cooled, it will reach saturation (100 percent relative humidity) at only 24°C. Further cooling will cause water to condense out. An air-conditioning system should therefore include a way to remove water vapor from the air being cooled to keep the relative humidity at a comfortable level.

The **dew point** of air with a certain water vapor content is the temperature at which the air would be saturated. Suppose the air somewhere contains 9.4 g/m^3 of water vapor. According to Fig. 14-6, air with this vapor content is saturated at 10°C, so the dew point of this air is 10°C. At temperatures above 10°C, the relative humidity of the air is below 100 percent; at 20°C, for instance, it would be 54 percent. If the temperature of the air drops to 10°C or less, however, some moisture will condense out. Dew point can be used as a guide to the possibility of fog. When the air temperature is well above the dew point, the relative humidity is low, and there is no danger of fog. When the air temperature is near the dew point and is decreasing, though, fog is likely.

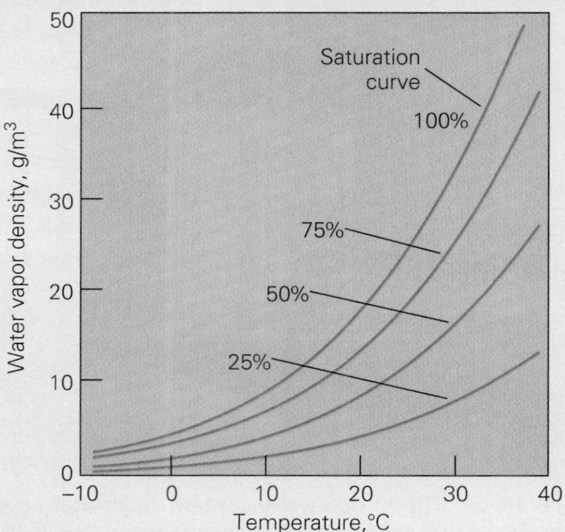

Figure 14-6 How the density of water vapor varies with temperature for various relative humidities. Each curve corresponds to a different relative humidity. The curve for 100% relative humidity represents saturation, when the air can hold no more water vapor.

the air becomes saturated and some of its water vapor condenses on dust specks or other microscopic particles into clouds of tiny water droplets or ice crystals, depending on the temperature, that are small enough to remain suspended aloft indefinitely. High clouds consist of ice crystals, low clouds of water droplets (Fig. 14-7).

The three chief processes in the atmosphere that can cause clouds to form are these.

1. A warm air mass moving horizontally meets a land barrier such as a mountain and rises (Figs. 14-8 and 14-9). Coastal mountains that lie in the paths of moisture-laden ocean winds may have permanent cloud caps over them.
2. An air mass is heated by contact with a warm part of the earth's surface. The air mass expands, and its buoyancy causes it to rise. This process, called **convection,** is discussed in Sec. 14.6 (see Fig. 14-16).

Cirrus clouds consist of ice crystals and occur at high altitudes. They are wispy or featherlike. *Stratus* clouds occur in broad, flat layers. The *cirrostratus* clouds shown combine features of both types. If such clouds grow thicker and darker, a storm is approaching.

Cumulus clouds, such as these, are dense and heaped upwards with flat bottoms and puffy tops. They are common features of fair weather and usually dissipate at night.

Clouds from which rain or snow fall have the word *nimbus* in their names. Shown are *cumulonimbus* clouds, which are dark and heavy with mountainous tops that constantly change shape. Strong winds, rain (sometimes hail), and thunderstorms are associated with them.

A "mackerel sky" like this contains small, round, high-altitude *cirrocumulus* clouds that happen to form patterns that suggest the scales of this fish. Such clouds indicate a change in weather with winds that fluctuate in speed and direction.

A cloud that occurs at a higher altitude than usual for its type is given the prefix *alto*, as in the case of the *altocumulus* clouds shown. Altocumulus clouds look lumpy, like cotton balls, and, if white against a blue sky, indicate fair weather. If they are dark and tightly packed, squalls may follow.

Nimbostratus clouds are low, thick, and relatively shapeless. Their presence is usually accompanied by steady rain or snow.

Figure 14-7 Six common cloud types.

Figure 14-8 Why mountains often have cloud caps over them.

Figure 14-9 Cloud cap over the mountainous island of Saba in the West Indies.

3. A warm air mass meets a cooler air mass and, being less dense, is forced upward over the cooler mass. This process is discussed in Sec. 14.8 (see Fig. 14-28).

Other sources of clouds include water vapor emitted by volcanoes and water vapor in the exhaust gases of aircraft engines. The latter condenses into ice crystals at high altitudes to form streaks of cirrus clouds ("contrails") that may persist for several hours.

Rain, Snow, Sleet, and Hail Rain falls when a cloud (or part of one) is cooled suddenly, usually by a rapid updraft, so that condensation is quick. Some of the water droplets in the cloud become larger than others and because air resistance affects them less, they move within the cloud faster than the smaller droplets. The smaller droplets tend to stick to the larger ones when they come in contact, and the result is larger and larger droplets. Finally the droplets become drops usually 1 to 2 mm in diameter (about the size of the letter "o" in this typeface), 100 times larger than typical cloud droplets. These drops are too heavy to stay aloft and the result is rain. Drops less than 0.5 mm in diameter fall as drizzle.

If a cloud is cold enough, it consists of ice crystals rather than water droplets, and these crystals can grow into snowflakes. As the photographs in Fig. 11-11 show, snowflakes have very open structures. As a result, 7 cm of snow is typically equivalent to only about 1 cm of rain.

Sleet consists of raindrops that have frozen on the way to the ground. **Hail** is associated with the violent up- and downdrafts in thunderclouds. When a raindrop rises high enough, it freezes; then when it falls it picks up more water, which freezes as the drop rises again; and so on. The result is a rocklike lump of ice layered like an onion that may be larger than a golfball (Fig. 14-10).

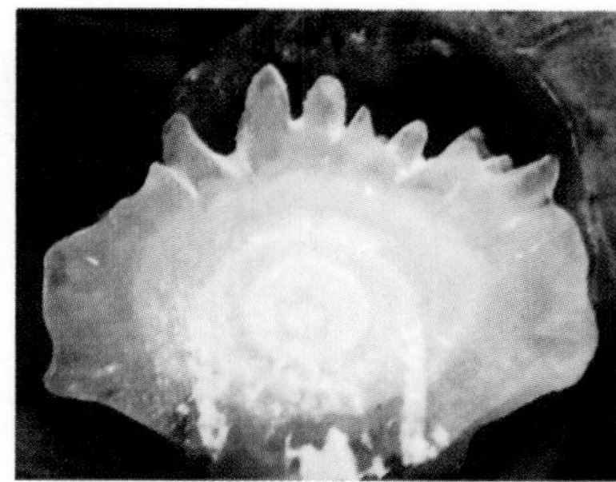

Figure 14-10 Cross section of a hailstone the size of a grapefruit that fell in Kansas in 1970.

Clouds are sometimes **seeded** with silver iodide to induce rain to fall from them. The crystal structure of silver iodide resembles that of ice; hence water molecules and droplets in a cloud can readily become attached to a silver iodide crystal. Such crystals are thus efficient condensation nuclei and so promote rain from a cloud. "Dry ice" (solid carbon dioxide) can also be used to seed clouds. However, the usual reason for not enough rain is not enough clouds of the right kind in the first place, so cloud seeding is seldom of much help in a drought.

Weather

Meteorology, the study of weather and weather patterns, is concerned with what we can think of as a vast air-conditioning system. Our spinning planet is heated strongly at the equator and weakly at the poles, and its water content is concentrated in the great ocean basins. From our point of view, it is the task of the atmosphere to redistribute this heat and moisture so that large areas of the land surface are habitable.

But air-conditioning by the atmosphere is far from perfect. It fails miserably in deserts, on mountaintops, in the polar regions. On sultry midsummer nights or on bitter January mornings we may question its efficiency even in our favored part of the world. Still, the atmosphere does succeed in making a surprisingly large part of the globe fit for people to live in.

Besides regulating air temperature and humidity, we expect the atmosphere to provide us with water in the form of rain or snow. The weather and climate of a region describe how effectively these things are done. **Weather** refers to the temperature, humidity, air pressure, cloudiness, and rainfall (or snowfall) at any given time. **Climate** is a summary of weather conditions over a period of years, including how temperature and rainfall vary with the seasons. For instance, a notable feature of the climate of North Dakota is its extreme warmth in summer and extreme cold in winter, whereas comfortable year-round temperatures with rainfall concentrated in the winter months characterize the climate of southern California. The speeds and directions of winds are often significant in describing weather and climate.

14.4 Atmospheric Energy

A Giant Greenhouse in the Sky

The energy that warms the air, evaporates water, and drives the winds comes to us from the sun. Solar energy arriving at the upper atmosphere

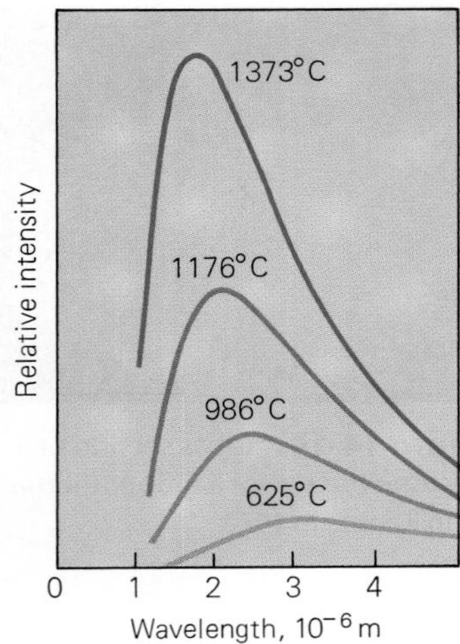

Figure 14-11 Every object gives off electromagnetic (em) radiation. These curves show how the intensity of the radiation varies with wavelength for objects at the temperatures indicated. More energy is given off, and the predominant wavelength decreases, as temperature increases. Em radiation from the sun (see Fig. 18-1) is mostly visible light; em radiation from the earth is mostly infrared light.

is called **insolation** (for *in*coming *sol*ar radi*ation*). Bright sunlight can bring energy at a rate of as much as 1.4 kW to each square meter on which it falls.

Every object gives off electromagnetic waves whose intensity and predominant wavelength depend on the temperature of the object (Fig. 14-11). The hotter the object, the more energy it emits and the shorter the average wavelength. The sun, whose surface temperature is about 5700°C, gives off a great deal of energy and its radiation is mainly visible light. The earth, whose surface temperature averages about 15°C, is a feebler source of energy, and its radiation is mainly in the long-wavelength infrared part of the spectrum to which the eye is not sensitive.

About 30 percent of the insolation is reflected directly back into space, chiefly by clouds (Fig. 14-12). The atmosphere absorbs perhaps 19 percent of the insolation, with ozone, water vapor, and water droplets in clouds taking up most of this amount. Slightly over half of the total insolation therefore reaches the earth's surface, where it is absorbed and becomes heat. A little of this heat is given to the atmosphere through contact with the warm surface, somewhat more by means of water evaporated from the oceans.

The warm earth also radiates energy back into the atmosphere, but the energy now is in the form of long-wavelength infrared radiation. These long waves are readily absorbed by atmospheric carbon dioxide and water vapor, whose molecules then transfer energy to the rest of the atmosphere. Thus a major source of atmospheric energy is radiation from the earth, not direct sunlight.

The way in which the atmosphere is heated from below rather than from above is often called the **greenhouse effect.** The interior of a greenhouse is warmer than the outside air because sunlight can enter through its windows, but the infrared radiation that the warm interior gives off cannot go through glass. As a result much of the incoming energy is trapped inside before being reemitted by the warm windows themselves. The atmosphere thus behaves like a giant greenhouse, with atmospheric gases that absorb infrared radiation acting like the windows of a greenhouse. Normally, incoming and outgoing energy flows are in balance, but an increasingly more efficient atmospheric greenhouse leads to more energy

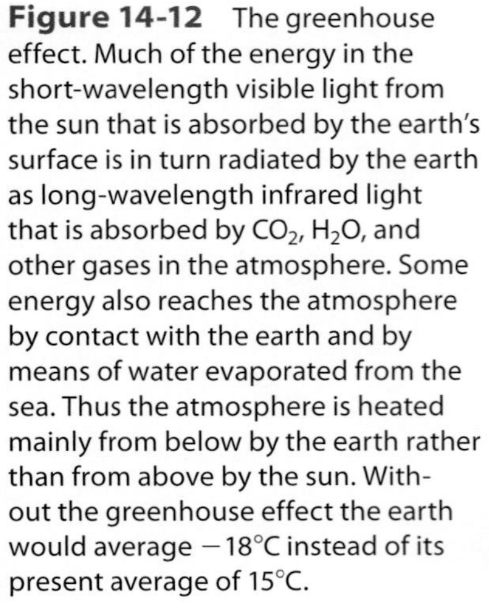

Figure 14-12 The greenhouse effect. Much of the energy in the short-wavelength visible light from the sun that is absorbed by the earth's surface is in turn radiated by the earth as long-wavelength infrared light that is absorbed by CO_2, H_2O, and other gases in the atmosphere. Some energy also reaches the atmosphere by contact with the earth and by means of water evaporated from the sea. Thus the atmosphere is heated mainly from below by the earth rather than from above by the sun. Without the greenhouse effect the earth would average −18°C instead of its present average of 15°C.

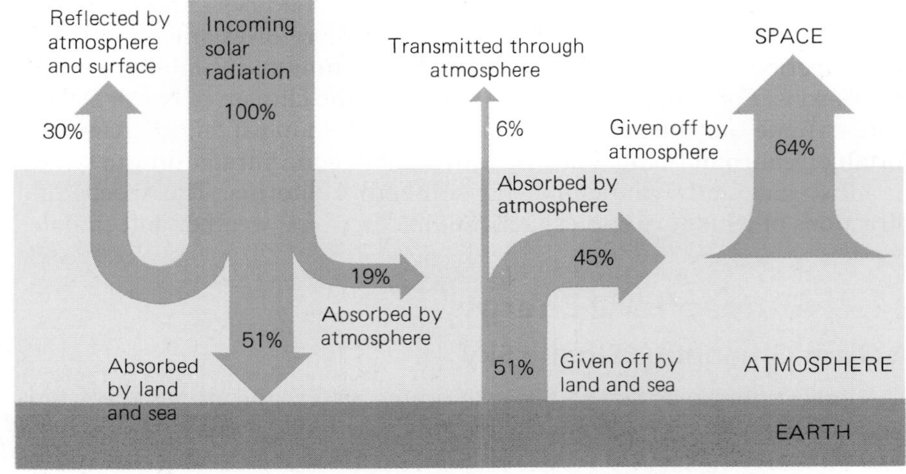

arriving than leaving. As a result the earth is warming up (see Sec. 4.3), but the enormous mass involved means that the earth's surface temperature responds only slowly to changes in energy supply. Eventually the two flows will be in balance again, though at a higher surface temperature.

Example 14.1

Air temperatures in the troposphere decrease with altitude whereas those in the stratosphere increase with altitude. Why do these variations occur?

Solution

The troposphere is heated largely through the absorption by carbon dioxide and water molecules of infrared radiation emitted by the earth's surface. Because it is heated from below, the temperature of the troposphere decreases with altitude.

The stratosphere is heated largely through the absorption by ozone molecules of ultraviolet radiation from the sun. Because it is heated from above, the temperature of the stratosphere increases with altitude.

Factors That Govern Air Temperatures If the earth had no atmosphere, it would grow intensely hot during the day and unbearably cold at night, as the airless moon does. The earth's atmosphere prevents these extremes. The constant movement of air around the world keeps daytime temperatures in any one place from climbing very high, and the ability of moist air to absorb the earth's infrared radiation prevents the rapid escape of heat by night.

How hot the atmosphere becomes over any particular region depends on a number of factors. Air near the equator is on the average much warmer than air near the poles because the sun's rays are more effective in heating the surface when they come from overhead than when they come at a slanting angle (Fig. 14-13). Air over a mountaintop may become warm at midday but cools quickly because it is thinner and contains less carbon dioxide and water vapor than air lower down. Clouds reflect sunlight, so a region with clouds overhead usually has lower air temperatures than a nearby region whose sky is clear.

Because the temperature of water changes more slowly than that of rocks and soil, the atmosphere near large bodies of water is usually cooler by day and warmer by night than the atmosphere over regions far from water. Desert regions commonly show abrupt changes in air temperature between day and night because so little water vapor is there to absorb infrared radiation. The atmospheric temperatures of some regions are influenced profoundly by winds and by ocean currents.

The earth's average temperature does not change very much with time, hence there must be an approximate balance between incoming and outgoing energy. That such a balance does indeed occur can be seen with the help of Fig. 14-14, which shows how the rates at which radiant energy enters and leaves the earth vary with latitude.

More energy arrives at the tropical regions than is lost there, and the opposite is true at the polar regions. Why then do not the tropics grow warmer and warmer while the poles grow colder and colder? The answer is to be found in the motions of air and water that shift energy from the regions of surplus to the regions of deficit. About 80 percent of the

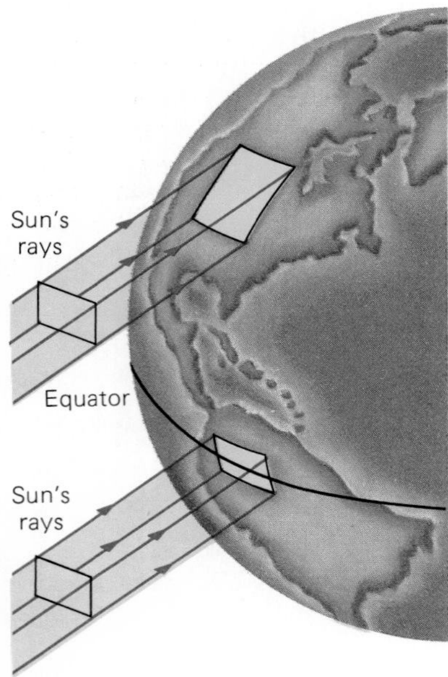

Figure 14-13 The equatorial regions of the earth are on the average warmer than the polar regions because at the equator the sun's rays are spread over a smaller surface.

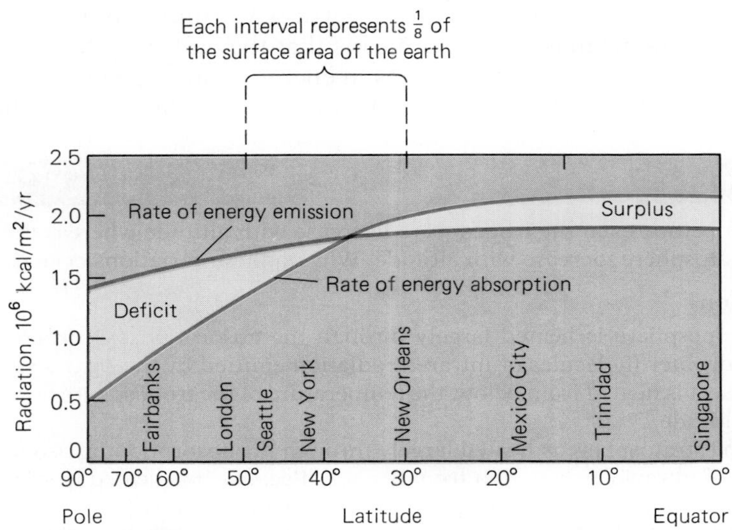

Figure 14-14 The annual balance between incoming solar radiation and outgoing radiation from the earth. More energy is gained than lost in the tropical regions, and more energy is lost than gained in the polar regions. The latitude scale is spaced so that equal horizontal distances on the graph correspond to equal areas of the earth's surface.

energy transport around the earth is carried by winds in the atmosphere, and the remainder is carried by ocean currents. We shall look at both of these mechanisms in the rest of this chapter.

14.5 The Seasons

They Are Due to the Tilt of the Earth's Axis

The average distance between the earth and the sun is about 150 million km. Because the earth's orbit is an ellipse rather than a circle, the earth-sun distance varies during the year from about 2.4 million km closer than the average to the same amount farther away. The earth is nearest the sun early in January and farthest from the sun early in July.

We might be tempted to attribute the seasons to the shape of the earth's orbit, especially if we happen to live in the southern hemisphere where January is a summer month and July a winter month. But this cannot be the reason, if only because the seasons are reversed in the northern hemisphere. In fact, the sunlight that reaches the earth varies in intensity by too little between the orbital extremes to give rise to the difference between summer and winter. After all, the earth's orbit differs from a perfect circle by only ±1.6 percent.

The 23.5° tilt of the earth's axis, not the shape of its orbit, is what causes the seasons. As a result of this tilt, for half of each year one hemisphere receives more direct sunlight than the other hemisphere, and in the other half of the year it receives less (Fig. 14-15). A beam of light that arrives at an angle to a surface delivers less energy per m^2 than does a similar beam that arrives perpendicularly, as we can see in Fig. 14-13.

The noon sun is at its highest in the sky in the northern hemisphere on about June 22 when the North Pole is tilted most toward the sun. The

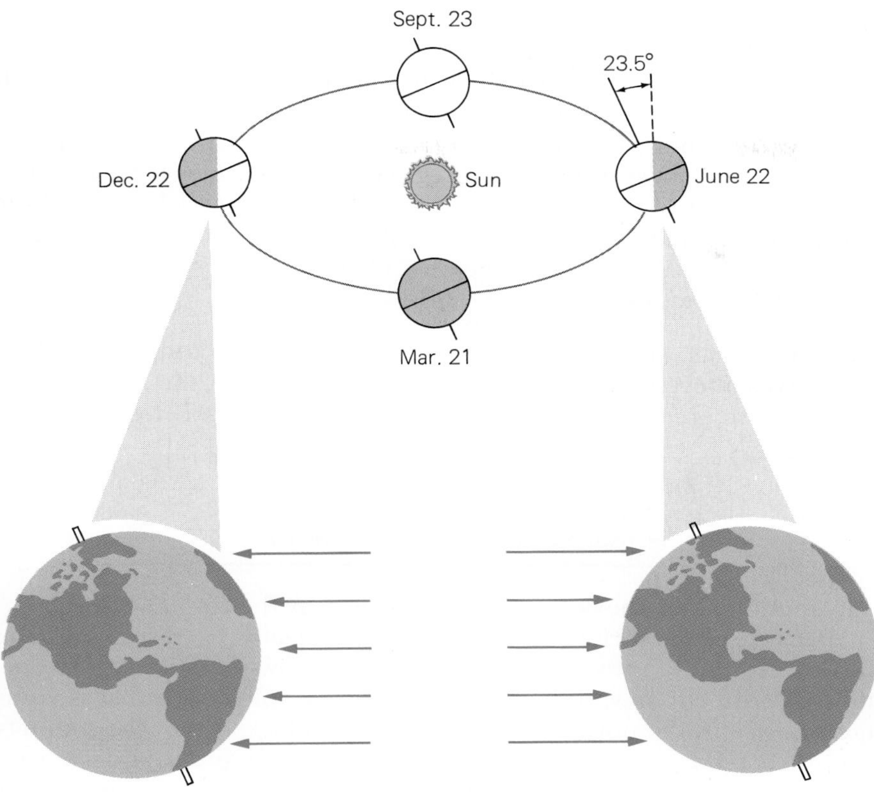

Figure 14-15 The seasons are caused by the tilt of the earth's axis together with its annual orbit around the sun. As a result, the daylight side of the northern hemisphere is tilted away from the sun in January, which means that sunlight strikes this hemisphere at a glancing angle and delivers less energy to a given area than in June. The seasons are reversed in the southern hemisphere.

Example 14.2

If the earth's axis were not tilted, would there still be seasons?

Solution

There would be seasons because the earth's orbit is an ellipse rather than a circle. As a result, the earth's distance from the sun varies during the year, but because the earth's orbit differs from a circle by very little, the seasons would be much less pronounced than the current seasons.

period of daylight in the northern hemisphere is longest on this date. The noon sun is at its lowest 6 months later, on or about December 22, when the North Pole is tilted most away from the sun. The period of daylight is then at its shortest; these times are called **solstices.** In the southern hemisphere the situation is, of course, reversed. On about March 21 and September 23 the sun is directly overhead at noon at the equator. The periods of daylight and darkness are then equal everywhere on the earth; these times are called **equinoxes.**

14.6 Winds

Currents of Air Driven by Temperature Differences

Winds are horizontal movements of air caused by pressure differences in the atmosphere. The greater the pressure difference between two regions,

Arctic and Antarctic Circles

On December 22, the shortest day of the year in the northern hemisphere, the 23°5 tilt of the earth's axis (see Fig.14-15) means that no sunlight reaches any point within 23.5° of the North Pole. The Arctic Circle is the boundary of this region of unbroken darkness. On the same day, which is the logest day of the year in the sourthern hemisphere, there are 24 hours of daylight at all points within 23°5 of the South Pole, and the Antarctic Circle is the boundary of this region of unbroken daylight. On June 22 the situations in the two hemispheres are reversed.

The latitude of the North Pole is 90°N, hence that of the Arctic Circle is 90°N − 23.5° = 66.5°N. The circle passes through northern Canada, Alaska, Siberia, Scandinavia, and southern Greenland. The latitude of the Antarctic Circle is similarly 66.5°S; it lies south of South America and Africa.

the faster the air between them moves. All pressure differences between places on the earth's surface can be traced to temperature differences.

When a region is warmer than its surroundings, the air above it is heated and expands (Fig. 14-16). The hot air rises, leaving behind a low-pressure zone into which cool air from the high-pressure neighborhood flows. The flow toward the heated region at low altitudes is balanced by an outward flow of the air that has risen. This air then cools and sinks to replace the air that has moved inward. Air movements of this kind, produced as the result of unequal heating of the earth's surface, are called **convection currents.**

Coriolis Effect The rotation of the earth affects the path of something that moves above its surface as this path is seen from the surface. This phenomenon is called the **coriolis effect,** and it has these results:

> In the northern hemisphere, a path that would be a straight line over a stationary earth instead is curved to the right. In the southern hemisphere the curvature is to the left.

Only motion along the equator is not affected by the coriolis effect.

Because of the coriolis effect, winds are deflected from straight paths into curved ones. Thus the air rushing into a low-pressure region does not move directly inward but instead follows a spiral path that is counterclockwise in the northern hemisphere and clockwise in the southern (Fig. 14-17). Examples of such spiral motion are, in order of decreasing size (but increasing violence), middle-latitude weather systems, hurricanes and other tropical storms, and tornados.

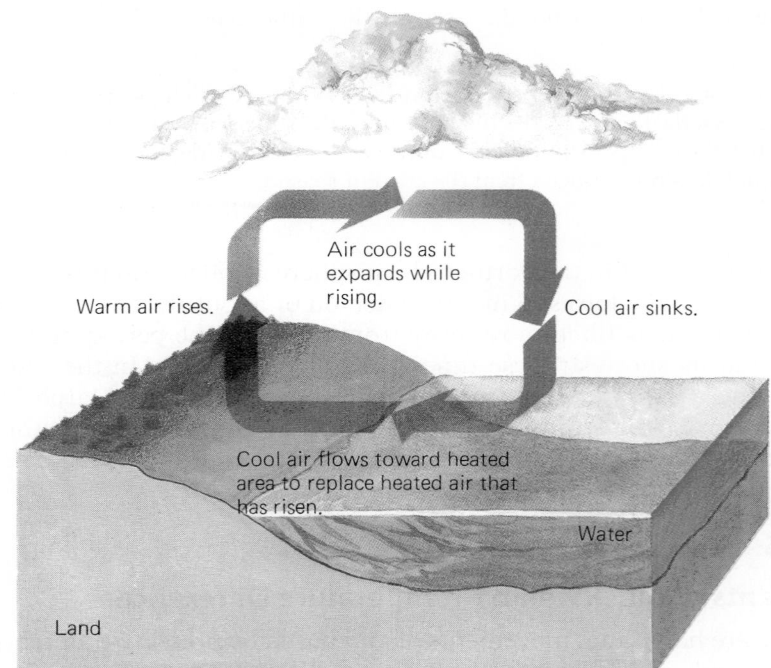

Figure 14-16 Convection currents are produced by unequal heating. The temperature of a land surface rises more rapidly in sunlight than the temperature of a water surface. The resulting convection produces the **sea breeze** found on sunny days near the shores of a body of water. At night, the land cools more than the sea, and the convection current is reversed to give a **land breeze** that blows offshore.

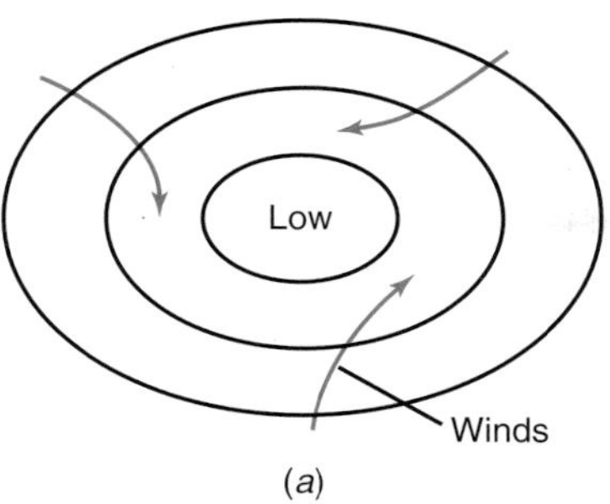

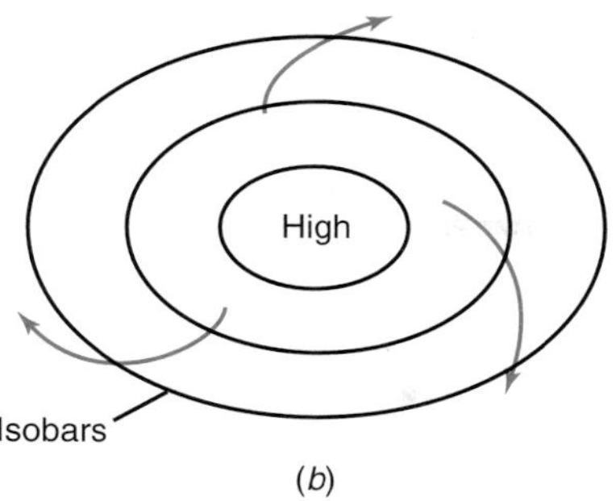

Figure 14-17 (*a*) Because of the coriolis effect, which is a consequence of the earth's rotation, winds in the northern hemisphere are deflected to the right. As a result, air does not flow directly toward the center of a low-pressure region but spirals inward in a counterclockwise direction. (*b*) Similarly, air flows away from the center of a high-pressure region in a clockwise spiral. In the southern hemisphere these directions are reversed. An **isobar** is a line of constant pressure on a weather map; it corresponds to a contour line of constant altitude on an ordinary map.

Monsoons

The seasonal winds called **monsoons** are large-scale sea and land breezes modified by the coriolis effect. During the summer, a continent is warmer than the oceans around it. The rising air over the continent creates a low-pressure region that pulls in moisture-laden sea breezes that bring rain. In winter the situation is reversed, with dry air moving seaward.

Monsoons are most pronounced in certain parts of Africa and Asia. Figure 14-18 shows how the summer and winter monsoons of India and southeast Asia arise. In summer, the motion of air around the low-pressure center on land is counterclockwise, and it is clockwise around the high-pressure center in the Indian Ocean. The result is a wet southwest monsoon that blows from May to September as in Fig. 14-18*a*. Every few years this monsoon is weaker than usual, less rain falls, crops suffer, and there may be widespread famine. From October to April dry winds blow from the northeast, as in Fig. 14-18*b*. The shifts in wind patterns usually take a few weeks but sometimes only a few days. About half the world's population depends on summer monsoons to provide the rain needed for agriculture. One of the reasons global warming is so serious is that, if it continues unabated, the Asian monsoons may no longer occur.

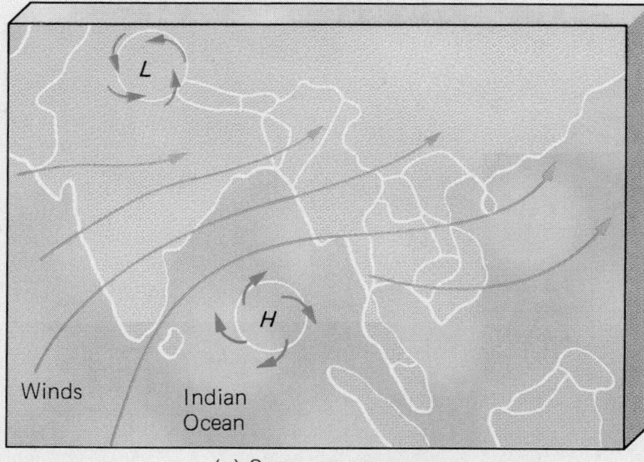

(*a*) Summer monsoon

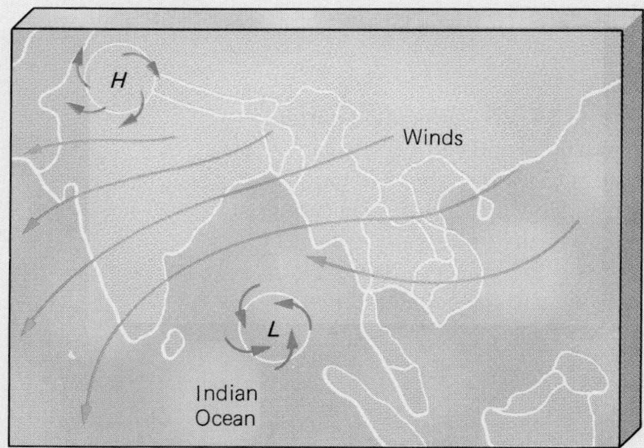

(*b*) Winter monsoon

Figure 14-18 (*a*) The summer monsoon of India and south Asia. Heating of the land produces a low-pressure region (*L*) centered inland and a high-pressure region (*H*) centered in the Indian Ocean that together cause southwest winds to occur. Rice cultivation in this part of the world depends on the warm, moist air brought by the summer monsoon. (*b*) The winter monsoon. Now the land is cooler than the ocean, so the low- and high-pressure regions are reversed to give dry northeast winds.

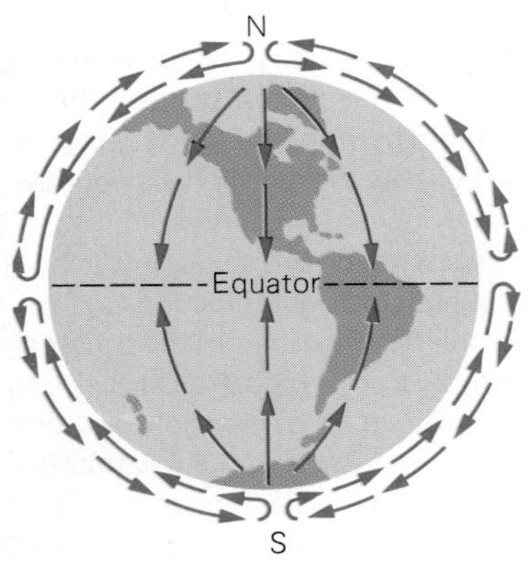

Figure 14-19 The convectional circulation that would occur if the earth did not rotate. The arrows in the center of the diagram indicate surface winds.

14.7 General Circulation of the Atmosphere

Alternate Belts of Wind and Calm

The earth is heated most at the equator and least at the poles. We therefore expect to find convection currents as part of the general atmospheric circulation.

Suppose for the moment that our planet did not rotate and that its surface was made up entirely of either land or water. On such an earth, air circulation would depend only on the difference in temperature between equator and poles. Air would rise along the heated equator, flow at high altitudes toward the poles, and at low altitudes return from the poles toward the equator (Fig. 14-19). We in the northern hemisphere would experience a steady north wind. (Winds are named for the direction they come from, so a north wind blows from the north.) Around the equator would be a belt of relatively low pressure, and near each pole a region of high pressure.

Because the earth does rotate, however, the north and south winds coming from the poles are deflected by the coriolis effect into large-scale eddies that lead to a generally eastward drift in the middle of each hemisphere and a westward drift in the tropics. The main features of the general circulation of the atmosphere are shown in Fig. 14-20.

The various wind zones were important to shipping in the days of sail, as their names indicate. Thus the steady easterlies on each side of the equator became known as the **trade winds** because they could be relied upon by sailing ships. The region of light, erratic wind along the equator, where the principal movement of air is upward, constitutes the **doldrums.** The **horse latitudes** that separate the trade winds in both hemispheres from the **prevailing westerlies** poleward of them are also regions of calm and light winds. Their name is supposed to have come from the practice of throwing overboard horses from sailing ships when the ships were becalmed and ran short of drinking water.

Jet Streams With increasing altitude the belts of westerly winds broaden until almost the entire flow of air is west to east at the top of the troposphere.

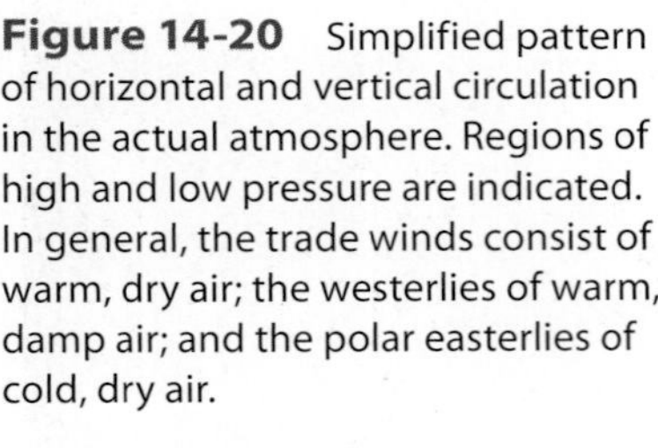

Figure 14-20 Simplified pattern of horizontal and vertical circulation in the actual atmosphere. Regions of high and low pressure are indicated. In general, the trade winds consist of warm, dry air; the westerlies of warm, damp air; and the polar easterlies of cold, dry air.

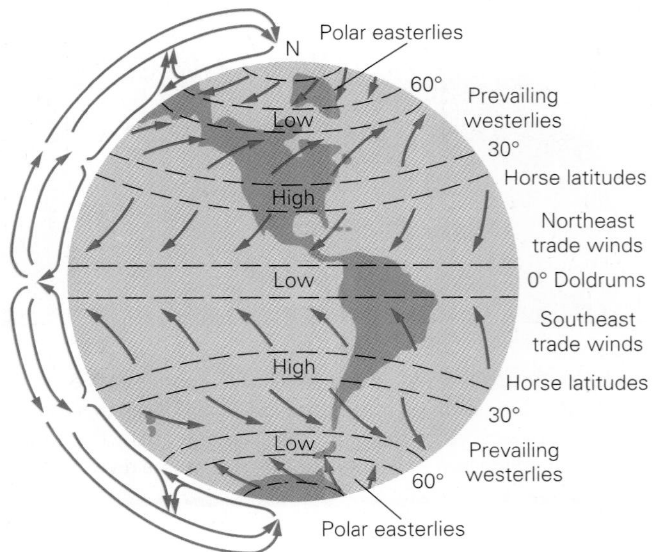

Example 14.3

The crew of a yacht is planning to sail from the east coast of the United States to England and later to sail home. What routes should the crew follow across the Atlantic on the passages out and back to have the courses more or less downwind as much of the time as possible?

Solution

To England, the crew should first sail northeast to reach the prevailing westerlies and then head eastward. Back to the United States, the crew should first sail south to the trade winds, then west, and finally north.

The westerly flow aloft is not uniform but contains narrow cores of high-speed winds called **jet streams,** usually only a few hundred kilometers wide but with winds of up to 500 km/h. The jet streams form zigzag patterns around the earth that change continuously and affect the variable weather of the middle latitudes by their influence on the paths of air masses closer to the surface. They are important to aircraft—an hour or more can be saved on a long west-east flight by using a route with a jet stream as a tail wind, and a corresponding addition to the flight time can be prevented in the other direction by avoiding a jet stream as much as possible.

At any given time the circulation near the earth's surface is more complicated than the pattern given in Fig. 14-20. An important factor is the presence of large seasonal low- and high-pressure regions caused by unequal heating due to the irregular distribution of landmasses and sea areas (Fig. 14-21). Smaller, short-lived cells also occur and profoundly affect local weather conditions, as we shall see next.

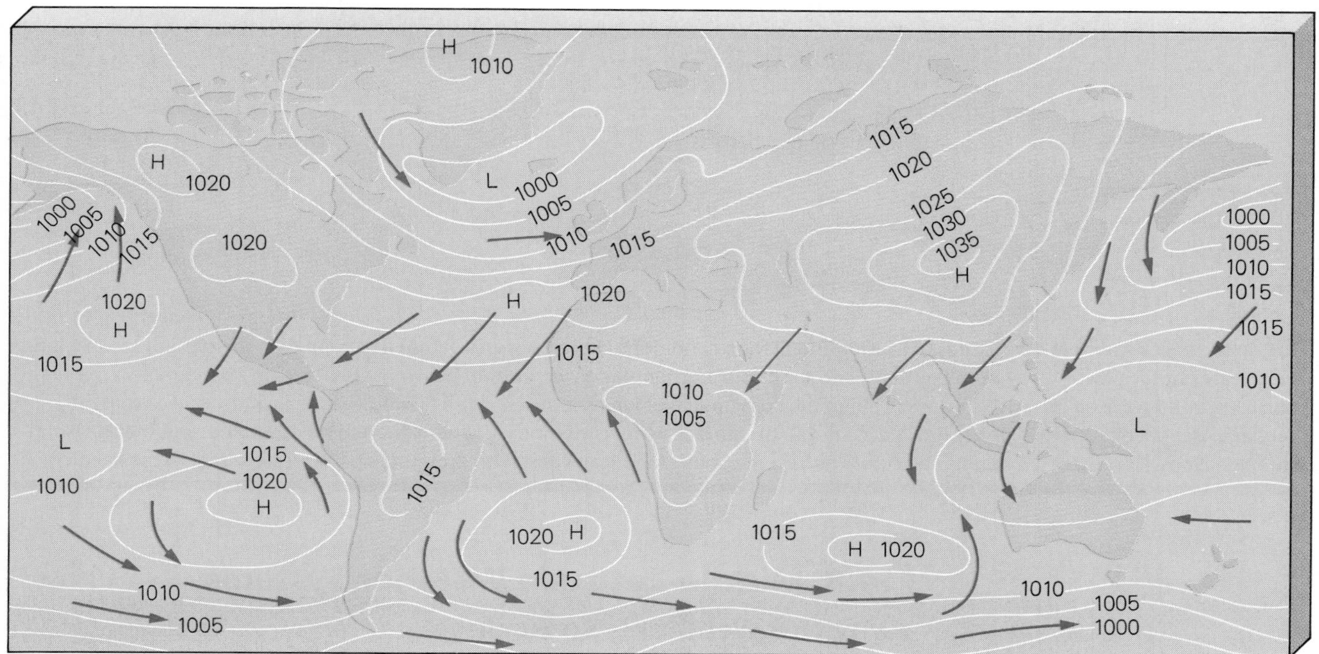

Figure 14-21 Average January sea-level pressures (in millibars) and wind patterns (red arrows). High- and low-pressure systems are indicated. Isobars connecting points of equal pressure are shown in white.

El Niño and La Niña

The surface winds of the tropical Pacific often blow westward from the Americas, as in the top part of Fig. 14-22. These are the easterly trade winds shown in Fig. 14-20. Together with the ocean currents they drive, the trade winds create a huge pool of warm water in the western side of the Pacific basin. The heated moist air above this pool rises to give an area of low pressure and brings abundant rain to Indonesia and northern Australia.

Every few years, however, a different pattern takes over. For reasons that are not completely understood, the trade winds weaken and the warm water pool shifts eastward. Now there is high pressure over the western Pacific and droughts parch Australia and East Asia. Low pressure over the eastern Pacific brings storms that lash California and torrential rains that soak Chile, Peru, and Ecuador on South America's west coast (Fig. 14-22). The upwelling of nutrient-rich cold water along this coast when the surface waters were blown westward stops and the coastal water becomes warm. As a result the fish population drops sharply in the eastern Pacific and the economies of Peru and Ecuador suffer accordingly. The entire phenomenon is called **El Niño,** Spanish for "the little boy," a reference to the infant Jesus, because the warm water appears around Christmas. The other pattern is called **La Niña,** "the little girl."

The influence of an El Niño can extend well beyond the Pacific, in part by disrupting the jet streams. The severe El Niño of 1982–1983 brought droughts that devastated agriculture in India, southern Africa, and Brazil, and abnormally heavy rains gave rise to floods in California. All the continents were affected in some way. In North America milder winters coincide with El Niños, and there are fewer tropical storms and hurricanes in the Atlantic.

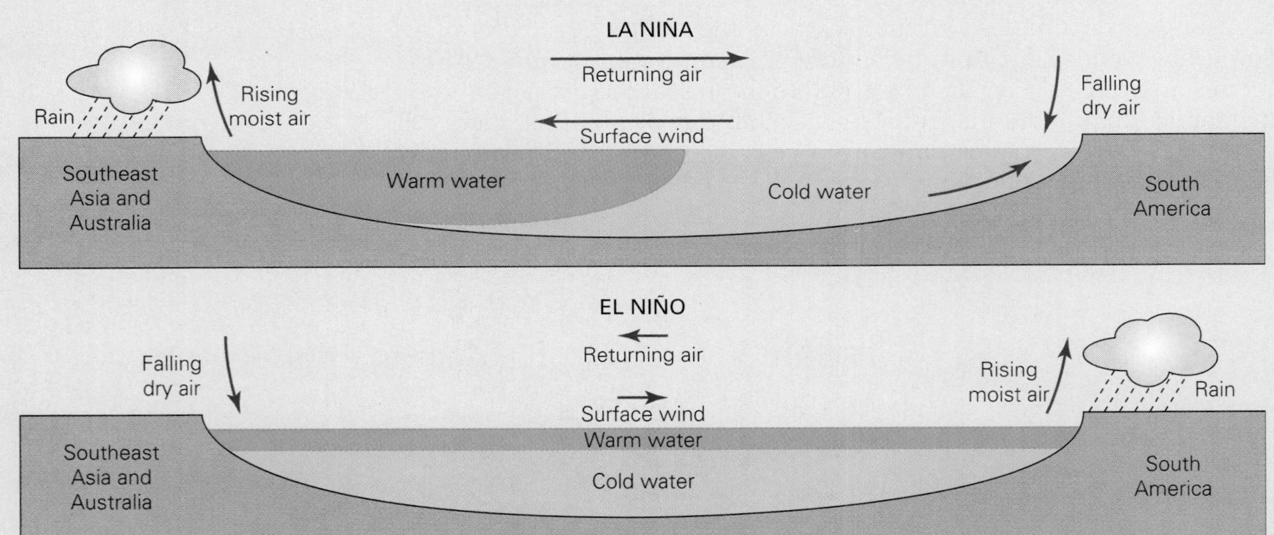

Figure 14-22 Simplified pictures of La Niña and El Niño conditions in the Pacific Ocean. In a La Niña year, the easterly trade winds push warm surface water away from South America and cold, nutrient-laden water from below rises to feed an abundant fish population. Descending dry air produces droughts along the coast, much of which is desert. In an El Niño year, the winds weaken and may vary in direction, or even reverse as shown. As a result, the surface water is warm everywhere, there is no upwelling, and the fish population declines. Now there are storms and heavy rain on the South American coast with droughts on the western side of the Pacific.

14.8 Middle-Latitude Weather Systems

Why Our Weather Is So Fickle

Day-to-day weather is more variable in the middle latitudes than anywhere else on earth. If we visit central Mexico or Hawaii, in the belt of the northeast trades, we find that one day follows another with hardly any change in temperature, humidity, or wind direction. On the other hand, in nearly all parts of the continental United States, drastic changes in weather are commonplace. The reason for this variability lies in the movement of warm and cold air masses and of storms derived from them through the belts of the westerlies.

In the northern part of the westerly belt an irregular boundary separates air moving generally northward from the horse latitudes and air moving southward from the polar regions. Great bodies of cold air at times sweep down over North America, and at other times warm air from the tropics extends far northward. The cold air is ultimately warmed and the warm air cooled, but a large volume of air can maintain nearly its original temperature and humidity for days or weeks.

These huge tongues of air, or isolated bodies of air detached from them, are the **air masses** of meteorology. The kind of air in an air mass depends on its source (Fig. 14-23). A mass formed over northern Canada is cold and dry, one from the North Atlantic or North Pacific is cold and humid, one from the Gulf of Mexico warm and humid, and so on. Weather prediction in the United States depends largely on following the movements of air masses from these various source areas.

Cyclones and Anticyclones Weather systems associated with air masses are usually several hundred to a thousand or more km across and move from west to east. A **cyclone** is an air mass in which the pressure is low at the center. As air rushes in toward the center of a cyclone, the

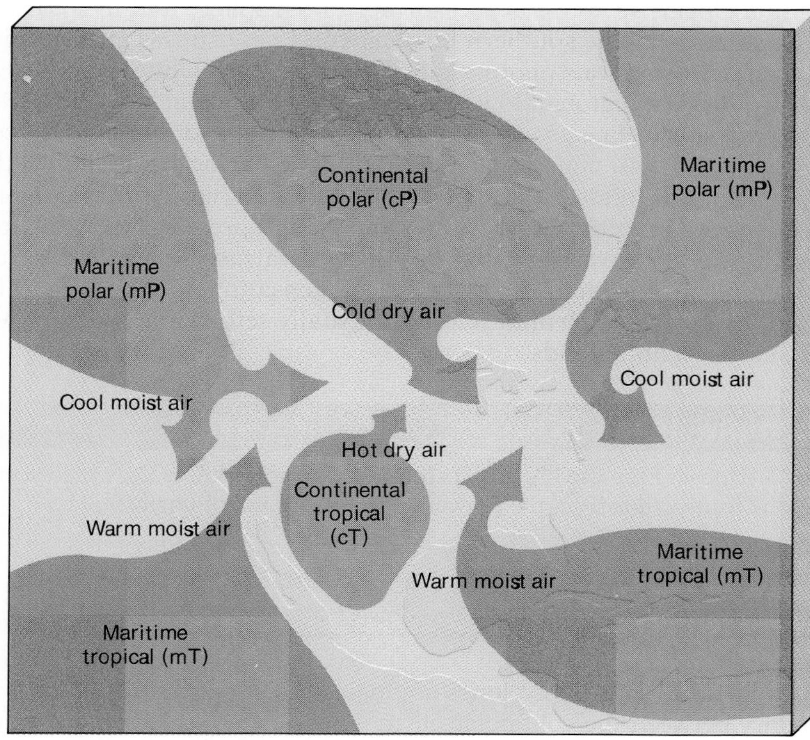

Figure 14-23 The air masses that affect weather in North America. The importance of the various air masses depends upon the season. In winter, for instance, the continental tropical air mass disappears and the continental polar air mass exerts its greatest influence.

Figure 14-24 Cyclonic weather systems are responsible for the variable weather of the middle latitudes. A typical system is 1500 km in diameter and moves eastward at 40 km/h; its characteristic winds usually do not exceed 65 km/h. The system here was centered about 2000 km north of Hawaii when it was photographed from a spacecraft.

moving air is deflected toward the right in the northern hemisphere and toward the left in the southern because of the coriolis effect (Fig. 14-17). As a result cyclonic winds blow in a counterclockwise spiral in the northern hemisphere and in a clockwise spiral in the southern hemisphere (Fig. 14-24).

An **anticyclone** is centered on a high-pressure region from which air moves outward. The coriolis effect therefore causes anticyclonic winds to blow in a clockwise spiral in the northern hemisphere and in a counterclockwise spiral in the southern hemisphere. These spirals are conspicuous in cloud formations photographed from earth satellites.

A cyclone is a region of low pressure, and air flowing into one rises in an upward spiral (Fig. 14-25). The rising air cools and its moisture content condenses into clouds. As a rule, cyclones bring unstable weather conditions with clouds, rain, strong shifting winds, and abrupt temperature changes. An anticyclone is a region of high pressure, and air flows out of it in a downward spiral. The descent warms the air and its relative humidity accordingly drops, hence condensation does not occur. The weather associated with anticyclones is usually settled and pleasant with clear skies and mild winds.

Example 14.4

Why are most of the world's deserts found in the horse latitudes, which separate the trade winds from the prevailing westerlies in both hemispheres?

Solution

As shown in Fig. 14-20, airflow in the horse latitudes is anticyclonic, which means that little rain falls there.

Fronts Middle-latitude cyclones originate at the **polar front,** which is the boundary between the cold polar air mass and the warmed air mass next to it. It is common for a kink to develop in this front with a wedge of warm

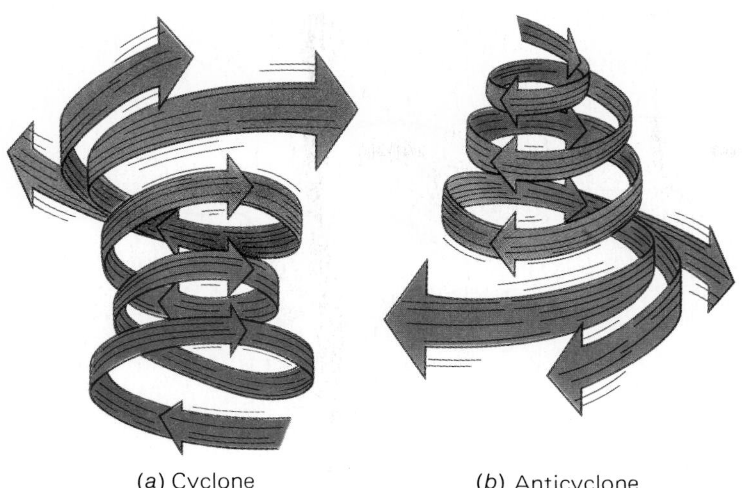

Figure 14-25 (*a*) In a cyclone in the northern hemisphere, the coriolis effect causes surface air to spiral counterclockwise as it moves toward the low-pressure center and rises. The air cools as it moves upward, which causes its moisture to condense out as clouds and rain. (*b*) In an anticyclone, cold air from above spirals outward from the high-pressure center as it sinks. The falling air warms, which decreases its relative humidity to give a clear sky. In the southern hemisphere the directions of spin are reversed.

Tornados

A tornado is a narrow but extremely violent cyclonic storm (Fig. 14-26). In the central United States, particularly in Kansas, Oklahoma, and Texas, about a thousand of them occur every year. Tornados typically form in the spring when warm, moist air moving north from the Gulf of Mexico collides with cold, dry air moving south from Canada (see Fig. 14-23). The warm air rises through the cold air to create a strong updraft that can power a severe thunderstorm. If the rising air in the thundercloud is swirling rapidly, the result may be a tornado. Waterspouts are weak tornados that form over open water.

Most tornados are about 50 m across, travel at about 50 km/h, and last for only a few minutes. Severe tornados may be over a kilometer across, travel at 100 km/h, and last for over an hour. Wind speeds up to 512 km/h (318 mi/h) have been recorded in especially violent tornados, and debris is sometimes found over 100 km from where the twister struck. In fact, much of the damage tornados cause is due to churned-up debris shredding whatever it encounters.

Figure 14-26 A tornado in Oklahoma. The spinning column of air spreads out as it rises because air pressure decreases with altitude.

air protruding into the cold air mass. This produces a low-pressure region that moves eastward as a cyclone. The eastern side of the warm-air wedge is a **warm front** since warm air moves in to replace cold air in its path; the western side is a **cold front** since cold air replaces warm air (Fig. 14-27).

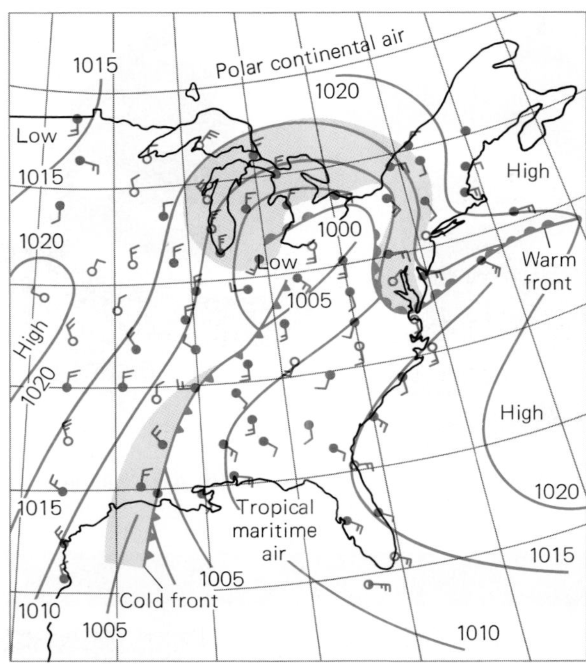

Figure 14-27 Weather maps show pressure patterns, winds, rain, and snow. This is a weather map of the eastern United States one April morning. A cold air mass on the west and north (polar continental air) is separated from a warm air mass (tropical maritime air) by a cold front extending from Louisiana to Michigan and by a warm front from Michigan to Virginia. Where the north end of the warm air mass lies between the two fronts a cyclone has formed, bringing rain (colored area) to the Great Lakes region. The unit of pressure in this map is the millibar. The small circles indicate clear skies; solid dots indicate cloudy skies. The small lines show wind direction, which is toward the circle or dot, and wind strength; the greater the number of tails, the faster the wind.

As warm air rises along an inclined front, it is cooled and part of its moisture condenses out. Clouds and rain are therefore associated with both kinds of fronts (Fig. 14-28). A cold front is generally steeper, since cold air is actively burrowing under warm air. The temperature difference is greater as well, so rainfall on a cold front is heavier and of shorter duration than on a warm front. A cold front with a large temperature difference is often marked by violent thundersqualls.

BIOGRAPHY Jacob A. B. Bjerknes (1897–1975)

The son and grandson of physics professors at Norway's University of Christiania (later Oslo, renamed after Norway became independent of Sweden in 1905), Bjerknes began his work in meteorology in collaboration with his father, Vilhelm. They established a network of observing stations all over Norway during World War I, and from the data they developed the concept of air masses that keep their identities for relatively long periods of time. They called the boundaries between air masses "fronts."

Bjerknes studied how fronts evolve and applied his findings to weather forecasting. In 1940 he joined the University of California at Los Angeles, where he extended his analysis of weather phenomena to include heat exchange between the atmosphere and the oceans. An especial interest was the phenomenon of El Niño. In 1952 Bjerknes pioneered the use of cloud photographs taken from rockets as an aid in forecasting. Today, of course, satellite-borne cameras are a routine tool of the meteorologist.

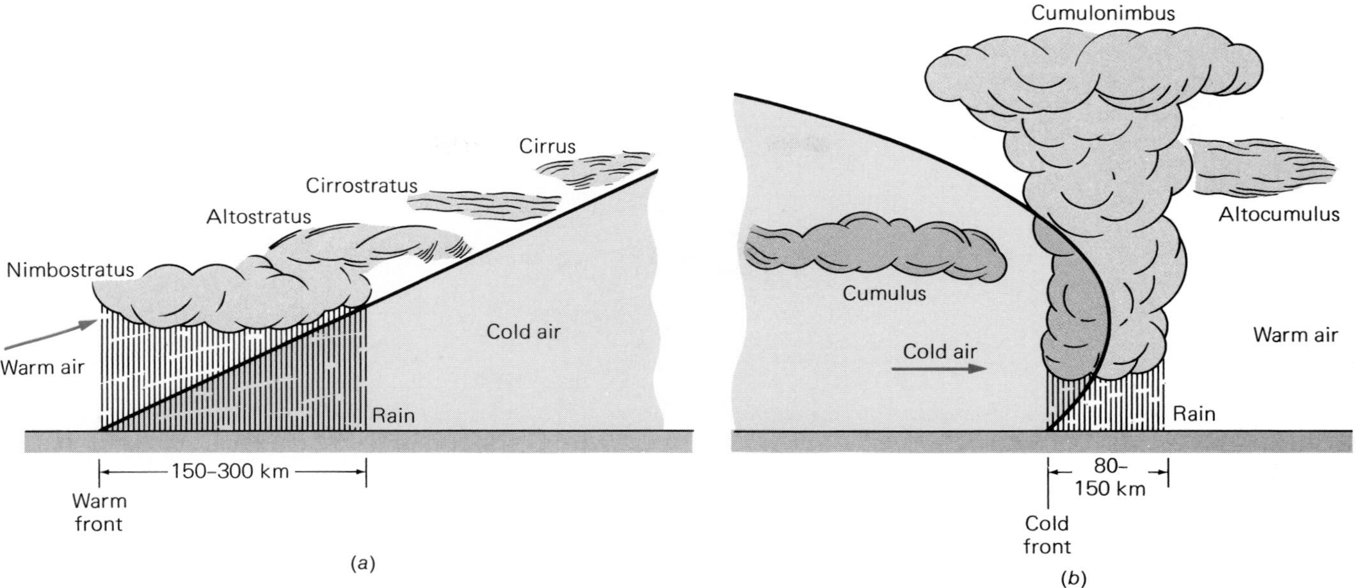

Figure 14-28 Cross-section diagrams of (*a*) a warm front and (*b*) a cold front. In each case the front is moving to the right. Photographs of the various cloud forms appear in Fig. 14-7. A warm front is the boundary between a warm, moist air mass that has moved (usually eastward) up and over a cool air mass ahead of it. The rising air cools, and its water vapor condenses into clouds and rain. A cold front is the boundary between a cool, relatively dry air mass that pushes warm, moist air ahead of it upward and out of its (usually eastward) path; it is steeper than a warm front. Again the rising warm air cools to produce clouds and rain, but now the weather change is more rapid and dramatic with stronger winds.

Weather Forecasting

In order to predict the weather for a certain region, it is necessary to know present and past conditions in detail over a much larger area. Data on air pressure, temperature, humidity, cloud cover, precipitation, and winds are collected every 3 hours at about 10,000 weather stations around the world and are available to forecasters everywhere in a standard code. Additional information is provided by weather satellites that look at the atmosphere from above and by balloon-borne instruments that make measurements at various altitudes. In both cases the observations are radioed to ground stations.

The first step in preparing a forecast is to draw a weather map, called a **synoptic** map, that shows the current situation. By comparing this map with the synoptic maps for the past day or two, a meteorologist can see how the various weather systems are developing and moving and may be able to spot new ones in their early stages. Then a **prognostic** map is prepared that shows for a future time the weather pattern the meteorologist thinks will have evolved from the current situation. The prognostic map indicates the weather that can be expected at that time.

Nowadays computers are used to digest the vast amounts of data that go into synoptic maps and also to help predict what will happen next. People are still essential, though: an educated guess by someone familiar with local weather may be more reliable than the calculations of a supercomputer. Forecasts for a day ahead are usually quite accurate, those for longer periods in the future less so. Even with all the resources available to meteorologists today, some phenomena, such as hurricanes, do not always behave as expected.

Are more accurate long-range forecasts a realistic prospect? The trouble is that the atmosphere is a very complex system and it interacts with land and sea in complicated ways. Although weather exhibits many regularities, even small differences in initial conditions may lead to entirely different outcomes. Because, to give a famous example, a tornado in Texas could, in principle, be triggered by the fluttering of a butterfly's wings in Brazil, some element of uncertainty will always exist in weather prediction.

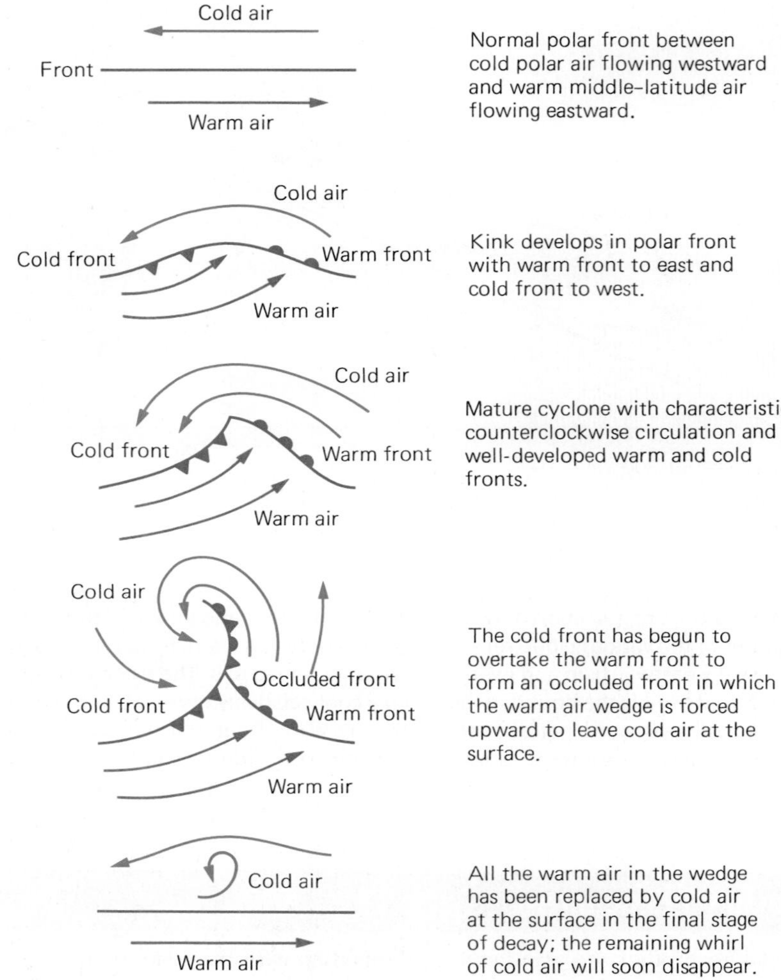

Figure 14-29 Life cycle of a middle-latitude cyclone in the northern hemisphere. Conventional weather-map symbols are used for cold, warm, and occluded fronts.

A warm front typically moves eastward at about 150 km/day. The cold front behind it moves faster, up to twice as fast, and eventually it overtakes the warm front to force the wedge of warm air upward (Fig. 14-29). The resulting **occluded front** is the last stage in the evolution of a cyclone, which soon afterward disappears. The total life span of a middle-latitude cyclone may be as little as a few hours or as much as a week, though the usual range is 3 to 5 days.

Climate

The climate of a region refers both to its average weather over a period of years and to the typical variations in its weather elements during each day and during each year. The most significant weather elements in determining climate are temperature and precipitation. Climates differ considerably around the world, ranging from the tropics, where there is no winter, to the polar regions, where summer is brief.

Tropical Cyclones

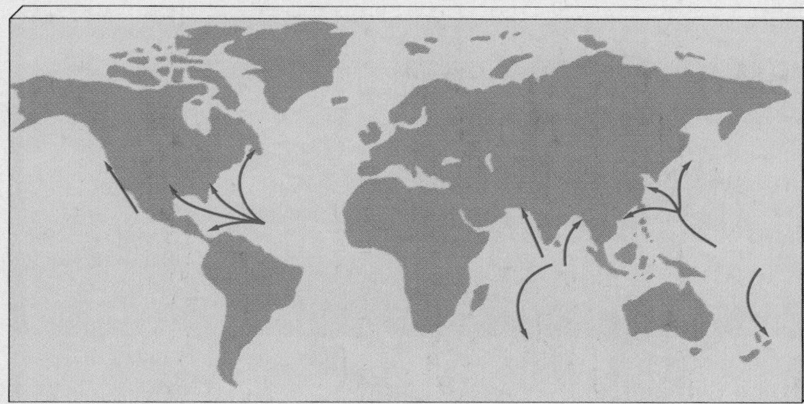

Figure 14-30 Typical storm tracks of tropical cyclones.

Compared with the cyclonic weather systems of the middle latitudes, tropical cyclones are less common—only 80 or 90 occur worldwide in an average year—but are much more violent with half of them having winds that exceed 120 km/h (75 mi/h). The hurricanes that several times a year batter the West Indies and the eastern coast of the United States are tropical cyclones, as are the typhoons of the western Pacific Ocean and the cyclones of the Indian Ocean (Fig. 14-30). A mature hurricane, one of the most powerful events on earth, has as much energy as all the electric energy used in the United States in 3 or 4 years.

A tropical cyclone begins as a small area of low pressure over warm (above 26.5°C) ocean water near the equator in late summer or early fall. Moist air flows into the depression and cools as it spirals upward (see Fig. 14-25*a*). The water vapor in this air condenses into cloud and rain droplets, a process that liberates heat; this is the reverse of the vaporization process described in Sec. 5.11, and it is what powers the cyclone. The heat given off causes the air pressure to drop further, which leads to more moisture from the ocean surface being sucked into the rising, spinning air column. This self-reinforcing sequence produces an increasingly intense storm that may end up as much as 2000 km across whose raging winds whirl around a calm "eye" of low pressure that is usually 20 to 100 km across (Fig. 14-31). The lowest surface pressure ever measured, 870 mb, occurred in the eye of a Pacific typhoon in 1979. Outside the eye rain falls in torrents. The most vigorous phase of these storms lasts for a few days to a week or so.

A typical North Atlantic hurricane originates near the bulge of Africa and gathers strength as it is swept westward by the trade winds. Then it turns north to swing around the permanent high-pressure region of the mid-Atlantic, and finally is blown eastward by the westerlies of the higher latitudes until it weakens and dies out without warm seawater to feed it energy. Hurricanes usually move at 15 to 50 km/h but may remain stationary for a day or two. Some of them, instead of staying at sea, curve across the eastern or southern United States, where their winds may do immense damage inland. Seawater surging ashore to batter low-lying coastal areas is even more destructive This was the case when Hurricane Katrina struck New Orleans in August 2005. The storm drove flood waters past inadequate earthen levees and concrete walls to devastate much of the low-lying city and leave hundreds of thousands of people homeless (Fig. 14-32).

Figure 14-31 This satellite photograph of Hurricane Katrina in the Gulf of Mexico was taken on August 28, 2005, when it had grown to category five with sustained winds of 260 km/h (160 mi/h), over twice the wind speeds of category one hurricanes.

(*Continued*)

Tropical Cyclones (continued)

Katrina was only one of the 26 tropical storms to brew in the North Atlantic in 2005, then the largest number in any year since records began in 1928. Fourteen of the storms grew into hurricanes. Although Katrina did the most damage, Wilma, which came in October, was the strongest hurricane on the books. Is global warming (see Sec. 4.3), which brings higher temperatures to ocean waters that then feed ever more energy to tropical storms, responsible for the rising frequency and violence of these storms? Or are the changes part of a natural variation that has been through previous decades-long cycles of more and less storm activity? The consensus among atmospheric scientists is that, while the growing number of storms per year probably cannot be blamed on global warming, their increasing ferocity probably can. Not good news for coastal dwellers.

Figure 14-32 New Orleans after Hurricane Katrina in 2005. The flood defenses were known not to be robust enough for such an assault; the savings were a fraction of the cost of the disaster, which had been predicted. And, of course, there is the question of how wise it was in the first place to develop so important a city nearly 10 m below sea level in a vulnerable location almost surrounded by water.

14.9 Tropical Climates

Hot and Wet or Hot and Dry

The equatorial belt of the doldrums, with its rapid evaporation and strong rising air currents, provides an ideal situation for abundant rain. Throughout the year the weather is hot with almost daily rains and light, changeable winds. The steaming rain forests of Africa, South America, and the East Indies occur where this belt crosses land (Fig. 14-33).

The horse latitudes, roughly 30° north and south of the equator, are also belts of little wind, but their climate is anything but humid. Air in these belts moves downward to become warmer and hence less saturated with water vapor. Thus the climate is dry, with clouds and rain only at long intervals. In a few areas, however, such as the Gulf Coast of the United States, the prevailing dryness is modified by moisture-laden winds of local origin.

Air returning to the warm equatorial belt from the horse latitudes tends to keep the little moisture it contains, except where a mountain range or strong convection currents force the air upward. Hence the trade-wind belts are for the most part dry regions. Seasonal shifts of the trade-wind belts give rainfall during part of the year to the equatorial margins of these belts. The poleward portions of the trade-wind belts, together with the adjacent horse latitudes, are the regions of the world's great deserts—the Sahara, the deserts of South Africa, the dry districts of Mexico and northern Chile, the dry interior of Australia.

14.10 Middle-Latitude Climates

Variety Is the Rule

The belts of prevailing westerlies generally have moderate average temperatures, although continental interiors show great variations with the seasons. On the other hand, oceanic islands and the western coasts of continents in these belts have even temperatures throughout the year.

In the northern hemisphere the huge landmasses of North America and Eurasia bring complications that are well illustrated by climates in the United States. Damp winds from the Pacific Ocean are forced upward by mountain ranges along the West Coast, and the western sides of the mountains therefore receive abundant rainfall. Once across the mountain barriers the westerlies have little remaining moisture, so that the region from the mountains east to the Great Plains is largely dry.

If the westerlies kept their direction as steadily as do the trade winds, dry conditions would continue across the continent to the East Coast. Instead, the cyclonic storms characteristic of this belt often bring moisture-laden air from the Gulf of Mexico and the Atlantic Ocean into the Mississippi Valley and the eastern states. Rainfall increases eastward across the country, becoming very abundant along the Gulf of Mexico.

Temperatures on the West Coast are conditioned by the prevailing wind from the ocean and change relatively little from season to season. In most other parts of the country, however, the difference between summer and winter is very marked. The fine climates of Florida and southern California owe their mildness to nearby warm oceans and to their locations near the junction of the belt of the westerlies and the horse latitudes.

In the bleak arctic and antarctic regions, summer warmth is a brief respite from the deep cold of the rest of the year. Moderate winds are the rule, although violent gales occur at times. The total amount of snow during the year is small simply because the low temperatures prevent the accumulation of much water vapor in the air.

Figure 14-33 Year-round high temperatures and abundant rainfall are characteristic of the equatorial regions and encourage plant growth. This rain forest is in Costa Rica.

Why Is the South Pole Colder Than the North Pole?

The main reason is the altitude difference between the North Pole region (a winter average of −30°C), which consists of sea ice floating on the Arctic Ocean, and the South Pole region (−60°C), which is over 2 km high on the world's loftiest continent. The water of the Arctic Ocean is also a more effective heat reservoir than the rock and ice of Antarctica. A third factor is the presence of tall mountain ranges in Antarctica that prevent warm air from the middle latitudes from reaching its interior.

14.11 Climate Change

An Icy Past, a Warm Future

Weather we expect to vary, both from day to day and from season to season. Nor are we surprised when one year has a colder winter or a drier summer than the one before. Less familiar are changes in climate. Even though climate represents averages in weather conditions over periods of, say, 20 or 30 years, there is plenty of evidence that it, too, is not constant but instead fluctuates markedly over long spans of time. The most dramatic fluctuations were the **ice ages** of the distant past.

The last ice age reached its peak about 20,000 years ago when huge ice sheets as much as 4 km thick covered much of Europe and North America (see Chap. 16). The vast amount of water locked up as ice lowered sea level nearly 100 m below what it is today. Then the ice began to melt and climates became progressively less severe; in a period of 12,000 years the average annual temperature of central Europe rose from −4°C to +9°C. By about 4000 B.C. average temperatures were a few degrees higher than those of today. A time of declining temperatures then set in, reaching a minimum in Europe between 900 and 500 B.C.

A gradual warming-up followed and came to a peak between A.D. 800 and 1200. So generally fine were climatic conditions then that the Vikings established flourishing colonies in Iceland and Greenland from which they went on to visit North America (Fig. 14-34). But then came cooler summers and exceptionally cold winters with extensive freezing of the Arctic Sea. The extreme weather from the fifteenth to the nineteenth centuries in Europe led it to be called the "Little Ice Age." Greenland became a less attractive place than formerly and the colony there disappeared; the coast of Iceland was surrounded by ice for several months per year (in contrast to a few weeks per year today); and glaciers advanced farther across mountain landscapes than ever before or since in recorded history. The average temperature in Europe then was around 1°C less than it is these days. This does not seem like much, but it made a considerable difference in climates.

Late in the nineteenth century a trend toward higher temperatures led to a marked shrinkage of the world's glaciers. In the first half of the twentieth century especially pronounced temperature increases took place whose most noticeable consequences were milder winters in the higher latitudes. These balmy conditions peaked about 1940, after which average worldwide temperatures did not change by much for 40 years. Then temperatures again started upward (see Sec. 4.3 and especially Fig. 4–6). The increase in average temperature has been 0.76°C since 1900.

Why Do Climates Change? A number of factors can influence climate, and it is not always clear which one is responsible for which change.

The most obvious factor is the sun's energy output. This is not constant but increases and decreases during the 11-year sunspot cycle (Sec. 18.6). Variations in a number of weather phenomena (for instance, in the paths of winter storms across the North Atlantic Ocean) parallel the sunspot cycle. Although the connections with the sunspot cycle seem to be real, nobody yet knows exactly how they come about. Longer-term rises and falls in solar radiation occur as well; the Little Ice Age may have been caused by one of them.

Climate and Collapse

The Viking Erik the Red established a colony in Greenland in the tenth century, which was a time of warm climate (Fig. 14–34). The colony, a few thousand strong, lasted for nearly 5 centuries until the Arctic climate deteriorated in the Little Ice Age. The colonists contributed to their fate: they cut down trees for fuel and construction until none were left, they used nonsustainable farming methods that led to soil erosion and a consequent shortage of food, and when starvation loomed, cultural taboos kept them from eating the abundant fish or learning from their Inuit ("Eskimo") neighbors how to hunt effectively. The Greenland Norse, already near the edge, could not survive the changing climate, unlike the more adaptable Inuit.

In his book *Collapse,* Jared Diamond examines the Greenland colony and other failed societies and then considers the "nagging thought: might such a fate eventually befall our own wealthy society?" His themes are "human environmental impact and climate change intersecting, environmental and population problems spilling over into warfare, the strengths but also the dangers of complex non-self-sufficient societies dependent on imports and exports, and societies collapsing swiftly after attaining peak population numbers and power."

Elizabeth Kolbert, another worried observer, has this to say about a worldwide peril in her book *Field Notes from a Catastrophe:* "As the effects of global warming become more and more difficult to ignore, will we react by finally fashioning a global response? Or will we retreat into ever narrower and more destructive forms of self-interest? It may seem impossible to imagine that a technologically advanced society could choose, in essence, to destroy itself, but that is what we are now in the process of doing."

Figure 14-34 The site of Erik the Red's Norse colony in Greenland, the world's largest island, which lies almost entirely above the Arctic Circle. His son Leif Eriksson sailed westward from here, landed in what is now Newfoundland, and become the first European to set foot in the Americas. Another Viking, Bjarni Herjolfsson, had reached the same region earlier but had not gone ashore.

The most promising explanation for the large-scale climatic changes, however, relates them to periodic changes that occur in the tilt of the earth's axis, the shape of its orbit, and the time of year when the earth is closest to the sun (Fig. 14-35). Seventy years ago Milutin Milankovitch (1879–1958), a Yugoslav mathematician and physicist, worked out how these changes might affect climate by altering the amount of sunlight (that is, insolation) received by the earth.

The differences in insolation on a global basis are small, 0.3 percent at most, but Milankovitch argued that what really counts is not the total insolation but the insolation in the polar regions in summer, which varies by up to 20 percent. Too little summer sunshine would not melt all the snow that fell during the winter before, and in time the accumulated snow would turn into great sheets of ice. In the southern hemisphere the ice would melt when it left the Antarctic continent and fell into the sea, but in the northern hemisphere the ice would move down across North America and Eurasia to produce an ice age.

Major ice ages have occurred every 100,000 years or so, with smaller cycles of cold and warm at closer intervals. The strongest evidence in favor of Milankovitch's hypothesis is that the periods of advance and retreat of the ice sheets are in accord with the various periods of the earth's orbital variations. Additional support comes from current theoretical models of the earth's climate, which respond to the known insolation changes with a prediction of regular ice ages.

Global Warming What about the warming trend of recent years? A few degrees more (which seem on the way) and we will be living in the warmest environment since humans evolved millions of years ago. It is clear today that the steady rise in CO_2 levels is increasing global temperatures (Sec. 4.4). Exactly how life on earth will ultimately be affected can only be guessed at, but it is sure to mean profound changes. We are

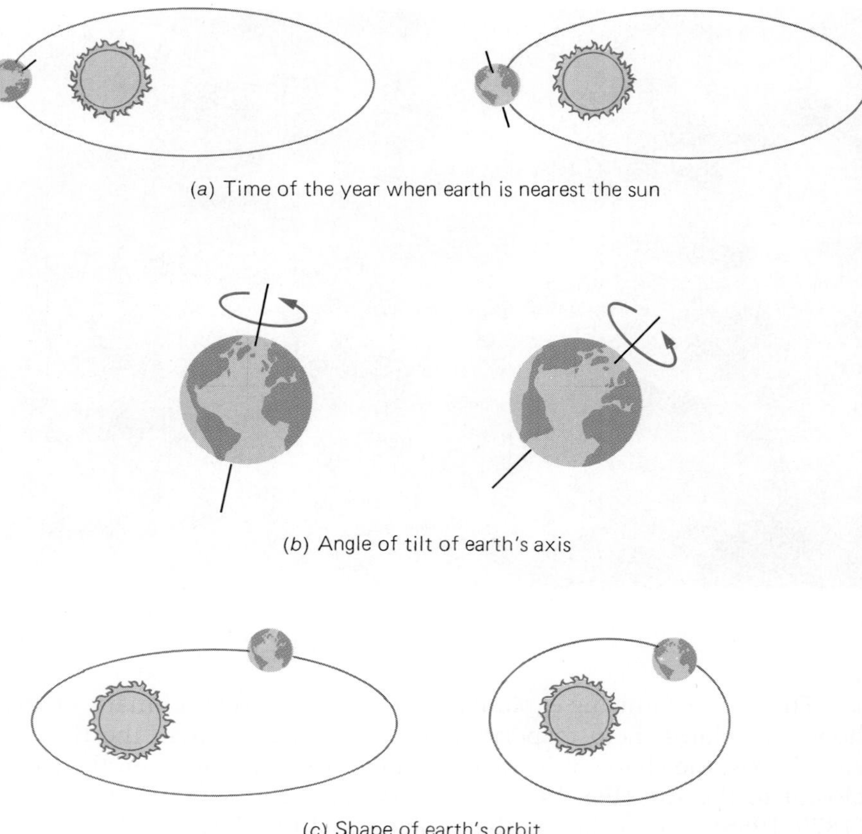

Figure 14-35 Three variations in the earth's motion that may be responsible for causing ice ages. (*a*) The time of year when the earth is nearest the sun varies with a period of about 23,000 years. (*b*) The angle of tilt of the earth's axis of rotation varies with a period of about 41,000 years. (*c*) The shape of the earth's elliptical orbit varies with a period of about 100,000 years. These variations have relatively little effect on the total sunlight reaching the earth but a considerable effect on the sunlight reaching the polar regions in summer. The ellipticity of the earth's orbit is vastly exaggerated here; the orbit is actually less than 2 percent away from a perfect circle.

in the midst of a gigantic experiment in climate modification, one that our descendants will surely regret. It is too late to stop the experiment because the persistence of greenhouse gases in the atmosphere means that global warming will continue for decades to come no matter what is done now. But a drastic cut in CO_2 emissions on a worldwide scale, though not easy to accomplish, may allow us to moderate the experiment before we go the way of the Greenland Vikings. How long do we have? One estimate gives us 10 to 20 years to take real action, but nobody really knows: all that is certain is, the sooner the better.

The Hydrosphere

Over 70 percent of the earth's surface is covered by the oceans and the seas that join with them. These waters are filled with plant and animal life—indeed, the early oceans were home to living things for more than a billion years before they began to move ashore. In addition, the oceans influence continental life in a variety of indirect ways. The oceans provide the reservoir from which water is evaporated into the atmosphere, later to fall as rain and snow on land. The oceans participate in the oxygen–carbon dioxide cycle both through the life they support and through the vast quantities of these gases dissolved in them. And the oceans help determine climates by their ability to absorb solar energy and transport it around the world.

14.12 Ocean Basins

Water, Water Everywhere

Each of the world's oceans lies in a vast basin bounded by continental landmasses. Typically an ocean bottom slopes gradually downward from the shore to a depth of 130 m or so before starting to drop more rapidly (Fig. 14-36). The average width of this **continental shelf** is 65 km, but it ranges from less than a kilometer (off the mountainous western coast of South America) to over 1000 km (off the low arctic coasts of the Eurasian landmass). The North, Irish, and Baltic Seas are part of the European continental shelf, while the Grand Banks off Newfoundland are part of the North American shelf. A sharp change in steepness marks the change from the continental shelf to the steeper **continental slope,** which after a fall of perhaps 2 km joins the **abyssal plain** of the ocean floor via the gentle **continental rise.**

The ocean basins average 3.7 km in depth, while the continents average only about 0.8 km in height above sea level. If the earth were smooth, it would be covered with a layer of water perhaps 2.4 km thick, but it seems likely that the oceans have always been confined to more or less distinct basins and presumably will continue to be. The waves that ruffle ocean surfaces are caused by winds that blow across them (Fig. 14-37).

The ocean floor, like the continents, has mountain ranges and valleys, volcanoes, and vast plains, many of them rivaling or exceeding in size their counterparts on land. The Hawaiian Islands, for instance, are

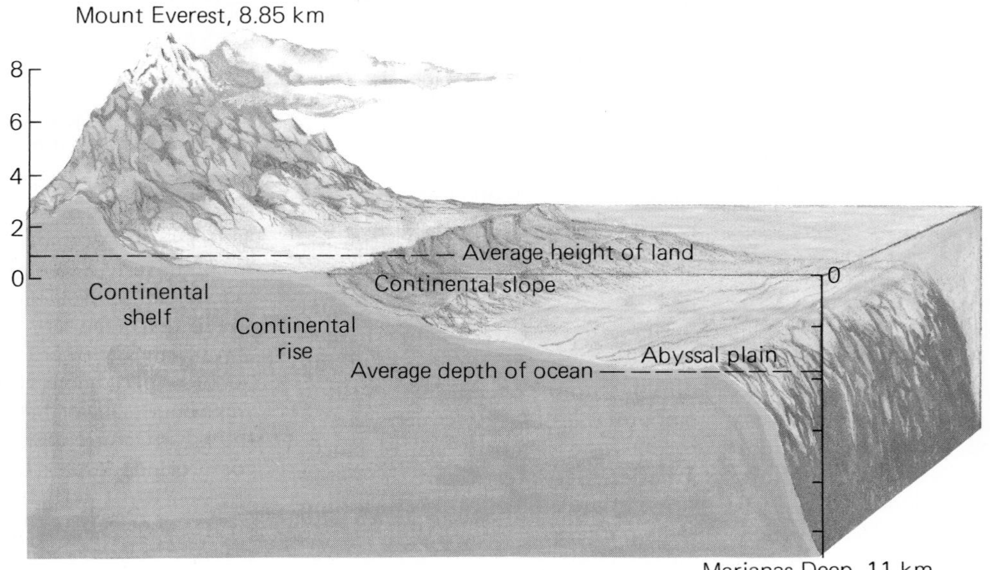

Figure 14-36 Profile of the earth's surface with heights above sea level and depths below it in km. In general, ocean depths are greater than land heights—the Marianas Deep can swallow Mt. Everest with over 2 km to spare. The vertical scale is greatly exaggerated here: the earth is actually quite smooth. Although the earth's circumference is about 40,000 km, the vertical distance from the deepest point in the oceans to the top of the highest mountain is less than 20 km. A car on a level road can easily go this far in 15 min.

Figure 14-37 Winds that blow over the surface of a body of water churn up waves that range from the ripples produced by a gentle breeze to the towering seas produced by a hurricane. The stronger the wind, the longer it blows, and the greater the distance over which the wind has been in contact with the water, the greater the average height of the waves that result. These three factors govern the amount of energy transferred to the water and thus govern how violent the resulting disturbance will be. At a given time and place in the ocean, wave heights vary considerably about their average. A third of the waves are normally higher than average, a tenth of them twice as high, and a very few three times as high.

Tsunamis

A **tsunami** consists of ocean waves usually caused by an earthquake on the seabed. In the open ocean tsunamis may be less than a meter high and, since their crests are many kilometers apart, ships at sea do not notice them and they lose little energy as they go. In deep water, tsunami speeds can exceed 700 km/h, as fast as a jet airliner. When the waves reach shallow water, they slow down, steepen to heights of as much as 30 m, and can go on to do immense damage on land (Fig. 14-38).

Because the location and magnitude of a major earthquake can be established soon after it occurs, even in midocean, warnings of a possible tsunami can usually be given a few hours before the first wave arrives. Tsunamis used to be called "tidal waves," but since they have nothing to do with tides, scientists now use the Japanese word for them, which means "harbor wave." Over 800 tsunamis have occurred since 1900, most of them in the Pacific Ocean and a few of which caused serious damage on shore. Reports of tsunamis go back to ancient Greece and Rome; one recorded in 365 A.D. killed thousands in the Egyptian port of Alexandria.

There is evidence that some tsunamis in the past were induced by large-scale landslides into the ocean. Of concern today is the instability of a volcano on La Palma, one of the Canary Islands in the Atlantic near Africa. It seems possible that its huge (half a trillion tons) western flank may someday split off and tumble into the ocean. The resulting tsunami might be 20 m high when it reaches the eastern coast of North America 8 h later where it would surge inland, devastating everything in its path.

Figure 14-38 During a severe earthquake—the most violent in 40 years—on December 26, 2004, off the northwestern corner of the Indonesian island of Sumatra a massive section of the earth's crust was abruptly thrust upward along a rupture hundreds of kilometers long. The result was a huge tsunami that spread across the Indian Ocean and smashed into the coasts of Indonesia, Sri Lanka, India, and Thailand. It reached East Africa 7 hours later. In all, at least 280,000 people, mainly in Indonesia and Sri Lanka, lost their lives and perhaps 2 million others were left homeless by waves up to 10 m high that surged ashore.

volcanoes that rise as much as 9000 m above the ocean floor, about half of their altitude being above sea level. Less conspicuous from the surface is the Mid-Atlantic Ridge, an immense submarine mountain range that extends from Iceland past the tip of South America before swinging into the Indian Ocean. Such islands as the Azores, Ascension Island, and Tristan da Cunha are all that protrude from the ocean of this ridge.

Much remains to be learned about the ocean floor—only 10 percent of its area has been surveyed well enough to reveal details smaller than 300 m. By contrast, 90 percent of the surface of Venus has been mapped to that extent by instruments carried on spacecraft.

Seawater and Freshwater By far the greatest part of the earth's surface water is in the form of seawater, as Fig. 14-5 shows. Much of the freshwater is stored as ice in the form of the glaciers and ice caps that cover one-tenth the land area of the earth. About 90 percent of this ice is located in the Antarctic ice cap, about 9 percent in the Greenland ice cap, and the remaining 1 percent in the various glaciers of the world. If suddenly melted, the ice would raise the sea level by perhaps 70 m. (By comparison, if all the water vapor in the atmosphere were condensed, the sea level would go up by only about 2.5 cm.)

Most of the earth's surface water probably appeared about 4 billion years ago when the young earth assumed its present internal structure. The water came from the rocks of the interior and took with it the same ions found in seawater today: the oceans have always been salty. Additional salts have been added to the oceans in the various ways illustrated in Fig. 11-33. However, seawater salinity has not changed very much because of various mechanisms that remove salt from the oceans. One of these mechanisms is quite direct: the loss of salt to the atmosphere when wind blows spray off wave tops. The resulting salt particles serve as nuclei for water molecules to stick to, and a substantial amount falls on land in rain and snow. Another mechanism that reduces salinity is the incorporation of various compounds in the shells of marine organisms, which end up in the sediments that coat the ocean floors.

Sea Ice

Even in the intense cold of the polar regions, sea ice is seldom more than a few meters deep. (Icebergs are huge masses of ice that have broken off from ice caps or glaciers that formed from accumulated snow on a land base.) There are several reasons why sea ice is so relatively thin. The chief one is that, as water under a surface layer of ice cools, it becomes denser and sinks, to be replaced by warmer water from underneath. Another reason is that ice itself acts as an insulator. Also, as seawater freezes, its salt content is not incorporated in the ice crystals but stays behind to increase the salinity of the remaining water, whose freezing point is accordingly lowered (see Sec. 11.4).

14.13 Ocean Currents

Four Great Whirlpools

The oceans affect climate in two ways. First, they act as heat reservoirs that moderate seasonal temperature extremes; more heat is stored in the upper few meters of the oceans than in the entire atmosphere. In spring and summer the oceans are cooler than the regions they border, since the insolation they receive is absorbed in a greater volume than on solid, opaque land. The heat stored in the ocean depths means that in fall and winter the oceans are warmer than nearby land areas. Heat flows readily between moving air and water. With enough of a temperature difference, the rate of energy transfer from warm water to cold air (or from warm air to cold water) can exceed the rate at which solar energy arrives at the earth.

Lacking such an adjacent heat reservoir, continental interiors experience lower winter temperatures and higher summer temperatures than those of coastal districts. In Canada, for instance, temperatures in the city of Victoria on the Pacific coast range from an average January minimum of 9°C to an average July maximum of 20°C, whereas in Winnipeg, in the interior, the corresponding figures are −13°C and 27°C.

The second way oceans influence climate is through surface currents, which are produced by the friction of wind on water. Such currents are much slower than movements in the atmosphere, with the fastest normal surface currents having speeds of about 10 km/h.

The wind-driven surface currents parallel to a large extent the major wind systems (Fig. 14-39). The northeast and southeast trade winds drive water before them westward along the equator, forming the **equatorial current.** In the Atlantic Ocean this current runs into South America and in the Pacific the current runs into the East Indies. At each of these contacts the current divides into two parts, one flowing south and the other north. Moving away from the equator along the continental edges, these currents then come under the influence of the westerlies, which drive them eastward across the oceans. Thus gigantic whirlpools are set up in both the Atlantic and the Pacific on either side of the equator. Many complexities are produced in the four great whirls by islands, continental projections, and undersea mountains and valleys.

The Gulf Stream The western side of the North Atlantic whirl—a warm current moving partly into the Gulf of Mexico, partly straight north along our southeastern coast—is the famous **Gulf Stream.** Forced away from the coast in the latitude of New Jersey by the westerlies, this current moves northeastward across the Atlantic (Fig. 14-40). The current

Figure 14-39 Principal ocean currents of the world. The swirling currents of the Pacific Ocean contain an estimated 100 million tons of rubbish, mainly plastic, that floats just below the surface and covers a total area twice that of the continental United States. The plastic debris will take hundreds of years to disintegrate.

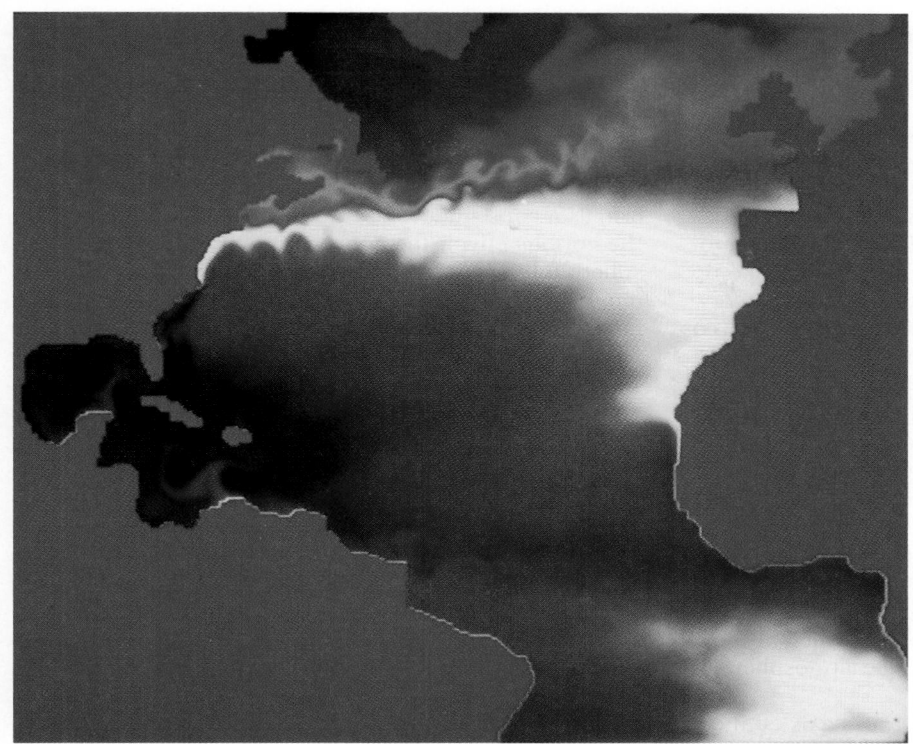

Figure 14-40 Surface temperatures of the North Atlantic Ocean are shown in this computer simulation. Temperatures range from about 28°C (dark red) in the tropics to 0°C (dark blue) near Greenland. The ribbon of dark red off the southern part of the American coast is the Gulf Stream, whose continuation across the Atlantic is responsible for the relatively warm water shown in yellow and green. Global warming is reducing the rate of flow of the Gulf Stream, a process that will eventually have the paradoxical effect of chilling the climate of northern Europe, perhaps by as much as 5°C.

splits on the European side into one part that moves south to complete the whirl and another part that continues northeastward past Great Britain and Norway into the Arctic Ocean. To compensate for the addition of water into the polar sea, the cold **Labrador Current** moves southward along the east coast of North America as far as New York. Down the west coast of North America moves the **California Current,** the southward-flowing eastern part of the North Pacific whirl.

Since ocean currents retain the temperatures of the latitudes from which they come for a long time, they exert a direct influence on the temperatures of neighboring lands. The influence is greatest, of course, where prevailing winds blow shoreward from the sea. The warm Gulf Stream has a far greater effect in tempering the climate of northwestern Europe than that of eastern United States, since prevailing winds in these latitudes are from the west. Cyclonic storms bring east winds to the Atlantic seaboard often enough, however, for the Gulf Stream to help raise temperatures in the South Atlantic states, and for the same reason the Labrador Current is in part responsible for the rigorous climate of New England and eastern Canada.

Thus the oceans, besides acting as water reservoirs for the earth's atmosphere, play a direct part in temperature control—both by preventing abrupt temperature changes in lands along their borders and by aiding the winds, through the motion of ocean currents, in their distribution of heat and cold over the surface of the earth.

Important Terms and Ideas

The earth's **atmosphere** is its envelope of air, which consists mainly of nitrogen and oxygen. The lowest part of the atmosphere is the **troposphere,** in which such weather phenomena as clouds and storms occur. Next comes the **stratosphere,** whose content of **ozone** (O_3) absorbs most of the ultraviolet radiation from the sun. Above the stratosphere is the **mesosphere** and still higher is the **thermosphere,** which contains most of the ions that make up the **ionosphere.** The ionosphere reflects radio waves.

Air at a given temperature is said to be **saturated** when it contains the maximum amount of water vapor possible without liquid water condensing out. The **relative humidity** of a volume of air is the ratio between the water vapor it contains and the amount that would be present at saturation. Low-altitude **clouds** consist of tiny water droplets; high-altitude clouds consist of ice crystals.

The energy that powers weather phenomena is **insolation,** which is **in**coming **sol**ar radi**ation.** Most of the insolation passes through the atmosphere and is absorbed by the earth's surface. The warm earth then radiates energy back into the atmosphere which absorbs some of it. This indirect heating of the atmosphere is called the **greenhouse effect.**

The **seasons** occur because the earth's axis is tilted, so that in half of each year one hemisphere receives more direct sunlight than the other hemisphere.

Convection currents are due to the uneven heating of a fluid: the warmer parts of the fluid expand and rise while the cooler parts sink. Such currents occur where air in the equatorial regions is heated strongly, expands and rises, and moves toward the poles. In the polar regions the air cools and sinks and then flows back on the surface toward the equator. In this way energy is shifted from the tropics to the higher latitudes.

The **coriolis effect** is the deflection of winds to the right in the northern hemisphere, to the left in the southern, as a consequence of the earth's rotation. Because of the coriolis effect, the convection currents in the atmosphere follow curved paths. A **cyclone** is a weather system centered on a low-pressure region. In the northern hemisphere the coriolis effect deflects winds moving inward in a cyclone into a counterclockwise spiral, in the southern hemisphere into a clockwise spiral. An **anticyclone** is a weather system centered on a high-pressure region; winds blowing outward from an anticyclone spiral clockwise in the northern hemisphere and counterclockwise in the southern hemisphere.

A **front** is the boundary between a mass of warm air and a mass of cold air; clouds and rain usually occur at a front.

Climate refers to averages in weather conditions over a period of years. The **ice ages** were times of severe cold in which ice sheets covered much of Europe and North America. Such long-term climatic changes are probably caused by changes in the earth's motions around the sun.

A **tsunami** consists of ocean waves caused by earthquakes on the seabed or by gigantic landslides into the ocean.

The **Gulf Stream** is a warm current that flows northeastward in the Atlantic Ocean. Westerly winds that blow across the Gulf Stream carry heat that moderates the climate of northwestern Europe.

Multiple Choice

1. Arrange the following gases in the order of their abundance in the earth's atmosphere:
a. oxygen **c.** nitrogen
b. carbon dioxide **d.** argon

2. Much of Tibet lies in altitudes above 5.5 km (18,000 ft). At such altitudes the Tibetans are above approximately
a. 10 percent of the atmosphere
b. 50 percent of the atmosphere
c. 90 percent of the atmosphere
d. 99 percent of the atmosphere

3. The region of the atmosphere closest to the earth's surface is the
a. thermosphere **c.** troposphere
b. ionosphere **d.** stratosphere

4. A characteristic feature of the stratosphere is the presence of
a. clouds **c.** ozone
b. dust **d.** ions

5. The atmosphere constituent chiefly responsible for absorbing ultraviolet radiation from the sun is
a. carbon dioxide **c.** ozone
b. water vapor **d.** helium

6. Chlorofluorocarbon (CFC) gases, which are widely used in refrigeration and in foam plastics, catalyze the breakdown in the upper atmosphere of
a. water vapor **c.** oxygen
b. carbon dioxide **d.** ozone

7. Completely dry air has a relative humidity of
a. 0 **c.** 50 percent
b. 1 percent **d.** 100 percent

8. The higher the temperature of a volume of air, the
a. more water vapor it can hold
b. less water vapor it can hold
c. greater its possible relative humidity
d. lower its possible relative humidity

9. When saturated air is cooled,
   a. it becomes able to take up more water vapor
   b. some of its water content condenses out
   c. the relative humidity goes down
   d. convection currents result

10. Clouds consist of
    a. water droplets at all altitudes
    b. ice crystals at all altitudes
    c. water droplets at low altitudes and ice crystals at high altitudes
    d. ice crystals at low altitudes and water droplets at high altitudes

11. If the atmosphere contained fewer salt crystals and dust particles than it now does,
    a. clouds would form less readily
    b. clouds would form more readily
    c. the formation of clouds would be unaffected
    d. snow would never fall

12. Clouds occur when moist air is cooled by
    a. expansion when it rises
    b. expansion when it falls
    c. compression when it rises
    d. compression when it falls

13. The atmosphere is heated chiefly by
    a. infrared radiation from the sun
    b. ultraviolet radiation from the sun
    c. infrared radiation from the earth
    d. ultraviolet radiation from the earth

14. The term *greenhouse effect* is used to describe the
    a. condensation of moisture to form dew
    b. conversion of carbon dioxide to oxygen by green plants
    c. heating of the atmosphere by direct solar radiation
    d. heating of the atmosphere by infrared radiation from the earth

15. Energy is transported from the tropics to the polar regions chiefly by
    a. winds c. carbon dioxide
    b. ocean currents d. ozone

16. The seasons occur as a result of
    a. variations in the sun's energy output
    b. variations in the distance between the earth and the sun
    c. variations in the orbital speed of the earth
    d. the tilt of the earth's axis

17. Because of the coriolis effect, a wind in the northern hemisphere is deflected
    a. upward c. toward the right
    b. downward d. toward the left

18. Because of the coriolis effect, a wind in the southern hemisphere is deflected
    a. upward c. toward the right
    b. downward d. toward the left

19. On a summer day sunlight warms coastal land until its temperature is higher than that of the adjacent sea. The result is a "sea breeze" that blows
    a. from sea to land
    b. from land to sea
    c. parallel to the shore
    d. any of these, depending on the part of the world

20. The flow of air in the upper atmosphere is largely from
    a. east to west
    b. west to east
    c. north to south
    d. south to north

21. The middle latitudes usually experience winds from the
    a. north c. east
    b. south d. west

22. The general direction of the trade winds in both hemispheres is from the
    a. north c. east
    b. south d. west

23. An airplane flies at the same speed relative to the air at a high altitude from New York to Paris and back.
    a. The New York to Paris flight takes less time.
    b. The Paris to New York flight takes less time.
    c. The two flights take the same time on the average.
    d. Any of these, depending on the season.

24. The trade-wind belts are regions of generally
    a. little rainfall c. low temperatures
    b. much rainfall d. westerly winds

25. A cyclone is a weather system centered about a
    a. region of low pressure
    b. region of high pressure
    c. hurricane
    d. cold front

26. The winds in an anticyclone
    a. blow directly toward its center
    b. spiral toward its center
    c. blow directly away from its center
    d. spiral away from its center

27. Unstable weather is associated with
    a. cyclones
    b. anticyclones
    c. trade winds
    d. the greenhouse effect

28. The most pleasant weather is found in
- **a.** a cold front
- **b.** a warm front
- **c.** a cyclone
- **d.** an anticyclone

29. The chief reason why the equatorial regions are warmer than the polar regions is that
- **a.** the equator is closer to the sun
- **b.** sunlight falls more nearly vertically on the equatorial regions
- **c.** sunlight is reflected by ice and snow in the polar regions
- **d.** there is more CO_2 in the air over the equatorial regions

30. The greatest seasonal variations in temperature occur in
- **a.** the west coasts of the continents
- **b.** the east coasts of the continents
- **c.** continental interiors
- **d.** isolated islands

31. Rain is most abundant in the
- **a.** prevailing westerlies of the middle latitudes
- **b.** horse latitudes between the middle latitudes and the trade-wind belts
- **c.** trade-wind belts
- **d.** doldrums of the equatorial regions

32. Ice ages
- **a.** cover the entire earth with a sheet of ice
- **b.** freeze all the oceans
- **c.** occurred seldom in the past
- **d.** occurred frequently in the past

33. The approximate percentage of the earth's surface covered by water is
- **a.** 10 percent
- **b.** 50 percent
- **c.** 70 percent
- **d.** 90 percent

34. Compared with the average height of the continents above sea level, the average depth of the ocean basins below sea level is
- **a.** smaller
- **b.** greater
- **c.** about the same
- **d.** sometimes smaller and sometimes greater, depending upon the tides

35. The deepest known point of the oceans is found in the
- **a.** Atlantic Ocean
- **b.** Pacific Ocean
- **c.** North Sea
- **d.** Panama Canal

36. The Hawaiian Islands are
- **a.** part of a sunken continent
- **b.** floating on the surface of the ocean
- **c.** located in shallow water
- **d.** volcanic peaks

37. Tsunamis are caused by
- **a.** monsoons
- **b.** typhoons
- **c.** icebergs
- **d.** undersea earthquakes

38. Most surface ocean currents are due to
- **a.** melting glaciers
- **b.** rivers
- **c.** winds
- **d.** differences in the altitude of the ocean surface

39. The Gulf Stream is a
- **a.** warm current in the North Atlantic
- **b.** cool current in the North Atlantic
- **c.** river that flows into the Gulf of Mexico
- **d.** warm wind blowing over the North Atlantic

40. The climate of northwestern Europe is greatly affected by the
- **a.** Gulf Stream
- **b.** abyssal zone
- **c.** Labrador Current
- **d.** trade winds

Exercises

14.1 Regions of the Atmosphere

1. What causes ionization to occur in the upper atmosphere?
2. Suppose you are climbing in an airplane that has no altimeter. How could you tell when you are approaching the top of the troposphere?
3. The tropopause, stratopause, and mesopause are respectively the upper boundaries of the troposphere, stratosphere, and mesosphere. What is characteristic of the air temperature at each of these boundaries?
4. What would happen if ozone were to disappear from the upper atmosphere? From the lower atmosphere?
5. Why are chlorofluorocarbon (CFC) gases, which are widely used in refrigeration and in foam plastics, harmful when released into the atmosphere?

14.2 Atmospheric Moisture

6. What does it mean to say that a certain volume of air has a relative humidity of 50 percent? Of 100 percent?

7. Under what circumstances, if any, can saturated air take up more water? Under what circumstances, if any, can some of the water content of saturated air condense out?
8. Why does the air in a heated room tend to be dry?
9. The air in a closed container is saturated with water vapor at 20°C. (a) What is the relative humidity? (b) What happens to the relative humidity if the temperature is reduced to 10°C? (c) If the temperature is increased to 30°C?
10. Why does dew form on the ground during clear, calm summer nights?
11. What does Fig. 14-5 tell us about the relative amounts of freshwater present on or near the earth's surface as groundwater, in lakes and rivers, in ice caps and glaciers, and as moisture in the atmosphere? How does the total amount of freshwater compare with the amount of seawater?

14.3 Clouds

12. What do high-altitude clouds consist of? Low-altitude clouds?
13. What initiates the fall of rain from a cloud? The fall of snow?
14. What is the origin of the streaks of cirrus cloud produced by high-altitude aircraft?

14.4 Atmospheric Energy

15. What is insolation? How does it affect the atmosphere?
16. (a) On a clear day, solar radiation is most intense at noon. Why? (b) The highest air temperatures occur a few hours later. Why?
17. What is the greenhouse effect and how is it related to the absorption of solar energy by the earth's atmosphere?
18. Compare the ways in which the troposphere and the stratosphere are heated.
19. If the earth's atmosphere were to disappear, what would happen to the rate at which the earth would radiate energy back into space?
20. Why does the average air temperature decrease going from the equator to the poles?
21. What are the two mechanisms by which energy of solar origin is transported around the earth? Which is more important?
22. Account for the abrupt changes in temperature between day and night in desert regions.

14.5 The Seasons

23. In the northern hemisphere, the longest day is in June and the shortest day is in December, but the warmest weather occurs in July and August and the coldest weather in January and February. What is the reason for these time lags?
24. If the earth's axis were tilted more than it is now, in what way, if any, would the seasons be affected?
25. The *Tropic of Cancer* is the most northerly latitude in the northern hemisphere at which the sun is ever directly overhead at noon. The *Tropic of Capricorn* is the corresponding latitude in the southern hemisphere. What are the latitudes of these tropics?

14.6 Winds

26. What is the basic cause of winds?
27. Distinguish between an isobar and a millibar.
28. A wind in the northern hemisphere starts to blow toward the equator. Toward what direction is the wind deflected by the coriolis effect? What about a wind in the southern hemisphere also starting to blow toward the equator?
29. On a summer night coastal land cools below the temperature of the adjacent sea. Does the resulting "land breeze" blow offshore or onshore? Does the coriolis effect influence the land breeze? If so, how?

14.7 General Circulation of the Atmosphere

30. Where in the atmosphere do the jet streams occur? What is their general direction?
31. An airplane flies at the same speed relative to the air at a high altitude from Chicago to London and then back to Chicago. How do the flight times for each leg compare?
32. The prevailing westerly winds of the middle latitudes of the northern hemisphere are generally weaker than those of the southern hemisphere. Can you think of why this is so?

14.8 Middle-Latitude Weather Systems

33. (a) What is the name of weather systems centered about regions of high pressure? (b) In what direction do winds in the northern hemisphere spiral around such a region? (c) In the southern hemisphere?
34. What is the name of weather systems centered about regions of low pressure? In what direction do winds in the northern hemisphere spiral around such a region? In the southern hemisphere?
35. How does the weather associated with a cyclone differ from that associated with an anticyclone?
36. Why are clouds and rain more likely to be associated with a cyclone than with an anticyclone?
37. When you face a wind associated with a cyclone in the northern hemisphere, in what approximate direction will the center of low pressure be? In what direction will the center of low pressure be if you do this in the southern hemisphere?

38. What is the approximate sequence of wind directions when the center of a cyclone passes north of an observer in the northem hemisphere?

39. What is the approximate sequence of wind directions when the center of an anticyclone passes south of an observer in the northern hemisphere?

40. What is the difference between the rainfall that accompanies the passage of a warm front and that which accompanies the passage of a cold front?

41. Cumulus clouds form when warm air rises vertically by convection, and stratus clouds form when a warm air mass moving horizontally encounters a cooler mass and is forced upward on top of the cooler mass. Which kind of clouds would you expect to be characteristic of a warm front?

42. Why can a hurricane be regarded as a heat engine?

14.9 Tropical Climates

43. The northeast and southeast trade winds meet in a belt called the doldrums. What is the characteristic climate of the doldrums and why does it occur?

44. A feature of most autumns in the northeastern United States is a period of mild, sunny weather called Indian summer, which occurs when an anticyclone happens to stall for a few days off the East Coast. Explain the connection.

14.11 Climate Change

45. The Milankovitch theory of ice ages relates them to variations in the tilt of the earth's axis, the shape of its orbit, and the time of year when the earth is closest to the sun. However, these variations affect the total amount of solar energy reaching the earth by no more than 0.3 percent. How did Milankovitch account for this apparent contradiction?

46. Water vapor is an important "greenhouse gas" in the atmosphere, yet much more attention is paid in this respect to carbon dioxide. Why do you think this is so?

47. What has kept certain regions of the earth from sharing fully in global warming?

48. From time to time a gigantic volcanic explosion sends a large amount of dust into the atmosphere, where it may remain for some years. How many consequences of such an event can you think of?

49. When did the Little Ice Age occur: several hundred years ago, several thousand years ago, several million years ago? By roughly how much did average European temperatures fall in the Little Ice Age?

14.12 Ocean Basins

50. How does the average depth of the ocean basins below sea level compare with the average height of the continents above sea level?

51. A wind begins to blow over the surface of a calm body of deep water. What factors govern the height of the waves that are produced?

52. Why is it believed that seawater has always been salty?

14.13 Ocean Currents

53. The salinity of seawater varies with location, but the relative proportions of the various ions in solution are almost exactly the same everywhere regardless of local circumstances. What is the significance of the latter observation?

54. The giant whirls of the oceans involve clockwise flows in the northern hemisphere and counterclockwise flows in the southern. Why?

55. In what two ways do the oceans influence climates on land?

56. England and Labrador are at about the same latitude on either side of the North Atlantic Ocean, but England is considerably warmer than Labrador on the average. Why?

57. The California Current along the California coast is cooler than the ocean to the west. How does this fact explain the numerous fogs on this coast?

58. The island of Oahu (one of the Hawaiian Islands) is at latitude 21°N and is crossed by a mountain range trending roughly northwest to southeast. Account for the more abundant rainfall on the northeastern side of the range.

59. Why does the equatorial current flow westward?

Answers to Multiple Choice

1. c, a, d, b	7. a	13. c	19. a	25. a	31. d	37. d
2. b	8. a	14. d	20. b	26. d	32. d	38. c
3. c	9. b	15. a	21. d	27. a	33. c	39. a
4. c	10. c	16. d	22. c	28. d	34. b	40. a
5. c	11. a	17. c	23. a	29. b	35. b	
6. d	12. a	18. d	24. a	30. c	36. d	

15

The Rock Cycle

Glacial valley in Alaska.

Goals

When you have finished this chapter you should be able to complete the goals ▶ given for each section below:

Rocks

15.1 Composition of the Crust
Oxygen and Silicon Are the Most Abundant Elements
- ▶ List in order of abundance the four chief elements in the earth's crust.
- ▶ Explain why the silicates can vary so much in composition and crystal structure.

15.2 Minerals
What Rocks Are Made Of
- ▶ Distinguish between rocks and minerals.
- ▶ Briefly describe quartz, feldspar, mica, the ferromagnesian minerals, the clay minerals, and calcite.

15.3 Igneous Rocks
Once Molten, Now Solid
- ▶ Distinguish among igneous, sedimentary, and metamorphic rocks.
- ▶ Compare the origins of the fine-grained and coarse-grained igneous rocks and give several examples of each type.

15.4 Sedimentary Rocks
Compacted Sediments or Precipitates from Solution
- ▶ Describe several fragmental sedimentary rocks.
- ▶ State the main characteristics of limestone and describe how it is formed.

15.5 Metamorphic Rocks
Formed from Other Rocks by Heat and/or Pressure
- ▶ Describe several metamorphic rocks and give their origins.

Within the Earth

15.6 Earthquakes
When Our Planet Trembles
- ▶ Distinguish among the four kinds of earthquake waves.

15.7 Structure of the Earth
Core, Mantle, and Crust
- ▶ Explain the evidence that suggests the division of the earth into core, mantle, and crust.

15.8 The Earth's Interior
A Mantle of Rock, a Core of Molten Iron
- ▶ Give several reasons for the belief that earth's core is largely molten iron.
- ▶ Identify the main source of heat that flows out of the earth's interior.

15.9 Geomagnetism
Electric Currents in the Core Seem to Be Responsible
- ▶ Compare the earth's magnetic field with the magnetic field of a bar magnet, and explain why no actual permanent magnet can give rise to the earth's field.

Erosion

15.10 Weathering
How Exposed Rocks Decay
- ▶ Describe the chemical and mechanical weathering of rocks.

15.11 Stream Erosion
Running Water Is the Chief Agent of Erosion
- ▶ Discuss the development of a valley carved by a river.

15.12 Glaciers
Rivers and Seas of Ice
- ▶ Discuss the development of a valley carved by a glacier.

15.13 Groundwater
Water, Water Everywhere (Almost)
- ▶ Define groundwater, saturated zone, water table, spring, and aquifer.

15.14 Sedimentation
What Becomes of the Debris of Erosion
- ▶ Discuss the deposition of stream and glacier sediments.
- ▶ Describe the processes by which sediments become rock.

Vulcanism

15.15 Volcanoes
Rivers of Lava, Clouds of Gas and Dust
- ▶ Describe the events that occur in a typical volcanic eruption.
- ▶ Identify the parts of the world where most volcanoes occur.

15.16 Intrusive Rocks
They Have Hardened Underground from Magma
- ▶ Describe the different kinds of intrusive bodies of igneous rock.

15.17 The Rock Cycle
Rocks Are Not Necessarily Forever
- ▶ Draw the rock cycle.

Soil, vegetation, and rock fragments form a thin surface layer on most land areas, but solid rock is always underneath. Rock underlies the sediments on the ocean floors as well. The deepest oil wells, which go down over 8 km, are drilled through rock similar to that at the surface. Some of the rock now out in the open was once buried several km inside the earth, and the material that makes up some volcanic rock probably rose in molten form from still greater depths, perhaps as much as 100 km down. These samples of rock from well below the earth's surface also turn out to be very much like rock that formed close to the surface.

Such direct observation tells us that the outer part of the earth, called its **crust,** is composed almost entirely of rock. However, the thickness of the crust is only 0.5 percent of the earth's 6400-km radius. There is no firsthand information about the rest of our planet, but its interior can be probed by indirect methods. After we have learned something about the structure of our planet and about the rocks that clothe it, we shall turn to the processes whose action has produced the landscapes around us.

Rocks

15.1 Composition of the Crust

Oxygen and Silicon Are the Most Abundant Elements

The average composition of the earth's crust is shown in Fig. 15-1. Only a few elements are abundant in the crust, while others are present in quite small amounts. Oxygen makes up nearly half the mass of the crust, most of it combined with silicon. Silicon and the two metals iron and aluminum account for three-fourths of the rest of the crust's mass. Lumped together in the 1.4 percent of "all others" are the carbon, hydrogen, and nitrogen present in all living things and such familiar metals as copper, lead, and silver.

Silicon never occurs by itself on the earth, but its compounds make up about 87 percent of the rock and soil of the earth's crust. In the chemistry of the earth, silicon has the same sort of central role that carbon has in the chemistry of living things.

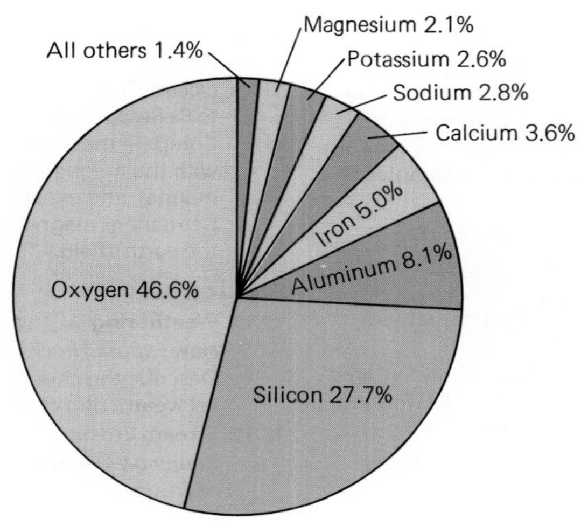

Figure 15-1 Average chemical composition of the earth's crust. Percentages are by mass. The total is not 100 percent because of rounding.

Nearly all the earth's silicon is combined either with oxygen in silicon dioxide (SiO_2), sometimes called **silica,** or with oxygen and one or more metals in the **silicates.** The differences in composition and structures of the silicates are reflected in a variety of colors, hardnesses, and crystal forms. The softness of talc and the hardness of zircon and beryl, the transparency of topaz and the deep color of garnet, the platy crystals of mica and the fibrous crystals of asbestos give some idea of the range of silicate properties. Glass is a mixture of silicates.

The basic structural unit of all silicates is the SiO_4^{4-} tetrahedron (a pyramid with a triangular base) shown in Fig. 15-2. In some silicates these units occur as single ions linked by positive metal ions. In more complex silicates the units form continuous chains, as in asbestos, or sheets, as in mica, with metal ions lying between them. Three-dimensional networks of SiO_4^{4-} units also occur. The number and variety of silicate minerals are due to the many different ways in which the basic SiO_4^{4-} unit can combine with metal ions to form stable crystal structures.

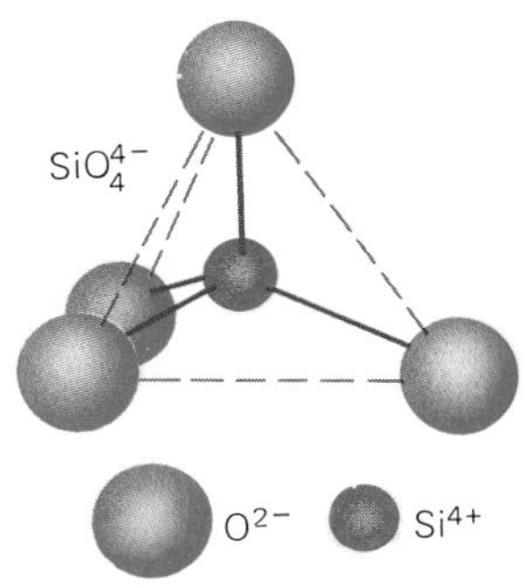

Figure 15-2 The silicon-oxygen tetrahedron is the fundamental unit in all silicate structures. Dashed lines show the tetrahedral form of SiO_4^{4-}; solid lines are bonds between the ions.

Asbestos

Asbestos is the name given to a group of fibrous minerals whose silicon-oxygen units are arranged in long chains. Because of its fibrous nature, mechanical strength, inertness, and resistance to fire, asbestos was once widely employed as reinforcement for building materials, as insulation in walls and around steam pipes, for fireproof theater curtains, in filters, in brake linings, and so on. The peak year for asbestos was 1973, when almost a million tons were used in the United States alone.

Unfortunately, asbestos fibers can cause serious lung and intestinal diseases such as cancer. For many years after their discovery nearly a century ago, the dangers of asbestos were ignored by both industry and government, sometimes even deliberately hidden by them. Hundreds of thousands of people around the world have died prematurely or will do so as a result of ingesting asbestos fibers at work or in their daily lives. One asbestos-related disease, mesothelioma, may take up to 50 years to show up after an exposure, after which it is usually fatal within 2 years. The eventual bill for compensation will total over \$100 billion.

When the perils of asbestos finally became widely known, its use became severely restricted and asbestos already in buildings such as schools was removed. There are two types of asbestos: the amphiboles, whose stiff, hard fibers are mainly responsible for the maladies associated with asbestos, and chrysotile, whose softer fibers are less likely to cause illness (Fig. 15-3). Chrysotile is still widely used in construction in a number of countries, notably Russia, China, and Brazil, and to a certain extent in the United States in certain applications where there is no adequate substitute. It is banned as a health hazard in the European Union.

Figure 15-3 Chrysotile asbestos.

15.2 Minerals

What Rocks Are Made Of

Most rocks are heterogeneous solids. The different homogeneous materials, called **minerals,** in a coarse-grained rock like granite are obvious to the eye. For a fine-grained rock, a microscope may be needed to distinguish the minerals that compose it.

A mineral is a crystalline inorganic solid found in nature that has a fairly specific chemical composition. Minerals are not usually identified by chemical names for two reasons. First, the same compound may occur in different forms: for instance, the minerals calcite and aragonite are both largely calcium carbonate ($CaCO_3$), but they differ in crystal form, hardness, density, and so on. Second, most minerals vary somewhat in composition from sample to sample, whereas the compositions of chemical compounds are invariable. About 4000 minerals are known, but the great majority are rare. The number of minerals important in ordinary rocks is so small that knowing something about only half a dozen minerals or mineral groups is enough for an introduction to geology.

Crystal Form and Cleavage In describing minerals, two important properties are **crystal form** and **cleavage.** Minerals are crystalline solids, which means that their atoms are arranged in lattice structures with definite geometric patterns. When a mineral hardens in a location where its crystals can grow freely, as in a cavity, perfect crystals are formed with smooth faces that meet each other at sharp angles. Each mineral has crystals of a distinctive shape, so that well-formed crystals make it easy to recognize a mineral (Fig. 15-4). Unfortunately good mineral crystals are rare, since neighboring crystals usually interfere with one another's growth.

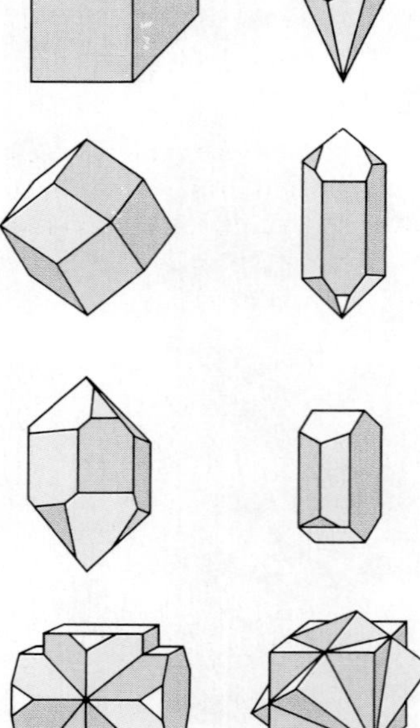

Figure 15-4 Some crystal forms found in minerals.

When well-developed crystals are not present, the characteristic crystal structure of a mineral may still reveal itself in the property called cleavage. This is the tendency of a substance to split along certain planes, which are determined by the arrangement of atoms in its lattice. When a mineral sample is struck with a hammer, its cleavage planes are revealed as the preferred directions in which it breaks.

Even in its original state, a mineral may show cleavage by flat, parallel faces and minute parallel cracks. The flat surfaces of mica flakes, for instance, and the ability of mica to be peeled apart in thin sheets show that this mineral has almost perfect cleavage in one direction (Fig. 15-5). Some minerals (for example, quartz) have no cleavage at all. When struck they shatter, like glass, along random curved surfaces.

Example 15.1

Opals are used in jewelry because of the beautiful play of colors that some specimens exhibit. An opal consists of SiO_2 and H_2O in various proportions linked together in microscopic crystals that form an amorphous material. Would you expect opals to exhibit cleavage?

Solution

Because an opal is amorphous, it has no regular structure and therefore cannot exhibit cleavage.

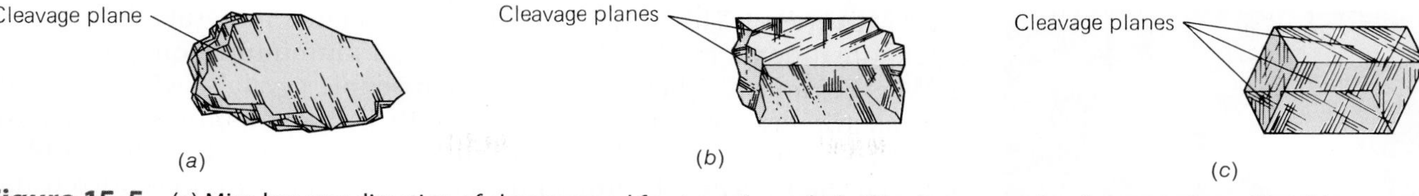

Figure 15-5 (*a*) Mica has one direction of cleavage and fractures irregularly if broken across its cleavage plane. (*b*) Feldspar has two perpendicular cleavage planes and fractures irregularly if broken across them. (*c*) Calcite has three directions of cleavage that are not perpendicular to each other.

Six Common Minerals Here are details of six common minerals and mineral groups.

Quartz Well-formed quartz (SiO_2) crystals are six-sided prisms and pyramids that show no cleavage (Fig. 15-6). They are colorless or milky (often gray, pink, or violet because of impurities), have a glassy luster, and are hard enough to scratch glass. They occur in many kinds of rock, sometimes appear as long, narrow deposits called **veins,** and often form assemblies of crystals inside cavities. Clear quartz (rock crystal) is used in jewelry and in optical instruments. Smoky quartz, rose quartz, and amethyst are colored varieties used in jewelry. Quartz sand is the chief raw material for making glass.

Figure 15-6 Quartz consists of silicon dioxide crystals and is found in many kinds of rock.

Feldspar This is the name of a group of minerals with very similar properties. The two classes of feldspar are a silicate of K and Al called orthoclase and a series of silicates of Na, Ca, and Al collectively called plagioclase. The crystals are rectangular, with blunt ends; they show good cleavage in two directions approximately at right angles (Fig. 15-7). They are sometimes clear; if not, their color is white or light shades of gray and pink. Feldspar is slightly harder than glass but not as hard as quartz. It is the most abundant single constituent of rocks, making up about 60 percent of the total weight of the earth's crust. Pure feldspar is used in the making of porcelain and as a mild abrasive.

Figure 15-7 Feldspar is the most abundant mineral in the earth's crust. Shown is a sample of orthoclase feldspar. The salmon-pink color is characteristic of this mineral.

Mica The two chief varieties of this familiar mineral are white mica—a silicate of H, K, and Al—and black mica—a silicate of H, K, Al, Mg, and Fe. Mica is easily recognized by its perfect and conspicuous cleavage in one plane (Fig. 15-8). It is a very soft mineral, only a trifle harder than a fingernail. Large sheets of white mica free from impurities are used as insulators in electrical equipment.

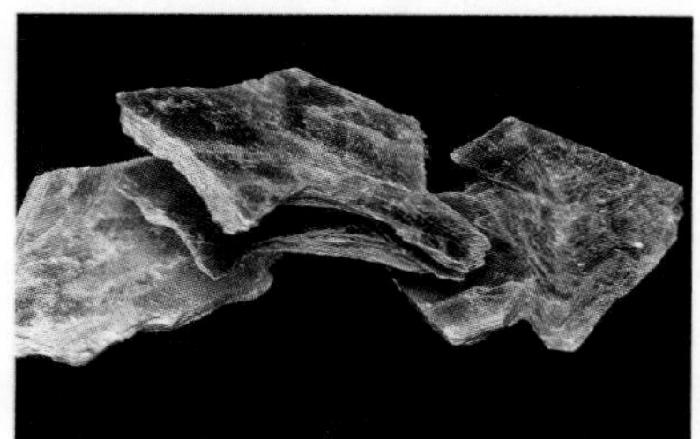

Figure 15-8 The micas are aluminum silicates having a sheet-type cleavage. As fine flakes, mica is the shiny mineral in some metamorphic rocks. Shown is white mica.

Ferromagnesian Minerals This name refers to a large group of minerals with varied properties. All of them are silicates of iron and magnesium, and most of them contain other metallic elements as well (for instance, calcium). Black mica belongs to this group; its composition includes H, K, and Al in addition to Fe and Mg. Nearly all ferromagnesian minerals are dark green to black (Fig. 15-9), but apart from composition and color the members of this group differ greatly from one another. The most abundant dark-colored constituents of common rocks belong to this group.

Figure 15-9 Olivine is an olive-green ferromagnesian mineral (a magnesium-iron silicate) that occurs mainly in igneous rocks.

Figure 15-10 Water molecules fit readily into the layered structures of clay minerals such as kaolin, shown here. When baked, wet clay loses its water content and becomes hard due to the formation of silicates that bind together the clay particles. The transformation of clay from a soft, easily shaped material into a rigid one is the basis of such ceramic products as bricks, pottery, and porcelain.

Figure 15-11 Calcite, the chief constituent of limestone and marble, consists of calcium carbonate crystals.

Clay Minerals This is a group of closely related minerals that are the chief constituents of clay. All are silicates of aluminum, some with a little Mg, Fe, and K. They consist of microscopic crystals, white or light-colored when pure, often discolored with iron compounds. They have a dull luster and are very soft, forming a smooth powder when rubbed between the fingers. Clay minerals have a low density and absorb water readily. They are distinguished from chalk by softness. Kaolin, one of the clay minerals, is an important ingredient in the manufacture of ceramics, paper, paint, and certain plastics (Fig. 15-10).

Calcite Calcite ($CaCO_3$) crystals are hexagonal, somewhat like those of quartz (Fig. 15-11). Unlike quartz they show perfect cleavage in three directions at angles of about 75°, so that fragments of calcite have a characteristic rhombic shape in which opposite sides are parallel. They are colorless or light in color, with a glassy luster. They are hard enough to scratch mica or a fingernail, but can be scratched by glass or by a knife blade. Like quartz, calcite is a common mineral of veins and crystal aggregates in cavities. It is the chief constituent of the common rocks limestone and marble and commercially serves as a source of lime for glass, mortar, and cement. Eggshells are mainly calcite, incidentally.

15.3 Igneous Rocks

Once Molten, Now Solid

There seems no limit to the variety of rocks on the earth's surface. We find coarse-grained rocks and fine-grained rocks, light rocks and heavy rocks, soft rocks and hard rocks, rocks of all sizes, shapes, and colors. But if we look closely, we will find order in this diversity, and a straightforward scheme for classifying rocks according to their origin has been developed.

1. **Igneous rocks** are rocks that have cooled from a molten state. The formation of some igneous rocks can actually be observed when molten lava cools on the side of a volcano. For others an igneous origin inside the earth is inferred from their composition and structure. Two-thirds of crustal rocks are igneous.
2. **Sedimentary rocks** consist of materials derived from other rocks and deposited by water, wind, or glacial ice. Some consist of separate rock fragments cemented together; others contain material precipitated from solution in water. Although sedimentary rocks make up only about 8 percent of the crust, three-quarters of surface rocks are of this kind.
3. **Metamorphic rocks** are igneous or sedimentary rocks that have been changed, or metamorphosed, by heat and pressure deep under the earth's surface. The changes may involve the formation of new minerals or simply the recrystallization of minerals already present.

Properties of Igneous Rocks The minerals in igneous rocks usually appear in the form of irregular grains that consist of interlocking crystals. This is to be expected when crystals grow together and interfere with one another's development. The principal minerals of these rocks contain silicon: quartz, feldspar, mica, and the ferromagnesians.

The siliceous liquids from which igneous rocks form are thick and viscous, much like molten glass. Sometimes, in fact, molten lava has the right composition and cools rapidly enough for crystals to have no time to develop. The result is a natural glass—the black, shiny rock called **obsidian** (Fig. 15-12).

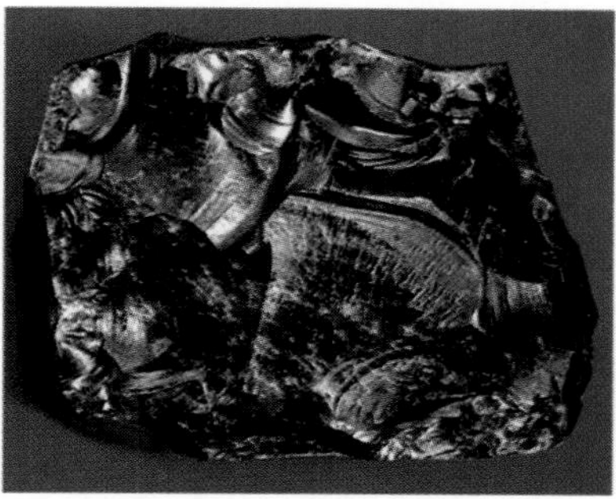

Figure 15-12 Obsidian is a glassy rock of volcanic origin.

More often cooling is slow enough to allow mineral crystals to form. If cooling is fairly rapid and if the molten material is highly viscous, the resulting rock may consist of tiny crystals or partly of crystals and partly of glass (Fig. 15-13). If cooling is very slow, mineral crystals grow large and a coarse-grained rock is formed (Fig. 15-14). Table 15-1 lists some common igneous rocks according to composition and grain size.

Figure 15-13 Basalt is a dark, fine-grained rock that emerged molten from the earth's interior. The ocean floors, under a thin sedimentary layer, consist of basalt, as do the volcanic islands of Hawaii and Iceland. Shown is pahoehoe basalt whose smooth, ropy surface resulted from the rapid cooling of a very fluid lava that came out of the opening in the foreground.

Grain size usually tells us not only the rate of cooling but also the environment in which a rock was cooled. Sufficiently fast cooling to give fine-grained rocks is common when molten lava reaches the earth's surface from a volcano and spreads out in a thin flow. Coarse-grained rocks, on the other hand, have cooled slowly, which must have occurred well beneath the surface. Such rocks are now exposed only because erosion has carried away the material that once covered them.

Figure 15-14 The faces of four American presidents (Washington, Jefferson, Theodore Roosevelt, and Lincoln) are carved in the granite of Mt. Rushmore, South Dakota. Granite is a coarse-grained igneous rock, generally light in color, in which quartz and feldspar are abundant. The faces of the feldspar crystals glisten, which gives granite an attractive appearance. Granite and similar rocks underlie the continents.

Table 15-1 Some Igneous Rocks

Mineral Composition	Coarse-Grained Rocks	Fine-Grained Rocks
Quartz Feldspar Ferromagnesian minerals	Granite	Rhyolite
No quartz Feldspar predominant Ferromagnesian minerals	Diorite	Andesite
No quartz Feldspar Ferromagnesian minerals predominant	Gabbro	Basalt

15.4 Sedimentary Rocks

Compacted Sediments or Precipitates from Solution

Sediments laid down by wind, water, or ice can become rock through the pressure of overlying deposits and by the gradual cementing of their grains with material deposited from underground water. The resulting rocks usually have distinct, somewhat rounded grains that have not grown together like the crystals of igneous rocks. A few sedimentary rocks, however, consist of intergrowing mineral grains that precipitated from solution in water.

Since sediments are normally deposited in layers, most sedimentary rocks have a banded appearance owing to slight differences in color or grain size from one layer to the next. Sedimentary rocks may often be recognized at a glance by the presence of fossils—remains of plants or animals buried with the sediments as they were laid down.

Figure 15-15 This conglomerate incorporates rounded volcanic and granitic rock fragments.

Sedimentary rocks may be divided into two groups according to the nature of the original sediments, as in Table 15-2. The three **fragmental rocks** are distinguished by their grain size. **Conglomerate** is cemented gravel whose fragments may have any composition and any size, from pebbles to boulders (Fig. 15-15). Conglomerate becomes **sandstone** as fragment size decreases. Sand grains may consist of many different minerals, but quartz is generally the most abundant. The hardness of sandstone and conglomerate depends largely on how well their grains are cemented together. Some varieties crumble easily. Others, especially those with silica as the cementing material, are among the toughest of

Table 15-2 Some Sedimentary Rocks

Group	Type	Constituents
Fragmental rocks	Conglomerate	Rock fragments
	Sandstone	Quartz usually most abundant
	Shale	Clay minerals
Chemical and biochemical precipitates	Chert	Microcrystalline quartz
	Limestone	Calcite

Figure 15-16 Shale is a soft sedimentary rock that has consolidated from mud deposits. These shale layers in the Hudson River Valley of New York were formed over 400 million years ago.

rocks. **Shale** is consolidated mud or silt; it is a soft rock usually composed of thin layers (Fig. 15-16).

Limestone is a fine-grained rock that consists chiefly of calcite. It may be formed either as a chemical precipitate or by the consolidation of shell fragments (Fig. 15-17). Small amounts of impurities may give limestone almost any color. **Chalk** is a loosely consolidated variety of limestone, often made up mainly of the shells of tiny one-celled animals.

Many Native American arrowheads are made of either the igneous rock obsidian or the sedimentary rock **chert** (microcrystalline quartz). Both rocks are hard and have sharp edges when broken. Two familiar varieties of chert are **flint** (Fig. 15-18) and **jasper.** Fragments of chert show the same sharp edges and smooth, concave surfaces as broken quartz or

Figure 15-17 Limestone deposit in western Texas. Some limestones originate as precipitates from solution; others have consolidated from the shells of marine organisms.

Coral Reefs

A common feature of shallow waters in the tropics is limestone coral reefs (Fig. 15-19). These reefs are produced by tiny creatures called coral polyps, which extract calcium carbonate from seawater to form hard external skeletons. When the polyps die, their skeletons remain behind, with new polyps growing on top. Coral reefs harbor plants and animals of many kinds, a profusion of life comparable with that in a rain forest.

The Great Barrier Reef off Australia's east coast is a series of coral reefs that extends for about 2000 km; it is the only structure of biological origin visible from space. Bermuda, the Bahamas, and the Florida Keys are coral islands and the Pacific Ocean contains many atolls, which are rings or horseshoes of coral reefs that enclose lagoons. Numerous coral reefs have been severely damaged in recent years, mainly by the use of dynamite to kill fish near them, by pollution of various kinds from agriculture on nearby shores, and by rising water temperatures.

In all, coral reefs cover about a million square kilometers of the earth's surface. The fossil record shows that reefbuilding organisms first appeared 225 million years ago. Limestone deposits that were once coral reefs are found in Wisconsin, Illinois, Indiana, and Texas.

Figure 15-19 A coral reef in Egypt.

Figure 15-18 Flint, a type of chert, consists largely of microcrystalline quartz and hence is hard and durable. Many Native American arrowheads, like this one, were made from flint.

obsidian, but the surfaces have a characteristic waxy luster. Chert may have almost any color; often a single specimen shows bands and pockets of several different colors. Not nearly as abundant as the other sedimentary rocks just described, chert is nevertheless common in pebble beds and gravel deposits because its hardness and resistance to chemical decay enable it to survive rough treatment from streams, waves, and glaciers.

Portland Cement

Portland cement, invented in 1824, is made by heating limestone and clay together in a rotating furnace at a high temperature until they partly melt. This produces various combinations of calcium, aluminum, and silicon oxides that are ground up together with additives that prevent the mixture from setting too quickly when water is added. The resulting Portland cement—so called because, when hard, it resembles a rock found near Portland, England—reacts with water to form a strong and durable adhesive that sets in a few hours and continues to harden over a period of weeks. Most Portland cement is mixed with water, sand, and gravel to form concrete, an artificial rock that is the most widely used building material, often reinforced with embedded steel rods.

Every year 2 billion tons of Portland cement are used worldwide, an average of over a quarter of a ton per person. China is the largest consumer of cement, next is India, then the United States. Cement plants account for 5 percent of all CO_2 emissions. About 300 kg of CO_2 are given off in making a ton of cement, some from the burning of fossil fuels for the required heat but most from the chemical reactions involved. For this reason improved methods of heating, though helpful, are unable to counterbalance the growing use of cement. However, varieties of cement and concrete are being developed that permit smaller amounts to be used in a given application, which may lead to reductions in CO_2 emissions from this source.

15.5 Metamorphic Rocks

Formed from Other Rocks by Heat and/or Pressure

The enormous pressures and high temperatures below the earth's surface can profoundly change sedimentary and igneous rocks that become deeply buried. Minerals stable at the surface are often unstable when crushed and baked and may react to form different substances. Other minerals keep their identities but their crystals increase in size. Hot liquids may add some new materials and dissolve out others. So many kinds of change are possible that no general rules can be set down to tell metamorphic rocks from others.

Many metamorphic rocks are characterized by a property called **foliation,** which means the arrangement of flat or elongated mineral grains in parallel layers. This effect is caused by extreme pressure in one direction, with the mineral grains growing out sideward as the rock is squeezed. Foliation gives a rock a banded or layered appearance, and when it is broken, the rock tends to split along the bands. Layering is also characteristic of sedimentary rocks, but in them the layering is caused by slight variations in color or grain size; layering in metamorphic rocks is due to the lining up of mineral grains.

Some Common Metamorphic Rocks The commoner metamorphic rocks may be classified according to the presence or absence of foliation, as in Table 15-3.

Slate is produced by the low-temperature metamorphism of the sedimentary rock shale, whose clay minerals form tiny flakes of mica. Although the individual flakes are too small to be seen, mica is responsible for the shiny surfaces seen whenever slate is split along its foliation (Fig. 15-20). Slate is harder than shale, finely foliated, and usually black or dark gray but sometimes light-colored.

Table 15-3 Some Metamorphic Rocks

Group	Type	Constituents	Origin
Foliated rocks	Slate	Mica and usually quartz, both in microscopic grains	Shale
	Schist	Mica and/or a ferromagnesian mineral, usually quartz also	Shale or fine-grained igneous rock
Foliated and banded rocks	Gneiss	Quartz, feldspar, mica	Various
Unfoliated rocks	Marble	Chiefly calcite	Limestone
	Quartzite	Chiefly quartz	Sandstone

Schist is formed from shale at higher temperatures than those that give slate, or from fine-grained igneous rocks. In it the mineral grains responsible for the foliation are large enough to be visible, giving the foliation surfaces a characteristic spangled appearance. Schist does not split as easily along the foliation as slate does, and its surfaces are rougher.

Gneiss is a coarse-grained rock formed under conditions of high temperature and pressure from almost any other rock except pure limestone and pure quartz sandstone. Its composition naturally depends on the nature of the original rock, but quartz, feldspar, and mica are the commonest minerals. In appearance gneiss resembles granite, except for its banding and foliation (Fig. 15-21).

The metamorphism of pure limestone and of pure quartz sandstone are relatively simple processes. Since each consists of a single mineral of simple composition, heat and pressure can produce no new substances but instead cause the growth and interlocking of crystals of calcite and quartz. Thus limestone becomes **marble** (Fig. 15-22), a rock composed of calcite in crystals large enough to be easily visible, and sandstone becomes the hard rock **quartzite.** Quartzite sometimes looks like sandstone, but its grains are so firmly intergrown that it splits across separate grains when broken to give smooth fracture surfaces in contrast to the rough surfaces of sandstone.

Figure 15-20 Slate results from the metamorphism of shale under pressure. This outcrop is in Antarctica.

Within the Earth

The earth's solid crust together with the atmosphere and oceans above is directly accessible to our instruments, and we may legitimately hope one day to understand their structures and behavior in detail. The interior of the earth, however, is beyond our direct reach. What we need to study it is some kind of indirect probe, and the waves sent out by earthquakes have turned out to be ideal for this purpose. Largely through the analysis of earthquake waves we now know a great deal about the earth's interior, which is hardly less remote than the most distant star, and we are continually learning more.

15.6 Earthquakes

When Our Planet Trembles

An **earthquake** consists of rapid vibrations of rock near the earth's surface. A single shock usually lasts no more than a few seconds, though

Figure 15-21 The metamorphic rock gneiss is foliated and has bands of different material.

Figure 15-22 The metamorphic rock marble is often used for statues, such as this larger-than-life one of David by Michelangelo in Florence, Italy.

The Richter Scale

Earthquake magnitudes are often expressed on the **Richter scale.** Each step of 1 on this scale represents an increase in vibration amplitude of a factor of 10 and an increase in energy release of a factor of about 30. Thus an earthquake of magnitude 5 produces vibrations 10 times larger than a quake of magnitude 4 and releases 30 times more energy. An earthquake of magnitude 0 is barely detectable; if the energy given off by such an earthquake could be concentrated, it would be just about enough to blow up a tree stump. An earthquake of magnitude 3 would be felt by people living near the location of the quake, and some damage to structures would occur when the magnitude is 4 or 5. Significant destruction is likely if the magnitude is 6 or more. The energy given off in a magnitude 9.5 earthquake, the strongest observed in the past century—it occurred off the Chilean coast (Fig. 15-23) and left 2 million people homeless—is perhaps 20 times the energy content of the world's yearly production of coal and oil. The Indonesian earthquake of 2004 that caused the tsunami described in Sec. 14.12 was nearly this severe; it moved every spot on the earth's surface by at least a centimeter, although so slowly far from the quake itself that the motion could not be felt. A magnitude 8 earthquake occurs somewhere an average of once a year, a magnitude 9 earthquake once every 30 or 40 years.

Figure 15-23 Some major earthquakes and their magnitudes on the Richter scale. The San Francisco and Tokyo earthquakes, together with the fires that followed them, led to the almost complete destruction of those cities. Over a quarter of a million people died in the Tangshan earthquake. The 2005 earthquake in Indonesia was an aftershock of the 2004 quake there.

severe quakes may last for as much as 3 min. Even in such brief times the damage done may be immense. Widespread fires often follow earthquakes in inhabited regions with broken water mains hindering their control, and landslides are common. Usually the first shock is the most severe, with weaker and weaker disturbances following from time to time for days or months afterward. A major earthquake may be felt over many thousands of square kilometers, but its destructiveness is limited to a much smaller area.

Of the million or so earthquakes per year strong enough to be noticed, only a few liberate enough energy to do serious damage (Fig. 15-24). About 15 really violent earthquakes occur each year on the average. When one of them happens to involve a densely populated urban area, the effects can be appalling: over 250,000 people died in the 1976 earthquake in Tangshan, China, as buildings collapsed around them (Fig. 15-25).

The great majority of earthquakes are caused by the sudden movement of large blocks of the earth's crust less than 70 km from the surface along fracture lines called **faults.** When the stresses that develop in the crust in a certain region become too great for the rock there to support, one side of a fault slips past the other side (Fig. 15-26). This movement causes vibrations that send out waves that may travel for long distances from their origin. Most earthquakes are over in a few seconds, but some last longer. For instance, the earthquake that caused the giant tsunami of 2004 took almost 10 minutes as the sides of a fault under the seabed off the Indonesian island of Sumatra shifted by 15 m over a distance of hundreds of kilometers.

Regions in which severe earthquakes are comparatively frequent include the mountain chains that fringe the Pacific and a broad belt that extends from the Mediterranean basin across southern Asia to China

Figure 15-24 Highway affected by the earthquake that shook Loma Prieta, California, and its vicinity in 1989. Horizontal vibrations of the ground are responsible for most of the damage an earthquake causes.

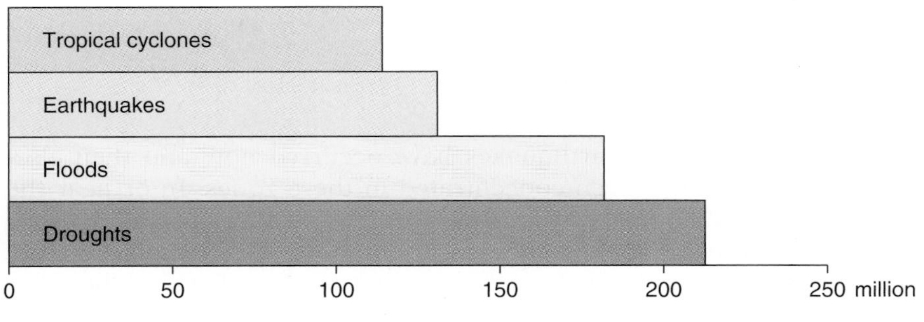

Figure 15-25 The average number of people affected by the 700 or so major natural disasters that occur each year.

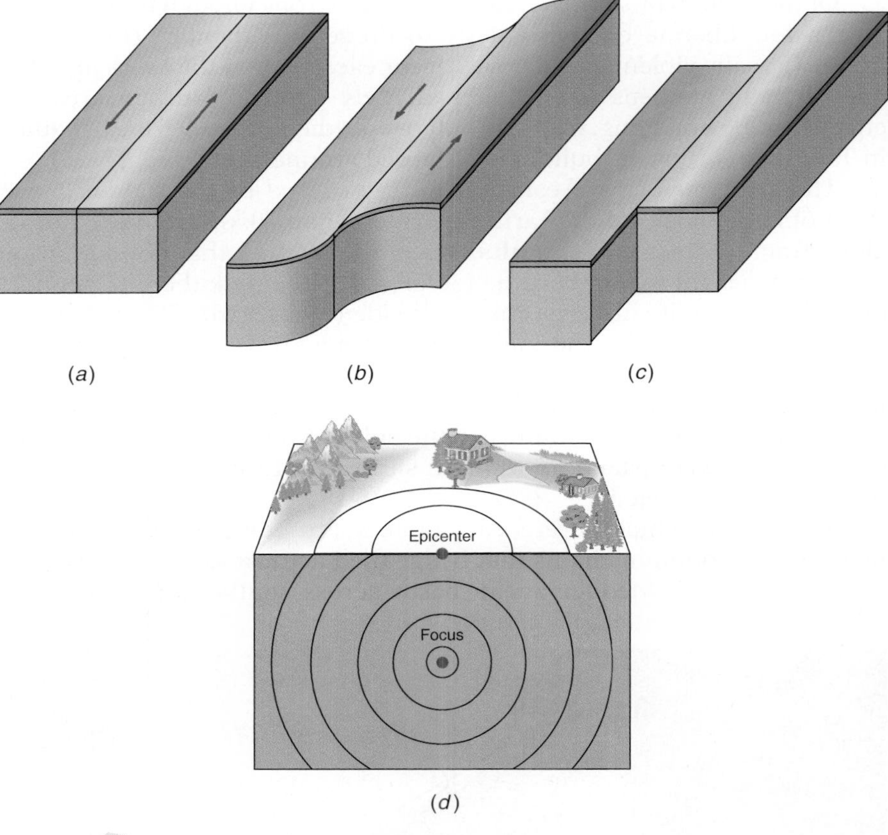

Figure 15-26 How an earthquake occurs. (*a*) The sides of a fault between crustal blocks have stuck together, preventing movement even though the forces (arrows) on them remain active. (*b*) As time goes on, stresses build up in the adjacent blocks, which deform as a result. (*c*) When the locked-in stresses become too great, the blocks suddenly shift to release them. The stored-up elastic energy powers the vibrations that constitute an earthquake. (*d*) The **focus** of an earthquake is the place where the crustal blocks moved; the **epicenter** of the quake is the point on the surface directly over the focus.

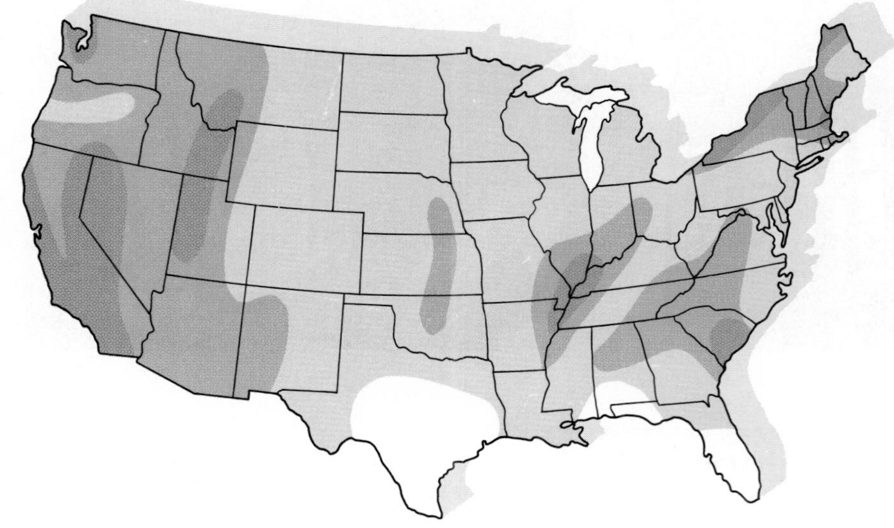

Figure 15-27 Earthquakes are expected during the next 100 years in the parts of the United States shown in color. The darker the color, the greater the likely damage. The earthquake probability is especially high in California because locked-in stresses have built up in many of the thousands of faults that riddle the region. The most severe earthquakes in U.S. history were a trio that occurred in the winter of 1811–1812 at New Madrid in southeastern Missouri, which is in the dark spot south of the Great Lakes in this map. The quakes were felt over most of the country east of the Rockies; the vibrations set church bells ringing as far away as Boston. The map does not show Hawaii or Alaska. In fact, the most earthquake-prone state is Alaska, which experiences a quake of magnitude 7 in most years.

(see Fig. 15-57). Major earthquakes have occurred now and then elsewhere, but most have been concentrated in these zones. In or near the earthquake belts lie most of the world's active volcanoes— which, as we shall see in Chap. 16, is no coincidence. Figure 15-27 is a map prepared by the U.S. Coast and Geodetic Survey that shows where earthquakes may be expected in the United States in the next 100 years.

A GEOPHYSICIST AT WORK

Andrea Donnellan
Jet Propulsion Laboratory and University of Southern California

I am a geophysicist who uses various types of space and earthbased data to monitor motions of the earth's crust and learn more about tectonic and earthquake processes. Traditional instruments such as seismometers measure brittle or elastic motion of the earth's crust resulting from earthquakes. However, the surface of the earth is constantly in motion as the tectonic plates move and grind past one another. Space-based observations have made it possible to measure that constant motion, giving us new insights into plate tectonics and earthquakes. We use the Global Positioning System (GPS) to precisely locate points on the ground to within a few millimeters of accuracy. By repeated measurements we can determine how the locations of the points change with time. We also use a technique called Interferometric Synthetic Aperture Radar (InSAR) to construct beautiful images of interference fringes that show how the surface of the earth is moving. We use computer programs to simulate and model the behavior of the earth's crust.

I've wanted to be a scientist for as long as I can remember, but I particularly enjoy being a geophysicist because my work encompasses a wide variety of skills ranging from technical fieldwork to mathematics and computer skills. In a typical year I might carry out fieldwork in Bolivia, California, Mongolia, or Antarctica; spend long productive hours at my computer; or meet with colleagues at various places around the world. I find that much of my current work involves collaboration with many individuals because we use so many different types of data and computer models. We are able to make much better progress on the earthquake problem by combining the expertise of many different people. For this reason, my QuakeSim project brings together several researchers from disciplines ranging from computer science to geology to construct computer simulations of earthquake processes. I am actively involved with the Southern California Earthquake Center and the Asia Pacific Economic Cooperation (APEC) on Earthquake Simulations. I am also acting as the project scientist for a mission called DESDynI (Deformation, Ecosystem Structure, and Dynamics of Ice) to study natural hazards, such as earthquakes and volcanoes, as well as ice sheets and the earth's forests.

We have learned some very interesting things about earthquakes by combining geological, GPS, and InSAR data with computer models. For example, the mountains near the 1994 Northridge, California, earthquake grew 40 cm taller as a result of the earthquake. We continued to make measurements of the region and have learned that the mountains continued to grow another 12 cm in the 2 years following the earthquake. Ninety percent of that motion occurred quietly; the aftershocks accounted for just 10 percent of the motion that we observed. Through computer models we were able to determine that the fault that broke in the earthquake continued to slip for those additional 2 years, redistributing stress on the fault. By studying other parts of the world we can learn more about how the tectonic plates behave, which improves our understanding of earthquakes. Spaceborne missions can systematically produce data and imagery from around the world, helping us use earth's many deforming regions and earthquakes to learn about the processes much more quickly.

Computer models are particularly useful to understanding tectonic and earthquake processes. The tectonic plates move very slowly and the complete earthquake cycle or process occurs over timescales of hundreds of thousands of years. By constructing models in the computer we simulate the whole process, in effect speeding it up so that we can study the entire system. This will help us understand how earthquakes occur and how earthquake faults interact. We also look for patterns in data using statistical and mathematical techniques. This type of analysis points out subtle features in data that even a trained eye might not observe.

I have found my career to be extremely rewarding. Being a scientist provides a satisfying balance of office work, fieldwork, and human interaction. I am a pilot and enjoy flying over and photographing faults. I returned to school to obtain a degree in computer science because much of my work is so computationally intensive. For 4 years my time was spent managing earth, planetary, and space scientists at JPL. I am now focusing on a mission to use InSAR and lidar (light detection and ranging) to study earthquakes, volcanoes, and other land surface processes. I am constantly being exposed to new science problems, which I find very exciting. My training and science research enables me to keep abreast of many fields of science and work with a variety of scientists.

Earthquakes give no obvious warning other than weak foreshocks that occasionally warn of larger quakes to come. Various subtle effects were at one time or another claimed to precede earthquakes, but none has proved a reliable predictor. Various animals, dogs and snakes in particular, sometimes behave strangely before earthquakes, but nobody knows exactly what they may be responding to. Despite intensive monitoring, the severe earthquakes that rocked California in 1989 and 1993 came as surprises. Possibly earthquakes are inherently unpredictable: once crustal stresses have built up somewhere, the situation may be so unstable that even a fairly minor event of some kind may be enough to trigger a quake without giving an unambiguous advance signal.

15.7 Structure of the Earth

Core, Mantle, and Crust

When an earthquake occurs at a fault in the crust, the rocks on both sides of the fault vibrate and send out waves that travel both through the earth's interior—"body waves"—and along its surface—"surface waves." The two kinds of body waves and the two kinds of surface waves are shown in Fig. 15-28.

An easy way to remember the difference between P and S waves is to think of P waves as "push-pull" vibrations and of S waves as "shakes." P waves are the fastest and so arrive first at a distant point when an earthquake occurs somewhere (Fig. 15-29). The S waves, which are slower, come next. The surface waves, which have to travel along the ground rather than the shorter distance through the earth, appear last. However, the surface waves may produce the strongest vibrations, particularly when the distance is no more than a few thousand kilometers. The Love surface waves are responsible for most earthquake damage to buildings and other structures.

The internal structure of the earth was discovered with the help of P and S earthquake waves. These waves do not travel in straight lines inside the earth (except directly downward toward the earth's center) because of refraction. As we learned in Chap. 7, refraction refers to the change in direction of a wave when its speed changes. The speeds of P and S waves change with depth in the earth in two ways, which have different effects on these waves.

One change in speed is a gradual increase with depth. This causes the P and S waves to travel in curved paths within the earth, as shown in Fig. 15-30. Figure 7-31 shows a similar effect in ocean waves as they approach the shore.

The other change in speed is more abrupt and occurs because the earth's interior consists of layers of different materials. When an earthquake body wave crosses the boundary between two layers, its speed changes, and the result is a sharp change in direction. Figure 7-34 shows a similar effect in light waves as they go from air into glass. As we see in Fig. 7-36, waves that move perpendicularly through a boundary are not deflected.

Shadow Zones Let us suppose an earthquake occurs somewhere. We consult various seismograph stations and find that most of them—never all—have detected P waves from this event. Curiously, the stations that

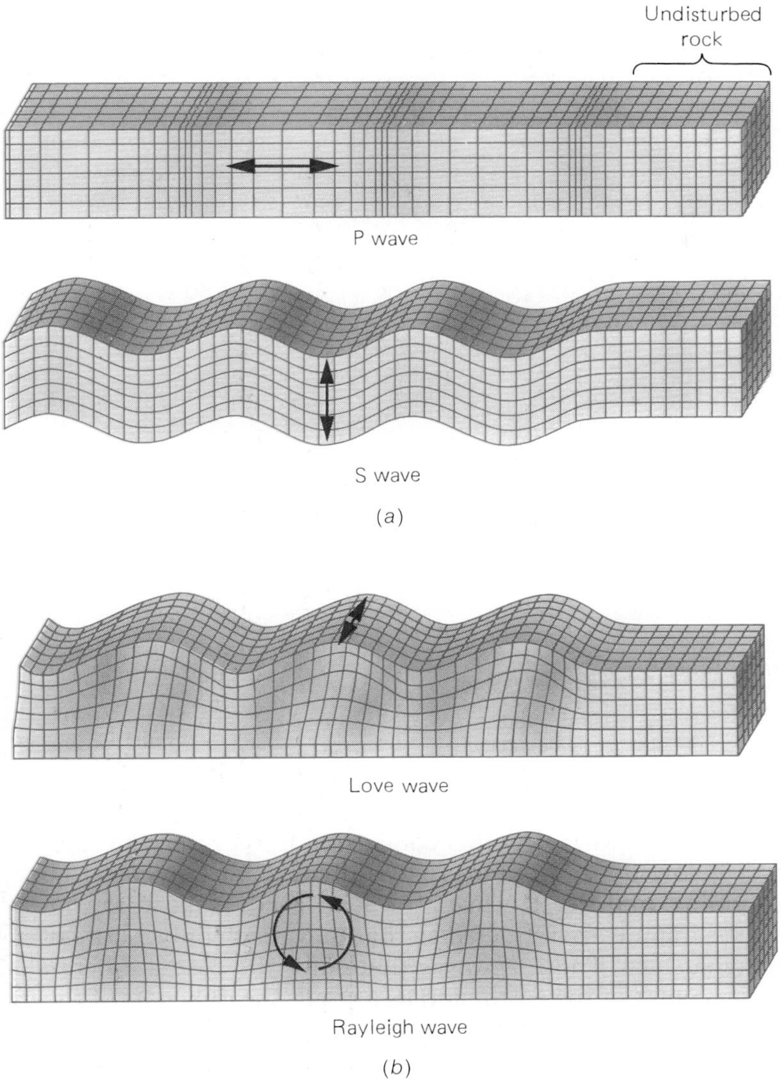

Figure 15-28 (*a*) Earthquake body waves travel through the earth's interior. P waves are longitudinal, like sound waves. S waves are transverse, like waves in a stretched string. P waves can move through a liquid; S waves cannot. (*b*) Earthquake surface waves travel on the earth's surface. Love waves are transverse, with their vibrations parallel to the surface. Rayleigh waves involve rotary motions, like the water waves of Fig. 7-2. All the waves shown here are moving from left to right. The arrows show the directions in which the rock particles move. (Love waves are named after the English mathematician Augustus Love and Rayleigh waves after the English physicist Lord Rayleigh.)

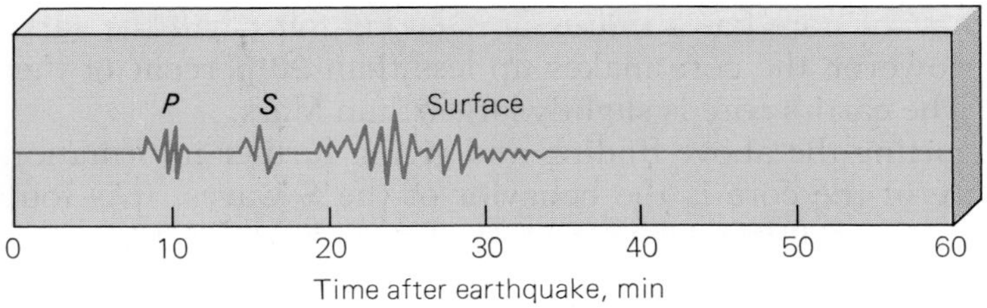

Figure 15-29 Earthquake waves are detected by instruments called **seismographs.** This is a record of waves from an earthquake that occurred about 5000 km from the location of the seismograph.

P waves did not reach lie along a large band on the side of the earth opposite the quake. We would find, if we looked at the records of other earthquakes, that no matter where they took place, similar **shadow zones** for their P waves existed. This is the chief clue that confirmed an early suspicion that the earth's interior is not uniform.

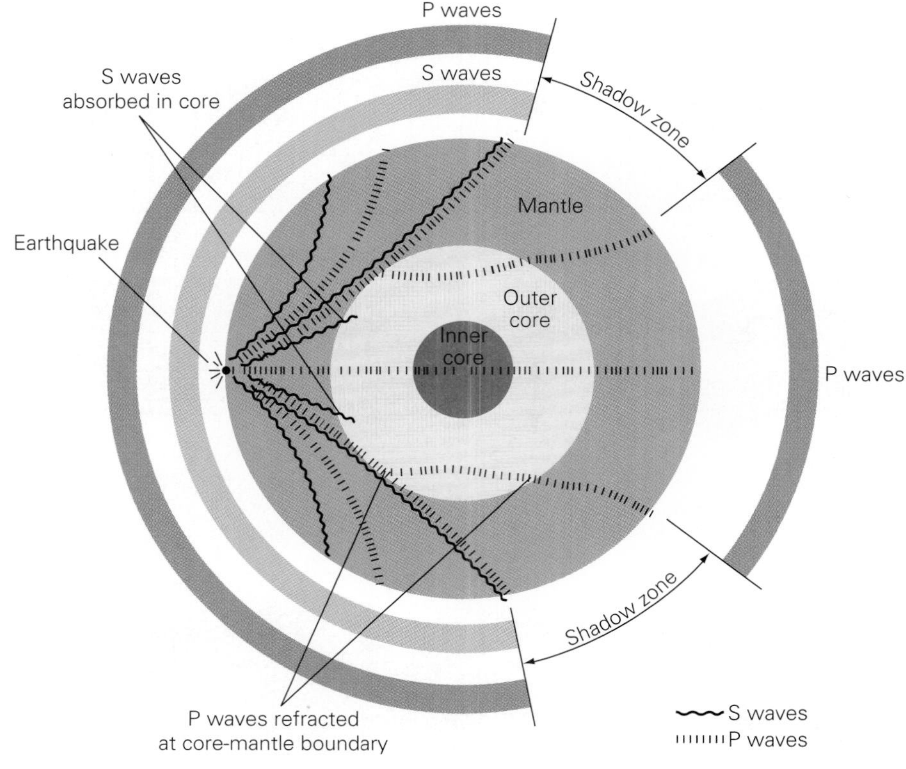

Figure 15-30 How earthquake waves travel through the earth. The existence of a shadow zone where neither P nor S waves arrive is evidence for a core. The inability of S waves to get through the core suggests that at least the outer part of it is liquid. The inner core is believed to consist of iron crystals.

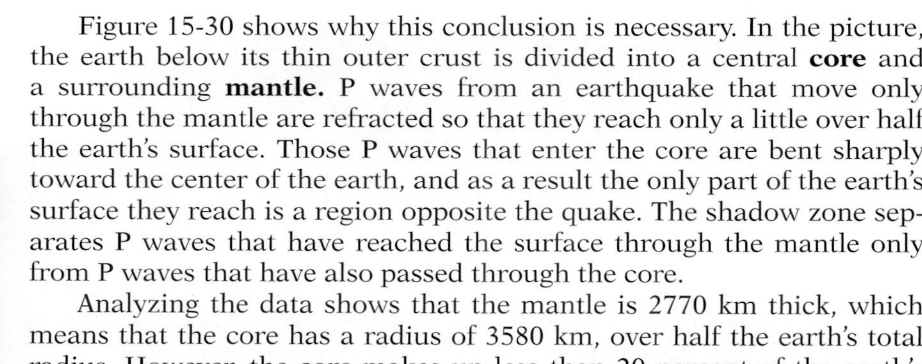

Figure 15-30 shows why this conclusion is necessary. In the picture, the earth below its thin outer crust is divided into a central **core** and a surrounding **mantle.** P waves from an earthquake that move only through the mantle are refracted so that they reach only a little over half the earth's surface. Those P waves that enter the core are bent sharply toward the center of the earth, and as a result the only part of the earth's surface they reach is a region opposite the quake. The shadow zone separates P waves that have reached the surface through the mantle only from P waves that have also passed through the core.

Analyzing the data shows that the mantle is 2770 km thick, which means that the core has a radius of 3580 km, over half the earth's total radius. However, the core makes up less than 20 percent of the earth's volume. The earth's core is slightly larger than Mars.

Supporting the above finding and giving further information about the nature of the core is the behavior of the S waves. It is found that these cannot get through the core at all (see Fig. 15-29). Because they are transverse, S waves cannot pass through a liquid, so the conclusion is that the earth's core is liquid! A liquid core accounts not only for the behavior of S waves but also for the marked changes in the speed of P waves when they enter and leave the core.

Figure 15-31 Inge Lehmann (1888–1993).

Sensitive seismographs have detected faint traces of P waves in the shadow zones, which suggested to Inge Lehmann (Fig. 15-31), a Danish geologist, that within the liquid core is a smaller, solid inner core. This idea is now well-established; the inner core is believed to have a radius

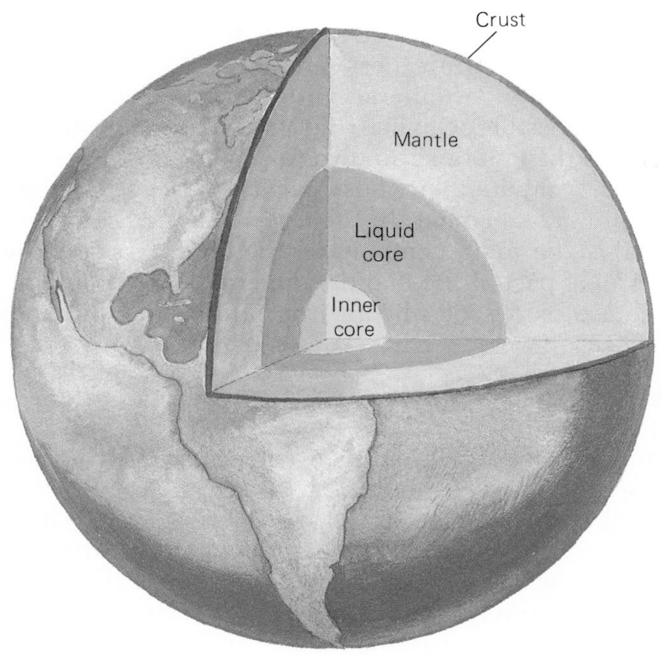

Figure 15-32 Structure of the earth. The mantle constitutes 80 percent of the earth's volume and about 67 percent of its mass.

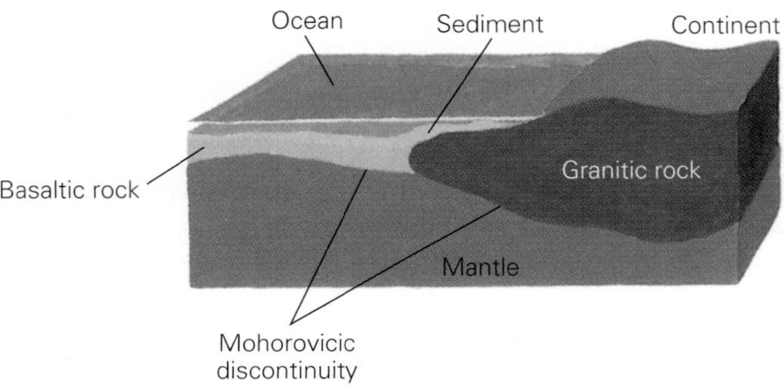

Figure 15-33 The Mohorovicic discontinuity separates the earth's crust from the mantle below it. The crust is thicker and has a different composition under the continents than under the oceans.

of 1215 km, almost 2/3 that of the moon. Thus the earth's interior has the onionlike structure shown in Fig. 15-32.

The Moho From observations of the waves from a 1909 earthquake it became clear that there is a distinct difference between the outer shell of the earth and the denser mantle below it. The surface between them is known as the **Mohorovicic discontinuity** (or just **Moho**), after its Croatian discoverer, and is considered to be the lower boundary of the crust. Under the oceans the crust is seldom much more than 5 km thick. Under the continents, though, the crust averages about 35 km, as much as 70 km under some mountain ranges (Fig. 15-33).

15.8 The Earth's Interior

A Mantle of Rock, a Core of Molten Iron

In the absence of a hole over 6000 km deep, anything said about the composition of the earth's interior must be a hypothesis, but a great deal of evidence supports the hypotheses that have been made.

In the case of the upper mantle, most studies point to igneous rocks composed mainly of ferromagnesian minerals. Deep in the mantle enormous pressures squeeze minerals into crystal structures that are the most compact possible. Olivine (Fig. 15-9), probably the most abundant mineral in the mantle, is known to occur in two forms, the normal one of the crust and another whose particles are packed together in an especially tight arrangement. In the upper mantle the more open variety is probably the main component, the denser variety lower down. In the innermost part of the mantle, minerals have probably separated into very dense oxides of silicon, iron, and magnesium that occupy even less space.

The Core Now we come to the liquid outer core. The average density of the earth is about twice the average density of the rocks at the earth's surface. The material of the mantle is only moderately denser than surface rocks, so that the core must be very heavy indeed.

Several clues point to iron as the logical candidate for the core material. It has almost the right density, it is liquid at the estimated pressure and temperature of the core, and it is abundant in the universe generally. Furthermore, iron is a good conductor of electricity, which is necessary in order to explain the earth's magnetism (Sec. 15.9). Because those meteorites that contain iron also contain some nickel, a reasonable conclusion is that there is nickel in the core also. As for the solid inner core, the kernel of the earth, many geophysicists believe it to be crystalline, because of the pressures there, and to consist of iron plus nickel and a lighter element as well, probably sulfur.

The Earth's Heat Below its surface the earth is extremely hot (Fig. 15-34). The rate at which heat flows outward from the interior is

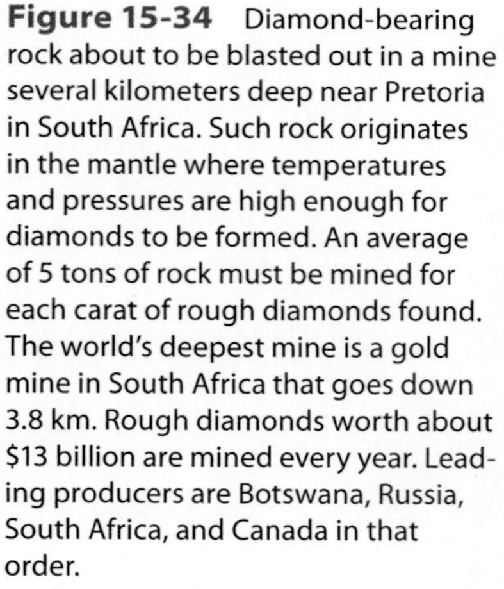

Figure 15-34 Diamond-bearing rock about to be blasted out in a mine several kilometers deep near Pretoria in South Africa. Such rock originates in the mantle where temperatures and pressures are high enough for diamonds to be formed. An average of 5 tons of rock must be mined for each carat of rough diamonds found. The world's deepest mine is a gold mine in South Africa that goes down 3.8 km. Rough diamonds worth about $13 billion are mined every year. Leading producers are Botswana, Russia, South Africa, and Canada in that order.

immense, about 100 times greater than the energy involved in such geological events as volcanoes and earthquakes. There is plenty of heat to spare to account for mountain building and other deformations that occur in the crust. In fact, the geological history of the earth is mainly a consequence of the steady heat streaming through its outer layers. Temperatures inside the earth are believed to go from perhaps 375°C at the top of the mantle to perhaps 5000°C in the core.

Part of the earth's heat is a relic of its early history, but most of it comes from radioactive uranium, thorium, and potassium isotopes. These isotopes are present in the minerals of the earth, and their radiations, which were discussed in Sec. 8.3, give up their energy to the minerals as they are absorbed. The earth is believed to have come into being 4.5 billion years ago as a cold clump of smaller bodies of metallic iron and silicate minerals that had been circling the sun. Heat due to radioactivity accumulated in the interior of the infant earth and in time caused partial melting. The influence of gravity then caused the iron to migrate inward to form the core while the lighter silicates rose to form the mantle and crust.

Today most of the earth's radioactivity is concentrated in the crust and upper mantle, where the heat it produces escapes through the surface and cannot collect to remelt the rest of the mantle or the inner core.

15.9 Geomagnetism

Electric Currents in the Core Seem to Be Responsible

Although the earliest known description of the compass and its use in navigation was published by Alexander Neckham in 1180, knowledge of the compass seems to have been widespread even further back in antiquity. Until 1600, however, it was believed that a force exerted by Polaris, the North Star, is what attracted magnetized needles. In that year Sir William Gilbert wrote of experiments he had performed with spherical pieces of lodestone, a naturally magnetized mineral. By comparing the direction of the magnetic force on an iron needle at various positions near a lodestone sphere with similar measurements made in other parts of the world by explorers, Gilbert concluded that the earth behaves like a giant magnet—*"magnus magnes ipse est globus terrestris."*

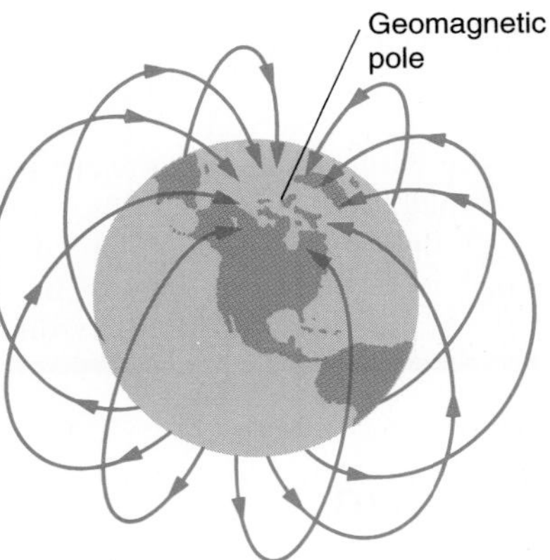

Figure 15-35 The earth's magnetic field originates in electric currents in its core of molten iron. The magnetic axis is tilted by 11° from the axis of rotation, so magnetic compasses do not point to true north.

The earth's magnetic field is much like the field that would come from a giant bar magnet located near the earth's center and tilted by 11° from the direction of the earth's axis of rotation (Fig. 15-35). No such magnet can possibly exist, since iron loses its magnetic properties above about 770°C and most of the earth's interior is much hotter than that. Instead, the magnetic field is believed to arise from electric currents generated in the swirling liquid iron of the core, a hypothesis supported by laboratory experiments. We recall from Fig. 6-33 that an electric current in the form of a loop has around it a magnetic field of the same kind as that of a bar magnet. Less than a billionth of the earth's rotational energy is needed to generate the observed magnetic field.

BIOGRAPHY William Gilbert (1544 –1603)

Gilbert practiced medicine in London and became the personal physician to Queen Elizabeth I in 1600. At about that time he published *De Magnete,* a book that described the results of his years of research on magnetism. An early advocate of experiment and observation in science—Galileo admired his work—Gilbert found that iron objects can be magnetized by rubbing them with a lodestone (an iron oxide mineral that is naturally magnetic); that heating an iron magnet causes it to lose its magnetism; that like magnetic poles repel and unlike poles attract (he introduced the word pole for the end of a magnet); and, most significant of all, that the earth itself is a giant magnet. In addition he showed that a number of popular beliefs were incorrect, for instance that garlic destroys magnetism and that magnetism cures headaches. Not until two centuries after *De Magnete* were any major additions made to our knowledge of magnetism.

Gilbert also studied static electricity, which the ancient Greeks had discovered by rubbing amber with silk, and found that other substances, too, can be charged by rubbing. Gilbert called all such substances *electrics,* after the Greek for "amber," and clearly distinguished between magnetic and electric attractions. In astronomy, Gilbert was the first prominent Englishman to support the copernican model of the universe, and he believed that the stars were at different distances from the earth, perhaps circled by habitable planets. The **gilbert** is a magnetic unit named after this remarkable man.

Field Reversals

When deposits of rock that contain magnetizable minerals are dated, specimens of different ages may have opposite magnetic polarities. The only explanation in most cases is that the earth's magnetic field has reversed itself periodically while these rocks were being formed. In the past 76 million years, 171 field reversals have apparently occurred. Such flip-flops fit in with the notion that the field is due to electric currents in the outer core, since changes may well take place in the patterns of flow in the liquid iron there from time to time. As we shall see in Sec. 16.6, these reversals helped to establish that the ocean floors are spreading, an important part of the plate tectonics picture of the earth that includes the drift of the continents.

The most recent field reversal occurred 780,000 years ago, nearly double the average interval between past reversals, so we seem to be overdue for another one. What will happen is that the field strength will decrease over a period of a few thousand years (a blink of the eye, geologically speaking) and then build up again with the field in the opposite direction. In the past hundred years the field has weakened by about 10 percent, which suggests that perhaps a reversal has already begun.

Erosion

It is not obvious that this solid earth under us, made up largely of hard, strong rock, is in a state of constant change. But rocks, hills, and mountains are permanent only by comparison with the brief span of human life, and the long history of the earth goes back not scores of years but billions of years. In this immense stretch of time continents have shifted across the globe, mountain ranges have been thrust upward and then leveled, and broad seas have appeared and disappeared.

All the processes by which rocks are worn down and by which the debris is carried away are included in the general term **erosion.** The underlying cause of erosion is gravity. Such agents of erosion as running water and glaciers derive their destructive energy from gravity, and gravity is responsible for the transport of the removed material.

15.10 Weathering

How Exposed Rocks Decay

We have all seen the rough, pitted surfaces of old stone buildings and monuments (Fig. 15-36). This kind of disintegration, brought about by rainwater and the gases in the air, is called **weathering.** Weathering contributes to erosion by preparing rocks for easy removal by the more active erosional agents, such as running water.

Some of the minerals in igneous and metamorphic rocks are especially susceptible to **chemical weathering,** since they were formed under conditions very different from those at the earth's surface. Ferromagnesian minerals are readily attacked by atmospheric oxygen, aided by carbon dioxide dissolved in water (which gives an acid solution) and by organic acids from decaying vegetation. This results in the formation of iron oxides, which give the red and brown colors that commonly appear as stains on the surface of rocks containing these minerals. Feldspars and other silicates containing aluminum are broken down to clay minerals. Among common sedimentary rocks limestone is most readily attacked by chemical weathering because calcite dissolves in weak acids (Fig. 15-37). Exposures of limestone can often be identified simply from the pitted surfaces and enlarged cracks that solution produces.

Quartz and white mica resist chemical attack and usually remain as loose grains when the rest of a rock is thoroughly decayed. Rocks consisting wholly of silica, like chert and most quartzites, are practically immune to chemical weathering.

Figure 15-36 After standing in the clean, dry air of Egypt for about 36 centuries, the carvings on Cleopatra's Needle were still sharp and clear. In 1881 the granite obelisk was moved to New York City, where the combination of climate and atmospheric pollution has almost erased the carvings. Acid rain running down the obelisk is the reason the damage increases toward the base.

Figure 15-37 Underground limestone gradually dissolves in acid water that seeps through it. When the limestone is near the surface, its disappearance can lead to a collapse of the ground above to leave a **sinkhole.** This sinkhole occurred in Winter Park, Florida. When limestone farther underground is dissolved, the result is a cave. Mammouth Cave in Kentucky is the world's longest; its various passages total 560 km.

Example 15.2

Both marble and slate are metamorphic rocks. Would you expect a marble tombstone or a slate one to be more resistant to chemical weathering?

Solution

When exposed to the atmosphere marble weathers fairly readily because its calcite content is soluble in rainwater that contains carbon dioxide. Slate consists largely of clay minerals that have metamorphosed to white mica, which is nearly as resistant as quartz to chemical weathering.

Mechanical weathering is often aided by chemical attack. Not only is the structure of a rock weakened by the decomposition of its minerals, but fragments of it are wedged apart because the chemical changes in a mineral grain usually result in an increased volume. The most effective process of mechanical disintegration that does not require chemical

Soil

Though the bulk of the earth's crust is solid rock, what we see on that part of the surface not covered by water is mainly soil with only occasional outcrops of bedrock. Soil originates in the weathering of rock, a complex disintegration process whose result is a coat of rock fragments and clay minerals mixed with varying amounts of organic matter.

Any type of rock may form the parent material of a soil. The particles of rock typically vary in size down to microscopic fineness and are intimately mixed with dark, partly decomposed plant debris called humus. The humus content decreases with depth (Fig. 15-38). A great many factors are involved in the production of soil, including microorganisms such as bacteria and fungi that are responsible for the decay of plant and animal residues and are important in maintaining the nitrogen content of soil. A significant fraction of the organic matter in soil, in fact, consists of the bodies, living and dead, of these microorganisms. Even so lowly a creature as the worm plays a vital role in mixing together the various soil constituents.

Some dust is always present in the lower atmosphere, but at various times and places gigantic dust storms of soil particles occur that can blot out daylight, make it hard to breathe, and damage crops on a large scale. In the 1930s such storms were a feature of the "dust bowl" in the Middle West of the United States. Today, on a much larger scale, they are taking place in northern China where they create dust clouds thousands of miles across that sometimes cross the Pacific to deposit dust on western North America. The chief factors responsible are the overcultivation of marginal soils, overgrazing of vegetation by sheep and goats, the cutting down of trees that anchor the soil, and the overpumping of groundwater that leaves the soil dry. Strong winds sweep millions of tons of loose soil into the air in late winter and early spring in the affected regions to produce the dust storms, which leave deserts behind. Harvests in China have been declining as a result of such ecological catastrophes, and its worried government is trying various schemes to keep the remaining soil in place.

The problem of soil erosion is not confined to China: erosion by water as well as by wind contributes to the loss of farmland soils in much of the world faster than new soil is formed. Iowa has lost about half its topsoil in the past 150 years.

Figure 15-38 Cross section of a typical soil. The darkening toward the top is due to the presence of humus.

action is the freezing of water in crevices. Just as water freezing in a car's engine on a cold night may split the engine block, so water freezing in tiny cracks is effective in disrupting rocks (Fig. 15-39). Plant roots help in rock disintegration by growing in cracks.

Weathering processes coat the naked rock of the earth's crust with a layer of debris made up largely of clay mixed with rock and mineral fragments. The upper part of the weathered layer, in which rock debris is mixed with decaying vegetable matter, is the **soil.** The formation of soil is an important result of weathering.

Figure 15-39 The expansion of water as it freezes into ice in cracks in rock has carved the sharp peak of the Matterhorn, a mountain on the border between Switzerland and Italy.

15.11 Stream Erosion

Running Water Is the Chief Agent of Erosion

By far the most important agent of erosion is the running water of streams and rivers. The work of glaciers, wind, and waves is impressive locally, but compared with running water, these other erosion agents play only minor roles in shaping the earth's landscapes. Even in deserts, mountainsides are carved with the unmistakable forms of stream-made valleys.

In a young landscape, streams and rivers are just starting their work and flow downhill swiftly with many rapids and waterfalls (Fig. 15-40). At first a river carves a narrow V-shaped channel (Fig. 15-41*a*). As time goes on, the channel deepens until the river's lower end is near the level of a nearby valley or perhaps nearly at sea level. As the rest of the channel becomes less

Figure 15-40 The running water of a stream may accomplish more erosion during a few hours of heavy rain than in months or years of normal flow. What does the work is not the moving water itself but the sand and pebbles hurled by the water at the sides and bed of its channel.

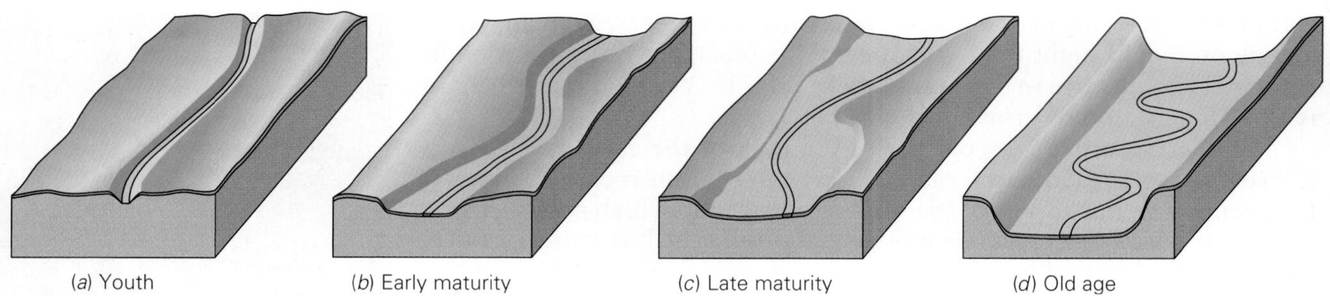

Figure 15-41 Successive stages in the development of a river valley.

steep, its sides begin to be cut away to broaden it (Fig. 15-41*b*). The eventual result is a **floodplain** with a flat floor (Fig. 15-41*c*). In dry weather the river wanders over its plain in a meandering course; in very wet weather it overflows its channel and spreads across the plain. The floodplain grows wider and wider, the river becomes more and more sluggish, and the sides of its valley become lower and lower (Fig. 15-41*d*).

During this development of the major valley, secondary streams extend their smaller valleys on either side. Soon a characteristic treelike pattern develops, separated from the patterns of adjacent rivers by sharp divides of high ground (Fig. 15-42). As floodplains widen along the main streams, divides are lowered by attack from the streams on either side. In the final stages of valley growth, when floodplains are wide and rivers broadly meandering, most of the divides are erased and those remaining are low and rounded (Fig. 15-43).

Actual landscapes seldom conform exactly to the simple valley shapes and patterns just described. One reason is the presence of rocks of different hardness: hard rocks usually remain as cliffs and high ridges, while the more easily eroded soft rocks wear away (Fig. 15-44). Many of the striking landforms produced by erosion are due simply to differences in resistance from one rock layer to the next.

Figure 15-42 The treelike drainage pattern of the Mississippi River covers more than 2.5 million square kilometers. Rivers to the west of the Continental Divide flow into the Pacific, those to its east flow into the Atlantic and the Gulf of Mexico.

Figure 15-43 The floodplain of the Sweetwater River in Wyoming. The wandering, shifting channel is typical of the old age of a river. The bends grow larger as their outer banks (where the water flows faster) are eroded, while sediments are deposited on the inner banks. Sometimes a bend becomes so extreme that its ends join together at a time of high water, which cuts a straight connection that leaves behind an oxbow lake. Such a lake can be seen to the right of the river in the middle of this photograph.

Whatever the valley shapes produced in various stages of landscape development, whatever different kinds of rock may be present, the tendency of stream erosion is to reduce the land surface to a flat plain almost at sea level. However, this outcome is seldom reached because, as we shall learn, other geologic processes continually counteract the effects of running water. There are very few regions in which geologic processes involving uplift do not occur at the same time as stream erosion. Thus most landscapes are the result of a complex of factors and reflect a balance among them rather than the action of stream erosion alone.

Figure 15-44 Parallel ridges and valleys produced by stream erosion in tilted layers of hard and soft rocks. Soft layers underlie the valleys; hard layers, the ridges. Landscapes and rock structures of this sort are typical of the Appalachian Mountains.

15.12 Glaciers

Rivers and Seas of Ice

In a cold climate with abundant snowfall, the snow of winter may not completely melt during the following summer, and a deposit of snow accumulates from year to year. Partial melting and the continual increase in pressure cause the lower part of a snow deposit to change gradually into ice. If the ice is sufficiently thick, its weight forces it slowly downhill. A moving mass of ice formed in this manner is called a **glacier.** About 10 percent of the earth's land area is today covered by glacial ice, which contains three-quarters of its freshwater.

Today's glaciers are of two main kinds:

1. **Valley glaciers**—found, for instance, in the Alps, on the Alaskan coast, in the western United States—are patches and tongues of ice lying in mountain valleys. These glaciers move slowly down their valleys and melt at their lower ends. The combination of downward movement and melting keeps their ends in roughly the same position from year to year. Valleys carved by glaciers have U-shaped cross sections instead of the V shapes produced by stream erosion (Fig. 15-45). Movement in the faster valley glaciers (a meter or more per day) is enough to keep their lower ends well below the timber line.
2. Glaciers of another type cover most of Greenland and Antarctica. These huge masses of ice thousands of square km in area that engulf

Figure 15-45 The U-shaped cross section of this valley in Alaska suggests that it was cut by the ancestor of today's glacier at a time when the climate there was colder than at present. Today's glaciers range in length up to the 109-km Hubbard Glacier in western Canada; all are smaller than in the past due to global warming.

hills as well as valleys are called **continental glaciers** or **ice caps.** They, too, move downhill, but the "hill" is the slope of their upper surfaces. An ice cap has the shape of a broad dome with its surface sloping outward from a thick central portion of greatest snow accumulation. Its motion is outward in all directions from its center. The icebergs of the polar seas are fragments that have broken off the edges of ice caps (see Fig. 4-7). Similar sheets of ice extended across Canada and northern Eurasia during the ice ages.

Glacial Erosion As a glacier moves, rock fragments held firmly by the ice at its bottom are dragged along. These fragments scrape and polish the underlying rock and are themselves ground down. Smooth, grooved rock surfaces and deposits of debris that contain boulders with flattened sides are common where the lower end of a valley glacier has melted back to reveal some of its bed. When such evidence of ice erosion is found far from present-day glaciers, we can infer that glaciers must have been active there in the past.

Glacial erosion is locally very impressive, particularly in high mountains. The amount of debris and the size of the boulders that a glacier can carry or push ahead of itself are often startling. But overall, the erosion accomplished by glaciers is small. Only rarely have they gouged rock surfaces deeply, and the amount of material they carry long distances is little compared with that carried by streams. Most glaciers of today are only feeble descendants of mighty ancestors, but even these ancestors succeeded only in modifying landscapes already shaped by running water.

15.13 Groundwater

Water, Water Everywhere (Almost)

Most of the water that falls as rain does not run off at once in streams but instead soaks into the ground. All water that thus penetrates the surface is called **groundwater.** There is more groundwater than all the freshwater of the world's lakes and rivers, though less than the water locked up in ice caps and glaciers (see Fig. 14-5).

The soil, the layer of weathered rock under it, and any porous rocks below act together as a sponge that can absorb huge quantities of water. During and just after a heavy rain all empty spaces in the sponge may be filled, and the ground is then said to be **saturated** with water. When the rain has stopped, water slowly drains away from hills into the adjacent valleys. A few days after a rain porous material in the upper part of a hill contains relatively little moisture, while that in the lower part may still be saturated. Another rain would raise the upper level of the saturated zone, prolonged drought would lower it. The fluctuating upper surface of the saturated zone is called the **water table.**

Beneath valleys the water table is usually closer to the surface than beneath nearby hills (Fig. 15-46). Groundwater in the saturated zone seeps slowly downward and sideward into streams, lakes, and swamps. The motion is rapid through coarse material like sand or gravel, slow through fine material like clay. It is this flow of groundwater that maintains streams when rain is not falling; a stream goes dry only when the water table drops below the level of its bed. A **spring** is formed where groundwater comes to the surface in a more or less definite channel.

An **aquifer** is a body of porous rock through which groundwater moves. Aquifers underlie more than half the area of the continental United States and supply about half its drinking water (Fig. 15-47).

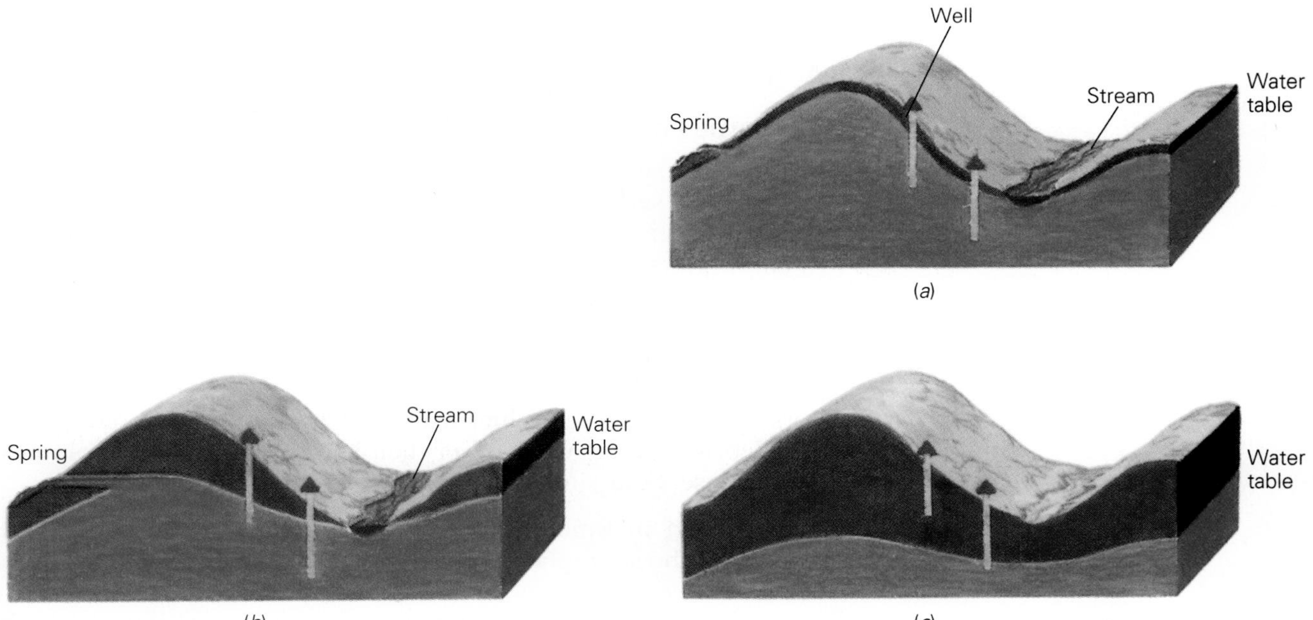

Figure 15-46 Cross sections through a landscape underlain by porous material. The position of the water table is shown (*a*) just after a heavy rain, (*b*) several days later, and (*c*) after a prolonged drought. The spring, the stream, and the upper well would be dry during the drought.

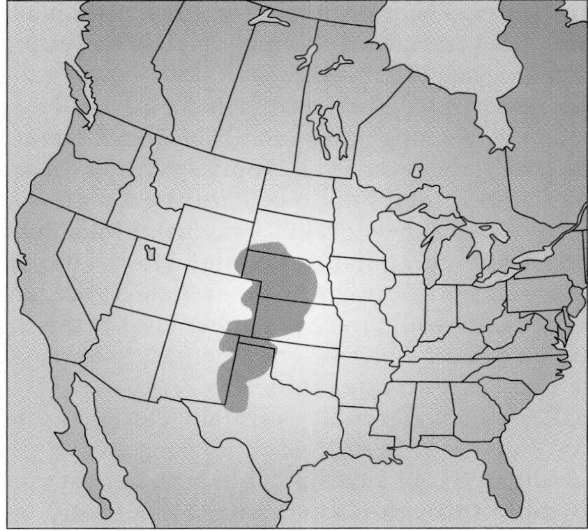

Figure 15-47 The Ogallala aquifer, the largest in the United States, supplies water for agricultural use in a heavily farmed region. It is shrinking and becoming contaminated by fertilizer runoff. Water from other aquifers throughout the world is also being withdrawn faster than natural processes can replenish it.

Because groundwater moves slowly, it can accomplish little mechanical wear, but its prolonged contact with rocks and soil allows it to dissolve much soluble material. Some of the dissolved material remains in solution (hence the "hardness" of water from many wells), and some is redeposited elsewhere. In regions underlain by limestone, the most soluble of ordinary rocks, caves are produced when water moving through tiny cracks enlarges the cracks by dissolving and removing the adjacent limestone.

15.14 Sedimentation

What Becomes of the Debris of Erosion

Most of the material transported by the agents of erosion is eventually deposited to form **sediments** of various sorts. The ultimate destination of erosional debris is the ocean, and the most widespread sediments collect in shallow parts of the ocean near continental edges. But much sedimentary material is carried to the sea in stages, deposited first in thick layers elsewhere—in lakes, in desert basins, in stream valleys. Each erosional agent has its own ways of giving up its load, and these ways leave their stamps on the deposits formed.

Rivers and streams, the chief agents of erosion, lose some of the abundant debris they carry whenever their speeds drop or their volumes of water decrease. Four sites of deposition are common:

1. Debris carried in time of flood is deposited in gravel banks and sandbars on the streambed when the swiftly flowing waters begin to recede.
2. The floodplain of a meandering river is a site of deposition whenever the river overflows its banks and loses speed as it spreads over the plain. In Egypt, for example, before construction of the Aswan Dam, the fertility of the soil was maintained for centuries by the deposit of black silt left each year when the Nile was in flood.

3. A common site of deposition, especially in the western United States, is the point where a stream emerges from a steep mountain valley and slows down as it flows onto a plain. Such a deposit, usually taking the form of a low cone pointing upstream, is called an **alluvial fan** (Fig. 15-48).
4. A similar deposit is formed when a stream's flow is stopped abruptly as the stream enters a lake or sea. This kind of deposit, built largely underwater and with a surface usually flatter than that of an alluvial fan, is called a **delta.**

Some of the material a glacier scrapes from its channel is spread as a layer of irregular thickness under the ice, and some is heaped up at the glacier's lower end where the ice melts. The pile of debris around the end and along the sides of the glacier, called a **moraine,** is left as a low ridge when the glacier melts back. Moraines in mountain valleys and in the North Central states are part of the evidence for a former wide extent of glaciation.

All the material deposited directly by ice goes by the name of **till,** an indiscriminate mixture of fine and coarse material. Till includes huge boulders that are often embedded in the fine, claylike material a glacier produces by its polishing action. Typically, most of the boulders are angular, but a few are rounded and show the flat scratched faces produced as they were dragged along the bed of the glacier.

Most important of the agents of deposition, because they handle by far the largest amount of sediment, are ocean currents. Currents deposit not only the materials eroded from coastlines by wave action but also the abundant debris brought to the ocean by rivers, wind, and glaciers. Visible deposits of waves and currents include beaches and sandbars, but the great bulk of the sediments brought to the ocean are laid down underwater (Fig. 15-49).

Figure 15-48 An alluvial fan consists of debris deposited by a mountain stream when it slowed down on reaching the plain.

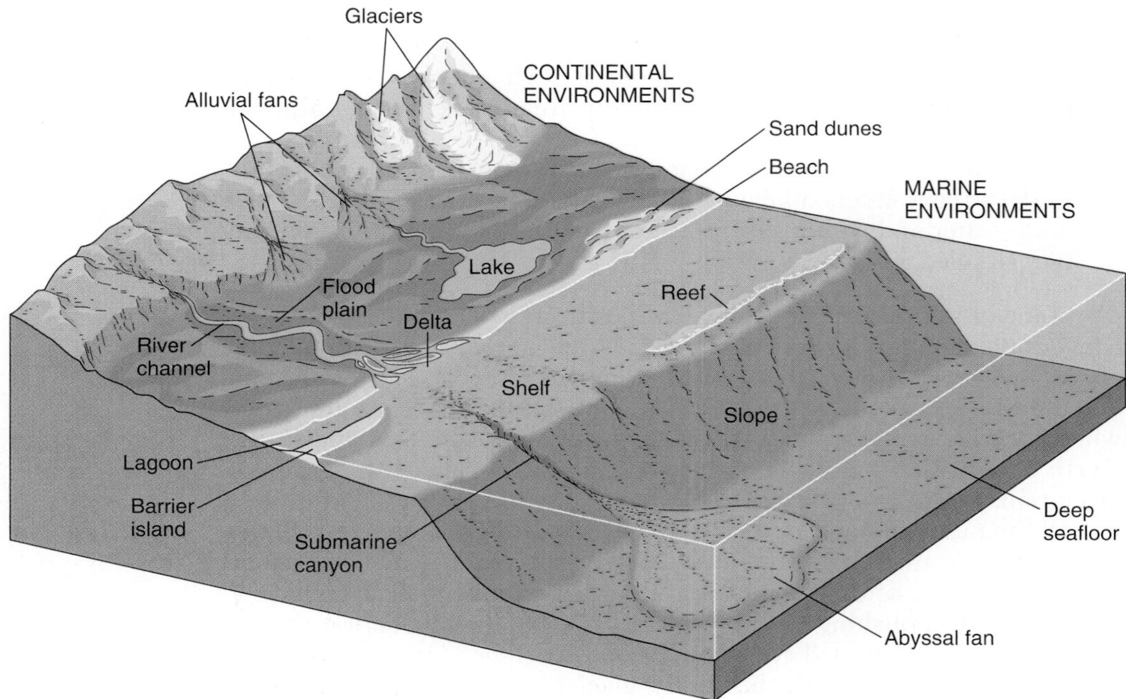

Figure 15-49 Erosion and sedimentation on land and sea.

Groundwater deposits material in the pore spaces of sediments, a process that helps to convert the sediments to rock. Much dissolved material precipitates in cracks to form veins, which are found in all kinds of rocks. Quartz and calcite are common in veins, and the ores of various metals are also found there. Spectacular examples of groundwater deposition are the **stalactites** that hang from the roofs of limestone caves, the **stalagmites** that rise from their floors (Fig. 15-50), and the colorful deposits often found around hot springs and geysers.

Example 15.3

Why are mineral deposits around hot springs thicker than those around ordinary springs?

Solution

Minerals are more soluble in hot water than in cold water, hence hot water contains a greater mineral load to deposit.

Lithification Sediments buried beneath later deposits are gradually hardened into rock, a process called **lithification.** Lithification can be a complex process, taking thousands or even millions of years. One important change in the sediment is compaction, the squeezing together of its grains under the pressure of overlying deposits. Some recrystallization may accompany compaction. The calcite crystals of limy sediments, in particular, grow larger and interlock with one another.

Figure 15-50 Carlsbad Caverns in New Mexico. Caves such as this occur when groundwater dissolves limestone and carries it away. The stalactites that hang like icicles from the roof of the cave and the stalagmite columns that grow from its floor consist of material that has precipitated from solution.

Figure 15-51 Petrified wood (this example is in Arizona) consists mainly of silica. Its beautiful colors come from other minerals deposited with the silica: those that contain iron give the reds and yellows, those that contain chromium give the greens, and those that contain cobalt give the blues.

Chemical changes brought about by circulating groundwater contribute to the hardening of many sediments. The grains of coarse sediments are cemented by material precipitated from solution in groundwater, and some sediments have much of their original material dissolved away and replaced by other substances. In petrified wood the original organic compounds have been removed, molecule by molecule, and replaced by silica. The whole process takes place so gradually that the finest details of the wood structure may be preserved (Fig. 15-51).

Sedimentary rocks are especially important in geology because they contain material that was deposited at or near the earth's surface and so record the changing surface conditions of past time. If we read the evidence with sufficient insight and imagination, we find before us a panorama of earth history: seas that once spread widely over the land, the advance and retreat of immense glaciers, the shifting sand dunes of ancient deserts (Fig. 15-52), and much more. Revealed are the living creatures that inhabited lands and seas of the past, for many sedimentary rocks contain fossil remains of plants and animals. Igneous and metamorphic rocks tell us by their structures something about conditions in the earth's interior; sedimentary rocks tell us about the history of surface landscapes.

Figure 15-52 These sandstone beds in Zion National Park, Utah, were once sand dunes that had been deposited by winds.

Vulcanism

Erosion and sedimentation are leveling processes through which the higher parts of the earth's surface are worn down and the lower parts are filled with the resulting debris. If their work could be completed, the continents would disappear and the earth would become a smooth globe covered with water. The fact that the continents still exist, not to mention mountain ranges on them, testifies to the action of two other processes:

1. Processes of **vulcanism,** which involve movements of molten rock
2. Processes of **diastrophism** (or **tectonism**), which involve movements of the solid materials of the crust

Vulcanism and diastrophism often occur together. The flow of molten rock may cause adjacent rock structures to be distorted, and it is common for major crustal shifts to be accompanied by volcanic eruptions and subsurface migrations of molten rock.

15.15 Volcanoes

Rivers of Lava, Clouds of Gas and Dust

A **volcano** is an opening in the earth's crust through which molten rock, usually called **magma** while underground and **lava** above ground, pours forth. Because lava accumulates near their openings, most volcanoes in the course of time build up mountains with a characteristic conical shape that steepens toward the top, with a small depression, or **crater,** at the summit. Lava escapes almost continuously from a few volcanoes, but the majority are active only at intervals.

A volcanic eruption is one of the most awesome spectacles on earth (Figs. 15-53 and 15-54). Earthquakes may provide warning of an eruption a few hours or a few days beforehand—minor shocks probably caused by

Example 15.4

An experiment is performed to determine the lowest temperature at which a certain magma can exist within the earth by melting a sample of rock that has hardened from this magma in a furnace. How meaningful are the results of this experiment?

Solution

The results mean little because the effect of the immense pressure within the earth on the melting point of the magma is not being taken into account.

the movement of gases and liquids underground. Eruptions follow a variety of patterns. Usually an explosion comes first, sending a great cloud of gases, dust, and rock fragments billowing from the crater. The exceptionally violent Tambora eruption in Indonesia in 1815, which killed 92,000 people, is thought to have blown over 100 km^3 of debris into the atmosphere. Enough stayed aloft to markedly reduce the sunlight reaching Europe and North America the following year—the "year without a summer." New England had blizzards in July and in Switzerland the 18-year-old Mary Shelley, kept indoors by constant rain, wrote *Frankenstein*. A number of airplanes have been damaged by flying into volcanic ash aloft. Most deaths from volcanoes are due not to lava flows but rather to clouds of gas and ash that sweep down from the craters and destroy everything in their paths. Sometimes a heavy rain washes ash and rock fragments down the slopes of a volcano; such a mudflow killed 25,000 people in 1985 in the town of Armero, Columbia.

Figure 15-53 The eruption of Mt. St. Helens on May 18, 1980, began with an explosion that devastated an area of over 500 km^2 in Washington State and sent a column of ash and smoke to a height of 20 km (see Fig. 15.54). Winds carried ash as far as Oklahoma. About 2.7 km^3 of volcanic rock, 0.5 km^3 of it molten, was expelled. The energy released was about 1.7×10^{18} J, equivalent to 400 million tons of TNT. Mt. St. Helens had also erupted a number of times in the past, leading Native Americans in the region to call it Coowit, the Lady of Fire.

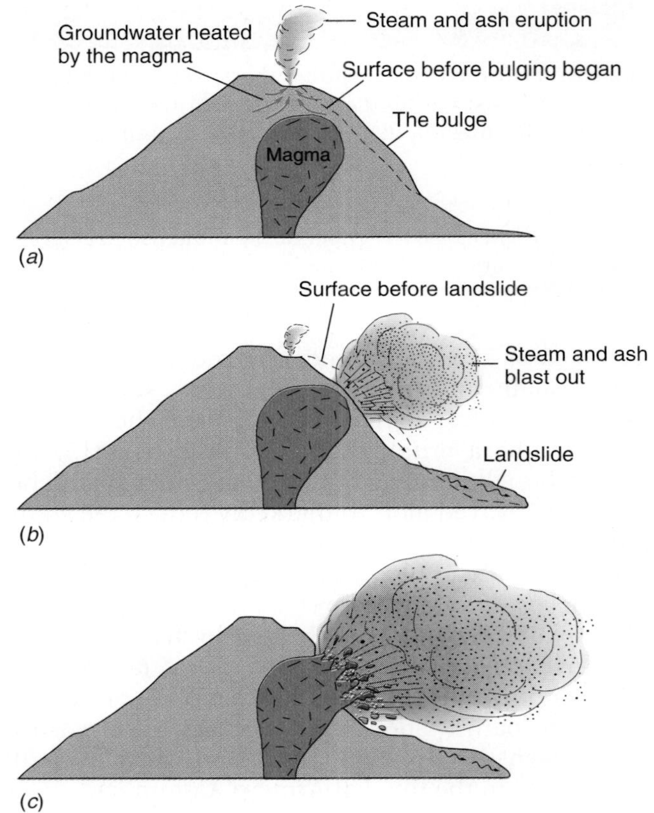

Figure 15-54 What happened at Mt. St. Helens in 1980. (*a*) A few weeks before the eruption, magma began welling up into the mountain and created a huge bulge (over 100 m outward) in its northern side. Small earthquakes and emissions of gas and ash occurred as the bulge formed. (*b*) Then the bulge broke loose and the rock that had sealed in the magma slid down the slope. (*c*) The sudden drop in pressure released the gases dissolved in the magma, and in the resulting explosion a froth of magma and gases poured out of the side of the mountain to devastate a wide area while dense clouds of ash shot upward for more than a day. Since the 1980 eruption magma has been flowing upward to the surface where it is oozing out to produce a lava dome. Because of this another violent eruption like that of 1980 seems unlikely, at least for some time to come.

Gas may continue to come out after the first explosion, and further bursts may recur at intervals. The cloud may persist for days or weeks with its lower part glowing red at night. Activity gradually slackens, and soon a tongue of white-hot (1100°C or so) lava spills over the edge of the crater or pours out of a fissure on the mountain slope. Other flows may follow the first, and explosions may continue, though they become weaker and weaker. Slowly the volcano quiets down, with only a small cloud of water vapor above to suggest its activity.

Volcanic Violence The chief factors that determine whether an eruption will be a quiet lava flow or a furious explosion are the **viscosity** of

Predicting an Eruption

Unlike earthquakes, which often occur with little or no warning, volcanoes commonly show signs that an eruption is coming hours or days before. Movements of gases and magma underground give rise to characteristic vibrations that can be heard as creaks, groans, and rumbles; sometimes small earthquakes are felt. Analyzing these vibrations gives clues as to when the eruption will begin, though not (yet?) how large it will be. Other signs of an impending eruption are changes in the shape of a volcano (see Fig. 15-54), sudden rises in its surface temperature, and the emission of volcanic gases such as sulfur dioxide. These effects can be detected by satellite measurements, which is often cheaper and easier than making observations on the volcanoes themselves. However, only a few satellites are suitably equipped, and these satellites survey the entire earth, so they only pass over a particular place every two weeks or so.

Figure 15-55 Rivers of red-hot lava pour into the Pacific Ocean from a volcano in Hawaii. The magma is relatively poor in silica, so the lava flows freely. Mauna Loa, also in Hawaii, is the largest volcano on earth; it rises over 15 km above the ocean floor. Mauna Loa's weight has depressed the ocean floor under it by about 8 km.

the magma and the amount of gas it contains. (The greater the viscosity of a liquid, the less freely it flows; honey is more viscous than water.) Magma is a complex mixture of various metal oxides and silica and usually has a lot of gas dissolved in it under pressure. Lavas vary in viscosity but commonly creep downhill slowly, like thick syrup or tar. The viscosity depends largely upon chemical composition: magmas with high percentages of silica are the most viscous.

Gas also affects viscosity: magmas with little gas in them are the most viscous. If the magma feeding a volcano happens to be rich in both gas and silica, the eruption will be explosive. A magma with modest gas and silica contents results in a quiet eruption (Fig. 15-55).

Volcanic gases include water vapor, carbon dioxide, nitrogen, hydrogen, and various sulfur compounds. The most prominent is water vapor. Some of it comes from groundwater heated by magma, some comes from the combination of hydrogen in the magma with atmospheric oxygen, and some was formerly incorporated in rocks deep in the crust and carried upward by the magma. Much of the water vapor condenses when it escapes to give rise to the torrential rains that often accompany eruptions.

Lava hardens into one or another of the various volcanic rocks (Fig. 15-56), which are all fine-grained because lava cools too rapidly for large mineral crystals to grow. Basalt is by far the most common volcanic rock and, when molten, is fluid enough to spread out over a wide area. The more viscous andesite usually produces steep, conical mountains and generally is associated with more rugged landscapes than basalt. Rhyolite, the most siliceous of ordinary lavas, forms small, thick flows and domes. Rhyolitic lava is sometimes so viscous and cools so rapidly that crystallization does not take place, leaving the natural glass obsidian.

Volcanic rocks of all kinds frequently have rounded holes due to gas bubbles that were trapped during the final stages of solidification. Viscous lavas may harden with so many cavities that the light, porous rock **pumice** results. Pumice is so light that it floats in water; it is sometimes used as a gentle abrasive.

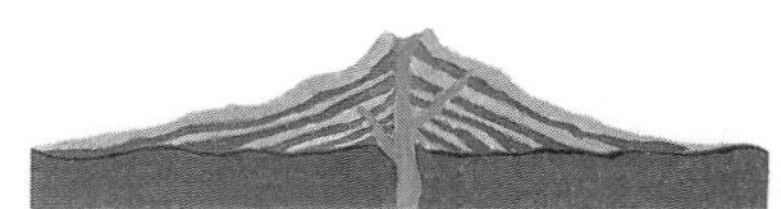

Figure 15-56 Cross section of a volcano. During explosive eruptions much liquid rock flies apart when it emerges. Deposits of the finer material may form the rock **tuff,** and deposits of the coarser material may form a kind of conglomerate called **volcanic breccia** whose fragments are angular rather than rounded. In the volcano shown, lava flows (solid color) alternate with beds of tuff and volcanic breccia.

Volcanic Regions About 530 volcanoes are active today around the borders of the Pacific Ocean (the "Ring of Fire"), on some of the Pacific Islands, in Iceland, in the Mediterranean, and in East Africa. Fifty or so of them erupt in an average year; see Fig. 15-57 and Table 15-4. One in

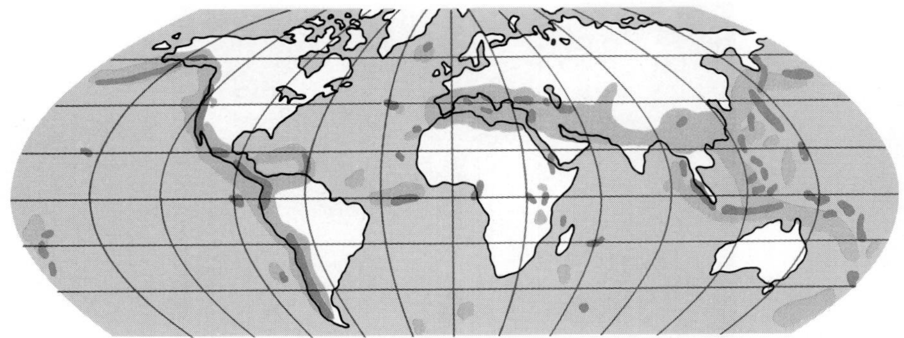

Figure 15-57 The principal earthquake (light color) and volcanic (dark color) regions of the world.

ten of the world's people lives near an active volcano. There are many thousands of undersea volcanoes.

In many other parts of the world volcanoes were active in earlier times. Where volcanoes became extinct in the recent geological past, their former splendor is suggested by isolated conical mountains, by solidified lava flows, and by hot springs, geysers, and steam vents. Some of the great mountains in the western United States are old volcanoes, scarred here and there with lava flows so recent that vegetation has not yet gained a foothold on them. In regions where volcanoes have been inactive for still longer, erosion may have removed all evidence of the original mountains and left only patches of volcanic rocks to indicate former igneous activity.

15.16 Intrusive Rocks

They Have Hardened Underground from Magma

Molten rock that rises through the earth's crust but does not reach the surface solidifies to form intrusive bodies (often called **plutons**) of various kinds. Because these bodies cool slowly, intrusive igneous

Table 15-4 Volcanoes Near Population Centers Around the World That Could Erupt at Any Time

Region	Country	Volcano
Western Pacific	Indonesia	Merapi
	Philippines	Taal
	Japan	Unzen
	Japan	Sakurajima
Hawaii	United States	Mauna Loa
Western North America	United States	Rainier
Central America	Mexico	Colima
	Guatemala	Santa Maria/Santiaguito
	Colombia	Galeras
Mediterranean	Italy	Vesuvius
	Italy	Etna
	Greece	Santorini
East Africa	Congo	Niragongo

rocks are coarser-grained than volcanic rocks that cool rapidly on the surface. We find intrusive rocks exposed only where erosion has uncovered them after they hardened.

The igneous origin of volcanic rocks is clear enough, for we can actually watch lava harden to solid rock. But no one has ever seen an intrusive rock like granite in a liquid state in nature. The belief that granite was once molten follows from indirect evidence such as the following:

1. Granite shows the same relations among its minerals that a volcanic rock shows. The separate grains are intergrown, and those with higher melting points show by their better crystal forms that they crystallized a little earlier than the others.
2. Some small intrusive formations show a continuous change between coarse granite and a rock indistinguishable from the volcanic rock rhyolite, whose igneous origin is clear.
3. Granite is found in masses that cut across layers of sedimentary rock and from which small irregular branches penetrate into the surrounding rocks. Sometimes blocks of the sedimentary rocks are found completely engulfed by the granite.
4. That granite was at a high enough temperature to be molten is shown by the baking and recrystallization of the rocks that it intrudes.

These four types of evidence apply equally well to the other intrusive rocks.

Plutons A **dike** is a wall of igneous rock that cuts across existing rock layers (Fig. 15-58). The largest dikes are hundreds of meters thick, but more often their thickness is between a few tenths of a meter and a few meters. The distinction between dikes and veins is that a dike is molten rock that has filled a fissure and solidified, whereas a vein consists of material deposited along a fissure from solution in water (Fig. 15-59).

Any kind of igneous rock may occur in a dike. Rapid cooling in small dikes may give rocks similar to those of volcanic origin, and slow cooling in larger dikes gives coarse-grained rocks. Dikes may cut any other kind of rock. They are often associated with volcanoes: some of the magma forces its way into cracks instead of moving upward through the central orifice. In regions of intrusive rocks, dikes are often found as offshoots of larger masses, as in Fig. 15-58. Also shown in the figure are **sills** and **laccoliths,** intrusive bodies that lie parallel to the strata in which they are found.

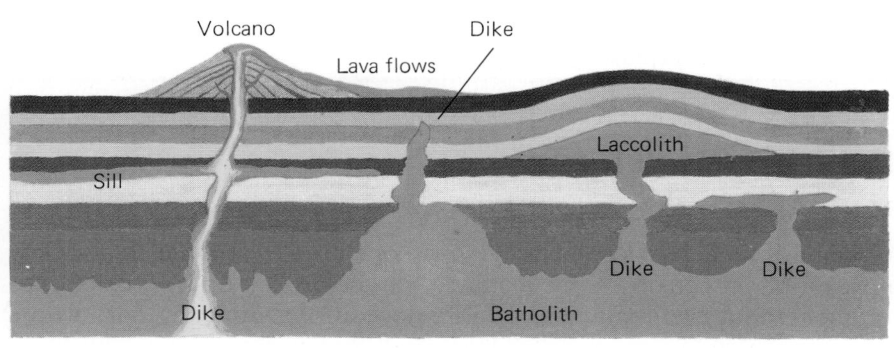

Figure 15-58 A batholith and associated dikes and sills; a laccolith and a volcano are also shown.

Figure 15-59 A dike of igneous rock intruded in sedimentary beds in Arizona's Grand Canyon. The beds on either side of the dike do not match, which suggests that the dike was intruded along a fault.

Batholiths are very large plutons that extend downward as much as several km. Visible exposures of batholiths cover hundreds of thousands of square km. The great batholith that forms the central part of the Sierra Nevada in California, for example, is about 800 km long and, in places, over 160 km wide. Granite is the principal rock in batholiths, although many have local patches of diorite and gabbro. Batholiths are always associated with mountain ranges, either mountains of the present or regions whose rock structure shows evidence of mountains in the distant past. The intrusion of a batholith is evidently one of the events that occurs in the process of mountain building.

Example 15.5

Suppose you find a nearly vertical contact between granite and sedimentary rocks with the sedimentary beds ending abruptly against the granite. How could you tell whether the granite had intruded into the sedimentary rocks or after solidifying had moved against the sedimentary rocks during a deformation of the crust?

Solution

If the granite had intruded the sedimentary rocks, they would show evidence of thermal metamorphism.

15.17 The Rock Cycle

Rocks Are Not Necessarily Forever

As we have seen, rocks can change from one kind to another in a variety of ways. An igneous rock, for instance, can be broken up by erosion into fragments that eventually end up in a deposit of sediments that in time lithifies into sedimentary rock. Heat and pressure can later transform

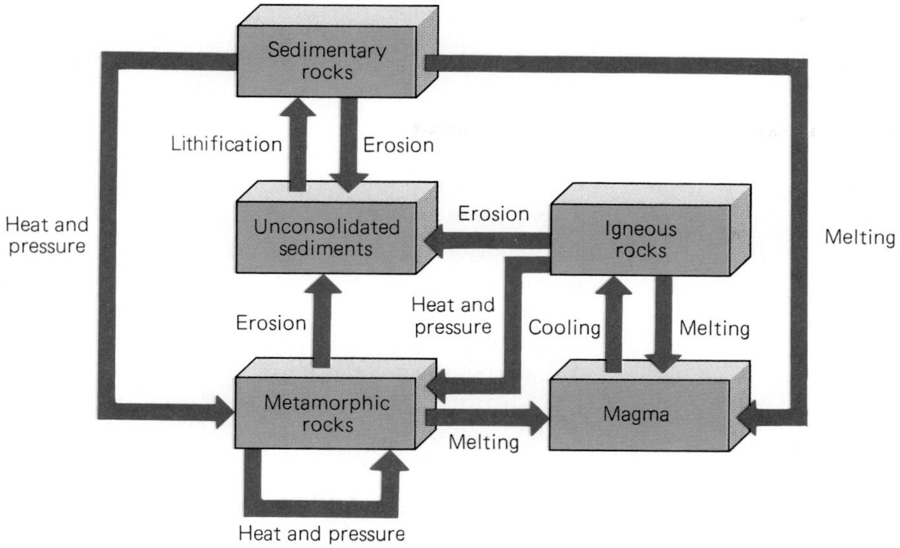

Figure 15-60 The rock cycle. Depending upon circumstances, different transformations are possible, including the conversion of one kind of metamorphic rock into another.

this rock into a metamorphic counterpart, which may later be melted underground into magma and still later harden into an igneous rock—probably not the same as the one the cycle began with, but perhaps a cousin to it. Other life histories are also possible, as shown in Fig. 15-60.

Important Terms and Ideas

The earth's outer shell of rock is called its **crust.** Oxygen and silicon are the most abundant elements in the crust. Crystalline **silicates** consist of continuous structures of oxygen and silicon ions, usually with metal ions as well. **Minerals** are the substances of which rocks are composed. The most abundant minerals are silicates; also common are carbonates and oxides. Quartz, feldspar, mica, calcite, and the ferromagnesian and clay minerals are six important kinds of minerals.

Igneous rocks (such as granite and basalt) have cooled from a molten state. **Sedimentary rocks** (such as sandstone and limestone) have consolidated from materials derived from the disintegration or solution of other rocks and deposited by water, wind, or glaciers. **Metamorphic rocks** (such as slate and marble) have been altered by heat and pressure beneath the earth's surface. **Soil** consists of rock fragments mixed with varying amounts of organic material.

Nearly all **earthquakes** are due to the sudden movement of solid rock along fracture surfaces called **faults** that are near the earth's surface. Earthquakes send out **body waves,** which pass through the earth's interior, and **surface waves,** which travel along the earth's surface. The two kinds of body waves are **P waves,** which are longitudinal, and **S waves,** which are transverse. The two kinds of surface waves are **Love waves,** which are transverse, and **Rayleigh waves,** which are rotary, like water waves.

The analysis of earthquake wave records shows that the earth's interior contains a **core,** probably consisting mainly of molten iron with a small solid inner core, and a solid **mantle** of ferromagnesian silicates. The temperature of the earth's interior increases with depth.

The **earth's magnetic field** resembles the magnetic field that would be produced by a huge bar magnet located near the earth's center. The field is due to electric currents in the molten iron core, and the direction of the field has reversed itself many times in the past.

The processes by which rocks are worn down and the debris carried away are included in the term **erosion.** The chief agent of erosion is running water, although **weathering** (the gradual disintegration of exposed rocks), **glaciers** (rivers and seas of ice formed from accumulated snow), and **groundwater** (subsurface water) also contribute. Most of the eroded material is deposited to form sediments, the bulk of which are laid down on ocean bottoms. **Lithification** refers to the gradual hardening into rock of sediments buried under later deposits.

A **volcano** is an opening in the earth's crust through which molten rock (called **magma** while underground, **lava** above ground) comes out. Intrusive bodies called **plutons** are formed by the solidification of magma under the surface.

Multiple Choice

1. The most abundant element in the earth's crust is
   a. oxygen
   b. nitrogen
   c. silicon
   d. carbon
2. The second most abundant element is
   a. iron
   b. silicon
   c. carbon
   d. aluminum
3. Minerals are
   a. silicon compounds
   b. common types of rock
   c. homogeneous solids of which rocks are composed
   d. always compounds
4. Cleavage, the tendency of certain minerals to split along certain planes, is conspicuous in
   a. quartz
   b. mica
   c. chalk
   d. chert
5. The most abundant mineral in the earth's crust is
   a. quartz
   b. mica
   c. feldspar
   d. calcite
6. A crystalline form of silicon dioxide (SiO_2) is
   a. quartz
   b. mica
   c. feldspar
   d. calcite
7. A mineral that is not a silicate is
   a. quartz
   b. mica
   c. feldspar
   d. calcite
8. The ferromagnesian minerals are usually
   a. transparent
   b. white or pink
   c. bluish
   d. dark green or black
9. Rocks that have been formed by cooling from a molten state are called
   a. igneous rocks
   b. sedimentary rocks
   c. metamorphic rocks
   d. precipitated rocks
10. Rocks that have been altered by heat and pressure beneath the earth's surface are called
    a. igneous rocks
    b. sedimentary rocks
    c. metamorphic rocks
    d. precipitated rocks
11. The majority of surface rocks are
    a. sedimentary
    b. metamorphic
    c. igneous
    d. found in the mantle as well
12. A general characteristic of rocks of volcanic origin is
    a. the presence of shale
    b. a light color
    c. coarse-grained structure
    d. fine-grained structure
13. Foliation occurs in
    a. sedimentary rocks
    b. metamorphic rocks
    c. igneous rocks
    d. all of the above
14. An example of a foliated rock is
    a. marble
    b. sandstone
    c. slate
    d. granite
15. Of the following rocks, the one that does not originate in sediments laid down by water, wind, or ice is
    a. marble
    b. conglomerate
    c. shale
    d. sandstone
16. Limestone may be metamorphosed into
    a. marble **c.** gneiss
    b. quartzite **d.** schist
17. Shale may be metamorphosed into
    a. marble **c.** slate
    b. sandstone **d.** granite
18. Mica is not found in
    a. slate
    b. schist
    c. gneiss
    d. marble
19. Fossils are most likely to be found in
    a. granite
    b. shale
    c. gneiss
    d. basalt

20. Most earthquakes are caused by
  a. volcanic eruptions
  b. landslides
  c. lightning strikes
  d. shifts of crustal rocks

21. Regions in which earthquakes are frequent are also regions in which
  a. the geomagnetic field is strong
  b. hurricanes are common
  c. volcanoes occur
  d. petroleum is found

22. Relative to an earthquake of magnitude 5 on the Richter scale, an earthquake of magnitude 6 releases
  a. 2 times more energy
  b. 10 times more energy
  c. 30 times more energy
  d. 100 times more energy

23. Earthquake P waves
  a. are longitudinal vibrations like sound waves
  b. are transverse vibrations like waves in a taut string
  c. cannot pass through the earth's core
  d. always travel in straight lines through the earth's interior

24. The earth's crust
  a. has very nearly the same thickness everywhere
  b. varies irregularly in thickness
  c. is always thinnest under the continents
  d. is always thinnest under the oceans

25. The part of the earth with the greatest volume is the
  a. inner core **c.** mantle
  b. outer core **d.** crust

26. The radius of the earth's core is roughly
  a. $\frac{1}{10}$ the earth's radius
  b. $\frac{1}{4}$ the earth's radius
  c. $\frac{1}{2}$ the earth's radius
  d. $\frac{3}{4}$ the earth's radius

27. The rocks of the mantle are believed to consist largely of
  a. feldspar
  b. quartz
  c. clay minerals
  d. ferromagnesian minerals

28. The reasons why the earth's core is believed to be largely molten iron do not include the
  a. magnetic properties of iron
  b. electrical conductivity of iron
  c. density of iron
  d. relative abundance of iron in the universe

29. Most of the heat of the earth's interior is believed to
  a. be left over from its formation
  b. come from radioactive materials
  c. be due to chemical reactions in the core
  d. be provided by solar radiation absorbed by the crust

30. The earth's magnetic field
  a. never changes
  b. has reversed itself many times
  c. is centered exactly at the earth's center
  d. originates in a permanently magnetized iron core

31. If we travel around the earth, we would find that the earth's magnetic field
  a. is the same in direction and strength
  b. is the same in direction but not in strength
  c. is the same in strength but not in direction
  d. varies in both direction and strength

32. A rock readily attacked by chemical weathering is
  a. limestone
  b. obsidian
  c. granite
  d. chert

33. The principal agent of erosion is
  a. groundwater **c.** ice
  b. running water **d.** wind

34. Which of the following is not produced by rivers?
  a. flood plains
  b. deltas
  c. alluvial fans
  d. moraines

35. In their early stages, river valleys exhibit characteristic
  a. U-shaped cross sections
  b. V-shaped cross sections
  c. water tables
  d. moraines

36. The last stage in the erosion of a river is the production of a (an)
  a. delta
  b. alluvial fan
  c. moraine
  d. floodplain

37. The approximate percentage of the earth's land area covered by ice today is
  a. 1 percent
  b. 2 percent
  c. 10 percent
  d. 30 percent

38. A fairly fast valley glacier might have a speed of
a. 1 m/h
b. 1 m/day
c. 1 m/month
d. 1 m/year

39. Most of the groundwater present in soil and underlying porous rocks comes from
a. streams and rivers
b. melting glaciers
c. springs
d. rain

40. A body of porous rock through which groundwater moves is called a (an)
a. water table
b. aquifer
c. moraine
d. batholith

41. The largest amounts of sediment are deposited
a. by glaciers
b. on river beds
c. on the ocean floors
d. by chemical precipitation

42. Minerals deposited by groundwater in rock fissures form
a. dikes
b. veins
c. sills
d. moraines

43. Most caves are produced by the solvent action of groundwater on
a. limestone
b. sandstone
c. granite
d. schist

44. The chief constituent of volcanic gases is
a. nitrogen
b. oxygen
c. carbon dioxide
d. water vapor

45. Molten rock underneath the earth's surface is called
a. magma
b. lava
c. obsidian
d. till

46. The most common volcanic rock is
a. granite
b. basalt
c. limestone
d. shale

47. The holes found in most volcanic rocks are due to
a. gases trapped in solidifying lava
b. erosion
c. marine organisms
d. rapid cooling

48. Active volcanoes are not found
a. in the West Indies
b. in the Mediterranean region
c. on the rim of the Pacific Ocean
d. in eastern Canada

49. A batholith is a
a. fissure from which groundwater emerges
b. natural rock pillar
c. large body of intrusive rock
d. volcanic cone

Exercises

15.1 Composition of the Crust
15.2 Minerals

1. What is the most abundant metal in the earth's crust? The next most abundant?
2. Do silicon compounds make up less than a quarter, between a quarter and a half, between half and three-quarters, or more than three-quarters of the mass of the crust?
3. What is the relationship between rocks and minerals?
4. What mineral is most abundant in the earth's crust? Does it make up more or less than half of the mass of the crust?
5. Both cleavage and crystal form are characteristic mineral properties. What is the difference between the two?
6. Graphite consists of layers of carbon atoms in hexagonal arrays (see Fig. 11-7), with each atom covalently bonded to three others. The layers are bonded together by van der Waals forces. Would you expect graphite to exhibit cleavage?
7. In the silicate minerals each Si^{4+} ion is always surrounded by four O^{2-} ions, yet no mineral has the formula SiO_4. Why not?
8. How could you distinguish calcite crystals from quartz crystals?

15.3 Igneous Rocks

9. Are the mineral grains in an igneous rock usually regular in form? What is the usual arrangement of the grains in an igneous rock?
10. Granite and rhyolite have similar compositions, but granite is coarse-grained whereas rhyolite is fine-grained. What does the difference in grain size indicate about the environments in which each rock formed?
11. Diorite is an igneous rock that has hardened slowly underground, and andesite, whose composition is similar, is an igneous rock that has hardened on the earth's surface. How can they be distinguished from one another?
12. Obsidian is a rock that resembles glass, in particular by sharing the property that its structure is closer to that of a liquid than to that of a crystalline solid. What does this observation suggest about the manner in which obsidian is formed?

15.4 Sedimentary Rocks

13. In what way does calcite differ from almost all other minerals? What rocks are largely calcite?
14. Of what rock do coral reefs consist?
15. What is the nature of chert and why is it so resistant to chemical and mechanical attack?

15.5 Metamorphic Rocks

16. What kind of rocks are most abundant in the earth's crust? On the earth's surface?
17. What happens to the density of a rock that undergoes metamorphism?
18. Why is gneiss the most abundant metamorphic rock?
19. The mineral grains of many metamorphic rocks are flat or elongated and occur in parallel layers. (a) What is this property called? (b) How does it originate?
20. Shale is a sedimentary rock that consolidated from mud deposits. What are the various metamorphic rocks that shale can become under progressively increasing temperature and pressure?
21. (a) What is the origin of limestone? (b) What rock is formed by the metamorphism of limestone? (c) What is the difference in structure that the metamorphism produces?
22. Distinguish between the foliation of a metamorphic rock and the stratification of a sedimentary rock.
23. Distinguish between quartz and quartzite.
24. How could you distinguish (a) chert from obsidian? (b) conglomerate from gneiss? (c) quartz from calcite?
25. How could you distinguish (a) granite from gabbro? (b) basalt from limestone? (c) schist from diorite?
26. Name the following rocks: (a) a rock consisting of intergrown crystals of quartz; (b) the rock resulting from the metamorphism of limestone; (c) an intrusive igneous rock with the same composition as andesite.
27. Name the following rocks: (a) a fine-grained, unfoliated rock with intergrowing crystals of quartz, feldspar, and black mica; (b) a finely foliated rock with microscopic crystals of quartz and white mica; (c) a fine-grained rock consisting principally of kaolin.

15.6 Earthquakes

28. What is the difference between the focus of an earthquake and its epicenter?
29. Each step of 1 on the Richter scale of earthquake magnitude represents an increase in vibration amplitude of a factor of 10. What is the approximate increase in the energy released?
30. What can be said about an earthquake whose magnitude is 0 on the Richter scale? Whose magnitude is 8 or more?

15.7 Structure of the Earth

31. Why is the mantle thought to be solid?
32. (a) Distinguish between earthquake P and S waves. (b) Which of them can pass through the mantle? (c) Through the core?
33. How does the radius of the earth's core compare with the total radius of the earth?
34. Where is the earth's crust thinnest? Where is it thickest?

15.8 The Earth's Interior

35. What evidence is there in favor of the idea that the earth's interior is very hot?
36. What is the source of the energy that powers most geological processes other than erosion?
37. (a) Why is it believed that the earth's outer core is a liquid? (b) Why is it believed that the liquid is mainly molten iron?

15.9 Geomagnetism

38. Why does a compass needle in most places not point due north?
39. Why is it unlikely that the earth's magnetic field originates in a huge bar magnet located in its interior?

15.10 Weathering

40. What is the most important mechanism of mechanical weathering?
41. Why are igneous and metamorphic rocks in general more susceptible to chemical weathering than sedimentary rocks?
42. Granite consists of feldspars, quartz, and ferromagnesian minerals. (a) What becomes of these minerals when granite undergoes weathering? (b) What kinds of sedimentary rocks can the weathering products form?
43. In what way is the weathering of rock important to life on earth?

15.11 Stream Erosion

44. What is the source of energy that makes possible the erosion of landscapes?
45. Is there a limit to the depth to which streams can erode a particular landscape? Is there a limit in the case of glaciers?
46. Why are streams and rivers so effective as agents of erosion on the earth's surface?

15.12 Glaciers

47. Under what circumstances does a glacier form?
48. Which is the more important agent of erosion, running water or glaciers? Why?
49. What agent of erosion produces valleys with a V-shaped cross section? A U-shaped cross section?
50. How is it possible for glaciers to wear down rocks that are harder than glacial ice?

15.13 Groundwater

51. What is a water table? An aquifer?
52. What is the immediate destination of most of the water that falls as rain on land?

15.14 Sedimentation

53. What is the eventual site of deposition of most sediments?
54. What kind of material is found in an alluvial fan? In a moraine?
55. Why are clay minerals and quartz particles abundant in sediments that have not been chemically deposited?
56. In sand derived from the attack of waves on granite, what mineral would you expect to be most abundant?
57. What is the probable origin of the following sedimentary rocks?
    a. A thick limestone
    b. A conglomerate with well-rounded boulders and numerous thin beds of sandy and clayey material
    c. A sandstone consisting of well-sorted, well-rounded grains of quartz

15.15 Volcanoes

58. What characteristic landscape features do active volcanoes produce? From what features could you conclude that volcanoes were once active in a region where eruptions have long since ceased?
59. What kinds of rocks are likely to be found in lava flows? What is the most common volcanic rock?
60. What factors determine the viscosity of a magma? What kinds of landscapes are produced by volcanoes whose lavas have relatively high and relatively low viscosities?
61. What is the cause of the holes found in many volcanic rocks?
62. What is the main constituent of volcanic gases?

15.16 Intrusive Rocks

63. (a) Why are metamorphic rocks often found near plutons? (b) Where would you expect to find the wider zone of thermal metamorphism, near a dike or near a batholith?
64. Distinguish between a dike and a vein.

Answers to Multiple Choice

1. a	8. d	15. a	22. c	29. b	36. d	43. a
2. b	9. a	16. a	23. a	30. b	37. c	44. d
3. c	10. c	17. c	24. d	31. d	38. b	45. a
4. b	11. a	18. d	25. c	32. a	39. d	46. b
5. c	12. d	19. b	26. c	33. b	40. b	47. a
6. a	13. b	20. d	27. d	34. d	41. c	48. d
7. d	14. c	21. c	28. a	35. b	42. b	49. c

16

The Evolving Earth

The Andes Mountains of South America are relatively young.

Goals

When you have finished this chapter you should be able to complete the goals ▶ given for each section below:

Tectonic Movement

16.1 Types of Deformation
Faults and Folds, Tilts and Warps, Rises and Falls
- ▶ Describe the various kinds of tectonic movement.

16.2 Mountain Building
The Rock Record Tells a Complicated Story
- ▶ List the main steps in the development of the great mountain ranges of the world.

16.3 Continental Drift
An Evolving Jigsaw Puzzle
- ▶ Outline the evolution of today's continents from the former supercontinent Pangaea.

Plate Tectonics

16.4 Lithosphere and Asthenosphere
A Hard Layer over a Soft Layer
- ▶ Distinguish between the lithosphere and the asthenosphere.

16.5 The Ocean Floors
The Youngest Part of the Crust

16.6 Ocean-Floor Spreading
Alternate Magnetization Is the Proof
- ▶ Explain how ocean-floor spreading accounts for present-day features of the ocean floors.

16.7 Plate Tectonics
How the Continents Drift
- ▶ Describe what happens at oceanic-continental, oceanic-oceanic, and continental-continental plate collisions.
- ▶ Outline the geological processes that occur in a subduction zone.
- ▶ Define transform fault and discuss the origin of the San Andreas Fault.

Methods of Historical Geology

16.8 Principle of Uniform Change
"No Vestige of a Beginning, No Prospect of an End"
- ▶ State what is meant by the principle of uniform change.
- ▶ Describe the basic ideas of the theory of evolution.

16.9 Rock Formations
History under Our Feet
- ▶ List the four basic principles of historical geology.

16.10 Radiometric Dating
A Clock Based on Radioactivity
- ▶ Describe how radiometric dating is used to find the ages of rocks.
- ▶ Describe how radiocarbon dating is used to find the ages of biological specimens.

16.11 Fossils
Tracing the History of Life
- ▶ Explain why animal fossils are more common than plant fossils, and account for the abundance of fossils on the floors of ancient shallow seas.
- ▶ Discuss the various ways in which fossils are useful in geology.

16.12 Geologic Time
Precambrian, Paleozoic, Mesozoic, Cenozoic
- ▶ Give the basis for the division of geological time into eras, periods, and epochs.
- ▶ List the four major divisions of geological time in order from past to present.

Earth History

16.13 Precambrian Time
Long Ago but Not Far Away

16.14 The Paleozoic Era
Plants and Animals in Variety and Abundance

16.15 Coal and Petroleum
Both Came from Once-Living Matter
- ▶ Describe the formation of coal and petroleum.

16.16 The Mesozoic Era
The Age of Reptiles
- ▶ Identify dinosaurs and discuss what may have led to their sudden disappearance.

16.17 The Cenozoic Era
The Age of Mammals
- ▶ Discuss the most recent ice age and its role in early human history.

16.18 Human History
We Are a Recent Species

Nothing about the earth is fixed, permanent, unchanging. What is today a great mountain that pierces the sky may in the future be nibbled down into a mere hill, while elsewhere an undersea mass of sediments may be thrust upward into a lofty plateau. How do we know that such things can happen? After all, though plenty of geologic activity is taking place around us, for the most part the pace is exceedingly slow. Only after millions of years can the processes now at work yield large-scale changes in the pattern of the continents and in their landscapes. What justifies the belief that the earth's crust never stops evolving is the record of the past, a record that can be read in the rocks of the present.

During the past few decades a major advance has occurred in our understanding of the large-scale forces that shape and reshape the earth's crust. The notion that the continents are slowly drifting relative to one another—a notion going back three-quarters of a century but largely scorned for most of that period—has turned out to be the only way to explain a variety of striking observations. These same observations also provide clues as to what makes the continents drift. So far-reaching are the implications of the new dynamic picture of the crust, and so suddenly did they come to light, that it is legitimate to speak of a revolution in geologic thought.

Tectonic Movement

Terra firma, the solid earth, is a symbol of stability and strength. On foundations of rock we anchor our buildings, our dams, our bridges. The massive rock of mountain ranges seems strong enough to withstand any force.

Yet even casual observation shows how naive such notions of the earth's stability are. High up on mountainsides we find shells of marine animals, shells that can be there only if rock formed beneath the sea has been lifted far above sea level. Sedimentary rocks, which must have been deposited originally in horizontal layers, are found tilted at steep angles or folded into arches and basins. Other layers have broken along cracks, and the fractured ends have moved apart. Gigantic forces must occur in the crust in order to lift, bend, and break even the strongest rocks. Such forces and the changes they cause are called **tectonic** (from the Greek word for "carpenter").

Figure 16-1 Reverse fault in sandstone near Klamath Falls, Oregon. A reverse fault occurs when a rock mass moves upward relative to adjacent rock.

16.1 Types of Deformation

Faults and Folds, Tilts and Warps, Rises and Falls

Cracks are found in rock formations of all kinds, some due to molten rock contracting as it cools and others due to mechanical stresses in the crust. A fracture surface along which motion has taken place is called a **fault.** In a rock outcrop a fault appears as a fairly straight line against which sedimentary layers and other structures end abruptly (Fig. 16-1). Near the fault, layers may be bent or crumpled, and along the fault streaks of finely powdered material may have developed from friction during movement. Three important kinds of fault are shown in Fig. 16-2.

Movement along faults usually takes place as a series of small, sudden displacements, with intervals of years or centuries between successive jerks. An immediate effect of displacement along a reverse fault or normal

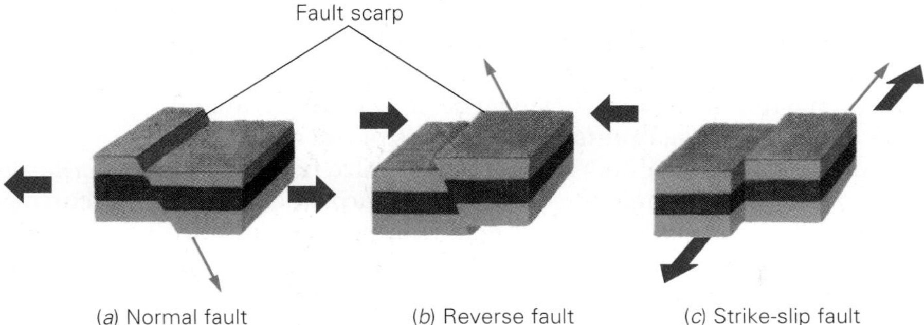

Figure 16-2 (*a*) A normal fault is an inclined surface along which a rock mass has slipped downward; it is the result of forces that tend to stretch the earth's crust. (*b*) A reverse fault is an inclined surface along which a rock mass has moved upward to override the neighboring mass; it is the result of forces that tend to compress the crust. (*c*) A strike-slip fault is a surface along which one rock has moved horizontally with respect to the other; it is the result of oppositely directed forces in the crust that do not act along the same line. Erosion modifies the fault scarps left by normal and reverse faults.

fault is a small cliff. Erosion then attacks the cliff and may level it before the next movement. If successive movements follow one another fast enough, erosion may not be able to keep up, with a high cliff as the result. Cliffs of this sort are called **fault scarps.** Good examples of scarps produced by normal faults are the steep mountain fronts of many of the desert ranges in Utah, Nevada, and eastern California. A more deeply eroded scarp produced by reverse faulting is the eastern front of the Rocky Mountains in Glacier National Park.

Folding takes place by slow, continuous movement, in contrast to the sudden displacements along faults (Figs. 16-3 and 16-4). Sometimes folding produces hills and depressions in the landscape directly, but more often erosion keeps pace with folding. Indirectly folds affect landforms

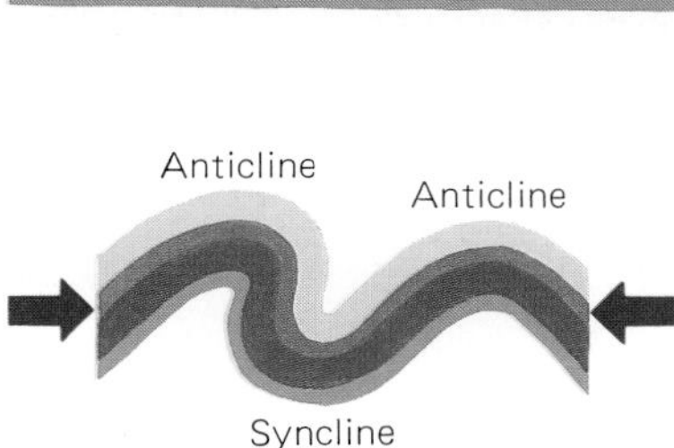

Figure 16-3 Cross section showing effect of folding in horizontal strata. Folds always shorten the crust and hence are produced by compressional forces. An anticline is an arch (a fold convex upward), and a syncline is a trough (a fold convex downward). In regions of intense folding, anticlines and synclines follow one another in long series.

Figure 16-4 Folded shale beds near Palmdale, California.

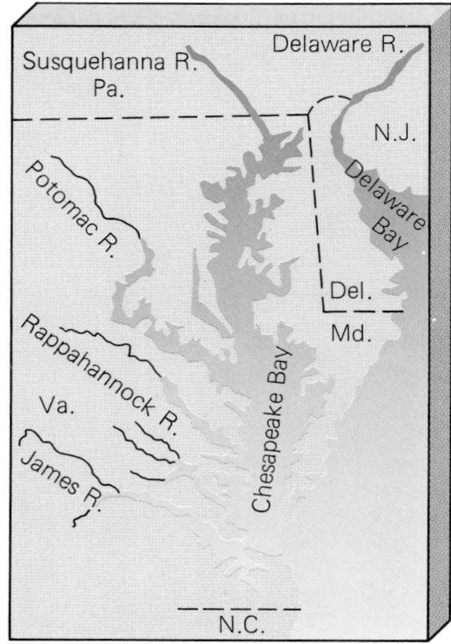

Figure 16-5 Drowned valleys on the Atlantic Coast of the United States.

by exposing tilted beds of varying degrees of resistance to the action of streams, so that long, parallel ridges and valleys develop, like those of the Appalachian Mountains (see Fig. 15-44). In these mountains, as in many others, the folding is very ancient. The present ridges are due to deep erosion after successive uplifts of the stumps of the old folds.

Large-scale crustal movements may involve whole continents or large parts of them, which may rise or fall, tilt or warp. Such events are occurring today as they have in the past. Coastal features in many parts of the world provide obvious evidence. For instance, a wave-cut cliff and terrace high above the present shore means that the coast there has been raised in fairly recent times, or stream erosion would have eaten these features away.

The sinking of land with respect to sea level is shown by long, narrow bays that fill the mouths of large stream valleys. A body of water like Chesapeake Bay (Fig. 16-5) could not be formed by wave erosion, since wave attack normally straightens a coastline rather than deeply indenting it. Instead, the shape of the bay suggests that it lies in a stream-carved valley whose lower part has been submerged beneath the sea. Elsewhere in the landscapes and sedimentary rocks of the United States a multitude of regional movements are recorded.

16.2 Mountain Building

The Rock Record Tells a Complicated Story

Mountains can form in a number of ways. Some mountains are accumulations of lava and fragmental material ejected by a volcano. Others are small blocks of the earth's crust raised along faults (Fig. 16-6). But the great mountain ranges of the earth, like the Appalachians, the Rockies, the Alps, and the Himalayas (where all 10 of the world's highest mountains are located), have a much longer and more complex history involving sedimentation, folding, faulting, igneous activity, repeated uplifts, and deep erosion. How

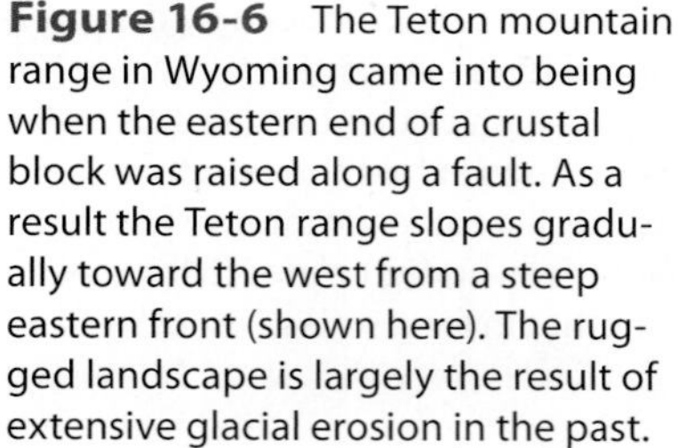

Figure 16-6 The Teton mountain range in Wyoming came into being when the eastern end of a crustal block was raised along a fault. As a result the Teton range slopes gradually toward the west from a steep eastern front (shown here). The rugged landscape is largely the result of extensive glacial erosion in the past.

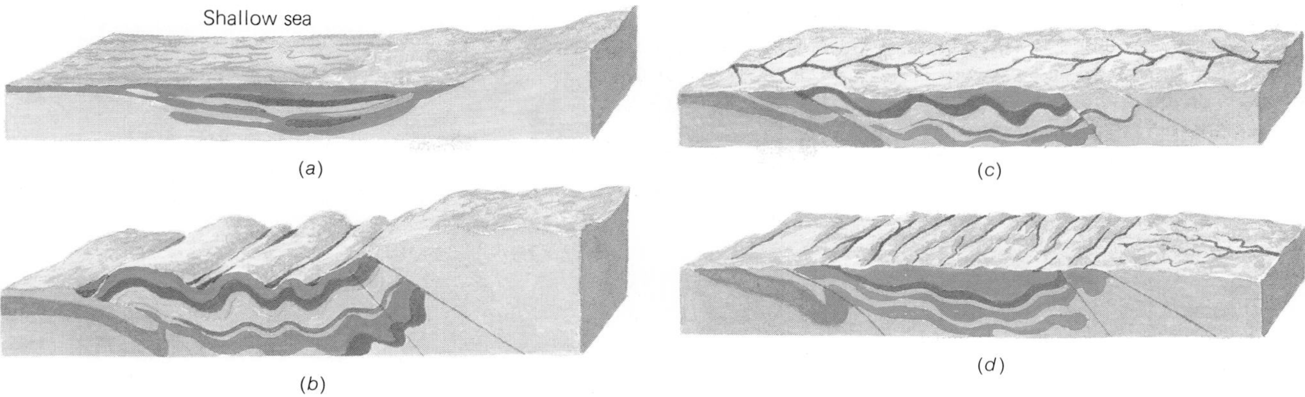

Figure 16-7 Successive stages in the evolution of the Appalachian Mountains. (*a*) Sediments accumulating in the Appalachian basin; (*b*) folding and reverse faulting of rocks in the basin; (*c*) original mountains worn down to a nearly level plain by stream erosion; (*d*) renewed erosion of the folded strata following vertical uplift, producing the parallel ridges and valleys of the present landscape.

the forces originate that push together and pull apart, lift and depress parts of the earth's crust are considered later in this chapter.

The layers of sedimentary rock in a mountain range are usually thicker than layers of similar age found under adjacent plains. Most of the layers in both mountains and plains, as shown by their sedimentary structures and their fossils, were formed from deposits that accumulated in shallow seas or on low-lying parts of the land—that is, on surfaces not far above or below sea level. As deposition continued, the surface must have been slowly sinking at the same time. The greater thickness of the strata in the present mountain area means that this part of the earth's surface was sinking more rapidly than nearby areas.

Besides their thickness, another conspicuous feature of the sedimentary rocks in major mountain ranges is their complex structure (Fig. 16-7). They are crumpled by intense folding and locally broken along huge reverse faults and minor normal faults. Thus in the development of a mountain range the next step must be a period in which the piled-up sediments are subjected to intense compressional forces that act horizontally. The compression raises the folded layers high above the sea, and erosion begins to wear down the exposed beds as folding continues. The appearance of most present-day mountain landscapes is not due directly to the compressions that folded and faulted their rocks but instead to erosion between periodic uplifts.

16.3 Continental Drift

An Evolving Jigsaw Puzzle

A glance at a map of the world suggests that at some time in the past the continents may have been joined together in one or two giant supercontinents. If the margins of the continents are taken to be on their continental slopes (see Fig. 14-36) at a depth of 900 m instead of their present sea-level

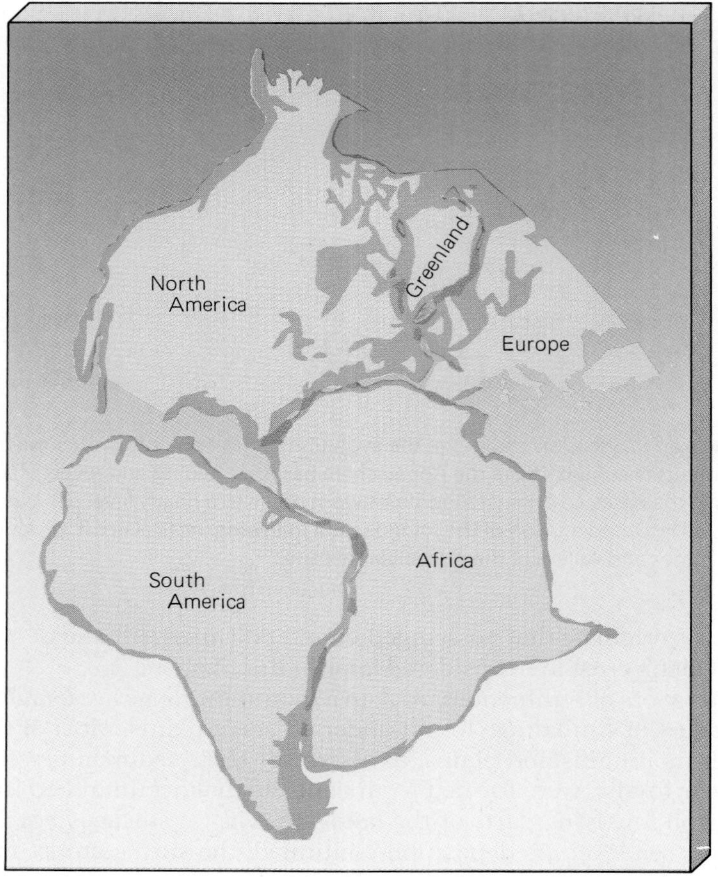

Figure 16-8 How some of the continents fit together. The boundary of each continent is taken at a depth of 900 m on its continental slope; the tan regions represent land above sea level at present, and the light orange regions represent submerged land on the continental shelf and slope. Overlaps are shown in dark orange and gaps in blue.

boundaries, the fit between North and South America, Africa, Greenland, and western Europe is remarkably exact, as Fig. 16-8 shows.

But merely matching up outlines of continents is not by itself proof that the continents have migrated around the globe. A detailed hypothesis of continental drift was proposed almost a century ago by Alfred Wegener, building on the work of others before him, who based his argument on biological and geologic evidence.

Wegener was troubled by the parallel evolution of living things. Going back through the ages, the fossil record shows that, until about 200 million years ago, whenever a new species appeared, it did so in many now-distant regions where suitable habitats existed. Evolution, in other words, proceeded at the same rate and in the same way in continents and oceans that today are widely separated. Only in the last 200 million years have plants and animals in the different continents developed in markedly different ways.

Wegener brought back an earlier idea that the continents were once all part of one huge landmass he called **Pangaea** ("all earth"). Then Pangaea broke up, and the continents slowly drifted to their present locations. This model found additional support in geologic data regarding prehistoric climates. Somewhat over 200 million years ago South Africa, India, Australia, and part of South America were burdened with great ice sheets, while at

BIOGRAPHY Alfred Wegener (1880–1930)

Wegener was a German meteorologist with a special interest in the Arctic. He participated in four expeditions to Greenland and died on the last of them. Like others before him, going back 3 centuries, Wegener was struck by the apparent fit of the continents shown in Fig. 16-8, but unlike them he went on to develop biological and geological arguments to support the notion that the continents had once been united. Wegener did not stop with tracing the subsequent movements of the continents but also considered other effects of their motions. He believed that mountain ranges were forced up by pressure on the leading edges of the drifting continents, with the Rockies and Andes as examples. The trailing edges of the drifting continents, in Wegener's view, left behind fragments that ended up as such island arcs as the West Indies.

Wegener first presented his ideas in 1912 and continued to refine them until his death. Some geologists were immediately attracted by the boldness and comprehensiveness of his scheme. Others were strongly opposed, at least partly because Wegener, a meteorologist, seemed to them an outsider heedlessly trying to overthrow established geological concepts. Doubters of Wegener explained the similarity of patterns of early life around the world by postulating a series of land bridges linking the continents. But no traces of such bridges were ever found. As for the mountains, it was widely believed that the earth was in the process of shrinking, with its surface becoming wrinkled the way a baked apple does as its water content evaporates. But the massive earth cannot contract as a baked apple does: the wrinkles of its continents can only have been produced by horizontal forces.

A more serious objection to continental drift was Wegener's inability to come up with a convincing mechanism for pushing the continents around. Today such a mechanism is known, and even though he got some details seriously wrong, Wegener now can be seen as a brilliant pioneer and not as the "pseudo-scientist" his opponents called him.

the same time a tropical rain forest covered North America, Europe, and China. At various other times, there was sufficient vegetation in Alaska and Antarctica for coal deposits to have resulted, and so currently frigid a place as Baffin Bay, which lies between Canada and Greenland, was a desert.

Wegener and his followers examined what was known about the climates of the distant past and tried to arrange the continents in each geologic period so that the glaciers of that period were near the poles and the hot regions were near the equator. Their efforts were, in general, quite successful, and in some cases startlingly so. Deposits of glacial debris and fossil remains of certain distinctive plant species follow each other in the same succession in Argentina, Brazil, South Africa, Antarctica, India, and Australia, for example. A recent discovery of this kind was the identification of a skull of the reptile Lystrosaurus in a sandstone layer in the Alexandra mountain range of Antarctica. This creature, which was about 90 cm long, flourished long ago in Africa.

Laurasia and Gondwana Today it seems clear that Pangaea did exist. About 200 million years ago it began to break apart into two supercontinents, **Laurasia** (which consisted of what is now North America, Greenland, and most of Eurasia) and **Gondwana** (South America, Africa,

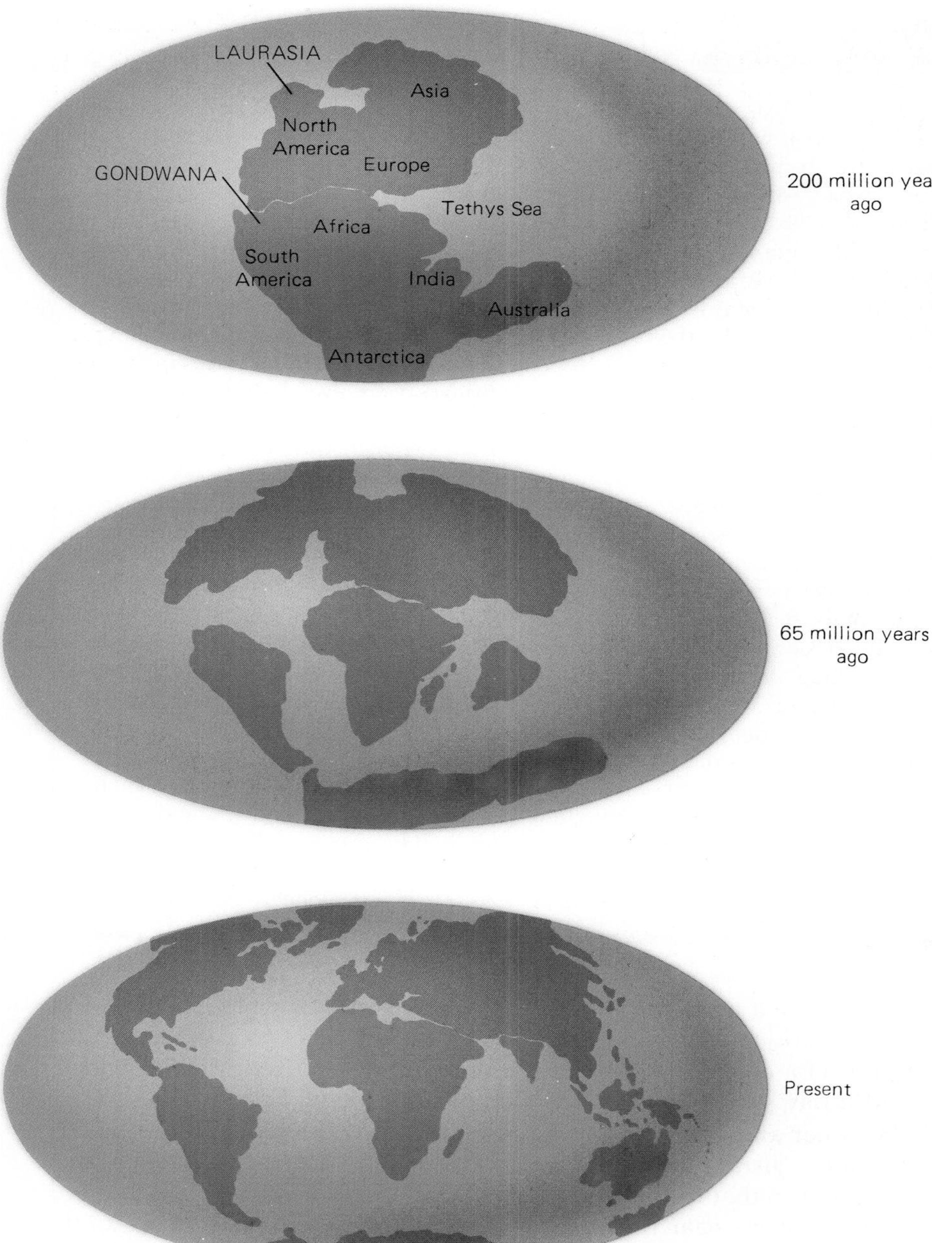

Figure 16-9 The landmasses of the earth as they may have appeared in the past and as they are today. The breakup of Pangaea into Laurasia and Gondwana began about 200 million years ago. There is evidence of even earlier continental drift, possibly as far back as 3.8 billion years ago.

Antarctica, India, and Australia). Laurasia and Gondwana were almost equal in size (Fig. 16-9). The separation of Pangaea into these supercontinents is supported by detailed geologic and biological evidence, for instance the differences between Laurasian and Gondwana fossils of the same age after the breakup.

Laurasia and Gondwana were separated by a body of water called the **Tethys Sea**. Today a little of the Tethys Sea survives as the Mediterranean, Caspian, and Black Seas. Its original extent can be gauged from the sediments that were subsequently uplifted to form the mountain ranges that stretch from Gibraltar eastward to the Pacific. The Pyrenees, Alps, and Caucasus of Europe, the Atlas Mountains of North Africa, and the Himalayas of Asia all were once part of the Tethys Sea.

Not long after Pangaea divided into Laurasia and Gondwana, the supercontinents themselves started to break up. The North Atlantic and Indian Oceans were the first to open, followed by the South Atlantic. Perhaps 80 million years ago Greenland began to move away from North America; 45 million years ago Australia split off from Antarctica and India finished its journey north to Asia; and 20 million years ago Arabia separated from Africa.

Plate Tectonics

"Continental drift" is not too accurate a description because the continents turn out to be merely passengers on a number of rigid rock plates that are continually moving across the face of the earth. Before we examine how this motion occurs and what its consequences are, we must first see why it is possible in the first place.

16.4 Lithosphere and Asthenosphere

A Hard Layer over a Soft Layer

At the heart of all current explanations for major crustal changes is the idea that the mantle of the earth is not a stiff structure, like the crust, but rather contains a layer near the top that is able to flow.

The crust and the outermost part of the mantle together make up a shell of hard rock 50 to 100 km thick called the **lithosphere.** The lithosphere

Why Continents Have Deep Roots

The notion of a relatively soft asthenosphere makes it possible to understand both why the continents are raised above the rest of the crust and why the crust is much thicker beneath them than beneath the oceans (see Fig. 15-33). Suppose that we place several blocks of wood of different sizes in a pool of water. The larger blocks float higher than the smaller while simultaneously extending down farther into the water (Fig. 16-10). Thus, if the lithosphere with its continents is imagined as floating on a denser asthenosphere, we can refer for guidance to the floating wooden blocks. This analogy suggests that exceptionally high regions—mountain ranges and plateaus—have roots that extend far downward. Such is actually the case; in fact, its discovery led the British scientist George Airy a century ago to propose the floating of the entire lithosphere.

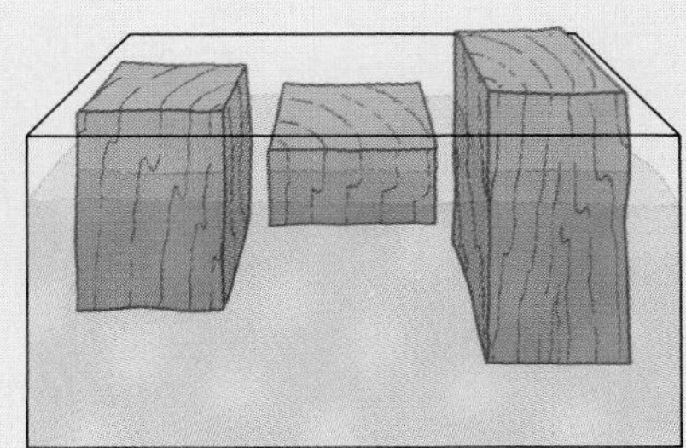

Figure 16-10 Large blocks of wood float higher and extend farther downward than smaller blocks of wood. This is why the thicker continental crust extends lower into the mantle than the oceanic crust, as shown in Fig. 15-33.

has no sharp boundary, as the crust does, but gradually turns into the softer **asthenosphere,** a region about 100 km thick. (*Lithos* means "rock" in Greek and *astheno* means "weak.") The asthenosphere is soft because its material is close to the melting point at the temperature and pressure found at such depths. Above the asthenosphere the temperature is too low and below it the pressure is too high for the material to deform easily.

The speeds of earthquake waves in the crust are different from their speeds in the mantle, which suggests that the two regions differ in their compositions or crystal structures or both. The difference between the lithosphere and the asthenosphere instead lies in their degrees of rigidity. In response to suddenly applied forces, the asthenosphere behaves like a solid and so can transmit the transverse S waves of earthquakes, for example. But when forces act on the asthenosphere over long periods of time, it responds like a thick, viscous fluid.

16.5 The Ocean Floors

The Youngest Part of the Crust

The mountains and valleys, plains and plateaus of the continents were mapped long ago, and few surprises are in store for future explorers. But the continents occupy less than 30 percent of the area of the earth's crust, while the rest lies hidden in darkness thousands of meters under the seas and oceans. Only in the past few decades have the floors of the oceans been studied and their physical characteristics determined. It is largely these findings that have clarified the evolution of the crust.

Methods used to investigate the ocean floors are not particularly subtle—the real problem is the vastness of the area to be covered. These days depths are charted by means of echo sounders. An instrument of this kind sends out a pulse of high-frequency sound waves and the time needed for the pulse to reach the sea floor, be reflected there, and then return to the surface is a measure of how deep the water is (Fig.7-13). A variant of this method reveals something of the structure of the sea floor itself. What is done is to set off an explosive charge in the water and study the returning echoes—one echo will come from the top of the sediment layer, and a later one from the hard rock underneath.

Samples of the sea floor can be obtained by dropping a hollow tube to the bottom on a long cable and then pulling it up filled with a core of the sediments into which it sank. Longer sediment cores can be obtained by drilling. These sediments can be examined later in the laboratory for their composition, their age, their fossil content, their magnetization, and so forth (Fig. 16-11). Another important technique is to tow a magnetometer behind a survey ship to obtain an idea of the direction and intensity of the magnetization of the rocks of the ocean floor over wide areas.

Four Important Findings Four findings about ocean floors have proved of crucial importance:

1. The ocean floors are, geologically speaking, very young. The oldest oceanic crust dates back only about 200 million years, in contrast to continental rocks that date back as much as 3800 million years.

Figure 16-11 Samples from a core of sediments extracted from the ocean floor are removed for examination. The record of the past is most complete in marine sediments.

Many parts of the ocean floor are much younger still, so that about one-third of the earth's surface has come into existence in $1\frac{1}{2}$ percent of the earth's history.

2. A worldwide system of narrow **ridges** and somewhat broader **rises** runs across the oceans (Fig. 16-12). An example is the Mid-Atlantic Ridge, which runs down the Atlantic Ocean from north to south. Iceland, the Azores, Ascension Island, and Tristan da Cunha are some of the higher peaks in this ridge. These ridges are offset at intervals by fracture zones that indicate sideways shifts of the ocean floors.
3. There is also a system of **trenches** several kilometers deep that rims much of the Pacific Ocean. These trenches parallel the belts in which most of today's earthquakes and volcanoes occur. Some of the trenches have **island arcs** on their landward sides that consist largely of volcanic mountains projecting above sea level.
4. The direction in which ocean-floor rocks are magnetized is the same along strips parallel to the midocean ridges, but the direction is reversed from strip to strip going away from a ridge on either side (Fig. 16-13).

16.6 Ocean-Floor Spreading

Alternate Magnetization Is the Proof

The first step toward understanding the above observations was taken in the early 1960s by the American geologists Harry H. Hess and Robert S. Dietz, who independently proposed that the ocean floors are spreading. (A similar hypothesis was put forward in 1928 by Arthur Holmes in England, but it remained practically unnoticed because supporting data were lacking.)

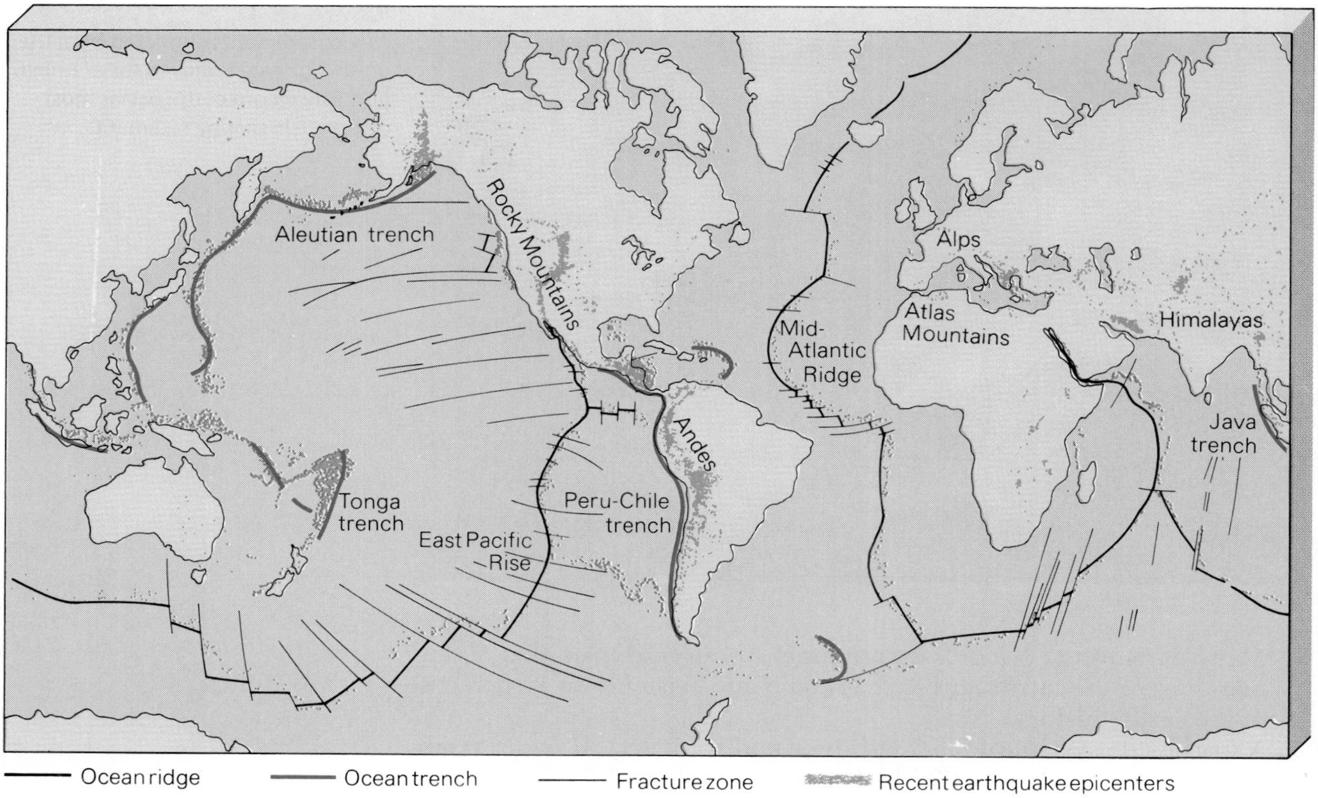

Figure 16-12 The worldwide system of oceanic ridges, rises, and trenches. The ridges and rises are offset by transverse fracture zones. The American oceanographers Marie Tharp and Bruce Heezen were pioneers in identifying these features of the ocean floors.

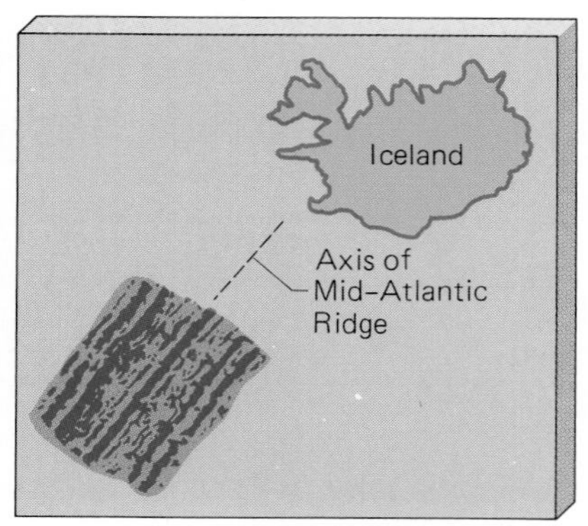

Figure 16-13 Pattern of magnetization along the Mid-Atlantic Ridge southwest of Iceland. Seafloor rocks whose directions of magnetization are the same as that of today's geomagnetic field are shown in dark blue; the intervening spaces represent rocks whose magnetization is in the opposite direction.

The basic idea of ocean-floor spreading is that molten rock is continually rising up along the midocean ridges (Figs. 16-14 and 16-15). The parts of the lithosphere on either side of a ridge move apart at speeds of a few centimeters per year—about as fast as fingernails grow—with the new material filling the gap as it hardens.

The fourth key observation mentioned earlier, which concerns the magnetization of rocks on either side of an ocean ridge, confirms the hypothesis of sea-floor spreading in a convincing way. As Fig. 16-13 shows in the case of a portion of the Mid-Atlantic Ridge southwest of Iceland, successive strips of rock lying parallel to the ridge are magnetized in alternate directions. To interpret this pattern, we draw upon the fact that the earth's magnetic field has periodically reversed itself many times in the past (see Sec. 15.9). What must have been happening is clear. As molten rock, unmagnetized in its liquid state, comes to the surface of the crust at a ridge, it hardens and the iron content of its minerals becomes magnetized in the same direction as that of the geomagnetic field at the time. When the direction of the geomagnetic field reverses, the new molten rock that cools then becomes magnetized in the opposite direction. Thus strips of alternate magnetization follow one another going away on both sides from a ridge. Iceland itself was formed less than 70 million years ago from magma rising through the rift in the Mid-Altantic Ridge.

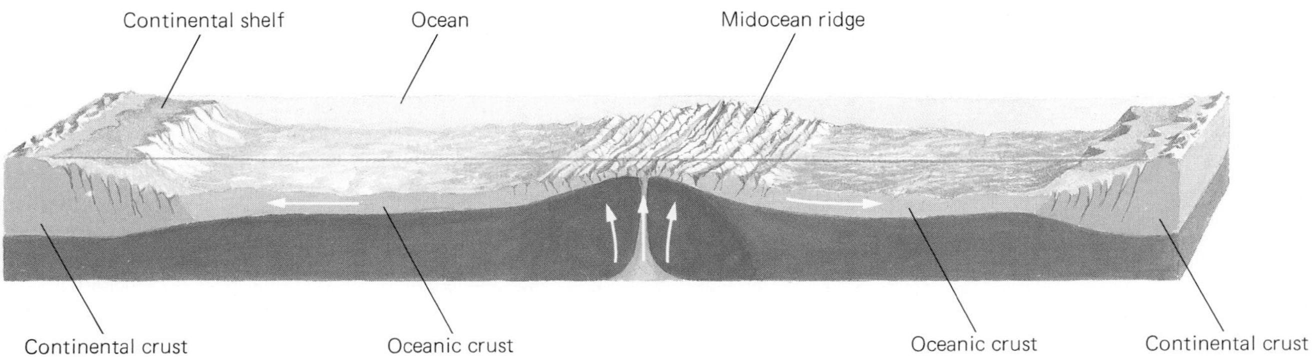

Figure 16-14 Ocean-floor spreading. A midocean ridge forms where molten rock rises from the asthenosphere along a rift in the ocean floor and pushes apart the lithosphere on both sides.

The reversals of the earth's magnetic field have been dated by measurements made on magnetized lava flows on land using methods based on radioactivity. This information can be used to find the ages of the magnetized strips of the sea floor, which gives the speeds of the lithospheric plates. Other ways to determine plate motions include using precise position data from the Global Positioning System (see Sec. 2.14). Typical speeds vary from 1 cm/year in the North Atlantic Ocean to 10 cm/year in the Pacific. The record is 15 cm/year near Easter Island—about as fast as the hair on your head grows.

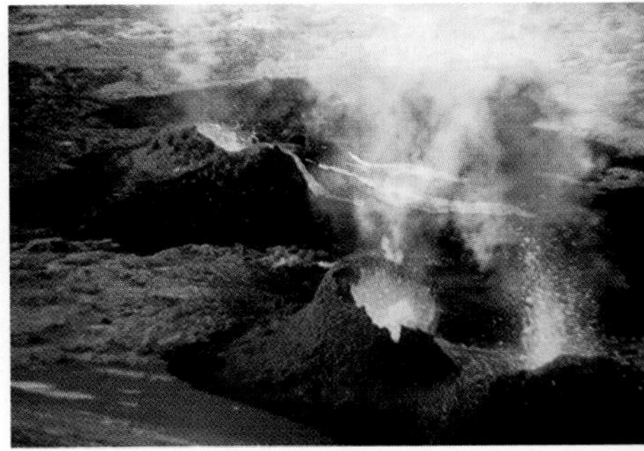

Figure 16-15 The island of Surtsey came into being in 1963 as a volcano that grew from the sea south of Iceland on the Mid-Atlantic Ridge. Molten rock rises to become new oceanic crust along this ridge, which forms the boundary between the North Atlantic and Eurasian plates. These plates are moving apart at about 1 cm per year.

16.7 Plate Tectonics

How the Continents Drift

Although the ocean floors are spreading at the midocean ridges, the earth as a whole does not expand. The spread of the ocean floors must therefore be balanced by other large-scale processes in the lithosphere. The study of these processes and their consequences has come to be known as **plate tectonics.**

The starting point of plate tectonics is the observation that the lithosphere is split not only along the ridges of the ocean floors but also along the trenches and fracture zones of Fig. 16-12. These cracks divide the lithosphere into seven huge **plates** and a number of smaller ones, all of which float on the plastic asthenosphere (Fig. 16-16).

New lithosphere is created where plates move apart at the midocean ridges, as we saw in Fig. 16-14. Where plates come together, on the other hand, lithosphere may be destroyed: the edge of one of the plates may slide under the edge of the other and partially melt upon reaching the hot asthenosphere. Figure 16-17 shows the three possible kinds of collision. To understand what is happening, we must keep in mind that continental crust (largely granitic rock) is less dense than oceanic crust (largely basaltic rock).

1. **Oceanic-continental plate collision.** When a plate whose edge is covered with oceanic crust moves against a plate whose edge is covered with continental crust, the denser oceanic slab slides underneath the continental slab (Fig. 16-17*a*). The region of contact is called a

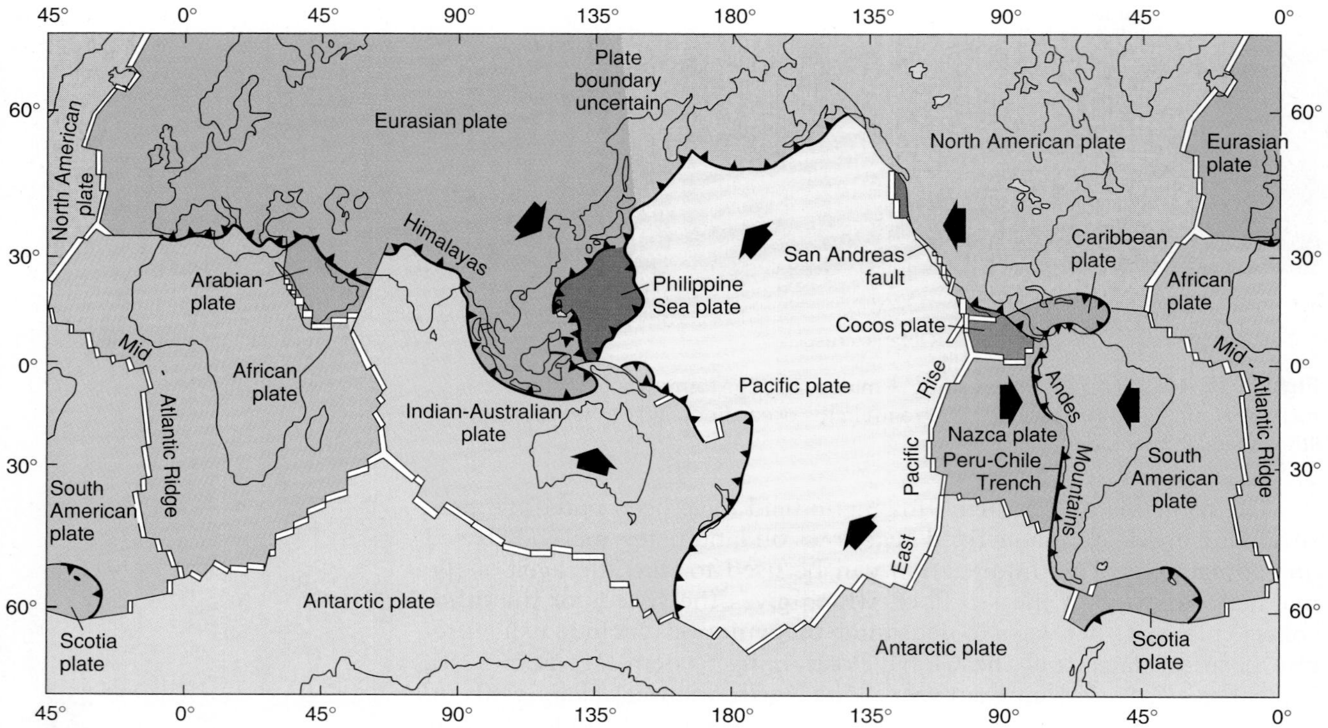

Figure 16-16 The chief lithospheric plates whose motion results in continental drift. The arrows show the directions of plate motions; the African plate is thought to be stationary. Diverging plate boundaries are shown as double lines. Converging boundaries are shown as heavy lines with triangles that point down subduction zones. Transform boundaries are shown as single lines. Because Japan lies on or near the intersections of four plates, one-tenth of the world's earthquakes rock its islands.

subduction zone, and a trench is formed there. Some of the oceanic slab melts in the asthenosphere, and magma from the lighter materials in the slab is carried upward by its buoyancy. The magma rises through cracks in the continental plate above to produce volcanoes at the surface and bodies of intrusive rock below the surface. An example of such a collision occurs at the western edge of South America, where the oceanic Nazca plate (which is moving eastward) meets the continental South American plate (which is moving westward). The result is a trench along the coasts of Peru and Chile and a range of volcanic mountains, the Andes (Figs. 16-18 and 16-19). Many of the world's 530 or so active volcanoes are found above subduction zones.

2. **Oceanic-oceanic plate collision.** When two plates whose edges are covered with oceanic crust collide, one of them slides underneath the other in a subduction zone (Fig. 16-17*b*). Volcanoes are again formed, this time on the ocean floor. Where these volcanoes are high enough to rise above the ocean, they appear as the chains of islands called **island arcs.** The island arcs that border the Asiatic side of the Pacific—the Aleutians, Japan, the Philippines, Indonesia, the Marianas—are believed to have come into being in this way. The West Indies in the Atlantic form another example of an island arc.

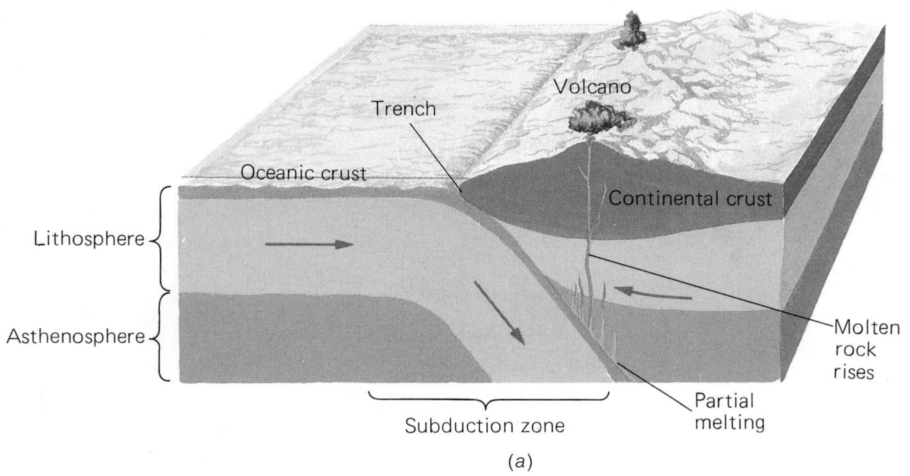

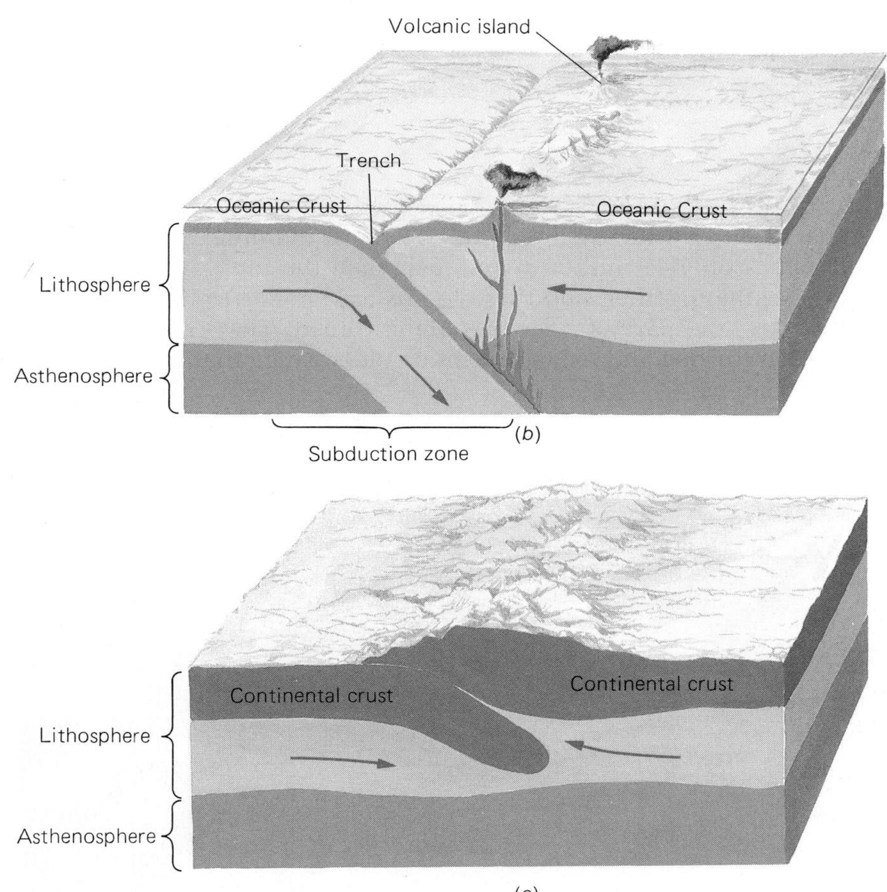

Figure 16-17 Three situations can occur when lithospheric plates come together. (*a*) Oceanic-continental plate collision. The Andes Mountains of South America are the result of such a collision. (*b*) Oceanic-oceanic plate collision. The islands of the West Indies originated in this way. (*c*) Continental-continental plate collision. A collision of this kind thrust up the Himalaya Mountains between India and the rest of Asia.

3. **Continental-continental plate collision.** Here both plates are too light relative to the underlying asthenosphere and too thick for either one to be forced under. The result is that the plate edges are pushed together and buckle, forming a mountain range (Fig. 16-17*c*). The massive Himalayas that divide India from the rest of Asia were

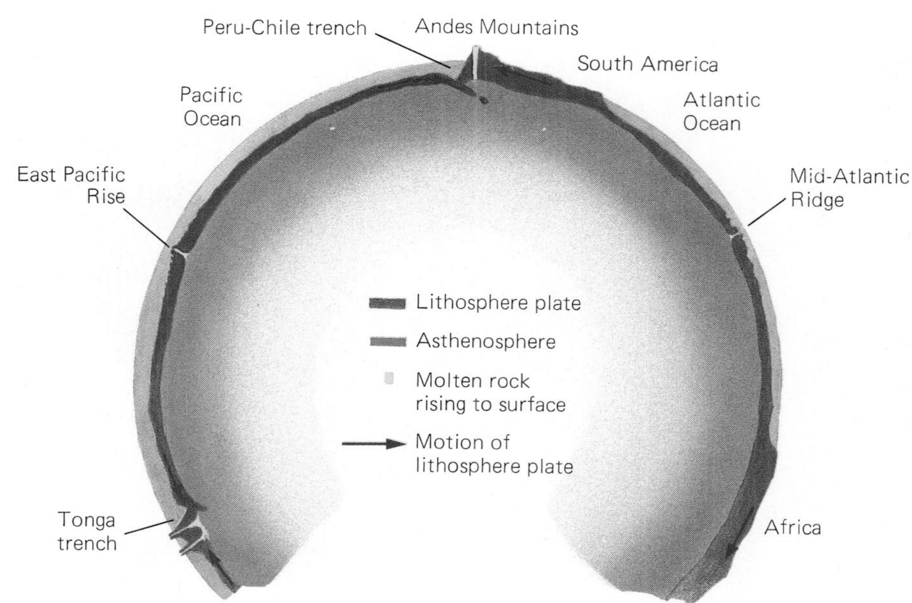

Figure 16-18 The sea floor spreads apart at midocean ridges where molten rock rises to the surface of the lithosphere. At a trench, one lithosphere plate is forced under another into the asthenosphere, where it melts. Mountain ranges, volcanoes, and island arcs are found where plates collide. The vertical scale is greatly exaggerated.

thrust upward in this manner. The geological evidence for this event is backed up by the fossil record. The oldest mammal fossils in India date back only 45 million years, just when the Indian and Eurasian plates came together, and these fossils are similar to those found in Mongolia, the part of Asia that India joined. The Ural Mountains between Europe and Asia are a much older range that formed in the same manner.

Because the earth's crust under the oceans is being continuously created at midocean ridges and destroyed at subduction zones, all of it is

Figure 16-19 The Andes are a young mountain range thrust upward along the western edge of South America where the eastward-moving Nazca plate is forced under the westward-moving Atlantic plate. These ruins in the Peruvian Andes are of the Inca city of Machu Picchu, which was built about 1450 and abandoned a century later at the time of the Spanish conquest of Peru.

Mount Everest

Mt. Everest, the highest mountain above sea level on earth, has an altitude of 8.85 km (29,028 ft) and is part of the Himalaya range in Asia (Fig. 16-20). (Measured from its base on the ocean floor, the Hawaiian mountain Mauna Kea is actually 1.4 km taller than Everest. Because the earth bulges at the equator as a consequence of its rotation, Mt. Chimborazo in Ecuador, which lies almost on the equator, is 2.2 km farther from the earth's center than Mt. Everest. However, Mt. Chimborazo, only 6.3 km above sea level, is not even the highest peak in the Andes.) The mountain was named after Sir George Everest, who supervised the Great Trigonometrical Survey of India, which mapped the subcontinent between 1823 and 1843.

The Himalayas were forced upward when the Indian plate plowed into the Eurasian plate about 45 million years ago. Before the collision the northward speed of the Indian plate was 15–20 cm/year. Afterward the plate slowed to around 5 cm/year as it advanced roughly 2000 km into the Eurasian plate. The Indian plate is still moving north, today at 2 cm/year, and the Himalayas are still rising. The world's 10 highest mountains are currently in the Himalayas, but the Andes are going up faster and some of their peaks may well look down on Everest in time.

Most of Mt. Everest consists of granite and metamorphic rocks with a topping of limestone that contains fossils of marine life from the Tethys Sea (see Fig. 16-9).

As we would expect, major earthquakes occur regularly along the still-active fault between the Indian and Eurasian plates. The most recent was the magnitude 7.6 quake of October 2005 that devastated Kashmir and adjacent northern Pakistan. About 87,000 people were killed, at least 130,000 were injured, and 2 million were left homeless as winter approached.

Figure 16-20 Mt. Everest lies on the border between Nepal and Tibet. Its summit was first reached in 1953 after many failed attempts.

relatively young. On the other hand, most continental crust is quite old, as much as 3.8 billion years old as compared with less than 180 million years for oceanic crust. Figure 16-21 shows the various geological processes that take place in the vicinity of a convergent plate boundary.

Example 16.1

The Andes are a young, still-growing mountain range on the west coast of South America. Why is there no similar mountain range on the east coast of South America?

Solution

The boundary of the western Atlantic plate is along the western edge of South America, and the Andes are the result of the collision between this plate, which is moving westward, and the eastern Pacific plate, which is moving eastward. The eastern edge of South America is not near a plate boundary so there is nothing to force a mountain range upward there.

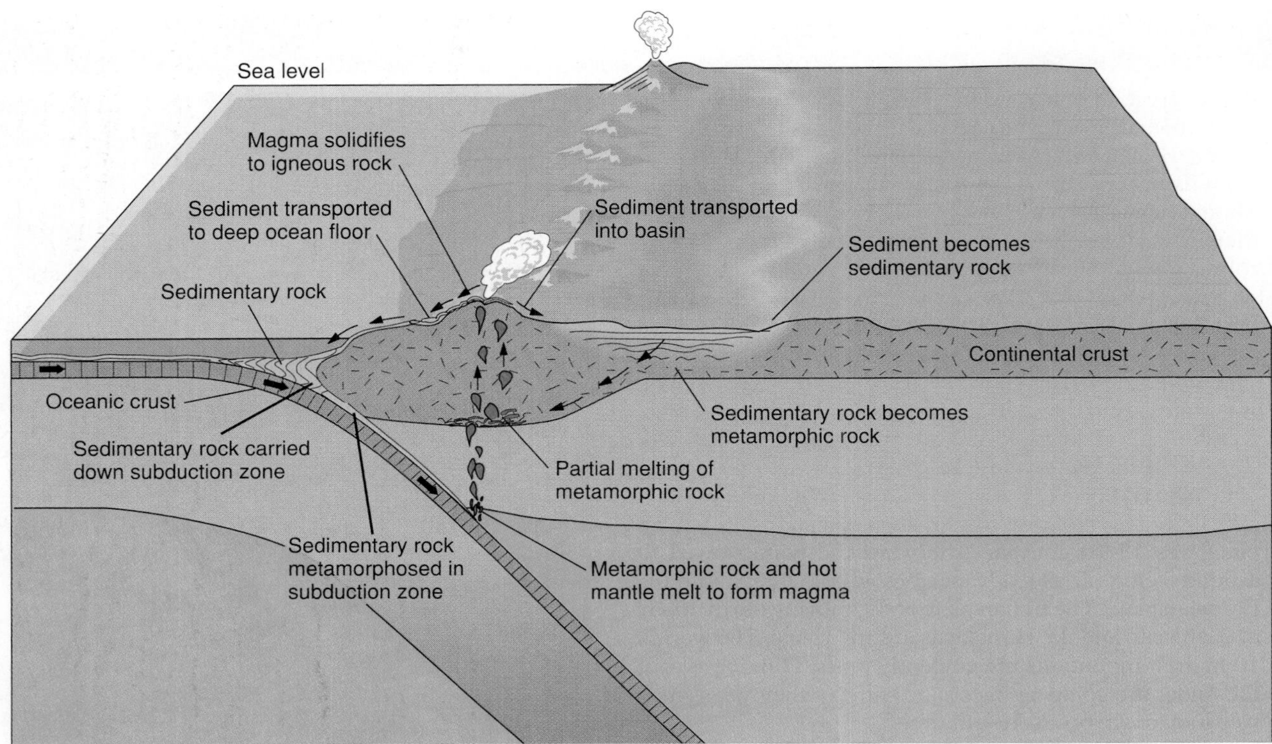

Figure 16-21 How the various transformations of the rock cycle shown in Fig. 15-60 occur where an oceanic plate collides with a continental plate.

Transform Faults Another type of plate boundary is a **transform fault**, which occurs where the edges of two plates slide past each other. (A transform fault is a large-scale version of the strike-slip fault shown in Fig. 16-2.) Earthquakes are common along transform faults, as we would expect. Since most transform faults are in ocean basins (see Fig. 16-16), the quakes associated with them usually are unnoticed except by geologists. A conspicuous exception is the 1200-km-long San Andreas Fault in California, which is one of the faults that lie along the boundary between the Pacific and North American plates (Figs. 16-23 and 16-24). Movement along the fault occurs continuously in some regions but elsewhere in sudden jerks that release accumulated stresses. The San Francisco earthquakes of 1906 and 1989 were caused by such abrupt slippages, and further earthquakes along the fault are sure to happen in the future.

How Tectonic Forces Arise What drives the lithosphere plates as they shape and reshape the earth's surface? Several plausible mechanisms have been put forward and it is possible that all contribute to some extent to plate motion. An obvious one is convection in the plastic asthenosphere due to uneven heating from the mantle below. Rock hot enough to flow gradually is supposed to rise at diverging plate edges (where mid-ocean ridges occur) and spread out horizontally while dragging along the plates above. Then, having cooled, the rock sinks at subduction zones. (Convection in the atmosphere is shown in Fig. 14-16.) Alternatively, the

The Hawaiian Islands

Many of the world's active volcanoes occur along the rifts between plates that are moving apart; an example is shown in Fig. 16-15. Another common location is above a subduction zone, as in Fig. 16-17*a* and *b*.

A third class of volcanoes are those produced by molten rock that comes to the surface above hot spots in the mantle. The Hawaiian Islands were formed as the Pacific plate moved northwestward at about 9 cm/year over such a hot spot (Fig. 16-22). From time to time a plume of molten rock erupted from this hot spot to create a volcano above it that rose above the ocean floor. These volcanoes, now the islands of the Hawaiian chain, became extinct after they passed the hot spot.

The only volcanic activity in that region today is on the island of Hawaii, whose southeastern part is over the hot spot, and on a young undersea volcano called Loihi that is 30 km southeast of Hawaii and feeding off the same plume. Eventually Loihi will grow into the next member of the Hawaiian island chain. There is evidence that the hot spot as well as the Pacific plate have been moving, which is why the islands do not lie along a straight line. About 20 other hot spots are active today around the world.

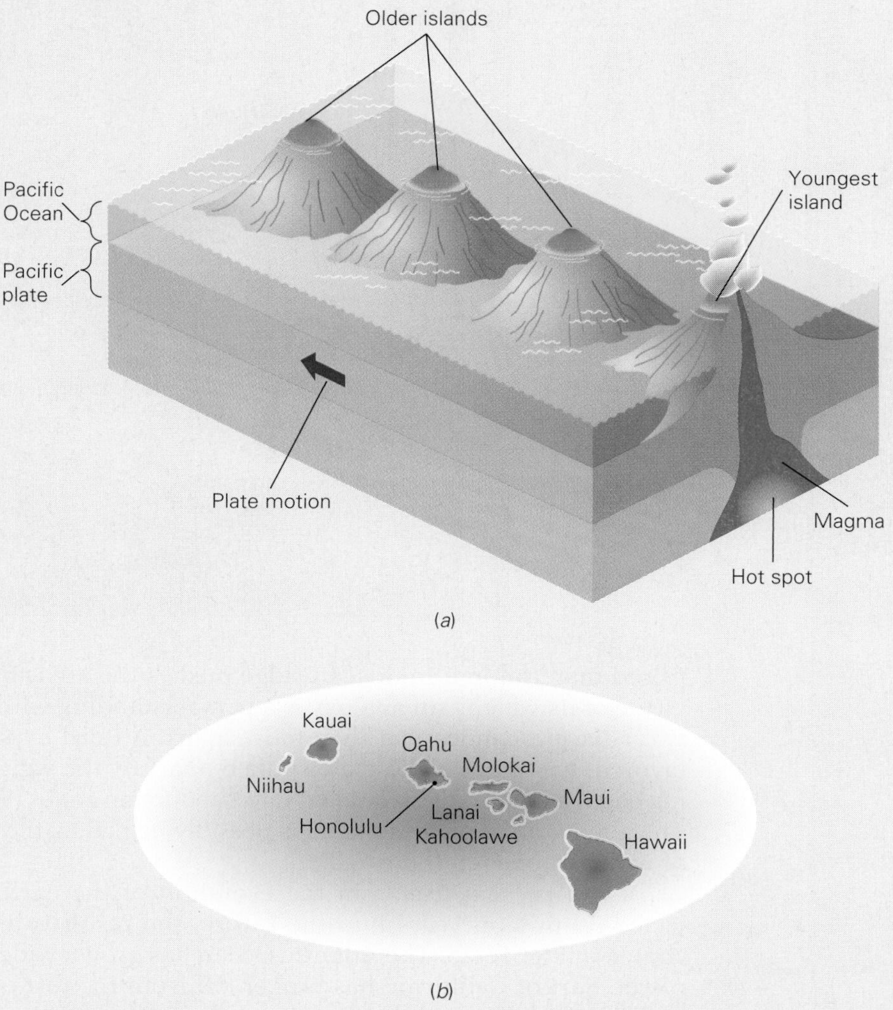

Figure 16-22 (*a*) Origin of the Hawaiian Islands. (*b*) The islands today.

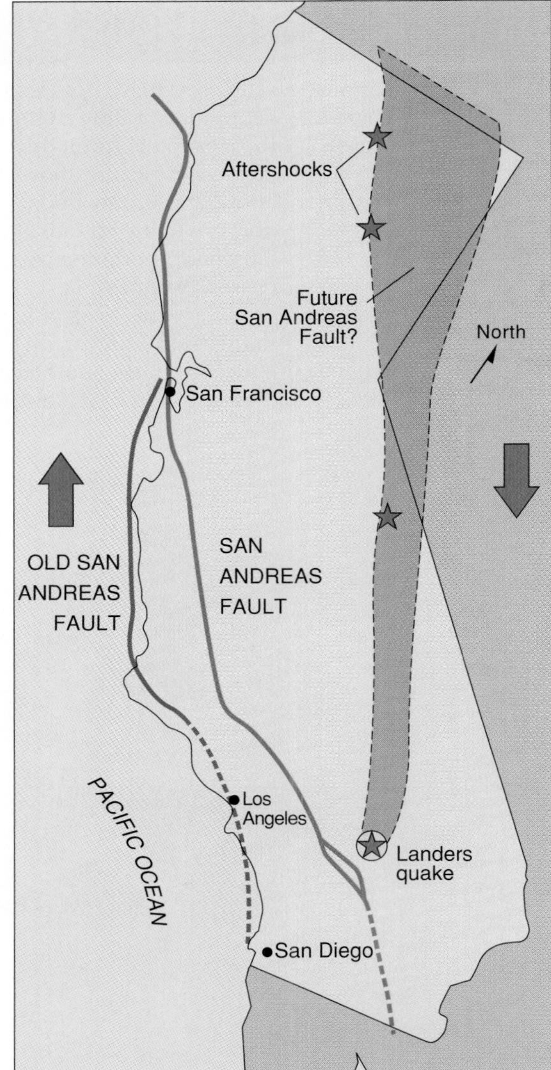

Figure 16-23 Until about 5 million years ago, motion along the transform boundary between the Pacific and North American plates occurred along a fault west of the present San Andreas Fault. Today the motion takes place along the latter fault at about 6 cm per year. The 1992 Landers earthquake had aftershocks along a line extending to the north, which suggests that perhaps in the future more of California will become attached to the Pacific plate and slide northwestward with it past the North American plate.

raised material in a midocean ridge pushes the adjacent plates apart by virtue of its weight; an analogy is a person standing with a foot in each of two adjacent canoes that then move apart. A third hypothesis, currently favored, has the oceanic plates pulled apart by the weights of their sinking edges as these edges descend into subduction zones (Fig. 16-17*a* and *b*). The ability of the plates to bend is possibly helped by the lubricating effect of water trapped in their rock.

When present trends in the evolution of the earth's crust are projected 30 million years into the future, the result is a picture like that shown in Fig. 16-25. The Atlantic Ocean has grown wider, the Pacific narrower. Part of California has broken off from the rest of North America, and the Arabian peninsula has been forced around to become an integral part of Asia. The islands of the West Indies have grown into a land bridge between the Americas, and the western Pacific islands have also increased markedly in extent.

(a)

(b)

Figure 16-24 (*a*) Part of the San Andreas Fault in California. The earthquake of October 17, 1989, that rocked San Francisco originated in a zone of this fault whose sides shifted by about 2 m relative to each other. The slippage occurred underground. (*b*) Geology is taken seriously in California.

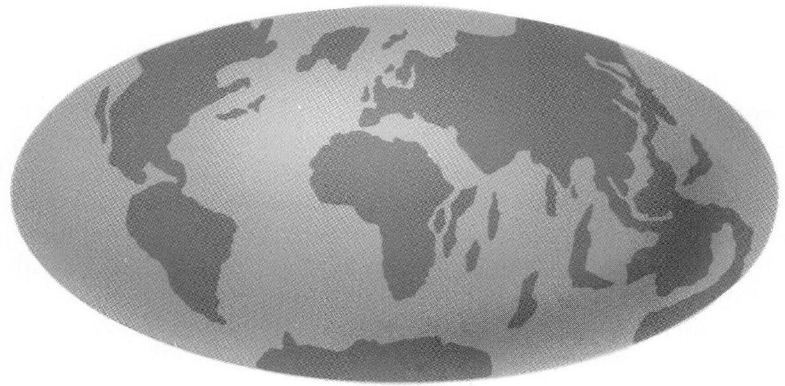

Figure 16-25 How the earth may appear 30 million years from now if present plate motions continue. Figure 16-9 shows the earth's landmasses as they were and as they are today to the same scale.

And after that? All that can be said is that the face of the earth will probably continue to change in the future, just as it has been changing as far back in the past as we have any evidence. One projection has all the continents merged again into a new Pangaea 250 million or so years from now. And then another breakup? Quite possibly.

Methods of Historical Geology

Two kinds of events in the history of the crust are significant. In one category are physical changes such as the drift of the continents, the upthrust and wearing away of mountains, the spread and retreat of glaciers and ice sheets. In the other are changes in the living things that populate the earth, from primitive one-celled organisms to the complex plants and animals of today. Although the rock record is less complete the farther back we go, it is still possible to trace the physical and biological evolution of the crust for billions of years into the past. Let us see how this is done.

16.8 Principle of Uniform Change

"No Vestige of a Beginning, No Prospect of an End"

The modern view of geologic evolution had its start two centuries ago. Before then a major obstacle to scientific thinking about the earth's history was the Bible. The Book of Genesis tells about the earth's beginning with beauty and simplicity. A seventeenth-century theologian, Bishop James Ussher, used the stories in the Bible to pinpoint the moment when our planet was created from a formless void: 9 o'clock in the morning of October 12, 4004 B.C. However, even the smallest knowledge of geology shows that the events recorded in today's landscapes cannot possibly have occurred in only a few thousand years. Early investigators nevertheless devoted their efforts to trying to fit what they found in the rocks around them into a literal interpretation of Genesis.

Even when freed from biblical shackles, geologists of the past usually went far astray in their arguments. One reason was their habit of generalizing from limited evidence. Another was their readiness to postulate tremendous events to explain particular findings. For instance, a comprehensive theory was formulated by Abraham Gottlob Werner on the basis of the geologic structures he found near his home in Freiberg, Germany. What he saw was granite overlain by folded, somewhat metamorphosed rocks, with these in turn overlain by flat sedimentary beds. Untraveled and deaf to the reports of others, Werner considered this sequence to be the same worldwide. Each of the three types of rocks, he presumed, was deposited by a universal ocean, with granite precipitating first and the upper beds last. All rocks, in Werner's system, were sedimentary rocks, and the geologic history of the earth consisted of three sudden precipitations from an ancient ocean that were followed by the disappearance of most of the water.

The French biologist Georges Cuvier greatly influenced geology at the beginning of the nineteenth century by studying **fossils,** which are the remains or traces of organisms preserved in rocks. In successive rock layers around Paris, Cuvier found distinct groups of animal fossils, different from one another and from the present animals of that region. He concluded that each group appeared on the earth as a result of a special creation and that each was destroyed by a universal disaster before the next creation. Thus Cuvier also regarded the earth's history as a succession of catastrophes that were separated by intervals of stable conditions.

BIOGRAPHY Charles Lyell (1797–1875)

A Scot, Lyell studied law at Oxford. Fascinated by geology, he traveled extensively and observed geologic changes then taking place—the shifting of streams, the building of deltas, the advances and retreats of shorelines, the outpouring of lava and ash from volcanoes. From rocks and landscapes he read the slow accumulation of these changes through the long past—rivers now cutting far below their former channels, coasts with remnants of old beaches high above the present shore, immense lava flows where no volcanoes exist today.

Lyell was the first to distinguish metamorphic rocks as a group, and so to understand clearly the cycle by which rocks are formed, destroyed, and re-formed. He continued the work of Cuvier and his English followers in tracing the succession of animals whose remains are entombed in rocks. Nowhere did he find signs of past changes brought about by agencies other than those in the world around him. Lyell reemphasized his predecessor Hutton's point that processes that seem unable to alter the landscape by very much may nevertheless have large-scale consequences if given enough time. If the earth is only a few thousand years old, catastrophism is indeed the only explanation, but if the earth is millions of years old, the slow pace of geologic change can explain what we see around us. Lyell's accumulation of evidence was overwhelming, and not long after his *Principles of Geology* was published, most geologists, initially skeptical, accepted his views.

Hutton, Lyell, and Darwin James Hutton, a Scot, based his thinking on a much larger body of observational evidence than Werner or Cuvier. By accepting that the earth is very old, Hutton found no need to invoke special mechanisms since he could account for what he saw in terms of processes under way in the present-day world: "In the phenomena of the earth," he concluded, "I see no vestige of a beginning, no prospect of an end."

Hutton's ideas, outlined in his 1785 book *Theory of the Earth,* were soon taken up by others who modified and extended them. Chief among Hutton's followers was Charles Lyell, whose goal was "to explain the former changes of the earth's surface by forces now in operation." This he was largely able to do, and his guiding idea became known as the **principle of uniform change**. Of course, the vigor of geologic processes has varied from time to time and from place to place, so that uniform is not the best word. However, with a few exceptions, such variations are quite different from the catastrophic happenings earlier thinkers were so fond of.

An important link in the chain of ideas that was forming was supplied by Lyell's friend Charles Darwin in 1859 (Fig. 16-26). Darwin's **theory of evolution** showed that changes in living things as well as those in the inorganic world of rocks could be explained in terms of processes operating all around us. The fossil groups found by Cuvier near Paris did not come from special creations but were stages in a continuous line of development. Lyell understood at once the significance of Darwin's work and became one of his most active supporters.

Figure 16-26 Charles Darwin (1809–1882).

How far back does the principle of uniform change hold? It must break down eventually because the earth in its early stages as a planet was certainly different from today's earth. All geology can say for certain

Evolution

Evolution is the "view of life" (in Darwin's words) that has turned out to be the only organizing principle from which all the different aspects of biology make sense. "I had two distinct objects in view: firstly to show that species had not been separately created, and secondly, that natural selection has been the chief agent of change." Starting from the simplest organisms, "endless forms most beautiful and most wonderful have been, and are being, evolved."

The ability to evolve can be considered one of the chief distinctions—perhaps the primary one—between life and nonlife.

Evolution operates through two mechanisms. The first is reproduction with variation: the members of each generation are not necessarily the same as each other or as their parents. Darwin did not know how variation comes about, but today we can trace it to changes (called mutations) in the DNA of which the genes of living things are composed.

The second mechanism is natural selection: the individual plants and animals that are best adapted to their environments are most likely to thrive and produce similarly successful offspring. In simple terms, "survival of the fittest." When living conditions alter drastically, which has happened often in the history of the earth, the pace of evolution is faster than during periods of stability. At such times previously dominant species (such as the dinosaurs) may even disappear entirely, while previously marginal species (such as the mammals) may become favored. Darwin's view was that "it is not the strongest of the species that survive, nor the most intelligent, but the ones most responsive to change."

It is sometimes said that evolution cannot have led to complex life forms because it involves chance events. However, although each individual mutation is indeed a chance event, whether the mutation survives as an element in the genetic makeup of the descendants of an organism does not depend on chance but on how well the descendants meet the challenges of life. Thus evolution is not random overall because it is directed by natural selection. The descendants of a particular organism may later follow different evolutionary paths, which leads to the diversity of living things, but this branching, too, is shaped by natural selection. In this way immense numbers of random events occurring over immense reaches of time have, under the nonrandom guidance of natural selection, produced a living world of variety and complexity. Only evolution can explain how this world came about in terms of natural processes, and it does this with complete success.

A huge body of evidence drawn from many disciplines supports the basic concepts of evolution. Thus the fossil record shows the development of new forms of life from older ones, with a time scale established by radiometric methods that is long enough (billions of years) for this development to have taken place. Attempts to discredit evolution always point to gaps that remain in the fossil record between some old and new forms of life. But these gaps are being steadily filled in with transitional fossils, confirming Darwin's vision. A notable example is the discovery in 2006 of fossils of a 2.75-m-long, 375-million-year-old creature that seems to be the missing link between fish and land-dwelling animals. The "fishapod"—pod comes from the Greek for foot—had anatomical features of both fish and land animals, for instance lungs as well as gills and fins that contained primitive limbs with wrists and fingers. From the transitional fishapods evolved amphibians, reptiles (such as dinosaurs; birds descended from them), and mammals (such as us).

The theory of evoution is a good example of the scientific method (Sec. 1.1) in action. Darwin formulated a hypothesis based on observations, further observations supported the hypothesis, and it was so successful in explaining a large variety of findings that it was accepted as a theory. And quite a theory it is: according to the National Academy of Sciences, the theory of evolution is no less than "the foundation of modern biology."

The distribution of life forms around the world shows the effects of evolution: Australia was long isolated from the other continents, and many plants and animals there are distinct from those elsewhere. The anatomies of different species often show great similarities (human arms have much in common with whale flippers), which suggests descent from a common ancestor. Certain body parts that today have no function can be traced back to forebears that needed them: an example is the coccyx at the base of the spine, the remnant of a tail. And, of course, even before Darwin farmers speeded up evolution by choosing plant and animal variations to breed that were especially desirable for their purposes. Even this very brief list shows the power of evolution to account for the history, variety, and relatedness of life on earth.

is that the oldest rocks now exposed contain a clear record of processes very similar to those of the present. Beyond that lies the realm of hypothesis, and in Chap. 19 we shall see how theories of the origin of the earth connect up with geologic history.

16.9 Rock Formations

History under Our Feet

The crustal events of the past are recorded in the rocks and landscapes of the present. It is usually possible to reconstruct these events in terms of processes still at work reshaping the face of the earth. Thus from moraines, lakes, and U-shaped valleys we learn of the spread and retreat of ancient glaciers. Wave-cut cliffs and terraces above the sea suggest recent elevation of the land. Hot springs and isolated, cone-shaped mountains signify past volcanic activity.

Other past events have left their traces as well. A geologist finds a bed of salt or gypsum buried beneath other strata, and he or she knows that the region must once have had a desert climate in which a lake or an arm of the sea evaporated. A layer of coal implies an ancient swamp in which partly decayed vegetation accumulated. A limestone bed with numerous fossils suggests a clear, shallow sea in which lived clams, snails, and other hard-shelled organisms (Fig. 16-27). As the long history is carried further and further back, the evidence becomes more shadowy and the geologist's reconstruction of the earth's surface similarly imprecise.

In trying to figure out how the earth's crust evolved, geology is faced with two fundamental problems: to arrange in order the events recorded in the rocks of a single outcrop or small region, and to correlate events in various regions of the world to give a connected history of the earth as a whole.

Reading the Rocks Some of the principles used by geologists to figure out the history of a small area are straightforward:

1. In a sequence of sedimentary rocks, the lowest bed is the oldest and the highest bed is the youngest. In Fig. 16-28 bed *A* must have been deposited before the others and bed *E* last.

Figure 16-27 Arizona's mile-deep Grand Canyon was carved from sedimentary layers, most of them deposited when the region was covered by shallow seas. Exposed rocks are mainly limestone, shale, and sandstone. Seventeen million years ago two streams, one in the west and the other in the east, began to cut channels which met 6 million years ago and became a single chasm through which the Colorado River flows today.

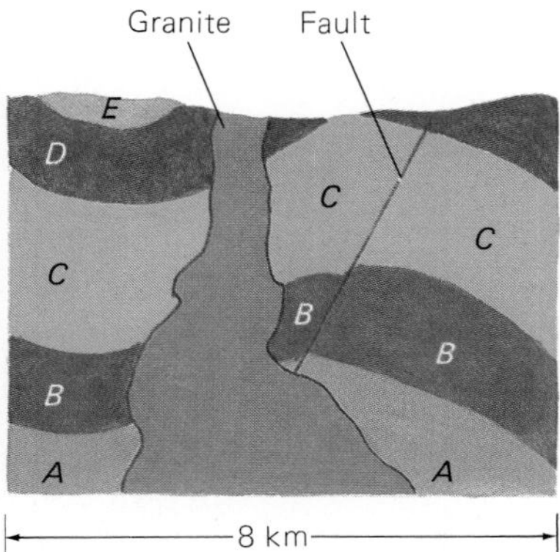

Figure 16-28 Schematic cross section showing folded sedimentary rocks that were displaced along a fault and then intruded by granite.

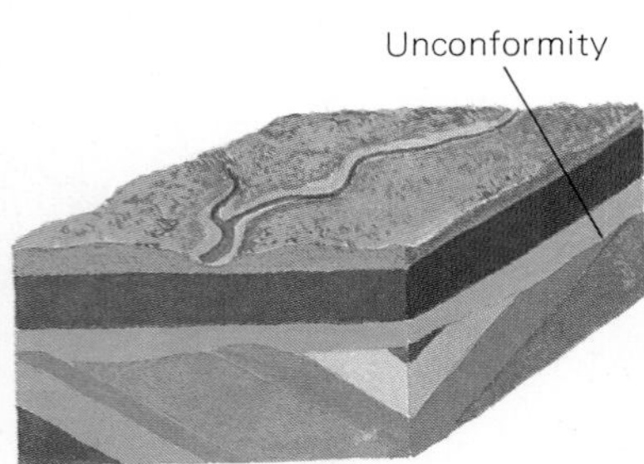

Figure 16-29 An unconformity is an irregular eroded surface that separates one set of rock layers from an earlier set. Shown is an unconformity above tilted lower layers; the layers above and below an unconformity can also be parallel.

2. Sedimentary beds were originally deposited in approximately horizontal layers.
3. Tectonic movement took place after the deposition of the youngest bed affected. Thus the layers of Fig. 16-28 were not folded until after bed *D* was laid down, and the fault must be younger than bed *C*.
4. An igneous rock is younger than the youngest bed it intrudes. The granite pluton shown in Fig. 16-28 is younger than bed *D*. (The age of an igneous rock refers to the time at which it solidified.)

Obvious as these rules are, much ingenuity may be needed to apply them to regions of heavily folded and faulted layers. The problem is especially difficult in regions where much of the rock structure is hidden by later sediments or vegetation.

A structure like that shown in Fig. 16-29 requires further attention. Here the lower, tilted beds are cut off by an uneven surface on which rest the upper horizontal beds. An irregular surface of this sort, separating two series of rocks, is called an **unconformity.**

An unconformity is a buried surface of erosion. It always involves at least four geologic events: the deposition of the oldest strata; tectonic movement that raises and perhaps tilts the existing strata; erosion of the elevated strata to produce an irregular surface; and finally a new period of deposition that buries the eroded surface (Fig. 16-30). Usually this last event involves the lowering of the eroded surface either beneath the sea or to a level where stream deposition can occur.

Example 16.2

In parts of Colorado and Wyoming, a long period during which marine sediments were deposited in a sinking basin ended with intense tectonic movement. A mountain range rose there that was later worn down by erosion to a nearly level plain. On parts of this plain stream and lake sediments were deposited. What rocks and rock structures would be expected in this region?

Solution

The original sedimentary deposits will have been converted into such metamorphic rocks as slate, schist, and marble, which will be exposed by the subsequent erosion. The outcrops of metamorphic rocks will be intruded by igneous rocks—granite batholiths and dikes of various kinds. Unconformities will occur where the later stream and lake sediments are deposited. Sandstone, shale, and conglomerate will predominate in the overlying sedimentary rocks.

16.10 Radiometric Dating

A Clock Based on Radioactivity

To trace the sequence of geologic events in various places is not enough. We also need to know which events happened at the same time in different places if we are to understand how the earth's crust evolved. We could get the data we need by following rock layers from one region to another around the globe, but in fact after a shorter or longer distance the layers are always either cut off by erosion or concealed by later deposits.

Table 16-1 Radionuclides Used for Dating Rocks

Parent Nuclide	Stable Daughter Nuclide	Half-Life, Billion Years
Potassium 40	Argon 40	1.3
Rubidium 87	Strontium 87	49
Thorium 232	Lead 208	14.1
Uranium 235	Lead 207	0.7
Uranium 238	Lead 206	4.5

Figure 16-30 This angular unconformity in Scotland consists of sandstone layers from the Devonian Period that overlie layers of Silurian rock that have been tilted almost vertically.

Two methods can be used to figure out the worldwide sequence of the events that have shaped the earth's surface: radioactive dating and fossil identification. As we shall find, each method has different advantages and disadvantages.

Methods based on radioactive decay make it possible to establish the ages of many rocks on an absolute rather than a relative time scale. Because the decay of any particular radionuclide proceeds at a steady rate, the ratio between the amounts of that nuclide and its eventual stable daughter in a rock sample indicates the age of the rock. (There may be several decays into intermediate daughter radionuclides before a stable daughter is reached.) The more there is of the daughter, the older the rock (Fig. 8-10). Radiometric dating can be accurate to ±2500 years per million years, but only certain rock specimens can be dated in this way.

Table 16-1 lists the radionuclides that have been found most useful in dating rocks. Age measurements of fossil-bearing rocks with the radioactive clock reveal that relatively large-brained humanlike creatures walked erect in Africa about 2 million years ago; that primitive mammals existed about 200 million years ago; that sea animals with hard shells first became abundant about 600 million years ago. The most ancient rocks whose ages have been established are found in northwestern Canada and are 4 billion years old; zircon samples have been discovered in Australia that seem to have crystallized even earlier.

Example 16.3

What are the chief assumptions made in the radiometric dating of the age of a rock?

Solution

(a) The rates of decay of the parent and any daughter radionuclides are constant, independent of time, temperature, and pressure.

(b) There has been no addition to or loss from the rock of any of the nuclides involved in the decay series from the radioactive parent to the eventual stable daughter.

Radiocarbon Another radionuclide permits the dating of more recent remains of living things. The carbon isotope ^{14}C, called **radiocarbon,** is beta radioactive with a half-life of 5700 years. Radiocarbon is produced in small quantities in the earth's atmosphere by the action of cosmic rays (discussed in Chap. 19) on nitrogen atoms, and the carbon dioxide of the

BIOGRAPHY Arthur Holmes (1890–1965)

Although Holmes was interested first in physics, he turned to geology because he felt it was more likely to lead to a good job. As it turned out, Holmes's great achievements in geology were largely due to his application of physics and chemistry to a subject that, in his student days, was much less of a science than he helped it to become. After graduating from the Imperial College of Science in London, he worked for a time in Africa and later for an oil company in what was then Burma. In 1924 he became head of the geology department at Durham University and in 1947 went to Edinburgh University where he remained until just before his death.

Holmes's main interest was petrology, the study of rocks, in particular igneous rocks and their origins, and he worked in this field with his wife Doris Reynolds, herself a distinguished geologist. However, his contributions to geology covered a wide span. He is best known for using radioactive methods to date rocks, which established a timescale for the geological record and provided a figure for the age of the earth based on measurements rather than on speculation: "It is perhaps a little delicate to ask of our Mother Earth her age, but Science admits no shame," he wrote. Holmes was an early champion of continental drift when it was a heretical idea, and his imaginative insight led him to anticipate some of the concepts of plate tectonics long before this theory became recognized as the key to understanding how the earth's surface evolves.

atmosphere accordingly contains a small proportion of radiocarbon. Green plants take in carbon dioxide in order to live, and so every plant contains radioactive carbon that it absorbs along with intake of ordinary carbon dioxide. Animals eat plants and thus become radioactive themselves; there are about 13 decays of ^{14}C per minute per gram of carbon in living tissue. As a result, every living thing on earth is slightly radioactive because of the radiocarbon it takes in. The mixing of radiocarbon with ordinary carbon is very efficient and so living plants and animals all have the same proportion of radiocarbon to ordinary carbon (^{12}C).

After death, however, the remains of living things no longer absorb radiocarbon, and the radiocarbon they contain keeps decaying away to nitrogen. After 5700 years, then, they have only half as much radiocarbon left—relative to their total carbon content—as they had as living matter, after 11,400 years only one-fourth as much, and so on. Determining the proportion of radiocarbon to ordinary carbon thus makes it possible to evaluate the ages of ancient objects and remains of organic origin (Figs. 16-31 and 16-32). This elegant method permits the dating of archeological specimens such as mummies, wooden implements, cloth, leather, charcoal from campfires, and similar remains of ancient civilizations that are between 1000 and 40,000 years old.

16.11 Fossils

Tracing the History of Life

Perhaps the most fascinating technique at the geologist's command for establishing relationships among rocks of different regions and for arranging beds in sequence makes use of **fossils.** The most common fossils, of

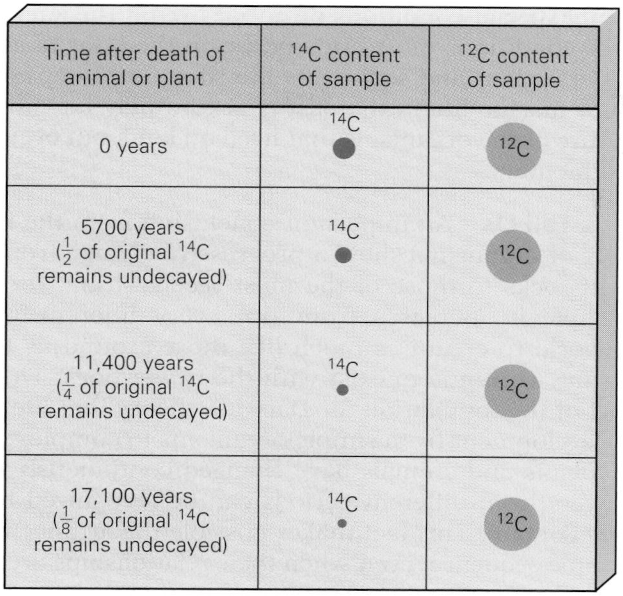

Time after death of animal or plant	^{14}C content of sample	^{12}C content of sample
0 years	^{14}C	^{12}C
5700 years ($\frac{1}{2}$ of original ^{14}C remains undecayed)	^{14}C	^{12}C
11,400 years ($\frac{1}{4}$ of original ^{14}C remains undecayed)	^{14}C	^{12}C
17,100 years ($\frac{1}{8}$ of original ^{14}C remains undecayed)	^{14}C	^{12}C

Figure 16-31 The principle of radiocarbon dating. The radioactive ^{14}C content of a sample of dead animal or plant tissue decreases steadily, while its ^{12}C content remains constant. Hence the ratio of ^{14}C to ^{12}C contents indicates the time that has elapsed since the death of an organism. The half-life of ^{14}C is 5700 years. (The relative proportions of ^{14}C and ^{12}C are greatly exaggerated here; there is actually very little ^{14}C in a carbon sample from animal or plant tissue.)

Figure 16-32 A mass spectrometer such as this one at the University of Minnesota can measure the proportion of ^{14}C atoms in carbon from a sample of plant or animal material. Because ^{14}C is radioactive whereas the more abundant ^{12}C isotope is not, their ratio indicates how long ago a plant or animal died.

course, are the hard parts of animals, such as shells, bones, and teeth (Fig. 16-33). On rare occasions an entire animal may be preserved: ancient insects have been trapped in amber, and immense woolly mammoths have been found frozen in the Arctic.

Plant fossils are relatively scarce since plants do not have durable hard parts. The structure of tree trunks is sometimes beautifully shown in petrified wood in which minerals carried by water have replaced the original plant tissue (see Fig. 15-51). The incomplete decay of buried leaves and wood fragments produces black, carbonaceous material that may keep the original organic structures—coal is a thick deposit of such material (Fig. 16-34). Occasionally fine sediments show impressions of delicate structures like leaves, feathers, and skin fragments. Some fossils are merely trails or footprints left in soft mud and covered by later sediments.

Conditions necessary for preservation have been much the same throughout geologic history. Chemical decay, bacteria, and scavengers have quickly disposed of most of the organisms that have lived on the

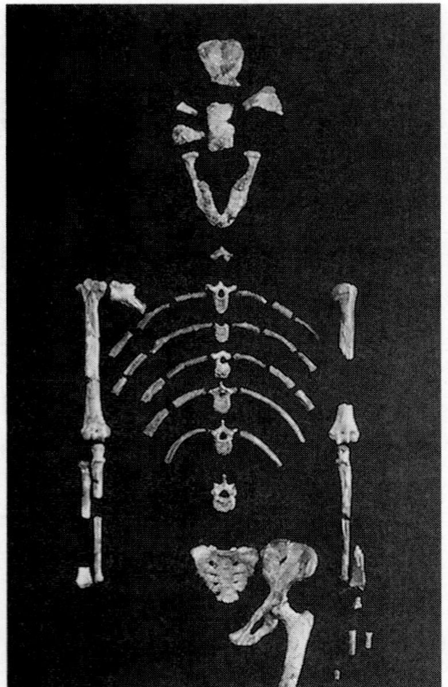

Figure 16-33 Fossil skeleton of "Lucy," a humanlike creature who lived 3 million years ago in what is today Ethiopia. She was between 1 and 1.5 m tall with long arms and curved fingers and toes that suggest she was at home both in trees and upright on the ground.

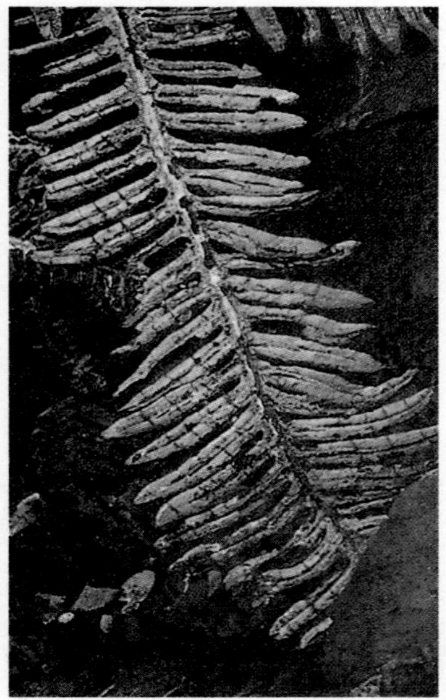

Figure 16-34 Fossil fern found in a coal deposit in Pennsylvania.

earth, and only special conditions of burial permit the survival of fossil groups. These conditions most often occur on the floor of shallow seas, where life is abundant and sediments are sometimes deposited rapidly. Our picture of marine life in the past is accordingly far more complete than our picture of the organisms that lived on land, but even the marine record is fragmentary.

What Fossils Tell Us An important conclusion from the fossil record is that groups of organisms show a progressive change from those buried in ancient rocks to those of the most recent strata. In general, the degree of complexity increases, from forms very different from those in the present world to creatures much like those around us today. Some older forms may continue to exist with the newer ones. These observations are part of the foundation for Darwin's theory that life has evolved by a steady development from simple organisms to complex ones.

Because plants and animals have changed continuously through the ages, rock layers from different periods can be recognized by the kinds of fossils they contain. This fact makes possible the arrangement of beds in a relative time sequence, even when their relationships are not directly apparent, and also provides a means of correlating the strata of different localities. If, for example, fossil snail shells and clamshells are found in a rock layer in New York that are similar to fossil shells from a layer in the Grand Canyon, the two layers are likely to be approximately the same age.

Fossils are useful not only in tracing the development of life and in correlating strata but in helping us to reconstruct the environment in which the organisms lived. Some creatures, like barnacles and scallops, live only in the sea, and it is likely that their close relatives in the past were similarly restricted to salt water. Other animals can exist only in freshwater. On land some organisms prefer desert climates, others cold climates, others warm and humid climates. Evidently many details about the conditions in which a rock was formed can be revealed by its fossil organisms.

16.12 Geologic Time

Precambrian, Paleozoic, Mesozoic, Cenozoic

Fossils enable us to arrange in sequence geologic events over the entire earth. Enough of these events can be dated accurately by radioactive methods for good estimates to be made for the dates of the others. The most recent 542 million years of the earth's history have been divided by geologists into three major divisions called **eras** (Table 16-2):

Cenozoic ("recent life") Era; began 66 million years ago
Mesozoic ("intermediate life") Era; began 251 million years ago and lasted 185 million years
Paleozoic ("old life") Era; began 542 million years ago and lasted 291 million years

Geologists have divided the 4 billion years before the Paleozoic Era—a shadowy span that covers over seven-eighths of the history of our planet—into just two major parts, the **Archaean** ("ancient") **Eon** and the later **Proterozoic** ("former life") **Eon,** that together make up **Precambrian time.** The dates of the various divisions of geologic time are

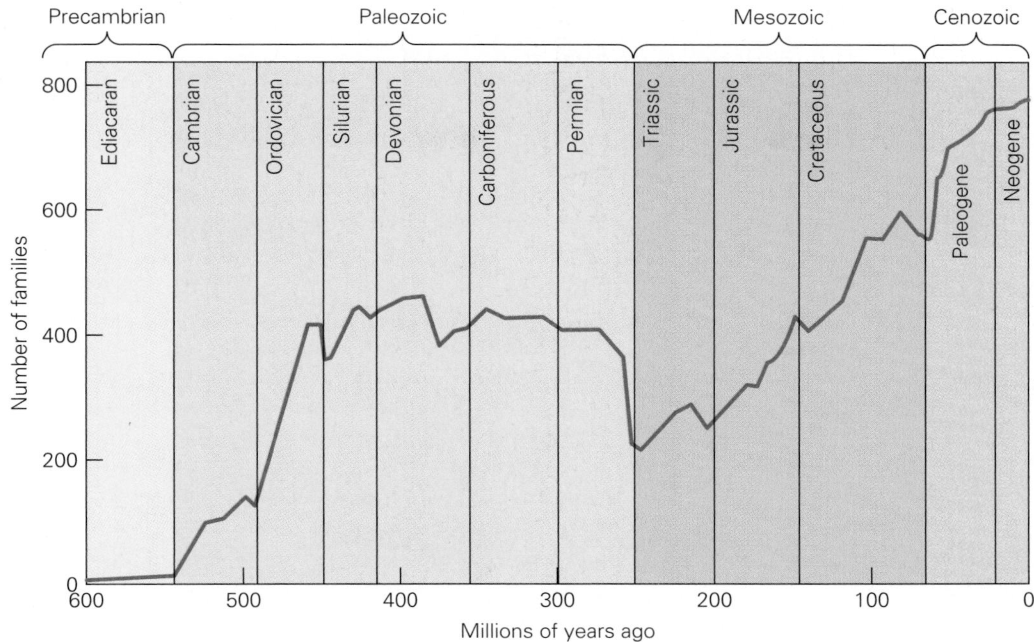

Figure 16-35 How the number of families of living things has varied in the past 600 million years. A family is a group of related species, and the greater the number of families, the greater the diversity of life. Each dip in the curve represents a biological extinction; there have been 20 major ones in all. The most severe extinction marks the end of the Paleozoic Era; the dinosaurs disappeared in the extinction at the end of the Mesozoic Era. Life is at its most diverse today, although human activity has begun to turn the graph downward.

changed from time to time as new evidence turns up; those in this book are the most recent ones.

The divisions of geologic time shown in Table 16-2 are based on dramatic changes that occurred from time to time in the earth's history—changes in landscapes, in climates, in types of organisms. In particular, the fossil record reveals a number of occasions, called **extinctions,** when animal and plant life became sharply reduced in both number and variety, to be followed in each case by the evolution of new forms (Fig. 16-35). Although it is hard to be precise, extinctions—with the exception of the one that ended the Mesozoic Era—seem to have taken place over periods of 100,000 years or more of environmental disturbances that involved drastic climate changes—a long time to us, but a relatively brief time in geology. Extinctions are used to divide geologic history into **periods.** During a typical period an expansion of living things is followed by an interval in which biological change is more gradual; then a time of more sudden change ends the period. Periods are similarly subdivided into shorter **epochs;** only the epochs of the Cenozoic Era are shown in Table 16-2.

The division into eras was based on exceptionally marked worldwide extinctions and subsequent expansions, and correlations were found between large-scale events in the earth's crust and the biological record. Not all these correlations have turned out to be as clear-cut as they once appeared, but the traditional organization of geologic time is still convenient and is universally used by scientists.

Table 16-2 Geologic Time. (The Earth Came into Existence About 4600 Million Years Ago)

Millions of Years Ago	Era	Period	Epoch	Duration, Million Years	The Biologic Record	The Geologic Record
66	Cenozoic	Neogene	Holocene	0.01	Humans become dominant	
251			Pleistocene	1.8	Rise of humans; large mammals abundant	Ice Age
			Pliocene	3.5	Flowering plants abundant; early humans	Atlantic Ocean widens
542			Miocene	18	Grasses abundant; rapid spread of grazing mammals	Alps and Himalayas form; Red Sea opens
		Paleogene	Oligocene	11	Apes and elephants appear	India collides with Asia
			Eocene	22	Primitive horses, camels, rhinoceroses; first grasses	Australia separates from Antarctica
			Paleocene	10	First large mammals and modern plants	Norwegian Sea and Baffin Bay open
	Mesozoic	Cretaceous		80	Spread of flowering plants; dinosaurs die out at end	Laurasia and Gondwana begin to break up
		Jurassic		54	First birds; dinosaurs at their peak	Laurasia separates from Gondwana
		Triassic		51	Dinosaurs and first mammals appear	Pangaea complete
	Paleozoic	Permian		48	Rise of reptiles; insects abundant	Laurasia and Gondwana come together to form Pangaea
2100		Carboniferous		60	Large nonflowering plants in enormous swamps; extensive forests; large insects and amphibians; sharks abundant	Coal being formed; Africa moves against Europe and North America
		Devonian		57	First forests and amphibians; fish abundant	Greenland and North America join Europe
		Silurian		28	First land plants and large coral reefs	
		Ordovician		44	First vertebrates (fish) appear	
	Precambrian time	Cambrian		54	Marine shelled invertebrates (earliest abundant fossils)	Early supercontinent breaks up
		Ediacaran		88	Various forms of simple multicellular life	
		Proterozoic Eon		1958	Before Ediacaran: complex one-celled organisms and algae colonies; atmospheric oxygen appears	Early supercontinent forms; "Snowball Earth" occurs
		Archaean Eon		2100	Primitive bacteria and stromatolites	Continents, oceans, and atmosphere form

Earth History

The earth came into being about 4.6 billion years ago. Current ideas on its origin are discussed in Chap. 19. With the oldest surface rocks the story of the earth leaves the realm of speculation. These rocks, which are found in northern Canada and date back over 4 billion years, consist of sedimentary and metamorphic rocks intruded by igneous rocks. This suggests that the cycle of erosion and renewal of continental crust was already well established as far back as the visible record of the earth's history extends.

16.13 Precambrian Time

Long Ago But Not Far Away

Nothing is known for certain about the locations of the continents in Precambrian time. The splitting of Pangaea into Laurasia and Gondwana and their subsequent breakup into today's continents began about 200 million years ago, a mere 4 percent of the earth's age. Plate movement seems to have been going on farther back in the past, with the Precambrian continents differently arranged across the face of the earth.

Although practically unaltered Precambrian sedimentary and igneous rocks are sometimes found, more often Precambrian rocks show considerable metamorphism and in many places have been greatly deformed. Both stream deposits and marine deposits are found in the sedimentary beds (Fig. 16-36). The volcanic rocks include all types, with basalt flows then as now the most common. Intrusive rocks are represented in great abundance and variety. Evidently geologic processes a billion years ago were not very different from those in the modern world. Precambrian rocks are exposed at the surface over a broad area covering most of eastern Canada and nearby parts of the United States.

Late in Precambrian time declines in the concentration of greenhouse gases in the atmosphere led to a series of at least three global ice ages—"Snowball Earth"—during which glaciers hundreds of meters thick covered land and sea everywhere. These big freezes occurred between 850 and 635 million years ago. Probably each ice age ended when volcanic eruptions released greenhouse gases sufficient for the earth's surface

Precambrian Time
PROTEROZOIC EON (2500–542)
Ediacaran (630–542)
various forms of simple multicellular life
Before Ediacaran: complex one-celled organisms and algae colonies; atmospheric oxygen appears
ARCHAEAN EON (4600–2500)
primitive bacteria and stromatolites
(dates are millions of years ago)

Figure 16-36 Sample of Precambrian rock from northern Canada that shows ripples and cracks characteristic of fine sediments deposited in shallow water and occasionally dried by exposure to the sun. Similar conditions occur along the shores of present-day lakes and seas.

to become warm again. Subsequent ice ages were never as extreme as these earliest ones.

Early Life Precisely when and where life began on earth nobody knows, because primitive organisms seldom leave traces. Nevertheless, old rocks that *could* show signs of life usually *do* show such signs. Apparently life began not long after the environment of the early earth allowed. In the Archaean Eon, evidence of some sort of primitive single-celled organisms, probably bacteria without nuclei, is found that dates back as far as 3.8 billion years ago. By 3.6 billion years ago cyanobacteria had developed and began to spread in the oceans. Joined together by sticky mucus into bluish-green mats, these bacteria could survive solar ultraviolet light deadly to most other life, and their photosynthetic activity started to provide the atmosphere with oxygen. (The original atmosphere was rich in CO_2 but lacked free O_2.) Besides producing O_2 the cyanobacteria caused the precipitation of calcium carbonate into characteristic structures of thin sheets of limestone called **stromatolites.** Stromatolites continue to be formed today in a few locations, notably in Australia.

In most of the Proterozoic Eon cyanobacteria were the dominant form of life and greatly increased the oxygen content of the atmosphere (Fig. 16-37). More complex single-celled organisms that contained nuclei arose, including algae that formed colonies in the oceans. When the last Snowball Earth ice age had thawed out, the bacteria that had survived were joined by a variety of simple multicellular organisms that began to develop. The newcomers did not have hard parts such as shells that could be preserved but did leave impressions in sediments that could be deciphered. The members of one group were apparently similar to modern coral polyps and jellyfish; the members of another had quiltlike structures up to a meter across but less than a centimeter thick that have no counterparts today. Evidence of such life forms in the Ediacara Hills of southern Australia led to the name Ediacaran Period for the time interval that bridged the end of Snowball Earth and the start of the Paleozoic Era with its abundant fossils. Ediacaran mementos have been found in many other parts of the world as well, for instance in Newfoundland where frondlike patterns were discovered in sediments laid down in this period. The Ediacaran Period was not formally recognized until 2004; the more recent geological divisions were finalized in 1891.

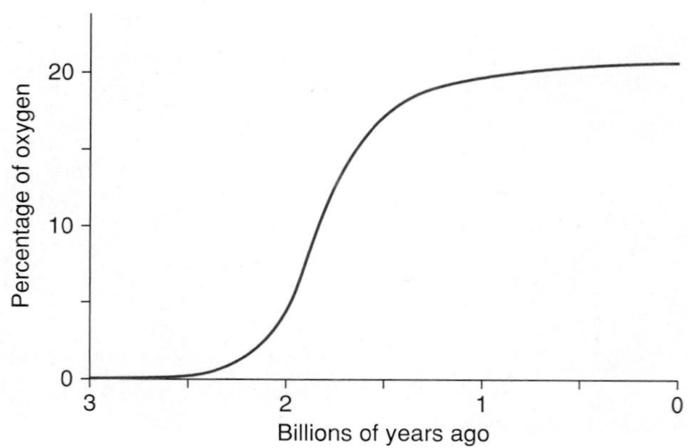

Figure 16-37 Free oxygen began to appear in the atmosphere about 2.5 billion years ago as the result of photosynthesis by primitive bacteria. The present concentration of O_2 in the air we breathe is nearly 21 percent. The oxygen-carbon dioxide cycle is shown in Fig. 13-22.

16.14 The Paleozoic Era

Plants and Animals in Variety and Abundance

The history of the Paleozoic Era is remarkably complete. No longer are there the doubts and vagueness that characterize Precambrian events, for Paleozoic strata are widely exposed, and their wealth of fossils makes possible correlation of rocks and events from one side of the earth to the other.

The oldest fossils of the Paleozoic Era are those of marine **invertebrates,** creatures without internal skeletons but with external shells of various kinds. All the major groups of the invertebrates are represented.

Paleozoic Era
Permian (299–251)
rise of reptiles, insects abundant
Carboniferous (359–299)
large nonflowering plants in enormous swamps; extensive forests; large insects and amphibians; sharks abundant
Devonian (416–359)
first forests and amphibians; fish abundant
Silurian (444–416)
first land plants and large coral reefs; first insects
Ordovician (488–444)
first vertebrates (fish) appear
Cambrian (542–488)
marine invertebrates with shells (earliest abundant fossils)
(dates are millions of years ago)

Trilobites

Trilobites were abundant in Paleozoic seas and oceans. Their fossils are common because they were among the first animals to have hard shells, which they shed periodically as they grew, like today's lobsters. Trilobites were the first creatures to have eyes, which appeared about 543 million years ago. These eyes had lenses of crystalline calcite, a feature not found elsewhere. Genetic evidence suggests that trilobite eyes could have been the ancestors of all other eyes, but it also seems possible that eyes evolved independently in as many as 65 different life forms.

Because trilobites evolved rapidly, fossils of their successive species are valuable markers that indicate the ages of rocks in which they are found. Trilobites ranged in length from a few millimeters to 70 cm (Fig. 16-38). Newly evolved predatory fish and squid may have been responsible for the disappearance of the trilobites near the end of the Paleozoic. Horseshoe crabs are the closest living relatives of trilobites. They are not true crabs but, like trilobites, are distant cousins of spiders and scorpions.

Figure 16-38 Trilobite fossil from the Paleozoic era.

Cambrian Explosion

Living things blossomed in the Paleozoic in an evolutionary riot that began in the Cambrian period—the "Cambrian explosion" of life. The ancestors of most modern life forms arose at this time. The first life was water-dwelling probably because there was not yet enough ozone in the atmosphere to screen out lethal ultraviolet light from the sun. The early Paleozoic was also a time when the earth's surface was being drastically reshaped; for instance, it was then that North America migrated from near the South Pole to the equator. Shifts such as this stressed living things through changes in climate and thus accelerated their rates of evolution.

The Great Dying

As Fig. 16-35 shows, the most severe mass extinction yet marks the divide between the Paleozoic and Mesozoic Eras. In all, 95 percent of the earth's plant and animal species vanished a quarter of a billion years ago. Not for millions of years was the normal complexity of ecosystems reestablished. What had happened?

Because the Great Dying (as it has been called) took place so long ago, geological processes have inexorably erased most potential evidence. But some clues remain, and they point to a series of immense volcanic eruptions that lasted half a million years and covered much of Siberia with lava. Vast clouds of ash and poisonous gases billowed into the atmosphere. A huge amount of carbon dioxide was also released, which caused global warming that raised world temperatures by around 6°C. Plant life died off on a large scale, which led oxygen levels to drop as the decay of the remains absorbed oxygen that could not be replaced by photosynthesis. Starved of oxygen, animal life suffered as well. Further harm was done by toxic hydrogen sulfide (H_2S) gas produced in the oceans by anaerobic bacteria (Sec. 11.9) which do not need oxygen. The H_2S that bubbled into the atmosphere was lethal to both plants and animals and also attacked the protective ozone layer, allowing solar ultraviolet radiation through to do more damage to living things. Eventually the volcanic activity subsided, conditions more favorable to living things returned, and the biological world began to flourish again, now with a different mix of life forms.

The Great Dying set the stage for the rise of the dinosaurs by leaving ecological niches vacant for them, just as the later extinction that ended the Mesozoic saw off the dinosaurs and allowed mammals (such as our ancestors) to thrive afterward.

Clams and snails increased in number and evolved considerably in the Paleozoic seas. Tiny coral polyps built widespread reefs in the middle Devonian Period. Starfish and sea urchins were not common, but some of their distant relatives that today are rare or extinct were more numerous then. Fishes were many and varied.

In late Paleozoic rocks there is for the first time much evidence of land-dwelling organisms. In the coal swamps of Carboniferous times grew dense forests of primitive plants—huge fernlike trees, enormous horsetails, primitive conifers. A modern person wandering through such a forest would find no bright-colored flowers, no grasses, few plants at all familiar except possibly some of the ferns and mosses. In and near these primeval forests lived a great variety of animals, including giant scorpions and the largest insects that ever lived, dragonflies whose wings spanned nearly a meter. The land-living vertebrates of the late Paleozoic are early members of our own family tree. In fact, the basic body plans that appeared in the Paleozoic have served for all animals ever since.

Fossil amphibians, oldest of the land vertebrates, appear first in Devonian rocks. These are relatives of modern frogs and salamanders, sluggish creatures that laid their eggs and spent the early part of their lives in water. Fossils of reptiles begin to appear in late Carboniferous rocks. These animals looked at first much like their amphibian ancestors but had the great advantage of being able to lay their eggs on land. The dry climate at the end of this era was hard on the amphibians, but the reptiles, not needing water to hatch their eggs, multiplied rapidly and developed many different species. During the Permian Period reptiles became the dominant creatures of the land.

The Paleozoic Era ended with a time of intense tectonic activity, which affected many parts of the world including the North American continent. The sediments that had accumulated for more than 300 million years in the

Appalachian trough were crumpled, fractured, and uplifted into a mountain chain that must have rivaled any modern range in height and grandeur. It is not hard to connect these dramatic geologic events with the crushing together of the continental masses that formed the supercontinent Pangaea at this time.

16.15 Coal and Petroleum

Both Came from Once-Living Matter

Coal is plentiful in Paleozoic rocks. Coal was formed from plant matter that accumulated under conditions where complete decay was prevented. A bed of coal nearly always implies an ancient swamp. Coal has been formed in swamps from the Devonian to the present day, but seldom have conditions been so favorable as in the Carboniferous Period. (The first part of this period is sometimes called the Pennsylvanian Period and the last part the Mississippian Period.) Apparently there were broad swamps almost at sea level that became periodically submerged so that partly decayed vegetation was covered with thin layers of marine sediments.

The formation of coal begins with the slow bacterial decay of the cellulose content of plants. Taking place underwater and in the absence of air, this decay results in a gradual removal of oxygen and hydrogen from the cellulose to leave a residue that is largely carbon. Also contributing to coal formation was the action of heat and pressure resulting from burial beneath later sediments.

The origin of petroleum is more obscure, for two reasons: fossils are not preserved in a fluid, and petroleum often migrates long distances from where it forms. Because petroleum hydrocarbons can be detected in modern marine sediments, because oils resembling petroleum can be prepared artificially from organic material, and because petroleum is associated with rocks formed from sediments deposited in shallow seas, there seems to be little doubt that marine life such as algae and plankton is the source of petroleum. Most petroleum apparently was formed more recently than coal: over half in the Cenozoic Era, about a quarter in the Mesozoic, and the rest in the Paleozoic.

Petroleum Formation Three steps seem to have been involved in producing petroleum. The first was bacterial decay in the absence of oxygen, an ideal site being the floor of a shallow sea. Then, as the organic debris was buried under later sediments, it was further modified by low-temperature chemical reactions. The final step was the "cracking" of complex hydrocarbons to straight-chain alkane hydrocarbons (see Chap. 13) under the influence of temperatures of 70°C to 130°C deep underground. If the temperatures became higher, the result was natural gas rather than oil.

Both gas and petroleum, like groundwater, can migrate freely through such porous rocks as loosely cemented sandstones and conglomerates. Wherever formed, they often find their way into porous beds, and it is from these beds that they are obtained by drilling. Since both gas and petroleum are lighter than water, they may be displaced by groundwater and so move upward to the surface to form oil seeps.

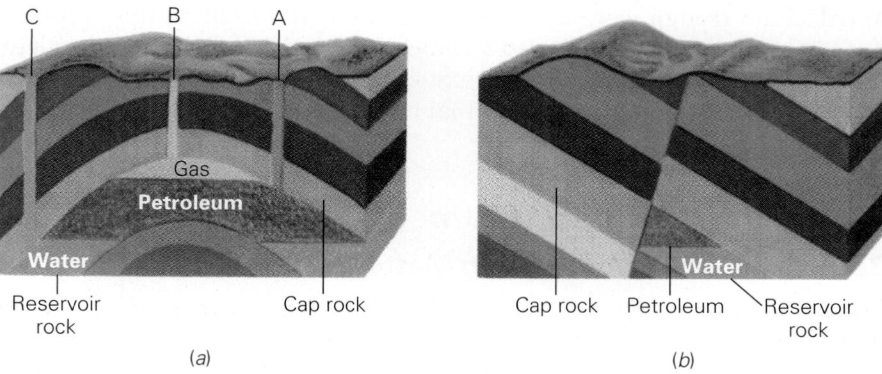

Figure 16-39 Two common types of structural traps in which petroleum accumulates: (*a*) a trap formed by an anticline (see Fig. 16-3); (*b*) a trap formed by a fault. In both cases petroleum in a porous reservoir rock (such as sandstone) is prevented from migrating upward by an impermeable cap rock (such as shale). A well drilled at A would strike petroleum, one drilled at B would strike gas, and one at C only water. About 80 percent of known petroleum deposits are found in anticline traps. Commercial oil deposits are typically 500–700 m deep; the deepest wells go down about 6 km.

Petroleum reservoirs consist of porous sandstones or carbonates capped by layers of impermeable clays or shales. The most common kinds of reservoir are shown in Fig. 16-39. The locations of the world's chief oil and gas deposits are shown in Fig. 4-15.

Mesozoic Era
Cretaceous (146–66)
spread of flowering plants; dinosaurs die out at end
Jurassic (200–146)
first birds; dinosaurs at their peak
Triassic (251–200)
dinosaurs and first mammals appear
(*dates are millions of years ago*)

16.16 The Mesozoic Era

The Age of Reptiles

The earliest Mesozoic sediments were laid down about 251 million years ago, a long time by ordinary reckoning. But the earth was already very old. Some 291 million years had elapsed since the beginning of the Paleozoic and $3\frac{1}{2}$ billion years since the oldest known rocks of the Precambrian. All the time that we include in the Mesozoic and Cenozoic Eras is only one-sixteenth of the history recorded in rocks of the earth's crust.

The Mesozoic Era saw Pangaea split into Laurasia and Gondwana, and this division was followed by their own breakup (see Fig. 16-9). Early in the Mesozoic Era North America began to part from Europe, and somewhat later, perhaps 120 million years ago, South America and Africa began to drift apart. By the end of the era Gondwana no longer existed. Australia, New Zealand, and India had all left Africa, though Arabia still remained attached. Africa itself was in the process of a shift northeastward, thus closing the western end of the Tethys Sea, while India, well on its way toward Asia, was moving into the eastern end. The Mid-Atlantic Ridge was already a prominent feature of the floor of the infant Atlantic Ocean.

At the end of the Paleozoic Era the world became hot and dry with widespread deserts and less plant life to provide oxygen to the atmosphere by photosynthesis. Reptiles, the dominant land animals, had evolved when the oxygen level was high, and most of them found it difficult to survive when the level fell. A small class of reptiles, however, happened to have especially efficient respiratory systems, and they took over under the new conditions. These were the **dinosaurs,** some of which developed into the largest land animals the earth has ever seen. (The word *dinosaur* comes from the Greek for "terrible lizard.") Birds are direct descendants of dinosaurs and share their remarkable respiratory systems, which allow geese to fly over the Himalayas at altitudes where the air is too thin for humans to live.

Some dinosaurs were carnivores with bodies designed for pursuing and eating other animals. Some were herbivores with jaws and digestive organs adapted for a vegetarian diet. Active species lived in open plains,

Dinosaurs and Pterosaurs

Dinosaur forms can be inferred by comparing their skeletons with the skeletons of living reptiles. Impressions of dinosaur skins sometimes occur in the rocks in which their bones are found, although the skin colors can only be guessed at. The dinosaurs shown in Fig. 16-40 were all plant-eaters except *Tyrannosaurus rex* ("king of the tyrant lizards") at upper right, which was about 3 m high at the hips, 11 m long, and had a mass of about 8 tons. The teeth of this creature were 15 cm long and the biting force of its jaws is estimated as at least 13.4 kilonewtons (almost 3 tons!) based on toothmarks found on fossil dinosaur bones discovered in Montana. With a top speed of perhaps 40 km/h, *T. Rex* could run faster than its herbivorous prey could flee. Thirty *T. Rex* skeletons have been identified thus far, the first in 1905. The largest modern land predator is the polar bear, whose mass is at most around 700 kg.

Pterosaurs (Greek for "winged lizards") took to the skies in the Mesozoic. The smallest were the size of robins, the largest were over 6 m long with wingspans of 12 m—the size of a small airplane. The toothless beaks of the latter monsters were 1.8 m long and tapered to narrow points. What did such a creature eat? Perhaps, like today's storks, it probed for tasty morsels on the bottoms of shallow lakes. Did pterosaurs build up speed to take off by running on the ground or by dropping off a tree or a cliff? In the air, did they soar on thermal updrafts like a glider or did they flap their wings? Nobody is sure about the answers; today's birds have masses of less than 15 kg, so we can't use them as guides. Dinosaurs, not pterosaurs, were the ancestors of modern birds. Crocodiles are the closest living relatives of birds.

Figure 16-40 A painting of some of the dinosaurs that lived in North America 65 to 70 million years ago at the end of the Mesozoic Era.

more sluggish ones in swamps. Some had bony armor for protection, others depended on speed to escape their enemies. The fastest dinosaurs could outrun today's sprinters, though not today's racehorses. The really large dinosaurs, though, moved sedately at a pace slower than that of a person walking. Not all the dinosaurs by any means were huge, but the biggest ones, called Futalognkosaurus, were 32 to 34 m (over 100 ft) long from head to tail; they lived in what is now northern Argentina 80 million years ago. Blue whales, the largest of today's animals and almost extinct due to overhunting, sometimes reach such lengths. The 100-ton mass of a blue whale is probably comparable to the mass of a Futalognkosaurus.

The largest modern land animal, the African elephant, weighs in at a mere 5–6 tons.

Meanwhile other land organisms were developing. Flowering plants appeared in mid-Mesozoic and with them a host of modern-looking insects suited for helping in the pollination of flowers. The first true birds, with feathered wings rather than membranes, arose from dinosaur ancestors in the Jurassic and were probably four-winged gliders. Birds with two flapping wings and lightweight skeletons, like modern ones, evolved later.

Sometime in the Triassic appeared the first **mammals,** creatures that probably descended from a group of small Permian reptiles. All during the Mesozoic mammals remained inconspicuous, seldom much larger than a cat. However, in several respects mammals represented an evolutionary advance over reptiles: they were warm-blooded (constant body temperatures), hence better able to cope with changes of temperature; they had bigger brains relative to their body size; and (except for the egg-laying spiny anteater and duck-billed platypus) they gave birth to well-developed live offspring which they cared for relatively long after birth, so some of the experience of one generation could be passed on to the next. Although most early mammals filled minor ecological niches, like rodents today, some were predators: the bones of a young dinosaur were found in the belly of a fossil opossum-sized mammal in China. One small mammal had flaps of skin that probably allowed it to glide through the air, like today's flying squirrels; bats arose much later.

What Happened to the Dinosaurs? For over 100 million years dinosaurs large and small roamed the earth. Then, 66 million years ago, at the height of their diversity and success, all of them disappeared. Not a single dinosaur fossil has ever been found in rocks formed after the end of the Mesozoic Era. The largest animals to survive were crocodiles, close cousins of dinosaurs. Dinosaurs were not the only victims: as many as 70 percent of the plant and animal species of the world were wiped out at about the same time. (Insects were more durable: most of the insect species present at the end of the Mesozoic are still with us. Today there are 200 million insects for each person on the earth.) This mass extinction is what divides the Mesozoic from the Cenozoic.

Throughout geological history extinctions of organisms have been by no means unusual. New species have evolved, flourished, and then died out as far back as the fossil record exists. There is no shortage of past events that could have led to these extinctions. The shifting of the continents and surges in volcanic activity associated with such shifts certainly led to repeated changes in climate, in sea level, in the amount of sunlight reaching the surface (which affects the photosynthesis that is at the base of the chain of life), in the carbon dioxide content of the atmosphere, and so forth. In particular, the end of the Mesozoic was a time of worldwide tectonic and volcanic activity, and it would have been surprising if a large-scale extinction had not occurred then. The late Mesozoic extinction is famous because the dinosaurs disappeared so suddenly and completely, but in fact, as Fig. 16-35 shows, other extinctions, notably the Great Dying that ended the Paleozoic, saw more species disappear.

Until 1979 there seemed no reason to suppose that the fate of the dinosaurs was unusual in any fundamental way. In that year came the discovery of exceptionally large traces of the element iridium in a thin clay layer in Gubbio, Italy, that separated marine limestones of the late Mesozoic from younger limestones. Iridium, a metal similar to platinum, is rare in the earth's crust but relatively abundant in the meteorites that bombard the earth from space. Other late-Mesozoic deposits rich in iridium were soon found elsewhere. The obvious inference was that an asteroid, perhaps 10 or 15 km across, struck the earth 65 million years ago. The impact could have sent up a vast cloud of debris that remained in the stratosphere for months, perhaps years, blocking sunlight and, together with widespread fires that destroyed habitats and spread out everywhere, wiped out much of the life on our planet.

Supporting the impact theory was the discovery in 1991 of the remains of a huge crater about 150 km across (as big as Maryland) that is centered on the edge of Mexico's Yucatán peninsula (Fig. 16-41). The crater's size and age are about right. Furthermore, the clay layer that marks the end of the Mesozoic seems to have originated in both ocean-floor and continental rocks that were pulverized by the impact, which fits in with the crater's location at the rim of an ocean basin.

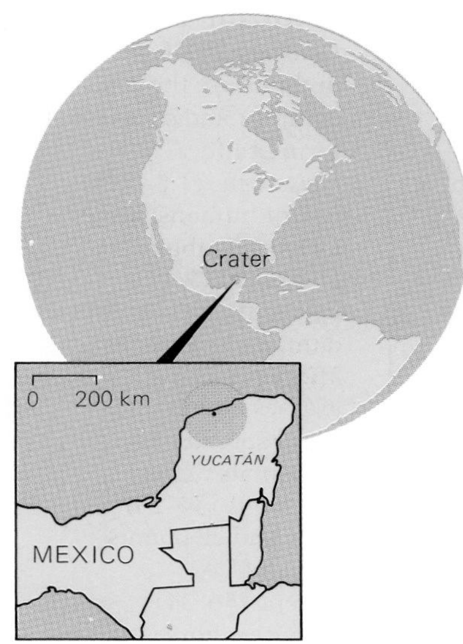

Figure 16-41 A gigantic crater whose traces were found at Chicxulub in Mexico's Yucatán peninsula may have been formed by the impact of the asteroid that led to the extinction of the dinosaurs and many other forms of life at the end of the Mesozoic Era.

There is another aspect to the disappearance of the dinosaurs. Intense volcanic activity took place at the end of the Mesozoic, leaving 300,000 km^2 of India covered with several km of lava (Fig. 16-42). Certainly the noxious gases and ash thrown into the atmosphere by the eruptions did life on our planet no good. Furthermore, iridium is abundant in the mantle and could have been brought to the surface in the volcanic magma. Was the asteroid impact or the volcanic activity the primary cause of the disappearance of the dinosaurs? Could the volcanic activity have been a consequence of the cosmic jolt in Yucatán? Did an original asteroid break up before reaching the earth, with one fragment going to Yucután and the other to India where it initiated the volcanic activity? Perhaps the two events are unrelated, although there is evidence that the volcanism began about the time the asteroid arrived. There are arguments for and against all these possibilities. In any case, were it not for the event or events involved, dinosaurs might well exist today. Would we? Nobody can say.

Example 16.4

The widespread biological extinctions of the past, for instance the drastic one in which the dinosaurs disappeared, lead to an interesting question. How can the events that caused them be squared with the principle of uniform change, according to which the earth's geologic history can be explained, in Lyell's words, by "forces now in operation"?

Solution

Certainly the event that caused not only the dinosaurs but a large proportion of the world's other living things to perish is a catastrophe in the usual sense of the word. But the ideas proposed to account for this catastrophe are all in accord with the laws of nature in today's universe. Such is not the case with Werner's idea that all rocks were formed in three precipitations from a universal ocean or with Cuvier's idea of the periodic destruction and re-creation of life by divine edict. Their catastrophes make no sense in terms of physics, chemistry, and biology. On the other hand, volcanoes and asteroids do exist, and even though they may affect the earth on a large scale only rarely, they still fit into the pattern of the principle of uniform change.

Figure 16-42 Volcanic eruptions at the end of the Mesozoic flooded a huge area of what is now India with basalt to form landscapes like this. Debris flung into the air by the eruptions could have partly blocked sunlight for a long period and thus contributed to the biological extinction at that time.

Cenozoic Era

Neogene:

Holocene (0.01–0) humans are the most complex and successful form of life
Pleistocene (1.8–0.01) rise of humans; large mammals abundant
Pliocene (5.3–1.8) flowering plants abundant; early humans
Miocene (23–5.3) grasses abundant; rapid spread of grazing mammals

Paleogene:

Oligocene (34–23) apes and elephants appear
Eocene (56–34) primitive horses, camels, rhinoceroses; first grasses
Paleocene (66–56) first large mammals and modern plants

(*dates are millions of years ago*)

16.17 The Cenozoic Era

The Age of Mammals

In many ways the Cenozoic Era of today has been different from preceding eras. During the Cenozoic the continents have stood for the most part well above sea level. No longer do shallow seas spread widely. In North America, beds originating from marine deposits are found only in narrow strips along the Pacific Coast and on the Atlantic Coast from New Jersey south to Yucatán. The thick Paleogene beds east and west of the Rocky Mountains are river, lake, and wind deposits made in continental basins. And climates during much of the Cenozoic have had a diversity like those of the present. The distribution of plants and animals shows that, instead of having widespread moderate climates like those of other eras, Cenozoic continents have had zones of distinct hot, cold, humid, and dry climates.

A characteristic of Cenozoic times has been widespread volcanic activity. From the Rockies to the Pacific Coast, lava flows and tuff beds testify to former volcanoes, some of which have only recently become extinct. Immense flows of basalt inundated an area of nearly a half-million km^2 in Oregon, Idaho, and Washington. Some of these flows today form the somber cliffs of the Columbia River Gorge.

Tectonic Activity The Cenozoic has also been a time of almost continuous tectonic disturbance, in contrast with the long periods of crustal stability in previous eras. Movements associated with the mountain-building episodes that divide the Cenozoic from the Mesozoic lasted well into the Paleogene. In mid-Paleogene the Alps and Carpathians of Europe and the Himalayas of Asia were folded and uplifted. Toward the end of the Paleogene the Cascade range of Washington and Oregon was formed, and other mountain ranges that have been active to the present day began to form around the border of the Pacific. Mountain ranges that had been folded earlier—the Appalachians, the Rockies, the Sierra Nevada—were repeatedly uplifted during the Cenozoic, and erosion following these uplifts has created their present topography.

It is not hard to associate this reshaping of continental landscapes with the spreading of sea floors and the grinding together of lithospheric plates that are still in action today. In the Cenozoic the continents continued their earlier drifts (see Fig. 16-9). In addition Greenland parted from Norway, Australia parted from Antarctica (New Zealand had done so earlier), and the Bay of Biscay opened up. More recently the Arabian peninsula broke off from Africa, the Gulf of California opened to separate Baja California from mainland Mexico, and Iceland rose above the surface of the Atlantic Ocean.

Mammals Thanks to their great adaptability, many mammals managed to survive the mass extinction that ended the Mesozoic Era. They continued an evolution and expansion that had begun during the Mesozoic. Carnivores like cats and wolves, armored beasts like rhinoceroses, agile creatures like deer and rabbits—ancestors of all these modern forms roamed the Paleogene landscape. A few mammals, like the whales and porpoises, took to life in the sea; another line, the bats, developed wings.

By the start of the Neogene Period, mammals began to dominate the earth as reptiles had before them. The most numerous of today's mammals are the rodents; a total of 4554 mammalian species are known. Side by side with the mammals developed modern birds and the trees of modern forests. As the end of the Tertiary approached, both the physical and the biological worlds assumed more and more closely their present aspects.

In the ice ages of the Pleistocene Epoch, great ice caps formed every 100,000 years or so in Canada and northern Europe, and valley glaciers advanced in high mountains elsewhere. (See Sec. 14.11.) Glacial deposits in North America show that ice spread outward from three centers of accumulation in Canada, as shown in Fig. 16-43. The changing climates proved a severe ordeal for living things. Nowadays mammals are still dominant, but in numbers of diversity of species they have declined markedly since the Pliocene.

The current interglacial period, now 10,000 years old, is known as the Holocene Epoch. This length of time is typical of interglacial periods in the past, which suggests another big chill may be about to begin. However, the interglacial period just before the most recent ice age lasted for an unusual 20,000 years, and it seems quite possible that the Holocene may follow suit. And there is the global warming discussed in Chap. 4. Although its duration thus far hardly registers on the geological clock, if the warming persists at its current rate, the ice-age cycle may well be disrupted, at least temporarily.

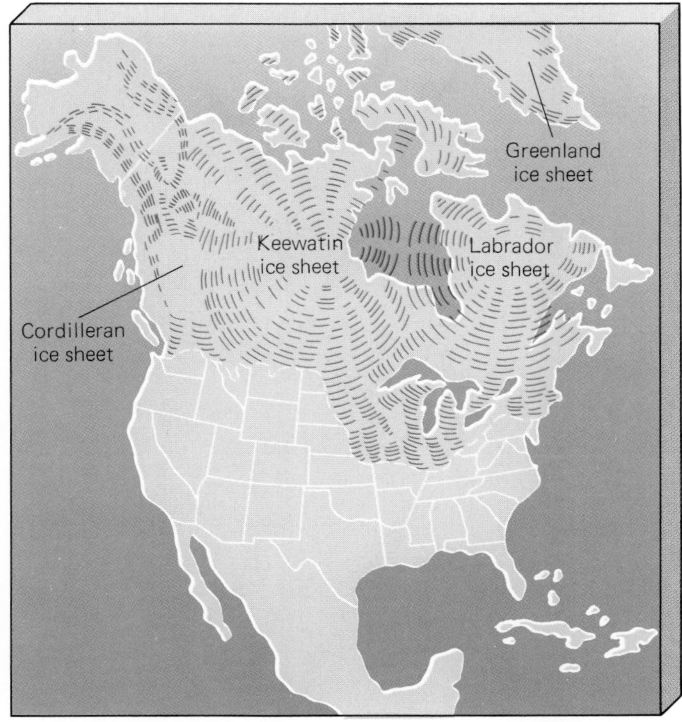

Figure 16-43 The maximum extent of Pleistocene glaciers, some over a kilometer thick, in North America. The major ice advances covered up to 30 percent of the earth's land surface about 20,000 years ago. Today we are in a warm interglacial period with about 10 percent under ice. Average worldwide temperatures when the last ice age was at its peak were 6°C colder than at present and sea level was more than 100 m lower.

Example 16.5

The Scandinavian landmass of northern Europe has been rising since the end of the most recent ice advance. The current rate is about 1 cm/year. What might be the reason for this rise?

Solution

When the thick sheet of ice that covered this region melted, the continental block there became lighter. The upward force of buoyancy thus had less weight to support, and the continental block therefore had a net upward force on it that led to its rise. Eventually the reduced buoyant force will equal the new weight of the continental block and the rise will stop.

16.18 Human History

We Are a Recent Species

According to the fossil record, our own species had its infancy in the Great Rift Valley of eastern Africa. Roughly 6.5 million years ago, biochemical evidence indicates that the descendants of a common ancestor split into two branches, one that evolved into some of today's apes and the other into today's humans (Fig. 16-44). The earliest of the humanlike creatures walked on two legs but still had small brains and eventually died out. There was not a continuous development from then on but rather a succession of perhaps 20 closely related species, usually overlapping, with each flourishing for a time and then either disappearing or turning into

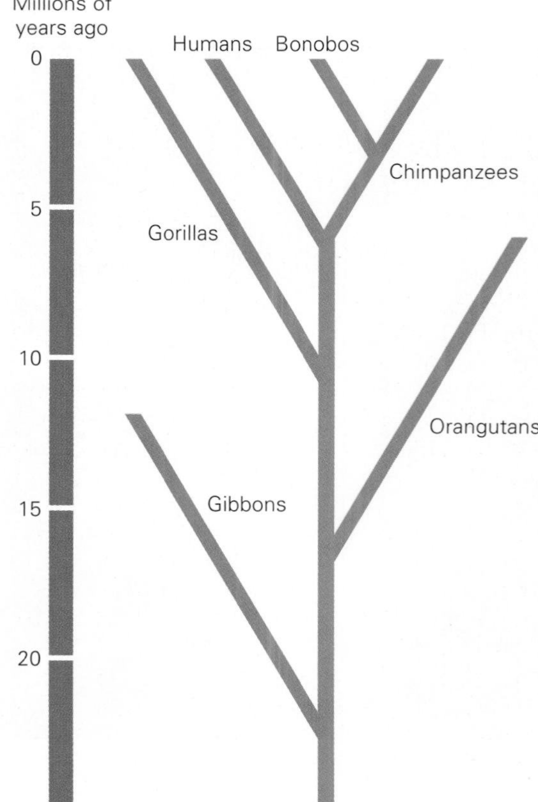

Figure 16-44 Evolutionary tree of humans and apes, based on DNA evidence. Monkeys split off earlier. Chimpanzees, bonobos, and humans have nearly all of their genes (and every bone in their bodies) in common. All other living things are also our relatives, if more distant. We share around 40 percent of our genes with fish and 25 percent with dandelions.

another. By 2 million years ago some early humans had larger brains than before and were making and using stone tools (hence their species name *Homo habilis,* "handy man"). Farther into the Pleistocene, around 200,000 years ago, modern humans (*Homo sapiens,* "wise man") emerged in Africa and about 65,000 years ago began to populate Europe and Asia. Their record is most complete in Europe, where stone implements, burial sites, drawings on cave walls (Fig. 16-45), and skeletal fragments give a fairly connected history.

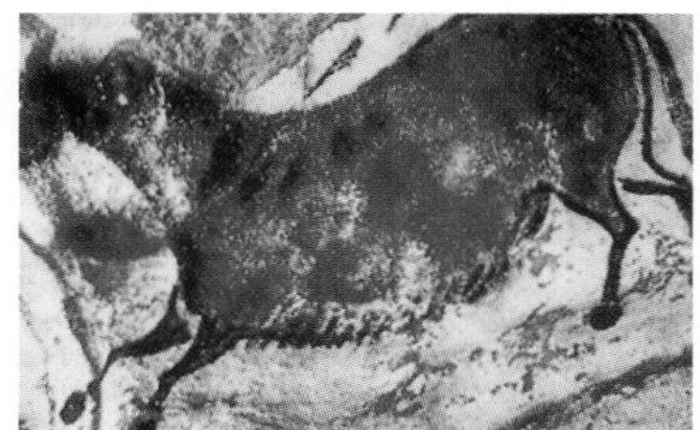

Figure 16-45 This painting is one of many that were made in a cave at Lascaux, France, during the most recent Ice Age 15,000 years ago.

Humans came to North America around 15,000 years ago, perhaps even earlier. Whether the earliest arrivals went from Siberia to Alaska over a land bridge that appeared when heavy glaciation lowered sea level (as later ones certainly did) or went from Europe to Canada via Greenland over frozen northern seas is unclear. They may have had the help of boats for part of either journey; Australia had been reached by watercraft well before then. Animal life in North America—which included woolly mammoths, lions, saber-toothed tigers, camels, giant ground sloths, and zebralike horses—was spectacularly varied and abundant at that time. Did hunting by the newcomers wipe out the large animals about 13,000 years ago or was it some environmental factor that did so? No one knows for sure, but both overkill and climatic change seem to have contributed. There were at least 10 million inhabitants of North America when the first Europeans came in the tenth century (see Fig. 14-34) and more than that in South America.

Evolution in humans is still continuing, as it is in other living things. Over 700 genes have been identified that have changed in the past 15,000 years. Most of these genes are involved in bone structure, skin color, digestion, brain function, and the senses of taste and smell. The time period covers the introduction and spread of farming, which changed the characteristics that promoted success in life from what they had been when people were migratory hunter-gatherers. An example is the emergence of the

Recent Cousins

Several hundred thousand years ago the Neanderthals, cousins of ours who were strongly built hunters with large brains (Fig. 16-46), diverged from the lineage that ultimately became modern humans and migrated from their original home in Africa to Europe and the Near East. (They are named after the German valley where their remains were first found in 1857.) About 35,000 years ago our direct ancestors began a major expansion north from Africa to Europe where they coexisted with the Neanderthals for 10,000 years or so. The Neanderthals then died out; exactly why is not known although there are plenty of ideas.

Was there any interbreeding? Apparently Neanderthals and humans were close enough genetically for this to have been possible, and several skeletons have been found that can be interpreted as hybrids. Traces of DNA from a number of Neanderthals have been recovered and when the difficult task of deciphering them is further advanced, we may know whether any Neanderthal genes found their way into our genome.

What may have been another cousin species, whose members were less than a meter tall with grapefruit-sized skulls, lived on the Indonesian island of Flores for thousands of years and used relatively advanced tools. They disappeared about 12,000 years ago after a major volcanic eruption on Flores. One hypothesis is that these little people in their isolation had evolved away from the lineage that led to us. Another is that their small stature was merely the result of a lack of iodine in their diets.

Figure 16-46 What a Neanderthal family might have looked like.

gene for lactase, the enzyme that permits adults to digest milk; today this gene is most common among people whose ancestors came from northern Europe, where cattle raising became widespread.

People and the Environment In the past, biological changes were brought about largely by changes in the physical environment, so the biological narrative follows the physical one through the ages. The modern world seems to be witnessing a reversal of this sequence. Not only are we able by virtue of our ingenuity to flourish in the environment that nature has provided, but we are able to alter that environment in a number of ways. These alterations have permitted a vast number of people to live relatively free (by historical standards) of starvation and disease. Today deaths due to ailments brought on by diets of plenty exceed those due to famine.

If continued into the future and swollen in scale by population growth, however, our present patterns of industry and agriculture menace the interplay between people and environment that has been so successful in the past. The indiscriminate use of pesticides and artificial fertilizers has already destroyed the ecological balance in many areas; numerous inland waters have been poisoned by industrial waste; noxious gases fill the air in densely populated regions; acid rain due to power-plant emissions has damaged forests and lakes on a large scale; vast areas of forest have had their trees cut and their soil washed away as a result; global warming due to carbon dioxide emissions is increasing; deposits of radioactive wastes lie waiting for chance catastrophe to disperse them; and so on. Although a lot has been done in developed countries to limit environmental damage, it is accelerating in the poorer ones as their people climb out of poverty. Most ominous of all is that, whatever we do in the future, a lot more of us will be doing it (see Sec. 4.1).

Will the next mass extinction be our fault? Thousands of years of hunting for food, skins, and furs have already caused numerous animal species to vanish. The ballooning human population and the accompanying expansion of agriculture have destroyed natural ecosystems everywhere. And global warming is altering the world's climates in ways that stress many life forms. It seems possible that, if these patterns of human behavior continue, their result will be a biological extinction large enough to show up on a graph like that of Fig. 16-35. Some biologists believe that two-thirds of present bird, mammal, and plant species will be gone by the year 2100. Will we be on the way out then? Humans are extremely adaptable, but there is surely a limit to how much we can abuse our planet and still flourish. How ironic if our mental capacities, which have enabled us to dominate life on earth despite other creatures being larger, stronger, faster, and with superhuman abilities such as flight, should be what brings about our decline.

Important Terms and Ideas

Movements of the solid materials of the earth's crust are called **tectonic.** The principal kinds of tectonic movement are faulting, folding, and regional uplift, sinking and tilting. A **fault** is a fracture surface along which motion has taken place. **Mountain building** begins with the deposition of sediments in a sinking area, followed by tectonic movement and erosion.

The crust and the upper part of the mantle make up a shell of hard rock called the **lithosphere.** Below the lithosphere is a layer of hot, soft rock called the **asthenosphere.** According to the theory of **plate tectonics,** the lithosphere is divided into seven huge **plates** and a number of smaller ones. The plates can move relative to one another in four ways: two plates can move apart with molten rock rising to form new ocean floor at the gap; one plate can slide under another and melt in a **subduction zone;** two plates can collide and buckle to form a mountain range; and adjacent plates can slide past each other along a **transform fault.**

Continental drift occurs because of plate motion. Today's continents were once part of two supercontinents called **Laurasia** (North America, Greenland, Europe, and most of Asia) and **Gondwana** (South America, Africa, Antarctica, India, and Australia) that were separated by the **Tethys Sea.**

According to the **principle of uniform change,** geologic processes in the past were the same as those in the present. An **unconformity,** which is a buried surface of erosion, indicates that tectonic uplift, erosion, and sedimentation have occurred in that order.

Radionuclides and their decay products in rocks make it possible to date geologic formations. The remains of living things can be dated with the help of **radiocarbon,** the radioactive carbon isotope ^{14}C. **Fossils,** the remains of organisms preserved in rocks, are useful in correlating strata, in tracing the development of living things, and in reconstructing ancient environments.

Geologic time is divided into **Precambrian time** and the **Paleozoic, Mesozoic,** and **Cenozoic Eras.** Eras are subdivided into periods and the periods into **epochs.** The divisions are marked by mass **extinctions** in which environmental changes cause the disappearance of many species of plants and animals. Dinosaurs were among the victims of the extinction at the end of the Mesozoic Era.

Multiple Choice

1. A crack in the earth's crust along which movement has taken place is called
 a. a fault
 b. a fold
 c. an earthquake
 d. a moraine

2. A long, narrow bay with an irregular outline, such as Chesapeake Bay, was formed by
 a. wave action
 b. stream erosion and later submersion
 c. glacier erosion and later submersion
 d. faulting

3. The rugged character of mountain landscapes is largely the result of
 a. folding
 b. faulting
 c. stream erosion
 d. glacier erosion

4. The Tethys Sea once separated
 a. North and South America
 b. South America and Africa
 c. Europe and Asia
 d. Laurasia and Gondwana

5. A mountain range that was not once part of the Tethys Sea is the
 a. Alps
 b. Pyrenees
 c. Andes
 d. Himalayas

6. North America, Greenland, and most of Eurasia once made up the supercontinent of
 a. Pangaea
 b. Laurasia
 c. Gondwana
 d. Atlantis

7. The thickness of the lithosphere varies between
 a. 2 and 5 km
 b. 5 and 35 km
 c. 50 and 100 km
 d. 500 and 1000 km

8. The layer of soft rock beneath the lithosphere is called the
 a. stratosphere
 b. thermosphere
 c. asthenosphere
 d. mantle

9. As compared with the continents, the ocean floors are
 a. much younger
 b. much older
 c. about the same age
 d. in some places older and in others younger

10. The ocean floor near a midocean ridge
 a. has the same constant magnetization on both sides
 b. is magnetized in one direction on one side and in the opposite direction on the other side
 c. has strips of alternate magnetization on both sides
 d. has no consistent pattern of magnetization

11. According to the hypothesis of sea-floor spreading, molten rock is rising up along the
 a. trenches that rim the Pacific Ocean
 b. ridges on midocean floors
 c. location of the Tethys Sea
 d. equator

12. The number of large plates into which the lithosphere is divided is
  a. 3
  b. 7
  c. 20
  d. 50

13. A typical speed for a lithospheric plate is
  a. 3 mm/year
  b. 3 cm/year
  c. 3 m/year
  d. 3 km/year

14. A region where an edge of a lithospheric plate slides under an edge of another plate is called a
  a. transform fault
  b. fault scarp
  c. moraine
  d. subduction zone

15. A subduction zone is never
  a. associated with a mountain range
  b. associated with an island chain
  c. located at a midocean ridge
  d. located at a continental margin

16. The Indian subcontinent
  a. was always part of Asia
  b. came into being 45 million years ago as the result of extensive vulcanism
  c. collided with Asia 45 million years ago
  d. began to move away from Asia 45 million years ago

17. The Andes seem to be the result of
  a. an oceanic plate sliding under a continental plate
  b. two continental plates colliding
  c. two oceanic plates colliding
  d. an enormous earthquake

18. If present trends continue,
  a. California will be detached from the rest of North America
  b. the Atlantic will become narrower
  c. The Pacific will become wider
  d. The West Indies will sink below the ocean surface

19. If the processes of plate tectonics were to stop acting, which of the following would be the last to cease?
  a. earthquakes
  b. volcanic eruptions
  c. erosion
  d. tsunamis

20. Scientific observation does not support
  a. the principle of uniform change
  b. Darwin's theory of evolution
  c. continental drift
  d. the origin of the earth in 4004 B.C.

21. An uneven surface on which a horizontal upper bed rests is called a (an)
  a. stratum
  b. fault
  c. dike
  d. unconformity

22. Radiocarbon dating is based upon the fact that
  a. ^{14}C is continually being formed in the remains of living things after their death
  b. ^{14}C is not radioactive
  c. the ^{14}C content of the remains of living things depends upon the time in the past when they came into being
  d. the ^{14}C content of the remains of living things depends upon the time in the past when they died

23. A suitable object for radiocarbon dating would be one whose age is
  a. 10,000 years
  b. 100,000 years
  c. 1 million years
  d. 10 million years

24. The oldest rocks that have been dated are about
  a. 4 million years old
  b. 4.5 million years old
  c. 1 billion years old
  d. 4 billion years old

25. Most fossils are found in
  a. volcanic rocks
  b. metamorphic rocks
  c. sedimentary rocks that formed from deposits on the floors of shallow seas
  d. sedimentary rocks that formed from stream deposits

26. The division of geologic time into eras and periods is based upon
  a. the coming of ice ages
  b. the disappearance of continents
  c. biological extinctions
  d. worldwide flooding

27. The earth was formed
  a. in 4004 B.C.
  b. about 2 million years ago
  c. about 4.6 billion years ago
  d. about 10 billion years ago

28. Arrange the following divisions of geologic time in their proper sequence, starting with the oldest:
  a. Paleozoic Era
  b. Precambrian time

c. Cenozoic Era
d. Mesozoic Era

29. Precambrian rocks are
a. never found
b. extremely rare
c. exposed in a number of regions
d. the most common rocks

30. In rocks of which one or more of the following eras does evidence of life appear?
a. Precambrian
b. Paleozoic
c. Mesozoic
d. Cenozoic

31. Ancient geologic processes as revealed in Precambrian strata were
a. primarily volcanic
b. primarily glacial
c. primarily erosion and sedimentation
d. of the same kinds as those of the present time

32. Large exposures of Precambrian rocks are to be found in
a. eastern Canada and adjacent parts of the United States
b. the Rocky Mountains
c. Texas and New Mexico
d. the Middle West

33. Coal is composed of
a. petrified wood
b. buried plant material that has partially decayed
c. buried animal material that has partially decayed
d. a variety of igneous rock

34. A bed of coal usually implies that the region was once a
a. desert
b. coniferous forest
c. swamp
d. river bed

35. Most coal deposits were formed in the
a. Cenozoic
b. Mesozoic
c. Paleozoic
d. Precambrian

36. Amphibians, fishes, and marine invertebrates were the dominant form of life in the
a. Cenozoic
b. Mesozoic
c. Paleozoic
d. Precambrian

37. The dinosaurs were
a. reptiles
b. primitive mammals
c. all carnivorous
d. still living when early humans appeared

38. Dinosaurs were abundant in the
a. Cenozoic
b. Mesozoic
c. Paleozoic
d. Precambrian

39. Dinosaurs were the dominant form of animal life for a period of about
a. 1000 years
b. 1 million years
c. 100 million years
d. 1 billion years

40. Unusually large traces of the rare element iridium are found in sediments deposited at the end of the
a. Cenozoic
b. Mesozoic
c. Paleozoic
d. Precambrian

41. The ancestors of the birds were
a. reptiles
b. mammals
c. amphibians
d. insects

42. Mammals were the dominant form of life in the
a. Cenozoic
b. Mesozoic
c. Paleozoic
d. Precambrian

43. Pangaea broke up into Laurasia and Gondwana in the
a. Cenozoic
b. Mesozoic
c. Paleozoic
d. Precambrian

44. Laurasia and Gondwana broke up in the
a. Cenozoic
b. Mesozoic
c. Paleozoic
d. Precambrian

45. The Cenozoic Era represents a period
a. of almost continuous tectonic activity
b. of relative stability, with erosion and sedimentation the chief geologic processes
c. of relatively uniform climate around the world
d. in which the reptile was the most advanced form of life

46. During the Ice Age
a. there was a single glacial advance
b. there were several glacial advances and retreats
c. the entire earth was covered with an ice sheet
d. all animal life perished and had to start over again afterward

Exercises

16.1 Types of Deformation

1. What landscape features are associated with faults?
2. List all the evidence you can for each of the following statements:
   a. Granite is an igneous rock.
   b. Mica schist is a rock that has been subjected to nonuniform pressure.
   c. Compressional forces exist in the earth's crust.
   d. Tectonic movement is going on at present.

16.2 Mountain Building

3. What geologic process is chiefly responsible for the landscape of a mountain range?
4. When stream erosion has been active for a long time in a region underlain by folded strata, what determines the position of the ridges and valleys?
5. Why is it believed that the region where the Rocky Mountains now stand was once near or below sea level?

16.3 Continental Drift

6. What kind of biological evidence supports the notion that all the continents were once part of a single supercontinent? What kind of climatological evidence supports the concept of continental drift?
7. The eastern coast of South America is a good fit against the western coast of Africa. What sort of evidence would you look for to confirm that the two continents were once part of the same landmass?
8. What body of water separated the ancient supercontinents of Laurasia and Gondwana? Are there any remnants of this body of water in existence today? If so, what are they?
9. What mountain ranges of today were once part of the Tethys Sea? What kind of evidence indicates that the region where these mountains are was once below sea level?

16.4 Lithosphere and Asthenosphere

10. Which is denser, the granitic rock of the continents or the basaltic rock of the ocean floors? Which extends deeper into the crust, the continents or the ocean floors?
11. (a) What is the difference between the earth's crust and its lithosphere? (b) How is it possible for a plastic asthenosphere to occur between a rigid lithosphere and a rigid mantle? (c) If the asthenosphere is plastic, how can transverse seismic waves travel through it?

16.5 The Ocean Floor

16.6 Ocean-Floor Spreading

12. North America, Greenland, and Eurasia fit quite well together in reconstructing Laurasia, but there is no space available for Iceland. Why is the omission of Iceland from Laurasia reasonable?
13. How do the ages of the ocean floors compare with the ages of continental rocks? What is the reason for the difference, if any?

16.7 Plate Tectonics

14. When continental drift was proposed almost a century ago, it was assumed that the continents move through soft ocean floors. Why is this hypothesis no longer considered valid? How does continental drift actually occur?
15. The energy source of erosional processes is the sun. Where does the energy involved in tectonic activity come from?
16. Where do subduction zones occur? What happens at them?
17. The Himalayas are the highest mountain range on earth. Why are they still rising?
18. How does the origin of the Himalayas differ from that of the oceanic mountains that constitute the Mid-Atlantic Ridge?
19. Which are younger, the Rocky Mountains or the Himalayas?
20. Is the Atlantic Ocean becoming narrower or wider? The Pacific Ocean?
21. The San Andreas Fault in California is a strike-slip fault that lies along the boundary between the Pacific and American plates. What does this indicate about the nature of the boundary?
22. In what geological zones are most volcanoes found?
23. Which plate collisions are responsible for creating island arcs such as the West Indies, the Aleutians, and Indonesia?
24. The distance between the continental shelves of the eastern coast of Greenland and the western coast of Norway is about 1300 km. If Greenland separated from Norway 65 million years ago and their respective plates have been moving apart ever since at the same rate, find the average speed of each plate.
25. The oldest sediments found on the floor of the South Atlantic Ocean, which are 1300 km west of the axis of the Mid-Atlantic Ridge, were deposited about 70 million years ago. What rate of plate movement does this finding suggest?

16.9 Rock Formations

26. In Fig. 16-47, beds *A* to *F* consist of sedimentary rocks formed from marine deposits and rocks *G* and *H* are granite. What sequence of events must have occurred in this region?

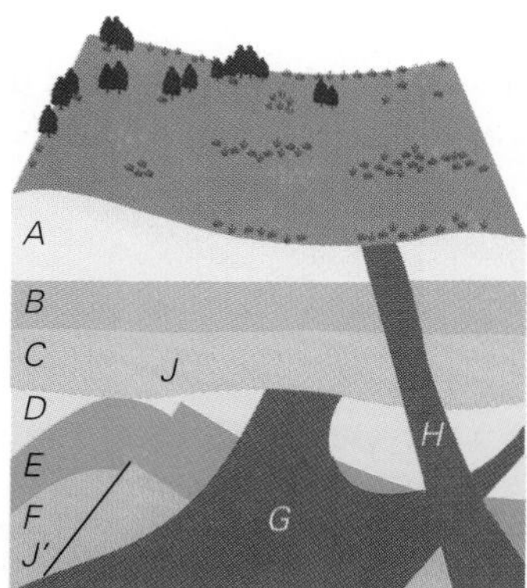

Figure 16-47

27. (a) What is an unconformity? (b) If one is shown in Fig. 16-47, where is it?
28. What is a fault? If one is shown in Fig. 16-47, where is it?

16.10 Radiometric Dating

29. What is the basis of the radiocarbon dating procedure?
30. The half-life of rubidium 87 is 47 billion years, and that of potassium 40 is 1.3 billion years. Would you expect the rubidium-strontium or potassium-argon method of radiometric dating to be more generally useful? Why?
31. The half-life of potassium 40 is 1.3 billion years. If a rock sample contains 0.010 percent of this radionuclide today, what was its percentage in the rock 3.9 billion years ago?

16.11 Fossils

32. Why are fossils still useful in dating rock formations despite the development of radioactive methods?
33. List as many different kinds of fossils as you can.
34. Why are most fossils found in beds that were once the floors of shallow seas?
35. Why are fossils never found in igneous rocks and only rarely in metamorphic rocks?

16.12 Geologic Time

36. What is the basis for the division of geologic time into eras and periods?
37. What is the oldest division of geologic time? In what division are we living today?
38. During what divisions of geologic time have living things existed on the earth?
39. The earth's history is sometimes divided into two **eons,** Cryptozoic ("hidden life") and Phanerozoic ("visible life"), with the first corresponding to Precambrian time and the second extending from the beginning of the Paleozoic Era to the present day. What do you think is the reason for this division?

16.13 Precambrian Time

40. The early atmosphere of the earth probably consisted of carbon dioxide, water vapor, and nitrogen, with little free oxygen. What is believed to be the source of oxygen in the present-day atmosphere? What bearing has this question on the relatively rapid development of varied and complex forms of life that marks the start of the Paleozoic Era?
41. Precambrian rocks include sedimentary, igneous, and metamorphic varieties. What does this suggest about the geologic activity in Precambrian time?
42. Precambrian rocks are exposed over a large part of eastern Canada. What does this suggest about the geologic history of this region since the end of Precambrian time?
43. What conspicuous difference is there between Precambrian sedimentary rocks and those of later eras?
44. What are the chief kinds of organisms that have left traces in Precambrian sedimentary rocks?

16.14 The Paleozoic Era

45. Paleozoic sedimentary rocks derived from marine deposits are widely distributed in all the continents. What does this indicate about the height of the continents relative to sea level in the Paleozoic Era?
46. Which of the following are found in Paleozoic rock formations? Dinosaur fossils; early human fossils; coal deposits; radiocarbon deposits.
47. Why is it believed that large parts of the United States were once covered by shallow seas?

16.15 Coal and Petroleum

48. Under what circumstances is coal formed?
49. What is believed to be the origin of petroleum? Of natural gas?

16.16 The Mesozoic Era

50. What are some of the chief differences between reptiles and mammals?
51. What kind of animals were the dinosaurs? Were they mostly small, mostly large, or were they of all sizes?

52. What is believed to be the reason or reasons for the disappearance of the dinosaurs? What is the evidence for this belief?

53. From what type of animal did birds evolve? Are modern birds closest anatomically to butterflies, bats, flying fish, or crocodiles?

16.17 The Cenozoic Era

54. About 200 million years ago today's continents were all part of the supercontinent Pangaea. During what geologic era did Pangaea break apart into Laurasia and Gondwana? During what era did Laurasia break up into North America, Greenland, and Eurasia?

55. The same reptiles were present on all continents during the Mesozoic Era, but the mammals of the Cenozoic Era are often different on different continents. Why?

56. In rocks of what era or eras would you expect to find fossils of (a) horses; (b) ferns; (c) clams; (d) insects; (e) apes?

57. What were the Ice Ages? When did they occur?

58. Minnesota has a great many shallow lakes. How do you think they originated?

Answers to Multiple Choice

1. a
2. b
3. c
4. d
5. c
6. b
7. c
8. c
9. a
10. c
11. b
12. b
13. b
14. d
15. c
16. c
17. a
18. a
19. c
20. d
21. d
22. d
23. a
24. d
25. c
26. c
27. c
28. b, a, d, c
29. c
30. a, b, c, d
31. d
32. a
33. b
34. c
35. c
36. c
37. a
38. b
39. c
40. b
41. a
42. a
43. b
44. b
45. a
46. b

17

The Solar System

Jupiter is the largest planet.

Goals

When you have finished this chapter you should be able to complete the goals ▶ given for each section below:

The Family of the Sun

17.1 The Solar System
Inner Planets Mostly Rock, Outer Planets Mostly Liquified Gas
- ▶ Distinguish between the rotation and revolution of a planet and state the two regularities of these motions shared by most planets and satellites.
- ▶ Identify the inner and outer planets and list the common properties of the members of each group.

17.2 Comets
Regular Visitors from Far Away in the Solar System
- ▶ Discuss the nature and appearance in the sky of comets.
- ▶ Give the two causes of the deflection of comet tails so they always point away from the sun.

17.3 Meteors
"Shooting Stars" Usually Smaller Than a Grain of Sand
- ▶ Distinguish among meteoroids, meteorites, and meteors and explain why meteor showers occur.

The Inner Planets

17.4 Mercury
It Always Appears near the Sun

17.5 Venus
Our Sister Planet
- ▶ Compare the surface features and atmosphere of Venus with those of the earth.

17.6 Mars
Small and Cold with a Varied Landscape

17.7 Is There Life on Mars?
Maybe, but Not Likely
- ▶ Discuss the possibility of life on Mars.

17.8 Asteroids
Millions of Tiny Planets between Mars and Jupiter
- ▶ State the nature and location in space of asteroids and describe the likelihood and danger of collisions with them.

The Outer Planets

17.9 Jupiter
Almost a Star

17.10 Saturn
Lord of the Rings
- ▶ Describe the compositions and structures of Jupiter and Saturn and the nature of Jupiter's Great Red Spot.
- ▶ Discuss the nature of Saturn's rings and why this was known before a spacecraft visit.
- ▶ Explain why several of the satellites of Jupiter and Saturn are of such interest to astronomers.

17.11 Uranus, Neptune, Pluto, and More
Far Out
- ▶ Describe Kuiper Belt objects and discuss why Pluto is considered one of them.

The Moon

17.12 Phases of the Moon
A Little Less Than a Month per Cycle
- ▶ Use a diagram to account for the phases of the moon.

17.13 Eclipses
Now You See It, Now You Don't
- ▶ Explain why eclipses of the sun and moon occur.

17.14 The Lunar Surface
Mountains and Maria, maybe Water but No Atmosphere
- ▶ Describe the surface features of the moon.

17.15 Evolution of the Lunar Landscape
A Violent Past, a Quiet Present
- ▶ Outline the evolution of the moon's surface.

17.16 Origin of the Moon
A Collision Was Probably Responsible
- ▶ Outline the probable origin of the moon.

Whether we look at them with the naked eye or with the help of the largest telescope, the stars are just points of light. On the other hand, most of the planets appear as disks in a telescope of even modest power. This does not mean, of course, that the planets are larger than the stars, only that the planets are much closer to us. The sun, like other stars, glows brightly because it is extremely hot. The planets are too cool to shine by themselves and we see them by the sunlight they reflect. The sun together with its accompanying planets, their satellites, and other smaller bodies make up the **solar system.** The members of the solar system dwell in emptiness and are separated by vast distances from everything else in the universe.

The Family of the Sun

Until the seventeenth century the solar system was thought to consist of only the five planets Mercury, Venus, Mars, Jupiter, and Saturn besides the sun, the earth, and the moon. In 1609, soon after having heard of the invention of the telescope in Holland, Galileo built one of his own. With his telescope, Galileo found four additional members of the solar system: the brighter of the moons (or **satellites**) that circle Jupiter. Since Galileo's time, improved telescopes have led to the discovery of many more members of the sun's family.

The list of planets, in order of average distance from the sun, comprises Mercury, Venus, Earth, Mars, Jupiter, Saturn, Uranus, and Neptune (Fig. 17-1). Pluto, once considered a planet, was reclassified in 2006 as a **dwarf planet** (Secs. 17.8 and 17.11), a new category with other members. All except Mercury and Venus have satellites. Thousands of small objects called **asteroids,** all less than 1000 km in diameter, follow separate orbits about the sun in the region between Mars and Jupiter. Similar objects, one larger than Pluto, are in orbit outside Pluto. Comets and meteors, in Galileo's time thought to be atmospheric phenomena, are now recognized as also belonging to the solar system.

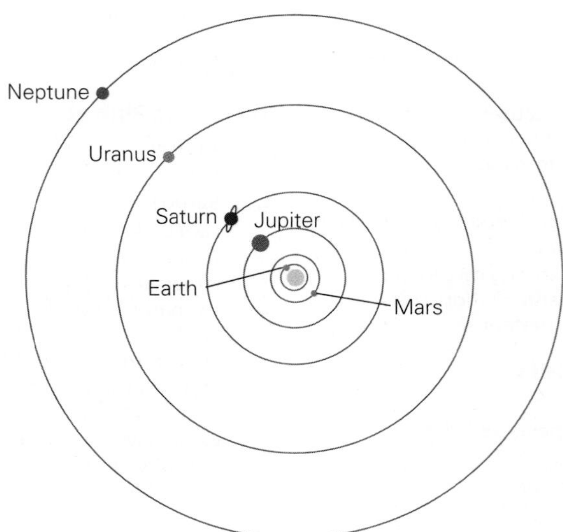

Figure 17-1 The solar system. The orbits of Mercury and Venus are too small to be shown on this scale. Diameters of sun and planets are exaggerated.

In recent years our knowledge of the solar system has been greatly increased by the voyages of spacecraft, most of them from the United States but with some notable European ones as well. There have been nearly 200 missions thus far, about two-thirds of them successful, with more in flight or planned. Spacecraft have landed on Venus and Mars, inspected comets and asteroids at close range, and astronauts have walked on the surface of the moon. So remarkable is modern technology that signals from the tiny 8-W transmitter on Pioneer 10 were still being picked up on the earth in 2003, 20 years after it left the solar system, over 30 years after it left Florida. The spacecraft was then so far away that its signals took more than 11 hours, traveling at the speed of light, to reach the earth. Effective contact was finally lost in that year.

Over 250 planets have been discovered that orbit stars other than the sun. These exoplanets are the subject of Sec. 19.11.

17.1 The Solar System

Inner Planets Mostly Rock, Outer Planets Mostly Liquified Gas

Not only is the solar system isolated in space, but also each of its principal members is far from the others. From the earth to our nearest neighbor, the moon, is about 384,000 km; from the earth to the sun is about 150 million km. It took the Apollo 11 spacecraft 3 days to reach the moon. Traveling at the same speed, Apollo 11 would take more than 3 years to reach the sun.

If we use a golf ball to represent the sun, a grain of sand 4 m away would represent the earth on the same scale. The moon would be a dust speck about 1 cm from the sand grain. The largest planet, Jupiter, would be a small pebble 18 m from the golf ball, and Neptune would be a still smaller pebble 120 m from the golf ball. The total mass of the planets is only 0.13 percent of the mass of the sun. The relative sizes of the planets are shown in Fig. 17-2.

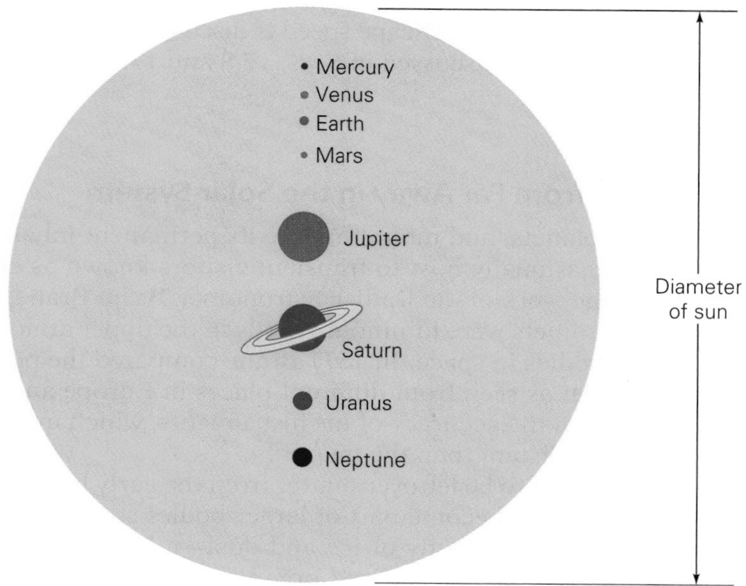

Figure 17-2 Relative sizes of planets and sun.

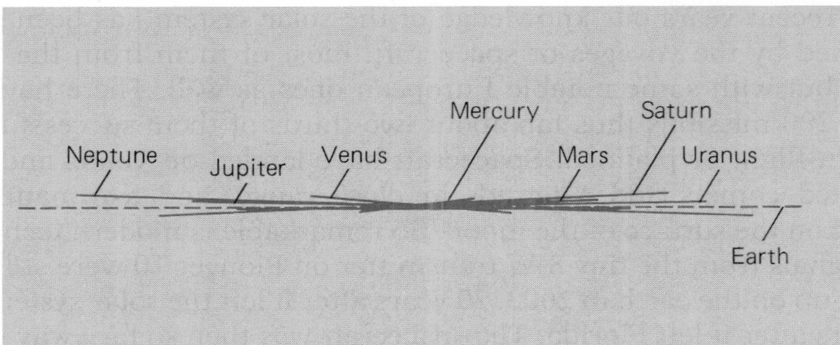

Figure 17-3 The orbits of the planets seen edgewise. The orbits all lie nearly in the same plane.

Planets **revolve** around the sun and **rotate** on their axes. Two aspects of these motions are notable:

1. Nearly all the revolutions and rotations are in the same direction. Only the rotation of Venus and the revolutions of a few minor satellites about their parent planets run contrary to the general motion. (Uranus is an exception of a different kind, since it rotates about an axis only 8° from the plane of its orbit.)
2. All the orbits lie nearly in the same plane (Fig. 17-3).

The principal data about the planets and their orbits are summarized in Table 17-1. The **inner planets** of Mercury, Venus, Earth, and Mars are relatively small, have similar densities, and are composed largely of rocky material. All have cores that probably consist mainly of iron. They rotate fairly slowly on their axes. Among them is only one satellite of any size, the moon; the two satellites of Mars are only a few km across.

The **outer planets** of Jupiter, Saturn, Uranus, and Neptune are large, not very dense compared with the earth (Saturn would float if placed in a large enough bathtub), and are composed largely of gases compressed to liquid form without iron cores. They rotate fairly rapidly on their axes. These giant planets have among them a total of over 130 satellites, a few large, most quite small. The inner planets have low escape speeds; the giant ones have high escape speeds. (Escape speed is discussed in Sec. 2.14.)

The dwarf planets are discussed in Secs. 17.8 and 17.11.

Where Comets Live

Most of the comets in the solar system, perhaps 10^{13} of them, make up the Oort Cloud, which lies about 50,000 times farther from the sun than the earth, a fifth of the distance to the nearest star. At that distance the sun is only as bright as Venus is in our sky. The cloud is named for the Dutch astronomer Jan Oort, who proposed its existence after studying the orbits of known comets. Because the Oort Cloud is so far from the sun, the orbits of its members are easily disturbed, for instance by stars that happen to pass not too far away, which occurs from time to time. Some of the changed orbits bring the corresponding comets to the inner solar system, where we are. The Hale-Bopp comet that decorated the sky in 1997 is believed to have leaked from the Oort Cloud.

A smaller number of comets form a swarm outside Neptune called the Kuiper Belt (named for the American astronomer Gerard Kuiper). Many objects in the Kuiper Belt—over 800 of them are known at present—are more like asteroids (see Sec. 17.8). At least five of these objects are huge, over 1000 km across. There may be many more like these far out in space and therefore hard to find.

17.2 Comets

Regular Visitors from Far Away in the Solar System

Besides the stars, planets, and moon that are its permanent inhabitants, the night sky is occasionally host to transient visitors known as **comets** (Fig. 17-4). Until the work of the Danish astronomer Tycho Brahe, it was not clear whether comets were luminous clouds in the upper atmosphere or were heavenly bodies in space. In 1577 Brahe compared the positions in the sky of a comet as seen from different places in Europe and found them the same within the accuracy of his instruments, which meant that the comet was more distant than the moon.

Comets are thought to be leftover matter from the early history of the solar system that did not become part of larger bodies such as the planets. A comet is composed chiefly of ice and dust—a "dirty snowball"—in some cases mixed with polymerized organic molecules. In addition,

Table 17-1 The Planets

Planet	Symbol	Mean Distance from Sun, Earth = 1[a]	Diameter, Thousands of km	Mass, Earth = 1[b]	Mean Density, Water = 1[c]	Surface Gravity, Earth = 1[d]	Escape Speed, km/s[e]	Period of Rotation on Axis	Period of Revolution around Sun	Eccentricity of Orbit[h]	Inclination of Orbit to Ecliptic[i]	Known Satellites[j]
Mercury	☿	0.39	4.9	0.055	5.4	0.38	4.3	59 days	88 days	0.21	7°00′	0
Venus	♀	0.72	12.1	0.82	5.25	0.90	10.4	243 days[f]	225 days	0.01	3°34′	0
Earth	⊕	1.00	12.7	1.00	5.52	1.00	11.2	24 h	365 days	0.02		1
Mars	♂	1.52	6.8	0.11	3.93	0.38	5.0	24.5 h	687 days	0.09	1°51′	2
Jupiter	♃	5.20	143	318	1.33	2.6	60	10 h	11.9 yr	0.05	1°18′	63
Saturn	♄	9.54	120	95	0.71	1.2	36	10 h	29.5 yr	0.06	2°29′	60
Uranus	♅	19.2	51	15	1.27	1.1	22	16 h[g]	84 yr	0.05	0°46′	27
Neptune	♆	30.1	50	17	1.70	1.2	24	16 h	165 yr	0.01	1°46′	13

[a] *The mean earth-sun distance is called the astronomical unit, where 1 AU = 1.496×10^{11} km.*
[b] *The earth's mass is 5.98×10^{24} kg.*
[c] *The density of water is 1 g/cm^3 = 10^3 kg/m^3.*
[d] *The acceleration of gravity at the earth's surface is 9.8 m/s^2.*
[e] *Speed needed for permanent escape from the planet's gravitational field.*
[f] *Venus rotates in the opposite direction from the other planets.*
[g] *The axis of rotation of Uranus is only 8° from the plane of its orbit.*
[h] *The difference between the minimum and maximum distances from the sun divided by the average distance.*
[i] *The ecliptic is the plane of the earth's orbit.*
[j] *Probably more small ones around Jupiter, Saturn, and Uranus.*

Figure 17-4 The comet Hale-Bopp, one of the largest ever to come near the earth, was a spectacular sight in the night sky for the first few months of 1997. Comets are named after their discoverers, here the professional astronomer Alan Hale and the amateur Thomas Bopp, who independently noticed a faintly glowing speck moving across the fields of their telescopes on the same night in July 1995. (About half a dozen comets are spotted every year by amateur astronomers.) The comet was then farther away than Jupiter. When it was closest to the earth, Hale-Bopp's tail stretched across 20° of the sky; for comparison, the apparent diameter of the full moon is 0.5°.

smaller amounts of various frozen gases are present, for instance carbon monoxide, carbon dioxide, methane, ammonia, and hydrogen cyanide.

A number of spacecraft have been sent out to intercept comets and find out more about them. On July 4, 2005, one of these spacecraft, NASA's Deep Impact, fired a 370-kg impactor into the path of the comet Tempel 1. The resulting collision produced a giant plume of debris and left a crater behind (Fig. 17-5). Analyzing the wealth of images and spectroscopic data sent back by the spacecraft as it flew past and by other spacecraft, as well as recorded by instruments on the earth, has provided detailed information on the comet's composition and structure. Apparently the comet is loosely held together by gravity rather than packed hard, and contains clay and carbonates, substances that need liquid water to form. But comets like Tempel 1 are thought to have originated from smaller bits of matter past Neptune's orbit where any water would be frozen. Unexpected findings like this are the rewards of space exploration, and solving such mysteries will add to our understanding of the universe we live in.

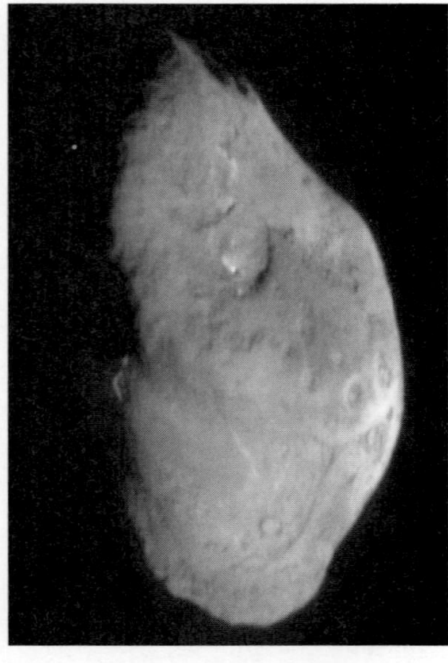

Figure 17-5 The NASA Deep Impact probe collided with the comet Tempel 1 in 2005 and left the crater just above the center of this photograph. The crater is 20 m deep.

In 2004 another NASA spacecraft, Stardust, passed close enough to the comet Wild 2 to take samples of its dust. The samples, parachuted down near Salt Lake City in 2006, contained minerals such as olivine (Fig. 15-9) that form only at high temperatures. But Wild 2, like Tempel 1, came from the Kuiper Belt beyond Neptune, where it is bitterly cold. This suggests that at least some of the material in both comets came into being closer to the sun before somehow ending up far out in the Kuiper Belt. In addition, organic compounds of various kinds were found, among them two that contain biologically usable nitrogen, which supports the idea (mentioned in Sec. 13.17) that the comets that rained down on the young earth may have had a role in the origin of life.

Comet Orbits The paths followed by comets that periodically pass near the earth are quite different from the nearly circular planetary orbits. A typical comet approaches the sun from far out in space beyond Pluto, swings around the sun, and then retreats. The orbit is a long, narrow ellipse, and the comet returns at regular intervals. Although most orbits are so large that their periods range up to millions of years, a few are smaller. Halley's comet, for instance, reappears every 76 years and has returned 28 times since the first sure record of its observation was made in 239 B.C. It was named after the English astronomer Edmund Halley, a contemporary of Newton, who in 1705 predicted that a comet last seen in 1682 would reappear in 1758, as it did. Halley's comet came within 93 million km of the earth in November 1985 on its most recent visit, when it was studied at close range by various spacecraft. The comet's solid nucleus was about 16 km long and 8 km across, with a mass of about 100 billion tons.

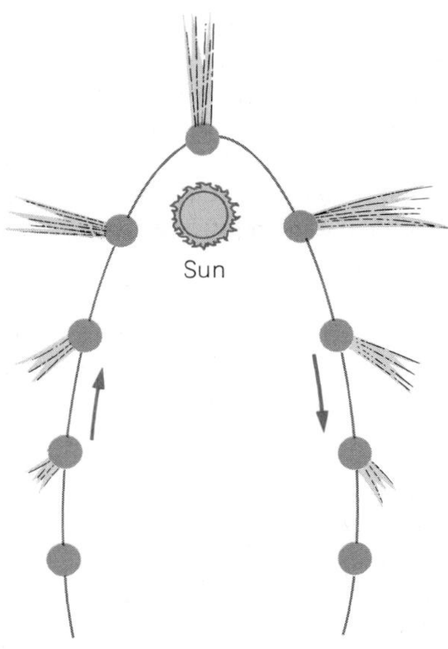

Figure 17-6 The tail of a comet always points away from the sun because of pressure from the sun's radiation and from the solar wind of ions. The tail is longest near the sun, when it is millions of km long, and seems to be absent far away from the sun.

Heads and Tails In the far reaches of the solar system comets are fairly small, only a few km across. Near the sun the frozen gases vaporize and take with them dust to form the huge but thin clouds that make up the comet heads and tails we see. Comets are visible only when close to the sun. This is partly because of sunlight scattered by cometary material but mainly because the gases are excited to luminescence by solar radiation.

Comets appear as small, hazy patches of light, often accompanied by long, filmy tails—hence their name, which comes from the Greek word for "hairy." Most comets are visible only with the help of a telescope, but from time to time one becomes conspicuous enough to be seen with the naked eye. Watched for a few weeks or months, a comet at first grows larger and its tail longer and more brilliant. Then it fades gradually, loses its tail, and eventually disappears.

A comet's tail always points away from the sun regardless of which way the comet is heading (Fig. 17-6). One reason is the **solar wind**, the stream of ions that constantly flows outward from the sun, which tends to sweep the comet's gases with it. Another reason is the pressure of solar radiation on the tail's dust particles.

Comet tails may stretch for millions of km across the sky. Near the sun, a comet loses material to space, and so is fainter each time it returns. Far from the sun, it contracts again into a relatively small body.

17.3 Meteors

"Shooting Stars" Usually Smaller Than a Grain of Sand

Meteoroids are small fragments of matter that the earth meets as it travels through space. Most meteoroids are smaller than grains of sand. The majority are believed to be the result of asteroid collisions; others are the debris of comets.

Moving swiftly through the atmosphere, meteoroids are heated rapidly by friction. Usually they burn up completely about 100 km above the earth, appearing as bright streaks in the sky—**meteors**, or "shooting stars" (Fig. 17-7). Sometimes, though, they are so large to begin with that a substantial portion may get through the atmosphere to the earth's surface (Fig. 17-8). The largest known fallen meteoroids, called **meteorites**, weigh many tons. The smallest meteoroids are so light that they float through

Figure 17-7 Time exposure showing a meteor streak in the night sky. Few meteoroids are larger than a grain of sand, and those that enter the earth's atmosphere usually burn up before they can reach the ground. On a clear dark night, 5 to 10 meteors per hour can be seen.

Figure 17-8 The Barringer Meteor Crater near Winslow, Arizona, 1200 m across and 175 m deep, was formed about 50,000 years ago. Its origin was ascribed to a meteoroid impact by the mining engineer D. M. Barringer in 1906. The meteoroid may have been 50 m across, weighed several hundred thousand tons, and traveled at over 40,000 km/h. It was completely destroyed upon impact.

the atmosphere without burning up. Many tons of these fine, dustlike micrometeoroids reach our planet daily.

A keen observer on an average clear night can spot 5 to 10 meteors an hour. Most of these meteors are random in occurrence and follow no particular pattern either in time or place in the sky. At several specific times of year, however, great meteor showers occur, with 50 or more meteors sometimes visible per hour that appear to come from the same part of the sky. The showers occur when the earth moves through a swarm of meteoroids that follow the same orbit about the sun.

When the meteoroids of a shower are spread out along their common orbit, the number of meteors seen is about the same each year. This is the case with the Perseid showers of mid-August. When the meteoroids are bunched together, the number of meteors seen varies from year to year. The Leonid showers of mid-November are an example, with intense displays every 33 years. Other conspicuous meteor showers are listed in Table 17-2. The Eta Aquarid and Orionid showers are due to dust particles from Halley's comet that are spread out in its orbit. The earth crosses this orbit twice a year with the showers as a result.

Meteoroid speeds range up to 72 km/s. This limit is significant, because a higher speed would imply an object arriving from outside the solar system. The conclusion is that all meteoroids, random as well as shower, are members of the solar system that follow regular orbits around the sun until they collide with the earth or another planet.

Table 17-2 Some Major Meteor Showers

Shower	Maximum Display	Hourly Rate
Quadrantid	Jan. 3	80
Eta Aquarid	May 4	60
Delta Aquarid	July 29	30
Perseid	Aug. 12	100
Orionid	Oct. 21	30
Leonid	Nov. 16	20
Geminid	Dec. 13	90

Meteorites Most meteorites that have been examined fall into two classes: stony meteorites, the great majority, which consist of silicate minerals much like those in common rocks of the earth's crust, and iron meteorites, which, like the earth's core, consist largely of iron with a small percentage of nickel (Fig. 17-9). A few meteorites are intermediate in character. All meteorites are sufficiently different from terrestrial rocks to be recognized as such, notably by smooth surface crusts, usually black, that formed when their surfaces melted on entry into the atmosphere and later hardened.

The best place to seek meteorites is the Antarctic, where conditions for preservation are ideal and on whose ice sheets scattered rocks are conspicuous in various regions. Most such rocks came from nearby mountains, but quite a few fell from the sky. Almost 20,000 meteorites have been collected thus far in the Antarctic in the last two decades, far more than have been found in the rest of the world in all the past. Stony meteorites classed as chondrites are the most common, and are not hard

Figure 17-9 Discovered in Greenland in 1894, this iron meteorite is 3.3 m long. The largest known meteorite has a mass of 55,000 kg and was found in Namibia. About 4 percent of meteorites are iron, 94 percent are stony, and the rest are stony-iron.

Visitors from the Moon and Mars

Not everything that falls on the earth is a meteoroid from space. In 1982 an unusual greenish brown rock about the size of a golf ball was found in Antarctica. The composition of this rock turned out to be identical with that of rocks brought back from the moon by Apollo 15 astronauts in 1971. A number of other similar moon rocks were found later, in Australia as well as in Antarctica. It seems possible that the impact of large meteoroids on the moon could have hurled the rocks out fast enough to escape the moon's gravitational pull.

About 30 other rocks found in Antarctica, India, Egypt, and France show signs of having come from Mars—the gas trapped in them has the same composition as the Martian atmosphere, for instance (Fig. 17-10). One such rock, the size and shape of a potato, seems to have been chipped off Mars and flung into space 15 million years ago by an asteroid collision. The rock then drifted through the solar system until it passed near enough to the earth about 13,000 years ago for the earth's gravity to pull it in. This much is well established; the time spans come from measurements of nuclides formed in the rock by cosmic ray bombardment in space and on the earth's surface. (Cosmic rays are the subject of Sec. 19.5.) Less certain are what seem to be traces of past life in the rock but which may have alternative explanations as well. Despite much study the origin of these traces remains uncertain.

Another meteorite thought to have come from Mars is called Nakhla. Apparently it hardened from lava about 1.3 billion years ago and 200 million years later was altered by contact with liquid water. An asteroid collision 11 million years ago blasted Nakhla into space and it wandered about the solar system until it fell in Egypt in 1911. Recent examination of Nakhla has turned up microscopic tunnels that resemble those made in terrestrial volcanic rocks by bacteria munching their way inside to obtain nutrients. Matter that contains carbon whose isotopic composition corresponds to a Martian origin was found in the tunnels, which, together with the evidence for contact with liquid water, adds up to a suspicion—no more yet—of at least former life on Mars.

Since rocks ejected from the moon and Mars are found on the earth, perhaps rocks from the earth now lie on the surfaces of the moon and Mars. Those reaching Mars long ago would have been slowed down by its dense early atmosphere and got through more or less intact. Finding ancient earth rocks on Mars would help in figuring out how the earth's crust developed and might also provide clues on the origin of earthly life, because the abundance ratios of the isotopes of some elements are affected by life processes.

Indeed, it is not at all impossible that microbes from the earth, protected from solar ultraviolet radiation by being embedded in a lump of ejected matter, could have ended up alive on Mars. After all, live bacteria have been found on the earth inside solid rock. If life is ever found to exist, or have existed, on Mars, one of the first questions is whether it is related to life on earth.

Figure 17-10 This meteorite, found in Antarctica in 1979, is one of a dozen or so believed to have come from Mars. Unlike other meteorites, which solidified about 4.5 billion years ago, these solidified much later, only 1.3 billion years ago, when Mars was still volcanically active. The meteorite here has a mass of 8 kg and is 15 cm long.

to identify because they are studded with tiny white mineral spheres ("chondrules"). The chrondules solidified from once-liquid rock 4.6 billion years ago when the solar system came into being.

The earth's neighbors in space—the moon, Mercury, Venus, Mars, and the satellites of Mars—are all deeply pitted with many huge meteor craters. Erosion on the earth has left traces of only 100 or so large craters, but there is no reason to think that a similar rain of giant meteoroids many km across did not fall long ago on our planet as well. By about 3 billion years ago the sizes and rate of arrival of the bombarding objects seem to have fallen to their present levels. This suggests that the planets and their satellites had by then finished sweeping up most of the larger debris left over after the solar system was formed.

The Inner Planets

17.4 Mercury

It Always Appears near the Sun

Mercury was named after the fleet-footed messenger of the gods in classical mythology because its position relative to the stars changes rapidly. Its symbol represents a winged helmet.

The innermost planet, Mercury always appears in the sky as a companion to the sun. Although as brilliant an object as Sirius, the brightest star, it is hard to see during the day because it is so near the sun; the best times for observation are near sunrise and sunset. To an observer on the earth, Mercury shows phases like those of the moon and Venus because its orbit lies within that of the earth. Mercury's apparent diameter is about 3 times greater when it is closest to the earth than when it is farthest away.

Mercury takes 59 of our days to make a complete turn on its axis and 88 of our days to circle the sun. As a result, a day on Mercury is 176 of our days.

The long days on Mercury and its closeness to the sun lead to high temperatures on the sunlit side, as much as 450°C (more than enough to melt lead). There is almost no atmosphere to transfer heat from the sunlit side to the dark side or to trap heat radiated by the surface. As a result the night temperature drops to less than −200°C just before sunrise. The only gases found near Mercury are the inert gases helium, argon, neon, and xenon, and only traces of them. Mercury is altogether an inhospitable place.

In 1974 the Mariner 10 spacecraft passed within a few hundred kilometers of Mercury and radioed back photographs (Fig. 17-11) as well as data of various kinds. Mercury turns out to have a surface pocked by meteoroid craters, much like the surface of the moon. Hills and valleys as well as craters abound in the rugged landscape. The youthful earth must also have suffered its share of assault by large meteoroids. The resulting craters on the earth, however, were later erased by the melting of the crust and by the erosive processes that continue to this day.

There is no evidence that Mercury ever melted, which presents a problem. Mercury seems to have a crust of silicate rocks whose density is much lower than the high average density of the planet; Mercury is the densest planet. Hence the interior probably has an iron-rich core like that of the earth but considerably larger in proportion to Mercury's size. Furthermore, Mercury has a magnetic field, which suggests that at least part of its interior is even now liquid. But if Mercury never completely melted after it had been formed, how did its heavy and light constituents separate, and why is the core molten today? Another puzzle.

Messenger

Mariner 10 has thus far been the only spacecraft to visit Mercury, about which we know less than any other planet. In 2004 a spacecraft appropriately called Messenger was sent to Mercury, which it will reach in March 2011 after a roundabout voyage to prevent the sun's gravity from accelerating it to too great a speed to permit its injection into an orbit around Mercury. The images radioed back by Mariner 10 only covered 45 percent of Mercury's surface at a resolution of about 1 km. Messenger will be able to return images of Mercury's entire surface at much higher resolution as well as make various measurements, which will vastly increase our knowledge of the planet and its history. Messenger will orbit Mercury for an earth year and then will crash into the planet.

What should the many geological features that Messenger will discover be called? The International Astronomical Union has prescribed that notable writers, musicians, and artists should be so honored; three prominent craters on Mercury are already named Shakespeare, Mozart, and Monet.

Figure 17-11 A mosaic of false color photographs of Mercury radioed back from the Mariner 10 spacecraft. The surface is heavily cratered as a result of meteoroid bombardment, probably several billion years ago when similar craters were formed on the moon.

Figure 17-12 The cloudy atmosphere of Venus as photographed by the Mariner 10 spacecraft on its way to Mercury. Like Mercury, Venus has no satellites.

17.5 Venus

Our Sister Planet

Venus is the brightest object in the sky apart from the sun and moon (Fig. 17-12). If we know where to look, we can sometimes see Venus during the day. Because its orbit, like that of Mercury, lies inside the earth's orbit, Venus never gets very far away in the sky from the sun and appears alternately as a "morning star" and an "evening star" (Fig. 17-13). Venus is usually farther from the sun than Mercury, however, and so is visible

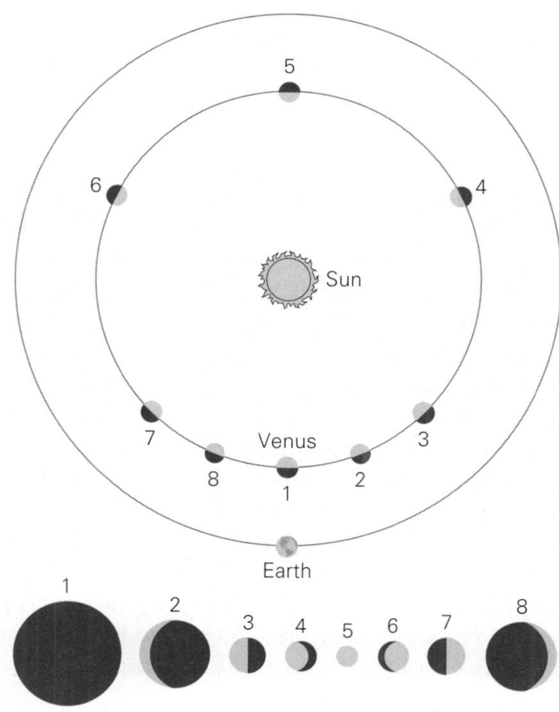

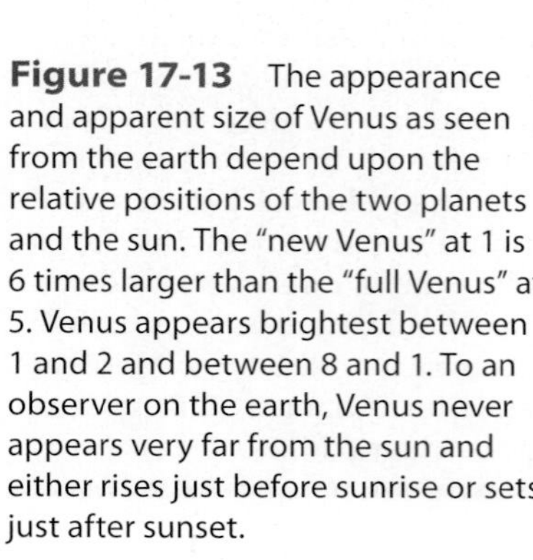

Figure 17-13 The appearance and apparent size of Venus as seen from the earth depend upon the relative positions of the two planets and the sun. The "new Venus" at 1 is 6 times larger than the "full Venus" at 5. Venus appears brightest between 1 and 2 and between 8 and 1. To an observer on the earth, Venus never appears very far from the sun and either rises just before sunrise or sets just after sunset.

The Phases of Venus

Galileo, the first person to study Venus with a telescope, found that it has phases like those of the moon and that its apparent size changes cyclically. He correctly interpreted this evidence as supporting the copernican thesis that Venus revolves around the sun. A good pair of binoculars can reveal the phases of Venus.

Figure 17-13 shows how these effects arise. A year on Venus is only 225 earth days long, so the positions of Venus and the earth relative to each other and to the sun change continually. When Venus is between the sun and the earth, as at 1, it appears dark to an observer on earth. If it could be seen, it would have its largest apparent diameter. At 2 and 8 Venus appears as a crescent, but because it is still fairly close to the earth, it is quite bright. At 3 and 7 we see half of Venus illuminated, and at 4 and 6 more than half. Although the sun gets in the way of our seeing the full Venus at 5, it is so far from the earth that it would not be especially brilliant even if it were visible.

for longer periods than Mercury. Venus was named after the Roman goddess of love and beauty and is represented by the traditional symbol of a mirror.

Venus has the distinction of spinning "backward" on its axis. That is, looking downward on its north pole, Venus rotates clockwise, whereas the earth and the other planets rotate counterclockwise. As a result the sun rises in the west on Venus, not in the east as on the earth. The rotation of our twisted sister is also extremely slow, so that a day on that planet represents 243 of our days.

In size and mass Venus is more like the earth than any other member of the sun's family. Mountains, craters, and fault-like cracks mar its surface (Fig. 17-14), which has two major highland regions, the larger with about the area of the United States. These "continents" cover only about 5 percent of Venus in contrast to the 30 percent extent of the earth's continents, and no water laps their edges. Some of the mountains of Venus are quite high, one of them topping Everest (Fig. 17-15). Evidence of volcanism is abundant, including volcanic peaks, extensive lava flows, and a huge volcanic crater about 100 km across. There are about 1600 large volcanoes, perhaps a million smaller ones. The amount of sulfur dioxide

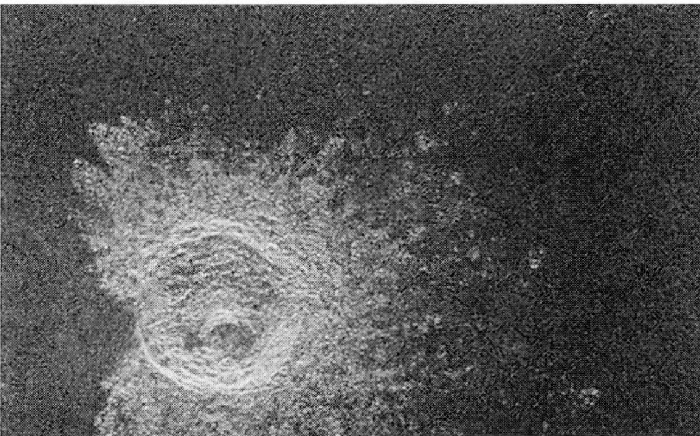

Figure 17-14 In 1990 the United States space probe Magellan began to circle Venus and to radio back radar images of its rugged surface. Features as small as 120 m across could be detected. The picture at left shows a ridge and valley network formed by intersecting faults that resembles the landscape of a region in the western United States. The picture at right shows an impact crater 12.5 km in diameter probably formed by a meteoroid moving at a shallow angle with the surface when it hit. The central peak in the crater consists of material that rebounded after the impact. Volcanic mountains and rivers and plains of hardened lava were also found. Some volcanic activity may continue today.

Figure 17-15 Radar images from the Magellan spacecraft were combined to give this false color perspective view of Maat Mons, at 8 km high the second tallest mountain on Venus. Lava flows extend for hundreds of kilometers across the fractured plains in the foreground. The mountains of Venus reflect radar signals much more strongly than ordinary rocks do. Perhaps metallic compounds emitted by its volcanoes were deposited on the mountains and turned into metal coatings, but nobody knows for sure.

in the Venusian atmosphere varies from time to time, which suggests that a few of the volcanoes are still active since this gas is belched out by volcanoes on the earth.

There is no evidence that the crust of Venus, like that of the earth, consists of huge shifting plates, which is odd since the two planets are otherwise so similar. Perhaps the absence of water on Venus is responsible: it is thought that water in the rock of the earth's plates might help the plates to bend and thus be tugged across the surface by the weight of their sinking edges (Sec. 16.7). Another mystery is why Venus, which is thought to have a molten iron core like that of the earth, has no magnetic field.

Meteoroids less than a kilometer or so across cannot make it through the dense atmosphere of Venus without burning up, but larger ones can and should have left far more impact craters on Venus's surface than the mere 968 that spacecraft have found. Without liquid water on Venus (it is too hot) there is no erosion to wear away craters, and the observed craters show little alteration by volcanic or tectonic activity. Something happened a half-billion years ago to blot out the impact craters that then existed, but what? The best guess is that immense lava flows at that time paved over Venus's surface one or more kilometers deep. Could such a lava inundation flood the earth in the future? Much of the earth's internal heat, which is what drives geologic processes, is spent in shaping and reshaping its lithosphere. This leaves a smaller proportion of the available energy for vulcanism, unlike the case of Venus where there is no plate activity. One less worry.

Life on Venus Is there life on Venus? Bernard de Fontanelle thought so when he wrote in 1686 that the inhabitants of Venus must "resemble the Moors of Granada; a small, dark people, burned by the sun, full of wit and fire, always in love, writing verse, fond of music, arranging festivals, dances, and tournaments every day."

Alas, the truth is less romantic. American and European spacecraft have passed close to Venus and Soviet ones have parachuted instruments to its surface. The information radioed back shows that the atmosphere of Venus is almost all carbon dioxide with a little nitrogen and traces of other gases. Lightning occurs there fairly often. The thick lemon-yellow clouds that permanently shroud Venus consist mainly of sulfuric acid droplets and are driven around the planet by strong east-to-west winds. Tornadoes whirl near its poles. At the surface, the pressure of the atmosphere is about 90 times the corresponding figure for the earth. On the earth the small amount of carbon dioxide in the atmosphere absorbs much of the radiation from the ground—the greenhouse effect. Venus, blanketed more effectively by far than the earth, has a surface temperature that averages 470°C, enough to melt lead (as in the case of Mercury's daylit surface). The existence of life on Venus seems impossible.

The surface of Venus is actually hot enough to glow a dull red, which can be seen at times on its night side through its clouds with even a small telescope. This glow was first observed by an Italian monk in 1643 when it was attributed to bonfires used to illuminate cities at night, or alternatively to firework displays set off by joyful inhabitants.

17.6 Mars

Small and Cold with a Varied Landscape

When it is at its closest to us, Mars is second only to Venus in brightness. Reddish orange in color, it has always been associated with violence and disaster (Fig. 17-16). The Romans named it after their god of war. The discoverer of its two satellites, the American astronomer Asaph Hall, continued the tradition by calling them Phobos (fear) and Deimos

The Moons of Mars

In what must be among the most bizarre coincidences in astronomy, the existence of two Martian satellites was conjectured in 1600 by Kepler and elaborated upon by Jonathan Swift in 1726 in *Gulliver's Travels*. Neither of them seems to have had the slightest justification for his belief. In the Voyage to Laputa, Swift has Gulliver learn not only that such satellites existed but that their periods of revolution were 10 h and $20\frac{1}{2}$ h, not far from their actual periods.

Figure 17-16 Mars at the beginning of summer in its northern hemisphere as seen from the Hubble Space Telescope. Iron oxides in the Martian soil are responsible for its red color. The ice cap around the north pole of Mars consists of water ice; the ice cap around its south pole consists of water ice and frozen carbon dioxide. Wispy white clouds of water ice cover about 1 percent of Mars on the average. Violent dust storms with winds of up to 150 km/h sometimes lash the entire planet and may last for months.

Figure 17-17 The two satellites of Mars are believed to be captured asteroids. Conceivably Phobos and Deimos long ago passed close enough to Mars to be slowed down by its atmosphere, which was denser in the past than it is today, and this allowed them to be trapped in orbits by Martian gravity. Phobos has a huge crater called Stickney visible at the left side of this photograph. The impact that created Stickney must have come close to shattering Phobos.

(terror) after the two sons of the Greek war god Ares. The symbol of Mars is a circle with an arrow, which is also the conventional male symbol.

The satellites of Mars were not discovered until 1877 because they are such tiny objects, far too insignificant to deserve the names they were given. Phobos, the inner one, is only 20 km high and 28 km across. Deimos, the outer one, is even smaller, 12 km high and 16 km across (Fig. 17-17). Shapes like these are possible only for astronomical bodies so small that the rigidity of their material is able to withstand the tendency of gravity to impose a spherical form, as discussed in Sec. 1.9.

Phobos is so close to Mars that its orbital period of 7 h 39 min is shorter than the length of the Martian day, and as a result it rises in the west and sets in the east, speeding across the Martian sky three times each day. The period of Deimos is 30 h 18 min, and it passes from east to west just as the earth's moon does. It seems likely that both satellites were originally asteroids that became trapped in orbits around Mars by its gravitational pull.

The diameter of Mars is slightly over half that of the earth, and its mass is 11 percent of the earth's mass. As a result the surface gravity on Mars is 38 percent of the earth's. An astronaut weighing 150 lb on the earth would weigh 57 lb on Mars. Although the overall density of Mars is 3.97 g/cm^3 as compared with 5.52 g/cm^3 for the earth, the difference is misleading because materials in the interior of the less massive Mars are not compressed as much as those in the earth's interior. Figured on a comparable basis, the two planets have about the same density, which suggests that Mars probably has about the same composition as the earth. Also like the earth, Mars seems to have a liquid core.

Example 17.1

If you were living on Mars, where in the sky would you expect to see the earth? When would be the best time to look?

Solution

Because the earth's orbit is inside the orbit of Mars, the earth would always appear near the sun in the sky, just as in the case of Venus as seen from the earth (see Fig. 17-13). The earth would best be seen just before sunrise or just after sunset, depending on the relative positions of Mars and the earth in their orbits. During the day the sky would be too bright to see the earth from Mars.

Geological Activity Mars today lacks the internal heat to power processes like those that shift and renew the earth's crustal plates or to drive a dynamo in its core to produce a magnetic field. But that may not have always been true: the Mars Global Surveyor, orbiting the planet, found a region where the Martian crust is magnetized in strips of alternating polarity. This also occurs in the earth's oceanic crust where molten rock comes to the surface, spreads out, and hardens while the geomagnetic field reverses itself periodically, thus locking the back-and-forth magnetic directions into the rock. The magnetized strips on Mars are about 200 km wide and up to 2000 km long on its southern highlands.

The first conclusion from the new findings is that, although Mars no longer has a magnetic field, it must once have had a strong one

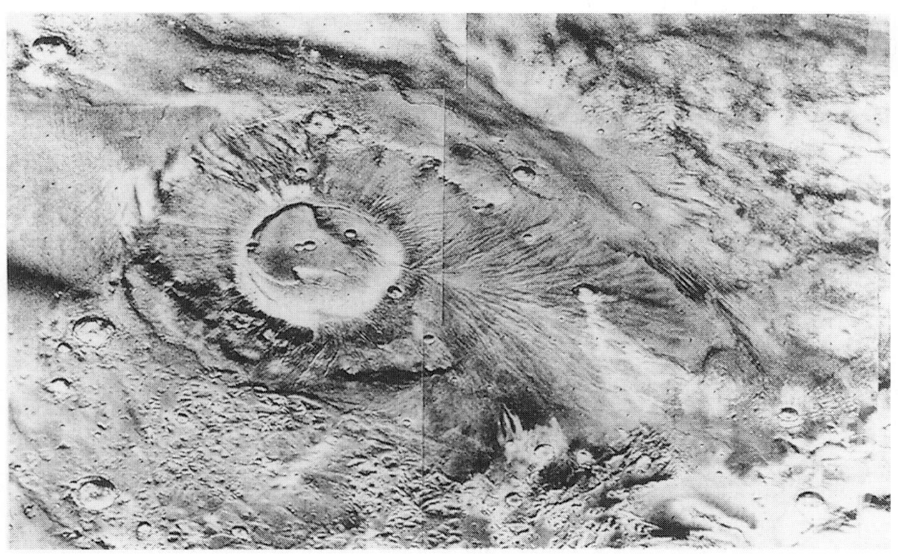

Figure 17-18 The crater of the Martian volcano Apollinaris Patera is 100 km across and has been scarred by running water. The lava flows that extend about 200 km southeast of the crater seem to have originated in a fissure. Although all are inactive today, some Martian volcanoes may just be dormant and may erupt again in the future. Another Martian volcano, Olympus Mons, is the largest in the solar system.

originating in a liquid iron core that reversed itself a number of times. Second, although the Martian surface today consists of an unbroken shell of rock, large-scale tectonic movements must have taken place there in the past. During Mars's first half-billion years, its interior evidently was sufficiently hot to support both a magnetic dynamo and plate-tectonic surface movement. Then, 4 billion years ago, the dynamo seems to have faded away, and not long afterward new crust stopped being formed.

Mars has nevertheless not been geologically dead since those early events. Its surface sports a host of intriguing features, some apparently of recent origin. The Martian landscape is extremely varied, with vast plains pitted with impact craters, regions broken up into irregular short ridges and depressions, a deep canyon as long as the distance from New York City to Los Angeles, deserts of windblown sand, and extinct volcanoes, some not very old and one of them three times the height of Everest (Fig. 17-18).

Water A lot of evidence points to running water in the past. Spacecraft (of the 38 missions sent thus far, 18 were successful) have sent back pictures of what seem to be dried-up river channels and drainage basins, dried-up lake beds, and structures that look like sedimentary deposits (Fig. 17-19). Seas and oceans were quite possibly parts of the Martian landscape. Eventually the water must have stopped flowing, though, because there are undisturbed meteoroid craters in areas that show signs of earlier carving by running water. The number of these craters suggests that no major erosion has taken place on Mars for a long time, probably billions of years.

Mars should have been cold and dry in the past, as it is now, and it has been a mystery how warm and wet periods occurred in its history. The discovery that the Martian river valleys and the craters made by giant asteroid impacts are about the same age suggests what may have happened. The kinetic energy of an asteroid of rock and ice 100 to 200 km across that smashed into Mars would have heated the planet's surface, and the asteroid's ice content would have become scalding rain that fell

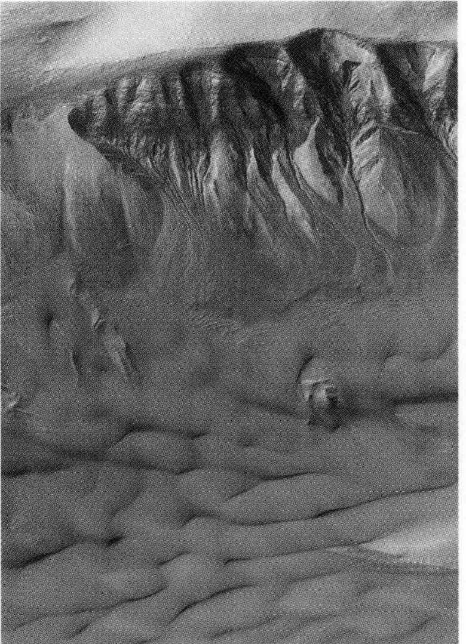

Figure 17-19 Gullies on the wall of a crater on Mars look just like gullies carved by streams of water on the earth.

Spirit and Opportunity

In 1997 the Pathfinder spacecraft landed on Mars and released the small six-wheeled Sojourner rover to survey the nearby Martian landscape. Much was learned, but the question of whether enough liquid water to support life had once existed on the planet remained without a definite answer.

Early in 2004 two larger rovers (Fig. 17-20) reached Mars to continue the search for traces of past water. Each rover had several cameras and a robot arm with an abrasion tool for boring into rock together with a microscope and devices for analyzing the composition of minerals that were encountered. The rovers could cover about 100 m per day and were designed to function for at least 3 months; in fact, they were still operating 3 years later. When the capsules that contained the rovers were separated from their spacecraft mother ships, their falls were at first slowed by drag in the Martian atmosphere. Next parachutes opened to reduce their speeds further, then rockets were fired to stop their descents just above the ground, and finally airbags cushioned their actual landings.

The first rover to arrive, Spirit, alit on the Connecticut-sized impact crater Gusev whose smooth floor suggests the sediment-covered bed of an ancient lake (Fig. 17-21). Gusev lies at the end of a 900-km valley that may have been carved by a river bringing water to the crater. The second rover, Opportunity, landed on Meridiani Planum, almost halfway around Mars from Gusev (Fig. 17-22). This is a plain the size of Oklahoma with deposits of gray hematite, an iron oxide usually formed by precipitation from water, which implies that water also was present here in the past. Because alternate explanations for the features of both Gusev and Meridiani Planum were conceivable, only actual visits could lead to definite conclusions, which is why these missions were launched.

In its travels, Opportunity found evidence almost everywhere it looked that indeed points to a former abundance of liquid water. For instance, a rock outcrop near the landing site seems to consist of fine layers of sedimentary rocks that apparently were disturbed by fluid flows in a characteristic way while hardening. Signs of the presence of sulfate, chloride, and bromide minerals were also found. Such minerals need the presence of water to form, either by evaporation from solution or by the alteration of existing minerals. An unexpected discovery was a scattering of small, round pebbles, which are thought to be iron-rich hematite grains that grew from water solution. The Meridiani site seems to be on the edge of what had been a vast salt sea.

At first, Spirit found nothing to suggest a moist past, but in June, a few km from its landing place, it, too, encountered rocks whose forms, mineralogy, and chemistry showed the influence of water. Later explorations turned up other water-affected rocks and minerals. All in all, the mission scientists are confident that Mars long ago had an environment of long duration that was warm and wet enough to have supported life of some kind. Whether it actually did, and if so what kind of life, remains for future visits of both orbiters and landers to establish.

for many years, caused flash floods, and fed the streams that carved the valleys seen today. Frozen groundwater would also have melted and come to the surface. After some thousands of years Mars would have cooled down and the water turned to ice although underground water could have remained liquid for much longer. There seem to have been 25 such impacts 3.5 billion years ago, 10 to 20 million years apart. Today there are enormous deposits of water ice in both polar regions of Mars.

17.7 Is There Life on Mars?

Maybe, But Not Likely

Mars rotates on its axis in a little over 24 h. Its revolution about the sun requires nearly 2 years, and its axis is inclined to the plane of its orbit at nearly the same angle as the earth's. Thus the Martian day and night have about the same lengths as ours, and the Martian seasons are 6 months long and at least as pronounced as ours. Mars has long fascinated astronomers and laypeople alike because it is in so many ways similar to the earth, which leads to the question of whether life of some kind is present there.

Although the conditions on Mars were once as hospitable to life as those on the earth, if life exists today it is adapted to an environment

Spirit and Opportunity (continued)

Figure 17-20 The rover Opportunity being tested before starting its 6-month voyage to Mars in July 2003 to search for evidence of surface water in the past. Its identical twin Spirit left a week earlier to study a different part of the planet.

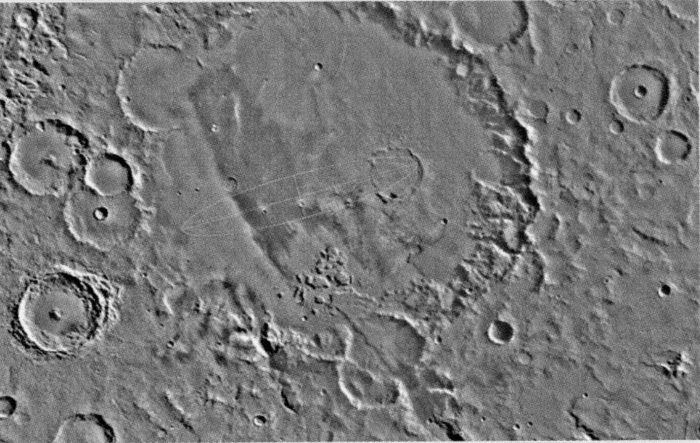

Figure 17-21 The bottom of the crater Gusev, the landing site of the Mars rover Spirit, may be a dried-up lake bed. The valley at lower right may have been cut by water flowing into the crater.

Figure 17-22 The rover Opportunity came across this 8-m-wide impact crater on its travels across the Martian landscape.

that would soon destroy most earthly organisms. For a start, Martian climates are severe by our standards. Over half again farther from the sun than the earth, Mars receives much less solar energy per m^2 than we do. Its atmosphere, largely carbon dioxide, is extremely thin—equivalent to the earth's atmosphere at an altitude of nearly 40 km—so little of the heat from the sun remains after nightfall. Daytime temperatures in summer at the equator rise to over 0°C, but at night drop to a chilly −23°C. The average surface temperature of the entire planet is about −55°C. The scanty Martian atmosphere is also unable to screen out harmful solar ultraviolet radiation, a function carried out in the earth's atmosphere by the ozone present at high altitudes (see Sec. 14.1). At times giant dust storms blow for months across the entire planet.

What about water, so essential for life? Nonliquid water is certainly present on Mars, a little as traces of water vapor in its atmosphere and much more frozen (along with carbon dioxide) in the ice caps of its polar regions and mixed with subsurface soil, perhaps like the permafrost of the earth's Arctic regions. As we saw, plenty of liquid water apparently washed across Mars long ago, but then disappeared. Until very recently, that was that: next to no liquid Martian water to sustain life.

Example 17.2

Why do temperatures on the surface of Mars vary more between day and night than do those on the earth's surface?

Solution

Because the atmosphere of Mars is much less dense than that of the earth, a greater proportion of the incoming solar radiation reaches the Martian surface during the day. In addition, the thinner atmosphere is less able to prevent infrared radiation from the Martian surface from escaping into space at night. As a result daytime temperatures are higher and nighttime temperatures are lower than they would be if Mars had an atmosphere like that of the earth.

Then high-resolution images from the Mars Global Surveyor revealed thousands of gullies whose form was exactly that of gullies cut by water gushing out of hillsides on earth. A typical Martian gully is 2 m deep, a few hundred meters long, and has a collapsed area on top where the water emerged with an apron of debris below. The most surprising aspect of these gullies is that they appear to be fresh, geologically speaking, having come into being "perhaps a million years ago, perhaps 10 thousand, perhaps yesterday," according to one of the scientists involved. The water would have soon evaporated or frozen, but not before creating the gullies. Some hypotheses of where the water came from start with deposits of underground ice and then propose ways it could melt and spill out from time to time, but perhaps melting snow was the source.

Evidence that some water flows still occur on Mars has been found by comparing photographs of Martian gullies taken in 1999 and 2000 with those taken of the same gullies in 2004 and 2005. The later pictures show what seem to be fresh mineral deposits on the sides of two gullies that were absent earlier. The light-colored deposits are thought not to be traces of dry dust that slid down the gullies because such dust elsewhere on Mars is always darker than its surroundings.

The fact that most terrestrial life requires a regular supply of liquid water and oxygen plus protection from solar ultraviolet radiation does not necessarily mean that life of some kind could not exist in their absence. The life processes of certain bacteria on the earth do not require oxygen, so an oxygen-containing atmosphere is not indispensable, at least for primitive forms of life. Conceivably there could have evolved on Mars organisms that can thrive on traces of water gleaned from the minerals in surface rocks. And shells of some sort might protect Martian life from ultraviolet radiation. Or life there could exist underground, its energy source heat from the interior rather than sunlight, as in some places on the earth. And extra heat may not be needed: bacteria have been found deep in the ice sheets of Greenland and Antarctica that seem to have survived for hundreds of thousands of years at temperatures as low as $-40°C$. One species remained alive at $-196°C$ when immersed in liquid nitrogen. Thin films of water often cling to the surfaces of mineral grains frozen inside ice deposits, and they could take care of the water needs of bacteria inside Mars.

Since conditions there long ago may have been comparable for a long period to those on the earth, life of some kind could have come into being on Mars. The loss of most of the carbon dioxide of its atmosphere,

Next Stop, Mars

Astronauts have already set foot on the moon and Mars is next, if all goes well, between 2030 and 2040. Before then much more will have been learned about the Red Planet.

Following the visit of Spirit and Opportunity, the NASA Mars Reconnaissance Orbiter began in 2006 to search for subsurface water, both liquid and frozen, together with studying surface minerals and atmospheric properties and behavior. An onboard camera can pick up details as small as a basketball on sites that might serve for future landings. The orbiter will also be able to relay signals to and from landers for the next 10 to 15 years.

In 2008 NASA's Phoenix lander descended in the far north of Mars to look for water and ice and for organic compounds characteristic of life in the terrain there. Phoenix also has microscopes to inspect soil samples for living organisms or their fossils. Phoenix is a fixed lander, but NASA's Mars Science Laboratory, planned for a 2011 launch and arrival a year later, will bring a rover much larger and more sophisticated than Spirit and Opportunity: a heavily laden SUV instead of golf carts. It will be able to travel at up to 90 m/h and will be powered by heat produced in the decay of plutonium (Sec. 8.11); it should operate for at least a full Martian year (1.9 earth years). Some of the devices for rock and soil analysis and for atmospheric studies will come from other countries, among them Russia, Spain, and Canada. In 2011 the European Space Agency (ESA) plans to send a rover about the size of Spirit and Opportunity but with more elaborate instruments focused on signs of life.

Still further ventures to Mars are between gleam-in-the-eye and actual-construction stages. The most eagerly anticipated of them are missions being developed by both NASA and ESA to bring back samples of Martian rocks and soil in the next 15 years or so.

vital for the greenhouse effect, and the disappearance of its surface water, some ending up frozen underground, were gradual, and it is not at all absurd to speculate that living things there could have adapted to the progressively harsher environment and have survived in some form to the present.

In 2004 three different studies, made using spectrometers on the earth and on a spacecraft orbiting Mars, found evidence for methane in several regions of the Martian atmosphere above icy ground. This was very exciting, because methane molecules are broken down in an average of a few hundred years by sunlight and so what was detected must have originated relatively recently. On the earth, nearly all the methane in the air is produced by plants and by bacteria, for instance those that rot plant matter and those that populate the digestive systems of cows. Nonbiological explanations are also possible, but the most obvious one, that the methane has a volcanic origin, is unlikely because if so it should be accompanied by sulfur dioxide, which is absent. The discovery of Martian methane gives even more immediacy to the search for definitive signs of life, past and present, on our neighbor in space: nobody laughs at the possibility any longer.

17.8 Asteroids

Millions of Tiny Planets between Mars and Jupiter

The asteroids are small, rocky objects, many of which circle the sun in a belt between Mars and Jupiter. A few pass inside the earth's orbit, at least one has an orbit entirely inside that of the earth, and another, Icarus, gets even closer to the sun than Mercury. Similar bodies are also found in the Kuiper Belt outside Neptune; see Sec. 17.11.

The largest asteroid, Ceres, is 950 km across, roughly the extent of Texas. It was the first to be discovered, on January 1, 1801, by the Italian astronomer Giuseppe Piazzi, who named it after the patron goddess

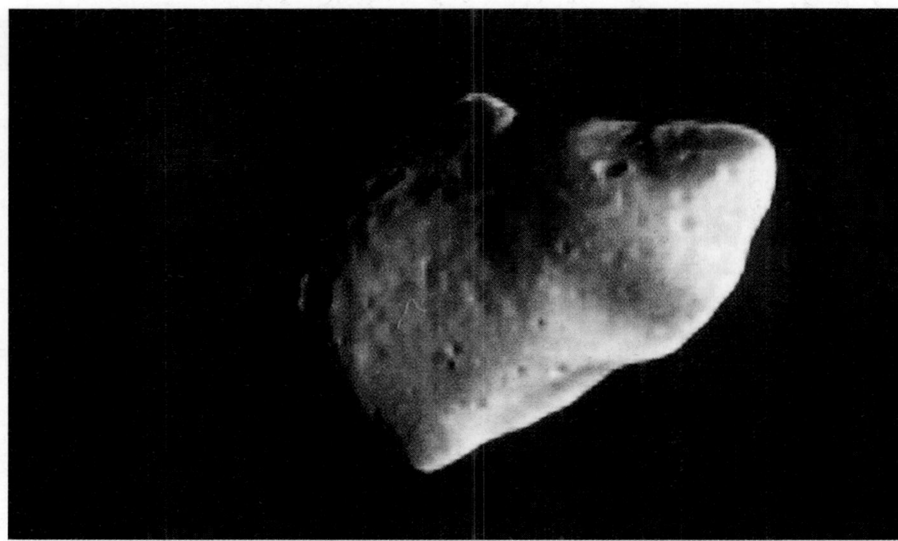

Figure 17-23 The first close-up photograph of an asteroid was taken in 1991 from the Galileo spacecraft. Called Gaspra, the asteroid is 19 km long and 12 km wide and is pocked with craters, some over a kilometer across. It rotates with a period of 7 h. Gaspra seems to have a magnetic field as strong as the earth's. Asteroids can be named after almost anybody except religious, political, or military leaders; one has been called Jerry Garcia after the former singer.

of his native Sicily. Ceres is today classified as a dwarf planet. For a long time asteroids were thought to be fragments from the breakup of a planet that was supposed to have once existed between Mars and Jupiter. Nowadays asteroids are considered to be bits of matter from the early solar system that never became part of a larger body because of the gravitational influence of the nearby giant planet Jupiter. Thus their study should tell us something about the raw materials from which the planets were formed 4.6 billion years ago.

The larger asteroids, several thousand in all with relatively few over 100 km across, have been tracked and named. Although there are millions more, the asteroid belt is so vast that they are usually far apart. But not always: most of the meteoroids that reach the earth are believed to be debris from asteroid collisions (the others come from the breakup of comets). Most asteroids are irregular in shape, too small for gravity to have pulled them into spheres (Fig. 17-23). At least one asteroid, Ida, has a satellite. Called Dactyl (after the Dactyli, mythical creatures supposed to live on Mount Ida in Greece), the moonlet is 1.4 km across.

Dawn

In 2007 NASA launched the spacecraft Dawn on a voyage to the asteroid Vesta and the dwarf planet Ceres. Dawn is expected to reach Vesta in 2011 to begin 6 months of studies from an orbit around it, and then go on to Ceres for 5 months of similar studies. Unlike previous spacecraft, once Dawn left the earth it was propelled by the reaction force (see Sec. 3.9) of xenon ions accelerated electrically to 34 km/s. The high ion speed means that very little xenon is needed—only 425 kg of xenon for the entire mission, far less than the many tons of chemical fuel needed by a conventional rocket engine. The reaction force on Dawn is less than the weight of a sheet of paper, but its effect grows steadily over time. In the 4 years of its flight to Vesta, Dawn's speed will have climbed to nearly 11 km/s, about the same as a conventional spacecraft although taking longer to go that fast.

The Asteroid Danger Because some asteroids have paths that intersect the earth's orbit, there is a chance that one of them might someday collide with the earth, which has occurred before. What would happen? Small asteroids would not threaten life on our planet. But a really big asteroid, of which there are plenty, is another story. As we saw in Sec. 16.16, an asteroid about 10 km across struck Mexico 65 million years ago, an event whose consequences may have led to the extinction of the dinosaurs as well as many other forms of life.

How often asteroids of various sizes collide with the earth can be estimated from their numbers in space and the distribution of crater sizes on the moon (where there is no erosion, unlike the earth). Figure 17-25 shows how the probable impact frequency varies with asteroid size. An asteroid 2 km across, which can be expected to strike the earth an average of every million years, could create catastrophe on a global scale. Its impact would release energy equivalent to 100,000 megatons of TNT—over 10,000 times

The Tunguska Event

On the morning of June 30, 1908, in the Tunguska River region north of Lake Baikal in Siberia, there was a brilliant flash in the sky and a great blast that devastated over 2000 km^2 of forest (Fig. 17-24). Here is how the event appeared to a Siberian farmer 200 km away: "When I sat down to have my breakfast beside my plow, I heard sudden bangs, as if from gunfire. My horse fell to its knees. From the north side above the forest a flame shot up. Then I saw that the fir forest had been bent over by the wind, as I thought of a hurricane. I seized hold of my plow with both hands so it would not be carried away. The wind was so strong it carried soil from the surface of the ground, and then the hurricane drove a wall of water up the Angora River."

Because no crater was found under the center of the blast, the most likely explanation for the Tunguska event seems to be the violent disintegration of an asteroid about 50 m across in the atmosphere at an altitude of 6 to 8 km. This origin is still a hypothesis because, although the breakup of so large an asteroid should have flung out thousands of tons of cosmic debris, 35 expeditions to Tunguska have turned up only relatively few iron and silicate particles of the expected kinds. Recently a deep lake 8 km from the point on the ground under the blast was found to have features that suggest an impact crater. It remains to be seen whether an asteroid remnant is buried under the lake.

Figure 17-24 The Tunguska event blew down thousands of trees over a large area in a radial pattern like the spokes of a wheel. It may have been caused by an asteroid exploding above the ground.

the energy of the largest hydrogen bomb ever tested—and leave a crater 30 km in diameter. Hundreds of asteroids of that size or larger have been detected that periodically pass near the earth's orbit, and there are no doubt thousands of smaller ones potentially able to cause serious damage in a collision. Of the asteroids that have been tracked, the one expected to come closest to the earth is about 250 m across, big enough to destroy a major city but likely to be a near miss. Current estimates indicate that in 2029 it will pass within 32,000 km of the earth's surface, which is inside the orbits of some artificial satellites and a close shave as astronomical

BIOGRAPHY Carl Sagan (1934–1996)

Born in Brooklyn, New York, Sagan studied physics and astronomy at Cornell and the University of Chicago. After a period at Harvard, he returned to Cornell, where he was a professor until his death. Sagan contributed to the understanding of many aspects of planetary science, for example the role of the greenhouse effect in heating the surface of Venus, how windblown dust causes seasonal changes in the appearance of Mars, and the nature of the organic molecules in the reddish clouds on Saturn's satellite Titan. His studies of the disastrous long-term effects of nuclear war on the earth were very influential. Sagan was involved in the research programs of the Mariner, Viking, Voyager, and Galileo spacecraft, and many of today's planetary scientists were his students or coworkers. An inspiring and proficient popularizer of science, his television series Cosmos was watched by 500 million people around the world, and the book based on the series became the best-selling science book ever published in English. Sagan was the author or coauthor of over 600 papers and articles and 20 books, one of which won a Pulitzer prize.

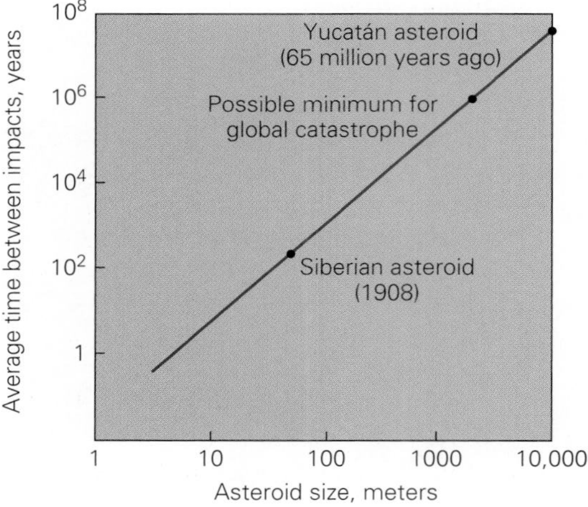

Figure 17-25 Large asteroids are much less likely to strike the earth than small ones, but those more than about 2 km across would devastate life on earth. (The scales are not linear.)

distances go. The asteroid would appear as a bright star drifting across the sky to nighttime viewers in Europe, Africa, and central Asia.

If a large asteroid headed for the earth were to be noticed in time, say 15 y ahead, a spacecraft could be sent to divert it by impact or explode a nuclear weapon on its surface to either divert it or blow it apart. However, if the asteroid is not a solid chunk of rock but a clump of cosmic rubble loosely held together mainly by gravity, as most asteroids seem to be, absorbing a shock would give it an unpredictable deflection; an explosion would also have an unpredictable result. So a better idea is probably to have a spacecraft spend a week or so nearby where it could exert enough of a gravitational tug—very little would be needed—to edge the asteroid into a safer orbit.

The Outer Planets

17.9 Jupiter

Almost a Star

The mammoth planet Jupiter, fittingly named after the most important of the Roman gods, is represented by a stylized lightning bolt because lightning was supposed to come from the god Jupiter. The symbolism has turned out to be appropriate since electric discharges regularly occur in Jupiter's atmosphere that produce bursts of radio waves detectable on the earth.

Jupiter, like Venus, is shrouded in clouds, but those of Jupiter are more spectacular (Fig. 17-26). The clouds are responsible for Jupiter's relatively high reflectivity, which together with its size makes it a bright object in the sky despite its distance from the sun. The clouds occur in alternate light and dark bands whose colors—mostly in the yellow-orange-red family—and markings change slightly from time to time.

Figure 17-26 Jupiter, the largest planet, has a mass $2\frac{1}{2}$ times that of all the other planets together. The Great Red Spot in its southern hemisphere is larger than the earth; it consists of swirling gases and varies in size and appearance.

One marking is particularly conspicuous, the Great Red Spot that today is about 25,000 km long and 10,000 km high, as large as two earths. Though its size, shape, and color are known to vary, the Spot itself does not seem to be a temporary phenomenon. It was definitely identified as long ago as 1831 and is probably the same marking described in 1660 by the French astronomer Jean Dominique Cassini. Although the nature and origin of the Great Red Spot remain uncertain, a plausible suggestion is that it is a kind of permanent hurricane into which energy is constantly fed by the Jovian wind system. The red color may be due to phosphorus from the breakup of the phosphorus compounds believed to contribute to the colors of the bands that circle the planet.

Cassini used the spot he found, whatever it was, to determine Jupiter's period of rotation. This is a little less than 10 h, which means that points on Jupiter's equator travel at the enormous speed of 45,000 km/h. (The earth's equatorial speed is only 1670 km/h.) Because of its rapid rotation, Jupiter bulges much more at the equator than the earth does.

Satellites Of Jupiter's 63 known satellites, the 4 that Galileo discovered in 1610 are easy to see with a pair of binoculars or a small telescope. The innermost, Io, is slightly larger than our moon and is spangled with active volcanoes. Indeed, no other body in the solar system is so volcanically active. Apparently the energy involved comes from the gravitational pulls of Jupiter and two of its other satellites, Ganymede and Europa, which knead Io like a lump of clay. Io flexes by nearly 100 m, which heats its interior and thereby powers over 100 active volcanoes that belch white-hot lava and jets of sulfurous gas. The most vigorous volcano, Loki, pours out more heat than all the earth's active volcanoes together.

The other large Jovian satellites are coated with ice: Europa, smaller than the moon; the giant Ganymede, larger than Mercury; and Callisto, a bit more modest in size (Fig. 17-27). All but Callisto show signs of geological activity. The rest of Jupiter's satellites are very small, only 2 to 250 km across, and most have large noncircular orbits that do not lie in the plane of Jupiter's orbit. Apparently these satellites were originally asteroids somehow captured by Jupiter as they circulated through the early solar system. A number of the irregular satellites revolve "backward" around

Figure 17-27 Jupiter and its four largest satellites, which were first seen and named by Galileo almost 400 years ago. In this composite image, Io, the innermost, is at the top, then in order of distance from Jupiter are Europa, Ganymede, and Callisto.

Jupiter, from east to west. Jupiter also has a ring, like those of Saturn but so faint that it can be detected only by spacecraft. The ring seems to consist of dust knocked off Jupiter's innermost satellites by micrometeoroid impacts and/or debris from volcanic eruptions on Io.

Structure Jupiter's volume is about 1300 times that of the earth, but its mass is only 318 times as great. The resulting low density—only a quarter that of the earth—means that Jupiter cannot be composed of rock and iron as is the earth. Like the other large planets—Saturn, Uranus, and Neptune—Jupiter must consist chiefly of hydrogen and helium, the two lightest elements.

Possibly Jupiter has a relatively small rocky inner core, perhaps the size of the earth, surrounded by a large outer core of liquid hydrogen under such enormous pressure that it behaves like a liquid metal. The interior is believed to be very hot, possibly 20,000°C. This is not hot enough for nuclear reactions to occur in its hydrogen content that would turn Jupiter into a star. However, if Jupiter's mass were perhaps 80 times

Europa

The Jovian satellite Europa (Fig. 17-28) is nearly the size of our moon but, unlike the moon, it is covered with a layer of water ice several kilometers thick. The ice is scarred by a network of curved, intersecting fracture lines, like a cracked eggshell (Fig. 17-29). A surface of this kind is to be expected if the ice is floating on a vast ocean of liquid water; currents in the ocean regularly break up the ice, and water wells up into the cracks and freezes there to give the patterns we see. Where does the heat come from that keeps the subsurface water liquid and powers its currents? Some may be the result of radioactivity in Europa's rocky core, but most arises from the continual squeezing and stretching due to tidal forces caused by the gravitational pulls of Jupiter and its other large satellites, as in the case of Io.

Given liquid water and heat, could life have arisen on Europa? After all, organisms have been discovered on earth that thrive in perpetual darkness with no need for sunlight, and living things have been found beneath Antarctic ice. Comet impacts on Europa may well have dumped on its surface billions of tons of such elements needed for life as carbon, nitrogen, and sulfur. Many scientists think that Europa, not Mars, is the most likely candidate for extraterrestrial life in the solar system today. NASA hopes to launch a spacecraft to visit Europa, Callisto, and Ganymede. If liquid water is indeed under Europa's ice coat, a lander may follow that will inspect the cracks and burrow through the ice to look for life.

Figure 17-28 Jupiter's satellite Europa.

Figure 17-29 Europa's shell of ice is covered with cracks and floats on an ocean of liquid water. This image, which covers an area 5 km wide, was radioed back by the spacecraft Galileo, which went into orbit around Jupiter late in 1995.

greater than it is, the internal temperatures would be high enough for Jupiter to be a miniature star (see Sec. 18.13). Surrounding the core is a dense layer of liquid hydrogen and helium that gradually turns into a gas with increasing distance. The outer part of Jupiter's atmosphere contains such gases as ammonia, methane, and water vapor as well as hydrogen and helium.

Spacecraft Visits United States spacecraft passed close to Jupiter after a number of journeys that lasted several years and covered over a billion km. Of the wealth of information radioed back, a few items are

Slingshot Effect

Spacecraft are often helped by the "slingshot effect" when they explore the solar system. When a spacecraft passes behind a planet (in terms of its orbital motion), the planet's gravitational pull accelerates the spacecraft. NASA's New Horizons probe, which was launched in 2006, received such a gravity assist when it passed by Jupiter 13 months later. The slingshot effect boosted the probe's speed by 2 m/s, which will save 3 years on the journey to Pluto, where it should arrive in 2015, and beyond.

especially notable. For example, Jupiter has a magnetic field many times stronger than the earth's. This field traps high-energy protons and electrons from the sun in belts that extend many Jovian radii outward. The earth has such belts too, but they are 10,000 times weaker than those of Jupiter. It is plausible that Jupiter's magnetic field is connected with the metallic nature of part of its volume, with the "metal" being highly compressed liquid hydrogen instead of the molten iron of the earth's core.

Another important finding confirmed that Jupiter radiates over twice as much energy as it receives from the sun. By contrast the atmospheres of Venus, Earth, and Mars are on average in balance, and radiate only as much energy as the sun provides. Apparently Jupiter is still cooling down, so that the extra heat has been left over from the planet's formation.

17.10 Saturn

Lord of the Rings

In its setting of brilliant rings, Saturn is the most striking and beautiful of the earth's kin (Fig. 17-30). Saturn, the Roman god of sowing seed and the father of Jupiter, was called Kronos in Greek mythology, from which the planet's symbol of a stylized K probably derives. The midwinter festivals of worship to Saturn, the saturnalia, were always splendid occasions of joy and revelry. Saturday is named after him.

Saturn resembles Jupiter in many respects, though smaller and less massive. Like Jupiter it consists largely of hydrogen and helium, radiates more heat than it receives from the sun, is flattened at the poles by rapid rotation, has a strong magnetic field and a dense atmosphere of hydrogen with some helium, and is surrounded by banded clouds in which gigantic thunderstorms occur. Auroras occur in the atmospheres of both planets. However, Saturn's core seems to have less liquid metallic hydrogen than

Figure 17-30 The rings of Saturn are not solid or gaseous but consist of small rock and ice fragments that orbit the planet. Galileo called Saturn "the planet with ears," but in 1655 the Dutch astronomer Christian Huygens recognized that the "ears" were actually rings. He also discovered Titan, Saturn's largest satellite.

Cassini and Huygens

In 2004 the Cassini spacecraft, a $3.3-billion American-European collaboration, reached Saturn after a 7-year journey. Cassini, 6.5 m long with the mass of a small elephant, carried a dozen instruments of various kinds to study Saturn, its rings, and several of its moons from orbit. Six months after its arrival Cassini sent off the smaller Huygens probe to Titan, Saturn's largest moon (Fig. 17-31), where it parachuted through the orange haze that hides its surface to land on soft ground that apparently is soaked with liquid methane. Huygens found a varied terrain with evidence of flowing hydrocarbon liquid (it is too cold there for liquid water) in the form of drainage channels and eroded plains. The presence of smooth, rounded pebbles supports this inference. Dark, featureless areas were seen that resemble lakes and seas with sharp boundaries that could be shorelines. Few craters scar Titan's landscape, unlike the landscapes of Mars and the moon, which confirms that erosion and deposition are continuing on Titan. Huygens had power enough to operate for only a short time on Titan, but Cassini flew close to it several dozen times. When all the images and data are analyzed, we will know much more about the still-active geology and chemistry of this remote world.

Cassini also inspected other of Saturn's satellites. One of them, the tiny (diameter 505 km) Enceladus, seems to have abundant liquid water below its surface that spews out in geysers of water vapor and ice crystals near its south pole. The vapor plumes show traces of organic compounds. Even though distant from the sun, Enceladus is warm enough for subsurface liquid water to be present, perhaps heated by the same tidal kneading that heats Jupiter's moon Europa. Given warmth, water, and organic compounds, did life of some kind appear? There is a long way to an answer, but right now there is nothing to rule it out.

Figure 17-31 Titan, Saturn's largest satellite.

Jupiter's core, perhaps with a large rocky kernel inside. Farther from the sun than Jupiter, Saturn is considerably colder, only about −180°C at the surface. Winds blow harder on Saturn, over 1800 km/h at its equator.

The satellites of Saturn—at least 60, probably with more small ones—range in size from 6 km across to the giant Titan, whose 5140 km diameter makes it only slightly smaller than Jupiter's Ganymede. An outer moon, Phoebe, orbits in the opposite direction to the others and may be a captured asteroid, as in the cases of many of Jupiter's outer satellites. Titan is the only satellite in the solar system with an atmosphere, which seems to consist largely of nitrogen with some methane and small amounts of other organic compounds. Titan's atmosphere is more like ours than that of any other body in the solar system. Reddish clouds of organic compounds float in this atmosphere and send showers of methane rain and perhaps snow from time to time to feed the liquid methane rivers and seas on Titan's surface. Methane apparently plays the same geological role on Titan that water does on the earth, although the gaseous methane in its atmosphere seems to come from its interior rather than from the evaporation of liquid methane on its surface. Titan and the earth are the only bodies in the solar system on which rain or snow of some kind falls to the ground.

Could life have developed on Titan? The answer seems to be that it might have had Titan been warmer. Titan's atmosphere, in particular, is similar to that of the earth before life emerged. However, chemical processes are slow at low temperatures. If life took a half billion years to come into being on the earth, a reasonable estimate, then the solar system is not old enough for this process to have occurred on the frigid (−179°C) surface of Titan. But conceivably heat from Titan's interior or from comet impacts produced local pools of liquid water that could persist under crusts of ice for long periods, which makes a search for life more reasonable. As far as astronomers can tell at present, conditions on Enceladus, a much smaller satellite of Saturn, may also be able to support life.

Example 17.3

You are in a spacecraft orbiting Saturn. Would you find that Jupiter shows phases like those of Venus? Would you find that Uranus does this?

Solution

Jupiter shows phases to you because its orbit is inside that of Saturn (see Fig. 17-13). Uranus does not show phases because its orbit is outside that of Saturn.

Saturn's Rings A number of rings surround Saturn at its equator, two of them bright and the others faint. The rings are inclined by 27° with respect to the plane of the earth's orbit, so we see them from different angles as Saturn proceeds in its leisurely $29\frac{1}{2}$-year tour around the sun (Fig. 17-32). Twice in each period of revolution the rings are edgewise to the earth, which happened last in 1996 and will happen again in 2011. In this orientation the rings are practically invisible, suggesting that they are very thin. In fact, they average only about 50 meters thick. Since the outer bright ring is 270,000 km in diameter, a sheet of paper is fat by comparison. In 2018 we will be able to see Saturn's rings in their fullest glory, tipped by 27° so that their upper (northern) surfaces face us.

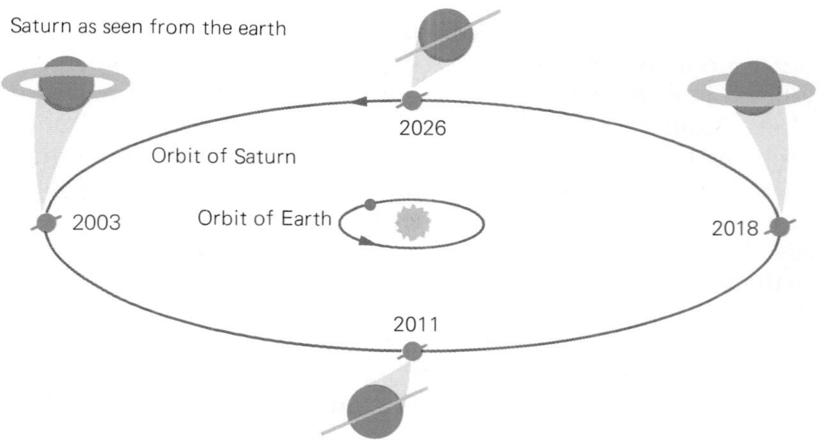

Figure 17-32 The appearance of Saturn's rings from the earth varies with Saturn's location in its orbit. Saturn's period of revolution is 29.5 years.

The rings are not the solid sheets they appear to be but instead consist of a multitude of small bodies, each of which revolves around Saturn as a miniature satellite. How do we know that Saturn's rings are not solid? One piece of evidence is that we can see stars through them. Another is that doppler-effect measurements show that the inner part of each ring moves faster than the outer part. This is in accord with Kepler's third law, which calls for a decrease in orbital speed with an increase in orbital size. (Thus Mercury's orbital speed is 48 km/s, and Pluto's is only 5 km/s.) If the rings were solid, their outer parts would have the highest speeds, contrary to what is found.

Nor can the rings be gaseous—they reflect sunlight and radar signals far too well. The only question left is the size and nature of the particles of which they consist. In 1980 and 1981 two Voyager spacecraft reached Saturn after 4-year journeys. They radioed back vast amounts of data on the planet, its satellites, and, of course, its rings. The findings about the rings confirmed what astronomers had suspected earlier: the particles are chunks of rock and ice that range in size from small stones to boulders. In addition, the rings we see from the earth are not uniform but are split into thousands of narrow ringlets. Two of the ringlets in one of the outer, faint rings seem to be braided around each other, a peculiarity that may result from the gravitational pulls of two small satellites whose orbits are just inside and just outside the ringlets.

Were Saturn's rings always there? Many—not all—astronomers think they are relatively young, perhaps fragments of a comet torn to shreds by Saturn's gravity or else the debris of a collision between an asteroid and one of Saturn's moons. Probably the rings are migrating inward and, in several hundred million years, will disappear into Saturn itself.

17.11 Uranus, Neptune, Pluto, and More

Far Out

Uranus, Neptune, and the dwarf planet Pluto owe their discovery to the telescope. Uranus was found in 1781 by the great English astronomer William Herschel. In Greek mythology, Uranus personifies the heavens and is the father of Saturn and the grandfather of Jupiter.

Herschel at first suspected Uranus to be a comet because, through his telescope, it appeared as a disk rather than as a point of light like a star. Observations made over a period of time showed its position to change relative to the stars, and its orbit, found from these data, revealed it to be a planet. Neptune was discovered in 1846 as the result of predictions based on its gravitational effects on the orbit of Uranus (see Chap. 1). Neptune was named after the Roman god of the sea.

Uranus has the distinction of rotating about an axis only 8° from the plane of its orbit, so we can think of it as spinning on its side. The most likely reason is a collision with a large object, perhaps the size of the earth, early in the history of the solar system. Because of this tilt, summer at a particular place on Uranus is a 42-year period of continuous sunshine, and winter is a 42-year period of continuous darkness. Uranus and Neptune are about the same size, smaller than Jupiter and Saturn but much larger than the other planets. Like Jupiter and Saturn they consist chiefly of hydrogen and helium.

BIOGRAPHY William Herschel (1738–1822)

Herschel was born in Germany (his name was originally Friedrich Wilhelm) and went to England in 1757 to avoid military service. In England Herschel prospered as an organist and music teacher, and composed 24 symphonies. He also found time for his hobby of astronomy. Good telescopes were expensive, so Herschel, with the help of his sister Caroline, built his own. By 1774 Herschel had made the finest telescope in the world, a reflector whose mirror he and his sister had ground themselves. With this telescope he systematically searched the sky, a labor rewarded in 1781 by the discovery of a seventh planet, Uranus. This created a great stir everywhere, both because Uranus was the first new planet to be identified for many centuries and because its distance from the sun was twice that of Saturn, thus doubling the size of the solar system. Six years later Herschel found 2 satellites circling Uranus and called them Titania and Oberon. (Today 27 are known, some very small.)

Caroline Herschel did important work herself, among other things discovering 8 comets. In her day—she lived from 1750 to 1848—women were unusual in science, but eventually her achievements were recognized with a gold medal from the Royal Astronomical Society in 1828 and another from the king of Prussia on her 96th birthday.

In 1788 Herschel married a rich widow, which enabled him to become a full-time observer of the heavens. The list of his achievements is unparalleled in astronomy. Among his other discoveries were binary (double) stars, and he established that each member of a pair revolved around the other. Using a thermometer to measure the heat produced, he studied the intensity of sunlight in various parts of its spectrum. To his surprise, the heating continued past the red end, which suggested to him that the sun emits invisible (today called infrared) radiation besides visible light. Most significantly, Herschel's studies of the Milky Way led him to conclude that it is a vast disk-shaped galaxy of stars, one of which is the sun. He speculated—correctly—that a number of fuzzy objects in the sky were other galaxies far away in space.

Herschel's renown became such that King George III provided the money for a huge new telescope with a mirror 1.2 m across. The first time the telescope was used, Herschel discovered two more satellites of Saturn to add to the five already known, and he was soon able to show that Saturn's rings are not stationary but revolve about their mother planet. He remained active almost until his death at 84 (the same span as the orbital period of Uranus). Herschel's work was carried on by his son John, who was a pioneer in photography (his word) and its application to astronomy.

The spacecraft Voyager 2 passed near Uranus in 1986 and near Neptune 3 years later (Fig. 17-33). Photographs radioed back to the earth show that both planets have several rings around them. Like the rings of Saturn, these rings consist of small particles but are too narrow to be seen from the earth. Uranus has 26 known satellites and Neptune has 13.

Voyager 2 found that Neptune's atmosphere is in much more violent motion than that of Uranus. Strong, turbulent winds blow across Neptune, and a huge hurricane can be seen that is nearly the width of the earth. From its appearance this storm is called the Great Dark Spot, and it is much like Jupiter's Great Red Spot (Fig. 17-33). A smaller hurricane is near Neptune's south pole. Both hurricanes are accompanied by white cirrus clouds of methane ice. Since Neptune is farther from the sun

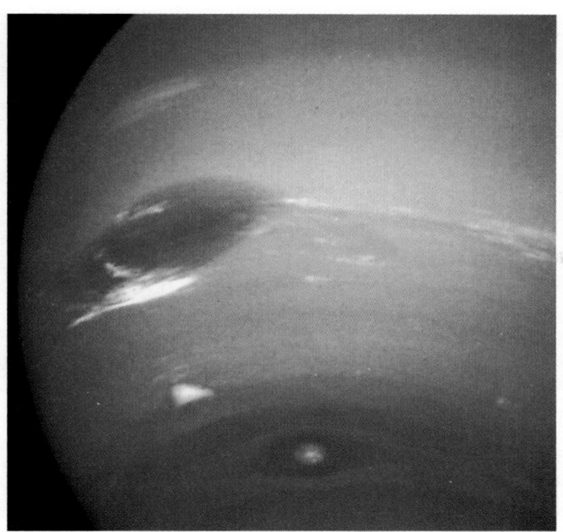

Figure 17-33 The spacecraft Voyager 2 left Florida in 1977 and arrived at Neptune 12 years later. This photograph radioed back by Voyager 2 shows Neptune's Great Dark Spot, which is almost as large as the earth. White patches are clouds that consist of methane ice crystals.

than Uranus and so receives less solar energy, why Neptune has stormier weather than Uranus is a puzzle.

Pluto Pluto was discovered in 1930 by the American astronomer Clyde Tombaugh, then 24 years old, who spent 10 months meticulously looking for a new planet whose existence was suspected on the basis of irregularities in the orbits of Uranus and Neptune. As it happens, the total mass of Pluto and its satellite is not enough to account for these irregularities, and Tombaugh seems to have found it purely by accident. Pluto was considered as a planet until 2006; it is now regarded as a dwarf planet. Pluto was named after the Greek god of the underworld, the home of the dead.

Pluto is so small, so far away, and so dimly lit that reliable information about it is hard to obtain. It is about two-thirds the size of our moon and seems to consist of rock and water ice with a thin atmosphere of

Dwarf Planets

Because Pluto is much smaller than the eight first-discovered planets (and indeed is smaller than seven of their satellites), is very different from the outer planets nearest it, and has an unusual orbit, many astronomers have never regarded it as a genuine planet. In 2006 the International Astronomical Union agreed that a planet must (1) orbit the sun, (2) have been forced by gravity into a spherical shape (see Sec. 1.9), and (3) be large enough to have swept up other objects from the neighborhood of its orbit. A new class of dwarf planets was established that need meet only the first two of these criteria. As for the third one, each of the eight "genuine" planets has a mass more than five thousand times the total mass of all the fragmentary bodies in its orbital vicinity; the masses are comparable for dwarf planets. The first dwarf planets to be nominated were Pluto, the asteroid Ceres, and a Kuiper Belt member called Eris, which is 5 percent larger and 27 percent more massive than Pluto and has a satellite, Dysnomia. (Eris is the Greek goddess of discord; Dysnomia is her daughter.) As mentioned in Sec. 17.2, the Kuiper Belt beyond Neptune is home to comets and to asteroids of various sizes, a number of them over 1000 km across. Probably many of the latter besides Eris will eventually be added to the list of dwarf planets.

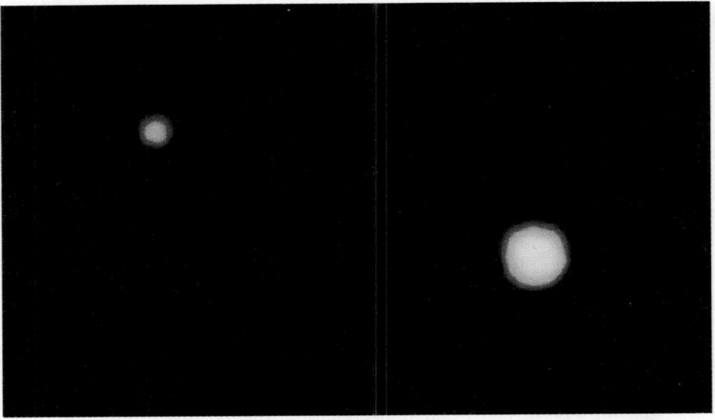

Figure 17-34 Pluto and its satellite Charon as seen by the Hubble Space Telescope. Pluto has two other, smaller satellites that are farther away.

nitrogen, carbon monoxide, and methane. Six other satellites in the solar system are also larger than Pluto. Pluto's average surface temperature is −230°C and it has prominent polar caps of frozen gas. It takes 248 of our years to orbit the sun. As in the case of Uranus, Pluto's axis is horizontal, that is, it lies near the plane of Pluto's orbit.

Pluto has a satellite, Charon, named for the ferryman who takes the dead across the river Styx to Pluto's domain (Fig. 17-34). Charon's diameter of 1200 km is a little over half that of Pluto, and it probably consists largely of ice with some rock as well. Charon is relatively close to Pluto and circles it every 6.4 days. Since Pluto also rotates with a 6.4-day period, Pluto and Charon always show the same sides to each other. There are signs that Charon, like Jupiter's satellite Europa, has an ocean of liquid water under a covering of ice. It is possible, again like Europa, that Charon is heated by tidal forces that knead its interior. Could there be underground life on Charon that draws its energy from this heat rather than from the extremely feeble sunlight? What little we know about Charon does not seem to rule it out. In 2005 the Hubble Space Telescope discovered two more moons that orbit Pluto. Both are over twice as far from Pluto as Charon and are quite small, not much over a tenth of Charon's diameter.

A spacecraft called New Horizons was launched in 2006 to inspect Pluto and its moons, which it should reach in 2015, the first spacecraft to do so. New Horizons will then go on past Pluto to examine Kuiper Belt objects, some comparable in size with Pluto. Clyde Tombaugh, who discovered Pluto, died in 1997 and New Horizons carries some of his ashes.

The Moon

The light that reaches us from the moon is reflected sunlight. Although closer to the earth than any other celestial body, the moon is nevertheless an average of 384,400 km away. Its diameter of 3476 km—a little more than a quarter of the earth's diameter—places it among the largest satellites in the solar system.

The moon circles the earth every $27\frac{1}{3}$ days and, like the earth, turns on its axis as it revolves. In the case of the moon, the rotation keeps pace exactly with the revolution, so the moon turns completely around only once during each circuit of the earth. This means that the same face of our satellite is always turned toward us and that the other side remains hidden from the earth (though not from spacecraft). The synchronization is a consequence of tidal bulges raised in the moon itself by the earth's gravitational pull, as described in Sec. 1.10 for the case of surface water on the earth attracted by the moon. Friction between the water bulges and the ocean bottoms is currently slowing down the earth's rotation on its axis. Friction involved in its tidal flexing similarly slowed the moon's spin in the past to the point where the tidal bulges are locked in place along a line between the moon and the earth. Most of the other large satellites in the solar system have their spin and orbital motions synchronized in the same way.

17.12 Phases of the Moon

A Little Less Than a Month per Cycle

The properties of the moon are given in Table 17-3. The $27\frac{1}{3}$-day period of the moon's orbit quoted above is the time needed for the moon to go through a complete circuit of the earth. *If* the earth did not move around the sun, we would see the moon in the same place in the sky relative to the stars at the same time of day every $27\frac{1}{3}$ days. But while the moon is circling the earth, the earth is carrying the moon with it in the earth's own motion around the sun. As a result, to us the moon's orbital period is increased to $29\frac{1}{2}$ days. *Relative to the stars,* the moon's period is $27\frac{1}{3}$ days; *relative to the sun,* the moon's period is $29\frac{1}{2}$ days. Since time on earth is

Example 17.4

If the moon circled Jupiter in an orbit the same size as its orbit around the earth, would its period of revolution be different from its period around the earth?

Solution

The period of the orbit around Jupiter would be shorter since the greater gravitational attraction of Jupiter would require a higher orbital speed for a stable orbit.

Table 17-3 The Moon (Percentages Are with Respect to the Same Property of the Earth)

Diameter	3476 km (27%)
Mass	7.35×10^{22} kg (1.2%)
Average density	3.3 g/cm^3 (60%)
Acceleration of gravity	1.7 m/s^2 (17%)
Escape speed	2.4 km/s (21%)
Average distance from the earth	384,400 km
Rotational period	27.3 days
Orbital period:	
Relative to the stars	27.3 days
Relative to the sun	29.5 days

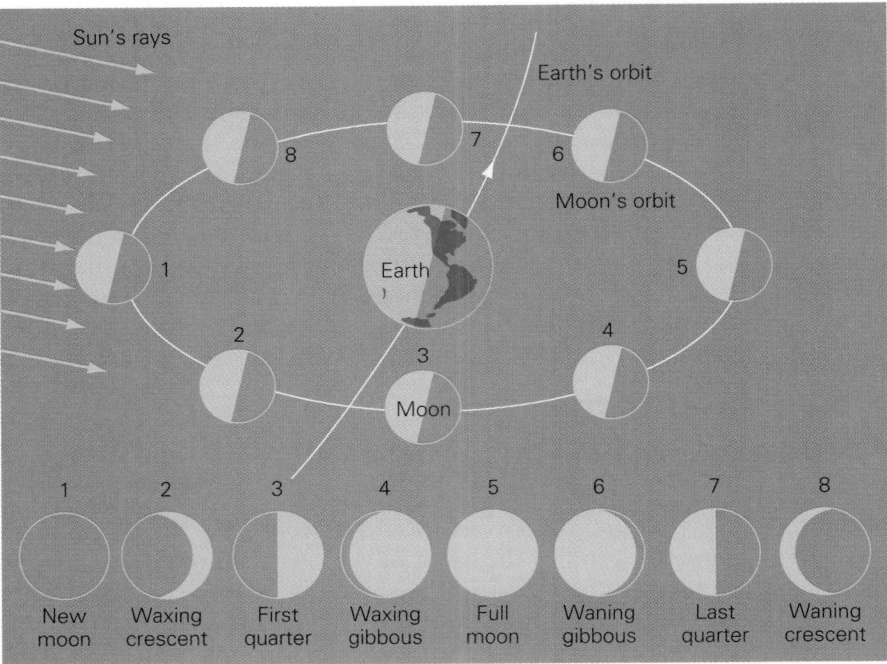

Figure 17-35 The origin of the moon's phases. As the moon revolves around the earth, we see it from different angles. When it is between us and the sun, we see only the dark side (new moon), and when it is on the opposite side of us from the sun, we see only the illuminated side (full moon). At other times we see parts of both sides. When the moon is approaching full moon, it is said to be **waxing;** when it is approaching new moon it is said to be **waning.** A way to remember what the moon looks like when it is waxing and waning is the rule of opposites: when the moon looks like a C, it is really Departing (waning); when it looks like a D, it is really Coming (waxing).

figured relative to the sun, we see the moon return to the same place in the sky at the same time of day every $29\frac{1}{2}$ days.

During each $29\frac{1}{2}$-day period the moon goes through its familiar cycle of **phases** (Fig. 17-35). First there is only a thin crescent in the western sky at sunset that soon falls below the horizon (Fig. 17-36). Each night afterward the illuminated part of the moon grows wider and moves eastward relative to the stars. After 2 weeks the moon has become full and rises in the east at sunset to light up the sky all night. Next the illuminated part of the moon grows narrower until, after 2 more weeks, it is a thin crescent that rises just before sunrise. Finally the moon disappears altogether for a few days before we see it again as a crescent at sunset.

These different aspects represent the amounts of the moon's illuminated surface visible to us in different parts of its orbit. When the moon is full, it is on the opposite side of the earth from the sun, so the side facing us is fully illuminated. In the "dark of the moon" (or new moon), the moon is moving approximately between us and the sun, so the side toward the earth is in shadow. But the shadow is not completely dark—there is a ghostly glow because of **earthshine,** sunlight reflected from the earth's surface that reaches the moon. In addition to clouds and surface ice, whitecaps on ocean waves contribute significantly to earthshine.

Figure 17-36 Waxing crescent moon 3 days after new moon.

17.13 Eclipses

Now You See It, Now You Don't

When the earth is between the sun and the moon (that is, at full moon), how can the sun illuminate the moon at all? Why doesn't the earth's shadow hide the moon completely? And when the moon passes between the sun and the earth, why isn't the sun hidden from our view?

Moonrise and Moonset

From Fig. 17-35 we can figure out the times of moonrise and moonset for each phase, keeping in mind that the earth spins counterclockwise when viewed from above the North Pole. Because a number of factors will be ignored, actual times of moonrise and moonset may vary by an hour or more from those we shall find.

Let us start with the new moon. Since the new moon is nearly in line with the sun, the new moon rises when the sun does at about 6 A.M. local time, is at its highest point at noon, and sets at about 6 P.M. The full moon is along the same earth-sun line but on the opposite side of the earth. The full moon therefore rises at 6 P.M. when the sun sets, is on the meridian at midnight, and sets at 6 A.M. A first-quarter moon is halfway between new moon and full moon; hence it rises at noon and sets at midnight. A last-quarter moon is halfway between full moon and new moon; hence it rises at midnight and sets at noon. The rises and sets of crescent and gibbous phases can be estimated in the same way (Table 17-4). For instance, a waxing crescent moon, which is at position 2 in Fig. 17-35, is intermediate between new moon and first quarter, and therefore rises at about 9 A.M. and sets at about 9 P.M.

Table 17-4 Approximate Times of Moonrise and Moonset

Phase	Rise Time	Set Time
New moon	6 A.M.	6 P.M.
Waxing crescent	9 A.M.	9 P.M.
First quarter	Noon	Midnight
Waxing gibbous	3 P.M.	3 A.M.
Full moon	6 P.M.	6 A.M.
Waning gibbous	9 P.M.	9 A.M.
Last quarter	Midnight	Noon
Waning crescent	3 A.M.	3 P.M.

The answers to these questions follow from the fact that the moon's orbit is tilted at an angle of 5.2° to the earth's orbit. Ordinarily the moon passes either slightly above or slightly below the direct line between sun and earth (Fig. 17-37). On the rare occasions when the moon does pass more or less directly before or behind the earth, an **eclipse** occurs—an eclipse of the moon when the earth obscures the moon, an eclipse of the sun when the moon's shadow touches the earth. The circular shape of the earth's shadow during a lunar eclipse is evidence for its spherical form, as the ancient Greek astronomers realized.

Total eclipses of the sun occur because, though the sun's diameter is about 400 times as great as that of the moon, the sun is also about 400 times as far away from the earth during much of its orbit. At these times the apparent diameters of both sun and moon are the same as seen from the earth, and total eclipses are possible (Fig. 17-38). Partial eclipses of

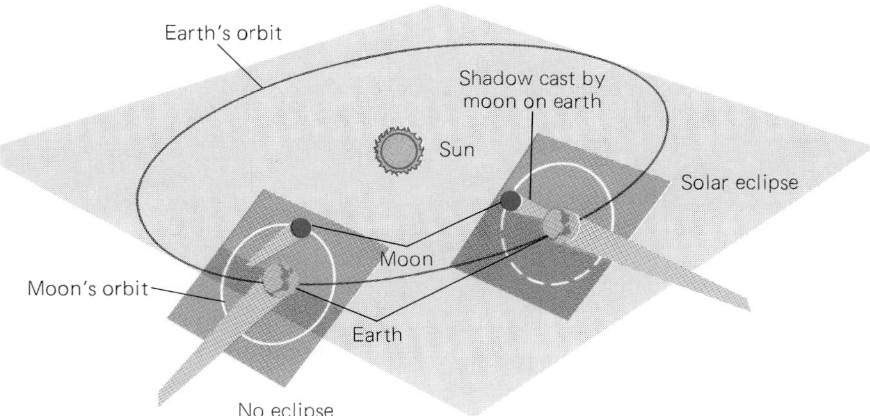

Figure 17-37 The orbit of the moon is tilted with respect to that of the earth. For this reason the moon normally passes above or below the direct line from the sun to the earth. Eclipses occur only on the rare occasions when the moon passes exactly between the earth and the sun (a solar eclipse, with the moon blocking out light from the sun) or exactly behind the earth (a lunar eclipse, with the earth blocking out light from the sun).

Future Eclipses

The friction of the earth's tidal bulges (see Sec. 1.10) on the ocean floors means that the bulges are always slightly ahead of the direct line between the centers of the earth and moon. The gravitational pull of the bulge nearest the moon is gradually speeding up the moon in its motion around the earth, which means that its orbit is increasing in size all the time. Radar measurements show that the moon is moving away from the earth at about 4 cm per year. As a result, in the distant future—perhaps 150 million years from now—the apparent size of the moon in the sky will be smaller than that of the sun and total solar eclipses will no longer take place, only annular eclipses.

Figure 17-38 In a total solar eclipse, the sun's disk is exactly obscured by the moon. The solar corona, a glowing gas cloud, can be seen during such an eclipse. Most total eclipses last less than 5 min. A total eclipse is visible somewhere on the earth about every year and a half, but there may be hundreds of years between successive eclipses at the same place.

the sun take place when the moon is not quite aligned with the sun, so that only part of the solar disk is obscured.

When the moon is farthest from the earth, its apparent diameter is less than that of the sun. The moon cannot then block the entire solar disk even when the moon lies directly between the sun and the earth. In this situation the result is an **annular** eclipse of the sun, with a ring of sunlight appearing around the rim of the moon. Table 17-5 lists the eclipses that will occur in the next few years.

17.14 The Lunar Surface

Mountains and Maria, maybe Water but No Atmosphere

To look at the moon is to wonder (Fig. 17-39). What is it made of? What is its origin? What is the nature of its landscape? What geologic processes occur on its surface and in its interior? Is there life on the moon? What is the ultimate destiny of the earth-moon system?

Until July 20, 1969, the study of the moon was more notable for the questions asked than for the answers available. On that day Neil Armstrong set foot on the moon, the first person ever to do so, after a 3-day voyage aboard the spacecraft Apollo 11 with two companions. Four days after that they returned to earth, bringing with them samples of the lunar surface. That historic expedition and the others that followed it (Fig. 17-40) answered a great many questions about our companion in space. But questions still remain, the chief one being: Where did the moon come from?

Table 17-5 Solar and Lunar Eclipses 2009–2013*

Date	Type	Where Visible
January 26, 2009	Annular solar	Africa, Antarctica, Asia, Australia
February 9, 2009	Partial lunar	Europe, Asia, Australia, Pacific, North America
July 7, 2009	Partial lunar	Australia, Pacific, Americas
July 22, 2009	Total solar	Asia, Pacific
August 6, 2009	Partial lunar	Americas, Europe, Africa, Asia
December 31, 2009	Partial lunar	Europe, Africa, Asia, Australia
January 15, 2010	Annular solar	Africa, Asia
June 26, 2010	Partial lunar	Asia, Australia, Pacific, Americas
July 11, 2010	Total solar	South America
December 21, 2010	Total lunar	Asia, Australia, Pacific, Americas, Europe
January 4, 2011	Partial solar	Europe, Africa, Asia
June 1, 2011	Partial solar	Asia, North America
June 15, 2011	Total lunar	South America, Europe, Asia, Australia
July 1, 2011	Partial solar	Indian Ocean
November 25, 2011	Partial solar	Africa, Antarctica, New Zealand
December 10, 2011	Total lunar	Europe, Africa, Asia, Australia
May 20, 2012	Annular solar	Asia, Pacific, North America
June 4, 2012	Partial lunar	Asia, Australia, Pacific, Americas
November 13, 2012	Total solar	Australia, Pacific, South America
November 28, 2012	Partial lunar	Europe, Africa, Asia, Australia, Pacific, North America
April 25, 2013	Partial lunar	Europe, Africa, Asia, Australia
May 10, 2013	Annular solar	Australia, Pacific
May 25, 2013	Partial lunar	Americas, Africa
October 18, 2013	Partial lunar	Americas, Europe, Africa, Asia
November 3, 2013	Partial solar	Americas, Europe, Africa

*Predictions by Fred Espenak, NASA/GSFC

Figure 17-39 The moon rotates on its axis at the same rate that it orbits the earth. As a result, the moon always presents this face to us. The dark areas are the maria, ancient lava flows that have been broken up by meteoroid bombardment.

A Base on the Moon

A life-support system that does not rely exclusively on supplies from the earth is necessary for the long-term lunar base proposed by the United States for 2025, which would follow preliminary visits starting in 2020. Some water seems to be present on the moon, though how much is uncertain. The water may have been brought to the moon by the comets and asteroids that pelted down upon it long ago. What is exciting about this water is not only that it might be used in the daily lives of visitors, who would anyway recycle most of their water, but that it could be separated into hydrogen and oxygen for rocket fuel, perhaps using solar energy. Then visitors to the moon would not have to carry with them fuel for their return to the earth, an enormous advantage. The combination of weak lunar gravity and a local fuel supply would make the moon a much better launching place for exploring the solar system than the earth itself. It remains to be seen how practical it will be to actually collect lunar ice, melt it, and convert it into rocket fuel.

Water is not the only lunar material that could help visitors. A cubic meter of the rock dust on the moon's surface contains, according to one researcher, the equivalent of two cheese sandwiches, two cola drinks, and two plums in such elements important to life as hydrogen, oxygen, carbon, nitrogen, and potassium. How to turn rock dust into lunch efficiently may well be worked out eventually, but a better bet for the first menus is something like nutritious and easily grown (though not necessarily very appetizing) blue-green algae.

Figure 17-40 Astronaut Charles M. Duke, Jr., collecting lunar samples during the Apollo 16 expedition in 1972. The crater at left is 40 m in diameter and 10 m deep. Behind it is the Lunar Roving Vehicle that permitted Duke and his fellow astronaut John W. Young to explore the lunar surface some distance from the landing craft. New astronaut visits to the moon are being planned for 2020. Possible projects for the first landings include installing an array of radiotelescopes on the dark side of the moon, studying x-rays from space, and accurately measuring the earth-moon distance.

The moon was not entirely a mystery even before the voyages of Apollo 11 and of the other manned spacecraft that followed it there. Even a small telescope reveals the chief features of the lunar landscape: wide plains, jagged mountain ranges, and innumerable craters of all sizes. Each mountain stands out in vivid clarity, with no clouds or haze to hide the smallest detail. Mountain shadows are black and sharp-edged. When the moon passes in front of a star, the star remains bright and clear up to the moon's very edge. From these observations we conclude that the moon has little or no atmosphere. Liquid surface water is likewise absent, as indicated by the complete lack of lakes, oceans, and rivers.

Example 17.5

Why does the moon not have an atmosphere although the earth does?

Solution

The moon's gravitational attraction is not great enough to prevent the escape of the rapidly moving gas molecules in an atmosphere. From Table 17-3 the escape speed of the moon is only 2.4 km/s, about a fifth of the escape speed of the more massive earth.

But there is still no substitute for direct observation and laboratory analysis. Each spacecraft that landed on the moon and returned to earth, whether piloted by human beings or not, has brought back information and samples of the greatest value. The lack of a protective atmosphere and of running water to erode away surface features means there is much to be learned on the moon about our common environment in space, both past and present. And from the composition and internal structure of the moon hints can be gleaned of its history, hints that bear upon the history of the earth as well. Thus the study of the moon is also a part of the study of the earth, doubly justifying the effort of its exploration.

Landscape Features With the help of no more than binoculars it is easy to distinguish the two main kinds of lunar landscape, the dark, relatively smooth **maria** (the singular is **mare**) and the lighter, ruggedly mountainous highlands. The mountains of the moon are thousands of meters high, which means that the moon's surface is about as irregular as the earth's.

Mare means "sea" in Latin, but the term is still used even though it has been known for a long time that these regions are not covered with water. The largest of the maria is Mare Imbrium, the Sea of Showers, which is over 1000 km across. The maria are circular depressions covered with dark, loosely packed material—not solid rock. They are not perfectly smooth but are marked by small craters, ridges, and cliffs. The maria consist of lava flows similar to basalt that have been broken up by meteoroid impacts. It is curious that nearly all the maria are on the lunar hemisphere that faces the earth (Fig. 17-41).

The lunar highlands are scarred by innumerable craters, some with mountain peaks at their centers (Fig. 17-42). Certain craters, such as Tycho and Copernicus, have conspicuous streaks of light-colored matter radiating outward. These **rays** may extend for hundreds or thousands of km, and seem to consist of lunar material sprayed outward after the meteoroid impacts that caused the craters. The impacts melted this material, and it cooled quickly in flight into glassy particles that reflect light well.

Figure 17-41 The far side of the moon, which always faces away from the earth, was photographed from the Galileo spacecraft in 1990. The dark markings are basaltic lava flows formed over 3 billion years ago. The dark region at lower left is a huge basin that was probably caused by the impact of a giant meteoroid.

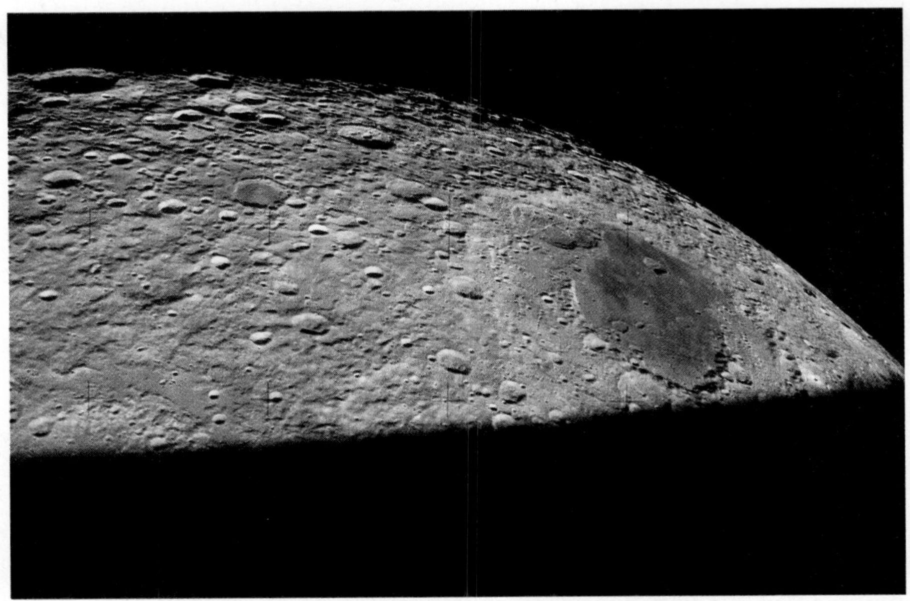

Figure 17-42 Cratered landscapes of the lunar highlands were caused by meteoroid impacts.

The **rilles** of the highlands are especially intriguing. These are narrow trenches up to 250 km long that look like dried-up riverbeds. The rilles were probably created by the collapse of subsurface channels through which molten rock once flowed from active volcanoes.

Instruments placed on the lunar surface show that moonquakes are few and mild compared with earthquakes. This implies that the interior of the moon is relatively cool, since heat from inside the earth powers its tectonic activity. As in the case of the earth, the data show the moon to consist of a rigid crust, a thick semisolid mantle, and a small, dense core. Evidence in lunar rocks of a magnetic field in the past suggests that the core then consisted of molten iron, as the earth's core does today. Later the core cooled, which stopped the motions that generated the magnetic field, though it may still be liquid. Because the moon's average density is only 60 percent of the earth's average density, the moon's core must be much smaller, relative to its size, than the earth's core.

17.15 Evolution of the Lunar Landscape

A Violent Past, a Quiet Present

The analysis of lunar rock and soil samples has led to a number of conclusions about the history of our satellite. On the basis of such samples the moon's story can be taken farther back than the earth's because, since the moon lacks an atmosphere, weathering did not occur there (Fig. 17-43). Even particles of pure iron have been found on the moon's surface, whereas only iron compounds occur on the earth's surface. Rocks have been found on the moon that radioactive dating reveals crystallized earlier than the most ancient terrestrial rocks, which are nearly 4 billion years old. Some lunar samples apparently solidified very soon after the solar system came into being. These rocks are older than any found on the earth.

As the molten outer part of the early moon gradually hardened into a light-colored crust, meteoroids of all sizes rained down. Some, as large as

Figure 17-43 This 2-cm rock fragment collected during the Apollo 11 expedition to the moon in 1969 resembles certain volcanic rocks on the earth, although it is different chemically and no weathering has taken place.

Rhode Island and better described as asteroids, smashed great basins hundreds of km across. About 4 billion years ago, as the bombardment moderated, radioactivity inside the moon produced enough heat to melt rock there. Lava that reached the surface flowed into the impact basins to form the dark maria that today cover about 20 percent of the moon's surface.

The youngest rocks found on the moon are 3 billion years old, so all igneous activity there must have stopped at that time. Meteoroids continued to crater the landscape and to pulverize surface rock into the powdery debris that today coats the moon.

17.16 Origin of the Moon

A Collision was Probably Responsible

Until recently theories of the origin of the moon fell into three categories (Fig. 17-44):

1. The moon was initially part of the earth and split off to become an independent body.
2. The moon was formed elsewhere in the solar system and was later captured by the earth's gravitational field.
3. The moon and the earth came into being together as a double-planet system.

Each of these approaches once seemed quite attractive, but strong arguments against all of them eventually appeared.

A fourth proposal is today widely accepted. Early in the history of the solar system, when the bits of matter in a particular region began to collect together, they could have produced more than one large body. Suppose another planet a little larger than Mars and with a composition slightly different from that of the infant earth developed nearby, and this planet crashed into the earth. The mantle of the other planet and some of the earth's would have been thrown off into orbit around the earth by the impact, a year or so later to form the moon. In this picture of the moon's

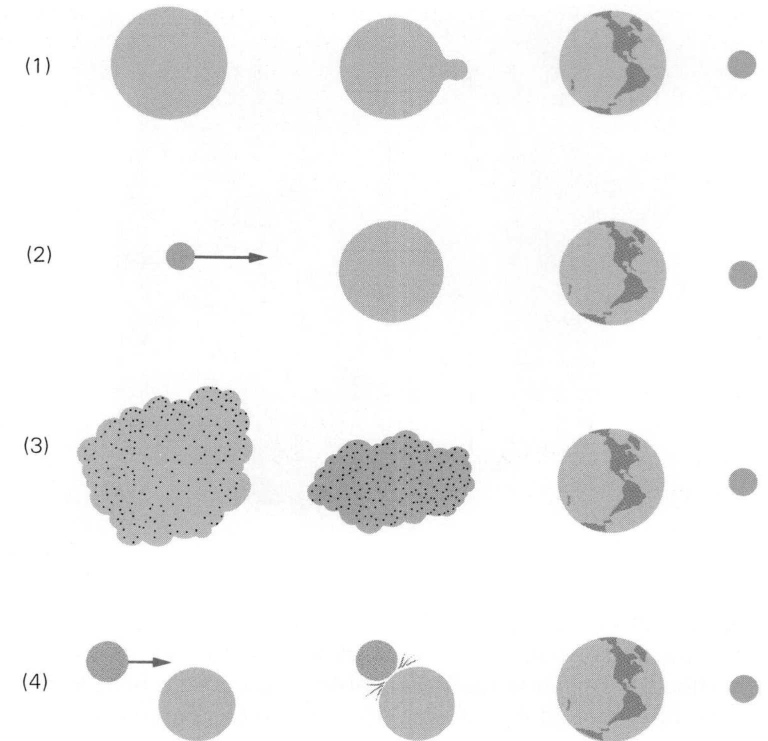

Figure 17-44 Four theories of the moon's origin: (1) The moon split away from the earth. (2) The moon was captured as it approached the earth from elsewhere. (3) The earth and moon were formed together from different clouds of particles. (4) Another early planet struck the earth and formed a larger earth plus the moon. The last theory seems the most plausible.

origin, the other planet's iron core was added to the earth's core. The orbital angular momentum that the moon has today can be accounted for if the collision were a glancing one rather than head-on.

Pulling so dramatic an event as a collision between two planets out of a hat always arouses suspicion among scientists. Why should we take this idea any more seriously than the discarded hypotheses mentioned earlier? The basic answer is that detailed computer simulations of such a collision and its aftermath fit well with what we know about the earth and moon and with what is strongly believed about the early history of the solar system. In addition, if similar collisions also occurred elsewhere in the solar system, then a number of other long-standing oddities can also be understood. For instance, we have already noted that a large-scale collision could have knocked Uranus into its sideways rotation, so it is possible that the moon's origin is just one example of a normal phenomenon.

Important Terms and Ideas

Comets and **meteoroids** are relatively small objects that pursue regular orbits in the solar system. Comets glow partly by the reflection of sunlight but mainly through the excitation of their gases by solar ultraviolet radiation. **Meteors** are the flashes of light that meteoroids produce when they enter the earth's atmosphere. Fallen meteoroids are called **meteorites.**

In order from the sun, the planets are **Mercury, Venus, Earth, Mars, Jupiter, Saturn, Uranus,** and **Neptune.** All but Mercury and Venus have satellites. There are also a number of **dwarf planets,** notably **Pluto.** In addition, thousands of **asteroids** are in orbits that lie between those of Mars and Jupiter and outside Neptune's orbit. All these bodies are visible by virtue of the sunlight they reflect.

The **inner planets** (Mercury, Venus, Earth, and Mars) are considerably smaller, less massive, and denser than the **outer planets,** which apparently consist largely of hydrogen, helium, and hydrogen compounds such as methane, ammonia, and water vapor. Jupiter, Saturn, Uranus, and Neptune are circled by **rings** of particles.

The **moon** shines by virtue of reflected sunlight. The **phases** of the moon occur because the area of its illuminated side visible to us varies with the position of the moon in its orbit. A **lunar eclipse** occurs when the earth's shadow obscures the moon, a **solar eclipse** when the moon obscures the sun.

The moon has neither an atmosphere nor liquid surface water. Although geologically inactive at present, its surface shows signs of having once melted and of having experienced many volcanic eruptions long ago. Meteoroid bombardment has been a more recent factor in shaping its landscapes.

Multiple Choice

1. Comets
- **a.** follow orbits around the earth
- **b.** follow orbits around the sun
- **c.** move randomly through the solar system
- **d.** are the tracks left in the atmosphere by meteoroids

2. Comets consist largely of
- **a.** hydrogen **c.** ice and dust
- **b.** iron **d.** rocks

3. The tail of a comet
- **a.** is constant in size
- **b.** always points away from its direction of motion
- **c.** always points toward the sun
- **d.** always points away from the sun

4. Meteors occur in the night sky
- **a.** only during spring and fall
- **b.** only during specific shower periods
- **c.** in approximately the same numbers every night
- **d.** every night, but with greater frequency during shower periods

5. On an average clear night, the number of meteors that can be seen per hour is
- **a.** 1 or 2 **c.** 100 to 200
- **b.** 5 to 10 **d.** 1000 to 2000

6. Meteoroids
- **a.** come from the sun
- **b.** come from the moon
- **c.** come from outside the solar system
- **d.** are members of the solar system

7. The majority of meteoroids consist of
- **a.** silicate minerals **c.** iron
- **b.** calcite **d.** nickel

8. The planets shine because they
- **a.** emit light **c.** reflect moonlight
- **b.** reflect sunlight **d.** reflect starlight

9. The planet nearest the sun is
- **a.** Mercury **c.** Saturn
- **b.** Venus **d.** Neptune

10. The smallest of the following planets is
- **a.** Mars **c.** Saturn
- **b.** Jupiter **d.** Uranus

11. The largest of the following planets is
- **a.** Mars **c.** Saturn
- **b.** Jupiter **d.** Uranus

12. The planet closest to the earth in size and mass is
- **a.** Mercury **c.** Mars
- **b.** Venus **d.** Uranus

13. The length of the year is shortest on
- **a.** Mercury **c.** Earth
- **b.** Venus **d.** Mars

14. A planet with virtually no atmosphere is
- **a.** Mercury **c.** Jupiter
- **b.** Mars **d.** Saturn

15. The planet that appears brightest in the sky is
- **a.** Venus **c.** Jupiter
- **b.** Mars **d.** Saturn

16. The planet whose surface most resembles that of the moon is
- **a.** Mercury **c.** Jupiter
- **b.** Venus **d.** Saturn

17. An astronaut would weigh least on the surface of
- **a.** Venus **c.** Mars
- **b.** Earth **d.** Jupiter

18. A dense atmosphere of carbon dioxide is found on
- **a.** Mercury **c.** Mars
- **b.** Venus **d.** the moon

19. A planet with no known satellites is
- **a.** Venus **c.** Saturn
- **b.** Jupiter **d.** Uranus

20. The surface of Titan, Saturn's largest moon,
- **a.** has many active volcanoes
- **b.** shows signs of erosion by flowing liquid
- **c.** is covered with liquid water
- **d.** is covered with water ice

21. The largest of Jupiter's satellites is about the size of
- **a.** a large ship **c.** Mercury
- **b.** the moon **d.** the earth

22. A planet that has phases like those of the moon is
- **a.** Venus **c.** Jupiter
- **b.** Mars **d.** Saturn

23. Compared with the earth, Mars has
- **a.** a denser atmosphere
- **b.** more surface water
- **c.** a lower average surface temperature
- **d.** a shorter year

24. The spacecraft that traveled to Mars did not find
- **a.** heavily cratered regions
- **b.** lava flows
- **c.** dust storms
- **d.** living things

25. The rings of Saturn
- **a.** are gas clouds
- **b.** are sheets of liquid
- **c.** are sheets of solid rock
- **d.** consist of separate particles

26. The planet with the most mass is
- **a.** Earth
- **b.** Jupiter
- **c.** Saturn
- **d.** Neptune

27. A planet that consists largely of hydrogen and helium is
- **a.** Mercury
- **b.** Venus
- **c.** Mars
- **d.** Jupiter

28. Most asteroids lie in a belt between
- **a.** Mercury and the sun
- **b.** Earth and Mars
- **c.** Earth and the moon
- **d.** Mars and Jupiter

29. The dwarf planets have diameters of about
- **a.** 1 km
- **b.** 10 km
- **c.** 1000 km
- **d.** that of the moon

30. The asteroids probably consist largely of
- **a.** ice
- **b.** frozen methane
- **c.** silicate minerals
- **d.** metallic iron

31. An asteroid 2 km or more across could cause a worldwide catastrophe. Such an asteroid is thought to collide with the earth an average of every
- **a.** thousand years
- **b.** million years
- **c.** billion years
- **d.** 10 billion years

32. Relative to the time the moon takes to circle the earth, the period of its rotation on its axis is
- **a.** shorter
- **b.** the same
- **c.** longer
- **d.** any of these, depending on the time of the year

33. The moon's diameter is about
- **a.** $\frac{1}{10}$ that of the earth
- **b.** $\frac{1}{4}$ that of the earth
- **c.** $\frac{1}{2}$ that of the earth
- **d.** $\frac{3}{4}$ that of the earth

34. At new moon, the moon is
- **a.** between the earth and the sun
- **b.** on the opposite side of the earth from the sun
- **c.** on the opposite side of the sun from the earth
- **d.** to the side of a line between the earth and the sun

35. An eclipse of the moon occurs when the
- **a.** moon passes directly between the earth and the sun
- **b.** sun passes directly between the earth and the moon
- **c.** earth passes directly between the sun and the moon
- **d.** moon's dark side faces the earth

36. Eclipses of the sun occur at
- **a.** new moon
- **b.** first or last quarter
- **c.** full moon
- **d.** anytime

37. Eclipses of the moon occur at
- **a.** new moon
- **b.** first or last quarter
- **c.** full moon
- **d.** anytime

38. The average density of the moon is
- **a.** lower than that of the earth
- **b.** about the same as that of the earth
- **c.** higher than that of the earth
- **d.** unknown

39. The moon's surface is
- **a.** perfectly smooth
- **b.** irregular but relatively smoother than the earth's surface
- **c.** about as irregular as the earth's surface
- **d.** much more irregular than the earth's surface

40. The moon's maria are
- **a.** bodies of water like the earth's oceans
- **b.** solid lava flows
- **c.** lava flows pulverized by meteoroid impacts
- **d.** hardened sediments

41. Most of the craters on the moon are probably the result of
- **a.** volcanic action
- **b.** meteoroid bombardment
- **c.** erosion
- **d.** collisions with asteroids

42. The moon's surface shows no signs of
- **a.** volcanic action
- **b.** meteoroid bombardment
- **c.** ever having melted
- **d.** glacier erosion

43. The time needed for the Apollo spacecraft to reach the moon was about
- **a.** 3 h
- **b.** 3 days
- **c.** 3 weeks
- **d.** 3 months

44. Relative to the oldest rocks that have been found on the earth, the oldest known lunar rocks are
- **a.** very much younger
- **b.** somewhat younger
- **c.** about the same age
- **d.** older

45. The interior of the moon is probably entirely or almost entirely
- **a.** liquid
- **b.** solid
- **c.** gaseous
- **d.** hollow

Exercises

17.1 The Solar System

1. Which planets are visible to the unaided eye?

2. (a) How is it possible to distinguish the planets from the stars by observations with the unaided eye? (b) By observations with a telescope?

3. Which is the largest planet? The smallest? Which planet is nearest the sun?

4. Why do the planets shine?

5. Which planets, if any, have no satellites?

6. Is the mass of the solar system concentrated in the sun or in the planets?

7. On which planets would a person weigh less than on the earth? On which planets would a person weigh more?

8. Suppose you were on Mars and watched the earth with the help of a telescope. What changes would you see in the earth's appearance as it moves around the sun in its orbit?

17.2 Comets

9. How do the orbits of comets compare in shape with the orbits of the planets?

10. Why do comets have tails only in the vicinity of the sun? Why do these tails always point away from the sun, even when the comet is receding from it?

11. When a comet is close enough to the sun to be seen from the earth, stars are visible through both the comet's head and tail. What does this imply about the danger to the earth from a collision with a comet?

17.3 Meteors

12. The Perseid meteor shower appears early every August. Does this mean that the orbits of the meteoroids in the Perseid swarm all have periods of exactly 1 year?

13. Over 90 percent of the meteorites found after a known fall are stony, yet most of the meteorites in museums are iron. Why do you think this is so?

14. If the earth had no atmosphere, would comets still be visible from its surface? Would meteors?

15. Why are most meteorites found in Antarctica?

16. Why are meteoroids believed to come from within the solar system?

17.4 Mercury

17. Why is it very unlikely that there is life on Mercury?

18. Mercury takes 59 of our days to turn completely on its axis, but the period of time between one sunrise and the next on Mercury is 176 of our days. What do you think is the reason for the difference?

17.5 Venus

19. Why are Mercury and Venus always seen either around sunrise or around sunset?

20. Venus is the brightest planet in the sky. How does its brightness compare with that of the brightest stars?

21. Why are Mercury and Venus the only planets that show phases like those of the moon?

22. Venus is brighter when it appears as a crescent than when we can see its full disk. Why?

23. Far fewer meteoroid craters of ancient origin have been observed on Venus than expected. Could running water have eroded the others? If not, what might be the reason?

24. Give two reasons why the surface of Venus is so much hotter than the earth's surface.

17.6 Mars

17.7 Is There Life on Mars?

25. Give three reasons why Venus is a brighter object in the sky than Mars.

26. Compare the likelihood that an astronaut on Mars will be struck by a meteoroid with its likelihood on earth.

27. Mars has surface features that seem to be the result of erosion by running water. Why does the presence of many meteoroid craters in some of these regions suggest that the running water disappeared there long ago?

28. What is the chief ingredient of the atmosphere of Mars?

29. Why does Mars appear red?

30. Are any volcanoes active on Mars today?

31. Why is ultraviolet radiation from the sun more of a hazard to life on Mars than on the earth?

17.8 Asteroids

32. Distinguish between asteroids and meteoroids.

33. What is believed to be the origin of the asteroids?

34. Why are few asteroids spherical, as planets are?

35. Is there any evidence that an asteroid ever collided with the earth? Could such a collision occur in the future?

17.9 Jupiter

36. Why is Jupiter thought not to consist mainly of rock with an iron core, as does the earth? Of what does it mainly consist?
37. What is believed to be the nature of the Great Red Spot on Jupiter?
38. Jupiter is relatively far from both the sun and the earth yet it can be seen in the sky with the unaided eye. Why is this possible?
39. The interior of Jupiter's satellite Io is thought to be heated through being flexed by the gravitational pulls of Jupiter and two other satellites. Why is Io's interior believed to be hot?

17.10 Saturn

40. What are the chief similarities between Jupiter and Saturn? The chief differences?
41. Why are Saturn's rings believed to consist of small particles rather than being solid sheets of matter or thin gas clouds?
42. Is it likely that Saturn's rings are permanent features?
43. Saturn's satellite Titan has an atmosphere. Do any of the other planetary satellites also have atmospheres?

17.11 Uranus, Neptune, Pluto, and More

44. Which planet resembles the earth most in size and mass? In surface conditions?
45. Is there any evidence that planets other than the earth today have crusts that consist of huge moving plates?
46. (a) Which planets besides Saturn have rings? (b) What is the nature of these rings?
47. What are thought to be the chief constituents of the giant planets Jupiter, Saturn, Uranus, and Neptune?
48. (a) What is the chief distinction between planets and dwarf planets? (b) What are the chief similarities? (c) What is the chief distinction between dwarf planets and planetary satellites? (d) Do any dwarf planets have satellites?
49. How does Pluto compare in size with the moon? With the satellites of the other planets?

17.12 Phases of the Moon

50. We always see the same hemisphere of the moon. Why?
51. What is wrong with the statement that the moon is more useful to us than the sun because the moon provides illumination at night when it is needed most?
52. The moon rises in the east at midnight on a certain night. Can it appear as a full moon?
53. Approximately how much time elapses between new moon and full moon?
54. Is the moon the largest satellite in the solar system? The smallest?
55. To what approximate length of time on the earth does the length of "day" at a given place on the moon correspond? The length of "night"?
56. Is the time the moon takes to complete an orbit around the earth relative to the sun the same as the moon's orbital period relative to the stars? If not, which period is longer and why is there a difference?

17.13 Eclipses

57. If the moon were smaller than it is, would total eclipses of the sun still occur? Would total eclipses of the moon still occur?
58. Eclipses of the sun and of the moon do not occur every month. Why not?
59. In what phase must the moon be at the time of a solar eclipse? At the time of a lunar eclipse?

17.14 The Lunar Surface

17.15 Evolution of the Lunar Surface

60. The moon's surface is about as irregular as that of the earth. What does this imply about temperatures in the moon's interior?
61. The moon's maria are dark, relatively smooth regions conspicuous even to the naked eye. What is their nature?
62. Moonquakes are weaker and occur much less often than earthquakes. What do these facts imply about temperatures in the moon's interior?
63. Why is it believed that the moon's interior is different in composition from the earth's interior?
64. Why is it believed that large-scale igneous activity ceased on the moon about 3 billion years ago?
65. What has been the chief influence that shaped the lunar landscape during the past 3 billion years?

Answers to Multiple Choice

1. b	**6.** d	**11.** b	**16.** a	**21.** c	**26.** b	**31.** b	**36.** a	**41.** b
2. c	**7.** a	**12.** b	**17.** c	**22.** a	**27.** d	**32.** b	**37.** c	**42.** d
3. d	**8.** b	**13.** a	**18.** b	**23.** c	**28.** d	**33.** b	**38.** a	**43.** b
4. d	**9.** a	**14.** a	**19.** a	**24.** d	**29.** c	**34.** a	**39.** c	**44.** d
5. b	**10.** a	**15.** a	**20.** b	**25.** d	**30.** c	**35.** c	**40.** c	**45.** b

18

The Stars

The Crab Nebula has a pulsar at its heart.

Goals

When you have finished this chapter you should be able to complete the goals ▶ given for each section below:

Tools of Astronomy

18.1 The Telescope
All Modern Ones Are Reflectors

18.2 The Spectrometer
Without It, Little Would Be Known about the Stars

18.3 Spectrum Analysis
Spectra Can Tell Us a Surprising Amount

▶ Describe how the spectrum of a star can provide information on the star's structure, temperature, composition, condition of matter, magnetism, and motion.

▶ Interpret in terms of stellar structure the observation that nearly all stars have absorption (dark line) spectra.

The Sun

18.4 Properties of the Sun
The Nearest Star

▶ State what is meant by the photosphere of the sun.

18.5 The Aurora
Fire in the Sky

▶ Describe the appearance in the sky and the origin of auroras.

18.6 Sunspots
They Come and Go in an 11-Year Cycle

▶ Discuss sunspots, the sunspot cycle, and some effects on the earth that are correlated with sunspot activity.

18.7 Solar Energy
It Comes from the Conversion of Hydrogen to Helium

▶ Explain why solar energy cannot come from combustion.

▶ Identify the basic process that gives rise to solar energy.

▶ Describe how the elements more massive than hydrogen are created and distributed throughout the universe.

The Stars

18.8 Stellar Distances
Not Easy to Measure

▶ Define light-year.

▶ Describe the parallax method of finding the distance to a star.

▶ Describe how the distance to a star can be found by comparing its apparent and intrinsic brightness.

18.9 Variable Stars
Stars Whose Brightness Changes, Usually in Regular Cycles

▶ Explain how Cepheid variable stars are used to find the distances of star groups.

18.10 Stellar Motions
The Stars Are Not Fixed in Space

18.11 Stellar Properties
Mass, Temperature, and Size

▶ Describe how the mass and size of a star can be found.

▶ Account for the relatively small range of stellar masses.

Life Histories of the Stars

18.12 H-R Diagram
Most Stars Belong to the Main Sequence

▶ State what is plotted on a Hertzsprung-Russell (H-R) diagram.

▶ Draw an H-R diagram and indicate the positions of main-sequence stars, red giants, and white dwarfs.

▶ Compare the properties of red giants and white dwarf stars.

18.13 Stellar Evolution
Life History of a Star

▶ Outline the life history of an average star like the sun.

▶ State what a brown dwarf is.

18.14 Supernovas
Exploding Stars

▶ Outline the life history of a very massive star.

▶ State what a supernova is.

18.15 Pulsars
Spinning Neutron Stars

▶ Define neutron star and pulsar and discuss the connection between them.

18.16 Black Holes
Even Light Cannot Escape from Them

▶ Describe what black holes are and explain how they can be detected.

The study of the stars began in earnest toward the end of the eighteenth century with the work of William Herschel. Herschel sought among the stars some kind of order, something as profound as the regularities Copernicus, Kepler, and Newton had found in the solar system. Like a pioneer in any other branch of science, Herschel began with observation and spent many years cataloging stars and measuring their apparent motions. From this study he was able to verify a structure for the universe that is not far from the one that today's astronomers believe to be correct.

Of the billions of stars in the universe, none (besides the sun) appears as more than a point of light to even the most powerful telescope. Less than a century ago, most scientists despaired of ever knowing the physical nature of the stars. Today, however, thanks to spectroscopic analysis, we not only have a great deal of detailed information on thousands of stars but also are able to trace the evolution of a star from its birth through maturity to its last agonies and eventual fate.

Tools of Astronomy

Light is the messenger that brings us information about the universe. Because the information arrives in code, so to speak, the astronomer must decipher it before being able to assess its significance. The tools of astronomy are devices that collect light and sort it into its component wavelengths, which are the elements of the code.

18.1 The Telescope

All Modern Ones Are Reflectors

In Herschel's time, as in ours, the telescope was the basic astronomical instrument. Much of his success was due to the improvements he introduced in telescope construction. Herschel was the first to build and use a large reflecting telescope, an instrument in which light is reflected from a concave mirror instead of being refracted through a lens (Fig. 18-1). A large lens tends to sag under its own weight, with the change in shape producing a distorted image, whereas even a sizable mirror can be adequately supported from behind. In addition, there is no problem of the dispersion of light of different wavelengths with a mirror (see Sec. 7.15).

Modern astronomical telescopes are reflectors. The 5-m-diameter Hale Telescope on Mt. Palomar, California, was the largest single-mirror reflector from 1948 to 1999, when the 8.3-m Subaru reflector was installed on Mauna Kea in Hawaii. All other modern large reflectors do not rely on single mirrors, which are limited in size by practical difficulties. Instead, a number of individual mirrors are linked to produce a single image. In one approach, hexagonal mirror segments are used to give a large collecting surface (Fig. 18-2). In another, separate circular mirrors are used. A telescope of this type with four 8.2-m mirrors has been installed on a Chilean mountain where the air is especially dry, stable, and clear; it should be able to observe an astronaut on the moon.

A still larger such telescope will have seven 8.4-m mirrors to give it more collecting area plus 10 times the resolving ability of the Hubble Space Telescope. Called the Giant Magellan Telescope (GMT), it will also be installed in Chile and—it is hoped—will start in 2015 to help solve such deep remaining astronomical mysteries as the natures of dark matter and dark energy, which we shall meet in Chap. 19.

Being considered for even farther in the future is OWL—the name comes both from the keen-eyed bird and from the telescope's *o*ver*w*helmingly *l*arge size—whose mirror, 100 m in diameter, would consist of 3048 adjustable hexagonal segments; its total weight would be 15,000 tons.

Twinkling stars, the result of turbulent motions in the atmosphere, delight poets but not astronomers. A cure for such flickering images, called adaptive optics, uses light from a bright "guide" star or from a reflected laser beam to detect the distortions, which are continuously corrected by adjusting a deformable mirror in the optical path of the telescope. Such a system can produce images as sharp as those of a telescope in space.

Size Matters In stellar astronomy the purpose of a big telescope is not magnification, for the stars are too distant to ever appear as more than points of light. One virtue of large mirrors and lenses lies in their light-gathering power, which enables more light from a given object to

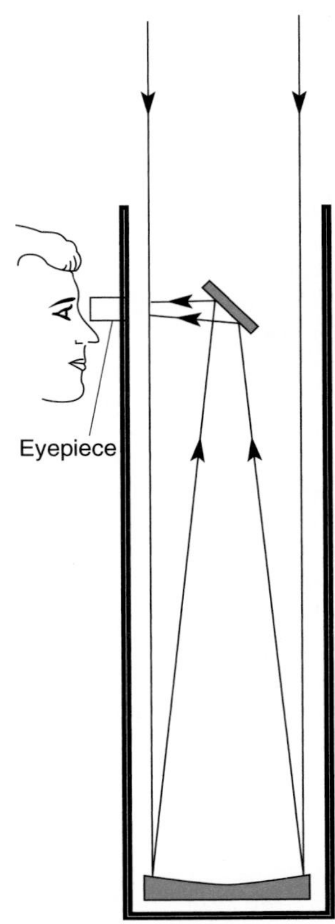

Figure 18-1 Simplified sketch of one type of reflecting telescope. The concave mirror produces a real image (see Chap. 7) of a distant object that is reflected outside the telescope by another mirror, normally for recording electronically or for spectroscopic analysis. When used in this way the telescope does not need additional optics—it acts as a giant telephoto lens. However, for direct viewing, a lens or set of lenses called an eyepiece is needed to give a virtual image whose object is the image produced by the mirror.

Figure 18-2 Each of the twin Keck telescopes atop Hawaii's Mauna Kea volcano has a mosaic mirror 10 m in diameter that consists of 36 hexagonal segments. The segments can be continuously adjusted to compensate for any distortions that may arise. Because the Keck telescopes are so much larger than the Hubble Space Telescope, they can see much fainter objects, but atmospheric turbulence reduces their resolving ability below that of the Hubble. Only four mirror segments were in place when this photograph was taken. A telescope with a still larger segmented mirror, 10.4 m across, has been installed on a peak on La Palma, one of Spain's Canary Islands.

Telescopes in Space

A telescope orbiting the earth in a satellite has a number of advantages over its earthbound cousins. An important one is that there is no atmosphere in the way to blur images and absorb various wavelengths of incoming radiation. Nor is there background radiation from the earth to swamp faint signals from space.

The first major telescope in orbit operating at visible wavelengths, the Hubble Space Telescope, named for the astronomer Edwin Hubble (biography in Sec. 19.6), has a mirror 2.4 m in diameter and was launched in 1990 (Fig. 18-3); it orbits the earth 5800 times each year. An unceasing stream of marvels has been revealed by the Hubble telescope, from close-ups of the planets to views of distant galaxies. For the cost of two B-1 bombers, it has transformed our knowledge of the astronomical universe.

Other telescopes in orbit are Chandra, which studies x-rays such as those emitted by the supernovas described in Sec. 18.14, and Spitzer, which operates in the infrared part of the spectrum. Both are named for astronomers, respectively Subramanyan Chandrasekhar and Lyman Spitzer, Jr.

Future space telescopes will be able to see yet farther into the depths of the universe. One of them, the James Webb Space Telescope (named for a past chief of NASA), is planned for launch in 2013. It will have a mirror 6.5 m in diameter that will give it 7 times the light-gathering ability of the Hubble telescope. To get the mirror into space, it will be made in 18 lightweight beryllium segments folded together that will unfurl when the telescope is in place. Each hexagonal segment will be very thin and have four actuators to control its shape precisely.

Because it orbits the earth, the Hubble telescope cannot make continuous observations of all parts of the sky. To avoid this limitation, the Webb telescope will hover in space at a place called a Lagrange point where the gravitational forces of the earth and sun balance out exactly. The Webb telescope will operate in the infrared because the expansion of the universe (Sec. 19.6) shifts the light from distant objects to this part of the spectrum through the doppler effect—the farther away an object is, the faster it appears to be receding from us, and the greater the shift. It is hoped that the new telescope will be able to study the formation of the earliest stars and galaxies. The telescope should also be able to detect Jupiter-sized planets that may orbit nearby stars.

Finding planets is also the task of the Corot Space Telescope, launched in 2007, whose 27-cm mirror will look at 120,000 stars and will be able to determine the sizes of any planets that are found. The Terrestrial Planet Finder, which NASA hopes to send off by 2011, will be able to detect signs of water, oxygen, and carbon dioxide, suggestive of the possibility of life, on any planets seen around other stars. Still other future space telescopes will study ultraviolet radiation, x-rays, and gamma rays from near and far sources.

be collected. Thus faint objects that would otherwise be invisible are revealed by a large telescope, and more light from brighter objects is available for study.

The second advantage of a large telescope is its ability to distinguish, or **resolve,** small details. As mentioned in Sec. 7.17, the diffraction of light waves causes every optical image to be blurred to a certain extent. The larger the lens or mirror, the less the blurring and the sharper the image.

Originally the light collected by a telescope went to an astronomer's eye, but later photographic plates were used and today electronic sensors are preferred. The latter methods have the advantage that they respond to the *total amount* of light that falls on them over the period of time they are exposed. The eye, on the other hand, responds only to the *brightness* of the light that reaches it. A telescope with a camera or sensor attached can be trained on the same area of the sky for hours or, if necessary, for several nights to detect objects too faint for the eye to pick up. Photographic plates (over 2 million in storage, some a century old) and, today, electronic methods of data storage provide permanent records that enable positions and properties of stars as they appear today to be compared with what they were years ago and with what they will be in years to come.

Figure 18-3 The Hubble Space Telescope was put in orbit in 1990 and is still in operation. Its record of discoveries is unsurpassed in variety and importance.

18.2 The Spectrometer

Without It, Little Would Be Known about the Stars

By itself, a telescope is of limited use in studying the stars. What is needed is a combination of a telescope and a spectrometer the same instrument that contributed so much to our knowledge of atomic structure (see Chap. 9). A spectrometer breaks light up into its separate wavelengths, as shown in Fig. 9-17. The resulting band of colors, with each wavelength separate from the others, is the spectrum that is recorded on a photographic plate or electronic medium.

The spectrum of a star does not seem impressive. If photographed in natural colors, it generally consists of a rainbow band crossed by a multitude of fine dark lines. Ordinarily color film is not used, so the spectrum shows simply black lines on a light gray background.

At first glance it does not seem that a few black lines on a photographic plate can get us very far in understanding the stars. But each of those lines has its own story to tell about how it was produced, and a specialist can piece together data from different lines into a comprehensive picture of a star. Some types of information obtainable from spectra are outlined in the next section.

A serious problem in astronomical spectroscopy is the absorption of light in the earth's atmosphere. Spectrometers mounted in sounding rockets, high-altitude balloons, satellites, and spacecraft are needed to study those parts of solar and stellar spectra that cannot reach the earth's surface.

18.3 Spectrum Analysis

Spectra Can Tell Us a Surprising Amount

Structure A spectrum of dark lines on a continuous colored background is an **absorption spectrum;** it is produced when light from a hot object passes through a cooler gas (see Sec. 9.8). Atoms and molecules of the gas absorb light of certain wavelengths and so leave narrow gaps in the band of color. Thus a star that has this kind of spectrum (and nearly all of them do) reveals at once something of its structure: it must have a hot, glowing interior surrounded by a relatively cool gaseous atmosphere.

Temperature From the continuous background of a star's spectrum, astronomers can find the temperature of its surface. What they need to know is where in the spectrum the star's radiation is brightest. Since the wavelength of maximum intensity decreases as the temperature rises, the point of maximum intensity in the spectrum is a measure of temperature (Fig. 18-4). Thus the hottest stars are blue-white (maximum intensity at the short-wavelength end of the spectrum), stars of intermediate temperature are orange-yellow, and the coolest visible stars are red. This relation holds for materials on the earth as well as for the stars, as we know from experience (see Fig. 5-5).

Composition Each element has a spectrum of lines with characteristic wavelengths. The elements present in a star's atmosphere can therefore be identified from the dark lines in its spectrum. In principle, all we have to do is measure the wavelength of each line in the spectrum and compare these wavelengths with those produced by various elements in the laboratory.

Condition of Matter In practice the identification of lines in a star's spectrum is not quite so easy. The wavelengths and intensities of the lines characteristic of a given element depend not only on the element but also on such conditions as temperature, pressure, and degree of ionization. These difficulties prove to be blessings in disguise, however, for once the lines are identified, they tell us not only which elements are present in

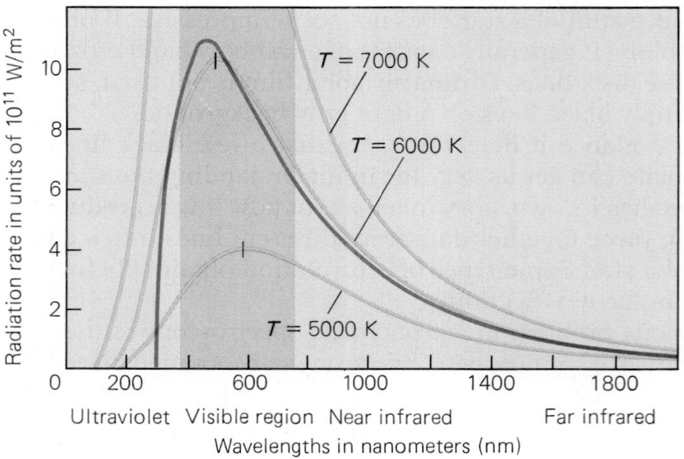

Figure 18-4 Relative intensity of the wavelengths of light emitted by bodies with the temperatures indicated. The wavelength of greatest intensity is shorter for hot bodies than for cooler ones. The red curve represents measurements of the sun's photosphere. (1 nm = 10^{-9} m)

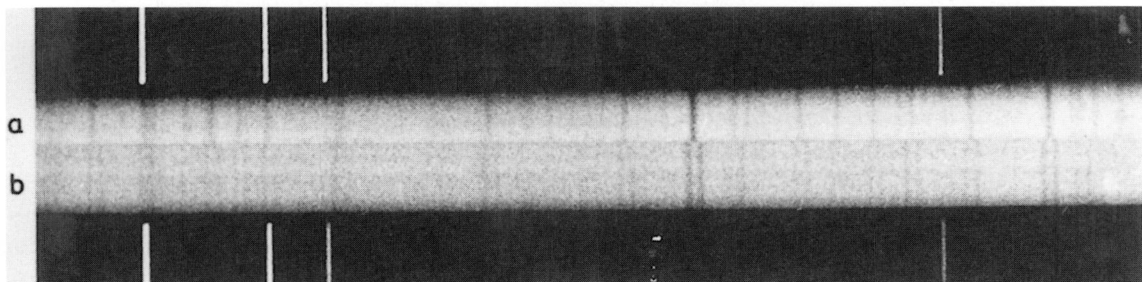

Figure 18-5 Spectra of the double star Mizar, which consists of two stars that circle each other, taken 2 days apart. In (*a*) the stars are in line with no motion toward or away from the earth, so their spectral lines are superimposed. In (*b*) one star is moving toward the earth and the other is moving away from the earth, so the spectral lines of the former are doppler shifted toward the blue end of the spectrum and those of the latter are shifted toward the red end.

a star's atmosphere but also something about the physical conditions in which the elements exist.

Chemical compounds also have spectral lines of recognizable wavelengths, so spectra provide a means of determining how much of the matter in a star's atmosphere is in the form of molecules rather than atoms.

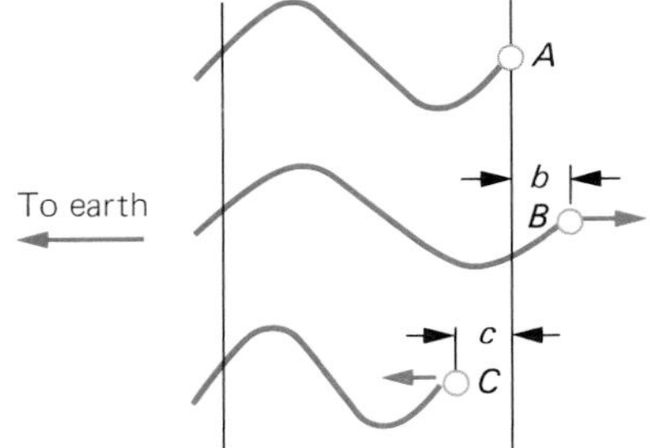

Figure 18-6 The doppler effect in stellar spectra. Star *A* is stationary with respect to the earth. Star *B* is moving away from the earth; it moves the distance *b* during the emission of one light wave, whose wavelength is therefore increased by *b*. Star *C* is approaching the earth; it moves the distance *c* during the emission of one wave, whose wavelength is thus decreased by *c*. Hence stars receding from the earth have spectral lines shifted toward the red (long-wavelength) end, while stars approaching the earth have spectral lines shifted toward the blue (short-wavelength) end.

Magnetic Fields The presence of a magnetic field causes individual energy levels within atoms to divide into several sublevels. When such atoms are excited and radiate, their spectral lines are accordingly split, each into a number of lines close to the original one. This phenomenon is called the **Zeeman effect** after its discoverer, the Dutch physicist Pieter Zeeman. With the help of the Zeeman effect the magnetic nature of sunspots has been established, and a large number of stars and clouds of matter in space have been discovered that appear to be strongly magnetized.

Motion

As we learned in Chap. 7, the doppler effect causes sounds produced by vehicles moving toward us to seem higher pitched than usual, whereas sounds produced by vehicles receding from us seem lower pitched than usual. Similarly a star moving toward the earth has a spectrum whose lines are shifted toward the blue (high-frequency) end, and a star moving away from the earth has a spectrum in which each line is shifted toward the red (low-frequency) end, as in Figs. 18-5 and 18-6. From the amount of the shift we can calculate the speed with which the star is approaching or receding.

The Sun

The sun is the glorious body that dominates the solar system, and the origin and destiny of the earth are closely connected with its life cycle. The astronomer has another reason for studying the sun closely, for it is in many ways a typical star, a rather ordinary member of the assembly of perhaps 10^{20} stars that make up the known universe. Thus the properties of the sun that we can observe by virtue of its relative closeness are interesting not only in themselves but also because they give us information about stars in general that would otherwise be out of reach.

18.4 Properties of the Sun

The Nearest Star

The sun's mass can be found from the characteristics of the earth's orbital motion around it. The result is a mass of 1.99×10^{30} kg, more than 300,000 times the earth's mass. The sun's radius of 6.96×10^{8} m can be established by simple geometry from the fact that its angular diameter as seen from the earth is 0.53°. The volume of the sun is such that 1,300,000 earths would fit into it.

The gases in the sun's interior are very hot, so they emit light copiously, and they are very dense, so the light is reabsorbed almost at once. Because both temperature and density fall off with distance from the sun's center, eventually there is a region in which the gases are still hot enough to radiate a great deal of light but not dense enough to prevent the light from escaping (Fig. 18-7). This region, called the **photosphere,** is what we see as the "surface" of the sun, although it has no sharp boundary. The temperature of the photosphere, 5800 K, is found in two ways: from the shape of its spectrum (see Fig. 18-4) and from the rate at which it gives off energy.

Of the thousands of lines in the sun's spectrum, about half have been traced to specific elements. The others probably come from highly excited energy levels or energy levels in ions rather than atoms. Lines of few compounds are found since the photosphere is hot enough to break up nearly all molecules.

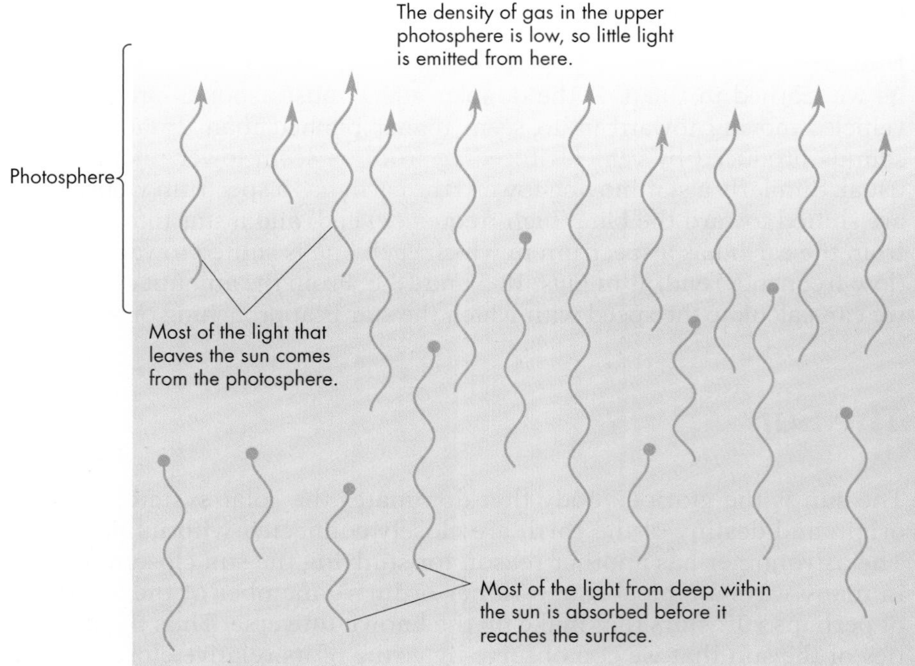

Figure 18-7 The light the sun emits comes from a thin outer layer called the photosphere. The photosphere is so thin (only 0.03 percent of the sun's radius) that it appears as a sharp surface.

Although conditions on the sun are very different from those on the earth, the basic matter of the two bodies appears to be the same. Even the relative amounts of different elements are similar, except that there is much more of the light elements hydrogen (72 percent by mass) and helium (27 percent) on the sun. At the relatively low temperatures here on the earth, most elements have combined to form compounds; in the sun's interior the elements occur as individual atoms or ions.

Above the Photosphere A rapidly thinning atmosphere, mainly of hydrogen and helium, extends past the photosphere. From this atmosphere great flamelike **prominences** sometimes project into space, much like sheets of gas standing on their sides (Fig. 18-8). Prominences occur in a variety of forms; a typical example is about 200,000 km long, 10,000 km wide, and 50,000 km high. Prominences are often associated with sunspots and, like sunspots, seem to have magnetic fields associated with them.

During a total eclipse of the sun, when the moon obscures the sun's disk completely (see Fig. 17-37), a wide halo of pearly light can be seen around the dark moon. This halo, or **corona,** may extend out as much as a solar diameter and seems to have a great number of fine lines extending outward from the sun immersed in its general luminosity. The corona consists of ionized atoms, mainly protons, and electrons in extremely rapid motion; its temperature can reach 2 million K.

Although the corona we see during eclipses is relatively near the sun, it is found in very diffuse form much farther out, well beyond the earth's orbit. The outward flow of ions (mostly protons) and electrons in this extension of the sun's atmosphere is the **solar wind,** which leaves the sun at about 400 km/s. The solar wind has been detected by spacecraft and helps deflect comet tails away from the sun as well as causing auroras in the earth's upper atmosphere.

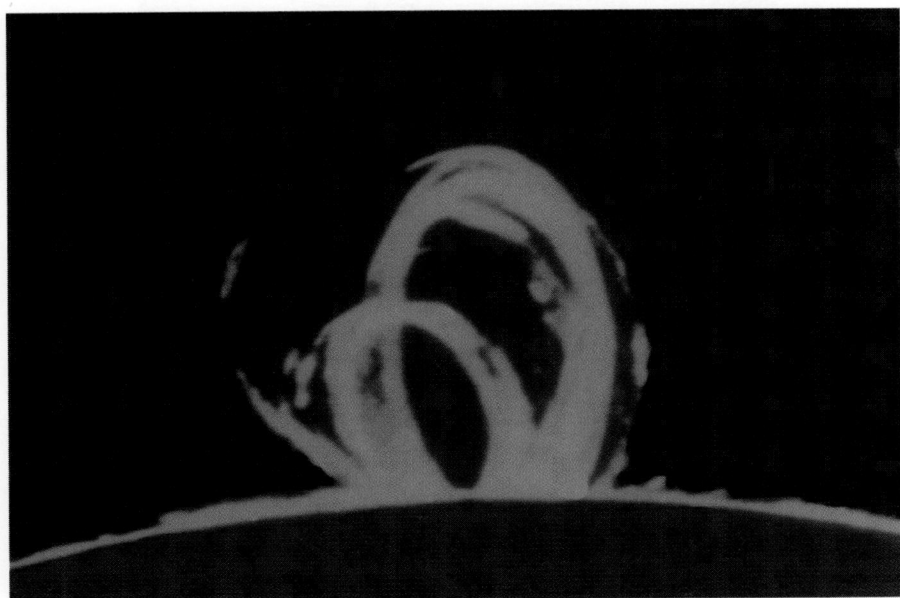

Figure 18-8 These solar prominences follow magnetic field lines.

18.5 The Aurora

Fire in the Sky

The aurora (or "northern lights") is one of nature's most awesome spectacles. In a typical auroral display, colored streamers seem to race across the sky, and glowing curtains of light pulsate as they change their shapes into weird forms and images. In the climax of the display the heavens seem on fire, with silent green and red flames dancing everywhere (Fig. 18-9). Then, after a while, the drama fades away, and only a faint reddish arc remains. Auroras are most common in the far north and far south. **Aurora borealis** is the name given (by Galileo) to this phenomenon in the northern hemisphere, and **aurora australis** in the southern. (Aurora was the Roman god of the dawn, boreas is Greek for "north wind," and australis is Greek for "of the south.")

Here is how an aurora that occurred in Russia in 1370 was described at the time: "During the autumn there were many signs in the sky. For many nights, people saw pillars in the sky and the sky itself was red, as if covered with blood. So red was the sky that even on the earth covered with snow all seemed red like blood, and this happened many times."

Auroras are caused by the solar wind. The protons and electrons take about a day to reach the earth. When they enter the upper atmosphere, they interact with the nitrogen and oxygen there so that light is given off. The process is similar to what occurs in a neon-filled glass tube when electricity is passed through it. The gas molecules are excited by charged particles moving near them, and this energy is then radiated as light in the characteristic wavelengths of the particular element. The green hues of an auroral display come from oxygen, the blues from nitrogen, and the reds from both oxygen and nitrogen. Auroras have also been observed in the atmospheres of Venus, Mars, Jupiter, and Saturn.

Figure 18-9 Aurora in Alaska. Auroras are caused by streams of fast protons and electrons from the sun that excite gases in the upper atmosphere to emit light.

The incoming streams of solar protons and electrons are affected by the earth's magnetic field in a complicated way. As a result most auroral displays occur in doughnutlike zones about 2000 km in diameter centered about the geomagnetic North and South Poles. Sometimes, though, the cloud of particles from the sun is so immense that auroras are visible elsewhere as well.

Even when auroras are not obvious as such, there is a faint glow in the night sky due to less concentrated streams of solar particles that interact with the upper atmosphere. The brightness of this **airglow** varies with solar activity, but it is always present to some extent (Fig. 18-10). Auroras and the airglow occur during the day as well as the night but are too dim to be visible in the daytime.

18.6 Sunspots

They Come and Go in an 11-Year Cycle

Dark patches called **sunspots** at times mar the intense luminosity of the sun's surface. Sunspots are cooler areas that appear dark only because we see them against a brighter background (Fig. 18-11). A spot whose temperature is 4500 K is hot enough to glow brilliantly but is considerably cooler than the rest of the solar surface, whose temperature is about 5800 K. They were first recorded by Chinese astronomers over 2000 years ago.

Sunspots change continually in form, each one growing rapidly and then shrinking, with lifetimes of from 2 to 3 days to more than a month. The largest sunspots are many thousands of km across, large enough to swallow several earths. Galileo noted in 1610 that they moved across the sun's disk, which he interpreted, as we do today, as a sign that the sun rotates on its axis. Solar rotation is confirmed by doppler shifts in the spectral lines of radiation from the edges of the sun's disk. The sun rotates faster at its equator, where a complete turn takes 27 days, than near its poles, where a complete turn takes about 31 days.

Sunspots generally appear in groups, each with a single large spot together with a number of smaller ones. Some groups contain as many as

Figure 18-10 This photograph of an aurora seen from above was taken from the American satellite Spacelab when it was halfway between Australia and Antarctica in 1985. The blue-green band and the tall red rays are the aurora. The brownish band along the earth's rim is the airglow, a faint luminescence of the atmosphere excited by streams of particles from the sun less concentrated than those responsible for the aurora.

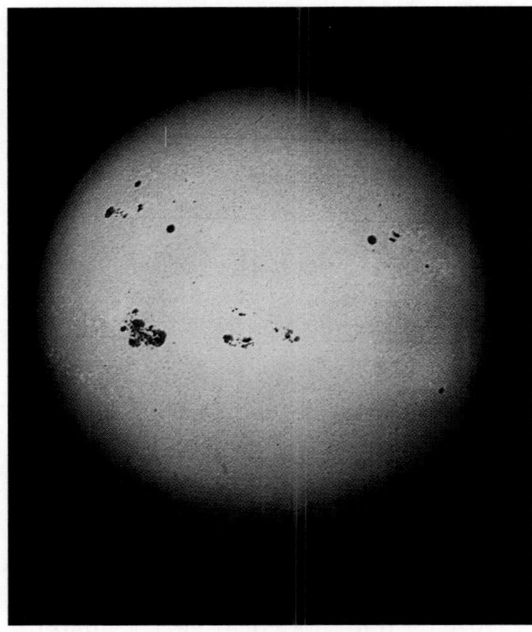

Figure 18-11 Sunspots appear dark because they are cooler than the rest of the solar surface, although quite hot themselves. Some sunspots are larger across than the earth. Most sunspots occur in groups.

80 separate spots. They tend to occur in two zones on either side of the solar equator and are rarely seen either near the equator or at latitudes on the sun higher than 35°. Strong magnetic fields are associated with sunspots, and there is little doubt these fields are involved in sunspot formation.

The number of spots on the sun increases and decreases with time in a regular cycle that covers about 11 years (Fig. 18-12). There is evidence that other stars also have cool spots, some of which come and go periodically like those on the sun. Many of these starspots are much larger than sunspots.

Sunspots and the Earth The sunspot cycle has aroused much interest because a number of effects observable on the earth—such as disturbances in its magnetic field, shortwave radio fadeouts, changes in cosmic-ray intensity, and unusual auroral activity—follow this cycle. It seems likely that the ionosphere changes that affect radio transmissions are due to intense bursts of ultraviolet and x-radiation that are more frequent during sunspot maximum. The magnetic, cosmic-ray, and auroral effects are due to vast streams of energetic protons and electrons that shoot out of the sun from the vicinity of sunspot groups to produce **solar storms** at the earth about 30 hours later.

Some aspects of weather and climate seem to be synchronized with sunspot activity. For instance, very few sunspots appeared between 1645 and 1715, a period during which temperatures worldwide were lower than usual—the "Little Ice Age" mentioned in Sec. 14.11 occurred at about that time. Apparently the events in the sun that cause sunspots are correlated with a slightly higher energy output. Even a small change in the sun's energy output would be enough to affect climates on the earth to the observed extent.

Solar Storms

Solar storms can have serious consequences. The fluctuating magnetic fields of a major storm in 1989 induced currents in electric transmission lines in eastern Canada strong enough to trip protective devices. This triggered power failures over a wide area by a domino effect: as one section of the electric grid became overloaded and shut down, its load was automatically shifted to another section that in turn became overloaded and shut down, and so on. Six million people lost electricity for 9 hours.

The particle streams of solar storms can affect spacecraft by damaging the solar cells that provide their energy, by altering control software in the spacecraft computers, and by producing flashes of light in the glass windows of star tracker navigation systems, which can cause the systems to lose their bearings. A satellite failure due to a solar storm in 1998 silenced most of the pagers in North America and in 2003 a large solar storm knocked two Japanese satellites out of service and affected communication and power grids around the world. The radiation in a severe solar storm may be strong enough to endanger astronauts in inadequately shielded spacecraft. Today as much as a day's warning of the arrival of a solar storm can be given by detectors in orbiting spacecraft.

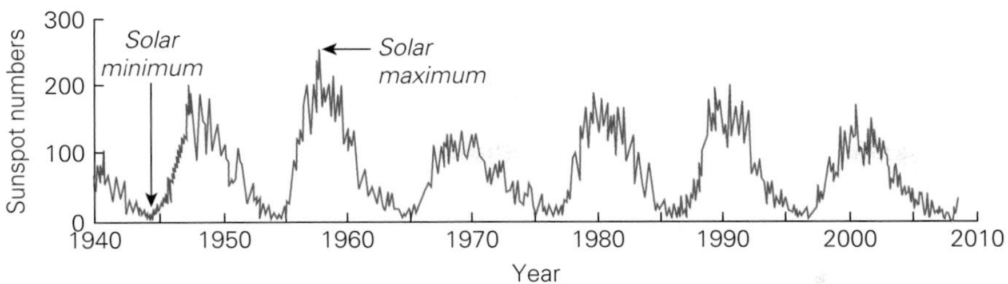

Figure 18-12 Sunspot numbers since 1940. Each cycle averages 11 years.

18.7 Solar Energy

It Comes from the Conversion of Hydrogen to Helium

Here on the earth, 150 million km from the sun, an area of 1 m^2 exposed to the vertical rays of the sun receives energy at a rate of about 1.4 kW. Adding up all the energy received over the earth's surface gives a staggering total, although this is only a tiny fraction of the sun's total radiation. And the sun has been emitting energy at this rate for billions of years. Where does all this energy come from?

We might be tempted to think of combustion, for fires give off what seems like a lot of heat and light. But a moment's thought shows that the sun is too hot to burn. Burning involves oxygen reacting with other substances to form compounds, but the sun is so hot that compounds cannot exist there except in its atmosphere. And even if burning were possible, the heat even the best fuels could give would be much too little to maintain the sun's temperatures.

Solar energy can come only from processes that take place inside the sun. Calculations based on reasonable assumptions lead to an estimate of 14 million K for the temperature and 1 billion atm for the pressure near the sun's center. The density of the matter there is nearly 10 times that of lead on the earth's surface. Under these conditions atoms of the lighter elements have lost all their electrons, and atoms of the heavier elements retain only their inmost electron shells. Thus matter in the sun's interior consists of atomic debris—free electrons and positive nuclei surrounded by a few electrons or none at all.

These atomic fragments move about far more rapidly than gas molecules at ordinary temperatures. Such speeds mean that two atomic nuclei may get close enough to each other—despite the repulsive electric force due to their positive charges—to react and form a single large nucleus. When this occurs among the light elements, the new nucleus has a little *less* mass than the combined masses of the reacting nuclei, as we saw in Chap. 8. The missing mass is converted to energy according to Einstein's formula $E_0 = mc^2$. So huge an amount of energy is given off in nuclear fusion reactions of this kind that there is no doubt they are responsible for solar energy.

Fusion Reactions in the Sun Most solar energy comes from the conversion of hydrogen into helium. This takes place both directly by collisions of hydrogen nuclei (protons) and indirectly by a series of steps in which carbon nuclei absorb a succession of hydrogen nuclei (Figs. 18-13 and 18-14). Each step can be duplicated in the laboratory and the energy released can be measured. The sun's interior is ideal for such energy-

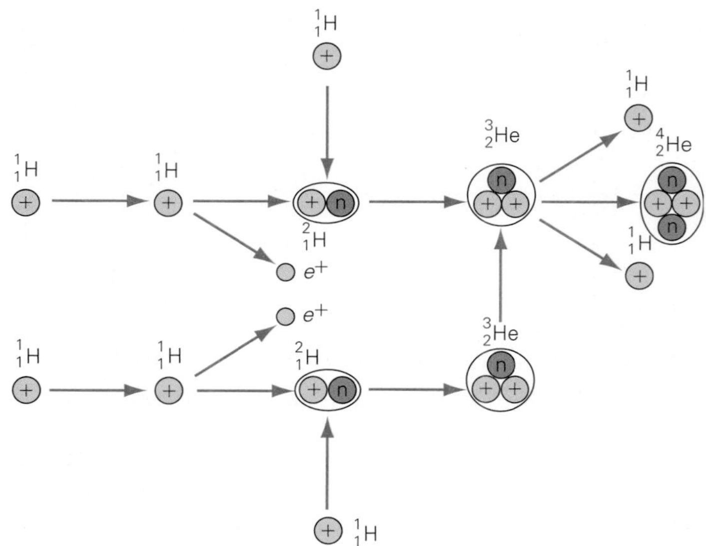

Figure 18-13 The proton-proton cycle. This is the chief nuclear reaction sequence that takes place in stars like the sun and cooler stars. Energy is given off at each step. The net result is the combination of four hydrogen nuclei to form a helium nucleus and two positrons. Two of the original six protons are left over and are available for further helium production.

producing events. For the entire process by either fusion mechanism, every kilogram of helium formed means that about 0.007 kg—about the mass of a tablespoon of water—of matter disappears. The corresponding energy release is

$$E_0 = mc^2 = (0.007\ \text{kg})(3 \times 10^8\ \text{m/s})^2 = 6.3 \times 10^{14}\ \text{J}$$

About 20 million kg of coal would have to be burned to obtain this amount of energy!

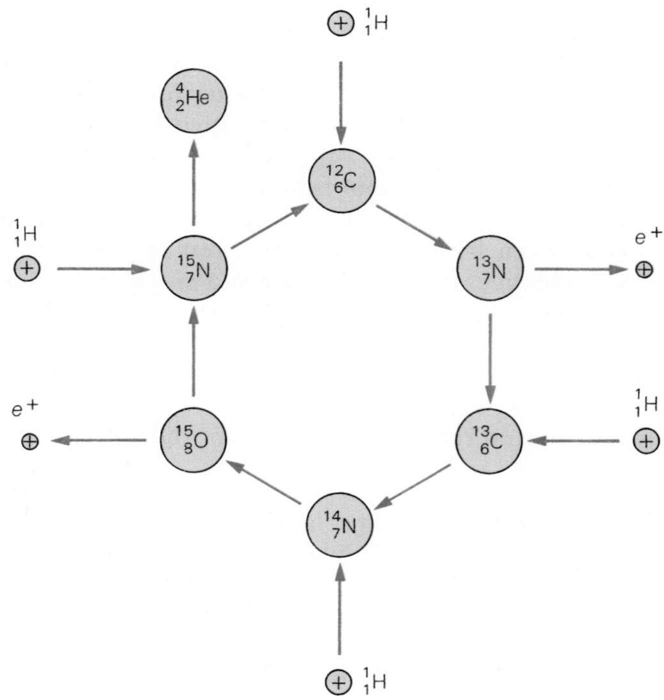

Figure 18-14 The carbon cycle also involves the combination of four hydrogen nuclei to form a helium nucleus with the evolution of energy. The $^{12}_6C$ nucleus is unchanged by the series of reactions and thus acts as a catalyst in the cycle. This cycle predominates in stars hotter than the sun.

BIOGRAPHY Hans Bethe (1906–2005)

After a distinguished early career at various German universities, when Hitler came to power Bethe moved first to England and then to the United States. He was professor of physics at Cornell University from 1937 to 1975 and remained active in research and public affairs even when officially retired. Notable among Bethe's many and varied contributions is his 1938 account of the sequences of nuclear reactions that power the sun and stars. During World War II he directed the theoretical physics division of the laboratory at Los Alamos, New Mexico where the atomic bomb was developed. A strong believer in nuclear energy—"it is more necessary now than ever before because of global warming"—Bethe was also an effective advocate of nuclear disarmament.

The relative likelihoods of the carbon and proton-proton cycles depend on temperature. In the sun and other stars like it, which have interior temperatures up to about 14 million K, the proton-proton cycle predominates. Most of the energy of hotter stars comes from the carbon cycle.

In every second the sun converts more than 4 billion kg of matter to energy, and it has enough hydrogen to be able to release energy at this rate for billions of years to come. In fact, the amount of matter lost in all of geologic history is not enough to have changed the sun's radiation appreciably. This confirms other evidence that the earth's surface temperature has not changed by very much during this period.

Origin of the Heavy Elements The reactions that convert hydrogen to helium are not the only ones that take place in the sun and other stars. With hydrogen and helium as raw materials and high temperatures and pressures to make things happen, most of the other elements are formed as well. The heaviest ones require still more extreme conditions to be produced, conditions that occur during the supernova explosions of heavy stars that are discussed later in this chapter. Such explosions also serve to scatter into space the various elements already created in the parent star.

Once scattered, these elements, heavy and light, mix with the hydrogen and helium of interstellar space and in turn become incorporated in new stars and their planets. We are all made of stardust.

Abundances of the Elements

There are a number of reasons why some elements are more common in the universe than others. To begin with, different nuclear reactions have different likelihoods of occurrence under comparable circumstances, which affects how often a given reaction sequence will take place inside stars. Another factor is the stability of a given nuclide. For instance, as we saw in Sec. 8.8, iron nuclei have high binding energies per nucleon and so are exceptionally stable. As a result, iron is a relatively abundant element. Also important is the number of different reaction sequences that can produce a given element. Only one sequence leads to a stable gold isotope, hence gold is relatively rare. Tin, by contrast, has 10 stable isotopes, and with so many possible routes to its creation is not at all scarce.

The Stars

18.8 Stellar Distances

Not Easy to Measure

Figure 18-15 Friedrich Bessel (1784–1840).

Aristotle pointed out long ago that, if the earth revolves around the sun, the stars should appear to shift in position, just as trees and buildings shift in position when we ride past them. Since he could detect no such shifts, Aristotle concluded that the earth must be stationary.

Another interpretation of the apparent lack of movement among the stars, suggested by some of the Greeks and later by Copernicus, is that the stars are simply too far away for such movement to be easily detected. An undoubted shift for one star was finally discovered in 1838 by the German astronomer Friedrich Bessel (Fig. 18-15), and others were found later. These shifts are so very small that the long failure to detect them is not surprising.

Bessel's discovery made possible the direct measurement of distances to the nearer stars. The method is simple (Fig. 1-13). The position of a star is determined twice, at times 6 months apart. From the measured change in the angle of the telescope, together with the fact that the telescope was moved by the 300-million-km diameter of the earth's orbit during the 6 months, the distance to the star can be calculated.

The **parallax,** as this apparent shift in position is called, is large enough to be measurable only for a few thousand of the nearer stars. The parallax of the closest star is equivalent to the diameter of a dime seen from a distance of 6 km. The distance from the earth to this star, Proxima Centauri, is about 4×10^{16} m.

A unit sometimes used to express stellar distances is the **light-year,** which is the distance light travels in a year and is equal to 9.46×10^{12} km. Thus Proxima Centauri is a little over 4 light-years away—which means that we see the star not as it is today but as it was 4 years ago. Only about 40 stars are within 16 light-years of the solar system. Such distances between stars are typical for much of the visible universe. Space is almost completely empty, far more empty even than the solar system with its tiny isolated planets.

Apparent and Intrinsic Brightnesses Parallax measurements are possible only for distances up to about 300 light-years. However, several indirect methods are available to find the distances of stars farther away than this. A very useful one is based on a comparison of the apparent and intrinsic brightnesses of stars.

The **apparent brightness** of a star is its brightness as we see it from the earth. This quantity expresses the amount of light that reaches us from the star. Its **intrinsic brightness,** on the other hand, is the true brightness of the star, a figure that depends upon the total amount of light it radiates into space.

The apparent brightness of a star depends on two things: its intrinsic brightness and its distance from us. A star that is actually very bright may appear faint because it is far away, and a star that is actually faint may have a high apparent brightness because it is close.

If both the apparent and the intrinsic brightness of a star are known, we can calculate its distance by finding out how far away an object with this intrinsic brightness must be located in order to send us the amount of light we observe. Such a calculation is not hard, and so we can find the distance to any star whose intrinsic brightness can be established.

A way to find the intrinsic brightness of a star was discovered by the American astronomer Walter Adams (Fig. 18-16). Studying the spectra of the nearer stars, for which intrinsic brightnesses are known, Adams observed that the spectra of stars with high intrinsic brightness showed certain relationships among the strengths of their lines. The spectra of stars with low intrinsic brightness showed somewhat different relationships. Adams thus could establish the intrinsic brightness of a star simply by looking at its spectrum. Assuming that the relationship holds for more distant stars, Adams then was able to use their spectra to find their intrinsic brightnesses and hence their distances. With this method stellar distances have been determined up to several thousand light-years.

Figure 18-16 Walter Adams (1876–1956).

18.9 Variable Stars

Stars Whose Brightness Changes, Usually in Regular Cycles

An extension of the brightness method for finding stellar distances is based upon the properties of a certain type of **variable star.** A variable star is one whose brightness varies continually (Fig. 18-17). Some variables show wholly irregular fluctuations, but most repeat a fairly definite cycle of change.

A typical variable grows brighter for a time, then fainter, then brighter once more, with irregular minor fluctuations during the cycle. Cycles range in length all the way from a few hours to several years. Maximum brightness for some variables is only slightly greater than minimum brightness, but for others it is several hundred times as great. Since the sun's radiation changes slightly during the sunspot cycle, we may think of it as a variable star with a very small range in brightness (a few percent at most) and a long period (11 years).

The light changes in some variable stars are easy to explain. These stars are actually double stars whose orbits we see edgewise, so that one member of each pair periodically gets in the way of the other. In other variables the appearance of numerous spots at regular intervals may be what is dimming their light. Still others seem to be pulsating, swelling

Figure 18-17 Superimposed and offset photographs taken at different times of the region of the sky in which the variable star WW Cygni appears. Only the brightness of this star has changed.

and shrinking so that their surface areas change periodically. Perhaps the irregular variables are passing through or behind ragged clouds of gas and dust that absorb some of their light.

Cepheid Variables A particular class of variable stars, called **Cepheid variables,** helps astronomers find out how far from us distant star groups that contain them are. Cepheid variables are bright yellow stars 5 to 10 times as massive as the sun that are in advanced stages of their lives; they are not very common. Their name comes from a typical example discovered in 1784 in the constellation Cepheus. Polaris, the North Star, is a Cepheid variable whose brightness varies by 10 percent with a period of 4 days.

Early in this century the American astronomer Henrietta Leavitt happened to be studying the Cepheid variables in a nearby galaxy called the Small Magellanic Cloud. She noticed that the brighter a Cepheid was, the longer its cycle took. Since all the stars in this galaxy are very nearly the same distance from the earth, Leavitt concluded that the average intrinsic brightness of a Cepheid anywhere in the sky can be found just by measuring its period. Comparing this calculated intrinsic brightness with the Cepheid's apparent brightness then gives its distance, as we know. This method can be used for greater distances than spectroscopic determinations because the period of a Cepheid can be determined even when it is very faint; the present record is 49 million light-years.

Unfortunately the individual stars in really distant galaxies appear smeared together, even in the largest telescopes, so the Cepheids they may contain cannot be picked out for analysis. More recently other methods have been developed for finding how far away such galaxies are. One method is based on the observation that the faster a galaxy rotates, the greater its intrinsic brightness. Nearby galaxies whose distances are known from the Cepheids in them are used to calibrate the rotation-brightness relationship.

BIOGRAPHY Henrietta Leavitt (1868–1921)

Leavitt was born in Massachusetts and studied astronomy at Radcliffe College. She then joined the Harvard College Observatory, initially as a volunteer, where she helped compile a photographic library of stars visible through the telescopes of the time. Perhaps her exceptional visual acuity arose as compensation for her deafness.

In 1907, while classifying the stars in this immense collection, Leavitt discovered that the average intrinsic brightness of a Cepheid variable is related to the period of its fluctuations. A bright Cepheid might take a month or two for each cycle, a faint Cepheid only a few days. Using Leavitt's finding, Harlow Shapley, who was then working at the Mt. Wilson Observatory in California, calibrated a distance scale based on Cepheid variables a few years later. Shapley used this cosmic yardstick to find the size of the Milky Way galaxy, the location of the sun relative to its center, and the distances to other galaxies. The Cepheid technique has ever since been an essential tool in astronomy; Edwin Hubble (see Sec. 19.6) used Cepheid data in his discovery that the universe is expanding.

18.10 Stellar Motions

The Stars Are Not Fixed in Space

As mentioned earlier, the speeds of stars that move toward or away from the earth can be found from the doppler shifts in their spectral lines. Motion across the line of sight can be followed by direct observation. The great distances of the stars make their apparent movements so slow that it is easy to think of them as being fixed in space. Nevertheless most stars are moving at speeds of several km per second relative to the earth (Fig. 18-18).

What about the sun? If the sun is moving toward a certain part of the sky, stars in that direction, on the average, should appear to be approaching us and to be spreading apart, just as trees in a forest seem to approach and spread apart when we drive toward them. Average stellar motions of this sort are indeed found near the constellation Cygnus, and in the opposite part of the sky stars are apparently receding and coming closer together. A study of these motions indicates that the sun and its family of planets are moving toward Cygnus at a speed of 200 to 300 km/s.

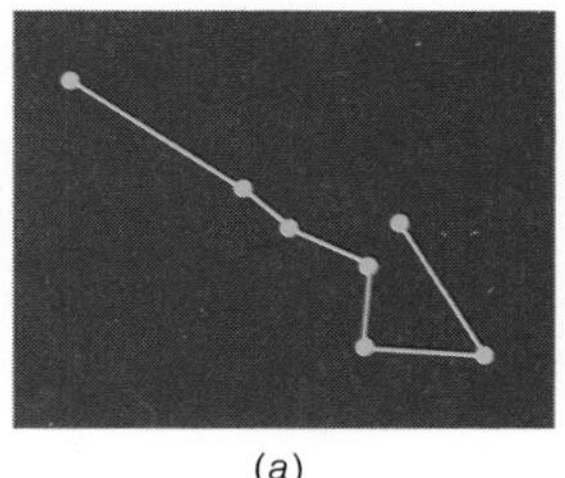

(*a*)

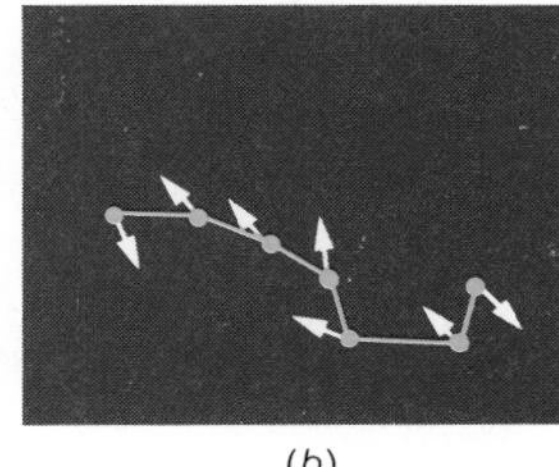

(*b*)

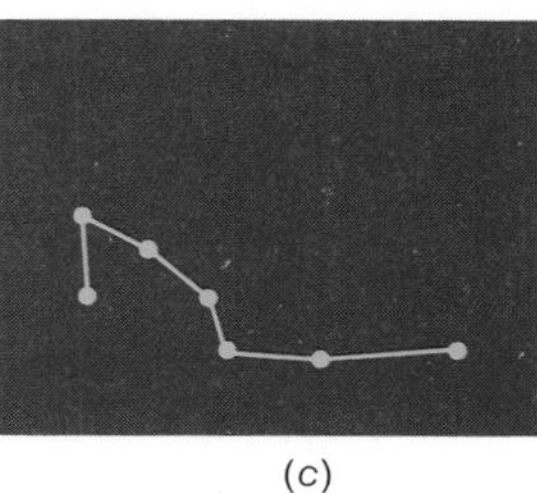

(*c*)

Figure 18-18 The Big Dipper (*a*) as it was 200,000 years ago, (*b*) as it is today (arrows show directions of motion of the stars), and (*c*) as it will be 200,000 years from now.

18.11 Stellar Properties

Mass, Temperature, and Size

We now turn to the properties of the stars. Many different types of stars are known, most of which fit into a pattern that can be understood in terms of a regular evolutionary sequence. Some stars, however, are still puzzles to the astronomer.

Mass The points of light that appear to the eye as single stars are often actually double, two stars close together. The members of such a star pair attract each other gravitationally, and each circles around the other (Fig. 18-19). From the characteristics of the orbits the masses of the stars can be calculated. Although this method is limited to such **binary stars,** they are common and stars of all kinds are found as members of such pairs. Three-star systems are also known; an example is Polaris, the North Star.

What the measurements show is that stellar masses range from $\frac{1}{40}$ to 150 times that of the sun—a smaller variation than planetary masses. It is not hard to see why normal stars should be limited in mass. In a body whose mass is smaller than a certain limit, gravity cannot squeeze its matter sufficiently to produce the temperatures needed for nuclear fusion reactions. At the other extreme, a very heavy star would become

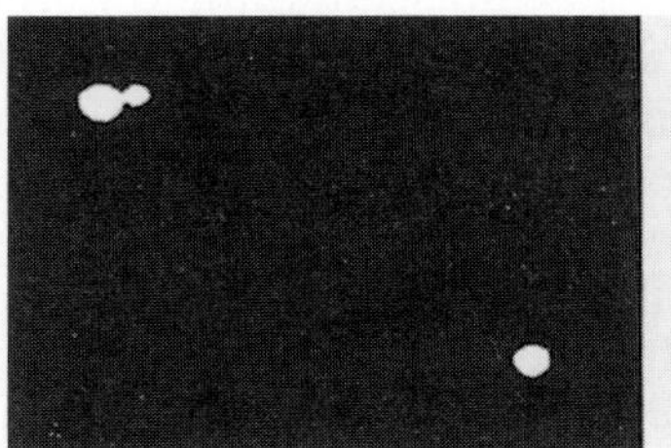  

Figure 18-19 The binary star shown here makes a complete rotation every 50 years.

so hot due to speeded-up nuclear reactions that gravity could not keep it together against the resulting outward pressure.

Temperature The temperature of a star is determined by finding the part of its spectrum in which the radiation is most intense (Fig. 18-4). This tells us the temperature of the star's surface—the photosphere from which radiation is emitted.

The surface temperatures of a few very hot stars range up to 40,000 K, but the great majority are between 3000 and 12,000 K. Probably many stars are cooler than 3000 K, which is near the boiling point of iron. However, unless they are relatively close to the earth, their radiation is then too feeble for us to detect. Like the sun, other stars must have enormously high internal temperatures to maintain their surface radiation. As mentioned earlier, the hottest stars are blue-white, those of intermediate temperature are orange-yellow, and the coolest are red.

Size If we know a star's surface temperature and its intrinsic brightness, we can find its size. The temperature tells us how much radiation is emitted from each square meter of the star's surface: the hotter it is, the more intense the radiation given off. The intrinsic brightness is a measure of the total radiation from the star's entire surface. We need only divide the total radiation by the radiation per square meter to find the number of square meters in the star's surface, and from this area the diameter and volume can be calculated.

There is also a more direct method of measuring stellar diameters based on the interference of light that can be used on the larger stars. Results obtained in this way agree with estimates from temperatures and intrinsic brightness.

The diameters of stars, unlike their masses, have an enormous range (Fig. 18-20). The smallest stars, composed almost entirely of neutrons, are only about 10 to 15 km across. The largest, like the giant red star Antares in the constellation Scorpio, have diameters over 500 times that of the sun. Antares is so huge that, if the sun were placed at its center, the four inner planets could pursue their normal orbits inside the star with plenty of room to spare.

Figure 18-20 The range of stellar sizes, from Antares (at the bottom) through the sun to a large white dwarf (black dot). Neutron stars are even smaller.

Giant stars like Antares have densities less than one-thousandth that of ordinary air—densities that correspond to a fairly good vacuum here on earth.

Life Histories of the Stars

Looking around us, we see human beings who seem quite different from one another: babies, children, young men and women, the middle-aged, the old. If we had just arrived from another world, we might think these kinds of people are all different species, not individuals of the same kind in different stages of development. Because stars have such long lives, it is easy for us to make the same mistake and think of them as belonging to separate categories. In fact, stars, like people, are born, mature, grow old, and die, so that the various kinds of stars we see fit into regular patterns of evolution.

18.12 H-R Diagram

Most Stars Belong to the Main Sequence

A century ago two astronomers, Ejnar Hertzsprung (Fig. 18-21) in Denmark and Henry Norris Russell (Fig. 18-22) in America, independently discovered that the intrinsic brightnesses of most stars are related to their temperatures. This relationship is shown in the graph of Fig. 18-23, which is called a **Hertzsprung-Russell** (or **H-R**) **diagram.** Each point on this graph represents a particular star.

About 90 percent of all stars belong to the **main sequence,** with most of the others in the **red giant** class at the upper right and in the **white dwarf** class at the lower left. The names giant and dwarf refer, as we might expect, to very large and very small stars respectively. The most abundant stars in the main sequence are the **red dwarfs** at its lower end, all of them too faint to be seen by the naked eye. Seventy percent of the members of our Milky Way galaxy of stars (see Sec. 19.1) are red dwarfs. Proxima Centauri, the nearest star to the sun, is a red dwarf.

Figure 18-21 Ejnar Hertzsprung (1873–1967).

Figure 18-22 Henry Norris Russell (1877–1957).

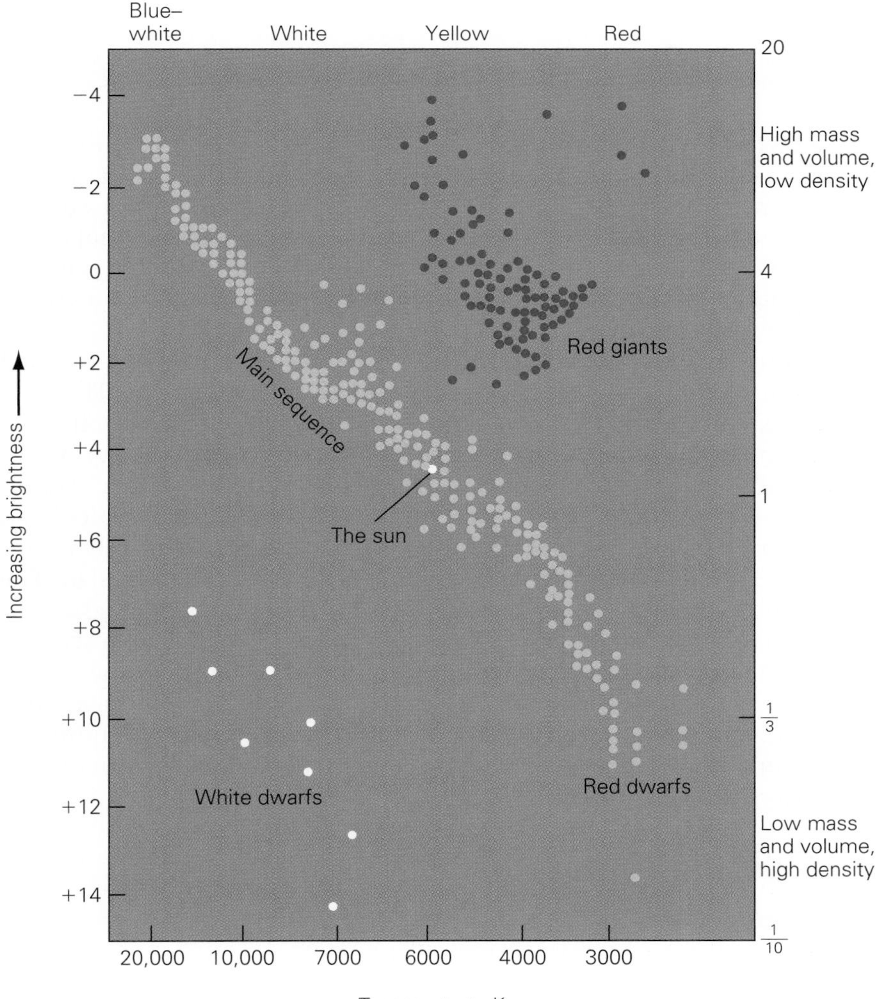

Figure 18-23 The Hertzsprung-Russell diagram plots stars according to temperature and intrinsic brightness. The numbers at the left express absolute magnitude (the astronomical measure of intrinsic brightness) with low numbers indicating bright stars and high numbers, faint stars. Masses at right correspond to main-sequence stars. Star colors are indicated at the top.

The position of a star on the H-R diagram is related to its physical properties. Stars at the upper end of the main sequence are large, hot, massive bodies. Stars at the lower end are small, dense, and reddish, cool enough so that chemical compounds form a considerable part of their atmosphere. In the middle are average stars like our sun, with moderate temperatures, densities, and masses and rather small diameters.

To the red giant class belong the huge, diffuse stars like Antares, with low densities and diameters often larger than that of the earth's orbit. Many of these stars have low surface temperatures, as their reddish color indicates, but their enormous surfaces make them very bright.

Example 18.1

A main-sequence star with the same intrinsic brightness as a white dwarf star is much redder than the dwarf. How does this observation indicate that the dwarf star is smaller than the main-sequence star?

Solution

The dwarf's color indicates that its temperature is higher than that of the main-sequence star. If both radiate energy at the same rate, the dwarf must be smaller since a hot object radiates energy at a greater rate per square meter than a cool one.

White Dwarfs The position of the white dwarfs in the H-R diagram reflects a combination of intensely hot surface and little total radiation. These properties suggest that such stars must be small, in fact comparable in size to the earth. However, their masses are all close to that of the sun, so that the density of a white dwarf is about 10^6 g/cm^3! A pinhead of such matter would weigh nearly a pound here on earth, and a cupful would weigh many tons.

Densities like this seem hard to believe, but they have been checked by enough methods to leave little doubt of their correctness. The only possible explanation is that atoms in these stars have collapsed. Instead of ordinary atoms with electrons relatively far from their nuclei, white dwarfs must have electrons and nuclei packed closely together. Matter in this state does not exist on earth, but its properties can be calculated from theories whose predictions have turned out correct in other situations. The greater the mass of a white dwarf, the smaller its size. The upper limit to the mass of a white dwarf is 1.4 times the sun's mass.

Only a few thousand white dwarfs are known. Their relative scarcity is more apparent than real, since they are so faint that only the nearer ones can be seen even in large telescopes. Enough of them have been found in recent years to suggest that the universe contains great numbers of these remarkable objects. Probably about 10 percent of the stars in the Milky Way are white dwarfs.

18.13 Stellar Evolution

Life History of a Star

The relationships revealed by the H-R diagram cannot have occurred through chance alone. Are the stars in different parts of the diagram perhaps in various stages of development? Does the mass of a star control

Figure 18-24 Stars form in gas clouds such as this one in the constellation Serpens. Clumps of matter in such a cloud that have a certain minimum mass are compressed by their own gravity to become hot and dense enough for energy-producing nuclear reactions to occur. This is how stars are born. Clumps too small for this to take place remain stillborn as "brown dwarfs"; no one knows how many brown dwarfs are present in the universe.

its temperature and the composition of its atmosphere? The answers to these questions seem to be yes, and the H-R diagram fits in well with modern ideas of the life history of a star.

Stars are believed to originate in gas clouds in space, clouds that consist largely of hydrogen. If a part of a gas cloud is dense enough, gravity will begin to pull it together into a still denser clump (Fig. 18-24). The contraction heats the clump, much as the gas in a tire pump is heated by compression, and the clump glows as a result. Such an infant star appears among the cooler giants in the H-R diagram.

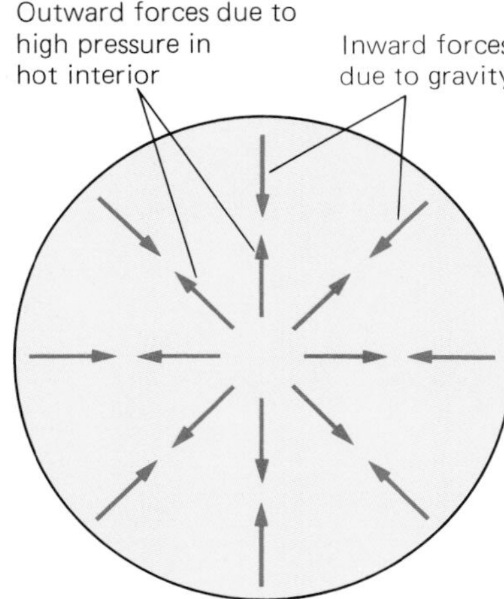

Figure 18-25 The tendency of a star to contract gravitationally is balanced by the tendency of its hot interior to expand.

Some thousands or millions of years later, the star's temperature will rise to the point where the nuclear reactions of Figs. 18-13 and 18-14 begin to occur, which convert its hydrogen into helium. The increase in temperature shifts the position of the star on the H-R diagram downward and to the left. From this time on the star's tendency to contract is opposed by the pressure of its hot interior (Fig. 18-25). Such a star is a stable member of the main sequence and maintains a constant size as long as its hydrogen supply holds out.

A star does not shine for some special reason—it shines because it has a certain mass and a certain composition.

The temperature a star reaches depends on its mass. Gravity in a large mass is more powerful than in a small mass and leads to more intense energy production to balance the resulting inward forces. For stars with abundant hydrogen, calculations show that the relationship between mass and temperature should be exactly that shown by stars in the H-R main sequence. The large, heavy stars at the upper end of the main sequence have high temperatures and shine brightly. The red dwarfs at the lower end are relatively cool and only faintly luminous. The sun's mass happens to be just right for life on earth, which is one reason why we are here: a mass 20 percent smaller and our planet would be colder than Mars, 20 percent greater and our planet would be hotter than Venus.

A heavy star consumes its hydrogen rapidly, so its lifetime in the main sequence is shorter than that of a less massive star. Accordingly, fewer

Brown Dwarfs

Brown dwarfs are lumps of matter between 10 and 75 times the mass of Jupiter that are born in the same way as ordinary stars by condensing under the influence of gravity from clouds of gas. With less mass they would not form lumps in this way, with more mass they would become hot enough to trigger nuclear reactions on a scale that would make them stars. So brown dwarfs are not actually stars, although their contraction (plus some help from a small amount of deuterium fusion in their first few million years, which can occur at a lower temperature than hydrogen fusion) does leave them barely warm enough for their glow to be detectable.

The feeble radiation from a brown dwarf is largely in the infrared part of the spectrum. Brown dwarfs are actually dark red in color, but the name red dwarf had already been given to true stars with less than half the sun's mass—which are a brighter red—by the time brown dwarfs were discovered in 1995. Because they do not generate energy after their initial deuterium is used up, brown dwarfs cool down after coming into being, unlike true stars, which may radiate for billions of years. Hundreds of brown dwarfs have been identified, and they may well be nearly as common as ordinary stars in our galaxy. The coolest known has a surface temperature of only about 400°C. A brown dwarf has been discovered that has a smaller companion in an orbit about the size of Pluto's.

How do brown dwarfs differ from planets? Of course, all planets orbit stars, but some brown dwarfs do, too. More of a distinction is that planets are richer in heavy elements than their parent stars, unlike brown dwarfs whose composition is that of a typical star. What really tells them apart is that, when it is first formed, some nuclear reactions (which soon stop) do take place in a brown dwarf, whereas planets do not have the mass needed to become hot enough for any nuclear reactions ever to occur. Once in existence, even though by different routes, there is not a lot to distinguish a really large planet from a really small brown dwarf.

than 1 percent of main-sequence stars are giants. The supply of hydrogen in a fairly modest star like our sun might last for 10 billion years. Probably the sun is now about halfway through this part of its life cycle. The heaviest stars may stay only as little as 10 million years in the main sequence and the lightest ones, the red dwarfs, may stay a trillion years.

Old Age When the hydrogen supply at last begins to run low in a star like the sun, the life of the star is by no means over but instead enters its most spectacular phase. Further gravitational contraction makes its core still hotter, and other nuclear reactions become possible—in particular those in which nuclei larger than helium are built up, for instance, the combination of three helium nuclei to form a carbon nucleus. The outer part of the star is heated and expands to as much as 100 times its former diameter. The expansion produces a cooling so that the result is a very large, cool star with a hot core. Such a bloated star is a red giant (Fig. 18-26). The shift from the main sequence to the upper right of the H-R diagram is relatively rapid. Energy is now being poured out at a great rate, so the star's life as a giant is much shorter than its stay in the main sequence.

Eventually the new energy-producing reactions run out of fuel too, and again the star shrinks, this time all the way down to the white dwarf state. A shell of gas from the outer part of the star goes out into space to form a bubblelike **planetary nebula** like that shown in Fig. 18-27. (Nebula is Latin for "cloud.") As a slowly contracting dwarf the star may glow for billions of years more with its energy now coming from the contraction, from nuclear reactions that involve elements heavier than helium, and from proton-proton reactions in a very diffuse outer atmosphere of hydrogen. Ultimately the star will grow dim and in time cease to radiate at all. It will now

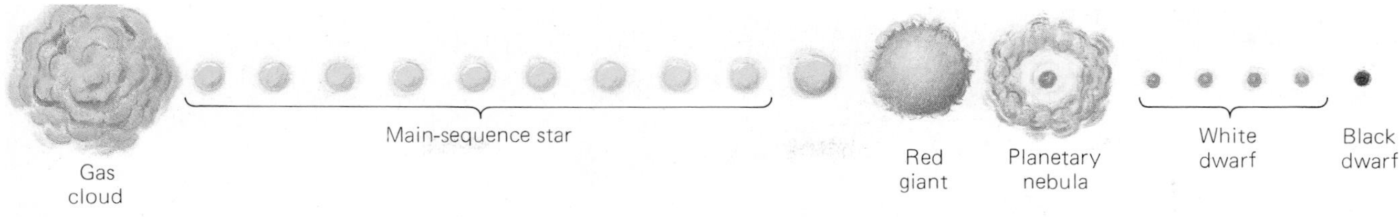

Figure 18-26 Life history of a star whose mass is near that of the sun. The sun is now about halfway through its estimated lifetime of 10 billion years. A planetary nebula contains roughly a third of the original star's mass.

be a **black dwarf,** a lifeless lump of matter. The universe is not thought to be old enough for any white dwarfs to have become black dwarfs as yet.

18.14 Supernovas

Exploding Stars

A heavy star—perhaps 9 to 25 times the sun's mass—toward the upper end of the main sequence has a rather different later history.

Such a star does not proceed in the usual way from red giant to white dwarf. Instead, after a relatively short lifetime of some millions of years, the star's great mass causes it to collapse abruptly when its fuel has run out and then to explode violently. The explosion, which takes less than a second, flings into space much of the star's mass. Such an event, which appears as one kind of **supernova,** is billions of times brighter than the original star ever was (Fig. 18-28). Nuclear reactions during the explosion create the heaviest elements which, together with the other elements formed during the earlier part of the star's life, are flung into space.

The other kind of supernova, equally bright, can occur in a binary system one of whose members is a white dwarf. Over time the dwarf gravitationally attracts gas from the other star. If the accumulated material brings the mass of the dwarf to 1.4 solar masses, the upper limit for

The Future of the Sun

For the sun, the star in which we have the most personal interest, we can expect about a billion more years of warming as the sun's temperature gradually increases. Life on earth will ultimately become impossible—not, as was once thought, because the sun will cool off but rather because it will grow too hot. In about 5 billion years the sun will expand into a red giant, at least as large around as the orbit of Venus, though perhaps not large enough to engulf the earth. Then, possibly 0.6 billion years later, its fuel gone, the sun will collapse into a white dwarf as its outer layers stream off as a planetary nebula (Fig. 18-27).

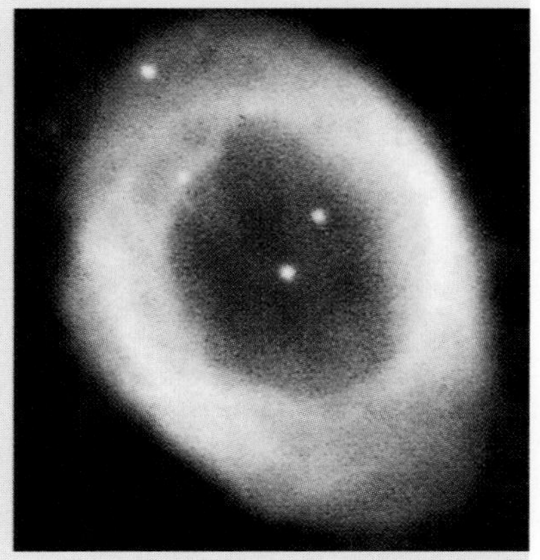

Figure 18-27 This planetary nebula in Lyra is a shell of gas moving outward from the star in the center, which is in the process of becoming a white dwarf. For a few thousand years this dying star is bright enough to heat the gas cloud so that it glows, but eventually the nebula will fade out and disappear into space. Not all planetary nebulas are spherical; some are egg-shaped, some have hourglass forms or appear as jets, still others are irregular.

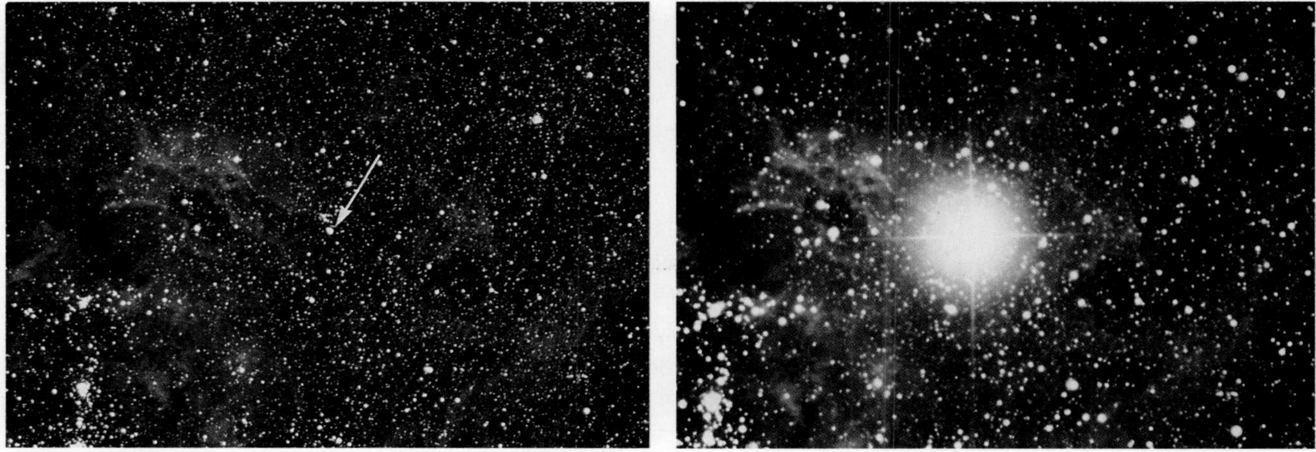

Figure 18-28 The explosion of this supernova in a nearby galaxy was detected on February 24, 1987. The arrow in the photograph at left indicates the star that became the supernova in the photograph at right. On the day before, a burst of neutrinos was recorded that corresponded to the emission of 10^{58} neutrinos from nuclear reactions in the core of the star as it collapsed.

a white dwarf to be stable, it collapses, which can trigger runaway fusion reactions in its carbon and oxygen content. What happens is a thermonuclear explosion that appears in the sky as a supernova. Whether such a system becomes a supernova depends on how fast matter accumulates on the dwarf: too slowly and not enough piles up, too rapidly and a feeble premature explosion is the result.

A supernova of either type may briefly outshine the entire galaxy to which it belongs. In a galaxy like ours about two supernovas normally occur every century, but by chance or for some unknown reason there have only been four—in 1006, 1054, 1572, and 1604—in our Milky Way galaxy in the past thousand years. A number of large stars in the Milky Way seem to be candidates to become supernovas in the future; one of them is Betelgeuse (Fig. 1-4) and another is Antares (Fig. 18-20), both red giants.

What is left after a supernova explosion is a dwarf star of extraordinary density. Its mass is typically about that of the sun but its diameter is only 10–15 km, the size of a city. The matter of such a star weighs billions of tons per teaspoonful. If the earth were this dense, it would fit into a large apartment house. Under the pressures inside such a star, the most stable form of matter is the neutron. Once the notion of **neutron stars** was purely speculative, but over a thousand have been identified thus far. Figure 18-29 shows the life history of a heavy star, and Fig. 18-30 shows how neutron stars and white dwarfs compare in size with the earth and the sun. Neutron stars range in mass from 1.4 up to about 2.7 times the sun's mass.

18.15 Pulsars

Spinning Neutron Stars

In 1967 unusual radio signals were picked up that came from a source in the direction of the constellation Vulpecula. They were found through meticulous work by Jocelyn Bell (now Jocelyn Bell Burnell; Fig. 18-31), at the time a graduate student at Cambridge University in England; her thesis advisor received a Nobel Prize for the discovery and for other

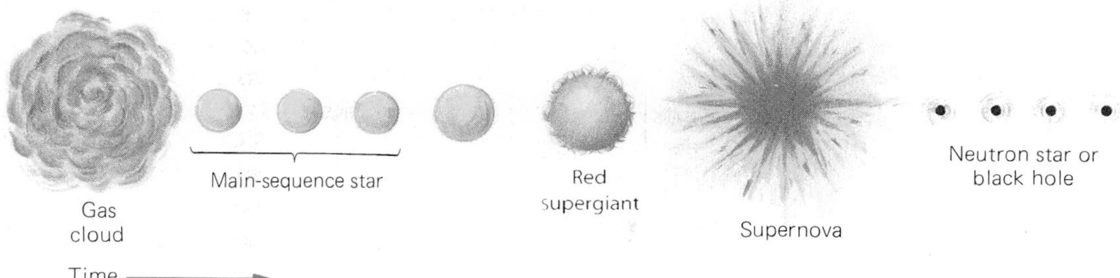

Figure 18-29 Life history of a star much heavier than the sun. Such a star spends less time in the main sequence than the star of Fig. 18-26. An example of a supergiant is Betelgeuse in the constellation Orion (see Fig. 1-4), which is farther across than the earth's orbit. In less than a million years Betelgeuse will have used up all its nuclear fuel supply and, after a sudden collapse, will explode into a supernova that will briefly rival the moon in brightness in the sky.

research in radioastronomy. The signals fluctuated with an extremely regular period, exactly 1.33730113 s. Since then over a thousand **pulsars** have been discovered with periods between 0.001 and 4 s. At first only radio emissions from pulsars were observed. Later, however, flashes of visible light were detected from several pulsars that were exactly synchronized with the radio signals.

The power output of a pulsar is about 10^{26} W, which is comparable with the total power output of the sun. So strong a source of energy cannot possibly be switched on and off in a fraction of a second. Instead it seems likely that pulsars are neutron stars that are spinning rapidly. Conceivably a pulsar has a strong magnetic field whose axis is at an angle to the axis of rotation, and this field traps tails of ionized gases that do the actual radiating. Whatever the mechanism, though, a pulsar is apparently like a lighthouse whose flashes are due to a rotating beam of light. The identification of pulsars with neutron stars is supported by evidence that the periods of pulsars are very gradually decreasing, which would be expected as they continue to lose energy (Fig. 18-32). About a dozen neutron stars, called **magnetars,** have been found with enormously strong magnetic fields, so strong that if one came closer to us than the moon, it might erase the data from every swipe card on earth.

The closest known pulsar is 280 light-years away. It is very dim, which suggests that there may be many—half a million?—other dim pulsars in our galaxy that are too faint to be seen because they are farther away.

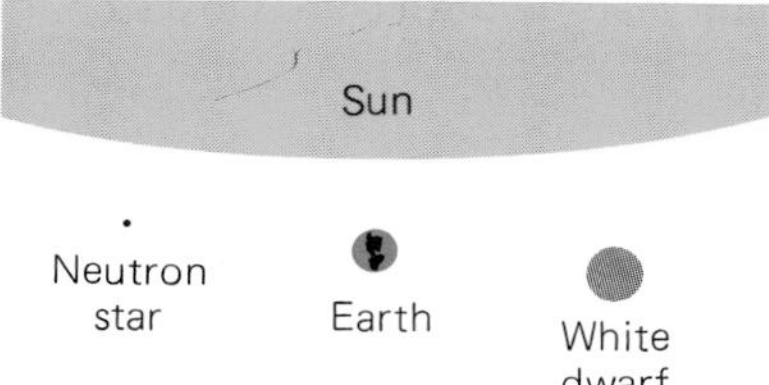

Figure 18-30 A comparison of a white dwarf and a neutron star with the sun and the earth.

18.16 Black Holes

Even Light Cannot Escape from Them

Does a neutron star represent the ultimate in compression? Apparently not. After it becomes a supernova, a very heavy star—25 to 40 times the sun's mass—leaves behind a remnant too massive to be stable as a neutron star. The remnant continues to contract until it is only a few km across. Such an object is called a **black hole** for a most interesting reason.

As we learned in Chap. 3, one of the results of Einstein's general theory of relativity is that light is affected by gravity. Thus starlight passing near the sun is bent by a small but measurable extent.

Figure 18-31 Jocelyn Bell Burnell (1943–).

Figure 18-32 This wispy object in space, called the Crab Nebula, has a pulsar at its heart. The nebula is the remnant of a supernova that was seen in A.D. 1054 and was visible in daylight. It has been expanding rapidly and glowing brightly ever since and is now 6 light-years across. The light and radio flashes from the Crab pulsar seem powerful enough to furnish the entire nebula with its energy. This pulsar flashes 30 times per second and is slowing down rapidly—1 part in 2400 per year. Both these observations are in accord with the time of its formation. A pulsar's lifetime is believed to be about 10 million years before it comes to a stop.

The more massive a star and the smaller it is, the stronger the pull of gravity at its surface and the higher the escape speed needed for something to leave the star. In the case of the earth, the escape speed is 11.2 km/s. In the case of the sun, 617 km/s. If the ratio M/R between the mass and radius of a star is large enough, the escape speed is more than the speed of light, and nothing, not even light, can ever get out (Fig. 18-33).

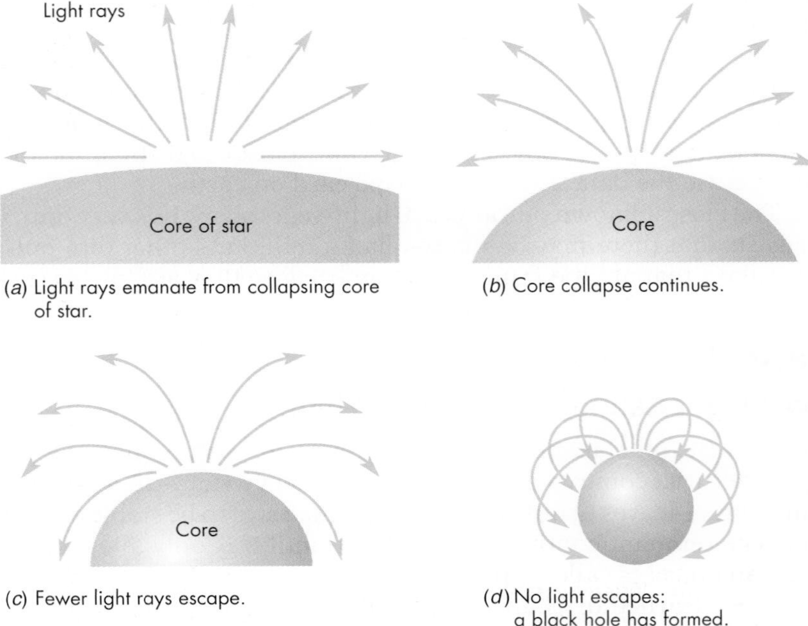

Figure 18-33 As the core of an old heavy star collapses, light that does not come out perpendicular to its surface is bent more and more strongly until it returns to the core. Finally, no light at all, regardless of direction, can leave the shrunken core, which is now a black hole.

The star cannot radiate and so is invisible, a black hole in space. Only very heavy stars end up as black holes; lighter ones eventually become white dwarfs (as in the case of the sun) or neutron stars.

Although anything passing close to a black hole will be gobbled up, never to return to the outside world, farther away the gravitational field of a black hole is the same as that of a star of the same mass. If the sun were to shrivel into a black hole (it is not nearly massive enough to do so), it would be 3 km across and the orbits of the planets would be unchanged. A black hole with the earth's mass would be the size of a grape.

Remarkably enough the concept of black holes was proposed as long ago as 1783 by John Mitchell in England, who called such an object a dark star. Not until the verification of Einstein's 1915 general theory of relativity (Sec. 3.13) was it established that light is indeed affected by gravity, however, so Mitchell's idea lay fallow until then.

Detecting Black Holes Since it is invisible, how can a black hole be detected? A black hole that is a member of a binary system (such double stars are quite common) will reveal its presence by its gravitational pull on the other star. In addition, the intense gravitational field of the black hole will suck matter from the other star, which will be compressed and heated before it falls inside to such high temperatures that x-rays will be emitted profusely (Fig. 18-34). When a binary system is discovered by the orbital motion of a star whose partner is invisible and x-rays are streaming out, the inference is that the partner is a black hole.

One of a number of invisible objects that astronomers believe on this basis to be black holes is known as Cygnus X-1, which is about 8000 light-years away. Its mass is 8.7 times that of the sun, and its radius may be only about 10 km. Enormous black holes whose masses are millions of times the solar mass are believed to be at the centers of galaxies of stars, as discussed in the next chapter.

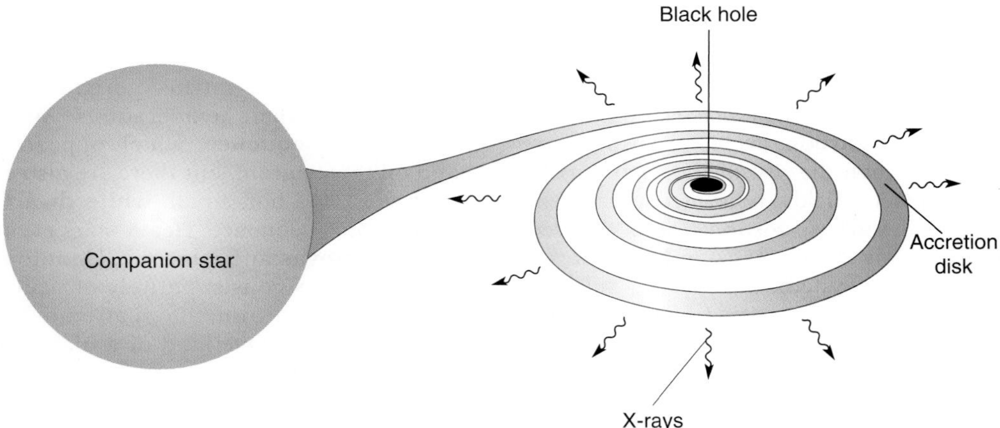

Figure 18-34 When matter from a companion star is pulled toward a black hole, it swirls into a spiral called an accretion disk on its way to being swallowed by the hole. Their compression heats the gases in the disk until they are hot enough to radiate x-rays.

Gamma-Ray Bursts

Every now and then short, intense bursts of gamma rays from space are detected by instruments at high altitudes and in satellites. There seem to be two possible explanations. One links the gamma-ray burst to the final convulsion of a **hypernova**, a super-supernova in which a star with more than 40 times the sun's mass collapses when its fuel is used up and then explodes so violently that neither a neutron star nor a black hole is left behind, as happens with less massive supernovas. In 2006 one such cosmic blast was observed when a distant star 150 times the sun's mass blew up and released a hundred times the energy of an ordinary supernova. Another monster star, called Eta Carinae and possibly the most massive star in our Milky Way galaxy, is only 7500 light-years away and is showing signs of instability.

The other possibility is the capture of a neutron star by a black hole, perhaps its partner in a binary system. In this case the burst would be very quick, less than 2 s in duration. In either case as much or more energy could be released as the sun gives off in its entire lifespan.

All the bursts thus far recorded came from distant galaxies and their energy was harmlessly absorbed in the atmosphere. But a hypernova within our galaxy, like Eta Carinae at some time in the future, might not let us off so easily. A flood of gamma rays that comes from nearby would still be mostly taken up in the atmosphere, but now the disruption of its oxygen and nitrogen molecules would occur on a much larger scale. In particular, there would be a lot of NO_2, a toxic component of smog that is brown and would darken the sky for as long as a year. The NO_2 would also destroy the ozone layer and thus allow harmful solar ultraviolet radiation to reach the earth's surface. Both effects would severely stress living things and might even lead to a mass biological extinction. In fact, the extinction that marked the end of the Ordovician Period (see Fig. 16-35) has characteristics that suggest a possible origin in such an event.

Important Terms and Ideas

A star's **spectrum** contains dark lines that correspond to particular frequencies of light absorbed by the ions, atoms, and molecules in the atmosphere around the star. These lines are superimposed on a bright background emitted by the star's surface. The spectrum and the background can be analyzed to provide information on the structure, temperature, composition, condition of matter, magnetism, and motion of the star.

The **sun** is a typical star whose temperature is 5800 K at the surface and perhaps 14 million K near the center. **Solar energy** comes from the conversion of hydrogen to helium, which can take place through both the **proton-proton** and **carbon cycles. Sunspots** are regions slightly cooler than the rest of the solar surface and have strong magnetic fields associated with them. Surrounding the sun is a diffuse **corona** of ions and electrons whose outward flow is the **solar wind** that extends through much of the solar system.

Auroras result from the excitation of gases in the earth's upper atmosphere by streams of protons and electrons from the sun.

The distance to a nearby star can be found from its apparent shift in position, or **parallax,** as the earth moves in its orbit. The distance to a star far away can be found by comparing its **intrinsic** and **apparent brightnesses,** which are its actual brightness and its brightness as seen from the earth. Stellar distances are often expressed in **light-years,** where a light-year is the distance light travels in a year.

Variable stars fluctuate continually in brightness. Notable are the **Cepheid variables,** whose intrinsic brightness and periods of variation are related, which permits their distances to be found. Other variable stars are actually pairs of stars that revolve about their common centers of gravity and periodically block each other's light.

The **Hertzsprung-Russell (H-R)** diagram is a graph on which the intrinsic brightnesses of stars are plotted versus their temperatures. Most stars belong to the **main sequence,** which appears as a diagonal band on the diagram, but there are also cool, large, **red giant** stars and hot, small, **white dwarfs** that lie outside the main sequence. Heavy stars eventually explode into **supernovas** and later subside into **neutron stars** whose interiors consist entirely of neutrons. A **pulsar** is a spinning neutron star that emits flashes of light and radio waves. One kind of a **black hole** is the collapsed core of an old heavy star that has contracted so much that its gravitational field is strong enough for the escape speed to be greater than the speed of light. Nothing, not even light, can escape from a black hole. Supermassive black holes lie at the centers of galaxies of stars.

A **brown dwarf** is a spherical lump of matter in space between 10 and 75 times the mass of Jupiter that is not heavy enough to trigger the nuclear reactions of a true star. It glows dark red because of heat left over from its gravitational contraction.

A **gamma-ray burst** is a brief but intense flood of gamma rays that reaches the earth from space. Suggested origins are an exceptionally powerful supernova explosion or the capture of a neutron star by a black hole.

Multiple Choice

1. Which one or more of the following are reasons why telescopes with large mirrors are useful in astronomy?
   a. Their ability to resolve objects close together in the sky
   b. Their ability to gather light
   c. Their magnifying ability
   d. Their ability to disperse light of different wavelengths

2. The fact that the spectra of most stars consist of dark lines on a bright background means that these stars
   a. have cool interiors surrounded by hot atmospheres
   b. have hot interiors surrounded by cool atmospheres
   c. have hot interiors surrounded by hot atmospheres
   d. have cool interiors surrounded by cool atmospheres

3. The examination of starlight with a spectrometer cannot provide information about an isolated star's
   a. temperature **c.** mass
   b. structure **d.** magnetic field

4. The sun is a (an)
   a. unusually small star
   b. unusually large star
   c. unusually hot star
   d. rather ordinary star

5. The surface temperature of the sun is approximately
   a. 600 K **c.** 6 million K
   b. 6000 K **d.** 14 million K

6. The sun's atmosphere
   a. extends out into the solar system
   b. consists mainly of oxygen and nitrogen
   c. consists of burning hydrogen
   d. is relatively cool

7. Auroras are caused by
   a. streams of colored gases from the sun
   b. streams of charged particles from the sun
   c. comets
   d. micrometeorites

8. Auroras occur mainly in the polar regions because of the effect of
   a. the moon
   b. low temperatures there
   c. the geomagnetic field
   d. the earth's equatorial bulge

9. Sunspots are
   a. dark clouds in the sun's atmosphere
   b. regions somewhat cooler than the rest of the sun's surface
   c. regions somewhat hotter than the rest of the sun's surface
   d. of unknown nature

10. The duration of the sunspot cycle is approximately
   a. 27 days **c.** 3 years
   b. 6 months **d.** 11 years

11. Sunspot activity does not affect
   a. shortwave radio communication
   b. the earth's magnetic field
   c. the aurora
   d. volcanic eruptions

12. The temperature of the sun's interior is believed to be about
   a. 600 K **c.** 14 million K
   b. 6000 K **d.** 14 billion K

13. Most of the matter in the sun is in the form of
   a. neutral atoms **c.** molecules
   b. ions **d.** liquids

14. The sun's energy comes from
   a. nuclear fission
   b. radioactivity
   c. the conversion of hydrogen to helium
   d. the conversion of helium to hydrogen

15. Proxima Centauri, the closest star to the sun, is at a distance of about
   a. 4 light-years
   b. 400 light-years
   c. 4 million light-years
   d. 4 billion light-years

16. If we know both the intrinsic and the apparent brightnesses of a star, we can find its
a. mass **c.** distance
b. temperature **d.** age

17. Cepheid variable stars are valuable to astronomers because
a. they are very brilliant, easily seen stars
b. their intrinsic brightnesses and periods of variation are directly related
c. their periods of variation are short
d. they all lie in the constellation Cepheus

18. In order to determine the mass of a star, it must
a. belong to the main sequence
b. be a Cepheid variable
c. be relatively close
d. be part of a binary-star system

19. The size of a star can be found from its
a. temperature and mass
b. temperature and intrinsic brightness
c. temperature and apparent brightness
d. mass and apparent brightness

20. The temperature of a star can be determined from
a. the wavelength at which its radiation is brightest
b. the wavelength at which its radiation is dimmest
c. its apparent brightness
d. the strength of its magnetic field

21. Variations among stars are least in their
a. distance from the sun **c.** size
b. mass **d.** temperature

22. The color of a relatively cool star is
a. white **c.** blue
b. yellow **d.** red

23. The reason stars more than about 100 times as massive as the sun are not found is that
a. they would split into double-star systems
b. they would be black holes from which no light can escape
c. the gravity of such a star would not hold it together against the pressure produced by nuclear reactions in its interior
d. the high internal pressures would prevent nuclear reactions from taking place

24. Stars belonging to the main sequence
a. have the same mass
b. have the same temperature
c. radiate at a steady rate
d. fluctuate markedly in brightness

25. The greater the mass of a main-sequence star, the
a. younger it is
b. longer it will remain part of the main sequence
c. less hydrogen it contains
d. hotter it is

26. The main sequence on the H-R diagram includes
a. the sun **c.** red giants
b. white dwarfs **d.** pulsars

27. A typical white dwarf star is about the size of
a. a large building **c.** the earth
b. the moon **d.** Jupiter

28. When the hydrogen supply in a typical main-sequence star begins to run out, other nuclear reactions occur and the star becomes a
a. red giant **c.** supernova
b. white dwarf **d.** neutron star

29. A star like the sun eventually becomes a
a. black hole **c.** supernova
b. white dwarf **d.** neutron star

30. A star that explodes as a supernova
a. has a mass much greater than that of the sun
b. has a mass much smaller than that of the sun
c. has a mass about equal to that of the sun
d. may have any mass

31. The brightest of the following types of stars is a
a. red giant **c.** supernova
b. white dwarf **d.** neutron star

32. The heaviest elements are created in
a. stellar interiors **c.** black holes
b. stellar atmospheres **d.** supernova explosions

33. A pulsar is not
a. the remnant of a supernova explosion
b. rotating rapidly
c. composed largely of neutrons
d. as large as the earth

34. A "black hole" appears black because
a. it is too cool to radiate
b. it is surrounded by an absorbing layer of gas
c. its gravitational field is too strong to permit light to escape
d. its magnetic field is too strong to permit light to escape

35. Black holes are remnants of
a. stars with small masses
b. stars with large masses
c. white dwarfs
d. black dwarfs

Exercises

18.1 The Telescope

1. Why are large telescopes valuable in astronomy?
2. A photograph of a star cluster shows many more stars than can be seen by looking by eye at the cluster through the same telescope. Why is there a difference?

18.2 The Spectrometer

18.3 Spectrum Analysis

3. Why do you think it is useful to put an astronomical telescope in a satellite?
4. What part of a star is responsible for the dark lines in its spectrum?
5. Suppose a star had a cool interior surrounded by a hot atmosphere. What kind of spectrum would it have?
6. Arrange the following types of stars in order of decreasing surface temperature: yellow stars, blue stars, white stars, red stars.
7. Suppose you examine the spectra of two stars receding from the earth and find that the lines in one are displaced farther toward the red end than those in the other. What conclusion can you draw?
8. A binary star consists of two stars that revolve around each other. Telescopes cannot resolve a binary star into its members, but an astronomer can recognize a binary star from its spectrum. What property of the spectrum is involved?
9. How can the composition of the sun be determined?
10. Helium was discovered in the sun before it was found on the earth (hence its name, which comes from *helios,* the Greek word for "sun"). How can this sequence have come about?

18.4 Properties of the Sun

11. What part of a star is responsible for the continuous bright background of its spectrum? What happens to the radiation produced underneath this part of the star?
12. Why is the sun's corona ordinarily not visible? How do we know it exists?
13. The sun's photosphere is at a temperature of 5800 K whereas the temperature of much of the corona exceeds 1,000,000 K. Why is the photosphere rather than the corona the source of most of the sun's radiation?
14. What element is most abundant on the sun? Next most abundant?

18.5 The Aurora

15. Suppose the earth's magnetic field were to disappear. What effect would this have on the aurora?

18.6 Sunspots

16. Why do sunspots appear dark if their temperatures are over 4500 K?
17. What aspect of sunspots changes during a sunspot cycle?
18. Give two methods for determining how fast the sun rotates on its axis. Is the rotation speed the same for the entire sun?
19. Intense auroral activity occurs in the earth's upper atmosphere about 30 h after a solar storm. What is the average speed of the ions emitted during the storm?

18.7 Solar Energy

20. Where are the lighter elements created? The heavier ones?
21. What aspect of the formation of helium from hydrogen results in the evolution of energy?
22. Why are conditions in the interior of a star favorable for nuclear fusion reactions?
23. According to the text, the sun's mass is about 2×10^{30} kg and it loses about 4×10^{9} kg per second as its hydrogen is converted into helium. Assuming that the sun has been radiating energy at the same rate as today during the 4.5-billion-year existence of the earth, what percentage of its mass has been lost in this period? What does this imply about the possibility that the intensity of solar radiation has been roughly constant during the life of the earth, as geological evidence suggests?

18.8 Stellar Distances

24. What information can be gained by comparing the intrinsic and apparent brightnesses of a star?

18.9 Variable Stars

25. Explain how the distance to a star cluster that contains Cepheid variables is determined.

18.11 Stellar Properties

26. How are stellar masses determined?
27. Which varies more, the masses of the stars or their sizes?

28. What data are needed to determine a star's average density? How would you expect the density to change from the surface layers to the interior?

29. Which stars do you think have the highest densities? The lowest?

30. A red star and a white star of the same apparent brightness are the same distance from the earth. Which is larger? Why?

31. How is a star's diameter estimated from measurements of temperature and intrinsic brightness?

32. The spectrum of a certain star shows a doppler shift that varies periodically from the red to the blue end of the spectrum. What kind of star is this?

18.12 H-R Diagram

18.13 Stellar Evolution

33. Must a star be spherical?

34. Why are relatively few stars very hot?

35. Is it possible for an object with the mass and composition of the sun to exist without radiating energy?

36. Why is the sun considered to be a star?

37. Why are most stars part of the main sequence on the H-R diagram?

38. Main-sequence stars are supposed to evolve into red giants, but relatively few stars lie between the main sequence and the group of red giants on the H-R diagram. Why?

39. The stars Betelgeuse and Deneb have similar intrinsic brightnesses but Betelgeuse is red and Deneb is white. (a) Which is larger? (b) Which has the greater density? (c) Which is hotter?

40. A giant star is much redder than a main-sequence star of the same intrinsic brightness. How does this observation indicate that the giant star is larger than the main-sequence star?

41. What are the chief characteristics of an average star in the upper left of the H-R diagram? In the lower left? In the upper right? In the middle of the main sequence?

42. Where would a star be located in the main sequence relative to the sun if its mass is less than half that of the sun? Would it remain in the main sequence for a shorter or longer time than the sun? Would it be cooler or hotter than the sun?

43. Sirius, the brightest star in the sky apart from the sun, is a blue-white star of great intrinsic brightness. (a) What does this suggest about its temperature? (b) About its average density? (c) About its position in the H-R diagram?

44. As a main-sequence star evolves, what happens to its position on the main sequence?

45. Why are there relatively few giant stars in the main sequence?

46. Into what kind of star will the sun eventually evolve?

47. In what part of its life cycle is a white dwarf star?

48. What happens to a very heavy star at the end of its period in the main sequence?

49. After a very long time, a white dwarf will cool down and become a black dwarf. What will the corresponding evolutionary path of the star be on the H-R diagram?

18.14 Supernovas

18.15 Pulsars

50. About how often do supernovas occur in the Milky Way galaxy?

51. What is left behind after a supernova explosion?

52. Which of the following types of star is the smallest? The largest? The most common? Neutron stars, white dwarfs, red dwarfs, black dwarfs.

53. (a) What is the characteristic behavior of a pulsar? (b) What is believed to be its nature? (c) From what does a pulsar originate?

18.16 Black Holes

54. How large are black holes? Can any star evolve into a black hole?

55. What prevents light from escaping from a black hole?

Answers to Multiple Choice

1. a, b	**6.** a	**11.** d	**16.** c	**21.** b	**26.** a	**31.** c
2. b	**7.** b	**12.** c	**17.** b	**22.** d	**27.** c	**32.** d
3. c	**8.** c	**13.** b	**18.** d	**23.** c	**28.** a	**33.** d
4. d	**9.** b	**14.** c	**19.** b	**24.** c	**29.** b	**34.** c
5. b	**10.** d	**15.** a	**20.** a	**25.** d	**30.** a	**35.** b

19

The Universe

Image of spiral galaxy of stars taken by Spitzer Space Telescope.

Goals

When you have finished this chapter you should be able to complete the goals ▶ given for each section below:

Galaxies

19.1 The Milky Way
A Spinning Disk of Stars
- ▶ Describe the Milky Way galaxy and indicate the sun's location in it.

19.2 Stellar Populations
A Clue to the History of Our Galaxy
- ▶ Compare Population I and II stars.

19.3 Radio Astronomy
Another Messenger from the Sky
- ▶ Explain what a radio telescope is.
- ▶ List the three ways in which radio waves from space are produced.

19.4 Galaxies
Island Universes of Stars
- ▶ Discuss the characteristics and distribution in space of galaxies.
- ▶ Explain what is meant by dark matter and describe the evidence for its existence.

19.5 Cosmic Rays
Atomic Nuclei Speeding through the Galaxy
- ▶ Distinguish between primary and secondary cosmic rays.
- ▶ Discuss the significance of cosmic rays in the evolution of the universe.

The Expanding Universe

19.6 Red Shifts
The Galaxies Are All Moving Away from One Another
- ▶ Explain what red shifts in galactic spectra indicate about the motions of galaxies.
- ▶ State Hubble's law and use it as evidence for the expansion of the universe.

19.7 Quasars
Brilliant, Tiny, and Far Away
- ▶ Outline the properties of quasars and what they suggest about the nature of these objects.

Evolution of the Universe

19.8 Dating the Universe
When Did the Big Bang Occur?
- ▶ Discuss the big bang theory of the origin of the universe.

19.9 After the Big Bang
A Glimmer of the Early Universe Still Exists
- ▶ Identify dark energy and explain why it is believed to exist.
- ▶ Outline the chief events after the big bang occurred.
- ▶ Explain the significance of the sea of radio waves that fills the universe.
- ▶ Discuss the various possibilities for the future of the universe, including the big crunch and the big rip.

19.10 Origin of the Solar System
A Gradual Process
- ▶ Outline the origin of the solar system.

Extraterrestrial Life

19.11 Exoplanets
Probably a Great Many
- ▶ Give the reasons why other planetary systems are hard to detect.

19.12 Interstellar Travel
Not Now, Probably Not Ever

19.13 Interstellar Communication
Why Not?
- ▶ Discuss the likelihoods of interstellar travel and communication.

Stars are not scattered at random throughout the universe but instead occur in immense swarms called **galaxies.** Each galaxy is separated from the others by vast reaches of nearly empty space. The stars of the galaxy to which our sun belongs appear in the sky as the Milky Way; the word *galaxy* comes from the Greek for "milk."

Of the many remarkable properties galaxies have, one stands out: they are moving apart from one another, so that the universe as a whole is expanding. If we project this expansion backward, we find that it began 13.7 billion years ago. Can it be that the entire universe was born in a cosmic **big bang** at that time and has been evolving ever since into the galaxies of today? As we shall see in this chapter, several lines of evidence support such a picture, which has been filled out in considerable detail.

Galaxies

A galaxy is a giant archipelago of stars. Just as studying the sun tells us a lot about stars in general, so studying our galaxy, the Milky Way, tells us a lot about galaxies in general.

19.1 The Milky Way

A Spinning Disk of Stars

The great band of misty light we see in the sky on a clear night is called the **Milky Way** and forms a continuous band around the heavens (Fig. 19-1). When we look at it with a telescope, as Galileo was the first to do in 1610, the Milky Way is an unforgettable sight. Instead of a dim glow, we now see countless individual stars, stars as numerous as sand grains on a beach. In other parts of the sky the telescope also reveals stars too faint for the naked eye, but nowhere else that many. Clearly the stars are not distributed evenly in space—a basic observation that implies much about the structure and evolution of the universe.

The appearance of the Milky Way tells us something about how the stars in our galaxy are arranged. Most of them are concentrated in a relatively

Figure 19-1 A mosaic of several photographs of the Milky Way between the constellations Cassiopeia and Sagittarius.

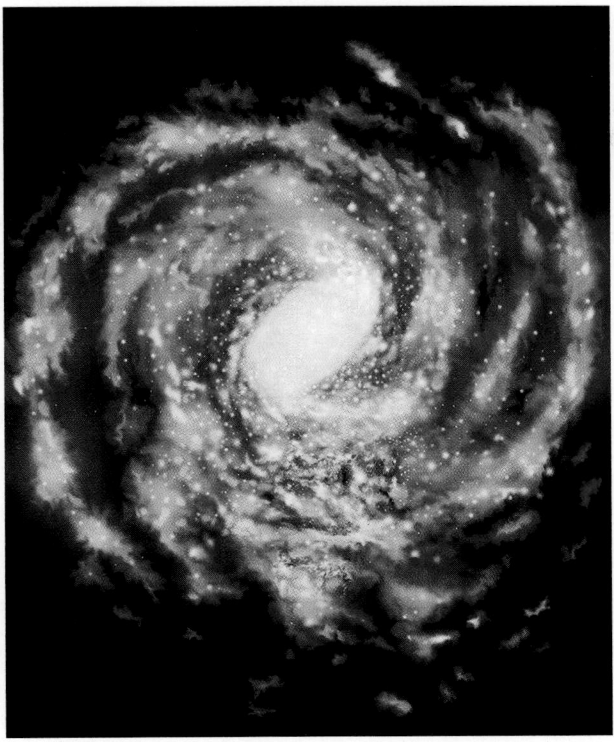

Figure 19-2 Computer simulation of our galaxy. The stars in the disk of the galaxy appear in the night sky as the Milky Way. The galaxy seems to have four main spiral arms and several smaller arms. The sun is a member of one of the latter and in this picture is located a little more than half the radius of the galaxy below its center. A massive black hole at the center of the galaxy holds it together as in the case of all galaxies. Every star we can see as an individual star without using a telescope is part of our galaxy.

thin disk with the sun near its central plane. When we look toward the rim of the disk, we see a great many stars, so many that they seem to form a continuous band of light. When we look above or below the disk, far fewer stars are to be seen. The disk of stars has a thicker central nucleus, so that it is shaped something like a fried egg with the outer stars concentrated in a number of spiral arms (Fig. 19-2).

The disk of the galaxy is roughly 130,000 light-years in diameter (there is no sharp boundary) and 10,000 light-years thick near the center. It is one of the larger galaxies of the universe with at least 200 billion stars. The sun is about 25,000 light-years from the center, which lies in the direction of the densest part of the Milky Way in the constellation Sagittarius. (For comparison, we are 8 light-minutes from the sun.) The stars of the galaxy are chiefly located in spiral arms that extend from the nucleus. Surrounded by so many similar bodies, the sun is not unusual in position, size, mass, or temperature. It is not even unusual in possessing a family of planets. There may well be billions of planets in our galaxy, at least some of them probably inhabited by some form of life.

The stars of our galaxy revolve about its center, which is what they must do if the galaxy is not to gradually collapse because of the gravitational attraction of its parts. (The planets do not fall into the sun because of their orbital motion, too.) The orbital speed of the sun and nearby stars around the galactic center is nearly 200 km/s. At this rate the sun makes a complete circuit once every 240 million years. Since its formation, the sun has revolved 20 times around the galactic center; the galaxy itself has completed at least twice that many turns since coming into being.

Galactic Nebulas

Galactic nebulas are irregular masses of diffuse material within our galaxy. Some appear as small glowing rings or disks surrounding stars, some take the form of lacy filaments, and many are wholly irregular in outline. The brightest, the Great Nebula in the constellation Orion (Fig. 19-3), is barely visible with the naked eye, but most of them are much fainter. These nebulas consist of gas and dust and shine only because they reflect light from nearby stars or are excited to luminescence by stellar radiation.

Clouds of gas and dust similar to galactic nebulas but without any luminosity sometimes reveal their presence as dark patches that obscure the light of stars beyond them (Fig. 19-4). Such dark nebulas may be fairly abundant but are difficult to find because bright stars shine through them except where they are especially dense. In fact, our entire galaxy is filled with rarefied nebular material with a density somewhere near one atom or molecule per cubic centimeter. The dark nebulas are only local concentrations of this interstellar matter. Empty space is not nearly as empty as it was once thought to be, but in most places the amount of interstellar material is so small that little starlight is absorbed.

Figure 19-3 The Great Nebula in Orion is a gas cloud excited to luminescence by hot stars in its center.

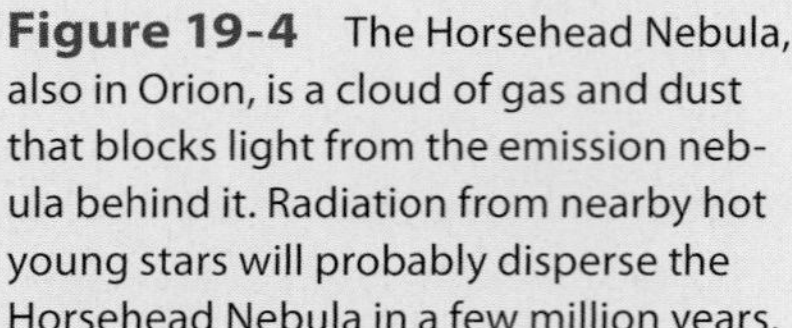

Figure 19-4 The Horsehead Nebula, also in Orion, is a cloud of gas and dust that blocks light from the emission nebula behind it. Radiation from nearby hot young stars will probably disperse the Horsehead Nebula in a few million years.

Its Heart Is a Black Hole Clouds of gas and dust surround the center of our galaxy and prevent us from seeing it. However, radio waves, infrared and ultraviolet light, x-rays, and gamma rays from near the center have been detected, and they suggest a monster black hole there that astronomers call Sagittarius A*. Its mass seems to be about 3.7 million times the sun's mass, and it sucks in material blown off by nearby supernova explosions. It is this material, compressed and thereby heated on its way to disappearing into the black hole, that gives off the observed radiations. Despite its giant mass, the black hole's diameter is probably less than 0.2 percent that of Mercury's orbit.

In support of the idea that a black hole is at the center of the Milky Way are the speeds of stars near the center. These stars move so fast that only a huge mass could keep them from flying off. Nothing but a black hole has the required mass (very large) and size (fairly small) to do the job. Evidence from other galaxies—there are many, as we shall see—suggests that such a black hole does not come first, gathering up stars later to form a galaxy around it, but instead the black hole and the galaxy develop together.

19.2 Stellar Populations

A Clue to the History of Our Galaxy

The stars in our galaxy fall into two categories. **Population I** stars are those in the central disk of the galaxy. These stars are of all ages, from those just coming into being to old ones that must have been formed early in the life of the galaxy.

Population II stars make up the **globular clusters** that surround the galaxy above and below its central disk. To the naked eye the largest of these assemblies of stars are just visible as faint patches of light. Through a telescope they are more spectacular, dense with stars near their centers and thinning out toward the edges (Fig. 19-5). About 150 of these clusters have been discovered, all orbiting the center of the galaxy.

In photographs of the globular cluster in the constellation Hercules, one of the largest, more than 50,000 stars have been counted. These are only the brightest stars, since the cluster is so far away that faint ones cannot be seen. The total number of stars may be over a million. The nearest clusters are about 20,000 light-years away from us and the farthest more than 100,000 light-years away. Light from the great Hercules cluster travels 33,000 years before reaching our eyes, so we see it as it appeared toward the end of the most recent ice age. The average distance between stars in a globular cluster is well under a light-year, which means they are much more closely packed than those near the sun. As a result the average time between the collisions of a given star with another star in such a cluster is perhaps 10,000 years; elsewhere in the galaxy the time is in the billions of years.

Figure 19-5 The globular cluster in Hercules contains perhaps a million stars.

The stars in the globular clusters of Population II are mostly very old, nearly as old as the galaxy. (Some are younger, having been born in collisions of old stars.) What seems to have happened is that all the matter of the galaxy was originally a spherical cloud of gas from which stars were forming. In time the cloud concentrated in a central disk, leaving behind as a sort of halo those stars that had already formed. New stars continued to come into being from the gas clouds of the disk, which is why the stars there are of all ages.

This picture is supported by the compositions of the stars of the two populations. Population II stars are extremely rich in hydrogen and helium. This follows from an early origin since the materials of the young universe are thought to have been these elements. Population I stars, on the other hand, contain heavier elements in some abundance. Such elements are produced in stars by the nuclear reactions that are responsible for stellar energy, and are thrown into space during the explosions that occur in the final stages of a star's active life. Compared with Population II stars, those of Population I are richer in heavy elements because they were formed from material that already contained such elements, material that was the debris of dying stars from Population II.

Although the globular clusters in our galaxy and in others consist of very old stars, new clusters can be formed when two galaxies collide and merge. In such an event, clusters come into being from concentrations of gas that occur, so the merged galaxy ends up with clusters both of new stars and of old ones that came from the original galaxies.

19.3 Radio Astronomy

Another Messenger from the Sky

A **radio telescope** is a directional antenna connected to a sensitive radio receiver. Usually a metal dish, like the concave mirror of a reflecting telescope, gathers radio waves over a large area and concentrates them on the antenna itself. With such an arrangement the direction from which a particular radio signal arrives can be established.

Radio waves from cosmic sources are quite weak when they reach the earth—signals from a cell phone on the moon would be among the strongest—so large antennas are needed to detect them. The largest steerable-dish radio telescope is located in Green Bank, West Virginia. Still larger is a fixed dish 305 m across in Arecibo, Puerto Rico, that consists of wire mesh fitted into a bowl-shaped hollow in the landscape (see Fig. 19-24). As the earth turns, a band in the sky can be surveyed; the great sensitivity of this telescope makes up for its restriction to this band.

A number of radio telescopes can be linked together electronically to precisely locate the source of radio waves by interference methods (Fig. 19-6). Under construction in a desert region of northern Chile is an array of 63 dish antennas. To come later is the Square Kilometer Array of thousands of radio dishes whose total collecting area will be a square kilometer—a million square meters. Half the dishes will be concentrated in a central area about 5 km across with the others scattered up to 3000 km away. The array will be so sensitive that it must have a remote location to prevent interference from terrestrial sources of radio waves; sites in Argentina, Australia, China, and South Africa have been proposed. The array, which is hoped to be operating by 2020, will be able to detect long-wavelength radiation from the youth of the universe, before it became transparent to light (see Sec. 19.9). Among its fields of inquiry will be the evolutions of galaxies and of cosmic magnetism and the nature of dark energy.

Figure 19-6 A radio telescope is a directional antenna designed to pick up radio waves from space. Shown are several of the 27 dish antennas, each 25 m in diameter, that make up the Very Large Array near Socorro, New Mexico. The data obtained from these antennas can be combined to reveal details about radio sources in space that would otherwise require a single dish many kilometers across (see Sec. 7.17).

Radio waves from space seem to originate in three ways. A common source is the random thermal motion of ions and electrons in a very hot gas, such as the atmosphere of a young star or the remnant thrown out by a supernova explosion. Another source consists of high-speed electrons that move in a magnetic field. The strongest radio sources, called **quasars,** are of this kind and may emit more energy as radio waves than as light waves. Quasars are discussed in Sec. 19.7.

Molecules in Space A third source of radio waves in space is hydrogen atoms and molecules of various kinds. These waves are spectral lines that happen to lie in the radio-frequency part of the spectrum rather than in the optical part. In particular, the hydrogen line whose wavelength is 21 cm has proved invaluable to astronomers because most interstellar material consists of cool hydrogen. Dark nebulas in our galaxy can be mapped accurately with the help of radio telescopes tuned to receive 21-cm radiation.

Traces of a surprisingly large number of chemical compounds—135 thus far—have been detected in galactic space by the radio waves their molecules emit when excited by collisions. Most of the molecules are fairly simple, such as carbon monoxide (CO), ammonia (NH_3), and water (H_2O); the hydroxide radical OH is also common. In addition some fairly complex organic molecules have been found, including formaldehyde (H_2CO), acetaldehyde (CH_3CHO), acrylonitrile (CH_2CHCN), and the alcohols methanol (CH_3OH) and ethanol (CH_3CH_2OH). Experiment and calculation have brought to light plausible reactions by which many of these compounds can be formed in space, but there is much still to be learned. The total mass of the atoms, molecules, and ions in the interstellar space of our galaxy is about a fifth the mass of its stars.

19.4 Galaxies

Island Universes of Stars

Our Milky Way galaxy is only one of perhaps 200 billion galaxies in the universe. Although galaxies have a variety of shapes, many appear as flat spirals with two curving arms that radiate from a bright nucleus. The telescope shows us spiral galaxies from different angles: some full face, some obliquely, and some edgewise, as in Figs. 19-7 and 19-8. Spiral galaxies contain from 1 to 100 billion stars. Our galaxy is so massive that 10 smaller galaxies revolve around it. Every galaxy that has been studied thus far has, like our Milky Way galaxy, a black hole at its core whose mass is millions of times that of

Figure 19-7 The Great Galaxy in Andromeda, which is 2.4 million light-years away, closely resembles our own Milky Way galaxy but is larger. The Andromeda galaxy contains about 100 billion stars, and its period of rotation is about 200 million years. It is the most distant object visible to the naked eye. The two bright objects nearby are dwarf galaxies held as satellites by the gravitational pull of their huge neighbor. The mass of the black hole at its centre is 30 million times the sun's mass, which is 8 times the mass of the black hole at the center of our galaxy.

Figure 19-8 The spiral galaxy in Coma Berenices appears edge-on to us.

the sun. The shape of a galaxy is not fixed but changes continually for complicated reasons, much as though waves were moving through it.

Other galaxies are so far away from ours that in early telescopes they were merely faint, fuzzy objects. William Herschel, the first to study these objects intensively, found evidence in support of an earlier guess that they were galaxies like the Milky Way, "island universes" in the sea of space. In his time, this was only a hypothesis, but today even galaxies too far away for their separate stars to be resolved have had their character revealed by spectroscopic studies. Doppler shifts have been detected in galaxies that show their stars to be revolving around the galactic centers, just as happens in our galaxy, with the inner ones moving faster than the outer ones (Fig. 19-9).

Most galaxies are concentrated in space in groups of up to a few hundred. Our galaxy is one of the three dozen or so members of the **Local Group.** The largest galaxy of this group is the one in the constellation Andromeda (Fig. 19-7), with our galaxy next in size. Groups of galaxies are further assembled by gravity into immense clusters that contain thousands of galaxies—and thousands of such clusters have been detected. On a still larger scale the clusters are clumped into superclusters.

Relative to their sizes, galaxies are closer together than individual stars. The distance from the sun to its nearest neighbor, Proxima Centauri, is about 30 million times the diameter of the sun, whereas the distance from our galaxy to its nearest neighbor, the Canis Major Dwarf, is only about a fifth of our galaxy's diameter. It is therefore not surprising that collisions between galaxies are far more common than collisions between stars, and even when galaxies collide, their stars rarely do. Our galaxy is in line for such a cataclysm in 2 or 3 billion years. At that time the Andromeda galaxy (Fig. 19.7), now headed our way, will collide with our galaxy. The meshing of the two galaxies will probably result in a new

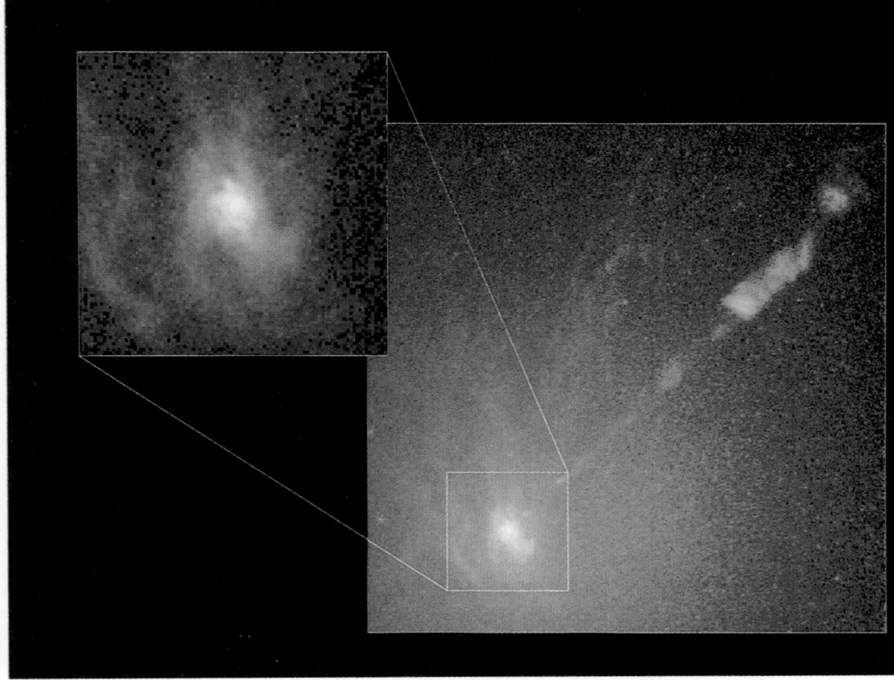

Figure 19-9 The central region of galaxy M87, which is 50 million light-years from the earth, as seen from the Hubble Space Telescope. A disk of hot gas orbits the center of this galaxy at a speed (750 km/s) that suggests the center is a black hole with a mass of 2 to 3 billion times that of the sun.

one whose form might be more like an elliptical blob than like the spiral shapes of the participants.

Dark Matter The stars together with the dust and gases among them that we observe in the sky are composed of the same protons, neutrons, and electrons of which the earth—and we ourselves—are composed. But there is very strong evidence that a large amount of invisible **dark matter** is also present in every galaxy. So much, in fact, that about 85 percent of the matter in the universe does not seem to either give off or absorb light. For instance, the outer stars in spiral galaxies rotate unexpectedly fast, which suggests that a lot of invisible matter must be part of each galaxy to keep it from flying apart. Similarly, the motion of individual galaxies within clusters of galaxies implies gravitational fields more powerful than the visible matter of the galaxies provides.

Measurements made using the Hubble Space Telescope have been able to map the distribution of dark matter over part of the universe. The results confirm that dark matter and visible matter usually occur together, so the galaxies we see are basically concentrations of dark matter that accumulated with visible matter in tow.

What can the dark matter be? An obvious candidate is ordinary matter in various forms. Examples are brown dwarfs, burned-out white dwarfs, and black holes. The snag here is that, if there were enough such objects in space to make up the missing mass, their number would be so great that they would have already been found. Neutrinos, too, apparently can be ruled out, both because their masses are too small and because they move too fast to remain inside galaxies.

There are other possibilities as well, but these involve hypothetical elementary particles with unusual properties. Until any of these particles is found, or some other explanation turns up, the nature of the dark matter in the universe will remain a major scientific mystery. About 20 different experiments are under way to directly detect dark matter.

The existence of dark matter—whatever it is—does not change the uniformity of material and structural pattern that runs through the vast array of stars and galaxies in the universe. The elements of the earth are the elements of the galaxies, the sun generates energy by a process repeated in billions of other stars, and the form of our galaxy appears

Example 19.1

In the past, substances such as phlogiston and caloric with unusual properties were suggested to explain otherwise mysterious phenomena but were later shown not to exist. Why should we take the idea of dark matter any more seriously?

Solution

In fact, some proposed new forms of matter with unusual properties were actually discovered later. An example is the neutrino, thought up in 1930 to help explain beta radioactivity but not experimentally confirmed for 23 years even though hundreds of billions of neutrinos pass through our bodies every second. Something strange is unquestionably going on in the gravitational behavior of galaxies, and the simplest explanation is dark matter. If there is no dark matter of some kind, then the laws of physics will have to be changed drastically despite their overwhelming success everywhere else, which seems less likely (though not impossible).

again and again in the rest of the universe. Enormous black holes have been found to lie at the hearts of a number of nearby galaxies, just as has been discovered in our galaxy, and strong evidence suggests that this is true in general. We can examine at first hand only a tiny fragment of the universe, yet so consistent is the whole that from this fragment we can extend our knowledge to as far as our instruments can reach.

19.5 Cosmic Rays

Atomic Nuclei Speeding through the Galaxy

The story of cosmic rays began early in this century, when it was discovered that the ionization in the atmosphere increased with altitude. At that time most scientists thought that the small number of ions always present in the air was due to radioactivity. After all, naturally radioactive substances such as radium and uranium are found everywhere on the earth. If this were the correct explanation, then when we go high into the atmosphere away from the earth and its content of radioactive materials, the proportion of ions that we find should drop. Instead it increases, as a number of balloon-borne experimenters learned between 1909 and 1914.

Figure 19-10 Victor Hess (1883–1964).

Finally Victor Hess (Fig. 19-10), an Austrian physicist, suggested the correct explanation. From somewhere *outside* the earth ionizing radiation is continually bombarding our atmosphere. This radiation was later called **cosmic radiation** because of its extraterrestrial origin.

Primary cosmic rays, which are the rays as they travel through space, are atomic nuclei, mainly protons, that move nearly as fast as light. Flashes of light that astronauts have reported seeing have been traced to cosmic-ray primaries that passed through the retinas of their eyes. The majority of primaries probably were shot out during supernova explosions in our galaxy and then were trapped there by magnetic fields. A few, though, have energies of more than 10^{20} eV, well over what accelerators on earth can produce (indeed, more than the energy of a golf ball in flight), which is truly enormous for atomic nuclei. Recent measurements of the arrival directions of these primaries show that almost all apparently come from the active cores of galaxies in which matter is swirling violently toward supermassive black holes at their hearts.

About as much energy reaches us in the form of cosmic rays as in the form of starlight, a hint of their significance. They seem to have been more than bystanders in the evolution of the universe. From what is known about the origin of the elements, there should be very much less of the light elements lithium, beryllium, and boron than the universe actually contains. Apparently the nuclei of the atoms of these elements are mostly—in the case of beryllium, nearly all—fragments of relatively abundant carbon and oxygen nuclei in space that were shattered by collisions with cosmic rays.

A Hazard for Astronauts

Cosmic rays pose a substantial cancer risk to astronauts on long voyages in space, such as the projected Mars missions. The shielding needed to protect astronauts is far too heavy (hundred of tons) for a spacecraft, and other conceivable solutions such as strong magnetic fields are also impractical. There may be no real solution to this problem, one of the reasons why many observers think robots would be a better choice as space explorers.

Secondary Cosmic Rays

Over a billion billion primaries arrive at the earth each second and carry with them energy equivalent to the output of a dozen large power plants. When a primary cosmic ray enters the earth's atmosphere, it disrupts atoms in its path to produce a shower of secondary particles. On the average more than one of these secondaries passes through each square centimeter at sea level per minute. Among the secondary particles are neutrons, which interact with nitrogen nuclei to produce the radioactive carbon 14 that is the basis of the radiocarbon dating method discussed in Sec. 16.10. Secondary cosmic rays cause some of the mutations in living things that are part of the process of evolution.

A really energetic primary cosmic ray produces an avalanche of as many as 100 billion secondaries that pepper 10 to 20 km^2 of the earth's surface. The Pierre Auger Observatory in western Argentina (named after a cosmic-ray pioneer) has 1600 12-ton detectors spread across 3000 km^2, about the area of Rhode Island, whose main purpose is to determine the directions in space from which the energetic primaries come. It is these data that led to the conclusion that such primaries come from the active cores of distant galaxies.

Example 19.2

Cosmic-ray intensity varies around the world in a manner that is correlated with the earth's magnetic field. (a) Why is such a correlation plausible? (b) Why do more cosmic rays reach the earth near the polar regions than near the equator?

Solution

(a) Primary cosmic rays are atomic nuclei and so are electrically charged particles, and moving charged particles experience forces in a magnetic field unless their motion is along the direction of the field. (b) A charged particle that approaches the equatorial regions moves perpendicularly to the earth's magnetic field there (see Fig 15-35), which means that the maximum deflecting force acts on them. A charged particle that moves toward the poles, on the other hand, is not deflected or is little deflected, since it moves parallel or nearly parallel to the magnetic field there.

The Expanding Universe

Now we turn to the evidence that led to the idea that the entire universe is growing larger and larger.

19.6 Red Shifts

The Galaxies Are All Moving Away from One Another

The spectra of galaxies share the curious feature that the lines in nearly all of them are shifted toward the red. Furthermore, the amount of shift increases with the distance of the galaxy from us. We can see this in Fig. 19-11, which shows two of the absorption lines of calcium in the spectra of several galaxies. Each galactic spectrum is shown between two comparison spectra, so that the shift of the lines toward the red (to the right in these pictures) is clear.

In Sec. 18.3 we saw that red shifts in stellar spectra result from motion away from the earth. We must therefore conclude that all the galaxies in the universe (except the few nearby ones in the Local Group) are receding from us. The recession speeds can be computed from the extent of the

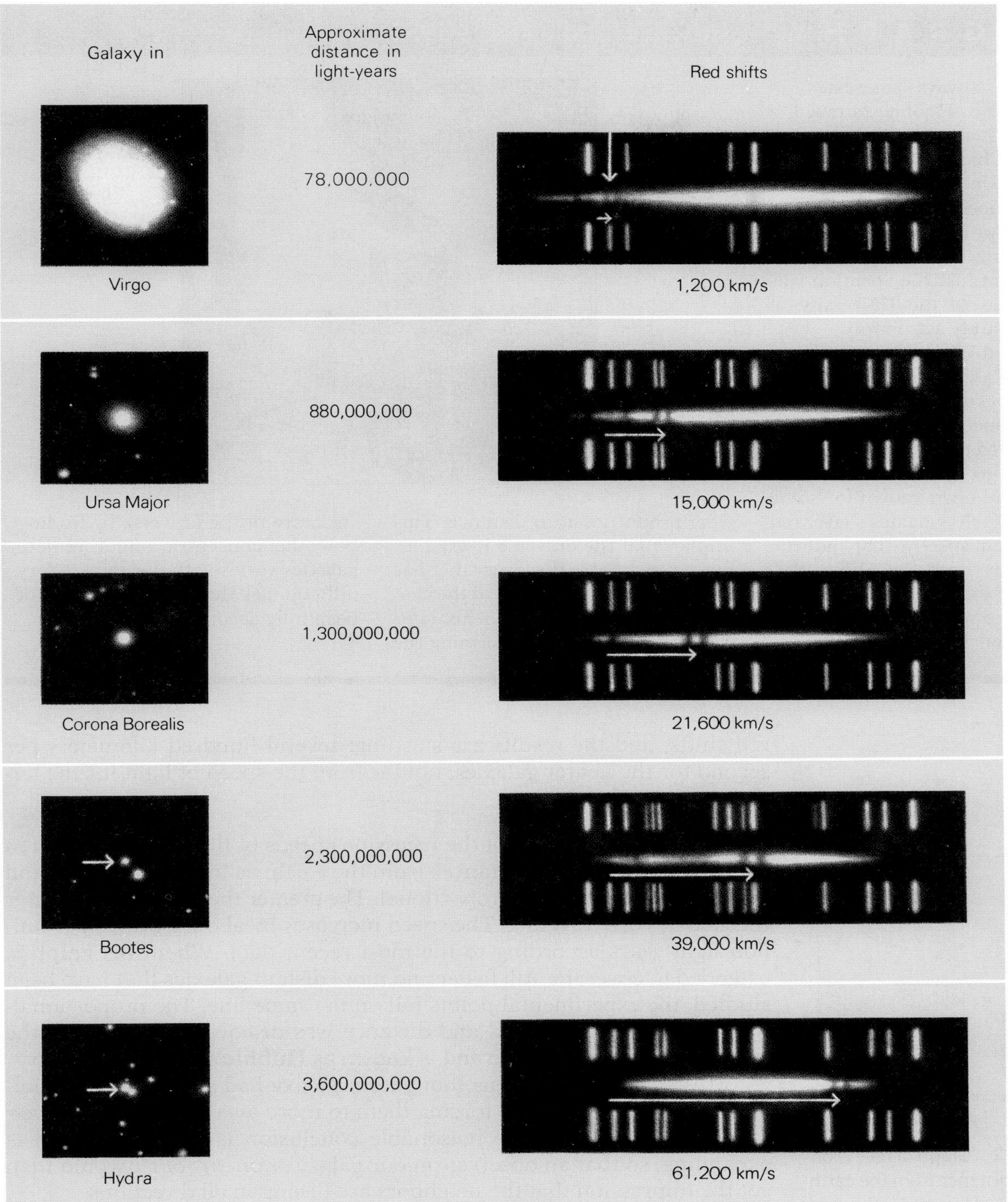

Figure 19-11 The red shifts in the spectral lines of distant galaxies increase with increasing distance. The indicated lines occur in the spectrum of calcium. Reference spectra taken in the laboratory are shown above and below each galactic spectrum. The red shifts come about because of the expansion of the universe. More recent studies indicate somewhat shorter distances than those shown here.

BIOGRAPHY Edwin Hubble (1889–1953)

Although always interested in astronomy, Hubble pursued a variety of other subjects at the University of Chicago. He then went to Oxford University in England, where he concentrated on law and Spanish. After two years of teaching at an Indiana high school, Hubble realized what his true vocation was and returned to the University of Chicago to study astronomy.

At Mount Wilson Observatory in Pasadena, California, Hubble made the first accurate measurements of the distances of spiral galaxies, which showed that they are far away in space from our own Milky Way galaxy. It had been known for some time that such galaxies have red shifts in their spectra that indicate motion away from the Milky Way, and Hubble joined his distance figures with the observed red shifts to conclude that the recession speeds were proportional to distance. This implies that the universe is expanding, a remarkable discovery that has led to the modern picture of the evolution of the universe. In his later work Hubble tried to determine the structure of the universe by finding how the concentration of remote galaxies varies with distance, a very difficult task that even today has not been fully accomplished.

red shifts, and the results are startling: several hundred kilometers per second for the nearer galaxies, not far from the speed of light for the farthest ones.

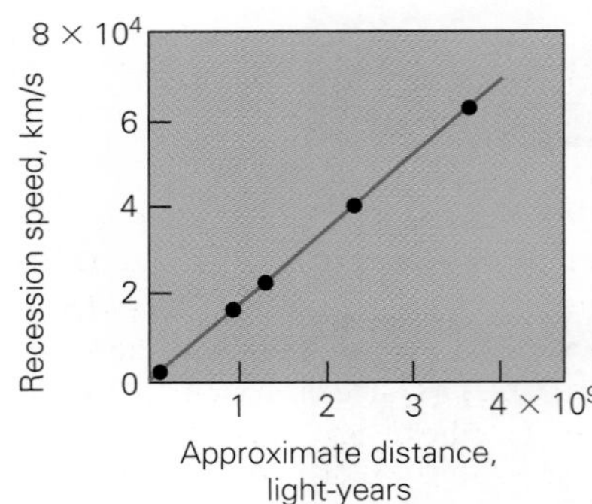

Figure 19-12 Graph of recession speed versus distance from the earth for the galaxies of Fig. 19-11. The straight line indicates that the universe is expanding at a uniform rate. Measurements made with the Hubble Space Telescope give this rate as 21 km/s per million light-years with an uncertainty of about 10 percent.

Hubble's Law If we plot the recession speeds of the galaxies shown in Fig. 19-11 versus their distances from the earth, as in Fig. 19-12, we find that these quantities are proportional. The greater the distance, the faster the galaxies are traveling. The speed increases by about 21 km/s per million light-years according to the most recent data. When this graph is extended to cover the still faster and more distant galaxies that have been studied, the experimental points fall on the same line. The proportionality between galactic speed and distance was discovered in 1929 by the astronomer Edwin Hubble and is known as **Hubble's law.**

At first it might seem as though our galaxy had some strange repulsion for all other galaxies, forcing them to move away from us with ever-increasing speed. A more reasonable conclusion is that space itself is expanding, so that an observer on our galaxy *or on any other* would then get the impression that the neighbors are fleeing in all directions.

The universe, in other words, is growing larger with the result that its component galaxies are moving ever farther apart (Fig. 19-13). The red shift in the light from a galaxy is determined by how much the universe has expanded since the galaxy emitted the light. As the light moves through the expanding universe, its wavelengths are stretched, so to speak, which corresponds to shifts toward the red (long-wavelength) end of the spectrum.

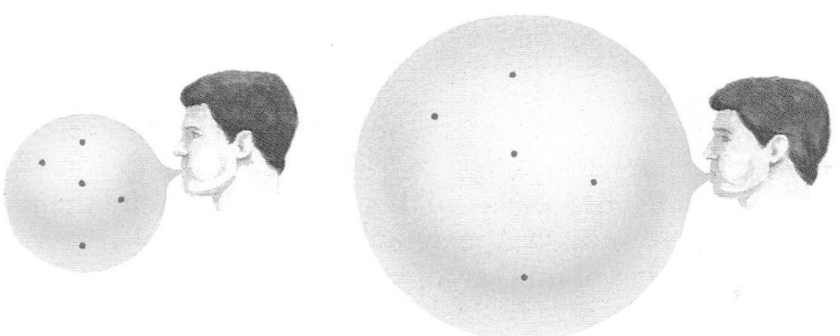

Figure 19-13 Two-dimensional analogy of the expanding universe. As the balloon is inflated, the spots on it become farther apart. A bug on the balloon would find that the farther away a spot is from it, the faster the spot seems to be moving away. This is true no matter where the bug is.

The concept of an expanding universe fits well into Einstein's general theory of relativity (Sec. 3.13), which we recall interprets gravitation as a warping of space and time.

Example 19.3

Collisions between galaxies are known to have occurred. How is this possible if the entire universe is expanding?

Solution

The expansion of the universe is an expansion of space itself. The galaxies of the universe move about independent of the general expansion and collisions occur from time to time just as they would if the universe were not expanding. As mentioned in Sec. 19.4, the Andromeda galaxy (Fig. 19-7) is headed toward our galaxy and will collide with it in 2 or 3 billion years.

19.7 Quasars

Brilliant, Tiny, and Far Away

The most remarkable objects in the sky whose spectra show red shifts are **quasars.** In a telescope, a quasar appears as a sharp point of light, just as a star does, not a fuzzy dot as it would if it were a distant galaxy. But unlike stars, the first quasars to be discovered, in the early 1960s, were powerful sources of radio waves (Fig. 19-14). Hence their name, a contraction of *quasi-stellar radio sources.* Many quasars were later found that do not give off radio waves but can be identified by their characteristic spectra.

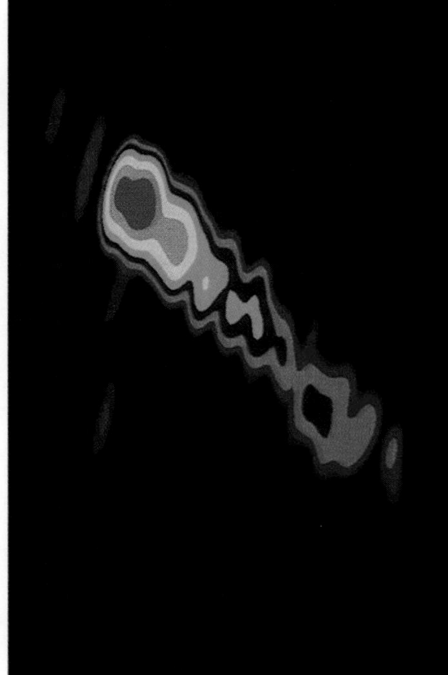

Figure 19-14 Quasars radiate many times more energy than ordinary galaxies like our own Milky Way, yet are far smaller in size. It is likely that quasars have black holes at their centers. The quasar shown here in a false-color radio image is 3 billion light-years away and emits energy at a rate of 10^{40} watts, equivalent to the output of 1000 ordinary galaxies. The jet of matter that extends from this quasar is 250,000 light-years long, although the quasar itself is only about 1 light-year across. The image of the quasar appears much larger here than it should because of diffraction (see Sec. 7.17).

Both the light and radio outputs of quasars fluctuate markedly, in some cases in less than a day. Thousands of quasars have been found, and there seem to be many more.

Quasar red shifts are usually large, corresponding to recession speeds of over three-quarters the speed of light. Such speeds mean that quasars are far away, so far away that the light reaching us from them was emitted early in the life of the universe. Because quasars are distant objects, the intensities of the light and radio waves we receive from them imply that they are giving off energy at colossal rates, billions or trillions of times more than ordinary stars. Where does the energy come from?

What Is a Quasar? Quasars vary in brightness too rapidly for them to be large objects. If a quasar were 100,000 light-years across, as a typical galaxy is, even a change in its energy output that took only a week to occur (if such a thing were possible) would appear to us as spaced out over 100,000 years. The observed short-period fluctuations point to a diameter for most quasars that is smaller than that of the solar system—but each quasar's energy output may be thousands of times the output of our entire galaxy.

Thus quasars must be small even though more powerful than any other energy sources in the universe. Apparently the heart of each quasar is a black hole whose mass is that of millions of suns. As nearby interstellar matter is pulled toward the black hole by its strong gravitational field, the matter is compressed and heated to temperatures such that radiation is given off in abundance. (Once inside the black hole, of course, the matter and its radiation vanish forever.) A diet of matter whose mass is that of a few suns per year seems to be enough to keep a quasar blazing away as observed.

Many astronomers believe that quasars are the cores of newly formed galaxies. Did all galaxies, including ours, once undergo a quasar phase? Nobody can say as yet, but it seems possible.

Evolution of the Universe

Everything we know about the universe points to its origin in an event—the **big bang**—billions of years ago in which space and time, matter and energy came into being. As the initial hot, dense universe expanded, very rapidly at first, its matter cooled and soon assumed its current forms. Local concentrations in the spreading matter then grew to become stars and galaxies. Today's physics can take us all the way back with some confidence to about 10^{-35} s after the big bang, and plausible hypotheses can get us even closer to the moment of birth, to the time when the entire universe was smaller than an atom. This is an amazing scientific accomplishment. Still ahead lies finding out what came before the big bang.

19.8 Dating the Universe

When Did the Big Bang Occur?

We can set a lower limit to the age of the universe by dating the oldest known objects it contains. These are globular clusters, which contain the first stars that formed, and the dimmest white dwarf stars that can be detected. Because a white dwarf is a dead star, it glows less and less brightly as it cools down, so the fainter it is, the farther back in time it came into being. The result for globular clusters is 12–13 billion years, for white dwarfs a little less.

An upper limit comes from using the Hubble's law figure of about 21 km/s per million light-years for the rate of expansion of the universe to calculate backward to a time when the galaxies were together in one place. Let us consider a galaxy 1 million light-years away from us, which is 9.5×10^{21} m. The galaxy's speed is 21 km/s $= 2.1 \times 10^4$ m/s, so it must have started its outward motion at a time

$$T = \frac{\text{distance}}{\text{speed}} = \frac{9.5 \times 10^{21}\ \text{m}}{2.1 \times 10^4\ \text{m/s}} = 4.5 \times 10^{17}\ \text{s}$$

Dark Energy

Dark matter is apparently not the only mystery ingredient in the universe. There is a variety of supernova, called Type 1a, which at its peak could outshine a billion suns. The brightness of such a supernova and the way in which it fades with time are closely related. By observing how the brightness of one of these supernovas changes over a period of weeks, its intrinsic brightness can be found, and comparing this with its observed brightness then gives its distance (Sec. 18.8). In 1998 astronomers discovered, to their surprise, that the remotest Type 1a supernovas were dimmer than their red shifts would predict. This means that the universe is expanding more rapidly today than it did in the past.

Several hundred Type 1a supernovas have now been studied, the farthest 7 billion light-years away. The results show that, until 5 billion years ago, the expansion of the universe was slowing down, as expected (see the discussion on this page), and then began to accelerate. The generally accepted conclusion is that some sort of influence is pushing the universe apart that at first could not counterbalance the pulling-together influence of gravity that acts to brake the expansion. As the universe grew larger, the matter in it spread out and the gravitational forces grew correspondingly weaker, which allowed the dispersive influence to take over and hasten the expansion.

Where does this cosmic repulsion come from? One suggestion is that Einstein's general theory of relativity does not correctly describe gravity on the largest scales of size, so that there is nothing to explain—the apparent repulsion is simply part of how gravity works. Most astronomers, however, favor the idea that energy of an unknown kind fills the universe and exerts an outward pressure on it. Although the nature of this **dark energy** is a puzzle, how much there is of it can be calculated. Dark energy apparently fills the universe with a uniform density that corresponds (by Einstein's formula $E_0 = mc^2$) to the equivalent in mass of a half-dozen hydrogen atoms per cubic meter. This does not seem like much, and indeed dark energy is of no importance in the solar system, but its effects on the evolution of the universe as a whole are significant.

Only 4 percent of the universe consists of the ordinary matter we are familiar with; 23 percent is the dark matter that galaxies contain; and no less than 73 percent is dark energy. Electromagnetic radiation—light, radio waves, x-rays—amount to only 0.005 percent (Fig. 19-15). Thus 96 percent of the universe awaits our understanding.

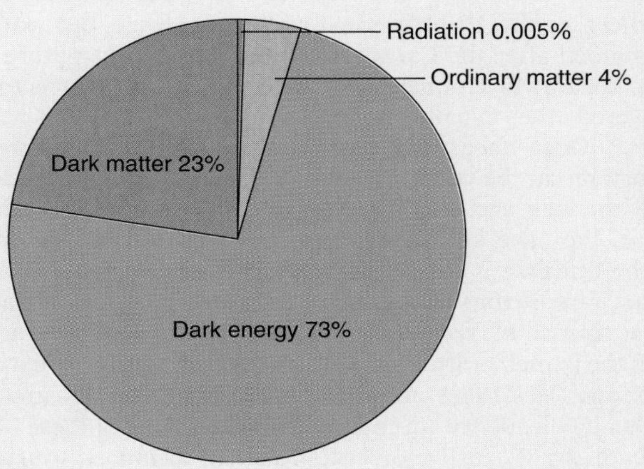

Figure 19-15 What the universe is apparently made of.

in the past. Since there are 3.2×10^7 s in a year, this would make the age of the universe 1.4×10^{10} years—14 billion years.

This calculation represents an upper limit because it assumes a constant rate of expansion. But the universe contains matter, and gravity must have been slowing the expansion down just as a ball thrown upward slows down as it rises. Since the expansion was once faster, the universe should be younger than 14 billion years. This conclusion holds even though recent evidence suggests that the expansion has actually been speeding up for the past 5 billion years (see sidebar *Dark Energy*).

In 2003 an entirely different method, based on studies of the cosmic background radiation (Sec. 19.9), was able to give a precise figure for the time of the big bang: 13.7 billion years ago, which agrees nicely with the discussion above.

AN ASTRONOMER AT WORK **Wendy Freedman**

Director, Carnegie Observatories, Pasadena, California.

When I was in the tenth grade, I had a physics teacher who, when explaining a difficult concept, would say to the class, "The girls don't have to listen to this." By eleventh grade, however, my physics teacher had an infectious love of the subject that inspired in me a desire to understand the universe, and helped propel me toward my current career. And my tenth grade teacher never realized that his comments only piqued my interest even more!

As a research astronomer, I have focused much of my recent career on measuring how fast the universe is expanding and determining what its fate will be. The expansion rate is characterized by something known to astronomers as the Hubble constant. It is named after the Carnegie astronomer, Dr. Edwin Hubble, who first discovered the expansion of our universe. Once measured, the Hubble constant can be used to calculate both the size and the age of our universe. We have known since 1929 that the universe is expanding, but the exact rate of that expansion remained a source of controversy, that is, until the launch of the Hubble Space Telescope (HST). For about a decade, I was a coleader of an international team of 27 astronomers who used HST to measure the distances to and speeds of other galaxies, galaxies not unlike our own Milky Way galaxy. At the end of that observing program we obtained a value of the Hubble constant that has also been recently and independently confirmed by another NASA mission called the Wilkinson Microwave Anisotropy Probe (WMAP). The age of the universe is now measured to be 13.7 billion years. But there is more to the story.

As we were evaluating the impact of our new measurements of the Hubble constant on the age of the universe, our calculations led to a paradox: the ages of stars in our Milky Way galaxy appeared to be older than the universe in which they were found. As one astronomer aptly put it, "You cannot be older than your mother." Clearly something was wrong. The resolution turned out not to be due to an error in our measurements but required a change in the cosmological model within which we astronomers had been interpreting these measurements.

Astronomers now believe that there is a new force of nature (the so-called dark energy) acting on very large scales and in a way that is now accelerating our universe. Physicists do not yet understand the origin of this new force but with time and more observing, nature will, I am sure, reveal her secrets to us. My current research interests have turned to understanding the nature of this new dark energy. Our Carnegie Supernova Project is aimed at making measurements of distant supernovas, very bright explosions resulting from the deaths of stars, which can be seen at vast distances. The study of supernovas led to the original discovery of dark energy. By making precise measurements of these objects over a range of distances, we can determine how the expansion of the universe changes over time, thereby allowing us to characterize the effects of dark energy on the expansion.

My research life has many facets, and I find my days extremely stimulating, interesting, and challenging. I am now the director of the Carnegie Observatories, which certainly keeps me busy. In a very exciting new venture with a consortium of institutions, we are embarking on a new project to build the world's largest telescope, a 24.5-meter reflector, named the Giant Magellan Telescope, to be located in the Chilean

Andes mountains in South America [see Sec. 17.1]. This telescope will control and focus seven 8.4-meter mirrors on a single mount, a bold technological challenge, holding the promise of entirely new scientific discoveries.

My own work ranges from traveling to telescopes to observe, analyzing data with computers, writing computer software programs, interacting with colleagues, attending scientific meetings and weekly seminars, giving talks on my work, writing papers, keeping abreast of new results and discoveries by reading papers from others in the field, and writing and reviewing proposals for grants and telescope time. Much of my daily interaction takes place via email. One of the things I like most about my work is the opportunity to continue to learn completely new things about our universe and the opportunity to participate in this fascinating scientific enterprise.

If there is one piece of advice I can give to young people, it is to find something that you love to do. There are very few things more rewarding in life than having a career that you enjoy.

19.9 After the Big Bang

A Glimmer of the Early Universe Still Exists

Figure 19-16 shows some stages in the history of the universe. Just after the big bang, the universe was a compact, intensely hot mixture of matter and energy. Particles and antiparticles were constantly annihilating each other to form photons of radiation, and just as often photons were materializing into particle-antiparticle pairs. The particles and antiparticles at this time were the quarks and leptons described in Sec. 8.15. Recreating these

Age 0

The big bang. The universe begins to expand, at first very rapidly.

Age 10^{-35} second

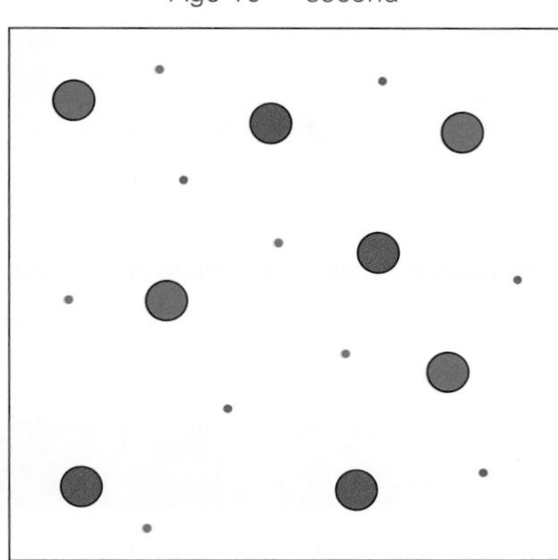

Quarks and leptons come into being. Particles slightly outnumber antiparticles. All the antiparticles are eventually annihilated to leave a matter universe with no antimatter.

Age 1 second

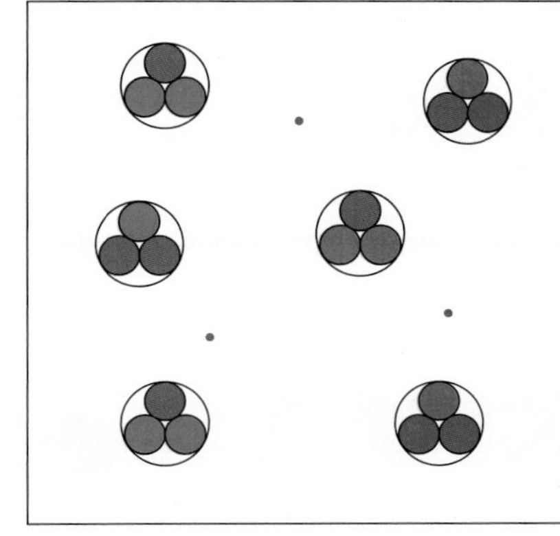

Quarks have joined to form neutrons and protons, some of which later react to form nuclei of light elements, mainly helium. There are as many electrons as protons, so the universe remains electrically neutral.

Age 3 minutes

Nuclear reactions stop. The hydrogen: helium ratio is 3:1 as in most of today's universe. The universe is still too hot (10^9 K) for atoms to exist.

Age 380,000 years

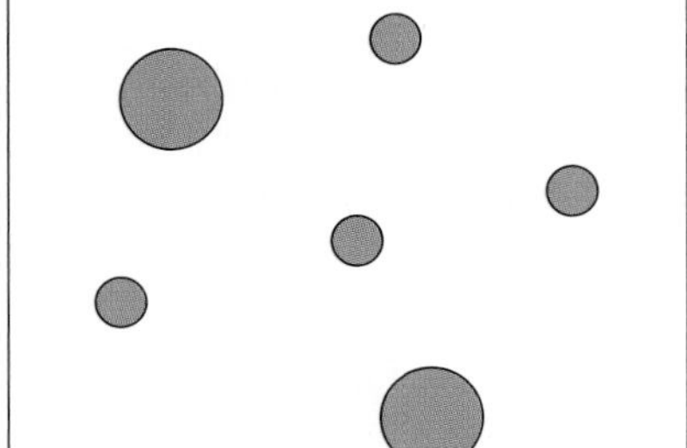

The universe has now cooled enough for nuclei and electrons to have joined to form neutral atoms. With few ions left to absorb it, radiation from this time was red-shifted by the expansion of the universe to become radio waves that fill the universe today.

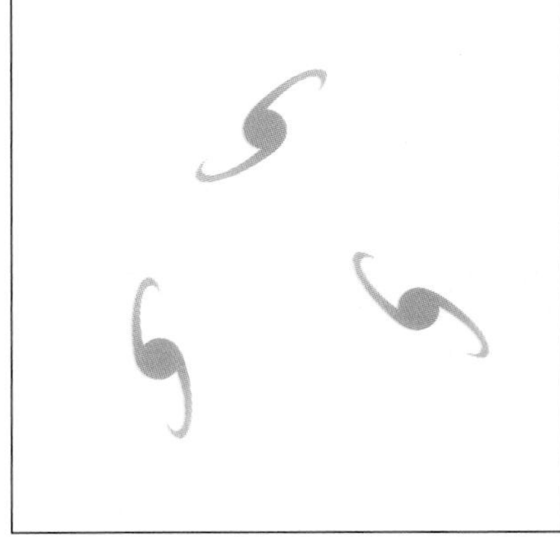

Under the influence of gravity, local concentrations of gas grew and began to condense into the first stars and galaxies when the universe was 400 million years old. After a billion years galaxies populated the universe, which continued to expand.

Figure 19-16 Snapshots from the youth of the universe.

Theory of Everything

Quantum theory tells us how matter and energy behave on the scale of size characteristic of molecules, atoms, protons, neutrons, and electrons, which are too small for us to see directly and where gravity has little influence. Einstein's general theory of relativity goes to the other extreme by accurately describing how matter and spacetime interact to produce gravitational effects on a larger scale all the way up to stars and galaxies. On this scale quantum effects, such as the wave-particle duality and the graininess of matter and energy, can be ignored. Between them the two pictures of how nature works might seem to include everything in the physical universe. However, the link between these pictures—there must be one—is not yet known, so there is no single Theory of Everything that would cover the overlap between the tiny (the province of quantum theory) and the massive (where general relativity reigns).

Where would we expect such an overlap? The most obvious place is the universe just after the big bang, when its entire mass occupied less space than an atom. Without a Theory of Everything nothing can be said about the first moments of the universe. Despite much effort, none of the approaches that have thus far been tried have led to testable predictions, so, as Wolfgang Pauli, a pioneer quantum physicist, said of another theory that lacked an anchor in reality, these approaches "aren't even wrong." But enough tantalizing hints have turned up to keep the search for a Theory of Everything alive.

A Map of the Cosmos

A number of surveys have measured galaxy distances by means of their red shifts to arrive at three-dimensional maps of parts of the sky. The most comprehensive survey thus far, made using an Australian telescope, examined no less than 106,688 galaxies found in 5 percent of the sky (Fig. 19-17). The farthest galaxies were 4 billion light-years away.

The results confirmed previous studies that found galaxies in groups, the groups in clusters, the clusters in superclusters, and sharp-edged "walls" of clusters separated by giant voids. What was new was that nothing bigger turned up: no structures were found more than about 300 million light-years across. Since the universe itself is around 10 billion light-years across, this means that the irregularities do not exceed 3 parts in 100. It seems that, overall, the universe is relatively uniform, which is a good fit with the big bang picture of its origin. On successively smaller scales of size lumpiness appears in the forms of aggregates of galaxies, galaxies themselves, and individual stars. Such lumpiness is to be expected as gravity concentrated density fluctuations in the early universe.

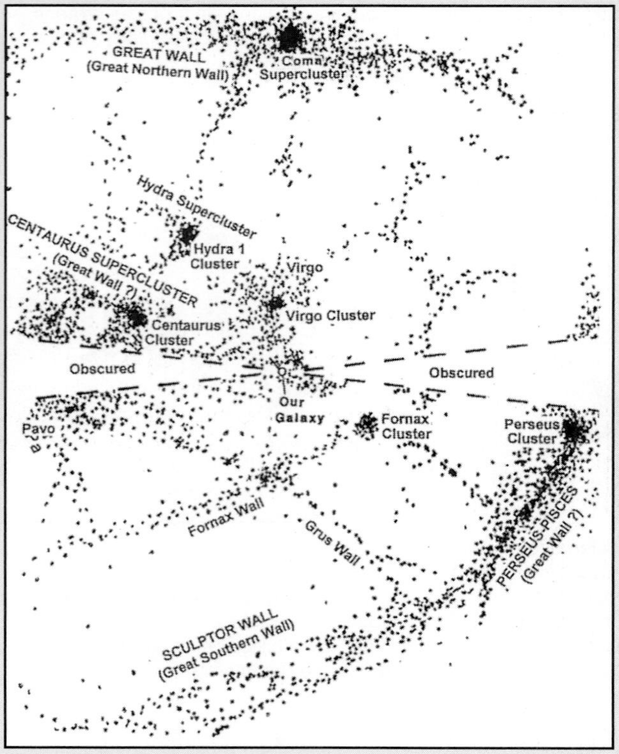

Figure 19-17 A map of the closest clusters and superclusters of galaxies. Dust in the Milky Way prevents light from the sectors marked "obscured" from reaching us. Because the map does not provide a three-dimensional picture, some features are not evident. For instance, the lines of galaxies called great walls actually extend as sheets of galaxies well above and below the map.

conditions is one of the reasons for the proposed International Linear Collider, which would be the most powerful particle accelerator ever built.

As the fireball expanded and cooled, the energies of the photons decreased. (We recall from Fig. 18-4 that the average wavelength of the radiation from a hot object increases as its temperature drops, which corresponds to a decrease in frequency and hence to a decrease in quantum energy hf.) Finally, shortly after the big bang, the photons had too little energy to create any more particle-antiparticle pairs. The annihilation of the existing particles and antiparticles continued, however. For a reason still unknown, particles outnumbered antiparticles by 30,000,001 to 30,000,000. When the annihilation was finished, some of the original particles were therefore left over, which ended up as the matter of today's universe.

When the universe was about a minute old, nuclear reactions began to form helium nuclei. Calculations show that the ratio of hydrogen nuclei (protons) to helium nuclei ought to have ended up as 3:1, the same ratio found in most of the universe today. Although nearly all the helium in the universe was formed in the first few minutes after the big bang, with some created later in the nuclear reactions that power stars, the helium on the earth mainly came from the alpha decay of radioactive elements in its interior.

After 380,000 years the universe was cool enough for electrons and nuclei to combine into atoms. Since photons interact strongly with charged particles but only weakly with neutral atoms, at this time matter and radiation were "decoupled" and the universe became transparent. The radiation that was left then continued to spread out with the rest of the universe, so that even today remnants of it must be everywhere.

Because the universe is expanding, an observer today would expect to find these remnants to have undergone a red shift to long wavelengths, in the range of radio microwaves. The radiation would not be easy to find, since it would be very weak. However, it ought to have two distinctive characteristics that would permit it to be identified: it should come almost equally strongly from all directions, and its spectrum should be the same as that which an object at about 2.7 K would radiate.

Remarkably enough, radiation of exactly this kind was discovered in 1965 (Figs. 19-18 and 19-19). It turns out to account for 99 percent of

Figure 19-18 Radio waves that originated early in the history of the universe were first detected by Arno Penzias and Robert Wilson as a persistent hiss in a sensitive microwave receiver attached to this 15-m-long horn antenna at Holmdel, New Jersey.

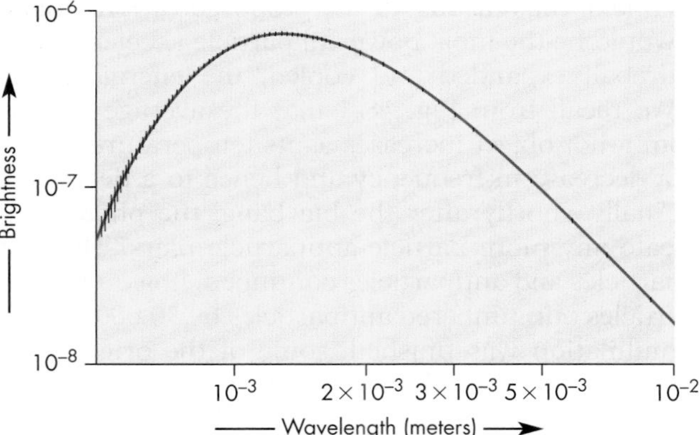

Figure 19-19 The spectrum of the cosmic background radiation corresponds to that of an object at a temperature of 2.7 K. The radiation originated when the universe became transparent at an age of about 300,000 years when its temperature was 3000 K. The waves were then lengthened by the expansion of the universe to their present wavelengths.

Whispers from the Past

As well as can be determined by instruments on the earth, the cosmic background radiation appears the same in all directions in the sky. This finding troubled astronomers, because only a completely smooth distribution of matter in the early universe could lead to a perfectly uniform afterglow—and a smooth distribution would leave no way to explain how such irregularities as galaxies could have developed.

To study the question further, NASA launched satellites in 1989 and 2001. Results from the first satellite showed that there are indeed very small variations in the radiation. These variations testified to the existence in the remote past of clumps of gas denser than the rest, clumps that were the seeds from which galaxies grew.

The more recent satellite, the Wilkinson Microwave Anisotropy Probe (WMAP), can pick up temperature differences down to a few millionths of a degree. The picture of the radiation it revealed is shown in Fig. 19-20. From the details of the temperature differences it was possible to establish, among other things, the precise age of the universe, its rate of expansion, how much dark matter it contains, and how much dark energy pervades it. The WMAP results thus not only independently confirmed the picture of the universe that earlier studies had suggested but also put much more reliable numbers into the picture.

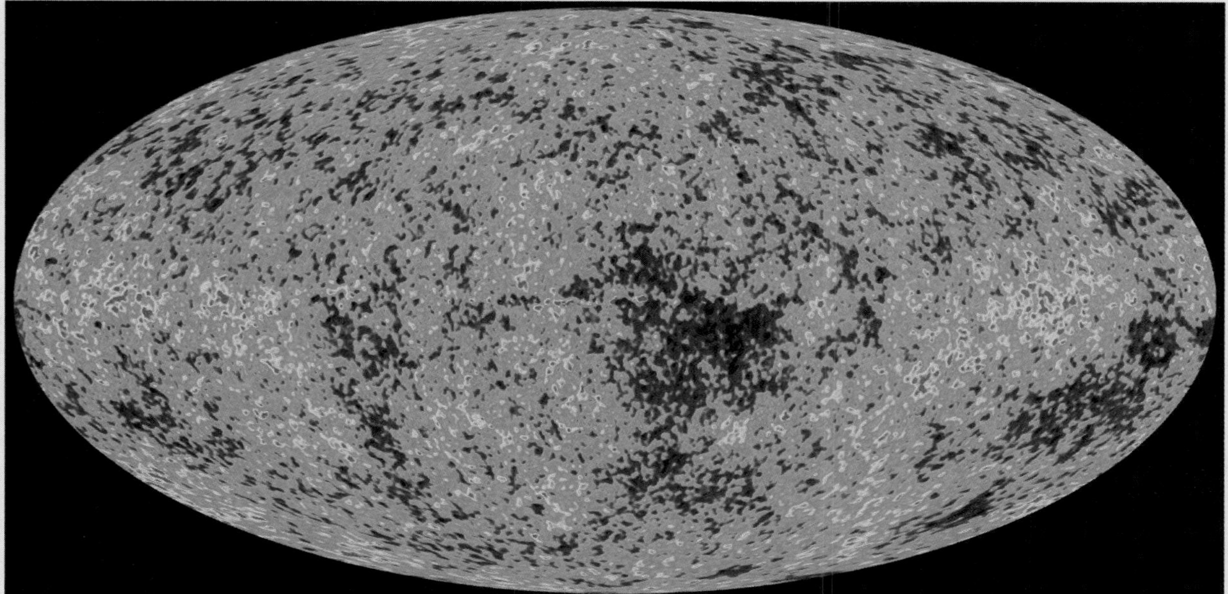

Figure 19-20 Temperature map of the early universe from WMAP data. The hottest regions are shown as red, the coolest as blue. A great deal of information about the universe and its history has been obtained from the details of this map.

the radiation in the universe—only 1 percent is starlight. The universe is bathed in a sea of radio waves whose ultimate source seems to have been the primeval fireball. Thus we have three observations that strongly support the big bang theory of the origin of the universe:

1. The expansion of the universe
2. The relative abundances of hydrogen and helium in the universe
3. The cosmic background radiation

The Future Before the discovery that the expansion of the universe is currently speeding up, it was not clear whether the expansion would continue forever (an **open universe**), or whether gravity would in time cause it to stop and then the universe would begin to collapse (a **closed universe**). If the universe is closed, all its matter and energy would eventually come together in a **big crunch.** Conceivably another big bang would follow, and then another cycle of expansion and contraction (Fig. 19-21). In this case the big bang of 13.7 billion years ago was only the latest in a series of big bangs that extends forever into the past and will continue forever into the future.

The key to knowing whether the universe is open or closed is how its average density of matter and energy compares with a certain critical value that depends on how fast the universe is expanding. Before evidence for dark energy came to light, although an open universe seemed more likely, there was still room for doubt. Now, even though the exact nature of dark energy remains to be found, the argument for an open universe seems very strong. Indeed, it has even been suggested that the push of dark energy may cause the universe to grow in size at an ever-increasing rate until a **big rip** tears it apart 20 or more billion years hence. Or, in another scenario whose likelihood is unknown, the influence of dark energy may fade and a big crunch does await.

19.10 Origin of the Solar System

A Gradual Process

By a billion years after the big bang much of the hydrogen and helium of the universe had accumulated in separate clouds under the influence of gravity. These clouds were the ancestors of today's galaxies. As the young

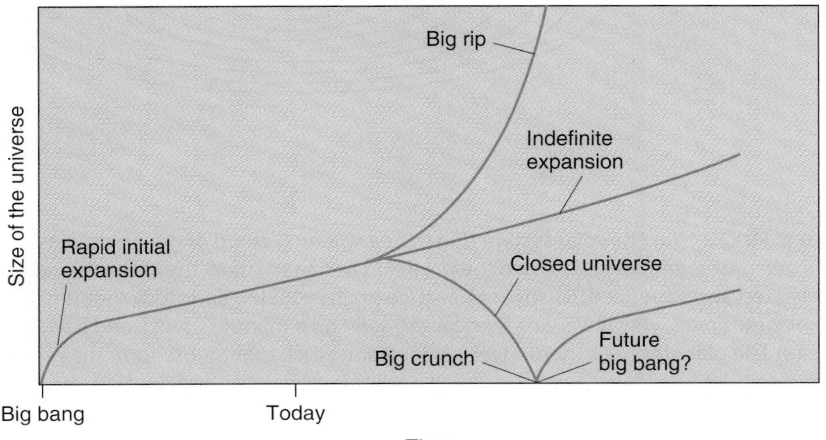

Figure 19-21 Three possible destinies for the universe.

galaxies contracted, local blobs of gas formed that became the first stars. Some of the early stars were very massive and went through their life cycles rapidly to end as supernovas. The explosions of these supernovas dumped elements heavier than hydrogen and helium into the remaining galactic gas. As a result the matter from which later stars condensed was a mixture of all the elements, not just hydrogen and helium. By the time the sun came into being, billions of years after the first generation of stars, the loose material of our galaxy contained between 1 and 2 percent of the heavier elements. These were in the form of small solid "dust" grains, some of ice, some of rock.

The swirling cloud that became the sun was originally much larger than the present solar system. As it shrank in the process of becoming a star, this **protosun** left behind a spinning disk of gas and dust (Fig 19-22). Bits of matter in the disk collided and stuck together to form larger and larger grains, perhaps the size of pebbles. In time these grains formed larger bodies called **planetesimals** a few kilometers across that ultimately joined to become the planets around 4.6 billion years ago. Near the protosun, which was heated by gravitational compression, it was too warm for gaseous elements to collect in any great amounts on the planetesimals, which is why the inner planets are rocky bodies. Farther out temperatures were lower, so the planetesimals there mirrored the composition of the original cloud by being mainly frozen gases.

As the planets were developing with their satellites around them, the protosun was gathering up more and more material. After about 30 million

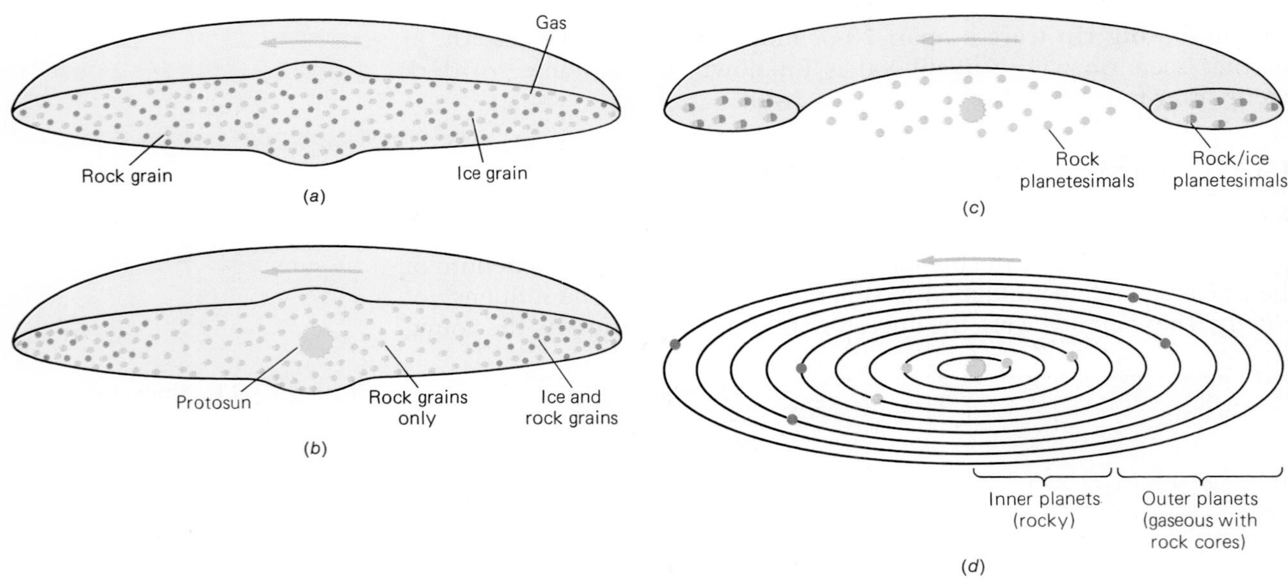

Figure 19-22 (*a*) The solar system began as a spinning cloud of gas and grains of ice (frozen gases) and rock. (*b*) When the protosun began to shine, the ice grains near it were heated and vaporized. (*c*) The rock and ice grains collided and stuck together to form planetesimals, which were rocky near the sun and a mixture of rock and ice farther away. (*d*) The planetesimals themselves collided and stuck together to form the planets. An intense solar wind blew away the gas and other material that had not become incorporated in the planets.

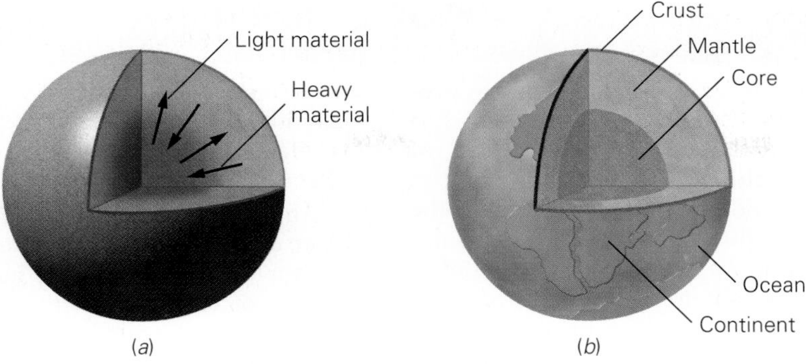

Figure 19-23 (*a*) When the earth was young, its interior melted. Under the influence of gravity, heavy materials then migrated inward to form the core and lighter ones were forced outward to form the mantle and crust. (*b*) Eventually the separation into core, mantle, and crust was complete, the outer parts cooled and hardened, and the early oceans and continents came into being.

years its internal temperature and pressure became sufficient for fusion reactions to occur in its hydrogen content, and the protosun became a star. Fast ions and electrons then began to stream out of the sun, much like today's solar wind but far more intense. This wind swept the solar system free of the gas and dust that had not yet been incorporated in the sun, the planets, and the satellites.

The infant planets were all heated by gravitational compression, which was naturally most effective in giant Jupiter and Saturn. Radioactivity was important as well in heating the inner planets. The earth became hot enough at this time to melt and separate into a dense iron core and a lighter rocky mantle (Fig. 19-23). The remaining planetesimals in the solar system bombarded the planets and satellites heavily, leaving craters on the solid ones that, in the case of the earth, erosion would later erase.

Extraterrestrial Life

The recipe for creating life begins with a warm rocky planet whose gravity is able to hold an atmosphere and whose distance from its parent star is such that it receives enough but not too much radiant energy. Then provide an atmosphere of certain quite common gases, allow the planet to cool down a bit so that water condenses out, and wait perhaps a billion years. This recipe was followed successfully on the earth. Let us now see how likely it is to have been followed elsewhere in the universe.

19.11 Exoplanets

Probably a Great Many

The modern view of the solar system is that it is a natural product of the evolution of the sun. Since the sun is by no means an exceptional star in any other respects, it is reasonable to suppose that other stars are also attended by planetary systems. Direct evidence in support of this idea was the discovery by the Hubble Space Telescope of disks of dust and gas around a number of newly formed stars, disks like the one pictured in Fig. 19-22*b*. In the region of the Orion Nebula that was studied, nearly all of the young stars (less than a million years old) there had such disks. Thus it may be that a star without planets is the exception, not the rule.

Looking for Planets

A few actual images of exoplanets—as yet only fuzzy dots—have been obtained by blocking out light from their parent stars by computer processing. However, the most successful quests for planets around sunlike stars make use of the fact that, as a planet orbits a star, its gravitational tug on the star causes the star to wobble back and forth. Unless the wobble is perpendicular to the line of sight, at times the star moves toward us, at other times away from us. These motions give rise to doppler shifts in the star's spectral lines that can be interpreted to reveal the mass of the planet and the size, shape, and period of its orbit. The sun's wobble as Jupiter circles it produces a doppler wavelength shift that would be just on the limit of detection.

Another method is to seek the slight dimming of the light from a star as a planet passes in front of it. An advantage of this procedure is that it can respond to smaller planets; the sun dims by 0.01 percent as the earth moves across its face, which could just barely be picked up from far away. NASA's Kepler satellite is planned to examine the light from 100,000 sunlike stars to look for brightness decreases due to such planet transits. Unfortunately both the wobble and dimming methods favor large planets revolving around small stars, whereas what we are really interested in are small earthlike planets revolving around sunlike stars.

An atmosphere around an exoplanet can sometimes be detected and analyzed if it passes across the line of sight from its parent star. If the spectrum of the star itself is subtracted from the combined spectra of the star and the planet, the gases in the planet's atmosphere, if there is one, can be identified. Space telescopes already in orbit have been used to look for exoplanet atmospheres, and two planned space observatories will continue the search.

Many of the 200 billion stars in the Milky Way are similar to the sun. Some fraction of these stars, even if not all, have planetary systems. And there are billions of other galaxies in the known universe. Even a conservative guess at the lower limit to the number of extrasolar planets, or **exoplanets,** in the universe yields billions of billions.

If planets are such common objects, why are we certain only of the eight in our own solar system plus not really a great many others? Three interrelated factors make it difficult to identify planets that belong to other stars. The first is distance: the closest star to us is thousands of times farther away than Pluto and other Kuiper Belt objects. The second is that planets shine by reflected light and not by themselves, which makes them faint compared with their parent stars. The third factor is size. Planets are necessarily small with relatively little mass, because if not, they would have become stars. The small size of a planet means that the light from its star dims by very little as it crosses the star's disk, and its small mass makes the wobbles it causes in the motion of the star hard to measure.

We Are Not Alone Despite these difficulties nearly 300 exoplanets, some in multiple-planet systems, have been unambiguously detected since the first was found in 1995, and more and more are turning up every year. So far they range from 0.02 times the earth's mass all the way up to 20 times Jupiter's mass. At least one has an atmosphere that contains water. Of stars like the sun that have been studied, 1 in 7 are known to have planets. And the smaller the planets, the more there seem to be. Although finding earth-sized planets is still extremely hard, it would therefore not be surprising if improved methods yield a big crop; there are indications that as many as 60 percent of sunlike stars have them.

From the point of view of life in the universe, just how many billions of planets there are is not really important. What is important is that a certain proportion of these planets surely meet, in the words of Harlow Shapley, "the happy requirement of suitable distance from the star, near-circular orbit, proper mass, salubrious atmosphere, and reasonable rotation period—all of which are necessary for life as we know it on earth." But not any location in our galaxy will do. Elements heavier than hydrogen and helium, which are needed for life, are most abundant toward the center of the galaxy. However, supernova explosions, which would destroy life on a planet anywhere near, are also more common near the center. A zone between 22,000 and 30,000 light-years from the center of the galaxy, about a tenth of the area of the galactic disk, seems ideal; our sun is in the middle of this zone. One of the smaller exoplanets that has been discovered, whose mass is five times that of the earth, also may occupy such a habitable zone around its parent star since its surface temperature is between 0°C and 40°C. This would suit Goldilocks: neither too hot nor too cold for life but just right.

Given the above conditions, life comes next. Would it necessarily resemble life on earth? This is an interesting question to which biology has an interesting answer. Terrestrial life has many examples of what is called convergent evolution: organisms that came from different genetic lines are found to have developed similar forms and functions as they respond to similar environmental pressures. Thus eyes may have independently evolved in possibly 65 different types of animals. Of course, extraterrestrial life is not likely to mirror life on earth exactly, but if a planet elsewhere is much like the earth, then the organisms that arise there may well be recognizable variations of organisms that now or in the past flourished here. Although the hypothesis of life on other worlds may not be directly verified in the near future, no serious arguments that dispute this hypothesis have been advanced. It seems likely that we are not alone in the universe.

19.12 Interstellar Travel

Not Now, Probably Not Ever

The great physicist Enrico Fermi was skeptical about widespread advanced life in the universe. If it existed, then "Why aren't the aliens here?" (Arthur C. Clarke's response was, "I'm sure the universe is full of intelligent life—it's just been too intelligent to come here.") The real answer is that the required interstellar journey is just not a practical proposition no matter how advanced a civilization might be. Astronauts have already visited the moon and spacecraft have traveled to all the other planets. If the money for such a project were available, as it may be in the next decade or two, there is no reason why people should not be able to set foot on Mars some years after a decision to do so is made and see for themselves whether life existed there in the past. But travel to the planetary systems of other stars is much more of an undertaking.

Suppose we wish to go to Proxima Centauri, the star nearest to the sun, to check on whether it has a planet that contains living things. This star is 4×10^{13} km away, a little over 4 light-years. The various Apollo spacecraft took 3 days to reach the moon, which is 100 million times closer than Proxima Centauri. To reach this star at the same rate, a similar spacecraft would therefore need 300 million days, nearly a million years. So existing technology is out of the question.

What about the future? According to Einstein's special theory of relativity, which has been found to be completely accurate in all its predictions, the ultimate limit to the speed that anything can have is the speed of light. This speed is about 300,000 km/s. At a speed 1 percent short of this, a spacecraft could reach Proxima Centauri in a few years, a reasonable enough period of time. However, the energy required would be fantastic, more than all the energy currently used in the entire world per year for each ton of spacecraft weight—and the spacecraft would have to weigh many tons. How can all this energy be produced? How can it be concentrated at one time and place? How can it be given to the spacecraft? And how could a like amount of energy be provided in the vicinity of Proxima Centauri to return the spacecraft to the earth?

These questions are not quibbles. They are fundamental to such an enterprise—and they have no answers at present. As for a space traveler visiting the stars that lie millions and billions of times farther away than Proxima Centauri, there does not seem to be any hope whatever.

The laws of physics hold everywhere in the universe, which means that the extreme unlikelihood of travel from the earth to an extrasolar planet is also true for travel from such a planet to the earth. It is easy to attribute ancient legends and archeological findings that are hard to explain in terms of known events to visitors from another world. However, all such attributions have turned out to lack any evidence to support them. A good story does not constitute proof of anything except the imagination of its author.

How Much Advanced Life?

The earth came into being about 4.6 billion years ago, and less than a billion years later life of some primitive sort had appeared. On a cosmic scale, this is pretty fast, and, together with other evidence, it strongly suggests that life could also have come into being on planets around other stars. But what kind of life? The next step, from one-celled microorganisms to multicellular organisms, took over 3 billion more years on the earth. And not until now, 4.6 billion years since the earth's origin, have living things on earth progressed to the point where interstellar communication can be taken seriously. Indeed, the development of an advanced civilization—as distinct from merely complex life forms—may require too many chance events to be inevitable as yet wherever life occurs. The dinosaurs were successful enough to be the dominant animals on earth for 140 million years (will we last that long?), but they never evolved to create an advanced civilization. If it were not for the asteroid (and/or lava flows) that drove them into extinction, we might well not be here today. So, although life is probably common in the universe, advanced civilizations may not be.

19.13 Interstellar Communication

Why Not?

Although travel to other worlds is almost surely impossible, by contrast communication with them seems quite feasible. The first serious proposal was made in 1820 by the German mathematician Karl Friedrich Gauss. Gauss suggested that trees be planted on the bare landscape of Siberia to form a right triangle large enough to be seen from the moon, which he thought might be inhabited. A better method for interstellar distances is to use radio waves.

A Search for Extraterrestrial Intelligence (SETI) is going on in which radio telescopes around the world seek messages from far away (Fig. 19-24). The program has been privately funded since 1993, when a Nevada senator, upset because until then "not a single Martian has said, 'take me to your leader,' and not a single flying saucer has applied for FAA approval," saw to it that NASA support was stopped; it was never restored. Part of the program is targeted on the 1000 nearest stars that resemble the sun. The entire sky is also scanned, though necessarily with less sensitivity. What is sought are patterns of emissions that are not the result of natural processes, such as pulsar bursts (Fig. 19-25). Around 5 million people around the world have volunteered their computers for use during

Figure 19-24 Among its other tasks, this extremely sensitive 305-m radio telescope built in a natural hollow at Arecibo in Puerto Rico is being used in the SETI (Search for Extraterrestrial Intelligence) program to monitor a total of 168 million radio channels for signals from space.

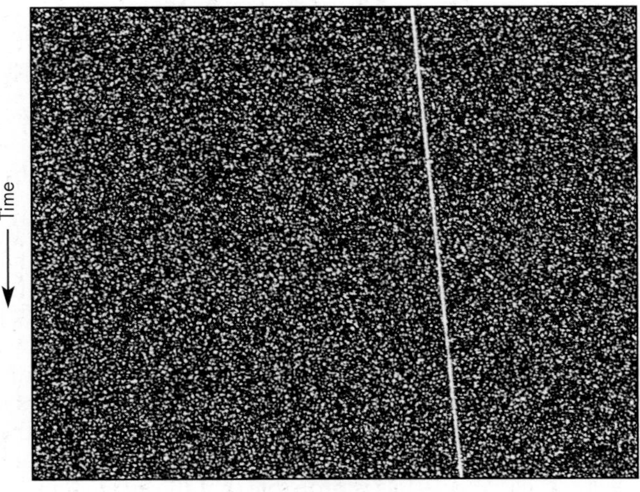

Figure 19-25 How a message from space might be detected. Each point on this plot corresponds to the frequency of a particular signal picked up by a radio telescope and the time at which it was received. The brighter the point, the stronger the signal. The random background is due to noise. The diagonal line records a signal sent out by the Pioneer 10 spacecraft when it was beyond the orbit of Pluto. The changing frequency of the signal is the result of the earth's rotation causing a changing doppler shift.

otherwise idle periods to process radio telescope data instead of reverting to screensavers. In a 2007 expansion of the program, 42 antenna dishes, each 5.5 m across, were connected to make the Allen Telescope Array in Hat Creek, California. The array may eventually grow to 350 dishes. The

Square Kilometer Array (Sec. 19.3), which is being planned, may also listen for messages from far away. Another approach recognizes that, if there is advanced life elsewhere, perhaps it, too, has invented lasers. Since April 2006, a telescope in Massachusetts has been scanning the sky for repetitive flashes of light that might be laser signals.

The SETI program and its predecessors have been on the lookout for 40 years, with no positive results. Although the early studies may not have been sufficiently sensitive or thorough, the latest equipment is able to respond to signals from up to 4000 light-years away sent out by transmitters no more powerful than transmitters we on earth could make today. Why the silence? To begin with, the search is far from complete. Second, maybe we are listening on the wrong frequencies. Third, for all our pride in our advanced civilization, we ourselves are not broadcasting messages that advertise our existence, and perhaps many of our counterparts elsewhere are similarly disinclined to do so. And so on. Even though SETI has thus far had no success, it is surely premature to consider the program a failure.

Indeed, what about sending out messages ourselves for other worlds to receive? The technology is already available in the radio band; in fact, a signal was deliberately sent into space in 1974 from the Arecibo radio telescope acting as a transmitter. In the optical band, a laser planned in the United States for other purposes would be powerful enough to shoot out pulses of light 3000 times brighter than the sun that could be detected up to perhaps 1000 light-years away. By virtue of the long distances involved, years at best, more likely centuries would be needed for an interchange of information to take place. But what an extraordinary, exhilarating prospect!

Important Terms and Ideas

Our **galaxy** is a huge, rotating disk-shaped group of stars that we see in the sky as the **Milky Way** from our location about two-thirds of the way out from the center. Most of the stars of the galaxy are found in two **spiral arms** that extend outward from the center. These stars are **Population I** stars and are of all ages. **Globular clusters** of mainly very old stars occur outside the central disk of the galaxy. These stars are **Population II** stars. **Spiral galaxies** are other collections of stars that resemble our galaxy. The universe apparently consists of widely separated galaxies of stars.

A **radio telescope** is a directional antenna connected to a sensitive radio receiver. Radio waves from space are produced by extremely hot gases, by fast electrons that move in magnetic fields, and by atoms and molecules excited to radiate. Especially notable sources are **quasars,** distant objects that emit both light and radio waves strongly and that may be powered by supermassive black holes at their centers.

Cosmic rays are atomic nuclei, mostly protons, that travel through the galaxy at speeds close to that of light. They probably were ejected during supernova explosions and are trapped in the galaxy by magnetic fields.

The spectral lines of distant galaxies show a doppler shift to the red arising from motion away from the earth. Since the speed of recession is observed to be proportional to distance, the **red shift** means that all the galaxies in the universe are moving away from one another. This **expansion of the universe** began 13.7 billion years ago and is currently accelerating.

The **big bang theory** holds that the universe originated in a great explosion 13.7 billion years ago. **Cosmic background radiation** left over from the early universe and doppler-shifted to radio frequencies has been detected. It is probable that the expansion of the universe will continue forever; if not, the universe will eventually begin to contract and will end up in a **big crunch** after which another cycle of expansion and contraction may occur. The solar system originated as part of an evolutionary process, and systems of **exoplanets** are probably quite common elsewhere in the universe.

There is apparently much more **dark matter** in galaxies than the luminous matter, such as stars, that we can see. The universe as a whole also contains a huge amount of **dark energy** whose nature, like that of dark matter, is unknown. Dark matter and dark energy together make up 96 percent of the stuff of the universe.

Multiple Choice

1. The stars in space are
   a. uniformly spread out
   b. distributed completely at random
   c. chiefly in the Milky Way
   d. mostly contained within widely separated galaxies
2. The Milky Way is
   a. a gas cloud in the solar system
   b. a gas cloud in the galaxy of which the sun is a member
   c. the galaxy of which the sun is a member
   d. a nearby galaxy
3. Relative to the center of our galaxy,
   a. its stars are stationary
   b. its stars move entirely at random
   c. its stars revolve
   d. Population I stars are stationary and Population II stars revolve
4. Our galaxy is approximately
   a. 4 light-years across
   b. 15,000 light-years across
   c. 130,000 light-years across
   d. 4 million light-years across
5. The number of stars in our galaxy is roughly
   a. 2000
   b. 2 million
   c. 2 billion
   d. 200 billion
6. Our galaxy has
   a. 1 main spiral arm
   b. 2 main spiral arms
   c. 3 main spiral arms
   d. 4 main spiral arms
7. Our galaxy is shaped roughly like a fried egg whose diameter is about
   a. twice its thickness
   b. 10 times its thickness
   c. 100 times its thickness
   d. 1 million times its thickness
8. Each cubic centimeter of space between the stars in our galaxy contains on the average about
   a. 1 atom or molecule
   b. 1 million atoms or molecules
   c. 1 mg of matter
   d. 1 g of matter
9. Evidence of various kinds suggests that at the center of our galaxy is a
   a. quasar
   b. pulsar
   c. neutron star
   d. black hole
10. Clouds of luminous gas and dust in our galaxy are called
    a. galactic clusters
    b. galactic nebulas
    c. quasars
    d. cosmic rays
11. The globular clusters of our galaxy
    a. contain around a hundred stars each
    b. contain around a million stars each
    c. contain around a billion stars each
    d. lie only in its central disk
12. The Population II stars in globular clusters are
    a. mostly very young
    b. mostly very old
    c. mostly about as old as the sun
    d. of all ages about equally
13. Compared with the Population II stars in globular clusters, the Population I stars in the galactic disk are
    a. richer in hydrogen and helium
    b. richer in the heavier elements
    c. older
    d. closer together
14. A radio telescope is basically a(an)
    a. device for magnifying radio waves
    b. telescope remotely controlled by radio
    c. directional antenna connected to a sensitive radio receiver
    d. optical telescope that uses electronic techniques to produce an image
15. Radio waves from space never originate in
    a. extremely hot gases
    b. fast electrons moving through magnetic fields
    c. molecules
    d. cosmic rays

16. Spiral galaxies
- **a.** are readily visible as such to the naked eye
- **b.** may be dark or bright
- **c.** are usually similar to our galaxy
- **d.** originate in supernova explosions

17. The stars in a galaxy are
- **a.** moving outward from its center
- **b.** moving inward toward its center
- **c.** revolving around its center
- **d.** stationary relative to its center

18. Cosmic rays carry energy to the earth at about the same rate as
- **a.** sunlight
- **b.** moonlight
- **c.** starlight
- **d.** meteoroids

19. Cosmic rays
- **a.** circulate freely through space
- **b.** are trapped in our galaxy by electric fields
- **c.** are trapped in our galaxy by magnetic fields
- **d.** are trapped in our galaxy by gravitational fields

20. Primary cosmic rays are composed largely of very fast
- **a.** protons
- **b.** neutrons
- **c.** electrons
- **d.** gamma rays

21. The origin of the most energetic cosmic rays is
- **a.** nuclear reactions in the sun
- **b.** solar flares
- **c.** the cosmic background radiation
- **d.** unknown

22. The red shift in the spectral lines of light reaching us from other galaxies implies that these galaxies
- **a.** are moving closer to one another
- **b.** are moving farther apart from one another
- **c.** are in rapid rotation
- **d.** consist predominantly of red giant stars

23. According to Einstein's general theory of relativity, the universe
- **a.** must be expanding
- **b.** must be contracting
- **c.** must be either expanding or contracting
- **d.** may be neither expanding nor contracting

24. Supernova explosions have no connection with
- **a.** the formation of heavy elements
- **b.** cosmic rays
- **c.** pulsars
- **d.** quasars

25. Current ideas suggest that what is responsible for the observed properties of a quasar is a massive
- **a.** neutron star
- **b.** black hole
- **c.** spiral galaxy
- **d.** star cluster

26. Quasars do not
- **a.** emit radio waves
- **b.** exhibit red shifts in their spectra
- **c.** vary in their output of radiation
- **d.** occur in the solar system

27. The term big bang refers to
- **a.** the origin of the universe
- **b.** the ultimate fate of the universe
- **c.** a supernova explosion
- **d.** the formation of a quasar

28. In the past several billion years the rate of expansion of the universe has
- **a.** been constant
- **b.** increased
- **c.** decreased
- **d.** alternately increased and decreased

29. The matter in the early universe that eventually condensed into galaxies and then into stars consisted of
- **a.** only hydrogen
- **b.** hydrogen and helium in an approximately 3:1 ratio
- **c.** hydrogen and helium in equal amounts
- **d.** hydrogen and helium in an approximately 1:3 ratio

30. The elements heavier than hydrogen and helium of which the planets are composed probably came from the
- **a.** sun
- **b.** debris of supernova explosions that occurred before the solar system came into being
- **c.** big bang
- **d.** big crunch

31. Soon after it came into being, the universe contained
- **a.** only matter
- **b.** only antimatter
- **c.** equal amounts of matter and antimatter
- **d.** slightly more matter than antimatter

32. Today the universe apparently contains
- **a.** only matter
- **b.** only antimatter
- **c.** equal amounts of matter and antimatter
- **d.** slightly more matter than antimatter

33. Radiation from the early history of the universe was stretched by the expansion of the universe until today it is in the form of
  a. x-rays
  b. ultraviolet waves
  c. infrared waves
  d. radio waves

34. Present evidence suggests that the major ingredient of the universe is in the form of
  a. dark matter
  b. dark energy
  c. cosmic rays
  d. black holes

35. It is least likely that the universe will
  a. continue to expand indefinitely
  b. expand at an ever-increasing rate until it is torn apart
  c. stop expanding and maintain a constant size
  d. collapse into a big crunch

36. The age of the universe is about
  a. 6000 years
  b. 13.7 million years
  c. 4.6 billion years
  d. 13.7 billion years

37. The age of the earth is about
  a. 6000 years
  b. 13.7 million years
  c. 4.6 billion years
  d. 13.7 billion years

38. It is likely that the planets, satellites, and other members of the solar system were formed
  a. together with the sun
  b. later than the sun from material it ejected
  c. later than the sun from material it captured from space
  d. elsewhere and were captured by the sun

39. Planets are always small compared with stars because otherwise
  a. the rotation of the planets would cause them to disintegrate
  b. the great mass of the planets would cause them to be pulled into their parent star
  c. the great mass of the planets would prevent them from being held in orbit and they would escape
  d. the planets would be stars themselves

40. The least likely reason why planetary systems have not been directly observed around stars other than the sun is that
  a. planets are small
  b. planets shine by reflected light
  c. planetary systems are rare
  d. other stars are far away

Exercises

19.1 The Milky Way

1. Are the stars uniformly distributed in space?
2. What is the Milky Way?
3. Why is the sun considered to be located in the central disk of our galaxy?
4. The earth undergoes four major motions through space. What are they?

19.2 Stellar Populations

5. What are the properties of globular clusters?
6. Both galactic nebulas and globular clusters occur in our galaxy. Distinguish between them.
7. Distinguish between Population I and Population II stars.
8. A number of elliptically shaped galaxies are known that seem to contain only Population II stars. Would you expect such galaxies to contain abundant gas and dust?
9. Where is most of the interstellar gas in our galaxy located? What is its chief constituent?

19.3 Radio Astronomy

10. Do radio telescopes magnify anything? If not, why are larger and larger ones being built?
11. List the three ways in which radio waves from space originate.
12. How does the mass of all the interstellar matter in our galaxy compare with the total mass of its stars?
13. Why are radio-astronomical studies of the distribution of hydrogen in the universe of greater interest than studies of the distribution of other elements?
14. What kind of evidence supports the belief that molecules of various kinds, including some fairly complex ones, exist in space?

19.4 Galaxies

15. Are most galaxies smaller, larger, or about the same size as our galaxy?
16. How can the rotation of a spiral galaxy be determined? Will this method work for all spiral galaxies?

17. (a) What is the observational evidence in favor of there being a great deal of dark matter in the universe? (b) Where is the dark matter located?
18. Black holes with masses of millions of times that of the sun are believed to lie at the hearts of which objects in space?

19.5 Cosmic Rays

19. In what way have cosmic rays affected the composition of the universe?
20. There is no day-night difference in cosmic-ray intensity. How does this observation bear on the possibility of a solar origin for cosmic rays?
21. What do you think ultimately becomes of the protons and neutrons knocked out of atmospheric atoms by cosmic rays?
22. Cosmic-ray primaries are mostly protons, but few protons are in the cosmic rays that reach the earth's surface. Why?
23. What traps cosmic rays in our galaxy?

19.6 Red Shifts

24. What is Hubble's law?
25. Why is the universe believed to be expanding?

19.7 Quasars

26. What is a quasar?
27. Why are quasars thought to be relatively small in size?
28. The spectra of quasars exhibit red shifts, never blue shifts. Why does this suggest that quasars are not members of our galaxy?

19.8 Dating the Universe

19.9 After the Big Bang

29. What is the observational evidence in favor of the big bang?
30. What is the evidence for the idea that dark energy fills the universe? Is the nature of dark energy known at present? How does the amount of dark energy compare with the amount of ordinary matter in the universe? With the amount of dark matter?
31. What are the origins of the helium found in the universe?
32. Suppose the spectral lines of distant galaxies were shifted toward the blue end. What would this suggest about the universe?
33. Distinguish between a closed and an open universe. How does the density of matter in the universe bear on the question of whether the universe is closed or open? What does present evidence suggest about the nature of the universe?

19.10 Origin of the Solar System

34. In what way or ways, if any, is the sun an unusual star?
35. Did the sun begin as a small body that grew to its present size, or as a large one that subsequently shrank?
36. The sun and the giant outer planets contain hydrogen and helium in abundance; the inner planets, very little. Why?
37. What heated all the planets as they were formed? Was there any other influence that helped heat the inner, rocky planets?
38. Did the planets come into being before, during, or after the formation of the sun?
39. Would you characterize the most likely next phase of the earth's history as fire or ice?

19.11 Exoplanets

19.12 Interstellar Travel

40. Why is it reasonable to suppose that many stars besides the sun have planets orbiting them? Why does it seem unlikely that spacecraft from the earth will ever visit other planetary systems?
41. Why is it hard to detect the planetary systems of other stars?

Answers to Multiple Choice

1. d	7. b	13. b	19. c	25. b	31. d	37. c
2. c	8. a	14. c	20. a	26. d	32. a	38. a
3. c	9. d	15. d	21. d	27. a	33. d	39. d
4. c	10. b	16. c	22. b	28. c	34. b	40. c
5. d	11. b	17. c	23. c	29. b	35. c	
6. d	12. b	18. c	24. d	30. b	36. d	

Math Refresher

A little basic mathematics is needed to appreciate much of physical science. This review is included mainly to help those readers whose mathematical skills have become rusty, but it is sufficiently self-contained to introduce such useful ideas as powers-of-10 notation to those who have not been exposed to them elsewhere.

Algebra

Algebra is the arithmetic of symbols that represent numbers. Instead of being limited to relationships among specific numbers, algebra can show more general relationships among quantities whose numerical values need not be known.

To give an example, in the theory of relativity it is shown that the "rest energy" of any object—that is, the energy it has due to its mass alone—is

$$E = mc^2$$

What this formula does is give a way to calculate the rest energy E in terms of mass m and speed of light c. The formula is not restricted to a particular object, but can be applied to any object whose mass is known. What is being shown is the way in which rest energy E varies with mass m *in general.*

If we are told only that the rest energy E of some object is 5 joules, we do not know upon what factors E depends or precisely how the value of E varies with those factors. (The joule is a unit of energy widely used in physics; it is equal to 1 kg · m^2/s^2.) The quantities E and m are **variables,** since they have no fixed values. On the other hand, c^2 is a **constant,** since it is the square of the speed of light c and the value of c—almost 300 million m/s, or about 186,000 mi/s—is the same everywhere in the universe. Thus the formula $E = mc^2$ tells us, in a simple and straightforward way, that the rest energy of something varies only with its mass and also how to find the numerical value of E if we are given the mass m of a particular object.

The convenience of algebra in science is increased by the use of standard symbols for constants of nature. Thus c always represents the speed of light, π always represents the ratio between the circumference and diameter of a circle, e always represents the electric charge of the electron, and so on.

Before we go further, it is worth reviewing how the arithmetical operations of addition, subtraction, multiplication, and division are expressed in algebra. Addition and subtraction are straightforward:

$$x + y = a$$

means that we obtain the sum a by adding the two quantities x and y together, while

$$x - y = b$$

means that we obtain the difference b when quantity y is subtracted from quantity x.

In algebraic multiplication, no special sign is ordinarily used, and the symbols of the quantities to be multiplied are merely written together. Thus these four expressions have the same meaning:

$$xy = c \qquad x(y) = c \qquad (x)(y) = c \qquad x \times y = c$$

When the quantity x is to be divided by y to yield the quotient e, we write

$$\frac{x}{y} = e$$

which can also be expressed as

$$x/y = e$$

whose meaning is the same.

If several operations are to be performed in a certain order, parentheses ( ), brackets [ ], and braces { } are used to indicate this order. For instance $a(x + y)$ means that we are first to add x and y together and then to multiply their sum $(x + y)$ by a. In essence $a(x + y)$ is an abbreviation for the same quantity written out in full:

$$a(x + y) = ax + ay$$

Let us find the value of

$$v = 5\left[\frac{(x - y)}{z}\right] + w$$

when $x = 15$, $y = 3$, $z = 4$, and $w = 10$. We proceed as follows:

1. Subtract y from x to give

$$x - y = 15 - 3 = 12$$

2. Divide $(x - y)$ by z to give

$$\frac{(x - y)}{z} = \frac{12}{4} = 3$$

3. Multiply $[(x - y)/z]$ by 5 to give

$$5\left[\frac{(x - y)}{z}\right] = 5 \times 3 = 15$$

4. Add w to $5[(x - y)/z]$ to give

$$v = 5\left[\frac{(x - y)}{z}\right] + w = 15 + 10 = 25$$

Positive and Negative Quantities

The rules for multiplying and dividing positive and negative quantities are simple. If the quantities are both positive or both negative, the result is positive; if one is positive and the other negative, the result is negative. In symbols,

$$(+a)(+b) = (-a)(-b) = +ab$$

$$\frac{+a}{+b} = \frac{-a}{-b} = +\frac{a}{b}$$

$$(-a)(+b) = (+a)(-b) = -ab$$

$$\frac{-a}{+b} = \frac{+a}{-b} = -\frac{a}{b}$$

Here are some examples:

$$(-3)(-5) = 15 \qquad \frac{-16}{-4} = 4$$

$$2(-4) = -8 \qquad \frac{10}{-5} = -2$$

$$(-12)6 = -72 \qquad \frac{-24}{4} = -6$$

To find the value of

$$w = \frac{xy}{x + y}$$

when $x = 5$ and $y = -6$, we begin by finding xy and $x + y$. These are

$$xy = (5)(-6) = -30$$
$$x + y = 5 + (-6) = 5 - 6 = -1$$

Hence

$$w = \frac{xy}{x + y} = \frac{-30}{-1} = 30$$

An example of the use of positive and negative quantities occurs in physics, where there are two kinds of electric charge. One kind is designated positive and the other negative. The force F that a charge Q_1 exerts on another charge Q_2 a distance r away is given by Coulomb's law as

$$F = K\frac{Q_1Q_2}{r^2}$$

where K is a universal constant. By convention, a positive value of F means a repulsion between the charges—the force tends to push Q_1 and Q_2 apart. A negative value of F means an attraction between the charges—the force tends to pull Q_1 and Q_2 together. A positive (=repulsive) force acts when *either* both charges are + *or* both are −: "like charges repel." When one charge is + and the other one −, the force is negative (=attractive): "opposite charges attract." Both the above observations about the types of force that occur together with the way in which the strength of F varies with the magnitudes of Q_1 and Q_2 and with their separation r are included in the simple formula $F = KQ_1Q_2/r^2$.

Exercises

A. Evaluate the following. The answers are given at the end of the Math Refresher.

1. $\frac{3(x + y)}{2}$ when $x = 5$ and $y = -2$

2. $\frac{1}{x - y} - \frac{1}{x + y}$ when $x = 3$ and $y = 2$

3. $\frac{4xy}{y + 3x} + 5$ when $x = 1$ and $y = -2$

4. $\frac{x + y}{2z} + \frac{z}{x - y}$ when $x = -2$, $y = 2$, and $z = 4$

5. $\frac{x + z}{y} + \frac{xy}{2}$ when $x = -2$, $y = 8$, and $z = 10$

6. $\frac{3(x + 7)}{y + 2}$ when $x = 3$ and $y = -6$

7. $\frac{5(3 - x)}{2(x + y)}$ when $x = -5$ and $y = 7$

Equations

An equation is a statement of equality: whatever is on the left-hand side of an equation is equal to whatever is on the right-hand side. An example of an arithmetical equation is

$$3 \times 9 + 8 = 35$$

since it contains only numbers. An example of an algebraic equation is

$$5x - 10 = 20$$

since it contains a symbol as well as numbers.

The symbols in an algebraic equation usually must have only certain values if the equality is to hold. To *solve* an equation is to find the possible values of these symbols. The solution of the equation $5x - 10 = 20$ is $x = 6$ since only when $x = 6$ is this equation a true statement:

$$\begin{aligned} 5x - 10 &= 20 \\ 5 \times 6 - 10 &= 20 \\ 30 - 10 &= 20 \\ 20 &= 20 \end{aligned}$$

The methods that can be used to solve an equation are based on this principle:

Any operation carried out on one side of an equation must be carried out on the other side as well.

Thus an equation remains valid when the same quantity is added to or subtracted from both sides or when the same quantity is used to multiply or divide both sides.

Two helpful rules follow from the above principle. The first is:

Any term on one side of an equation may be shifted to the other side by changing its sign.

To check this rule, let us consider the equation

$$a + b = c$$

If we subtract b from each side of the equation, we obtain

$$\begin{aligned} a + b - b &= c - \mathrm{b} \\ a &= c - b \end{aligned}$$

Thus b has disappeared from the left-hand side and $-b$ is now on the right-hand side. Similarly, if

$$a - d = e$$

then it is true that

$$a = e + d$$

The second rule is:

A quantity that multiplies one side of an equation may be shifted so as to divide the other side, and vice versa.

To check this rule, let us consider the equation

$$ab = c$$

If we divide both sides of the equation by b, we obtain

$$\frac{ab}{b} = \frac{c}{b}$$
$$a = \frac{c}{b}$$

Thus b, which was a multiplier on the left-hand side, is now a divisor on the right-hand side. Similarly, if

$$\frac{a}{d} = e$$

then it is true that

$$a = ed$$

Let us use the above rules to solve $5x - 10 = 20$ for the value of x. What we want is to have just x on the left-hand side of the equation. The first step is to shift the -10 to the right-hand side, where it becomes $+10$:

$$5x - 10 = 20$$
$$5x = 20 + 10 = 30$$

Now we shift the 5 so that it divides the right-hand side:

$$5x = 30$$
$$x = \frac{30}{5} = 6$$

The solution is $x = 6$.

When each side of an equation consists of a fraction, all we need do is **cross multiply** to remove the fractions:

$$\frac{a}{b} = \frac{c}{d} \qquad \frac{a}{b} \times \frac{c}{d} \qquad ad = bc$$

For practice, let us solve the equation

$$\frac{5}{a+2} = \frac{3}{a-2}$$

for the value of a. We proceed as follows:

Cross multiply to give	$5(a - 2) = 3(a + 2)$
Multiply out both sides to give	$5a - 10 = 3a + 6$
Shift the -10 and the $3a$ to give	$5a - 3a = 6 + 10$
Carry out the indicated addition and subtraction to give	$2a = 16$
Divide both sides by 2 to give	$a = 8$

Exercises

B. Solve each of the following equations for the value of x:

1. $3x + 7 = 13$

2. $5x - 8 = 17$

3. $2(x + 5) = 6$

4. $7x - 10 = 0.5$

5. $\frac{x+7}{6} = x + 2$

6. $\frac{4x - 35}{3} = 9(1 - x)$

7. $\frac{3x - 42}{9} = 2(7 - x)$

8. $\frac{1}{x+1} = \frac{1}{2x-1}$

9. $\frac{3}{x-1} = \frac{5}{x+1}$

10. $\frac{1}{3x+4} = \frac{2}{x+8}$

Exponents

There is a convenient shorthand way to express a quantity that is to be multiplied by itself one or more times. In this scheme a superscript number called an **exponent** is used to show how many times the multiplication is to be carried out, as follows:

$$a = a^1$$
$$a \times a = a^2$$
$$a \times a \times a = a^3$$
$$a \times a \times a \times a = a^4$$

and so on. The quantity a^2 is read "a squared" because it is equal to the area of a square whose sides are a long. The quantity a^3 is read as "a cubed" because it is equal to the volume of a cube whose edges are a long. Past an exponent of 3 we read a^n as "a to the nth power," so that a^5 is "a to the fifth power."

Suppose we have a quantity raised to some power, say a^n, that is to be multiplied by the same quantity raised to another power, say a^m. In this event the result is that quantity raised to a power equal to the sum of the original exponents:

$$a^n \times a^m = a^n a^m = a^{n+m}$$

To convince ourselves that this is true, we can work out $a^3 \times a^4$:

$$(a \times a \times a)(a \times a \times a \times a) = a \times a \times a \times a \times a \times a \times a$$
$$a^3 a^4 = a^7$$

Because the process of multiplication is basically one of repeated addition,

$$(a^n)^m = a^{nm}$$

where $(a^n)^m$ means that a^n is to be multiplied by itself the number of times indicated by the exponent m. Thus

$$(a^2)^4 = a^{2\times 4} = a^8$$

because

$$(a^2)^4 = a^2 \times a^2 \times a^2 \times a^2 = a^{2+2+2+2} = a^8$$

Reciprocal quantities are expressed according to the above scheme but with negative exponents:

$$\frac{1}{a} = a^{-1} \qquad \frac{1}{a^2} = a^{-2} \qquad \frac{1}{a^3} = a^{-3} \qquad \frac{1}{a^4} = a^{-4}$$

Roots

When the **square root** of a quantity is multiplied by itself, the product is equal to the quantity. The usual symbol for the square root of a quantity a is $\sqrt{a}$. Thus

$$\sqrt{a} \times \sqrt{a} = a$$

Here are some examples of square roots:

$\sqrt{1} = 1$	because	$1 \times 1 = 1$
$\sqrt{4} = 2$	because	$2 \times 2 = 4$
$\sqrt{9} = 3$	because	$3 \times 3 = 9$
$\sqrt{100} = 10$	because	$10 \times 10 = 100$

$$\sqrt{30.25} = 5.5 \qquad \text{because} \qquad 5.5 \times 5.5 = 30.25$$
$$\sqrt{16B^2} = 4B \qquad \text{because} \qquad 4B \times 4B = 16B^2$$

In the case of a number smaller than 1, the square root is larger than the number itself:

$$\sqrt{0.01} = 0.1 \qquad \text{because} \qquad 0.1 \times 0.1 = 0.01$$
$$\sqrt{0.25} = 0.5 \qquad \text{because} \qquad 0.5 \times 0.5 = 0.25$$

Similarly, multiplying the **cube root** $\sqrt[3]{a}$ of a quantity a by itself twice yields the quantity:

$$\sqrt[3]{a} \times \sqrt[3]{a} \times \sqrt[3]{a} = a$$

An expression of the form $\sqrt[n]{a}$ is read as "the nth root of a"; for instance, $\sqrt[4]{16}$ is "the fourth root of 16" and is equal to 2 because $2 \times 2 \times 2 \times 2 = 16$.

Although procedures exist for finding square and cube roots arithmetically, in practice electronic calculators or printed tables are normally used nowadays.

Here is an example of how a square root arises naturally in physics. Let us solve for r the equation

$$F = K\frac{Q_1Q_2}{r^2}$$

which expresses Coulomb's law of electric force. What we do is this:

Multiply both sides by r^2 to give $\qquad Fr^2 = KQ_1Q_2$

Divide both sides by F to give $\qquad r^2 = \frac{KQ_1Q_2}{F}$

Take square root of both sides to give $\qquad r = \sqrt{\frac{KQ_1Q_2}{F}}$

In algebra, a fractional exponent is used to indicate a root of a quantity. In terms of exponents we would write the square root of a as

$$\sqrt{a} = a^{1/2}$$

because

$$a^{1/2} \times a^{1/2} = (a^{1/2})^2 = a^{2\times 1/2} = a^1 = a$$

In a similar way the "cube root" of a, which is $\sqrt[3]{a}$, is indicated by the exponent $\frac{1}{3}$ because

$$a^{1/3} \times a^{1/3} \times a^{1/3} = (a^{1/3})^3 = a^1 = a$$

In general, the nth root of any quantity is indicated by the exponent $1/n$:

$$\sqrt[n]{a} = a^{1/n}$$

A few examples will indicate how fractional exponents fit into the general pattern of exponential notation:

$$(a^6)^{1/2} = a^{(1/2)\times 6} = a^3$$
$$(a^{1/2})^6 = a^{6\times 1/2} = a^3$$
$$(a^3)^{-1/3} = a^{(-1/3)\times 3} = a^{-1}$$
$$a^6 a^{1/2} = a^{6+1/2} = a^{6\,1/2}$$

Powers of 10

There is a convenient and widely used method for expressing very large and very small numbers that makes use of powers of 10. Any number in decimal form can be written as a number between 1 and 10 multiplied by some power of 10. The power of 10 is positive for numbers larger than 10 and negative for numbers smaller than 1. Positive powers of 10 follow this pattern:

$$
\begin{aligned}
10^0 &= 1 &&= 1 \text{ with decimal point moved 0 places} \\
10^1 &= 10 &&= 1 \text{ with decimal point moved 1 place to the right} \\
10^2 &= 100 &&= 1 \text{ with decimal point moved 2 places to the right} \\
10^3 &= 1000 &&= 1 \text{ with decimal point moved 3 places to the right} \\
10^4 &= 10{,}000 &&= 1 \text{ with decimal point moved 4 places to the right} \\
10^5 &= 100{,}000 &&= 1 \text{ with decimal point moved 5 places to the right} \\
10^6 &= 1{,}000{,}000 &&= 1 \text{ with decimal point moved 6 places to the right}
\end{aligned}
$$

and so on. The exponent of 10 in each case indicates the number of places through which the decimal point is moved to the right from 1.00000 . . . . Equivalently, the exponent gives the number of zeroes that follow the 1.

Negative powers of 10 follow a similar pattern:

$$
\begin{aligned}
10^0 &= 1 &&= 1 \text{ with decimal point moved 0 places} \\
10^{-1} &= 0.1 &&= 1 \text{ with decimal point moved 1 place to the left} \\
10^{-2} &= 0.01 &&= 1 \text{ with decimal point moved 2 places to the left} \\
10^{-3} &= 0.001 &&= 1 \text{ with decimal point moved 3 places to the left} \\
10^{-4} &= 0.000{,}1 &&= 1 \text{ with decimal point moved 4 places to the left} \\
10^{-5} &= 0.000{,}01 &&= 1 \text{ with decimal point moved 5 places to the left} \\
10^{-6} &= 0.000{,}001 &&= 1 \text{ with decimal point moved 6 places to the left}
\end{aligned}
$$

and so on. Here the exponent of 10 in each case shows the number of places through which the decimal point is moved to the left from 1. The number of zeroes between the decimal point and the 1 is one less than the exponent, that is, $n - 1$.

Here are some examples of powers-of-10 notation:

$$
\begin{aligned}
8000 &= 8 \times 1000 = 8 \times 10^3 \\
347 &= 3.47 \times 100 = 3.47 \times 10^2 \\
8{,}700{,}000 &= 8.7 \times 1{,}000{,}000 = 8.7 \times 10^6 \\
0.22 &= 2.2 \times 0.1 = 2.2 \times 10^{-1} \\
0.000{,}035 &= 3.5 \times 0.000{,}01 = 3.5 \times 10^5
\end{aligned}
$$

An advantage of powers-of-10 notation is that it makes calculations involving large and small numbers easier to carry out. The rules for working with exponents that were reviewed in the previous section hold for exponents of 10. We have here

Multiplication: $10^n \times 10^m = 10^{n+m}$

Division: $\dfrac{10^n}{10^m} = 10^{n-m}$

Raising to power: $(10^n)^m = 10^{nm}$

Taking a root: $(10^n)^{1/m} = 10^{n/m}$

An example will show how a calculation involving powers of 10 is worked out:

$$\frac{460 \times 0.000{,}03 \times 100{,}000}{9000 \times 0.006{,}2} = \frac{(4.6 \times 10^2) \times (3 \times 10^{-5}) \times (10^5)}{(9 \times 10^3) \times (6.2 \times 10^{-3})}$$

$$= \frac{4.6 \times 3}{9 \times 6.2} \times \frac{10^2 \times 10^{-5} \times 10^5}{10^3 \times 10^{-3}}$$

$$= 0.25 \times \frac{10^{2-5+5}}{10^{3-3}} = 0.25 \times \frac{10^2}{10^0}$$

$$= 25$$

Another virtue of this notation is that it permits us to express the accuracy with which a quantity is known in a clear way. The speed of light in free space c is often given as simply 3×10^8 m/s. If c were written out as 300,000,000 m/s we might be tempted to think the speed is precisely equal to this number, right down to the last zero. Actually, the speed of light is 299,792,458 m/s. For our purposes we do not need this much detail. By writing just $c = 3 \times 10^8$ we automatically indicate both how large the number is (the 10^8 tells how many decimal places are present) and how precise the quoted figure is (the single digit 3 means that c is closer to 3×10^8 than it is to either 2×10^8 or 4×10^8 m/s). If we wanted more precision, we could write $c = 2.998 \times 10^8$ m/s. Again how large c is and how precise the quoted figure is are both obvious at a glance.

To be sure, sometimes one or more zeroes in a number are meaningful in their own right and not solely decimal-point indicators. In the case of the speed of light, we can legitimately state that, to three-digit accuracy

$$c = 3.00 \times 10^8 \text{ m/s}$$

since c is closer to this figure than to 2.99×10^8 or 3.01×10^8 m/s. In the last sample calculation, the quantity $(4.6 \times 3)/(9 \times 6.2)$ actually equals 0.2473118. . . . It is rounded off to 0.25 because the result of a calculation may have no more significant digits than those in the least precise of the numbers that went into it.

Exercises

C. Express the following numbers in powers-of-10 notation:

1. 720 = **2.** 890,000 =

3. 0.02 = **4.** 0.000,062 =

5. 3.6 = **6.** 0.4 =

7. 49,527 = **8.** 0.002,943 =

9. 0.0014 = **10.** 49,000,000,000 =

11. 0.000,000,011 = **12.** 1.4763 =

D. Express the following numbers in decimal notation:

1. $3 \times 10^{-4} =$ **2.** $7.5 \times 10^3 =$

3. $8.126 \times 10^{-5} =$ **4.** $1.01 \times 10^8 =$

5. $5 \times 10^2 =$ **6.** $3.2 \times 10^{-2} =$

7. $4.32145 \times 10^3 =$ **8.** $6 \times 10^6 =$

9. $5.7 \times 10^0 =$ **10.** $6.9 \times 10^{-5} =$

E. Evaluate the following in powers-of-10 notation:

1. $\dfrac{30 \times 80{,}000{,}000{,}000}{0.0004} =$

2. $\dfrac{30{,}000 \times 0.000{,}000{,}6}{1000 \times 0.02} =$

3. $\dfrac{0.0001}{60{,}000 \times 200} =$

4. $5000 \times 0.005 =$

5. $\dfrac{5000}{0.005} =$

6. $\dfrac{200 \times 0.000{,}04}{400{,}000} =$

7. $\dfrac{0.002 \times 0.000{,}000{,}05}{0.000{,}004} =$

8. $\dfrac{500{,}000 \times 18{,}000}{9{,}000{,}000} =$

Answers

A.
1. 4.5
2. 0.8
3. −3
4. −1
5. 7
6. −7.5
7. 10

B.
1. 2
2. 5
3. −2
4. 1.5
5. −1
6. 2
7. 8
8. 2
9. 4
10. 0

C.
1. 7.2×10^2
2. 8.9×10^5
3. 2×10^{-2}
4. 6.2×10^{-5}
5. 3.6×10^0
6. 4×10^{-1}
7. 4.9527×10^4
8. 2.943×10^{-3}
9. 1.4×10^{-3}
10. 4.9×10^{10}
11. 1.1×10^{-8}
12. 1.4763×10^0

D.
1. 0.0003
2. 7500
3. 0.000,081,26
4. 101,000,000
5. 500
6. 0.032
7. 4321.45
8. 6,000,000
9. 5.7
10. 0.000,069

E.
1. 6×10^{15}
2. 9×10^{-4}
3. 8.3×10^{-11}
4. 2.5×10^1
5. 10^6
6. 2×10^{-8}
7. 2.5×10^{-5}
8. 10^3

The Elements

Atomic Number	Element	Symbol	Atomic Mass*	Atomic Number	Element	Symbol	Atomic Mass*	Atomic Number	Element	Symbol	Atomic Mass*
1	Hydrogen	H	1.008	38	Strontium	Sr	87.62	75	Rhenium	Re	186.2
2	Helium	He	4.003	39	Yttrium	Y	88.91	76	Osmium	Os	190.2
3	Lithium	Li	6.941	40	Zirconium	Zr	91.22	77	Iridium	Ir	192.2
4	Beryllium	Be	9.012	41	Niobium	Nb	92.91	78	Platinum	Pt	195.1
5	Boron	B	10.81	42	Molybdenum	Mo	95.94	79	Gold	Au	197.0
6	Carbon	C	12.01	43	Technetium	Tc	(97)	80	Mercury	Hg	200.6
7	Nitrogen	N	14.01	44	Ruthenium	Ru	101.1	81	Thallium	Ti	204.4
8	Oxygen	O	16.00	45	Rhodium	Rh	102.9	82	Lead	Pb	207.2
9	Fluorine	F	19.00	46	Palladium	Pd	106.4	83	Bismuth	Bi	209.0
10	Neon	Ne	20.18	47	Silver	Ag	107.9	84	Polonium	Po	(209)
11	Sodium	Na	22.99	48	Cadmium	Cd	112.4	85	Astatine	At	(210)
12	Magnesium	Mg	24.31	49	Indium	In	114.8	86	Radon	Rn	(222)
13	Aluminum	Al	26.98	50	Tin	Sn	118.7	87	Francium	Fr	(223)
14	Silicon	Si	28.09	51	Antimony	Sb	121.8	88	Radium	Ra	226.0
15	Phosphorus	P	30.97	52	Tellurium	Te	127.6	89	Actinium	Ac	(227)
16	Sulfur	S	32.06	53	Iodine	I	126.9	90	Thorium	Th	232.0
17	Chlorine	Cl	35.45	54	Xenon	Xe	131.3	91	Protactinium	Pa	231.0
18	Argon	Ar	39.95	55	Cesium	Cs	132.9	92	Uranium	U	238.0
19	Potassium	K	39.10	56	Barium	Ba	137.3	93	Neptunium	Np	(237)
20	Calcium	Ca	40.08	57	Lanthanum	La	138.9	94	Plutonium	Pu	(244)
21	Scandium	Sc	44.96	58	Cerium	Ce	140.1	95	Americium	Am	(243)
22	Titanium	Ti	47.90	59	Praseodymium	Pr	140.9	96	Curium	Cm	(247)
23	Vanadium	V	50.94	60	Neodymium	Nd	144.2	97	Berkelium	Bk	(247)
24	Chromium	Cr	52.00	61	Promethium	Pm	(145)	98	Californium	Cf	(251)
25	Manganese	Mn	54.94	62	Samarium	Sm	150.4	99	Einsteinium	Es	(252)
26	Iron	Fe	55.85	63	Europium	Eu	152.0	100	Fermium	Fm	(257)
27	Cobalt	Co	58.93	64	Gadolinium	Gd	157.3	101	Mendelevium	Md	(260)
28	Nickel	Ni	58.70	65	Terbium	Tb	158.9	102	Nobelium	No	(259)
29	Copper	Cu	63.54	66	Dysprosium	Dy	162.5	103	Lawrencium	Lw	(262)
30	Zinc	Zn	65.38	67	Holmium	Ho	164.9	104	Rutherfordium	Rf	(261)
31	Gallium	Ga	69.72	68	Erbium	Er	167.3	105	Dubnium	Db	(262)
32	Germanium	Ge	72.59	69	Thulium	Tm	168.9	106	Seaborgium	Sg	(263)
33	Arsenic	As	74.92	70	Ytterbium	Yb	173.0	107	Bohrium	Bh	(262)
34	Selenium	Se	78.96	71	Lutetium	Lu	175.0	108	Hassium	Hs	(265)
35	Bromine	Br	79.90	72	Hafnium	Hf	178.5	109	Meitnerium	Mt	(266)
36	Krypton	Kr	83.80	73	Tantalum	Ta	180.9	**			
37	Rubidium	Rb	85.47	74	Tungsten	W	183.9				

*Masses in parentheses are those of the most stable isotopes of the elements.

**Elements with atomic numbers 113–116 and 118 have been created in the laboratory in small quantities but not yet named.

Answers to Odd-Numbered Exercises

Chapter 1

1. None. The scientific method is based on observation and experiment together with reasoning from their results.

3. When first proposed, a scientific interpretation is called a hypothesis. If the hypothesis states a regularity or relationship, it is called a law after it has been verified by further observation and experiment. If the hypothesis uses general considerations to account for specific findings, it is called a theory after it has been verified.

5. Science is a reliable guide to the natural world precisely because it is self-correcting by being open to change in the light of new evidence.

7. At the North or South Pole.

9. The moon's apparent diameter remains constant, and its eastward motion through the sky is uniform.

11. In the ptolemaic model, the sun, moon, and other planets all revolve around the earth. In the copernican model, the moon revolves around the earth, and the earth and the other planets revolve around the sun with the earth rotating daily on its axis. There is direct observational evidence that supports the copernican model.

13. Ptolemaic system: (a) the sun circles the earth every day; (b) the stars circle the earth in a little less than a day, so that the difference between the speeds of the sun and stars causes the sun to move eastward relative to the stars; (c) the moon circles the earth.

Copernican system: (a) the earth turns on its axis once a day; (b) the earth moves around the sun once a year; (c) the moon circles the earth as the earth moves around the sun.

15. Yes, because $T_e^2/R_e^3 = T_a^2/R_a^3$.

17. The tidal range is a maximum at spring tides, a minimum at neap tides. At spring tides, the sun, moon, and earth are in a straight line. At neap tides, the moon-earth line is perpendicular to the sun-earth line.

19. The moon revolves around the earth in the same direction as the earth's rotation. Hence the moon is in the same place relative to a point on the earth's surface about 50 minutes later every day.

21. $10^6/10^{-6} = 10^{12}$.

23. (405 mi)(1.61 km/mi) = 652 km.

25. 10.8 ft^2.

Chapter 2

1. $d = vt = 960$ m for the round trip, hence the distance to the cliff is $d/2 = 480$ m.

3. $v = d_1/t_1 = 3.14$ m/s; $t_2 = d_2/v = 1.6$ s.

5. $t = d/v = 500$ s $= 8.3$ min.

7. Three.

9. $d^2 = (70\text{ m})^2 + (40\text{ m})^2 = 6500\text{ m}^2$, $d = \sqrt{d^2} = 81$ m.

11. Yes.

13. $a = (80\text{ km/h})/20\text{ s} = (4\text{ km/h})/\text{s}$; hence with $v_0 = 80$ km/h and $v = 130$ km/h, $t = (v - v_0)/a = 12.5$ s.

15. $a = v_1/t = 4$ m/s, $v_2 = v_1 + at = 60$ m/s.

17. (a) $a = (v_2 - v_1)/t = 6$ s.
(b) $d = v_1 t + (1/2)at^2 = 105$ m.

19. Average acceleration is $a = 2d/t^2 = 2\text{ m/s}^2$ and final speed is $v = at = 40$ m/s.

21. The apple will remain stationary with respect to the barrel, since both are falling with the same acceleration.

23. When thrown; at the top of the path.

25. (a) Yes. (b) The stone thrown horizontally is moving faster because its velocity has a horizontal as well as a vertical component.

27. (a) The same time. When dropped, the coin has the same initial speed as the elevator, so this constant speed does not affect the motion of the coin relative to the elevator. (b) More time. The downward acceleration of the elevator is not shared by the coin; hence the coin's acceleration relative to the elevator is less than g by the amount of the elevator's acceleration. (c) The same time for the reason given in (a). (d) Less time. The upward acceleration of the elevator is not shared by the coin, hence the coin's acceleration relative to the elevator is more than g by the amount of the elevator's acceleration.

29. The dive lasts $t = -\sqrt{2h/g} = 2.6$ s and the person enters the water at $v = gt = 25$ m/s.

31. (a) $v_2 = v_1 + gt = +1.1$ m/s, which is upward.
(b) Now $v_2 = -13.6$ m/s, which is downward.

33. $v = v_0 + gt = 21.8$ m/s.

35. (a) $v_0 = 30$ m/s and at the highest point $v = 0$. Since $v = v_0 - gt, t = (v_0 - v)/g = 3.06$ s. (b) The time of fall equals the time of rise, so total time is (2)(3.06 s) = 6.12 s. (c) 30 m/s.

37. (a) Since $h = (1/2)gt^2$, $t = \sqrt{2h/g} = 15.6$ s. (b) The horizontal speed of the wiper is $v_h = 100$ m/s, so $d = v_h t = 1.56$ km. (c) The vertical speed of the wiper is $v_v = gt = 153$ m/s. Hence $v = \sqrt{v_v^2 + v_h^2} = 182$ m/s.

39. (a) $h = (1/2)gt^2 = 123$ m. (b) $d = v_h t = 1$ km.
(c) The vertical speed of the bullet is $v_v = gt = 49$ m/s. Hence $v = \sqrt{v_v^2 + v_h^2} = 206$ m/s.

41. When the wrench is dropped, it is moving at the same speed as the boat and in the same direction. The downward speed of the wrench increases as it falls, but its horizontal speed remains the same as that of the boat. Therefore the wrench hits the deck directly below the point from which it was dropped. Only if the boat were accelerated would it be possible for the wrench to miss the deck.

43. $F = ma = m(v/t), t = mv/F = 16.8$ s.

45. $a = F/m = 3.33\text{ m/s}^2; t = v/a = 7.2$ s.

47. $a = v/t = 1.5\text{ m/s}^2$; $F = ma = 120$ N.

49. $m = F_1/a_1 = 4$ kg; $F_2 = ma_2 = 4$ N; $F_3 = ma_3 = 40$ N.

51. The mass of an object is a measure of its inertia, which is the resistance it offers to any change in its state of rest or of motion. The object's weight is the gravitational force with which the earth attracts it.

53. The downward gravitational force is balanced by an equal upward force of the ground on him. The sum of these two forces is zero, hence Albert does not move.

55. g(moon) = (85/490) g(earth) = 1.7 m/s^2.

57. $v_0 = -15$ m/s and $v = 20$ m/s, so $a = (v - v_0)/t = (20$ m/s$) - (-15$ m/s$)/0.005$ s $= (35$ m/s$)/0.005$ s $= 7000$ m/s^2 and $F = ma = 420$ N.

59. The force is her weight of mg plus her mass times her upward acceleration; hence $F = mg + ma = 708$ N.

61. Action and reaction forces always act on different bodies.

63. (a) The reaction force is the upward force exerted by the table on the book; without this force the book would fall to the floor. (b) The reaction force is the gravitational pull the book exerts on the earth.

65. A propeller works by pushing backward on the air whose reaction force in turn pushes the propeller itself and the airplane it is attached to forward. No air, no reaction force, so the idea is no good.

67. (a) $F = ma = 270$ N. (b) Yes; 270 N backward.

69. Under no circumstances.

71. $F = mv^2/r = 12.5$ N.

73. $v = \sqrt{Fr/m} = 6.3$ m/s.

75. $Fc = mv^2/r, r = mv^2/Fc = 60$ m.

77. $F = mg + mv^2/r, r = mv^2/(F - mg) = 637$ m.

79. Sprinters could not improve their time in the 100-m dash on the moon because their masses are the same there as on the earth. With the force that their legs can exert also unchanged, their acceleration will be the same, and hence their motion will not differ from that on the earth.

81. At an altitude of two earth's radii, the mass would be the same, but the weight would be $\frac{1}{9}$ as great because the distance from the earth's center is three times as great and gravitational force varies as $1/r^2$.

83. The sun's gravitational pull on the earth varies during the year since the distance from the earth to the sun varies.

85. The earth must travel faster when it is nearest the sun in order to counteract the greater gravitational force of the sun.

87. $F = Gm_1m_2/r^2 = 7.4 \times 10^{-8}$ N. Since 1 g $= 10^{-3}$ kg, 1g of anything weighs $w = mg = 9.8 \times 10^{-3}$ N, which is more than 10^5 (a hundred thousand) times as great as the gravitational force of the lead on the cheese.

89. No, because they and the airplane are "falling" at the same rate.

91. A satellite must have the minimum speed $v = \sqrt{rg}$. For Jupiter, $r = 0.715 \times 10^8$ m and $g = (2.6)(9.8$ m/s$^2) = 1.6$ m/s^2, so $v = 4.27 \times 10^4$ m/s.

Chapter 3

1. Yes, because all changes require that work be done.

3. $W = Fd = 400$ J. The crate's mass does not matter.

5. $W = 0$ because the earth moves perpendicular to the direction of the force acting on it from the sun.

7. $m = W/gh = (490$ J$)/(9.8$ m/s$^2)(10$ m$) = 5$ kg.

9. $P = W/t = Fd/t, F = Pt/d = 14.4$ kN.

11. $P = W/t = mgh/t = 65$ W since 10 h = 36,000 s.

13. $W = Fd = Pt, F = Pt/d = P/v = 0.21$ kN. The horse's mass does not matter.

15. Efficiency $= (mgh/t)/P_{\text{input}} = 0.78 = 78$ percent.

17. $d = $ KE$/F = 5$ m.

19. Less work is needed.

21. $Fd = \frac{1}{2}mv^2$, $F = mv^2/2d = 0.96$ kN.

23. $P = (\text{KE}_2 - \text{KE}_1)/t, t = (\text{KE}_2 - \text{KE}_1)/P = 7.5$ s.

25. (a) PE increases. (b) PE decreases.

27. When it is farthest from the sun; when it is closest to the sun. The work needed to pull a planet away from the sun to a given distance increases with the distance, so the planet's PE is greatest the farthest it is from the sun. The gravitational force of the sun on a planet is greatest when it is closest to the sun; hence its speed is also greatest there in order that gravitational and centripetal forces be in balance.

29. Ball A reaches the bottom first because all of its original PE becomes KE of downward motion. Part of B's original PE becomes KE of its rotation, so there is less PE available to become KE of downward motion. As a result B moves more slowly than A and reaches the bottom first.

31. The difference in height is $h = 2r$ so the KE at the bottom is greater by $mgh = 2mgr$. If the speed at the top is v_1 and at the bottom is v_2, then $(1/2)mv_2^2 = (1/2)mv_1^2 + 2mgr$, $v_2 = \sqrt{v_1^2 + 4gr} = 6.35$ m/s. We did not have to know the yo-yo's mass.

33. $mv_2^2 = mv_1^2 + mgh$, $v_2 = \sqrt{v_1^2 + 2gh} = 16$ m/s. The skier's KE at the bottom is equal to her KE at the top plus the change in her PE.

35. Most of the work done in hammering the nail is dissipated as heat owing to friction between the nail and the wood.

37. $t = d/v = 1250$ s, $W = Pt = 1.75 \times 10^6$ J; hence 1.75×10^6 J$/(4 \times 10^4$ J/g$) = 44$ g of fat is metabolized.

39. (a) $W = Fd = 1760$ J. (b) PE $= mgh = 1568$ J. (c) 192 J was lost as heat due to friction in the pulleys.

41. Their speeds are the same. The golf ball, which has the greater mass, has the greater KE, PE, and momentum.

43. Most of the momentum is given to the earth through the road and the rest is given to the air the car was moving through.

45. Since $2 \times \left(\frac{1}{2}mv^2\right) = \frac{1}{2}m(\sqrt{2}v)^2$, the speed increases by $\sqrt{2}$ and so the momentum increases by $\sqrt{2}$ as well.

47. (a) When $m = M$. (b) When m is less than M. (c) When m is greater than M.

49. (a) The speed decreases as rainwater collects in the truck, since the total momentum must remain constant. (b) The reduced speed is unchanged because the water that leaked out carried with it the momentum it has gained.

51. $v_1 = m_2v_2/m_1 = 0.7$ m/s.

53. Let Δ stand for "change in." Then $m_1\Delta v_1 = m_2\Delta v_2$, $\Delta v_2 = \Delta v_1 (m_1/m_2) = 17$ m/s; $vf = v_1 - \Delta v = 28$ m/s.

55. (a) $m_1v_1 + m_2 v_2 = (m_1 + m_2)v_3, v_3 = 2.8$ m/s. (b) Initial KE $= (1/2)m_1v_1^2 + (1/2)m_2v_2^2 = 440$ J and final KE $= (1/2)(m_1 + m_2)v_3^2 = 392$ J. Hence the KE lost is 48 kJ.

57. The resulting water will add to the oceans, and more of the earth's mass will be distant from its axis than before. Conservation of angular momentum requires that the earth in that event spin more slowly to compensate, and the day will be longer.

59. As in the case of a skater pulling in her arms (see Sec. 4.10), conservation of angular momentum required that the earth spin faster when it became smaller. (This has shortened the length of the day by about 3 microseconds.)

61. The predictions of relativity apply at much higher speeds than those in everyday life, and these predictions are supported by experimental evidence.
63. The rod appears longest to the stationary observer.
65. The KE of a moving object increases with its speed until it would be infinite at the speed of light. Since nothing can have an infinite amount of KE, the speed of light is the ultimate speed limit.
67. Mass must be considered as a form of energy.
69. $m = E/c^2 = 3.35 \times 10^5$ J/$(3 \times 10^8$ m/s$)^2 = 3.7 \times 10^{-12}$ kg.
71. The total rest energy of 1 kJ of anything is $mc^2 = 9 \times 10^{16}$ J. Here $(5.4 \times 10^6)/(9 \times 10^{16}) = 6 \times 10^{-11}$, which is 6×10^{-9} percent.

Chapter 4

1. (a) The world's population is increasing and becoming more prosperous, so it will need more energy. (b) Most of today's energy comes from fossil fuels, whose supply will eventually run out. (c) Burning fossil fuels produces CO_2, the leading cause of global warming.
3. Over four times (actually 4.4 times).
5. Food and wood: photosynthesis in plants is powered by sunlight. Water power: sunlight evaporates water that later falls as rain and snow on high ground. Wind power: unequal heating of the earth's surface causes motions in the atmosphere. Fossil fuels: they come from the remains of living things, which contain energy ultimately derived from sunlight through photosynthesis.
7. Nearly half the world's population.
9. Seawater and the earth's surface absorb solar radiation more efficiently than ice, which is very reflective, so global warming and further melting will persist.
11. Millions of years ago.
13. Sunlight that strikes the earth's surface heats it, and the surface itself then radiates infrared light. The atmosphere allows visible light from the sun to pass through it, but various gases in it absorb infrared light and it is this absorbed infrared light that heats the atmosphere.
15. It is absorbed by oceans, soils, and forests.
17. Coal contributes the most CO_2, then oil, and last natural gas.
19. Transportation.
21. Build engines that burn fuel more efficiently; use catalytic converters; use additives to increase fuel efficiency.
23. For the same amount of energy produced, coal is cheaper than natural gas.
25. Coal.
27. One-half.
29. Coal is almost pure carbon whereas oil and natural gas also contain hydrogen whose burning produces water rather than carbon dioxide.
31. (a) In fission, a large atomic nucleus splits into two smaller ones. In fusion, two small nuclei join to form a larger one. (b) Both processes give off large amounts of energy.
33. Heat from the nuclear fissions in the reactor is used to produce steam that runs turbines connected to electric generators.
35. Safe operation; abundant, cheap fuel; little radioactive waste; no greenhouse gas emissions.
37. Nuclear fusion is the energy source of the sun and stars.
39. Hydropower.
41. A photovoltaic cell converts energy in sunlight directly to electric energy. It is also called a solar cell.
43. Average output per turbine = (0.4)(2 M) = 0.8 MW, hence (500 MW)/(0.8 MW/turbine) = 625 turbines.
45. Geothermal energy is a continuous source that does not vary with time of day or weather conditions.
47. (a) On the earth, hydrogen is found only combined with other elements from which it must first be separated. (b) At ordinary temperatures and pressures, hydrogen is a gas, which makes it difficult to transport and store.
49. (a) Corn ethanol is expensive; so much energy is needed in all the steps involved that there is little or no energy gain; on an overall basis, CO_2 emissions are not reduced; agricultural land is diverted from food production at a time of population growth. (b) The needed technology has not yet been perfected.
51. Seconds per year = (365 days)(24 h/day)(3600 s/h) = 3.15×10^7 s, hence $P = E/t = 2.1 \times 10^3$ W = 2.1 kW.
53. (a) An overall cap on emissions is set and companies are given or buy at auction permits to emit CO_2 whose total equals the cap. Companies with unused permits can sell them to companies that need additional ones. (b) Tax CO_2 emissions, so that products and services with high CO_2 emissions are more expensive than comparable ones using better technology instead of cheaper as at present.

Chapter 5

1. The metal expands more than the glass when heated, thereby loosening the lid.
3. White; red.
5. $T_F = \frac{9}{5}(37°) + 32° = 98.6°F$.
7. $T_C = \frac{5}{9}(T_F - 32°) = -80°C$.
9. A piece of ice at 0°C is more effective in cooling a drink than the same weight of water at 0°C because of the heat of fusion that must be added to the ice before it melts. Hence the ice will absorb more heat from the drink than the cold water.
11. Aluminum, because it has the highest specific heat.
13. Metals are better conductors of heat than wood and therefore conduct heat away from your hand more rapidly.
15. The temperature difference is 80°C and 4.19 kJ/kg is needed per 1°C change in temperature. Hence E = (4.19 kJ/kg · °C)(80°C)(0.2 kg) = 67 kJ.
17. Since 4.19 kJ increases the temperature of 1 kg of water by 1°C, 12,000 kJ increases the temperature of 60 kg of water by (12,000 kJ)/(4.19 kJ/kg · °C)(60 kg) = 48°C.
19. $Q = mc\Delta T$, $\Delta T = Q/mc$ = 67°C.
21. The stone's energy is mgh = 9800 J = 9.8 kJ. Since 4.19 kJ increases the temperature of 1 kg of water by 1°C and here there is 10^4 kg of water, the temperature rise is (9.8 kJ)/(4.19kJ/kg · °C)(10^4 kg) = 2.3×10^{-4} °C.
23. $m = dV = dLWH$ = 78 kg.
25. The density of the bracelet is $d = m/V$ = 12.5 g/cm^3 = 1.25×10^4 kg/m^3. Since the density of gold is 1.9×10^4 kg/m^3, the bracelet cannot be pure gold.
27. (a) $V = m/d$ = 0.055 m^3 = 55 L. (b) $V = m/d$ = 140 m^3.

29. (a) $d = m/V = 3m/4\pi r^3 = 5.52 \times 10^3$ kg/m. (b) The interior of the earth must consist of denser materials than those at the surface; no.

31. The person reduces the pressure in the straw by sucking on it, and atmospheric pressure then forces the liquid upward.

33. Since the heights of the water columns are the same, the pressures are the same.

35. $p = F/A = 1.5 \times 10^5$ N/m^2 = 150 kPa.

37. $A = F/p = mg/p = 0.0588$ m^2; 1/4 of this is 147 cm^2.

39. (a) $F = pA = 0.12$ N. (b) An external pressure exerted on a fluid is transmitted uniformly throughout the fluid.

41. (a) Because pressure in a fluid increases with depth, the upward force on the bottom of a submerged object is greater than the downward force on its top. (b) If the buoyant force exceeds the object's weight, it will float.

43. The block will stay where it is because there is no water underneath it to furnish a buoyant force.

45. (a) The weight of the ship is unchanged, hence the volume of water it displaces is unchanged. (b) Because the volume of the ship's hull is now greater, the height of its deck above water increases.

47. The water level is unchanged.

49. The boat is more stable in freshwater because the buoyancy of its keel is less there.

51. The anchor's volume is $V = m/d = 0.0128$ m^2. The weight of water the anchor displaces is $w = mg = Vdg = 129$ N, which is the buoyant force on the anchor. Hence $F = mg - F$(buoyant) = 851 N.

53. The raft's volume is $V = (L)(W)(H) = 1.8$ m^3. The maximum upward force the raft can exert is (buoyant force − weight of raft) = d(water)$gV - d$(balsa)gV = 15.3 kN. Each person weighs mg = 637 N so 24 people can be supported.

55. (a) The mass of 200 L = 0.2 m^3 of seawater is $m = dV$ = 206 kg, so the empty tank can support up to 206 kg. The empty tank will therefore float. (b) The mass of 0.2 m^3 of freshwater is 200 kg, so the mass of the filled tank is 236 kg and it will sink. (c) The mass of 0.2 m^3 of gasoline is 136 kg, so the mass of the filled tank is now 172 kg and it will float.

57. (a) $V_2 = p_1V_1/p_2 = 333$ cm^3. (b) $T_1 = 273$ K, $T_2 = 546$ K, $V_2 = V_1T_2/T_1 = 2000$ cm^3.

59. The volume of oxygen at atmospheric pressure is V(atm) = V(cyl)p(cyl)/p(atm) = 6.73 m^3 so its mass is $m = dV = 9.4$ kg.

61. $T_1 = 293$ K, $T_2 = T_1V_2/V_1 = 586$ K = 313°C.

63. $T_1 = 293$ K, $T_2 = p_2T_1/p_1 = 410$ K = 137°C.

65. No. The only temperature scale on which such a comparison might make sense is the absolute temperature scale.

67. The molecules themselves occupy volume.

69. Lower.

71. The thermal energy of a solid resides in oscillations of its particles about fixed positions.

73. T = 27°C = 300 K. Since the average energy is proportional to the average temperature, doubling the energy doubles the temperature to 600 K = 327°C.

75. At the same temperature the average KE of the hydrogen and CO_2 molecules is the same. Hence $m_Hv_H^2 = m_{CO_2}v_{CO_2}^2$, $v_{CO_2} = v_H\sqrt{m_H/m_{CO_2}} = 0.34$ km/s.

77. The water molecules with the highest energies leave the liquid surface, and blowing them away prevents them from returning to the liquid. The remaining liquid contains the less energetic molecules, which means it is cooler.

79. (a) Increase the liquid temperature, which increases the average molecular KE while leaving unchanged the intermolecular attractive forces. (b) Reduce the pressure above the liquid, which decreases the likelihood that vapor molecules will return to the liquid after colliding with air molecules. (c) Arrange for a current of air to blow over the liquid surface, which will remove vapor molecules before they can return to the liquid. (d) Increase the area of the liquid surface, which brings more liquid molecules to the surface, where it is easiest for them to leave.

81. To heat the water to 100°C the heat needed is (4.19 kJ/kg · °C)(1 kg)(80°C) = 335 kJ.
Thus the water is heated to 100°C with 165 kJ left over. The mass of water turned into steam is (165 kJ)/(2260 kJ/kg) = 0.073 kg = 73 g.

83. $Q = m(c\Delta T$ + heat of vaporization) = 561 kJ. Since $Q = Pt$, $t = Q/P = 468$ s = 7.8 min.

85. (a) Higher than 50°C. (b) Lower than 50°C. (c) Higher than 50°C. (d) 50°C.

87. There is no low-temperature reservoir for it to use.

89. The kitchen will warm up because the refrigerator gives off more heat than it absorbs. Leaving its door open means that it will run continuously and hence add even more heat to the kitchen.

91. No.

93. Eff(max) = $1 - T_2/T_1 = 0.65$ = 65 percent; hence the actual efficiency is 62 percent of the maximum possible.

95. Eff(max) = $1 - T_{cold}/T_{hot} = 0.073$ = 7.3 percent. The hot and cold reservoirs are enormous.

97. The entropy decrease involved in the evolution of today's animals is more than balanced by the entropy increase that occurred as the food the living things took in, digested, and metabolized lost its initial order. The food itself also represented an increase in entropy, as mentioned in Sec. 5.16.

Chapter 6

1. Experiments show that every charge is either attracted or repelled by a + or a − charge; a charge that is attracted by a + charge is always repelled by a − charge, and vice versa; every charge obeys Coulomb's law when brought near a known charge; and so forth. Since all electrical phenomena can be accounted for on the basis of two kinds of charge only, there is no reason to suppose any other kind of charge exists.

3. (a) Since the ball is positively charged, it has fewer electrons than are needed for neutrality. (b) $N = Q/e = 6.25 \times 10^6$ electrons.

5. Protons in its nucleus provide the positive charge of an atom, and the electrons that surround the nucleus provide the negative charge.

7. The number of protons is $N = m/m_p = 6.0 \times 10^{23}$ protons, so $Q = Ne = 1.6 \times 10^4$ C.

9. No; yes.

11. $F = Ke^2/R^2 = 4.2 \times 10^{-8}$ N.

13. $F = Kq_1q_2/r^2 = 5.4 \times 10^{-7}$ N.

15. $F_2/F_1 = (R_1/R_2)^2$, $R_2 = R_1\sqrt{F_1/F_2} = 20$ mm.

17. (a) $F = Kq_1q_2/r^2 = 180$ N. (b) When the charges are brought into contact, the 1×10^{-5} C positive charge cancels out this amount of negative charge to leave -1×10^{-5} C. The latter charge is equally divided between the spheres, so when they are separated, the force between them is $F = (9 \times 10^9)(5 \times 10^{-6})(5 \times 10^{-6})/(0.1)^2$ N = 23 N and is now repulsive.
19. When the forces balance, $Gm_em_m/d^2 = KQ^2/d^2$ and so $Q = \sqrt{Gm_em_m/K} = 5.7 \times 10^{13}$ C.
21. A current would flow briefly and then stop as charge builds up on the lightbulb and prevents further charge from being transferred.
23. The need for extremely low temperatures.
25. $W = QV = 4.32 \times 10^5$ J.
27. $W = QV = 8 \times 10^8$ J.
29. $I = V/R = 10$ A, so this must be the minimum rating of the fuse.
31. $V = IR = 75$ V.
33. The circuit elements might get too hot and start a fire or even melt. The criterion is therefore how much heat the circuit can tolerate without hazard. In the case of house wiring, this depends upon the thickness of the wire used and the nature of its insulation.
35. (a) In series, the batteries are connected in line with the + terminal of one battery attached to the − terminal of the next. The voltages of the batteries add, so the voltage of the combination is greater than that of the individual batteries. (b) In parallel, batteries with the same voltage are connected with all their + terminals attached together and all their − terminals attached together. The voltage of the combination is the same as that of each battery, but the combination can provide more current than each battery could by itself.
37. (a) $I = P/V = 0.625$ A. (b) $R = V/I = 192\ \Omega$. (c) $P = 75$ W.
39. $I = P/V = 1.7$ A.
41. $P = IV = 3600$ W; 36 bulbs.
43. $W = Pt = IVt = 2.7$ kWh = 9.7 MJ.
45. (a) $I = P/V = 7.4 \times 10^{-5}$ A. (b) $Q = 1.5$ C/s $\times$ 3600 s = 5400 C; $W = QV = 7290$ J; $t = W/P = 7.29 \times 10^3$ J/10^{-4} W = 7.29×10^7 s = 844 days = 2.3 years since 1 day = 86,400 s and 1 year = 365 days.
47. In an hour, the resistance element gives off $W = Pt$ = (2000 W)(3600 s) = 7200 kJ. To raise the temperature of 1 kg of water through 60°C requires (4.19 kJ/kg · °C)(60°C) = 251.4 kJ; hence (7200 kJ)/(251.4 kJ/kg) = 28.6 kg of water will be so heated in an hour.
49. Either pole of the magnet attracts opposite poles within the iron, leading to a net attractive force since these poles line up facing the external magnet.
51. At a given point a test body free to move travels along a field line, by definition, and it can travel in only one direction at that point.
53. Direction *D*.
55. Attract; repel.
57. 0° or 180°; 90°.
59. To the left.
61. The magnetic field through the loop undergoes both an increase and a decrease in each half of a complete rotation.
63. When the generator is connected to an outside circuit, work must be done to turn its shaft as the current produced delivers energy to the circuit.
65. The changing magnetic field produced by an alternating current.
67. When the connection is made, a momentary current will occur in the secondary winding as the current in the primary builds up to its final value. Afterward, since the primary current will be constant and hence its magnetic field will not change, there will be no current in the secondary.
69. (a) $I_2 = (N_1/N_2)I_1 = 1.5$ A; 60 Hz. (b) No current will flow in the secondary coil.
71. (a) $N_1/N_2 = V_1/V_2$ = (5000 V)/(240 V) = 20.8. (b) $P = IV$ so $I_2 = P/V_2$ = (10,000 W)/(240 V) = 41.7 A.

Chapter 7

1. (a) In a longitudinal wave, the particles of the medium vibrate parallel to the direction in which the wave is moving. In a transverse wave, the particles vibrate perpendicularly to this direction. (b) No; water waves are a combination of both.
3. $f = 1/(1.2 \text{ s}) = 0.83$ Hz, $v = f\lambda = 5$ m/s.
5. $v = f/\lambda = 0.3$ m/s; $t = d/v = 3$ s.
7. Sound travels fastest in solids because their constituent particles are more tightly bound together than those of liquids and gases. Sound travels slowest in gases because their molecules interact only during random collisions.
9. The moon has no atmosphere to transmit sound waves.
11. The distance of the man from the spike is $d = v_st_1 = 686$ m, where $v_s = 343$ m/s is the speed of sound in air. Hence the speed of sound v in the rail is $v = d/t_2$ = 686 m/0.14 s = 4900 m/s.
13. $f = v/\lambda$ = (343 m/s)/0.25 m = 1372 Hz.
15. 10^4 = 10,000 times stronger.
17. In all kinds of waves.
19. A shift toward the red end of the spectrum means a shift to lower frequencies, which corresponds to motion away from the earth.
21. When light is absorbed, the absorbing material is heated.
23. The electric and magnetic fields of an electromagnetic wave are perpendicular to each other and to the direction of propagation.
25. The sun does not lose momentum due to the light it emits because momentum is a vector quantity and the sun radiates equally in all directions.
27. Red; violet; violet; red.
29. Sound takes $t = d/v$ = 30 m/(343 m/s) = 0.087 s to travel 30 m, but radio waves take only 5×10^6m/(3×10^8 m/s) = 0.017 s to travel 5000 km.
31. (a) $f = 1/T = 10^9$ Hz. (b) $\lambda = c/f = 0.3$ m. (c) Microwaves.
33. Speed, frequency, wavelength.
35. The mirror must be half your own height. It does not matter how far away you are.
37. No, because light cannot have a speed greater than c.
39. The object appears to have its normal horizontal dimensions but to be farther above the water surface than it actually is.
41. Path *C*.

43. $h' = h/n = 30.5$ mm.
45. A real image is formed by light rays that pass through it. A screen at the location of the image would show the image. A virtual image is formed by the backward extension of light rays that were diverted by reflection or refraction. The rays that seem to come from it do not actually pass through a virtual image, and it cannot appear on a screen but can be seen by the eye.
47. In both cases a ray that passes through the center of the lens is not deviated.
49. The glass tube magnifies the mercury column, which is therefore narrower than it appears to be.
51. Path *B*.
53. 30 cm behind the lens; 12 cm high; erect; virtual.
55. 37.5 cm in front of the lens; 6 cm high; inverted; real.
57. 21.4 cm in front of the lens; 17 mm high; inverted; real.
59. Nearsightedness; astigmatism.
61. Violet light has the lower speed in glass.
63. (a) The stars would be red on a black field; instead of having stripes, the rest of the flag would be solid red. (b) Instead of white stars in a blue field, the upper left-hand part of the flag would be solid blue; the rest of the flag would have blue and black stripes.
65. Blue light is scattered most, red light is scattered least.
67. (a) All. (b) All except polarization.
69. The wavelengths in visible light are very small relative to the size of a building, whereas those in radio waves are more nearly comparable.
71. $d_o = 1.22\lambda L/D = 41$ m.

Chapter 8

1. Nearly all the mass of an atom is located in its small central nucleus. Its positive charge is also in the nucleus. The electrons that carry the atom's negative charge move about the nucleus a relatively large distance away.
3. The isotopes have the same atomic number and hence the same electron structure, therefore they have the same chemical behavior. They have different numbers of neutrons and hence different atomic masses.
5. 8p, 10n; 12p, 14n; 26p, 31n; 47p, 62n.
7. The limited range of the strong interaction whereas the electric proton-proton repulsion has a long range.
9. (a) A ${}^{4}_{2}$He nucleus, which consists of two protons and two neutrons; an electron or positron; a high-frequency electromagnetic wave. (b) Gamma rays are the most penetrating in general, alpha particles, the least.
11. Z decreases by 2; A decreases by 4.
13. (a) A nucleus emits an electron when it contains too many neutrons to be stable, and it emits a positron when it contains too many protons. (b) ${}^{14}_{8}$O emits a positron and ${}^{19}_{8}$O emits an electron.
15. $84 - 2 = 82$; $210 - 4 = 206$.
17. 92; 233; uranium.
19. ${}^{207}_{82}$Pb.
21. It remains the same.
23. If $\frac{1}{4}$ of the original amount remains after 10 years, the half-life must be 5 years since $\frac{1}{4} = (\frac{1}{2})(\frac{1}{2})$.
25. 0.5 kg; 0.125 kg; radon and helium.
27. $\text{KE} = \frac{1}{2}mv^2 = (\frac{1}{2})(9.1 \times 10^{-31}\ \text{kg})(10^6\ \text{m/s})^2 = 4.55 \times 10^{-19}$ J. Since 1 eV = 1.6×10^{-19} J, this energy is equal to 2.84 eV.
29. 1 eV = 1.6×10^{-19} J, so 26 eV = 4.16×10^{-18} J. Since KE $= \frac{1}{2}mv^2$, $v = \sqrt{2\text{KE}/m} = 3.02 \times 10^6$ m/s.
31. The energy needed to remove an electron is very much less.
33. The binding energy per nucleon is greatest for nuclei of intermediate size.
35. The mass equivalent of the total binding energy is $m_E = 0.529$ u. Hence $m = 26m_H + 30m_n - m_E = 55.937$ u.
37. $2m_H + 2m_n = 4.0330$ u and so the mass difference is 0.0304 u = 28.3 MeV. There are four nucleons in ${}^{4}_{2}$He, hence the binding energy per nucleon is 7.1 MeV.
39. $m_p + m_e = 1.0078$ u. The difference between this and m_n is 0.009 u = 0.8 MeV, which is less than the observed binding energies per nucleon in stable nuclei. Hence neutrons do not decay inside nonradioactive nuclei.
41. Collisions with the nuclei of the moderator slow the fast neutrons produced in fission. This is desirable because ^{235}U undergoes fission more readily when struck by slow neutrons than by fast ones; hence the presence of a moderator promotes a chain reaction when this nuclide is the fuel. In addition, ^{238}U absorbs fast neutrons without undergoing fission but has little tendency to absorb slow neutrons.
43. (a) $m = Pt/c^2 = 9.6 \times 10^{-4}$ kg = 0.96 g. (b) Fissions per second = (total power)/(energy per fission) = 3.1×10^{19} per second.
45. The chief difference is that in fission heavy nuclei split into lighter ones, whereas in fusion light nuclei join to form heavier ones. The chief similarity is that in both processes mass is converted into energy.
47. (a) Their electric charges have opposite signs. (b) They annihilate each other with the lost mass becoming gamma rays or particle-antiparticle pairs.
49. The protons and electrons would form hydrogen atoms. With no strong interaction to hold them together, nuclei other than a single proton could not exist. The neutrons would decay into protons and electrons, which would also join to become hydrogen atoms.
51. Both are neutral electrically. The neutron is associated with both the strong and weak nuclear interactions, the neutrino with the weak interaction only. Both have mass, but that of the neutrino is very much smaller. Both have antiparticles. The neutrino is stable; the neutron beta-decays in free space into a proton, an electron, and a neutrino.
53. Leptons are not subject to the strong interaction and are point particles with no detectable size. Hadrons are subject to the strong interaction, have definite sizes, and apparently consist of various combinations of quarks.
55. No; it is possible that there is a reason why quarks cannot exist except in combination with each other as hadrons.

Chapter 9

1. Electrons have mass, while photons do not. Electrons have charge, while photons do not. Electrons may be stationary or move with speeds of up to almost the speed of light, while photons always travel with the speed of light. Electrons are constituents of ordinary matter, while photons are

not. The energy of a photon depends upon its frequency, while that of an electron depends upon its speed.

3. Interference, diffraction, and polarization phenomena and agreement with the electromagnetic theory of light argue for a wave nature. The photoelectric effect and the nature of line spectra argue for a particle nature.

5. Even a faint light involves a great many photons.

7. The photon energies are greater; their speeds and number per second are unchanged.

9. $hf = 1.3 \times 10^{-17}$ J; $hf = 1.3 \times 10^{-28}$ J.

11. 3 photons.

13. The frequency corresponding to a wavelength of 5.5×10^{-7} m is $f = c/\lambda = (3 \times 10^8 \text{ m/s})/5.5 \times 10^{-7} \text{ m} = 5.45 \times 10^{-14} \text{ s}^{-1}$, so $hf = (6.63 \times 10^{-34} \text{J} \cdot \text{s})(5.45 \times 10^{14} \text{ s}^{-1}) = 3.61 \times 10^{-19}$J. Hence $(1400 \text{ J/m}^2 \cdot \text{s})/(3.61 \times 10^{-19} \text{ J/photon}) = 3.9 \times 10^{21}$ photons/m$^2 \cdot$ s reach the earth from the sun.

15. Energy per pulse $= Pt = 0.01$ J; energy per photon $= hf = hc/\lambda = 3.1 \times 10^{-19}$ J/photon; photons per pulse = $(0.01 \text{ J})/(3.1 \times 10^{-19} \text{ J/photon}) = 3.2 \times 10^{16}$ photons.

17. (a) $E = hf$, so $f = E/h = 4 \times 10^{-19} \text{ J}/6.63 \times 10^{-34} \text{ J} \cdot \text{s} = 6.03 \times 10^{14} \text{ s}^{-1} = 6.03 \times 10^{14}$ Hz. (b) $f = c/\lambda = (3 \times 10^8 \text{ m/s})/2 \times 10^{-7} \text{ m} = 1.5 \times 10^{15}$ Hz. The energy of a photon of light of this frequency is $E = hf = (6.63 \times 10^{-34} \text{ J} \cdot \text{s})(1.5 \times 10^{15} \text{ s}^{-1}) = 9.95 \times 10^{-19}$ J, so the maximum energy of the photoelectrons is $(9.95 - 4) \times 10^{-19} \text{ J} = 6.0 \times 10^{-19}$ J.

19. KE $= QV = eV = hf$, $f = eV/h = 2.4 \times 10^{-18}$ Hz.

21. No.

23. The proton's KE may be less than, equal to, or more than the photon energy, depending upon what the wavelength is.

25. (a) Same speed. (b) The wave packet is slower.

27. $\lambda = h/mv = 3.6 \times 10^{-11}$ m. This wavelength is comparable with atomic dimensions; hence the wave character of the electron will affect any interactions it has with atoms in its path.

29. $\lambda = h/mv = 2.6 \times 10^{-11}$ m. This is 15 times smaller than the molecule's diameter and no wave behavior would be found.

31. Since 1eV $= 1.6 \times 10^{-19}$ J, the energy of each electron is 6.4×10^{-15} J and from KE $= \frac{1}{2}mv^2$ its speed is $v = \sqrt{\dfrac{2 \times 6.4 \times 10^{-15} \text{ J}}{9.1 \times 10^{-31} \text{ kg}}} = 1.2 \times 10^8$ m/s

The corresponding de Broglie wavelength is $\lambda = h/mv = 6.63 \times 10^{-34} \text{ J} \cdot \text{s}/(9.1 \times 10^{-31} \text{ kg} \times 1.19 \times 10^8 \text{ m/s}) = 6.1 \times 10^{-10}$ m.

33. The dimensions and momenta of all objects other than atomic particles are so large that the uncertainties in their positions and momenta are too small in comparison to be detectable.

35. (a) A continuous emission spectrum. (b) An emission line spectrum. (c) An absorption line spectrum.

37. A negative total energy signifies that the electron is bound to the nucleus. The KE of the electron is, of course, a positive quantity; the PE of the electron is sufficiently negative to make the total energy negative.

39. (a) The state of lowest energy. (b) $n = 1$.

41. The doppler effect shifts the frequencies in the emitted light to higher and lower frequencies, depending on whether the motion is toward or away from the observer, by amounts that depend on the speeds of the atoms and their directions. As a result the spectral lines are slightly fuzzy instead of perfectly sharp.

43. The centripetal force mv^2/r on the electron is provided by the electrical attraction Ke^2/r^2 of the hydrogen nucleus. Hence $mv^2/r = Ke^2/r^2$ and $v = \sqrt{Ke^2/mr}$. In the $n = 1$ orbit, $r = 5.3 \times 10^{-11}$ m; so $v = \sqrt{\dfrac{9 \times 10^9(1.6 \times 10^{-19})^2}{9.1 \times 10^{-31} \times 5.3 \times 10^{-11}}}$ m/s $= 2.2 \times 10^6$ m/s

45. (a) $\lambda = h/mv = 3.68 \times 10^{-63}$ m.
(b) n = circumference/$\lambda = 2.55 \times 10^{74}$. (c) No. The de Broglie wavelength is so small compared with the orbit circumference that no quantum effects could occur.

47. The light waves from a laser are coherent; that is, they are exactly in step with one another.

49. It is an excited state whose lifetime is longer than that of a normal excited state.

51. The results of quantum mechanics are in better quantitative agreement with experiment than those of the Bohr theory and can be applied to a greater variety of situations.

53. A high value of ψ^2 signifies a high probability of finding the particle; a low value of ψ^2 signifies a low probability.

55. Under all circumstances.

Chapter 10

1. (a) The change from water to ice is a physical change because chemically the substance remains the same; the only differences between ice and water are in their physical properties. (b) The change from iron to rust is a chemical change because the chemical compositions of the two substances are different.

3. Although heating is a physical process, it can produce chemical changes such as the decomposition of mercuric oxide into its constituent elements mercury and oxygen.

5. Elements: mercury, liquid hydrogen. Compounds: alcohol, pure water. Solutions: seawater, beer.

7. Hydrogen; oxygen.

9. These formulas represent the ratios in which the atoms of the various elements are present in the respective compounds. They do not provide information on the structures of the individual molecules or crystals, on the physical properties of the compounds, or on how to prepare the compounds.

11. Iron has the characteristic metallic luster, conducts heat and electricity well, combines directly with oxygen, and liberates hydrogen from dilute acids. Sulfur has no metallic luster and is a poor conductor of heat and electricity, combines readily with metals, does not react with dilute acids, and forms an acid when its oxide is dissolved in water.

13. Sodium is a very active metal, and so it combines readily, whereas platinum is highly inactive and therefore does not tend to combine at all.

15. Vertical column.

17. In each group of the periodic table, going from top to bottom means increasing metallic activity. Hence in Group 1 potassium is more active than sodium and in Group 2 calcium is more active than magnesium. In each period of the table, going from left to right means decreasing metallic activity. Hence in Period 3 sodium is more active than magnesium, and in Period 4 potassium is more active than calcium. In a more complex activity series, cesium would appear as more active than potassium.

19. (a) Each outer shell has a single electron outside filled inner shells. (b) Each outer shell lacks an electron of being filled. (c) Each outer shell is filled.

21. Both F and Cl atoms lack one electron of having closed outer shells.

23. Electrons are liberated from metals illuminated by light more easily than from nonmetals because the outer electrons of metal atoms are less tightly bound, which is also the reason they tend to form positive ions. Electrons are most readily liberated from metals in Group 1 of the periodic table.

25. A chlorine ion has a closed outer shell, whereas a chlorine atom lacks an electron of having a closed outer shell.

27. The outermost electron shell of an atom determines its chemical activity, hence the similarity of the outermost shells of the transition elements means that their chemical behavior must also be similar.

29. Selenium is a nonmetal because it requires only two electrons to complete its outer shell.

31. $+2e$; relatively easy.

33. The attractive force of the two protons in an H_2 molecule is greater than the attractive force of the single proton in an H atom. The mutual repulsion of the two electrons in H_2 means that they tend to be on opposite ends of the molecule, and so this repulsive force is smaller than the increased attractive force of the two protons on each electron.

35. Inert gas atoms contain only closed outer shells and so they cannot accommodate other electrons, as would occur in covalent bonding.

37. $+1$; -1; -2.

39. BaI_2; NH_4ClO_3; $SnCrO_4$; $LiPO_4$.

41. 14; 8.

43. Barium hydride; lithium phosphate; lead(II) oxide; copper(II) bromide; potassium hydroxide.

45. a, c, d.

47. (a) 2; (b) 3; (c) 8; (d) 7.

49. $2Na + 2H_2O \rightarrow 2NaOH + H_2$

51. $2Al + 3Cl_2 \rightarrow 2AlCl_3$

53. $2C_2H_6 + 7O_2 \rightarrow 4CO_2 + 6H_2O$.

55. $2C_4H_{10} + 13O_2 \rightarrow 8CO_2 + 10H_2O$.

57. $Fe_2O_3 + 3CO \rightarrow 2Fe + 3CO_2$.

Chapter 11

1. (a) Ionic, NaCl; covalent, diamond; van der Waals, ice; metallic, copper. (b) In each case the bonding is due to electric forces. (c) ions; atoms; molecules; ions.

3. Ionic crystals are more common.

5. By heating it gradually; if it is glass, it will sag slowly, but this will not occur if it is a crystalline solid.

7. These forces are too weak to hold inert gas atoms together to form molecules against the forces exerted during collisions in the gaseous state.

9. Ca^2+, F^-; K^+, I^-.

11. The solubility of a gas in a liquid decreases with increasing temperature.

13. An unsaturated solution contains less solute than the solvent can dissolve under its conditions of temperature and pressure. A saturated solution contains the maximum solute that the solvent can normally dissolve under these conditions. A supersaturated solution contains more solute than the solvent can normally dissolve under these conditions, but such a solution is unstable and the excess solute will leave the solution if it is disturbed.

15. (a) Add some additional sugar and see if it dissolves. (b) Cool the solution and see if any sugar crystallizes out.

17. Let the solutions cool. Because the solubility of KNO_3 falls faster with temperature than that of $NaNO_3$, more KNO_3 will precipitate out of solution than $NaNO_3$.

19. A solution of an electrolyte conducts electricity, a solution of a nonelectrolyte does not.

21. (a) AgCl would be precipitated from the solution with Cl^- ions; $AgNO_3$ is soluble. (b) AgCl would be precipitated from the solution with Ag^+ ions; NaCl is soluble.

23. Water; $Ca^{2+} + 2HCO_3^- \rightarrow CaCO_3 + CO_2 + H_2O$.

25. 3 percent; two-thirds.

27. Pure acids in the liquid state consist of neutral covalently bonded molecules. They form H^+ ions by reaction with water.

29. HBr is a strong acid like HCl because Br is below Cl in the periodic table, and thus its acid would have similar properties, in particular complete dissociation.

31. There is no difference.

33. 3; 10.

35. Water dissociates to a very small extent into H^+ and OH^- ions; hence it is both a weak acid and a weak base.

37. The ionic equation is $K^+ + OH^- + H^- + NO_3^- \rightarrow H_2O + K^+ + NO_3^-$. The actual chemical change is the combination of H^+ and OH^- to form H_2O.

39. (a) Calcium chloride, $CaCl_2$. (b) $2HCl + Ca(OH)_2 \rightarrow CaCl_2 + 2H_2O$.

41. Potassium acetate, $KC_2H_3O_2$;
$HC_2H_3O_2 + KOH \rightarrow K_2H_3O_2 + H_2O$.

43. $2NH_3 + H_2SO_4 \rightarrow (NH_4)_2SO_4$.

45. When $AlCl_3$ is dissolved in water, it first dissociates into Al^{3+} and Cl^- ions: $AlCl_3 \rightarrow Al^{3+} + 3Cl^-$. Some of the Al^{3+} ions then react with water to form $AlOH^{2+}$ and H^+ ions: $Al^{3+} + H_2O \rightarrow AlOH^{2+} + H^+$. Hence the solution contains Al^{3-}, $AlOH^{2+}$, H^+, and Cl^- ions and is acidic.

Chapter 12

1. His emphasis on accurate measurements of mass in chemical processes.

3. Air provides the oxygen needed for combustion.

5. The number of molecules and the average molecular energies are the same. The mole of oxygen molecules has more mass than the mole of hydrogen molecules.

7. Each mole of $MgAl_2O_4$ contains 2 mol of Al, hence 5 mol contains 10 mol of Al.

9. $\frac{1}{2}$ mol of N_2 and $1\frac{1}{2}$ mol of H_2.

11. The atomic mass of U is 238 g/mol, so 10 mol has a mass of 2.38 kg. The number of atoms is (10 mol) (6.02×10^{23} atoms/mol) = 6.02×10^{24} atoms since 1 formula unit = 1 atom of U here.

13. 2Fe = 111.70 u and 3O = 48.00 u, so the mass of 1 mole of Fe_2O_3 is 159.70 g and the mass of 2 moles is 319.40 g.

15. The formula mass of glucose is 180.16 g/mol and 500 kg = 5×10^5 g, hence the number of moles is 5×10^5 g/(180.16 g/mol) = 2775 mol.

17. Urea, 47 percent; ammonium nitrate, 35 percent; ammonium sulfate, 21 percent.

19. (a) One mol of Na and 1 mol of Cl combine to form 1 mol of NaCl, so the number of moles is the same for each substance. The number of moles of Na in 50 g is 50 g/(22.99 g/mol) = 2.17 mol. The mass of 2.17 mol of Cl is (2.17 mol)(35.46 g/mol) = 77 g. (b) The mass of 2.17 mol of NaCl is 50 g + 77 g = 127 g.

21. The number of moles in 200 g of K is 200 g/(39.1 g/mol) = 5.12 mol. In K_2S there is 1 mol of S for each 2 mol of K, hence the number of moles of S is half the number of moles of K or 2.56 mol. The corresponding mass of S is (2.56 mol)(32.06 g/mol) = 82 g.

23. In an atomic bomb explosion, the liberated energy comes from rearrangements of particles within atomic nuclei, whereas in a dynamite explosion the liberated energy comes from rearrangements within the electron clouds of atoms.

25. Exothermic; *a, b, e, f.*

27. Before the initial substances can react, energy is needed to break some of the bonds holding their respective constituent atoms to one another in order for these atoms to reassemble themselves to form new substances.

29. The higher the temperature, the greater the number of activated molecules.

31. Grind the salt into a fine powder; increase the water temperature; stir the water.

33. (a) The explosion of dynamite; the precipitation of AgCl when solutions containing Ag^+ and Cl^- are mixed. (b) The rusting of iron; the formation of ammonia gas from a solution of NH_4OH.

35. At room temperature few of the molecules will have energies as great as the activation energy, and since only these few molecules can react, the process is a slow one.

37. Most are reversible.

39. The number of molecules is the same on both sides of the equation.

41. Decrease the yield because the reaction is exothermic; increase the yield because in the reaction three molecules combine to form only two; use a catalyst.

43. $N_2 + O_2 + \text{energy} \leftrightarrow 2NO$; backward reaction favored; no effect; backward reaction favored; forward reaction favored; no effect.

45. Na; Al; I^-; Cl; Hg^{2+}.

47. (a) Hydrogen is given off.
(b) $2Li + 2H_2O \rightarrow 2LiOH + H_2$. (c) Lithium is oxidized and hydrogen is reduced.

49. Positive electrode: Cl_2; negative electrode: H_2.

51. Silver.

53. The water content of the electrolyte undergoes electrolysis with the production of H_2 and O_2 gases.

Chapter 13

1. There are more carbon compounds than compounds of any other element because of the ability of carbon atoms to form covalent bonds with one another.

3. Covalent bonds that consist of shared electron pairs.

5. Gasoline is not a compound but is a mixture of different alkenes.

7. The isomers have different molecular structures.

9. Four electrons.

11. Two bonds, so that the structure of CO_2 is O= C= O.

13. Such molecules are extremely reactive.

15. Bromopropane has two isomers:

```
    H   H   Br              H   Br  H
    |   |   |               |   |   |
H — C — C — C — H       H — C — C — C — H
    |   |   |               |   |   |
    H   H   H               H   H   H
```

17. The middle C atom has 5 bonds instead of the correct 4.

19. There is no way these atoms can be arranged so that each carbon atom participates in four covalent bonds and each hydrogen atom participates in one covalent bond.

21. Yes.

23. (a) The two straight-chain isomers are

```
  H   H   H   H                 H   H   H   H
  |   |   |   |                 |   |   |   |
H—C = C — C — C—H   and   H—C — C = C — C—H
          |   |                 |           |
          H   H                 H           H
```

What might seem to be a third isomer,

```
    H   H   H   H
    |   |   |   |
H — C — C — C = C — H
    |   |
    H   H
```

is really the first of the above reversed, which is not a true difference.

(b)

```
    H       H
    |       |
H — C — C = C — H
    |   |
    H   |
        |
    H — C — H
        |
        H
```

25. The molecules of aromatic compounds contain rings of six carbon atoms; the molecules of aliphatic compounds do not.

27. All the bonds are single. The structural formula is

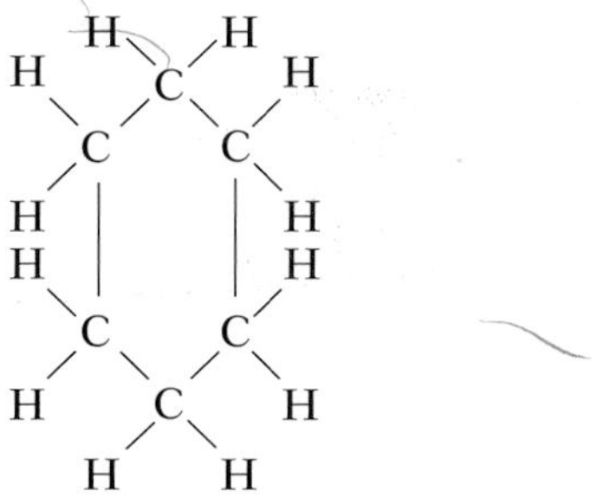

29. Carbon dioxide.

31. The compound is an aldehyde, namely proprionaldehyde.

33. Esters are nonelectrolytes, while salts in solution are electrolytes. Salts (such as sodium chloride) are crystals in their pure state, while the simpler esters (such as methyl acetate) are liquids or gases.

35. The compound has the structure of ethene with three of the H atoms replaced by Cl atoms.

37.

```
     H   H   H                      H   OH    H
     |   |   |                      |   |     |
 H — C — C — C — OH   and   H — C — C ——— C — H
     |   |   |                      |   |     |
     H   H   H                      H   H     H
```

39. (a)

```
     H   H   H   H   H                  H   H   H   H
     |   |   |   |   |                  |   |   |   |
 H — C — C — C — C — C — H          H — C — C — C — C — H
     |   |   |   |   |                  |   |   |   |
     H   H   H   H   H                  H   H   C   H
                                               /|\
       normal pentane                         H H H
                                            isopentane

                  H  H  H
                   \ | /
                     C
         H           |           H
           \         |         /
       H — C    —    C    —    C — H
           /         |         \
         H           C           H
                   / | \
                  H  H  H
                neopentane
```

(b) Normal pentane. (c) Neopentane.

41. In plants, carbohydrates are obtained by photosynthesis from CO_2 and H_2O; in animals, they are obtained by eating plants or foods derived from plants. In both plants and animals, fats are synthesized from carbohydrates.

43. Photosynthesis.

45. Thermal insulation and protection from mechanical injury.

47. Amino acids; some are synthesized by the body, others must be present in food.

49. (a) In a mutation, one or more of the letters in the genetic code expressed in the DNA of the gene is altered. (b) If a mutation occurs in the DNA of a reproductive cell of an organism, the descendants of the organism may be different in some way. Thus the occurrence of mutations is an essential part of the continual evolution of living things.

Chapter 14

1. Ultraviolet and x-rays from the sun.

3. At the tropopause the temperature has decreased to a minimum and is about to increase; at the stratopause it has increased to a maximum and is about to decrease; at the mesopause it has decreased to a minimum and is about to increase.

5. The CFC gases catalyze the breakdown of ozone in the upper atmosphere.

7. When the air is heated; when the air is cooled.

9. (a) 100 percent. (b) The air remains saturated and so the relative humidity remains 100 percent, while the excess water vapor condenses out. (c) The relative humidity decreases.

11. In order of decreasing volume of freshwater: ice caps and glaciers; groundwater; lakes and rivers; atmospheric moisture. The volume of seawater is many times greater than that of freshwater.

13. (a) The sudden cooling of the cloud (or part of it), usually by a rapid updraft, causes condensation of water vapor into raindrops that become heavy enough to fall. (b) If the cloud consists of ice crystals, its sudden cooling results in the formation of snowflakes that become heavy enough to fall.

15. (a) Incoming solar radiation. (b) It heats the atmosphere both directly and, through the greenhouse effect, indirectly.

17. The interior of a greenhouse is warmer than the outside air because sunlight can enter through its windows, but the infrared radiation that the warm interior gives off cannot escape through them. The carbon dioxide and water vapor contents of the atmosphere act as a trap of this kind for the earth as a whole. The atmosphere is transparent to visible light, which is absorbed by the earth's surface. The temperature of the surface is thereby increased, which in turn increases the rate at which it emits infrared radiation. The carbon dioxide and water vapor in the atmosphere absorb the infrared radiation, which leads to a warming of the lower atmosphere.

19. The rate of energy radiation into space would not change.

21. Winds and ocean currents carry energy around the earth in the forms of warm air and warm water, respectively. Winds are more effective in energy transport than ocean currents.

23. A large amount of heat must be absorbed or lost by a region of the earth's surface before it reaches its final temperature when the rate of arrival of solar energy changes. Since the difference between the rates of energy absorption and energy loss is always small compared with the heat content of the earth's surface, the temperature of the surface cannot change rapidly enough to keep pace with changes in the rate at which solar energy arrives; hence the time lags in seasonal weather conditions.

25. On June 22 the North Pole is tilted closest to the sun and hence is the day of maximum sunlight in the northern hemisphere. Because of the 23.5° tilt of the earth's axis, the noon sun is directly overhead 23.5° north of the equator. Hence

the latitude of the *Tropic of Cancer* is 23.5°N. Similarly the latitude of the *Tropic of Capricorn* is 23.5°S; the South Pole is tilted closest to the sun on December 22, when the noon sun is directly overhead at this latitude.

27. An isobar is a line on a weather map that joins points that have the same atmospheric pressure. A millibar (mb) is a unit of pressure equal to 100 Pa; average sea-level atmospheric pressure is 1013 mb.

29. In a land breeze, cool air over the land flows offshore to replace warm air that rises over the sea. In the northern hemisphere, the land breeze is deflected to the right from being directly offshore; in the southern hemisphere, to the left.

31. The Chicago-London flight takes less time because high-altitude winds at that latitude are westerly.

33. (a) Anticyclones. (b) The winds spiral clockwise outward from its center. (c) The winds spiral counterclockwise outward from its center.

35. Anticyclonic weather is generally steady with relatively constant temperature, clear skies, and light winds. Cyclonic weather is unsettled with rapid changes in temperature that accompany the passages of cold and warm fronts, cloudy skies, rain, and fairly strong, shifting winds.

37. Cyclonic winds in the northern hemisphere are counterclockwise; hence when you face the wind, the center of low pressure will be on your right. Cyclonic winds in the southern hemisphere are clockwise; hence when you face the wind, the center of low pressure will be on your left.

39. Northwest → west → southwest.

41. Stratus clouds are characteristic of a warm front.

43. The doldrums are at the equator, so it is quite warm there with considerable evaporation of water and thus high humidity. The airflow is largely upward, so surface winds are light and erratic. The rising currents of moist air lead to considerable rainfall.

45. Milankovitch pointed out that what is important is the amount of solar energy reaching the polar regions in summer, which varies by up to 20 percent. Too little summer sunshine would not melt all the snow that fell in the winter before, and this snow would accumulate and turn into great sheets of ice that would move equatorward to produce an ice age.

47. The presence in the atmosphere of sulfate particles from the burning of fossil fuels partly blocks sunlight in these regions.

49. Several hundred years ago; about 1°C.

51. (a) The greater the wind speed, the higher the waves. (b) The longer the period of time during which the wind blows, the higher the waves. (c) The greater the distance (fetch) over which the wind blows across the water, the higher the waves. Each of the above factors ceases to have a strong effect on wave height after a certain point; for example, after a day or two the waves will have reached very nearly the maximum height possible for the wind speed and fetch of a given situation.

53. Because their waters must be thoroughly mixed in the course of time to obtain a uniform composition of ions, the seas and oceans of the world cannot be static bodies but must exhibit large-scale currents, both vertical and horizontal.

55. (a) Ocean currents transport warm and cold water around the world, and this water heats or cools air that blows over it before the air reaches land. (b) The oceans act as heat reservoirs that help prevent sudden temperature changes on land near their shores.

57. When warm moist air from the west blows over the colder California Current, its temperature drops and moisture from the now supersaturated air condenses into tiny droplets to form a fog.

59. It is driven by the northeasterly and southeasterly trade winds on either side of the equator.

Chapter 15

1. Aluminum; iron.

3. A rock is an aggregate of grains of one or more minerals.

5. Crystal form refers to the shape of a crystal, which is determined by the pattern in which its constituent particles are linked together. Cleavage refers to the tendency, if any, of a crystal to break apart in a regular way, which is determined by the presence of weak bonds in certain directions in its structure.

7. An isolated SiO_4^{4-} tetrahedron has a net charge of $-4e$, so an assembly of such tetrahedra is electrically impossible. Minerals that contain isolated SiO_4^{4-} tetrahedra also have positive metal ions in their crystal structures that bond the tetrahedra together, producing electrical neutrality.

9. Igneous rocks consist of random arrangements of irregular mineral grains that have grown together.

11. Diorite is a coarse-grained rock and andesite is a fine-grained rock.

13. Calcite does not contain silicon. Limestone and marble are largely calcite.

15. Chert consists largely of microscopic quartz crystals and hence is hard and durable.

17. The density increases because the pressures under which metamorphism occurs lead to more compact rearrangements of the atoms in the various minerals.

19. (a) Foliation. (b) Foliation results from the growth of platy or needlelike crystals along planes of movement in a rock produced by directed pressure (stress).

21. (a) Limestone is produced by the consolidation of shell fragments and also by precipitation of calcium carbonate from solution. (b) Marble. (c) The grains become larger.

23. Quartz is a mineral whose chemical composition is SiO. Quartzite is a hard rock formed by the metamorphosis of sandstone; it consists mainly of quartz.

25. (a) Granite, which contains only a small proportion of ferromagnesian material, is light in color, whereas gabbro, with much ferromagnesian material, is dark. (b) Limestone reacts readily with an acid, unlike basalt. (c) Schist is foliated, whereas diorite is not.

27. (a) Rhyolite. (b) Slate. (c) Shale.

29. A factor of 30.

31. The mantle is believed to be solid because earthquake S waves can pass through it.

33. The core's radius is about half the earth's radius.

35. (a) Measurements made in mines and wells indicate that temperature increases with depth. (b) Molten rock from the interior emerges from volcanoes. (c) The outer core is liquid, which means that it must be at a high temperature.

37. (a) Earthquake S waves cannot go through the outer core. (b) Iron has the right density, it is liquid at the temperature and pressure of the core, it is abundant in the universe generally, and it is a good conductor of electricity so currents in it can produce the geomagnetic field.

39. Ferromagnetic materials lose their magnetic properties at high temperatures, and sufficiently high temperatures exist throughout all of the earth's interior except near the surface of the crust to cause such a loss. Also, both the direction and strength of the field are observed to vary, and in fact the field has reversed its direction many times in the past, which cannot be reconciled with the notion of a permanent magnet in the interior.

41. Igneous and metamorphic rocks are formed under conditions of heat and pressure very different from those at the earth's surface, and minerals stable under the former conditions are not necessarily stable under the latter conditions. Most sedimentary rocks, on the other hand, consist of rock debris that has already undergone chemical weathering and so are relatively resistant to further attack. The chief exception is limestone, which is soluble in water that contains carbon dioxide.

43. The rock debris produced by weathering is the principal constituent of soil.

45. The maximum depth to which streams can erode a landscape is sea level since streams flow downhill into the seas. Glacial erosion is not limited in this way, and glaciers can wear away landscapes to depths well below sea level.

47. A glacier forms when the average annual snowfall in a region exceeds the annual loss by evaporation and melting.

49. Streams; glaciers.

51. (a) A water table is the upper surface of an underground zone that is saturated with groundwater. (b) An aquifer is a body of porous rock through which groundwater can move.

53. The ocean floors.

55. Quartz is resistant to chemical attack and so survives weathering and erosion. Feldspar, the most common mineral, is converted into clay minerals by the carbonic acid of surface waters.

57. (a) Precipitate from groundwater. (b) Stream deposits. (c) Sand dunes.

59. Basalt, rhyolite, andesite, obsidian; basalt.

61. Such holes were produced by bubbles of gas trapped in lava as it solidified.

63. (a) The intruded magma that solidifies to form a pluton is very hot, and thus nearby rocks often undergo thermal metamorphism. (b) Near a batholith, because of the greater heat that had to be dissipated in its cooling.

Chapter 16

1. Both normal and thrust faults produce cliffs called fault scarps. A strike-slip fault is often marked by a rift, which is a trench or valley caused by erosion of the disintegrated rock produced during the faulting.

3. Erosion.

5. The Rocky Mountains contain thick layers of sedimentary rocks that can have been formed only from sediments deposited over a long period of time; hence the region must once have been relatively low lying in order that rivers and streams containing erosional debris would have flowed into it.

7. If South America and Africa were once joined together, there should be similar geologic formations and fossils of the same kinds at corresponding locations along their respective east and west coasts. This is indeed found for material deposited up to about 100 million years ago, which is when these continents must have begun to separate.

9. (a) The Pyrenees, Alps, and Caucasus of Europe; the Atlas Mountains of North Africa; and the Himalayas of Asia. (b) Thick deposits of sedimentary rocks; fossils of sea creatures.

11. (a) The crust is distinguished from the mantle beneath it by a sharp difference in seismic-wave velocity, which suggests a difference in the composition of the minerals involved or in their crystal structures or in both. The lithosphere is distinguished from the asthenosphere beneath it by a difference in their behavior under stress: the lithosphere is rigid whereas the asthenosphere is capable of plastic flow. (b) The asthenosphere is plastic because its material is close to its melting point under the conditions of temperature and pressure found in that region of the mantle. Above the asthenosphere the temperature is too low and below it the pressure is too high for the material of the mantle to be plastic. (c) When a large stress is applied over a long period of time, the asthenosphere gradually flows in response to it. When brief, relatively small forces are applied, as is the case with seismic waves, the asthenosphere is rigid enough to transmit them as a solid does.

13. The ocean floors are relatively recent in origin; the oldest sediments date back only about 135 million years. Continental rocks, in contrast, date back more than 4000 million years. The reason is that, owing to their low density and consequent buoyancy, the continental blocks are not forced down into the mantle in subduction zones but remain as permanent features of the lithosphere plates they are part of. The ocean floors, on the other hand, are continually being destroyed in such zones as new ocean floors are deposited at midocean ridges.

15. The earth's interior.

17. The Indian plate is still moving northward into the Eurasian plate and thereby is continuing to force the Himalayas upward.

19. The Himalayas.

21. There is relative motion between the two plates along their boundary; the Pacific plate is moving northwestward relative to the American plate.

23. Oceanic-oceanic plate collisions.

25. 1.9 cm/year.

27. (a) An unconformity is an eroded surface buried under rocks that were subsequently deposited. (b) An unconformity is at *J*.

29. The ratio between the radiocarbon and ordinary carbon contents of all living things is the same. When a plant or animal dies, its radiocarbon content decreases at a fixed rate. Hence the ratio between the radiocarbon and ordinary carbon contents of an ancient specimen of organic origin will reveal its age.

31. 3.9 billion years is 3 half-lives, so the amount of potassium 40 that remains is $(\frac{1}{2})(\frac{1}{2})(\frac{1}{2}) = \frac{1}{8}$ of the original amount, which must have been 8(0.010 percent) = 0.080 percent.

33. Actual plant or animal tissues, usually of a hard nature such as teeth, bones, hair, and shells. Entire insects have been found preserved in amber. Plant tissues that have become coal through partial decay but that retain their original forms. Tissues that have been replaced by material (such as silica) deposited from groundwater; petrified wood is an example. Sometimes a porous tissue such as bone will have its pore spaces filled with a deposited mineral. Impressions that remain in a rock of plant or animal structures that have themselves disappeared. Footprints, wormholes, or other cavities produced by animals in soft ground that have later filled with a different material and so can be distinguished today.

35. Igneous rocks have hardened from a molten state, and no fossil could survive such temperatures. Metamorphic rocks have been altered under conditions of heat and pressure severe enough to distort or destroy most fossils.

37. Precambrian time; the Cenozoic Era.

39. Abundant fossils exist in rocks belonging to the Phanerozoic Eon, which permits tracing the evolution of living things during this span of time. Few fossils exist from the Cryptozoic Eon, making it difficult to determine the forms of life that were present then and how they developed.

41. Precambrian geologic activity must have been similar to that of today.

43. Precambrian sedimentary rocks contain few fossils, whereas later sedimentary rocks usually contain abundant fossils.

45. Much of the area of the continents must have been near or below sea level during at least part of the Paleozoic since shallow seas must have been widespread on their surfaces.

47. Sedimentary rocks are found in many parts of the United States that contain the fossil shells of marine organisms.

49. Petroleum is thought to have originated in the remains of marine animals and plants that became buried under sedimentary deposits. After bacterial decay in the absence of oxygen, low-temperature chemical reactions produced further modifications. Then complex hydrocarbons were "cracked" under the influence of temperatures of 70 to 130°C to the straight-chain alkane hydrocarbons found in petroleum. When the temperatures were higher, the result was the smaller alkanes that make up natural gas.

51. Reptiles; all sizes.

53. Dinosaurs; crocodiles.

55. During the Mesozoic Era today's continents were joined together so the animal populations (which were largely reptiles) could move freely among them. During the Cenozoic Era the continents were split apart, and the evolution of some of the mammals that replaced the reptiles proceeded differently on the various landmasses.

57. The ice ages involved the formation of ice sheets that covered large areas of the earth's surface. In the most recent of the ice ages there were four major episodes during which ice advanced across the continents, separated by interglacial periods during which the ice retreated poleward. The glacial advances took place during the past 2 million years, that is, the Pleistocene Epoch of the Quaternary Period of the Cenozoic Era. In the latest of the glacial episodes, ice covered much of Canada and northeastern United States and began to recede only about 20,000 years ago.

Chapter 17

1. Mercury, Venus, Mars, Jupiter, Saturn.

3. Jupiter; Mercury; Mercury.

5. Mercury and Venus.

7. Less: Mercury, Venus, Mars. More: Jupiter, Saturn, Uranus, Neptune.

9. All are ellipses but those of comets are long and narrow whereas those of the planets are more nearly circular.

11. The density of a comet is extremely low when it is in the vicinity of the earth, and in a collision most or all of the comet material would simply be absorbed in the upper atmosphere.

13. Stony meteorites resemble ordinary rocks, whereas iron ones are conspicuously different; also, stony meteorites are more readily eroded than iron ones.

15. Conditions for both preservation and recognition are best there.

17. The sunlit side of Mercury is too hot and its dark side is too cold for life to exist. Also, Mercury has only a trace of an atmosphere, and it contains only inert gases.

19. Mercury and Venus are closer to the sun than the earth is, hence an observer on the earth always sees them in the vicinity of the sun. When one of them is east of the sun, it disappears below the horizon after the sun and is visible in the early evening; when it is west of the sun, it rises above the horizon before the sun and is visible in the early morning.

21. Their orbits are inside the earth's orbit (see Fig. 17-13).

23. (a) Venus is too hot for liquid water to have existed on its surface. (b) Lava flows may have flooded the surface and covered most of the ancient craters.

25. Venus is closer to the sun than Mars and hence receives more sunlight to reflect. It is larger than Mars, so there is more reflecting area. Venus is surrounded by clouds whereas Mars has none, and these clouds are better reflectors of sunlight than the Martian surface (the white polar ice caps on Mars are too small to make much difference in this respect).

27. Because running water would fill craters with sediments and level their raised rims, the presence of many meteoroid craters means that there has been no running water for a long time on such parts of the surface of Mars.

29. Iron oxides in its soil.

31. In contrast to the earth's atmosphere, the Martian atmosphere is very thin and consists largely of carbon dioxide, so there is little oxygen to form ozone that would absorb the ultraviolet radiation.

33. Asteroids are believed to be matter left over from the early solar system that never became part of a large body because of the gravitational influence of nearby Jupiter.

35. (a) An asteroid apparently struck the earth in what is now Mexico 65 million years ago; see Sec. 16.16. (b) Yes.

37. It consists of swirling gases and resembles a permanent hurricane in Jupiter's atmosphere. The red color may be due to phosphorus.

39. Io has many active volcanoes.
41. We can see stars through the rings; the inner part of each ring moves faster than the outer part; they reflect sunlight and radar signals too well to be gaseous. Finally, spacecraft have sent back photographs of the rings showing their structure.
43. No
45. No.
47. Hydrogen and helium.
49. The moon and six other satellites are larger than Pluto.
51. Not only does the sun provide daylight, but the light of the moon is reflected sunlight.
53. Two weeks.
55. Two weeks; two weeks.
57. No; yes.
59. New moon; full moon.
61. The maria are approximately circular depressions covered with pulverized rocks. They are apparently lava flows that were broken up by meteorite impacts.
63. The average density of the moon is much less than that of the earth. Part of the reason is the smaller total mass of the moon, which reduces the pressures in its interior. However, this factor is not enough to account for the large difference in the densities. Hence the moon must have a different composition from that of the earth, perhaps by virtue of a smaller proportion of iron.
65. Meteoroid impacts.

Chapter 18

1. The larger the telescope, the more light it can gather (thereby revealing faint objects in the sky) and the sharper the images it can produce (thereby resolving—that is, separating—objects that are close together).
3. The earth's atmosphere absorbs light of various frequencies, notably in the ultraviolet. The entire spectrum can be received by a telescope in orbit outside the atmosphere. Such a telescope is also unaffected by clouds and by the scattering of light by atmospheric dust.
5. An emission (bright-line) spectrum.
7. The star whose spectral lines are displaced farther to the red is moving away from the earth faster than the other star.
9. The presence of the spectral lines of a particular element in the solar spectrum means that this element must be present in the sun.
11. (a) The photosphere. (b) The radiation is absorbed by the gases inside the star, which reradiate to produce a steady outward flow of energy. The photosphere is hot enough to radiate but the gases above it are not dense enough to absorb all the light it emits. (c) The atmosphere of a star above its photosphere.
13. The density of the corona is extremely low, so the energy it radiates is very small despite its high temperature.
15. If the earth's magnetic field disappeared, auroras would be less frequent than at present since charged particles passing near the earth would not be deviated toward it by the magnetic field. Also, there would be no tendency for auroras to occur in definite zones centered on the geomagnetic poles.
17. Their number increases and then decreases.
19. $v = d/t = 1.4 \times 10^6$ m/s.
21. A helium nucleus has less mass than the total mass of the four hydrogen nuclei (protons) that combine to form it, and the "missing" mass appears as energy.
23. Assuming that the mass lost is small compared with the sun's original mass, (mass lost)/(original mass) = (mass lost)/(current mass) = 2.8×10^{-4} = 0.028 percent. This is so little that it is entirely possible that the sun's radiation rate has not changed by very much during the life of the earth.
25. One would begin by measuring the apparent brightnesses and periods of the Cepheid variables. From the known relationship between the period of a Cepheid and its intrinsic brightness, the latter can be computed, and a comparison of the intrinsic and apparent brightnesses then yields the distance of the star and, hence, of the cluster.
27. Stellar sizes vary more than stellar masses.
29. Black holes, neutron stars, and white dwarfs have the highest densities and giant stars the lowest.
31. The surface temperature of a star determines the radiation it emits per unit area, while its intrinsic brightness is a measure of its total radiation; knowing both quantities permits computing the star's surface area and hence its diameter.
33. Stars must be spheres or nearly so because if parts of their surfaces were at different distances from their centers, the resulting pressure differences due to gravity would cause the material of the stars to flow until they had spherical shapes. See Sec. 1.9.
35. Such an object must contract owing to gravity, which causes both a rise in temperature and an increase in density. As a result the hydrogen present begins to react to form helium with the release of considerable energy. Thus any object with the mass and composition of the sun must radiate energy like the sun.
37. A star on the main sequence is in an equilibrium condition with its tendency to expand owing to high temperature exactly balanced by its tendency to contract gravitationally. The condition lasts until the star's hydrogen content decreases beyond a certain proportion, which requires a relatively long time compared with its earlier and later phases. Therefore most stars are members of the main sequence simply because this is the longest stage in a star's evolution.
39. Betelgeuse; Deneb; Deneb.
41. It is large, heavy, hot, and bright, with prominent hydrogen and helium lines in its spectrum. It is small, exceedingly dense, very hot, and dim. It is huge, diffuse, cool, and bright. It is moderately small with moderate temperature, density, and mass, with a spectrum in which lines of metallic elements are prominent.
43. (a) It is very hot. (b) Its average density is low. (c) Upper end of main sequence.
45. Heavy stars use up their hydrogen rapidly, so their lifetimes in the main sequence are shorter than those of smaller stars.
47. Near the end of its life cycle.
49. Diagonally downward (since the star's luminosity will decrease) and to the right (since its temperature will decrease).
51. A neutron star or, if the supernova is very massive, a black hole.

53. (a) A pulsar emits bursts of radio waves at regular intervals. (b) Pulsars are believed to be very small, dense stars that consist almost entirely of neutrons. (c) Pulsars are believed to originate in supernova explosions.

55. Its strong gravitational field.

Chapter 19

1. No. Stars are concentrated in galaxies that are relatively far apart from one another.

3. The Milky Way is composed of stars in the spiral arms of our galaxy and so defines its central disk. Since the earth is close to the plane of the Milky Way, the sun must be part of the central disk of the galaxy.

5. A typical globular cluster is an assembly of hundreds of thousands of Population II stars that are relatively close together. They are found in all galaxies; in spiral galaxies, they are mostly located in the corona outside the central disk and move at high speeds about the galactic center. Since globular clusters are much smaller than galaxies and are always found as members of them, they cannot be considered as being themselves galaxies.

7. Population I stars are found in the spiral arms of spiral galaxies and are of all ages, including very young stars. Population II stars are found outside the arms of spiral galaxies and most are very old.

9. Most of the interstellar gas is located in the spiral arms of the galaxy, and its chief constituent is hydrogen.

11. Random thermal motion of ions and electrons in a very hot gas; fast electrons moving in a magnetic field; spectral lines of atoms and molecules.

13. Hydrogen is by far the most abundant element in the universe.

15. Most galaxies are smaller than our galaxy, which is one of the largest known.

17. (a) Two clues are the high speeds of outer stars in spiral galaxies and the motions of individual galaxies in clusters of galaxies. (b) Dark matter is located in all galaxies.

19. Collisions with primary cosmic rays broke up some of such relatively abundant nuclei in space as those of carbon and oxygen into the nuclei of lighter elements, notably lithium, beryllium, and boron, that otherwise would be rarer in the universe than they are observed to be.

21. The protons pick up electrons and become hydrogen atoms, while most of the neutrons are absorbed by carbon nuclei to form radiocarbon. Some neutrons escape from the earth entirely and decay into protons and electrons in space.

23. Magnetic fields.

25. Red shifts in the spectra of galaxies indicate that they are all moving apart.

27. The radiations from quasars vary too rapidly for them to be large in size.

29. The uniform expansion of the universe; the relative abundances of hydrogen and helium in the universe; the cosmic background radiation.

31. Most helium was formed in nuclear reactions soon after the big bang, and some was formed later in nuclear reactions inside stars. A little came from the alpha decay of radioactive nuclei.

33. (a) A closed universe will eventually stop expanding and then will contract to a big crunch; an open universe will expand forever. (b) Knowing the density of matter in the universe would indicate whether the universe is open (low density) or closed (high density). (c) Apparently the universe is open.

35. The young sun was much larger than it is today, perhaps as far across as the entire present solar system.

37. Gravitational contraction; radioactivity.

39. Fire, in the sense that the earth will be strongly heated when the sun swells into a red giant.

41. Other stars are all very far away; planets are small in size; planets are dim objects because they shine by reflected light.

Glossary

A

Absolute zero is the lowest temperature possible, corresponding to no random molecular movement. This temperature is −273°C. The *absolute temperature scale* gives temperatures in °C above absolute zero, denoted K. Thus the freezing point of water is 273 K.

The **acceleration** of an object is the rate of change of its velocity. An object is accelerated when its speed changes, when its direction of motion changes, or when both change.

The **acceleration of gravity** is the downward acceleration of an object in free fall near the earth's surface. Its value is 9.8 m/s^2.

An **acid** is a substance whose molecules contain hydrogen and whose water solution contains hydrogen ions. A *strong acid* dissociates completely into ions when dissolved, a *weak acid* only partly.

The **activation energy** of a reaction is the energy that must be supplied initially for the reaction to start.

An **alcohol** is a hydrocarbon derivative in which one or more hydrogen atoms have been replaced by OH groups.

The **alkali metals** are a family of soft, light, extremely active metals with similar chemical properties. The alkali metals are lithium, sodium, potassium, rubidium, cesium, and francium in order of atomic number.

An **alluvial fan** is a deposit of sediments where a stream emerges from a steep mountain valley and flows onto a plain.

An **alpha particle** is the nucleus of a helium atom. It consists of two protons and two neutrons and is emitted in certain radioactive decays.

An **amorphous solid** is one whose constituent particles show no regularity of arrangement.

The **amplitude** of a wave is the maximum value of whatever quantity is periodically varying.

The **angular momentum** of a rotating object is a measure of its tendency to continue to spin. Angular momentum is a vector quantity. The angular momentum of an isolated object or system of objects is *conserved* (remains unchanged).

The **antiparticle** of a particle has the same mass and general behavior, but has a charge of opposite sign and differs in certain other respects. A particle and its antiparticle can *annihilate* each other with their masses turning entirely into energy. In the opposite process of *pair production* a particle-antiparticle pair materializes from energy. *Antimatter* consists of antiparticles.

An **aquifer** is a body of porous rock through which groundwater moves.

Archimedes' principle states that the buoyant force on an object immersed in a fluid is equal to the weight of the fluid displaced by the object.

An **asteroid** is one of many relatively small bodies that revolve around the sun in orbits that lie between those of Mars and Jupiter.

The **asthenosphere** is a layer of rock capable of plastic deformation that is just below the lithosphere in the earth's mantle.

The earth's **atmosphere** is its gaseous envelope of air. The four regions of the atmosphere are, from the earth's surface upward, the *troposphere,* the *stratosphere,* the *mesosphere,* and the *thermosphere.*

Atmospheric pressure is the force with which the atmosphere presses down upon each unit area at the earth's surface. Its value is normally about 1.013×10^5 Pa.

An **atom** is the ultimate particle of any element. It consists of a *nucleus,* composed of neutrons and protons, and a number of *electrons* that move about the nucleus relatively far away.

The **atomic mass** of an element is the average mass of its atoms expressed in atomic mass units (u).

The **atomic number** of an element is the number of electrons in each of its atoms or, equivalently, the number of protons in its nucleus. An element is a substance all of whose atoms have the same atomic number.

An **aurora** is a display of changing colored patterns of light that appear in the sky, particularly at high latitudes. The aurora arises from the excitation of atmospheric gases at high altitudes by streams of protons and electrons from the sun.

Avogadro's number is the number of atoms in a mole of any element; it is also the number of formula units in a mole of any compound.

B

A **base** is a substance whose molecules contain OH groups and whose water solutions contain OH^- ions. A *strong base* dissociates completely into ions when dissolved, *weak base* only partly.

A **batholith** is a very large body of intrusive rock, mainly granite, that extends downward as much as 10 km.

A **benzene ring** consists of six carbon atoms linked together in a flat hexagon.

The **big bang theory** holds that the universe originated 13.7 billion years ago in the explosion of a hot, dense aggregate of matter. If the explosion was violent enough, as seems likely, the expansion of the universe will continue forever; if not, the universe will eventually begin to contract and will end up in a *big crunch* after which another cycle of expansion and contraction may occur.

The **binding energy** of a nucleus is the energy equivalent of the difference between its mass and the sum of the masses of the nucleons it is composed of. The *binding energy per nucleon* is least for very light and very heavy nuclei. Hence the *fusion* of very light nuclei to form heavier ones and the *fission* of very heavy nuclei to form lighter ones are both processes that give off energy.

One kind of **black hole** is the collapsed core of an old heavy star that has contracted to so small a size that its gravitational field prevents the escape of anything, including light. Supermassive black holes lie at the centers of galaxies of stars.

In the **Bohr model of the atom,** electrons are supposed to move around nuclei in circular orbits of definite size. When an electron jumps from one orbit to another, a photon of light is either emitted or absorbed whose energy corresponds to the difference in the electron's energy in the two orbits.

Boyle's law states that the volume of a gas is inversely proportional to its pressure when the temperature is held constant.

A **brown dwarf** is a spherical lump of matter in space between 10 and 75 times the mass of Jupiter that is not heavy enough to trigger the nuclear reactions of a true star. It glows dark red because of heat left over from its gravitational contraction.

Brownian movement is the irregular motion of suspended particles due to molecular bombardment.

The upward **buoyant force** on an object immersed in a fluid is equal to the weight of the fluid displaced by the object and is due to the increase in pressure with depth in the fluid.

C

In a **cap-and-trade** system for controlling CO_2 emissions, an overall cap on them is set for a region. Companies there are given or buy at auction permits to emit CO_2 whose total equals the cap. Companies that do not use their full quotas can sell the leftover permits to companies that exceed their quotas.

A **carbohydrate** is a compound of carbon, hydrogen, and oxygen whose molecules contain two atoms of hydrogen for each one of oxygen. Carbohydrates are manufactured in green plants from water and carbon dioxide in the process of photosynthesis. Sugars, starches, chitin, and cellulose are carbohydrates.

A **catalyst** is a substance that can alter the rate of a chemical reaction without itself being permanently changed by the reaction.

The **Cenozoic Era** refers to the past 66 million years of the earth's history; the term means "recent life."

The **centripetal force** on an object moving in a curved path is the inward force that must be exerted to produce this motion. It always acts toward the center of curvature of the object's path.

Charles's law states that the volume of a gas is proportional to its absolute temperature when the pressure is held constant.

In a **chemical reaction,** the atoms of the reacting substances (*reactants*) combine differently to form new substances; for instance, in the reaction of hydrogen and oxygen to form water, H_2 and O_2 molecules become H_2O molecules.

Chlorophyll is the catalyst that makes possible the reaction of water and carbon dioxide in green plants to produce carbohydrates. The reaction is called *photosynthesis.*

Cleavage is the tendency of a substance to split along certain planes determined by the arrangement of particles in its crystal lattice.

Climate refers to average weather conditions in a region over a period of years.

A **cloud** is composed of water droplets or ice crystals small enough to remain suspended aloft.

Coal is a fuel, largely carbon, that was formed chiefly from plant material that accumulated under conditions preventing complete decay.

Coherent light consists of light waves of the same frequency that are all in step with one another; in *incoherent light* the individual waves are not in step.

In a **colloid,** tiny particles, droplets, or bubbles of one or more substances are uniformly dispersed in another substance to form an intimate mixture that is not quite a solution. Emulsions, sols, gels, and aerosols are colloids.

Combustion is the rapid combination of oxygen with another substance during which heat and light are given off.

Comets are objects, largely ice, dust, and frozen gases, that follow regular orbits in the solar system and appear as small, hazy patches of light accompanied by long, filmy tails when near the sun.

A **compound** is a homogeneous combination of elements in definite proportions. The properties of a compound are different from those of its constituent elements.

A **conductor** is a substance through which electric current can flow readily.

A **constellation** is a group of stars whose arrangement in the sky suggests a particular pattern.

According to the theory of **continental drift,** today's continents were once part of two primeval supercontinents called Laurasia (North America, Greenland, Europe, and most of Asia) and Gondwana (South America, Africa, Antarctica, India, and Australia), which were separated by the Tethys Sea. The continents are still in motion.

Convection currents result from the uneven heating of a fluid. The warmer parts of the fluid expand and rise because of their buoyancy, while the cooler parts sink.

In the **copernican system** the earth and the other planets revolve around the sun, the moon revolves around the earth, and the earth rotates on an axis passing through Polaris.

The earth's **core** is a spherical region around the center of the earth whose radius is about 3600 km. It probably consists mainly of molten iron and has a small solid inner core.

The **coriolis effect** is the deflection of winds to the right in the northern hemisphere, to the left in the southern as a consequence of the earth's rotation.

The sun's **corona** is a vast cloud of extremely hot, rarefied gas that surrounds the sun. It is visible during solar eclipses.

The **cosmic background radiation** is electromagnetic radiation left over from the early universe that has been doppler-shifted to radio frequencies. Its spectrum corresponds to that of an object at a temperature of 2.7 K.

Cosmic rays are atomic nuclei, mostly protons, that travel through the galaxy at speeds close to that of light. They probably were ejected during supernova explosions and are trapped in the galaxy by magnetic fields.

Coulomb's law states that the force between two electric charges is directly proportional to both charges and inversely proportional to the square of the distance between them. The force is repulsive when the charges have the same sign, attractive when they have different signs.

In a **covalent bond** between atoms, the atoms share one or more electron pairs. The atoms in a molecule are held together by covalent bonds. A *covalent crystal* consists of atoms that share electron pairs with their neighbors.

The **crust** of the earth is its outer shell of rock. The crust averages about 35 km in thickness under the continents and about 5 km under the oceans.

A **crystalline solid** is one whose atoms or molecules are arranged in a definite pattern.

An electric **current** is a flow of charge from one place to another. A direct current is one that always flows in one direction; an alternating current periodically reverses its direction of flow. The unit of current is the *ampere* (A), which is equal to the flow of 1 coulomb per second (C/s).

Cyclones are weather systems centered about regions of low pressure. In the northern hemisphere cyclones are characterized by counterclockwise winds, in the southern hemisphere by clockwise winds. *Anticyclones* are weather systems centered about regions of high pressure; the characteristic winds of anticyclones are opposite in direction to those of cyclones.

D

The universe is pervaded with **dark energy** that is accelerating its expansion. The nature of dark energy is unknown. The universe consists of 73 percent dark energy, 23 percent dark matter, and only 4 percent of ordinary matter.

There is apparently much more **dark matter** in the universe than the luminous matter, such as stars, that we can see. The nature of the dark matter is not known; it may consist of hypothetical elementary particles yet to be discovered.

The matter waves that correspond to a moving object have a **de Broglie** wavelength inversely proportional to its momentum.

The *law of* **definite proportions** states that the elements that make up a chemical compound are always combined in the same definite proportions by mass.

A **delta** is a deposit of sediments where a stream flows into a lake or the sea.

The **density** of a substance is its mass per unit volume.

Diffraction refers to the ability of waves to bend around the edge of an obstacle in their path.

A **dike** is a wall-like mass of igneous rock that cuts across existing rock layers.

An ionic compound **dissociates** into free ions when dissolved in water.

The **doppler effect** is the change in perceived frequency of a wave due to relative motion between its source and an observer. Light and sound waves exhibit the doppler effect.

E

An **earthquake** consists of rapid vibratory motions of rock near the earth's surface usually caused by the sudden movement of rock along a fault. Earthquake **P** (for primary) waves are longitudinal oscillations in the solid earth like those in sound waves; earthquake **S** (for secondary) waves are transverse oscillations in the earth like those in a stretched string; earthquake surface waves are oscillations of the earth's surface. The *focus* of an earthquake is the place where the rock shifted; the *epicenter* is the point on the surface directly above the focus. Each step of 1 in the earthquake *Richter scale* represents a change by a factor of 10 in vibration amplitude and a change by a factor of about 30 in energy release.

A lunar **eclipse** occurs when the earth's shadow obscures the moon. A *solar eclipse* occurs when the moon obscures the sun.

Electric charge is a fundamental property of certain of the elementary particles of which all matter is composed. There are two kinds of charge, *positive* and *negative;* charges of like sign repel, unlike charges attract. The unit of charge is the *coulomb* (C). The principle of *conservation of charge* states that the net electric charge in an isolated system remains constant.

In an **electrochemical cell** such as a battery, electric current is produced by an oxidation-reduction reaction whose two half-reactions take place at different locations.

Electrolysis refers to the liberation of free elements from a liquid when an electric current passes through it.

An **electrolyte** is a substance that separates into free ions when dissolved in water.

An **electromagnet** is a current-carrying coil of wire with an iron core to increase its magnetic field.

Electromagnetic induction refers to the production of a current in a wire when there is relative motion between the wire and a magnetic field.

Electromagnetic waves are coupled periodic electric and magnetic disturbances that spread out from accelerated electric charges. Gamma rays, x-rays, ultraviolet radiation, visible light, infrared radiation, millimeter waves, microwaves, and radio waves are all electromagnetic waves that differ only in their frequencies. Electromagnetic waves all travel in a vacuum with the *speed of light.*

An **electron** is a tiny, negatively charged particle found in matter.

The **electronvolt** (eV) is a unit of energy equal to 1.6×10^{-19} J, which is the amount of energy acquired by an electron accelerated by a potential difference of 1 volt. A *MeV* is 10^6 eV and a *GeV* is 10^9 eV.

An **element** is a substance all of whose atoms have the same number of protons in their nuclei. Elements can neither be decomposed nor transformed into one another by ordinary chemical or physical means. There is a limited number of elements, and all other substances are combinations of them in various proportions.

An **elementary particle** cannot be separated into other particles. A *lepton* is an elementary particle that is not affected by the strong interaction and has no internal structure; the electron is a lepton. A *hadron* is an elementary particle that is affected by the strong interaction and is composed of *quarks,* particles with electric charges of $\pm\frac{1}{3}e$ or $\pm\frac{2}{3}e$ that have not been experimentally isolated as yet. Protons and neutrons are hadrons.

An **endothermic reaction** is one that must be supplied with energy in order to occur.

Energy is the property something has that enables it to do work. The unit of energy is the *joule* (J). The basic forms of energy are kinetic energy, potential energy, and rest energy. The *law of conservation of energy* states that energy can be neither created nor destroyed, although it may change from one form to another (including mass).

The **energy levels** of an atom are those specific energies its electrons can have. In the *ground state* of the atom, all its electrons are in their lowest possible energy levels. In the *excited states* of the atom, one or more of its electrons are in higher energy levels.

Entropy is a measure of the disorder of the particles that make up a body of matter. In a system of any kind isolated from the rest of the universe, entropy cannot decrease.

A chemical **equation** expresses the result of a chemical change. When the equation is *balanced,* the number of each kind of atom is the same on both sides of the equation.

A chemical **equilibrium** occurs when a chemical reaction and its reverse reaction both take place at the same rate.

At the vernal and autumnal **equinoxes** the sun is directly overhead at noon at the equator. The periods of daylight and darkness on these days are the same everywhere on the earth.

Erosion refers to all processes by which rock is disintegrated and worn away and its debris removed.

The **escape speed** is the minimum speed an object needs to permanently escape from the gravitational pull of a particular astronomical body.

An **ester** is an organic compound formed when an alcohol reacts with an acid.

According to Darwin's *theory of* **evolution,** which is one of the basic principles of biology and is supported by all known evidence, living things evolve through the operation of two mechanisms. One is reproduction with variation, so that the members of each generation may not be identical to their parents. The second is natural selection: the individual organisms best adapted to their environments are most likely to survive and produce offspring.

The Pauli **exclusion principle** states that no more than one electron in an atom can have a particular set of quantum numbers.

An **exoplanet** is a planet in orbit around a star other than the sun.

An **exothermic reaction** is one that liberates energy.

The **expanding universe** refers to the fact that all of the galaxies visible to us seem to be rapidly moving apart from one another. The evidence for the recession is the observed *red shift* in galactic spectra, which is a shift of their spectral lines toward the red end of the spectrum that is interpreted as a doppler effect.

The divisions of geological time are marked by mass **extinctions** in which environmental changes cause the disappearance of many species of plants and animals. Dinosaurs were among the victims of the extinction at the end of the Mesozoic Era.

F

A **fault** is a rock fracture along which movement has occurred.

The **ferromagnesian minerals** are a group of minerals that are silicates of iron and magnesium and green to black in color.

A **floodplain** is the wide, flat floor of a river valley produced by the sidewise cutting of the stream when its slope has become too gradual for further downcutting.

Foliation refers to the alignment of flat or elongated mineral grains characteristic of many metamorphic rocks.

A **force** is any influence that can cause an object to be accelerated. The unit of force is the *newton* (N).

A **force field** is a region of altered space (for example, around a mass, an electric charge, a magnet) that exerts a force on appropriate objects in that region.

The **formula mass** of a substance is the sum of the atomic numbers of the elements it contains, each multiplied by the number of times it appears in the formula of the substance.

The **fossil fuels** coal, oil, and natural gas were formed by the partial decay of the remains of plants and marine organisms that lived millions of years ago.

Fossils are the remains or traces of organisms preserved in rocks.

When we say something is moving, we mean that its position relative to something else—the **frame of reference**—is changing. The choice of an appropriate frame of reference depends on the situation.

The **frequency** of a wave is the number of waves that pass a given point per second. The unit of frequency is the *hertz* (Hz). The frequency of a sound wave is its *pitch*.

A **frontal surface** is the surface separating a warm and a cold air mass; a *front* is where this surface touches the ground. A *cold front* generally involves a cold air mass moving approximately eastward. A *warm front* generally involves a warm air mass moving approximately eastward.

In a **fuel cell,** the electrons transferred in an oxidation-reduction reaction pass through an external circuit as they go from one reactant to the other. These moving electrons constitute an electric current. Because the reactants are fed in and the products removed continuously, a fuel cell can provide current indefinitely, unlike having to be replaced or recharged as would a battery.

A **functional group** is a group of atoms in an organic molecule whose presence determines its chemical behavior to a large extent. An example is the OH group that characterizes the alcohols.

The four **fundamental interactions,** in order of decreasing strength, are the strong, electromagnetic, weak, and gravitational. They give rise to all the physical processes in the universe. The electromagnetic and weak interactions, and probably the strong as well, are closely related.

G

Our **galaxy** is a huge, rotating, disk-shaped group of stars that we see as the Milky Way from our location about two-thirds of the way out from the center. *Spiral galaxies* are other collections of stars that resemble our galaxy. The universe apparently consists of widely separated galaxies of stars.

Gamma rays are very high frequency electromagnetic waves.

A **gamma-ray burst** is a brief but intense flood of gamma rays that reaches the earth from space. Suggested origins are an exceptionally powerful supernova explosion or the capture of a neutron star by a black hole.

Geologic time is divided into *Precambrian time* and the *Paleozoic, Mesozoic,* and *Cenozoic eras* of earth history. Eras are subdivided into *periods* and the periods into *epochs*.

Geothermal energy comes from the heat of the earth's interior.

A **glacier** is a large mass of ice formed from compacted snow that gradually moves downhill.

Globular clusters are roughly spherical assemblies of very old stars outside the plane of our galaxy but associated with it.

Gondwana was the ancient supercontinent which later split up to form what are today South America, Africa, Antarctica, India, and Australia.

Newton's law of **gravity** states that every particle in the universe attracts every other particle with a force that is directly proportional to both of their masses and inversely proportional to the square of the distance between them.

The **greenhouse effect** refers to the process by which a greenhouse is heated: sunlight can enter through its windows, but the infrared radiation the warm interior gives off is absorbed by glass, so the incoming energy is trapped. The earth's atmosphere is heated in a similar way by absorbing infrared radiation from the warm earth.

Greenhouse gases are gases that absorb infrared radiation; the chief ones in the atmosphere are carbon dioxide (CO_2), methane, nitrous oxide (N_2O), and a group of gases used in refrigeration called CFCs and HCFCs.

Groundwater is rainwater that has soaked into the ground. The *water table* is the upper surface of that part of the ground whose pore spaces are saturated with water. A *spring* is a channel through which groundwater comes to the surface.

H

The **half-life** of a radioactive nuclide is the period of time required for one-half of an original sample to decay.

The **halogens** are a family of highly active nonmetals with similar chemical properties. The halogens are fluorine, chloride, bromine, iodine, and astatine in order of atomic number.

Heat is energy of random molecular motion. The heat that a body possesses depends upon its temperature, its mass, and the kind of material of which it is composed. The unit of heat is the joule (J).

A **heat engine** is a device that converts some of the heat that flows from a hot reservoir to a cold reservoir into mechanical energy or work.

The **heat of fusion** of a substance is the amount of heat needed to change 1 kg of the substance from solid to liquid at its freezing point. It is also the heat that 1 kg of the substance gives up when it changes from liquid to solid at its freezing point.

The **heat of vaporization** of a substance is the amount of heat needed to change 1 kg of the substance from liquid to vapor at its boiling point. It is also the heat that 1 kg of the substance gives up when it changes from vapor to liquid at its boiling point.

Heat transfer can take place by *conduction,* in which heat is carried from one place to another by molecular collisions; by *convection,* in which the transport is by the motion of a volume of hot fluid (liquid or gas); or by *radiation,* in which the transport is by means of electromagnetic waves.

The **Hertzsprung-Russell diagram** is a graph on which are plotted the intrinsic brightness and spectral type of individual stars. Most stars belong to the *main sequence,* which follows a diagonal line on the diagram, but there are also small, hot *white dwarf* stars and large, cool *red giant* stars occupying other positions not on the main sequence. *Red dwarfs* at its lower end are the most abundant stars in the main sequence.

Humidity refers to the moisture content of air. *Relative humidity* is the ratio between the amount of moisture in a volume of air and the maximum amount of moisture that volume of air can hold when completely saturated. It is usually expressed as a percentage.

A **hydrocarbon** is an organic compound containing only carbon and hydrogen. An *unsaturated hydrocarbon* is one whose molecules contain more than one bond (that is, shared electron pair) between adjacent carbon atoms. In a *saturated hydrocarbon* there is only one bond between adjacent carbon atoms.

The **hydronium ion** consists of a water molecule with a hydrogen ion attached. Its symbol is H_3O^+. (Sometimes more than one water molecule may be attached to a single hydrogen ion; the notion of the hydronium ion is for convenience only.)

A **hypothesis** is an interpretation of scientific results as first proposed. After thorough checking, it becomes a *law* if it states a regularity of relationship, or a *theory* if it uses general considerations to account for specific phenomena. (These terms are often used in senses slightly different from the ones indicated, but the definitions stated here correspond to their usual meanings.)

I

The **ice ages** were times of severe cold in which ice sheets covered much of Europe and North America.

An **ideal gas** is one that obeys the formula pV/T = constant, which is a combination of Boyle's and Charles's laws. The behavior of actual gases corresponds approximately to that of an ideal gas.

Igneous rocks are rocks that have been formed from a molten state by cooling.

A *real* **image** of an object is formed by light rays that pass through it; the image would therefore appear on a screen. A *virtual image* can only be seen by the eye because the light rays that seem to come from the image do not actually pass through it.

The **inert gases** are a family of almost totally inactive elements consisting of helium, neon, argon, krypton, xenon, and radon in order of atomic number.

Inertia is the apparent resistance an object offers to any change in its state of rest or of motion.

Insolation is an acronym for *incoming solar radiation.* Insolation is the energy that powers weather phenomena.

An **insulator** is a substance through which electric charges can move only with difficulty.

Interference occurs when two or more waves of the same kind pass the same point in space at the same time. If the waves are "in step" with each other, their amplitudes add together to produce a strong wave; this situation is called *constructive interference.* If the waves are "out of step" with each other, their amplitudes tend to cancel out and the resulting wave is weaker; this situation is called *destructive interference.*

Intrusive rocks are igneous rocks that flowed into regions below the surface already occupied by other rocks and gradually hardened there.

An **ion** is an atom or molecule that has an electric charge because it has gained or lost one or more electrons. The process of forming ions is called *ionization.*

In an **ionic bond,** electrons are transferred from one atom to another and the resulting ions then attract each other. An *ionic crystal* consists of individual ions in an array in which attractive and repulsive forces balance out.

The **ionosphere** is a region in the upper atmosphere that contains layers of ions. The ionosphere reflects radio waves and so makes possible long-range radio communication.

Isomers are compounds whose molecules contain the same atoms but in different arrangements.

The **isotopes** of an element have atoms with the same atomic number but different atomic masses; their nuclei contain the same number of protons but different numbers of neutrons.

K

Kepler's laws of planetary motion state that (1) the paths of the planets around the sun are ellipses, (2) the planets move so that their radius vectors sweep out equal areas in equal times, and (3) the ratio between the square of the time required by a planet to make a complete revolution around the sun and the cube of its average distance from the sun is a constant for all the planets.

Kinetic energy is the energy a body has by virtue of its motion. The kinetic energy of a moving body is equal to $\frac{1}{2}mv^2$, one-half the product of its mass and the square of its speed.

According to the **kinetic theory of matter,** all matter consists of tiny individual molecules that are in constant random motion. Heat is the kinetic energy of these random molecular motions.

L

A **laser** is a device that produces an intense beam of monochromatic, coherent light from the cooperative radiation of excited atoms. (The waves in a coherent beam are all in step with one another.)

Laurasia was the ancient supercontinent which later split up into what are today the continents North America, Greenland, Europe, and most of Asia.

Lava is molten rock on the earth's surface.

A **lens** is a piece of glass or other transparent material shaped to produce an image by refracting light that comes from an object. A *converging* lens brings parallel light to a single point at a distance called the *focal length* of the lens. A *diverging lens* spreads out parallel light so that it seems to come from a point behind the lens.

Light consists of electromagnetic waves in a range of frequencies to which the eye is sensitive. *White light* is a mixture of different frequencies, each of which produces the visual sensation of a particular color.

A **light-year** is the distance light travels in 1 year; it is about 9.46×10^{12} km.

Lines of force are imaginary lines used to describe a field of force. They are closest together where the field is strongest, farthest apart where the field is weakest.

Fats, oils, waxes, and sterols are **lipids,** which are synthesized in plants and animals from carbohydrates.

Lithification refers to the gradual hardening into rock of sediments buried under later deposits.

The **lithosphere** is the earth's outer shell of rigid rock. It consists of the crust and the outermost part of the mantle.

M

Magma is molten rock below the earth's surface.

A **magnet** is an object that can attract iron objects. When freely suspended, the *north pole* of a magnet points north while its *south pole* points south. Like poles repel each other; unlike poles attract.

A **magnetic field** is present around every moving charge and electric current. A magnetic field exerts a sideways force on any other moving charge or electric current in its presence.

Magnetic forces are exerted by moving charges (such as electric currents) on one another. The magnetic force between two moving charges is a modification of the electric force between them due to their motion; it is a relativistic effect.

The earth's **mantle** is the solid part of the earth between the core and the crust; it is about 2800 km thick.

The **mass** of a body is the property of matter that manifests itself as inertia. It may be thought of as the quantity of matter in the body. The unit of mass is the *kilogram* (kg).

The **mass number** of an atomic nucleus is the number of *nucleons* (neutrons and protons) it contains.

A **matter wave** is associated with rapidly moving objects, whose behavior in certain respects resembles wave behavior. The matter waves associated with a moving object are in the form of a group, or packet, of waves which travel with the same speed as the object. The quantity that varies in a matter wave is called the object's *wave function,* and its square is the object's *probability density.* The greater the probability density at a certain place and time, the greater the likelihood of finding the object there at that time.

The **Mesozoic Era** is the period in the earth's history from 251 million years ago to 66 million years ago; the term means "intermediate life."

The **metallic bond** that holds metal atoms together arises from a "gas" of electrons that can move freely through a solid metal. These electrons are also responsible for the ability of metals to conduct heat and electricity well.

Metalloids (also called semimetals) are elements whose properties are intermediate between metals and nonmetals.

Metals possess a characteristic sheen (metallic luster) and are good conductors of heat and electricity. They combine with nonmetals more readily than with one another. A metal atom has one or several electrons outside closed shells or subshells and combines chemically by losing these electrons to nonmetal atoms.

Metamorphic rocks are rocks that have been altered by heat and pressure deep under the earth's surface.

Meteoroids are pieces of matter moving through the solar system; *meteors* are the flashes of light they produce when entering the earth's atmosphere; and *meteorites* are the remains of meteoroids that reach the ground.

Methane, the main constituent of natural gas, is a compound of carbon and hydrogen with the chemical formula CH_4.

Minerals are the separate homogeneous substances of which rocks are composed. The most abundant minerals are silicates; also common are carbonates and oxides.

A **mole** of a substance is that amount of it whose mass in grams is equal to its formula mass expressed in atomic mass units. A mole of anything contains the same number of formula units as a mole of anything else; this number is called *Avogadro's number.* Because of the way the mole is defined, a chemical equation can be interpreted in terms of moles as well as in terms of formula units such as atoms, molecules, or ions.

A **molecule** is an electrically neutral combination of two or more atoms held together strongly enough to be experimentally observable as a particle.

In a **molecular crystal,** van der Waals forces bond the molecules together to form a solid.

The linear **momentum** of a body is the product of its mass and its velocity. The law of *conservation of momentum* states that, when several objects interact with one another (for instance, in an explosion or a collision), if outside forces do not act upon them, the total momentum of all the objects before they interact is exactly the same as their total momentum afterward. (See also **angular momentum.**)

A **moraine** is the pile of debris around the end of a glacier left as a low, hummocky ridge when the glacier melts back. The deposited material is called *till.*

Newton's *first law of* **motion** states that every object continues in its state of rest or of uniform motion in a straight line if no force acts upon it. The *second law of motion* states that when a net force F acts on an object of mass m, the object is given an acceleration of F/m in the same direction as that of the force. The *third law of motion* states that when an object exerts a force on another object, the second object exerts an equal but opposite force on the first. Thus for every *action force* there is an equal but opposite *reaction force;* the two forces act on different objects.

N

The **neutralization** of an acid by a base in water solution is a reaction between hydrogen and hydroxide ions to form water.

A **neutrino** is an uncharged particle with very little mass that is emitted during the beta decay of a nucleus. It interacts only feebly with matter. Slightly different neutrinos are emitted in the decay of certain unstable elementary particles.

The **neutron** is an electrically neutral elementary particle whose mass is slightly more than that of the proton. Neutrons are present in all atomic nuclei except that of ordinary hydrogen.

A **neutron star** is an extremely small star, smaller than a white dwarf, which is composed almost entirely of neutrons.

Nonmetals have an extreme range of physical properties; in the solid state they are usually lusterless and brittle and are poor conductors of heat and electricity. Some nonmetals form no compounds whatever; the others combine more readily with active metals than with one another. A nonmetal atom lacks one or several electrons of having closed shells or subshells and combines chemically by picking up electrons from metal atoms or by sharing electrons with other nonmetal atoms.

Nuclear fission occurs when a large nucleus splits into two or more smaller nuclei. Considerable energy is given off each time fission occurs.

Nuclear fusion occurs when two small nuclei unite to form a large one. Considerable energy is given off each time fusion occurs.

A **nuclear reactor** is a device in which fissions occur at a controlled rate.

Nucleic acid molecules consist of long chains of *nucleotides* whose precise sequence governs the structure and functioning of cells and organisms. *DNA* has the form of a double helix and carries the genetic code. One form of the simpler *RNA* acts as a messenger in protein synthesis.

The **nucleus** of an atom is its small, heavy core, containing all the atom's positive charge and most of its mass. The nucleus of ordinary hydrogen consists of a single proton; nuclei of other atoms consist of protons and neutrons, which are jointly called *nucleons.*

All the atoms of a **nuclide** have the same atomic and mass numbers.

O

Ohm's law states that the current in a metallic conductor is equal to the potential difference between the ends of the conductor divided by its resistance; symbolically, $I = V/R$.

Organic chemistry refers to the chemistry of carbon compounds.

Oxidation is the chemical combination of a substance with oxygen. More generally, an element is *oxidized* when its atoms lose electrons and *reduced* when its atoms gain electrons.

Ozone is a form of oxygen whose molecules consist of three oxygen atoms each. Ozone is present in the upper atmosphere where it absorbs solar ultraviolet radiation.

P

Pair production is the materialization of a particle-antiparticle pair (for instance, an electron and a positron) when a sufficiently energetic gamma ray passes near an atomic nucleus.

The **Paleozoic Era** is the period in the earth's history from 542 million years ago to 251 million years ago; the term means "ancient life."

Pangaea was the ancient landmass from which all the continents originated.

The **parallax** of a star is the apparent shift in its position relative to more distant stars as the earth revolves in its orbit.

The **period** of a wave is the time needed for a complete wave (crest + trough or compression + rarefaction) to pass a given point.

Mendeleev's **periodic law** in modern form states that if the elements are listed in the order of their atomic numbers, elements with similar properties recur at regular intervals. Such similar elements make up *groups;* the halogens are an example. A tabular arrangement of the elements showing this recurrence of properties is called a *periodic table.* In a periodic table, a *period* is a horizontal row of elements with different properties; a group appears as a vertical column.

Petroleum is a naturally occurring liquid mixture of hydrocarbons believed to have originated from plant and animal matter buried long ago.

The **pH** of a solution expresses its degree of acidity or basicity in terms of its hydrogen-ion concentration. A pH of 7 signifies a neutral solution, a smaller pH than 7 signifies an acidic solution, and a higher pH than 7 signifies a basic solution.

The **phases of the moon** occur because the area of the illuminated side of the moon visible to us varies with the position of the moon in its orbit.

The **photoelectric effect** is the emission of electrons from a metal surface when light shines on it.

Electromagnetic waves transport energy in tiny bursts called **photons** that resemble particles in many respects. The energy of a photon is related to the frequency f of the corresponding wave by $E = hf$, where h is *Planck's constant*.

The **photosphere** of the sun is a thin outer layer from which most of the light the sun radiates comes from.

Photosynthesis is the reaction between water and carbon dioxide to produce carbohydrates that occurs in green plants with the help of sunlight and the catalyst chlorophyll.

A **photovoltaic cell,** also called a **solar cell,** converts the energy in sunlight directly to electric energy.

A **planet** is a large, spherical satellite of the sun that appears in the sky as a bright object whose position changes relative to the stars. In order from the sun the eight planets are Mercury, Venus, Earth, Mars, Jupiter, Saturn, Uranus, and Neptune. A **dwarf planet** is a satellite of the sun much smaller than any of the eight regular planets but large enough for gravity to have given it a spherical shape.

According to the theory of **plate tectonics,** the lithosphere of the earth is divided into seven huge *plates* and a number of smaller ones. Continental drift occurs because the plates can move relative to one another.

A **pluton** is a body of intrusive rock formed by the solidification of magma under the surface.

A **polar molecule** is one that behaves as if it were negatively charged at one end and positively charged at the other. A *polar liquid* is a liquid whose molecules are polar whereas a *nonpolar liquid* has molecules whose charge is symmetrically arranged.

A **polymer** is a long chain of simple molecules, called *monomers,* that have joined together, usually under the influence of heat and catalysts.

A **positron** is a positively charged electron.

The **potential difference** (or **voltage**) between two points is the work needed to take a positive electric charge of 1 C from one of the points to the other. The unit of potential difference is the *volt* (V), equal to 1 J/C.

Potential energy is the energy an object has by virtue of its position. The gravitational potential energy of an object of mass m that is a height h above some reference level is $PE = mgh$, where g is the acceleration of gravity. Since mg is the object's weight w, an alternative formula is $PE = wh$.

Power is the rate at which work is being done. The unit of power is the *watt* (W), equal to 1 J/s.

Precambrian time is the name given to the period in the earth's history preceding the Paleozoic Era, which began 542 million years ago.

A **precipitate** is an insoluble solid that forms as a result of a chemical reaction in solution.

The **pressure** on a surface is the perpendicular force per unit area that acts on it. The unit of pressure is the *pascal* (Pa), equal to 1 N/m^2. Another common pressure unit is the *bar,* equal to 10^5 Pa, which is very close to atmospheric pressure.

Proteins are the chief constituents of living matter and consist of long chains of *amino acid* molecules. The sequence of amino acids in a protein molecule together with the form of the molecule determines its biological role.

The **proton** is a positively charged elementary particle found in all atomic nuclei.

The **ptolemaic system** is an incorrect hypothesis of the astronomical universe in which the earth is at the center with all of the other celestial bodies revolving around it in more or less complex orbits.

A **pulsar** is a star that emits extremely regular flashes of light and radio waves. Pulsars are believed to be neutron stars that are spinning rapidly.

The **Pythagorean theorem** states that the sum of the squares of the short sides of a right triangle is equal to the square of its *hypotenuse* (longest side).

Q

Quantum mechanics is a theory of atomic phenomena based on the wave nature of moving things. Newtonian mechanics is an approximate version of quantum mechanics valid only on a relatively large scale of size.

Four **quantum numbers** are needed to specify completely the physical state of an atomic electron. These are the *principal quantum number, n,* which governs the energy of the electron; the *orbital quantum number, l,* which determines the magnitude of the electron's angular momentum; the *magnetic quantum number, m_l,* which determines the direction of the electron's angular momentum; and the *spin magnetic quantum number, m_s,* which determines the orientation of the electron's spin.

According to the **quantum theory of light,** light travels in tiny bursts (or *quanta*) of energy called *photons.* The higher the frequency of the light, the more energy each of its photons has.

A **quasar** is a distant celestial object that emits both light and radio waves strongly and that may be powered by a super-massive black hole at its center.

R

In **radioactive decay,** certain atomic nuclei spontaneously emit *alpha particles* (helium nuclei), *beta particles* (electrons or positrons), or *gamma rays* (high-frequency electromagnetic waves).

Radiocarbon dating is a procedure for establishing the approximate age of once-living matter on the basis of the relative amount of radioactive carbon it contains.

A **radio telescope** is a directional antenna connected to a sensitive radio receiver. Radio waves from space are produced by extremely hot gases, by fast electrons that move in magnetic fields, and by atoms and molecules excited to radiate.

Reflection occurs when a wave bounces off a surface. Light is reflected by a mirror, sound by a wall. In **internal reflection,** light that arrives at a medium of lower index of refraction (for instance, glass to air) at a large enough angle is reflected back.

A **reformer** is a device that extracts hydrogen, usually for a fuel cell, from fuels such as natural gas, ethanol, or gasoline.

Refraction occurs when a wave changes direction on passing from one medium to another in which its speed is different. The *index of refraction* of a medium is the ratio between the speed of light in free space and its speed in the medium.

A **refrigerator** uses mechanical energy to transfer heat from a cold reservoir to a hot one, a path opposite to the normal direction of heat flow. It is the reverse of a heat engine

Einstein's *general theory of* **relativity** interprets gravity as a warping of space and time; this theory correctly predicted that the path of a light ray is bent when it passes through a gravitational field. The *special theory of relativity* deals with situations in which there is relative motion between an observer and what is being observed. Compared with their measurements when there is no relative motion, lengths are shorter, time intervals are longer, and kinetic energies are greater. Special relativity accounts for the relationships between electricity and magnetism. Another conclusion of special relativity is that nothing can travel faster than the speed of light.

Electrical **resistance** is a measure of the difficulty electric current has in passing through a certain body of matter. The unit of resistance is the *ohm* (Ω, the Greek capital letter *omega*), equal to 1 V/A.

The **resolving power** of an optical system is a measure of its ability to produce distinct images of nearby objects, which is limited by diffraction.

Resonance occurs when periodic impulses are given to an object at a frequency equal to one of its natural frequencies of vibration.

The **rest energy** of an object is the energy it possesses by virtue of its mass alone. An object's rest energy is given by the product of its mass measured when it is at rest and the square of the speed of light, namely m_0c^2.

The **right-hand rule** states that when a current-carrying wire is grasped so that the thumb of the right hand points in the direction of the current (assumed to flow from + to −), the fingers of that hand point in the direction of the magnetic field around the current.

S

A **salt** is one of a class of ionic compounds most of which are crystalline solids at ordinary temperatures and most of which consist of a metal combined with one or more nonmetals. Any salt can be formed by mixing the appropriate acid and base and evaporating the solution to dryness.

A **satellite** is, in general, an astronomical body revolving about some other body. Usually the term refers to the satellites of the planets; thus the moon is a satellite of the earth.

The molecules of a **saturated organic compound** contain only single carbon-carbon bonds; those of an *unsaturated compound* contain one or more double or triple bonds.

A **scalar quantity** has magnitude only. Mass and speed are examples.

The **scientific method** of studying nature can be thought of as consisting of four steps: (1) formulating a problem; (2) observation and experiment; (3) interpretation of the results; (4) testing the interpretation by further observation and experiment.

A **scientific theory** is a framework of ideas and relationships that ties together a variety of observations and experimental findings and permits as-yet unknown phenomena and connections to be predicted.

Sedimentary rocks consist of materials derived from other rocks that have been decomposed by water, wind, or glacial ice. They may be fragments cemented together or material precipitated from water solution.

A **seismograph** is a sensitive instrument designed to detect earthquake waves.

Semimetals (also called metalloids) are elements whose properties are intermediate between metals and nonmetals.

Sequestration is a method of carbon capture and storage that involves pumping CO_2 emitted by a power plant into an underground reservoir.

An electron **shell** in an atom consists of all the electrons having the same principal quantum number n. When a particular shell contains all the electrons possible it is called a *closed shell.* An electron *subshell* in an atom consists of all the electrons having both the same principal quantum number n and the same orbital quantum number l. When a particular subshell contains all the electrons possible it is called a *closed subshell.* Closed shells and subshells are exceptionally stable.

The **significant figures** in a number are its accurately known digits. When numbers are combined arithmetically, the result has as many significant figures as those in the number with the fewest of them.

A **sill** is a sheetlike mass of igneous rock lying parallel to preexisting rock layers.

Soil is a mixture of rock fragments, clay minerals, and organic matter. The organic matter consists largely of partially decomposed plant debris called *humus.*

A **solar prominence** is a large sheet of luminous gas projecting from the solar surface.

The **solar system** consists of all the objects that orbit the sun, namely the planets and their satellites together with the asteroids, meteoroids, and comets.

The **solar wind** consists of streams of ions that constantly flow outward from the sun. It deflects comet tails and causes auroras in the earth's upper atmosphere.

The *summer* **solstice** occurs when the North Pole is tilted toward the sun; the period of daylight in the northern hemisphere is longest on this day. The *winter solstice* occurs when the North Pole is tilted away from the sun; the period of daylight in the northern hemisphere is shortest on this day.

A **solution** is a homogeneous mixture of elements or compounds without any fixed proportions. The substance present in larger amount is the *solvent,* the other is the *solute.* When a solid or gas is dissolved in a liquid, the liquid is always considered the solvent. The *solubility* of a substance is the maximum amount that can be dissolved in a given quantity of solvent at a given temperature. A *saturated solution* is one that contains the maximum amount of solute that can be absorbed at a given temperature.

The **specific heat** of a substance is the amount of heat needed to change the temperature of 1 kg of the substance by 1°C.

A **spectrum** is the band of different colors produced when a light beam passes through a glass prism or is diffracted by a device called a *grating.* An *emission spectrum* is one produced by a light source alone. It may be a *continuous spectrum* with all colors present, or a *bright-line spectrum,* in which only a few specific frequencies characteristic of the source appear. An *absorption spectrum* is one produced when light from a glowing source passes through a cool gas. It is also called a *dark-line spectrum* because it appears as a continuous band of colors crossed by dark lines corresponding to characteristic frequencies absorbed by the gas.

The **speed** of a moving object is the rate at which it covers distance. The unit of speed is the m/s. The *terminal speed* of a falling body is the speed at which its weight is balanced by the force of air resistance, so it cannot fall any faster. This limit arises because air resistance increases with speed.

The **spin** of an elementary particle such as an electron refers to its intrinsic angular momentum, which is the same for all particles of a given kind. Electron spin plays an important part in atomic and molecule structures.

A **star** is a large, self-luminous body of gas held together gravitationally that obtains its energy from nuclear fusion reactions in its interior.

Stellar brightness. The *apparent* brightness of a star is its brightness as seen from the earth; the *intrinsic* brightness of a star is a measure of the total amount of light it radiates into space.

The **structural formula** of a compound is a diagram that shows the bonds between the atoms in its molecules.

In a **subduction zone,** the edge of a lithospheric plate slides under the edge of an adjacent plate.

Sublimation is the direct conversion of a substance from the solid to the vapor state without it first becoming a liquid.

A **sunspot** is a dark marking on the solar surface. Sunspots range up to some thousands of km across, last from several days to over a month, and have temperatures as much as 1000°C cooler than the rest of the solar surface. The *sunspot cycle* is a regular variation in the number and size of sunspots whose period is about 11 years.

Superconductivity refers to the loss of all electrical resistance by certain materials at very low temperatures.

A **supernova** is a heavy star that explodes with spectacular brightness and emits vast amounts of material into space.

T

Temperature is that property of a body of matter that gives rise to sensations of hot and cold; it is a measure of average molecular kinetic energy.

The *first law of* **thermodynamics** is the law of conservation of energy. The *second law of thermodynamics* states that some of the heat input to a heat engine must be wasted in order for the engine to operate.

A **thermometer** is a device for measuring temperature. In the *Celsius scale,* the freezing point of water is given the value 0°C and the boiling point of water the value of 100°C. In the *Fahrenheit scale* these temperatures are given the values of 32°F and 212°F.

The **tides** are twice-daily rises and falls of the ocean surface. They are due to the different attractive forces exerted by the moon and, to a smaller extent, the sun on different parts of the earth. *Spring tides* have a large range between high and low water. They occur when the sun and moon are in line with the earth and thus add together their tide-producing actions. *Neap tides* have small ranges and occur when the sun and moon are 90° apart relative to the earth and thus their tide-producing actions tend to cancel each other out.

Till is the unsorted material deposited by a glacier and left behind when the glacier recedes.

A **transform fault** occurs where the edges of two lithosphere plates slide past each other.

A **transformer** is a device that transfers electric energy in the form of alternating current from one coil to another by means of electromagnetic induction.

A **tsunami** consists of ocean waves caused by earthquakes on the seabed or by gigantic landslides into the ocean.

U

The **uncertainty principle** states that it is impossible at the same time to determine accurate values for the position and momentum of a particle. Hence, in dealing with electrons within atoms, all we can consider are probabilities rather than specific positions and states of motion.

An **unconformity** is an uneven surface separating two series of rock beds. It is a buried surface of erosion that involves at least four geologic events: formation of the older rocks; diastrophic uplift of these rocks; erosion; and the deposit of sediments on the eroded landscape.

According to the *principle of* **uniform change,** the geologic processes that shaped the earth's surface in the past are the same as those active today.

V

Van der Waals forces arise from the electrical attraction between nonuniform charge distributions in atoms and molecules. They enable atoms and molecules to form solids without sharing or transferring electrons.

A **variable star** is one whose brightness changes regularly. The intrinsic brightness of a *Cepheid variable* is related to its period of variation, which permits the distance of the star, and hence the distance of the galaxy it is in, to be determined.

A **vector** is an arrowed line whose length is proportional to the magnitude of some quantity and whose direction is that of the quantity. A *vector quantity* is a quantity that has both magnitude and direction.

Veins consist of minerals deposited from solutions along cracks in rocks.

The **velocity** of an object specifies both its speed and the direction of its motion. A car has a speed of 30 m/s; its velocity is 30 m/s to the northwest.

Viscosity is the resistance of fluids to flowing motion; liquids are more viscous than gases.

A **volcano** is an opening in the earth's crust through which molten rock (called *magma* while underground, *lava* above ground) comes out, usually at intervals rather than continuously.

W

The **wavelength** of a wave is the distance between adjacent crests, in the case of transverse waves, or between adjacent compressions, in the case of longitudinal waves.

Waves carry energy from one place to another by a series of periodic motions of the individual particles of the medium in which the waves occur. (Electromagnetic waves are an exception.) There is no net transfer of matter in wave motion. In a *longitudinal wave* the particles of the medium vibrate back and forth in the direction in which the waves travel. In a *transverse wave* the particles vibrate from side to side perpendicular to the wave direction. Sound waves are longitudinal; waves in a stretched string are transverse; water waves are a combination of both since water molecules move in circular orbits when a wave passes.

Weather refers to the state of the atmosphere in a particular place at a particular time, whereas *climate* refers to the weather trends in a region through the years.

Weathering is the surface disintegration of rock by chemical decay and mechanical processes such as the freezing of water in crevices.

The **weight** of an object is the force with which gravity pulls it toward the earth. The weight of an object of mass m is mg, where g is the acceleration of gravity.

Work is a measure of the change, in a general sense, a force causes when it acts upon something. The work done by a force acting on an object is the product of the magnitude of the force and the distance through which the object moves while the force acts on it. If the direction of the force is not the same as the direction of motion, the projection of the force in the direction of motion must be used. The unit of work is the *joule* (J).

X

X-rays are high-frequency electromagnetic waves given off when matter is struck by fast electrons.

Photo Credits

Chapter 1

Figure 1-2: NASA/JPL/Malin Space Systems; 1-6: Corbis; p. 13: © Pixtal/age Fotostock; 1-9: The Bancroft Library, University of California, Berkeley; p. 15 & 20: © Pixtal/age Fotostock; 1-16: NASA; 1-21(both): © Bill Brooks/Alamy

Chapter 2

Opener: NASA; 2-1: The Image Works; 2-7: Courtesy Mitsubishi Estate, New York; 2-10: Royalty Free/Corbis; p. 42(both): McGraw-Hill College Division Photo Research Library; 2-17: Guy Sauvage/Photo Researchers, Inc.; 2-21: Jean Marc Barey/Photo Researchers, Inc.; 2-25 & 2-31: Bob Daemmrich/The Image Works; 2-35: © Dimitri Iundt/TempSport/Corbis; 2-40: NASA/Johnson Space Center; 2-42: © Comstock/PunchStock; 2-44: NASA

Chapter 3

Opener: Getty Images; 3-2: Royalty Free/Corbis; 3-12: E.R. Degginger/Color-Pic; 3-16: Philip Mehta/Contact Press Images; 3-17: © Photodisc/Vol. 27; 3-18: Culver Pictures; 3-19: Corbis; 3-25: © Duomo/Corbis; 3-27: NASA; 3-28: Pictorial Parade/Archive/Hulton/Getty; 3-31: E.R. Degginger/Color-Pic; 3-32: Brock May/Photo Researchers, Inc.; p. 94: Library of Congress Prints and Photographs Division (LC-USZ62-60242)

Chapter 4

Opener: © Bill Ross/Corbis 4-7: U.S. Geological Survey; 4-16: American Petroleum Institute; 4-17: © Photodisc/Getty Images; 4-18: Toyota Motor Sales, USA, Inc.; 4-19: Walter S. Siler/Index Stock Imagery; 4-20: © Digital Vision/Punch Stock; 4-21: Phil Degginger/Color-Pic; 4-22: Images courtesy of BP PLC, 2008; 4-23: Electric Power Cooperative; 4-24: © Dr. Parvinder Sethi; 4-26: Courtesy of the U.S. Department of Energy; 4-27: New York Power Authority; 4-28: Photo by M. Smith, U.S. Geological Survey; 4-29: AstroPower; 4-30: Stone/Getty Images; 4-31: Peter Menzel/Stock, Boston; 4-32: © Still Pictures; 4-33: Courtesy Fuel Cells 2000; 4-34: GreenFuel Technologies Corporation; 4-35: David Young-Wolff/Photo Edit, Inc.

Chapter 5

Opener: Royalty Free/Corbis; 5-2: The McGraw-Hill Companies/John A. Karachewski, photographer; 5-6: Stockbyte/Punchstock Images; 5-10: Dennis Stock/Index Stock Imagery; 5-15: Royalty Free/Corbis; 5-16: Courtesy John Deere; 5-18: Timothy O'Keefe/Bruce Coleman, Inc.; 5-20: Tom Stewart/Corbis/Stock Market; p. 168: © Pixtal/age Fotostock; 5-23: George Hall/Woodfin Camp & Associates; 5-24: AIP Emilio Segre Visual Archives, Zeleny Collection; 5-33: SuperStock; 5-38: Terry Wild Studio; 5-39: Gary Kessler; 5-43: Paul Silverman/Fundamental Photographs; 5-50: Will & Deni McIntyre/Photo Researchers, Inc.; p. 188: Corbis

Chapter 6

Opener: © Comstock/JupiterImages; 6-2: Leif Skoogers/Woodfin Camp & Associates; p. 201: © Pixtal/age Fotostock; 6-9: Corbis; 6-11: Tom Way; p. 211: AIP Emilio Segre Visual Archives, W.F. Meggers Gallery of Nobel Laureates; 6-15: AIP Emilio Segre Visual Archives; 6-17: © Stockbyte/PunchStock; 6-19: AIP Emilio Segre Visuals Archives, E. Scott Barr Collection; 6-23: Antman/The Image Works; 6-25: The McGraw-Hill Companies, Inc./Jacques Cornell, photographer; 6-30: Corbis; p. 225: © Pixtal/age Fotostock; 6-36: E.R. Degginger/Color-Pic; 6-39: Greg David; 6-41: Courtesy General Electric; p. 231: AIP Emilio Segre Visual Archives; 6-44: © Brand X Pictures/PunchStock; 6-49: Steve Cole/Getty Images

Chapter 7

Opener: PhotoLink/Getty Images; 7-1: S. Cazenave/Vandystadt/Photo Researchers, Inc.; 7-10: © Goodshoot/PunchStock; 7-11: Corbis; 7-14: Jim Wehtje/Getty Images; 7-17: David R. Frazier/Photo Researchers, Inc.; 7-19: Andy Sacks/Stone/Getty; 7-20: Bob Daemmrich/Stock, Boston; p. 256: Cavendish Laboratory, University of Cambridge; 7-23: Deutsches Museum, courtesy AIP Emilio Segre Visual Archives; 7-25: Topham/The Image Works; 7-27(both): Images courtesy of Furuno USA, Inc.; 7-32: Courtesy John Deere; 7-33: Berenice Abbott; 7-38: Runk/Schoenberger/Grant Heilman Photography; 7-40: Courtesy Zeiss; 7-49: Andrew Wood/Photo Researchers, Inc.; 7-54: Bill Ross/Woodfin Camp & Associates; 7-55: © Steve Hamblin/Alamy; 7-56: Royalty Free/Corbis; 7-61: Courtesy Noise Cancellation Technologies; 7-62: Berenice Abbott; p. 279: Corbis; 7-64: Photo Researchers, Inc.

Chapter 8

Opener: SuperStock; p. 291: Wiedergaberecht nicht bein Deutschen Museum; p. 294: Radium Institute, courtesy AIP Emilio Segre Visuals Archives; 8-6: CNRI/Science Photo Library/Photo Researchers, Inc; p. 306: Helmholtz-Zentrum Berlin Bildarchiv; 8-16: Argonne National Laboratory managed and operated by UChicago Argonne, LLC, for the U.S. Department of Energy under Contract No. DE-AC02-06CH11357; 8-18: Nuclear Energy Institute; 8-22: Courtesy Princeton University; p. 315: © Bettmann/Corbis; 8-25: Catherine Puedras/Science Library/Photo Researchers, Inc.; 8-27: Courtesy Brookhaven National Laboratory; 8-30: ©CERN Geneva; p. 321: Michelangelo D'Agostino

Chapter 9

Opener: SuperStock; 9-3: Corbis; 9-4: AP Wide World; 9-6: iStockphoto; 9-8: AIP Emilio Segre Visuals Archives, Lande Collection; 9-11: AP Wide World Photo; 9-10: Royalty Free/Corbis; p. 335: AIP Emilio Segre Visuals Archives, Physics Today Collection; 9-12: Steve Allen/Getty Images; 9-13: Courtesy of the Center for Disease Control; p. 339: AIP Emilio Segre Visuals Archives, W.F. Meggers Gallery of Nobel Laureates, Gift of William Numeroff; 9-16: Royalty Free/Corbis; p. 342: Niels Bohr Archive, courtesy AIP Emilio Segre Visuals Archives; 9-30: Courtesy Spectra Physics; 9-33: Kim Steele/Getty Images; p. 350: Photograph by Francis Simon, courtesy AIP Emilio Segre Visuals Archives; 9-37: Royalty Free/Corbis

Chapter 10

Opener: © Pepiera Tom/Iconotec.com; p. 365: Engraving by William Henry Worthington, published 1823, from the 1814 painting by William Allen, original in the National Portrait Gallery, London; 10-7 & 10-8: Royalty Free/Corbis; 10-9: Mitchell Funk; p. 371: Corbis; p. 380: Thomas Hollyman/Photo Researchers, Inc.; 10-19: Gerd Ludwig/Woodfin Camp & Associates; 10-25: Courtesy Bernzomatic

Chapter 11

Opener: Alfred Pasieka/Science Photo Library/Photo Researchers, Inc.; 11-1: E.R. Degginger/Color-Pic; 11-3: Courtesy of Jenifer Glynn/U.S. National Library of Medicine; 11-2: Courtesy Biochemistry Dept., King's College, London; 11-8: ©Vladpans/eStock Photo; 11-10: E.R. Degginger/Color-Pic; 11-11: International Science & Technology Magazine; 11-15: Iconica/Getty; 11-17: Digital Vision/Getty Images; 11-19: Eastcott/Momatiuk/The Image Works; 11-31: Dennis MacDonald/Index Stock Imagery; p. 409: Courtesy Department Library Services American Museum of Natural History, photo by Clyde Fisher; 11-35: Courtesy Betz Dearborn; 11-37: Photograph courtesy USGS Photo Library, Denver, CO; 11-38: Photograph by Belinda Rain, courtesy of EPA/National Archives; 11-40: The McGraw-Hill Companies, Inc./Stephen Frisch, photographer

Chapter 12

Opener: © Digital Vision/PunchStock; p. 430: Corbis; 12-2: © Pixtal/age Fotostock; 12-4: © Comstock/Corbis; 12-5: Harald Sund; 12-10: Corbis; p. 444: Judith Iriarte-Gross; 12-15: Paul Silverman/Fundamental Photographs; 12-17: Courtesy Weirton Steel Corporation; 12-21: Ballard Power Systems

Chapter 13

Opener: Courtesy A. Beiser, Photo by Rick Tomlinson; 13-3: © Chad McDermott/Alamy; 13-4: Justin Sullivan/Getty Images; 13-5: U.S.

Fish & Wildlife Service; 13-9: Joseph Nettis/Stock, Boston; p. 471: Corbis; 13-11 & 13-12: Corbis; 13-14: Gary Kessler; 13-15: Courtesy Regal Ware; 13-17: The McGraw-Hill Companies/Joe Franek, photographer; 13-18: Courtesy A. Beiser, Photo by Rick Tomlinson; 13-19: Ulrike Welsch/Photo Researchers, Inc.; 13-20: © Goodshoot/PunchStock; 13-23: Courtesy American Red Cross; p. 488: Bettmann/Corbis; 13-26: Grant Heilman Photography; 13-28: Camera Press; 13-29: M. Freeman/PhotoLink/Getty Images; 13-31: © Dr. Parvinder Sethi

Chapter 14

Opener: NASA; p. 501: Corbis; 14-3: Kent Knudson/PhotoLink/Getty Images; 14-7(all) & 14.9: Cloud Charts, Inc.; 14-10: NCAR; 14-24: NASA; 14-26: Royalty Free/Corbis; p. 522: Eugene M. Rasmusson; 14-31: NOAA; 14-32: Photo Courtesy of U.S. Army/U.S. Coast Guard photograph by Petty Officer 2nd Class Kyle Niemi; 14-33: © Dr. Parvinder Sethi; 14-34: Lionel Delevingne/Stock Boston; 14-37: Stuart Westmoreland/ Photo Researchers, Inc.; 14-38: AP Wide World Photo; 14-40: National Center for Atmospheric Research/National Science Foundation, Produced by William Holland/NCAR

Chapter 15

Opener: Alan Sussman/The Image Works; 15-3: © Dr. Parvinder Sethi; 15-4: Copyright © Bruce Molnia, Terra Photographics; 15-6: Siede Preis/ Getty Images; 15-7: © Dr. Parvinder Sethi; 15-8: © The McGraw-Hill Companies, Inc./Bob Coyle, photographer; 15-9 to 15-12: E.R. Degginger/ Color-Pic; 15-13: © Dr. Parvinder Sethi; 15-14: C. Borland/PhotoLink/ Getty Images; 15-15: The McGraw-Hill Companies/John A. Karachewski, photographer; 15-16: John Buitenkant/Photo Researchers, Inc.; 15-17: E.R. Degginger/Color-Pic; 15-18: Spike Mafford/Getty Images; 15-19: Digital Vision/Getty Images; 15-20: Photo by P.D. Rowley, U.S. Geological Society; 15-21: The McGraw-Hill Companies/John A. Karachewski, photographer; 15-22: Philip Coblentz/Brand X Pictures/PictureQuest; 15-24: Photograph by D. Keefer, from U.S. Geological Survey Open-File Report 89-687; p. 555: Andrea Donnellan; 15-31: Courtesy of Dr. Soren Gregersen; 15-34: Stock Photo/The Web Stock House; p. 562: Corbis; 15-36: Runk/Schoenberger/Grant Heilman Photography; 15-37: Leif Skoogfors/Woodfin Camp & Associates; 15-38: Courtesy USDA; 15-39: Ann Purcell/Photo Researchers, Inc.; 15-40: Jeff Gnass; 15-43: Photo by W.R. Hansen, U.S. Geological Survey; 15-45: Alan Sussman/The Image Works; 15-48: Marli Bryant Miller/Earthlens; 15-50: National Park Service; 15-51: Henry Tomasz/Index Stock Imagery; 15-52: Robert Glusic/Getty Images; 15-53: U.S. Geological Service; 15-55: Dr. Parvinder Sethi; 15-59: Asa C. Thoresen/Photo Researchers, Inc.

Chapter 16

Opener: C.C. Plummer; 16-1: Marli Bryant Miller/Earthlens; 16-4: Copyright © Bruce Molnia, Terra Photographics; 16-6: Bryant Miller/ Earthlens; p. 593: Corbis; 16-11: Woods Hole Oceanographic Institution; 16-15: Ragnar Larusson/Photo Researchers, Inc.; 16-19: Royalty Free/ Corbis; 16-20: Dinodia Picture Agency/The Image Works; 16-24a: © Brand X Pictures/PunchStock; 16-24b: Getty Images; p. 609: Courtesy Department Library Services, American Museum of Natural History; 16-26: Courtesy of the National Library of Medicine; 16-27: Joe Sohm/ The Image Works; 16-30: Marli Bryant Miller/Earthlens; p. 614: By kind permission of the Department of Geological Sciences, University of Durham; 16-32: The McGraw-Hill Companies/Joe Franek, photographer; 16-33: Courtesy of the Cleveland Museum of Natural History; 16-34: E.R. Degginger/Color-Pic; 16-36: Kenneth W. Fink/Bruce Coleman, Inc.; 16-38: Steve Cole/Getty Images; 16-40: The National History Museum, London; 16-42: photo courtesy of Anne-Lise Chenet; 16-47: Courtesy NASA/JPL-Caltech; 16-45: Corbis

Chapter 17

Opener: Stock Trek/Getty Images; 17-4: Photo by Wally Pacholka; 17-5: NASA Jet Propulsion Laboratory (NASA-JPL); 17-7: Courtesy of California Institute of Technology/Hale Observatories; 17-8: © U.S. Geological Survey; 17-9: Jonathan Blair/Woodfin Camp & Associates; 17-10 to 17-12: NASA; 17-14(both): Jet Propulsion Laboratory; 17-15: NASA; 17-16: ESA/ ISO, CEA SACLAY & ISOCAM CONSORTIUM; 17-17: ESA/DLR/FU Berlin (G. Neukum); 17-18: NASA; 17-19: NASA/JPL/MSSS; 17-20 to 17-23: NASA/JPL; 17-24: Corbis; p. 662: Photo by Michael Okoniewski, Copyright © 1994; 17-26: StockTrek/Getty Images; 17-27 to 17-29: NASA/JPL; 17-30: NASA; 17-31: NASA/JPL/University of Arizona; p. 670: Corbis; 17-33: NASA; 17-34 & 17-36: Courtesy California Institute of Technology/Hale Observatories; 17-38: National Optical Astronomy Observatory; 17-39: Courtesy California Institute of Technology/Hale Observatories; 17-40 to 17-43: NASA

Chapter 18

Opener: NASA/Johnson Space Center; 18-2: Roger Ressmeyer/Corbis; 18-3: Lowell Observatory Archives; 18-5: Courtesy California Institute of Technology/Hale Observatories; 18-8: Our Universe/JPL/NASA; 18-9: Michael Giannechini/Photo Researchers, Inc.; 18-10: NASA/The Image Works; 18-11: Courtesy California Institute of Technology/Hale Observatories; p. 701: Mark Marten/Photo Researchers, Inc.; 18-15: Courtesy Department Library Services, American Museum of Natural History; 18-16 & 18-17: Courtesy Yerkes Observatory; p. 704: AIP Emilio Segre Visuals Archives; 18-19(all): Courtesy California Institute of Technology/Hale Observatories; 18-21: Courtesy Yerkes Observatory; 18-22: AIP Emilio Segre Visuals Archives; 18-24: Photo Researchers, Inc.; 18-27: Courtesy of The Observatories of the Carnegie Institution of Washington; 18-28(both): © Anglo-Australian Telescope, Photo by David Malin; 18-31: The Image Works; 18-32: Courtesy UC Regents/Lick Observatory

Chapter 19

Opener: NASA/JPL-Caltech/S. Willner (Harvard-Smithsonian Center for Astrophysics); 19-1: Courtesy California Institute of Technology/ Hale Observatories; 19-2: Take 27 Ltd.; 19-3: Hayden Planetarium of the American Museum of Natural History; 19-4: © Anglo-Australian Telescope, Photo by David Malin; 19-5: Courtesy California Institute of Technology/Hale Observatories; 19-6: David Parker/Science Photo Library/Photo Researchers, Inc.; 19-7: Courtesy California Institute of Technology/Hale Observatories; 19-8: Photo Researchers, Inc.; 19-9: NASA; 19-10 & p. 734: Corbis; 19-14: Dr. Stephen Unwin/Science Photo Library/Photo Researchers, Inc.; p. 738: Andrea Donnellan, Ph.D.; 19-18: Courtesy Bell Labs; 19-20:/WMAP Science Team; 19-24: NASA

Index

Conversion Factors

1 meter (m) = 100 cm = 39.4 in. = 3.28 ft

1 centimeter (cm) = 10 millimeters (mm) = 0.394 in.

1 kilometer (km) = 1,000 m = 0.621 mi

1 foot (ft) = 12 in. = 0.305 m

1 inch (in.) = 0.0833 ft = 2.54 cm

1 mile (mi) = 5,280 ft = 1.61 km

1 liter = 1,000 cm^3 = 10^{-3} m^3 = 1.056 quart

1 day = 86,400 s = 2.74×10^{-3} year

1 year = 3.16×10^7 s = 365 days

1 m/s = 3.28 ft/s = 2.24 mi/h = 3.60 km/h

1 ft/s = 0.305 m/s = 0.682 mi/h = 1.10 km/h

1 mi/h = 1.47 ft/s = 0.447 m/s = 1.61 km/h

1 kilogram (kg) = 1,000 grams (g)

(Note: kg corresponds to 2.21 lb in the sense that the weight of 1 kg is 2.21 lb.)

1 atomic mass unit (u) = 1.66×10^{-27} kg
= 1.49×10^{-10} J
= 931 MeV

1 newton (N) = 0.221 lb

1 pound (lb) = 4.45 N

1 joule (J) = 2.39×10^{-4} kcal
= 6.24×10^{18} eV

1 kWh = 3.6 GJ

1 kilocalorie = 4,185 J = 3,089 ft·lb

1 electron volt (eV) = 10^{-6} MeV = 10^{-9} GeV
= 1.60×10^{-19} J
= 1.18×10^{-19} ft·lb = 3.83×10^{-23} kcal

1 watt (W) = 1 J/s

1 kilowatt (kW) = 1,000 W = 1.34 hp

1 horsepower (hp) = 746 W

1 pascal (Pa) = 1 N/m^2

1 atmosphere of pressure (atm) = 1.013×10^5 Pa
= 14.7 lb/in.^{2}

1 bar = 10^5 Pa

$$°C = \frac{5}{9}(°F - 32°)$$

$$°F = \frac{9}{5}°C + 32°$$

$$K = °C + 273$$

Powers of Ten

10^{-10} = 0.000,000,000,1	10^0 = 1
10^{-9} = 0.000,000,001	10^1 = 10
10^{-8} = 0.000,000,01	10^2 = 100
10^{-7} = 0.000,000,1	10^3 = 1000
10^{-6} = 0.000,001	10^4 = 10,000
10^{-5} = 0.000,01	10^5 = 100,000
10^{-4} = 0.000,1	10^6 = 1,000,000
10^{-3} = 0.001	10^7 = 10,000,000
10^{-2} = 0.01	10^8 = 100,000,000
10^{-1} = 0.1	10^9 = 1,000,000,000
10^0 = 1	10^{10} = 10,000,000,000

Multipliers for SI Units

a	atto-	10^{-18}	da	deka-	10^1
f	femto-	10^{-15}	h	hecto-	10^2
p	pico-	10^{-12}	k	kilo-	10^3
n	nano-	10^{-9}	M	mega-	10^6
μ	micro-	10^{-6}	G	giga-	10^9
m	milli-	10^{-3}	T	tera-	10^{12}
c	centi-	10^{-2}	P	peta-	10^{15}
d	deci-	10^{-1}	E	exa-	10^{18}

Physical and Chemical Constants

Speed of light in vacuum	c	3.00×10^8 m/s
Charge on electron	e	1.60×10^{-19} C
Gravitational constant	G	6.67×10^{-11} N·m^2/kg^2
Acceleration of gravity at earth's surface	g	9.81 m/s^2
Planck's constant	h	6.63×10^{-34} J·s
Coulomb constant	k	8.99×10^9 N·m^2/C^2
Electron rest mass	m_e	9.11×10^{-31} kg
Neutron rest mass	m_n	1.675×10^{-27} kg
Proton rest mass	m_p	1.673×10^{-27} kg
Avogadro's number	N_o	6.02×10^{23} formula units/mole